WHAT IS A QUANTUM FIELD THEORY?

Quantum field theory (QFT) is one of the great achievements of physics, of profound interest to mathematicians. Most pedagogical texts on QFT are geared toward budding professional physicists, however, whereas mathematical accounts are abstract and difficult to relate to the physics. This book bridges the gap. While the treatment is rigorous whenever possible, the accent is not on formality but on explaining what the physicists do and why, using precise mathematical language. In particular, it covers in detail the mysterious procedure of renormalization. Written for readers with a mathematical background but no previous knowledge of physics and largely self-contained, it presents both basic physical ideas from special relativity and quantum mechanics and advanced mathematical concepts in complete detail. It will be of interest to mathematicians wanting to learn about QFT and, with nearly 300 exercises, also to physics students seeking greater rigor than they typically find in their courses.

MICHEL TALAGRAND is the recipient of the Loève Prize (1995), the Fermat Prize (1997), and the Shaw Prize (2019). He was a plenary speaker at the International Congress of Mathematicians and is currently a member of the *Académie des sciences* (Paris). He has written several books in probability theory and well over 200 research papers.

"This book accomplishes the impossible task: It explains to a mathematician, in a language that a mathematician can understand, what is meant by a quantum field theory from a physicist's point of view. The author is completely and brutally honest in his goal to truly explain the physics rather than filtering out only the mathematics, but is at the same time as mathematically lucid as one can be with this topic. It is a great book by a great mathematician."

- Sourav Chatterjee, *Stanford University*

"Talagrand has done an admirable job of making the difficult subject of quantum field theory as concrete and understandable as possible. The book progresses slowly and carefully but still covers an enormous amount of material, culminating in a detailed treatment of renormalization. Although no one can make the subject truly easy, Talagrand has made every effort to assist the reader on a rewarding journey though the world of quantum fields."

- Brian Hall, *University of Notre Dame*

"A presentation of the fundamental ideas of quantum field theory in a manner that is both accessible and mathematically accurate seems like an impossible dream. Well, not anymore! This book goes from basic notions to advanced topics with patience and care. It is an absolute delight to anyone looking for a friendly introduction to the beauty of QFT and its mysteries."

- Shahar Mendelson, *Australian National University*

"I have been motivated to try and learn about quantum field theories for some time but struggled to find a presentation in a language that I as a mathematician could understand. This book was perfect for me: I was able to make progress without any initial preparation and felt very comfortable and reassured by the style of exposition."

- Ellen Powell, *Durham University*

"In addition to its success as a physical theory, quantum field theory has been a continuous source of inspiration for mathematics. However, mathematicians trying to understand quantum field theory must contend with the fact that some of the most important computations in the theory have no rigorous justification. This has been a considerable obstacle to communication between mathematicians and physicists. It is why, despite many fruitful interactions, only very few people would claim to be well versed in both disciplines at the highest level.

There have been many attempts to bridge this gap, each emphasizing different aspects of quantum field theory. Treatments aimed at a mathematical audience often deploy sophisticated mathematics. Michel Talagrand takes a decidedly elementary approach to answering the question in the title of his book, assuming little more than basic analysis. In addition to learning what quantum field theory is, the reader will encounter in this book beautiful mathematics that is hard to find anywhere else in such clear pedagogical form, notably the discussion of representations of the Poincaré group and the BPHZ Theorem. The book is especially timely given the recent resurgence of ideas from quantum field theory in probability and partial differential equations. It is sure to remain a reference for many decades."

- Philippe Sosoe, *Cornell University*

WHAT IS A QUANTUM FIELD THEORY?

A First Introduction for Mathematicians

MICHEL TALAGRAND

CAMBRIDGE
UNIVERSITY PRESS

CAMBRIDGE
UNIVERSITY PRESS

University Printing House, Cambridge CB2 8BS, United Kingdom

One Liberty Plaza, 20th Floor, New York, NY 10006, USA

477 Williamstown Road, Port Melbourne, VIC 3207, Australia

314–321, 3rd Floor, Plot 3, Splendor Forum, Jasola District Centre, New Delhi – 110025, India

103 Penang Road, #05–06/07, Visioncrest Commercial, Singapore 238467

Cambridge University Press is part of the University of Cambridge.

It furthers the University's mission by disseminating knowledge in the pursuit of education, learning, and research at the highest international levels of excellence.

www.cambridge.org
Information on this title: www.cambridge.org/9781316510278
DOI: 10.1017/9781108225144

First published 2022

A catalogue record for this publication is available from the British Library

Library of Congress Cataloging-in-Publication Data
Names: Talagrand, Michel, 1952– author.
Title: What is a quantum field theory? : a first introduction for mathematicians / Michel Talagrand.
Description: First edition. | New York : Cambridge University Press, 2021. | Includes bibliographical references and index.
Identifiers: LCCN 2021020786 (print) | LCCN 2021020787 (ebook) | ISBN 9781316510278 (hardback) | ISBN 9781108225144 (epub)
Subjects: LCSH: Quantum field theory. | BISAC: SCIENCE / Physics / Mathematical & Computational
Classification: LCC QC174.45 .T35 2021 (print) | LCC QC174.45 (ebook) | DDC 530.14/3–dc23
LC record available at https://lccn.loc.gov/2021020786
LC ebook record available at https://lccn.loc.gov/2021020787

ISBN 978-1-316-51027-8 Hardback

If all mathematics were to disappear, physics would be set back exactly one week.

Richard Feynman

Physics should be made as simple as possible, but not simpler.

Albert Einstein

The career of a young theoretical physicist consists of treating the harmonic oscillator in ever-increasing levels of abstraction.

Sydney Coleman

Contents

Introduction

As a teenager in the sixties reading scientific magazines, countless articles alerted me to "the infinities plaguing the theory of Quantum Mechanics". Reaching 60 after a busy mathematician's life, I decided that it was now or never for me to really understand the subject.[1] The project started for my own enjoyment, before turning into the hardest of my scientific life. I faced many difficulties, the most important being the lack of a suitable introductory text. These notes try to mend that issue.

I knew no physics to speak of, but it was not particularly difficult to get a basic grasp of topics such as Classical Mechanics, Electromagnetism, Special and even General Relativity. They are friendly for mathematicians as they can be made rigorous to our liking.

Quantum Mechanics was a different challenge. Quite naturally, I looked first for books written by mathematicians for mathematicians. By a stroke of bad luck, the first book I tried was to me a shining example of everything one should *not* do when writing a book. Being a mathematician does not mean that I absolutely need to hear about "a one-dimensional central extension of V by a Lie algebra 2-cocycle" just to learn the Heisenberg commutation relations. Moreover, while there is no question that mastery of some high level form of Classical Mechanics will help reaching a deeper understanding of Quantum Mechanics, Poisson Manifolds and Symplectic Geometry are not absolute prerequisites to get started. Other books written by mathematicians are less misguided, but seem to cover mostly topics which barely overlap those in physicists' textbooks with similar titles. To top it all, I was buried by the worst advice I ever received, to learn the topic from Dirac's book itself! The well-known obstacle of the difference of language and culture between mathematics and physics is all too real. To a mathematician's eye, some physics textbooks are chock-full of somewhat imprecise statements made about rather ill-defined quantities. It is not rare that these statements are simply untrue if taken at face value. Moreover arguments full of implicit assumptions are presented in the most authoritative manner. Looking at elementary textbooks can be an even harder challenge. These often use simple-minded approaches

[1] Of course, the realization that I could no longer do research at the level that I wanted to played a major part in this decision.

which need not be fully correct, or try to help the reader with analogies which might be superficial and misleading.[2]

Luckily, in 2012 I ran into the preliminary version of Brian Hall's *Quantum Theory for Mathematicians* [40], which made me feel proud again for mathematicians. I learnt from this book many times faster than from any other place. The "magic recipe" for this was so obvious that it sounds trivial when you spell it out: *explain the theory in complete detail, starting from the very basics, and in a language the reader can understand.* Simple enough, but very difficult to put into practice, as it requires a lot of *humility* from the author, and humility is not the most common quality among mathematicians. I don't pretend to be able to emulate Brian's style, but I really tried and his book has had a considerable influence on mine.

After getting some (still very limited) understanding of Quantum Mechanics came the real challenge: Quantum Field Theory. I first looked at books written by mathematicians for mathematicians. These were obviously not designed as first texts or for ease of reading. Moreover, they focus on building rigorous theories. As of today these attempts seem to be of modest interest for most physicists.[3] More promising seemed studying Gerald Folland's heroic attempt [31] to make Quantum Field Theory accessible. His invaluable contribution is to explain what the physicists do rather than limiting the topic to the (rather small) mathematically sound part of the theory. His book is packed with an unbelievable amount of information, and, if you are stuck with minimum luggage on a desert island, this is a fantastic value. Unfortunately, as a consequence of its neutron-star density, I found it also much harder to read than I would have liked, even in sections dealing with well-established or, worse, elementary mathematics. Sadly, this book has no real competitors and cannot be dispensed with, except by readers able to understand physics textbooks.[4] No doubt my difficulties are due to my own shortcomings, but still, it was while reading Weinberg's treatise [87] that I finally understood what induced representations are, and this is not the way it should have been. So, as the days laboring through Folland's book turned into weeks, into many months, I felt the need to explain his material to myself, and to write the text from which I would have liked to learn the first steps in this area.

In the rest of the introduction I describe what I attempt to do and why. I try to provide an easily accessible introduction to some ideas of Quantum Field Theory for a reader well versed in undergraduate mathematics, but not necessarily knowing any physics beyond the high-school level or any graduate mathematics.

I must be clear about a fundamental point. A striking feature of Quantum Field theory is that it is not mathematically complete. This is what makes it so challenging for mathematicians. Numerous bright people have tried for a long time to make this topic rigorous and have yet to fully succeed. I have nothing new to offer in this direction. This book contains

[2] I am certainly not the first mathematician to be appalled by the way physics students get treated, but it seems futile to discuss this matter further.

[3] This being said, the multiple author treatise [9] is a magnificent piece of work. It is overwhelming at first, but most rewarding once you get into it.

[4] See our Reading Suggestions for physics textbooks on page 732.

statements that nobody knows how to mathematically prove. Still, I try to explain some basic facts *using mathematical language*. I acknowledge right away that familiarity with this language and the suffering I underwent to understand the present material are my only credentials for this task.

My main concern has been to spare the reader some of the difficulties from which I have very much suffered reading others' books (while of course I fear introducing new ones), and I will comment on some of these.

First, there is no doubt that the search for generality and for the "proper setting" of a theory is a source of immense progress in mathematics, but it may become a kind of disease among professional mathematicians.[5] They delight in the "second cohomology group of the Lie algebra" but do not explain why the important theorem holds when the Lie group is $\mathbb{R}$. I feel that in an introductory work generality should be indulged in only when it is useful beyond doubt. As a specific example, I see no need to mention cotangent bundles to explain basic mechanics in Euclidean space, but the use of tensor products *does* clarify Fock spaces. Rather than pursuing generality, I find more instructive to explain in complete detail simple facts and situations, especially when they are not immediately found in the literature.[6] I strive not to refer the reader to extensive specialized works, which may have different notation, and may not have been written to be easily accessible. Of course, other very different approaches are also possible [62].

Second, as an extremely ignorant person, I have suffered from the fact that many textbooks assume much known by the reader, such as "standard graduate courses in mathematics". For mathematics (except on very few occasions), I have assumed nothing that I did not learn in my first three years of university in Lyon, 1969–1972: almost all the more advanced mathematics I need are built from the ground up. For physics, I have tried to assume basically nothing known beyond high-school level.

Third, it was hard at first to recognize that different authors are treating in fact the same material, but each in his own way. I have tried different ways to explain the material which confused me the most.

Fourth, when doing physics, and taking steps of dubious mathematical value (which is often required to proceed) I have tried to be very candid about it and to explain clearly what the problems are.[7]

Fifth, and most importantly, I believe that brevity is not such a desirable goal that it should be reached at the reader's expense. The goal of a textbook is to communicate ideas, not to encrypt them in the shortest possible way (however beautifully this is done). Reading an introductory textbook such as this one *should simply not be a research project*. This book is long because:

5 I personally found that the only rather accessible article of the volume [18] is due to Edward Witten! But of course this volume is not designed as an introductory course.

6 This attitude is motivated by the fact that whatever small successes I had in my own mathematical research were always based on a thorough understanding of very simple structures.

7 I do not believe that authoritative statements, use of implicit assumptions, or contrived arguments as to which obvious problems are not real help the readers.

- *It starts with basic material.*
- *The proofs are very detailed.*[8]

As G. Folland appropriately points out [31], readers of notes such as these are likely to be "tourists": they do not look to acquire professional expertise in the topic. He and I disagree in that I don't think most tourists enjoy extreme sports. My overwhelming concern has been to make the book easy to read, by providing everywhere as high a level of detail as I could manage, and by avoiding anything really complicated until the last chapters. Tourists may not enjoy extreme sports, but they might enjoy leisurely sight-seeing. A number of appendices strive to provide an accessible introduction to a number of rewarding topics which complement the main story.[9]

It seemed most useful not to duplicate what is done everywhere else. First, I acknowledge the fundamental importance of historical perspective in understanding a topic, but I make no attempt whatsoever in this direction: there is no point repeating in a clumsy way what is said excellently elsewhere. Besides, one might gain by presenting early certain central and simple ideas even if they came later in the development of the theory.[10] I am in good company here, see Weinberg's treatise [87]. Second, I concentrate on the points I had the most difficulty understanding, and I treat these in considerable detail, trying also to explain how these points are presented in physics textbooks. The subtitle of this book, a first introduction for mathematicians, *does not mean* that it intends to be the fastest possible introduction to the topic, but rather that the reader is not assumed to know anything whatsoever about it. A bare-bones treatment (covering far more material than I do) has already been written [23], and I am aiming here at a more fulfilling level of understanding.[11] I felt it useful to cover some of the fundamental structures in sufficient detail to provide a solid understanding. One of my most glaring shortcomings is the inability to make sense of a mathematical statement unless I have taken it apart to the very last bolt and reconstructed it entirely. I have tried to do just that for these fundamental structures. This often takes several times longer than in standard textbooks. Obviously in these the reader is expected to produce whatever efforts are required to become a professional and master the field, while I try to be much less demanding. On the other hand, some fundamentally important topics, which I found easier to learn, get only succinct coverage here when detailed understanding is not indispensable.

[8] Yes, for *every* topic, I do *whatever it takes* to complete all details.

[9] The choice of these topics is highly personal and reflects both my interests and the history of my learning of this topic. There are points which I felt I just *had to* understand, but which I am certain many readers will feel comfortable to accept without proof.

[10] Furthermore, it *does help* to entirely forget about some early missteps of the theories.

[11] In practice, I have pursued every thread of thought until I felt I reached some real understanding of it, or until I was forced to accept that some kind of miracle is taking place.

Quantum Field Theory is difficult and voluminous. It involves a great many deep ideas. Not all of them are extremely difficult, but the overall quantity is staggering. Being simultaneously detailed and thorough limits the number of topics I might cover, and difficult choices had to be made. Roger Penrose [63, page 657], characterizes Quantum Field Theory as "this magnificent, profound, difficult, sometimes phenomenonally accurate, and yet often tantalizingly inconsistent scheme of things". As the contribution of the present work is only to describe in a self-contained manner some simple facts in mathematical language, it seemed appropriate that it covers mostly topics where mathematical language brings benefits well beyond a simple matter of notation. This is no longer the case when one ventures in the "inconsistent scheme of things" part of the theory. Still, I briefly enter this territory, in order to provide at least some form of answer to the question which serves as title, and some glimpse at physics methods. I deliberately choose to explain these as clearly as possible on the simplest possible "toy models" without any attempt to study realistic models (where the principles are similar but obscured by many accessory complications). I do not attempt what is already done in so many places, such as describing the tremendous successes of Quantum Electrodynamics. You will not learn here any real physics or what is the Standard Model.[12] Rather, I try to prepare the reader to study books which cover these topics, among which I first recommend Folland's book [31].[13] I have also decided to stay entirely away from path integrals (also known as Feynman's integrals). There is no doubt that from the point of view of physics, path integrals are the correct approach, but they are not well defined mathematically, and I do not see what yet another heuristic discussion of them would bring.

On the other hand, the study of renormalization, the method to circumvent the dreadful infinities, receives far more attention than it does in standard textbooks. The procedure itself is rigorous. It involves only rather elementary (but magnificently clever) mathematics. There seems to exist no other detailed source than the original papers or specialized monographs such as [52] or [70].[14] I prove in full detail the possibility of renormalization at all orders of the so-called ϕ_4^4 and ϕ_6^3 theories, cases of somewhat generic difficulty.

No magic wand will make Quantum Field Theory really easy and some effort will be required from the reader. My goal has been to make this effort easier for the reader than it was for me without sacrificing the potential for enjoyment and enrichment this fascinating topic offers.

A number of people helped me while I wrote this book. Shahar Mendelson had a transformative impact, and there are simply no words by which I can express my gratitude for the time and energy he invested in what was for him a pure labor of love. He and Roy Wilsker rescued me many times from the brink of disaster. Gerald Folland spent considerable time and displayed infinite patience in trying to explain some of the most delicate points of his book [31]. Ellen Powell relentlessly asked for clarification of many imprecise statements.

12 Nor will you meet any idea that was not well formed fifty years ago.

13 However demanding, this book is simply brilliant at a number of places, and I see no purpose in repeating the parts which I cannot improve in my own eyes.

14 The major textbooks do not enter into details on this topic, but only illustrate some ideas on examples, "referring the serious reader to the study of the original papers" see e.g. [58, page 157]. This is even the case of books which primarily deal with renormalization, such as [1, 15].

Guilherme Henrique de Paula Reis and Shuta Nakajima read in detail every single proof, however technical, and contributed in a major way to removing numerous obscurities and outright mistakes.[15] Comments of Sourav Chatterjee, Carlos Guesdes, Brian Hall, Amey Joshi, Bernard Lirola, Patrick Lopatto, Hengrui Luo, David Saunders, Krzysztof Smuteck, Phil Sosoe, Zenyuang Zhang and others on the successive versions had a major impact. It was a pure delight to work with my editor Diana Gillooly.[16] I express my gratitude to them all.

Part of the work involved in this project was performed while the author was employed by C.N.R.S and the author is grateful for this support.

15 I am solely responsible for the remaining ones.
16 Unless of course the matter at hand was the length of the book.

Part I

Basics

Our broad goal is to construct models explaining the interactions of elementary particles. The multiple ingredients and fundamental ideas are introduced in succession in the main chapters. We will then discover the bad news: even in simple models, the predictions are non-sensical, being expressed by diverging integrals. Still, a process called renormalization allows finite physical predictions from these infinities. Such is the program of the main chapters. Along the way, a wealth of interesting questions of mathematics will be encountered. Many of these are explored in the appendices, which also contain more advanced material on certain matters of physics.* A minimum reading program does not require you to ever look at an appendix. The goal of such a program would be the central Chapter 13, which explains how the infinities occur and hints at how to remove them. The chapter deals only with spinless particles, so it can be reached without learning any of the material related to spin (which is nonetheless of great mathematical interest), and entirely ignoring Part II of the book. This minimum program can then be extended according to the reader's main interests.

The reader is assumed to be a mathematician. As far as knowledge is concerned, this means little more than the expectation that she will not faint at reading the words "group" and "Hilbert space". As far as attitude is concerned this means that she is expected to be more interested in trying to understand what is going on and why certain assumptions and methods are reasonable rather than in the details of specific computations that can be measured against experiments.

Here seems to be a good place to comment on the level of mathematical rigor of the book. A very significant part (particularly in the appendices) is just standard, fully rigorous mathematics, since after all this is what this author is best qualified to write. Another significant part is "basically rigorous" in the sense that it could (probably) be made rigorous but the complication of doing so would obscure the main purpose, and whenever the choice arises, we choose clarity over formality. The rest of the book takes place outside mathematics, in the physicist's fairyland where one manipulates objects whose very existence is rather

* Even though we have aimed at high standards of detail and clarity throughout, some of the appendices might be a bit more demanding mathematically than the main text.

dubious.† Roughly speaking this is the case of Part III, whereas most of Parts I, II and IV are either fully rigorous or basically rigorous. We have tried to help the reader figure out the level of rigor of various sections, but of course the boundaries of "basically rigorous" are somewhat fuzzy. Let us stress, however, that Part IV, treating renormalization (the art to make sense of certain diverging integrals) is fully rigorous.

There are many possible plans one could make to treat the present subject. One could choose to follow the historical development. One could choose the "logical order": explain the basics of Classical Mechanics, Quantum Mechanics and Special Relativity before attacking Quantum Field Theory. We have chosen a less-traveled route, which might be more appealing to mathematicians.

First, let us ask why one might be interested in this topic. There are obvious reasons, of course. Any learned person should want to know something about a theory, which is certainly one of the greatest intellectual achievements of all times, and which has been so well confirmed by experiment with the discovery of the Higgs boson.‡ But more specifically, why might mathematicians be interested in Quantum Field Theory? For the *very* ambitious, there is the fact that putting the theory on a firm mathematical footing remains a formidable challenge. More modestly, an aesthetically appealing aspect of the theory is that so much of it is determined just by the requirement that it should be consistent with Special Relativity, i.e. invariant under Lorentz transformations. Our presentation emphasizes this aspect, culminating in Chapter 10 with simple arguments (using no more than elementary linear algebra) that allow the discovery of the basic free fields. Prior to this chapter, only the minimum is said about Quantum Mechanics, delaying the description of fundamental practical matters such as perturbation theory until they are actually needed, and until motivation has been gained.

Probably the reader would like to have a more precise road map of what lies ahead, but whatever precise information we might try to give at this stage would only make little sense to a reader not knowing the basic concepts of Quantum Mechanics. Consequently, a technical overview of the first few chapters is delayed until the beginning of Chapter 3.

† And one even "proves theorems" about them!

‡ Or, maybe more accurately, of a boson that looked like a Higgs boson at the time of this writing.

1

Preliminaries

As the present project started as an attempt at rewriting Folland's textbook [31], we begin in the same direction, attempting though to give more details.

1.1 Dimension

The numerical value of many physical quantities depends on the unit one chooses to measure them. My height is 1.8 m, or 180 cm, or 1.90×10^{-16} light-years. The use of light-years here as a unit is weird, but not so much more than the use of centimeters to measure distances at the scale of a nucleon as many textbooks do. (A nucleon has a size of about 10^{-15} m = 10^{-13} cm.) Tradition unfortunately has more weight than rationality in these matters.

The concept of "physical dimension" (which definitely differs from dimension in the mathematical sense) expresses how the numerical value of a physical quantity depends on the units you choose to measure it. A distance has dimension $[l]$ where l stands of course for length. If you increase the unit of length by a factor 100, the corresponding measure *decreases* by a factor 100: 100 cm = 1 m. Then a surface has dimension $[l^2]$: $(100)^2\,\mathrm{cm}^2 = 1\,\mathrm{m}^2$. A volume has dimension $[l^3]$: $1\,\mathrm{km}^3 = (10^3)^3\,\mathrm{m}^3 = 10^9\,\mathrm{m}^3$. The unit of time can be chosen independently from the unit of length. Time has dimension $[t]$, so speed, which is a distance divided by a time, has dimension $[lt^{-1}]$. Thus 1 m/s = 3,600 m/h = 3.6 km/h. Acceleration, which is a change of speed divided by a time, has dimension $[lt^{-2}]$. It is of course a convention to choose time and length as fundamental quantities. One could make other choices, such as choosing time and speed as fundamental quantities. This is indeed basically what is actually done. Since 1983, in the international system the speed of light is *defined* to be *exactly*

$$c = 299{,}792{,}458 \text{ m/s} \tag{1.1}$$

and this serves as a definition of the meter given the unit of time.[1]

A formula in physics *must give a correct result independently of the system of units used.* This is a strong constraint. This is why it often makes sense to multiply or divide quantities of different dimensions, but it *never* makes sense to add them. As we learn in kindergarten,

[1] The reason for this definition is that the speed of light is a fundamental constant of Nature, and that it makes little sense to have its value change as the accuracy of measurements improves.

you do not add pears with bananas. Furthermore, when a quantity occurs in a formula as the argument of, say, an exponential, it must be *dimensionless,* i.e. its value must be independent of the unit system. To understand a formula in physics it always helps to check that it makes sense with respect to dimension, a task we will perform many times.

The unit of mass can be chosen independently of the units of length and time. Momentum,[2] the product of a mass and a speed, then has dimension $[lt^{-1}m]$, and angular momentum, the product of a momentum and a distance, has dimension $[l^2t^{-1}m]$. Energy has the dimension of a mass times the square of a speed, that is $[l^2t^{-2}m]$. Less known is the *action* which occurs in Lagrangian Mechanics as the integral over a time interval of a quantity with the dimension of energy, and thus has the same dimension $[l^2t^{-1}m]$ as the angular momentum.

A fundamental constant of Nature is Planck's constant h, which represents the basic quantum of action (and in particular has the dimension of an action). In physical equations it often occurs in combinations with a factor of $1/2\pi$, so one defines the reduced Planck constant

$$\hbar = \frac{h}{2\pi}, \tag{1.2}$$

whose value is about[3]

$$\hbar = 1.0546 \times 10^{-34}\ \mathrm{J \cdot s}. \tag{1.3}$$

This is small, as becomes more apparent if this value is expressed in units more related to the microscopic world[4] $\hbar \simeq 6.6 \times 10^{-16}\ \mathrm{eV \cdot s}$. It is important to note that energy times time, momentum times length and angular momentum all have the same dimension as $\hbar$ so that their quotients by $\hbar$ are dimensionless. These quotients will occur in countless formulas.

Exercise 1.1.1 The Planck-Einstein relation gives the energy E of a photon of frequency ν as $E = h\nu$. Check that this formula makes sense with respect to dimension.

Exercise 1.1.2 The de Broglie momentum–wavelength relation states that to a particle of momentum p is associated a wavelength $\lambda = h/p$. Check that this formula makes sense with respect to dimension.

1.2 Notation

Since to enjoy this topic one has to read the work of physicists, it is best to adopt their notation from the beginning. Complex numbers play a central role, and the conjugate of a complex number a is denoted by a^*. Even some of the best authors let the reader decide

[2] Please do not worry if you do not have a real feeling for the concepts of momentum, energy, etc. (despite the fact that you should experience them every day). This is not going to be an obstacle.

[3] An action has the dimension of an energy times a time. In the International System of Units, the unit of energy is the joule J and the unit of time is the second s, so action is measured in $\mathrm{J \cdot s}$. From May 2019, the value of h is *defined* to be exactly $6.62607015 \times 10^{-34}\mathrm{J \cdot s}$.

[4] A joule is a huge energy at the microscopic scale. A more appropriate unit of energy at this scale is the electron-volt eV, the energy acquired by an electron going through a difference of potential of one volt.

whether i denotes a complex number with $i^2 = -1$ or an integer index. Since this requires no extra work, the complex number will be denoted by i, so that $\mathrm{i}^* = -\mathrm{i}$.

When working with complex Hilbert spaces we adopt the convention that the inner product $(\cdot,\cdot)$ is anti-linear in the *first* variable (while often mathematicians use the convention that it is anti-linear in the second variable). One says that the inner product is *sesqui-linear*. That is, as another example of our notation for complex conjugation, we write

$$(ax,y) = a^*(x,y)$$

for any vectors x, y and any complex number a. Moreover

$$(y,x) = (x,y)^*. \tag{1.4}$$

The norm $\|x\|$ of a vector x is given by $\|x\|^2 = (x,x)$, and we recall the Cauchy-Schwarz inequality

$$|(x,y)|^2 \le \|x\|^2\|y\|^2,$$

where $|a|$ denotes the modulus of the complex number a. A basic example of a complex Hilbert space[5] is the space $\mathbb{C}^n$, where the inner product is defined by $(x,y) = \sum_{i\le n} x_i^* y_i$, with the obvious notation $x = (x_i)_{i\le n}$. Another very important example is the space $L^2(\mathbb{R})$ of complex-valued functions f on the real line for which $\int_{\mathbb{R}} |f|^2 \mathrm{d}x = \int_{\mathbb{R}} |f(x)|^2 \mathrm{d}x < \infty$, where $|f(x)|$ denotes the modulus of $f(x)$. The inner product is then given by $(f,g) = \int_{\mathbb{R}} f^* g \mathrm{d}x$. A physicist would actually write

$$(f,g) = \int_{-\infty}^{\infty} \mathrm{d}^1 x f(x)^* g(x), \tag{1.5}$$

where the superscript 1 refers to the fact that one integrates for a one-dimensional measure. The reason for which the $\mathrm{d}^1 x$ is put before the function to integrate is that this makes the formula easier to parse when there are multiple integrals. We will use this convention systematically. We will not however mention the dimension in which we integrate when this dimension is equal to one.

An *operator* A on a finite-dimensional Hilbert space $\mathcal{H}$ is simply a linear map $\mathcal{H} \to \mathcal{H}$. Its *adjoint* $A^\dagger$ is defined by

$$(A^\dagger(x), y) = (x, A(y)), \tag{1.6}$$

for all vectors x, y. (Mathematicians would use the notation A^* rather than $A^\dagger$.)

Exercise 1.2.1 (a) If A is an operator and α a number, prove the formula

$$(\alpha A)^\dagger = \alpha^* A^\dagger.$$

(b) If A and B are operators, prove that $(AB)^\dagger = B^\dagger A^\dagger$.

An operator (still on a finite-dimensional space) is called *Hermitian* if $A = A^\dagger$. A Hermitian operator A has the crucial property that if a subspace F is such that $A(F) \subset F$ then the orthogonal complement $F^\perp$ of F is also such that $A(F^\perp) \subset F^\perp$.

[5] In this work we consider only finite-dimensional or, more generally, *separable* Hilbert spaces.

Exercise 1.2.2 Deduce this from the fact that $(A(x),y) = (x,A(y))$ for all x,y.

As a consequence, a Hermitian operator has a simple structure: there exists an orthonormal basis of eigenvectors.[6]

Exercise 1.2.3 Give a complete proof of this fact by induction over the dimension of the space.

Moreover the eigenvalues are real since $(x,A(x))$ is real for all x. Indeed,

$$(x,A(x)) = (A(x),x)^* = (x,A(x))^*,$$

where the second equality uses that A is Hermitian. If $A(x) = \lambda x$ then $(x,A(x)) = \lambda(x,x)$, so that $\lambda(x,x) = \lambda^*(x,x)$ and $\lambda = \lambda^*$. It is this property of having a basis of eigenvectors, with real eigenvalues, which makes the class of Hermitian operators so important.

A few times we will need the notion of *anti-linear* operator. Such a map T does satisfy $T(x+y) = T(x) + T(y)$ but for a scalar a we have $T(ax) = a^*T(x)$.

We should also mention that in physics, vectors and inner products are denoted differently, using Dirac's ubiquitous notation, which we will explain later. In most situations however we prefer to use standard mathematical notation, and it is unlikely that anybody reading this will mind.

1.3 Distributions

Laurent Schwartz invented the theory of distributions to give a rigorous meaning to many formal calculations of physicists. The theory of distributions is a fully rigorous part of mathematical analysis. In the main text however we will use only the very basics of this theory at a purely informal level. In Appendix L the reader may find an introduction to rigorous methods.

We will consider distributions on $\mathbb{R}^n$ but here we assume $n = 1$. The central object is the space $\mathcal{S} = \mathcal{S}(\mathbb{R})$ of *rapidly decreasing functions*, called also *test functions* or *Schwartz functions*. A complex-valued[7] function ζ on $\mathbb{R}$ is a test function if it has derivatives of all orders and if for any integers $k,n \geq 0$ one has[8]

$$\sup_x |x^n \zeta^{(k)}(x)| < \infty. \tag{1.7}$$

A distribution is simply a linear functional (which also satisfies certain regularity conditions which will not concern us before we reach Appendix L, as they will be satisfied in all the examples we will consider). That is, a distribution Φ is a complex linear map from $\mathcal{S}$ to $\mathbb{C}$, and for each test function ζ the number $\Phi(\zeta)$ makes sense. Such a distribution should actually be called a *tempered* distribution, but we will simply say "distribution" since we will

[6] Later we will meet a far-reaching extension of this fact, the spectral theorem for self-adjoint operators.

[7] A test function is typically complex-valued. At times, to avoid complications created by anti-linear operators, we also consider real-valued test functions. The space of such functions is denoted by $\mathcal{S}_{\mathbb{R}}$. Please try to remember that the subscript $\mathbb{R}$ on a space of test functions means that we consider only real-valued functions.

[8] The property is often described in words as follows: As $|x| \to \infty$, the function and each of its derivatives decrease faster than $|x|^{-n}$ for each n. This is what motivates the name "rapidly decreasing function".

hardly consider any other type of distribution. Tempered distributions are also known under the name of *generalized functions.* This name has the advantage of explaining the point of the theory of distributions: it generalizes the theory of functions. Indeed, a sufficiently well-behaved function[9] f defines a distribution (= generalized function) Φ_f by the formula

$$\Phi_f(\zeta) = \int \mathrm{d}x\, \zeta(x) f(x). \tag{1.8}$$

Throughout the book *we maintain the convention that when the domain of integration is not mentioned, this domain is the whole space.* Thus (1.8) means $\Phi_f(\zeta) = \int_{\mathbb{R}} \mathrm{d}x f(x)\zeta(x) = \int_{-\infty}^{\infty} \mathrm{d}x f(x)\zeta(x)$.

As we are going to see, distributions can be strange animals. On the other hand, distributions given by a formula such as (1.8) are much better behaved. The short way to describe the situation where the distribution Φ is of the type Φ_f as in (1.8) is to simply say that Φ *is a function.*

In general a distribution is certainly not given by a formula of the type (1.8). However, when dealing with distributions we will maintain the *notational fiction* that they are functions, i.e. for a test function ζ we will write

$$\int \mathrm{d}x\, \zeta(x)\Phi(x) := \Phi(\zeta). \tag{1.9}$$

The use of the symbol := here is to stress that the right-hand side is a definition of the left-hand side, so that you may be reassured that your memory did not fail and that there is no point in looking back for the definition of the left-hand side. Equation (1.9) indeed defines the left-hand side, since the symbol $\Phi(x)$ is a *notation*, and a priori really *makes no sense whatsoever by itself.* It is only the integral $\int \mathrm{d}x\, \zeta(x)\Phi(x)$ which makes sense for any test function, and the value of this integral is (by definition) the quantity $\Phi(\zeta)$, as expressed in (1.9).[10] The central objects of this work, quantum fields, have precisely the previous property. The value of a quantum field cannot be specified at any given point. This value makes sense only "when it is integrated against a test function", or *smeared* in physics-type language.

Even for distributions, it is however sometimes possible to give a meaning to the quantity $\Phi(x)$ for certain values of x. Given an open interval I of the real line, we say that a distribution *is a function on I* if there exists a well-behaved function f on I such that $\Phi(\zeta) = \int \mathrm{d}x\, \zeta(x) f(x)$ whenever the test function ζ has compact support contained in I.[11] It is then reasonable to define $\Phi(x) = f(x)$ for $x \in I$. However, unless Φ is a function, it is not possible to assign a meaning to the symbol $\Phi(x)$ for each value of x.

9 The meaning of the expression "well-behaved" varies depending on the context. In the present case, f needs to be locally integrable and "not grow too fast at infinity".

10 The notational fiction (1.9) is however quite useful when one likes to be informal, as we simply do not distinguish distributions from functions until this leads us into trouble.

11 When $I = \mathbb{R}$ this looks different from the definition (1.8) because then it is not required that ζ has compact support. However, using the regularity properties that are part of the definition of a distribution, and on which we do not dwell here one may show that these definitions coincide, see Appendix L.

Distributions can be added, or multiplied by a scalar, but in general *cannot be multiplied.*[12] The great appeal of distributions is that they can always be differentiated. The derivative of a distribution Φ is the distribution *defined* by the formula

$$\Phi'(\zeta) := -\Phi(\zeta') \tag{1.10}$$

for every test function ζ. The reason behind this definition is best understood by integrating by parts when Φ is given by the formula (1.8) for a well-behaved function f. Then $\Phi'(\zeta) = \int \mathrm{d}x f'(x)\zeta(x)$.

Exercise 1.3.1 Convince yourself from the preceding definition that "if a distribution is actually a nice function, its derivatives as a function and as a distribution coincide". Hint: Recalling (1.8), prove that $(\Phi_f)' = \Phi_{f'}$.

Pretending that Φ' is also a function, we write $\Phi'(\zeta)$ as $\int \mathrm{d}x\zeta(x)\Phi'(x)$, and equally shamelessly we write (1.10) as

$$\int \mathrm{d}x\, \zeta(x)\Phi'(x) := -\int \mathrm{d}x\, \zeta'(x)\Phi(x). \tag{1.11}$$

In several dimensions, the class of test functions is defined as the class of infinitely differentiable functions such that the product of any partial derivative (of any order) and any polynomial in the variables is bounded. The reader may refer to Section L.1 for more about test functions.

1.4 The Delta Function

Besides reviewing the delta function, this section introduces the idea of a smooth cutoff, how to get rid of the troublesome part of an integral.

Mathematically, the delta "function" δ is simply the distribution given by

$$\delta(\zeta) = \zeta(0) \tag{1.12}$$

for any test function $\zeta \in \mathcal{S}$. Pretending that the delta "function" is actually a true function we will shamelessly write (1.12) as

$$\zeta(0) = \int \mathrm{d}x\zeta(x)\delta(x). \tag{1.13}$$

The name "delta function" is historical. Physicists have been using this object long before distributions were invented.

Exercise 1.4.1 (a) Convince yourself that it makes perfect sense to say that $\delta(x) = 0$ if $x \neq 0$.
(b) Make sure that you understand that despite the terminology, the delta function δ is not a function in the mathematical sense and that the quantity $\delta(0)$ *makes no sense*.

12 One of the reasons for the dreaded infinities which will occur later is that we will have no other choice to proceed than pretending we can multiply certain distributions.

(c) Convince yourself from (1.13) that, in the words of physicists, "the delta function δ is the function of x which is equal to zero for $x \neq 0$ and to infinity for $x = 0$, but in such a way that its integral is 1".
(d) Convince yourself that the *derivative* of the delta function δ, i.e. the distribution δ' given by $\delta'(\zeta) = -\zeta'(0)$ "does not look at all like a function".

For $a \neq 0$ let us *define* $\delta(ax)$ by

$$\int \mathrm{d}x\zeta(x)\delta(ax) := \frac{1}{|a|}\int \mathrm{d}x\zeta(x/a)\delta(x) = \frac{1}{|a|}\zeta(0), \tag{1.14}$$

so that

$$\delta(ax) = \frac{1}{|a|}\delta(x), \tag{1.15}$$

and in particular $\delta(-x) = \delta(x)$.[13]

Proper mathematical terminology requires one to say "the function f" but sooner or later one always says "the function $f(x)$" to carry at the same time the information that the variable is called x. In the same manner we will use expressions such as "the distribution $\Phi(x)$" which *should not* be interpreted as meaning that the quantity $\Phi(x)$ makes sense for a given x.

We will often write the quantity $\delta(x-y)$. It can be seen as a "function" of x depending on the parameter y. This "function" makes sense only when integrated in x against a test function ζ, and one has

$$\zeta(y) = \int \mathrm{d}x\zeta(x)\delta(x-y). \tag{1.16}$$

It can also be seen as a "function" of y depending on the parameter x, and one has

$$\zeta(x) = \int \mathrm{d}y\zeta(y)\delta(x-y). \tag{1.17}$$

Exercise 1.4.2 The quantity $\delta(x-y)$ can also be seen as a distribution Φ in the variables x, y. For a test function $\xi(x,y)$ one has

$$\iint \mathrm{d}x\mathrm{d}y\xi(x,y)\delta(x-y) := \Phi(\xi) := \int \mathrm{d}x\xi(x,x).$$

Convince yourself that this is consistent with (1.16) and (1.17).

Note also that $\delta(x-y) = \delta(y-x)$. We will shift freely between the previous meanings of the quantity $\delta(x-y)$. More generally we will stay very informal. Everything we say at this stage could be made rigorous, but this is not our objective.[14]

[13] Please be prepared: soon we will start manipulating delta functions as if they were functions, for example taking for granted the first equality in (1.14).
[14] It makes no sense to carefully climb molehills when the Himalayas are waiting for us.

Exercise 1.4.3 For a test function ξ convince yourself of the formula (various versions of which we will use many times)

$$\int \mathrm{d}z \delta(x-z)\delta(z-y)\xi(z) = \delta(x-y)\xi(x). \tag{1.18}$$

We will make massive use of the formula[15]

$$\delta(x) = \frac{1}{2\pi}\int \mathrm{d}y \exp(\mathrm{i}xy), \tag{1.19}$$

and we first need to make sense of it. In studying physics, one must keep in mind at all times that the goal is to make predictions about the behavior of the physical world. This is difficult enough. To study the physical world, one makes models of it. One should as far as possible concentrate on problems that arise from the physical world, and stay away from the problems that arise not from the physical world, but from the models we made of it. We may never know for sure whether the physical world is finite or not, but certainly events located very far from our experiments are unlikely to affect very much their outcome, so their inclusion in our model is an idealization, and in (1.19) points very far from the origin should be discounted, for example by a factor $\exp(-ay^2)$ for a very small $a > 0$. This is an example of what is called a "smooth cutoff". Thus, rather than (1.19) we mean

$$\delta(x) = \lim_{a\to 0}\frac{1}{2\pi}\int \mathrm{d}y \exp(\mathrm{i}xy - ay^2). \tag{1.20}$$

Here the limit is "in the sense of distributions". *By definition of convergence in the sense of distributions*, this means that for every test function ζ,

$$\zeta(0) = \int \mathrm{d}x \zeta(x)\delta(x) = \lim_{a\to 0}\int \mathrm{d}x \zeta(x)\psi_a(x), \tag{1.21}$$

where

$$\psi_a(x) := \frac{1}{2\pi}\int \mathrm{d}y \exp(\mathrm{i}xy - ay^2) = \frac{1}{2\sqrt{a\pi}}\exp(-x^2/4a),$$

the second equality resulting from the computation of the Gaussian integral, see Lemma C.3.3. Making the change of variables $x = \sqrt{a}y$, (1.21) becomes

$$\zeta(0) = \lim_{a\to 0}\int \mathrm{d}y \zeta(y\sqrt{a})\psi_1(y),$$

which holds by dominated convergence since ψ_1 has integral 1 and ζ is uniformly bounded.[16]

A regularization procedure as in (1.20) can make sense only if it is robust enough. You might have chosen an origin different from mine, but this does not matter since for all b it follows from (1.20) that we actually have

$$\delta(x) = \lim_{a\to 0}\frac{1}{2\pi}\int \mathrm{d}y \exp(\mathrm{i}xy - a(y-b)^2).$$

15 $\exp(x)$ is just another notation for e^x.

16 Certainly the reader has observed the fundamental idea there: a family of functions of integral 1, which peaks more and more narrowly around zero converges to the delta function in the sense of distributions, consistently with Exercise 1.4.1 (a).

Exercise 1.4.4 It would seem at first that the substitution $z = y - b$ in the right-hand side brings a factor $\exp(-\mathrm{i}xb)$. Why is there no such factor? Hint: Where is the delta function different from zero?

We will also need the obvious three-dimensional generalization of (1.19). In this case, not only might you and I have chosen different origins, we might also move at relativistic speed with respect to each other (and consequently we may not agree on the way we measure distances). There is however little point in investigating which specific regularization schemes would take care of this: as we will see later in this section, far more general regularization schemes work.

1.5 The Fourier Transform

Besides reviewing some basic facts about Fourier transforms, this section provides the first example of certain calculations common in physics.

The Fourier transform will play a fundamental role.[17] Let us temporarily denote by $\mathcal{F}_m$ the Fourier transform, that is

$$\mathcal{F}_m(f)(x) = \frac{1}{\sqrt{2\pi}} \int \mathrm{d}y \exp(-\mathrm{i}xy) f(y), \tag{1.22}$$

where the subscript m reminds you that this is the way mathematicians like to define it (whereas our choice of normalization will be different). The right-hand side is defined for f integrable, and in particular for a Schwartz function $f \in \mathcal{S}$. Using integration by parts in the first equality, and differentiation under the integral sign in the second one, we obtain the fundamental facts that for any test function f,

$$\mathcal{F}_m(f')(x) = \mathrm{i}x\mathcal{F}_m(f)(x)\ ;\ \mathcal{F}_m(xf) = \mathrm{i}\mathcal{F}_m(f)', \tag{1.23}$$

where we abuse notation by denoting by xf the function $x \mapsto xf(x)$. An essential fact is that the Fourier transform of a test function is a test function. The details of the proof are a bit tedious, and are given in Section L.1.[18]

The Plancherel formula is the equality

$$(\mathcal{F}_m(f), \mathcal{F}_m(g)) = (f, g), \tag{1.24}$$

for $f, g \in \mathcal{S}$, where $(f,g) = \int \mathrm{d}x f(x)^* g(x)$. It is very instructive to "prove" this formula the way a physicist would, since this is a very simplified occurrence of the type of computations that are ubiquitous in Quantum Field Theory:

$$\begin{aligned}(\mathcal{F}_m(f), \mathcal{F}_m(g)) &= \frac{1}{2\pi} \int \mathrm{d}x \Big(\int \mathrm{d}y_1 \exp(-\mathrm{i}xy_1) f(y_1) \Big)^* \int \mathrm{d}y_2 \exp(-\mathrm{i}xy_2) g(y_2) \\ &= \frac{1}{2\pi} \iint \mathrm{d}y_1 \mathrm{d}y_2 f(y_1)^* g(y_2) \int \mathrm{d}x \exp(\mathrm{i}x(y_1 - y_2))\end{aligned}$$

[17] As will be explained later, it provides a natural correspondence between the "position representation" and the "momentum representation".

[18] The essential point is that iteration of the previous relations and Plancherel's formula show that if f is a test function then for each $n, k \in \mathbb{N}$ the function $x^n \mathcal{F}_m(f)^{(k)}(x)$ belongs to L^2.

$$
\begin{aligned}
&= \iint \mathrm{d}y_1 \mathrm{d}y_2 f(y_1)^* g(y_2)\delta(y_1 - y_2) \\
&= \int \mathrm{d}y_2 f(y_2)^* g(y_2) \\
&= (f,g), \qquad (1.25)
\end{aligned}
$$

where we have used (1.19) in the third line and have integrated first in y_1 in the fourth line. Although this type of manipulation might look scary at first to a mathematician, it suffices in fact to insert a factor $\exp(-ax^2)$ in the first line and let $a \to 0$ to make the argument rigorous using (1.20).

As a consequence of Plancherel's formula, for $f \in \mathcal{S}$, the Fourier transform of f has the same L^2 norm as f, i.e. $\mathcal{F}_m$ is an isometry when $\mathcal{S}$ is provided with the L^2 norm. Since $\mathcal{S}$ is dense in L^2 for this norm, an elementary result asserts that we may extend by continuity the Fourier transform as a linear map from L^2 to itself and this extension still satisfies (1.24). Please observe that it is by no means obvious that the right-hand side of the formula (1.22) is well-defined when $f \in L^2$.

One of the miracles of the Fourier transform is that it can be inverted by a formula very similar to (1.22):

$$
\mathcal{F}_m^{-1}(g)(y) = \frac{1}{\sqrt{2\pi}} \int \mathrm{d}x \exp(\mathrm{i}xy) g(x). \qquad (1.26)
$$

To justify the notation $\mathcal{F}_m^{-1}$ we observe that, using (1.26) for $g = \mathcal{F}_m(f)$, and using again (1.19),

$$
\begin{aligned}
\mathcal{F}_m^{-1}\left(\mathcal{F}_m(f)\right)(y) &= \frac{1}{2\pi} \int \mathrm{d}x \exp(\mathrm{i}xy) \int \mathrm{d}z \exp(-\mathrm{i}xz) f(z) \\
&= \frac{1}{2\pi} \int \mathrm{d}z f(z) \int \mathrm{d}x \exp(\mathrm{i}x(y-z)) \\
&= \int \mathrm{d}z f(z)\delta(y-z) \\
&= f(y). \qquad (1.27)
\end{aligned}
$$

This again can be made rigorous just as (1.25). Let us now look back at (1.21), which we write, using Fubini's theorem,

$$
\zeta(0) = \lim_{a \to 0} \frac{1}{2\pi} \int \mathrm{d}y \exp(-ay^2) \left(\int \mathrm{d}x \exp(\mathrm{i}xy)\zeta(x) \right),
$$

and by dominated convergence we obtain

$$
\zeta(0) = \frac{1}{\sqrt{2\pi}} \int \mathrm{d}y\, \mathcal{F}_m^{-1}(\zeta)(y). \qquad (1.28)
$$

Incidentally, it is now quite obvious that the regularization scheme in (1.20) is very robust. To see this, let us investigate for which regularizing families of functions ψ_a, we have in the sense of distributions

$$
\delta(x) = \lim_{a \to 0} \frac{1}{2\pi} \int \mathrm{d}y \exp(\mathrm{i}xy)\psi_a(y).
$$

This means that for any test function $\zeta \in \mathcal{S}$ it holds that

$$\zeta(0) = \lim_{a\to 0} \frac{1}{2\pi} \iint \mathrm{d}x\mathrm{d}y\, \zeta(x)\exp(\mathrm{i}xy)\psi_a(y). \tag{1.29}$$

The right-hand side is

$$\lim_{a\to 0} \frac{1}{\sqrt{2\pi}} \int \mathrm{d}y\, \mathcal{F}_m^{-1}(\zeta)(y)\,\psi_a(y) = \lim_{a\to 0} \int \mathrm{d}y\, \theta(y)\psi_a(y),$$

where $\theta := (2\pi)^{-1/2}\mathcal{F}_m^{-1}(\zeta)$. Since θ is a test function, and since its integral is $\zeta(0)$ by (1.28), (1.29) holds true whenever ψ_a converges to the constant function 1 in the sense of distributions, that is

$$\lim_{a\to 0} \int \mathrm{d}y\, \eta(y)\psi_a(y) = \int \mathrm{d}y\, \eta(y)$$

for each test function η. This is the case for example (using dominated convergence) when $\psi_a(y) = \psi(ay)$ where $\psi \in \mathcal{S}$ satisfies $\psi(0) = 1$, and in particular when $\psi(x) = \exp(-x^2)$ as in (1.20).

Let us also mention that it is possible to define a notion of "Fourier transform of a distribution", and once this is done (1.28) is equivalent to the statement "the delta function is the Fourier transform of the constant function $1/\sqrt{2\pi}$", which is a more elaborate way to describe the way we made sense of (1.19).

Mathematicians love the symmetry between (1.22) and (1.26), but in physics it is better, thinking that x has the dimension of a length and p of a momentum to define the Fourier transform of a function f as

$$\hat{f}(p) = \int \mathrm{d}x\, \exp(-\mathrm{i}xp/\hbar)f(x) = \sqrt{2\pi}\mathcal{F}_m(f)(p/\hbar) \tag{1.30}$$

and the inverse Fourier transform as

$$\check{\xi}(x) = \int \frac{\mathrm{d}p}{2\pi\hbar} \exp(\mathrm{i}xp/\hbar)\xi(p). \tag{1.31}$$

This makes sense because the quantity $xp/\hbar$ is dimensionless. The Plancherel formula then becomes

$$\int \mathrm{d}x\, |f(x)|^2 = \int \frac{\mathrm{d}p}{2\pi\hbar} |\hat{f}(p)|^2. \tag{1.32}$$

Exercise 1.5.1 Make sure you understand (1.31) by writing out all details. The factor $2\pi\hbar$ will occur constantly.[19]

There are obvious multidimensional versions of these formulas. There is one factor $2\pi\hbar$ per dimension in the analog of (1.31). When integrating in p we will always include these factors.

Exercise 1.5.2 Write the multidimensional versions of these formulas.

[19] According to (1.2), we have $2\pi\hbar = h$ but we write all formulas in term of $\hbar$.

The formula (1.23) now becomes

$$\widehat{-i\hbar\frac{\mathrm{d}f}{\mathrm{d}x}}(p) = -i\hbar\widehat{\frac{\mathrm{d}f}{\mathrm{d}x}}(p) = p\hat{f}(p). \tag{1.33}$$

The operator $-i\hbar\mathrm{d}/\mathrm{d}x$ is fundamental in Quantum Mechanics, and (1.33) means that it is much simpler to express on $\hat{f}$ than on f: applying this operator to the function f simply amounts to multiplying the Fourier transform of f by p.

Key facts to remember:

- One should always check that an equation in physics makes sense from the point of view of physical dimension.
- The complex conjugate of a complex number a is denoted by a^* and the adjoint of an operator A by $A^\dagger$.
- Distributions generalize functions but their value is not defined at every point and they make sense only when integrated against a test function. The "delta function" is not a function!
- The ubiquitous Fourier transform is not exactly defined as it would be in mathematics.

2

Basics of Non-relativistic Quantum Mechanics

Quantum Field Theory may be described as an attempt to bring together non-relativistic Quantum Mechanics and Special Relativity. As a first task one has to understand some basics of Quantum Mechanics. In this chapter we review the absolute minimum required. We make no attempt to state axioms in a formal and complete way, since the goal, as so often in this book,[1] is to explain things clearly rather than formally. If you have never heard of Quantum Mechanics, it is unlikely that these basic elements will suffice. Despite the fact that if you pick a textbook at random from your library shelf, it is likely to vampirize your time, there *do exist* good, well-written textbooks, and there is no point in duplicating too much of them here. On the mathematical side, the reader-friendly book by Hall [40] provides a good introduction. On the physics side, Cohen-Tannoudji, Diu, and Laloë's treatise [11] is a masterpiece of pedagogy. These textbooks contain far more material than will actually be needed, but they are excellent pieces to read more if you need to do so.

Different approaches may be tried to provide a sketch of Quantum Mechanics. One could cover right away all the needed topics systematically. The reader may then face neural overload (as I experience myself when being confronted with such texts). Another approach, which we try here, is to introduce every element only when and as needed, and to develop even simple and basic ideas only when required. That may ease the mental overload, but it may then take longer for the reader to form a global mental picture of what is going on. If this is your experience, try Hall's textbook [40] for help.

Another point must be stressed. We do our best to present the basics of Quantum Mechanics in a non-pedantic manner, but still in mathematical language (at the expense, of course, of not always giving complete arguments). It is far from obvious that the principles we will explain have any relevance to the physical world. Yet it is their real justification, and one can only marvel at the depth of insight of the great minds who discovered them. To explain this relevance, one must lay out examples of concrete situations, where the predictions of the model can be compared with lab experiments. Needless to say, this author is not the best qualified to do this, but many such examples are covered in a very progressive manner in Cohen-Tannoudji, Diu, and Laloë's book [11].

More basic facts about Quantum Mechanics are covered at the beginning of Chapter 11 and require only the material of the present chapter as a prerequisite.

[1] The appendices are more often fully mathematical.

2.1 Basic Setting

We must first stress that one does not "prove" the basic concepts of Quantum Mechanics, or of any physical theory. One builds models, and *the ultimate test of the validity of such models is whether they predict correctly the results of experiments.* Still, one strives for mathematical consistency and elegance.[2]

The purpose of Mechanics is to describe the state of mechanical systems and to determine their time-evolution. Consider, for example, one of the simplest mechanical systems: a massive dimensionless point. Its state at a given time is described by its position and velocity. That this, indeed, is the correct way to specify even such a simple system is by itself a deep fact: it is the position and the velocity (and not, say, the acceleration) at a given time that determine the future motion of the point. This fact is delicate enough that apparently it is not understood by the general public, which seems to still believe that an astronaut stepping out of the International Space Station (ISS) will start falling toward Earth.[3]

As we pointed out, it is not easy to relate the principles of Quantum Mechanics to actual physical experiments. On the positive side, this means that there is no real loss in starting to learn them even with little knowledge of Classical Mechanics. A very brief introduction to Classical Mechanics will be given in Sections 6.4 and 6.5.

Principle 1[4] *The state of a physical system is described by a unit vector in a complex*[5] *Hilbert space* $\mathcal{H}$.

This Hilbert space is called the *state space*, and the unit vector is called the *state vector*. The state space will always be either finite-dimensional or separable (i.e. admitting a countable orthonormal basis). As in the case of the lowly classical massive point, the correct description of the state of a system encompasses a huge amount of wisdom. The fact that it is done by a vector in Hilbert space allows some of the most surprising features of Quantum Mechanics. *It makes physical sense to consider linear combinations of different states.*[6] The principles of Quantum Mechanics, and this one in particular, really do not appeal to our everyday intuition. There seems to be no remedy to this situation.

This is a good place to start explaining Dirac's notation, which is nearly ubiquitous in physics textbooks. Dirac's notation (also called bra–ket notation) is not really appropriate for writing mathematical arguments, so our use of it will be somewhat limited. Nonetheless, it must be learned in order to read physics literature. In this notation, vectors (i.e. elements of $\mathcal{H}$) are denoted by $|\alpha\rangle, |\beta\rangle$, and so on. These are called "ket vectors". Let us stress that

[2] This is a critical philosophical difference between mathematicians and physicists: mathematicians have logical systems that they can use to have confidence in systems, and, conversely, experience has made them cautious about vagueness and lack of logical rigor. Physicists, who can never "prove" that something is physically true, look to experimental results (and, occasionally, beauty) to provide confidence in their assumptions or to tell them to go back to the drawing board.

[3] Of course, the ISS too is falling toward Earth, since otherwise, its motion would be in a straight line. What is meant here is that the astronaut falls at a *faster rate*.

[4] Particularly important principles are outlined for the reader's convenience but there is no claim that these form a complete or independent set of axioms. I may use at times the word "axiom" rather than principle.

[5] All Hilbert spaces considered in this work are complex Hilbert spaces. According to the quote of Richard Feynman that opens this book, he apparently considered complex numbers as a part of physics rather than a part of mathematics. That complex numbers play such a major part in our models of the world is a rather awesome fact.

[6] This allows, in particular, for interferences.

in this notation, α is *not* an element of $\mathcal{H}$; it is a *label*. It is the whole symbol $|\alpha\rangle$ that is an element of $\mathcal{H}$. For example, we may very well denote by $\{|i\rangle; 1 \le i \le n\}$ a basis of $\mathcal{H}$, where n is the dimension of $\mathcal{H}$. The notation $|i\rangle$ for what a mathematician would write, say, e_i, looks strange at first until one realizes that the real information is carried by the index i, and that the letter e is just a support for this index and is no longer needed in the Dirac notation, where the support for the index is now the symbol $|\cdot\rangle$.

On the other hand, any element x of $\mathcal{H}$ induces a linear functional $y \mapsto (x, y)$ on $\mathcal{H}$. When $x = |\alpha\rangle$, this linear functional is denoted by $\langle\alpha|$ and is called a "bra vector". So $|\alpha\rangle$ is an element of $\mathcal{H}$, while $\langle\alpha|$ is a linear functional on $\mathcal{H}$. The value of the functional $\langle\alpha|$ on the vector $|\beta\rangle$ is denoted by $\langle\alpha|\beta\rangle$ (rather than $\langle\alpha||\beta\rangle$). This is simply the inner product of the vector $|\alpha\rangle$ and the vector $|\beta\rangle$. Thus, corresponding to (1.4) we have

$$\langle\alpha|\beta\rangle = \langle\beta|\alpha\rangle^*. \tag{2.1}$$

Let us observe in particular that $\langle\alpha|\alpha\rangle$ is the square of the norm of $|\alpha\rangle$.

An essential ingredient of any physical theory is the concept of an *observable*, that is, a quantity that can, at least in principle, be measured by an apparatus in an experiment, and the goal of the theory is to predict this measured value. The concept of observable is particularly important in Quantum Mechanics. In this theory, many quantities are *not* observable, and this leads to all kinds of nonsense if forgotten. In the basic example, the position of the point should be such an observable.

To explain more about the principles of Quantum Mechanics, in the first stage, we assume that the state space is *finite-dimensional*.[7]

> **Principle 2** *To each observable $\mathcal{O}$, there corresponds a Hermitian operator A on the state space $\mathcal{H}$.*

The statement is not very precise yet, as it does not say how we use the operator A to measure the observable, but we will come to that soon.

A fundamental fact about operators is that they may not commute. In general, $AB \neq BA$, and this lack of commutativity has immeasurable consequences. As a measure of the "lack of commutativity," we define the *commutator*

$$[A, B] := AB - BA, \tag{2.2}$$

of two operators, which will be of constant use. It should be stressed that operators are a complicated business. To a large extent, this explains the intricacies of Quantum Mechanics.

In the case of $\mathcal{H} = \mathbb{C}^2$, the simplest situation of interest,[8] the space of Hermitian operators is a four-dimensional real vector space. A natural basis consists of the following four matrices:

$$\sigma_0 = I = \begin{pmatrix} 1 & 0 \\ 0 & 1 \end{pmatrix}; \sigma_1 = \begin{pmatrix} 0 & 1 \\ 1 & 0 \end{pmatrix}; \sigma_2 = \begin{pmatrix} 0 & -\mathrm{i} \\ \mathrm{i} & 0 \end{pmatrix}; \sigma_3 = \begin{pmatrix} 1 & 0 \\ 0 & -1 \end{pmatrix}. \tag{2.3}$$

[7] When that is not the case, we have to find the proper generalization of the notion of Hermitian operator, which we will explain later.

[8] Nature has made considerable use of this case, although this cannot be explained yet. In words you might have heard, $\mathbb{C}^2$ is the appropriate state space to describe the spin of particles of spin 1/2.

The matrices σ_1, σ_2 and σ_3 are called the *Pauli matrices* and are of fundamental importance. The reader may try to compute the commutators of these matrices to become convinced that lack of commutativity is already in full force. One should also observe that σ_3 is diagonal, so it is a bit simpler than σ_1 and σ_2. As will become gradually clear, the reason for this choice is that in our laboratories, the vertical direction is privileged.

In Dirac's notation the action of the operator A on the vector $|\alpha\rangle$ is denoted by $A|\alpha\rangle$. In mathematical notation the action of the operator A on the vector $x \in \mathcal{H}$ is denoted by $A(x)$ or sometimes Ax. Physicists like to write Ax, and my own training is to write $A(x)$. This schizophrenic pressure prevents me from being consistent.

Since A is Hermitian, it has an orthonormal basis of eigenvectors, which we denote by $|i\rangle$, so that $A|i\rangle = \lambda_i|i\rangle$ for some real numbers λ_i (which may or may not be different as i varies). Here i ranges from 1 to the dimension n of $\mathcal{H}$. Since $\{|i\rangle; i \le n\}$ is an orthonormal basis, we have

$$|\alpha\rangle = \sum_{i \le n} \langle i|\alpha\rangle |i\rangle, \tag{2.4}$$

because $\langle i|\alpha\rangle$ equals the inner product of the vectors $|i\rangle$ and $|\alpha\rangle$. In mathematical notation one would write $x = \sum_{i \le n}(e_i, x)e_i$. It is pleasant to rewrite (2.4) as

$$|\alpha\rangle = \sum_{i \le n} |i\rangle\langle i|\alpha\rangle,$$

or even

$$1 = \sum_{i \le n} |i\rangle\langle i|, \tag{2.5}$$

where 1 is the identity operator.[9] Physicists call such a relation a *resolution of the identity* or a *completeness relation*, because the system of vectors $\{|i\rangle; i \le n\}$ is complete enough to reconstitute the whole space.

Exercise 2.1.1 If $|\alpha\rangle$ is a unit vector, describe the operator $|\alpha\rangle\langle\alpha|$.

The next two principles relate the operator associated to an observable with the actual measurements of this observable.

Principle 3 *When measuring (by an experiment) the value of the observable $\mathcal{O}$ for a system in state $|\alpha\rangle$, the value obtained is always one of the eigenvalues λ_i of the Hermitian operator A associated to $\mathcal{O}$.*

At this stage we can only give trivial examples of physical systems, but let us try to consider a system consisting of one single point, which can be at any of n different locations $(\lambda_i)_{i \le n}$ on the real line. (In the next few pages, this example will be referred to as the *basic example.*) In this basic example, the position of the point should be such an observable. According to Principle 3, it seems sensible that the state space is just $\mathbb{C}^n$, and that the

9 This relation holds because by (2.4) it holds whenever applied to a vector $|\alpha\rangle$.

operator corresponding to the position of the point on the real line is such that for $x = (x_i)_{i\le n} \in \mathbb{C}^n$, one has $A(x) = (\lambda_i x_i)_{i\le n}$ (where $(\lambda_i)_{i\le n}$ are the n possible positions of the point on the real line).[10] Equivalently, each vector $e_i = |i\rangle$ of the canonical basis is an eigenvector of eigenvalue λ_i, $A(e_i) = \lambda_i e_i$ or, in Dirac's notation, $A|i\rangle = \lambda_i |i\rangle$.

The amazing thing (amazing at least compared to our macroscopic experience) is that repeating the experiment on identically prepared systems usually gives *different values*, however carefully these systems have been prepared.

Principle 4 *If the system is in state $|\alpha\rangle$ (so that $\langle\alpha|\alpha\rangle = 1$), then the* probability *of measuring the value λ_i for the observable $\mathcal{O}$ is*

$$\sum_{j;\lambda_j=\lambda_i} |\langle\alpha|j\rangle|^2. \tag{2.6}$$

Here, *probability* means that if you repeat the experiment many times on identically prepared systems, the frequency of measuring the value λ_i is about the value of the right-hand side (with an accuracy that increases with the number of experiments).

Let us examine Principle 4 in the case of the basic example. If the system is in state $|\alpha\rangle = x = (x_i)_{i\le n} \in \mathbb{C}^n$ (with $\sum_{j\le n} |x_j|^2 = 1$), then $x_i = (e_i, x) = \langle i|\alpha\rangle$, so that $|x_i|^2$ is the probability of measuring the system at position λ_i.

As explained, the state of a physical system is described by a unit vector of $\mathcal{H}$, and conversely, every unit vector of $\mathcal{H}$ describes a state of the system.[11] Very often we will consider non-zero vectors that need not be unit vectors. It is very convenient, as we will do without further warning, *to consider that such a vector x describes the same state of the system as the unit vector $x/\|x\|$*. With this convention, when the state $|\alpha\rangle$ is not normalized, the expression (2.6) has to be divided by $\langle\alpha|\alpha\rangle$. Please note at this stage that even when we use this convention, the state $|\alpha\rangle$ of a system is a *non-zero vector*.

The statement "the state of a system is described by a unit vector in a complex Hilbert space $\mathcal{H}$" has to be qualified in a precise way. In Classical Mechanics, if you know the state of the system, you can, in principle, know its exact state at any future time and predict the outcome of any future measurement. Quantum Mechanics allows no such prediction. It does not predict with certainty the outcome of a given experiment. What it does is to make *statistical predictions* about the results of many identical experiments. The technically correct way to express this is to say that "an ensemble of identical systems is described by a unit vector in a Hilbert space".[12] For simplicity we will keep saying that "a system is described by a given state vector", but this *does not* mean, even for simple systems consisting of one single particle, that from this state vector one could (even in principle) determine quantities such as the position or the momentum of the particle. What Quantum Mechanics does is to predict (statistically) the results you will get when you *measure* this position or this momentum. In fact in Quantum Mechanics it *does not make sense* to talk about the position of the particle or its momentum except at a time when you measure it. It is a misconception

10 For what else could A depend on?

11 The correspondence is not one-to-one, however. If a is a complex number of modulus 1, x and ax describe the same state.

12 One may try to think of a state as specifying a preparation process. Repeating the same experiment on similarly prepared systems yields results that are only statistically predictable.

to think that in Quantum Mechanics, systems are in a kind of "undetermined state". The state vector should be thought of (at least according to the standard interpretation) as a *complete description of everything that can possibly be known* about the system. (It of course touches philosophical issues to decide whether this state vector is "a complete description of reality".) It is *the result of measurements on the system* that are not determined, not the state of the system itself.

Going back to Principle 4, we are guaranteed to get the value λ_i out of our measurement only in the very special case where $|\langle\alpha|j\rangle| = 0$ for $\lambda_j \neq \lambda_i$, that is, when $|\alpha\rangle$ is an eigenvector of eigenvalue λ_i. In the case of the basic example this means that only one of the components x_i of x is not zero.

Let us examine what happens if we proceed twice in succession to the measurement of $\mathcal{O}$ on the same system. We consider here the idealized situation where the second measurement takes place "immediately after" the first measurement. (What this means at the experimental level will be discussed a few pages later.) Then both measurements must yield the same result.[13] Thus, at the time of the second measurement, the system must be such that any measurement of $\mathcal{O}$ yields a given value (the value obtained in the first measurement) so that the state of the system must be an eigenvector of A. Consequently, if before the first measurement the state of the system was not an eigenvector of A, this first measurement has *changed* the state of the system. In the case of the basic example, this means that measuring position changes the state x of the system as soon as at least two components of x are not zero. *Measurements typically cannot be performed without changing the state of the system.*

In order to form a clearer picture of the situation, let us anticipate Section 2.15. In the absence of measurements, the time-evolution of the state of a system is *entirely deterministic*. The measurement is not producing a temporary disturbance of the system. Rather, the measurement sends the system on a new course for ever after. Which new course exactly depends on the result of the measurement, a result that can be predicted only statistically.

Much ink (and ingenuity) has been expended debating the fact that randomness appears to be intrinsic to Nature at the most fundamental level. As a probabilist, living a full century after this discovery, I must confess that I do not find it more startling than the older discovery that Earth is orbiting the Sun and not the other way around, so I will say no more about this. Far more annoying is the fact that the change of state occurring in a measurement process (which is often called "the collapse of the wave function") is hard to describe within the setting of Quantum Mechanics. In the preceding formulation of the basic principles, we implicitly assumed that Quantum Mechanics applies only to the microscopic world, while the experimental apparatus, as well as the physicist performing the experiment are *outside* its domain of validity. But where, then, is the boundary of the domain of validity? It would be more satisfactory that Quantum Mechanics should itself be able to describe the measuring apparatus as well as the observer performing the experiment. This raises issues that are not fully resolved to this day. For a simple discussion of these issues, we refer to Isham's book [43], and for a more detailed (yet accessible) account to Weinberg's book [88]. In fact there remain a number of other conceptual difficulties within Quantum Mechanics

[13] At least that is the way things behave. Physics would be harder otherwise.

(see e.g. [20]). The legendary question by A. Einstein to N. Bohr offers a glimpse of this: Do you really believe that the moon is not there when nobody looks at it? (According to the standard interpretation of Quantum Mechanics, it makes no sense to assert that the moon is there when nobody looks at it.[14])

In Dirac's notation, $\langle\beta|A|\alpha\rangle$ denotes the inner product of the vectors $|\beta\rangle$ and $A|\alpha\rangle$. To compute this quantity one may use (2.4), but this computation gives us an opportunity to illustrate one of the strengths of Dirac's notation, the use of (2.5) to obtain the following:

$$\langle\beta|A|\alpha\rangle = \sum_{i,j\le n} \langle\beta|i\rangle\langle i|A|j\rangle\langle j|\alpha\rangle = \sum_{i,j\le n} \langle\beta|i\rangle\langle\alpha|j\rangle^*\langle i|A|j\rangle. \tag{2.7}$$

This type of manipulation using completeness relations is of constant use in physics.[15]

A consequence of Principle 4 and (2.6) is that for a system in state $|\alpha\rangle$, the expected value,[16] or average value, of the measurement of the observable $\mathcal{O}$ is

$$\sum_{i\le n} |\langle\alpha|i\rangle|^2\lambda_i = \langle\alpha|A|\alpha\rangle, \tag{2.8}$$

where the equality is obtained using (2.7) together with the fact that $\langle i|A|j\rangle = \langle i|j\rangle\lambda_j = \lambda_i$ if $i = j$ and zero otherwise. To lighten our terminology, we will use expressions such as "average value of the observable" or even "average of the observable" as shorthand for "average value of the measurement of the observable". Similarly we will say "expected value of the observable" for "expected value of the measurement of the observable". Let us note that the expected value $\langle\alpha|A|\alpha\rangle$ of the observable $\mathcal{O}$ does not determine the probability law of Principle 4.

We obviously get the same result in (2.8) if we replace $|\alpha\rangle$ by $a|\alpha\rangle$ where the complex number a is of modulus 1, $|a| = 1$. Thus as far as predicting the result of the measurement of $\mathcal{O}$ (which is the purpose of the model) is concerned, all these vectors represent the same state, and one should say that the state of the system is actually represented by a *unitary ray*, the set of vectors $a|\alpha\rangle$ for $|a| = 1$ and a certain unit vector $|\alpha\rangle$. (The name *unitary ray* is traditional, even though this set is a circle!)

2.2 Measuring Two Different Observables on the Same System

Suppose now that we consider a second observable $\mathcal{O}'$ with a corresponding Hermitian operator B. Then if a system in state $|\alpha\rangle$ is such that it will always yield the same value

[14] One should say that the problem here is formulated in a rather outrageous way to make the point clear. The universe seems to have done fine prior to our existing! The problem is nonetheless real. On the one hand, one may argue that Quantum Mechanics applies only to the microscopic world. Then a measurement process will be anything that interacts with a macroscopic object such as a photosensitive chemical in a photographic emulsion and has nothing to do with a conscious observer. On the other hand, if one refuses this arbitrary and ill-defined boundary between macroscopic and microscopic worlds, the quantum realm extends all the way to the consciousness of the observer. It is then very difficult to escape the conclusion that this consciousness plays a role, and matters become very murky. The theory of *decoherence* tries to address these issues.

[15] Including cases where the mathematical justification is not ironclad.

[16] The measured value of the observable O in state $|\alpha\rangle$ is a random variable, and as such has an expected value. When we repeat the experiment many times and average the corresponding measurements, the quantity we obtain is near this expected value. This is why both names are used.

when $\mathcal{O}$ is measured, and also when $\mathcal{O}'$ is measured, then $|\alpha\rangle$ must be an eigenvector of both A and B. Then $[A, B]|\alpha\rangle = 0$. When no such $|\alpha\rangle$ exists,[17] there does not exist a state for which both $\mathcal{O}$ and $\mathcal{O}'$ can be measured with certainty. It is simply *impossible* to ever know at the same time the values of both $\mathcal{O}$ and $\mathcal{O}'$ for any state of the system. It is fallacious to think that to know them both one just has to measure $\mathcal{O}$ and then $\mathcal{O}'$. After the measurement of $\mathcal{O}'$ has taken place you no longer know the value of $\mathcal{O}$. If you measure $\mathcal{O}$, then measure $\mathcal{O}'$, and "immediately after" measure $\mathcal{O}$ again, the result of this second measurement of $\mathcal{O}$ *will* sometimes be different from the result of the first measurement. This is because, as we explained earlier, the measurement of $\mathcal{O}'$ *has changed* the state of the system. Right after the measurement of $\mathcal{O}'$, the state of the system is an eigenvector of B, and by hypothesis, this eigenvector of B is *not* an eigenvector of A, so that in this state of the system, the result of the measurement of A cannot be predicted with certainty.

To prove at the experimental level that the difference between the first and the second measurements of $\mathcal{O}$ is really due to the measurement of $\mathcal{O}'$, but not to the evolution of the system, one cannot of course perform the measurements of $\mathcal{O}$, $\mathcal{O}'$, and then $\mathcal{O}$ again "immediately after each other". One simply arranges that the time between the measurements is very short compared with the timescale at which the system evolves. This is most easily done when the spontaneous evolution of the system is very slow, as in the case of the spin of an electron in Earth's very weak magnetic field. The actual spectacular physical experiment, where $\mathcal{O}$ and $\mathcal{O}'$ are, for example, the x and y components of the spin of an electron, is described in many introductory textbooks. (Generally speaking, the predictions of Quantum Mechanics are extremely well confirmed by experiments.)

Exercise 2.2.1 Consider two observables $\mathcal{O}$ and $\mathcal{O}'$ and the corresponding Hermitian operators A and A'. Assume for simplicity that all the eigenvalues of A are distinct, and all the eigenvalues of A' are distinct. Prove that if A and A' commute, the probability to measure a certain value λ of $\mathcal{O}$ for a system in a given state $|\alpha\rangle$ remains the same if we measure $\mathcal{O}'$ before we measure $\mathcal{O}$. Prove that if A and A' do not commute, there are certain states $|\alpha\rangle$ for which this is not true. Hint: A and A' commute if and only if they have a common basis of eigenvectors.

2.3 Uncertainty

Heisenberg's uncertainty principle, which we will study in the present section, is related to but different from the phenomenon of the previous section. If the system is in state x with $(x, [A, B]x) \neq 0$, one cannot measure both observables $\mathcal{O}$ and $\mathcal{O}'$ with arbitrary accuracy. We are not talking here of successive measurements on the same experiment, where the first measurement changes the state of the system. We are talking of measurements on different experiments. One repeats the experiment many times, each time measuring *either $\mathcal{O}$ or $\mathcal{O}'$*. If the results of measuring $\mathcal{O}$ are concentrated in a small interval, then the results of measuring $\mathcal{O}'$ must spread out. In the important special case where $[A, B]$ is a multiple of the

[17] Recall here that since $|\alpha\rangle$ is the state of the system, it must be non-zero. An important particular case is when $[A, B]$ is a multiple of the identity.

identity, whatever the state of the system, you can *never* measure both $\mathcal{O}$ and $\mathcal{O}'$ with arbitrary accuracy. A quantitative version of this statement is given in Proposition 2.3.2.

Definition 2.3.1 Consider an observable $\mathcal{O}$ with associated Hermitian operator A. The *uncertainty* $\Delta_x A \geq 0$ of $\mathcal{O}$ in the state $x \in \mathcal{H}$ is given by

$$(\Delta_x A)^2 := (x, A^2 x) - (x, Ax)^2. \tag{2.9}$$

For instance, in the case of the basic example, the uncertainty of the position in state x is $\sqrt{\sum_{i\leq n} \lambda_i^2 |x_i|^2 - (\sum_{i\leq n} \lambda_i |x_i|^2)^2}$. To make sense of (2.9), one may observe that $(\Delta_x A)^2$ is just the variance of the probability distribution of Principle 4, or in other words, $\Delta_x A$ is the standard deviation of this probability distribution. The physical content of this definition should be stressed. When you make a measurement of the observable $\mathcal{O}$ for a system in state x, you get a *random* result, and $\Delta_x A$ is the standard deviation of this random result.[18]

To explain (2.9) in more mathematical terms, this quantity measures "the square-deviation of $\mathcal{O}$ from its average in state x". Indeed, denoting by $\mathbf{1}$ the identity operator of $\mathcal{H}$ (and since x is of norm 1),

$$(\Delta_x A)^2 = (x, A'^2 x), \tag{2.10}$$

where the Hermitian operator $A' := A - (x, Ax)\mathbf{1}$ is "the deviation of $\mathcal{O}$ from its average in state x". When x is an eigenvector of A, obviously $(\Delta_x A)^2 = 0$. Conversely, when $(\Delta_x A)^2 = 0$, since

$$(x, A'^2 x) = (A'x, A'x) = \|A'x\|^2,$$

then $A'x = 0$ so that x is an eigenvector of A. Therefore, $(\Delta_x A)^2 = 0$ if and only if x is an eigenvector of A, i.e. if and only if the measurement of $\mathcal{O}$ in state x offers no uncertainty, in accord with our calling $\Delta_x A$ the "uncertainty of $\mathcal{O}$ in state x".

Observe finally that $(x, A'x) = 0$, so that (2.10) means

$$(\Delta_x A)^2 = (\Delta_x A')^2. \tag{2.11}$$

Proposition 2.3.2 ([71]) *For two Hermitian operators A and B,*

$$\Delta_x A \Delta_x B \geq \frac{1}{2} |(x, [A, B]x)|. \tag{2.12}$$

When the right-hand side is not zero, this gives a quantitative lower bound on the product of the uncertainty of measuring $\mathcal{O}$ and $\mathcal{O}'$ in state x, a quantitative version of Heisenberg's uncertainty principle. We will soon meet fundamental operators for which $[A, B]$ is a non-zero multiple of the identity and the right-hand side is never zero.

Proof Defining B' in the obvious manner, using (2.11), and observing that $[A, B] = [A', B']$ because $\mathbf{1}$ commutes with everything, we see that it suffices to prove (2.12) for A' and B' or, equivalently, in the case where $(x, Ax) = (x, Bx) = 0$. In that case,

[18] Standard deviation is a statistical measure of how "spread out" a random variable is. Intuitively this means here that the outcomes of different experiments are likely to differ by values of order $\Delta_x A$.

$(\Delta_x A)^2 = (x, A^2 x)$, and similarly for B. The key fact is that, using symmetry (and denoting by $\mathrm{Im}\, z$ the imaginary part of the complex number z),

$$(x,[A,B]x) = (x,ABx) - (x,BAx) = (Ax,Bx) - (Bx,Ax) \\ = (Ax,Bx) - (Ax,Bx)^* = 2\mathrm{i}\mathrm{Im}\,(Ax,Bx). \tag{2.13}$$

Therefore, using the Cauchy-Schwarz inequality,

$$|(x,[A,B]x)| \leq 2|(Ax,Bx)| \leq 2\|Ax\|\|Bx\|. \tag{2.14}$$

Now, since $\|Ax\|^2 = (Ax,Ax) = (x,A^2x) = (\Delta_x A)^2$ and similarly for B, (2.14) proves (2.12). □

2.4 Finite versus Continuous Models

Mathematicians are trained to think of physical space as $\mathbb{R}^3$. But our continuous model of physical space as $\mathbb{R}^3$ is of course an idealization, both at the scale of the very large and at the scale of the very small. This idealization has proved to be very powerful, but in the case of Quantum Field Theory, it creates multiple problems, and in particular the infamous infinities (in the form of diverging integrals).

Can we dispense with continuous models and their analytical problems? A physical measurement is made through a device with finite accuracy, and this measurement is no different from the same measurement rounded to the last significant digit. The result of the measurement is also bounded,[19] so it may yield only finitely many possible values, and we might be able to study physics using only finite-dimensional Hilbert spaces (of huge dimension).

There is a fundamental reason why we stubbornly keep infinite models. Probably the most important guiding principle in finding good models is that a proper theory should be Lorentz invariant,[20] reflecting the fact that physics should be the same for all inertial observers[21] (who undergo no acceleration). There is no way this can be implemented in a finite model, say one which replaces the continuous model of physical space by a finite grid. Lorentz invariance can be recovered from a finite model only "in the infinite limit". Further, there is no canonical choice for such a finite model, so that one has to show that the results obtained by a finite approximation are indeed essentially independent of how this approximation is performed. Heuristically, this is plausible but it is quite another matter to really prove independence. In fact, it can be argued that settling this question is of the same order of difficulty as constructing a continuous model which in some sense would be the limit of the finite model as the grid becomes finer. In the case of Quantum Field Theory, this is a highly non-trivial task. Most importantly, considering finite models does not really solve anything. The infinities reappear in the guise of quantities that blow up as the grid becomes finer, and it is very hard to make sense of this behavior.

19 For example, as of today, no object at distance from us greater than a few billion light-years has been observed. If you feel like arguing that this bound is not large enough, then try to argue that the bound of 10^{10000} light-years is not large enough.

20 See Chapter 4.

21 This fact is extremely well-established experimentally.

For these reasons we uncomfortably but realistically consider continuous models, even though they are not really well defined. Since nobody really knows how to solve the analytical difficulties related to these models, there is little point in working toward a partial solution to these difficulties, and our efforts in this direction will be minimal. It is embarrassing (in particular for this author) to write quantities that do not really make sense, but they would make sense if we replaced the continuous model by a suitable finite approximation.[22] The simple device of "putting the universe in a box" which we study later in Section 3.9 goes a long way in this direction.

2.5 Position State Space for a Particle

In this section we analyze how the previous machinery works to describe a very simple system, a massive point that can be located anywhere on the real line. This provides a first concrete example, and at the same time allows us to discuss the intricacies of considering infinite-dimensional state spaces. Almost nothing of what we explained in the finite-dimensional case will carry on exactly the same, but a suitable infinite-dimensional reinterpretation of the concepts will basically suffice.

The state space $\mathcal{H}$ is the space $L^2 = L^2(\mathbb{R}, dx)$ of complex-valued square-integrable functions[23] on the real line.[24] An element of $\mathcal{H}$ is thus a complex-valued function[25] f on $\mathbb{R}$. The traditional terminology is to call this function the *wave function*. A wave function of norm 1 therefore describes the possible state of a massive point, which for simplicity we will call a *particle*.

The basic idea is that the position of a particle in state f is not really determined, but that the function $|f|^2$ represents the probability density to find this particle at a given location. This statement will eventually appear as the proper interpretation of (2.6) in the present "continuous case". To develop this idea, consider an interval I of $\mathbb{R}$, and the operator 1_I defined by $1_I(f)(x) = f(x)$ if $x \in I$ and $1_I(f)(x) = 0$ if $x \notin I$. This operator is bounded since $\|1_I(f)\| \leq \|f\|$. After we develop the right generalization of Hermitian operators in infinite dimensions, it will become apparent that this operator corresponds to an observable, and the average value of this observable on the state f is

$$(f, 1_I(f)) = \int_I dx |f(x)|^2,$$

which is the probability to find the particle in the set I. It is worth repeating the fundamental fact: When you actually measure whether the particle is in I or not, you get a yes/no answer. But you are certain to find the particle in I only if its state vector f is an eigenvector of 1_I

22 As far as we know, there is no other way to give a precise meaning to these quantities. Physicists seriously trying to explain why what they do makes sense use exactly this procedure, see e.g. Duncan's book [24].

23 In L^2, functions which differ on a set of measure zero are identified, but this is never an issue here.

24 You may also wonder why we choose such a state space. It is quite natural in view of our "basic example" of a point taking n possible positions on the real line. The true answer that this is the fruitful model which will become only gradually apparent.

25 For clarity we often denote functions in position state space by Latin letters and functions on momentum state space by Greek letters.

of eigenvalue 1, i.e. $f(x) = 0$ for $x \notin I$, and you are certain not to find it in I only when f is an eigenvector of 1_I of eigenvalue 0, i.e. $f(x) = 0$ for $x \in I$.[26]

In the present setting, the position of the particle is an observable so that it corresponds to a "Hermitian operator" X, which will be called the *position operator*. It is not difficult to guess what the operator X should be. If indeed $|f|^2$ represents the probability density that the particle is at a given location, its average position is given by

$$\int \mathrm{d}x x|f(x)|^2 = \int \mathrm{d}x f(x)^*(xf(x)). \tag{2.15}$$

This quantity should be $(f, X(f))$, so that X should be the operator which sends f to the function xf (where of course xf is the function $x \mapsto xf(x)$).

This operator X is not defined everywhere, because the function xf may not be square-integrable. It is defined only for the functions f such that $xf \in L^2$. Accordingly we need to introduce the concept of *unbounded operators.*

Definition 2.5.1 An unbounded operator A on a Hilbert space $\mathcal{H}$ is a linear map from a dense linear subspace $\mathcal{D}(A)$ into $\mathcal{H}$. The space $\mathcal{D}(A)$ on which A is defined is called the *domain* of A.

We will say "operator" rather than "unbounded operator", and when the operator is bounded, i.e. $\|A(y)\| \leq C\|y\|$ for some constant C and all $y \in \mathcal{H}$ we will say "bounded operator".[27] The reason for this terminology is that most of the operators we shall use are unbounded. Not all unbounded operators are of interest. The important class, the so-called "self-adjoint operators", is the proper generalization of the Hermitian operators, and we start to introduce the necessary concepts to define it. These concepts are important in the rigorous study of Quantum Mechanics, but they will be rather marginal for us. If you find their study too demanding, you should jump right after Exercise 2.5.20.

Let us recall from Section 1.2 that in finite dimension the adjoint $A^\dagger$ of an operator A is given by (1.6), i.e. $(A^\dagger(y), x) = (y, A(x))$ for all $x, y \in \mathcal{H}$. We would like to use the same definition (1.6) in infinite dimensions. Certainly this formula can make sense only for $x \in \mathcal{D}(A)$. Looking at the left-hand side, we see that when this formula holds, we have

$$|(y, A(x))| = |(A^\dagger(y), x)| \leq \|A^\dagger(y)\|\|x\|,$$

so that $A^\dagger(y)$ may exist only when $y \in \mathcal{H}$ such that the linear map $x \mapsto (y, A(x))$, which is defined on $\mathcal{D}(A)$, is also bounded; $|(y, A(x))| \leq C\|x\|$ for a constant C independent of x.

These observations point the way to the proper definition of the adjoint of an unbounded operator. We define the subspace $\mathcal{D}(A^\dagger)$ as the set of $y \in \mathcal{H}$ for which there exists a constant C independent of x such that $|(y, A(x))| \leq C\|x\|$ for all $x \in \mathcal{D}(A)$. For such a y, by

[26] Generally speaking, it is not always obvious to understand the meaning of the observable corresponding to a given operator.

[27] The reader may of course wonder why an operator which is not defined everywhere is called an "unbounded operator". The reason is that it can be shown that an operator which is defined everywhere and is not extremely pathological is automatically bounded. It is obvious that the position operator cannot be bounded, because if $\|X(f)\| \leq C\|f\|$, then $(f, X(f)) \leq C$ for $\|f\| = 1$. This is absurd because $(f, X(f))$ is the average of the measured position of the particle, and there exist of course particles that are arbitrarily far away.

continuity, the map $x \mapsto (y, A(x))$ from $\mathcal{D}(A)$ to $\mathbb{C}$ extends to the whole of $\mathcal{H}$. Therefore, by the Riesz representation theorem,[28] there exists an element $A^\dagger(y)$ of $\mathcal{H}$ such that

$$\forall x \in \mathcal{D}(A)\,,\ (A^\dagger(y), x) = (y, A(x)), \tag{2.16}$$

and $A^\dagger(y)$ is uniquely determined by this condition since $\mathcal{D}(A)$ is dense.

Definition 2.5.2 When $\mathcal{D}(A^\dagger)$ is dense, the operator $A^\dagger$ with domain $\mathcal{D}(A^\dagger)$ defined by (2.16) is called the *adjoint* of A.

Observe that when A is a bounded operator we have $\mathcal{D}(A^\dagger) = \mathcal{H}$.

Definition 2.5.3 An operator A is called *symmetric* if

$$\forall x, y \in \mathcal{D}(A)\,,\ (A(y), x) = (y, A(x)). \tag{2.17}$$

Exercise 2.5.4 Consider the space $\mathcal{D}$ of continuously differentiable functions f on $[0,1]$. Consider the operator A on $L^2([0,1])$ with domain $\mathcal{D}$ defined by $A(f) = \mathrm{i}f'$. Is A symmetric? What happens if one considers instead the domain $\mathcal{D}_\alpha = \{f \in \mathcal{D};\ f(1) = \alpha f(0)\}$, where α is a complex number of modulus 1?

The following should be obvious.

Lemma 2.5.5 *If A is symmetric, then $\mathcal{D}(A) \subset \mathcal{D}(A^\dagger)$*[29] *and $A^\dagger(y) = A(y)$ for $y \in \mathcal{D}(A)$.*

Definition 2.5.6 An operator is called *self-adjoint* if it is symmetric and if $\mathcal{D}(A^\dagger) = \mathcal{D}(A)$.

If we remember that in finite dimensions an operator A is Hermitian if and only if $A = A^\dagger$, we see that the previous definition exactly generalizes this definition to the infinite-dimensional case: A is self-adjoint if it is equal to its adjoint, i.e. $A^\dagger$ has the same domain as A and coincides with A on this domain. The fundamental importance of self-adjoint operators in Quantum Mechanics is that they correspond to observables. In infinite dimensions physicists often call an operator Hermitian when in fact they mean self-adjoint (so, you will hardly ever see the word "self-adjoint" in physics textbooks).

One may stress the point of Definition 2.5.6 as follows. If we *decrease* $\mathcal{D}(A)$ we *increase* $\mathcal{D}(A^\dagger)$. That is if $\mathcal{D}' \subset \mathcal{D}(A)$ and we denote by B the operator with domain $\mathcal{D}'$, given by $B(x) = A(x)$ for $x \in \mathcal{D}'$, then the domain of $B^\dagger$ contains the domain of $A^\dagger$. This should be obvious from the definitions. Thus in a sense a self-adjoint operator is a symmetric operator such that its domain *is as large as possible*.

Exercise 2.5.7 Prove that if the operator A is symmetric and if $A^\dagger$ is symmetric, then $A^\dagger$ is self-adjoint.

[28] The Riesz representation theorem states that given a bounded linear functional φ on $\mathcal{H}$, that is $|\varphi(x)| \le C\|x\|$ for all x, there exists $y \in \mathcal{H}$ such that $\varphi(x) = (y, x)$.

[29] In particular, $\mathcal{D}(A^\dagger)$ is dense and the adjoint is well defined.

The following exercise is of fundamental importance to get a feeling for self-adjoint operators.

Exercise 2.5.8 Consider the Hilbert space $\mathcal{H}$ of sequences $x = (x_n)_{n\geq 0}$ with norm $\|x\|^2 = \sum_{n\geq 0} |x_n|^2$. Consider a sequence $(\lambda_n)_{n\geq 0}$ of complex numbers.
(a) Prove that the formula

$$A((x_n)_{n\geq 0}) = (\lambda_n x_n)_{n\geq 0}$$

defines an operator on the domain

$$\mathcal{D}(A) = \Big\{x = (x_n)_{n\geq 0} \in \mathcal{H}\ ;\ \sum_{n\geq 0} |\lambda_n x_n|^2 < \infty\Big\}.$$

(b) Prove that A is symmetric if and only if $\lambda_n \in \mathbb{R}$ for each $n \geq 0$.
(c) Prove that the adjoint of $A^\dagger$ of A has the domain $\mathcal{D}(A^\dagger) = \mathcal{D}(A)$ and is given by the formula

$$A^\dagger((x_n)_{n\geq 0}) = (\lambda_n^* x_n)_{n\geq 0}.$$

(d) Prove that A is self-adjoint when $\lambda_n \in \mathbb{R}$ for each n.

Exercise 2.5.9 Consider a sequence $(\mathcal{H}_n)_{n\geq 0}$ of Hilbert spaces and their direct sum $\bigoplus_{n\geq 0} \mathcal{H}_n$. Thus an element $x \in \bigoplus_{n\geq 0} \mathcal{H}_n$ is a sequence $(x_n)_{n\geq 0}$ where $x_n \in \mathcal{H}_n$, with norm $\|x\|^2 := \sum_{n\geq 0} \|x_n\|^2$. Consider for each $n \geq 0$ a bounded operator A_n from $\mathcal{H}_n$ to $\mathcal{H}_{n+1}$ and the adjoint operator B_n from $\mathcal{H}_{n+1}$ to $\mathcal{H}_n$, that is the operator such that $(x, A_n(y)) = (B_n(x), y)$ for $x \in \mathcal{H}_{n+1}$ and $y \in \mathcal{H}_n$. Consider the following domain:

$$\mathcal{D}(A) = \Big\{x = (x_n)_{n\geq 0}\ ;\ \sum_{n\geq 0} \|x_n\|^2 < \infty\ ,\ \sum_{n\geq 0} \|A_n(x_n)\|^2 < \infty\Big\},$$

and define $\mathcal{D}(B)$ similarly. Then the operator A: $(x_n)_{n\geq 0} \mapsto (0, A_0(x_0), A_1(x_1), \ldots)$ is defined on the domain $\mathcal{D}(A)$. Define the operator B similarly and prove that A and B are adjoint of each other.

Exercise 2.5.10 Recall that 1 denotes the identity operator. Prove that a symmetric operator A is self-adjoint if $A + \mathrm{i}1$ and $A - \mathrm{i}1$ are onto.[30] Caution: This takes only four lines but is very tricky.

Exercise 2.5.11 If A is as in Exercise 2.5.8 and symmetric, prove that $A + \mathrm{i}1$ and $A - \mathrm{i}1$ are onto.

Exercise 2.5.12 Consider the space $\mathcal{D}$ of continuously differentiable functions on the unit circle and the operator on L^2 (of the unit circle) with domain $\mathcal{D}$ given for $f \in \mathcal{D}$ by $A(f) = \mathrm{i}f'$. Prove that A is symmetric. Prove that if f is in the domain of $A^\dagger$ one may find a continuous function g such that $f = g$ a.e. Hint: This is difficult if you do not use Fourier series.

30 The converse is proved in Theorem J.1.

When we define an operator A by a concrete formula, in typical cases this formula shows that the operator is defined on a dense subspace $\mathcal{D}(A)$, but this subspace is often too small for A to be self-adjoint. When A is the restriction to $\mathcal{D}(A)$ of a self-adjoint operator defined on the larger domain, there is a simple "automatic procedure" to construct this larger domain, based on the following elementary fact.

Definition 2.5.13 The *graph* of the operator A is the set of pairs $(u, v) \in \mathcal{H} \times \mathcal{H}$ with $u \in \mathcal{D}(A)$ and $v = A(u)$.

Lemma 2.5.14 *If A is an unbounded operator on a Hilbert space $\mathcal{H}$, the graph of $A^\dagger$ is a closed subset of $\mathcal{H} \times \mathcal{H}$.*

Proof By definition of the adjoint we have $z \in \mathcal{D}(A^\dagger)$ and $y = A^\dagger(z)$ if and only if for each $x \in \mathcal{D}(A)$ it holds that $(y, x) = (z, A(x))$. Given x, the set of pairs $(z, y) \subset \mathcal{H} \times \mathcal{H}$ which satisfy this condition defines a closed subset of $\mathcal{H} \times \mathcal{H}$. Thus the graph of $A^\dagger$ is the intersection of a family of closed sets and therefore is closed. □

Thus, if A is self-adjoint, its graph is a closed subset of $\mathcal{H} \times \mathcal{H}$. This leads to the following definition:

Definition 2.5.15 An operator A is *essentially self-adjoint* if the closure in $\mathcal{H} \times \mathcal{H}$ of the graph of A is the graph of a self-adjoint operator.

Thus the "automatic procedure" to check if an operator A can be "made self-adjoint" by properly extending its domain of definition is to compute the closure of the graph of A and see if it is the graph of a self-adjoint operator. This method is much used in serious rigorous texts, but we choose here not to concentrate on "technicalities".

Exercise 2.5.16 Consider the subspace $\mathcal{L}_0$ of $L^2 = L^2([0, 1])$ consisting of functions of integral zero. Consider the map $T: L^2 \to L^2$ given by $T(f)(x) = \int_0^x \mathrm{d}t f(t)$. Prove that it is one-to-one. Let $\mathcal{D}_0 = T(\mathcal{L}_0)$. Prove that the operator $A = \mathrm{i}T^{-1}$ with domain $\mathcal{D}_0$ is symmetric, that its graph is closed but that it is not self-adjoint.

One rarely uses the actual details of the domain of a self-adjoint operator, and it usually suffices to know that a given operator is essentially self-adjoint. In some sense, self-adjoint operators have a simple structure (while operators which are simply symmetric may have all kinds of pathologies). The most spectacular way to reveal this structure is called the "spectral theory for self-adjoint operators". It is a generalization of the fact that in finite dimensions a Hermitian operator has an orthogonal basis of eigenvectors. This theory is of fundamental importance in Quantum Mechanics. As it is a favorite topic for authors attempting a rigorous presentation of the subject, many proofs are available, and those presented in the textbook of Brian Hall [40] are particularly detailed. The spectral theorem is arguably of less fundamental importance in Quantum Field Theory, for the simple reason that the central parts of Quantum Field Theory are too far from mathematical rigor to really take advantage of this result. While the spectral theory solves some of the difficulties in constructing a suitable continuous model, this is of limited help as other difficulties remain unsolved to this day.

For this reason, and in order to keep the present volume to a reasonable size, we will not enter spectral theory here, which is extensively covered in the treatise by Reed and Simon [68]. We limit ourselves to an intuitive understanding of some of the main conclusions of this theory. For this we return to the "multiplication by x" operator of (2.15), which is obviously self-adjoint since it is symmetric and both its domain and the domain of its adjoint consist of the functions f such that $xf \in L^2$.

Exercise 2.5.17 Prove this last statement concerning the domain of the adjoint. Hint: A function f belongs to this domain if the integral $\int \mathrm{d}x f(x) x g(x)$ exists whenever $g \in L^2$.

It is important to observe that this operator *does not* have any eigenvector (although a function with support near a given point x should be thought of as an approximate eigenvector with eigenvalue x).[31] A more general example of a self-adjoint operator is obtained by considering the "same" "multiplication by x" operator on the space $L^2 = L^2(\mathbb{R}, \mathrm{d}\mu, \mathbb{C})$ of functions which are square-integrable for a more general measure $\mathrm{d}\mu$ than Lebesgue's measure.[32] The measure $\mathrm{d}\mu$ is always assumed to give finite mass to bounded sets.

Exercise 2.5.18 Prove that the "multiplication by x" operator has an eigenvector with eigenvalue a if and only if the measure $\mathrm{d}\mu$ gives positive mass to the point a.

In the previous example the eigenvalues have "multiplicity 1". A more general example consists of the "multiplication by x" operator on the space $L^2(\mathbb{R}^2, \mathrm{d}\mu, \mathbb{C})$, where the coordinates of a point in $\mathbb{R}^2$ are denoted by (x, y). The good news is that the most general self-adjoint operator is just a tad more complicated than this operator.

Exercise 2.5.19 Prove that this operator has an eigenvector with eigenvalue a if and only if the measure $\mathrm{d}\mu$ gives positive mass to the set $\{a\} \times \mathbb{R}$. Prove that this eigenvalue has finite multiplicity n if and only if the restriction of μ to this set is supported by a set of cardinality n but not by a set of cardinality $n-1$.

Thus in this example, the multiplicity of an eigenvalue can be finite or infinite. Spectral theory then asserts that a "generic" self-adjoint A operator is not more complicated than this, in a sense that we will develop in Section 2.6. When facing a statement on unbounded operators, one should first try to understand what it means for the "multiplication by x" operator.

It is very easy to write all kinds of nonsense if one ignores the fact that unbounded operators are *not* defined everywhere (but only on their domain $\mathcal{D}$). Many such examples are given in this book by Hall [40]. Interestingly enough, in a random sample of physics textbooks on the shelves of my university's library, it turned out that only a very small fraction even mentions potential problems on this matter. Rather, one pretends that every self-adjoint operator[33] is defined everywhere and diagonalizable. Of course, it is more

[31] Thus in general a self-adjoint operator does not have eigenvectors.
[32] Mathematicians love to call Lebesgue's measure the standard volume measure on $\mathbb{R}^n$.
[33] Words like "self-adjoint" themselves are rarely to be found in these books.

admissible in physics than in mathematics to forge ahead and ignore all kinds of potential problems until a looming disaster imposes more caution.

Exercise 2.5.20 Convince yourself that if $|\gamma\rangle = A|\beta\rangle$ then $\langle\gamma| = \langle\beta|A^\dagger$, where $\langle\beta|A^\dagger$ denotes the composition of the operator $A^\dagger$ from $\mathcal{H}$ to itself and the operator $\langle\beta|$ from $\mathcal{H}$ to $\mathbb{R}$.

It would have been an embarrassment not to state Definitions 2.5.1–2.5.6 in this book, but as we manage to mostly skirt spectral theory they will play a rather small part in the sequel.

Going back to our model for a particle on the line, the state space L^2 on which the position operator is represented by the "multiplication by x" operator X will be used many times. We will call it the "position state space" since "the position of the particle" is simply related to the element of L^2 (the wave function) which represents the state of the particle.

A second fundamental operator is the *momentum operator*[34] P defined by

$$P(f) = -\mathrm{i}\hbar\frac{\mathrm{d}f}{\mathrm{d}x}. \tag{2.18}$$

Its domain consists of the functions f which are differentiable almost everywhere and such that their differential belongs to L^2. It is not at all intuitive that such an operator measures the momentum of a particle, and it is best for us to take this fact for granted as one of those fundamental principles of Quantum Mechanics whose success is every day confirmed by experience.[35] The momentum operator will also appear very naturally when we study Stone's theorem in Section 2.14.

Exercise 2.5.21 Prove that the momentum operator commutes with the translation operators given by $T(f)(x) = f(x+a)$ when $a \in \mathbb{R}$ and explain the physical meaning of this relation.

The momentum operator does not have any eigenvectors. Such an eigenvector would be a function $f(x) = a\exp(\mathrm{i}px/\hbar)$ but such a function is not square-integrable for $a \neq 0$.

A fundamental property is that the position operator and the momentum operator do not commute, but rather satisfy the following so-called *canonical commutation relation*:

$$[X, P] = \mathrm{i}\hbar 1. \tag{2.19}$$

Generally speaking, the domain of the commutator of two unbounded operators might be very small.[36] In the present case, the commutator is defined on well-behaved functions, and in particular on Schwartz functions. Formula (2.19) is a consequence of the formula $(xf)' - xf' = f$.

[34] One may regret that the letter P has not been reserved for the position operator, but in physics it is irrevocably associated to momentum. The letter P comes from the Latin *pulsus*. The occurrence of a "momentum operator", rather than a "velocity operator" is less surprising if one knows that in Classical Mechanics momentum, the product of the mass of the particle and its velocity, is indeed more fundamental than velocity.

[35] At a very intrinsic level, (2.18) has to do with the wave-particle duality. As we shall see soon, particles are also waves. On page 40 we will give arguments making this formula less mysterious.

[36] It might happen that this domain consists only of the 0 vector, even when the domains of the two operators are dense.

It is a good time to mention that $\mathbf{1}$ will always in our notation denote the identity operator of the Hilbert space under study (unless $\mathcal{H} = \mathbb{C}^2$ in which case it will be denoted by I). Many textbooks make the more sweeping convention of simply omitting such an operator and write simply $[X, P] = i\hbar$. Also, the desire to respect alphabetic order makes us often write (2.19) as $[P, X] = -i\hbar\mathbf{1}$.

Exercise 2.5.22 Consider two operators A, B which are assumed to be defined everywhere and satisfy $[A, B] = \mathbf{1}$. Prove by induction on n that $[A, B^n] = nB^{n-1}$. Prove that A and B cannot both be bounded.[37] Hint: Arguing by contradiction, prove that $n\|B^{n-1}\| \le 2\|A\|\|B\|\|B^{n-1}\|$.

The momentum operator looks very different from the position operator, but in some sense it is a very similar operator, but "in a very different position". This will be explained in Section 2.7.

2.6 Unitary Operators

We introduce now unitary operators between Hilbert spaces. It is a fundamental notion in at least two respects:

- Mathematically it provides (as explained in this section) a way to recognize whether two different models "are in fact the same".
- Physically, countless processes are represented by unitary operators. Why this is the case is explained at the beginning of Section 2.10.

Definition 2.6.1 A linear operator U between Hilbert spaces is called *unitary* if it is one-to-one, *onto*, and preserves the norm, $\|U(x)\| = \|x\|$.

Unitary transformations are in a sense the "natural class of isomorphisms between Hilbert spaces". The "polarization identity" $\|x + y\|^2 = \|x\|^2 + \|y\|^2 + 2\mathrm{Re}(x, y)$ shows that a unitary operator preserves the inner product,[38]

$$(U(x), U(y)) = (x, y). \tag{2.20}$$

It is almost obvious that the set $\mathcal{U}(\mathcal{H})$ of unitary operators on a Hilbert space $\mathcal{H}$ forms a group.

Next we reformulate the condition that an operator on a Hilbert space $\mathcal{H}$ is unitary using the notion of adjoint operator. As already noted, for a bounded operator A one has $\mathcal{D}(A) = \mathcal{D}(A^\dagger) = \mathcal{H}$ and

$$\forall x, y \in \mathcal{H}\ ;\ (A^\dagger(x), y) = (x, A(y)). \tag{2.21}$$

A unitary operator on a Hilbert space is bounded, so its adjoint is defined everywhere. Furthermore by (2.21) one has $(U^\dagger U(x), y) = (U(x), U(y))$ and (2.20) is equivalent to

37 We recall that an operator A is *bounded* if for some constant C and all x we have $\|A(x)\| \le C\|x\|$. The operator norm $\|A\|$ of A is then the smallest possible value of C.

38 One has to be careful that an operator preserving the inner product is necessarily one-to-one, but need not be onto.

$U^\dagger U = 1$. Consequently, an operator on a complex Hilbert space is unitary if and only if it is invertible and

$$U^{-1} = U^\dagger. \tag{2.22}$$

Thus a unitary operator also satisfies $UU^\dagger = 1$.

The following trivial fact is stressed because of its considerable importance.

Lemma 2.6.2 *A unitary operator U from a Hilbert space $\mathcal{H}$ to a Hilbert space $\mathcal{H}'$ induces a map*

$$A \mapsto A' = UAU^{-1}$$

from the operators on $\mathcal{H}$ to the operators on $\mathcal{H}'$.

Of course here $\mathcal{D}(A') = U(\mathcal{D}(A))$. The philosophy is that "A' is a copy of A so it has the same properties as A". For example, if A is self-adjoint, so is A'.

If you use the Hilbert space $\mathcal{H}$, the vector x to model a state of a physical system and the operator A to represent an observable, while I use $\mathcal{H}'$, Ux and A' respectively, we will get the same predictions. Such models are called *unitarily equivalent*. This seems to be a trivial observation, but the point is that even if the models are unitarily equivalent, one might be easier to use than the other (because unitary transformations are far from being trivial). Using a basis of eigenvectors of A surely makes computations related to A simpler. Quite accurately, one may think of unitary transformations simply as changes of basis.

2.7 Momentum State Space for a Particle

To put the ideas of the previous section to use we go back to the model of a massive particle on the line which we studied in Section 2.5. Let us consider $\mathcal{H}' = L^2(\mathbb{R}, \mathrm{d}p/(2\pi\hbar))$, where the notation $\mathrm{d}p/(2\pi\hbar)$ means that we include the factor $1/(2\pi\hbar)$ whenever we integrate in p. Consider the Fourier transform $U: f \mapsto \hat{f}$ of (1.30) from $\mathcal{H}$ to $\mathcal{H}'$. It is a unitary operator since it preserves the scalar product by (1.32) and since it has an inverse, the inverse Fourier transform $\varphi \mapsto \check{\varphi}$. The state which was represented by $f \in \mathcal{H}$ is now represented by $\varphi = \hat{f} \in \mathcal{H}'$. As in Lemma 2.6.2 the Fourier transform U transports an operator A on $\mathcal{H}$ to the operator $A' = UAU^{-1}$ on $\mathcal{H}'$ given by $A'(\varphi) = \widehat{A(\check{\varphi})}$. Using (1.33) for $f = \check{\varphi}$ yields (provided φ is well behaved)

$$P'(\varphi)(p) = p\varphi(p) \tag{2.23}$$

and, similarly,

$$X'(\varphi) = \mathrm{i}\hbar \frac{\mathrm{d}\varphi}{\mathrm{d}p}. \tag{2.24}$$

(The plus sign here is not surprising since (2.19) implies that $[X', P'] = \mathrm{i}\hbar 1$.) It is now the momentum operator which looks simple and the position operator which looks complicated. Just as in position state space $|f|^2$ represented the probability density of the location of the particle in state f, we can now argue that $|\varphi|^2$ represents the probability density of the momentum of the particle (when the basic measure is $\mathrm{d}p/2\pi\hbar$). For this reason we will call this space $\mathcal{H}'$ the *momentum state space*. Taking the Fourier transform is the standard way to analyze a wave. Thus it can be said that using momentum state space *amounts to thinking of a particle as wave*. This is the *wave-particle duality*.[39]

So, the probability density of the location of the particle is $|f|^2$, and the probability density of its momentum is $|\varphi|^2$, where φ is the Fourier transform of f. One shows in Fourier analysis that it never happens that both a function and its Fourier transform are sharply concentrated around a single value. This is closely related to Heisenberg's uncertainty principle, which here asserts according to (2.19) that it is impossible that both position and momentum have a sharply defined value. Momentum state space is of fundamental importance (much more so than position state space) because the kinetic energy of a particle depends simply on its momentum (and its mass). The fun however is that particles interact not only according to their momenta, but also according to their respective positions, and this largely contributes to make the universe quite complicated and interesting.

The self-adjoint momentum operator (2.18) on position state space does not look at all like an operator of "multiplication by a function". Yet, after Fourier transform, on momentum state space, this operator is just "multiplication by p". This is a particularly striking instance of the general fact mentioned on page 36: given a self-adjoint operator A on a Hilbert space $\mathcal{H}$ one may find a unitary map U (to a certain L^2 space) such that the operator UAU^{-1} is a "multiplication by a function" operator, refer to Hall's textbook [40, Theorem 10.9].

Exercise 2.7.1 Consider a self-adjoint operator which in momentum state space is given by "multiplication by a function $g(p)$". Prove that this operator commutes with the translations of Exercise 2.5.21.

It can be shown that Exercise 2.7.1 has a converse: operators commuting with the translations are of the from "multiplication by a function $g(p)$". Thus the momentum operator must be of this type. The formula (2.23) corresponds to the simplest case of interest of such an operator.

2.8 Dirac's Formalism

It is regrettable that neither the position nor the momentum operator has a basis of eigenvectors, for this would indeed be very convenient. Paul Dirac invented a remarkable formalism to deal with this problem. It is used in almost every physics textbook, where it is typically considered as self-evident. We try to explain some of the basic features and meaning of this formalism here, striving as usual to explain what this means, but with no serious attempt to make matters rigorous.[40]

39 Not everybody uses the name "momentum state space".
40 The so-called theory of "rigged Hilbert spaces" can be used to make matters rigorous, but we have chosen to stay on the heuristic level.

Dirac's formalism works beautifully. It allows a great economy of thought and notation. It is however unfriendly to mathematicians, and the mathematically inclined reader must brace for a kind of cold shower, as things are likely to look horrendously confusing at first. Let us stress that it is not necessary to master this formalism to follow most of the rest of this book. Still, we will use it at times, if only to formulate results in the same language as they are found in physics textbooks. The reader who finds the present section overwhelming is encouraged to move on and to come back when the need arises.

As a consequence of the use of Dirac's formalism, if one looks at a physics textbook discussing (say) particles on the real line, one may find it difficult to recognize any of the previous material. First, one is likely to find very early the sentence "let $|x\rangle$ denote the state of a particle located at $x\ldots$". An element of the position state space $\mathcal{H} = L^2(\mathbb{R}, \mathrm{d}x)$ that could be said to be located at x would have to be an eigenvector of the position operator X with eigenvalue x and these do not exist. The expression $|x\rangle$ can however be given a meaning in a kind of "distributional sense". It makes sense only when integrated against a Schwartz function f, that is for such a function the integral $\int \mathrm{d}x|x\rangle f(x)$ makes sense as an element of $\mathcal{H}$. The value of this integral is simply the function f seen as an element of $\mathcal{H}$. Quite naturally, we denote by

$$|f\rangle \text{ the function } f \in \mathcal{S} \text{ seen as an element of } \mathcal{H}, \tag{2.25}$$

so that

$$|f\rangle = \int \mathrm{d}x\, |x\rangle f(x). \tag{2.26}$$

This should certainly remind us of a basis expansion $f = \sum_i (f, e_i)e_i = \sum_i e_i(f, e_i)$, and physicists think of $|x\rangle$ as a "continuous basis". In this manner we have given a meaning to the quantity $|x\rangle$ as "an element of $\mathcal{H}$ in the distributional sense". The principle at work here is important enough to be stated clearly:[41]

To make sense of a formula written in Dirac's formalism, try to integrate it against one or several Schwartz functions. (2.27)

We will however not stop at considering Dirac formalism as just a way to encode complicated formulas in a simpler manner. Rather, we will try to free our imagination and to give a "mathematical" interpretation of the objects appearing in this formalism, at least at the level of intuition. Matters will get slippery, since to really make mathematical sense[42] of what we shall do would require heavy formalism which we want to avoid. Still, in this way we will develop a way to look at things which *in the long run* helps intuition.[43] Another point which must be stressed is that our manipulations of quantities involving Dirac's formalism, as e.g. in the rest of this section, never pretend to "prove" anything, and simply check on important instances that the formalism is self-consistent.

[41] Thus, you will know what to try first when meeting exercises asking you to make sense of formulas written in this formalism!

[42] However hard this may be for a mathematician, one has to accept that certain things make some kind of sense even if it is not mathematical.

[43] Please trust me on that assertion, even if you suffer now.

Let us then look at $|x\rangle$ again. If we recall (1.16), that is

$$f(y) = \int \mathrm{d}x f(x)\delta(x-y) = \int \mathrm{d}x f(x)\delta_x(y)$$

where $\delta_x(y) = \delta(x-y) = \delta(y-x)$, so that $f = \int \mathrm{d}x f(x)\delta_x$. Comparing with (2.26), we see that in position state space $|x\rangle$ "corresponds to the Dirac function δ_x",[44] even though $\mathcal{H}$ consists of square-integrable functions, which is not the case of "the function δ". This matches very well with the fact that an eigenvector of the position operator of eigenvalue x must take value zero at points different from x.

Let us next try to give a meaning to the quantity $\langle x|$. As a consequence of (2.26), given a test function g we should have

$$\langle g| = \int \mathrm{d}x g(x)^* \langle x|,$$

so that $\langle g|f\rangle = \int \mathrm{d}x g(x)^* \langle x|f\rangle$. Comparing with the formula

$$\langle g|f\rangle = \int \mathrm{d}x g(x)^* f(x) \tag{2.28}$$

we conclude that

$$\langle x|f\rangle = f(x). \tag{2.29}$$

This formula is not really defined for all $f \in L^2 = \mathcal{H}$, but it certainly makes sense for $f \in \mathcal{S}$. This formula also makes $\langle x|$ appear as a linear functional defined on a subspace of $\mathcal{H}$, and in particular on $\mathcal{S}$.[45] The formula (2.29) is also consistent with the interpretation that "$|x\rangle$ corresponds to the function δ_x".

Using (2.29) we reformulate (2.26) as follows: for $f \in \mathcal{S}$,

$$|f\rangle = \int \mathrm{d}x |x\rangle\langle x|f\rangle, \tag{2.30}$$

which looks much better than (2.26) (but means the same). It can be summarized in the formula

$$\int \mathrm{d}x |x\rangle\langle x| = 1, \tag{2.31}$$

which is a kind of continuous version of (2.5). Physicists call (2.31) a "completeness relation", meaning that there are enough elements in the continuous basis $|x\rangle$ to capture the whole space. They are well aware that $|x\rangle \notin \mathcal{H}$, but pretend nonetheless that this is the case and that $|x\rangle$ is an eigenvector of the position operator X with eigenvalue x. They maintain this position until disaster strikes, and then they call the attention of the reader to the fact that "the state $|x\rangle$ is unphysical" or sometimes, that it is "improper".

44 You might have observed at this stage that one has sometimes to use somewhat ill-defined expressions such as this one. As far as possible they are always between quotes.

45 Of course the function $f \mapsto f(x)$ is not continuous in the L^2 norm, i.e. there does not exist a number C such that $|f(x)| \le C\|f\|_2$ for all $f \in \mathcal{S}$. This corresponds to the fact that the functional $\langle x|$ is not defined on all of L^2.

Now, how can we define $\langle y|x\rangle$ for two such "states" x, y? Applying the functional $\langle y|$ to both sides of (2.26), we should get

$$f(y) = \langle y|f\rangle = \int \mathrm{d}x \langle y|x\rangle f(x), \tag{2.32}$$

and since $f(y) = \int \mathrm{d}x f(x)\delta(x-y)$ from (1.16) we can summarize this by the formula

$$\langle y|x\rangle = \delta(x-y)\ (= \delta(y-x)). \tag{2.33}$$

It is important for the sequel that the first part of the next exercise be very clear to you.

Exercise 2.8.1 (a) Convince yourself that (2.33) is *nothing more* than a compact way to write the formula (2.28).

(b) Convince yourself that (2.33) is consistent with the interpretation of $|x\rangle$ as the Dirac function δ_x, that is that the following somehow makes sense:

$$\langle y|x\rangle = \int \mathrm{d}z \delta_y(z)\delta_x(z) = \delta(y-x). \tag{2.34}$$

Given x prove this equality as an equality of distributions in y, that is prove that for any test function f one has

$$\int \mathrm{d}y f(y) \int \mathrm{d}z \delta_y(z)\delta_x(z) = \int \mathrm{d}y \delta(y-x) f(y),$$

by proving that both sides are equal to $f(x)$. (The formula (2.34) can also be seen as a result on convolution of measures.)

Even though (2.33) sounds like a triviality, this is not the case because it defines a kind of normalization of the "states" $|x\rangle$. (Elements of $\mathcal{H}$ can be normalized to norm 1, but $|x\rangle$ is not such an element.) This is obscured here by the simplicity of the situation, but the proper normalization would be less apparent, if, as will occur later, rather than L^2 we would use the Hilbert space with squared norm $\int \mathrm{d}x |f(x)|^2 w(x)$ for a weight function $w(x)$. We will come back to this later in Section 4.10.

Besides the "states" $|x\rangle$, physics textbooks consider for $p \in \mathbb{R}$ the "states $|p\rangle$ of momentum p". Such a state would have to be an eigenvector of the momentum operator P, and the corresponding differential equation yields the solution $a \exp(\mathrm{i}px/\hbar)$, which does not belong to $\mathcal{H}$ for $a \neq 0$. Nonetheless, thinking of $|p\rangle$ as the function $x \mapsto \exp(\mathrm{i}px/\hbar)$ and thus of $\langle p|$ as the function $x \mapsto \exp(-\mathrm{i}px/\hbar)$, at least for $f \in \mathcal{S}$ it makes perfect sense to consider (using again the notation (2.25))

$$\langle p|f\rangle := \int \mathrm{d}x \exp(-\mathrm{i}px/\hbar) f(x) = \hat{f}(p), \tag{2.35}$$

which turns out to be the Fourier transform of f. Also, since $|x\rangle$ corresponds to the case $f = \delta_x$, we get

$$\langle p|x\rangle = \exp(-\mathrm{i}px/\hbar).$$

Moreover, recalling (1.19) and (1.15),

$$\langle p|p'\rangle = \int dx \exp(i(p'-p)x/\hbar) = 2\pi\delta((p'-p)/\hbar) = 2\pi\hbar\delta(p'-p).$$

We observe here the factor $2\pi\hbar$ on the right-hand side. Generally speaking delta functions on momentum space go with a factor $2\pi\hbar$ (or more precisely one such factor for each dimension) because we have chosen our normalization in order that the proper measure on this space is $dp/2\pi\hbar$, and thus the integral of $2\pi\hbar\delta$ is still 1.

We also have the completeness relation

$$\int \frac{dp}{2\pi\hbar}|p\rangle\langle p| = 1, \tag{2.36}$$

which is a concise way to express that for $f \in \mathcal{S}$ one has

$$|f\rangle = \int \frac{dp}{2\pi\hbar}|p\rangle\langle p|f\rangle. \tag{2.37}$$

Indeed, $\langle p|f\rangle = \hat{f}(p)$, and recalling that $|p\rangle$ is just the function $x \mapsto \exp(ipx/\hbar)$, so that $\langle x|p\rangle = \exp(ipx/\hbar)$ the required equality

$$\langle x|f\rangle = f(x) = \int \frac{dp}{2\pi\hbar} \exp(ipx/\hbar)\hat{f}(p)$$

is just the Fourier inversion formula (1.31).

Exercise 2.8.2 Convince yourself that in momentum state space, "$|p\rangle$ corresponds to the function $2\pi\hbar\delta_p$".

However, this is not the end of the story. Physicists think of the state space as some unspecified abstract Hilbert space, and the state of the system is represented by a vector $|\alpha\rangle$ "independent of any specific representation". Using the "position state space" amounts to describing the state $|\alpha\rangle$ by the function[46] $x \mapsto \langle x|\alpha\rangle$, while using "momentum state space" amounts to describing it by the function $p \mapsto \langle p|\alpha\rangle$. These are related by the completeness relations (2.31) and (2.36)

$$\langle p|\alpha\rangle = \int dx \langle p|x\rangle\langle x|\alpha\rangle = \int dx \exp(-ipx/\hbar)\langle x|\alpha\rangle \tag{2.38}$$

$$\langle x|\alpha\rangle = \int \frac{dp}{2\pi\hbar}\langle x|p\rangle\langle p|\alpha\rangle = \int \frac{dp}{2\pi\hbar} \exp(ipx/\hbar)\langle p|\alpha\rangle \tag{2.39}$$

which simply express that these functions are Fourier and inverse Fourier transforms of each other. We have checked that these claims are indeed correct when one uses the position state space, with the proper interpretation of these formulas. This formalism works beautifully in many more situations. Once it has been developed, it never becomes necessary to use a specific Hilbert space as a state space, which is the way the vast majority of physics

[46] For physicists this function is always well-defined, so please accept their point of view here, as we do not go into technicalities.

textbooks proceed. As we strive to formulate results in a more precise mathematical language we shall mostly not use this point of view.

Let us pursue a bit by a sample computation in the physicist's way [73]. If $\mathcal{O}$ is an observable, why not consider the matrix[47] $\langle x|\mathcal{O}|y\rangle$ for $x, y \in \mathbb{R}$? Let us compute this matrix when $\mathcal{O}$ is the momentum operator (2.18). First, using the completeness relation (2.36),

$$\langle x|P|y\rangle = \int \frac{\mathrm{d}p}{2\pi\hbar} \langle x|P|p\rangle\langle p|y\rangle. \tag{2.40}$$

Now, since we are arguing as physicists, we use that "$|p\rangle$ is an eigenvector for P, of eigenvalue p", so that $P|p\rangle = p|p\rangle$, and thus

$$\langle x|P|p\rangle\langle p|y\rangle = p\langle x|p\rangle\langle p|y\rangle = p\exp(\mathrm{i}p(x-y)/\hbar).$$

Now, recalling (1.19) we have

$$\int \frac{\mathrm{d}p}{2\pi\hbar} \exp(\mathrm{i}p(x-y)/\hbar) = \delta(x-y)$$

and thus (differentiating with respect to x under the integral sign),

$$\int \frac{\mathrm{d}p}{2\pi\hbar} p\exp(\mathrm{i}p(x-y)/\hbar) = -\mathrm{i}\hbar\delta'(x-y),$$

and the above relations yield the formal expression

$$\langle x|P|y\rangle = -\mathrm{i}\hbar\delta'(x-y) = \mathrm{i}\hbar\delta'(y-x). \tag{2.41}$$

In the present case, while the previous computation might seem mysterious to a mathematician, it is easy to understand the result. The quantity $\langle x|P|y\rangle$ can be defined in the distributional sense, requiring that for $f, g \in \mathcal{S}$ one should have

$$\iint \mathrm{d}x\mathrm{d}y g(x)^* f(y)\langle x|P|y\rangle = \langle g|P|f\rangle, \tag{2.42}$$

while, by definition of P, writing f' the derivative of f, $Pf(x) = -\mathrm{i}\hbar f'(x)$, so that

$$\langle g|P|f\rangle = -\mathrm{i}\hbar \int \mathrm{d}x g(x)^* f'(x).$$

Now, using successively (1.16) and (1.11),

$$f'(x) = \int \mathrm{d}y f'(y)\delta(y-x) = -\int \mathrm{d}y f(y)\delta'(y-x),$$

so that one may write

$$-\mathrm{i}\hbar \int \mathrm{d}x g^*(x) f'(x) = \iint \mathrm{d}x\mathrm{d}y\, g(x)^* f(y)(\mathrm{i}\hbar\delta'(y-x)).$$

Thus the right-hand side above equals the left-hand side of (2.42), and this is the meaning of (2.41).

47 The name "matrix" is Physics terminology.

2.9 Why Are Unitary Transformations Ubiquitous?

Let us go back to the setting of a general Hilbert space, whose elements are denoted $x, y, z, \ldots$. To each observable is associated a self-adjoint operator. Conversely, to each self-adjoint operator is associated an observable (although it is another matter in concrete situations to design an experiment that actually measures it). We now describe a fundamental class of such observables. Given $x \in \mathcal{H}$ of norm 1, $(x,x) = 1$, the projector $P_x(y) := (x,y)x$ is Hermitian since

$$(z, P_x(y)) = (z,x)(x,y) = (x,z)^*(x,y) = ((x,z)x, y) = (P_x(z), y).$$

Exercise 2.9.1 In Dirac's formalism for a norm-1 vector $|\alpha\rangle$ one writes the projector $P_\alpha = |\alpha\rangle\langle\alpha|$. Why does it seem to require no proof that P_α is Hermitian?

Thus the operator P_x corresponds to an observable $\mathcal{O}$. The possible values of $\mathcal{O}$ are the eigenvalues of P, namely 0 and 1. We may describe $\mathcal{O}$ as asking the question: Is the state of the system equal to x?[48] The average value of this operator in state y is given by $(y, P_x(y)) = (y,x)(x,y) = |(x,y)|^2$ and is the probability to obtain the answer "yes" to your question. It is called the *transition probability* between x and y.[49] In physics, the inner product (x,y) is often called an *amplitude*, so that *the transition probability is the square of the modulus of the amplitude*.[50] The transition probability does not change if one multiplies x and y by complex numbers of modulus 1, as expected from the fact that this multiplication does not change the state represented by either x or y.

We should expect that *any transformation*[51] *which preserves the physical properties of a system preserves the transition probabilities*. Transition probabilities are preserved by unitary transformations, since for such a transformation $|(Ux, Uy)|^2 = |(x,y)|^2$. They are also preserved under *anti-unitary* transformations, i.e. anti-linear operators which preserve the inner product. Conversely, which are the transformations of $\mathcal{H}$ which preserve transition probabilities? A deep theorem of Eugene Wigner [92, appendix to Chapter 20] shows that this is the case *only* for unitary and anti-unitary transformations.[52] A fundamental consequence is that any transformation which preserves the physics of the system corresponds to a unitary or anti-unitary transformation. As we explain in the next section, there are many of these, corresponding to the symmetries of Nature. The symmetries of Nature of a certain type naturally form a group, bringing us to group theory. Furthermore time-evolution will also be represented by a unitary transformation.

[48] This *does not* mean that there is any way to "determine" the state y of the system. You may only perform a measurement that will give you a yes/no answer to the question: Is the system in state x? This experiment changes the state of the system.

[49] The reason for this name is that indeed this is the probability that the system exhibits a transition from state y to state x when you measure the observable $\mathcal{O}$.

[50] It would help to remember now the general principle that the square of the modulus of an amplitude represents a probability.

[51] The next section will make clear what is meant exactly by "transformation of the system".

[52] Anti-unitary transformations are much less used than unitary ones. They occur naturally in the study of time-reversal, a topic we do not treat at all.

2.10 Unitary Representations of Groups

Certain types of invariance in Nature are among the most important guiding principles in developing physical theories about the real world. This will be a recurring theme in this book. It forces us to choose models which satisfy certain symmetries and this implies extremely strict restrictions on the possible forms of physical theories.

In this section we start to use this principle in the simplest case, translation invariance. In physics each observer uses a *reference frame* to describe the positions of points in space (or in space-time). These reference frames need not have the same origin, may use different privileged directions and may even move with respect to each other.[53] Here we just consider the situation of different origins. If you study the motion of an object using a different origin for your reference frame than mine, we may disagree on the coordinates of the object, but we should agree that it follows the same laws of physics. Mathematically, the space $\mathbb{R}^3$ acts on itself by translations, and we examine first the effect of these translations at a purely classical level. Suppose that the system you are studying is translated by a vector $\boldsymbol{a}$.[54] The object at position $\boldsymbol{x}$ that you were studying has now been moved to position $\boldsymbol{x}+\boldsymbol{a}$. Say that you use a function f on $\mathbb{R}^3$ to measure e.g. the electrical potential at a point of space. Before you translate the system, the value $f(\boldsymbol{x})$ measures the potential at the point $\boldsymbol{x}$. After the translation this value of the potential occurs at the point $\boldsymbol{x}+\boldsymbol{a}$. Thus the value of the new function $U(\boldsymbol{a})(f)$ you use to measure the potential at this point $\boldsymbol{x}+\boldsymbol{a}$ equals the value of the old function f at the point $\boldsymbol{x}$:

$$U(\boldsymbol{a})(f)(\boldsymbol{x}+\boldsymbol{a}) = f(\boldsymbol{x}),$$

i.e.

$$U(\boldsymbol{a})(f)(\boldsymbol{x}) = f(\boldsymbol{x}-\boldsymbol{a}). \tag{2.43}$$

Observe the all-important minus sign and note the fundamental property $U(\boldsymbol{a}+\boldsymbol{b}) = U(\boldsymbol{a})U(\boldsymbol{b})$.

Suppose now that more generally we study a system whose state is described by a vector x in a Hilbert space $\mathcal{H}$. If the system is translated by a vector $\boldsymbol{a}$ we expect that the system will be described by a new state $U(\boldsymbol{a})(x)$. This new description should not change the physics. The transition probability between x and y should be the same as the transition probability between $U(\boldsymbol{a})(x)$ and $U(\boldsymbol{a})(y)$, i.e. $|(x,y)|^2 = |(U(\boldsymbol{a})(x), U(\boldsymbol{a})(y))|^2$. According to the discussion of the previous section, $U(\boldsymbol{a})$ is either unitary or anti-unitary. Moreover, it is obvious that $U(\boldsymbol{0})$ should be the identity. What becomes very interesting is when we perform two such translations in succession, first by a vector $\boldsymbol{a}$ and then by a vector $\boldsymbol{b}$. The state x is transformed first in $U(\boldsymbol{a})(x)$ and then in $U(\boldsymbol{b})(U(\boldsymbol{a})(x))$. This also amounts to perform the translation by the vector $\boldsymbol{a}+\boldsymbol{b}$, and this transforms the state x into the state $U(\boldsymbol{a}+\boldsymbol{b})(x)$. Therefore $U(\boldsymbol{b})U(\boldsymbol{a})$ and $U(\boldsymbol{a}+\boldsymbol{b})$ should represent the same transformation of the system. We cannot conclude that these transformations are equal, because states are

[53] If you sit in a train, you will typically think of the train as fixed and the landscape as moving.

[54] Equivalently, assume that this system remains fixed, but that you translate your reference frame by a vector $-\boldsymbol{a}$. Translating the system by $\boldsymbol{a}$ is called an *active transformation* whereas translating your reference frame by a vector $-\boldsymbol{a}$ is called a *passive transformation*.

really represented by whole rays rather than points, and operators of the form $c\mathbf{1}$, where c is of modulus 1 and $\mathbf{1}$ is the identity operator, are unitary, but do not change the ray structure. Rather, we expect that

$$U(\boldsymbol{a}+\boldsymbol{b}) = r(\boldsymbol{a},\boldsymbol{b})U(\boldsymbol{a})U(\boldsymbol{b}), \tag{2.44}$$

where $r(\boldsymbol{a},\boldsymbol{b})$ is a complex number of modulus 1. In particular

$$U(\boldsymbol{a}) = r(\boldsymbol{a}/2,\boldsymbol{a}/2)U(\boldsymbol{a}/2)U(\boldsymbol{a}/2).$$

Since the composition of two anti-unitary transformations is unitary, we have shown that $U(\boldsymbol{a})$ is always unitary. This motivates the following definition.

Definition 2.10.1 A map U which associates to each element a in a group G a unitary operator $U(a)$ on $\mathcal{H}$ is called a *projective unitary representation*[55] if for each $a,b \in G$ it satisfies

$$U(ab) = r(a,b)U(a)U(b) \tag{2.45}$$

where the complex number $r(a,b)$ is of modulus 1.

In this definition G denotes a group that is not necessarily commutative, so as customary the group operation is denoted in multiplicative form. In the case $G = \mathbb{R}^3$, (2.45) specializes to (2.44). We may then rephrase the argument which opens this section:

Any quantum system comes equipped with a projective unitary representation of the group of translations. This representation describes how translations affect the state of the system. (2.46)

The word "quantum" in the expression "quantum system" simply stresses the fact that the system is studied through Quantum Mechanics. The same argument shows also that the Euclidean group (generated by translations and rotations) has a projective unitary representation in any quantum system.

Exercise 2.10.2 Using that $(ab)c = a(bc)$ prove that the function r of (2.45) satisfies the identity

$$r(a,bc)r(b,c) = r(ab,c)r(a,b).$$

Our Definition 2.10.1 of projective representations is perfectly adapted to Quantum Mechanics, and follows the way physicists actually think about these objects. Mathematicians however like to look at these somewhat differently. This point of view brings conceptual clarification at the expense of an extra (thin) layer of abstraction, and is explained in Section 2.12. The mathematically inclined reader may like to look at this material now.

[55] These representations are also called *ray representations*.

Definition 2.10.3 A map U which associates to each element a in a group G a unitary operator $U(a)$ on $\mathcal{H}$ is called a *unitary representation* if it satisfies

$$U(ab) = U(a)U(b) \tag{2.47}$$

for $a, b \in G$.

Thus, a unitary representation is simply a projective unitary representation for which $r(a,b) \equiv 1$ for $a, b \in G$.

2.11 Projective versus True Unitary Representations

Let us start the discussion of the concepts involved in Definitions 2.10.1 and 2.10.3. The word "unitary" refers of course to the fact that each of the operators $U(a)$ is unitary. Unless mentioned otherwise, all representations are unitary, so that we shall nearly always omit the word "unitary", and the expressions "representation" and "projective representations" have to be understood by default as "unitary representation" and "projective unitary representations".

To insist that a representation satisfies $r(a,b) = 1$ for all a, b we will sometimes say *true* representation, even though throughout the book, the word "representation" means "true representation". When we consider a representation that is only a projective representation we will always say so explicitly. It is most important to understand the relationship between representations and projective representations.

- The concept of "representation" is *far more restrictive* than the concept of "projective representation".
- From the point of view of mathematics, the nice objects are representations. The study of group representations is a vast subject in mathematics.
- From the point of view of Quantum Mechanics, the natural objects are projective representations.

The following explains an important relationship between representations and projective representations.

Definition 2.11.1 Given a true representation V of G, and for $a \in G$ a number $\lambda(a)$ of modulus 1, the formula

$$U(a) := \lambda(a)V(a) \tag{2.48}$$

defines a projective representation, since (2.45) holds for the function

$$r(a,b) = \lambda(ab)/(\lambda(a)\lambda(b)). \tag{2.49}$$

When this is the case we will say that the projective representation U *arises* from the true representation V.

More generally, there is an important idea behind this definition: two projective representations U, U' such for each $a \in G$ one has $U(a) = \lambda(a)U'(a)$ for some complex number $\lambda(a)$ with $|\lambda(a)| = 1$ are to be thought of as "the same projective representation".

Given a projective representation, the immediate question is whether it arises from a true representation as in Definition 2.11.1. The point here is that even though projective representations are natural for Quantum Mechanics, true representations are much nicer mathematical objects, and it is easier to calculate with them. It would be sad to go through the complications of dealing with a projective representation just because we cannot recognize that this representation is really a true representation in disguise as given by (2.48).

Lemma 2.11.2 *Consider a projective representation U and assume that we can find a function $\lambda(a)$ with $|\lambda(a)| = 1$ such that (2.49) holds. Then U arises from a true representation.*

Proof It is immediate to check that $V(a) = \lambda(a)^{-1}U(a)$ is a true representation, and U arises from V. □

Thus, to prove that a projective representation arises from a true representation, given the function r it "is sufficient" to find a function λ which satisfies (2.49). This might be difficult if r is complicated.

We shall study later a projective representation that does not arise from a true representation, and which will be of constant use. That example motivates the fundamental program of investigating in detail how true and projective representations are related, and what obstacles may prevent a projective representation from arising from a true representation. We shall say a few words about this in Appendix A. We do not pursue this goal in the main text, since our priority is not to understand all possible projective representations, but rather to describe as simply as possible those which are the most important in our topic.

2.12 Mathematicians Look at Projective Representations

This material is not needed to follow the main story. It assumes that you know some very basic group theory. A map U from G to the group $\mathcal{U}(\mathcal{H})$ of unitary transformations of $\mathcal{H}$ is a true representation if and only if it is a group homomorphism. The group $\mathcal{U}(\mathcal{H})$ has a remarkable subgroup, the subgroup consisting of the transformations $\lambda 1$ with $|\lambda| = 1$. Let us denote by $\mathcal{U}_p(\mathcal{H})$ the quotient of $\mathcal{U}(\mathcal{H})$ by this subgroup, and by Φ the quotient map $\mathcal{U}(\mathcal{H}) \to \mathcal{U}_p(\mathcal{H})$. Thus the elements of $\mathcal{U}_p(\mathcal{H})$ are unitary operators "up to a phase", i.e. up to a multiplicative constant of modulus 1. It is immediate to check that a map U from a group G into $\mathcal{U}(\mathcal{H})$ is a projective representation in the sense of Definition 2.10.1 if and only $\Phi \circ U$ is a group homomorphism from G to $\mathcal{U}_p(\mathcal{H})$. The important object is thus the map $\Phi \circ U$. Accordingly, mathematicians *define a projective representation as a group homomorphism from G to $\mathcal{U}_p(\mathcal{H})$*. This formalizes the idea that two projective representations U and U' such that $U(a) = \lambda(a)U'(a)$ "are the same projective representation" (because this is the case if and only if $\Phi \circ U = \Phi \circ U'$). Another benefit of this approach is that it becomes natural to define "continuous projective representations", a topic which is investigated in Section A.2.

In mathematical language, the fundamental question, is, given a projective representation U of G, that is a group homomorphism from G to $\mathcal{U}_p(\mathcal{H})$, whether there exists a true representation V, that is a group homomorphism from G to $\mathcal{U}(\mathcal{H})$, such that $U = \Phi \circ V$.

2.13 Projective Representations of ℝ

We do not investigate in detail how true and projective representations are related in general, but we examine this question in the centrally important case $G = \mathbb{R}$. However, we must first discuss a technical question. In the cases of greatest interest, G is a topological group, and to avoid pathologies, one requires also a mild continuity assumption.

Definition 2.13.1 The map $a \mapsto U(a)$ which associates to each element a of G a unitary operator $U(a)$ is called *strongly continuous* if for each $x \in \mathcal{H}$ the map $a \mapsto U(a)(x)$ from G to $\mathcal{H}$ is continuous.

The topology on $\mathcal{H}$ is the topology induced by its norm, so the condition of strong continuity means that for each $x \in \mathcal{H}$ the norm $\|U(a)(x) - U(a_0)(x)\|$ goes to 0 as $a \to a_0$. Despite the adjective "strong", this condition is much weaker than the continuity of the map $a \mapsto U(a)$ in the operator norm.

A simple but instructive example of a representation is the case where $G = \mathbb{R}$, $\mathcal{H} = L^2(\mathbb{R})$ and $U(a)(f) \in L^2(\mathbb{R})$ is the function $w \mapsto f(w - a)$. The map $a \mapsto U(a)$ is not continuous when the space of unitary operators is provided with the topology induced by the operator norm but it is strongly continuous (as one sees by approximating f with a continuous function of bounded support).

When the map $a \mapsto U(a)$ is strongly continuous, then for $x, y \in \mathcal{H}$ the map $a \mapsto (x, U(a)(y))$ is continuous. This apparently weaker condition is equivalent to strong continuity. To prove this, assume the weaker condition. Then as $a \to a_0$, $(U(a)(x), U(a_0)(x))$ tends to the square of the norm of $U(a_0)(x)$, and since both vectors $U(a)(x)$ and $U(a_0)(x)$ have the same norm they become close to each other (as follows from the relation $\|u - v\|^2 = \|u\|^2 + \|v\|^2 - 2\mathrm{Re}(u, v)$).

Theorem 2.13.2 *A strongly continuous projective unitary representation of* $\mathbb{R}$ *arises from a true representation. That is, for such a projective representation* $U(u)$ *we have* $U(u) = \lambda(u)V(u)$ *where* $V(u)$ *is a true representation.*

Physicist's proof of Theorem 2.13.2 Since $U(u)$ is unitary, it is invertible. Since $U(0) = r(0,0)U(0)^2$, it holds that $r(0,0)U(0) = 1$. Since U is unitary, we have $|r(0,0)| = 1$, and $U'(u) := r(0,0)U(u)$ is a strongly continuous projective representation with $U'(0) = 1$. To prove that U arises from a true representation, it is sufficient to prove that U' arises from a true representation. Thus we may assume that $U(0) = 1$. A physicist will by default assume that for u small,[56] one has

$$U(u) = 1 + uiH + O(u^2), \tag{2.50}$$

for a certain operator H. Since $U(u)$ is unitary,

$$1 = U(u)U(u)^\dagger = 1 + ui(H - H^\dagger) + O(u^2),$$

56 We will comment a bit later on this.

so that $H - H^\dagger = 0$, i.e. H is Hermitian. Let us then set $V(u) = \exp(\mathrm{i}uH)$. This operator is unitary since

$$V(u)^\dagger = \exp(-\mathrm{i}uH^\dagger) = \exp(-\mathrm{i}uH) = V(u)^{-1},$$

and we thus define a true unitary representation. (If you find the first equality mysterious, please think of the finite-dimensional case, where the exponential is given by the usual power series.) Next, we prove that $U(u) = \lambda(u)V(u)$ where $\lambda(u)$ is of modulus 1. For this we fix u and we observe that for an integer n (and because u/n is small for n large),

$$U(u/n) = 1 + \mathrm{i}\frac{u}{n}H + O(n^{-2}) = V(u/n) + O(n^{-2}),$$

so that one should have

$$U(u/n)^n = V(u) + O(n^{-1}).$$

Since U is a projective representation, $U(u/n)^n = \lambda(n)U(u)$ where $|\lambda(n)| = 1$, and thus $U(u) = \lambda(n)^{-1}V(u) + O(n^{-1})$. Considering a subsequence (n_k) such that $\lambda(n_k)$ converges as $k \to \infty$ and letting $k \to \infty$ we conclude that $U(u) = \lambda V(u)$ where $|\lambda| = 1$.[57] □

Even in finite dimension the previous proof assumes more about U than what the theorem states. The relation (2.50) holds if and only if the map $u \mapsto U(u)$ is differentiable at $u = 0$ and this is not one of the hypotheses of the theorem. There is *no reason whatsoever* why a continuous projective representation should be differentiable since we may obtain a projective representation from a true representation using (2.48) for a nasty function λ. Furthermore, the previous argument fails in infinite dimension, because (2.50) cannot possibly be true. Even for a true representation there exist in general many vectors x for which the map $u \mapsto U(u)(x)$ is not differentiable at 0. A mathematician's proof of Theorem 2.13.2 may be found in Section A.1.

We may also define the concept of (not necessarily unitary) projective representation by keeping the relation (2.45) but dropping the requirement that the operators are unitary. Similarly we may define the concept of (not necessarily unitary) representation. When we consider representations that are not unitary we will say so explicitly.

2.14 One-parameter Unitary Groups and Stone's Theorem

A (strongly continuous) one-parameter unitary group is simply a (strongly continuous) unitary representation of $\mathbb{R}$, that is a map which associates to $t \in \mathbb{R}$ a unitary operator $U(t)$ on Hilbert space $\mathcal{H}$ in such a manner that

$$U(s)U(t) = U(s+t), \tag{2.51}$$

and (the continuity condition)

$$\forall x, y \in \mathcal{H}, \ \lim_{t \to 0}(x, U(t)(y)) = (x, y). \tag{2.52}$$

57 In Quantum Mechanics a *phase* is a complex number of modulus 1. To lighten terminology we will say that "$U(u) = V(u)$ up to a phase".

The archetypical example is the operator $U(t)$ on $L^2(\mathbb{R})$ given for $f \in L^2$ and $w \in \mathbb{R}$ by

$$U(t)(f)(w) = \exp(\mathrm{i}tw/\hbar)f(w). \tag{2.53}$$

This is simply the operator "multiplication by the function $\exp(\mathrm{i}t \cdot /\hbar)$." Another example is the operator $V(t)$ on $L^2(\mathbb{R})$ given for $f \in L^2$ and $w \in \mathbb{R}$ by

$$V(t)(f)(w) = f(w+t). \tag{2.54}$$

In both cases it is a nice exercise of elementary analysis to prove that these operators are strongly continuous. These one-parameter groups are closely related by the Fourier transform. Indeed,

$$\widehat{V(t)(f)} = U(t)\hat{f}. \tag{2.55}$$

Exercise 2.14.1 Make sure you understand every detail of the proof of the important formula (2.55).

Theorem 2.14.2 (Stone's theorem) *There is a one-to-one correspondence between the strongly continuous one-parameter unitary groups on a Hilbert space $\mathcal{H}$ and the self-adjoint operators on $\mathcal{H}$. Given the unitary group U, the corresponding self-adjoint operator A is called the* infinitesimal generator *of U. It is defined by the formula*[58]

$$A(x) = \lim_{t \to 0} \frac{\hbar}{\mathrm{i}t}(U(t)(x) - x). \tag{2.56}$$

and its domain $\mathcal{D} = \mathcal{D}(A)$ is the set of x for which the previous limit exists.

The most important part of Stone's theorem is the statement that the formula (2.56) defines a self-adjoint operator. Recalling that the domain of a self-adjoint operator "is large", the mathematical content of this part of Stone's theorem can be understood as a differentiability statement: there are many points x for which the map $t \mapsto U(t)(x)$ is differentiable at $t = 0$.

The impact on Quantum Mechanics of this part of Stone's theorem is simply staggering. Each time we are given a one-parameter unitary group, we can construct a self-adjoint operator, which means that there is a new fundamental quantity with physical meaning which we can measure. One-parameter unitary groups occur e.g. as follows:

- From the operation of space translation in a given direction (to be detailed later in the present section). The corresponding operator is called the *momentum operator in that direction.*
- From the operation of rotation around a given axis. The corresponding operator is called *angular momentum about this axis.*[59]
- From time-evolution, with the construction of the fundamental Hamiltonian operator, as studied in the next section.

[58] The use of the factor i in the denominator of (2.56) is precisely to ensure that A is self-adjoint (as opposed to skew-adjoint i.e. $A^\dagger = -A$) since self-adjoint operators are the class of interest in physics. Conventions such as this one often differ between mathematicians and physicists.

[59] The theory of angular momentum occupies a large part in Quantum Mechanics textbooks. We say a few words about it in Section 8.7.

Given the self-adjoint operator A, the converse part of Stone's theorem constructs a corresponding one-parameter group denoted by

$$U(t) = \exp(itA/\hbar). \tag{2.57}$$

It is easy to understand what this means when $\mathcal{H}$ is finite-dimensional (so that A is then simply a Hermitian operator). Then the exponential of an operator is simply given by the usual power series, and it should then be obvious that $\exp(B)^\dagger = \exp(B^\dagger)$. Since A is Hermitian ($A = A^\dagger$), it holds that $U(t)^\dagger = \exp(-itA/\hbar) = U(-t) = U(t)^{-1}$ so that $U(t)$ is indeed unitary. It should also be obvious that when A and B *commute* one has $\exp(A+B) = \exp(A)\exp(B)$ and this implies the formula $U(s+t) = U(s)U(t)$. If one has developed spectral theory, it is easy to understand (2.57). A self-adjoint operator A is essentially a "multiplication by x operator" and $\exp(itA/\hbar)$ is simply "multiplication by $\exp(itx/\hbar)$", see Exercise 2.14.3. In Appendix B we provide a self-contained construction of the operator $\exp(itA/\hbar)$ which does not use spectral theory.

The following exercise should not be missed, as it provides some intuition for Stone's theorem. It requires some fluency with integration theory.

Exercise 2.14.3 (a) Consider a real-valued measurable function h on $\mathbb{R}$. Prove that the operators $U(t)$ defined by $U(t)(f) = \exp(ith/\hbar)f$ form a strongly continuous one-parameter group of unitary operators of $L^2 = L^2(\mathbb{R}, dx, \mathbb{C})$.
(b) Prove that the operator A of (2.56) has domain $\mathcal{D} = \{f \in L^2; hf \in L^2\}$ and that for $f \in \mathcal{D}$ one has $A(f) = hf$.
(c) Consider the self-adjoint operator A "multiplication by the function h" given by $A(f) = hf$ for $f \in \mathcal{D} = \{f \in L^2; hf \in L^2\}$. Prove that the corresponding one-parameter group is the group $U(t)$ considered in (a).

Let us now suppose that we are given a one-parameter unitary group $U(t)$. We no longer assume that $\mathcal{H}$ is finite-dimensional, and we proceed to prove the truly remarkable fact that the formula (2.56) defines a self-adjoint operator. The proof takes about three pages, and is rather rewarding. The underlying ideas however are not needed for the sequel, so this material may be skipped without harm.

Lemma 2.14.4 *The operator A is symmetric, that is $(A(y),x) = (y,A(x))$ for $x,y \in \mathcal{D}$.*

Proof We observe the obvious but important fact that

$$(U(t)(y) - y, x) = (y, U(-t)(x) - x) \tag{2.58}$$

so that

$$\left(\frac{\hbar}{it}(U(t)(y) - y), x\right) = \left(y, \frac{\hbar}{-it}(U(-t)(x) - x)\right), \tag{2.59}$$

and letting $t \to 0$ gives the result. □

Lemma 2.14.5 *The domain $\mathcal{D}$ where the limit (2.56) exists is a dense subspace of $\mathcal{H}$.*

Assuming this lemma for a moment, to help the reader form an understanding of the situation, we complete the proof of Theorem 2.14.2 when $\mathcal{H}$ is finite-dimensional (the general case requires extra work). Then a dense subspace of $\mathcal{H}$ can be only $\mathcal{H}$ itself, and since A is symmetric it is Hermitian. Moreover, taking the derivative of $t \mapsto U(t+s) = U(t)U(s)$ at $t = 0$ yields $U'(s) = (\mathrm{i}/\hbar)AU(s)$ hence the formula $U(t) = \exp(\mathrm{i}tA/\hbar)$. In the finite-dimensional case this formula makes it obvious that there is a one-to-one correspondence between A and U.

Our proof of Stone's theorem is rather elementary, but it requires the concept of integration of a continuous Hilbert-space-valued function $h(s)$ with compact support. This integral is defined in a quite straightforward manner by approximation by step functions. We shall use that[60]

$$\Big\| \int \mathrm{d}s h(s) \Big\| \le \int \mathrm{d}s \|h(s)\|, \tag{2.60}$$

and that for a bounded linear operator B (possibly valued in another Hilbert space) one has

$$B\Big(\int \mathrm{d}s h(s)\Big) = \int \mathrm{d}s B(h(s)). \tag{2.61}$$

Proof of Lemma 2.14.5 Consider an infinitely differentiable function f with compact support, an element $x \in \mathcal{H}$ and define

$$y := \int \mathrm{d}s U(s)(x) f(s), \tag{2.62}$$

the integral of the Hilbert-space-valued function $h(s) = U(s)(x)f(s)$. We prove first that $y \in \mathcal{D}$. First we note the formula

$$\begin{aligned} U(t)(y) &= \int \mathrm{d}s U(t)U(s)(x)f(s) = \int \mathrm{d}s U(t+s)(x)f(s) \\ &= \int \mathrm{d}s U(s)(x) f(s-t), \end{aligned} \tag{2.63}$$

using (2.61) in the first equality and change of variables in the last one. The dependence on t in the right-hand side is then through the differentiable function $f(s-t)$, and we are in a situation where we may differentiate under the integral sign. It is then straightforward to see that $y \in \mathcal{D}(A)$ and $A(y) = \mathrm{i}\hbar \int \mathrm{d}s U(s)(x) f'(s)$.

Next, we prove that the elements of the type (2.62) are dense in $\mathcal{H}$. Given $x \in \mathcal{H}$ we construct f such that the element (2.62) of $\mathcal{D}$ is arbitrarily close to x. Since U is strongly continuous, for each $\varepsilon > 0$ there exists $\alpha > 0$ such that $\|U(t)(x) - x\| \le \varepsilon$ for $|t| < \alpha$. Choosing $f \ge 0$ supported in $[-\alpha, \alpha]$ and of integral 1 we obtain, using (2.60),

$$\|y - x\| = \Big\| \int \mathrm{d}s U(s)(x) f(s) - \int \mathrm{d}s x f(s) \Big\| \le \int \mathrm{d}s f(s) \|U(s)(x) - x\| \le \varepsilon.$$

□

Our next observation is that $\mathcal{D}$ is stable under the action of the operators $U(s)$.

60 As usual when the domain of integration is not specified, the integration is over the whole space, here $\mathbb{R}$.

Lemma 2.14.6 *For each $s \in \mathbb{R}$ and each $x \in \mathcal{D}$ we have $U(s)(x) \in \mathcal{D}$ and $AU(s)(x) = U(s)A(x)$. Furthermore for $x \in \mathcal{D}$ the function $\varphi(s) = U(s)(x)$ is differentiable and*

$$\varphi'(s) = \frac{\mathrm{i}}{\hbar}U(s)A(x) = \frac{\mathrm{i}}{\hbar}AU(s)(x). \tag{2.64}$$

Proof Let s and x be as above and $y = \varphi(s) = U(s)(x)$. We first note the formula

$$\varphi(t+s) - \varphi(s) = U(t)(y) - y = U(s)(U(t)(x) - x).$$

Taking the derivative in t at $t = 0$ in the equality $\varphi(t+s)-\varphi(s) = U(s)(U(t)(x)-x)$ implies that φ is differentiable and $\varphi'(s) = (\mathrm{i}/\hbar)U(s)A(x)$, whereas taking the same derivative in the equality $\varphi(t+s) - \varphi(s) = U(t)(y) - y$ implies that $y \in \mathcal{D}(A)$ and $\varphi'(s) = (\mathrm{i}/\hbar)A(y)$. □

Now comes the key idea of the proof. For $x \in \mathcal{D}$ there is an explicit formula for $U(s)(x) - x$, and moreover this formula demonstrates that $x \in \mathcal{D}$. This allows a complete understanding of $\mathcal{D}$.

Lemma 2.14.7 *If $x \in \mathcal{D}$ then for all s we have*

$$U(s)(x) - x = \frac{\mathrm{i}}{\hbar}\int_0^s \mathrm{d}v U(v)A(x) = \frac{\mathrm{i}}{\hbar}\int_0^s \mathrm{d}v A(U(v)(x)). \tag{2.65}$$

Conversely, if

$$U(s)(x) - x = \frac{\mathrm{i}}{\hbar}\int_0^s \mathrm{d}v U(v)(z) \tag{2.66}$$

for all $s \in \mathbb{R}$ and some $z \in \mathcal{H}$ then $x \in \mathcal{D}$ and $A(x) = z$.

Proof First, (2.64) implies (2.65). Conversely, if (2.66) holds then

$$\frac{U(t)(x) - x}{t} - \frac{\mathrm{i}}{\hbar}z = \frac{\mathrm{i}}{\hbar t}\int_0^t \mathrm{d}v(U(v)(z) - z)$$

and the right-hand side goes to 0 as $t \to 0$ because $v \mapsto U(v)$ is strongly continuous, so that $v \mapsto U(v)(z)$ is continuous. □

Here starts the tricky part: we have to prove that A is self-adjoint, that is (since we know that A is symmetric) that every element of the domain of $A^\dagger$ is an element of $\mathcal{D}$. So, let us consider an element y in the domain of $A^\dagger$, with the goal of proving that $y \in \mathcal{D}$. By definition of the domain of $A^\dagger$, there exists $z \in \mathcal{H}$ for which

$$\forall x \in \mathcal{D},\ (y, A(x)) = (z, x). \tag{2.67}$$

We will prove that

$$U(s)(y) - y = \frac{\mathrm{i}}{\hbar}\int_0^s \mathrm{d}v U(v)(z), \tag{2.68}$$

which implies that $y \in \mathcal{D}$ according to the second part of Lemma 2.14.7. To prove (2.68) the idea is to prove that both sides have the same inner product with every element x in the dense space $\mathcal{D}$. Considering such x, let us recall that by (2.58) we have

$$(U(s)(y) - y, x) = (y, U(t)(x) - x), \tag{2.69}$$

for $t = -s$. Now, (2.65) implies

$$U(t)(x) - x = \frac{\mathrm{i}}{\hbar}\int_0^t \mathrm{d}v A(U(v)(x)),$$

so that, using (2.61) for the operator B: $x \mapsto (y,x)$ in the first line, and (2.67) in the second line,

$$\begin{aligned}(y, U(t)(x) - x) &= \frac{\mathrm{i}}{\hbar}\int_0^t \mathrm{d}v(y, A(U(v)(x)))\\ &= \frac{\mathrm{i}}{\hbar}\int_0^t \mathrm{d}v(z, U(v)(x))\\ &= \frac{\mathrm{i}}{\hbar}\int_0^t \mathrm{d}v(U(-v)(z), x)\\ &= \Big(-\frac{\mathrm{i}}{\hbar}\int_0^t \mathrm{d}v U(-v)(z), x\Big).\end{aligned}$$

Using (2.69), and since $\mathcal{D}$ is dense, we obtain

$$U(s)(y) - y = -\frac{\mathrm{i}}{\hbar}\int_0^{-s} \mathrm{d}v U(-v)(z).$$

Changing v into $-v$ in the integral implies (2.68) and we have proved one direction in Stone's theorem. We recall that a self-contained proof of the converse can be found in Appendix B.

Let us note the following easy but important fact.

Proposition 2.14.8 *We have $x \in \mathcal{D}(A)$ and $Ax = \lambda x$ if and only if $U(t)(x) = \exp(\mathrm{i}t\lambda/\hbar)x$.*

Proof It is obvious from (2.56) that if $U(t)(x) = \exp(\mathrm{i}t\lambda/\hbar)x$ then $x \in \mathcal{D}(A)$ and $Ax = \lambda x$. To prove the converse we observe that the first equality in (2.65) implies $\mathrm{d}U(s)(x)/\mathrm{d}s = \mathrm{i}\lambda U(s)(x)/\hbar$ and $U(0)(x) = x$. □

Let us investigate the contents of Stone's theorem in the case of the one-parameter groups (2.53) and (2.54) given by

$$U(t)(f)(w) = \exp(\mathrm{i}tw/\hbar)f(w)\;;\; V(t)(f)(w) = f(w+t).$$

As the state space is now $L^2(\mathbb{R})$ its generic element is denoted f, while x and y now denote elements of $\mathbb{R}$. It should be obvious that the infinitesimal generator of the group (2.53) is the "multiplication by x" operator X. The infinitesimal generator P of the group (2.54) is given by

$$P(f)(x) = \lim_{t\to 0} \frac{\hbar}{\mathrm{i}t}(f(x+t) - f(x)) = -\mathrm{i}\hbar \frac{\partial}{\partial x} f(x),$$

and this is the momentum operator $P = -\mathrm{i}\hbar\partial/\partial x$. As the group is really natural, it is less surprising that its infinitesimal generator should be fundamental. It is quite instructive in this case to look at the relation (2.57) which here becomes

$$V(t) = \exp(\mathrm{i}tP/\hbar) = \exp(t\partial/\partial x). \tag{2.70}$$

Applied on a well-behaved function this is Taylor's formula:

$$f(x+t) = \sum_{n\geq 0} \frac{t^n}{n!}\left(\frac{\partial}{\partial x}\right)^n (f)(x) = \exp(\mathrm{i}tP/\hbar)(f)(x). \tag{2.71}$$

It bears repeating the considerable importance of Stone's theorem as a means of *constructing* self-adjoint operators: each time we construct a one-parameter unitary group we construct such an operator. Here is a fundamental example.

Definition 2.14.9 Assume that the one-parameter group $U_i(s)$ describes the action of translation of physical space by s in the ith direction. Then the momentum operator P_i in the same direction is *defined* by

$$U_i(s) = \exp(-\mathrm{i}sP_i/\hbar). \tag{2.72}$$

At first it sounds strange to define the momentum operator in this way because we might feel that we know it from the physics of the model, but at a deeper level the important relation is (2.72). We will soon meet situations where it is the unitary group $U_i(t)$ rather than P_i which is naturally constructed.[61]

To understand the reason behind the definition (2.72), and in particular the reason behind the minus sign in the exponent, let us look at a special case. We go back to the relation (2.43) (in one dimension), where f is not a potential, but a wave function in position state space, measuring the probability density of the presence of a particle. Then the action $U(s)$ of translation of physical space by s on the wave function f is given by

$$U(s)(f)(x) = f(x-s) = \exp(-\mathrm{i}sP/\hbar)(f)(x),$$

where we use (2.71) in the second equality. This is indeed a special case of (2.72).

Let us recall that the operators P and X satisfy the canonical commutation relation (2.19):

$$[X,P] = \mathrm{i}\hbar 1. \tag{2.73}$$

It is absolutely fundamental that in a precise sense there is a *unique* pair of "nice" operators which satisfy it. "Nice" operators satisfy a slightly stronger form of the relation (2.73) which we explain now. First, recalling the operators U, V of (2.53) and (2.54), we observe that

[61] This is also related to a fundamental principle of Classical Mechanics: Noether's theorem relating symmetries of Lagrangians with invariants of the corresponding theory.

$$U(t)\big(V(s)(f)\big)(x) = \exp(\mathrm{i}tx/\hbar)f(x+s),$$

while

$$V(s)\big(U(t)(f)\big)(x) = \exp(\mathrm{i}t(x+s)/\hbar)f(x+s).$$

Since this relation holds whatever the function f we have

$$V(s)U(t) = \exp(\mathrm{i}ts/\hbar)U(t)V(s). \tag{2.74}$$

This relation is called the *integrated version of the canonical commutation relation* (2.73).

We now proceed to show that (2.74) formally implies (2.73). Taking the derivative of (2.74) at zero in t gives, since $U'(0) = \mathrm{i}X/\hbar$,

$$\frac{\mathrm{i}}{\hbar}V(s)X = \frac{\mathrm{i}s}{\hbar}V(s) + \frac{\mathrm{i}}{\hbar}XV(s).$$

Since $V'(0) = \mathrm{i}P/\hbar$, taking now the derivative at $s = 0$ gives the required relation $-PX/\hbar^2 = \mathrm{i}1/\hbar - XP/\hbar^2$. This argument is formal: since X is not bounded it is far from obvious that the function $s \mapsto XV(s)$ has a derivative, but nonetheless it can be shown that the argument is correct when the operators are applied to a nice function, say in the Schwartz space $\mathcal{S}$.

On the other hand, there are examples of operators satisfying (2.73) but not (2.74). To see this, consider $\mathcal{H} = L^2([0,1])$, and for $x \in [0,1]$ and $s \in \mathbb{R}$, denote by $w_s(x)$ the value of $x+s$ modulo 1. Consider $U(t)$ given by (2.53), and define $V(s)$ by $V(s)(f)(x) = f(w_s(x))$. Then since $V(1)$ is the identity, (2.74) is certainly not satisfied. The infinitesimal generators X and P of these one-parameter groups are respectively given by $X(f)(x) = xf(x)$ and $P(f) = -\mathrm{i}\hbar\,\mathrm{d}f/\mathrm{d}x$. The operator X is now bounded, and these operators satisfy (2.73) on the domain of P.

One should say that (2.74) is the fruitful form of the commutation relation (2.73), as it rules out pathologies such as in the previous example. The Stone and von Neumann theorem (which is stated precisely in Theorem C.3.1 in Appendix C) asserts that *up to unitary equivalence there is only one minimal way to realize the commutation relation (2.74).* In particular, once we reached the understanding that the commutation relation (2.74) is central, we know that our model for a particle is unique and canonical. The study of (2.74) is pursued in Appendix C.

2.15 Time-evolution

Consider a physical system, the state of which is described by a vector in $\mathcal{H}$.

Principle 5 *If the system does not change with time (not in the sense that it does not evolve, but in the sense that it is not subjected to variable external influences), its evolution between time t_0 and time t_1 is described by a unitary operator*[62] $U(t_1,t_0)$.

[62] Popular accounts of advanced physics concerning the properties of black holes often discuss the mysterious fact that in certain approaches "information is lost". It was a revelation when Peter Woit explained to me that this just amounts to the fact that unitarity of time-evolution fails, indeed a revolutionary idea.

This operator depends only on $t_1 - t_0$, reflecting the fact that the laws of physics are believed not to change with time.[63]

Please observe the notation: the evolution $U(t_1,t_0)$ is from t_0 to t_1. The reason for this notation is that the evolution of the system from t_0 to t_1 and then from t_1 to t_2 is represented by $U(t_2,t_1)U(t_1,t_0)$, which also represents the same evolution as $U(t_2,t_0)$ so that as in (2.45) these two operators should differ only by a phase, $U(t_2,t_1)U(t_1,t_0) = cU(t_2,t_0)$ for some c of modulus 1. Thus $U(t_2-t_1,0)U(t_1-t_0,0) = cU(t_2-t_0,0)$. This means that $U(t) := U(t,0)$ should be a projective representation of $\mathbb{R}$ in $\mathcal{H}$, and on physical grounds it should be continuous in some sense. As shown in Section 2.13 this projective representation arises from a true representation, so we can as well assume that it already is a true representation, and Stone's theorem describes these. Therefore, there exists a self-adjoint operator H on $\mathcal{H}$ such that

$$U(t) = \exp(-\mathrm{i}tH/\hbar). \tag{2.75}$$

Probably it is worth making explicit the following fundamental point:

The time-evolution of a quantum system is entirely deterministic.

The minus sign in (2.75) is conventional. The reason for this convention will appear in Section 9.7. Since $\hbar$ has the dimension of an energy times a time, H has the dimension of an energy. It is called the *Hamiltonian* of the system. Although this is certainly not obvious[64]

The Hamiltonian should be thought of as representing the energy of the system.

A consequence of (2.75) is that if ψ belongs to the domain of H, then, by the second part of (2.64), $\psi(t) := U(t)(\psi)$ satisfies the equation

$$\psi'(t) = -\frac{\mathrm{i}}{\hbar}H(\psi(t)). \tag{2.76}$$

This is the fundamental *Schrödinger equation*. The case one sees first when opening a textbook on Quantum Mechanics concerns a free (i.e. submitted to no influence) particle of mass m on the line. In this case the Hilbert space is $L^2(\mathbb{R},\mathrm{d}x)$. We denote an element of L^2 by f, while x denotes a point of $\mathbb{R}$. Then

$$H = \frac{p^2}{2m}, \tag{2.77}$$

or, equivalently,

$$H(f) = -\frac{\hbar^2}{2m}\frac{\mathrm{d}^2 f}{\mathrm{d}x^2}.$$

It is absolutely not obvious or intuitive yet that this is the correct formula, although one should note that if $p = mv$ is the momentum of a classical particle, then $p^2/2m = mv^2/2$

[63] Thus, during a given interval of time, the time-evolution of the system has to be the same, independently of the moment at which the time-evolution started.

[64] We will take this for granted. It is difficult to justify this without some notions of Classical Mechanics.

is its kinetic energy, so that (2.77) is consistent with the fact that the Hamiltonian should represent the energy of the system.

The state of the particle at time t is described by the wave function $f(\cdot,t) \in L^2$. It satisfies the equation (2.76) which here is

$$\frac{\partial f(x,t)}{\partial t} = \frac{\mathrm{i}\hbar}{2m}\frac{\partial^2 f(x,t)}{\partial x^2}. \tag{2.78}$$

Free particles are not very exciting. More interesting is the case where there is a potential V, so that V is a function on $\mathbb{R}$. The Hamiltonian is then

$$H(f) = -\frac{\hbar^2}{2m}\frac{\mathrm{d}^2 f}{\mathrm{d}x^2} + Vf, \tag{2.79}$$

where Vf is the multiplication of f by the function V. The wave function then satisfies the relation

$$\frac{\partial f(x,t)}{\partial t} = \frac{\mathrm{i}\hbar}{2m}\frac{\partial^2 f(x,t)}{\partial x^2} - \frac{\mathrm{i}}{\hbar}V(x)f(x,t), \tag{2.80}$$

which is called Schrödinger equation as well. Countlessly many examples of this situation (and higher-dimensional cases) can be found in Quantum Mechanics textbooks. The fundamental case of quadratic potentials will be examined in Section 2.18. Mathematically, the challenge is to show that the Hamiltonian (2.79), which is defined, say, on functions in the Schwartz space is actually a self-adjoint operator when considered on the proper domain. Stone's theorem tells us that it makes sense to consider the time-evolution of the system as provided by (2.75) if and only if this is the case. It bears repeating that it makes *no sense* to speak of time-evolution unless H is self-adjoint. It is in this kind of situation that the notion of essentially self-adjoint operators of Definition 2.5.15 is useful, as it provides a method to construct the domain of H. It is an important topic to prove that certain classes of operators are self-adjoint and many results exist in that direction.

Going back to the general setting, the fundamental problem is to compute the time-evolution, that is to compute $\psi(t)$ knowing $\psi(0)$. This is very easy when $\psi(0)$ is an eigenvector of H, $H(\psi(0)) = E\psi(0)$ because then

$$\psi(t) = \exp(-\mathrm{i}tE/\hbar)\psi(0) \tag{2.81}$$

solves the equation (2.76). The eigenvalue equation

$$H(\psi) = E\psi \tag{2.82}$$

is called the *time-independent Schrödinger equation.* If H has a basis of eigenvectors, we can then describe the time-evolution of the system by approximating any vector by a combination of such vectors. In other words, when H is diagonalizable, it is easy to compute $\exp(-\mathrm{i}tH/\hbar)$. Therefore once we have found a basis of eigenvectors of H we have reached as complete an understanding of time-evolution as is possible. Alas, more often than not it is impossible to solve exactly the time-independent Schrödinger equation, so that one needs to develop approximation methods, another important topic for practitioners.

Another issue is that H may not have a basis of eigenvectors, as is typically the case for the Hamiltonian in equation (2.80). There is a simple reason for that, which we explain now. Under (2.81), $\psi(t)$ is of the type $\alpha\psi(0)$ where α is a complex number of modulus 1, and as we have seen at the end of Section 2.1 it represents the same physical state, so that the measurements at time t do not differ from the measurements at time zero. Let us describe without any proof the case of a potential $V \leq 0$, which goes to zero fast enough at infinity (and for which $|V|$ is "not too large"). Then there is a sequence (possibly finite or even empty) of eigenstates with negative eigenvalues which correspond to "bound states". The name "bound state" reflects the fact that a particle in this state is "bound" by the potential, it does not go away. If one measures the position of the particle at any time, there only a small probability that it is far away. The corresponding eigenstates do not however span the whole state space. The orthogonal complement $\mathcal{H}_{\text{scat}}$ of their span consists of "scattering states", which one should picture as particles coming from infinity, interacting with the potential and going to infinity again. The space $\mathcal{H}_{\text{scat}}$ contains no eigenvector. This is not surprising: the results we obtain when measuring the position of such a particle at time t certainly depend on t, so the time-evolution cannot be of the type 2.81.

2.16 Schrödinger and Heisenberg Pictures

The previous description of an evolving state $\psi(t)$ and of time-independent operators is called the Schrödinger picture.

We have seen that we may be able to improve matters by re-shuffling the state space using a unitary transformation. A fundamental idea is to try this using a *time-dependent* unitary transformation $V(t)$, replacing the state ψ by $V(t)(\psi)$, and replacing the operator A by $A(t) = V(t)AV(t)^{-1}$. Here we use a first simple implementation, with $V(t) = \exp(\mathrm{i}tH/\hbar) = U(t)^{-1}$. This is called the *Heisenberg picture*. In the Heisenberg picture the state $\psi(t)$ is replaced by $V(t)(\psi(t)) = U(t)^{-1}U(t)(\psi) = \psi$, so that *states do not change with time*. On the other hand, an operator A is replaced by the operator

$$A(t) = U(t)^{-1}AU(t) = \exp(\mathrm{i}tH/\hbar)A\exp(-\mathrm{i}tH/\hbar). \tag{2.83}$$

Suppose that at time $t = 0$ the system is in state ψ. Then, in the Schrödinger picture, the average value of A at time t is given by

$$(\psi(t), A\psi(t)) = (U(t)\psi, AU(t)\psi) = (\psi, U(t)^{-1}AU(t)\psi) = (\psi, A(t)\psi), \tag{2.84}$$

where the last expression is the same quantity in the Heisenberg picture. Thus these two pictures are fortunately consistent with each other.

We may wonder why there is anything to gain by moving from the Schrödinger picture, where the simpler objects, the states, evolve, but where the complicated objects, the operators, are constant, to the Heisenberg picture where the simpler objects are constant, but the complicated ones evolve. One reason is that while it is correct in principle that the states, which are simply vectors of $\mathcal{H}$ are simpler objects than operators, it will often happen that the operators of interest have a simple description, while the states have a very complicated one. Another reason will be given at the end of the present section.

The evolution (2.83) results in the equation[65]

$$\dot{A}(t) := \frac{\mathrm{d}A(t)}{\mathrm{d}t} = \frac{\mathrm{i}}{\hbar}[H, A(t)]. \tag{2.85}$$

Thus when A and H commute, $A(t) = A$ does not depend on time. *Operators which commute with the Hamiltonian have the same expression in the Schrödinger and the Heisenberg picture.* In particular if this is the case the average observed value of A on a given state does not depend on time. This is of course fundamental. In particular since H commutes with itself, its average value does not change over time. As H represents the energy of the system, this is the *principle of conservation of energy.*[66]

At this point the reader would certainly like to see concrete examples. However the time-evolution of an operator in the Heisenberg picture depends not only on the operator but also on the Hamiltonian, and we should not expect any simple way to write this time-evolution unless we understand the Hamiltonian well. The only simple Hamiltonian we have written so far is given by $H = P^2/(2m)$. It commutes with P, so that $P(t) = P$. To compute $X(t)$, it is convenient to work in momentum state space (and in this case the operator is denoted $X'(t)$ as in (2.24) and the generic element of the state space is denoted by φ), because then H is just "multiplication by $p^2/(2m)$" and $\exp(\mathrm{i}tH/\hbar)$ is multiplication by $\exp(\mathrm{i}tp^2/(2m\hbar))$. To compute $X'(t)(\varphi)$, according to (2.83), we multiply φ by the function $p \mapsto \exp(-\mathrm{i}tp^2/(2m\hbar))$, we apply $X'(0)$, and we multiply again by the function $\exp(\mathrm{i}tp^2/(2m\hbar))$. According to (2.24), applying $X'(0)$ is simply taking $\mathrm{i}\hbar$ times the derivative in p, so that finally

$$X'(t)(\varphi) = X'(0)(\varphi) + \frac{pt}{m}\varphi, \tag{2.86}$$

where the last term is the application of the operator "multiplication by pt/m" to the function φ. More examples will be given in Section 2.18.

Exercise 2.16.1 Convince yourself that when the support of φ is close to a given value p, the last term in (2.86) represents the shift in position due to the velocity p/m of the "particle represented by the state φ".

The following observation is also very important. In non-relativistic Quantum Mechanics, position is given by an operator, while time is a *label*. So in this approach, position and time are of a different nature, in contradiction with the theory of Special Relativity. In the Heisenberg picture, Hermitian operators are labeled with t when they represent a measurement at time t. In Quantum Field Theory, there will be no position operator. Position will be demoted to a simple label and Hermitian operators will be labeled by points in space-time. Such an operator should be thought of as making a measurement at the corresponding point of space-time. The Heisenberg picture is important because it allows in this manner to put space and time on an equal footing.

[65] Informality starts to creep in, as it is not clear in which domain the following equality should hold.

[66] A fundamental theorem of Classical Mechanics, called Noether's theorem, relates symmetries of a system with invariant quantities. It brings clearly forward the fact that conservation of energy is a consequence of the fact that the laws of physics are time-invariant. If you plan to build a perpetual motion machine, you will save much effort by studying this theorem first.

2.17 A First Contact with Creation and Annihilation Operators

High-energy interacting particles create other particles, and a relativistic theory must consider multiparticle systems, where the number of particles may vary.

Let us describe what is probably the simplest example of a multiparticle system.[67] The "particles" are as simple as possible.[68]

Consider a separable Hilbert space with an orthonormal basis $(e_n)_{n\geq 0}$. The idea is that the state of the system is described by e_n when the system consists of n particles. The important structure consists of the operators a and $a^\dagger$ defined on the domain

$$\mathcal{D} = \Big\{ \sum_{n\geq 0} \alpha_n e_n \; ; \; \sum_{n\geq 0} n|\alpha_n|^2 < \infty \Big\} \tag{2.87}$$

by

$$a(e_n) = \sqrt{n}\, e_{n-1} \; ; \; a^\dagger(e_n) = \sqrt{n+1}\, e_{n+1}. \tag{2.88}$$

The definition of $a(e_n)$ is to be understood as $a(e_0) = 0$ when $n = 0$. The reason for the factors $\sqrt{n}$ and $\sqrt{n+1}$ is not intuitive, and will become clear only gradually.

The notation is consistent, since for each n, m,

$$(e_n, a(e_m)) = \sqrt{m}\delta_n^{m-1} = \sqrt{m}\delta_{n+1}^m = (a^\dagger(e_n), e_m), \tag{2.89}$$

where δ_n^m is the Kronecker symbol (equal to 1 if $n = m$ and to 0 otherwise).

Exercise 2.17.1 Prove that $a^\dagger$ is the adjoint of a. Prove in particular that if $|(y, a(x))| \leq C\|x\|$ for $x \in \mathcal{D}$ then $y \in \mathcal{D}$.

Exercise 2.17.2 Prove that for each $\lambda \in \mathbb{C}$ the operator a has an eigenvector with eigenvalue λ. Can this happen for a symmetric operator?

It should be obvious from (2.88) that

$$a^\dagger a(e_n) = n e_n \; ; \; a a^\dagger(e_n) = (n+1)e_n. \tag{2.90}$$

Let us then consider the self-adjoint operator[69]

$$N := a^\dagger a. \tag{2.91}$$

Thus $N(e_n) = n e_n$. Since a system in state e_n has n particles, the observable corresponding to this operator is "the number of particles". The operator N is therefore called the *number operator*.

As another consequence of (2.90),

$$[a, a^\dagger](e_n) = e_n, \tag{2.92}$$

[67] I will not apologize for being the only author in the known universe to present this material before introducing the harmonic oscillator rather than the other way around. My point is to emphasize that for the purposes of Quantum Field Theory the idea of creation and annihilation operators is *absolutely central*.

[68] I will write the word particle without quotes, but as we will see later these are not really particles in the usual sense.

[69] It is formally obvious that N is self-adjoint using the formula $(AB)^\dagger = B^\dagger A^\dagger$, but of course it requires a little proof to show that N is self-adjoint with domain $\{y \; ; \; \sum_{n\geq 0} n^2 |y_n|^2 < \infty\}$.

which one can write as

$$[a,a^\dagger] = 1, \tag{2.93}$$

at least on the domain $\mathcal{D}$.[70] It is not surprising that such a simple basic relation will turn out to be fundamentally important. In fact, at least formally, the operators

$$P := \mathrm{i}\sqrt{\frac{\hbar}{2}}(a^\dagger - a)\ ;\ X := \sqrt{\frac{\hbar}{2}}(a^\dagger + a) \tag{2.94}$$

are self-adjoint and satisfy the canonical commutation relation $[X,P] = \mathrm{i}\hbar 1$, so in a sense (2.93) is just another way to look at this relation. The operators (2.94) are called X and P as in Section 2.5 because (barring pathologies) by the Stone–von Neumann theorem mentioned at the end of Section 2.14, the pair (P,X) is (up to a unitary transformation) characterized by the canonical commutation relation.[71]

Since the relation (2.93) is just a reformulation of the canonical commutation relation, the Stone–von Neumann theorem implies that, barring pathologies,

There is a unique *minimal operator a which satisfies (2.93).*

Minimal here means that there is no non-trivial subspace that is invariant under both a and $a^\dagger$. Unique means that two such operators are related by a unitary transformation. The intuitive reason for this result is clear: one can find an orthonormal basis (e_n) in which the operators a and $a^\dagger$ are given by (2.88).[72] As the operator a satisfying (2.93) is essentially unique, there is no point to specify it.

In the sequel we will write countless formulas involving an unspecified operator a satisfying (2.93).

Let us look again at the operators a and $a^\dagger$. By our convention, the non-unit vector $\sqrt{n+1}e_{n+1}$ represents the same state of the system as the unit vector e_{n+1}, i.e. the state where $n+1$ particles are present. Thus $a^\dagger$ transforms the n-particle state e_n into the $(n+1)$-particle state $\sqrt{n+1}e_{n+1}$. It "creates a particle" and is therefore called the *creation operator*. For similar reasons the operator a is called the *annihilation operator* as it destroys a particle. One should carefully distinguish e_0, the unit vector describing the state where no particles are present, from the element zero of the state space, *which does not represent any state of the system*, and one should remember the relation $a(e_0) = 0$.

We comment now on the square root factors in (2.88). These factors are related to a very deep law of Nature: bosons (in a very simple form here) like to huddle together (and, among other things, this is what makes the laser possible). You may have heard of bosons and fermions in connection with spin, but this is a different and even deeper story. It will become clearer only gradually why our quanta of excitation here should be called bosons. Here we simply explain why they like to huddle together by making a sample computation. Consider the state $e_0 + e_n$, or if one wants to stick with unit vectors, the state $(e_0 + e_n)/\sqrt{2}$.

70 We remind the reader that when writing the commutator of unbounded operators, one always must inquire what is its domain of definition.

71 In fact, the work of the next section will prove that there exists an orthonormal sequence (e_n) in $L^2(\mathbb{R})$ such that the operators (2.94) are exactly the operators of Section 2.5.

72 As exemplified in the next section it *does* require work to find the proper vectors e_n!

It is not an eigenvector of the number operator N, so it does not have a well-defined number of particles. Using the operator (2.91) to count the number of particles of this state, (2.6) shows that we will obtain either 0 or n each with probability $1/2$. Let us now create a new particle by applying the operator $a^\dagger$ to this state. We obtain the vector $e_1+\sqrt{n+1}e_{n+1}$. The corresponding state does not have a well-defined number of particles either. If we normalize this vector to unity and use (2.6), we find that when measuring the number of particles, the probability to observe 1 particle is $1/(1+(n+1))$, much smaller than the probability $(n+1)/(1+(n+1))$ of observing $n+1$ particles. Informally, the new particle loves to join the existing n particles. Bosons are gregarious.

2.18 The Harmonic Oscillator

The fundamental structure outlined in the previous section is connected to an equally fundamental system, the *harmonic oscillator*. A classical one-dimensional harmonic oscillator of angular frequency[73] ω consists of a point of mass m on the real line which is pulled back to the origin with a force $m\omega^2$ times the distance to the origin. The quantum version of this system is the space $\mathcal{H} = L^2(\mathbb{R})$ with Hamiltonian

$$H := \frac{1}{2m}(P^2 + \omega^2 m^2 X^2), \tag{2.95}$$

where P and X are respectively the momentum and the position operators of Section 2.5. This is the Hamiltonian (2.79) in the case where $V(x) = m\omega^2 x^2/2$. That this formula provides a quantized version of the classical harmonic oscillator is not obvious at all.[74] We will explain in Section 6.6 the systematic procedure of "canonical quantization" to discover formulas such as (2.77) or (2.95). This procedure is by no means a proof of anything, and the resulting formulas are justified only by the fact that they provide a fruitful model. So there is little harm to accept at this stage that the formula (2.95) is indeed fundamental. We have not proved yet that this formula defines a self-adjoint operator, but this is a consequence of the analysis below.

Exercise 2.18.1 Prove that a symmetric operator which admits an orthonormal basis of eigenvectors is self-adjoint. Hint: If we denote by (e_n) an orthonormal basis of eigenvectors, and λ_n the eigenvalue of e_n, the natural domain $\mathcal{D}$ of the operator is

$$\Big\{x = \sum_n x_n e_n\ ;\ \sum_n (1+|\lambda_n|^2)|x_n|^2 < \infty\Big\}.$$

The program for this section is first to find a basis of eigenvectors for the Hamiltonian (2.95), and then to examine how some classical quantities transform under quantization.

[73] For a periodic function such as $\exp(i\omega t)$, the quantity ω is called the *angular frequency*, whereas the frequency, i.e. the reciprocal of the period, equals $\omega/2\pi$.

[74] On the other hand we should certainly expect that quadratic potentials will be important, as they are among the simplest non-linear potentials.

To start the study of this Hamiltonian, we recall that P and X satisfy the commutation relation $[X,P] = \mathrm{i}\hbar 1$. Let us now consider

$$a = \alpha X + \mathrm{i}\beta P \tag{2.96}$$

where

$$\alpha = \sqrt{\frac{\omega m}{2\hbar}}\ ;\ \beta = \sqrt{\frac{1}{2\omega m\hbar}},$$

so that $\alpha\beta = 1/(2\hbar)$ and (at least formally)

$$a^\dagger = \alpha X - \mathrm{i}\beta P, \tag{2.97}$$

and thus

$$[a,a^\dagger] = -\mathrm{i}\alpha\beta[X,P] + \mathrm{i}\alpha\beta[P,X] = 2\alpha\beta\hbar 1 = 1, \tag{2.98}$$

the fundamental relation (2.93). Let us also observe that $\omega m\hbar$ has the dimension of the square of a momentum, and that $\hbar/\omega m$ has the dimension of the square of a length, so that a and $a^\dagger$ are dimensionless.[75] We observe that

$$P = \mathrm{i}\sqrt{\frac{\omega m\hbar}{2}}(a^\dagger - a), \tag{2.99}$$

$$X = \sqrt{\frac{\hbar}{2\omega m}}(a^\dagger + a), \tag{2.100}$$

and then

$$\begin{aligned} H &= \frac{1}{2m}(P^2 + \omega^2 m^2 X^2) = \frac{\omega\hbar}{4}(-(a-a^\dagger)^2 + (a+a^\dagger)^2) \\ &= \frac{\omega\hbar}{2}(a^\dagger a + aa^\dagger) = \omega\hbar\,(a^\dagger a + \frac{1}{2}1). \end{aligned} \tag{2.101}$$

This is basically the number operator (2.91), so we may guess that a basis of eigenvectors for the Hamiltonian is simply a basis of vectors e_n as in (2.88).

Theorem 2.18.2 *The Hamiltonian (2.95) admits an orthogonal basis of eigenvectors e_n with corresponding eigenvalues $(n + 1/2)\omega\hbar$. These eigenvectors satisfy the relations (2.88).*

The heuristic argument found in the vast majority of physics textbooks proceeds as follows. First, one computes

$$[H,a] = \omega\hbar(a^\dagger a^2 - aa^\dagger a) = \omega\hbar(a^\dagger a^2 - (a^\dagger a + 1)a) = -\omega\hbar a, \tag{2.102}$$

and, similarly, $[H,a^\dagger] = \omega\hbar a^\dagger$. If φ is an eigenvector for H, $H(\varphi) = \lambda\varphi$, then

$$Ha(\varphi) = aH(\varphi) - \omega\hbar a(\varphi) = (\lambda - \omega\hbar)a(\varphi),$$

75 In the sense that they do not need to be rescaled if we change our system of units.

so that either $a(\varphi)$ is an eigenvector of eigenvalue $\lambda - \omega\hbar$ or it is 0. Iterating this statement, for each k, either $a^k(\varphi) = 0$ or else $a^k(\varphi)$ is an eigenvector of eigenvalue $\lambda - k\omega\hbar$. On the other hand, (2.101) implies that

$$(H(\varphi),\varphi) = \omega\hbar\big((a^\dagger a(\varphi),\varphi) + \frac{1}{2}(\varphi,\varphi)\big) = \omega\hbar\big((a(\varphi),a(\varphi)) + \frac{1}{2}(\varphi,\varphi)\big) \geq 0,$$

so that all the eigenvalues of H are ≥ 0.

Let us then start with an eigenvector φ of eigenvalue λ. As we have seen, for each k, either $a^k(\varphi) = 0$ or else $a^k(\varphi)$ is an eigenvector of eigenvalue $\lambda - k\omega\hbar$. Since all eigenvalues are ≥ 0, there exists a largest k for which $\psi := a^k(\varphi) \neq 0$ and therefore $a(\psi) = 0$. Since $a(\psi) = 0$, it is obvious from (2.101) that ψ is an eigenvector with eigenvalue $\omega\hbar/2$. Consequently the original eigenvalue λ is of the type $\lambda = \omega\hbar(k + 1/2)$. One can then continue the analysis as will be detailed below, using now that since $[H,a^\dagger] = \omega\hbar a^\dagger$, if φ is an eigenvector of eigenvalue λ, then $a^\dagger(\varphi)$ is an eigenvector of eigenvalue $\lambda + \omega\hbar$.

The problem with the previous argument is that we implicitly assume that the eigenvector φ belongs to the domain of a ($a^\dagger$), in line with physicists' approach which assumes by default that all operators are defined everywhere.

Proof of Theorem 2.18.2 The secret of a rigorous argument is to construct only Schwartz functions on which P and X, and hence a and $a^\dagger$ are well-defined. The first step is to look for a norm-1 vector e_0 which satisfies $a(e_0) = 0$, because we have observed that such a vector will be an eigenvector, presumably with the smallest possible eigenvalue. Recalling the definition (2.96) of a this amounts to solving the differential equation $\omega m x\varphi + \hbar\varphi' = 0$, so that we can choose e_0 given by the function

$$\varphi_0(x) = \theta \exp(-\omega m x^2/2\hbar) \tag{2.103}$$

for an appropriate value of θ $(= (m\omega/\pi\hbar)^{1/4})$ which makes it of norm 1. One then defines recursively the vectors e_n by the second part of (2.88), i.e. $e_{n+1} = a^\dagger(e_n)/\sqrt{n+1}$ and one uses (2.92) to obtain

$$\sqrt{n+1}a(e_{n+1}) = aa^\dagger(e_n) = a^\dagger a(e_n) + e_n.$$

It follows recursively from this equality that $a(e_n) = \sqrt{n}e_{n-1}$, the first part of (2.88), because

$$a^\dagger a(e_n) + e_n = \sqrt{n}a^\dagger(e_{n-1}) + e_n = (n+1)e_n.$$

Moreover

$$\sqrt{n+1}(e_{n+1},e_{n+1}) = (e_{n+1},a^\dagger(e_n)) = (a(e_{n+1}),e_n) = \sqrt{n+1}(e_n,e_n),$$

so that recursively the norm of e_n is 1. Furthermore, (2.101) shows that e_n is an eigenvector of H of eigenvalue $(n + 1/2)\omega\hbar$. Thus the vectors e_n are orthogonal to each other since they are eigenvectors of the Hamiltonian with different eigenvalues. Finally, one sees by induction that e_n is a function of the type $P_n(x)\varphi_0(x)$ where P_n is a polynomial of degree n.[76] Thus the closed linear span $\mathcal{G}$ of the vectors e_n contains all the functions which are the product of a polynomial by φ_0. Thus $\mathcal{G}$ contains all the functions $\exp(\mathrm{i}tx)\varphi_0(x)$, as is seen by approximating the exponential by the first terms of its power series. Moreover, by computing averages $\int \mathrm{d}t\psi(t)\exp(\mathrm{i}tx)\varphi_0(x)$ of the functions $\exp(\mathrm{i}tx)\varphi_0(x)$, $\mathcal{G}$ contains all functions $\varphi(x)\varphi_0(x)$ where φ is the Fourier transform of a Schwartz function. Since every Schwartz function is the Fourier transform of a Schwartz function (its inverse Fourier transform), $\mathcal{G}$ contains all functions $\varphi(x)\varphi_0(x)$ where φ is a Schwartz function. Since the previous set of functions is dense we have $\mathcal{G} = \mathcal{H}$ and the vectors e_n form an orthonormal basis of L^2. □

Exercise 2.18.3 Prove in detail that the closure of the space of functions of the type $P(x)\varphi_0(x)$ where P is a polynomial contains the functions $\exp(\mathrm{i}tx)\varphi_0(x)$ for $t \in \mathbb{R}$. Hint: Use the dominated convergence theorem.

Since e_n is an eigenvector for the Hamiltonian H with eigenvalue $\omega\hbar(n + 1/2)$ it represents a state of energy $\omega\hbar(n + 1/2)$. We will describe the situation by saying that this state has "n quanta of excitation".[77] Let us pause a moment here. The classical harmonic oscillator is a continuous system, which can have any positive energy. On the other hand, the possible values of the energy of the quantum harmonic oscillator form a discrete set. This is quite a general phenomenon. It motivates the "quantum" in the name Quantum Mechanics.

The state with lowest energy has no quantum of excitation. It is important that it *does not* have zero energy, but rather energy $\omega\hbar/2$. To understand why the energy of a state cannot be zero, going back to (2.95), the harmonic oscillator models a particle on the line in a certain potential which is zero only at the origin. To have zero energy the particle would need to have both kinetic and potential energy zero, that is to be still at the origin. Thus we would exactly know both its momentum and its position. This is however impossible by Heisenberg's uncertainty principle.

If we think of a quantum of excitation as a (very simplified) "particle", the harmonic oscillator is a model where the number of particles can take any value $n \geq 0$. The average number of particles is given by the number operator N. Since obviously N and H commute, "the average number of particles does not change with time". Moreover, it can be said that the "particles" do not interact with each other (as will become gradually apparent).

Later, in Section 3.4 we will study the following problem.[78] Given a space $\mathcal{H}$ which describes the possible states of a single particle, what is the state space of the system where any number of the same particle type may be present? In a sense the harmonic oscillator is the simplest example of this situation. The individual particle is just a quantum of excitation[79] and the harmonic oscillator is a state space where any number of particles of this type may be present. This explains why the harmonic oscillator will be ubiquitous in the sequel. More precisely, systems unitarily equivalent to it will be ubiquitous. The particular sequence of energy eigenfunctions in $L^2(\mathbb{R})$ which we constructed is not important,

[76] The relationship between these polynomials and the classical Hermite polynomials is explained in Section C.2.
[77] Do not look for some deep meaning here, this is just a convenient way to describe the situation.
[78] Specifically on page 84.
[79] We should think of a quantum of excitation as having state space $\mathbb{C}$.

compared to the pair of operators $a, a^\dagger$ satisfying the relations (2.88) in an orthonormal basis e_n, together with a Hamiltonian of the form $\hbar\omega(a^\dagger a + (1/2)1)$, for which e_n is an eigenvector of eigenvalue $\hbar\omega(n+1/2)$.

The free quantum field, a basic object of our future studies is in a sense a sum[80] of non-interacting[81] harmonic oscillators (in the previous sense).[82] Some mysterious looking formulas become transparent when one realizes that they reduce to calculations on a single harmonic oscillator.

We perform below some simple computations for later use, that also have some intrinsic interest. Indeed, when we quantize a classical system, we find a quantized version of all the classical quantities attached to the system, and we expect that they satisfy the same relations as the corresponding classical quantities. It is a good consistency check to prove some of these relations. Changing focus, it is in general difficult to explain how the classical behavior of a system emerges from quantum behavior. This can be done simply for the harmonic oscillator (see Section C.2).

We now compute how the various operators connected with the harmonic oscillator evolve in the Heisenberg picture. We recall that the evolution of an operator A is given by (2.83) or, equivalently, (2.85).

Exercise 2.18.4 Prove directly from the definition that

$$a(t) = \exp(-\mathrm{i}t\omega)a. \tag{2.104}$$

Alternatively (2.104) follows from (2.85) and (2.102). Of course, one has

$$a^\dagger(t) = \exp(\mathrm{i}t\omega)a^\dagger. \tag{2.105}$$

One can then time-evolve the operators P and X of (2.99) and (2.100):

$$P(t) = \mathrm{i}\sqrt{\frac{\omega m\hbar}{2}}(a^\dagger\exp(\mathrm{i}t\omega) - a\exp(-\mathrm{i}t\omega)) \tag{2.106}$$

$$X(t) = \sqrt{\frac{\hbar}{2\omega m}}(a^\dagger\exp(\mathrm{i}t\omega) + a\exp(-\mathrm{i}t\omega)), \tag{2.107}$$

yielding the relation

$$\dot{X}(t) := \frac{\mathrm{d}X(t)}{\mathrm{d}t} = \frac{1}{m}P(t). \tag{2.108}$$

This echoes the classical relation that the momentum is the mass times the time-derivative of the position. As a consequence of (2.108),

$$\left[\dot{X}(t), X(t)\right] = -\frac{\mathrm{i}\hbar}{m}1. \tag{2.109}$$

80 This statement is correct when we put the universe in a box, otherwise it is a kind of continuous sum.
81 I will sometimes use the word "independent" as synonymous to "non-interacting".
82 I wasted an infinitely long time trying to figure out what the authors of many elementary textbooks mean when they say "one replaces each point of space-time by a harmonic oscillator".

Let us note also that, reproducing the computation of (2.101) we have

$$H = \frac{1}{2m}\big(P(t)^2 + \omega^2 m^2 X(t)^2\big). \tag{2.110}$$

This is expected since H does not evolve with time as it commutes with itself, so that (2.110) also follows from time-evolving (2.95), i.e. multiplication by $\exp(\mathrm{i}H/\hbar)$ to the left and $\exp(-\mathrm{i}H/\hbar)$ to the right.

Let us also note for further use that combining with (2.108) this yields

$$H = \frac{m}{2}\big(\dot{X}(t)^2 + \omega^2 X(t)^2\big). \tag{2.111}$$

This equation has an important content. Under the Hamiltonian of the harmonic oscillator, the time-evolution of the position operator is given by (2.107), and conversely, the time-evolution (2.107) determines the Hamiltonian.

Key ideas to remember:[83]

- One does not "prove" the basic principles of Quantum Mechanics. The ultimate test for a model is the agreement of its predictions with experiments.
- In Quantum Mechanics the state of a physical system is represented by a vector in a complex Hilbert space, the state space $\mathcal{H}$. This vector encompasses everything we can possibly know about the system.
- When the state space $\mathcal{H}$ is finite-dimensional, observables correspond to Hermitian operators on $\mathcal{H}$.
- The predictions of Quantum Mechanics are only statistical in nature.
- It is typically not possible to know the values of two different observables at the same time.
- Measurements typically change the state of the system.
- The theory of self-adjoint operators generalizes the theory of Hermitian operators to infinite-dimensional state spaces.
- Massive particles can also be seen as waves. Position state space is the appropriate way to describe a massive particle when it is thought of as a particle, and momentum state space when it is thought of as a wave.
- The position operator X and the momentum operator P of a particle in one dimension satisfy the canonical commutation relation $[X,P] = \mathrm{i}\hbar 1$.
- Fear not Dirac formalism, but make sense of it by integration against test functions.
- Unitary operators are the natural class of isomorphisms between complex Hilbert spaces.
- Projective unitary representations are ubiquitous, reflecting the symmetries of the universe, but mathematicians prefer true unitary representations.
- Stone's theorem provides a one-to-one correspondence between self-adjoint operators and strongly continuous unitary representations of $\mathbb{R}$.

[83] If you have any doubt about what the following succinct statements mean, please review the corresponding section.

- The time-evolution of a quantum system is entirely deterministic. It is governed by a special self-adjoint operator, the Hamiltonian, which should be thought of as representing the energy of the system.
- The Schrödinger picture and the Heisenberg picture are equivalent ways to describe the time-evolution of a quantum system.
- The harmonic oscillator is a ubiquitous quantum system. It can be thought of as describing a multiparticle system. The creation and annihilation operators on the state space are a fundamental structure associated to this system.

3

Non-relativistic Quantum Fields

We will learn in Section 3.1 how to describe a multiparticle system, and to construct Hamiltonians describing the evolution of particles that do not interact. Their total number remains constant. This is not necessarily the case for interacting particles, a phenomenon that has its roots in the equivalence of mass and energy. For example an electron may annihilate with a positron, leaving (at low kinetic energy) only energy in the form of two photons.[1] To describe interacting particles we must consider situations where the number of particles may not be fixed. This is the purpose of Section 3.4.

Interesting physics comes from interactions, and our long-term goal is to construct Hamiltonians which model these interactions, and to study the corresponding time-evolution. It turns out that the need to make the theory compatible with Special Relativity puts very stringent requirements on the possible form of the Hamiltonians, and that their construction is a highly non-trivial task. They will be built out of simpler objects, quantum fields, which are operator-valued distributions. In Section 3.10 we will construct a first toy example of such a field, not trying yet to make it compatible with Special Relativity. It will then be a long road to learn how our theory should be consistent with Special Relativity, what we should choose as models for a particle, and what kind of quantum fields are consistent with Special Relativity. All these are necessary steps before any sensible Hamiltonian can be written out.

3.1 Tensor Products

The present section is standard material, but our presentation attempts to balance rigor and readability.

Principle 6 *If the states of two systems $\mathcal{S}_1$ and $\mathcal{S}_2$ are represented by the unitary rays in two Hilbert spaces $\mathcal{H}_1$ and $\mathcal{H}_2$ respectively, the appropriate Hilbert space to represent the system consisting of the union of $\mathcal{S}_1$ and $\mathcal{S}_2$ is the tensor product $\mathcal{H}_1 \otimes \mathcal{H}_2$.*

[1] The reason why there cannot be a single photon produced is the conservation of momentum. At high energy other particles may be created.

Our first task is to describe this space.[2] A mathematician would love to see an "intrinsic" definition of this tensor product, a definition that does not use bases or a special representation of these Hilbert spaces. This can be done elegantly as in e.g. Dimock's book [23]. We shall not enjoy this piece of abstraction and we shall go the ugly way.

If $(e_n)_{n\geq 1}$ is an orthonormal basis of $\mathcal{H}_1$ and $(f_n)_{n\geq 1}$ is an orthonormal basis of $\mathcal{H}_2$ then the vectors $e_n \otimes f_m$ constitute an orthonormal basis of $\mathcal{H}_1 \otimes \mathcal{H}_2$, which is thus the set of vectors of the type $\sum_{n,m\geq 1} a_{n,m} e_n \otimes f_m$ where the complex numbers $a_{n,m}$ satisfy $\sum_{n,m\geq 1} |a_{n,m}|^2 < \infty$. Here the quantity $e_n \otimes f_m$ is just a notation, which is motivated by the fact that for $x = \sum_{n\geq 1} \alpha_n e_n \in \mathcal{H}_1$ and $y = \sum_{n\geq 1} \beta_n f_n \in \mathcal{H}_2$ one defines $x \otimes y \in \mathcal{H}_1 \otimes \mathcal{H}_2$ by

$$x \otimes y = \sum_{m,n\geq 1} \alpha_n \beta_m e_n \otimes f_m. \tag{3.1}$$

When either $\mathcal{H}_1$ or $\mathcal{H}_2$, or both, are finite-dimensional, the definition is modified in the obvious manner.

Exercise 3.1.1 When $\mathcal{H}_1$ and $\mathcal{H}_2$ are finite-dimensional, what is the dimension of $\mathcal{H}_1 \otimes \mathcal{H}_2$? How does it compare with the dimension of the usual product $\mathcal{H}_1 \times \mathcal{H}_2$?

When either $\mathcal{H}_1$ or $\mathcal{H}_2$ is infinite-dimensional, $\mathcal{H}_1 \otimes \mathcal{H}_2$ is an infinite-dimensional Hilbert space. The important structure is the bilinear form from $\mathcal{H}_1 \times \mathcal{H}_2$ into this space given by (3.1).

Recalling that (x, y) denotes the inner product in a Hilbert space we observe the formula

$$(x \otimes y, x' \otimes y') = (x, x')(y, y'), \tag{3.2}$$

which is a straightforward consequence of the fact that the basis $e_n \otimes f_m$ is orthonormal.[3]

The problem with our definition of the tensor product is that one is supposed to check that "it does not depend on the choice of the orthonormal basis", a tedious task that joins similar tasks under the carpet.[4] The good news is that all the identifications one may wish for are true. If both $\mathcal{H}_1$ and $\mathcal{H}_2$ are the space of square-integrable functions on $\mathbb{R}^3$, then $\mathcal{H}_1 \otimes \mathcal{H}_2$ is the space of square-integrable functions on $\mathbb{R}^6$.[5] This fits very well with the Dirac formalism: If $|\boldsymbol{x}\rangle$ denotes the Dirac function at $\boldsymbol{x}$, (so that these generalized vectors provide a generalized basis of $\mathcal{H}_1$, and similarly for $|\boldsymbol{y}\rangle$, then $|\boldsymbol{x}\rangle|\boldsymbol{y}\rangle$ denotes the Dirac function at the point $(\boldsymbol{x}, \boldsymbol{y}) \in \mathbb{R}^6$, and these generalized vectors provide a generalized basis of $\mathcal{H}_1 \otimes \mathcal{H}_2$. Furthermore, if $f \in \mathcal{H}_1$ and $g \in \mathcal{H}_2$ then $f \otimes g$ identifies with the function $(\boldsymbol{x}, \boldsymbol{y}) \mapsto f(\boldsymbol{x})g(\boldsymbol{y})$ on $\mathbb{R}^6$.

As another example, assume that $\mathcal{H}_1$ is the space of square-integrable functions over $\mathbb{R}^3$ and that $\mathcal{H}_2$ has dimension 2. Let (e_1, e_2) be an orthogonal basis of $\mathcal{H}_2$. Then each element $f \in \mathcal{H}_1 \otimes \mathcal{H}_2$ is of the type $f_1 \otimes e_1 + f_2 \otimes e_2$ where $f_1, f_2 \in \mathcal{H}_1$. That is, $\mathcal{H}_1 \otimes \mathcal{H}_2$ identifies with the space of pairs (f_1, f_2) of functions on $\mathbb{R}^3$. According to (3.2) the norm

[2] It is not indicated in the notation that we consider the tensor product of $\mathcal{H}_1$ and $\mathcal{H}_2$ "as complex Hilbert spaces", because we will never consider other types of tensor products.

[3] This may also serve as a basis-independent definition of the inner product on $\mathcal{H}_1 \otimes \mathcal{H}_2$, but it is a non-trivial task.

[4] Viewing a change of basis as a unitary transformation, this is essentially the same task as checking (3.4) below.

[5] This is of course a special case of a general theorem on the tensor product of two L^2 spaces.

is given by $\|(f_1, f_2)\|^2 = \int d^3\boldsymbol{x}(|f_1(\boldsymbol{x})|^2 + |f_2(\boldsymbol{x})|^2)$. If $\mathcal{H}_1$ is used to describe the position of a particle, and $\mathcal{H}_2$ is used to describe some kind of internal state (spin, as we will study later) then $\mathcal{H}_1 \otimes \mathcal{H}_2$ can be used to describe both the position of the particle together with its internal state. In the case of the electron, this pair of functions is called a *Pauli spinor*, and provides a (non-relativistic) description of a particle of spin $1/2$.

The axiom that tensor products correctly describe systems composed of several elements might look innocuous enough, but in fact it has stupendous consequences. To get a glimpse at these, let us consider a system consisting of two components, with the state of the first component being described by a vector in a two-dimensional space $\mathcal{H}_1$, and the state of the second component by a vector in another two-dimensional space $\mathcal{H}_2$. Let e_1, e_2 be the basis of $\mathcal{H}_1$, and f_1, f_2 the basis of $\mathcal{H}_2$. Then $\mathcal{H}_1 \otimes \mathcal{H}_2$ contains elements such as

$$x = \frac{1}{\sqrt{2}}(e_1 \otimes f_2 - e_2 \otimes f_1). \tag{3.3}$$

A state like x is "entangled".[6] The idea of this name is that the "properties of the two components of the system cannot be separated from each other". This is true in a very deep sense. It has been possible to prepare entangled states consisting of two photons *moving in opposite directions*, and to design and carry out (fantastically difficult) experiments whose results cannot be explained if one assumes that each of the photons has well-defined properties independent of the other photon.[7] A very accessible discussion can be found e.g. in Isham's textbook [43, Section 9.3]. Quantum Mechanics forces us to revise the notion, based on our macroscopic experience, that a system is made from smaller, well-localized parts. This is a real conceptual revolution.

There is no doubt that entangled states exist at the quantum level. If, however, one assumes that the rules of Quantum Mechanics extend[8] to the macroscopic world, entangled states are at the root of some thorny issues (which are not fully resolved to this day) such as the famous Schrödinger's cat paradox and its refinement called *Wigner's friend*.

The notion of tensor product generalizes in a straightforward manner to several components. A simple but important result concerns extension of operators to such a tensor product. Consider Hilbert spaces $\mathcal{H}_1, \ldots, \mathcal{H}_n$ and their tensor product $\bigotimes_{k\leq n} \mathcal{H}_k$. Consider for $k \leq n$ a unitary operator U_k on $\mathcal{H}_k$. Then there is a unitary operator U on $\bigotimes_{k\leq n} \mathcal{H}_k$ such that if $x_i \in \mathcal{H}_i$ for $i \leq n$,

$$U(x_1 \otimes \cdots \otimes x_n) = U_1(x_1) \otimes \cdots \otimes U_n(x_n). \tag{3.4}$$

This is straightforward to check but uninspiring.

When each $U_k = U_k(t) = \exp(\mathrm{i}tA_k)$ is a strongly continuous unitary group, then (3.4) defines a strongly continuous unitary group of operators $U(t)$. Let us call A the infinitesimal generator of this semigroup, as provided by Stone's theorem. It is almost obvious that if x_k belongs to the domain of A_k for each $k \leq n$ then $\bigotimes_{k\leq n} x_k$ belongs to the domain of A, and moreover

6 The technical definition of an entangled state is that it is not of the type $e \otimes f$ for $e \in \mathcal{H}_1$ and $f \in \mathcal{H}_2$.
7 The key term here is: violation of the Bell inequalities.
8 And if they do not, where is the boundary?

$$A(x_1 \otimes \cdots \otimes x_n) = A_1(x_1) \otimes x_2 \otimes \cdots \otimes x_n + x_1 \otimes A_2(x_2) \otimes \cdots \otimes x_n + \cdots + x_1 \otimes \cdots \otimes A_n(x_n). \tag{3.5}$$

Exercise 3.1.2 Prove the previous statements in every detail.

As a special case assume that for each $k \le n$ we are given a time-independent Hamiltonian H_k on $\mathcal{H}_k$, and the corresponding time-evolution $U_k(t)$. The corresponding strongly continuous unitary group $U(t)$ represents a certain evolution of the joint "n-particle system". We will *define* the expression "the particles do not interact with each other" as meaning that the time-evolution of the n-particle system is described by $U(t)$. The formula corresponding to (3.5) giving the Hamiltonian H of the n-particle system simply expresses that the energy of this system is the sum of the individual energies of each particle.

Let us now examine the case where $\mathcal{H}_k = L^2(\mathbb{R})$ for $k \le n$, and where $H_k = \hbar\omega_k(a^\dagger a + (1/2)1)$ is the Hamiltonian of a harmonic oscillator as in (2.101). Then (when the particles do not interact with each other) the Hamiltonian of the n-particle system is given by

$$H = \sum_{k \le n} \hbar\omega_k \Big(a_k^\dagger a_k + \frac{1}{2}1\Big). \tag{3.6}$$

Here, $a_k = \bigotimes_{\ell \le n} a_{\ell,n}$ where $a_{\ell,n} = 1$ for $\ell \ne k$ and $a_{k,n} = a$. The operators a_k satisfy $[a_k, a_k^\dagger] = 1$ whereas for $k \ne \ell$, a_k and $a_k^\dagger$ commute with a_ℓ and $a_\ell^\dagger$. The formula (3.6) represents the "sum of the energies of independent harmonic oscillators" and will occur many times in the sequel. In particular each of the individual terms of this formula can be described as "the energy of a harmonic oscillator".[9]

Non-interacting particles are not so interesting, it is the interaction that contains the physics. To obtain models where these particles interact, one then adds an "interaction term" to the Hamiltonian, a topic to which we will return later.

3.2 Symmetric Tensors

There is a very important twist to the notion of tensor product when one considers systems composed of several, say n *identical* particles. Identical particles are indistinguishable from each other, even in principle. If an electron in motion scatters on an electron at rest, two moving electrons come out of the experiment and there is no way telling which of them was the electron at rest (and it can be argued that the question may not even make sense). This has to be built in the model. To this aim, we consider a Hilbert space $\mathcal{H}$ with a basis $(e_i)_{i \ge 1}$, a given integer n and we denote by $\mathcal{H}_n$ the tensor product of n copies of it, in the above sense. We observe that a permutation σ of $\{1, 2, \ldots, n\}$ induces a transformation of $\mathcal{H}_n$, simply by transforming the basis element $\bigotimes_{k \le n} e_{i_k}$ into $\bigotimes_{k \le n} e_{i_{\sigma(k)}}$. If an element x of $\mathcal{H}_n$ describes the state of a system consisting of n identical particles, its image under this transformation describes the same particle system, so that it must be of the type λx for

9 And this despite the fact that the operator a_k is not unitarily equivalent to the operator a. We leave as a teaser to the reader to show that there are many non-trivial subspaces of $\mathcal{H}$ which are invariant under a_k and $a_k^\dagger$.

$\lambda \in \mathbb{C}$. Since the transform of x has the same norm as x then $|\lambda| = 1$, that is the transform of x differs from x only by a phase.[10]

In Part I of this book we consider the simplest case where the transform of each state x is x itself. Particles with this property are called *bosons*.[11] These are not the most common and interesting particles, but must be understood first. Later on, we will meet *fermions*, which comprise most of the important particles (and in particular electrons). Let us denote by S_n the group of permutations of $\{1,\ldots,n\}$.

Principle 7 Consider the state space $\mathcal{H}$ of a boson, provided with a basis $(e_i)_{i\geq 1}$. Then the state space which represents a system of n such identical bosons[12] is

$$\mathcal{H}_{n,s} = \Big\{ \sum_{i_1,\ldots,i_n} \alpha_{i_1,i_2,\ldots,i_n} e_{i_1} \otimes \cdots \otimes e_{i_n} \in \mathcal{H}_n \ ; \ \forall \sigma \in \mathsf{S}_n \, , \ \alpha_{i_1,\ldots,i_n} = \alpha_{i_{\sigma(1)},\ldots,i_{\sigma(n)}} \Big\}. \tag{3.7}$$

The subscript s stands for "symmetric". For example, the tensor $e_1 \otimes e_2 + e_2 \otimes e_1$ is symmetric but the tensor $e_1 \otimes e_2 - e_2 \otimes e_1$ is not (it is "anti-symmetric"). The norm on $\mathcal{H}_{n,s}$ is the obvious quadratic norm. We observe that $\mathcal{H}$ is the 1-particle space,

$$\mathcal{H} = \mathcal{H}_{1,s}.$$

We may also consider an element α of $\mathcal{H}_{n,s}$ as just a family of numbers $(\alpha_{i_1,\ldots,i_n})_{i_1,\ldots,i_n \geq 1}$ with $\alpha_{i_1,\ldots,i_n} = \alpha_{i_{\sigma(1)},\ldots,i_{\sigma(n)}}$ for each $\sigma \in \mathsf{S}_n$. For obvious reasons we call such a family a *tensor*. The squared norm of such a tensor is simply $\sum_{i_1,\ldots,i_n} |\alpha_{i_1,\ldots,i_n}|^2$.

Exercise 3.2.1 The previous description assumes that $\mathcal{H}$ is infinite-dimensional. Convince yourself that when $\mathcal{H}$ is one-dimensional, this is also the case for $\mathcal{H}_{n,s}$.

Computations in $\mathcal{H}_{n,s}$ simplify considerably if one uses a proper basis. You will understand much better what happens if you try to solve the next exercise before reading the general construction.

Exercise 3.2.2 Prove that the elements of the type $e_j \otimes e_j$ and $(e_{j_1} \otimes e_{j_2} + e_{j_2} \otimes e_{j_1})/\sqrt{2}$ for $j_1 \neq j_2$ form an orthonormal basis of $\mathcal{H}_{2,s}$. Try to extend the construction to the case $n = 3$. Hint: You may look at (3.11) below.

To define the appropriate basis of $\mathcal{H}_{n,s}$, let us consider an integer sequence $(n_k)_{k\geq 1}$ with $n_k \geq 0$ and $\sum_{k\geq 1} n_k = n$. We consider the element $f(n_1,\ldots,n_k,\ldots)$ of $\mathcal{H}_{n,s}$ defined as follows: It is the element $\alpha = (\alpha_{i_1,\ldots,i_n})_{i_1,\ldots,i_n \geq 1}$ of $\mathcal{H}_{n,s}$ such that $\alpha_{i_1,\ldots,i_n} = 0$ unless for each $k \geq 1$ exactly n_k of the indices $i_1,\ldots,i_n$ are equal to k, in which case $\alpha_{i_1,\ldots,i_n} = 1$ (see (3.9) below if this definition does not appeal to you). Thus for example $f(2,0,\ldots) = e_1 \otimes e_1$,

[10] Let us recall that two vectors x, y differ by a phase if $x = \lambda y$ where $|\lambda| = 1$. More generally a phase is a complex number of modulus 1.

[11] There are different types of bosons, say with different masses, etc. and these types can be distinguished from one another. It is the bosons of *the same type* which cannot be distinguished from one another.

[12] So, an n-particle state is simply a non-zero element of $\mathcal{H}_{n,s}$.

$f(1,1,0,\dots) = e_1 \otimes e_2 + e_2 \otimes e_1$ and $f(1,2,0,\dots) = e_1 \otimes e_2 \otimes e_2 + e_2 \otimes e_1 \otimes e_2 + e_2 \otimes e_2 \otimes e_1$. There are $n! / \prod_{k \geq 1} n_k!$ coefficients $\alpha_{i_1,\dots,i_n}$ which are equal to 1 and the others are 0, so that

$$\|f(n_1,\dots,n_k,\dots)\|_{n,s} = \sqrt{\frac{n!}{\prod_k n_k!}}. \tag{3.8}$$

A synthetic expression can be given as follows: If $i_1,\dots,i_n$ is a sequence such that exactly n_k of these indices equal k, then

$$f(n_1,\dots,n_k,\dots) = \frac{1}{\prod_{k \geq 1} n_k!} \sum_{\sigma \in \mathsf{S}_n} e_{i_{\sigma(1)}} \otimes \cdots \otimes e_{i_{\sigma(n)}}. \tag{3.9}$$

Using Dirac's notation, for a sequence $(n_i)_{i \geq 1}$ of integers with $\sum_{i \geq 1} n_i = n$ we define

$$|n_1,n_2,\dots,n_k,\dots\rangle := \sqrt{\frac{\prod_{k \geq 1} n_k!}{n!}} f(n_1,n_2,\dots,n_k,\dots), \tag{3.10}$$

and one easily checks that these elements form an orthonormal basis of $\mathcal{H}_{n,s}$. For example,

$$|2,1,0,0,\dots\rangle = \frac{1}{\sqrt{3}}(e_1 \otimes e_1 \otimes e_2 + e_1 \otimes e_2 \otimes e_1 + e_2 \otimes e_1 \otimes e_1). \tag{3.11}$$

This basis is very convenient. It has a simple physical meaning. The basis vectors e_k represent different states of a single particle, labeled $1,\dots,k,\dots$. In the n-particle state $|n_1,n_2,\dots,n_k,\dots\rangle$ exactly n_k of the n particles are in the state k.

> **Exercise 3.2.3** (a) Write the complete proof that $\mathcal{H}_{n,s}$ is the closed linear span of the elements $|n_1,n_2,\dots,n_k,\dots\rangle$ with $\sum_{i \geq 1} n_i = n$.
> (b) What happens when $\mathcal{H}$ is one-dimensional?

3.3 Creation and Annihilation Operators

We define and study the operators $A_n(\xi)$ and $A_n^\dagger(\eta)$, which will play a crucial role in the next section. It is convenient here to define $\mathcal{H}_{0,s} := \mathbb{C}$. For $n \geq 1$ and $\xi \in \mathcal{H}$ the operator $A_n(\xi)\colon \mathcal{H}_{n,s} \to \mathcal{H}_{n-1,s}$ transforms an n-particle state into an $(n-1)$-particle state, and for this reason is called an *annihilation operator*. For $n \geq 0$ and $\eta \in \mathcal{H}$ the operator $A_n^\dagger(\eta)$: $\mathcal{H}_{n,s} \to \mathcal{H}_{n+1,s}$ transforms an n-particle state into an $(n+1)$-particle state, and is called a *creation operator*.

In order to avoid writing formulas which are too abstract, we pick an orthonormal basis $(e_i)_{i \geq 1}$ of $\mathcal{H}$, so that an element η of $\mathcal{H}$ identifies with a sequence $(\eta_i)_{i \geq 1}$.[13] Similarly we think of an element α of $\mathcal{H}_{n,s}$ as a symmetric tensor $(\alpha_{i_1,\dots,i_n})_{i_1,\dots,i_n \geq 1}$.

Let us then introduce an important notation: Given a sequence $i_1,\dots,i_{n+1}$ of length $n+1$ we denote by $i_1,\dots,\hat{i}_k,\dots i_{n+1}$ the sequence of length n where the term i_k is omitted.

[13] Thus we assume that $\mathcal{H}$ is infinite-dimensional, leaving the finite-dimensional case to the reader.

Thus, for example, $1, 2, \hat{3}, 4$ is the sequence $1, 2, 4$. Given $\eta \in \mathcal{H}$, $n \geq 0$ and $\alpha \in \mathcal{H}_{n,s}$ we define[14]

$$A_n^\dagger(\eta)(\alpha)_{i_1,\ldots,i_{n+1}} := \frac{1}{\sqrt{n+1}} \sum_{\ell \leq n+1} \eta_{i_\ell} \alpha_{i_1,\ldots,\widehat{i_\ell},\ldots,i_{n+1}}. \tag{3.12}$$

This makes sense since the sequence $i_1, \ldots, \widehat{i_\ell}, \ldots, i_{n+1}$ is of length n and $\alpha \in \mathcal{H}_{n,s}$ has n indices. For example,

$$A_2^\dagger(\eta)(\alpha)_{i_1,i_2,i_3} = \frac{1}{\sqrt{3}}(\eta_{i_1}\alpha_{i_2,i_3} + \eta_{i_2}\alpha_{i_1,i_3} + \eta_{i_3}\alpha_{i_1,i_2}).$$

To see that the left-hand side of (3.12) is also symmetric in $i_1, \ldots, i_{n+1}$, look at what happens when you exchange i_1 and i_2: the first two terms of the summation get exchanged, while all the other ones do not change since α is symmetric in its indices. Defining the operator $C_\ell(\eta)$ by

$$C_\ell(\eta)(\alpha)_{i_1,\ldots,i_{n+1}} := \eta_{i_\ell}\alpha_{i_1,\ldots,\widehat{i_\ell},\ldots,i_{n+1}},$$

one has

$$\|C_\ell(\eta)(\alpha)\|^2 = \sum_{i_1,\ldots,i_{n+1} \geq 1} |\eta_{i_\ell}|^2 |\alpha_{i_1,\ldots,\widehat{i_\ell},\ldots,i_{n+1}}|^2 = \|\eta\|^2 \|\alpha\|^2$$

and since (3.12) implies $A_n^\dagger(\eta)(\alpha) = (n+1)^{-1/2} \sum_{\ell \leq n+1} C_\ell(\eta)(\alpha)$ it holds that

$$\|A_n^\dagger(\eta)(\alpha)\| \leq \frac{1}{\sqrt{n+1}} \sum_{\ell \leq n+1} \|C_\ell(\eta)(\alpha)\| \leq \sqrt{n+1}\|\eta\|\|\alpha\|, \tag{3.13}$$

and indeed $A_n^\dagger(\eta)(\alpha) \in \mathcal{H}_{n+1,s}$.

Consider next $\xi \in \mathcal{H}$ and β in $\mathcal{H}_{n,s}$. We define $A_n(\xi)(\beta) \in \mathcal{H}_{n-1,s}$ by

$$A_n(\xi)(\beta)_{i_1,\ldots,i_{n-1}} := \sqrt{n} \sum_{i \geq 1} \xi_i^* \beta_{i_1,\ldots,i_{n-1},i}. \tag{3.14}$$

It is obvious that this quantity is symmetric in $i_1, \ldots, i_{n-1}$. Let us note a striking feature here: the dependence of $A_n(\xi)$ in ξ is *anti*-linear. Also, using the Cauchy-Schwarz inequality,

$$\Big|\sum_{i \geq 1} \xi_i^* \beta_{i_1,\ldots,i_{n-1},i}\Big|^2 \leq \sum_{i \geq 1} |\xi_i|^2 \sum_{i \geq 1} |\beta_{i_1,\ldots,i_{n-1},i}|^2$$

we obtain

$$\|A_n(\xi)(\beta)\|^2 = \sum_{i_1,\ldots,i_{n-1} \geq 1} |A_n(\xi)(\beta)_{i_1,\ldots,i_{n-1}}|^2 \leq n \sum_{i \geq 1} |\xi_i|^2 \sum_{i_1,\ldots,i_{n-1},i \geq 1} |\beta_{i_1,\ldots,i_{n-1},i}|^2, \tag{3.15}$$

[14] This formula might be hard to understand right now but it will become transparent very soon.

so that

$$\|A_n(\xi)(\beta)\| \le \sqrt{n}\|\xi\|\|\beta\|, \tag{3.16}$$

and indeed $A_n(\xi)(\beta) \in \mathcal{H}_{n-1,s}$.

The meaning of the formulas (3.14) and (3.12) will become much clearer using the basis (3.10) of the spaces $\mathcal{H}_{n,s}$. Recalling the quantity $f(n_1,\ldots,n_k,\ldots)$ defined on page 78, let us first prove that

$$A_n(e_k)(f(n_1,\ldots,n_k,\ldots)) = \sqrt{n} f(n_1,\ldots,n_{k-1},n_k-1,n_{k+1},\ldots), \tag{3.17}$$

where it is understood that the right-hand side is zero if $n_k = 0$. For this we use (3.14) for $\beta = f(n_1,\ldots,n_k,\ldots)$ and for $\xi = e_k$, so that $\xi = \sum \xi_i e_i$ where $\xi_i = \delta_k^i$, for the usual Kronecker symbol δ_k^i. Consequently the terms in the summation in the right-hand side of (3.14) are zero unless $i = k$, so this relation becomes

$$A_n(\xi)(\beta)_{i_1,\ldots,i_{n-1}} = \sqrt{n}\beta_{i_1,\ldots,i_{n-1},k}.$$

Since $\beta = f(n_1,\ldots,n_k,\ldots)$, by definition of this quantity, the right-hand side is zero unless among the indices $i_1,\ldots,i_{n-1}$ there are exactly n_ℓ indices which are equal to ℓ for $\ell \neq k$, and exactly $n_k - 1$ indices which are equal to k. In this case it is equal to $\sqrt{n}$, and this proves (3.17).

Next we prove that

$$A_n^\dagger(e_k)(f(n_1,\ldots,n_k,\ldots)) = \frac{n_k+1}{\sqrt{n+1}} f(n_1,\ldots,n_{k-1},n_k+1,n_{k+1},\ldots). \tag{3.18}$$

For this we now appeal to (3.12), with $\eta = e_k$, and $\alpha = f(n_1,\ldots,n_k,\ldots)$, that is

$$A_n^\dagger(e_k)(\alpha)_{i_1,\ldots,i_{n+1}} = \frac{1}{\sqrt{n+1}} \sum_{\ell \le n+1} \delta_{i_\ell}^k \alpha_{i_1,\ldots,\widehat{i_\ell},\ldots,i_{n+1}}.$$

Then the terms in the summation are zero unless $i_\ell = k$, and unless, among the indices $i_1,\ldots,i_{\ell-1},i_{\ell+1},\ldots,i_{n+1}$, for each k' the value k' occurs exactly $n_{k'}$ times. That is, the summation is zero unless this occurs for at least one value of ℓ. But then among the indices $i_1,\ldots,i_{n+1}$ the value k' must occur $n_{k'}$ times, except for $k' = k$ where it must occur $n_k + 1$ times. Furthermore the summation is then equal to $n_k + 1$, which proves (3.18).

Using (3.10) the formulas (3.18) and (3.17) imply respectively[15]

$$A_n^\dagger(e_k)|n_1,n_2,\ldots,n_k,\ldots\rangle = \sqrt{n_k+1}|n_1,n_2,\ldots,n_k+1,\ldots\rangle, \tag{3.19}$$

and

$$A_n(e_k)|n_1,n_2,\ldots,n_k,\ldots\rangle = \sqrt{n_k}|n_1,n_2,\ldots,n_k-1,\ldots\rangle. \tag{3.20}$$

Exercise 3.3.1 Given an operator $A\colon \mathcal{H}_1 \to \mathcal{H}_2$ between Hilbert spaces, the formula (1.6) defines its adjoint from $\mathcal{H}_2$ to $\mathcal{H}_1$. Use the previous formulas to prove that $A_n^\dagger(\gamma)$ and $A_{n+1}(\gamma)$ are adjoint of each other.

[15] You may wonder why I did not use these formulas to *define* A and $A^\dagger$. The problem is that then it takes some work to prove (3.13) and (3.16), which are needed to prove that $A_n(\xi)(\beta)$ and $A^\dagger(\eta)(\alpha)$ are well defined.

Given $\gamma_1,\ldots,\gamma_n \in \mathcal{H}$ we define the element $\varphi_n(\gamma_1,\ldots,\gamma_n) \in \mathcal{H}_{n,s}$ by

$$\varphi_n(\gamma_1,\ldots,\gamma_n) = \frac{1}{\sqrt{n!}} \sum_{\sigma \in \mathsf{S}_n} \gamma_{\sigma(1)} \otimes \cdots \otimes \gamma_{\sigma(n)}, \tag{3.21}$$

so that e.g. $\varphi_2(e_1,e_2) = (e_1 \otimes e_2 + e_2 \otimes e_1)/\sqrt{2}$ and $\varphi_2(e_1,e_1) = \sqrt{2}e_1 \otimes e_1$. The normalization factor $\sqrt{n!}$ is to ensure that $\varphi_n(\gamma_1,\ldots,\gamma_n)$ is of norm 1 when the sequence (γ_i) is orthonormal. These elements will be given a natural interpretation in the next section.

Proposition 3.3.2 *We have*

$$A_n^\dagger(\eta)(\varphi_n(\gamma_1,\ldots,\gamma_n)) = \varphi_{n+1}(\gamma_1,\ldots,\gamma_n,\eta), \tag{3.22}$$

$$A_n(\xi)(\varphi_n(\gamma_1,\ldots,\gamma_n)) = \sum_{\ell \le n} (\xi,\gamma_\ell)\varphi_{n-1}(\gamma_1,\ldots,\widehat{\gamma}_\ell,\ldots,\gamma_n), \tag{3.23}$$

and these formulas entirely determine the operators $A_n^\dagger(\eta)$ and $A_n(\xi)$.

As a consequence our construction does not depend on the choice of the orthogonal basis of $\mathcal{H}$.

Proof Using linearity, anti-linearity and continuity it suffices to prove these formulas when $\xi,\eta,\gamma_1,\ldots,\gamma_n$ are basis vectors. Consider indices $i_1,\ldots,i_n$ and let n_k be the number of indices $i_1,\ldots,i_n$ equal to k. Then (3.9) implies

$$\varphi_n(e_{i_1},\ldots,e_{i_n}) = \frac{\prod_{k\ge 1} n_k!}{\sqrt{n!}} f(n_1,\ldots,n_k,\ldots) = \sqrt{\prod_{k \ge 1} n_k!}\,|n_1,\ldots,n_k,\ldots\rangle,$$

and the required formulas are a simple consequence of (3.19) and (3.20). The last assertion is obvious since the elements of the type $\varphi_n(\gamma_1,\ldots,\gamma_n)$ span $\mathcal{H}_{n,s}$. □

Exercise 3.3.3 Assume that $\mathcal{H}$ is the space of square-integrable functions on $\mathbb{R}$. Prove that $\mathcal{H}_{n,s}$ identifies with the class of square-integrable functions f on $\mathbb{R}^n$ which are symmetric in their arguments. (For simplicity we will simply call such a function symmetric.) Hint: Use an orthonormal basis of $\mathcal{H}$. We denote by $x_1,\ldots,x_n,y$ real numbers, so that e.g. $(x_1,\ldots,x_{n-1},y) \in \mathbb{R}^n$ and for $f \in \mathcal{H}_{n,s}$ the quantity $f(x_1,\ldots,x_{n-1},y)$ denotes the value of the function f at this point of $\mathbb{R}^n$.

Given $n \ge 1$, $f \in \mathcal{H}_{n,s}$ and $g \in \mathcal{H}$ prove that the value $A_n(g)(f)(x_1,\ldots,x_{n-1})$ at the point $(x_1,\ldots,x_{n-1}) \in \mathbb{R}^{n-1}$ is given by

$$A_n(g)(f)(x_1,\ldots,x_{n-1}) = \sqrt{n}\int \mathrm{d}y g(y)^* f(x_1,\ldots,x_{n-1},y). \tag{3.24}$$

Given f in $\mathcal{H}_{n,s}$ and $h \in \mathcal{H}$ prove that

$$A_n^\dagger(h)(f)(x_1,\ldots,x_{n+1}) = \frac{1}{\sqrt{n+1}} \sum_{k \le n+1} h(x_k) f(x_1,\ldots,\widehat{x}_k,\ldots,x_{n+1}). \tag{3.25}$$

For $g,h \in \mathcal{H}$ and $f \in \mathcal{H}_{n,s}$ prove that $A_{n-1}^\dagger(h)A_n(g)(f) \in \mathcal{H}_{n,s}$ is n times the symmetrized version of the function $\int \mathrm{d}y g(y)^* h(x_1) f(y,x_2,\ldots,x_n)$.

3.4 Boson Fock Space

A relativistically correct version of Quantum Mechanics must describe systems with a variable number of particles, because the equivalence of mass and energy allows creation and destruction of particles. Let us assume that the space $\mathcal{H}$ describes a single particle. We have constructed in (3.7) the space $\mathcal{H}_{n,s}$ which describes a collection of n identical particles. The boson Fock space will simply be the direct sum of these spaces (in the sense of Hilbert space) as $n \geq 0$ and will describe collections of any number of identical particles.[16] We do not yet incorporate any idea from Special Relativity. The construction of the boson Fock space is almost trivial. The non-trivial structure of importance is a special family of operators described in Theorem 3.4.2.

For $n = 0$ we define $\mathcal{H}_{0,s} = \mathbb{C}$, and we denote by $e_\emptyset$ its basis element (e.g. the number 1). The element $e_\emptyset$ represents the state where *no particles are present*, that is, the vacuum. It is of course of fundamental importance. Then we define

$$\mathcal{B}_0 = \bigoplus_{n \geq 0} \mathcal{H}_{n,s}, \tag{3.26}$$

the *algebraic* sum of the spaces $\mathcal{H}_{n,s}$, where again $\mathcal{H}_{n,s}$ is the space defined in (3.7). By definition of the algebraic sum, any element α of $\mathcal{B}_0$ is a sequence $\alpha = (\alpha(n))_{n \geq 0}$ with $\alpha(n) \in \mathcal{H}_{n,s}$ and $\alpha(n) = 0$ for n large enough. Let us denote by $(\cdot,\cdot)_n$ the inner product on $\mathcal{H}_{n,s}$. Consider $\alpha(n), \beta(n) \in \mathcal{H}_{n,s}$ and $\alpha = (\alpha(n))_{n \geq 0}$, $\beta = (\beta(n))_{n \geq 0}$. We define

$$(\alpha, \beta) := \sum_{n \geq 0} (\alpha(n), \beta(n))_n. \tag{3.27}$$

The boson Fock space $\mathcal{B}$ is the space of sequences $(\alpha(n))_{n \geq 0}$ such that $\alpha(n) \in \mathcal{H}_{n,s}$ and

$$\|(\alpha(n))_{n \geq 0}\|^2 := \sum_{n \geq 0} \|\alpha(n)\|^2 < \infty,$$

where $\|\alpha(n)\|$ is the norm in $\mathcal{H}_{n,s}$. We will hardly ever need to write down elements of $\mathcal{B}$ which are not in $\mathcal{B}_0$.

We will somewhat abuse notation by considering each $\mathcal{H}_{n,s}$, and in particular $\mathcal{H} = \mathcal{H}_{1,s}$, as a subspace of $\mathcal{B}_0$. Again, $\mathcal{H}_{n,s}$ represents the n-particle states. Given ξ, η in $\mathcal{H}$ we recall the operators $A_n(\xi)$ and $A_n^\dagger(\eta)$ of the previous section.

Definition 3.4.1 (The creation and annihilation operators) Given ξ, η in $\mathcal{H}$ we define the operators $A^\dagger(\eta)$ and $A(\xi)$ on $\mathcal{B}_0$ as follows. If $\alpha \in \mathbb{C} = \mathcal{H}_{0,s}$ we set

$$A^\dagger(\eta)(\alpha) = \alpha\eta \; ; \; A(\xi)(\alpha) = 0$$

and for $n \geq 1, \alpha \in \mathcal{H}_{n,s}$ we set

$$A^\dagger(\eta)(\alpha) = A_n^\dagger(\eta)(\alpha) \; ; \; A(\xi)(\alpha) = A_n(\xi)(\alpha). \tag{3.28}$$

[16] In old textbooks this construction is called *second quantization*.

Theorem 3.4.2 *The operators $A(\xi)$ and $A^\dagger(\eta)$ on $\mathcal{B}_0$ have the following properties.*

$$\text{For } n \geq 1\,,\ A(\xi) \text{ maps } \mathcal{H}_{n,s} \text{ into } \mathcal{H}_{n-1,s}. \tag{3.29}$$

$$\text{For } n \geq 0\,,\ A^\dagger(\eta) \text{ maps } \mathcal{H}_{n,s} \text{ into } \mathcal{H}_{n+1,s}. \tag{3.30}$$

$$A(\xi)(e_\emptyset) = 0\ ;\ A^\dagger(\eta)(e_\emptyset) = \eta \in \mathcal{H} = \mathcal{H}_{1,s} \subset \mathcal{B}_0. \tag{3.31}$$

$$\forall \gamma \in \mathcal{H}\,,\ \forall \alpha,\beta \in \mathcal{B}_0\,,\ (A^\dagger(\gamma)(\alpha),\beta) = (\alpha, A(\gamma)(\beta)). \tag{3.32}$$

$$\forall \xi,\eta \in \mathcal{H}\,,\ [A(\xi), A^\dagger(\eta)] = (\xi,\eta)1. \tag{3.33}$$

$$\forall \xi,\eta \in \mathcal{H}\,,\ [A(\xi), A(\eta)] = [A^\dagger(\xi), A^\dagger(\eta)] = 0. \tag{3.34}$$

$$\text{The map } \eta \mapsto A^\dagger(\eta) \text{ is linear. The map } \xi \mapsto A(\xi) \text{ is anti-linear.} \tag{3.35}$$

Property (3.29) shows that $A(\xi)$ decreases the number of particles by 1, and for this reason it is called an *annihilation operator*.[17] Similarly, from (3.30), $A^\dagger(\eta)$ increases the number of particles by 1 and is called a *creation operator*. Let us stress the meaning of (3.31): $\mathcal{B}$ contains a copy of $\mathcal{H}$. This copy is obtained by applying the creation operators $A^\dagger(\eta)$ to the vacuum $e_\emptyset$. Property (3.32) *does not* say that $A^\dagger(\gamma)$ is the adjoint $A(\gamma)^\dagger$ of $A(\gamma)$. To reach such a conclusion we would have to define $A(\gamma)$ and $A^\dagger(\gamma)$ on a proper domain. This is a question of mathematical interest which is examined in Exercise 3.4.3. However defining $A(\xi)$ and $A^\dagger(\eta)$ on $\mathcal{B}_0$ is sufficient for our needs.

The operator-valued maps $\xi \mapsto A(\xi)$ and $\eta \mapsto A^\dagger(\eta)$ appearing in Theorem 3.4.2 are of fundamental importance for the rest of the book. These are the key ingredients from which we may construct quantum fields, i.e. operator-valued maps. We stress the content of (3.35): The map $A^\dagger$ is linear whereas the map A is *anti*-linear.

We may now consider the vectors $|n_1,\ldots,n_k,\ldots\rangle$ of (3.10) for any sequence (n_k) such that finitely many of these integers are not zero. These constitute an orthogonal basis of $\mathcal{B}$. Let us write $a_k = A(e_k)$ and $a_k^\dagger = A^\dagger(e_k)$. Then (3.19) and (3.20) become respectively

$$a_k^\dagger|n_1,n_2,\ldots,n_k,\ldots\rangle = \sqrt{n_k+1}|n_1,n_2,\ldots,n_k+1,\ldots\rangle, \tag{3.36}$$

and

$$a_k|n_1,n_2,\ldots,n_k,\ldots\rangle = \sqrt{n_k}|n_1,n_2,\ldots,n_k-1,\ldots\rangle, \tag{3.37}$$

to be compared with (2.88).

Proof of Theorem 3.4.2 Only (3.32) to (3.34) are not obvious at this stage. Using linearity and continuity as expressed by (3.16) and (3.13), it suffices to prove these formulas when ξ and η are basis vectors of $\mathcal{H}$ and when both α and β are of the type $|n_1,n_2,\ldots,n_k,\ldots\rangle$. But in that case they are straightforward consequences of (3.36) and (3.37). □

[17] From now on this name is reserved for the operator $A(\xi)$ on the boson Fock space and will never be used for the operator $A_n(\xi)$ of the previous section.

The following very simple exercise explains in which domain one should define $A(\xi)$ and $A^\dagger(\xi)$ so that they are adjoints of each other.

Exercise 3.4.3 Figure out the relevance of Exercise 2.5.9 to the situation of Theorem 3.4.2.

Exercise 3.4.4 Find a direct proof of (3.32) to (3.34) from (3.14) and (3.12). (This is somewhat challenging.)

The creation and annihilation operators greatly simplify the manipulation of the elements of the boson Fock space. Indeed, as a consequence of (3.22) for any n one has

$$A^\dagger(\eta_1)\cdots A^\dagger(\eta_n)(e_\emptyset) = \varphi_n(\eta_1,\ldots,\eta_n). \tag{3.38}$$

In words: Repeated applications of n creation operators $A^\dagger(\eta_k)$ to the vacuum (the state $e_\emptyset$) results in a state consisting of n particles in the corresponding states. This will be used many times. Informally, one may say that $\varphi_n(\eta_1,\ldots,\eta_n)$ "represents n particles in the states $\eta_1,\ldots,\eta_n$".

It can be very challenging for a mathematician to understand how physics textbooks usually describe the boson Fock space. The elements of this space are always written as $A^\dagger(\eta_1)\cdots A^\dagger(\eta_n)(e_\emptyset)$, or in Dirac's notation $A^\dagger(\eta_1)\cdots A^\dagger(\eta_n)|0\rangle$. One is not told on which space the creation operators act, as it is understood that they act "on the state space". Furthermore, η_1, etc. are usually not true states, but "improper states" e.g. "states with a definite momentum" such as the improper states $|p\rangle$ we met in Section 2.8.

The previous formulas need to be modified in the obvious manner when $\mathcal{H}$ is finite-dimensional. The case $\mathcal{H} = \mathbb{C}$ is of importance. In this case, let us denote by e_1 a basis of $\mathcal{H}$ (e.g. the number 1). Then $|n\rangle = e_1 \otimes \cdots \otimes e_1 \in \mathcal{H}_{n,s}$ is a unit vector, which is a basis of $\mathcal{H}_{n,s}$. The vectors $|n\rangle$ ($n \geq 0$) form an orthonormal basis of the boson Fock space. The operators $a = A(e_1)$ and $a^\dagger = A^\dagger(e_1)$ satisfy the relations $a|0\rangle = 0$, $a|n\rangle = \sqrt{n}|n-1\rangle$ for $n \geq 1$ and $a^\dagger|n\rangle = \sqrt{n+1}|n+1\rangle$ for $n \geq 0$. Up to the name for the basis, these are the relations (2.88). With respect to these natural creation and annihilation operators, the boson Fock space of $\mathcal{H} = \mathbb{C}$ has exactly the same structure as the harmonic oscillator with respect to its own natural creation and annihilation operators. Since we think of the boson Fock space as describing a multiparticle system, this confirms the idea that the harmonic oscillator is in a sense a multiparticle system, and it is the simplest possible because $\mathcal{H} = \mathbb{C}$ is the simplest possible.[18]

Conversely the boson Fock space of a one-dimensional Hilbert space naturally appears as the state space of the harmonic oscillator of Hamiltonian $\hbar\omega(a^\dagger a + (1/2)\mathbf{1})$.

3.5 Unitary Evolution in the Boson Fock Space

We now explain how time-evolution taking place in $\mathcal{H}$ can be extended in a canonical way to the whole of $\mathcal{B}$, describing the evolution of non-interacting particles. (Using the larger space

[18] The particles there are the "quanta of excitation" of page 69.

$\mathcal{B}$ however allows one to construct more complicated Hamiltonians which make different particles interact.) For this consider a unitary operator U on $\mathcal{H}$ and as explained at the end of Section 3.1 for $n \geq 1$ the unitary operator U_n on $\mathcal{H}^{\otimes n}$ such that

$$\forall x_1, \ldots, x_n \in \mathcal{H}\,,\ U_n(x_1 \otimes \cdots \otimes x_n) = \bigotimes_{i \leq n} U(x_i). \tag{3.39}$$

It is straightforward to check that $U_n(\mathcal{H}_{n,s}) \subset \mathcal{H}_{n,s}$, see also Exercise 3.5.1. Thus the restriction of U_n to $\mathcal{H}_{n,s}$ is a unitary operator on $\mathcal{H}_{n,s}$. Therefore, there exists a unitary operator $U_{\mathcal{B}}$ on $\mathcal{B}$ which coincides with U_n on $\mathcal{H}_{n,s}$. For $n = 0$ we define U_0 as the identity on $\mathcal{H}_{0,s} = \mathbb{C}$.

Exercise 3.5.1 With the notation (3.21), prove that

$$U_n(\varphi_n(\gamma_1, \ldots, \gamma_n)) = \varphi_n(U(\gamma_1), \ldots, U(\gamma_n)). \tag{3.40}$$

In particular, when $U(t)$ is the unitary group of time-evolution under a Hamiltonian H, then $U_{\mathcal{B}}(t)$ describes the time-evolution of the multiparticle system when the particles do not interact with each other. In fact, one may *define* the statement "the particles do not interact with each other" as meaning that the time-evolution of the multiparticle system is of the type $U_{\mathcal{B}}$ for a certain time-evolution U. That this definition makes sense is apparent from the formula (3.41) below. In this manner, to a Hamiltonian H on $\mathcal{H}$ corresponds a Hamiltonian $H_{\mathcal{B}}$ on $\mathcal{B}$. As follows from (3.5), $H_{\mathcal{B}}$ satisfies the relation

$$H_{\mathcal{B}}\Big(\sum_{\sigma \in S_n} x_{\sigma(1)} \otimes \cdots \otimes x_{\sigma(n)}\Big)$$
$$= \sum_{k \leq n, \sigma \in S_n} x_{\sigma(1)} \otimes \cdots \otimes x_{\sigma(k-1)} \otimes H(x_{\sigma(k)}) \otimes x_{\sigma(k+1)} \otimes \cdots \otimes x_{\sigma(n)}. \tag{3.41}$$

The physical meaning of this relation is again that the energy of a non-interacting multiparticle system is the sum of the energies of the individual particles (even though these cannot be distinguished from each other).

More generally given any self-adjoint operator H on $\mathcal{H}$ there is a canonical extension $H_{\mathcal{B}}$ of H to the boson Fock space for the simple reason that, mathematically, every self-adjoint operator is a Hamiltonian by the converse of Stone's theorem, with associated time-evolution $U(t) = \exp(-itH/\hbar)$.[19] The formula (3.41) still holds in that case. If you find this argument horrifying, just repeat it without the word "Hamiltonian". For a self-adjoint operator H, consider the one-parameter group $U(t) = \exp(itA/\hbar)$, construct from this group a one-parameter group $U_{\mathcal{B}}(t)$ on $\mathcal{H}_{\mathcal{B}}$. Then $H_{\mathcal{B}}$ is the infinitesimal generator of this group.

The following result is simple yet fundamental.

Proposition 3.5.2 *Consider a unitary operator U from a Hilbert space $\mathcal{H}'$ to a Hilbert space $\mathcal{H}$. It has a canonical extension W from the boson Fock space $\mathcal{B}'$ of $\mathcal{H}'$ to the*

[19] The existence of this time-evolution certainly does not imply that it makes physical sense!

boson Fock space $\mathcal{B}$ of $\mathcal{H}$, which sends $\mathcal{B}'_0$ to $\mathcal{B}_0$. For $\xi,\eta \in \mathcal{H}'$ denote by $A'(\xi)$ and $A'^{\dagger}(\eta)$ the annihilation and creation operators of Theorem 3.4.2 relative to $\mathcal{H}'$. Then for any vectors $\xi,\eta \in \mathcal{H}'$ it holds that

$$WA'(\xi)W^{-1} = A(U(\xi))\ ;\ WA'^{\dagger}(\eta)W^{-1} = A^{\dagger}(U(\eta)). \tag{3.42}$$

This equality holds as an equality of operators on $\mathcal{B}_0$.

Proof To construct W we proceed just as in the case where $\mathcal{H} = \mathcal{H}'$. To prove the rest we lighten notation and we assume without real loss of generality that $\mathcal{H} = \mathcal{H}'$. Then $W = U_{\mathcal{B}}$. To prove that, say, $U_{\mathcal{B}}A^{\dagger}(\eta) = A^{\dagger}(U(\eta))U_{\mathcal{B}}$ we compute, using (3.40) in the third line:

$$\begin{aligned} A^{\dagger}(U(\eta))U_{\mathcal{B}}(\varphi_n(\gamma_1,\dots,\gamma_n)) &= A^{\dagger}(U(\eta))\varphi_n(U(\gamma_1),\dots,U(\gamma_n)) \\ &= \varphi_{n+1}(U(\eta),U(\gamma_1),\dots,U(\gamma_n)) \\ &= U_{\mathcal{B}}\varphi_{n+1}(\eta,\gamma_1,\dots,\gamma_n) \\ &= U_{\mathcal{B}}A^{\dagger}(\eta)\varphi_n(\gamma_1,\cdots,\gamma_n). \end{aligned}$$

□

3.6 Boson Fock Space and Collections of Harmonic Oscillators

Hopefully the ideas expressed in this section will gradually become clearer, but right now the reader may find them rather murky. We have no choice though. As explained toward the end of the section, Nature apparently uses these ideas. The plan is to meditate about the formulas (3.36) and (3.37). These support the physicists' view that, when $\mathcal{H}$ is infinite-dimensional then the boson Fock space $\mathcal{B}$ represents in some sense an infinite collection of harmonic oscillators. This representation is closely related to the traditional way quantum fields are constructed, although for a mathematician the direct construction of the boson Fock space is likely to be much clearer. A remarkable property is that the choice of any orthogonal basis of $\mathcal{H}$ induces such a representation as an infinite collection of harmonic oscillators. It is a bit like the fact that on $\mathbb{R}^n$ the probability with a density proportional to $\exp(-\|x\|^2)$ is a product probability with respect to *any* choice of an orthogonal basis. The reader may review Section 2.18 now.

Let us start with a one-dimensional Hilbert space $\mathcal{H}$. As explained at the end of Section 3.4, its boson Fock space, equipped with its corresponding creation and annihilation operators can be naturally thought of as the state space of a harmonic oscillator.[20] Picking a basis $(e_k)_{k\geq 1}$ in a standard Hilbert space makes it appear as a sum of one-dimensional Hilbert spaces, and each of them corresponds to an independent harmonic oscillator. In some canonical sense, an infinite number of independent harmonic oscillators live in the boson Fock space of a standard Hilbert space. Elementary physics textbooks, which do not want to mention tensor products and other high mathematics often attempt to explain boson Fock space using risqué metaphors[21] such as "replacing each point of space by a harmonic

[20] This requires choosing a value of the frequency ω, and we will see later how to do it.
[21] Trying to make sense of these was certainly the hardest obstacle for this author to get started, although one may argue that this again falls under the heading "cultural differences".

oscillator". In the same questionable style, one could say that "constructing the boson Fock space amounts to replacing each dimension by a harmonic oscillator".

To be more specific, assume that there exists a Hamiltonian H on a standard Hilbert space $\mathcal{H}$ such that (e_k) are eigenvectors of H with $H(e_k) = \hbar\omega_k e_k$ for certain numbers ω_k (to make this close to the situation of the harmonic oscillator). As we have explained at the end of Section 3.5, the Hamiltonian H gives rise to a Hamiltonian $H_{\mathcal{B}}$ on $\mathcal{B}$, and (3.41) shows that

$$H_{\mathcal{B}}|n_1,n_2,\ldots,n_k,\ldots\rangle = \hbar\Big(\sum_k n_k\omega_k\Big)|n_1,n_2,\ldots,n_k,\ldots\rangle, \tag{3.43}$$

i.e. $|n_1,\ldots,n_k,\ldots\rangle$ is an eigenvector of eigenvalue $\hbar\sum_k n_k\omega_k$, in accordance with the idea that the energy of a non-interacting multiparticle system is the sum of the energies of the individual particles.

Comparing with (3.36) and (3.37) gives

$$H_{\mathcal{B}} = \sum_k \hbar\omega_k a_k^\dagger a_k. \tag{3.44}$$

Thus $H_{\mathcal{B}} = \sum_k H_k$ where $H_k = \hbar\omega_k a_k^\dagger a_k$, where $[a_k, a_k^\dagger] = 1$. Comparing with (3.6) we find it plausible to interpret $H_k + \hbar\omega_k 1/2$ as the Hamiltonian of a harmonic oscillator. Adding or removing a constant term i.e. a multiple of the identity operator[22] in the Hamiltonian is irrelevant to the evolution of the system, so that one could say that within a constant term the Hamiltonian (3.44) is the sum of the Hamiltonians of independent harmonic oscillators,

$$H_{\mathcal{B}} = \sum_k \hbar\omega_k\Big(a_k^\dagger a_k + \frac{1}{2}1\Big). \tag{3.45}$$

A slight problem is the infinite sum in (3.45), so the constant term $(\hbar/2)\sum_k \omega_k 1$ is (often) infinite. But, as we already explained, it is only in our idealization that the sum is infinite, so that this infinite constant term is no disaster. Even a very large constant term has no influence: in the lab we can measure only *differences* in energy between two situations.[23]

The stupendous fact is that this constant term, which corresponds to the energy of the vacuum (no particle present), is apparently *present in Nature*, as is shown by the Casimir effect,[24] a simple case of which we will investigate in Chapter 7.[25]

The formula (3.44) involves the choice of a basis of eigenvectors for the Hamiltonian H. When several eigenvectors correspond to the same eigenvalue, the choice of this basis is by no means unique. Still, the Hamiltonian $H_{\mathcal{B}}$ depends only on H and not on the choice of the basis of eigenvectors for H, as we will verify now. To see this let us consider another

22 Such a quantity is called a c-number in physics, where c stands for classical, as opposed to q-number, which is operator-like.

23 This term is however a huge problem when one tries to reconcile Quantum Field Theory with General Relativity, where what matters is the actual value of the energy-momentum density, not the differences. This is the "cosmological constant problem", and it is huge indeed: the energy density of the vacuum as predicted by Quantum Field Theory is larger than the measured value by 120 orders of magnitude, easily the worse prediction ever made by physics.

24 However, refer to Jaffe's article [44] for a different view.

25 The existence of this effect seems to indicate that the correct Hamiltonian is not (3.44) but (3.45).

orthogonal basis of eigenvectors (f_ℓ) with $e_k = \sum_\ell c_k^\ell f_\ell$. To simplify notation we assume that all the ω_k are equal. The operator U such that $U(f_k) = e_k$ is unitary, thus $U^{-1} = U^\dagger$ is unitary. In particular $g_\ell := U^\dagger(f_\ell) = \sum_k (c_k^\ell)^* f_k$ is an orthogonal basis, so that $\sum_k c_k^\ell (c_k^{\ell'})^* = \delta_\ell^{\ell'}$. Thus

$$\sum_k a_k^\dagger a_k = \sum_k A^\dagger(e_k)A(e_k) = \sum_k A^\dagger\Big(\sum_\ell c_k^\ell f_\ell\Big)A\Big(\sum_{\ell'} c_k^{\ell'} f_{\ell'}\Big)$$

$$= \sum_k \sum_{\ell,\ell'} c_k^\ell c_k^{\ell'*} A^\dagger(f_\ell)A(f_{\ell'}) = \sum_\ell A^\dagger(f_\ell)A(f_\ell). \tag{3.46}$$

Exercise 3.6.1 Still assuming that H has a basis of eigenvectors, carry out the case where ω_k is not independent of k. Hint: How do you relate two different bases of eigenvectors?

3.7 Explicit Formulas: Position Space

In this section we start to manipulate the boson Fock space in a simple setting, learning how to write basic formulas. We recall the maps A and $A^\dagger$ of Definition 3.4.1.

We have shown in Section 3.5 how a self-adjoint operator (and in particular a Hamiltonian) on $\mathcal{H}$ can be canonically extended to $\mathcal{B}$. Here we explain how the physicists write the formula for such an extension in the typical but crucial case, where $\mathcal{H}$ is $L^2(\mathbb{R})$, modeling the position of a single particle. Here and at so many other places we will use the concept of an *operator-valued distribution*. This is simply a linear map that associates an operator to each test function (putting again under the rug some rather technical regularity properties that this map should satisfy, about which the reader may learn on page 707). We are going to define the maps a and $a^\dagger$, which are *operator-valued distributions*, taking their values in the operators on $\mathcal{B}_0$. Denoting as usual distributions as functions, we will write the quantities $a(x)$ and $a^\dagger(x)$ (where x is a point of $\mathbb{R}$) even though the quantities $a(x)$ and $a^\dagger(x)$ *do not* make sense by themselves, but only when they are integrated against a function $f \in \mathcal{S}$. The result of such an integration is an *operator*. The *defining* formulas are then

$$\int \mathrm{d}x f(x)^* a(x) := A(f)\,; \quad \int \mathrm{d}x f(x) a^\dagger(x) := A^\dagger(f). \tag{3.47}$$

The reason for the term $f(x)^*$ is that the map $f \mapsto A(f)$ is anti-linear rather than linear. Pretending that we can use (3.47) for the "function" $f = \delta_y$ we obtain the identity $a(y) = A(\delta_y)$. Again, this is *formal* because $\delta_y \notin \mathcal{S}$. Furthermore $A(f)$ is defined only for $f \in \mathcal{H}$ and certainly $\delta_y \notin \mathcal{H}$. So, all of this is a matter of *notation* and has little intrinsic content. Similarly we may pretend that $a^\dagger(x) = A^\dagger(\delta_x)$.

Now, using (3.33) in the first equality,

$$[A(f), A^\dagger(g)] = (f,g)\mathbf{1} = \mathbf{1}\int \mathrm{d}x f(x)^* g(x) = \mathbf{1}\iint \mathrm{d}x\mathrm{d}y f(x)^* g(y)\delta(x-y). \tag{3.48}$$

Comparing with the formal expression (obtained by using (3.47) in the left-hand side below)

$$[A(f), A^\dagger(g)] = \iint \mathrm{d}x\mathrm{d}y f(x)^* g(y)[a(x), a^\dagger(y)],$$

one is lead to write[26]

$$[a(x), a^\dagger(y)] = \delta(x-y)\mathbf{1}. \tag{3.49}$$

This looks very impressive, but is *simply a formal way* to express the relation $[A(f), A^\dagger(g)] = (f,g)\mathbf{1}$.

Exercise 3.7.1 Consider a function $\xi \in \mathcal{S}^2$. Prove that it makes sense to define the operator $S := \iint \mathrm{d}x\mathrm{d}y\xi(x,y)a^\dagger(x)a(y)$ as follows. When applied to a function $f \in \mathcal{H}_{n,s}$ (that is, to a square-integrable function f on $\mathbb{R}^n$ which is symmetric in its arguments) then $S(f)$ is n times the symmetrized version of the function $\int \mathrm{d}y\xi(x_1,y)f(y,x_2,\ldots,x_n)$

In general one cannot multiply distributions, so it is not clear what products such as $a(x)^2$ or $a^\dagger(x)a(x)$ mean. Our next task is to show that the product $a^\dagger(x)a(x)$ *can* be defined in a sensible way, as an operator-valued distribution. (The importance of this specific quantity will become clear only as we write some of the many formulas where it occurs.) To guess what this definition should be, we make a *formal* calculation where we have x as a fixed parameter. Consider an element f of $\mathcal{H}_{n,s}$, which is simply a square-integrable function f on $\mathbb{R}^n$, symmetric in its arguments. Then, since $a(x) = A(\delta_x)$, from (3.24) we get that the function $g := a(x)(f)$ is given by

$$g(x_1,\ldots,x_{n-1}) = \sqrt{n} f(x_1,\ldots,x_{n-1},x). \tag{3.50}$$

Since $a^\dagger(x) = A^\dagger(\delta_x)$, (3.25) implies that the function $a^\dagger(x)a(x)(f) = a^\dagger(x)(g)$ is given by

$$\begin{aligned} a^\dagger(x)(g)(x_1,\ldots,x_n) &= \frac{1}{\sqrt{n}} \sum_{k\le n} \delta_x(x_k) g(x_1,\ldots,\widehat{x_k},\ldots,x_n) \\ &= \sum_{k\le n} \delta_x(x_k) f(x_1,\ldots,\widehat{x_k},\ldots,x_n,x). \end{aligned} \tag{3.51}$$

Since f is symmetric in its arguments, and since $\delta_x(x_k) = 0$ unless $x = x_k$ we have[27]

$$\delta_x(x_k) f(x_1,\ldots,\widehat{x_k},\ldots,x_n,x) = \delta_x(x_k) f(x_1,\ldots,x_n)$$

and the previous formula yields

$$a^\dagger(x)a(x)(f)(x_1,\ldots,x_n) = \Big(\sum_{k\le n} \delta_x(x_k)\Big) f(x_1,\ldots,x_n). \tag{3.52}$$

26 One has to be aware of the dangers of writing distributions as functions. It makes no sense whatsoever to set $y = x$ in the formula (3.49). The quantity $\delta(0)$ does not make sense, and as we are going to see soon, the operator $a(x)a^\dagger(x)$ cannot be meaningfully defined, so that the bracket $[a(x), a^\dagger(x)]$ does not make sense either.

27 More rigorously, for a test function g, $\delta_x(a)g(x) = \delta(x-a)g(x) = g(a)\delta(x-a) = g(a)\delta_x(a)$.

Since for $\psi \in \mathcal{S}$ it holds that $\int \mathrm{d}x \delta_x(x_k)\psi(x) = \psi(x_k)$, it makes sense to define the operator $D := \int \mathrm{d}x \psi(x) a^\dagger(x) a(x)$ by

$$D(f)(x_1,\ldots,x_n) = \sum_{k \le n} \psi(x_k) f(x_1,\ldots,x_n). \tag{3.53}$$

Using this formula for a test function ψ defines $a^\dagger(x)a(x)$ as an operator-valued distribution, but the formula makes sense on far weaker conditions on ψ, as we will extensively use.

Exercise 3.7.2 Proceeding as in the proof of (3.53), show that for a smooth function $V(x,y)$ it makes sense to define the Hamiltonian

$$H_V := \iint \mathrm{d}x\mathrm{d}y V(x,y) a^\dagger(x) a^\dagger(y) a(x) a(y) \tag{3.54}$$

by

$$H_V(f)(x_1,\ldots,x_n) = \sum_{k \ne \ell} V(x_k,x_\ell) f(x_1,\ldots,x_n)$$

whenever $f \in \mathcal{H}_{n,s}$.

Exercise 3.7.3 For a smooth function $V(x,y)$, how do you define $\iint \mathrm{d}x\mathrm{d}y V(x,y)$ $a^\dagger(x)a(y)$?

Consider now a symmetric function f of n variables which is of the type

$$f(x_1,\ldots,x_n) = \sum_{\sigma \in \mathsf{S}_n} f_{\sigma(1)}(x_1) \cdots f_{\sigma(n)}(x_n), \tag{3.55}$$

where $f_1,\ldots,f_n \in \mathcal{H}$. Consider a self-adjoint operator H on $\mathcal{H}$, and its canonical extension $H_{\mathcal{B}}$ to the boson Fock space. Then (3.41) shows that

$$H_{\mathcal{B}}(f)(x_1,\ldots,x_n) = \sum_{k \le n} \sum_{\sigma \in \mathsf{S}_n} g_{\sigma,k}(x_1,\ldots,x_n), \tag{3.56}$$

where

$$g_{\sigma,k}(x_1,\ldots,x_n) = f_{\sigma(1)}(x_1) \cdots f_{\sigma(k-1)}(x_{k-1}) H(f_{\sigma(k)})(x_k) f_{\sigma(k+1)}(x_{k+1}) \cdots f_{\sigma(n)}(x_n).$$

In the case where H is the operator "multiplication by a function ψ", then $H(f_{\sigma(k)})(x_k) = \psi(x_k) f_{\sigma(k)}(x_k)$ and (3.56) implies that $H_{\mathcal{B}}(f)(x_1,\ldots,x_n)$ is given by the right-hand side of (3.53). The linear span of the functions of the type (3.55) is a very large space (in particular it is dense in $\mathcal{H}_{n,s}$), so at the heuristic level we obtain the following important formula (which we will use many times). If H is the operator "multiplication by the function ψ" then $H_{\mathcal{B}}$ is given by the formal but correct expression[28]

$$H_{\mathcal{B}} = \int \mathrm{d}x \psi(x) a^\dagger(x) a(x). \tag{3.57}$$

28 To go beyond the heuristic level we would have to explain how (3.57) defines a self-adjoint operator.

Let us recall that as explained on page 31 the operator on $\mathcal{H}$ given by multiplication of a function by the indicator of C (i.e. the function which equals 1 on C and zero elsewhere) "measures whether the particle belongs to C". As a particular case of (3.57), for a set C, the operator $\int_C \mathrm{d}x a^\dagger(x)a(x)$ on $\mathcal{B}$ then "counts the number of particles in C". As a supplementary check of this property, consider the case where f is obtained by symmetrization of a product $f_1 \times \ldots \times f_n$ where $f_1, \ldots, f_k$ are zero on the complement of C and $f_{k+1}, \ldots, f_n$ are zero on C. Then f is an eigenvector of this operator with eigenvalue k.

In particular

$$N := \int \mathrm{d}x a^\dagger(x)a(x),$$

is the "number operator" which counts the total number of particles. Using (3.53) for the function $\psi(x) \equiv 1$ we obtain that the space $\mathcal{H}_{n,s}$ is an eigenspace of this operator, with eigenvalue n. Of course, a state has a well-defined number of particles only if it belongs to one of the spaces $\mathcal{H}_{n,s}$.

Assuming now that $f = f(x_1, x_2, \ldots, x_n)$ is smooth and symmetric, from (3.50) we get

$$\frac{\mathrm{d}^2}{\mathrm{d}x^2} a(x)(f)(x_1, \ldots, x_{n-1}) = \sqrt{n} \frac{\partial^2}{\partial x^2} f(x_1, \ldots, x_{n-1}, x),$$

and proceeding as in (3.53) we may write

$$\left(\int \mathrm{d}x a^\dagger(x) \frac{\mathrm{d}^2}{\mathrm{d}x^2} a(x) \right)(f)(x_1, \ldots, x_n) = \sum_{k \le n} \frac{\partial^2}{\partial x_k^2} f(x_1, \ldots, x_n).$$

For a particle of mass m in a potential V in one dimension, recall from (2.79) that the Hamiltonian is given by

$$H(f) = -\frac{\hbar^2}{2m} \frac{\mathrm{d}^2 f}{\mathrm{d}x^2} + Vf.$$

The previous considerations lead to the formal expression

$$H_{\mathcal{B}} = \int \mathrm{d}x a^\dagger(x) \left(-\frac{\hbar^2}{2m} \frac{\mathrm{d}^2}{\mathrm{d}x^2} + V(x) \right) a(x) \tag{3.58}$$

for the extension of this Hamiltonian to the multiparticle system.

Exercise 3.7.4 Try to make sense of the following Hamiltonian (which is often found in textbooks)

$$H = \int \mathrm{d}x a^\dagger(x) \left(-\frac{\hbar^2}{2m} \frac{\mathrm{d}^2}{\mathrm{d}x^2} \right) a(x) + \frac{1}{2} \iint \mathrm{d}x \mathrm{d}y a^\dagger(x) a^\dagger(y) V(x,y) a(x) a(y).$$

What is the meaning of these two terms?[29] Hint: Use (3.58) and Exercise 3.7.2.

[29] The last integrand is written exactly in this form because it also makes sense if V is an operator, not just a scalar.

Having succeeded to give a meaning (in the distributional sense) to the product $a^\dagger(x)a(x)$ we should not jump to the conclusion that we are going to be as lucky each time we try to multiply distributions. In our forthcoming formal computations we will meet many expressions which are ill-defined. One of them is the product $a(x)a^\dagger(x)$. This is to be expected in view of the formula

$$a(x)a^\dagger(y) = a^\dagger(x)a(y) + \delta(x-y),$$

since by taking $y = x$ one obtains the infinite term $\delta(0)$, and if one tries as above a formal computation to guess a possible meaning of $a(x)a^\dagger(x)$ (even in the distributional sense) the same obstacle arises.

Another quantity that creates problems is $a^\dagger(x)^2$. From our definition of creation operators, it is obvious that $A^\dagger(f)^2(e_\emptyset)(x_1,x_2) = \sqrt{2}f(x_1)f(x_2)$, so that formally, using this relation for $f = \delta_x$ we obtain $a^\dagger(x)^2(e_\emptyset)(x_1,x_2) = \sqrt{2}\delta_x(x_1)\delta_x(x_2)$. It is not obvious how to make sense of this even as a distribution since even an integral of such quantities does not look at all like a function in $L^2(\mathbb{R}^2)$ because it takes non-zero values only on the diagonal of $\mathbb{R}^2$.

3.8 Explicit Formulas: Momentum Space

In the previous section we have worked in position state space in order to present the simplest possible formulas. It is however in momentum state space that similar formulas will be actually used in the sequel. The adaptation of the formalism is completely straightforward, but, even though it will not be used for a while, it is probably better explained now while the previous material is fresh in the reader's mind.

Let us consider now the Hilbert space $\mathcal{H}' = L^2(\mathbb{R}^3, \mathrm{d}^3\boldsymbol{p}/(2\pi\hbar)^3)$.[30] We denote by $\mathcal{B}'$ the boson Fock space of $\mathcal{H}'$. We denote B and $B^\dagger$ the corresponding maps as in Definition 3.4.1. We may then consider the "operators" $b(\boldsymbol{p})$ and $b^\dagger(\boldsymbol{p})$ which are the operator-valued distributions given for $\xi \in \mathcal{S}^3 (= \mathcal{S}(\mathbb{R}^3)$, the space of test functions on $\mathbb{R}^3)$ by

$$B(\xi) = \int \frac{\mathrm{d}^3\boldsymbol{p}}{(2\pi\hbar)^3}\xi(\boldsymbol{p})^* b(\boldsymbol{p})\,;\; B^\dagger(\xi) = \int \frac{\mathrm{d}^3\boldsymbol{p}}{(2\pi\hbar)^3}\xi(\boldsymbol{p}) b^\dagger(\boldsymbol{p}). \tag{3.59}$$

These are the same formulas as in (3.47), but where now the underlying space is $\mathbb{R}^3$ with the measure $\mathrm{d}^3\boldsymbol{p}/(2\pi\hbar)^3$ rather than $\mathbb{R}$ with the measure $\mathrm{d}x$. The "operators" $b(\boldsymbol{p})$ and $b^\dagger(\boldsymbol{p})$ satisfy the commutation relations

$$[b(\boldsymbol{p}), b^\dagger(\boldsymbol{p}')] = (2\pi\hbar)^3\delta^{(3)}(\boldsymbol{p}-\boldsymbol{p}')1. \tag{3.60}$$

This is the version of (3.49) which is appropriate here since in momentum space delta functions come with a factor $(2\pi\hbar)^3$.[31] The operators $b(\boldsymbol{p}), b^\dagger(\boldsymbol{p})$ also satisfy

$$[b(\boldsymbol{p}), b(\boldsymbol{p}')] = [b^\dagger(\boldsymbol{p}), b^\dagger(\boldsymbol{p}')] = 0. \tag{3.61}$$

[30] The reason for the notation $\mathcal{H}'$ will be apparent later: we keep the name $\mathcal{H}$ for a more important space.

[31] To ensure that $\int \frac{\mathrm{d}^3\boldsymbol{p}}{(2\pi\hbar)^3}(2\pi\hbar)^3\delta^{(3)}(\boldsymbol{p}-\boldsymbol{p}')\xi(\boldsymbol{p}) = \xi(\boldsymbol{p}')$.

Exercise 3.8.1 Check (3.60) in complete detail.

Exercise 3.8.2 Recalling that $|0\rangle$ represents the vacuum in Dirac's notation, what is the meaning of $b^\dagger(\boldsymbol{p})|0\rangle$ and $b^\dagger(\boldsymbol{p})b^\dagger(\boldsymbol{p}')|0\rangle$?

Exercise 3.8.3 Convince yourself that (heuristically) "$b(\boldsymbol{p})$ is of dimension $[l^{3/2}]$".

Finally, let us note the following version of (3.57). Consider a real-valued function ψ on $\mathbb{R}^3$ and the operator H "multiplication by the function $\psi(\boldsymbol{p})$" on $\mathcal{H}'$. Then its extension $H_{\mathcal{B}'}$ to $\mathcal{B}'$ is formally given by the formula

$$H_{\mathcal{B}'} = \int \frac{\mathrm{d}^3\boldsymbol{p}}{(2\pi\hbar)^3}\psi(\boldsymbol{p})b^\dagger(\boldsymbol{p})b(\boldsymbol{p}). \tag{3.62}$$

3.9 Universe in a Box

We have already faced some of the analytical difficulties and complications inherent to a model of a particle on the entire real line. This type of complication will become much more serious later on. On the other hand, as we mentioned in Section 2.4, replacing the whole space by a finite set would be technically much simpler, but raises other tricky issues. In this section we describe a compromise between these two extreme attitudes, which we will use numerous times in the sequel, starting with the next section.

When we model an experiment in our lab, what is the relevance of the values the wave function takes at places further away than 10^{10000} light-years (assuming that such places exist)? Probably very little, so we may realistically use a finite "box" rather than $\mathbb{R}^3$ to model physical space. For this we consider a (very large) number L and the box $B = [-L/2, L/2]^3$ which is thus supposed to represent the whole physical space. We should think of L as much larger than any distance we can reach. We denote by L^2_B the corresponding L^2 space. This is now our position state space, and as traditional its elements are called *wave functions*. As L is extremely large, it seems reasonable to assume that the wave functions of anything of interest to us will be indistinguishable from zero long before the boundary of B is reached.[32] Our choice of a basis for L^2_B is dictated by mathematical convenience and the behavior of this basis on the boundary of B is unimportant.

Consider the set $\mathcal{K}$ of elements $\boldsymbol{k} \in \mathbb{R}^3$ such that $L\boldsymbol{k}/2\pi\hbar \in \mathbb{Z}^3$, i.e. all the coordinates of $\boldsymbol{k}$ are of the type $2\pi n\hbar/L$ for $n \in \mathbb{Z}$.[33] Consider for $\boldsymbol{k} \in \mathcal{K}$ the functions

$$f_{\boldsymbol{k}}(\boldsymbol{x}) = \frac{1}{L^{3/2}}\exp(\mathrm{i}\boldsymbol{x}\cdot\boldsymbol{k}/\hbar). \tag{3.63}$$

They form an orthonormal basis of L^2_B, and $f_{\boldsymbol{k}}$ is an eigenvector for each of the three momenta operators $-\mathrm{i}\hbar\partial/\partial x_j$. Any function f has an expansion along this basis,

$$f = \sum_{\boldsymbol{k}\in\mathcal{K}} (f_{\boldsymbol{k}}, f) f_{\boldsymbol{k}}, \tag{3.64}$$

[32] Of course we choose the origin of the box close to our location.

[33] In physics, $\boldsymbol{k}$ is the traditional notation for the wave vector. This is just an unfortunate coincidence, wave vectors are not considered in this text.

and we should think of the family of numbers $(f_{\boldsymbol{k}}, f)$ for $\boldsymbol{k} \in \mathcal{K}$ as the representation of f in momentum state space. Indeed applying the momentum operator $-\mathrm{i}\hbar\partial/\partial x_j$ simply amounts to multiplying the coefficient $(f_{\boldsymbol{k}}, f)$ by k_j. As a consequence

Momentum state space becomes a set of functions on a discrete set.

This is a significant technical simplification. Now,

$$(f_{\boldsymbol{k}}, f) = \frac{1}{L^{3/2}} \int_B \mathrm{d}^3 \boldsymbol{y} \exp(-\mathrm{i}\boldsymbol{k} \cdot \boldsymbol{y}/\hbar) f(\boldsymbol{y}) = \frac{1}{L^{3/2}} \hat{f}_B(\boldsymbol{k}), \tag{3.65}$$

where f_B denotes the function on $\mathbb{R}^3$ which is equal to f on B and zero otherwise and $\hat{f}_B$ denotes the Fourier transform of f_B. Thus (3.64) becomes

$$f(\boldsymbol{x}) = \frac{1}{L^3} \sum_{\boldsymbol{k} \in \mathcal{K}} \exp(\mathrm{i}\boldsymbol{x} \cdot \boldsymbol{k}/\hbar) \hat{f}_B(\boldsymbol{k}). \tag{3.66}$$

We may think of B as divided into small boxes $B_{\boldsymbol{k}}$ of side $2\pi\hbar/L$ centered on the elements $\boldsymbol{k}$ of $\mathcal{K}$. The integral of the function $\boldsymbol{p} \mapsto \exp(\mathrm{i}\boldsymbol{x} \cdot \boldsymbol{p}/\hbar)\hat{f}_B(\boldsymbol{p})$ on such a little box $B_{\boldsymbol{k}}$ is about $(2\pi\hbar/L)^3$ times its value at $\boldsymbol{k}$, so that letting $L \to \infty$ in (3.66) one recovers the inverse Fourier transform formula (1.31).[34] Also, using (3.64) at $\boldsymbol{x} = 0$ gives $f(0) = \sum_{\boldsymbol{k} \in \mathcal{K}} (f_{\boldsymbol{k}}, f)$ i.e.

$$\delta^{(3)}(\boldsymbol{x}) = \frac{1}{L^3} \sum_{\boldsymbol{k} \in \mathcal{K}} \exp(-\mathrm{i}\boldsymbol{x} \cdot \boldsymbol{k}/\hbar) = \frac{1}{L^{3/2}} \sum_{\boldsymbol{k} \in \mathcal{K}} f_{\boldsymbol{k}}(\boldsymbol{x}), \tag{3.67}$$

a discrete version of (1.19).

Exercise 3.9.1 If f_n is an orthonormal basis of $L^2(\mathbb{R})$ many physics textbooks state the formula

$$\sum_{n \geq 1} f_n(x)^* f_n(y) = \delta(x - y). \tag{3.68}$$

Try to make sense of this by integrating both sides against a test function. How does this formula relate to (3.67)?

3.10 Quantum Fields: Quantizing Spaces of Functions

A Quantum Field is simply an operator-valued distribution[35] *defined on space-time.*

In this section we construct our first quantum field (although, technically, the creation and annihilation operators $a^\dagger(x)$ and $a(x)$ of Section 3.7 are already quantum fields).

Let us consider $L > 0$ and the interval $B = [-L/2, L/2]$, of length L.[36] Our basic object is a function u: $B \to \mathbb{R}$ with $u(-L/2) = u(L/2)$, a condition which is called a *periodic*

[34] It is possible to make the argument completely rigorous.

[35] An operator-valued distribution is a linear map from the space of test functions to a given space of operators. Intuitively, the value of this map is not defined at every point but only when smeared by a test function.

[36] We are putting here in practice the method of the previous section. Our real interest is functions defined on the whole real line, but we replace the whole line by a very large box to ensure that momentum space becomes discrete.

boundary condition. This function also depends on $t \in \mathbb{R}$, so we denote it $u(t,x)$.[37] It models the y-displacement of a homogeneous elastic string located in the strip $B \times \mathbb{R}$, so that at time t the string occupies the points $(x, u(t,x))$ for $-L/2 \le x \le L/2$. Here "homogeneous" means that the mass density and the elasticity of the string do not depend on the point of the string. An elastic string is subject to the following equation of motion

$$\frac{\partial^2 u}{\partial t^2} = \alpha^2 \frac{\partial^2 u}{\partial x^2}, \tag{3.69}$$

for some constant α. The physics of this equation is easy to understand. An elastic string likes to be in a straight line. When it is bent, elasticity creates a force proportional to $\partial^2 u/\partial x^2$ at the point x and Newton's law of motion implies that the acceleration of this point is given by (3.69). We will consider the following more general equation of motion

$$\frac{\partial^2 u}{\partial t^2} = \alpha^2 \frac{\partial^2 u}{\partial x^2} - \beta u, \tag{3.70}$$

where $\beta \ge 0$. In addition to the elasticity of the string which tends to make it straight, each point $(x, u(t,x))$ is submitted to a force of intensity proportional to $|u(t,x)|$ which tends to pull it back to the point $(x, 0)$. One can picture many very thin elastic bands, each joining a point $(x, 0)$ to a point $(x, u(t,x))$ doing this pulling back.

The string obeying the equation of motion (3.70) is a mechanical system, far more complicated but essentially of the same nature as the simple free massive point we have already studied. Our purpose is to quantize this system. Quantization is *not* a mathematical procedure, because Quantum Mechanics is not a consequence of Classical Mechanics. In quantizing a system one always has to make assumptions, and these are justified only by the fact that the resulting model is successful. Later, in Chapter 6 we will learn systematic methods of describing mechanical systems and quantizing them. As we have yet to develop these tools, we will make some educated guesses.

Differential equations such as (3.70) can be solved using the Fourier transform in a discrete form. We denote by $\mathcal{K}$ the set of points $k \in \mathbb{R}$ such that $Lk/2\pi\hbar \in \mathbb{Z}$, so the elements k of $\mathcal{K}$ are of the type $k = 2\pi n\hbar/L$ for $n \in \mathbb{Z}$. For $k \in \mathcal{K}$ we consider the functions

$$f_k(x) = \frac{1}{\sqrt{L}} \exp(\mathrm{i}xk/\hbar).$$

The restriction $k \in \mathcal{K}$ exactly says that f_k satisfies the periodic boundary condition. These functions form an orthonormal basis of $L_B^2 := L^2(B, \mathrm{d}x)$, and we can expand

$$u(t,x) = \sum_{k \in \mathcal{K}} u_k(t) f_k(x). \tag{3.71}$$

The important feature is that

$$f_k''(x) = -k^2/\hbar^2 f_k(x),$$

37 It is the value of $u(t,x)$ which is of interest now and which we try to quantize. One should think of x and t as *labels*, not variables.

so that (with some handwaving) the equation (3.70) is equivalent to the fact that for each k we have

$$\ddot{u}_k(t) := \frac{\mathrm{d}^2 u_k(t)}{\mathrm{d}t^2} = -\omega_k^2 u_k(t) := -(\alpha^2 k^2/\hbar^2 + \beta)u_k(t). \tag{3.72}$$

Assuming for simplicity $\beta > 0$, we have $\omega_k^2 > 0$ for each k and (3.72) is the equation of motion of a classical harmonic oscillator of angular frequency ω_k.[38] Thus the number $u_k(t)$ describes the position at time t of a classical harmonic oscillator of angular frequency ω_k. The idea we follow is very simple: we replace this number by the operator which quantizes the position at time t of a harmonic oscillator with the same frequency.[39]

Unfortunately we are going to get the wrong result if we proceed exactly in this direction. The function u is real-valued, and one should use real-valued functions to expand it. Such functions are given e.g. by $g_0 = f_0 = 1/\sqrt{L}$ and

$$g_k(x) = \begin{cases} \sqrt{2/L}\cos(kx/\hbar) & \text{if } k > 0 \\ \sqrt{2/L}\sin(kx/\hbar) & \text{if } k < 0. \end{cases} \tag{3.73}$$

We can then expand $u(t,x) = \sum_{k\in\mathcal{K}} v_k(t)g_k(x)$, and now the functions $v_k(t)$ are real-valued. Since $g_k'' = -k^2/\hbar^2 g_k$, we see as before that the function $v_k(t)$ satisfies $\ddot{v}_k(t) = -\omega_k^2 v_k(t)$.

We learned that a quantum system is governed by a Hamiltonian, which in turn produces a time-evolution of the operators in the Heisenberg picture. For the time being however we will forget all these sophisticated considerations and we use a very simple-minded quantization procedure. We replace each of the numbers $v_k(t)$ by the operator quantizing at time t the position of a harmonic oscillator of angular frequency ω_k. Let us recall the formula (2.107) for this operator:

$$X(t) = \sqrt{\frac{\hbar}{2\omega_k m}}(a^\dagger \exp(\mathrm{i}t\omega_k) + a\exp(-\mathrm{i}t\omega_k)), \tag{3.74}$$

where the operator a satisfies $[a,a^\dagger] = 1$. In the case of (2.107) the number m was the mass of the harmonic oscillator but in our case nothing tells us how to choose it. It is apparent however that the choice of m is simply a normalization condition (provided the choice is independent of k). Since at this stage we have no clue how to choose m, we put this issue aside and we set $\gamma = \sqrt{\hbar/2m}$, a normalization constant.

Let us then for each $k \in \mathcal{K}$ consider operators $c(k)$ with $[c(k), c^\dagger(k)] = 1$, such that these operators commute for different values of k: $[c(k), c(k')] = 0$, etc. The short-hand way to describe this situation is the sentence

The operators $c(k)$ satisfy the relation $[c(k), c^\dagger(k)] = 1$,
all the other commutators being zero.

We start to think as physicists here, not telling what the state space is. This attitude is based on the assumption that in some sense, the choice of the state space does not matter, but only

[38] When $\beta = 0$, for $k = 0$, the equation is $\ddot{u}_0(t) = 0$. This simply expresses that the center of mass of the string may follow a uniform motion, and one should add the constraint $u_0 = 0$. Besides this point, there is not much to change to the analysis.

[39] We start to make sense of the physicists' metaphor of "attaching a harmonic oscillator to each point of space", although here "space" means the set $\mathcal{K}$ which we use to describe momentum (rather than position).

the commutation relations between the operators matter.[40] In Exercise 3.10.2 we provide an actual construction of such operators.

We then replace the function $v_k(t)$ by the operator

$$\frac{\gamma}{\sqrt{\omega_k}}(c(t,k)+c^\dagger(t,k)), \tag{3.75}$$

where $c(t,k) = c(k)\exp(-it\omega_k)$ and $c^\dagger(t,k) = c^\dagger(k)\exp(it\omega_k)$. This provides a time-dependent operator which is a candidate for quantization of the function $u(t,x)$:

$$\varphi(t,x) = \sum_{k\in\mathcal{K}} \frac{\gamma}{\sqrt{\omega_k}}(c(t,k)+c^\dagger(t,k))g_k(x). \tag{3.76}$$

Exercise 3.10.1 Convince yourself that we would have gotten a wrong result if we had not reduced to real-valued functions. Hint: We would not get a self-adjoint field.[41]

The formula (3.76) is potentially our first quantum field. Unfortunately it is not clear what this infinite sum might mean. As the next exercise explores, it is actually not possible to define a bona fide operator by the previous formula.

Exercise 3.10.2 To explore the meaning of (3.76) we need a concrete realization of the operators $c(k)$. For this we consider a standard (infinite-dimensional and separable) Hilbert space $\mathcal{H}$ and a basis $(e_k)_{k\in\mathcal{K}}$. We then take $c(k) = A(e_k)$ and $c^\dagger(k) = A^\dagger(e_k)$, which are operators on $\mathcal{B}$. We recall the element $e_\emptyset$ of $\mathcal{B}$, which is about the simplest non-zero element of the space.
(a) Prove that if a sequence $(\alpha_k)_{k\in\mathcal{K}}$ satisfies $\sum_{k\in\mathcal{K}}|\alpha_k|^2 = \infty$ then it is not possible to define sensibly the value of the operator

$$\sum_{k\in\mathcal{K}} \alpha_k(c(k)+c^\dagger(k)) \tag{3.77}$$

on the vector $e_\emptyset$.
(b) Prove on the other hand that if $\sum_{k\in\mathcal{K}}|\alpha_k|^2 < \infty$ then the operator (3.77) may be defined with a dense domain. Hint: Prove that this operator is well-defined on each vector of the type $c^\dagger(i_1)\cdots c^\dagger(i_n)(e_\emptyset)$.

The way to make sense of the formula (3.76) is as an operator-valued distribution. Let us here define a test function as an infinitely differentiable function ξ such that each derivative $\xi^{(n)}$ satisfies the boundary condition $\xi^{(n)}(-L/2) = \xi^{(n)}(L/2)$. Given a test function ξ the sequence $(\int_B \mathrm{d}x g_k(x)\xi(x))_k$ decreases fast (as is seen by integration by parts), and we may then define the integral $\int_B \mathrm{d}x\varphi(t,x)\xi(x)$ in the obvious manner, as being the operator

$$\int_B \mathrm{d}x\varphi(t,x)\xi(x) = \sum_{k\in\mathcal{K}} \frac{\gamma}{\sqrt{\omega_k}}\Big(\int_B \mathrm{d}x\xi(x)g_k(x)\Big)(c(t,k)+c^\dagger(t,k)). \tag{3.78}$$

[40] However, as we shall eventually see, in the case we are considering here with an infinity of operators, the situation is trickier.
[41] In Quantum Mechanics, self-adjoint operators correspond to real numbers.

The quantum field (3.76) is of considerable importance. It is the version in "1 + 1 dimensions" (that is, when physical space is one-dimensional) of one of the most important quantum fields, to be studied in great detail in Chapter 5. It will be more profitable to study it after we learn more, but we can already understand though that nothing can be difficult, since we are basically studying independent harmonic oscillators, and whatever formula we prove will ultimately reduce to proving something about a single harmonic oscillator.

It is appropriate to think of the (ill-defined) operator $\varphi(t,x)$ as measuring the position of the string at the point x of space and at time t.

The purpose of the forthcoming exercise is to show that the operator-valued distribution $\varphi(t,x)$ satisfies the differential equation (3.70). A complete proof of a similar result is given later in Proposition 6.1.3.

Exercise 3.10.3 The operator-valued distribution $\partial^2\varphi(t,x)/\partial x^2$ is defined by

$$\int_B \mathrm{d}x \frac{\partial^2\varphi(t,x)}{\partial x^2}\xi(x) = \int_B \mathrm{d}x\varphi(t,x)\xi''(x).$$

Defining the derivative of $\varphi(t,x)$ with respect to t in the obvious manner, prove that the operator-valued distribution $\partial^2\varphi/\partial t^2 - \alpha^2\partial^2\varphi/\partial x^2 + \beta\varphi$ is zero by computing its value against a test function ξ.

There is however one unpleasant task ahead of us, which will not get any easier if we delay it, so we confront it now over the next page or so. It is to manipulate the formula (3.76) to express the quantum field using the functions f_k rather than g_k. For this we use that for $k > 0$ we have $g_k = (f_k + f_{-k})/\sqrt{2}$ whereas $g_k = \mathrm{i}(f_{-k} - f_k)/\sqrt{2}$ for $k < 0$. Fixing the value of $k > 0$ we consider the contributions of the values k and $-k$ to the sum (3.76), which, since $\omega_k = \omega_{-k}$, is

$$\frac{\gamma}{\sqrt{2\omega_k}}\Big(f_k(x)\big(c(t,k) + c^\dagger(t,k) + \mathrm{i}c(t,-k) + \mathrm{i}c^\dagger(t,-k)\big) + f_{-k}(x)\big(c(t,k) + c^\dagger(t,k) - \mathrm{i}c(t,-k) - \mathrm{i}c^\dagger(t,-k)\big)\Big). \tag{3.79}$$

Let us then define

$$a(k) := \frac{1}{\sqrt{2}}(c(k) + \mathrm{i}c(-k))\ ;\ a(-k) := \frac{1}{\sqrt{2}}(c(k) - \mathrm{i}c(-k)), \tag{3.80}$$

and $a(t,k) := a(k)\exp(-\mathrm{i}\omega_k t)$, $a(t,-k) := a(-k)\exp(-\mathrm{i}\omega_k t)$, so that the quantity (3.79) equals

$$\frac{\gamma}{\sqrt{\omega_k}}\Big(f_k(x)\big(a(t,k) + a^\dagger(t,-k)\big) + f_{-k}(x)\big(a(t,-k) + a^\dagger(t,k)\big)\Big).$$

Writing down what this means we get

$$\frac{\gamma}{\sqrt{\omega_k}}\sum_{\ell\in\{-k,k\}} a(t,\ell)f_\ell(x) + a^\dagger(t,\ell)f_{-\ell}(x).$$

It is straightforward to check that

$$[a(k), a^\dagger(k)] = [a(-k), a^\dagger(-k)] = 1, \tag{3.81}$$

with all the other commutators zero.

Exercise 3.10.4 Check explicitly that $[a(k), a^\dagger(-k)] = 0$.

Setting $a(0) = c(0)$, the formula (3.76) then becomes

$$\varphi(t,x) = \frac{1}{\sqrt{L}} \sum_{k \in \mathcal{K}} \frac{\gamma}{\sqrt{\omega_k}} \big(\exp(\mathrm{i}(-t\omega_k + xk/\hbar))a(k) + \exp(\mathrm{i}(t\omega_k - xk/\hbar))a^\dagger(k)\big), \tag{3.82}$$

where the operators $a(k)$ satisfy the relations (3.81), with all the other commutators being zero. This formula (or, rather, more advanced versions of it) is going to stay with us a long time, although we will understand it only gradually. Let us try to remember that the mysterious factor $1/\sqrt{\omega_k}$ *arises from* (3.74) *concerning a single position operator.*

Looking at (3.82) we notice the operators

$$\exp(-\mathrm{i}t\omega_k)a(k)\ ;\ \exp(\mathrm{i}t\omega_k)a^\dagger(k),$$

which of course remind us of (2.104) and (2.105), and which express that $a(t,k)$ and $a^\dagger(t,k)$ are the time-evolved operators of $a(k)$ and $a^\dagger(k)$ under the Hamiltonian $\hbar\omega_k a^\dagger(k)a(k)$. Since there is such a contribution for each value of k we expect that the Hamiltonian

$$\sum_k \hbar\omega_k a^\dagger(k)a(k) \tag{3.83}$$

will be relevant, and that in some sense the operator $\varphi(t,x)$ is the Heisenberg time-evolution of the operator $\varphi(0,x)$ under this Hamiltonian.[42]

We will pursue these matters in due time, but for now we start with simpler considerations which will already bring forward the thorny issues plaguing the topic. The formula (3.83) attaches to each point of $\mathcal{K}$ a harmonic oscillator. As $L \to \infty$, the points of $\mathcal{K}$ become closer to each other, and we would like to give a meaning to the procedure "attach to each point p of $\mathbb{R}$ the position operator of a harmonic oscillator of angular frequency $\omega(p)$". We assume $\omega(p)$ is a smooth function of p with $\omega(p) \geq \alpha > 0$.[43] We define the operator-valued distributions $a(p)$ and $a^\dagger(p)$ as in Section 3.8, but in dimension 1 rather than 3, and the commutation relations are

$$[a(p), a^\dagger(p')] = 2\pi\hbar\delta(p - p')1.$$

We expect the continuous limit of the formula (3.83) to be[44]

$$H = \hbar \int \frac{\mathrm{d}p}{2\pi\hbar}\omega(p)a^\dagger(p)a(p), \tag{3.84}$$

42 Not accounting for an unimportant constant term, (3.83) is the Hamiltonian $\sum_k \hbar\omega_k(a^\dagger(k)a(k) + (1/2)1)$ obtained "by attaching non-interacting harmonic oscillators to each point of space". The occurrence of this Hamiltonian is completely expected. It represents the sum of the energies of independent harmonic oscillators, and such oscillators occur in (3.72). In his textbook [93], Zee uses for a quantum field the metaphor of a mattress with quantized springs attached.

43 We are not really interested in generality, but no other property of ω matters.

44 It would be possible to derive the correct formulas by analyzing how one should go from the discrete case to the continuum limit, but as this requires a bit of work, we will *guess* them.

which, as we have explained, is a well-defined operator on the boson Fock space of L^2. Presumably, we then have a continuous version

$$H = \int \frac{\mathrm{d}p}{2\pi\hbar}\frac{1}{2m}(\Psi(p)^2 + \omega(p)^2 m^2 \Phi(p)^2) \tag{3.85}$$

of the formula (2.101), where $\Phi(p)$ and $\Psi(p)$ respectively give "the position and the momentum of the harmonic oscillator located at p". We may guess from (2.100) that the operator-valued distribution $\Phi(p)$"giving the position of the harmonic oscillator at p" is given by the formula (keeping again from now on the explicit value of m):

$$\Phi(p) = \sqrt{\frac{\hbar}{2m\omega(p)}}(a(p) + a^\dagger(p)). \tag{3.86}$$

The formula (3.86) is to be taken in the sense of distributions. It means that for a test function $\xi \in \mathcal{S}$ one has

$$\Phi(\xi) = \sqrt{\frac{\hbar}{2m}}\Big(A\Big(\frac{\xi^*}{\sqrt{\omega}}\Big) + A^\dagger\Big(\frac{\xi}{\sqrt{\omega}}\Big)\Big), \tag{3.87}$$

where $\xi/\sqrt{\omega}$ is the test function[45] given by $\xi/\sqrt{\omega}(p) = \xi(p)/\sqrt{\omega(p)}$. As in (3.86) we may write an operator-valued distribution "expressing the momentum of the harmonic oscillator at p":

$$\Psi(p) = \mathrm{i}\sqrt{\frac{\omega(p)m\hbar}{2}}(a^\dagger(p) - a(p)), \tag{3.88}$$

which now means that for a test function ξ one has

$$\Psi(\xi) = \mathrm{i}\sqrt{\frac{m\hbar}{2}}(A^\dagger(\xi\sqrt{\omega}) - A(\xi^*\sqrt{\omega})).$$

Plugging (3.86) and (3.88) into (3.85), performing the straightforward calculation (taking into account that a and $a^\dagger$ do not commute) we get

$$H = \frac{\hbar}{2}\int \frac{\mathrm{d}p}{2\pi\hbar}\omega(p)\big(a^\dagger(p)a(p) + a(p)a^\dagger(p)\big). \tag{3.89}$$

The first problem is that the "straightforward calculation" involved considering $\Phi(p)^2$ and the undefined terms $a^\dagger(p)^2$ and $a(p)^2$. *By a kind of miracle these ill-defined quantities cancel algebraically*. The second problem is that (3.89) contains the undefined term $a(p)a^\dagger(p)$. For this term we brazenly use that since $[a(p), a^\dagger(p)] = 2\pi\hbar\delta(p - p')1$, we have $a(p)a^\dagger(p) = a^\dagger(p)a(p) + 2\pi\hbar\delta(0)1 = a^\dagger(p)a(p) + \text{constant}1$, so that within an irrelevant multiple of the identity (3.89) coincides with (3.84).

The situation is unpleasant but not dramatic. The problems connected to the continuous limit are in a sense an artifact of our continuous model and one should not worry too much about them. "Putting the universe in a box", that is replacing physical space by a finite rectangular box as we did at the beginning of the present section removes most of

45 This is a consequence of our hypotheses on the function $\omega(p)$.

these problems. We may argue that it is reasonable to view formal computations done in the continuous limit as simply a convenient way to encode computations done in the more reliable discrete approximation. Therefore these computations should be trusted unless proved otherwise. There will be many of them.

Now, what about the value of m? We better leave this question for a future time when we have figured some rationale for the normalization of this quantum field.

Key ideas to remember:

- If $\mathcal{H}$ is the state space appropriate to describe a single particle, a certain subspace of the n-fold tensor product $\mathcal{H} \otimes \cdots \otimes \mathcal{H}$ is the appropriate state space to describe a system of n indistinguishable particles. For bosons, this is the subspace $\mathcal{H}_{n,s}$ of symmetric tensors.
- The boson Fock space, the direct sum of the $\mathcal{H}_{n,s}$ is the appropriate state space to describe a system of any number of the same type of bosons.
- The fundamental structure of the boson Fock space is the construction for $\xi \in \mathcal{H}$ of the creation operator $A^\dagger(\xi)$ and the annihilation operator $A(\xi)$ which are both unbounded operators on the Fock space. They satisfy the relation $[A(\xi), A^\dagger(\eta)] = (\xi, \eta)1$.
- Putting the universe in a box often makes our formalism less prickly.
- Quantum Fields are operator-valued distributions.

4

The Lorentz Group and the Poincaré Group

It is an extremely well-established experimental fact that the speed of light c is the same for all "inertial observers" (those who do not undergo accelerations). This is what has led to the definition (1.1). The analysis of the consequences of this remarkable fact has forced a complete revision of Newton's ideas. Each observer uses a copy of $\mathbb{R}^3$ to describe the location of points and a copy of $\mathbb{R}$ to describe time. Space and time are however not different entities but are different aspects of one single entity, space-time.[1] Different inertial observers may use different coordinates to describe the points of space-time, but these coordinates must be related in a way that preserves the speed of light. To formulate this mathematically, the Euclidean inner product on $\mathbb{R}^4$ has to be replaced by another bilinear form, the Lorentz bilinear form, and the change of coordinates between observers must preserve the Lorentz bilinear form. These transformations form a group, the Lorentz group. To a large extent the mathematics of Special Relativity reduce to the study of this group. In the present chapter we introduce the Lorentz group and a related group, the Poincaré group, the group of transformations of space-time generated by the Lorentz group and the translations.

The present chapter is essentially independent of the previous ones.

4.1 Notation and Basics

Special Relativity uses standard notation, which we introduce now. Points in $\mathbb{R}^3$ are denoted in **boldface** to distinguish them from points in space-time. The components of a point $x \in \mathbb{R}^4$ are denoted by (x^0, x^1, x^2, x^3). The idea is that $x = (x^0, x^1, x^2, x^3)$ describes the point $\boldsymbol{x} = (x^1, x^2, x^3)$ of space at time $t = x^0/c$, and we often write $x = (x^0, \boldsymbol{x})$. Thus x^0/c has the dimension of a time, and each coordinate x^ν has the dimension of a length.

We consider the 4×4 matrix $\eta_{\mu\nu}$, which is diagonal with $\eta_{00} = 1$ and $\eta_{ii} = -1$. Here we follow the convention that Greek indices range from 0 to 3 while Latin indices range from 1 to 3, so that the condition $\eta_{ii} = -1$ means $\eta_{11} = \eta_{22} = \eta_{33} = -1$. Let us note that

$$\eta_{\mu\nu} = \eta_{\nu\mu}.$$

[1] The reader with no exposure to the delightful theory of Special Relativity may get minimal help by looking at Appendix E, but is strongly advised to open a real textbook such as that of Taylor and Wheeler [81].

On $\mathbb{R}^4$ we consider the following bilinear form, called the *Lorentz bilinear form*,

$$(x, y) = \eta_{\mu\nu} x^\mu y^\nu. \tag{4.1}$$

This expression follows the convention that repeated indices *in the same term, one of which is a subscript and one of which is a superscript* are summed, so that

$$(x, y) = \sum_{0 \le \mu, \nu \le 3} \eta_{\mu\nu} x^\mu y^\nu = x^0 y^0 - x^1 y^1 - x^2 y^2 - x^3 y^3.$$

We do not use a special notation for the Lorentz bilinear form, since it can hardly be confused with the Euclidean dot product which we will not use in this setting. Still, we shall denote $\mathbb{R}^4$ equipped with the Lorentz bilinear form by $\mathbb{R}^{1,3}$ as a reminder that our version of the Lorentz bilinear form uses one + sign and three − signs.[2] We note that $(x, y) = (y, x)$. The convention that repeated indices are summed will be in force *every time* we deal with $\mathbb{R}^{1,3}$. The space $\mathbb{R}^{1,3}$ will be called *Minkowski space.*

A *Lorentz transformation* L is a linear transformation on $\mathbb{R}^{1,3}$ which preserves the bilinear form (4.1), that is

$$(L(x), L(y)) = (x, y). \tag{4.2}$$

Definition 4.1.1 For a four-vector $x = (x^0, \boldsymbol{x})$ we use the notation

$$x^2 := (x, x)\ ;\ \boldsymbol{x}^2 = \sum_{1 \le i \le 3} (x^i)^2\ ;\ |\boldsymbol{x}| = \sqrt{\boldsymbol{x}^2}. \tag{4.3}$$

In principle the notation x^2 is ambiguous as this could also be the third component of x, but in practice there is no problem. Thus

$$x^2 = (x^0)^2 - \boldsymbol{x}^2.$$

A Lorentz transformation is necessarily injective since if $x \neq 0$ there exists y with $(x, y) \neq 0$, so that $(L(x), L(y)) = (x, y) \neq 0$ and $L(x) \neq 0$. It is then bijective since $\mathbb{R}^{1,3}$ is finite-dimensional. It is straightforward to check that the Lorentz transformations form a group, denoted by $O(1, 3)$, the numbers 1 and 3 referring again to one + and the three − signs among the numbers $\eta_{\mu\mu}$. As a consequence of (4.2) let us note for further use that

$$(L(x), y) = (x, L^{-1}(y))\ ;\ (x, L(y)) = (L^{-1}(x), y). \tag{4.4}$$

Let us spend a few lines to explain the physical ideas here. The identity $4(x, y) = (x + y, x + y) - (x - y, x - y)$ shows that a linear map L is a Lorentz transformation if and only if it preserves the quantity $x^2 = (x, x)$.[3] The set of $x \in \mathbb{R}^{1,3}$ for which $(x, x) = 0$ is called the *light cone of the origin*. We will often omit the words "of the origin" as we consider no other light cone. The trajectory in space-time of a point moving in a straight line at the speed of light and departing (respectively reaching) the origin at time zero in the

[2] The present convention goes under various names, such as *particle's physicist's sign convention*. Relativists mostly use three + signs and one − sign.

[3] The name "Lorentz norm" here would be improper because (x, x) can be negative.

direction of the unit vector $\boldsymbol{u}$ is the set of points of the form $(tc, tc\boldsymbol{u})$ for $t \geq 0$ (respectively $t \leq 0$).[4] The light cone is exactly the union of such trajectories. Thus, the basic assumption of Special Relativity, that light travels at speed c for any inertial observer, implies that the light cone should be the same for two such observers. The natural way to ensure this is that the transformation L which relates the way two inertial observers refer to points of space-time preserves the Lorentz bilinear form.

Exercise 4.1.2 Prove that if a linear transformation A of $\mathbb{R}^{1,3}$ satisfies $(x,x) = 0 \Rightarrow (A(x), A(x)) = 0$ then A is a multiple of a Lorentz transformation. Hint: Assuming $A \neq 0$, reduce to the case where A fixes $e_0 = (1,0,0,0)$. This exercise will be easier is you study up to Section 4.3 first.

The matrix of a linear transformation L on $\mathbb{R}^{1,3}$ will be denoted by $(L^\mu{}_\nu)$. The rule is that μ, the index on the left, represents the row and ν, the index on the right, represents the column. Thus (thinking of $x = (x^\nu)$ as a column vector),

$$L(x)^\mu = L^\mu{}_\nu x^\nu \Big(= \sum_{0 \leq \nu \leq 3} L^\mu{}_\nu x^\nu \Big). \tag{4.5}$$

With our present row and column convention, the rule for matrix multiplication is then

$$(LM)^\mu{}_\nu = L^\mu{}_\rho M^\rho{}_\nu. \tag{4.6}$$

When dealing with matrices that do not represent Lorentz transformations, the index on the left still represents the row and the index on the right still represents the column, but both indices are lower indices. The matrix multiplication rule then takes the form

$$(AB)_{i,j} = \sum_k A_{i,k} B_{k,j}. \tag{4.7}$$

The map L is a Lorentz transformation if and only if

$$\eta_{\mu\nu} L^\mu{}_\lambda x^\lambda L^\nu{}_{\lambda'} y^{\lambda'} = (L(x), L(y)) = (x,y) = \eta_{\lambda\lambda'} x^\lambda y^{\lambda'}, \tag{4.8}$$

for all $x, y \in \mathbb{R}^{1,3}$, or, equivalently,[5]

$$\eta_{\mu\nu} L^\mu{}_\lambda L^\nu{}_{\lambda'} = \eta_{\lambda\lambda'}. \tag{4.9}$$

Writing all indices as lower indices as in the case (4.7) this means that

$$L^T \Xi L = \Xi, \tag{4.10}$$

where Ξ is the matrix $(\eta_{\mu\nu})$ and where L^T is the transpose of L.[6] Taking determinants proves that $(\det L)^2 = 1$, and hence the important fact that $\det L = \pm 1$. Since $\Xi = \Xi^{-1}$, taking the inverse of (4.10) and replacing L by L^{-1} proves that $L^T \in O(1,3)$.

[4] This simply means that at time $t = (tc)/c$ the position of the point is $tc\boldsymbol{u}$, at distance $|tc|$ from the origin, in the direction of the unit vector $\boldsymbol{u}$ when $t > 0$ and in the opposite direction when t is negative.

[5] It is an automatic part of the formalism that λ and λ' are fixed indices in (4.9) but are summed in (4.8). To prove the equivalence, multiply (4.9) by $x^\lambda y^{\lambda'}$ and sum over λ and λ'. In the reverse direction, choose x, y in the canonical basis.

[6] We abuse notation by not distinguishing between L and its matrix.

The map $L \mapsto \det L$ is a group homomorphism from $O(1,3)$ to $\{-1,1\}$. A second less obvious such homomorphism is given by $L \mapsto \operatorname{sign} L^0{}_0$. To prove this we have to show that given two Lorentz transformations L and M the quantity

$$(LM)^0{}_0 = L^0{}_\nu M^\nu{}_0 = L^0{}_0 M^0{}_0 + L^0{}_i M^i{}_0$$

has the same sign as $L^0{}_0 M^0{}_0$. It suffices for this to prove that

$$|L^0{}_i M^i{}_0| < |L^0{}_0 M^0{}_0|. \tag{4.11}$$

Using (4.9) for $\lambda = \lambda' = 0$ yields in particular

$$(L^0{}_0)^2 = 1 + \sum_{1 \le i \le 3} (L^i{}_0)^2. \tag{4.12}$$

We observe for further use that $|L^0{}_0| \geq 1$. Now, (4.12) used for M implies $(M^0{}_0)^2 = \sum_{1\le i\le 3}(M^i{}_0)^2 + 1$ whereas the same relation used for L^T implies

$$(L^0{}_0)^2 = \sum_{1 \le i \le 3} (L^0{}_i)^2 + 1. \tag{4.13}$$

Thus (4.11) follows, since, using the Cauchy-Schwarz inequality, we have

$$(L^0{}_i M^i{}_0)^2 \le \sum_{i \le 3} (L^0{}_i)^2 \sum_{i \le 3} (M^i{}_0)^2 = ((L^0{}_0)^2 - 1)((M^0{}_0)^2 - 1) < (L^0{}_0 M^0{}_0)^2.$$

The transformations such that $L^0{}_0 > 0$ are by far the most important since they preserve the direction in which time flows. They are called *orthochronous*[7] *transformations*. The two homomorphisms from $O(1,3)$ into $\{-1,1\}$ show that $O(1,3)$ is made up of four pieces (determined by the four possible choices for the sign of $L^0{}_0$ and of $\det L$). We will denote

$$SO^\uparrow(1,3) = \{L \in O(1,3)\ ;\ L^0{}_0 > 0\ ,\ \det L = 1\}$$

the most important of these pieces. It will soon be apparent that $SO^\uparrow(1,3)$ is connected. The proper terminology is to call $SO^\uparrow(1,3)$ "the restricted Lorentz group", although we will abuse this terminology and often call it "the Lorentz group". There is little risk of confusion since we rarely use the group $O(1,3)$ itself. From this point on let us agree that when we say "Lorentz transformation", unless otherwise specified, we *always mean an element of* $SO^\uparrow(1,3)$, an orthochronous transformation of determinant one. There are two important elements of $O(1,3) \setminus SO^\uparrow(1,3)$, the parity transformation $P\colon (x^0, \boldsymbol{x}) \to (x^0, -\boldsymbol{x})$, and time reversal $(x^0, \boldsymbol{x}) \to (-x^0, \boldsymbol{x})$. We will not study time reversal, but the parity transformation will be of considerable importance.

Lemma 4.1.3 *A Lorentz transformation*[8] *sends the set* $\bar{V}_+ = \{x^2 = (x,x) \geq 0,\ x^0 \geq 0\}$ *to itself.*

[7] From the Greek ortho=proper, and chrono=time.
[8] That is, in our terminology, an element of $SO^\uparrow(1,3)$. Time reversal, which is not orthochronous, does not satisfy the conclusion of the lemma.

Proof Consider $L \in SO^{\uparrow}(1,3)$ and $x \in \bar{V}_+$. Then $L(x)^2 = x^2 \geq 0$ and

$$L(x)^0 = L^0{}_\mu x^\mu = L^0{}_0 x^0 + L^0{}_i x^i,$$

where $L^0_{\ 0} > 0$, $x^0 \geq 0$, $|L^0{}_i x^i| \leq L^0{}_0 x^0$ using the Cauchy-Schwarz inequality, (4.13), and $\sum_{i\leq 3}(x^i)^2 \leq (x^0)^2$. Thus $L(x)^0 \geq 0$. □

There is some fundamental physics going on in this lemma. Suppose I am located at the origin and I flip a coin at time zero, and that you are located at $\boldsymbol{x}$ and try to record at time x^0/c the result of this coin flipping. This may be done as follows if $x^2 > 0$: I send you the appropriate message using a light ray starting from the origin at time zero. It reaches you at time $|\boldsymbol{x}|/c$. Since $x^2 > 0$ it holds that $|\boldsymbol{x}|/c \leq x^0/c$, so that you have received the message before you have to record the result of the coin flipping. In more general terms what happens at $\boldsymbol{x}$ at time x^0/c may be influenced by what happens at the origin at time zero: there is a *causality relation*, and it must hold for all observers, as the previous lemma states.[9] The situation is very different when $x^2 < 0$. You are too far away for me to inform you in due time of the coin flipping result. Indeed, when we know more about the Lorentz group is will be easy to show (in Exercise 4.3.2) that when $x^2 < 0$ there exists a Lorentz transformation L for which $L(x)^0 < 0$. In physical terms, for some observers the recording event takes place *before* the coin is flipped![10]

Definition 4.1.4 Two points x, y of $\mathbb{R}^{1,3}$ are *causally separated* if $(x - y)^2 < 0$.

The previous discussion has led us to the following important statement, to be remembered.

If two points $x, y \in \mathbb{R}^{1,3}$ are causally separated what happens at x cannot influence what happens at y. (4.14)

The next page or so is devoted to playing with a very nice notational formalism, which finds its roots in the theory of General Relativity, where it is universally used to manage tensorial calculus. Not every author uses this formalism in the present setting (this is a kind of overkill). Whether I use it or not, this will conflict with some of the textbooks the reader might open, so I follow my own taste. The formalism will not be used much in the sequel, besides some notation such as (4.17). It is not necessary to master it, except for readers wishing to study Section D.12, Section E.2 or Appendix G, where we perform some computations on tensors that exemplify the remarkable power of this formalism.

We define $\eta^{\mu\nu} := \eta_{\mu\nu}$ and observe that

$$\eta^{\mu\nu}\eta_{\lambda\nu}\Big(= \sum_{0\leq\nu\leq 3} \eta^{\mu\nu}\eta_{\lambda\nu}\Big) = \delta^\mu_\lambda, \tag{4.15}$$

9 The set $\bar{V}_+$ is called the causal future of the origin.

10 Let us now enjoy getting really confused together. In Quantum Field Theory, (at least in the flowery description of physicists) "virtual particles" may travel faster than light. A virtual electron may be created at the origin at time zero and be annihilated at time x^0/c and location $\boldsymbol{x}$ where $x^2 < 0$. How can we make sense that for some observers the annihilation takes place *before* the creation? Elementary, my dear Watson: these observers see a virtual *anti-electron* created first and then annihilated later.

where δ^μ_λ is the Kronecker symbol. We can use the $\eta^{\nu\mu}$ to *raise* and the $\eta_{\nu\mu}$ to *lower* indices, as follows:

$$x_\nu := \eta_{\nu\mu} x^\mu. \tag{4.16}$$

This *simply amounts to changing the sign* of the indices μ as they are raised or lowered when $1 \le \mu \le 3$, e.g. $x^0 = x_0$ and $x_1 = -x^1$. Nonetheless this notation is very convenient. Thus we can define

$$\eta^\mu{}_\lambda := \eta^{\mu\nu}\eta_{\nu\lambda}.$$

The summation in the right-hand side can be viewed either as lowering the index λ in $\eta^{\mu\lambda}$ or as raising the index μ in $\eta_{\mu\lambda}$. Then (4.15) simply means that $\eta^\mu{}_\lambda = \delta^\mu_\lambda$. With the convention of raising and lowering indices, (4.1) takes the elegant form

$$(x, y) = x_\mu y^\mu = x^\mu y_\mu. \tag{4.17}$$

The end of this section is for fun and will be used only in Sections D.12 and E.2. The indices in a matrix coefficient $L^\mu{}_\nu$ can be raised or lowered as any other index, but then one should respect the difference between the left and the right index.[11] For example we can write (4.5) as $L(x)_\mu = L_{\mu\nu} x^\nu$. To see this we simply write

$$L(x)_\mu = \eta_{\mu\nu} L(x)^\nu = \eta_{\mu\nu} L^\nu{}_\lambda x^\lambda = L_{\mu\lambda} x^\lambda.$$

Exercise 4.1.5 Prove that $L_{\mu\nu} x^\nu = L_\mu{}^\nu x_\nu$.

The matrix multiplication formula (4.6) can be written in several different forms such as $(LM)_{\mu\nu} = L_\mu{}^\rho M_{\rho\nu}$, as is seen by writing

$$(LM)_{\mu\nu} = \eta_{\mu\lambda}(LM)^\lambda{}_\nu = \eta_{\mu\lambda} L^\lambda{}_\tau M^\tau{}_\nu = L_{\mu\tau} M^\tau{}_\nu = L_\mu{}^\rho \eta_{\rho\tau} M^\tau{}_\nu = L_\mu{}^\rho M_{\rho\nu}.$$

Moreover (4.9) can be written as $L_{\nu\lambda} L^\nu{}_{\lambda'} = \eta_{\lambda\lambda'}$ or even (raising λ in the previous equality and recalling that $\eta^\lambda{}_{\lambda'} = \delta^\lambda_{\lambda'}$) in the more appealing form

$$L_\nu{}^\lambda L^\nu{}_{\lambda'} = \delta^\lambda_{\lambda'}. \tag{4.18}$$

Exercise 4.1.6 Make sure you understand how to derive the previous equation.

Denoting by M the transformation such that $M^\lambda{}_\nu = L_\nu{}^\lambda$, this implies $M^\lambda{}_\nu L^\nu{}_{\lambda'} = \delta^\lambda_{\lambda'}$ so that $M = L^{-1}$. That is, to go from L to L^{-1} one simply exchanges right and left indices. Furthermore, $M^\lambda{}_\nu = L_\nu{}^\lambda = \eta_{\nu a} L^a{}_b \eta^{b\lambda}$, and viewing the right-hand side as a product of three 4×4 matrices gives $\det M = \det L$. Therefore since $M = L^{-1}$ we obtain again that $(\det L)^2 = 1$.

4.2 Rotations

The linear transformations of $\mathbb{R}^n$ which preserve the Euclidean norm (the length) of a vector form a group, the orthogonal group $O(n)$. A transformation preserves the length of

[11] This is precisely why we state the rule that "the left index represents the row and the right index represents the column". Raising or lowering indices does not affect it.

a vector if and only if its matrix M in an orthonormal basis satisfies $M^T M = MM^T = 1$, the identity matrix. Consequently $\det M = \pm 1$. The subgroup $SO(n)$ of transformations of determinant 1 is called the *special orthogonal group.*

Exercise 4.2.1 If you are not familiar with the previous results, prove them by copying arguments of the previous section.

The group $SO(2)$ is the familiar group of matrices of the form,

$$R_\theta = \begin{pmatrix} \cos\theta & -\sin\theta \\ \sin\theta & \cos\theta \end{pmatrix},$$

where θ is called the *angle of the rotation*. The group $SO(3)$ will be fundamental. It is simply called the *group of rotations*. Given the matrix M of a rotation,

$$\det(M - 1) = \det M \det(1 - M^{-1}) = \det(1 - M^T) = \det(1 - M) = -\det(M - 1),$$

using in the last equality that the dimension of these matrices is odd. Thus $\det(M - 1) = 0$ and there are invariant vectors under a rotation, which constitute the axis of the rotation. The plane perpendicular to the axis of rotation is invariant under the rotation. Rotations are studied in more detail in Appendix D.

Given $T \in O(3)$ we may define an operator T' on $\mathbb{R}^{1,3}$ by

$$T' : (x^0, \boldsymbol{x}) \mapsto (x^0, T(\boldsymbol{x})). \tag{4.19}$$

It is obvious that T' is a Lorentz transformation, and that $\det T' = \det T$. Moreover, T' leaves the vector $e_0 := (1,0,0,0)$ invariant. Conversely, if a Lorentz transformation S leaves e_0 invariant, it maps the set $\{(0,\boldsymbol{x}); \boldsymbol{x} \in \mathbb{R}^3\}$ to itself, because this is exactly the set of vectors x for which $(e_0, x) = x^0 = 0$. Thus one may define a linear map T on $\mathbb{R}^3$ by $S(0,\boldsymbol{x}) = (0, T(\boldsymbol{x}))$. It is clear that T is an isometry, so that if $\det S = 1$, T is a rotation. For this reason we will call a Lorentz transformation a *rotation* if it fixes e_0 and has determinant 1.

4.3 Pure Boosts

In this section we try to completely clarify the nature of some basic Lorentz transformations called *pure boosts.*

A rotation does not mix space and time coordinates, but a general Lorentz transformation does, and pure boosts are the simplest type of transformations mixing space and time coordinates. As we will see a general Lorentz transformation can then entirely be described by applying first a rotation and then a pure boost. The crucial point of mixing time and space coordinates is that in (4.1) the signs are not the same for the time coordinate and the space coordinates. To understand the consequences of this, let us first study the linear transformations V of $\mathbb{R}^2$ which preserve the bilinear form $B(x,y) := x_1 y_1 - x_2 y_2$, i.e. $B(V(x), V(y)) = B(x,y)$. The condition $B(V(e_1), V(e_1)) = 1$ implies that $V(e_1) = (a,b)$ where $a^2 - b^2 = 1$. The condition $B(V(e_1), V(e_2)) = 0$ implies that $V(e_2) = (\lambda b, \lambda a)$

for some $\lambda \in \mathbb{R}$. The condition $B(V(e_2), V(e_2)) = -1$ implies $\lambda^2 = 1$. If one moreover requires that $\det V = 1$, the matrix of V then has to be of the type

$$\begin{pmatrix} a & b \\ b & a \end{pmatrix}.$$

We will consider only the case $a > 0$. Then we can set $a = \cosh s, b = \sinh s$, and the previous matrix becomes

$$M_s = \begin{pmatrix} \cosh s & \sinh s \\ \sinh s & \cosh s \end{pmatrix}. \tag{4.20}$$

This surely reminds one of a rotation matrix. The point of the parameterization is the remarkable formula $M_{s+t} = M_s M_t$.

Given $s \in \mathbb{R}$ we can then define the boost B^s as the transformation which operates by the matrix (4.20) on the coordinates (x^0, x^3) while leaving the coordinates (x^1, x^2) unchanged:

$$B^s := (x^0, x^1, x^2, x^3) \mapsto (x^0 \cosh s + x^3 \sinh s, x^1, x^2, x^0 \sinh s + x^3 \cosh s). \tag{4.21}$$

It is called a (pure) boost along the x^3 axis. It is straightforward that $B^s \in SO^{\uparrow}(1,3)$, and $B^s(e_0) = (\cosh s, 0, 0, \sinh s)$.

Observe that when $r := (\cosh s, 0, 0, \sinh s)$, (4.21) can be written as

$$B^s((x^0, \boldsymbol{x})) = (x^0 r^0 + \boldsymbol{r} \cdot \boldsymbol{x}, x^0 \boldsymbol{r} + r^0 \boldsymbol{x}_{\parallel} + \boldsymbol{x}_{\perp}), \tag{4.22}$$

where for $\boldsymbol{x} \in \mathbb{R}^3$ we denote by $\boldsymbol{x}_{\parallel}$ and $\boldsymbol{x}_{\perp}$ its components parallel and perpendicular to $\boldsymbol{r}$. This formula motivates the following.

Definition 4.3.1 Given $r = (r^0, \boldsymbol{r}) \in \mathbb{R}^{1,3}$ with $r^2 := r_\mu r^\mu = (r^0)^2 - \boldsymbol{r}^2 = 1$ and $r^0 > 0$ we define the *pure boost* B_r by (4.22).

Thus for $r = (\cosh s, 0, 0, \sinh s)$ we have $B_r = B^s$ and in particular when $r = e_0$ the pure boost B_r is the identity. Otherwise, this is simply the transformation which acts as identity on the space $\{(0, \boldsymbol{x}); \boldsymbol{x} \cdot \boldsymbol{r} = 0\}$, while (as will become apparent from (4.23)) acting by a transformation of the type (4.20) on the two-dimensional space generated by e_0 and r.

Exercise 4.3.2 Prove that if $x^2 < 0$ there is a pure boost B such that $B(x)^0 = 0$ and a pure boost such that $B(x)^0 < 0$.

To understand the physical meaning of a pure boost, consider an observer (called Alice) using a reference frame such that when Alice measures the position of any space-time point by x, I measure the position of the same point by $B_r(x)$. In particular we both use the same origin 0 of space-time. Imagine a clock being located at the origin Alice uses for space, and marking the time Alice uses.[12] Then when Alice's clock marks time t mine marks time[13] $r^0 t$. Moreover Alice's clock is then located at the point of coordinates $ct\boldsymbol{r}$ in my reference

[12] Thus, this clock is stationary for Alice.
[13] This is the phenomenon of time-contraction, which is briefly discussed in Appendix E.

frame. That is, in my reference frame, Alice's clock moves away at speed $c|\boldsymbol{r}|/r^0$ in the direction of $\boldsymbol{r}$.[14]

Lemma 4.3.3 *(a) We have* $B_r(e_0) = r$.
(b) If S is a rotation, then

$$B_{S(r)} = SB_rS^{-1}. \tag{4.23}$$

(c) $B_r \in SO^{\uparrow}(1,3)$.

Proof Taking $x^0 = 1$ and $\boldsymbol{x} = 0$ in (4.22) yields $B_r(e_0) = (r^0, \boldsymbol{r}) = r$. To check formula (4.23) we write it as $B_{S(r)}S = SB_r$. It should then be obvious from (4.22), since when S is a rotation of $\mathbb{R}^3$, $S(\boldsymbol{x}_\parallel)$ and $S(\boldsymbol{x}_\perp)$ are the components of $S(\boldsymbol{x})$ which are parallel and perpendicular to $S(\boldsymbol{r})$. Finally consider $r \in \mathbb{R}^{1,3}$ with $r^2 = 1$ and $r^0 > 0$. Then we can find a rotation S for which $q := S^{-1}(r)$ is such that $\boldsymbol{q}$ is parallel to the third axis. Then $B_q \in SO^{\uparrow}(1,3)$ since it is of the type (4.21) and therefore $B_r = B_{S(q)} \in SO^{\uparrow}(1,3)$ by (4.23). □

The content of (4.23) is that "the geometry of a boost is the same as the geometry of a boost of the type B^s".

Observing that

$$r^0\boldsymbol{x}_\parallel + \boldsymbol{x}_\perp = (r^0 - 1)\boldsymbol{x}_\parallel + \boldsymbol{x} = \frac{(r^0)^2 - 1}{1 + r^0}\boldsymbol{x}_\parallel + \boldsymbol{x} = \frac{\boldsymbol{r}^2}{1 + r^0}\boldsymbol{x}_\parallel + \boldsymbol{x},$$

and that $\boldsymbol{x}_\parallel = (\boldsymbol{r} \cdot \boldsymbol{x})\boldsymbol{r}/\boldsymbol{r}^2$ we obtain from (4.22) the following formula: the transformation $B = B_r$ is given by

$$B((x^0, \boldsymbol{x})) = (x^0r^0 + \boldsymbol{r} \cdot \boldsymbol{x}, x^0\boldsymbol{r} + \frac{(\boldsymbol{r} \cdot \boldsymbol{x})\boldsymbol{r}}{1 + r^0} + \boldsymbol{x}), \tag{4.24}$$

or, equivalently,

$$B^0{}_0 = r^0\ ;\ B^0{}_j = B^j{}_0 = r^j\ ;\ B^i{}_j = \delta^i_j + \frac{r^ir^j}{1 + r^0}. \tag{4.25}$$

Exercise 4.3.4 Consider $r = (r^0, \boldsymbol{r}) \in \mathbb{R}^{1,3}$ with $r^2 = r_\mu r^\mu = 1$ and $r^0 > 0$. Let $s > 0$ with $r^0 = \cosh s$, so that $|\boldsymbol{r}| = \sinh s$. Thus there exists a rotation R which transforms $B^s(e_0) = (\cosh s, 0, 0, \sinh s)$ into r. Prove that the operator

$$RB^sR^{-1} \tag{4.26}$$

is the same for each rotation R which transforms $B^s(e_0) = (\cosh s, 0, 0, \sinh s)$ into r. Hint: Prove first that if R is a rotation which fixes the third axis it commutes with B^s. This provides an alternate definition of $B_r := RB^sR^{-1}$ for any such R. Prove (4.23) from this definition.

[14] Generally speaking, any two inertial observers can choose their respective coordinate systems in such a way that the change between these coordinate systems is represented by a pure boost.

Let us collect two simple facts for further use. You are not expected to understand right now the relevance of these facts. We denote by P the *parity operator*

$$P(r^0, \boldsymbol{r}) = (r^0, -\boldsymbol{r}) \tag{4.27}$$

1. First,

$$B_r^{-1} = B_{Pr}. \tag{4.28}$$

When $\boldsymbol{r}$ is along the third axis, this results from the fact that $(B^s)^{-1} = B^{-s}$. The general case follows by writing $r = S(r')$ where S is a rotation, where $\boldsymbol{r}'$ is along the third axis, and using (4.23).

2. Consider a Lorentz transformation $A \in SO^{\uparrow}(1,3)$, and let $r = A(e_0)$. Then $B_r^{-1}A(e_0) = B_r^{-1}(r) = e_0$. That is, $B_r^{-1}A$ fixes e_0, so that it is a rotation because it is of determinant 1. Consequently, $A = B_r R$ where R is a rotation. This is a kind of polar decomposition. It makes it obvious that one can move continuously from the identity to A (because this is the case both for B_r and R), so that $SO^{\uparrow}(1,3)$ is connected. Matters are tricky however. The product of two pure boosts B_r, B_q is not in general a pure boost (unless $\boldsymbol{r}$ and $\boldsymbol{q}$ are parallel).

The vast majority of textbooks use Lie algebras to study the Lorentz group (and for many other purposes). While we choose a different presentation, it is nonetheless highly recommended to learn at least the idea of this fundamental approach, if only to understand the literature. We present some elements of this theory in Appendix D, and we suggest that the reader starts the study of this appendix now.

4.4 The Mass Shell and Its Invariant Measure

In basic Quantum Mechanics we have used "position space" and "momentum space", and similarly Minkowski space comes in two versions, a "position space" version, the points x of which represent a location in space-time, $x = (x^0, x^1, x^2, x^3)$, $x^0 = ct$, and are measured in units of length, and a "momentum space" version, the points $p = (p^0, p^1, p^2, p^3)$ of which represent the "four-momentum", also called the "energy-momentum" of a point particle where the coordinates are in units of momentum.[15] For x in "position space" and p in "momentum space" the quantity $(x,p)/\hbar$ is dimensionless.

We will accept the basic fact that a particle of four-momentum $p = (p^0, \boldsymbol{p})$ has energy $E = cp^0$ and momentum $\boldsymbol{p}$.[16] We will also accept the fundamental relation

$$p^2 = p_\nu p^\nu = m^2c^2, \tag{4.29}$$

where m is the *rest mass* of the particle. The words "rest mass" do not refer to any new concept. When $m > 0$, the rest mass of an object is the usual mass as we experience it at small velocity in non-relativistic situations.[17] We will often simply say "mass" rather than

15 We will not attempt to use notation that would distinguish between these two versions of $\mathbb{R}^{1,3}$.
16 Thus, p gives us both the energy and the momentum of the particle, hence the name energy-momentum.
17 The situation is different when $m = 0$. As we will see soon, an object of mass zero is never at rest.

"rest mass".[18] Equivalently, the energy E and the momentum $\boldsymbol{p}$ of a particle of mass m are connected by the relation

$$\frac{E^2}{c^2} - \boldsymbol{p}^2 = m^2c^2, \tag{4.30}$$

or, equivalently, $E^2 = c^2\boldsymbol{p}^2 + m^2c^4$. The case $\boldsymbol{p} = 0$ should be familiar to the reader. The case $m = 0$ is Einstein's equation $E = c|\boldsymbol{p}|$ for photons.

For $m \geq 0$ we define the set

$$X_m := \{p \in \mathbb{R}^{1,3}\ ;\ p^2 = m^2c^2\ ,\ p^0 \geq 0\},$$

which is called the "mass shell" in physics. This is the set of points of the type $p = (\omega_{\boldsymbol{p}}, \boldsymbol{p})$ where

$$\omega_{\boldsymbol{p}} := \sqrt{m^2c^2 + \boldsymbol{p}^2}. \tag{4.31}$$

The reason for the name is that the hyperboloid $p^2 = m^2c^2$ somewhat looks like a shell. The mass shell is to Minkowski space what a sphere is to Euclidean space.

Let us first assume that $m > 0$. Then X_m consists of the possible values of a four-vector p when it represents the four-momentum of a particle of mass m, and it is the orbit of the point[19] $p^* = (mc,0,0,0)$ under the action of $SO^{\uparrow}(1,3)$. Indeed, if $L \in SO^{\uparrow}(1,3)$, then $p = L(p^*)$ satisfies $p^2 = (p^*)^2 = m^2c^2$ because $L \in O(1,3)$ and satisfies $p^0 > 0$ because $L^0{}_0 > 0$ since $L \in SO^{\uparrow}(1,3)$. Conversely, as we will use many times, given $p \in X_m$ there is a unique pure boost B such that $B(p^*) = p$, namely, recalling the notation of Section 4.3, $B = B_{p/(mc)}$. When $m = 0$ the situation is slightly different, since $(0,0,0,0) \in X_0$ is invariant under the action of $SO^{\uparrow}(1,3)$. One may check that $X_0 \setminus \{(0,0,0,0)\}$ is the orbit under $SO^{\uparrow}(1,3)$ of the point $p^* = (1,0,0,1)$, so that it represents all the possible values of the four-momentum of a massless particle. For any observer the four-momentum of the particle belongs to $X_0 \setminus \{(0,0,0,0)\}$ i.e. $(p^0)^2 = |\boldsymbol{p}|^2 \neq 0$ so its momentum is not zero: a massless particle is never observed at rest.

Exercise 4.4.1 Check the previous claim.

It is fundamentally important that on X_m there exists a measure that is invariant under the action of the Lorentz group (as stated precisely in (4.37)). This measure is to Minkowski space what the uniform measure on the sphere is to Euclidean space. This measure will be central in much of the further constructions. It is defined only up to a multiplicative factor. An explicit formula for an invariant measure can be found by considering for $\varepsilon > 0$ the set

$$X_{m,\varepsilon} = \{p\ ;\ m^2c^2 < p^2 < m^2c^2 + \varepsilon\ ;\ p^0 \geq 0\}. \tag{4.32}$$

This set is invariant under the action of $SO^{\uparrow}(1,3)$ as follows from Lemma 4.1.3. The four-dimensional volume measure $\mathrm{d}^4p/(2\pi\hbar)^4$ on $\mathbb{R}^{1,3}$ is also invariant under this action because Lorentz transformations have determinant 1. Thus the restriction $\mathrm{d}\lambda_{m,\varepsilon}$ of this measure to

[18] The reason for the name "rest mass" is to prevent any confusion with the quantity m' of (E.4).
[19] Here the $*$ is just a notation and has nothing to do with complex conjugation, as p is not a number but a four-vector.

$X_{m,\varepsilon}$ is Lorentz invariant. One can then obtain a Lorentz invariant measure λ_m on X_m by taking a limit as $\varepsilon \to 0$ of $\varphi(\varepsilon)\mathrm{d}\lambda_{m,\varepsilon}$ where the normalization factor $\varphi(\varepsilon)$ is chosen to make things converge. To apply this procedure we consider a smooth function f and compute the integral

$$\int_{X_{m,\varepsilon}} \frac{\mathrm{d}^4 p}{(2\pi\hbar)^4} f(p) = \int_{m^2c^2<p^2<m^2c^2+\varepsilon} \frac{\mathrm{d}^4 p}{(2\pi\hbar)^4} f(p).$$

For this, we integrate in p^0 first. Given $\boldsymbol{p}$, the range of p^0 is

$$\sqrt{m^2c^2+\boldsymbol{p}^2} < p^0 < \sqrt{m^2c^2+\boldsymbol{p}^2+\varepsilon},$$

i.e. $h_{\boldsymbol{p}}(m^2c^2) < p^0 < h_{\boldsymbol{p}}(m^2c^2+\varepsilon)$ where $h_{\boldsymbol{p}}(x) := \sqrt{x+\boldsymbol{p}^2}$. Recalling the quantity $\omega_{\boldsymbol{p}} = h_{\boldsymbol{p}}(m^2c^2)$ of (4.31) and using the approximation $h_{\boldsymbol{p}}(m^2c^2+\varepsilon) \simeq h_{\boldsymbol{p}}(m^2c^2) + \varepsilon h'_{\boldsymbol{p}}(m^2c^2)$, at the first-order in ε the range of p^0 is the interval $(\omega_{\boldsymbol{p}}, \omega_{\boldsymbol{p}} + \varepsilon/2\omega_{\boldsymbol{p}})$. Therefore

$$\int_{X_{m,\varepsilon}} \frac{\mathrm{d}^4 p}{(2\pi\hbar)^4} f(p) \simeq \frac{\varepsilon}{4\pi\hbar} \int \frac{\mathrm{d}^3 \boldsymbol{p}}{(2\pi\hbar)^3\omega_{\boldsymbol{p}}} f((\omega_{\boldsymbol{p}}, \boldsymbol{p})). \tag{4.33}$$

Here we have kept a factor $(2\pi\hbar)^{-3}$ with $\mathrm{d}^3\boldsymbol{p}$ by consistency with our previous choices. Integrals over the mass shell are ubiquitous, so it helps to simplify notation. Given the value of m and given $\boldsymbol{p}$ we write

$$p = (\omega_{\boldsymbol{p}}, \boldsymbol{p}). \tag{4.34}$$

This notation will be of constant use and should be learned now. We will write (4.33) as

$$\lim_{\varepsilon\to 0+} \frac{2\pi\hbar}{\varepsilon} \int_{X_{m,\varepsilon}} \frac{\mathrm{d}^4 p}{(2\pi\hbar)^4} f(p) = \int \frac{\mathrm{d}^3 \boldsymbol{p}}{(2\pi\hbar)^3 2\omega_{\boldsymbol{p}}} f(p), \tag{4.35}$$

it being understood that $f(p)$ is a function of $\boldsymbol{p}$ through the relation (4.34).

Thus we have shown that the measure

$$\int_{X_m} \mathrm{d}\lambda_m(p) f(p) := \int \frac{\mathrm{d}^3 \boldsymbol{p}}{(2\pi\hbar)^3 2\omega_{\boldsymbol{p}}} f(p) \tag{4.36}$$

is a Lorentz invariant measure on X_m because it is a limit of Lorentz invariant measures. Again, this choice is determined only up to a multiplicative constant, and in particular the factor 2 in the denominator does not have any intrinsic value.

Exercise 4.4.2 How would you describe λ_m in the region where $|\boldsymbol{p}|$ is much smaller than mc?

For notational purposes it is convenient to think of $\mathrm{d}\lambda_m$ not as a measure on X_m, but as a measure on the whole of $\mathbb{R}^{1,3}$, which gives mass zero to the complement of X_m. For this reason we will always write $\int \mathrm{d}\lambda_m f$ rather than $\int_{X_m} \mathrm{d}\lambda_m f$. The measure $\mathrm{d}\lambda_m$ is Lorentz invariant, in the sense that for any Lorentz transformation $L \in SO^{\uparrow}(1,3)$ one has

$$\int \mathrm{d}\lambda_m(p) f(L(p)) = \int \mathrm{d}\lambda_m(p) f(p). \tag{4.37}$$

It is not invariant under the full Lorentz group, as it is not invariant under the map $(p^0,\boldsymbol{p}) \mapsto (-p^0,\boldsymbol{p})$. To obtain invariance under the full Lorentz group we would need to add a copy of $\mathrm{d}\lambda_m$ on the set

$$\{p\ ;\ p^2 = m^2c^2\ ;\ p^0 < 0\}. \tag{4.38}$$

In physics textbooks it is usually argued that a Lorentz invariant measure is given by the formula

$$\int \frac{\mathrm{d}^4 p}{(2\pi\hbar)^4}\Big((2\pi\hbar)\delta(p^2 - m^2c^2)\Big) f(p).$$

This is seen by the change of variable $p = L(p')$ since $\mathrm{d}^4 p$ is invariant under this change of variables because Lorentz transformations have determinant 1 and since $L(p)^2 = p^2$. Actually you may not always find this exact formula, because authors use a system of units in which $\hbar = 1 = c$, and may use a different normalization for the π factor. The previous measure is invariant under the full Lorentz group, so that it will also have a part on the set (4.38), and to obtain a measure on X_m we consider instead

$$\int_{p^0\geq 0} \frac{\mathrm{d}^4 p}{(2\pi\hbar)^4}\Big((2\pi\hbar)\delta(p^2 - m^2c^2)\Big) f(p). \tag{4.39}$$

To compute the above expression, one integrates in p^0 first. This brings in the function $\delta(\varphi(p^0))$ where $\varphi(x) = x^2 - \boldsymbol{p}^2 - m^2c^2$. Now, one has the useful formula

$$\delta(\varphi(x)) = \sum_{\{y\ ;\ \varphi(y)=0\}} \frac{1}{|\varphi'(y)|}\delta_y(x), \tag{4.40}$$

where in the right-hand side we have as usual $\delta_y(x) = \delta(x-y)$. This formula holds whenever the set $\{y\ ;\ \varphi(y) = 0\}$ is finite and $\varphi'(y) \neq 0$ at each point of this set. This formula is actually the *definition* of the left-hand side. This definition is obviously consistent with the idea that the delta function is the limit of the true functions $(2\varepsilon)^{-1}\mathbf{1}_{[-\varepsilon,\varepsilon]}$.[20] In the present case, $\varphi^{-1}(0) = \{\omega_{\boldsymbol{p}}, -\omega_{\boldsymbol{p}}\}$ and $\varphi'(x) = 2x$. Here the restriction to the set $p^0 \geq 0$ eliminates the contribution of $-\omega_{\boldsymbol{p}}$, yielding again the formula (4.36).

Despite the fact that the factor 2 in the denominator of (4.36) has *no intrinsic significance* and *is an artifact of the method of derivation*, the majority of textbooks religiously carry this factor 2, giving the impression that it is canonical and important. *This is simply not true.* Despite this, as it goes nowhere to try to write notes where the formulas differ from those of the literature, I will use the standard formula (4.36) in the sequel.[21]

We conclude this section with some heuristic considerations which may help our intuition. If they do not help your intuition or you find them confusing, please ignore them, there is nothing of central importance here. Let us recall that, in the words of physicists, the Dirac delta function is the function which is equal to zero outside the origin, and which is infinity at the origin, in such a way that its integral is 1.

[20] The quantity $\delta(\varphi(x))$ is *not* defined for any smooth function φ. In particular, the expression $\delta(x^2)$ makes no sense.

[21] You probably wonder why I make so much fuss about this factor 2. This will be explained when we discuss (6.62) on page 166.

Definition 4.4.3 (Heuristic!) Given a point $p \in X_m$ we heuristically consider the "function" $\delta_{m,p}$ on X_m which is zero outside p, infinity at p, in such a way that its integral *with respect to* $\mathrm{d}\lambda_m$ is equal to one.

The reason for the "m" in the notation $\delta_{m,p}$ is to emphasize the fact that this is a function on X_m. Thus, for any $\xi \in \mathcal{S}^4$, the space of Schwartz functions[22] on $\mathbb{R}^{1,3}$, we have

$$\int \mathrm{d}\lambda_m(p')\delta_{m,p}(p')\xi(p') = \xi(p), \tag{4.41}$$

which is a rigorous definition of $\delta_{m,p}$ as a "distribution on X_m".

Exercise 4.4.4 Consider the canonical unitary map

$$J : \mathcal{H}' = L^2(\mathbb{R}^3, (2\pi\hbar)^{-3}\mathrm{d}^3\boldsymbol{p}) \to \mathcal{H} = L^2(X_m, \mathrm{d}\lambda_m)$$

given by $J(f)(p) = \sqrt{2\omega_{\boldsymbol{p}}}f(\boldsymbol{p})$ for $p = (\omega_{\boldsymbol{p}}, \boldsymbol{p})$. Pretending that $\delta^{(3)}_{\boldsymbol{p}} \in \mathcal{H}$ and $\delta_{m,p} \in \mathcal{H}'$, convince yourself of the formula

$$\delta_{m,p} = (2\pi\hbar)^3\sqrt{2\omega_{\boldsymbol{p}}}J(\delta^{(3)}_{\boldsymbol{p}}). \tag{4.42}$$

Hint: All there is to it is that $\sqrt{2\omega_{\boldsymbol{p}}}\sqrt{2\omega_{\boldsymbol{p}}} = 2\omega_{\boldsymbol{p}}$!

Exercise 4.4.5 If you know a bit of measure theory, convince yourself that heuristically: "The function $\delta_{m,p}$ is the density with respect to $\mathrm{d}\lambda_m$ of the Dirac measure δ_p".

4.5 More about Unitary Representations

Before continuing, we need to learn more about representation theory.[23] In the present section we investigate some basic properties of unitary representations at a very elementary level.

Consider a unitary representation U of a group G on a Hilbert space $\mathcal{H}$ and a unitary operator W on $\mathcal{H}$. Then one can construct a new representation U_W by the formula

$$U_W(a) = W^{-1}U(a)W. \tag{4.43}$$

This construction however is not very interesting because in some sense U_W is the "same representation" as U, as they differ only by a "change of basis" in $\mathcal{H}$. More generally, the following concept crystallizes the idea that two unitary representations are "the same".

Definition 4.5.1 Two unitary representations U, U' of a group G in Hilbert spaces $\mathcal{H}$ and $\mathcal{H}'$ respectively are called *unitarily equivalent* if there is a unitary map W from $\mathcal{H}$ to $\mathcal{H}'$ such that

$$\forall a \in G\ ;\ U(a) = W^{-1}U'(a)W. \tag{4.44}$$

22 We denote by $\mathcal{S}^4$ both the space of Schwartz functions on $\mathbb{R}^4$ and on $\mathbb{R}^{1,3}$ since the Lorentz bilinear norm is irrelevant to the definition of this space.

23 A standard reference is Serre's book [74].

The relation (4.44) can be equivalently written as

$$WU(a) = U'(a)W.$$

$$\begin{array}{ccc} \mathcal{H} & \xrightarrow{U(a)} & \mathcal{H} \\ \downarrow W & & \downarrow W \\ \mathcal{H}' & \xrightarrow{U'(a)} & \mathcal{H}' \end{array}$$

For this reason the operator W is called an *intertwining operator* between the two representations. Thus, an intertwining operator between two representations witnesses that they are unitarily equivalent. It may happen that two unitary representations are unitarily equivalent in a really non-trivial manner. For example, the two representations of $\mathbb{R}$ on $L^2(\mathbb{R})$ given by (2.53) and (2.54) are unitarily equivalent since (2.55) shows that the Fourier transform is an intertwining map between them.

Exercise 4.5.2 Assuming $\mathcal{H}$ and $\mathcal{H}'$ to be finite-dimensional, prove that the unitary representations U and U' are unitarily equivalent if and only if there exists an orthonormal basis of $\mathcal{H}$ and an orthonormal basis of $\mathcal{H}'$ such that in these respective bases the matrices of $U(a)$ and $U'(a)$ are identical for each $a \in G$.

Definition 4.5.3 Given a (not necessarily unitary) representation U of the group G, a subspace[24] $\mathcal{G}$ of $\mathcal{H}$ is called *invariant* if $U(a)\mathcal{G} \subset \mathcal{G}$ for each $a \in G$.

Let us denote by $V(a)$ the restriction of $U(a)$ to $\mathcal{G}$. It should be obvious that V defines a representation of G on $\mathcal{G}$. Moreover if U is unitary, this is also the case for V as indeed $V(a)$ is norm-preserving and invertible.

A simple but fundamental property of *unitary* representations is that if $\mathcal{G}$ is an invariant subspace of $\mathcal{H}$ then its orthogonal complement

$$\mathcal{G}^{\perp} := \{y \in \mathcal{H} \,;\, \forall x \in \mathcal{G}\,,\, (y,x) = 0\}$$

is also invariant. This follows immediately from the equalities[25]

$$(U(a)(y),x) = (y, U(a)^{\dagger}(x)) = (y, U(a)^{-1}(x)) = (y, U(a^{-1})(x)).$$

Consequently, if a unitary representation admits a non-trivial invariant subspace $\mathcal{G}$ then it is built out of two simpler representations, namely its restrictions to $\mathcal{G}$ and to $\mathcal{G}^{\perp}$. This leads to the following.

Definition 4.5.4 A representation is called *irreducible* if it admits no non-trivial invariant subspace.

This definition also holds for representations that are not necessarily unitary, and we will need some of these. (Note however that when a representation is not unitary, it need

[24] Throughout the book, concerning Hilbert spaces, the word "subspace" is used as a short-hand for "closed linear subspace".

[25] Since we left Minkowski's space, it should be obvious that $(\cdot,\cdot)$ denotes the inner product, not the Lorentz bilinear form.

not be true that the orthogonal complement of an invariant subspace is also invariant.) An irreducible representation cannot be readily understood in terms of simpler pieces.

Let us now recall a simple fact about Hilbert spaces, the proof of which can be found in many elementary textbooks.

Lemma 4.5.5 *A closed subspace $\mathcal{H}_0$ of a Hilbert space $\mathcal{H}$ coincides with $\mathcal{H}$ if and only if $\mathcal{H}_0^\perp = \{0\}$, i.e. if zero is the only element of $\mathcal{H}$ which is orthogonal to each element of $\mathcal{H}_0$. In particular, given a subset C of $\mathcal{H}$, its closed linear span, that is the smallest closed linear subspace which contains C, coincides with $\mathcal{H}$ if and only if zero is the only element of $\mathcal{H}$ which is orthogonal to each element of C.*

The following result is almost obvious, but nonetheless of constant use.

Lemma 4.5.6 *A representation U is irreducible if and only if whenever x is a non-zero vector, the closed linear span of all the $U(a)(x)$ for $a \in G$ is all of $\mathcal{H}$, or, equivalently, if and only if zero is the only vector y which satisfies $(y, U(a)(x)) = 0$ for all $a \in G$.*

Proof Given $x \in \mathcal{H}$ the closed linear span of all the $U(a)(x)$ for $a \in G$ is an invariant subspace. Thus if U is irreducible, when x is not zero, this span is $\mathcal{H}$. Conversely, an invariant subspace that contains a non-zero vector x also contains the closed linear span of all the $U(a)(x)$ for $a \in G$. The last equivalence follows from Lemma 4.5.5. □

The following simple result is also very useful.

Lemma 4.5.7 (Schur's lemma) *Consider an irreducible representation U on a* finite-dimensional *space. Then an operator V which commutes with all the operators $U(a)$ is a multiple of identity.*

Proof Since we are working with a *complex* finite-dimensional Hilbert space, V has an eigenvalue. The corresponding eigenspace $\mathcal{G}$ is invariant under all the operators $U(a)$, because $VU(a)(x) = U(a)V(x) = \lambda U(a)(x)$ when $V(x) = \lambda x$. Since $\mathcal{G}$ is not zero it must be equal to $\mathcal{H}$ so that V is a multiple of the identity. □

To say that a representation takes place in a finite-dimensional space we will simply say that it is a *finite-dimensional representation*

Corollary 4.5.8 *The only irreducible finite-dimensional representations*[26] *of a commutative group are one-dimensional.*

Proof Given $b \in G$, the operator $U(b)$ commutes with every operator $U(a)$ because $U(b)U(a) = U(ba) = U(ab) = U(a)U(b)$ so it is a multiple of the identity. Every one-dimensional subspace is then an invariant subspace so that if the representation is irreducible it has to be of dimension 1. □

The following exercise presents a sort of converse to Schur's lemma, which does not require the representation to be finite-dimensional.

[26] Of course we assume that the representation is not trivial, it is not of dimension zero.

Exercise 4.5.9 Prove that if the only bounded operators V on $\mathcal{H}$ which commute with all operators $U(a)$ of a unitary representation U are multiples of the identity, then the representation U is irreducible.

For a mathematician a fundamental problem is to classify all the unitary representations of a given group, up to unitary equivalence. There is a vast literature on this subject, with some very accessible texts, such as that of Hall [39], and in Appendix D we present in a self-contained manner some results which are used in most physics textbooks.

4.6 Group Actions and Representations

We first put forward a general principle to construct representations. Let us say that a group G *acts* on a space X if there is a map $(a,x) \mapsto a \cdot x$ from $G \times X$ to X with the following properties:

$$1 \cdot x = x \; ; \; a \cdot (b \cdot x) = (ab) \cdot x \tag{4.45}$$

for all a,b in G, all x in X, and where 1 denotes the unit of G. Then the map $T_a \colon x \mapsto a \cdot x$ is a transformation of X and the map $a \mapsto T_a$ is a group homomorphism, T_1 is the identity and $T_a T_b = T_{ab}$. A fundamental fact is that for any set Y, G then acts naturally on the set $\mathcal{F}(X,Y)$ of functions from X to Y. The action $U(a)(f)$ of a on the function f is given by the formula

$$U(a)(f)(x) = f(a^{-1} \cdot x). \tag{4.46}$$

The reason for the all important term a^{-1} rather than a is the following computation:

$$\begin{aligned} U(a)[U(b)(f)](x) &= U(b)(f)(a^{-1} \cdot x) = f(b^{-1} \cdot (a^{-1} \cdot x)) \\ &= f((b^{-1}a^{-1}) \cdot x) = f((ab)^{-1} \cdot x) \\ &= U(ab)(f)(x), \end{aligned} \tag{4.47}$$

so that we have the essential formula

$$U(a)U(b) = U(ab). \tag{4.48}$$

Exercise 4.6.1 Assume that G also acts on Y and define now $U(a)(f)(x) = a \cdot f(a^{-1} \cdot x)$. Prove that (4.48) still holds true.[27]

In practice we will consider cases where there is a topology, both on X and G, and the action will always be continuous. A case of special interest is when there is a measure $d\mu$ on X which is invariant under the action of G, that is every measurable set A has the same measure as its image by T_a for each a. Then the restriction of $U(a)$ to $L^2(X, d\mu)$ is unitary, so that U defines a unitary representation of G in $L^2(X, d\mu)$:

[27] Writing $a \cdot f$ for $U(a)(f)$ one may motivate this formula by $(a \cdot f)(a \cdot x) = a \cdot f(x)$. The transform of f applied to the transform of x is the transform of $f(x)$.

$$(U(a)(f), U(a)(g)) = \int_X \mathrm{d}\mu(x) f(a^{-1} \cdot x)^* g(a^{-1} \cdot x)$$
$$= \int_X \mathrm{d}\mu(x) f(x)^* g(x) = (f, g).$$

As a prime example of this situation, the Lorentz group acts on the mass shell X_m on which the invariant measure $\mathrm{d}\lambda_m$ lives. Thus we obtain a unitary representation U of the Lorentz group on the space $\mathcal{H} = L^2(X_m, \mathrm{d}\lambda_m)$ by the formula

$$U(L)(\xi)(p) = \xi(L^{-1}(p)). \tag{4.49}$$

Physicists use a very different language to describe this representation and more complicated ones. We illustrate this vivid difference by transcribing such a description from a physics textbook (taking into account different normalization and notation). Even if you studied Dirac's formalism in Section 2.8, it may feel to you like an alien language, but we will learn more about it in Section 4.10. Here it goes:

Consider ket vectors $|\boldsymbol{p}\rangle$ normalized by

$$\langle \boldsymbol{p}'|\boldsymbol{p}\rangle = (2\pi\hbar)^3 \delta^{(3)}(\boldsymbol{p}' - \boldsymbol{p}), \tag{4.50}$$

and let us define the Lorentz-normalized kets $|p\rangle$ by

$$|p\rangle = \sqrt{2\omega_{\boldsymbol{p}}}|\boldsymbol{p}\rangle, \tag{4.51}$$

where $\omega_{\boldsymbol{p}} = \sqrt{m^2c^2 + \boldsymbol{p}^2}$. Then the new normalization equation is

$$\langle p'|p\rangle = 2\omega_{\boldsymbol{p}}(2\pi\hbar)^3 \delta^{(3)}(\boldsymbol{p} - \boldsymbol{p}'), \tag{4.52}$$

and the completeness relation based on the Lorentz invariant measure becomes

$$1 = \int \frac{\mathrm{d}^3\boldsymbol{p}}{(2\pi\hbar)^3 2\omega_{\boldsymbol{p}}} |p\rangle\langle p|. \tag{4.53}$$

With these Lorentz-normalized states, we can define a unitary representation of the Lorentz group simply:

Theorem. *If we define $U(L)$ by $U(L)|p\rangle = |L(p)\rangle$, then U is a unitary representation of the Lorentz group.*

Proof To prove that U is unitary, we use the completeness relation

$$U(L)U(L)^\dagger = U(L)1U(L)^\dagger$$
$$= \int \frac{\mathrm{d}^3\boldsymbol{p}}{(2\pi\hbar)^3 2\omega_{\boldsymbol{p}}} U(L)|p\rangle\langle p|U(L)^\dagger$$
$$= \int \frac{\mathrm{d}^3\boldsymbol{p}}{(2\pi\hbar)^3 2\omega_{\boldsymbol{p}}} |L(p)\rangle\langle L(p)| = 1,$$

using the result of Exercise 2.5.20, (4.53) and Lorentz invariance in the last line. □

The transcription ends here.[28] One can only admire the conciseness and power of Dirac's formalism, which physicists consider as self-evident. At the intuitive level the situation is clear: the state $|p\rangle$ represents the ideal case of a function ξ which takes non-zero values only at the point p of X_m. We will study this idea in more detail in Section 4.10.

If you are faced on your own with the task of deciphering the meaning of the previous transcription, it may immediately occur to you that the state space has to be $L^2(X_m, \mathrm{d}\lambda_m)$, for what else could it be? Otherwise you may try to apply the rule of "integrating against a test function" to make sense of Dirac's formalism.[29] Here the test function is $\xi \in \mathcal{S}^4$ and since $p \in X_m$ it is natural to define

$$|\xi\rangle := \int \mathrm{d}\lambda_m(p)\xi(p)|p\rangle,$$

so that, making a change of variable in the last equality,

$$U(L)|\xi\rangle = \int \mathrm{d}\lambda_m(p)\xi(p)U(L)|p\rangle = \int \mathrm{d}\lambda_m(p)\xi(p)|L(p)\rangle = \int \mathrm{d}\lambda_m(p)\xi(L^{-1}(p))|p\rangle,$$

i.e. $U(L)|\xi\rangle = |U(L)(\xi)\rangle$ where $U(L)(\xi)$ is as in (4.49). At this point it is even more obvious that the state space is indeed $L^2(X_m, \mathrm{d}\lambda_m)$, on which we have been operating by the formula (4.49), and that for $\xi \in \mathcal{S}^4$ the element $|\xi\rangle$ is simply ξ seen as an element of $L^2(X_m, \mathrm{d}\lambda_m)$. As a further check of this property, for another test function η,

$$\langle\eta|\xi\rangle = \int \mathrm{d}\lambda_m(p')\mathrm{d}\lambda_m(p)\eta(p')^*\xi(p)\langle p'|p\rangle.$$

Going back to the definition of $\mathrm{d}\lambda_m(p')$ and using (4.52) one reaches the expected formula $\langle\eta|\xi\rangle = \int \mathrm{d}\lambda_m(p)\eta(p)^*\xi(p)$.

4.7 Quantum Mechanics, Special Relativity and the Poincaré Group

Quantum Mechanics in its standard formulation is not compatible with Special Relativity and it is very difficult to reconcile these two theories. Early attempts in this direction go under the name of "Relativistic Quantum Mechanics". This theory, which is the subject of numerous textbooks, runs into severe inconsistencies and we will not try to describe it.[30]

Modern physics postulates that the laws of physics should be invariant under space-time translations, and under Lorentz transformations, because these represent symmetries of space-time. They should then be invariant under the action of the group generated by these transformations, the *Poincaré Group* (also called the extended Lorentz group by some

[28] Not the least remarkable feature of the preceding text is that it occurs on page 7 of a textbook whose title contains the words "for mathematicians". I could say that I felt sad not to qualify as a mathematician, but I already had trouble on page 4 of the same book.

[29] Everything is easy if you have already guessed the meaning of the states $|p\rangle$ and $|\boldsymbol{p}\rangle$, which will be explored in detail in Section 4.10. Here, however, you pretend that you have not guessed it.

[30] Quantum Field Theory is at present the most successful attempt at reconciling these two theories, even though it suffers from many problems of its own.

authors). Thus the Poincaré group $\mathcal{P}$ is the group of transformations of $\mathbb{R}^{1,3}$ of the type $x \mapsto a + A(x)$ where[31] $(a, A) \in \mathbb{R}^{1,3} \times SO^{\uparrow}(1,3)$. Performing first the transformation $x \mapsto b + B(x)$ and then the transformation $x \mapsto a + A(x)$ gives the transformation $x \mapsto a + A(b) + AB(x)$. Thus, the composition law of the Poincaré group is given by

$$(a, A)(b, B) = (a + A(b), AB). \tag{4.54}$$

The group constructed this way from $\mathbb{R}^{1,3}$ and $SO^{\uparrow}(1,3)$ is called their semidirect product and denoted $\mathbb{R}^{1,3} \rtimes SO^{\uparrow}(1,3)$.

Exercise 4.7.1 Prove the formula $(a, A)^{-1} = (-A^{-1}(a), A^{-1})$.

Exercise 4.7.2 Convince yourself that you have known semidirect products since high school! Two geometrical objects in the Euclidean plane are "congruent" if one can be transformed into the other by an element of $\mathbb{R}^2 \rtimes SO(2)$, the semidirect product of the group of translations and the group of rotations.

It seems necessary to elaborate a bit on the fundamental idea that the laws of physics should be invariant under the action of the Poincaré group. This idea is the keystone of Special Relativity and of our approach to Quantum Field Theory. Either at the classical or the quantum level, we describe a physical object by a point of a "state space" which describes its state. For example, at the classical level, we describe the electromagnetic field by a function from $\mathbb{R}^{1,3} \to \mathbb{R}^4$: at each point of space-time we give the value of the electrostatic potential and of the three components of the magnetic field. Different observers use different coordinate systems on the state space. Mathematically this means that there is a group homomorphism from the Poincaré group to a group of transformations of the state space, reflecting this change of coordinates, and these transformations must preserve all the laws of physics. For example the electromagnetic field satisfies the Maxwell equations[32] and it was known even before the invention of Special Relativity that these equations are invariant under Lorentz transformations.[33] An equation *cannot* be physically correct unless it displays the proper invariance. This gives us a fundamental method to discover fruitful equations/objects: *look for the simplest possible objects with the correct invariance properties.* It is quite an awesome fact that many of our most fascinating discoveries will occur simply by following this path.[34]

Non-relativistic Quantum Mechanics is *not a part* of Quantum Field Theory. Large parts of Quantum Mechanics have to be abandoned when trying to make this theory consistent with Special Relativity, but we will accept that the basic tenet remains true: The correct description of the state of a physical system is a non-zero vector in a complex Hilbert space.

31 Please observe the change of notation here. A Lorentz transformation will be very often denoted by one of the letters A, B, C, simply because visually the symbols (a, A) look better to me than the symbols (a, L).

32 These are a highlight of nineteenth-century physics.

33 A brief account of the Maxwell equations is given in Section E.2.

34 Nature however does not always use the simplest possible structures. For example, the Klein-Gordon field, to be discussed later, is much simpler than the electromagnetic field, but is not observed.

Reflecting the fact/belief that the Poincaré group is a group of symmetries of Nature, we will assume that[35]

Every quantum system comes equipped with a projective representation of the Poincaré group. (4.55)

This representation will always play an essential part in the study of the system.[36] Moreover, elementary systems, i.e. systems which cannot be broken into smaller subpieces, should correspond to "minimal" representations, i.e. irreducible ones. Elementary particles are such elementary systems.

To each elementary particle one may associate an irreducible representation of the Poincaré group. (4.56)

Here, elementary particle means "elementary at the level of the properties under study".[37] Let us study the idea some more. The state space of our model will describe multiparticle systems,[38] but will contain subspaces that correspond to single-particle states. The restriction of the action of the Poincaré group to such a subspace will be an irreducible representation of this group.[39]

In Chapter 9 we shall describe in detail the most important representations of the Poincaré group. In the next section we study the simplest non-trivial representation of that kind. It is fundamental for the sequel, not because it corresponds to the most important particles, but because it is at the basis of the toy models for interaction that we shall study.

4.8 A Fundamental Representation of the Poincaré Group

The Poincaré group $\mathcal{P}$ is a group of transformations of $\mathbb{R}^{1,3}$ and these transformations preserve the volume measure. Therefore we obtain a unitary representation of $\mathcal{P}$ on $L^2(\mathbb{R}^{1,3}, \mathrm{d}^4x)$ by the formula (4.46):

$$V(a,A)(f)(x) = f(A^{-1}(x-a)), \tag{4.57}$$

where we abuse notation by writing $V(a,A)$ rather than $V((a,A))$, and where we have used that $A^{-1}(x-a)$ is the result of the action of $(a,A)^{-1}$ on the point x, as is shown by the formula of Exercise 4.7.1. Let us note in particular that $V(a,1)$ *represents the action on* $L^2(\mathbb{R}^{1,3}, \mathrm{d}^4x)$ *of translating space-time by* a.

35 What we mean here is that the representation is such that *the action of time-translations describes the evolution of the system with time*. This assumption is not valid for quantum systems studied in non-relativistic Quantum Mechanics.

36 The physical meaning of this representation is that it describes "changes of coordinates" in the state space by which different observers relate their descriptions of the system under study.

37 Two bound quarks might be considered as forming an elementary particle for certain purposes, but this will hardly ever be the case for two unbound nucleons.

38 It does not seem possible, as is attempted by "Relativistic Quantum Mechanics" to build a coherent theory of 1-particle systems. It does not make much sense either, as high-energy physics witnesses creation and annihilation of particles.

39 This fundamental idea is due to Eugene Wigner. Then (4.56) begs for a converse. Given an irreducible representation of the Poincaré group, may one construct a consistent model in which this representation is associated to an elementary particle? This question is the object of Chapter 10.

As it turns out, this representation is far from being irreducible. This will become apparent after we use the Fourier transform. In Minkowski's space $\mathbb{R}^{1,3}$ we define the Fourier transform[40] as

$$\hat{f}(p) = \int \mathrm{d}^4x \exp(\mathrm{i}(x,p)/\hbar) f(x) = \int \mathrm{d}^4x \exp(\mathrm{i}(x^0 p^0 - \boldsymbol{x}\cdot\boldsymbol{p})/\hbar) f(x). \tag{4.58}$$

Thus the ordinary sign convention is reversed for the time coordinate x^0. The inverse Fourier transform is given by

$$\check{\xi}(x) = \int \frac{\mathrm{d}^4p}{(2\pi\hbar)^4} \exp(-\mathrm{i}(x,p)/\hbar)\xi(p). \tag{4.59}$$

Lemma 4.8.1 *For $f \in L^2(\mathbb{R}^{1,3}, \mathrm{d}^4x)$ we have*

$$\widehat{V(a,A)(f)}(p) = \exp(\mathrm{i}(a,p)/\hbar)\hat{f}(A^{-1}(p)). \tag{4.60}$$

Proof This is of course straightforward. By continuity in the L^2 norm we assume that f is a test function. This ensures that all the integrals we write make sense. The left-hand side is

$$\int \mathrm{d}^4x \exp(\mathrm{i}(x,p)/\hbar) f(A^{-1}(x-a)).$$

Making the change of variables $x = a + A(y)$, and since the measure d^4x on $\mathbb{R}^{1,3}$ is invariant under this transformation, this is

$$\exp(\mathrm{i}(a,p)/\hbar) \int \mathrm{d}^4y \exp(\mathrm{i}(A(y),p)/\hbar) f(y),$$

which is the right-hand side of (4.60) since $(A(y),p) = (y, A^{-1}(p))$. □

Since the Fourier transform is a unitary operator, V is unitarily equivalent to the representation $\hat{V}$ on $L^2(\mathbb{R}^{1,3}, \mathrm{d}^4p/(2\pi\hbar)^4)$ given by the formula

$$\hat{V}(a,A)(\varphi)(p) = \exp(\mathrm{i}(a,p)/\hbar)\varphi(A^{-1}(p)).$$

This representation is not irreducible because the sets $X_{m,\varepsilon}$ of (4.32) are invariant under the action of $SO^{\uparrow}(1,3)$ so that the subspace $L^2(X_{m,\varepsilon}, \mathrm{d}^4p/(2\pi\hbar)^4)$ of L^2 consisting of functions supported by $X_{m,\varepsilon}$ is an invariant subspace of the representation.

We may however use (4.60) to guess the following formula.

Theorem 4.8.2 *Consider $m \geq 0$. Then the operators $U(a,A)$ defined for $(a,A) \in \mathcal{P}$ and $\varphi \in L^2(X_m, \mathrm{d}\lambda_m)$ by*

$$U(a,A)(\varphi)(p) = \exp(\mathrm{i}(a,p)/\hbar)\varphi(A^{-1}(p)) \tag{4.61}$$

form a unitary representation of $\mathcal{P}$ in $L^2(X_m, \mathrm{d}\lambda_m)$.

We again abuse notation by writing $U(a,A)$ rather than $U((a,A))$. This representation has a chance to be irreducible because the action of $SO^{\uparrow}(1,3)$ on X_m is transitive: Given p_1 and p_2 in X_m one may find $A \in SO^{\uparrow}(1,3)$ with $A(p_1) = p_2$.

40 The two versions of Minkowski space are involved here. The Fourier transform sends $L^2(\mathbb{R}^{1,3}, \mathrm{d}^4x)$ to $L^2(\mathbb{R}^{1,3}, \mathrm{d}^4p/(2\pi\hbar)^4)$.

Proof The operator $U(a,A)$ is unitary since the exponential has modulus 1 and $\mathrm{d}\lambda_m$ is invariant under the transformation $p \mapsto A^{-1}(p)$. To prove the representation property, we compute $U(a,A)U(b,B)(\psi)$ for a given function ψ. For this we apply (4.61) to the function

$$\varphi(p) := U(b,B)(\psi)(p) = \exp(\mathrm{i}(b,p)/\hbar)\psi(B^{-1}(p)),$$

and we obtain

$$\exp(\mathrm{i}(a,p)/\hbar)\exp(\mathrm{i}(b,A^{-1}(p))/\hbar)\psi(B^{-1}(A^{-1}(p))). \tag{4.62}$$

Now, since $A \in SO^{\uparrow}(1,3)$, for any $x,y \in \mathbb{R}^{1,3}$ we have $(A(x),A(y)) = (x,y)$ and thus $(b,A^{-1}(p)) = (A(b),p)$. Moreover $B^{-1}(A^{-1}(p)) = (AB)^{-1}(p)$. Therefore the quantity (4.62) equals

$$\exp(\mathrm{i}(a + A(b),p)/\hbar)\psi((AB)^{-1}(p)) = U((a,A)(b,B))(\psi)(p). \qquad \square$$

The following exercise prepares you for a key point of the proof of the next result. You should skip it if you know too little measure theory.

Exercise 4.8.3 (a) Consider two measurable functions $\varphi, \psi \geq 0$ on $\mathbb{R}$. Prove that if for each $t \in \mathbb{R}$ one has $\int \mathrm{d}x\varphi(x)\psi(x+t) = 0$ then either $\varphi = 0$ a.e. or $\psi = 0$ a.e. Hint: Prove that $\int \mathrm{d}t\mathrm{d}x\varphi(x)\psi(x+t) = \int \mathrm{d}x\varphi(x)\int \mathrm{d}t\psi(t)$.
(b) There is probability $\mathrm{d}R$ (called a Haar measure[41]) on the group of rotations $SO(3)$ with the property that given any measurable function θ on $\mathbb{S}$ and any $T \in SO(3)$ one has $\int \mathrm{d}R\theta(RT) = \int \mathrm{d}R\theta(R)$ (left-invariance). Prove that given a function φ on the sphere $\mathbb{S}$ and any $x \in \mathbb{S}$ one has $\int \mathrm{d}R\varphi(Rx) = \int \mathrm{d}\mu(y)\varphi(y)$, where $\mathrm{d}\mu$ is the uniform measure on $\mathbb{S}$.
(c) Given two measurable functions $\varphi, \psi \geq 0$ on $\mathbb{S}$, prove that if $\int \mathrm{d}\mu\varphi(x)\psi(R(x)) = 0$ for each rotation R, then either $\varphi = 0$ a.e. or $\psi = 0$ a.e. Hint: Prove that

$$\int \mathrm{d}R \int \mathrm{d}\mu(x)\varphi(x)\psi(R(x)) = \int \mathrm{d}\mu(x)\varphi(x) \int \mathrm{d}\mu(x)\psi(x).$$

Proposition 4.8.4 *The representation of Theorem 4.8.2 is irreducible.*

Proof To avoid measure-theoretic technicalities we give only a sketch of the proof. Consider a non-zero function φ in $L^2(X_m, \mathrm{d}\lambda_m)$. We want to show that the closed linear span of all the functions $U(a,A)(\varphi)$ is $L^2(X_m, \mathrm{d}\lambda_m)$. Consider a function ψ which is orthogonal to each of these functions. The goal is to prove that ψ is zero almost everywhere (a.e.), which completes the proof according to Lemma 4.5.5. Thus, for each $a \in \mathbb{R}^{1,3}$ and each $A \in SO^{\uparrow}(1,3)$ it holds that

$$\int \mathrm{d}\lambda_m(p)\psi(p)^* \exp(\mathrm{i}(a,p)/\hbar)\varphi(A^{-1}(p)) = 0. \tag{4.63}$$

Any Schwartz function $\xi \in S^4$ is the Fourier transform of another Schwartz function f (namely, its inverse Fourier transform), i.e. $\xi(p) = \int \mathrm{d}^4a \exp(\mathrm{i}(a,p))f(a)$ and thus

41 Numerous books present the construction of the Haar measure.

$$\int d\lambda_m(p)\xi(p)\psi(p)^*\varphi(A^{-1}(p))$$
$$= \int d^4 a f(a) \int d\lambda_m(p)\psi(p)^* \exp(i(a,p)/\hbar)\varphi(A^{-1}(p)) = 0.$$

Since the Schwartz functions ξ are dense in $L^2(X_m, d\lambda_m)$, for each A the function $p \mapsto \psi(p)^*\varphi(A^{-1}(p))$ has to be zero almost everywhere for $d\lambda_m$, i.e. (replacing A by A^{-1})

$$\int d\lambda_m(p)|\psi(p)||\varphi(A(p))| = 0. \quad (4.64)$$

From this we have to deduce that $\psi = 0$ a.e. The situation is pretty similar to that of Exercise 4.8.3 (c), but $SO^{\uparrow}(1,3)$ is not compact. Still there exists a left-invariant measure (a Haar measure) dA on $SO^{\uparrow}(1,3)$, and one can show as in the exercise that for a certain constant C, any $\varphi \in L^1(X_m, d\lambda_m)$ and any $p \in X_m$ one has $\int dA\varphi(A(p)) = C\int d\lambda_m(p')\varphi(p')$. Then by exchanging the order of integration one obtains

$$\int dA \int d\lambda_m(p)|\psi(p)||\varphi(A(p))| = C \int d\lambda_m(p)|\varphi(p)| \int d\lambda_m(p)|\psi(p)|,$$

so that under (4.64) the left-hand side is zero and then $\psi = 0$ a.e. □

Exercise 4.8.5 Given $\alpha > 0$ prove that the formula

$$U(a,A)(\varphi)(p) = \exp(\alpha i(a,p))\varphi(A^{-1}(p))$$

defines a unitary representation of $\mathcal{P}$ in $L^2(X_m, d\lambda_m)$. Do we obtain new representations this way? Hint: For $\beta > 0$ the representations corresponding to (α, m) and $(\alpha\beta, m\beta)$ are unitarily equivalent.

It is essential for the sequel to understand well the representation of Theorem 4.8.2, as it is the basis of our toy models.

4.9 Particles and Representations

Consider a particle that corresponds (in the sense of (4.56)) the representation of Theorem 4.8.2. What are its properties? In this section we make the case that, not surprisingly, this particle has mass m. The key here is the formula (4.57): $V(a,1)$ corresponds to a space-time translation by a. Since $U(a,A)$ is in a sense the restriction of (a representation unitarily equivalent to) $V(a,A)$ to an invariant subspace,[42]

We interpret $U(a,1)$ as a space-time translation by a. (4.65)

[42] One can view (in a sense) $U(a,A)$ as operating on the space of functions on $\mathbb{R}^{1,3}$ whose Fourier transform is supported by X_m. These functions can be characterized by a certain differential equation, or "wave equation", the Klein-Gordon equation which will play an important role later. This approach is very popular in physics. My problem here is that it is not clear what is the norm which makes the representation unitary. I never ran into a single textbook describing such a norm, or even mentioning the need for it, and I am unable to explain what is the mental picture physicists form at this point. More generally more complicated representations of the Poincaré group can be defined on more complicated spaces of functions satisfying certain wave equations.

Let us stress right away the shift from non-relativistic Quantum Mechanics. There, we specify the Hamiltonian, which in turn determines the time-evolution of the system. Here we specify the representation $U(a,1)$ which in particular *entirely describes the time-evolution of the system,* and should be thought of as replacing the equations of evolution. Nonetheless the relation between the Hamiltonian and the time-evolution remains the same through Stone's theorem (as is recalled below).

Our goal is to use (4.65) to compute the Hamiltonian and the momentum operators. For this we recall that given a function h on X_m we may define the operator "multiplication by h" on $L^2(X_m, \mathrm{d}\lambda_m)$. It is simply the operator given by $T(f) = fh$ when this makes sense, i.e. $fh \in L^2(X_m, \mathrm{d}\lambda_m)$. When h is real-valued, the corresponding operator is self-adjoint and the corresponding observable is simply "the value of h".[43] The four operators we are trying to compute will be of this type. The following exercise is an almost obvious variation of Exercise 2.14.3. Its result will be used several times.

> **Exercise 4.9.1** Consider a real-valued function h on X_m. Define $U(t)$ as the operator "multiplication by the function $\exp(\mathrm{i}th/\hbar)$". These operators form a one-parameter unitary group. Prove that the infinitesimal generator A of this group (in the sense of (2.56)) is the operator "multiplication by the function h".

Let us recall the definition (2.72) of the momentum operator. If $U_i(s)$ is the one-parameter group of transformations of the state space which describes the operation "translation of physical space by s along the ith direction" then $U_i(s) = \exp(-\mathrm{i}sP_i/\hbar)$ where P_i is the momentum operator in the same direction. Thus if we denote by $U(a,A)$ the action of the Poincaré group on the state space, the momentum operator P_1 in the direction of the first coordinate satisfies

$$\exp(-\mathrm{i}sP_1/\hbar) = U(a(s),1), \tag{4.66}$$

where $a(s) = (0,s,0,0)$. In the case of the representation of Theorem 4.8.2, $U(a,1)$ is the operator "multiplication by $\exp(\mathrm{i}(a,p)/\hbar)$". Since $(a(s),p) = -sp^1$, $U(a(s),1)$ is the operator "multiplication by $\exp(-\mathrm{i}sp^1/\hbar)$". Using (4.66) we obtain that $\exp(-\mathrm{i}sP_1)$ is the operator "multiplication by $\exp(-\mathrm{i}sp^1/\hbar)$", from which Exercise 4.9.1 shows that P_1, the momentum operator in the direction of the first coordinate, is the operator "multiplication by p^1". This makes physical sense because this operator is precisely the self-adjoint operator corresponding to the observable p^1. It should be that way in order that in our representation the quantity p has indeed, as the name intended, the signification of the momentum of the particle.

Let us now compute the Hamiltonian H. Setting $b(t) = (ct,0,0,0)$ then $U(b(t),1)$ represents translation of space-time by t along the time coordinate. The tricky point is that this represents *time-evolution by* $-t$. To understand this, let us recall from the beginning of Section 2.10 that space-time translation by a means that the object at space-time position x has been moved to position $x + a$. If my twentieth birthday, which was supposed to take

[43] To make sense of this, recall that in position state space, position is given by the "multiplication by x operator", as we learned in Section 2.5, whereas (2.23) teaches us that in momentum state space, momentum is given by the "multiplication by p operator".

place now, has been moved to next year, I am now only 19 years old: I became one year younger. Now, by the convention (2.75), time-evolution by time $-t$ is given by $\exp(itH/\hbar)$, so that as in (4.66) we have

$$\exp(itH/\hbar) = U(b(t), 1).$$

Since $(b(t), p) = ctp^0$, $U(b(t), 1)$ is the operator "multiplication by $\exp(ictp^0/\hbar)$", so that $\exp(itH/\hbar)$ is the operator "multiplication by $\exp(ictp^0/\hbar)$" and therefore the Hamiltonian H is just "multiplication by the function cp^0".[44] This operator corresponds to the observable cp^0. Things again make physical sense: As always the Hamiltonian should correspond to the energy of the system, and cp^0 is precisely the energy of the particle. To make sure that the important facts above are remembered, let us state them again.

Proposition 4.9.2 *The Hamiltonian for a particle described by the representation of Theorem 4.8.2 is the operator "multiplication by the function cp^0" and the momentum operator in the direction of e_j is the operator "multiplication by the function p^j".*

The operator $H^2/c^2 - \mathbf{P}^2 = H^2/c^2 - \sum_{1\le i\le 3} P_i^2$ is then the operator "multiplication by the function $(p^0)^2 - \boldsymbol{p}^2 = p^2$" i.e. multiplication by m^2c^2 since $p^2 = m^2c^2$ for $p \in X_m$. We have obtained the identity

$$\frac{H^2}{c^2} - \mathbf{P}^2 = m^2c^2 1.$$

This expresses a relation between energy and momentum which is a "quantized version" of the relation $E^2/c^2 - \boldsymbol{p}^2 = m^2c^2$ of (4.30).[45] We have thus argued that our mathematical object corresponds to a *particle of mass m*.[46] When $m > 0$ we will say that the particle is *massive* whereas when $m = 0$ it is *massless*.

We will later study more complicated representations which are fundamentally important in physics. These representations will describe particles which as here have mass $m \geq 0$. These particles will also have another characteristic, "spin". These include the most important particles with which the universe is built. These representations will however not be used in the simple toy models that we consider, so if your goal is simply to understand these and the dreaded infinities that plague them, you may safely skip the entire Part II of the book.

It turns out that the representation of Theorem 4.8.2 corresponds to the case where the spin is zero, so that the corresponding particle will be described as "spinless".[47]

In Section 4.11, we will argue as physicists that the only representation of the Poincaré group associated to a massive spinless particle is the representation of Theorem 4.8.2. The considerations of Section 4.10 are not really needed to understand these arguments.

[44] Signs match nicely, which is of course the purpose of the convention that $\exp(itH/\hbar)$ represents time-evolution by time $-t$.

[45] The expression "quantized version" means that each of the classical quantities E, $\boldsymbol{p}^2$ has been replaced by a self-adjoint operator measuring it.

[46] Mathematical objects and physical objects belong to different realms. At some point we have to argue that a certain mathematical model is appropriate to describe a certain physical object, and like here this step takes place outside mathematics.

[47] You need to study Part II to understand the technical meaning of this expression, but you will do fine just by knowing that this means that the corresponding representation is that of Theorem 4.8.2.

4.10 The States $|\boldsymbol{p}\rangle$ and $|p\rangle$

Many of our later results will be stated in physics language. This section attempts to go beyond intuition and to give mathematical precision to the idea of the "states" $|p\rangle$ of Section 4.6 which are ubiquitous in this language. I see *no other way* to do this than to enter into the following considerations. As with everything having to do with Dirac formalism, this has the potential to be lethally confusing for a mathematician, so you should not hesitate to skip the present section if it is not your cup of tea, and you will do just fine by staying at the intuitive level as the physicists do. There is nothing deep going on here, only formal manipulations checking that the formalism is self-consistent. It is advised to review Section 2.8 at this point.

Let us first review the meaning of the elements $|\boldsymbol{p}\rangle$. They are defined only in the distributional sense. If $\tau \in \mathcal{S}^3 := \mathcal{S}(\mathbb{R}^3)$ is a Schwartz function, and $|\tau\rangle$ denotes the state obtained by seeing τ as an element of the state space $\mathcal{H}' = L^2(\mathbb{R}^3, \mathrm{d}^3\boldsymbol{p}/(2\pi\hbar)^3)$, then

$$\int \frac{\mathrm{d}^3\boldsymbol{p}}{(2\pi\hbar)^3}\tau(\boldsymbol{p})|\boldsymbol{p}\rangle = |\tau\rangle. \tag{4.67}$$

Given another Schwartz function μ and $|\mu\rangle$ the corresponding state, applying $\langle\mu|$ to the above relation (4.67) we then have

$$\int \frac{\mathrm{d}^3\boldsymbol{p}}{(2\pi\hbar)^3}\tau(\boldsymbol{p})\langle\mu|\boldsymbol{p}\rangle = \langle\mu|\tau\rangle = \int \frac{\mathrm{d}^3\boldsymbol{p}}{(2\pi\hbar)^3}\mu(\boldsymbol{p})^*\tau(\boldsymbol{p}), \tag{4.68}$$

which leads to the heuristic formula $\mu(\boldsymbol{p})^* = \langle\mu|\boldsymbol{p}\rangle$ i.e. $\langle\boldsymbol{p}|\mu\rangle = \mu(\boldsymbol{p})$, and thus also to the interpretation "$|\boldsymbol{p}\rangle = (2\pi\hbar)^3\delta^{(3)}_{\boldsymbol{p}}$ seen as an element of $\mathcal{H}'$".

Suppose now that we want to use $\mathcal{H} = L^2(X_m, \mathrm{d}\lambda_m)$ as a state space. We can then define states $|p\rangle$ in the distributional sense: for each $\xi \in \mathcal{S}^4$ we have

$$\int \mathrm{d}\lambda_m(p)\xi(p)|p\rangle = |\xi\rangle, \tag{4.69}$$

where $|\xi\rangle$ is the state obtained by considering ξ as an element of $\mathcal{H}$, and just as before we obtain the heuristic formula $\langle p|\xi\rangle = \xi(p)$. Let us also note for further use that

$$\int \mathrm{d}\lambda_m(p)|p\rangle\langle p| = 1. \tag{4.70}$$

Indeed, since $\langle p|\xi\rangle = \xi(p)$, for any $\xi \in \mathcal{S}^4$ we have

$$\Big(\int \mathrm{d}\lambda_m(p)|p\rangle\langle p|\Big)|\xi\rangle = \int \mathrm{d}\lambda_m(p)|p\rangle\xi(p) = |\xi\rangle,$$

using (4.69) in the last equality, and $\mathcal{S}^4$ is dense in $\mathcal{H}$.

To relate the states $|\boldsymbol{p}\rangle$ and $|p\rangle$, consider the canonical isometry J from $\mathcal{H}'$ to $\mathcal{H}$ given by $J(\tau)(p) = \sqrt{2\omega_{\boldsymbol{p}}}\tau(\boldsymbol{p})$ where $p = (\omega_{\boldsymbol{p}}, \boldsymbol{p})$. We apply J to both sides of (4.67) to obtain

$$\int \frac{\mathrm{d}^3\boldsymbol{p}}{(2\pi\hbar)^3 2\omega_{\boldsymbol{p}}}(\sqrt{2\omega_{\boldsymbol{p}}}\tau(\boldsymbol{p}))(J(\sqrt{2\omega_{\boldsymbol{p}}}|\boldsymbol{p}\rangle)) = J|\tau\rangle. \tag{4.71}$$

The function $\xi := J(\tau)$ is given by $\xi(p) = \sqrt{2\omega_{\boldsymbol{p}}}\tau(\boldsymbol{p})$ where $p = (\omega_{\boldsymbol{p}}, \boldsymbol{p})$, and $J|\tau\rangle = |\xi\rangle$, so that (4.71) reads

$$\int \mathrm{d}\lambda_m(p)\xi(p)J(\sqrt{2\omega_{\boldsymbol{p}}}|\boldsymbol{p}\rangle) = |\xi\rangle,$$

and comparing with (4.69) we conclude (heuristically) that $|p\rangle = J(\sqrt{2\omega_{\boldsymbol{p}}}|\boldsymbol{p}\rangle)$, which a physicist writes $|p\rangle = \sqrt{2\omega_{\boldsymbol{p}}}|\boldsymbol{p}\rangle$ as in (4.51) since for her the spaces $\mathcal{H}'$ and $\mathcal{H}$ are identical, being the unspecified "state space".

Proposition 4.10.1 *The states $|p\rangle$ are Lorentz invariant. That is, $U(A)|p\rangle = |A(p)\rangle$ where U is the representation of the Lorentz group defined in (4.49).*

Proof Such a statement makes sense only when integrated against a function $\xi \in \mathcal{S}^4$, that is we have to prove the equality

$$\int \mathrm{d}\lambda_m(p)\xi(p)U(A)|p\rangle = \int \mathrm{d}\lambda_m(p)\xi(p)|A(p)\rangle. \tag{4.72}$$

When $\xi \in \mathcal{S}^4$, $|\xi\rangle$ is just the function ξ seen as an element of $\mathcal{H}$, and we define accordingly $U(A)|\xi\rangle = |U(A)(\xi)\rangle$. Then[48]

$$\int \mathrm{d}\lambda_m(p)\xi(p)U(A)|p\rangle = U(A)\int \mathrm{d}\lambda_m(p)\xi(p))|p\rangle = U(A)|\xi\rangle = |U(A)(\xi)\rangle.$$

On the other hand, by change of variable, we have $\int \mathrm{d}\lambda_m(p)\xi(p)|A(p)\rangle = \int \mathrm{d}\lambda_m(p)\xi(A^{-1}(p))|p\rangle = \int \mathrm{d}\lambda_m(p)U(A)(\xi)(p)|p\rangle = |U(A)(\xi)\rangle$. The equality (4.72) is proved. □

As recalled above, using the state space $\mathcal{H}' = L^2(\mathbb{R}^3, (2\pi\hbar)^{-3}\mathrm{d}^3\boldsymbol{p})$ we can heuristically think of $|\boldsymbol{p}\rangle$ as the state which corresponds to the function $(2\pi\hbar)^3\delta^{(3)}_{\boldsymbol{p}}$ if we pretend that this function belongs to $\mathcal{H}'$.

Exercise 4.10.2 Convince yourself in a similar manner that $|p\rangle$ corresponds to the delta function $\delta_{m,p}$ *seen as an element of* $L^2(X_m, \mathrm{d}\lambda_m)$. Hint: Integrate against a test function.

Exercise 4.10.3 Make sense of the formula

$$\langle p'|p\rangle = \delta_{m,p}(p') = \delta_{m,p'}(p).$$

4.11 The Physicists' Way

In the present section we explain how a physicist might describe the action of the Poincaré group on a massive spinless particle, and how this relates to the previous mathematical considerations. Here "spinless" means by definition that the state of the particle is completely described by its momentum $\boldsymbol{p}$, or equivalently by its four-momentum $p = (\omega_{\boldsymbol{p}}, \boldsymbol{p})$. Our arguments are heuristic and do not constitute a mathematical proof. On the other hand, they

48 This manipulation really *defines* $U(A)|p\rangle$.

go far beyond just rediscovering the formula (4.61): they argue that there is *no other possible action* of the Poincaré group on the state space than the action given by (4.61).

In this section we argue as physicists. We entirely forget about the considerations of the previous section (which concern only mathematicians). We consider as self-evident the existence of states of given four-momentum.[49] A state $|\alpha\rangle$ of four-momentum p is by definition such that

$$\forall a \in \mathbb{R}^{1,3}\ ,\ U(a,1)|\alpha\rangle = \exp(\mathrm{i}(a,p)/\hbar)|\alpha\rangle. \tag{4.73}$$

The hypothesis that the particle is spinless, i.e. that the state of the particle is entirely described by its four-momentum means that given p the space of solutions of the previous equation is of dimension 1. In particular two states (i.e. unit vectors) satisfying this relation differ only by a phase.

We start with a general fact, valid with any unitary representation of the Poincaré group on any state space.

Lemma 4.11.1 *Consider a state $|\alpha\rangle$ of four-momentum p. Then $U(0,A)|\alpha\rangle$ has four-momentum $A(p)$.*

Proof For $a \in \mathbb{R}^{1,3}$ and $A \in SO^{\uparrow}(1,3)$ we have the relation

$$(a,A) = (a,1)(0,A) = (0,A)(A^{-1}(a),1)$$

so that since U is a representation

$$U(a,A) = U(a,1)U(0,A) = U(0,A)U(A^{-1}(a),1). \tag{4.74}$$

Using the second equality of (4.74) in the first line and (4.73) in the second line, we obtain

$$\begin{aligned} U(a,1)U(0,A)|\alpha\rangle &= U(0,A)U(A^{-1}(a),1)|\alpha\rangle \\ &= \exp(\mathrm{i}(A^{-1}(a),p)/\hbar)U(0,A)|\alpha\rangle \\ &= \exp(\mathrm{i}(a,A(p))/\hbar)U(0,A)|\alpha\rangle, \end{aligned} \tag{4.75}$$

which proves the claim by (4.73). □

Let us go back to the study of the massive spinless particle. When the particle is at rest, it has four-momentum[50] $p^* := (mc,0,0,0) = mce_0$. Recalling that a state of given momentum is unique only up to a phase, let us fix once for all a state $|p^*\rangle$ of momentum p^* (in the sense of (4.73)). Let us recall that $A \in SO^{\uparrow}(1,3)$ is a rotation if and only if $A(e_0) = e_0$, or, equivalently, if and only if $A(p^*) = p^*$. Thus if R is a rotation, Lemma 4.11.1 implies that $U(0,R)|p^*\rangle$ has four-momentum p^*, so that this state differs from $|p^*\rangle$ only by a phase, i.e. $U(0,R)|p^*\rangle = \varphi(R)|p^*\rangle$. The map φ: $SO(3) \to \mathbb{C}$ satisfies $\varphi(1) = 1$ and $\varphi(R)\varphi(R') = \varphi(RR')$. We will show later, in Lemma 8.5.2 that this implies that φ takes only the value 1. Thus we have shown that

$$R \text{ rotation } \Rightarrow U(0,R)|p^*\rangle = |p^*\rangle. \tag{4.76}$$

[49] For a mathematician, these "states" exist only in the distributional sense, as was explored in the previous section, but here we pretend as physicists do that they are bona fide states.

[50] The $*$ is just a notation that has nothing to do with complex conjugation!

Given $A, B \in SO^{\uparrow}(1,3)$ with $A(p^*) = B(p^*)$ then $B^{-1}A(p^*) = p^*$, i.e. $B^{-1}A$ is a rotation, so that $U(0, B^{-1}A)|p^*\rangle = |p^*\rangle$ and since $U(0, B^{-1}A) = U(0, B^{-1})U(0, A)$ this means that $U(0, A)|p^*\rangle = U(0, B)|p^*\rangle$. Given a four-momentum p we may then define the state $|p\rangle$ as being $U(0, A)|p^*\rangle$ where A is any element of $SO^{\uparrow}(1,3)$ for which $A(p^*) = p$. This notation makes sense because according to Lemma 4.11.1, the state $|A(p^*)\rangle$ has four-momentum $A(p^*) = p$.

Now we investigate how $U(0, A)$ operates on $|p\rangle$. Consider $B \in SO^{\uparrow}(1,3)$ with $p = B(p^*)$ so $|p\rangle = U(0, B)|p^*\rangle$ and

$$U(0, A)|p\rangle = U(0, A)U(0, B)|p^*\rangle = U(0, AB)|p^*\rangle = |AB(p^*)\rangle = |A(p)\rangle.$$

Since $U(a, A) = U(a, 1)U(0, A)$ and since $|A(p)\rangle$ has four-momentum $A(p)$ we have finally obtained that

$$U(a, A)|p\rangle = \exp(\mathrm{i}(a, A(p))/\hbar)|A(p)\rangle. \tag{4.77}$$

The previous formula describes the action of $U(a, A)$ on a "continuous basis" and hence on every state. The analysis at the end of Section 4.6 should convince us that (4.77) is nothing else than the formula (4.61), but the next exercise argues again that this is the case.

Exercise 4.11.2 Given a Schwartz function $\xi \in \mathcal{S}^4$, consider as in the previous section the state $|\xi\rangle := \int d\lambda_m(p)\xi(p)|p\rangle$. Arguing at the level of rigor of a physicist, deduce from (4.77) that $U(a, A)|\xi\rangle = |\eta\rangle$ where $\eta(p) = \exp(\mathrm{i}(a, p)/\hbar)\xi(A^{-1}(p))$.

Key ideas to remember:

- "Space" and "time" are not separate realities and each depends on the observer. Only space-time is defined independently of the observer.
- To reflect that the speed of light does not depend on the inertial observer we provide space-time with the Lorentz inner product. It is then denoted by $\mathbb{R}^{1,3}$.
- Physics appears to respect causality. When two points x, y of space-time are causally separated, $(x - y)^2 < 0$, what happens at one point cannot influence what happens at the other.
- The Lorentz group consists of the linear transformations of space-time which respect the Lorentz bilinear form.
- The four-momentum p of a particle of rest mass m satisfies $p^2 = m^2c^2$. The set of all possible such momenta is the mass shell X_m, on which exists an invariant measure $d\lambda_m$.
- The Poincaré group $\mathcal{P}$ is a fundamental group of symmetries of space-time which leaves physics invariant.
- An action of a group on a measure space is often the road to the construction of unitary representations of this group. This leads us to discover a fundamental irreducible unitary representation of $\mathcal{P}$ which corresponds to a particle of mass m.
- The simplest mathematical structures which satisfy the correct invariance properties are often physically relevant.

5

The Massive Scalar Free Field

The massive scalar free field is an absolutely central object in Quantum Field Theory.[1] It is both canonical and natural. It is, in a sense, the simplest Lorentz invariant quantum field which respects causality.[2] In this chapter we describe this object and we prove its most striking properties. We then reconstruct the concrete formulas used to describe it in physics textbooks.

A vast majority of these books follow a different path to motivate the massive scalar free field (a notable exception being Weinberg's book [87]). Their approach is closer to both the historical development and the physical insight which motivates the definition of the free field. Understanding the details of this approach requires learning some Classical Mechanics, which is not really needed for what follows here. For this reason it will be presented separately, in, Chapter 6.

5.1 Intrinsic Definition

We recall that in Minkowski's space $\mathbb{R}^{1,3}$ we define the Fourier transform and its inverse by the formulas (4.58) and (4.59). Let us consider $m > 0$, the invariant measure $\mathrm{d}\lambda_m$ on X_m, and the Hilbert space $\mathcal{H} = L^2(X_m, \mathrm{d}\lambda_m)$. We consider the corresponding boson Fock space $\mathcal{B}$ and its subspace $\mathcal{B}_0$, as constructed in Section 3.4. We recall the maps A and $A^\dagger$ of Definition 3.4.1. If it helps you to have a concrete realization of these objects, you may think of $\mathcal{B}_0 = \bigoplus_{n\geq 0} \mathcal{H}_{n,s}$, and where $\mathcal{H}_{n,s}$ is the space of functions in $L^2(X_m^n, \mathrm{d}\lambda_m^n)$ which are symmetric in their arguments, the operators A and $A^\dagger$ being given by an obvious transcription of the formulas of Exercise 3.3.3. The space of Schwartz test functions on $\mathbb{R}^{1,3}$ is denoted by $\mathcal{S}^4$.[3]

For $f \in \mathcal{S}^4$, its Fourier transform $\hat{f}$ (given by (4.58)) also belongs to $\mathcal{S}^4$. We may then see it as an element of $\mathcal{H} = L^2(X_m, \mathrm{d}\lambda_m)$ by restricting[4] $\hat{f}$ to X_m.

1 At least from a theoretical point of view. It plays a far lesser role in the description of Nature.

2 As will become apparent much later, the quanta of this field are spinless neutral massive particles. The most important such particle is the Higgs boson. The word "scalar" reflects the fact that the quanta are spinless. It also reflects the (related) fact that the field has a single component, in contrast to the several component fields studied later in Chapter 10.

3 The notation $\mathcal{S}^{1,3}$ would not make real sense since the Lorentz structure is irrelevant in the definition of this space.

4 Equivalently, we may think of $\mathrm{d}\lambda_m$ as a measure on $\mathbb{R}^{1,3}$, a point of view which we adopt in our notation by writing $\int \mathrm{d}\lambda_m(p)\xi(p)$ rather than $\int_{X_m} \mathrm{d}\lambda_m(p)\xi(p)$.

Definition 5.1.1 The massive[5] scalar free quantum field is the operator-valued distribution φ defined for $f \in \mathcal{S}^4$ by

$$\varphi(f) = \frac{1}{\sqrt{c}}(A(\widehat{f^*}) + A^\dagger(\hat{f})), \tag{5.1}$$

where c is the speed of light and where the right-hand side is seen as an operator from $\mathcal{B}_0$ to $\mathcal{B}_0$.

The relevance and the meaning of this formula are not obvious now, but will become gradually clear. The reason for the occurrence of the term f^* is that A is anti-linear. Equivalently, it suffices to consider (5.1) for $f \in \mathcal{S}^4_{\mathbb{R}}$, the space of *real-valued* test functions on $\mathbb{R}^4$.

Exercise 5.1.2 Prove that indeed if one defines $\varphi(f) = (A(\hat{f}) + A^\dagger(\hat{f}))/\sqrt{c}$ when f is real-valued, then when f is complex-valued, $\varphi(f)$ is given by the formula (5.1). Prove (formally) that for $f \in \mathcal{S}^4$ we have $\varphi(f)^\dagger = \varphi(f^*)$. In particular for $f \in \mathcal{S}^4_{\mathbb{R}}$, $\varphi(f)$ is self-adjoint.[6]

The invariant measure $d\lambda_m$ is defined intrinsically only up to a multiplicative constant, but in (5.1) we have in mind the formula (4.36) for $d\lambda_m$. It may be better anyway for the time being to think of the quantum field as being intrinsically defined only up to a multiplicative constant. The choice of a particular multiplicative constant is largely a matter of convention. We will explain in due time the inessential reason why we have chosen here to insert the multiplicative factor $1/\sqrt{c}$.

Let us investigate the basic properties of the quantum field (5.1), and first of all how it behaves with respect to the action of the Poincaré group $\mathcal{P}$.[7] We have no less than three different actions of the Poincaré group to consider here.

- First we define the operation $V(c,C)$ of $\mathcal{P}$ on $\mathcal{S}^4$ by (4.57), i.e.

$$V(c,C)(f)(x) = f(C^{-1}(x-c)). \tag{5.2}$$

- On the other hand, there is also the action $U(c,C)$ of $\mathcal{P}$ on $\mathcal{H} = L^2(X_m, d\lambda_m)$ given by (4.61), i.e. for a function ξ on X_m,

$$U(c,C)(\xi)(p) = \exp(i(c,p)/\hbar)\xi(C^{-1}(p)). \tag{5.3}$$

  These actions cannot be confused with each other as they do not operate on the same space. The action (5.2) operates on $\mathcal{S}^4$ and the action (5.3) operates on $\mathcal{H}$. We recall that by (4.60) we have

$$\widehat{V(c,C)(f)}(p) = \exp(i(c,p)/\hbar)\hat{f}(C^{-1}(p)) = U(c,C)(\hat{f})(p). \tag{5.4}$$

- The action $U(c,C)$ on $\mathcal{H}$ induces a corresponding action $U_{\mathcal{B}}(c,C)$ on the Fock space $\mathcal{B}$ (see Section 3.5). That is, $U_{\mathcal{B}}(c,C)$ is the canonical extension to $\mathcal{B}$ of the unitary operator $U(c,C)$ on $\mathcal{H}$. It sends $\mathcal{B}_0$ to itself.

[5] This field depends on a parameter $m > 0$ which is implicit in the notation.
[6] It is not very intuitive what the corresponding observable is, although mathematically the situation is rather similar to (2.100).
[7] Consistency with the action of the Poincaré group is probably the strongest guiding principle of our theories.

We now state the fundamental invariance property of φ. As we shall see, this property determines the structure of φ to a large extent.

Proposition 5.1.3 *For each $(c,C) \in \mathcal{P}$ and each $f \in \mathcal{S}^4$ we have*

$$U_{\mathcal{B}}(c,C) \circ \varphi(f) \circ U_{\mathcal{B}}(c,C)^{-1} = \varphi(V(c,C)f), \tag{5.5}$$

where for legibility we denote by $\circ$ the composition of operators on $\mathcal{B}_0$.

$$\begin{array}{ccc}
\mathcal{B}_0 & \xrightarrow{\varphi(f)} & \mathcal{B}_0 \\
{\scriptstyle U_{\mathcal{B}}(c,C)^{-1}}\uparrow & & \downarrow{\scriptstyle U_{\mathcal{B}}(c,C)} \\
\mathcal{B}_0 & \xrightarrow{\varphi(V(c,C)f)} & \mathcal{B}_0
\end{array}$$

Proof To avoid a conflict of notation in this proof we denote the speed of light by c. It suffices to check (5.5) when f is real-valued. Using (5.1) and Proposition 3.5.2 the left-hand side of (5.5) is

$$\mathsf{c}^{-1/2} U_{\mathcal{B}}(c,C) \circ (A(\hat{f}) + A^{\dagger}(\hat{f})) \circ U_{\mathcal{B}}(c,C)^{-1} = \mathsf{c}^{-1/2}(A(U(c,C)\hat{f}) + A^{\dagger}(U(c,C)\hat{f})).$$

Since $U(c,C)\hat{f} = \widehat{V(c,C)(f)}$ by (5.4) this is the right-hand side of (5.5). □

Pretending that φ is an operator-valued function so that $\varphi(f) = \int \mathrm{d}^4 x \varphi(x) f(x)$, then, since $\mathrm{d}^4 x$ is invariant under the action of $\mathcal{P}$,

$$\varphi(V(c,C)(f)) = \int \mathrm{d}^4 x \varphi(x) f(C^{-1}(x-c)) = \int \mathrm{d}^4 x \varphi(c + C(x)) f(x),$$

and (5.5) takes the form

$$\varphi(c + C(x)) = U_{\mathcal{B}}(c,C) \circ \varphi(x) \circ U_{\mathcal{B}}(c,C)^{-1}. \tag{5.6}$$

An intuitive way to understand the importance of this property is to think of $V(c,C)$ as a change of coordinates in space-time. A point y in space-time is then described by the point x such that $y = c + C(x)$. As a function of this new coordinate, the field is then $\varphi_{\text{new}}(x) := \varphi(c + C(x))$. The corresponding change of coordinates in $\mathcal{H}$ is given by $U(c,C)$, in $\mathcal{B}$ is given by $U_{\mathcal{B}}(c,C)$ and in the space of operators on $\mathcal{B}$ is given by the map $X \mapsto U_{\mathcal{B}}(c,C) \circ X \circ U_{\mathcal{B}}(c,C)^{-1}$. The content of (5.6) is then that when we compute the value of the field φ as a function of the new coordinate x the value we obtain is given by the new coordinates of $\varphi(x)$, i.e. $\varphi_{\text{new}}(x) = (\varphi(x))_{\text{new}}$.[8] Similar considerations will be of central importance in Chapter 10.

Let us now substantiate the claim that the covariance[9] property (5.5) largely determines the structure of the quantum field. Let us try to *guess* the form of a field that would satisfy (5.5), with no attempt at rigor. Presumably such a field has a creation and an annihilation

[8] At some intuitive level, this means that the free quantum field exists independently of the system of coordinates we use to describe it.

[9] In a relativistic setting, I will use the word *covariance* with the informal meaning of "behaving properly with respect to the action of Lorentz transformations".

part.[10] Let us try first to find for $x \in \mathbb{R}^{1,3}$ a *creation* operator $\varphi(x)$ which satisfies the covariance property (5.6). Then, if $\varphi(x)$ is a creation operator, it should be of the type $A^\dagger(F_x)$, where $F_x \in L^2(X_m, \mathrm{d}\lambda_m)$, so we may think of F_x as a function of p. Let us then investigate the consequences of (5.6). On the one hand, since $\varphi(x) = A^\dagger(F_x)$, we have

$$\varphi(c + C(x)) = A^\dagger(F_{c+C(x)}).$$

On the other hand, using Proposition 3.5.2 in the second equality,

$$\begin{aligned} U_{\mathcal{B}}(c,C) \circ \varphi(x) \circ U_{\mathcal{B}}(c,C)^{-1} &= U_{\mathcal{B}}(c,C) \circ A^\dagger(F_x) \circ U_{\mathcal{B}}(c,C)^{-1} \\ &= A^\dagger(U(c,C)F_x), \end{aligned}$$

so that in order to satisfy (5.6) the function F_x must satisfy

$$F_{c+C(x)}(p) = (U(c,C)F_x)(p) = \exp(\mathrm{i}(c,p)/\hbar)F_x(C^{-1}(p)). \tag{5.7}$$

Taking $x = 0$ and $c = 0$ yields $F_0(p) = F_0(C^{-1}(p))$ so that F_0 is constant. Taking $x = 0$ yields $F_c(p) = \lambda \exp(\mathrm{i}(c,p)/\hbar)$ for some constant λ. Thus we must have $\varphi(x) = \lambda A^\dagger(\exp(\mathrm{i}(x,p)/\hbar))$. This formula is rather formal, since $\exp(\mathrm{i}(x,p)/\hbar)$ is not an element of $L^2(X_m, \mathrm{d}\lambda_m)$. Its rigorous version is simply that $\varphi(f) = \lambda A^\dagger(\hat{f})$ for $f \in \mathcal{S}^4$, as is shown by the following formal calculation:

$$\begin{aligned} \varphi(f) &= \int \mathrm{d}^4x f(x)\varphi(x) = \lambda \int \mathrm{d}^4x f(x) A^\dagger(\exp(\mathrm{i}(x,p)/\hbar)) \\ &= \lambda A^\dagger\Big(\int \mathrm{d}^4x f(x)\exp((\mathrm{i}(x,p)/\hbar))\Big) = \lambda A^\dagger(\hat{f}). \end{aligned} \tag{5.8}$$

Assuming now that φ is the sum of a creation and an annihilation part, and proceeding in the same fashion as before for the annihilation part, we find that for some τ we must have

$$\varphi(f) = \lambda A(\widehat{f^*}) + \tau A^\dagger(\hat{f}). \tag{5.9}$$

Thus the covariance condition (5.5) goes a long way toward determining the form of the free field, but does not completely specify it, since (5.1) corresponds to the case $\lambda = \tau$ in (5.9).

It turns out that the condition $\lambda = \tau$ is related to another fundamental property of the quantum field of Definition 5.1.1, which we explore now.

Definition 5.1.4 We say that $f, g \in \mathcal{S}^4$ are *causally separated* if they have compact support and if for x in the support of f and y in the support of g we have $(x-y)^2 < 0$.

That is, each point x of the support of f is causally separated from each point y of the support of g, and according to (4.14) what happens at x cannot influence what happens at y. We will prove later the following second fundamental property of the field (5.1).

[10] This is a very non-trivial assumption. One may wonder whether there exist entirely different structures of interest, although the fact that the fertile imagination of physicists has not discovered them is not a good sign. One may also wonder whether one could not use a "different version" of creation operators, and this indeed is possible as we will see after we study fermion Fock space.

Theorem 5.1.5 (Microcausality) *If f and g in $\mathcal{S}^4$ are causally separated then the massive scalar field of (5.1) satisfies*

$$[\varphi(f),\varphi(g)] = 0. \tag{5.10}$$

Let us explain the fundamental physical idea behind this result. Pretending that φ is a function this means that if x and y are causally separated the operators $\varphi(x)$ and $\varphi(y)$ commute. These operators are self-adjoint, so that they correspond in principle to quantities that can be measured. Also, as supported by (5.6), $\varphi(x)$ "corresponds to a measurement at the space-time point x". The fact that no influence travels at a speed faster than light is well established at the macroscopic scale and there is no reason to believe that it fails at the microscopic scale. To respect causality, according to (4.14), the measurement at x cannot influence the measurement at y, and the natural condition to ensure that is the commutation of the operators $\varphi(x)$ and $\varphi(y)$ (see Exercise 2.2.1). Accordingly condition (5.10) is called *microcausality*. It is believed to be essential in building theories that preserve causality.

The field (5.9) need not satisfy (5.10). It is the desire to satisfy this property that motivates the choice $\lambda = \tau$ in (5.9).[11]

As a consequence one may say that the massive scalar free field is determined up to a multiplicative constant by the physically transparent requirements of covariance and microcausality. Its definition is "intrinsic". The same requirements, when properly generalized, will lead us in Chapter 10 to the discovery of all the fundamental quantum fields.

As Theorem 5.1.5 belongs to an important circle of ideas we explore it in great detail. We first define an important technical tool.

Definition 5.1.6 The tempered distribution Δ_0 is given by

$$\Delta_0(f) = \int \mathrm{d}\lambda_m(p)\hat{f}(p) \tag{5.11}$$

for any $f \in \mathcal{S}^4$.

In fact Δ_0 also depends on m but the value of m remains implicit. This distribution is essentially the inverse Fourier transform of $\mathrm{d}\lambda_m$.[12] Of course, we will pretend that Δ_0 is a function, $\Delta_0(f) = \int \mathrm{d}^4x\,\Delta_0(x)f(x)$.

Lemma 5.1.7 *There exists an infinitely differentiable function J on $\mathbb{R}^+$ such that if for $x^2 = x_\mu x^\mu < 0$ we define*

$$I(x) = J(\sqrt{-x^2}), \tag{5.12}$$

whenever f is a test function with compact support in the region $x^2 < 0$ we have $\int \mathrm{d}^4x\,\Delta_0(x)f(x) = \int \mathrm{d}^4x\,I(x)f(x)$.

[11] It can be shown that the field (5.9) satisfies (5.10) if and only if $|\lambda|^2 = |\tau|^2$, and that furthermore this case does not construct new fields compared to the case $\tau = \lambda$.

[12] Our arguments are self-contained and you need no previous knowledge of the Fourier transform of distributions or measures to follow our discussion. If you already know what these are, I should add that in our setting there are annoying issues concerning how one should define the Fourier transform of distributions so that it coincides with the Fourier transform of functions when the distributions are actually functions, but these hardly concern us since we nearly always pretend that distributions are functions. These issues are examined in Section L.4, to which the reader is referred to understand why we use the word "nearly" in the previous sentence.

Note that in particular we have $I(x) = I(-x)$. The statement of the lemma is partially described by the sentence "the distribution Δ_0 is a function in the region $x^2 < 0$". In some sense we have

$$I(x) = \int \mathrm{d}\lambda_m(p) \exp(\mathrm{i}(x,p)/\hbar) = \int \frac{\mathrm{d}^3\boldsymbol{p}}{(2\pi\hbar)^3 2\omega_{\boldsymbol{p}}} \exp(\mathrm{i}(x^0\omega_{\boldsymbol{p}} - \boldsymbol{x}\cdot\boldsymbol{p})/\hbar). \tag{5.13}$$

The problem is that this integral is not absolutely convergent and it is not obvious what it means.[13] The explicit expression we obtain below for the function J shows that it is related to certain special functions (Bessel functions) but this is irrelevant for our purposes.

Proof The proof is a non-trivial piece of mathematical analysis. It can be skipped without harm, as the ideas will not be used again. For $\varepsilon > 0$ the integral

$$\int \mathrm{d}\lambda_m(p)\mathrm{d}^4x f(x) \exp(\mathrm{i}(x,p)/\hbar) \exp(-\varepsilon\omega_{\boldsymbol{p}}) \tag{5.14}$$

is absolutely convergent, because $\int \mathrm{d}^4x|f(x)| < \infty$ and $\int \mathrm{d}\lambda_m(p)\exp(-\varepsilon\omega_{\boldsymbol{p}}) < \infty$ (since $\exp(-\varepsilon\omega_{\boldsymbol{p}}) < \exp(-\varepsilon|\boldsymbol{p}|)$).

We set

$$I_\varepsilon(x) = \int \mathrm{d}\lambda_m(p) \exp(\mathrm{i}(x,p)/\hbar) \exp(-\varepsilon\omega_{\boldsymbol{p}}). \tag{5.15}$$

Since the integral (5.14) is absolutely convergent, Fubini's theorem[14] asserts that we may compute it by integrating first either in p or in x, yielding the identity

$$\int \mathrm{d}^4x I_\varepsilon(x) f(x) = \int \mathrm{d}\lambda_m(p) \exp(-\varepsilon\omega_{\boldsymbol{p}}) \hat{f}(p). \tag{5.16}$$

As $\varepsilon \to 0^+$ the right-hand side of (5.16) converges to $\Delta_0(f) = \int \mathrm{d}^4x \Delta_0(x) f(x)$, so the issue is to compute $\lim_{\varepsilon\to 0^+} I_\varepsilon(x)$ and to prove uniform convergence of this limit over compact subsets of the region $x^2 < 0$. We reformulate the definition of $I_\varepsilon(x)$:

$$I_\varepsilon(x) = \int \frac{\mathrm{d}^3\boldsymbol{p}}{(2\pi\hbar)^3 2\omega_{\boldsymbol{p}}} \exp(-\mathrm{i}(x^0\omega_{\boldsymbol{p}} - \boldsymbol{x}\cdot\boldsymbol{p})/\hbar - \varepsilon\omega_{\boldsymbol{p}}).$$

We assume $x^2 < 0$. To compute $I_\varepsilon(x)$ we may assume by rotational invariance that $\boldsymbol{x}$ is in the direction of e_3 (so that $\boldsymbol{x} = |\boldsymbol{x}|e_3$) and going to cylindrical coordinates, that is setting $p^1 = r\cos\theta$, $p^2 = r\sin\theta$, $p^3 = z$ and setting $\omega := \sqrt{r^2 + z^2 + m^2c^2}$ we obtain

$$I_\varepsilon(x) = \int_0^\infty \mathrm{d}r \int_{-\infty}^\infty \mathrm{d}z \frac{r}{2(2\pi)^2\hbar^3\omega} \exp(-\mathrm{i}(x^0\omega - |\boldsymbol{x}|z)/\hbar - \varepsilon\omega).$$

13 It can never be stressed enough that considering any integral whatsoever which does not converge absolutely is intrinsically a delicate endeavor, as the value given to the integral *may depend on its definition*. For example an integral in two variables may be given different values depending on the order of integration chosen. Integrals can be formally manipulated, with the result of these manipulations often reaching the correct answer. A rigorous justification for these manipulations is a different and often harder matter. Correctness of the result is the relevant criterion in physics, and most authors of physics textbooks show very little interest in justifying the way they manipulate integrals, or even in defining the values of these integrals.

14 This theorem is reviewed on page 475.

We set $d := 1/(2(2\pi)^2\hbar^3)$. Now comes the key idea: to make the change of variable $z := \sqrt{r^2+m^2c^2}\sinh u$, so that $\omega = \sqrt{r^2+m^2c^2}\cosh u$, $\mathrm{d}z = \omega \mathrm{d}u$ and

$$I_\varepsilon(x) = d\int_0^\infty r\mathrm{d}r \int_{-\infty}^\infty \mathrm{d}u \exp\Big(\sqrt{r^2+m^2c^2}\big(-\mathrm{i}(x^0\cosh u - |\boldsymbol{x}|\sinh u)/\hbar - \varepsilon\cosh u\big)\Big). \tag{5.17}$$

Consider $b > 0$ with $b^2 = -x^2 = |\boldsymbol{x}|^2 - (x^0)^2$. Define the parameter τ by $x^0 = b\sinh\tau$ and $|\boldsymbol{x}| = b\cosh\tau$, so that $x^0\cosh u - |\boldsymbol{x}|\sinh u = b\sinh(\tau - u)$. Making the change of variable $u \to u + \tau$ yields

$$I_\varepsilon(x) = d\int_0^\infty r\mathrm{d}r \int_{-\infty}^\infty \mathrm{d}u \exp\Big(\sqrt{r^2+m^2c^2}\big(-(\mathrm{i}b\sinh u)/\hbar - \varepsilon\cosh(u+\tau)\big)\Big). \tag{5.18}$$

Let us set $\alpha := b\sqrt{r^2+m^2c^2}/\hbar$ and $f_{\varepsilon,r}(u) := \exp(-\varepsilon\sqrt{r^2+m^2c^2}\cosh(u+\tau))$, so that

$$I_\varepsilon(x) = d\int_0^\infty r\mathrm{d}r \int_{-\infty}^\infty \mathrm{d}u f_{\varepsilon,r}(u)\exp(-\mathrm{i}\alpha\sinh u). \tag{5.19}$$

The important feature of this integral is the oscillatory nature of the exponential term. To bring it forward we will use that by integration by parts, for a nice function $f(u)$,

$$\int_{-\infty}^\infty \mathrm{d}u f(u)\exp(\mathrm{i}\alpha\sinh u) = \frac{1}{\mathrm{i}\alpha}\int_{-\infty}^\infty \mathrm{d}u T(f)(u)\exp(-\mathrm{i}\alpha\sinh u), \tag{5.20}$$

where

$$T(f)(u) = \frac{\mathrm{d}}{\mathrm{d}u}\frac{f(u)}{\cosh u}.$$

Using this formula three times we obtain

$$I_\varepsilon(x) = J_\varepsilon(b) := d\int_0^\infty \frac{r\mathrm{d}r}{(\mathrm{i}\alpha)^3}\int_{-\infty}^\infty \mathrm{d}u T^3(f_{\varepsilon,r})(u)\exp(-\mathrm{i}\alpha\sinh(u)).$$

Using (a) of Exercise 5.1.8, the integral

$$J(b) := J_0(b) = d\int_0^\infty \frac{r\mathrm{d}r}{(\mathrm{i}\alpha)^3}\int_{-\infty}^\infty \mathrm{d}u T^3(1)(u)\exp(-\mathrm{i}\alpha\sinh(u)) \tag{5.21}$$

is well defined because the factor α^{-3} ensures the convergence in r of the outer integral. Using (b) of Exercise 5.1.8 and dominated convergence we obtain that $J_\varepsilon(b) \to J_0(b)$ as $\varepsilon \to 0$. This convergence is obviously uniform over the regions $b \geq b_0 > 0$ as desired.

Finally we have to show that $J(b)$ is an infinitely differentiable function of b. For this we again use (5.20) to prove that differentiating in b in the integrand of this right-hand side still gives an integrable function. □

Exercise 5.1.8 (a) Prove that $\int_{-\infty}^\infty \mathrm{d}u\,|T^3(1)(u)| < \infty$.
(b) Prove that $g_\varepsilon(r) := \int_{-\infty}^\infty \mathrm{d}u\,|T^3(1 - f_{\varepsilon,r})(u)|$ is bounded and converges pointwise to zero as $\varepsilon \to 0$.

Exercise 5.1.9 Extend the result of Lemma 5.1.7 to the regions $x^2 > 0, x^0 > 0$ and $x^2 > 0, x^0 < 0$. Hint: Integrate now in spherical coordinates.[15]

Proof of Theorem 5.1.5 It suffices to consider the case where f and g are real-valued. Let us pay a lot of attention to the following calculation, which will occur many times. It follows from the definition of φ that

$$c[\varphi(f), \varphi(g)] = [A(\hat{f}) + A^\dagger(\hat{f}), A(\hat{g}) + A^\dagger(\hat{g})],$$

and then from (3.33) and (3.34) that

$$c[\varphi(f), \varphi(g)] = \big((\hat{f}, \hat{g}) - (\hat{g}, \hat{f})\big)\mathbf{1}. \tag{5.22}$$

Now,

$$(\hat{f}, \hat{g}) = \int \mathrm{d}\lambda_m(p)\hat{f}(p)^*\hat{g}(p) \tag{5.23}$$

whereas, making the change of variable $y = x + a$ in the second line,

$$\begin{aligned} \hat{f}(p)^*\hat{g}(p) &= \iint \mathrm{d}^4x\mathrm{d}^4y \exp(\mathrm{i}(y - x, p)/\hbar) f(x)g(y) \\ &= \iint \mathrm{d}^4x\mathrm{d}^4a \exp(\mathrm{i}(a, p)/\hbar) f(x)g(x + a) \\ &= \hat{h}(p), \end{aligned} \tag{5.24}$$

where $h(a) = \int \mathrm{d}^4x f(x)g(a + x)$. Going back to (5.23) this means that $(\hat{f}, \hat{g}) = \int \mathrm{d}^4x \Delta_0(x)h(x)$. Now, the hypothesis that the supports of f and g are causally separated implies that h has compact support, which is contained in the set $\{x^2 < 0\}$, so that as proved in Lemma 5.1.7 the previous integral is $\int \mathrm{d}^4x I(x)h(x)$. Finally, exchanging f and g changes $h(x)$ into $h(-x)$ as is obvious from the definition of h. Moreover $I(x) = I(-x)$ because $I(x)$ depends only on x^2. Thus exchanging f and g does not change the value of $(\hat{f}, \hat{g})$, i.e. $(\hat{f}, \hat{g}) = (\hat{g}, \hat{f})$ and this finishes the proof. □

Combining (5.23) and the first line of (5.24) we obtain

$$\begin{aligned} (\hat{f}, \hat{g}) &= \iiint \mathrm{d}^4x\mathrm{d}^4y\mathrm{d}\lambda_m(p) \exp(\mathrm{i}(y - x, p)/\hbar) f(x)g(y) \\ &= \iint \mathrm{d}^4x\mathrm{d}^4y \Delta_0(y - x) f(x)g(y). \end{aligned} \tag{5.25}$$

(The reader who finds this manipulation scary might learn in Appendix M how it could be done more rigorously.) If we pretend that distributions are functions, we may write (5.22) as

$$c[\varphi(x), \varphi(y)] = (\Delta_0(y - x) - \Delta_0(x - y))\mathbf{1}. \tag{5.26}$$

This is a more precise statement than (5.10). Still, the proof of Theorem 5.1.5 is needed to really address the fine points concerning distributions. In the sequel we will be rather

[15] This is very clever. If I were you, I would not even try. But you may enjoy studying the solution.

informal, and we will simply say that Theorem 5.1.5 follows from (5.26) and the fact that $I(x) = I(-x)$ when $x^2 < 0$.

It is possible to formulate the previous results differently, as in e.g. Dimock's textbook [23], without ever mentioning distributions. This is the purpose of the next exercise.

Exercise 5.1.10 (a) Prove that for $f, g \in \mathcal{S}^4_{\mathbb{R}}$ we have

$$c[\varphi(f), \varphi(g)] = (f, Eg) = \int \mathrm{d}^4x f(x) Eg(x)$$

where

$$Eg(x) = \int\int \mathrm{d}^4y \mathrm{d}\lambda_m(p)\big((\exp(\mathrm{i}(y-x,p)/\hbar) - \exp(\mathrm{i}(x-y),p)/\hbar)\big)g(y).$$

(b) Prove that $Eg(x) = 0$ if $\{x\}$ and the support of g are causally separated. Hint: This is really challenging. You may study Dimock's textbook [23, page 125] for help.

5.2 Explicit Formulas

In this section we justify formulas for the free massive scalar field as a physicist might write them. Unfortunately the way this is usually done does not respect the most important property (5.5) of the field, and we refer the reader to Section 5.4 for this matter.

The basic idea is to "express φ in terms of the 'operators' $b(\boldsymbol{p})$ and $b^\dagger(\boldsymbol{p})$ of Section 3.8". The problem is that $\varphi(x)$ as defined by (5.1) operates on the boson Fock space $\mathcal{B}$ of $\mathcal{H} = L^2(X_m, \mathrm{d}\lambda_m)$ whereas the operators of Section 3.8 operate on the boson Fock space $\mathcal{B}'$ of $\mathcal{H}' = L^2(\mathbb{R}^3, \mathrm{d}\boldsymbol{p}^3/(2\pi\hbar)^3)$. So either one has to transport φ to a unitarily equivalent version operating on $\mathcal{B}'$ or one has to transport the operators of Section 3.8 to a unitarily equivalent version operating on $\mathcal{B}$. We chose the second solution, which is preferable in the long run.

Our purpose is to now define operators which will turn out to be the "transported version" of the operators of Section 3.8 on the boson Fock space of $\mathcal{H}$. Given a function $\xi \in \mathcal{H}' = L^2(\mathbb{R}^3, \mathrm{d}\boldsymbol{p}^3/(2\pi\hbar)^3)$ we define the function $J\xi \in \mathcal{H} = L^2(X_m, \mathrm{d}\lambda_m)$ by $J\xi(p) = \sqrt{2\omega_{\boldsymbol{p}}}\xi(\boldsymbol{p})$ where as usual

$$p = (\omega_{\boldsymbol{p}}, \boldsymbol{p})\ ;\ \omega_{\boldsymbol{p}}^2 = c^2m^2 + \boldsymbol{p}^2, \tag{5.27}$$

and we observe that J is an isometry. We then define two operator-valued distributions $a(\boldsymbol{p})$ and $a^\dagger(\boldsymbol{p})$ as follows: for each test function $\xi \in \mathcal{S}^3$,

$$A(J\xi) = \int \frac{\mathrm{d}^3\boldsymbol{p}}{(2\pi\hbar)^3}\xi(\boldsymbol{p})^* a(\boldsymbol{p})\ ;\ A^\dagger(J\xi) = \int \frac{\mathrm{d}^3\boldsymbol{p}}{(2\pi\hbar)^3}\xi(\boldsymbol{p}) a^\dagger(\boldsymbol{p}). \tag{5.28}$$

These are distributions on $\mathbb{R}^3$ valued into the operators on $\mathcal{B}_0$. Since J is an isometry,

$$[A(J\xi), A^\dagger(J\eta)] = (J\xi, J\eta)1 = \int \mathrm{d}\lambda_m(p)(J\xi(p))^* J\eta(p)1 = \int \frac{\mathrm{d}^3\boldsymbol{p}}{(2\pi\hbar)^3}\xi(\boldsymbol{p})^*\eta(\boldsymbol{p})1,$$

and just as in Section 3.8, we obtain the commutation relations $[a(\boldsymbol{p}),a^\dagger(\boldsymbol{p}')] = (2\pi\hbar)^3\delta^{(3)}(\boldsymbol{p}-\boldsymbol{p}')1$ and $[a(\boldsymbol{p}),a(\boldsymbol{p}')] = [a^\dagger(\boldsymbol{p}),a^\dagger(\boldsymbol{p}')] = 0$ corresponding to (3.60) and (3.61).

Next we relate these operators to the operators $b(\boldsymbol{p})$ and $b^\dagger(\boldsymbol{p})$ of (3.59).

Lemma 5.2.1 *Consider the canonical extension W of J from $\mathcal{B}'$ to $\mathcal{B}$. Then*

$$a(\boldsymbol{p}) = Wb(\boldsymbol{p})W^{-1}\ ;\ a^\dagger(\boldsymbol{p}) = Wb^\dagger(\boldsymbol{p})W^{-1}. \tag{5.29}$$

Proof From the first part of (3.59) and the first part of (3.42) we deduce that

$$\int \frac{\mathrm{d}^3\boldsymbol{p}}{(2\pi\hbar)^3}\xi(\boldsymbol{p})^* Wb(\boldsymbol{p})W^{-1} = WB(\xi)W^{-1} = A(J\xi)$$

and we compare with (5.28). □

Going back to the problem of an explicit expression for φ we use the first relation in (5.28) for the function $\xi \in \mathcal{S}^3$ given by $\xi(\boldsymbol{p}) = \widehat{f^*}(p)/\sqrt{2\omega_{\boldsymbol{p}}}$ so that $J\xi = \widehat{f^*}$; and we use similarly the second relation of (5.28) for the function $\xi \in \mathcal{S}^3$ such that $J\xi = \hat{f}$ to obtain

$$\varphi(f) = \int \frac{\mathrm{d}^3\boldsymbol{p}}{(2\pi\hbar)^3\sqrt{2c\omega_{\boldsymbol{p}}}}(\widehat{f^*}(p)^* a(\boldsymbol{p}) + \hat{f}(p)a^\dagger(\boldsymbol{p})),$$

and replacing $\hat{f}$ and $\hat{f}^*$ by their values this is[16]

$$\int \mathrm{d}^4x f(x) \int \frac{\mathrm{d}^3\boldsymbol{p}}{(2\pi\hbar)^3\sqrt{2c\omega_{\boldsymbol{p}}}}(e^{-\mathrm{i}(x,p)/\hbar}a(\boldsymbol{p}) + e^{\mathrm{i}(x,p)/\hbar}a^\dagger(\boldsymbol{p})).$$

Now, since $\varphi(f) = \int \mathrm{d}^4x f(x)\varphi(x)$ we have discovered the fundamental formula:

$$\varphi(x) = \int \frac{\mathrm{d}^3\boldsymbol{p}}{(2\pi\hbar)^3\sqrt{2c\omega_{\boldsymbol{p}}}}(e^{-\mathrm{i}(x,p)/\hbar}a(\boldsymbol{p}) + e^{\mathrm{i}(x,p)/\hbar}a^\dagger(\boldsymbol{p})). \tag{5.30}$$

Observe that the only dependence on m is through $\omega_{\boldsymbol{p}}$.

There is a mathematically far more elegant way to write this formula, see (5.42). Mathematical elegance would dictate to use the formula (5.42) throughout the book, but it makes little sense to write a text which would be hard to connect to the existing literature.

Pretending from (5.30) that the operator $\varphi(x)$ is well defined, many introductory textbooks state that the quantum field (5.30) "creates and destroys particles at the space-time point x" see e.g. Peskin and Schroeder's book [64, page 24]. This statement should be taken on an intuitive rather than a technical level. In relativistic Quantum Mechanics the very existence of an "exact" position operator is questionable, as explained in Appendix F. It does not make sense to think of a particle (such as a particle created by the field $\varphi(x)$) as having a very precise location.

At a later stage we will need "a discrete version" of the previous formulas and now is a good time to examine it. We follow the procedure of Section 3.9 of "putting the spatial universe in a box", i.e. replacing physical space $\mathbb{R}^3$ by the box $[-L/2, L/2]^3$. Each point

16 The reason for which we write here $e^{-\mathrm{i}(x,p)/\hbar}$ rather than our usual $\exp(-\mathrm{i}(x,p)/\hbar)$ is that this is much closer to the notation used in physics textbooks.

$\boldsymbol{k} \in \mathcal{K}$ replaces in momentum space a small box of side $2\pi\hbar/L$, and of volume L^{-3} for the natural measure $\mathrm{d}^3\boldsymbol{p}/(2\pi\hbar)^3$, and the discrete version of (5.30) is

$$\varphi(x) = \frac{1}{L^3}\sum_{\boldsymbol{k}\in\mathcal{K}} \frac{1}{\sqrt{2c\omega_{\boldsymbol{k}}}}\Big(\exp(\mathrm{i}(-x^0\omega_{\boldsymbol{k}} + \boldsymbol{x}\cdot\boldsymbol{k})/\hbar)a(\boldsymbol{k}) + \exp(\mathrm{i}(x^0\omega_{\boldsymbol{k}} - \boldsymbol{x}\cdot\boldsymbol{k})/\hbar)a^\dagger(\boldsymbol{k})\Big), \tag{5.31}$$

where $\omega_{\boldsymbol{k}} = \sqrt{\boldsymbol{k}^2 + c^2m^2}$ and where (3.60) is replaced by

$$[a(\boldsymbol{k}), a^\dagger(\boldsymbol{k}')] = L^3\delta_{\boldsymbol{k}}^{\boldsymbol{k}'}1. \tag{5.32}$$

Calling now a what is $a/L^{3/2}$ in (5.31) this formula can also be written as

$$\varphi(x) = \frac{1}{L^{3/2}}\sum_{\boldsymbol{k}\in\mathcal{K}} \frac{1}{\sqrt{2c\omega_{\boldsymbol{k}}}}\Big(\exp(\mathrm{i}(-x^0\omega_{\boldsymbol{k}} + \boldsymbol{x}\cdot\boldsymbol{k})/\hbar)a(\boldsymbol{k}) + \exp(\mathrm{i}(x^0\omega_{\boldsymbol{k}} - \boldsymbol{x}\cdot\boldsymbol{k})/\hbar)a^\dagger(\boldsymbol{k})\Big), \tag{5.33}$$

where (5.32) is replaced by

$$[a(\boldsymbol{k}), a^\dagger(\boldsymbol{k}')] = \delta_{\boldsymbol{k}}^{\boldsymbol{k}'}1. \tag{5.34}$$

Exercise 5.2.2 Prove that $L^{3/2}\sqrt{2c\omega_{\boldsymbol{p}}}$ has dimension $[l^{5/2}m^{1/2}t^{-1}]$ so that in a sense

$$\varphi \text{ has dimension } [l^{-5/2}m^{-1/2}t], \tag{5.35}$$

not such an intuitive result. Convince yourself that "the dimension of φ" in the continuous case is the same.

The meaning of the infinite summation in (5.31) is problematic, but matters are much better when one integrates again a "test function" i.e. an infinitely differentiable function,[17] because the Fourier coefficients of such a function converge rapidly to zero. That is, even the discrete formula (5.31) should be understood in the sense of distributions.

5.3 Time-evolution

The unitary representation $U(a,A)$ of (5.3) of the Poincaré group on $\mathcal{H}$ specifies the action of time-translations, so it specifies a time-evolution. The corresponding Hamiltonian is simply "multiplication by the function cp^0" as we learned in Proposition 4.9.2. The canonical extension $U_{\mathcal{B}}(a,A)$ of $U(a,A)$ to $\mathcal{B}$ defines a unitary representation of the Poincaré group on $\mathcal{B}$, so it specifies a time-evolution that is governed by a Hamiltonian $H_{\mathcal{B}}$. The relevance of the following formula will become apparent only gradually.

Proposition 5.3.1 *The time-evolution in the boson Fock space $\mathcal{B}$ is governed by the Hamiltonian formally given by the formula*

$$H_{\mathcal{B}} = \int \frac{\mathrm{d}^3\boldsymbol{p}}{(2\pi\hbar)^3} c\omega_{\boldsymbol{p}} a^\dagger(\boldsymbol{p})a(\boldsymbol{p}). \tag{5.36}$$

17 With all its derivatives satisfying the periodic boundary conditions.

Proof One transports the formula (3.62) from $\mathcal{B}'$ to $\mathcal{B}$ by multiplying to the left by W and to the right by W^{-1} and uses (5.29). □

Pretending that φ is a function, consider the special case

$$\varphi(d+x) = U_{\mathcal{B}}(d,1) \circ \varphi(x) \circ U_{\mathcal{B}}(d,1)^{-1} \tag{5.37}$$

of (5.6), and specialize further to the case $d = (ct,0,0,0)$. Then, as explained on page 126, $U(d,1)^{-1}$ is the time-evolution $\exp(-\mathrm{i}tH/\hbar)$ during a time t and consequently $U_{\mathcal{B}}(d,1)^{-1}$ is the time-evolution $\exp(-\mathrm{i}tH_{\mathcal{B}}/\hbar)$ during the same time t. The right-hand side of (5.37) is then the time-evolution of $\varphi(x)$ in the Heisenberg picture. We will make this idea precise in Theorem 6.11.1.

5.4 Lorentz Invariant Formulas

There are many different ways to write the formula (5.30) depending on the way the operators $a(\boldsymbol{p})$ and $a^\dagger(\boldsymbol{p})$ are defined, i.e. which factors are put inside the definition of these operators, and which factors are left in the integral. Besides the unimportant numerical factors, there are really two different philosophies in normalizing the operators $a(\boldsymbol{p})$. The first one leads to formula (5.30). We have chosen this approach to stay consistent with Folland's book [31] and many other textbooks. A drawback of this formula is that it does not display properly the invariance with respect to Lorentz transformations, namely (5.6) fails even when $c = 0$.

The second philosophy is to bring Lorentz invariance forward, and to write instead

$$\sqrt{c}\varphi(x) = \int \frac{\mathrm{d}^3\boldsymbol{p}}{(2\pi\hbar)^3 2\omega_{\boldsymbol{p}}} \left(e^{-\mathrm{i}(x,p)/\hbar}\sqrt{2\omega_{\boldsymbol{p}}}a(\boldsymbol{p}) + e^{\mathrm{i}(x,p)/\hbar}\sqrt{2\omega_{\boldsymbol{p}}}a^\dagger(\boldsymbol{p})\right). \tag{5.38}$$

The purpose of the present section is to show that the previous formula makes it obvious that φ is Lorentz invariant, see (5.42). It brings forward the operators

$$a(p) := \sqrt{2\omega_{\boldsymbol{p}}}a(\boldsymbol{p})\ ;\ a^\dagger(p) := \sqrt{2\omega_{\boldsymbol{p}}}a^\dagger(\boldsymbol{p}). \tag{5.39}$$

These operators should be thought of as a normalization of the operators $a(\boldsymbol{p})$ and $a^\dagger(\boldsymbol{p})$ respecting Lorentz invariance.[18]

Exercise 5.4.1 Manipulate the definitions to show that for each function $\xi \in \mathcal{S}^4$ we have

$$\int \mathrm{d}\lambda_m(p)\xi(p)^* a(p) = A(\xi), \tag{5.40}$$

where in the right-hand side ξ is seen as an element of $L^2(X_m, \mathrm{d}\lambda_m)$.

Next we explore the Lorentz invariance properties of the operators $a(p)$. This simple material is better left as an exercise.

[18] For this reason, we will meet these operators again in Section 13.11.

Exercise 5.4.2 (a) Let us recall the action $U_{\mathcal{B}}(c,C)$ of the Poincaré group on the boson Fock space $\mathcal{B}$ of $L^2(X_m, d\lambda_m)$. Prove the following formula:

$$U_{\mathcal{B}}(c,C) \circ a(p) \circ U_{\mathcal{B}}(c,C)^{-1} = \exp(-i(c, C(p))/\hbar) a(C(p)). \tag{5.41}$$

(b) To make sure you understand the formula (5.41), define

$$W(c,C)(a(p)) = \exp(-i(c,C(p))/\hbar) a(C(p))$$

and check directly that

$$W(b,B)W(c,C)(a(p)) = W((b,B)(c,C))(a(p)).$$

(c) In which way is the formula (5.41) different from the formula (4.61)?
(d) Show that (5.38) can be written as

$$\sqrt{c}\varphi(x) = \int d\lambda_m(p)\big(\exp(-i(x,p)/\hbar)a(p) + \exp(i(x,p)/\hbar)a^\dagger(p)\big). \tag{5.42}$$

(e) Explain why the property (5.6) simply follows from the previous formula.

We explore next how the operators $a(\boldsymbol{p})$ themselves transform under Lorentz transformations. The following exercise confirms that for invariance properties, it is $a(p) = \sqrt{2\omega_{\boldsymbol{p}}} a(\boldsymbol{p})$ rather than $a(\boldsymbol{p})$ which transforms nicely.

Exercise 5.4.3 To each $\boldsymbol{p} \in \mathbb{R}^3$ we can associate $p = (\omega_{\boldsymbol{p}}, \boldsymbol{p}) \in X_m$. Given $C \in SO^\uparrow(1,3)$ let us define $\hat{C}(\boldsymbol{p})$ by $C(p) = (\omega, \hat{C}(\boldsymbol{p}))$ for a certain ω (which is of course $\omega_{\hat{C}(\boldsymbol{p})}$). Prove the formula

$$U_{\mathcal{B}}(c,C) \circ a(\boldsymbol{p}) \circ U_{\mathcal{B}}(c,C)^{-1} = \exp(-i(c,C(p))/\hbar)\sqrt{\frac{\omega_{\hat{C}(\boldsymbol{p})}}{\omega_{\boldsymbol{p}}}}\, a(\hat{C}(\boldsymbol{p})), \tag{5.43}$$

and compare it with (5.41).

The next exercise is tailored to mathematical intuition, and makes implicit use of the idea of "distribution on X_m". You should not be worried if your way to think about these objects is different from this abstraction.

Exercise 5.4.4 Convince yourself that heuristically (5.40) means $a(p) = A(\delta_{m,p})$ where $\delta_{m,p}$ is the "function" of $L^2(X_m, d\lambda_m)$ of Definition 4.4.3.

Key ideas to remember:

- The quantum free field depends on a parameter $m \geq 0$. It is an operator-valued distribution on space-time. It is valued on the set of unbounded operators on the boson Fock space of a particle of mass m.
- The quantum free field behaves remarkably under the action of the Poincaré group.
- It respects causality through the microcausality condition.
- To a large extent it is determined by the previous two properties.

6

Quantization

The purpose of this chapter is to make the free quantum field appear as a quantization of the Klein-Gordon field, the classical field[1] that satisfies the fundamental Klein-Gordon equation.[2]

We will pursue this goal with increasingly sophisticated approaches, but in its simpler version the quantization procedure was already performed in Section 3.10. The differential equation (3.70) of that section is simply the version of the Klein-Gordon equation "in 1+1 dimensions", when physical space is one-dimensional, and it is a simple matter to extend the procedure to the case of "1+3 dimensions".

In order to go beyond the naive procedure of Section 3.10, one must develop rationales for the quantization procedure. For this one has to discuss two issues:

- How does one describe classically a mechanical system?
- What is the natural way to quantize such a system?

These vast questions have been the objects of considerable work. To the first one, we will give very limited answers, through the shortest possible introduction to Lagrangian and Hamiltonian Mechanics. The second question, how to quantize a classical system, is in general genuinely difficult: Quantum Mechanics is not a consequence of Classical Mechanics. We will explore in considerable detail several routes one may follow to quantize the Klein-Gordon field. Each of these encompasses a different aspect of the wisdom developed by generations of physicists. This may of course feel overwhelming[3] to the reader with no previous exposure to Classical or Quantum Mechanics.[4] Arguments of this type will not be used in the sequel and may be skipped. It is, however, important to go over the short

[1] "Classical" here means "non-quantum". This field is a creature of Classical Mechanics. It is not observed in Nature.

[2] The Klein-Gordon equation was first proposed as a relativistically correct wave equation for a single particle but, for a variety of reasons, cannot be considered as such. The procedure of quantizing the Klein-Gordon field is sometimes called "second quantization" in older textbooks, despite the fact that the Klein-Gordon equation is not a quantization of anything. The name "second quantization" is also sometimes applied to the procedure of going from a state space to its boson Fock space. The unfortunate and confusing name "second quantization" is a remainder of a period at the birth of the theory where the wave function itself was thought of as a field that could be quantized. This non-sense can still be found in rather recent textbooks.

[3] The reason we treat the subject in such detail is that different physics textbooks give different arguments which may be hard at first to relate to each other.

[4] A further source of difficulty might be that many of the arguments are entirely heuristic and far from rigorous mathematics.

Section 6.10. We show there that, given a classical system there may exist a very large number of non-unitarily equivalent corresponding quantum systems.

Having made the free quantum field appear as the quantization of the Klein-Gordon field, we examine in the last section how to relate the Hamiltonians at the classical and quantum level.

6.1 The Klein-Gordon Equation

Lorentz invariance being a crucial property we adopt the notation used in Chapter 4. We denote $(x^\nu) = (x^0, x^1, x^2, x^3)$ the generic point of Minkowski space $\mathbb{R}^{1,3}$ and we introduce the notation

$$\partial_\nu = \frac{\partial}{\partial x^\nu}. \tag{6.1}$$

We have argued that a differential equation can be of physical interest only if it is "Lorentz invariant". To find such equations let us try to invent a linear differential operator Δ for a function u on $\mathbb{R}^{1,3}$ which would be "Lorentz invariant" in the sense that it commutes with all the operators $V(c,C)$ of (5.2) i.e.[5]

$$V(c,C)\Delta(u) = \Delta(V(c,C)(u)). \tag{6.2}$$

Two immediate choices which come to mind are $\Delta(u) = u$ and

$$\Delta(u) = \eta^{\mu\nu}\partial_\mu\partial_\nu u,$$

where, as in Chapter 4, $\eta^{\mu\nu}$ is such that $\eta^{00} = 1$ and $\eta^{11} = \eta^{22} = \eta^{33} = -1$, and where, as is usual in relativistic notation, there is a summation over repeated indices.

Exercise 6.1.1 Check that these operators are Lorentz invariant.

Out of these two operators we form the equation

$$m^2c^2u + \hbar^2\eta^{\mu\nu}\partial_\mu\partial_\nu u = 0, \tag{6.3}$$

known as the *Klein-Gordon equation*.[6] As for the coefficients m^2c^2 and $\hbar^2$ let us say simply that they are constrained by dimensional considerations.

Exercise 6.1.2 Perform dimensional analysis to prove that the two terms in (6.3) have the same dimension.[7]

It should be obvious that if u is a solution of the Klein-Gordon equation and if $(c,C) \in \mathcal{P}$ then the function $x \mapsto V(c,C)u(x) = u(C^{-1}(x-c))$ is also a solution of this equation. In a sense the Klein-Gordon equation is the simplest Lorentz invariant differential equation,

[5] It would be more appropriate to call such an operator "Poincaré invariant", but translation invariance is already a basic idea of Classical Mechanics and the word "Lorentz invariant" focuses on the new feature.

[6] One way to discover this equation is to start with the relativistic relation $E^2 = (mc^2)^2 + (pc)^2$ between energy and momentum. Using the standard ansatz $E \to i\hbar\partial/\partial t = ci\hbar\partial/\partial_0$ and $p \to -i\hbar\nabla$ yields the Klein-Gordon equation.

[7] A more clever way to reach this conclusion is that m^2c^2 is a momentum squared, and if you know that $\hbar$ has the dimension of a length times a momentum, it is plain that the second term has the dimension of a momentum squared too.

and as such is very important: All the quantum fields we will write satisfy this equation in the sense of distributions.[8]

The Klein-Gordon equation can be written in an even prettier form if one uses our convention to raise and lower indices using the matrix $\eta^{\mu\nu}$, defining

$$\partial^\nu = \eta^{\mu\nu}\partial_\mu, \tag{6.4}$$

which simply amounts to putting a minus sign if $\nu = 1,2,3$. It then takes the form[9]

$$m^2c^2u + \hbar^2\partial^\nu\partial_\nu u = 0. \tag{6.5}$$

Much of the rest of this chapter is devoted to arguing, following several different routes, that the quantum field (5.30) is "a quantization of the Klein-Gordon field".[10] The first step in this direction is that the quantum field itself satisfies the Klein-Gordon equation.

Proposition 6.1.3 *The quantum field (5.1) satisfies the Klein-Gordon equation as an operator-valued distribution.*

Proof This sounds very impressive, but it is a very weak statement. Recalling the definition (1.10) of the derivative of a distribution, the statement means that for each $f \in \mathcal{S}^4_{\mathbb{R}}$ we have $\varphi(g) = 0$ where $g = \hbar^2\eta^{\mu\nu}\partial_\mu\partial_\nu f + m^2c^2 f$. Now, using (1.33) (or more accurately, the version of this result for the Fourier transform as defined in (4.58)),

$$\hat{g}(p) = (-\eta^{\mu\nu}p_\mu p_\nu + m^2c^2)\hat{f}(p) = (m^2c^2 - p^2)\hat{f}(p)$$

is identically zero on $X_m = \{p^2 = m^2c^2\}$. Thus it is zero as an element of $L^2(X_m, \mathrm{d}\lambda_m)$ and $A(\hat{g}) = A^\dagger(\hat{g}) = 0$ so that $\varphi(g) = 0$ by (5.1). □

A general principle is at work here. A tempered distribution satisfies the Klein-Gordon equation if and only if its Fourier transform vanishes outside the set $\{p^2 = m^2c^2\}$ (which is the union of X_m and its symmetric set with respect to the origin).

6.2 Naive Quantization of the Klein-Gordon Field

In this section we "quantize the solutions of the Klein-Gordon equation" from the most naive point of view. We are interested here in *real-valued* solutions (since a complex-valued solution can be considered as consisting of two independent real-valued solutions). Let us start with some general observations concerning the solutions of the Klein-Gordon equation. This is a second-order differential equation, and we expect from the elementary theory of

[8] It bears repeating that despite the fundamental importance of the Klein-Gordon equation, Nature has used it sparingly. A classical field obeying this equation has not been observed.

[9] The operator $\partial^\mu\partial_\mu$ is sometimes called the d'Alembertian and is then denoted by $\Box$,

$$\Box = \partial_0^2 - \sum_{1\le k\le 3}\partial_k^2 = \frac{\partial^2}{c^2(\partial t)^2} - \sum_{1\le k\le 3}\frac{\partial^2}{(\partial x^k)^2}$$

and the Klein-Gordon equation is then written $\hbar^2\Box u + m^2c^2u = 0$.

[10] The precise meaning of the term "Klein-Gordon field" will be explained in the next section.

such equations that if for a given value of x^0 we know both functions $\boldsymbol{x} \mapsto u(x^0, \boldsymbol{x})$ and $\boldsymbol{x} \mapsto \partial_0 u(x^0, \boldsymbol{x})$ then u is completely determined.[11] A nice solution is one for which at any time x^0 these two functions are small (e.g. Schwartz functions) in the spatial variables. We intend first to look at such functions.

In our presentation we replace the space $\mathbb{R}^3$ of spatial coordinates by a torus, i.e. we replace the spatial universe by the finite box $B = [-L/2, L/2]^3$ of Section 3.9, using periodic boundary conditions, simply because this makes the manipulations somewhat less formal than if we stay in the continuous setting. We recall the functions $f_{\boldsymbol{k}}(\boldsymbol{x})$ of (3.63):

$$f_{\boldsymbol{k}}(\boldsymbol{x}) = \frac{1}{L^{3/2}} \exp(\mathrm{i}\boldsymbol{x} \cdot \boldsymbol{k}/\hbar). \tag{3.63}$$

From (3.65) we obtain the Fourier expansion of the function $\boldsymbol{x} \mapsto u(x^0, \boldsymbol{x})$:

$$u(x) = \sum_{\boldsymbol{k} \in \mathcal{K}} u_{\boldsymbol{k}}(x^0) f_{\boldsymbol{k}}(\boldsymbol{x}) = \frac{1}{L^{3/2}} \sum_{\boldsymbol{k} \in \mathcal{K}} u_{\boldsymbol{k}}(x^0) \exp(\mathrm{i}\boldsymbol{x} \cdot \boldsymbol{k}/\hbar). \tag{6.6}$$

The function $f_{\boldsymbol{k}}$ is an eigenvector of the Laplacian, of eigenvalue $-\boldsymbol{k}^2/\hbar^2$, i.e.

$$\sum_{1 \le \nu \le 3} \frac{\partial^2 f_{\boldsymbol{k}}}{(\partial x^\nu)^2} = -\frac{\boldsymbol{k}^2}{\hbar^2} f_{\boldsymbol{k}}. \tag{6.7}$$

Consequently we expect that, denoting by $'$ differentiation with respect to x^0,

$$m^2 c^2 u(x) + \hbar^2 \partial^\nu \partial_\nu u(x) = \sum_{\boldsymbol{k} \in \mathcal{K}} \big(\hbar^2 u''_{\boldsymbol{k}}(x^0) + (m^2 c^2 + \boldsymbol{k}^2) u_{\boldsymbol{k}}(x^0)\big) f_{\boldsymbol{k}}(\boldsymbol{x}), \tag{6.8}$$

so that u will satisfy the Klein-Gordon equation if and only if for each $\boldsymbol{k}$ we have

$$\hbar^2 u''_{\boldsymbol{k}}(x^0) + (m^2 c^2 + \boldsymbol{k}^2) u_{\boldsymbol{k}}(x^0) = 0. \tag{6.9}$$

Defining $v_{\boldsymbol{k}}(t) := u_{\boldsymbol{k}}(ct)$ this amounts to saying that for each $\boldsymbol{k}$ we have

$$v''_{\boldsymbol{k}} + \tilde{\omega}_{\boldsymbol{k}}^2 v_{\boldsymbol{k}} = 0, \tag{6.10}$$

where we have defined

$$\tilde{\omega}_{\boldsymbol{k}} := \frac{c}{\hbar} \sqrt{c^2 m^2 + \boldsymbol{k}^2}. \tag{6.11}$$

The purpose of the tilde is to distinguish this angular frequency from the momentum $\omega_{\boldsymbol{k}} = \sqrt{c^2 m^2 + \boldsymbol{k}^2}$. The reader should keep in mind the equality $\hbar \tilde{\omega}_{\boldsymbol{k}} = c\omega_{\boldsymbol{k}}$.

At this stage it is obvious that we are just following the same steps as in Section 3.10. (The reader will pay attention to the fact that the number ω_k of Section 3.10 corresponds to $\tilde{\omega}_{\boldsymbol{k}}$ and not to $\omega_{\boldsymbol{k}}$.) To get the correct result we have to use real-valued functions. For this we fix a set $\mathcal{K}' \subset \mathcal{K} \setminus \{0\}$ such that for $\boldsymbol{k} \in \mathcal{K} \setminus \{0\}$ exactly one of the points $\boldsymbol{k}, -\boldsymbol{k}$ belongs to $\mathcal{K}'$. We define $g_0(\boldsymbol{x}) = L^{-3/2}$ and

$$g_{\boldsymbol{k}}(\boldsymbol{x}) = \begin{cases} \sqrt{2}/L^{3/2} \cos(\boldsymbol{k} \cdot \boldsymbol{x}/\hbar) & \text{if } \boldsymbol{k} \in \mathcal{K}', \\ \sqrt{2}/L^{3/2} \sin(\boldsymbol{k} \cdot \boldsymbol{x}/\hbar) & \text{if } -\boldsymbol{k} \in \mathcal{K}'. \end{cases} \tag{6.12}$$

11 Here we consider this equation as describing the time-evolution of a function on $\mathbb{R}^3$. If "position" and "velocity" are known at a given time, the entire trajectory is determined.

The fruitful expansion is then

$$u(x^0,\boldsymbol{x}) = \sum_{\boldsymbol{k}\in\mathcal{K}'} w_{\boldsymbol{k}}(x^0) g_{\boldsymbol{k}}(\boldsymbol{x}). \tag{6.13}$$

Since $g_{\boldsymbol{k}}$ is also an eigenvector of the Laplacian with the same eigenvalue as $f_{\boldsymbol{k}}$, the functions $v_{\boldsymbol{k}}(t) := w_{\boldsymbol{k}}(ct)$ are now real-valued and satisfy the equation (6.10). Our naive quantization method replaces the real number $v_{\boldsymbol{k}}(t)$ which expresses the position of a harmonic oscillator at time t by the operator representing the position at time t of a harmonic oscillator with the same frequency.

Let us recall once again the formula (2.107) which gives *at time t* the operator $X(t)$ quantizing the position of a harmonic oscillator of angular frequency ω and "mass" m:

$$X(t) = \sqrt{\frac{\hbar}{2\omega m}}(a^\dagger \exp(\mathrm{i}t\omega) + a\exp(-\mathrm{i}t\omega)), \tag{6.14}$$

where $[a,a^\dagger] = 1$. It is apparent here that we must take $\omega = \tilde{\omega}_{\boldsymbol{k}}$ and that the choice of m is simply a normalization condition (provided this choice is independent of $\boldsymbol{k}$). Also, let us stress the quantities of (6.14) have the dimension $[l]$ because this operator represents a position, and for this m must have the dimension of a mass. We do not know the dimension of the Klein-Gordon operator yet, but it may well be different (and it will be different). Thus for our choice of m we are not limited to quantities that have the dimension of a mass.

Exercise 6.2.1 Would we get a more general result by replacing t in the right-hand side of (6.14) by $t - t_0$?

We pluck out of thin air the choice

$$m = \hbar^2,$$

for which we will find a better motivation when we use a similar but more elaborate method later. Let us also observe that since $x^0 = ct$ we have

$$\exp(\mathrm{i}t\tilde{\omega}_{\boldsymbol{k}}) = \exp(\mathrm{i}x^0\tilde{\omega}_{\boldsymbol{k}}/c) = \exp(\mathrm{i}x^0\omega_{\boldsymbol{k}}/\hbar). \tag{6.15}$$

Thus quantization will replace the real numbers $v_{\boldsymbol{k}}(t) = w_{\boldsymbol{k}}(x^0)$ by the operators

$$\frac{1}{\sqrt{2\hbar\tilde{\omega}_{\boldsymbol{k}}}}(c(x^0,\boldsymbol{k}) + c^\dagger(x^0,\boldsymbol{k})) \tag{6.16}$$

where $c(x^0,\boldsymbol{k}) = \exp(-\mathrm{i}x^0\omega_{\boldsymbol{k}}/\hbar)c(\boldsymbol{k})$ (etc.) and the operators $c(\boldsymbol{k}), c^\dagger(\boldsymbol{k})$ satisfy

$$[c(\boldsymbol{k}), c^\dagger(\boldsymbol{k})] = 1, \tag{6.17}$$

with all the other commutators zero.

We have succeeded in quantizing the Klein-Gordon field, but we would like to express the result using the functions $f_{\boldsymbol{k}}$ rather than $g_{\boldsymbol{k}}$, and we turn to this now. This can be done in a completely elementary manner exactly as in Section 3.10, but instead we reproduce a brilliant abstract argument given by Folland [31]. The functions $g_{\boldsymbol{k}}$ can be written as $g_{\boldsymbol{k}} = \sum_{\boldsymbol{\ell}} \alpha_{\boldsymbol{k},\boldsymbol{\ell}} f_{\boldsymbol{\ell}}$ for a certain (infinite) orthogonal matrix $\alpha_{\boldsymbol{k},\boldsymbol{\ell}}$. Since the function $g_{\boldsymbol{k}}$ is real-valued, we also have $g_{\boldsymbol{k}} = \sum_{\boldsymbol{\ell}} \alpha^*_{\boldsymbol{k},\boldsymbol{\ell}} f^*_{\boldsymbol{\ell}}$. Consider now a Hilbert space $\mathcal{H}$ and for $e \in \mathcal{H}$ the corresponding operators $A(e)$ and $A^\dagger(e)$ on the boson Fock space of $\mathcal{H}$. Consider an orthogonal sequence $(e_{\boldsymbol{k}})$ of $\mathcal{H}$. We can make the choice

$$c(\boldsymbol{k}) = A(e_{\boldsymbol{k}})\ ;\ c^\dagger(\boldsymbol{k}) = A^\dagger(e_{\boldsymbol{k}}),$$

since these operators satisfy appropriate commutation relations. Replacing the numbers $w_{\boldsymbol{k}}(x^0)$ by the operators (6.16) leads us to quantize the function $u(x)$ by $\varphi(x)$ given by

$$\sum_{\boldsymbol{k}\in\mathcal{K}} \frac{1}{\sqrt{2c\omega_{\boldsymbol{k}}}}\Big(\exp(-\mathrm{i}x^0\omega_{\boldsymbol{k}}/\hbar)\sum_{\ell\in\mathcal{K}}\alpha_{\boldsymbol{k},\ell} f_\ell A(e_{\boldsymbol{k}}) + \exp(\mathrm{i}x^0\omega_{\boldsymbol{k}}/\hbar)\sum_{\ell\in\mathcal{K}}\alpha^*_{\boldsymbol{k},\ell} f^*_\ell A^\dagger(e_{\boldsymbol{k}})\Big).$$

Now the coefficients $\alpha_{\boldsymbol{k},\ell}$ are not zero only when $\omega_{\boldsymbol{k}} = \omega_\ell$, so that

$$\exp(-\mathrm{i}x^0\omega_{\boldsymbol{k}}/\hbar)\alpha_{\boldsymbol{k},\ell} = \exp(-\mathrm{i}x^0\omega_\ell/\hbar)\alpha_{\boldsymbol{k},\ell}$$

and the previous quantity is

$$\sum_{\ell\in\mathcal{K}} \frac{1}{\sqrt{2c\omega_\ell}}\Big(\exp(-\mathrm{i}x^0\omega_\ell/\hbar) f_\ell A(e'_\ell) + \exp(\mathrm{i}x^0\omega_\ell/\hbar) f^*_\ell A^\dagger(e'_\ell)\Big),$$

where $e'_\ell = \sum_{\boldsymbol{k}} \alpha^*_{\boldsymbol{k},\ell} e_{\boldsymbol{k}}$ is a new orthogonal sequence in $\mathcal{H}$ (since the adjoint of the orthogonal matrix $(\alpha_{\boldsymbol{k},\ell})$ is orthogonal). In other words, the quantization $\varphi(x)$ of $u(x)$ is given by the formula

$$\frac{1}{L^{3/2}}\sum_{\boldsymbol{k}\in\mathcal{K}} \frac{1}{\sqrt{2c\omega_{\boldsymbol{k}}}}\big(\exp(\mathrm{i}(-x^0\omega_{\boldsymbol{k}} + \boldsymbol{x}\cdot\boldsymbol{k})/\hbar)a(\boldsymbol{k}) + \exp(\mathrm{i}(x^0\omega_{\boldsymbol{k}} - \boldsymbol{x}\cdot\boldsymbol{k})/\hbar)a^\dagger(\boldsymbol{k})\big), \tag{6.18}$$

where the operators $a(\boldsymbol{k})$ satisfy

$$[a(\boldsymbol{k}), a^\dagger(\boldsymbol{k})] = 1, \tag{6.19}$$

with all the other commutators equal to zero. Taking into account the normalization (6.19), this is exactly the formula (5.33).

6.3 Road Map

There was of course quite a lot of guesswork in the derivation of the formula (6.18). In particular the method provides no support for the fact that we have chosen $m = \hbar^2$ rather than, say, m depending on $\omega_{\boldsymbol{k}}$. Our goal now is to build a more solid framework leading to the same result with less guesswork. For this we will have to learn some elements of Classical Mechanics.

Probably the single most important idea of Classical Mechanics is that the future behavior of a mechanical system is entirely determined by the positions and the velocities of its different parts at a given time (the so-called *initial conditions*). This determination occurs through a system of differential equations, known as the *equations of motion*. How, then, does one produce these equations of motion and how does one study them? Classical Mechanics has built a remarkable body of work around this question.

In the next two sections we will briefly introduce Lagrangian and Hamiltonian Mechanics. Both use one single function (called respectively the Lagrangian and the Hamiltonian) to generate the equations of motion. Hamiltonian mechanics is a powerful reformulation of Lagrangian Mechanics, and has largely superseded it in Classical Mechanics. For our purposes however the situation is far less clear. Lagrangian Mechanics smoothly blends with Special Relativity to produce Lorentz invariant formulas. This is much less the case for Hamiltonian Mechanics where time does play a very special role.

Our quantization procedures will not use Lagrangians. We still discuss Lagrangian Mechanics for two reasons: first, it is closely connected to Hamiltonian Mechanics. Second, and even more importantly, the standard route (which we will not travel) followed in many Quantum Field Theory textbooks is to systematically investigate simple Lagrangians to discover many quantum fields.

The structure of Hamiltonian Mechanics is almost unbelievably simple:

- The state of the system is described by a single point in a space called the phase space.[12]
- The time-evolution of the system is governed by the Hamiltonian, which is a function on the phase space, through Hamilton's equation of motion, which are very simple first-order differential equations.

For a system with n degrees of freedom in the simplest cases the state space is $\mathbb{R}^n \times \mathbb{R}^n$, and the Hamiltonian function is thus a real-valued function on this space. In Section 6.6 we describe a simple procedure, called canonical quantization, to construct a Quantum Hamiltonian from this classical Hamiltonian function. This procedure is by no means universally valid, but it dates from the origins of Quantum Mechanics and has been very successful in simple cases. In Section 6.7 we make a big jump: We pretend we can use this method to quantize the solutions of the Klein-Gordon equation. In the end, this procedure is really *the same* as the procedure in Section 6.2, but instead of being based on our amateurish guesses (which could easily have gone wrong had we not known the correct answer beforehand), it is now backed by a systematic procedure with lots of successes to its credit and the formidable authority of the great physicists of the past.[13]

6.4 Lagrangian Mechanics

In discussing Classical Mechanics we will adopt the traditional notation to denote time derivatives by a dot, so that

$$\dot{x}(t) := \frac{\mathrm{d}x(t)}{\mathrm{d}t} \ ; \ \ddot{x}(t) := \frac{\mathrm{d}^2x(t)}{\mathrm{d}t^2}.$$

First, consider a system consisting of a single time-dependent point $x(t)$ in $\mathbb{R}^n$. In this case the *Lagrangian* is a real-valued function $L = L(x,v)$ on $\mathbb{R}^n \times \mathbb{R}^n$ (which has the dimension of an energy). The reason for the choice of notation is that x represents a position and v represents a velocity. Lagrangian Mechanics, asserts that the motion $x(t)$ between times t_0 and t_1 is stationary[14] for the *action*

[12] In a sense, phase space in Hamiltonian mechanics corresponds to the state space of Quantum Mechanics.

[13] However great, the physicists of the past did err at times. It is quite remarkable that some of these mistakes find their ways in elementary physics textbooks a full two generations after they took place, greatly contributing to the difficulty of self-education.

[14] The meaning of this word is explained below.

$$S := \int_{t_0}^{t_1} \mathrm{d}t\, L(x(t), \dot{x}(t)). \tag{6.20}$$

This is known as *Hamilton's principle of least action* because the actual motion often minimizes S. Since L has the dimension of an energy, S has the dimension of energy times time i.e. of an action.[15]

Saying that S is stationary means (leaving technicalities aside here and later) that if we consider a family of motions $x(t,\varepsilon)$ for $t_0 \le t \le t_1$ and $\varepsilon \in \mathbb{R}$, with

$$\forall t_0 \le t \le t_1 \;,\; x(t,0) = x(t) \;;\; \forall \varepsilon \;,\; x(t_0,\varepsilon) = x(t_0) \;,\; x(t_1,\varepsilon) = x(t_1), \tag{6.21}$$

then the derivative at $\varepsilon = 0$ of the function

$$S(\varepsilon) = \int_{t_0}^{t_1} \mathrm{d}t\, L\Big(x(t,\varepsilon), \frac{\mathrm{d}x(t,\varepsilon)}{\mathrm{d}t}\Big) \tag{6.22}$$

is zero.

Assuming $x(t,\varepsilon)$ to be nice, let $w(t)$ be the derivative of $\varepsilon \mapsto x(t,\varepsilon)$ at $\varepsilon = 0$. One gets

$$S'(0) = \int_{t_0}^{t_1} \mathrm{d}t \sum_{k \le n} \Big(w_k(t) \frac{\partial L}{\partial x_k}(x(t), \dot{x}(t)) + \dot{w}_k(t) \frac{\partial L}{\partial v_k}(x(t), \dot{x}(t)) \Big),$$

where we have used that the derivative at $\varepsilon = 0$ of $\dot{x}_k(t,\varepsilon)$ is $\dot{w}_k(t)$. Observing that $w(t_0) = w(t_1) = 0$ from the second part of (6.21) one may integrate by parts to get

$$S'(0) = \int_{t_0}^{t_1} \mathrm{d}t \sum_{k \le n} w_k(t) \Big(\frac{\partial L}{\partial x_k}(x(t), \dot{x}(t)) - \frac{\mathrm{d}}{\mathrm{d}t} \frac{\partial L}{\partial v_k}(x(t), \dot{x}(t)) \Big).$$

Saying that this is zero whatever the choice of the w_k with $w_k(t_0) = w_k(t_1) = 0$ yields the *Euler-Lagrange equations*

$$\frac{\partial L}{\partial x_k}(x(t), \dot{x}(t)) = \frac{\mathrm{d}}{\mathrm{d}t} \frac{\partial L}{\partial v_k}(x(t), \dot{x}(t)), \tag{6.23}$$

which are the equations of motion we are looking for. Writing $\nabla_x L$ for the gradient of the function $x \mapsto L(x,v)$ and $\nabla_v L$ for the gradient of the function $v \mapsto L(x,v)$, the Lagrange equations can be written in condensed form

$$\nabla_x L = \frac{\mathrm{d}}{\mathrm{d}t} \nabla_v L. \tag{6.24}$$

In the previous argument to obtain the Lagrange equations it suffices to consider variations of the type $x(t,\varepsilon) = x(t) + \varepsilon w(t)$, where $w(t_0) = w(t_1) = 0$. The way this argument is written in the calculus of variations and in physics textbooks is to consider an "infinitesimal variation" δx of the motion (so here $\delta x = \varepsilon w(t)$ for infinitesimally small ε), and, observing that the variation $\delta(\dot{x})$ of $\dot{x}$ is the time-derivative $\dot{\delta x}$ of the variation δx of x, to write the first-order corresponding change in the action as

[15] This result motivates the definition of dimension of an action on page 10.

$$\delta S = \int_{t_0}^{t_1} \mathrm{d}t \sum_{k \le n} \left(\delta x_k \frac{\partial L}{\partial x_k} + \delta \dot{x}_k \frac{\partial L}{\partial v_k} \right)$$

and then to integrate by parts as above. The argument is informal but sound.[16]

Let us observe that if we multiply the Lagrangian by a constant factor this does not change the Euler-Lagrange equations of motion. However it will turn out from other considerations that this unspecified "multiplicative constant" in the Lagrangian is in fact unique if one wants to get the correct expression of the energy of the system. This will become apparent in the next section.

The rule of the game is that one has to pick the right Lagrangian function to get the correct equations of motion. There is no magic in this process, although one might wonder why the method is so successful.[17] To get a feeling for what happens, consider the Lagrangian

$$L(x,v) = \frac{m}{2} v \cdot v - V(x), \tag{6.25}$$

where $\cdot$ denotes the dot product in $\mathbb{R}^n$ and V is a "potential function". Then $\nabla_v L = mv$, so that (6.24) becomes

$$m\ddot{x} = -\nabla_x V, \tag{6.26}$$

which is Newton's law of motion for a particle of mass m submitted to the force $-\nabla_x V$ created by the potential V.

Exercise 6.4.1 Consider the case $n = 1$, $V(x) = m\omega^2 x^2/2$. This is the classical harmonic oscillator. Show that the solution of the Lagrange equations giving the actual motion is not always a minimum of the action. Hint: Try $t_0 = 0$, $t_1 = 2\pi/\omega$, $x(t_0) = x(t_1) = 1$.

The previous considerations are adapted to the study of the motion of a time-dependent point $x(t) = (x_k(t))_{k \le n} \in \mathbb{R}^n$, representing a system with n degrees of freedom, one for each component. We now make a mind-boggling shift, as our goal is to study a *classical field*,[18] that is a time-dependent function $u(t, \cdot)$ on $\mathbb{R}^3$.

The value $u(t, \boldsymbol{x})$ of the field at each point $\boldsymbol{x}$ should be thought of as an independent degree of freedom. This is a "continuous version" of the previous case in a rather natural way, but a headache is created by the clash of notations. The variable, which was called x before is now called u, and the "index", which refers to the corresponding degree of freedom of the variable, and which was called k before, is now called $\boldsymbol{x}$.

[16] If you do not understand now the point of this comment, you will when meeting unsound arguments!

[17] The process of picking the right Lagrangian is helped by Noether's theorem, a fundamental link between the Lagrangian and quantities conserved during the evolution of the system. Knowing what quantities are conserved often allows one to guess the right Lagrangian. The approach of mechanics through Lagrangians is put forward in particular in the classical treatise of Landau and Lifshitz.

[18] The study of classical fields is an important part of Classical Mechanics. The most important classical field is the electromagnetic field. The discovery of the "concept of fields" is a triumph of nineteenth century physics, whose crowning achievement is the work of J. Maxwell.

We want to find interesting equations of motion for the field using an appropriate Lagrangian. If we mimic the previous situation, we expect this Lagrangian to depend on two functions $u\colon \boldsymbol{x} \mapsto u(\boldsymbol{x})$ and $v\colon \boldsymbol{x} \mapsto v(\boldsymbol{x})$. A fruitful class of Lagrangians will be obtained using a *Lagrangian density* $\mathcal{L}$, which here is a function on $\mathbb{R} \times \mathbb{R} \times \mathbb{R}^3$, by the formula

$$L(u,v) = \int \mathrm{d}^3\boldsymbol{x}\, \mathcal{L}(u, v, \partial_{\boldsymbol{x}} u), \tag{6.27}$$

where of course $\partial_{\boldsymbol{x}} u$ denotes the gradient of the function u, and where on the right-hand side we write u for $u(\boldsymbol{x})$, v for $v(\boldsymbol{x})$, etc.[19] There is actually no mathematical reason why the Lagrangian density should not also depend on higher derivatives of the functions u and v, but the present situation covers all the cases of interest to us.[20] Assume now that $u(t,\boldsymbol{x})$ is a time-dependent function of $\boldsymbol{x}$, i.e. a function on $\mathbb{R}^4$. The action between times t_0 and t_1 is then given by

$$S = \int_{t_0}^{t_1} \mathrm{d}t\, L(u, \partial_t u) = \int_{t_0}^{t_1} \mathrm{d}t \int \mathrm{d}^3\boldsymbol{x}\, \mathcal{L}(u, \partial_t u, \partial_{\boldsymbol{x}} u),$$

where in the expression $L(u, \partial_t u)$, u and $\partial_t u$ are seen as time-dependent functions of $\boldsymbol{x}$, and where in the last integral u stands for $u(t,\boldsymbol{x})$, etc.

In the case $t_0 = -\infty, t_1 = \infty$ this becomes

$$S = \iint \mathrm{d}t\,\mathrm{d}^3\boldsymbol{x}\, \mathcal{L}(u, \partial_t u, \partial_{\boldsymbol{x}} u).$$

There is a pleasant feature here: Time plays the same role as spatial coordinates, and this setting will naturally lead to Lorentz invariant equations.[21] Consider a function u on Minkowski space $\mathbb{R}^{1,3}$. Recalling the notation ∂_ν of (6.1) we are interested in functions u which are stationary for the quantity

$$S = \frac{1}{c} \int \mathrm{d}^4x\, \mathcal{L}(u, \partial_\nu u), \tag{6.28}$$

where $\mathcal{L}(u, \partial_\nu u)$ is shorthand for $\mathcal{L}(u, \partial_0 u, \partial_1 u, \partial_2 u, \partial_3 u)$. The purpose of the factor $1/c$ is to keep the correct dimension since $\mathrm{d}x^0 = c\mathrm{d}t$. Here $\mathcal{L}$ is a function $\mathcal{L}(u, v_0, v_1, v_2, v_3)$ on $\mathbb{R}^5$. The traditional notation[22] is to denote by $\partial\mathcal{L}/\partial(\partial_\nu u)$ the partial derivative of $\mathcal{L}$ with respect to v_ν:

$$\frac{\partial \mathcal{L}}{\partial(\partial_\nu u)} := \frac{\partial \mathcal{L}}{\partial v_\nu}.$$

19 You may wonder why suddenly we allow the Lagrangian density to depend on the gradient of u. This is completely natural in view of the fact that the interesting equations of motion for a classical field are often differential equations and that we want the Lagrangian to lead us to these.

20 Higher derivatives actually often lead to problematic physical theories. So do "non-local" terms such as products of values of u at different points.

21 Despite the fact that Lagrangian mechanics was invented much before Special Relativity, the two theories blend very well.

22 I cannot change this questionable notation which is almost universally used. However weird it might feel at the beginning, it is just fine once one gets used to it.

This amounts to thinking of u and $\partial_\nu u$ as independent variables, which is certainly not the case.[23] For an infinitesimal variation δu of u, to first-order the variation of S is therefore given by

$$c\delta S = \int \mathrm{d}^4x\Big(\delta u\frac{\partial \mathcal{L}}{\partial u} + \partial_\nu(\delta u)\frac{\partial \mathcal{L}}{\partial(\partial_\nu u)}\Big), \tag{6.29}$$

where as usual there is a summation over the repeated indices ν, and where we have used that $\partial_\nu(\delta u) = \delta(\partial_\nu u)$. Integrating the last term by parts (assuming that everything vanishes fast enough at infinity) and using that the resulting integral must vanish for any choice of δu we obtain the following form of the Euler-Lagrange equations:

$$\frac{\partial \mathcal{L}}{\partial u} = \partial_\nu \frac{\partial \mathcal{L}}{\partial(\partial_\nu u)}. \tag{6.30}$$

In order to obtain a Lorentz invariant theory one should look for Lagrangian densities which are "Lorentz invariant" in some sense. Informally, the condition is that after making a Lorentz change of variables the form of the Lagrangian remains the same.[24] A more formal description is to consider the operator $\mathbf{L}$ which associates to a function u on $\mathbb{R}^{1,3}$ the function $\mathbf{L}(u)$ on $\mathbb{R}^{1,3}$ given by $\mathbf{L}(u)(x) = \mathcal{L}((u(x), \partial_\nu u(x))$ and to request that this operator $\mathbf{L}$ commutes with the operators $V(c,C)$ as in (6.2). Among the simplest Lagrangian densities with this property are u^2 and $\eta^{\mu\nu}\partial_\mu u \partial_\nu u$.

Exercise 6.4.2 Recall that $\mathcal{L}$ is a function on $\mathbb{R}^5$, which we now view as a function on $\mathbb{R} \times \mathbb{R}^{1,3}$. Then if the function $\mathcal{L}$ is such that for each Lorentz transformation L, every $u \in \mathbb{R}$ and every $w \in \mathbb{R}^{1,3}$ we have $\mathcal{L}(u,w) = \mathcal{L}(u, Lw)$ the corresponding Lagrangian density is Lorentz invariant in the previous sense. Hint: Almost obvious using the chain rule.

Let us then combine the densities u^2 and $\eta^{\mu\nu}\partial_\mu u \partial_\nu u$ to form the following new Lagrangian density

$$\mathcal{L}(u, \partial_\nu u) = \frac{1}{2}(\hbar^2 c^2 \eta^{\mu,\nu}\partial_\mu u \partial_\nu u - m^2 c^4 u^2), \tag{6.31}$$

where m is a positive mass. The coefficients here are constrained by dimensional considerations: The two terms of the equation must have the same dimension.

Exercise 6.4.3 Check that this is the case.

The Lagrangian density (6.31) is given to you by an ukase, but there are not that many simple Lorentz invariant Lagrangian densities that one can form. A standard procedure in Quantum Field Theory is to perform a systematic exploration of these Lagrangian densities to discover various fields. It is used by many authors but we shall follow another route.[25]

[23] One can also think of this as a shorthand for numbering the variables in $\mathcal{L} = \mathcal{L}(slot1, slot2, \ldots)$ so that $\partial\mathcal{L}/\partial(\partial x^1)$ means $\partial\mathcal{L}/\partial(slot2)$.

[24] So that two physicists using different coordinate systems can agree on the laws of physics.

[25] With some experience one can understand a lot about a quantum field theory just by looking at the corresponding classical Lagrangian. This is brilliantly used in the last chapters of Folland's book [31].

Let us have a closer look at the density (6.31):

$$\mathcal{L}(u,\partial_\nu u) = \frac{1}{2}\hbar^2(\partial_t u)^2 - \frac{1}{2}\Big(\hbar^2 c^2 \sum_{1\le i\le 3}(\partial_i u)^2 + m^2c^4u^2\Big). \tag{6.32}$$

This is of the same type as (6.25). The term in $(\partial_t u)^2$ corresponds to (a density of) kinetic energy and the rest to (a density of) potential energy. The potential energy increases as the field gets away from zero and also as it exhibits space variations.

For the Lagrangian density (6.31) we have

$$\frac{\partial\mathcal{L}}{\partial u} = -m^2c^4u\ ;\ \frac{\partial\mathcal{L}}{\partial(\partial_\nu u)} = \hbar^2c^2\eta^{\mu\nu}\partial_\mu u = \hbar^2c^2\partial^\nu u,$$

and the Euler-Lagrange equation (6.30) is precisely the Klein-Gordon equation $m^2c^2u + \hbar^2\partial^\mu\partial_\mu u = 0$.

Fixing the Lagrangian density fixes the dimension of u. The proper dimension of the Lagrangian density (6.31) is such that the quantity (6.28) should have the dimension of an action. That is, after the density is integrated over space-time, (an operation which adds a factor $[l^3t]$ to the dimension), it has the dimension of an action, i.e. the density should have dimension $[ml^{-1}t^{-2}]$. This determines the dimension of u: the dimension of $m^2c^4u^2$ has to be $[ml^{-1}t^{-2}]$. One finds the rather curious result that u has dimension $[l^{-5/2}m^{-1/2}t]$, which is the dimension of the quantum field which we found in (5.35). *It is to match these dimensions that we inserted the factor* $1/\sqrt{c}$ *in (5.1).*

Textbooks treating the present topics usually choose a system of units in which $\hbar = 1$ and $c = 1$. This simplifies formulas by removing the need to write such factors, and considerably simplifies dimensional analysis since length and time can then be expressed in units of mass, and only the way the formulas depend on mass then matters. This approach is better suited to a trained physicist who has already mastered dimensional analysis, which may not be the case of most readers here. In retrospect however one can understand another reason why textbooks proceed this way: The results obtained by the full-dimensional analysis we perform here are not very enlightening.

6.5 From Lagrangian Mechanics to Hamiltonian Mechanics

Hamiltonian Mechanics is a profound reformulation of Lagrangian Mechanics.[26] It will lead us to natural methods of quantizing classical systems through the identification of a special function, the Hamiltonian function, which governs the time-evolution of the system and represents the energy of this system. The state of the mechanical system is described by a single point in "phase space", and the evolution of the system i.e. the motion of this point is described by a first-order system of differential equations[27] determined by the Hamiltonian function. Achieving this simplicity will however require some abstraction.

[26] Furthermore, it is as exquisitely beautiful as it is powerful.
[27] Hamilton's equations of motion, to be described below.

Let us go back to a system with n degrees of freedom described by a Lagrangian $L(x, v)$. Let us introduce the following notation

$$p_k = \frac{\partial L}{\partial v_k}, \tag{6.33}$$

which is called the *conjugate momentum* to the variable x_k. In the fundamental case (6.25), the conjugate momentum is just the usual momentum $p = mv$. The following gives a hint of what is coming.

Lemma 6.5.1 *The quantity*

$$\sum_{k\le n} \dot{x}_k(t)p_k(x(t),\dot{x}(t)) - L(x(t),\dot{x}(t)) \tag{6.34}$$

is a constant of motion.

The proof is immediate: One computes the time derivative of this quantity,

$$\sum_{k\le n}(\ddot{x}_k p_k + \dot{x}_k \dot{p}_k) - \sum_{k\le n}\Big(\dot{x}_k\frac{\partial L}{\partial x_k} + \ddot{x}_k\frac{\partial L}{\partial v_k}\Big),$$

and writing the Euler-Lagrange equation as $\dot{p}_k = \partial L/\partial x_k$ one sees that it is zero. The secret is that the quantity (6.34) represents the total energy of the system (and this is how one determines the unknown multiplicative constant in the Lagrangian, since the energy of the system is known "on physical grounds").

Suppose now that we can invert the equations (6.33), that is, setting $p = (p_k)_{k\le n}$ we can compute v_k as a function of p and x,

$$v_k = v_k(x, p). \tag{6.35}$$

This is not possible in general but it can be done in many interesting situations. For example, in the case (6.25) we have $p_k = mv_k$, so that $v_k = p_k/m$. In the case (6.35) we can define for $x, p \in \mathbb{R}^n$ the quantity $v(x, p) = (v_k(x, p))_{k\le n}$ and the *Hamiltonian function*

$$H(x, p) = \sum_{k\le n} v_k(x, p)p_k - L(x, v(x, p)). \tag{6.36}$$

The important point here is that *we consider now x and p as new variables.*[28] The space of the pairs (x, p) is called *phase space*. Here it is simply a fancy name for $\mathbb{R}^n \times \mathbb{R}^n$, but the deep underlying idea is that each point in the phase space specifies a set of initial conditions of the system, here the position and the momentum.

For example in the case (6.25)

$$H(x, p) = \frac{1}{2m}p \cdot p + V(x), \tag{6.37}$$

describes the energy of the system, the sum of the kinetic and potential energy. It is a general fact that the Hamiltonian function of the system describes its energy. We will often say

28 In particular we completely forget about the relation (6.33), and x and p are now seen as "independent".

"Hamiltonian" rather than "Hamiltonian function". It should always be clear from the context whether we mean the Hamiltonian function of Classical Mechanics or the Hamiltonian operator of Quantum Mechanics.

Lemma 6.5.2 *We have*

$$\frac{\partial H}{\partial x_k}(x,p) = -\frac{\partial L}{\partial x_k}(x, v(x,p)) \; ; \; \frac{\partial H}{\partial p_k}(x,p) = v_k(x,p). \tag{6.38}$$

Proof Starting from (6.36) one simply computes these partial derivatives according to the chain rule, and one observes that the other terms cancel thanks to (6.33), i.e. in shorthand notation

$$\frac{\partial H}{\partial x_k} = \sum_{\ell \le n} \frac{\partial v_\ell}{\partial x_k} p_\ell - \sum_{\ell \le n} \frac{\partial v_\ell}{\partial x_k}\frac{\partial L}{\partial v_\ell} - \frac{\partial L}{\partial x_k}$$

and

$$\frac{\partial H}{\partial p_k} = v_k + \sum_{\ell \le n} \frac{\partial v_\ell}{\partial p_k} p_\ell - \sum_{\ell \le n} \frac{\partial v_\ell}{\partial p_k}\frac{\partial L}{\partial v_\ell}.$$

□

The motion of a point $x(t)$ is described in (x,v) coordinates by $(x(t),\dot{x}(t)) = (x(t),v(t))$ and in phase space by $(x(t),p(t))$ where

$$p_k(t) = \frac{\partial L}{\partial v_k}(x(t),\dot{x}(t)).$$

The Euler-Lagrange equation (6.23) then reads

$$\dot{p}_k(t) = \frac{\partial L}{\partial x_k}(x(t),\dot{x}(t)).$$

Since $\dot{x}(t) = v(t) = v(x(t),p(t))$ by (6.35), according to the first part of (6.38) the previous equation becomes

$$\dot{p}_k(t) = -\frac{\partial H}{\partial x_k}(x(t),p(t)), \tag{6.39}$$

while the second part of (6.38) yields

$$\dot{x}_k(t) = \frac{\partial H}{\partial p_k}(x(t),p(t)). \tag{6.40}$$

Equations (6.39) and (6.40) together are called *Hamilton's equations of motion*. They are written in shorthand as

$$\dot{p}_k = -\frac{\partial H}{\partial x_k} \; ; \; \dot{x}_k = \frac{\partial H}{\partial p_k}. \tag{6.41}$$

The evolution of the system is then represented by a trajectory in phase space given by Hamilton's equations of motion (6.41): Knowing the function H, and the initial conditions at a given time we can in principle compute the position in phase space for all times. For example in the case (6.25) these equations are

$$\dot{p}_k = -\frac{\partial V}{\partial x_k} \; ; \; \dot{x}_k = \frac{p_k}{m} \tag{6.42}$$

which together give

$$m\ddot{x}_k = -\frac{\partial V}{\partial x_k}, \tag{6.43}$$

Newton's equation of motion of a point of mass m submitted to a conservative force arising from the potential V. Let us also observe that Lemma 6.5.1 is now immediate, since

$$\dot{H} = \sum_{k\le n}\Big(\dot{x}_k\frac{\partial H}{\partial x_k} + \dot{p}_k\frac{\partial H}{\partial p_k}\Big) = 0$$

from Hamilton's equation of motion.

One reason for which Hamilton's formulation of mechanics is both successful and beautiful is that there is a large class of transformations of phase space for which Hamilton's equations are invariant. These transformations are called *canonical transformations*. The beauty of this approach is that it forgets that the phase space is made up of pairs (x, p), but is thought of as a single object. That is, canonical transformations can mix momentum and position variables, and provide a very powerful tool. A few more words about Hamiltonian Mechanics can be found in Section H.5.

To be useful, the preceding construction has to be independent of arbitrary choices. The next exercise checks this at a very elementary level, by investigating the effect of a change of orthonormal[29] basis in $\mathbb{R}^n$.

Exercise 6.5.3 For a rotation R we consider a new variable $\bar{x}$ with $x = R(\bar{x})$. Then $v = R(\bar{v})$, and expressing the Lagrangian in terms of these new variables we obtain the new Lagrangian $\bar{L}(\bar{x}, \bar{v}) = L(R(\bar{x}), R(\bar{v}))$. Prove that computing the Hamiltonian $\bar{H}$ in the new coordinates yields

$$\bar{H}(\bar{x}, \bar{p}) = H(R(\bar{x}), R(\bar{p})). \tag{6.44}$$

This important fact shows that our construction of the Hamiltonian function behaves as expected with respect to change of basis, i.e. in other words that it "does not depend on the choice of orthonormal basis".

Exercise 6.5.4 Prove that the change of variable $x = R(\bar{x})$ amounts to a change of variable $(x, p) = (R(\bar{x}), R(\bar{p}))$ in phase space. Prove that this transformation is a canonical transformation (which however does not mix position and momentum variables).

Let us now investigate the "continuous case" where rather than having only n variables $(x_k)_{k\le n}$ depending on time we have a continuous family $u(t, \boldsymbol{x})$ of such variables, where $\boldsymbol{x}$ plays the role of the index, and where we use a Lagrangian density $\mathcal{L}(u, v, \partial_{\boldsymbol{x}} u)$ as in (6.27), where $\mathcal{L}$ is a function $\mathcal{L}(u, v, v^1, v^2, v^3)$. (Note that the second variable is called v rather than v^0.) Then we can mimic (6.33) and define (using again the traditional notation)

$$\pi = \frac{\partial \mathcal{L}}{\partial(\partial_t u)} := \frac{\partial \mathcal{L}}{\partial v}. \tag{6.45}$$

29 The possibility of making such a change of basis will be crucial to diagonalize a certain quadratic form.

When (6.45) allows us to compute $v = v(u,\pi,v^1,v^2,v^3)$ as a function of u,π,v^1,v^2,v^3, the continuous version of (6.36) is then to define the Hamiltonian density[30]

$$\mathcal{H}(u,\pi) = \pi v - \mathcal{L}(u,v,\partial_x u). \tag{6.46}$$

Here $u = u(\boldsymbol{x})$ and $\pi = \pi(\boldsymbol{x})$ are two functions, and, setting

$$v(\boldsymbol{x}) = v(u(\boldsymbol{x}),\pi(\boldsymbol{x}),\partial_1 u(\boldsymbol{x}),\partial_2 u(\boldsymbol{x}),\partial_3 u(\boldsymbol{x})),$$

the formula (6.46) means

$$\mathcal{H}(u,\pi)(\boldsymbol{x}) = \pi(\boldsymbol{x})v(\boldsymbol{x}) - \mathcal{L}(u(\boldsymbol{x}),v(\boldsymbol{x}),\partial_1 u(\boldsymbol{x}),\partial_2 u(\boldsymbol{x}),\partial_3 u(\boldsymbol{x})). \tag{6.47}$$

We then define the Hamiltonian as the integral $\int \mathrm{d}^3\boldsymbol{x}\,\mathcal{H}(u,\pi)(\boldsymbol{x})$, and it should represent the energy of the system "in position u and at momentum π". Thus we have to think of the phase space as the space of pairs of functions (u,π), and of the Hamiltonian as a functional of these pairs.[31]

In the relativistic notation the Lagrangian density $\mathcal{L}(u,\partial_\nu u)$ is written in terms of

$$\partial_0 u = \frac{\partial u}{\partial(ct)} = c^{-1}\frac{\partial u}{\partial t} = c^{-1}\partial_t u \tag{6.48}$$

instead of $\partial_t u$. (The reader should remember this factor c^{-1} relating the derivatives with respect to x^0 and t.) Let us examine what happens in the case of the Lagrangian density (6.31):

$$\begin{aligned}\mathcal{L}(u,\partial_\nu u) &= \frac{1}{2}(\hbar^2c^2\eta^{\mu\nu}\partial_\mu u\partial_\nu u - m^2c^4u^2)\\ &= \frac{1}{2}\hbar^2(\partial_t u)^2 - \frac{1}{2}\hbar^2c^2\sum_{1\le i\le 3}(\partial_i u)^2 - \frac{1}{2}m^2c^4u^2,\end{aligned} \tag{6.49}$$

that is the case where

$$\mathcal{L}(u,v,v^1,v^2,v^3) = \frac{1}{2}\hbar^2v^2 - \frac{1}{2}\hbar^2c^2\sum_{1\le i\le 3}(v^i)^2 - \frac{1}{2}m^2c^4u^2.$$

Then $\pi = \partial\mathcal{L}/\partial v = \hbar^2 v$ so that $v = \pi/\hbar^2$ and (6.47) yields

$$\mathcal{H}(u,\pi) = \frac{1}{2\hbar^2}\pi^2 + \frac{1}{2}\hbar^2c^2\sum_{1\le i\le 3}(\partial_i u)^2 + \frac{1}{2}m^2c^4u^2. \tag{6.50}$$

We should think of the first term as representing kinetic energy and of the second and third terms as representing potential energy. It is the desire for the natural factor 1/2 in the potential energy which motivates the factor 1/2 in the Lagrangian. Moreover it is very

[30] We observe that in (6.45) and (6.46) *time plays a special role*, so this construction does not respect Lorentz invariance. Whereas in Classical Mechanics the Hamiltonian approach is unquestionably the most powerful, in the relativistic case the Lagrangian approach is far more appropriate to bring forward Lorentz invariance.
[31] In physics one traditionally calls a *functional* a function whose arguments are themselves functions.

nice that the potential energy term is a quadratic function of u, about the simplest type you can imagine.[32]

An obvious question is how to extend Hamilton's equations of motion (6.39) and (6.40) to the present continuous setting. The question is examined in Appendix H.

6.6 Canonical Quantization and Quadratic Potentials

Before we attempt to quantize an infinite-dimensional system such as the one described by (6.50), let us examine how we might quantize a one-dimensional system. The classical Hamiltonian being a function of position x and momentum p, the most obvious way to quantize it is "to replace x and p by the corresponding position and momentum operators X and P constructed in Section 2.5". This simple procedure called "canonical quantization" in physics, goes back to the very beginnings of Quantum Mechanics. It bears repeating an important feature. The number x, which describes the position of the system, is replaced by an operator (measuring the position of that system).

The operators X and P do not commute, but rather satisfy the commutation relation $[X,P] = i\hbar 1$. Canonical quantization runs into difficulties when the Hamiltonian is a bit complicated. For example, how do we handle a term such as x^2p^2? Do we replace it by X^2P^2, or by $XPXP$? It is known that there is no universally consistent scheme to proceed. These are so-called *ordering ambiguities*.[33] Fortunately, we will have to deal only with Hamiltonians of the form (6.37) for which this difficulty does not arise. Moreover our potentials are quadratic, which makes matters even simpler.[34]

Let us then look at the one-dimensional case (6.37) for the quadratic potential $V(x) = \kappa x^2/2$. This is the Hamiltonian of a harmonic oscillator. The equation of motion (6.43) becomes $m\ddot{x} = -\kappa x$, describing an oscillatory motion of angular frequency $\omega = \sqrt{\kappa/m}$. Therefore it is a good idea to write $\kappa = m\omega^2$ and the Hamiltonian as

$$H(x,p) = \frac{1}{2m}p^2 + \frac{m\omega^2}{2}x^2. \tag{6.51}$$

Here, replacing p by P and x by X leads to the quantized Hamiltonian

$$H := \frac{1}{2m}P^2 + \frac{m\omega^2}{2}X^2, \tag{6.52}$$

which we have already met in Section 2.10. Under this Hamiltonian the position operator at time t is given by

$$X(t) := \sqrt{\frac{\hbar}{2\omega m}}(\exp(i\omega t)a^\dagger + \exp(-i\omega t)a), \tag{2.100}$$

[32] Furthermore, quadratic potentials are intrinsically connected to harmonic oscillators.

[33] In this case (one-dimensional), this obstruction can be circumvented for observables which are at most cubic in x and p. Inconsistencies are shown to arise in general via the Groenewold–van Hove's theorem.

[34] The process of canonical quantization, in general, depends on the choice of canonical coordinates used to describe the corresponding classical system. Classical theories related by canonical transformations may correspond to unitarily inequivalent quantum theories. This led to the theory of *geometric quantization* which tries to get rid of the coordinates using only the natural geometric structure on the phase space.

where a is (the essentially unique) operator such that $[a,a^\dagger] = 1$. As a consequence of (2.111), specifying this time-evolution exactly amounts to specifying the Hamiltonian (6.52).

Now, how do we proceed for an n-dimensional system, for example a point in $\mathbb{R}^n$? The straightforward generalization of the procedure of Section 2.5 is to use $L^2(\mathbb{R}^n)$ as position state space, to replace the coordinates x_k by the operators X_k given by $X_k(f)(x) = x_k f(x)$ and the momenta p_k by the operators $P_k = -i\hbar\partial/\partial x_k$. The resulting operators X_k and P_k commute between themselves, while

$$[X_k, P_\ell] = i\hbar\delta_k^\ell 1. \tag{6.53}$$

In the case of the Hamiltonian (6.37) the quantized Hamiltonian is simply

$$H_n := \frac{1}{2m}\sum_{k\le n} P_k^2 + V, \tag{6.54}$$

where V denotes the "multiplication by the function $V(x)$" operator on $L^2(\mathbb{R}^n)$.

Exercise 6.6.1 Show that this procedure does not depend on our initial choice of orthogonal basis.

The relations (6.53) are preserved by a change of orthonormal basis. A deep result, the Stone–von Neumann Theorem, which we prove in Theorem C.3.1, asserts (modulo technical assumptions) that such a system of relations is unique up to unitary equivalence, providing a considerable generalization of the result of Exercise 6.6.1. It then makes sense to call this quantization procedure "canonical" since (at least in the absence of ordering ambiguities) it is the one single sensible way to quantize the system.[35]

Let us then assume, still in dimension n, that the potential V is given by a quadratic (positive definite) form. Then we can perform a change of orthogonal basis so that it becomes of the type

$$V(x) = \frac{m}{2}\sum_{k\le n}\omega_k^2 x_k^2,$$

and the Hamiltonian becomes

$$H(x,p) = \sum_{k\le n}\frac{1}{2m}p_k^2 + \frac{m\omega_k^2}{2}x_k^2. \tag{6.55}$$

The Hamiltonian function appears now as a sum of terms, each being the Hamiltonian of a harmonic oscillator, and these oscillators do not interact.[36] This is as simple as a multidimensional system can be. The quantized Hamiltonian may be written as $\sum_{k\le n}\omega_k\hbar(a_k^\dagger a_k + (1/2)1)$ where the operators a_k satisfy the commutation relations $[a_k, a_k^\dagger] = 1$, all the other commutators being equal to zero. Let us consider the operators

[35] The linguistic problem that a mathematician may face is that physicists still call "canonical quantization" similar procedures in situations where they are far less unique. We come back to this point later.

[36] Since the energy of the system is the sum of the energy of the components.

$$X_k(t) = \sqrt{\frac{\hbar}{2\omega_k m}}(\exp(\mathrm{i}\omega_k t)a_k^\dagger + \exp(-\mathrm{i}\omega_k t)a_k).$$

In the quantization procedure, the "position $(x_k)_{k\le n}$ of the point" of classical mechanics is replaced by the "position operator $(X_k(0))_{k\le n}$". The time-evolved position operator is the sequence $(X_k(t))_{k\le n}$.

6.7 Quantization through the Hamiltonian

Our goal in this section is to quantize the Klein-Gordon field by pretending that the previous method also applies to this infinite-dimensional situation. The reader is warned though that the manipulations are pretty formal. Intuitively, after a proper change of basis, the system appears as a continuous sum of non-interacting harmonic oscillators. The proper change of basis is realized by the Fourier transform, and it brings us to a situation similar to that of Section 3.10. However in order not to have to deal with mathematically tricky continuous sums, as in Section 6.2 we replace the space $\mathbb{R}^3$ of spatial coordinates by a torus, i.e. the spatial universe by the finite box $B = [-L/2, L/2]^3$ of Section 3.9 and use periodic boundary conditions. The phase space consists of pairs of functions (u,π) on B.[37] We will parametrize functions by their Fourier transform. Recalling the functions $g_{\boldsymbol{k}}(\boldsymbol{x})$ of (6.12), we write

$$u(\boldsymbol{x}) = \sum_{\boldsymbol{k}\in\mathcal{K}'} b_{\boldsymbol{k}} g_{\boldsymbol{k}}(\boldsymbol{x})\ ;\ \pi(\boldsymbol{x}) = \sum_{\boldsymbol{k}\in\mathcal{K}'} c_{\boldsymbol{k}} g_{\boldsymbol{k}}(\boldsymbol{x}), \tag{6.56}$$

with $b_{\boldsymbol{k}}, c_{\boldsymbol{k}} \in \mathbb{R}$. Thus we describe the phase space as a set of families $(c_{\boldsymbol{k}}, b_{\boldsymbol{k}})_{\boldsymbol{k}\in\mathcal{K}'}$. Recalling the Hamiltonian density (6.50), to compute the Hamiltonian $\int_B \mathrm{d}^3\boldsymbol{x}\mathcal{H}(u,\pi)(\boldsymbol{x})$ we replace u and π by their Fourier expansions as above. Using integration by parts,

$$\int_B \mathrm{d}^3\boldsymbol{x} \sum_{1\le\nu\le 3} \partial_\nu u(\boldsymbol{x})\partial_\nu v(\boldsymbol{x}) = -\int_B \mathrm{d}^3\boldsymbol{x} u(\boldsymbol{x}) \sum_{1\le\nu\le 3} \frac{\partial^2 v}{(\partial x_\nu)^2}(\boldsymbol{x}), \tag{6.57}$$

and that moreover the function $g_{\boldsymbol{k}}$ is an eigenvector of the Laplacian, of eigenvalue $-\boldsymbol{k}^2/\hbar^2$, as expressed in (6.7). Since the functions $g_{\boldsymbol{k}}$ form an orthogonal basis, it is straightforward to formally express the Hamiltonian as

$$\int_B \mathrm{d}^3\boldsymbol{x}\mathcal{H}(u,\pi)(\boldsymbol{x}) = \frac{1}{2}\sum_{\boldsymbol{k}\in\mathcal{K}'} \left(\frac{1}{\hbar^2}c_{\boldsymbol{k}}^2 + (m^2c^4 + \boldsymbol{k}^2c^2)b_{\boldsymbol{k}}^2\right), \tag{6.58}$$

although the meaning of the infinite sum is unclear. (Problems created by such infinite sums are discussed in the next section.)

Exercise 6.7.1 Complete the previous computation.

[37] One great advantage of doing physics rather than mathematics is that I do not have to tell you what regularity property I assume for these functions!

Recalling the definition (6.11) of $\tilde{\omega}_{\boldsymbol{k}}$, the formula (6.58) decomposes the Hamiltonian as the sum

$$\sum_{\boldsymbol{k}\in\mathcal{K}'}\Big(\frac{1}{2\hbar^2}c_{\boldsymbol{k}}^2+\frac{1}{2}\hbar^2\tilde{\omega}_{\boldsymbol{k}}^2 b_{\boldsymbol{k}}^2\Big). \tag{6.59}$$

Each term is the Hamiltonian of a harmonic oscillator.[38] This is exactly an infinite-dimensional version of (6.55), the "position" of the "point" being described by the "coordinates" $b_{\boldsymbol{k}}$. To quantize the system we then replace each of the $b_{\boldsymbol{k}}$ by a suitable "position operator", and the object which quantizes "the point" is the sequence of operators measuring the coordinates of this point. In this quantization procedure, the "point", i.e. the classical field $u(\boldsymbol{x}) = \sum_{\boldsymbol{k}\in\mathcal{K}'} b_{\boldsymbol{k}} g_{\boldsymbol{k}}(\boldsymbol{x})$ is then replaced by "an operator measuring the position of the point". It appears as an operator-valued function $\psi(\boldsymbol{x})$ (although further analysis reveals that it is better to think of it as a distribution) because this is what one obtains by replacing the numbers $b_{\boldsymbol{k}}$ by operators in the previous formula. The intuitive meaning of the operator-valued function ψ is that the operator $\int \mathrm{d}^3\boldsymbol{x}\,\psi(\boldsymbol{x})g_{\boldsymbol{k}}(\boldsymbol{x})$ measures the component of the quantized field on the basis element $g_{\boldsymbol{k}}$. When bringing in time-evolution, we are then led to quantize the time-evolved operator corresponding to $b_{\boldsymbol{k}}$ at time t by the formula (6.16), after which one repeats the steps of Section 6.2, and one lets the size of the box go to infinity to guess the limiting continuous formulas for the quantum field.

Despite the considerable amount of hand-waving, our quantization procedure is now arguably related to a successful and general quantization machinery rather than being largely guesswork. Furthermore *this approach motivates the use of* $m = \hbar^2$, by comparing the terms of (6.59) with (6.51).

6.8 Ultraviolet Divergences

Let us now comment on the problem posed by the infinite sum in (6.58). Such infinite sums create all kinds of problems. Here the meaning of the infinite sum itself is unclear. Later we will meet similar problems in the form of diverging series (or diverging integrals when we do not put the universe in a box). Such problems go under the name of *ultraviolet divergences*. The reason for the name is that ultraviolet light has high energy and short wavelength. High energy corresponds to high values of $\boldsymbol{k}$. Large energies correspond to small physical distances. This important physical principle can be understood remembering Einstein's relation $E = h\nu$ between the energy E of a photon and its frequency ν since high frequencies correspond to short wavelength and short distances.

We must keep in mind that the use of $\mathcal{K}$ is just an idealization. There is no reason to believe that our continuous model of space-time applies to arbitrarily small scales. The large values of $\boldsymbol{k}$ should somehow be discounted, although we do not know exactly how this should be done. We may hope however to obtain results that do not depend on the specific method to discount these values. A first example of this phenomenon is given in Chapter 7 with the

[38] In the end, we have simply argued again, in a vastly more elaborate way, that the Klein-Gordon field is just a sum of independent harmonic oscillators.

study of the Casimir effect. We will brilliantly pass this first test, but matters are not always that simple. The brutal methods used in physics, such as *ultraviolet cutoff* reflect well the desperate situation: Only the terms for which $\boldsymbol{k}^2$ is less than a certain number are retained in the sum.

It could shed some light to briefly discuss the current conventional wisdom in physics regarding these matters. What is the structure of space-time at a very short scale is naturally an extremely difficult problem, of which we do not have the solution: This solution would probably require building a quantum theory of gravity. A simpler question is: At about what scale should the continuous model break down? A reasonable guess can be made using the "naturalness principle". This principle states that in a natural system of units, a physical quantity should always be of order 1: It is not difficult to find natural mechanisms that will produce, say a factor 2π but much harder to think of a natural mechanism that will produce a factor 10^{-12}. Thus according to this principle the scale at which the continuous model breaks down should be of order 1 in a natural system of units. A natural system of units was invented by Max Planck. In this system the speed of light, Planck's constant and Newton's gravitational constant are all set to 1. This determines units of length, time and mass, known as Planck's length, Planck's time and Plank's mass, respectively. According to this wisdom, the scale at which the continuous model of space-time should fail must be about Planck's length, about 10^{-35} m.[39] To probe space at this scale, we would need accelerators about 10^{15} times more powerful than what is available today, so experimentation is unlikely in the foreseeable future.

Another major idea is that the description of reality at a given scale need not really depend on the detail of what happens at very much smaller scales. For example continuous models of matter ignoring its atomic structure are very successful at describing the vast majority of mechanical phenomenon at our scale. Atoms are only 10^{10} times smaller than we are, while the Planck length is $(10^{10})^2$ times smaller than the nucleons. Thus there is a *lot* of room for this change of scale to hide the detailed effects of quantum gravity in particle physics,[40] and we should be able to build a theory of particle physics despite our lack of understanding of quantum gravity.

6.9 Quantization through Equal-time Commutation Relations

In the present section we attack the problem of quantizing the Klein-Gordon field from another angle. This approach too is rooted in the canonical quantization of Section 6.6, but it does not directly use Hamiltonians and is somewhat different from our two previous (nearly-identical) methods.

As already explained, when one quantizes a system representing one single particle in n-dimensional space, a successful scheme is to represent the coordinates of the particle

[39] Planck length is *unimaginably small*. It is 10^{20} times smaller than the diameter of a nucleon. If we scale the nucleon to the size of our galaxy, Planck length is scaled to just about 1 cm.

[40] This idea is at the heart of the renormalization group approach.

and its momenta along the axes by $2n$ operators $X_1, \ldots, X_n, P_1, \ldots, P_n$, which satisfy the commutation relations

$$[P_k, X_{k'}] = -i\hbar\delta_k^{k'}1, \tag{6.60}$$

and all the other commutators equal to zero.

We can then time-evolve these operators according to (2.82) and reach the more impressive looking but equivalent relations

$$[P_k(t), X_{k'}(t)] = -i\hbar\delta_k^{k'}1, \tag{6.61}$$

which are called for obvious reasons the "equal time commutation relations". If one believes that the functions u and π on $\mathbb{R}^3$ in (6.49) have the same structure, in the sense that each $u(\boldsymbol{x})$ is a coordinate and each $\pi(\boldsymbol{x})$ the corresponding momentum, it is natural in quantization to try to obtain a continuous generalization of the relations (6.61). In this section we are going to prove first that the quantum field (5.30) satisfies exactly these relations. We will then show how the procedure can be reversed, so that starting with these relations one can discover the formula (5.30) (which is a standard way to find these formulas in the literature). Thus, we go back to the setting of Chapter 5.[41]

Proposition 6.9.1 (The equal time commutation relations) *The quantum field $\varphi(x)$ given by (5.30) and the quantity $\pi(x) = c\hbar^2\partial_0\varphi(x)$ satisfy the relations*[42]

$$[\pi(ct,\boldsymbol{x}), \varphi(ct,\boldsymbol{y})] = -i\hbar\delta^{(3)}(\boldsymbol{x}-\boldsymbol{y})1, \tag{6.62}$$

$$[\varphi(ct,\boldsymbol{x}), \varphi(ct,\boldsymbol{y})] = 0, \tag{6.63}$$

$$[\pi(ct,\boldsymbol{x}), \pi(ct,\boldsymbol{y})] = 0. \tag{6.64}$$

It requires a little bit of work to give a precise meaning to these relations, but for the time being we discuss them at a formal level. First, the quantity π is just the "conjugate momentum" defined by the formula (6.45) for the Lagrangian (6.49), which makes its occurrence rather natural.

The numerical factor $i\hbar$ in the relation (6.62) is absolutely what is expected here. Let us review then how we reached it. We defined the invariant measure by the formula (4.36) and the free field by the formula (5.1). Lo and behold! We heard the music of the spheres and we obtained the relation (6.62) with exactly the expected factor $i\hbar$! Magic seems to be involved here because given a number C, changing φ into $C\varphi$ changes the left-hand side of (6.62) by a factor C^2 and destroys this relation. Thus the "extraneous" factor 2 in the formula (4.36) plays a crucial role in the miracle of (6.62). I could not see more than a coincidence in this "miracle".[43] Anyway, we have found a way which is in a sense "canonical" to normalize the

41 So that in this section $\mathcal{H}$ denotes the state space and not a Hamiltonian density!

42 Here and in the sequel we make the customary abuse of notation: We denote by $\varphi(ct,\boldsymbol{x})$ the quantum field $\varphi(x)$ given by (5.30) when $x = (ct,\boldsymbol{x})$.

43 And here again I would have saved much trouble if any of the authors of the main textbooks had been kind enough to make this clear.

quantum field, which was intrinsically defined up to this point only up to a multiplicative constant, by imposing the relations (6.62).[44]

There is an obvious relationship between Proposition 6.9.1 and (5.10), at least at the formal level. This becomes apparent if we write the latter as $[\varphi(x), \varphi(y)] = 0$ whenever $(x-y)^2 < 0$, so that this implies (6.63) for $\boldsymbol{x} \neq \boldsymbol{y}$. Furthermore, differentiating in the time component of x implies $[\pi(x), \varphi(y)] = 0$ in the same domain $(x-y)^2 < 0$ (observe that this does not contradict (6.62)). Differentiating in the time component of y then yields $[\pi(x), \pi(y)] = 0$ in the same domain. Conversely, using Exercise 4.3.2, (6.63) together with Lorentz invariance imply (5.10). If we are willing to go further in the way of formal manipulations, (6.62) itself is a consequence of (5.26). Indeed, differentiating (5.26) with respect to y_0 yields

$$[\varphi(x), \pi(y)] = \hbar^2(\partial_0 \Delta_0(y-x) + \partial_0 \Delta_0(x-y))1. \tag{6.65}$$

Also, following (5.13), we should have

$$\partial_0 \Delta_0(x) = \frac{\mathrm{i}}{2\hbar} \int \frac{\mathrm{d}^3 \boldsymbol{p}}{(2\pi\hbar)^3} \exp(\mathrm{i}(x,p)/\hbar)$$

equals $(\mathrm{i}/2\hbar)\delta^{(3)}(\boldsymbol{x})$ for $x^0 = 0$. Therefore (6.65) implies that if $x^0 = y^0$ then $[\varphi(x), \pi(y)] = \mathrm{i}\hbar\delta^{(3)}(\boldsymbol{x} - \boldsymbol{y})$.

Let us now try to give a precise meaning to the statements of Proposition 6.9.1. The problem is that it does not in general make sense to "fix the value of t" in a distribution on $\mathbb{R}^{1,3}$ any more than it makes sense to think of the value of a distribution as existing at every point.

Exercise 6.9.2 Convince yourself of the previous statement in the case of the distribution $\delta^{(4)}$.

We are going to show however that it makes sense, at a given value of t, to consider φ as a distribution in $\boldsymbol{x}$, as is already apparent for the formula (5.30). Thus the previous formal arguments are essentially correct. To formulate things precisely, for a function $f \in \mathcal{S}^3$ and $t \in \mathbb{R}$ let us denote

$$\mathcal{F}(f)(\boldsymbol{p}) = \int \mathrm{d}^3\boldsymbol{x} \exp(-\mathrm{i}\boldsymbol{x}\cdot\boldsymbol{p}/\hbar) f(\boldsymbol{x})\ ;\ \mathcal{F}_t(f)(p) = \exp(\mathrm{i}ctp^0/\hbar)\mathcal{F}(f)(\boldsymbol{p}).$$

Thus, up to a numerical factor, $\mathcal{F}$ is the three-dimensional Fourier transform. Let us stress that we view $\mathcal{F}_t(f)$ as a function of $p \in X_m$ and that $\mathcal{F}_t(f) \in \mathcal{H} = L^2(X_m, \mathrm{d}\lambda_m)$. We can then define the quantity $\varphi(ct, \boldsymbol{x})$ as a distribution in $\boldsymbol{x}$, depending on the parameter ct by the formula

$$\int \mathrm{d}^3\boldsymbol{x}\varphi(ct,\boldsymbol{x}) f(\boldsymbol{x}) = \frac{1}{\sqrt{c}}(A(\mathcal{F}_t(f^*)) + A^\dagger(\mathcal{F}_t(f))). \tag{6.66}$$

[44] Let us point out though a small caveat. The relation $\pi(ct,\boldsymbol{x}) = \hbar^2 \partial_t \varphi(t,\boldsymbol{x})$ is determined by the choice of Hamiltonian, so the normalization of the quantum field we obtain by this process is just as "canonical" as is the choice of the Hamiltonian density (6.50). Whereas there do exist particles that are described by the quantized version of the Klein-Gordon field, the classical Klein-Gordon field itself is not observed in Nature, so on which grounds does one tell what its energy should be? But nonetheless (6.62) is a convenient, simple and natural way to fix the otherwise undetermined multiplicative constant in the quantum field.

As a further check that this makes sense, if we consider $f \in \mathcal{S}^4$ and for $t \in \mathbb{R}$ we denote f_t the function in $\mathcal{S}^3$ given by $f_t(\boldsymbol{x}) = f(ct,\boldsymbol{x})$, then

$$\hat{f}(p) = \int c\,\mathrm{d}t\,\mathcal{F}_t(f_t)(p), \tag{6.67}$$

and thus (6.66) implies, as it should, that

$$\int c\,\mathrm{d}t\Big(\int \mathrm{d}^3\boldsymbol{x}\,\varphi(ct,\boldsymbol{x})f(ct,\boldsymbol{x})\Big) = \varphi(f) = \frac{1}{\sqrt{c}}(A(\widehat{f^*}) + A^\dagger(\hat{f})).$$

So now $\varphi(ct,\boldsymbol{x})$ is well defined for each t, and (6.66) implies that its derivative in t is also well defined. We interpret $\pi(ct,\boldsymbol{x})$ as $\hbar^2\partial_t\varphi(ct,\boldsymbol{x}) = c\hbar^2\partial_0\varphi(ct,\boldsymbol{x})$. The statement of Proposition 6.9.1 is now well defined.

Proof of Proposition 6.9.1 Let us first consider (6.62). This is equivalent to saying that for two functions $f,g \in \mathcal{S}^3_{\mathbb{R}}$, we have

$$\Big[\int \mathrm{d}^3\boldsymbol{x}\,\pi(ct,\boldsymbol{x})f(\boldsymbol{x}), \int \mathrm{d}^3\boldsymbol{x}\,\varphi(ct,\boldsymbol{x})g(\boldsymbol{x})\Big] = -\mathrm{i}\hbar\int \mathrm{d}^3\boldsymbol{x}\,f(\boldsymbol{x})g(\boldsymbol{x})1. \tag{6.68}$$

To prove this we first note that

$$\int \mathrm{d}^3\boldsymbol{x}\,\pi(ct,\boldsymbol{x})f(\boldsymbol{x}) = \hbar^2\int \mathrm{d}^3\boldsymbol{x}\,\partial_t\varphi(ct,\boldsymbol{x})f(\boldsymbol{x}) = \frac{\hbar^2}{\sqrt{c}}(A(\partial_t\mathcal{F}_t(f)) + A^\dagger(\partial_t\mathcal{F}_t(f))),$$

and we recall that $[A(u),A^\dagger(v)] = (u,v)1$ where $(\cdot,\cdot)$ denotes the inner product in $L^2(X_m,\mathrm{d}\lambda_m)$. Since $\partial_t\mathcal{F}_t(f)(p) = icp^0\hbar^{-1}\mathcal{F}_t(f)(p)$, taking into account the factor $1/\sqrt{c}$ in (6.66), the left-hand side of (6.68) is exactly

$$\hbar\Big((\mathrm{i}p^0\mathcal{F}_t(f),\mathcal{F}_t(g)) - (\mathcal{F}_t(f),\mathrm{i}p^0\mathcal{F}_t(g))\Big)1 = -2\hbar\mathrm{i}(p^0\mathcal{F}_t(f),\mathcal{F}_t(g))1,$$

where we have used that $p^0 \in \mathbb{R}$. Now since on X_m we have $p^0 = \omega_{\boldsymbol{p}}$,

$$(p^0\mathcal{F}_t(f),\mathcal{F}_t(g)) = \int \frac{\mathrm{d}^3\boldsymbol{p}}{2(2\pi\hbar)^3}\mathcal{F}(f)(\boldsymbol{p})^*\mathcal{F}(g)(\boldsymbol{p}).$$

Furthermore, just as in (5.24) we have $\mathcal{F}(f)^*\mathcal{F}(g) = \mathcal{F}(h)$ where $h(\boldsymbol{a}) = \int \mathrm{d}^3\boldsymbol{x}\,f(\boldsymbol{x})g(\boldsymbol{a}+\boldsymbol{x})$. An appropriate version of (1.28) shows that

$$2(p^0\mathcal{F}_t(f),\mathcal{F}_t(g)) = \int \frac{\mathrm{d}^3\boldsymbol{p}}{(2\pi\hbar)^3}\mathcal{F}(h)(\boldsymbol{p}) = h(\boldsymbol{0})$$

and this finishes the proof of (6.68). To prove (6.63) we write

$$\begin{aligned} c\Big[\int \mathrm{d}\boldsymbol{x}\,\varphi(ct,\boldsymbol{x})f(\boldsymbol{x}), \int \mathrm{d}\boldsymbol{x}\,\varphi(ct,\boldsymbol{x})g(\boldsymbol{x})\Big] &= (\mathcal{F}_t(f),\mathcal{F}_t(g)) - (\mathcal{F}_t(g),\mathcal{F}_t(f)) \\ &= 2\mathrm{i}\,\mathrm{Im}\,(\mathcal{F}_t(f),\mathcal{F}_t(g)). \end{aligned}$$

Since h is real-valued it holds that $\mathcal{F}(h)(-\boldsymbol{p}) = \mathcal{F}(h)(\boldsymbol{p})^*$. Therefore the change of variable $\boldsymbol{p} \to -\boldsymbol{p}$ shows that

$$(\mathcal{F}_t(f), \mathcal{F}_t(g)) = \int \frac{\mathrm{d}^3\boldsymbol{p}}{(2\pi\hbar)^3 2\omega_{\boldsymbol{p}}} \mathcal{F}(h)(\boldsymbol{p})$$

is a real number and this proves (6.63). The proof of (6.64) is similar. □

Alternative "Proof" of Proposition 6.9.1 Let us sketch how to obtain (6.62) by the type of formal calculation ubiquitous in this topic (and seemingly unavoidable). As we pointed out, it seems mostly safe to believe in such computations on the grounds that they encode a computation in the discrete situation. Here we encourage the reader to repeat the same computation in the discrete setting using the formula (5.31). We recall that

$$\varphi(x) = \int \frac{\mathrm{d}^3\boldsymbol{p}}{(2\pi\hbar)^3\sqrt{2c\,\omega_{\boldsymbol{p}}}}\big(\exp(-\mathrm{i}(x,p)/\hbar)a(\boldsymbol{p}) + \exp(\mathrm{i}(x,p)/\hbar)a^\dagger(\boldsymbol{p})\big) \tag{6.69}$$

and, consequently,

$$\pi(ct,\boldsymbol{x}) = \mathrm{i}\hbar \int \frac{\mathrm{d}^3\boldsymbol{p}}{(2\pi\hbar)^3}\sqrt{\frac{c\,\omega_{\boldsymbol{p}}}{2}}\big(-\exp(-\mathrm{i}(x,p)/\hbar)a(\boldsymbol{p}) + \exp(\mathrm{i}(x,p)/\hbar)a^\dagger(\boldsymbol{p})\big), \tag{6.70}$$

where $x = (ct, \boldsymbol{x})$. Then, taking into account that $[a(\boldsymbol{p}), a(\boldsymbol{q})] = 0$, etc., we get

$$[\pi(ct,\boldsymbol{x}), \varphi(ct,\boldsymbol{y})] = \frac{\mathrm{i}\hbar}{2}(\mathrm{I} + \mathrm{II})$$

where, setting $y = (ct, \boldsymbol{y})$,

$$\mathrm{I} = -\int \frac{\mathrm{d}^3\boldsymbol{p}_1}{(2\pi\hbar)^3}\frac{\mathrm{d}^3\boldsymbol{p}_2}{(2\pi\hbar)^3}\sqrt{\frac{\omega_{\boldsymbol{p}_2}}{\omega_{\boldsymbol{p}_1}}}\exp\big(\mathrm{i}(y,p_2)/\hbar - \mathrm{i}(x,p_1)/\hbar\big)[a(\boldsymbol{p}_1), a^\dagger(\boldsymbol{p}_2)]$$

and

$$\mathrm{II} = \int \frac{\mathrm{d}^3\boldsymbol{p}_1}{(2\pi\hbar)^3}\frac{\mathrm{d}^3\boldsymbol{p}_2}{(2\pi\hbar)^3}\sqrt{\frac{\omega_{\boldsymbol{p}_2}}{\omega_{\boldsymbol{p}_1}}}\exp\big(\mathrm{i}(x,p_1)/\hbar - \mathrm{i}(y,p_2)/\hbar\big)[a^\dagger(\boldsymbol{p}_1), a(\boldsymbol{p}_2)].$$

Now, since $[a(\boldsymbol{p}_1), a^\dagger(\boldsymbol{p}_2)] = (2\pi\hbar)^3\delta^{(3)}(\boldsymbol{p}_1 - \boldsymbol{p}_2)1$, integrating in $\boldsymbol{p}_1$ first yields, using that x and y have the same time component, and remembering that $(y,p) = y^0p^0 - \boldsymbol{y}\cdot\boldsymbol{p}$,

$$\mathrm{I} = -\int \frac{\mathrm{d}^3\boldsymbol{p}}{(2\pi\hbar)^3}\exp(-\mathrm{i}(\boldsymbol{y}-\boldsymbol{x})\cdot\boldsymbol{p})/\hbar)1 = -\delta^{(3)}(\boldsymbol{x}-\boldsymbol{y})1$$

using (1.19). The term II is similarly found to be $-\delta^{(3)}(\boldsymbol{y}-\boldsymbol{x})1 = -\delta^{(3)}(\boldsymbol{x}-\boldsymbol{y})1$. □

Proposition 6.9.1 may serve as a third approach to the quantization of the Klein-Gordon field, which is found in many physics textbooks. One *postulates* that a suitable quantization of this field will be an operator-valued "function" φ with the following properties:

- It will satisfy the Klein-Gordon equation.
- It will satisfy the equal-time commutation relations.

One *has to* postulate something about the quantization one is looking for. Postulating as here that things work the same as in the familiar settings should be the first attempt.

Assuming then that the field satisfies the Klein-Gordon equation and the equal-time commutation relations, let us try to find a formula for it. Our derivation is heuristic: distributions are treated as functions, all integrals exist, etc.[45]

The quantum field satisfies the Klein-Gordon equation i.e. $\hbar^2\partial^\mu\partial_\mu\varphi + m^2c^2\varphi = 0$. Taking the Fourier transform we obtain $(m^2c^2 - p^2)\hat{\varphi}(p) = 0$. Thus $\hat{\varphi}$ is supported by the set $\{p; p^2 = m^2c^2\} = X_m \cup -X_m$. We then express φ as the inverse Fourier transform of $\hat{\varphi}$, i.e. as an integral over $X_m \cup -X_m$. We bring back the part of the integral on $-X_m$ to X_m by the transformation $p \to -p$, and (crossing our fingers) we obtain an expression

$$\varphi(x) = \int \mathrm{d}\lambda_m(p)(\exp(-\mathrm{i}(x,p)/\hbar)b(p) + \exp(\mathrm{i}(x,p)/\hbar)c(p)), \tag{6.71}$$

where $b(p)$ and $c(p)$ are operator-valued functions. Since we are quantizing a real-valued vector field, and that real numbers correspond to Hermitian operators, we look for φ to be Hermitian, and this implies that $c(p) = b(p)^\dagger$. Let us write[46] $a(\boldsymbol{p}) = b(p)/\sqrt{2c\,\omega_{\boldsymbol{p}}}$ where $p = (\omega_{\boldsymbol{p}}, \boldsymbol{p})$, so that (6.71) implies (6.69) and in particular

$$\begin{aligned}\varphi(0,\boldsymbol{x}) &= \int \frac{\mathrm{d}^3\boldsymbol{p}}{(2\pi\hbar)^3\sqrt{2c\,\omega_{\boldsymbol{p}}}}(\exp(\mathrm{i}\boldsymbol{x}\cdot\boldsymbol{p}/\hbar)a(\boldsymbol{p}) + \exp(-\mathrm{i}\boldsymbol{x}\cdot\boldsymbol{p}/\hbar)a^\dagger(\boldsymbol{p}))\\ &= \int \frac{\mathrm{d}^3\boldsymbol{p}}{(2\pi\hbar)^3}\exp(\mathrm{i}\boldsymbol{x}\cdot\boldsymbol{p}/\hbar)d(\boldsymbol{p}),\end{aligned} \tag{6.72}$$

where

$$d(\boldsymbol{p}) = \frac{1}{\sqrt{2c\,\omega_{\boldsymbol{p}}}}(a(\boldsymbol{p}) + a^\dagger(-\boldsymbol{p})). \tag{6.73}$$

That is, $\varphi(0,\boldsymbol{x})$ is the inverse Fourier transform of $d(\boldsymbol{p})$, so that $d(\boldsymbol{p})$ is the Fourier transform of $\varphi(0,\boldsymbol{x})$:

$$d(\boldsymbol{p}) = \int \mathrm{d}^3\boldsymbol{x}\exp(-\mathrm{i}\boldsymbol{x}\cdot\boldsymbol{p}/\hbar)\varphi(0,\boldsymbol{x}). \tag{6.74}$$

Starting from (6.70) and setting

$$f(\boldsymbol{p}) = \frac{\mathrm{i}\hbar\sqrt{c\,\omega_{\boldsymbol{p}}}}{\sqrt{2}}(-a(\boldsymbol{p}) + a^\dagger(-\boldsymbol{p})) \tag{6.75}$$

we obtain in the same manner

$$f(\boldsymbol{p}) = \int \mathrm{d}^3\boldsymbol{x}\exp(-\mathrm{i}\boldsymbol{x}\cdot\boldsymbol{p}/\hbar)\pi(0,\boldsymbol{x}). \tag{6.76}$$

[45] A special case of the "canonical quantization method" is described here, but the method is quite general. It allows the discovery of many important quantum fields. Some books such as that of Greiner and Reinhardt [35] go as far as "discovering" the boson Fock space by considering the Schrödinger equation (2.80) as the equation of a classical field, and applying the previous method to it. This may not contribute to conceptual clarity but motivates the name of "second quantization" for the construction of the boson Fock space.

[46] It is perfectly possible to write the computation below in terms of $b(p)$, but for consistency we follow our policy to express the quantum field in term of the operators $a(\boldsymbol{p})$.

Consequently

$$[f(\boldsymbol{p}_1), d(\boldsymbol{p}_2)] = \int \mathrm{d}^3\boldsymbol{x}\mathrm{d}^3\boldsymbol{y} \exp(-\mathrm{i}(\boldsymbol{x}\cdot\boldsymbol{p}_1 + \boldsymbol{y}\cdot\boldsymbol{p}_2)/\hbar)[\pi(0,\boldsymbol{x}), \varphi(0,\boldsymbol{y})]$$

$$= -\mathrm{i}\hbar \int \mathrm{d}^3\boldsymbol{x}\mathrm{d}^3\boldsymbol{y} \exp(-\mathrm{i}(\boldsymbol{x}\cdot\boldsymbol{p}_1 + \boldsymbol{y}\cdot\boldsymbol{p}_2)/\hbar)\delta^{(3)}(\boldsymbol{x}-\boldsymbol{y})1$$

$$= -\mathrm{i}\hbar \int \mathrm{d}^3\boldsymbol{x} \exp(-\mathrm{i}\boldsymbol{x}\cdot(\boldsymbol{p}_1 + \boldsymbol{p}_2)/\hbar)1$$

$$= -\mathrm{i}\hbar(2\pi\hbar)^3\delta^{(3)}(\boldsymbol{p}_1 + \boldsymbol{p}_2)1. \tag{6.77}$$

Moreover, the same calculation implies $[d(\boldsymbol{p}_1), d(\boldsymbol{p}_2)] = [f(\boldsymbol{p}_1), f(\boldsymbol{p}_2)] = 0$ and from these we may compute the commutators between the a and the $a^\dagger$ by simple algebra, using (6.73) and (6.75):

$$[a(\boldsymbol{p}), a^\dagger(\boldsymbol{p}')] = (2\pi)^3\delta^3(\boldsymbol{p}-\boldsymbol{p}')1, \tag{6.78}$$

with all the other commutators being zero.

In physics textbooks the previous argument is presented by obtaining first the explicit formula (6.79) for $a(\boldsymbol{p})$ and a similar formula for $a^\dagger(\boldsymbol{p})$, from which one can compute the commutator of these operators.[47]

Exercise 6.9.3 Let us introduce the notation

$$A \overset{\leftrightarrow}{\partial_t} B := A\partial_t B - B\partial_t A.$$

(a) Prove that if A and B are functions that are solutions of the Klein-Gordon equation and which, for any given t, go fast enough to zero as $\boldsymbol{x}$ goes to infinity, the quantity $\int \mathrm{d}^3\boldsymbol{x} A \overset{\leftrightarrow}{\partial_t} B$ (where the integral is computed at a given value of x^0) is independent of this value.

(b) Consider a function A on $\mathbb{R}^4$ such that at given x^0 the function $\boldsymbol{x} \mapsto A(x^0, \boldsymbol{x})$ is a test function. Consider a distribution B which is well defined at every given value of x_0. Prove that if A and B satisfy the Klein-Gordon equation then the quantity $\int \mathrm{d}^3\boldsymbol{x} A \overset{\leftrightarrow}{\partial_t} B$, where the integral is computed at a given value of x^0, is independent of this value..

Exercise 6.9.4 Prove formally that

$$\frac{\mathrm{i}}{\hbar}\sqrt{2c\,\omega_{\boldsymbol{p}}}a(\boldsymbol{p}) = \int \mathrm{d}^3\boldsymbol{x}\varphi(x) \overset{\leftrightarrow}{\partial_t} \exp(\mathrm{i}(x,p)/\hbar). \tag{6.79}$$

Argue that the right-hand side is independent of x^0.

47 These formulas are an overkill here, but they are useful when one studies fields that are not necessarily free fields. The operator $a(\boldsymbol{p})$ defined by the formula (6.79) then depends also on t, and can be used to study the asymptotic properties of the field.

Exercise 6.9.5 The goal of this exercise is to better understand the mysterious looking quantity $\int d^3\boldsymbol{x} A \overleftrightarrow{\partial_t} B$ when A, B are solutions of the Klein-Gordon equation. As in (6.71) the function A should be of the type

$$A(x) = \int d\lambda_m(p)\big(\exp(i(x,p)/\hbar)f^+(p) + \exp(-i(x,p)/\hbar)f^-(p)\big).$$

for two functions f^+ and f^- on X_m. Using a similar formula for B, with corresponding functions g^+, g^-, compute $\int d^3\boldsymbol{x} A \overleftrightarrow{\partial_t} B$ as an expression involving the functions f^+, f^-, g^+, g^-.

6.10 Caveat

We have seen several arguments, each showing that a natural way to quantize the Klein-Gordon field yields the scalar free field, building a solid case that there is a unique way to quantize this field. Unfortunately, you have been robbed blind. The Stone–von Neuman theorem which we will prove in Appendix C asserts that when we have *finitely many* canonical commutation relations, they have an essentially unique solution (in the sense that any two solutions are unitarily equivalent). But this fails completely when one considers an *infinity* of canonical commutation relations: There are many non-equivalent ways to implement them. i.e. there are many non-unitarily equivalent sequences of operators which satisfy them. Why this is true is also explained in detail in Appendix C. The commutation relations might be called canonical, but the sequences of operators which satisfy them are not. In presence of an infinite number of canonical commutation relations, it makes very little sense to talk of canonical quantization. The free field (5.1) satisfying the fundamental property (5.5) is the "correct" quantization of the Klein-Gordon field, and it simply *cannot be discovered by any of the previous methods.*[48] In particular, despite the commutation relations (13.191) the field given by the first line in (6.72) has no reason to be unitarily equivalent to the free field (5.1).

To give a simple example of two sequences of operators satisfying the canonical commutation relations which are not unitarily equivalent, consider such a sequence $d(\boldsymbol{k})$, so that $[d(\boldsymbol{k}), d^\dagger(\boldsymbol{k})] = 1$, with all other commutators equal zero. Then, for two real numbers α, β with $\alpha^2 - \beta^2 = 1$, the operators $c(\boldsymbol{k}) = \alpha d(\boldsymbol{k}) + \beta d^\dagger(\boldsymbol{k})$ and $c^\dagger(\boldsymbol{k}) = \alpha d^\dagger(\boldsymbol{k}) + \beta d(\boldsymbol{k})$ also satisfy the canonical commutation relations. However it can be proved that there is no unitary operator which transforms $d(\boldsymbol{k})$ into $c(\boldsymbol{k})$ when $\beta \neq 0$, see Appendix C.

There is a simple reason why the canonical commutation relations occupy a far more central place in many physics textbooks than the theoretically more fulfilling description of the free field used in Chapter 5. As will become apparent in Chapter 13, when computing quantities which can actually be measured in experiments, the commutation relations between the operators $a(\boldsymbol{p})$ and $a^\dagger(\boldsymbol{p})$ in the formula (5.30) are all that matters.[49]

[48] This entire section is again motivated by the very long time during which I was confused by certain physics textbooks.

[49] Unfortunately, it will be shown in Section 13.4 that the whole setting in which one performs these practical computations is mathematically unsound, so that our clean theoretical understanding of the free field will not be very useful at that stage.

6.11 Hamiltonian

In Section 5.3 we heuristically discussed the following result, which we now prove rigorously.

Theorem 6.11.1 *Given any* $f \in \mathcal{S}^3$ *the operator* $\int \mathrm{d}^3\boldsymbol{x}\varphi(ct,\boldsymbol{x})f(\boldsymbol{x})$ *is the time-evolution of the operator* $\int \mathrm{d}^3\boldsymbol{x}\varphi(0,\boldsymbol{x})f(\boldsymbol{x})$ *in the Heisenberg picture corresponding to the Hamiltonian* $H_{\mathcal{B}}$ *given by*

$$H_{\mathcal{B}} = \int \frac{\mathrm{d}^3\boldsymbol{p}}{(2\pi\hbar)^3} c\omega_{\boldsymbol{p}} a^\dagger(\boldsymbol{p})a(\boldsymbol{p}). \tag{5.36}$$

We will summarize this statement by the sentence "The field $\varphi(ct,\cdot)$ is the time evolution of the field $\varphi(0,\cdot)$".

Proof According to Proposition 5.3.1, $\exp(-\mathrm{i}tH_{\mathcal{B}}/\hbar)$ is the time-evolution in $\mathcal{B}$, so it is the canonical extension to $\mathcal{B}$ of the time-evolution $\exp(-\mathrm{i}tH/\hbar)$ in $\mathcal{H}$. Therefore using Proposition 3.5.2 we obtain

$$\exp(\mathrm{i}tH_{\mathcal{B}}/\hbar)\big(A(\mathcal{F}_0(f)) + A^\dagger(\mathcal{F}_0(f)\big)\exp(-\mathrm{i}tH_{\mathcal{B}}/\hbar) = A(g) + A^\dagger(g), \tag{6.80}$$

where $g := \exp(\mathrm{i}tH/\hbar)\mathcal{F}_0(f) = \mathcal{F}_t(f)$ since $\exp(\mathrm{i}tH/\hbar)$ is the "multiplication by $\exp(\mathrm{i}ct\omega_{\boldsymbol{p}}/\hbar)$" operator. This is the result according to (6.66). □

In the preceding sections we have argued different ways that the free quantum field is the quantization of the Klein-Gordon field, whose time-evolution is governed by the Hamiltonian density (6.50)

$$\mathcal{H}(u,\pi) = \frac{1}{2\hbar^2}\pi^2 + \frac{1}{2}\hbar^2c^2 \sum_{1\le\nu\le 3}(\partial_\nu u)^2 + \frac{1}{2}m^2c^4u^2. \tag{6.50}$$

For things to make sense, we expect that if in this expression we replace $u(x)$ by the quantum field $\varphi(x)$ and $\pi(x)$ by $\hbar^2\partial_t\varphi(x)$, we somehow "get a Hamiltonian density governing the time-evolution of the free field". In other words, we expect that in some sense the quantity

$$\int \mathrm{d}^3\boldsymbol{x}\mathcal{H}(\varphi(x),\hbar^2\partial_t\varphi(x)) \tag{6.81}$$

equals the Hamiltonian (5.36). Here, $x = (ct,\boldsymbol{x})$, and we expect the result to be independent of t.

Our next goal is to perform this computation, showing the "equality" of the quantities (5.36) and (6.81), provided we discard the usual extraneous infinite term created when replacing terms such as $a(\boldsymbol{p})a^\dagger(\boldsymbol{p}')$ by $a^\dagger(\boldsymbol{p}')a(\boldsymbol{p}) + \delta^{(3)}(\boldsymbol{p}-\boldsymbol{p}')1$ when $\boldsymbol{p} = \boldsymbol{p}'$. The computation involves as in Section 3.10 algebraic cancellation of terms that are ill-defined,[50] so it *is* formal, but we stick to our belief that such a formal computation simply encodes a correct computation in the discretized situation. We start by computing the contribution of the last term of (6.50), that is we compute

[50] A diplomatic way to say that they are not defined at all.

$$\int \mathrm{d}^3\boldsymbol{x}\varphi(x)^2 = \int \mathrm{d}^3\boldsymbol{x}\frac{\mathrm{d}^3\boldsymbol{p}}{(2\pi\hbar)^3}\frac{\mathrm{d}^3\boldsymbol{p}'}{(2\pi\hbar)^3}\frac{1}{2c\sqrt{\omega_{\boldsymbol{p}}\omega_{\boldsymbol{p}'}}}U(\boldsymbol{p})U(\boldsymbol{p}') \tag{6.82}$$

where

$$U(\boldsymbol{p}) = \exp(-\mathrm{i}(x,p)/\hbar)a(\boldsymbol{p}) + \exp(\mathrm{i}(x,p)/\hbar)a^\dagger(\boldsymbol{p}).$$

For this we first integrate in $\boldsymbol{x}$, using that for $r \in \mathbb{R}^{1,3}$, according to (1.19),

$$\int \mathrm{d}^3\boldsymbol{x}\exp(\mathrm{i}(x,r)/\hbar) = \exp(\mathrm{i}x^0r^0/\hbar)(2\pi\hbar)^3\delta^{(3)}(\boldsymbol{r}).$$

We use this relation four times, with $r = \pm p \pm p'$. Since $x^0\omega_{\boldsymbol{p}} = ct\omega_{\boldsymbol{p}}$ and $\omega_{-\boldsymbol{p}} = \omega_{\boldsymbol{p}}$ we find that the quantity (6.82) is

$$\int \frac{\mathrm{d}^3\boldsymbol{p}}{(2\pi\hbar)^3 2c\omega_{\boldsymbol{p}}}\Big(\exp(-2\mathrm{i}\omega_{\boldsymbol{p}}ct/\hbar)a(\boldsymbol{p})a(-\boldsymbol{p}) + a(\boldsymbol{p})a^\dagger(\boldsymbol{p})$$
$$+ a^\dagger(\boldsymbol{p})a(\boldsymbol{p}) + \exp(2\mathrm{i}\omega_{\boldsymbol{p}}ct/\hbar)a^\dagger(\boldsymbol{p})a^\dagger(-\boldsymbol{p})\Big).$$

The contributions of the three other terms of (6.50) are computed similarly. Grouping them the terms depending on t cancel out and replacing the terms $a(\boldsymbol{p})a^\dagger(\boldsymbol{p})$ by $a^\dagger(\boldsymbol{p})a(\boldsymbol{p})$ gives the expression (5.36).

Exercise 6.11.2 Complete the previous computation.

Even when one follows carefully every step it is hard to understand why the previous computation works at all. We next explain that there is no magic here. Since the system is in a sense a sum of independent harmonic oscillators, we expect that in the end the whole computation reduces to a statement *about a single harmonic oscillator*, namely the formula (2.111). Let us demonstrate this by moving to the discrete setting, when the spatial universe is replaced by a box. To perform the calculation, we first observe that if the operator-valued functions $u(\boldsymbol{x})$ and $\pi(\boldsymbol{x})$ satisfy

$$u(\boldsymbol{x}) = \sum_{\boldsymbol{k}\in\mathcal{K}} b_{\boldsymbol{k}} f_{\boldsymbol{k}}(\boldsymbol{x})\ ;\ \pi(\boldsymbol{x}) = \sum_{\boldsymbol{k}\in\mathcal{K}} c_{\boldsymbol{k}} f_{\boldsymbol{k}}(\boldsymbol{x}) \tag{6.83}$$

where $b_{-\boldsymbol{k}} = b_{\boldsymbol{k}}^\dagger$ and $c_{-\boldsymbol{k}} = c_{\boldsymbol{k}}^\dagger$ then the same computation leading to (6.58) yields

$$\int_B \mathrm{d}^3\boldsymbol{x}\mathcal{H}(u,\pi)(\boldsymbol{x}) = \sum_{\boldsymbol{k}\in\mathcal{K}}\Big(\frac{1}{2\hbar^2}c_{\boldsymbol{k}}c_{\boldsymbol{k}}^\dagger + \frac{1}{2}\hbar^2\tilde{\omega}_{\boldsymbol{k}}^2 b_{\boldsymbol{k}}b_{\boldsymbol{k}}^\dagger\Big). \tag{6.84}$$

Furthermore it follows from (6.18) that $u(\boldsymbol{x}) = \varphi(0,\boldsymbol{x})$ and $\pi(\boldsymbol{x}) = \hbar^2\partial_t\varphi(0,\boldsymbol{x})$ satisfy (6.83) for

$$b_{\boldsymbol{k}} = \frac{1}{\sqrt{2c\omega_{\boldsymbol{k}}}}(a(\boldsymbol{k}) + a(-\boldsymbol{k})^\dagger)\ ;\ c_{\boldsymbol{k}} = \mathrm{i}\hbar\sqrt{\frac{c\omega_{\boldsymbol{k}}}{2}}(-a(\boldsymbol{k}) + a(-\boldsymbol{k})^\dagger).$$

For these values the right-hand side of (6.84) then equals

$$\sum_{k\in\mathcal{K}} \frac{c\omega_k}{4}\big((a(\boldsymbol{k})+a(-\boldsymbol{k})^\dagger)(a(\boldsymbol{k})^\dagger+a(-\boldsymbol{k})+(-a(\boldsymbol{k})+a(-\boldsymbol{k})^\dagger)(-a(\boldsymbol{k})^\dagger+a(-\boldsymbol{k}))\big)$$

$$=\sum_{k\in\mathcal{K}} \frac{c\omega_k}{2}(a(\boldsymbol{k})a^\dagger(\boldsymbol{k})+a(-\boldsymbol{k})^\dagger a(-\boldsymbol{k}) = \sum_{k\in\mathcal{K}} \frac{c\omega_k}{2}(a(\boldsymbol{k})a^\dagger(\boldsymbol{k})+a(\boldsymbol{k})^\dagger a(\boldsymbol{k})), \quad (6.85)$$

and since $a(\boldsymbol{k})a^\dagger(\boldsymbol{k}) = a(\boldsymbol{k})^\dagger a(\boldsymbol{k})+1$, ignoring this constant term 1, the previous expression is just the Hamiltonian (5.36) when the universe has been put in a box, and we have obtained a much shorter proof of the desired equality of (6.81) and (5.36).[51]

There is a systematic way to get rid of the infinite terms which occur in the previous computations.[52] These terms occur when replacing the expression $a(\boldsymbol{p})a^\dagger(\boldsymbol{p})$ by $a^\dagger(\boldsymbol{p})a(\boldsymbol{p}) + \delta^{(3)}(\boldsymbol{0})1$. They would not occur if in the expression $\mathcal{H}(\varphi(x),\hbar^2\partial_t\varphi(x))$ we had decided beforehand to write all creation operators $a^\dagger$ to the left of the annihilation operators a. Such an operation is called *normal ordering*. The normal-ordering of the expression C is denoted by $:C:$. To illustrate this procedure by a simple example,

$$c\,:\varphi(f)\varphi(g):= A^\dagger(\widehat{f^*})A^\dagger(\widehat{g^*})+A^\dagger(\widehat{f^*})A(\widehat{g})+A^\dagger(\widehat{g^*})A(\widehat{f})+A(\widehat{f})A(\widehat{g}).$$

If we revisit our computation that the quantity (6.81) equals the Hamiltonian (5.36), we can then state the following.

Proposition 6.11.3 *The following expression is a Hamiltonian density governing the time-evolution of the free field:*

$$:\frac{\hbar^2}{2}\Big((\partial_t\varphi(x))^2+c^2\sum_{1\le\nu\le 3}(\partial_\nu\varphi(x))^2\Big)+\frac{1}{2}m^2c^4\varphi(x)^2: \,. \quad (6.86)$$

This formula represents the Hamiltonian density as a function of a field that itself time-evolves according to the Hamiltonian. It is tempting to use the same procedure to "define" Hamiltonians where the density is expressed as in (6.86) as a function of φ itself, but with an extra "interaction term". For such Hamiltonians nice formulas such as (5.36) do not exist. This procedure is commonly used in physics, but the results are heuristic, and the existence of the resulting objects has not been proved, and is arguably dubious.

Key ideas to remember:

- Lagrangian Mechanics naturally leads us to construct classical systems with Lorentz invariant behavior.
- Hamiltonian Mechanics is a powerful reformulation of Lagrangian Mechanics, but it singles out the role of time.
- Various paths lead to consider the free quantum field as a quantization of the Klein-Gordon equation.
- None of these methods actually leads uniquely to the correct quantization, due to the fact that many essentially different sequences of operators satisfy the canonical commutation relations when there is an infinity of them.

[51] Note that when using the normalization (6.19) there is no factor L^{-3}.
[52] In this case, it is easy to get rid of the infinities. We will not be so lucky in the future.

7

The Casimir Effect

The Casimir effect can be roughly described as follows. If two well-polished conducting plates are held parallel very close to each other in a good vacuum, there is an attractive force between them.[1] Let us stress that there is *no electrical charge whatsoever* involved here, and that such a force is entirely unexpected in the classical view of the world. It has actually been measured by experimentalists.[2] At a crude level, the explanation is that between the two plates, there is less room than in unbounded space for the "vacuum energy", and the closer the plates the more so. Bringing the plates closer decreases the amount of energy in the vacuum between them, and this translates into an attractive force.

7.1 Vacuum Energy

The real-world Casimir effect is not due to the scalar field we have studied so far, but due to the electromagnetic field. However, we will follow a very enlightening discussion in Zees's textbook [93] on the simplest possible case, where the electromagnetic field is replaced by the massless scalar field, that is the case $m = 0$ of the field we have been studying, and where physical space is one-dimensional. We consider two "plates", at positions 0 and $d > 0$. We will assume that the scalar field is zero on each conducting plate i.e. it lives in the interval $[0, d]$.[3]

The first goal is to "compute" the "vacuum energy" in the interval $[0, d]$. We reproduce in the case $m = 0$ the analysis of Section 6.7, replacing the box $B = [-L/2, L/2]^3$ with the interval $[0, d]$. If $(f_k)_{k\geq 1}$ is a basis of $L^2([0, d])$ consisting of eigenvectors of the Laplacian of eigenvalue $-a_k^2$ (i.e. $f_k'' = -a_k^2 f_k$) we were led in that section to attach to each f_k a "mode of excitation" of the scalar field represented by a harmonic oscillator of ground state energy $\hbar c|a_k|/2$.[4] The vacuum energy is then the lowest possible energy, the sum $\sum_k \hbar c|a_k|/2$ of

[1] A force besides the gravitational attraction of course.
[2] This is a difficult experiment because the force is rather weak.
[3] This assumption is motivated by analogy with the electromagnetic field which behaves just that way.
[4] Let us get some details on how to reach this formula. We found a ground state energy of $\hbar\tilde{\omega}_{\boldsymbol{k}}/2$ where $\tilde{\omega}_{\boldsymbol{k}}$ is given by (6.11) when $m = 0$, i.e. $\tilde{\omega}_{\boldsymbol{k}} = c\sqrt{\boldsymbol{k}^2/\hbar}$ but the eigenvalue $\boldsymbol{k}^2/\hbar^2$ corresponds to a_k^2.

the ground state energies of all these harmonic oscillators.[5] Now a suitable basis is provided by the functions[6] $\sin(\pi nx/d)$ for $n \geq 1$, giving a total vacuum energy[7]

$$\sum_{n\geq 1} \frac{n\pi\hbar c}{2d} = \frac{\pi\hbar c}{2} \sum_{n\geq 1} \frac{n}{d}. \tag{7.1}$$

This formula cannot be correct since it gives an infinite value. Furthermore, our model is not reliable for high-energy contributions.[8] The goal of the chapter is to make sense of (7.1).

7.2 Regularization

The Casimir force is essentially an energy difference. The infinity in (7.1) is coming from the high-frequency modes which, from a physical point of view, will just go through the conducting plates.[9] For these waves, the existence of the conducting plate at position d does not matter, and so the high modes contribute equally on either side of the plates, that is, they are not contributing to the force we want to calculate. To make a sense of this difference of two infinities, we introduce a damping factor on the high-energy terms such that the finite part of the difference is brought to light, and hope that in the end it does not depend on the procedure.

So, let us introduce an unspecified "damping function φ" and rather than (7.1) use for the vacuum energy the quantity

$$E_\alpha(d) := \frac{\pi\hbar c}{2} \sum_{n\geq 1} \frac{n}{d}\varphi\Big(\frac{\alpha n}{d}\Big), \tag{7.2}$$

where $\alpha > 0$ is a small parameter. We assume very little about the function φ, namely that it is smooth enough and decreases fast enough at infinity so that the series of (7.2) makes sense. We also want

$$\varphi(0) = 1, \tag{7.3}$$

so that when α gets small, the first terms of (7.2) resemble the first terms of (7.1).

Consider now a third conducting plate at position $D \gg d$. The point of this device is that we do not have to consider the even harder problem of making sense of the vacuum energy

5 I must admit that I do not find it very easy to believe in the validity of such a procedure, but, indeed the Casimir effect is for real.

6 It is rather obvious that these functions are orthogonal to each other, but less obvious that they span $L^2([0,d])$. To prove this we have to show that a function f which is orthogonal to each of the functions $\sin(\pi nx/d)$ is 0. Observe then that the odd function g on $[-d,d]$ such that $g(x) = f(x)$ for $x \in [0,d]$ is orthogonal to each function $\sin(\pi nx/d)$ and $\cos(\pi nx/d)$ and hence 0.

7 There are some subtle issues here. In order to make real sense of the notion of "an orthonormal basis of eigenvectors of the Laplacian" one has to define the Laplacian as a self-adjoint operator. There are multiple ways to do this. The choice of the functions $\sin(\pi nx/d)$ corresponds to the "Dirichlet boundary conditions" $f(0) = f(d) = 0$ which are appropriate here since the field is zero on both plates. The "periodic boundary conditions" which we have been often using would give the basis $\sin(2\pi nx/d)$ and $\cos(2\pi nx/d)$. Interestingly these two choices give the same answer for $\sum |a_k|$ but we do not know of a general result to that effect.

8 High-frequency waves have a short wave length, and the structure of space at extremely small scale is unknown, see Section 6.8.

9 High-frequency waves have a short wave length, and when this wave length is much shorter than the inter-atomic distances in matter, "to them matter just looks empty". Our bodies stop visible light but not X-rays.

of an unbounded interval. The vacuum energy in the interval $[d, D]$ is $E_\alpha(D-d)$, so that the total energy in both intervals is $E_\alpha(d)+E_\alpha(D-d)$, and the force on the plate at d is[10]

$$-E'_\alpha(d)+E'_\alpha(D-d). \tag{7.4}$$

Proposition 7.2.1

$$\lim_{\alpha\to 0}(-E'_\alpha(d)+E'_\alpha(D-d)) = -\frac{\pi\hbar c}{24}\Big(\frac{1}{d^2}-\frac{1}{(D-d)^2}\Big). \tag{7.5}$$

The amazing part of this result is that the limit does not depend on the choice of the regularizing function, and we can believe that we have succeeded in calculating the Casimir force to be about $-\pi\hbar c/(24d^2)$ when $d \ll D$. Similar formulas in the more realistic case of the electromagnetic field are confirmed by experiment.

To start the proof of Proposition 7.2.1 we observe that

$$-\frac{2}{\pi\hbar c}E'_\alpha(d) = \frac{1}{\alpha d}\sum_{n\geq 1}\psi\Big(\frac{\alpha n}{d}\Big), \tag{7.6}$$

where $\psi(x) = x\varphi(x)+x^2\varphi'(x)$. Thus we are led to estimate the sum $\sum_{n\geq 1}\psi(\varepsilon n)$, where $\varepsilon = \alpha/d$. Estimating such sums is the object of the famous Euler-Maclaurin summation formula. In the present case, this formula reduces to the application of the following simple lemma.

Lemma 7.2.2 *If ψ is three times differentiable, then*

$$\int_a^b \mathrm{d}x\,\psi(x) = \frac{b-a}{2}(\psi(b)+\psi(a)) - \frac{(b-a)^2}{12}(\psi'(b)-\psi'(a))+(b-a)^3\mathcal{R}, \tag{7.7}$$

where

$$|\mathcal{R}| \leq L \sup_{a\leq x\leq b}|\psi''(x)| \tag{7.8}$$

and where L is a numerical constant.[11]

Proof Let us be quick and dirty. It is straightforward to check that no remainder is necessary when ψ is a polynomial of degree ≤ 2. Subtracting such a polynomial from ψ we may assume that $\psi(a)=\psi'(a)=0$, and then all the terms are bounded by a constant times $(b-a)^3$ times the maximum of the second derivative of ψ. □

Using this for $a = n\varepsilon$, $b = (n+1)\varepsilon$, the remainder is bounded by

$$L\varepsilon^3 \sup_{n\varepsilon\leq x\leq(n+1)\varepsilon}|\psi''(x)|.$$

[10] The force is minus the gradient of the potential energy.
[11] If it is your game you may like to find the best possible constant in (7.8) but we do not care about this.

Summing over $n \geq 0$ and using that $\varphi(0) = 1$ we obtain that (provided ψ'' decreases fast enough)

$$\sum_{n\geq 1} \psi(\varepsilon n) = \frac{1}{\varepsilon}\int_0^\infty \mathrm{d}x\, \psi(x) - \varepsilon\frac{\psi'(0)}{12} + O(\varepsilon^2). \tag{7.9}$$

Now $\psi'(0) = \varphi(0) = 1$, so that going back to (7.6) gives

$$-\frac{2}{\pi\hbar c}E'_\alpha(d) = \frac{1}{\alpha^2}\int_0^\infty \mathrm{d}x\, \psi(x) - \frac{1}{12d^2} + O(\alpha). \tag{7.10}$$

This proves (7.4) as the "infinities", i.e. the term in $1/\alpha^2$ cancels when removing the corresponding term for $D - d$ rather than d.

The previous argument is nit-picking. It is much simpler to apply the well-known identity[12](see e.g. the youtube video of Brady Haran [96])

$$\sum_{n\geq 1} n = -\frac{1}{12}$$

to (7.1) and to differentiate in d. I am not kidding. I have seen books on string theory that take this identity as a starting point in the most serious manner.

Key ideas to remember:

- We tamed our first really serious infinities with flying colors!
- It looks crazy to "associate the ground state energy of a harmonic oscillator to each mode of vibration" but the Casimir force is for real.

[12] This identity is not as crazy as it sounds. The zeta function $\zeta(s) = \sum_{n\geq 1} 1/n^s$ analytically continues to the value $\zeta(-1) = -1/12$.

Part II

Spin

It sounds paradoxical at first to have an entire part of the book called "spin" with the single Section 8.7 offering a "brief introduction to spin".

Let us recall from Definition 2.10.1 that a projective unitary representation U associates to each element a in a group G a unitary operator $U(a)$ such that for $a,b \in G$ we have $U(ab) = r(a,b)U(a)U(b)$ where $|r(a,b)| = 1$. When $r(a,b) \equiv 1$, U is called a *(true) unitary representation*. The reader should also review Section 4.5, and in particular the notion of unitarily equivalent representations.

We argued in Section 2.10 that, as a consequence of the fact that the group of translations represents a group of symmetries of Euclidean space, any quantum system on $\mathbb{R}^3$ should come equipped with a projective unitary representation of $\mathbb{R}^3$. As the group of rotations $SO(3)$ represents another group of symmetries of Euclidean space, we expect *that every quantum system comes equipped with a strongly continuous* projective unitary representation U of $SO(3)$*. It will be useful in the sequel to consider the following very simple example of this situation. Suppose that we study a system whose state vector in position state space is a function f on $\mathbb{R}^3$, and that we rotate this system by a rotation R. Thinking that the value of the wave function at a point is related to the probability density of finding the particle at that point, it should be clear that the wave function of the system after the rotation should be $f_{\text{new}}(\boldsymbol{x}) = f(R^{-1}(\boldsymbol{x}))$. That is, the "new" wave function at a point $\boldsymbol{x}$ is the "old" wave function at the point $R^{-1}(\boldsymbol{x})$ before the rotation has taken place. As we have seen this formula defines a representation of $SO(3)$

$$U(R)(f)(\boldsymbol{x}) = f(R^{-1}(\boldsymbol{x})), \tag{II.1}$$

in the position state space $L^2(\mathbb{R}^3)$. For further considerations, it is more convenient to move to momentum state space $L^2(\mathbb{R}^3, \mathrm{d}^3p/(2\pi\hbar)^3)$. Taking the Fourier transform, it is straightforward to check that the action takes the same form as in (II.1), for a function φ,

$$U(R)(\varphi)(\boldsymbol{p}) = \varphi(R^{-1}(\boldsymbol{p})). \tag{II.2}$$

* Some kind of continuity seems physically obvious, and strong continuity is the fruitful notion.

We can expect that the simplest projective representations of $SO(3)$, that is the finite-dimensional irreducible unitary projective representations of $SO(3)$, will be especially important. It turns out that for each integer $j \geq 0$ there is an essentially unique such representation of dimension $j+1$. Furthermore, for j odd, these representations definitely do not arise from true representations. This has mind-boggling consequences, such as the fact that if one performs a rotation of angle 2π around a given axis on a quantum system equipped with such a representation, it may change state. Even more remarkable is that Nature has made extensive use of this structure. These representations are the correct tools to describe a certain "internal state" of particles, which has no classical equivalent† and which is called the *spin* of the particle. The case $j = 1$ (called spin 1/2) is of particular importance since we are built from electrons and nucleons of spin 1/2. This book being already several books in one, it is simply not possible to study spin in detail here (although we do cover a lot of the underlying mathematics). We will say just enough about spin to argue in the next chapter that a certain fundamental family of representations of the Poincaré group may be interpreted as "describing massive particles with spin". These representations of the Poincaré group are in turn the foundations for the construction of free fields in Chapter 10 corresponding to massive particles with spin (which are themselves basic to understanding the interactions of such massive particles). In some sense, this entire part of the book takes its roots in the notion of spin, and it is only natural that it should be named after it. The material presented in this part is both fundamental and beautiful. It will however not be used in our study of interactions, which we restrict to the simplest case of spinless particles, so that the reader interested only in understanding the basic problems of interaction theory may proceed directly to Part III.

† One may argue that it resembles angular momentum.

8

Representations of the Orthogonal and the Lorentz Group

Mathematically, some miracle happens: The complication of having to consider projective representations goes away when $SO(3)$ is replaced by a certain group of complex 2×2 matrices, the special unitary group $SU(2)$. The special orthogonal group $SO(3)$ is the quotient of $SU(2)$ by a two-to-one group homomorphism.[1] Thus any projective unitary representation of $SO(3)$ gives rise to a projective[2] representation of $SU(2)$. The miracle is that all the strongly continuous projective representations of $SU(2)$ arise from true representations in the sense of Definition 2.11.1. In some sense, $SU(2)$ is the "correct version" of $SO(3)$. A similar situation occurs with the Lorentz group $SO^{\uparrow}(1,3)$, for which "the correct version" is $SL(2,\mathbb{C})$, the group of 2×2 matrices of determinant 1.

In the present chapter we develop three related but distinct threads of ideas. In the next section, we define the groups $SU(2)$ and $SL(2,\mathbb{C})$ and list simple properties. In Section 8.2 we provide a first construction of the fundamental family of representations of $SU(2)$. In Section 8.3 we show how to use tensor products to construct new representations of a group out of known ones and we provide a second construction of the fundamental family of representations of $SU(2)$. These two sections are largely independent of the rest of the material in this chapter. In Section 8.4 we start in a new direction. We construct the fundamental two-to-one homomorphism κ from $SL(2,\mathbb{C})$ to $SO^{\uparrow}(1,3)$, and in Section 8.5 from this homomorphism we construct projective representations of $SO(3)$ on $\mathbb{C}^2$ which are intrinsically not true representations. In Section 8.7 we relate projective representations of $SO(3)$ and true representations of $SU(2)$ and we link these representations to the idea of spin. The last two sections, starting with Section 8.9 develop the third direction: How to incorporate parity as an additional symmetry of the theory, and how this leads to the discovery of Dirac matrices.

8.1 The Groups $SU(2)$ and $SL(2,\mathbb{C})$

These two groups of 2×2 matrices are fundamental and deserve to be looked at in great detail. Their intimate connection with the group of rotations and the Lorentz group will be

[1] Here two-to-one means that this homomorphism is onto, and that the inverse image of each element consists of two elements.

[2] Let us recall here that the word "unitary" is not always repeated, as all our representations are understood to be unitary unless otherwise specified.

described only in Section 8.4, so for the time being we take their importance for granted. We describe a few elementary but basic properties.

Definition 8.1.1 The group $SL(2,\mathbb{C})$ is the group of 2×2 complex matrices of determinant 1.

In this terminology L stands for "linear" and S for "special", i.e. of determinant 1. Of course 2 stands for the dimension and $\mathbb{C}$ for the fact that we deal with complex matrices. This group is connected but is not compact, as is seen by considering the elements

$$\begin{pmatrix} a & 0 \\ 0 & 1/a \end{pmatrix}$$

which do not have a limit as $a \to \infty$. As a consequence, it is known on general grounds that it cannot have a non-trivial continuous finite-dimensional unitary representation. A direct proof of this fact follows from the description of all the continuous finite-dimensional representations given in Appendix D. Consequently, when we discuss representations of $SL(2,\mathbb{C})$, it is always understood that *these are non-unitary representations.*

Given a 2×2 matrix A, the matrix A^* called the *conjugate* of A is obtained from A by replacing each entry by its complex conjugate. The matrix A^T is the transpose of A. Thus

$$A = \begin{pmatrix} \alpha & \beta \\ \gamma & \delta \end{pmatrix} \Rightarrow A^* = \begin{pmatrix} \alpha^* & \beta^* \\ \gamma^* & \delta^* \end{pmatrix} \; ; \; A^T = \begin{pmatrix} \alpha & \gamma \\ \beta & \delta \end{pmatrix}.$$

We also consider $A^\dagger$, the conjugate-transpose of A, i.e.

$$A = \begin{pmatrix} \alpha & \beta \\ \gamma & \delta \end{pmatrix} \Rightarrow A^\dagger = \begin{pmatrix} \alpha^* & \gamma^* \\ \beta^* & \delta^* \end{pmatrix}.$$

The name of conjugate-transpose for $A^\dagger$ is justified by $A^\dagger = (A^T)^* = (A^*)^T$. To understand the notation, let us observe that to each 2×2 complex matrix A corresponds an operator on $\mathbb{C}^2$: Identifying the elements of $\mathbb{C}^2$ with column vectors, it is the operator $x \mapsto Ax$ where Ax is the column vector product of the matrix A and the column vector x. Under this natural identification, the operator corresponding to the matrix $A^\dagger$ is the adjoint of the operator corresponding to the matrix A. Here physics notation clashes with standard mathematical notation and the reader should be careful to distinguish between the conjugate matrix A^* and the transpose-conjugate matrix $A^\dagger$.

Definition 8.1.2 Given a group G of $n \times n$ matrices the *defining representation of G* is the representation of G on $\mathbb{C}^n$ obtained by considering as above each element of G as an operator on $\mathbb{C}^n$.

The following rather trivial fact will turn out to be quite important.

Lemma 8.1.3 *The map* $A \mapsto A^{\dagger -1}$ *is a group homomorphism of* $SL(2,\mathbb{C})$, *that is* $(AB)^{\dagger -1} = A^{\dagger -1} B^{\dagger -1}$.

Proof Each of the operations $A \mapsto A^\dagger$ and $A \mapsto A^{-1}$ reverses the order of multiplication, e.g. $A^\dagger B^\dagger = (BA)^\dagger$. □

We can immediately write no less than four different (non-unitary) representations of dimension 2 of $SL(2,\mathbb{C})$.[3] They are given respectively by

$$\pi(A) = A\ ;\ \pi(A) = A^{\dagger-1}\ ;\ \pi(A) = A^*\ ;\ \pi(A) = (A^T)^{-1}.$$

Each of these formulas defines a representation because $\pi(AB) = \pi(A)\pi(B)$.

The first two representations above, $\pi(A) = A$ and $\pi(A) = A^{\dagger-1}$ are of fundamental importance. They are not equivalent in the sense that there exists no element $B \in SL(2,\mathbb{C})$ for which $BAB^{-1} = A^{\dagger-1}$ for all A. This will be obvious in a moment. The following says that, on the other hand, the representation $\pi(A) = A^*$ is equivalent to the representation $\pi(A) = A^{\dagger-1}$.

Lemma 8.1.4 *Consider the matrix*

$$J = \begin{pmatrix} 0 & 1 \\ -1 & 0 \end{pmatrix}. \tag{8.1}$$

Consider $C \in SL(2,\mathbb{C})$. Then denoting by C^* the conjugate matrix of C, where all entries have been replaced by their complex conjugate, we have $J^{-1}C^*J = C^{\dagger-1}$.

Proof For $C \in SL(2,\mathbb{C})$ we have

$$C = \begin{pmatrix} \alpha & \beta \\ \gamma & \delta \end{pmatrix}$$

with $\alpha\delta - \beta\gamma = 1$, so that

$$C^{-1} = \begin{pmatrix} \delta & -\beta \\ -\gamma & \alpha \end{pmatrix};\ C^{\dagger-1} = \begin{pmatrix} \delta^* & -\gamma^* \\ -\beta^* & \alpha^* \end{pmatrix}, \tag{8.2}$$

while, by direct computation, and since $J^{-1} = -J$,

$$J^{-1}C^*J = \begin{pmatrix} 0 & -1 \\ 1 & 0 \end{pmatrix}\begin{pmatrix} \alpha^* & \beta^* \\ \gamma^* & \delta^* \end{pmatrix}\begin{pmatrix} 0 & 1 \\ -1 & 0 \end{pmatrix} = \begin{pmatrix} \delta^* & -\gamma^* \\ -\beta^* & \alpha^* \end{pmatrix}.$$ □

It is rather obvious that the representations $\pi(A) = A$ and $\pi(A) = A^*$ are not equivalent. Indeed there cannot exist a matrix B such that $BAB^{-1} = A^*$ because the matrix BAB^{-1} has the same eigenvalues as the matrix A, while this is not in general the case for A^*. Consequently the representations $\pi(A) = A$ and $\pi(A) = A^{\dagger-1}$ are not equivalent either.

Exercise 8.1.5 Show that the representation $\pi(A) = (A^T)^{-1}$ is equivalent to the representation $\pi(A) = A$.

Definition 8.1.6 The group $SU(2)$ is the group of 2×2 complex unitary matrices A (i.e. such that $AA^\dagger$ is the identity) of determinant 1.

Thus $SU(2)$ is a subgroup of $SL(2,\mathbb{C})$. It is exactly the subset of $SL(2,\mathbb{C})$ consisting of matrices A for which $A = A^{\dagger-1}$. We can view $SU(2)$ as the group of unitary operators

[3] We will eventually learn how to describe all finite-dimensional representations of $SL(2,\mathbb{C})$. There are no other non-equivalent representations of dimension two than the ones we describe here.

of determinant 1 on a two-dimensional Hilbert space $\mathcal{H}$. As in the case of $SL(2,\mathbb{C})$, the elements of $SU(2)$ act on $\mathcal{H}$ in a canonical manner: If we represent the elements of $\mathcal{H}$ as column matrices, this operation is matrix multiplication. In contrast with $SL(2,\mathbb{C})$ we do not get a new representation when we make A act by $A^{\dagger-1}$ on $\mathbb{C}^2$ because $A^{\dagger-1} = A$. Also in contrast with $SL(2,\mathbb{C})$, $SU(2)$ is a compact group.

When one needs to do explicit computations, the following is useful to know.

Lemma 8.1.7 *An element of $SU(2)$ is of the type*

$$U = \begin{pmatrix} \alpha & \beta \\ -\beta^* & \alpha^* \end{pmatrix} \tag{8.3}$$

for complex numbers α, β with $|\alpha|^2 + |\beta|^2 = 1$.

Proof The rows of the matrix must be orthogonal of length 1, so that given the first row (α, β) the second row must be of the type $(-c\beta^*, c\alpha^*)$ and computing the determinant of the matrix yields $c = 1$. □

Let us call a *curve* a continuous map φ from $[0,1]$ to $SL(2,\mathbb{C})$. We say that the curve φ is a *loop* if $\varphi(0) = \varphi(1)$.

Corollary 8.1.8 *The group $SU(2)$ is simply connected i.e. all loops can be continuously contracted to a point. In fact each loop can be contracted to the unit I. That is, given a loop φ, there is a continuous map f from $[0,1]^2$ to $SU(2)$ such that $f(0,t) = I$ and $f(1,t) = \varphi(t)$.*

Proof It follows from Lemma 8.1.7 that topologically $SU(2)$ is the three-sphere in $\mathbb{C}^2 \simeq \mathbb{R}^4$. □

Lemma 8.1.9 *The group $SL(2,\mathbb{C})$ is simply connected.*

Proof For $C \in SL(2,\mathbb{C})$ the matrix $CC^\dagger$ is positive Hermitian i.e. its eigenvalues are ≥ 0, because $(x, CC^\dagger x) = \|C^\dagger x\|^2 \geq 0$ for any x. Then there is a basis on which $CC^\dagger$ is diagonal. This makes it obvious that there exists a positive Hermitian matrix B such that $B^2 = CC^\dagger$. Then $B^{-1}CC^\dagger B^{\dagger-1} = I$, so that $B^{-1}C \in SU(2)$. Thus $C = BA$ where $A \in SU(2)$ and B is positive Hermitian. It is easy to show that B is determined by the relation $BB^\dagger = CC^\dagger$ (see Exercise 8.1.10) and that the map $C \mapsto (B,A)$ is continuous. Consider a loop $C(t)$ in $SL(2,\mathbb{C})$. Using the unique decomposition $C(t) = B(t)A(t)$ as above, then $A(t)$ defines a loop in $SU(2)$ and $B(t)$ defines a loop in the space of positive Hermitian matrices of determinant 1. Since we have proved that we can contract the curve $A(t)$ to the unit I, it suffices to show that the curve $B(t)$ can be contracted to a constant loop.[4] For $0 \leq t, u \leq 1$ let us define $B(t,u)$ as follows. Since $B(t)$ is positive Hermitian, there is an orthonormal basis such that the basis vectors are eigenvectors of $B(t)$ with eigenvalues $\lambda(t) \geq 1$ and $1/\lambda(t)$. (This basis is essentially uniquely determined unless $B(t) = I$.) In the same orthogonal

[4] We are sketching here the idea of homotopy. The reader can find more details in a basic topology book such as that of Munkres [57].

basis the eigenvalues of $B(t,u)$ are $\lambda(t,u)$ and $1/\lambda(t,u)$ where $\lambda(t,u) = 1 + u(\lambda(t) - 1)$. It is then obvious that the map $(t,u) \to B(t,u)$ is continuous, whereas $B(t,1) = B(t)$ and $B(t,0) = I$. □

Exercise 8.1.10 Prove that a positive Hermitian matrix has a unique positive Hermitian square root. Prove that the map that sends a positive Hermitian matrix A to its positive Hermitian square root is continuous.

8.2 A Fundamental Family of Representations of $SU(2)$

For each $j \geq 0$ there exists an irreducible unitary representation π_j of $SU(2)$ of dimension $j+1$, and this representation is unique up to unitary equivalence. The representations π_j being fundamental objects, we shall look at them from several angles. On the other hand, the uniqueness up to unitary equivalence is less fundamental for our main story, so its proof is relegated to Appendix D.

In the present section we will follow the idea of Section 4.6 to construct the representations π_j. This presentation appeals to mathematicians, although physicists usually prefer to use tensors to describe this representation, the topic of the next section.

Lemma 8.2.1 *The elements of $SU(2)$ act on the unit sphere $\mathbb{S} = \{z_1, z_2; |z_1|^2 + |z_2|^2 = 1\}$ of $\mathbb{C}^2$. The canonical probability $\mathrm{d}\mu$ on $\mathbb{S}$ is invariant under this action.*

Proof The elements of $SU(2)$ are unitary operators which act on the two-dimensional Hilbert space $\mathbb{C}^2$. Since they preserve the norm, they leave the unit sphere $\mathbb{S}$ of $\mathbb{C}^2$ invariant. Thus $SU(2)$ acts on $\mathbb{S}$. Since $\mathbb{S}$ identifies with the unit sphere of $\mathbb{R}^4$, it has a canonical probability measure $\mathrm{d}\mu$, which is simply the normalized surface measure of the unit sphere of $\mathbb{R}^4$. It is invariant under the action of any element of $SU(2)$. This is obvious if one looks at what happens in a basis where this element of $SU(2)$ is diagonal. Indeed the transformations $(z_1, z_2) \to (az_1, z_2)$ where $|a| = 1$ preserve $\mathrm{d}\mu$ and the action of a diagonal element of $SU(2)$ is the composition of two such operations (one for each coordinate). □

Consequently, as explained in Section 4.6, this action of $SU(2)$ on $\mathbb{S}$ induces a unitary representation of $SU(2)$ on $L^2(\mathbb{S}, \mathrm{d}\mu)$. This representation is however not irreducible. Indeed for every $j \geq 0$ the space of (restrictions to $\mathbb{S}$ of) homogeneous polynomials of degree j in the coordinates is invariant under this action. Such a polynomial is of the form

$$f(z_1, z_2) = \sum_{0 \leq i \leq j} c_i z_1^i z_2^{j-i}, \tag{8.4}$$

so that these polynomials form a space $\mathcal{H}_j$ of dimension $j+1$. Through a brief but violent struggle with elementary integrals (using Gaussian integrals to compute integrals of polynomials on the sphere in $\mathbb{R}^4$) one may prove that the monomials $z_1^i z_2^{j-i}$ form an orthogonal basis in $L^2(\mathbb{S}, \mathrm{d}\mu)$, and that

$$\|z_1^i z_2^{j-i}\|^2 = \frac{1}{(j+1)\binom{j}{i}}. \tag{8.5}$$

Exercise 8.2.2 The identity $|z_1|^2 + |z_2|^2 = 1$ holds on $\mathbb{S}$. Raise this identity to the power j to check the formula

$$\sum_{0 \le i \le j} \binom{j}{i} \|z_1^i z_2^{j-i}\|^2 = 1.$$

Let us rephrase the definition of this representation.

Definition 8.2.3 The representation π_j of $SU(2)$ on the space $\mathcal{H}_j$ of homogeneous polynomials of degree j on the unit sphere of $\mathbb{C}^2$ is given for $A \in SU(2)$ by

$$\pi_j(A)(f)(z_1, z_2) = f(z_1', z_2') \tag{8.6}$$

where

$$\begin{pmatrix} z_1' \\ z_2' \end{pmatrix} = A^{-1} \begin{pmatrix} z_1 \\ z_2 \end{pmatrix} = A^{\dagger} \begin{pmatrix} z_1 \\ z_2 \end{pmatrix}. \tag{8.7}$$

Let us stress again that in (8.7) (as in (4.46)) appears A^{-1} rather than A. Let us also observe the following immediate fact, where I denotes the 2×2 identity matrix,

$$\pi_j(-I)(f) = (-1)^j f. \tag{8.8}$$

Exercise 8.2.4 For $j = 1$, prove that the representation π_1 is equivalent to the representation $\pi(A) = A$.

Proposition 8.2.5 *The representation π_j is irreducible.*

Consider the element f of $\mathcal{H}_j$ given by (8.4), and assume that $f \neq 0$. Let us denote by $\mathcal{G}$ the closed linear span of the vectors $\pi_j(A)(f)$ for $A \in SU(2)$. The goal is to show that $\mathcal{G} = \mathcal{H}_j$ (after which the conclusion follows by Lemma 4.5.6). We will reach it in several steps.

Lemma 8.2.6 *If $\mathcal{G}$ contains the polynomial $\sum_i c_i z_1^i z_2^{j-i}$ then it contains each monomial $c_k z_1^k z_2^{j-k}$. In particular if $c_k \neq 0$ the space $\mathcal{G}$ contains the monomial $z_1^k z_2^{j-k}$.*

Proof Consider the case where A is diagonal, with diagonal elements a, a^{-1}, $|a| = 1$. Then

$$\pi_j(A) f(z_1, z_2) = \sum_{0 \le i \le j} c_i a^{-i} z_1^i a^{j-i} z_2^{j-i}. \tag{8.9}$$

Thus, multiplying the above right-hand side by a^{2k-j}, we obtain that for any $0 \le k \le j$ the space $\mathcal{G}$ contains the polynomials

$$g(z_1, z_2) = \sum_{0 \le i \le j} c_i a^{2k-2i} z_1^i z_2^{j-i}.$$

Taking $a = \exp(i\theta)$, averaging over $0 \le \theta \le 2\pi$ and using that $\int_0^{2\pi} d\theta \exp(in\theta) = 0$ for $n \neq 0$ yields that $\mathcal{G}$ contains the monomial $c_k z_1^k z_2^{j-k}$. □

Lemma 8.2.7 *If $\mathcal{G}$ contains a monomial $z_1^k z_2^{j-k}$ then for each $\theta \in \mathbb{R}$ it contains the polynomial*

$$(z_1 \cos\theta + z_2 \sin\theta)^k(-z_1 \sin\theta + z_2 \cos\theta)^{j-k}. \tag{8.10}$$

Proof Consider the case where A is the rotation of angle θ. □

Proof of Proposition 8.2.5 As a consequence of Lemma 8.2.6, $\mathcal{G}$ contains a monomial $z_1^k z_2^{j-k}$. Therefore it contains the polynomial (8.10) for each θ. When $\sin\theta\cos\theta \neq 0$, the coefficient of z_1^j in this polynomial is not zero. According to Lemma 8.2.6, $\mathcal{G}$ contains the monomial z_1^j. It then contains the polynomial (8.10) in the case $k = 0$. When $\sin\theta\cos\theta \neq 0$, the coefficient of each monomial in this polynomial is $\neq 0$. Then $\mathcal{G}$ contains each such monomial by Lemma 8.2.6, and therefore coincides with $\mathcal{H}_j$. □

The previous proof is the simplest "by hand" proof I could figure out. The alternate proof below is an elementary transcription of the "standard" proof using Lie algebras, about which the reader can learn in Appendix D.

Magic Proof of Proposition 8.2.5 Consider a subspace $\mathcal{G}$ of $\mathcal{H}_j$ which is invariant under each map $\pi_j(A)$ for $A \in SU(2)$. If σ is a traceless 2×2 Hermitian matrix, then $\exp it\sigma \in SU(2)$, so that given $f \in \mathcal{G}$ then $\pi_j(\exp it\sigma)(f) \in \mathcal{G}$. The derivative in t of this t-dependent element of $\mathcal{G}$ also belongs to $\mathcal{G}$:

$$-\frac{1}{\mathrm{i}}\frac{\mathrm{d}}{\mathrm{d}t}\pi_j(\exp it\sigma)(f) \in \mathcal{G}. \tag{8.11}$$

Now, $\pi_j(\exp it\sigma)(f)(z_1, z_2) = f(z_1(t), z_2(t))$ where

$$\begin{pmatrix} z_1(t) \\ z_2(t) \end{pmatrix} = \exp(-it\sigma)\begin{pmatrix} z_1 \\ z_2 \end{pmatrix}.$$

Using then (8.11) for $t = 0$ we obtain that for $f \in \mathcal{G}$, $\dot{z}_1\frac{\partial f}{\partial z_1} + \dot{z}_2\frac{\partial f}{\partial z_2} \in \mathcal{G}$ where

$$\begin{pmatrix} \dot{z}_1 \\ \dot{z}_2 \end{pmatrix} = \sigma\begin{pmatrix} z_1 \\ z_2 \end{pmatrix}.$$

Here σ can be any of the Pauli matrices $\sigma_1, \sigma_2, \sigma_3$, or even any linear combination with complex coefficients of such matrices. Using for σ the matrices $\sigma_1 \pm i\sigma_2$ one finds that $\mathcal{G}$ is invariant under the operators $z_2\frac{\partial}{\partial z_1}$ and $z_1\frac{\partial}{\partial z_2}$. We now show that this implies that $\mathcal{G} = \mathcal{H}_j$ whenever $\mathcal{G}$ is not empty. Indeed, consider $f(z_1, z_2) = \sum_{0 \le i \le j} c_i z_1^i z_2^{j-i} \neq 0 \in \mathcal{G}$. Consider the largest integer k for which $c_k \neq 0$. Applying k times the operator $z_2\frac{\partial}{\partial z_1}$ shows that the monomial z_2^j belongs to $\mathcal{G}$, and applying i times the operator $z_1\frac{\partial}{\partial z_2}$ to this monomial shows that the monomial $z_1^i z_2^{j-i}$ belongs to $\mathcal{G}$. □

Let us note that π_j extends to a representation of $SL(2,\mathbb{C})$ by using the very same formulas (8.6) and (8.7). Of course this representation is not unitary.

In physics one often makes use of special properties of the representation π_j. These are investigated in the next exercise.

Exercise 8.2.8 Prove the following. Consider the orthonormal basis $f_i = \alpha_i z_1^i z_2^{j-i}$ for $0 \le i \le j$ where $\alpha_i > 0$ is a normalization constant. Prove that $\alpha_i = \alpha_{j-i}$. Prove that the matrix J of Lemma 8.1.4 satisfies

$$\pi_j(J)(f_i) = (-1)^i f_{j-i} \; ; \; \pi_j(J^{-1})(f_i) = (-1)^{j-i} f_{j-i}. \tag{8.12}$$

For $C \in SU(2)$ denote by $\pi_j(C)_{k,\ell}$ the matrix of $\pi_j(C)$ in this basis. Then

$$\pi_j(C)^*_{k,\ell} = \pi_j(C^*)_{k,\ell} = (-1)^{k+\ell}\pi_j(C)_{j-k,j-\ell}. \tag{8.13}$$

8.3 Tensor Products of Representations

Tensor products provide a powerful method to construct new representations from old ones. The basic idea is that if for $j = 1,2$, U_j is a representation of a group G on $\mathcal{H}_j$ then $U_1 \otimes U_2(A) := U_1(A) \otimes U_2(A)$ is a new representation of G, this time on $\mathcal{H}_1 \otimes \mathcal{H}_2$. When U_1 and U_2 are unitary, so is the new representation. As will be obvious soon, even when U_1 and U_2 are irreducible, this need not be the case of $U_1 \otimes U_2$. However when we are working with unitary representations, $\mathcal{H}_1 \otimes \mathcal{H}_2$ can be decomposed in an orthogonal sum of irreducible invariant subspaces for $U_1 \otimes U_2$ and it is a very interesting problem to figure out which irreducible representations occur this way. In the case where $G = SU(2)$ the problem is known under the name *addition of angular momentum*. It is treated in detail in most Quantum Mechanics textbooks, and in Proposition D.7.6. Typically inside $\mathcal{H}_1 \otimes \mathcal{H}_2$ one can find irreducible invariant subspaces on which $U_1 \otimes U_2$ is "larger" than either U_1 or U_2.

Let us put this idea of taking tensor products to work in the simplest possible manner. Consider the tensor product $\mathcal{H} = (\mathbb{C}^2)^{\otimes j}$ of j copies of $\mathbb{C}^2$. Thus, if e_1, e_2 denote the standard basis vectors of $\mathbb{C}^2$, an orthonormal basis of this space consists of the 2^j vectors $e_{n_1} \otimes e_{n_2} \otimes \cdots \otimes e_{n_j}$ where $n_i \in \{1,2\}$ for $1 \le i \le j$, and a generic element of the tensor product is a sum

$$x = \sum x_{n_1,\ldots,n_j} e_{n_1} \otimes e_{n_2} \otimes \cdots \otimes e_{n_j}$$

where the summation is over all $n_i \in \{1,2\}$ for $1 \le i \le j$. Using simpler notation, we write $x = (x_{n_1,\ldots,n_j})$ where it is understood that all indices are equal to 1 or 2. Each element A of $SU(2)$ can be seen as a unitary operator on $\mathbb{C}^2$ and the operator $U(A) := A^{\otimes j}$ is therefore unitary on $\mathcal{H}$. Thus $A \mapsto U(A)$ defines a unitary representation of $SU(2)$. In concrete terms, if $A_{n,k}$ denotes the entries of the matrix A, with $n \in \{1,2\}$ indicating the row and $k \in \{1,2\}$ indicating the column, $U(A)$ transforms the tensor $(x_{n_1,\ldots,n_j})$ into the tensor

$$\Big(\sum_{k_1,\ldots,k_j} A_{n_1,k_1} \cdots A_{n_j,k_j} x_{k_1,\ldots,k_j}\Big). \tag{8.14}$$

This representation is however not irreducible because it leaves invariant the subspace S_j of symmetric tensors, those for which $x_{n_1,\ldots,n_j}$ is invariant under any permutation of the indices. For such a symmetric tensor, the value of $x_{n_1,\ldots,n_j}$ depends only on how many of the indices are equal to 1 (the rest being equal to 2) so $\mathcal{S}_j$ is a space of dimension $j+1$. Let us denote by Π_j the restriction of the representation $A \mapsto U(A)$ to $\mathcal{S}_j$, so that

$$\Pi_j(A)((x_{n_1,\ldots,n_j})) = \Big(\sum_{k_1,\ldots,k_j} A_{n_1,k_1} \cdots A_{n_j,k_j} x_{k_1,\ldots,k_j} \Big). \tag{8.15}$$

Proposition 8.3.1 *The representation Π_j of $SU(2)$ in $\mathcal{S}_j$ is unitarily equivalent to the representation π_j of $SU(2)$ in the space of polynomials described in Definition 8.2.3.*

Proof Let us recall the space $\mathcal{H}_j$ of polynomials of the form (8.4). We define the map $W: \mathcal{S}_j \to \mathcal{H}_j$ by

$$W((x_{n_1,\ldots,n_j})) = \sum_{n_1,\ldots,n_j} x_{n_1,\ldots,n_j} z_{n_1} z_{n_2} \cdots z_{n_j}, \tag{8.16}$$

where the summation is over $n_1,\ldots,n_j \in \{1,2\}$. First we prove that $\sqrt{j+1}W$ is unitary. For this let c_i be the common value of the numbers $x_{n_1,\ldots,n_j}$ when exactly i of the indices n_ℓ are equal to 1, so that the right-hand side of (8.16) equals

$$\sum_{0 \le i \le j} \binom{j}{i} c_i z_1^i z_2^{j-i}$$

and according to (8.5) the square of the norm of this element is

$$\sum_{0 \le i \le j} \binom{j}{i}^2 |c_i|^2 \|z_1^i z_2^{j-i}\|^2 = \frac{1}{j+1} \sum_{0 \le i \le j} \binom{j}{i} |c_i|^2 = \frac{1}{j+1} \|(x_{n_1},\ldots,x_{n_j})\|^2.$$

We reformulate (8.7) as $z'_n = \sum_{k=1,2} A^*_{k,n} z_k$ and we compute

$$\begin{aligned}
\pi_j(A) \circ W((x_{n_1,\ldots,n_j})) &= \sum_{n_1,\ldots,n_j} x_{n_1,\ldots,n_j} z'_{n_1} z'_{n_2} \cdots z'_{n_j} \\
&= \sum_{k_1,\ldots,k_j} \Big(\sum_{n_1,\ldots,n_j} A^*_{k_1,n_1} \cdots A^*_{k_j,n_j} x_{n_1,\ldots,n_j} \Big) z_{k_1} \cdots z_{k_j} \\
&= \sum_{n_1,\ldots,n_j} \Big(\sum_{k_1,\ldots,k_j} A^*_{n_1,k_1} \cdots A^*_{n_j,k_j} x_{k_1,\ldots,k_j} \Big) z_{n_1} \cdots z_{n_j}.
\end{aligned}$$

Comparing with the transformation rule (8.14) this means that

$$\pi_j(A) \circ W = W \circ \Pi_j(A^*),$$

and we have shown that the representations $A \mapsto \pi_j(A)$ and $A \mapsto \Pi_j(A^*)$ are unitarily equivalent.

Finally, since by Lemma 8.1.4 we have $J^{-1}A^*J = A$ for $A \in SU(2)$, it holds that $A^{\otimes j} = V^{-1}(A^*)^{\otimes j} V$ for $V = J^{\otimes j}$ and thus the representations $A \mapsto \Pi_j(A^*)$ and $A \mapsto \Pi_j(A)$ are unitarily equivalent. Therefore π_j and Π_j are also unitarily equivalent. □

We now use the same idea to describe some representations of $SL(2,\mathbb{C})$. A key feature here is that the map $A \mapsto A^{\dagger-1}$ is a representation of $SL(2,\mathbb{C})$, and that it may also be "tensorized". Consider the space $\mathcal{S}_{j,\ell}$ of tensors

$$x = (x_{n_1,\ldots,n_j,m_1,\ldots,m_\ell})$$

which are symmetric in both $n_1,\ldots,n_j$ and $m_1,\ldots,m_\ell$,[5] a space of dimension $(j+1)(\ell+1)$.

Definition 8.3.2 The (j,ℓ) representation[6] of $SL(2,\mathbb{C})$ is the representation on $\mathcal{S}_{j,\ell}$ such that the action of $A \in SL(2,\mathbb{C})$ on $\mathcal{S}_{j,\ell}$ is given by the restriction to this space of the operator $A^{\otimes j} \bigotimes (A^{\dagger-1})^{\otimes \ell}$.

In Appendix D we prove that this representation is irreducible, and that all irreducible finite-dimensional representations of $SL(2,\mathbb{C})$ are obtained this way. We will use only the $(1,0)$, $(0,1)$ and $(1,1)$ representations. The representation $(1,1)$ of $SL(2,\mathbb{C})$ lives on $\mathbb{C}^4$. We will gradually understand its importance.

8.4 $SL(2,\mathbb{C})$ as a Universal Cover of the Lorentz Group

The group $SL(2,\mathbb{C})$ is closely related to the Lorentz group $SO^{\uparrow}(1,3)$ and the group $SU(2)$ to the group $SO(3)$ of rotations. We thoroughly investigate this relationship. The importance of the present section cannot be overestimated. It is crucial for all the matters concerning representations of the Lorentz group and the group of rotations. The meaning of the title of this section will be explained only later.

Let us denote by $\mathcal{H}$ the space of 2×2 Hermitian matrices. The basic observation is that if $M \in \mathcal{H}$ and $A \in SL(2,\mathbb{C})$ then the matrix $AMA^\dagger$ is Hermitian. Thus each $A \in SL(2,\mathbb{C})$ induces a linear map $\kappa(A)\colon \mathcal{H} \to \mathcal{H}$ given by

$$\kappa(A)(M) = AMA^\dagger. \tag{8.17}$$

Moreover it is straightforward to check that this is a group homomorphism i.e. $\kappa(A)\kappa(B) = \kappa(AB)$. As it turns out, $\mathcal{H}$ can be naturally identified with $\mathbb{R}^{1,3}$, and under this identification we are going to prove that $\kappa(A)$ is a Lorentz transformation. In this manner we define a group homomorphism κ from $SL(2,\mathbb{C})$ to $SO^{\uparrow}(1,3)$.

To make matters explicit, we observe that the space $\mathcal{H}$ of 2×2 Hermitian matrices is of dimension four as a *real* vector space. A natural basis consists of the following four matrices:

$$\sigma_0 = I = \begin{pmatrix} 1 & 0 \\ 0 & 1 \end{pmatrix} ; \sigma_1 = \begin{pmatrix} 0 & 1 \\ 1 & 0 \end{pmatrix} ; \sigma_2 = \begin{pmatrix} 0 & -\mathrm{i} \\ \mathrm{i} & 0 \end{pmatrix} ; \sigma_3 = \begin{pmatrix} 1 & 0 \\ 0 & -1 \end{pmatrix}. \tag{8.18}$$

We can then identify $\mathcal{H}$ with $\mathbb{R}^{1,3}$ using the correspondence

$$x \in \mathbb{R}^{1,3} \leftrightarrow M(x) := x^\mu \sigma_\mu = \begin{pmatrix} x^0 + x^3 & x^1 - \mathrm{i}x^2 \\ x^1 + \mathrm{i}x^2 & x^0 - x^3 \end{pmatrix}, \tag{8.19}$$

5 That is, the value of $x_{n_1,\ldots,n_j,m_1,\ldots,m_\ell}$ does not change under any permutation of $n_1,\ldots,n_j$ or any permutation of $m_1,\ldots,m_\ell$.

6 This representation is usually denoted $(j/2,\ell/2)$, and we do not follow the common usage here.

where we use the standard notation $x^\mu\sigma_\mu = \sum_{0\le\mu\le 4} x^\mu\sigma_\mu$. This notation will be used many times and should be learned now. The key fact is that

$$\det M(x) = (x^0)^2-(x^1)^2-(x^2)^2-(x^3)^2 = x_\mu x^\mu = x^2 = (x,x). \tag{8.20}$$

For every $A \in SL(2,\mathbb{C})$ and every $x \in \mathbb{R}^{1,3}$ the matrix $AM(x)A^\dagger$ is Hermitian, so there exists a unique element $\kappa(A)(x) \in \mathbb{R}^{1,3}$ such that

$$M(\kappa(A)(x)) = AM(x)A^\dagger, \tag{8.21}$$

which is just the way to rewrite (8.17) under our identification of $\mathcal{H}$ and $\mathbb{R}^{1,3}$. Moreover, since $\det A = 1$, we have $\det M(\kappa(A)(x)) = \det M(x)$, which means from (8.20) that $\kappa(A)$ preserves the quantity (x,x) i.e. $\kappa(A) \in O(1,3)$, that is, κ is a group homomorphism from $SL(2,\mathbb{C})$ into $O(1,3)$. The notation $\kappa(A)$ will be massively used in the sequel and should be memorized at this point. It is easy and very instructive to compute $\kappa(A)$ when A is diagonal. In that case, since $A \in SL(2,\mathbb{C})$ (and since any non-zero complex number is of the type $\exp a$ for some $a \in \mathbb{C}$), it is of the type

$$A = \begin{pmatrix} \exp a & 0 \\ 0 & \exp(-a) \end{pmatrix} = \exp(a\sigma_3),$$

where $a \in \mathbb{C}$ and where the exponential of the matrix is computed as the sum of the usual power series (or should simply be considered here as a convenient notation). The following is obtained by a few lines of straightforward computation but is nonetheless very important. For $s \in \mathbb{R}$,

$$\kappa(\exp s\sigma_3/2) = B^s, \tag{8.22}$$

where B^s is the boost considered in Section 4.3. Moreover, for $\theta \in \mathbb{R}$,

$$\kappa(\exp(-\mathrm{i}\theta\sigma_3/2)) = R_\theta, \tag{8.23}$$

where R_θ is the rotation of angle θ around the third axis of $\mathbb{R}^3$ seen as a Lorentz transformation in Section 4.2.

Exercise 8.4.1 Make sure you carry out the computations (8.22) and especially (8.23) to understand the confusing change of sign here.

This doubling of the angle is related to the fact that there are two factors A in the right-hand side of (8.21). What happens for other directions than the third axis is investigated in Appendix D.

Since $SL(2,\mathbb{C})$ is connected, the image of this group under κ is contained in the connected component of the identity of $O(1,3)$, i.e. $SO^\uparrow(1,3)$. After studying Appendix D it will be obvious to the reader that this image is the whole of $SO^\uparrow(1,3)$.

Exercise 8.4.2 Find a direct proof of this fact. Hint: Prove that a group of rotations that contains all the rotations around the third axis and at least another rotation is the whole of $SO(3)$. To prove that $\kappa(SU(2))$ contains a rotation that is not around the third axis, you may use direct computation or Lemma 8.4.3.

Lemma 8.4.3 *If $\kappa(A)$ is the identity then $A = \pm I$. Thus if $\kappa(A) = \kappa(B)$ then either $A = B$ or $A = -B$.*

Therefore the inverse image of any element of $SO^{\uparrow}(1,3)$ is determined up to a sign, and thus κ is a two-to-one homomorphism from $SL(2,\mathbb{C})$ onto $SO^{\uparrow}(1,3)$.

Proof If $\kappa(A)$ is the identity then in particular since $M(e_0) = I$ we have $AA^\dagger = I$ so $A \in SU(2)$. Thus $A = UBU^{-1}$ where $B \in SU(2)$ is diagonal and $U \in SU(2)$. Thus B is of the type $B = \exp(i\theta\sigma_3)$. Since $\kappa(U)\kappa(B)\kappa(U)^{-1} = I$ then $\kappa(B) = I$, and by (8.23) θ is a multiple of π so $B = \pm I$. □

An element of $SO^{\uparrow}(1,3)$ can be seen as a real 4×4 matrix, and thus as an operator on $\mathbb{C}^4$. In this manner the map κ can be thought of as a representation of $SL(2,\mathbb{C})$ of dimension four.

Exercise 8.4.4 If you keep reading you will eventually find a proof that this representation is equivalent to the representation (1,1) of Definition 8.3.2, but it would be much more rewarding to figure this out yourself. This is not easy. Hint: You may try to read Proposition D.12.2.

Let us recall that we say that an element of $SO^{\uparrow}(1,3)$ is a rotation if it is of the type $(x^0, \boldsymbol{x}) \to (x^0, T(\boldsymbol{x}))$ where $T \in SO(3)$. In this manner we identify $SO(3)$ as a subgroup of $SO^{\uparrow}(1,3)$. As we have observed an element of $SO^{\uparrow}(1,3)$ is a rotation if and only if it fixes $e_0 = (1,0,0,0)$, and since $M(e_0) = I$ this means that $\kappa(A)$ is a rotation if and only if $AA^\dagger = I$, i.e. if and only if $A \in SU(2)$. Thus the restriction of κ to $SU(2)$ is valued in $SO(3)$, and is also a two-to-one homomorphism. This map is also of considerable importance.[7] In summary:

The map κ is a two-to-one homomorphism from $SL(2,\mathbb{C})$ to $SO^{\uparrow}(1,3)$. Its restriction to $SU(2)$ is a two-to-one homomorphism from $SU(2)$ to $SO(3)$.[8]

Exercise 8.4.5 The restriction of κ to $SU(2)$ is valued in $SO(3)$. Thinking of an element of $SO(3)$ as a unitary transformation, this restriction can be seen as a unitary representation of $SU(2)$. Prove that it is unitarily equivalent to the representation π_2 of Definition 8.2.3. Warning: This is not so easy.

We end this section by investigating in more detail the map κ for further use. Let us say that a matrix is *positive definite Hermitian* if it is Hermitian and its eigenvalues are positive.

Lemma 8.4.6 *If $V \in SL(2,\mathbb{C})$ is positive definite Hermitian then $\kappa(V)$ is a pure boost. Moreover each pure boost is of the type $\kappa(V)$ for a unique positive definite Hermitian matrix.*

Proof If V is positive definite Hermitian it can be diagonalized in an orthonormal basis, so that $V = UCU^{-1}$, where U is unitary and C is diagonal with positive eigenvalues.

[7] It is unfortunate that many physics textbooks do not distinguish between $SL(2,\mathbb{C})$ and $SO^{\uparrow}(1,3)$ or between $SU(2)$ and $SO(3)$. They even sometimes go as far as not distinguishing between a Lie group and its Lie algebra.

[8] The reason for the name "universal cover" will be explained in the next section.

Thus $\kappa(V) = \kappa(U)\kappa(C)\kappa(U)^{-1}$. Since C is diagonal with positive eigenvalues, it is of the type $\exp(s\sigma_3/2)$, and we have then seen earlier that $\kappa(C)$ is the boost B^s. Since $R = \kappa(U)$ is a rotation, we know from Lemma 4.3.3 (b) that $\kappa(V) = RB^sR^{-1}$ is a pure boost. Conversely, given a boost B, to prove the existence of V, we reduce to the case where $B = B^s$ and we take $V = \exp(s\sigma_3/2)$. The uniqueness follows from the fact that if $\kappa(W) = B$ then $W = \pm V$, and that $-V$ is negative definite. □

Exercise 8.4.7 For an operator L on $\mathbb{R}^{1,3}$ we denote by L^T the operator whose matrix is the transpose of the matrix of L. Prove that

$$\kappa(A^\dagger) = \kappa(A)^T. \tag{8.24}$$

Hint: Consider first the cases where A is positive definite Hermitian and where A is unitary. Prove then that each element of $SL(2,\mathbb{C})$ is the product of two such matrices.

8.5 An Intrinsically Projective Representation

There is a fundamental difference between true and projective representations, which we illustrate in this section. Recalling first that each element of $SL(2,\mathbb{C})$ can be thought of as a linear transformation of $\mathbb{C}^2$ (a matrix acting on $\mathbb{C}^2$ in the natural manner) we will prove that if we consider a function[9] $\rho : SO^{\uparrow}(1,3) \to SL(2,\mathbb{C})$ such that for each $C \in SO^{\uparrow}(1,3)$ it holds that

$$\kappa(\rho(C)) = C, \tag{8.25}$$

then the restriction of ρ to $SO(3)$ is a projective unitary representation of $SO(3)$ which does not arise from a true representation (whereas ρ itself is a non-unitary projective representation of $SL(2,\mathbb{C})$ which does not arise from a true representation).[10]

We will show that a function ρ satisfying (8.25) can never be really satisfactory,[11] and that even its restriction to $SO(3)$ cannot be satisfactory either.

We observe first that

$$C \in SO(3) \Rightarrow \rho(C) \in SU(2), \tag{8.26}$$

because $A \in SU(2)$ if and only if $\kappa(A) \in SO(3)$.

Let us consider first a toy situation. The map $z \mapsto z^2$ from the group $\mathbb{U}$ of complex numbers of modulus 1 to itself is a two-to-one group homomorphism, which somehow corresponds to κ in the present situation. It is not possible to find a map θ from $\mathbb{U}$ to itself such that $\theta(z)^2 = z$ in a continuous and single-valued way. More accurately, this can be done locally but not on all $\mathbb{U}$. (The reader may try to show that it is not possible to find θ

[9] ρ is effectively a procedure by which, for each $C \in SO^{\uparrow}(1,3)$ we choose one element A of $SL(2,\mathbb{C})$ such that $\kappa(A) = C$. In mathematics, such a function is called a section of the map κ.

[10] If one adopts the abstract point of view of Section 2.12 "all the possible choices of ρ correspond to a single projective representation of $SO^{\uparrow}(1,3)$". To see this, we observe that the only multiples of the identity that belong to $SL(2,\mathbb{C})$ are $\pm I$. Using Lemma (8.4.3) shows that the group $SL(2,\mathbb{C})_p$, the quotient of $SL(2,\mathbb{C})$ by the subgroup of multiples of the identity, i.e. by subgroup $\{I, -I\}$, identifies with $SO^{\uparrow}(1,3)$, and the canonical map from $SL(2,\mathbb{C})$ to $SL(2,\mathbb{C})_p$ identifies with κ. The composition of κ and ρ does not depend on ρ: it is the identity of $SO^{\uparrow}(1,3)$.

[11] As is made precise in the last sentence of the statement of Theorem 8.5.1.

that is continuous.) Here one may construct ρ nicely in a neighborhood of the identity of $SO(3)$ but it is impossible to do this globally.

To investigate this, let us assume that we have chosen for each $C \in SO^{\uparrow}(1,3)$ a $\rho(C) \in SL(2,\mathbb{C})$ satisfying (8.25). Using this relation for $C = \kappa(A)$, $A \in SL(2,\mathbb{C})$ shows that $\kappa[\rho(\kappa(A))] = \kappa(A)$ so that

$$\rho(\kappa(A)) \in \{\pm A\}. \tag{8.27}$$

Thus, the "choice" of $\rho(C)$ is really only up to a sign. We will lighten notation by writing (8.27) as $\rho(\kappa(A)) = \pm A$.

Theorem 8.5.1 *For C_1, C_2 in $SO^{\uparrow}(1,3)$ we have*

$$\rho(C_1C_2) = \pm\rho(C_1)\rho(C_2). \tag{8.28}$$

The restriction of ρ to $SO(3)$ is a projective unitary representation of $SO(3)$ on $\mathbb{C}^2$. This projective representation is not continuous for the usual topology (the topology on 3×3 matrices) and does not arise from a true representation.[12]

However careless we are in constructing ρ, it satisfies (8.28), which implies that it is a (non-unitary) projective representation of $SO^{\uparrow}(1,3)$ on $\mathbb{C}^2$. Yet, however clever we are, we cannot do much better than this. Any such ρ provides an example of a projective representation which is genuinely not a true representation. The reason why ρ is not continuous is that the usual topology is not the natural topology to consider here. For the natural topology, described in Section A.2, ρ is continuous.[13]

Proof Given $C_1, C_2 \in SO^{\uparrow}(1,3)$, we have

$$\kappa(\rho(C_1C_2)) = C_1C_2 = \kappa(\rho(C_1))\kappa(\rho(C_2)) = \kappa(\rho(C_1)\rho(C_2)),$$

and this proves (8.28). For $C \in SO(3)$, $\rho(C) \in SU(2)$ is an element of the group of unitary transformations of $\mathbb{C}^2$ and consequently (8.28) proves that the restriction of ρ to $SO(3)$ is a projective unitary representation. Next we prove that ρ is not continuous. Let us recall that $\kappa(\exp(-i\theta\sigma_3/2)) = R_\theta$, the rotation of angle θ around the third axis. Hence

$$\rho(R_\theta) = \rho(\kappa(\exp(-i\theta\sigma_3/2))) = \pm\exp(-i\theta\sigma_3/2) = \exp(-i\varphi(\theta)\sigma_3)$$

where $\varphi(\theta) \in \{\theta/2, \theta/2+\pi\}$. Increasing θ from 0 to 2π we see that if φ is continuous then if, say, $\varphi(0) = 0$ then $\varphi(2\pi) = \pi$ contradicting the fact that $\varphi(0) = \varphi(2\pi)$ since $R_{2\pi} = R_0$. Therefore φ is not continuous, and as θ moves from 0 to 2π, $\rho(R_\theta)$ must exhibit at some point a brutal change of sign.

Next if ρ arises from a true representation we can find $\lambda(C) \in \mathbb{U}$ such that $\rho'(C) := \lambda(C)\rho(C)$ satisfies

$$\rho'(C_1)\rho'(C_2) = \rho'(C_1C_2). \tag{8.29}$$

12 The same results are true if one does not restrict ρ to $SO(3)$.

13 In all the results stated about the structure of continuous projective representations, such as Proposition 8.6.2, continuity refers to the natural topology as in Section A.2. These technical considerations are not essential.

Applying κ to both sides we obtain $\lambda(C_1)\lambda(C_2)C_1C_2 = \lambda(C_1C_2)C_1C_2$, so that $\lambda(C_1)\lambda(C_2) = \lambda(C_1C_2)$. As is shown in Lemma 8.5.2, this proves that $\lambda \equiv 1$, so that to prove that (8.29) cannot hold it suffices to prove that ρ itself is not a true representation. Now $\rho(R_\pi) = \exp(-\mathrm{i}\theta\sigma_3/2)$ where $\theta \in \{\pi, 3\pi\}$, so that $\rho(R_\pi)^2 = -I$. Then we cannot have both $\rho(R_\pi)^2 = \rho(R_\pi^2)$ and $\rho(R_\pi^2) = \rho(1) = I$. □

Lemma 8.5.2 *If a map $\varphi\colon SO(3) \to \mathbb{C}$ satisfies $\varphi(1) = 1$ and $\varphi(C_1C_2) = \varphi(C_1)\varphi(C_2)$ for all choices of C_1, C_2 then φ takes only the value 1.*

In particular the one-dimensional representations of $SO(3)$ are trivial.

Proof If you have heard of the theory of Lie algebras, this should be obvious when you assume that φ is differentiable. The derivative φ' of φ is a Lie algebra homomorphism into a commutative Lie algebra, so that everything that is a bracket is sent to zero, and the brackets generate the Lie algebra of $SO(3)$ as is shown in particular by the commutation relations between the standard generators.

A direct proof is hardly more difficult, and has the advantage not to assume any kind of regularity on φ. Given $A \in SO(3)$, $\varphi(A)\varphi(A^{-1}) = \varphi(1) = 1$ so that $\varphi(A^{-1}) = \varphi(A)^{-1}$. Given $A, B \in SO(3)$, then $\varphi(ABA^{-1}B^{-1}) = 1$, so it suffices to show that every element of $SO(3)$ is of this type. The key observation is that if A and C are two rotations of the same angle, but not necessarily of the same axis, then we can find a rotation B for which $A = BCB^{-1}$. This applies in particular to the case $C = A^{-1}$ which is a rotation of the same angle as A but around the opposite axis. Then $ABA^{-1}B^{-1} = A^2$ which finishes the proof since every rotation is a square. □

Let us look deeper into the topological structure of the two-to-one map $A \mapsto \kappa(A)$ from $SU(2)$ to $SO(3)$. Given a continuous map φ from $[0,1]$ to $SO(3)$ (i.e. a curve in $SO(3)$) we can lift it into a continuous map ψ from $[0,1]$ to $SU(2)$ such that $\kappa(\psi(t)) = \varphi(t)$. This is possible because locally κ is a homeomorphism: There exists a neighborhood W of the unit in $SU(2)$ such that κ is a homeomorphism between W and $\kappa(W)$. However if φ is a loop, i.e. $\varphi(0) = \varphi(1)$ this need not be the case for ψ. There are exactly two different types of loops φ in $SO(3)$, those for which their lifting[14] is also a loop, and those for which it is not. Two loops of a different type cannot be continuously deformed into each other, but two loops of the same type can (as the reader will show using that $SU(2)$ is simply connected). The words "universal cover" in the title of the previous section reflect the fact that there is a covering $A \mapsto \kappa(A)$ from the simply connected group $SU(2)$ to $SO(3)$, and that all curves in $SO(3)$ can be lifted to curves in $SU(2)$.[15]

The following exercise brings forward a rather extraordinary fact. In a sense, while a rotation of angle 2π around a given axis is by no means the identity, a rotation of angle 4π cannot be distinguished from the identity.

[14] There are exactly two ways to continuously lift a curve φ, and they are related by a change of sign.
[15] It is also possible to show that $SL(2,\mathbb{C})$ is simply connected, so that "$SL(2,\mathbb{C})$ is the universal cover of $SO^{\uparrow}(1,3)$".

Exercise 8.5.3 Let $R_\theta \in SO(3)$ be the rotation around the third axis by an angle θ. Prove by an explicit construction that the loop $\varphi(\theta) = R_{4\pi\theta}, 0 \le \theta \le 1$ can be continuously contracted to the trivial loop. (This is quite challenging.)

Even though the choice $\rho(C)$ of one of the two elements $\pm A$ such that $\kappa(\pm A) = C$ cannot be done in a really nice manner, it still can be of some use as shown in the following exercise, where I denotes the 2×2 identity matrix.

Exercise 8.5.4 Consider ρ as above and a representation π of $SU(2)$ and the map $\pi'\colon A \mapsto \pi(\rho(A))$. Prove that if $\pi(-I)$ is the identity then π' is a representation of $SO(3)$. Prove that if $\pi(-I)$ is minus the identity then π' is a projective representation of $SO(3)$.

We may apply in particular the result of Exercise 8.5.4 to $\pi = \pi_j$ since (8.8) implies that $\pi_j(-I)$ is $(-1)^j$ times the identity.

Exercise 8.5.5 Using Exercise 8.5.4 for π_2 we construct a representation of $SO(3)$ of dimension 3. Try to figure out what it is. (It is easy to guess the result, but difficult to prove it, unless you are familiar with Appendix D.)

Exercise 8.5.6 (a) If π is an irreducible representation of $SU(2)$ prove that $\pi(-I)$ is always $\pm$ the identity. Hint: Use Schur's lemma.
(b) With the notation of Exercise 8.5.4 prove that if π is irreducible then π' is always a projective representation.

In some sense Exercise 8.5.4 produces all the irreducible projective unitary representations of $SO(3)$. A formal proof of this fact is given in Exercise 8.6.3.

Exercise 8.5.7 Since κ is a two-to-one homomorphism, for each $C \in SO^{\uparrow}(1,3)$ the set $\kappa^{-1}(C) = \{A \in SL(2,\mathbb{C}), \kappa(A) = C\}$ has two elements. Prove that

$$\kappa^{-1}(C)\kappa^{-1}(B) = \kappa^{-1}(CB), \tag{8.30}$$

where on the left the product of two subsets F, G of $SL(2,\mathbb{C})$ is defined as $FG = \{AB; A \in F, B \in G\}$. The map $C \to \kappa^{-1}(C)$ is called in many textbooks a *double-valued representation.*

One never sees any of the considerations of the present section in textbooks of physics. The reason is explained in detail in Appendix D. Physicists define "the action of rotations on $\mathbb{C}^2$", in a manner which we explain now in the case of rotations around the third axis. The action on $\mathbb{C}^2$ of such a rotation R_θ of angle θ is that of the matrix

$$C(\theta) := \exp(-\mathrm{i}\theta\sigma_3/2). \tag{8.31}$$

Formula (8.31) makes sense because (8.23) implies that $\kappa(C(\theta)) = R_\theta$. The problem however is that $C(\theta)$ is not determined by R_θ because changing θ into $\theta + 2\pi$ does not change R_θ but changes the sign of $C(\theta)$. As physicists say, there is a "sign ambiguity", so in our language they define a "double-valued representation". For the purposes of Quantum Mechanics, the sign ambiguity is irrelevant since projective representations are as good as true representations.

8.6 Deprojectivization

The following is a special case of a fundamental result of Bargmann [4] which is discussed further in Section A.1.

Theorem 8.6.1 *A strongly continuous unitary projective representation of $SU(2)$ arises from a true representation in the sense of Definition 2.11.1.*

The proof in the case of finite-dimensional representations is easier than in the general case, and is given in Section A.3.

Corollary 8.6.2 *Consider a strongly continuous unitary projective representation π' of $SO(3)$. Then there exists a true unitary representation π of $SU(2)$ and for $A \in SU(2)$ a complex number $\lambda(A)$ of modulus 1 such that*

$$\forall A \in SU(2)\ ;\ \pi'(\kappa(A)) = \lambda(A)\pi(A). \tag{8.32}$$

Proof Apply Theorem 8.6.1 to the projective representation $A \to \pi'(\kappa(A))$ of $SU(2)$. □

Exercise 8.6.3 Prove that all strongly continuous unitary projective representations of $SO(3)$ arise from a projective representation of $SU(2)$ constructed through the procedure of Exercise 8.5.4.

8.7 A Brief Introduction to Spin

In this section and the next we work *within non-relativistic* Quantum Mechanics.[16]

On page 181 we argued that, as a consequence of the fact that $SO(3)$ is a group of symmetries of Euclidean space, every quantum system should come equipped with a strongly continuous projective unitary representation of $SO(3)$. In view of Corollary 8.6.2, we expect that:

Every quantum system comes equipped with a strongly continuous unitary representation of $SU(2)$.

For example, in the case where $\mathcal{H} = L^2 = L^2(\mathbb{R}^3, \mathrm{d}^3\boldsymbol{p}/(2\pi\hbar)^3)$, the representation corresponding to (II.2) is given for $A \in SU(2)$ by

$$U(A)(\varphi)(\boldsymbol{p}) = \varphi(\kappa(A^{-1})(\boldsymbol{p})). \tag{8.33}$$

We somewhat abuse notation here since $\kappa(A^{-1}) \in SO^{\uparrow}(1,3)$ operates on $\mathbb{R}^4$ rather than $\mathbb{R}^3$, but we have seen that when $A \in SU(2)$ the operator $\kappa(A^{-1})$ can also be considered as a rotation. Let us stress that (8.33) is really nothing more than (II.2). We simply have replaced the group $SO(3)$ by "its nicer version $SU(2)$".

The momentum state space $\mathcal{H}$ above is appropriate to describe (at the level of non-relativistic Quantum Mechanics) particles whose state is entirely described by their momenta. But it is a fundamental experimental fact that the state of many particles (say, electrons) is *not* described only by their momenta.[17] Besides the momenta, many

16 So, we forget that we ever heard of the Lorentz and Poincaré group.
17 That is, the result of certain experiments cannot, even statistically, be predicted when knowing only the momentum of the particle.

particles have "internal degrees of freedom" which *have no classical equivalent.* The use of the momentum state space $L^2 = L^2(\mathbb{R}^3, d^3p/(2\pi\hbar)^3)$ is not appropriate to describe such particles, as it really describes only their momenta, but not their "internal degrees of freedom".

Let us recall the fundamental representation π_j of Section 8.2 and its state space $\mathcal{H}_j$. The successful proposal to describe particles, including their "internal degrees of freedom", is to use the state space $L^2 \otimes \mathcal{H}_j$, *equipped with the representation* $U \otimes \pi_j$ of $SU(2)$, where U is given by (8.33). We then state informally the following definition.

Definition 8.7.1 A particle whose state space is $L^2 \otimes \mathcal{H}_j$ equipped with the representation $U \otimes \pi_j$ of $SU(2)$ above is said to be of spin $j/2$.[18]

Thus the possible values of the spin of a particle are $0, 1/2, 1$, etc., and it is traditional to say that spins are "half-integers". Note that in this definition *spin is a property of the particle.*

If we recall Principle 6 on page 73 the use of the tensor product $L^2 \otimes \mathcal{H}_j$ is completely natural: The momentum state space L^2 is appropriate to describe the momentum of the particle, and $\mathcal{H}_j$ is the appropriate "state space" to describe the "internal degrees of freedom". Actually, the success of the model based on this tensor product can be interpreted as showing that momentum and "internal degrees of freedom" are of an entirely different nature.[19]

Let us now have a closer look at the representation $U \otimes \pi_j$. The tensor product $L^2 \otimes \mathcal{H}_j$ identifies with the space $L^2(\mathbb{R}^3, \mathcal{H}_j, d^3p/(2\pi\hbar)^3)$ of $\mathcal{H}_j$-valued functions on $\mathbb{R}^3$ and with this identification the representation $U \otimes \pi_j$ is given by the formula

$$(U \otimes \pi_j)(A)(\varphi)(\boldsymbol{p}) = \pi_j(A)(\varphi(\kappa(A^{-1})(\boldsymbol{p}))), \tag{8.34}$$

with the same abuse of notation as in (8.33). Indeed, it suffices to prove this formula when φ is of the type $\varphi(\boldsymbol{p}) = \psi(\boldsymbol{p})\eta$ for $\eta \in \mathcal{H}_j$ and $\psi \in L^2$, and it becomes obvious. There are two different ways in which A appears in the previous formula. The term $\kappa(A^{-1})$ has nothing to do with spin, and reflects that the momentum of the particle has changed due to the rotation. It is the term $\pi_j(A)$ which reflects the spin.[20]

8.8 Spin as an Observable

Intuitively, particles with spin exhibit a kind of "internal angular momentum". To discuss this we outline in the briefest manner the theory of angular momentum. This theory is of fundamental importance and is extensively developed in any book on Quantum Mechanics, but we do not need to enter its details here.

[18] This is an oversimplification, as it ignores the case where other quantum numbers might be needed to describe the internal state of the particle.

[19] Furthermore, the terminology "*internal* degrees of freedom" reflects the heuristic idea that whereas the four-momentum of the particle is easily visible from outside, the "internal state" of the particle is more hidden and reflects what happens "inside" the particle.

[20] Quantum Mechanics describes the situation by stating that total angular momentum is the sum of orbital angular momentum, corresponding to the term $\kappa(A^{-1})(\boldsymbol{p})$, and of spin angular momentum, corresponding to the term $\pi_j(A)$. Spin angular momentum is described in the next section.

There are many one-parameter subgroups of $SU(2)$, for example the group consisting of the matrices $\exp(-\mathrm{i}\theta\sigma_3/2)$. The special importance of this group is that it intuitively corresponds to the rotations around the z-axis, since $\kappa(\exp(-\mathrm{i}\theta\sigma_3/2))$ is such a rotation, of angle θ. Given a unitary representation Π of $SU(2)$, the formula $V(\theta) = \Pi(\exp(-\mathrm{i}\theta\sigma_3/2))$ defines a one-parameter unitary group. According to Stone's theorem, the operator

$$J_3(x) := \lim_{\theta\to 0} \frac{\hbar}{\mathrm{i}\theta}(V(\theta)(x) - x) \tag{8.35}$$

is self-adjoint. This operator is called the *angular momentum operator with respect to the z axis.*[21]

Exercise 8.8.1 Compute the operator J_3 in the case of the representation (8.33).

When $\Pi = \pi_j$ let us denote by S_3[22] the operator defined by the formula (8.35). Since the space $\mathcal{H}_j$ is of dimension $j+1$, S_3 can have at most $j+1$ eigenvalues. On the other hand, the relation (8.9) witnesses the state $z_1^k z_2^{j-k}$ being an eigenvector of eigenvalue $\hbar(j-2k)/2$ of S_3, so that $\hbar j/2, \hbar(j/2-1), \ldots, -\hbar j/2$ are exactly the eigenvalues of S_3.

In the setting of the previous section, one says that the operator $\mathbf{1}\otimes S_3$ *measures the spin of the particle along the z-axis*. Observe here the traditional but very confusing terminology:

- The word "spin" is used on the one hand to describe a fixed property of the particle, that its internal degrees of freedom are described by the state space $\mathcal{H}_j$ equipped with the representation π_j.
- On the other hand, for such a particle, "spin" is also something that you may measure along any given direction.

The possible values of the spin (as an observable) of a particle of spin $j/2$ (as a property of the particle) in any given direction are $\hbar j/2, \hbar(j/2-1), \ldots, -\hbar j/2$.

Exercise 8.8.2 Assume that a particle of spin $j/2$ has spin $\hbar(j/2-k)$ in the z-direction, in the sense that the measurement of this spin will always yield the value $j/2-k$. In the ideal case where the wave function $L^2 \to \mathcal{H}_j$ is zero unless $\boldsymbol{p}$ is in the positive z-direction, show that the action of a rotation of angle θ in the z direction "multiplies the wave function by a phase $\exp(-\mathrm{i}\theta(j/2-k))$". (Here a rotation of angle 4π is different from a rotation of angle 2π!)

To understand how one should define spin, and more generally angular momentum, with respect to another axis, you may study the first three sections of Appendix D.

8.9 Parity and the Double Cover $SL^+(2,\mathbb{C})$ of $O^+(1,3)$

The group $SO^{\uparrow}(1,3)$ represents a fundamental group of symmetries of space-time. Another fundamental symmetry that does not belong to $SO^{\uparrow}(1,3)$ is the parity operator P given by

[21] This is the reason for the index 3 in the notation J_3.
[22] S stands for spin.

(4.27) i.e. $P(x^0,\boldsymbol{x}) = (x^0, -\boldsymbol{x})$. An important difference of nature between P and the elements of $SO^{\uparrow}(1,3)$ is that P is a "discrete symmetry" which cannot be reached continuously from the identity. Physically, we can (at least in principle) operate on $\mathbb{R}^{1,3}$ by $SO^{\uparrow}(1,3)$ but not by P: We cannot have an object "go through the mirror" and become its parity image. Still it is of great interest to consider theories that have a symmetry corresponding to this operator. Including parity in our group of symmetries leads us to the following definition.

Definition 8.9.1 We denote by $O^+(1,3)$ the subgroup of $O(1,3)$ generated by $SO^{\uparrow}(1,3)$ and P, or, equivalently, the subgroup of $O(1,3)$ consisting of orthochronous transformations.[23]

The double cover $SL(2,\mathbb{C})$ of $SO^{\uparrow}(1,3)$ will be of considerable importance in the next chapter. To include parity in our theories, the double cover $SL^+(2,\mathbb{C})$ of $O^+(1,3)$ is convenient. It is to $O^+(1,3)$ what $SL(2,\mathbb{C})$ is to $SO^{\uparrow}(1,3)$. In the present section we construct it in the most elementary manner possible. Benefits of this approach include:

- Trying to find representations of $SL^+(2,\mathbb{C})$ will lead us in the most natural way to the Dirac matrices.
- In the next chapter, to study the projective representation of the Poincaré group $\mathcal{P} = \mathbb{R}^{1,3} \rtimes SO^{\uparrow}(1,3)$ we will be led to study instead the true representations of $\mathcal{P}^* := \mathbb{R}^{1,3} \rtimes SL(2,\mathbb{C})$, and replacing $SL(2,\mathbb{C})$ by $SL^+(2,\mathbb{C})$ is the natural way to introduce parity into the theory.

Since $O^+(1,3)$ is generated by $SO^{\uparrow}(1,3)$ and P, we guess that $SL^+(2,\mathbb{C})$ will be generated by $SL(2,\mathbb{C})$ and a new element P'. We then just have to guess what the multiplication rules should be, and check that our guess works. The single non-trivial fact that lets us guess these rules is as follows.

Lemma 8.9.2 *For $A \in SL(2,\mathbb{C})$ we have*

$$P\kappa(A)P = \kappa(A^{\dagger -1}). \tag{8.36}$$

In (8.19) we could have replaced $M(x)$ by $M(P(x))$ which also satisfies (8.20); and in the right-hand side of (8.21) we could have replaced A by $A^{\dagger -1}$. The lemma states that these two changes do exactly the same thing.

Proof We first observe the identity

$$M(P(x))M(x) = \det M(x)I, \tag{8.37}$$

where I is the 2×2 identity matrix, so that when $M(x)$ is invertible,

$$M(P(x)) = (\det M(x))M(x)^{-1}. \tag{8.38}$$

On the other hand, it follows from (8.21) that

$$M(\kappa(A)(x))^{-1} = A^{\dagger -1}M(x)^{-1}A^{-1}. \tag{8.39}$$

[23] As defined in Section 4.1 these are the Lorentz transformations L with $L^0_0 > 0$, i.e. which preserve the flow of time.

Replacing x by $\kappa(A)(x)$ in (8.38), using (8.39) and since $\det M(\kappa(A)(x)) = \det M(x)$ we obtain, using again (8.38) in the last inequality,

$$M(P\kappa(A)(x)) = \det M(x) A^{\dagger -1} M(x)^{-1} A^{-1} = A^{\dagger -1} M(P(x)) A^{-1}. \tag{8.40}$$

This equality holds whenever $M(x)$ is invertible so that it holds everywhere by continuity. The right-hand side is $M(\kappa(A^{\dagger -1})P(x))$, so that $P\kappa(A)(x) = \kappa(A^{\dagger -1})P(x)$ for each x i.e. $P\kappa(A) = \kappa(A^{\dagger -1})P$. □

It is then not very difficult to invent the following, denoting by I the identity 2×2 matrix.

Definition 8.9.3 The group $SL^+(2,\mathbb{C})$ is obtained by adding a new element P' to $SL(2,\mathbb{C})$ with the multiplication rules

$$P'I = P' \,;\; P'AP' = A^{\dagger -1} \tag{8.41}$$

whenever $A \in SL(2,\mathbb{C})$. (Thus in particular, taking $A = I$ we have $(P')^2 = I$.) The homomorphism κ is extended to $SL^+(2,\mathbb{C})$ by defining[24]

$$\kappa(P'A) = P\kappa(A). \tag{8.42}$$

More specifically, the elements of $SL^+(2,\mathbb{C})$ are either of the type A or the type $P'A$ for $A \in SL(2,\mathbb{C})$ and the way to multiply these elements is entirely determined by the rules (8.41), for example $A(P'B) = P'A^{\dagger -1}B$ and $(P'A)(P'B) = A^{\dagger -1}B$. It is straightforward to verify (by considering cases) that we do define a group $SL^+(2,\mathbb{C})$ in this manner, for which κ is a two-to-one homomorphism onto $O^+(1,3)$. For example, to prove that κ is a homomorphism, $\kappa(UV) = \kappa(U)\kappa(V)$, there are four cases to consider, depending on whether U and V are of the type A or $P'A$ for $A \in SL(2,\mathbb{C})$. We write simply

$$\kappa(P'AP'B) = \kappa(A^{\dagger -1}B) = \kappa(A^{\dagger -1})\kappa(B) = P\kappa(A)P\kappa(B) = \kappa(P'A)\kappa(P'B)$$

and

$$\kappa(AP'B) = \kappa(P'A^{\dagger -1}B) = P\kappa(A^{\dagger -1})\kappa(B) = \kappa(A)P\kappa(B) = \kappa(A)\kappa(P'B),$$

the other two cases being even easier.

The following exercise should not be missed. It provides a concrete realization of $SL^+(2,\mathbb{C})$. In this exercise 4×4 matrices are written by 2×2 blocks.

Exercise 8.9.4 For $A \in SL(2,\mathbb{C})$ consider the matrix

$$\begin{pmatrix} A^{\dagger -1} & 0 \\ 0 & A \end{pmatrix}$$

and the matrix

$$\begin{pmatrix} 0 & I \\ I & 0 \end{pmatrix},$$

where I is the 2×2 identity matrix. Show that the group generated by these matrices is isomorphic to $SL^+(2,\mathbb{C})$.

24 This extension is not unique. Another extension κ' is given by the formula $\kappa'(P'A) = -P\kappa(A)$.

8.10 The Parity Operator and the Dirac Matrices

The short story of the present section is that when trying to find representations of $SL^+(2,\mathbb{C})$, the fundamental Dirac matrices appear rather immediately. This was one of our motivations in introducing the group $SL^+(2,\mathbb{C})$.

The key property of a representation S of $SL^+(2,\mathbb{C})$ is that since $A^{\dagger-1} = P'AP'$ it holds that for each $A \in SL(2,\mathbb{C})$ one has

$$S(P')S(A)S(P') = S(A^{\dagger-1}). \tag{8.43}$$

It is quite easy to discover[25] the following method of construction: Starting with a representation θ of $SL(2,\mathbb{C})$ on a space $\mathcal{G}$, consider the representation S on $\mathcal{H} = \mathcal{G} \oplus \mathcal{G}$ given by $S(A)(x,y) = (\theta(A^{\dagger-1})(x), \theta(A)(y))$ and $S(P')(x,y) = (y,x)$. The simplest case of this construction is when θ is the defining representation by 2×2 matrices, i.e. $\theta(A) = A$. Writing S in matrix form yields the formulas

$$S(A) := \begin{pmatrix} A^{\dagger-1} & 0 \\ 0 & A \end{pmatrix}, \tag{8.44}$$

where the right-hand side is a 4×4 matrix decomposed in blocks of size 2×2, and

$$S(P') = \begin{pmatrix} 0 & I \\ I & 0 \end{pmatrix}, \tag{8.45}$$

where I is the 2×2 identity matrix. For $A \in SL(2,\mathbb{C})$ we define $S(P'A) = S(P')S(A)$. Checking that S defines a representation is as straightforward as (and very similar to) checking that the rules of Definition 8.9.3 define a group.

The copy of $\mathbb{C}^4$ on which the representation S of $SL^+(2,\mathbb{C})$ given by (8.44) and (8.45) acts is called the space of *Dirac spinors*. It seems a pure coincidence that the space of Dirac spinors has the same dimension four as space-time. One should carefully distinguish the following two (very different) representations of $SL^+(2,\mathbb{C})$ of dimension four:

- The extension of κ constructed in the previous section.
- The representation in the space of Dirac spinors given by (8.44) and (8.45).

These representations are very different because the restriction of the first one to $SL(2,\mathbb{C})$ is irreducible, which is not the case for the second one. At the end of the section the reader will find exercises analyzing the irreducible representations of $SL^+(2,\mathbb{C})$.

Let us have a closer look at the fundamentally important structure occurring here. For $x \in \mathbb{R}^{1,3}$ we define the following 4×4 matrix:

$$\gamma(x) := \begin{pmatrix} 0 & M(P(x)) \\ M(x) & 0 \end{pmatrix}. \tag{8.46}$$

Then the two equations (8.21) and (8.40), i.e.

$$M(\kappa(A)(x)) = AM(x)A^\dagger \text{ and } M(P\kappa(A)(x)) = A^{\dagger-1}M(P(x))A^{-1}$$

[25] At the end of this section we present as an exercise the analysis of the representations of $SL^+(2,\mathbb{C})$, and this analysis leads to the method we describe here. But it is not hard to just *guess* this method.

are equivalent to the relation

$$\gamma(\kappa(A)(x)) = S(A)\gamma(x)S(A)^{-1}. \tag{8.47}$$

Moreover since $\kappa(P')(x) = P(x)$ we see from (8.46) and the definition of $S(P')$ that (8.47) holds for $A = P'$. It then follows that (8.47) *holds for each* $A \in SL^+(2,\mathbb{C})$. The matrices $\gamma(x)$ enjoy a number of remarkable properties. For example, using (8.37) it is immediate to see that $\gamma(x)^2 = \det M(x)1 = x^2 1$, so that replacing x by $x+y$ (the usual polarization procedure) yields the identity

$$\{\gamma(x),\gamma(y)\} := \gamma(x)\gamma(y) + \gamma(y)\gamma(x) = 2(x,y)1. \tag{8.48}$$

Incidentally, given two operators A, B, the quantity

$$\{A,B\} = AB + BA$$

is called the *anti-commutator* of these operators and will be of much use later.

Let us define $\gamma_\mu = \gamma(e_\mu)$ where $(e_\mu)_{\mu=0,1,2,3}$ is the standard basis of $\mathbb{R}^{1,3}$, so that, recalling (8.18) and (8.19),

$$\gamma_0 = \begin{pmatrix} 0 & \sigma_0 \\ \sigma_0 & 0 \end{pmatrix};\ \gamma_1 = \begin{pmatrix} 0 & -\sigma_1 \\ \sigma_1 & 0 \end{pmatrix};\ \gamma_2 = \begin{pmatrix} 0 & -\sigma_2 \\ \sigma_2 & 0 \end{pmatrix};\ \gamma_3 = \begin{pmatrix} 0 & -\sigma_3 \\ \sigma_3 & 0 \end{pmatrix}. \tag{8.49}$$

The matrices[26] γ_μ are called the *Dirac matrices in the Weyl or chiral presentation*. They satisfy the property

$$\{\gamma_\mu,\gamma_\nu\} = 2\eta_{\mu\nu}1. \tag{8.50}$$

It is this property (rather than the specific form of the matrices) that is really important. Given matrices γ_μ satisfying this property, so do the matrices $L\gamma_\mu L^{-1}$ where L is any invertible matrix.[27] Different choices of L give different sets of matrices. One then informally talks of "different presentations" of the Dirac matrices. There exist other useful presentations.

Since $\kappa(A)(x) = \kappa(A)(x)^\nu e_\nu = \kappa(A)^\nu{}_\mu x^\mu e_\nu$, we have $\gamma(\kappa(A)(x)) = \kappa(A)^\nu{}_\mu x^\mu \gamma_\nu$ while $\gamma(x) = \gamma_\mu x^\mu$ so we obtain from (8.47) that $\kappa(A)^\nu{}_\mu x^\mu \gamma_\nu = S(A)\gamma_\mu x^\mu S(A^{-1})$ and therefore the fundamental relation:

$$\kappa(A)^\nu{}_\mu \gamma_\nu = S(A)\gamma_\mu S(A)^{-1}, \tag{8.51}$$

which holds whenever $A \in SL^+(2,\mathbb{C})$. This is the way the transformation property (8.47) is expressed in physics textbooks.

The following provides another description of the group $SL^+(2,\mathbb{C})$.

Exercise 8.10.1 Consider the set $Pin(1,3)$ of 4×4 invertible matrices L for which there exists a 4×4 matrix $\theta(L)$ such that

$$L\gamma_\mu L^{-1} = \theta(L)^\nu{}_\mu \gamma_\nu.$$

[26] In case you have seen other presentations of this material, please observe that γ_μ is not to be confused with $\gamma^\mu = \eta^{\mu\nu}\gamma_\nu$.

[27] It is true that any matrices which satisfy (8.50) are of the type $L\gamma_\mu L^{-1}$ for a certain invertible matrix L.

(a) Prove that this relation is equivalent to $\gamma(\theta(L)(x)) = L\gamma(x)L^{-1}$.
(b) Prove that $Pin(1,3)$ is a group, and that the map $L \to \theta(L)$ is a group homomorphism from $Pin(1,3)$ to $O(1,3)$.
(c) Prove that the map $L \to \theta(L)$ from $Pin(1,3)$ to $O(1,3)$ is two-to-one. Hint: This is a bit challenging. One has to prove that if L commutes with all the γ matrices, it is $\pm$ the identity. Prove first that the linear span of all the products of γ matrices consists of all 4×4 matrices. This is standard material found in many books.
(d) We can define a group homomorphism S from $SL^+(2,\mathbb{C})$ into the group of 4×4 matrices by the formulas (8.44) and (8.45). Prove that S is valued into $Pin(1,3)$. Prove that $O^+(1,3) = \theta(S(SL^+(2,\mathbb{C})))$.
(e) Using (c) and (d) prove that S is an isomorphism from $SL^+(2,\mathbb{C})$ onto the subgroup $\theta^{-1}(O^+(1,3))$ of $Pin(1,3)$. This provides an alternate description of $SL^+(2,\mathbb{C})$.
(f) Compute $\theta(T)$ when T is the matrix

$$T = \begin{pmatrix} 0 & I \\ -I & 0 \end{pmatrix}.$$

Show that $\theta(T)$ can be interpreted as "time-reversal".
(g) We provided a construction of $SL^+(2,\mathbb{C})$ by "adding a new element P' to $SL(2,\mathbb{C})$". In the same manner "add a new element T' to $SL^+(2,\mathbb{C})$" to obtain a group isomorphic to $Pin(1,3)$. What are the multiplication rules in this group?

In the next two exercises we analyze the finite-dimensional irreducible representations of $SL^+(2,\mathbb{C})$. Consider such a finite-dimensional representation S.

Exercise 8.10.2 Assume that the restriction of S to $SL(2,\mathbb{C})$ is irreducible.
(a) Prove that the representation $A \mapsto S(A^{\dagger -1})$ of $SL(2,\mathbb{C})$ must be equivalent to S. Hint: Use (8.43).
(b) Using the results of Section D.10 describe all the irreducible representations of $SL(2,\mathbb{C})$ with the previous property.
(c) For these representations construct representations of $SL^+(2,\mathbb{C})$ whose restriction to $SL(2,\mathbb{C})$ is the given representation.

Exercise 8.10.3 If you succeeded in solving Exercise 8.4.4 and (8.10.2), you may try to prove the following more precise result. The extension κ to $SL^+(2,\mathbb{C})$ is unitarily equivalent to the representation S of $SL^+(2,\mathbb{C})$ on $\mathbb{C}^2 \otimes \mathbb{C}^2$ such that for $A \in SL(2,\mathbb{C})$ we have $S(A) = A \otimes A^{\dagger -1}$ while $S(A)(P')$ is the map that sends the tensor x_{n_1,n_2} to the tensor $-x_{n_2,n_1}$. Hint: All the ingredients are contained in the proof of Proposition D.12.2 and the solution of Exercise D.12.3, but it still requires some patience to write this down.

Exercise 8.10.4 Assume now that the restriction of S to $SL(2,\mathbb{C})$ is not irreducible, so that there exists a proper subspace $\mathcal{G}$ of the space $\mathcal{H}$ on which lives the representation which is invariant under all the maps $S(A)$, $A \in SL(2,\mathbb{C})$.
(a) Prove that the space $\mathcal{G}' = S(P')(\mathcal{G})$ is invariant under all the operators $S(A)$ for $A \in SL(2,\mathbb{C})$.

(b) Prove that $\mathcal{G} \cap \mathcal{G}' = \{0\}$. Hint: Prove that the space $\mathcal{G} \cap \mathcal{G}'$ is invariant under S, and use that S is irreducible.
(c) Prove that $\mathcal{H} = \mathcal{G} \oplus \mathcal{G}'$.
(d) Prove that S is equivalent to a representation of $SL^+(2,\mathbb{C})$ on $\mathcal{G} \oplus \mathcal{G}$ of the type $S(A)(x,y) = (\theta(A^{\dagger -1})(x), \theta(A)(y))$ for $A \in SL(2,\mathbb{C})$ and $S(P')(x,y) = (y,x)$ for a certain irreducible representation θ of $SL(2,\mathbb{C})$.

Key ideas to remember:

- The group $SU(2)$ "is a better version of $SO(3)$" and the group $SL(2,\mathbb{C})$ "is a better version of $SO^{\uparrow}(1,3)$". There is a two-to-one homomorphism κ from $SL(2,\mathbb{C})$ to $SO^{\uparrow}(1,3)$.
- For each $j \geq 0$ there exists an irreducible unitary representation π_j of $SU(2)$ of dimension $j+1$. This representation is unique up to unitary equivalence.
- Projective unitary representations of $SU(2)$ arise from true representations but this is not the case for $SO(3)$.
- The word "spin" has two very different meanings. On the one hand, "having spin $j/2$" for $j \geq 0$ is an intrinsic property of a particle. On the other hand "spin" is a kind of angular momentum whose value can be measured in any direction. For a particle of spin $j/2$, the possible values of this measurement are $-\hbar j/2, -\hbar j/2+1, \dots, \hbar j/2$.
- The group $SL(2,\mathbb{C})$ is a subgroup of $SL^+(2,\mathbb{C})$ which now contains a parity operator. The search for representations of $SL^+(2,\mathbb{C})$ leads to the Dirac matrices.

9

Representations of the Poincaré Group

Let us recall that the Poincaré group $\mathcal{P}$ is the group of transformations of $\mathbb{R}^{1,3}$ of the type $x \mapsto a+A(x)$ where $(a, A) \in \mathbb{R}^{1,3} \times SO^{\uparrow}(1,3)$. It is the semidirect product $\mathbb{R}^{1,3} \rtimes SO^{\uparrow}(1,3)$ and the composition law is given by

$$(a, A)(b, B) = (a + A(b), AB). \tag{9.1}$$

As we started explaining in Section 4.7, a fundamental idea of Eugene Wigner is that to each elementary particle corresponds an irreducible projective unitary representation of the Poincaré group, so these are of central importance in Quantum Field Theory. In a truly impressive work, Wigner [90] classified these representations.

Even though to each elementary particle is attached a projective unitary representation of the Poincaré group, this does not mean that all these representations correspond to observed particles. Those which do are of course the most important. To each representation corresponding to a particle is attached a number $m \geq 0$ which physically corresponds to the mass of the particle. The particle is called massive if $m > 0$ and massless if $m = 0$. Wigner reached the following very striking conclusion: For a massive particle, besides the mass, there exists exactly one other characteristic (namely, its spin) which determines how the Poincaré group acts on it.

Learning the present topic proved particularly arduous. A given representation of the Poincaré group can be described by many different formulas. A common method for authors to skirt the difficulty of the present topic and shorten their presentation at the reader's expense is to state formulas without giving any clue of what they might mean, how one might discover them and how one might decide if a different formula would describe the same representation.[1] On the other hand, the well-developed mathematical theory of group representations bears on this topic and some mathematicians as Folland [31] might choose to directly throw it at the unprepared reader.[2]

To avoid the various pitfalls ubiquitous in the literature, we have treated the subject in great detail. Rather than displaying our dizzying mastery of abstract mathematics, our emphasis is on various descriptions of the representations corresponding to actual particles, and in clarifying some points directly motivated by physics which are hardly ever treated either in mathematics or physics textbooks.

[1] See e.g. the formula (C.12) in Araki's book [2].
[2] I personally wasted a very long time there.

9.1 The Physicists' Way

The physicists' view of this topic is pragmatic and efficient.[3] Following their methods, we will easily discover the fundamental tool of *induced representations.*[4] The present section continues Section 4.11, which the reader should review now. The goal is to determine the action $U(a, A)$ of a Poincaré transformation (a, A) on the state space of a massive particle of mass $m > 0$. The arguments are entirely heuristic, and for simplicity we assume that U is a true representation.

At rest, the four-momentum of the particle is $p^* := (mc, 0, 0, 0)$. We will use in a crucial way that a Lorentz transformation is a rotation if and only if it fixes p^*.[5] As we discussed in Section 8.7, the four-momentum of a particle is not enough to completely describe it. The particle has "internal degrees of freedom", which are described by a vector u in a finite-dimensional Hilbert space $\mathcal{V}$. We may then describe the state of a particle at rest by $|p^*, u\rangle$. Since this state has four-momentum p^* we expect from (4.73) that[6]

$$U(a, 1)|p^*, u\rangle = \exp(\mathrm{i}(a, p^*)/\hbar)|p^*, u\rangle. \tag{9.2}$$

The space $\mathcal{V}$ comes equipped with a certain irreducible unitary projective representation V of $SO(3)$, which for simplicity we assume to be a true representation.[7] Since a rotation R seen as a Lorentz transformation fixes p^* we expect that for such a rotation

$$U(0, R)|p^*, u\rangle = |p^*, V(R)(u)\rangle. \tag{9.3}$$

Combining (9.2) with (9.3) we have determined the action $U(a, R)$ of any element $(a, R) \in \mathbb{R}^{1,3} \rtimes SO(3)$ on a state $|p^*, u\rangle$:

$$U(a, R)|p^*, u\rangle = \exp(\mathrm{i}(a, p^*)/\hbar)|p^*, V(R)(u)\rangle. \tag{9.4}$$

Let us then look at the situation in abstract terms. Presumably the states of the type $|p^*, u\rangle$ form a copy of $\mathcal{V}$, and (9.4) defines an irreducible unitary representation of $H := \mathbb{R}^{1,3} \rtimes SO(3)$ on this space. How can we "extend" this representation to a larger space and to the larger group $H' := \mathbb{R}^{1,3} \rtimes SO^{\uparrow}(1,3)\ (= \mathcal{P})$? The answer to that question is known under very general conditions on the groups H and H'. This is the general "theory of induced representations". Rather than getting buried into abstract mathematics we however keep thinking as physicists. This will lead us to the solution of this problem in our specific case of interest.

Let us *define* the state $|p, u\rangle$ by

$$|p, u\rangle := U(0, D_p)|p^*, u\rangle, \tag{9.5}$$

[3] It is based on a large number of implicit assumptions, which are considered as self-evident, since they are well supported by observation. We shall not try to present them in the form of axioms.

[4] What we call an induced representation, for lack of a better term, does not exactly coincide with the use of this name in the "general theory of induced representations". As we do not enter that theory there is however no risk of confusion.

[5] Using terminology to be introduced later, the group of rotations is the *little group* of p^*, see Definition 9.3.1.

[6] Please remember that here we are thinking like physicists, who consider this relation as self-evident. If we think like mathematicians, we may turn to the analysis of Section 4.9 to see that (9.2) is really the *definition* of the statement that the state $|p^*, u\rangle$ has four-momentum p^*.

[7] We know what these are: the representations π_j on the spaces $\mathcal{H}_j$. The reason for using generic names rather than the specific names $\mathcal{H}_j$ and π_j is simply that the forthcoming argument does not use this fact.

where D_p is the pure boost which sends p^* to p.[8] As a mathematician, you may wonder what is the precise meaning of this formula, but a physicist will find this perfectly clear. Since $|p^*,u\rangle$ has four-momentum p^*, Lemma 4.11.1 shows that $|p,u\rangle$ has four-momentum $D_p p^* = p$.

Next we find how Lorentz transformations act on these states. Consider $A \in SO^{\uparrow}(1,3)$. The key observation is that since $R := D_{Ap}^{-1} A D_p$ leaves p^* invariant it is a rotation. Since $AD_p = D_{Ap}R$ and U is a representation, it holds that $U(0,A)U(0,D_p) = U(0,D_{Ap})U(0,R)$. Thus, using (9.5), and then using (9.3) in the third equality,

$$\begin{aligned} U(0,A)|p,u\rangle = U(0,A)U(0,D_p)|p^*,u\rangle &= U(0,D_{Ap})U(0,R)|p^*,u\rangle \\ &= U(0,D_{Ap})|p^*,V(R)(u)\rangle = |Ap,V(R)(u)\rangle, \end{aligned}$$

and thus (and this is the key point)

$$U(0,A)|p,u\rangle = |Ap, V(D_{Ap}^{-1} A D_p)(u)\rangle.$$

Finally, since $U(a,A) = U(a,1)U(0,A)$ we obtain the required formula:

$$U(a,A)|p,u\rangle = \exp(\mathrm{i}(a,Ap)/\hbar)|Ap, V(D_{Ap}^{-1} A D_p)(u)\rangle. \tag{9.6}$$

The representation U *is entirely determined* by V. We have discovered the *little group method*. As we will detail later, an irreducible unitary representation of $\mathcal{P}$ is determined by an irreducible unitary representation of the subgroup of $SO^{\uparrow}(1,3)$ consisting of elements which fix a certain element of $\mathbb{R}^{1,3}$. Conversely to each irreducible unitary representation of this subgroup is associated an irreducible unitary representation of $\mathcal{P}$.

If the reader has understood the relationship between relations (4.61) and (4.77) she will be able to guess how we should interpret the formula (9.6): The Poincaré group acts on the space[9] $L^2(X_m,\mathcal{V},\mathrm{d}\lambda_m)$ by the formula[10]

$$\tilde{U}(a,A)(\varphi)(p) := \exp(\mathrm{i}(a,p)/\hbar)V(D_p^{-1} A D_{A^{-1}(p)})(\varphi(A^{-1}(p))). \tag{9.7}$$

Let us derive this formula more precisely. Recalling the formula (4.69), the states $|p,u\rangle$ are defined in the distributional sense by

$$\varphi = \int \mathrm{d}\lambda_m(p)\xi(p)|p,u\rangle, \tag{9.8}$$

where u is a given point of $\mathcal{V}$, where $\xi \in \mathcal{S}^4$ and where φ is the element of $L^2(X_m,\mathcal{V},\mathrm{d}\lambda_m)$ given by $\varphi(p) = \xi(p)u$. Using that for a complex number α we have $(\alpha\xi(p))u = \xi(p)(\alpha u)$ we obtain that for any ξ we have $\int \mathrm{d}\lambda_m(p)\alpha\xi(p)|p,u\rangle = \int \mathrm{d}\lambda_m(p)\xi(p)|p,\alpha u\rangle$. Thus $\alpha|p,u\rangle = |p,\alpha u\rangle$. In particular $\xi(p)|p,u\rangle = |p,\xi(p)u\rangle = |p,\varphi(p)\rangle$ so that (9.8) becomes

$$\varphi = \int \mathrm{d}\lambda_m(p)|p,\varphi(p)\rangle. \tag{9.9}$$

8 Thus, with the notation of Definition 4.3.1, $D_p = B_{p/mc}$. The reason for the change of notation will appear soon.

9 Since the physicists never tell us what the state space is, we have to guess both the state space and the action $U(a,A)$ on it.

10 Which is obtained from (9.6) by changing p to $A^{-1}(p)$.

This formula will then hold for any nice function $X_m \to \mathcal{V}$ because such a function can be well approximated by a finite sum of functions of the type $p \to \xi(p)u$. Using (9.6) and making the change of variable $p \to A^{-1}(p)$ in the third line,

$$\begin{aligned}\tilde{U}(a,A)(\varphi) &= \int \mathrm{d}\lambda_m(p) U(a,A)|p,\varphi(p)\rangle \\ &= \int \mathrm{d}\lambda_m(p) \exp(\mathrm{i}(a,Ap)/\hbar)|Ap, V(D_{Ap}^{-1} A D_p)(\varphi(p))\rangle \\ &= \int \mathrm{d}\lambda_m(p) \exp(\mathrm{i}(a,p)/\hbar)|p, V(D_p^{-1} A D_{A^{-1}(p)})(\varphi(A^{-1}(p)))\rangle \\ &= \int \mathrm{d}\lambda_m(p)|p, \exp(\mathrm{i}(a,p)/\hbar) V(D_p^{-1} A D_{A^{-1}(p)})(\varphi(A^{-1}(p)))\rangle,\end{aligned}$$

where we have used again in the last line that for a complex number α we have $\alpha|p,u\rangle = |p,\alpha u\rangle$. Comparing with (9.9) this yields (9.7).

We have completed our task in this section, which was to *guess* the formula (9.7). In Section 9.4 we will *prove* that this formula actually defines a unitary representation of the Poincaré group.

Exercise 9.1.1 Pretending as usual that the distribution $\delta_{m,p'}$ is a function, given $u \in \mathcal{V}$, apply (9.7) to the "function" $\varphi(p) = u\delta_{m,p'}(p)$ to recover (9.6).

9.2 The Group $\mathcal{P}^*$

In the previous section we assumed that $U(a,A)$ is a representation, but we are actually interested in projective representations. We have learned in Section 8.7 that projective representations of $SO(3)$ are simply related to true representations of $SU(2)$. A similar phenomenon occurs here. Projective representations of $\mathcal{P}$ are closely related to true representations of the group $\mathcal{P}^* := \mathbb{R}^{1,3} \rtimes SL(2,\mathbb{C})$, where the group law is given by

$$(a,A)(b,B) = (a + \kappa(A)(b), AB).$$

Here κ is the fundamental two-to-one homomorphism (8.17) from $SL(2,\mathbb{C})$ to $SO^{\uparrow}(1,3)$ given by the formula

$$M(\kappa(A)(x)) = AM(x)A^{\dagger}$$

for the matrices $M(x)$ of (8.19) i.e.

$$M(x) := x^{\mu}\sigma_{\mu} = \begin{pmatrix} x^0 + x^3 & x^1 - \mathrm{i}x^2 \\ x^1 + \mathrm{i}x^2 & x^0 - x^3 \end{pmatrix}.$$

To simplify notation in the forthcoming relatively complicated formulas, for $A \in SL(2,\mathbb{C})$ we will write

$$A(b) := \kappa(A)(b). \tag{9.10}$$

This *important* notation is used in the rest of the book and must be learned now. It is *essential* not to forget it, as the following exercise shows.

Exercise 9.2.1 Is it true that $(-A)(b) = -A(b)$?

With the convention (9.10), the group law of $\mathcal{P}^*$ takes exactly the same form (4.54) as the group law of $\mathcal{P}$.

Even though we do not plan to classify all the projective unitary representations of $\mathcal{P}$, it is reassuring that the theorem of Bargmann [4] (of which we already stated a special case in Corollary 8.6.2) ensures that each such representation arises from a true unitary representation of $\mathcal{P}^*$. The precise statement is as follows.

Theorem 9.2.2 *Given a strongly continuous unitary projective representation π' of $\mathcal{P}$ there exists a strongly continuous unitary representation π of $\mathcal{P}^*$ such that for $(c,C) \in \mathcal{P}^*$ we have*

$$\pi'(c,\kappa(C)) = \lambda(c,C)\pi(c,C), \tag{9.11}$$

where $\lambda(c,C)$ is a complex number of modulus 1.

Exercise 9.2.3 Consider ρ as in Section 8.5, that is ρ is such $\kappa(\rho(A)) = A$.
(a) Prove that given an irreducible representation π of $\mathcal{P}^*$, the formula

$$\pi'(a,A) = \pi(a,\rho(A)) \tag{9.12}$$

defines a projective representation of $\mathcal{P}$.
(b) Prove that if π and π' are as in the theorem,

$$\pi'(a,A) = \lambda'(a,A)\pi(a,\rho(A))$$

where $\lambda'(a,A)$ is a number of modulus 1.

9.3 Road Map

According to Theorem 9.2.2 we may now focus on the study of the strongly continuous unitary representations of $\mathcal{P}^*$. The two most obvious questions are as follows:

- Which are the physically important representations of $\mathcal{P}^*$ and how do we describe them?
- What explains the structure of the representations of $\mathcal{P}^*$?

The short answer to the second question is that the semidirect product structure $\mathcal{P}^* = \mathbb{R}^{1,3} \rtimes SL(2,\mathbb{C})$ is essential. To explain why, in Section A.5 we provide a complete description of the unitary representation of certain finite semidirect products. It may greatly help the reader to grasp at least the content of the main conclusions of that section in order to form the overall picture of what is happening here. Since, however, the primary goal of this book is to make the link between mathematical and physical ideas we focus in the main text on answering the first question.

9.3.1 How to Construct Representations?

Definition 9.3.1 Given any $p^* \in \mathbb{R}^{1,3}$, the group

$$G_{p^*} := \{A \in SL(2,\mathbb{C})\ ;\ A(p^*)(= \kappa(A)(p^*)) = p^*\},$$

is called[11] the *little group* of p^*.

The *central point* of the whole representation theory of $\mathcal{P}^*$ is that to each point p^* of $\mathbb{R}^{1,3}$ and each unitary irreducible representation V of its little group G_{p^*} in a finite-dimensional space $\mathcal{V}$, one can associate an irreducible unitary representation of $\mathcal{P}^*$ by a construction directly inspired of the method of Section 9.1. It will be called the representation of $\mathcal{P}^*$ *induced* by V. In the physically relevant cases the method is described in Theorem 9.4.2,[12] and as far as the problem of constructing the required representations is concerned *we are done.*

Eugene Wigner proved that every irreducible unitary representation of $\mathcal{P}^*$ arises by the previous method. His proof relies on the structure of $\mathcal{P}^*$ as a semidirect product. It is too technical to be given here, but some of the essential ideas are explained in Section A.5.

In mathematical terms, a representation V of G_{p^*} in a space $\mathcal{V}$ determines a unitary representation of $H := \mathbb{R}^{1,3} \rtimes G_{p^*}$ on $\mathcal{V}$ by the formula $U(a,A)(u) = \exp(\mathrm{i}(a,p^*)/\hbar)V(A)(u)$, and the problem is to "extend" it to the larger group $H' = \mathcal{P}^*$. As we mentioned, the solution of this problem is known in great generality through the general "theory of induced representations", and the special structure of H and $H' = \mathcal{P}^*$ as semidirect products is inessential here. We do not enter the general theory of induced representations, but instead perform all constructions in the specific cases needed, even if these constructions could be performed in a far more general setting. In Section A.4 we have however covered the general theory of induced representations in the technically simpler situation where all groups are finite. Studying that section should be most instructive for the abstract-minded reader, as several of the key ideas there parallel those of the main text.

9.3.2 Surviving the Formulas

The induced representations of $\mathcal{P}^*$ will be basically given by the complicated formula (9.7). The most urgent task is to find a simpler way to describe these representations. We will pursue that goal in two somewhat different directions:

- In Section 9.5 we provide a general method to describe induced representations by simpler formulas. A given representation can be described in many different ways, and we learn how to recognize what is what.
- In Section 9.8 we learn how to describe induced representations in an abstract, but powerful and intrinsic manner.

11 This is physicists' terminology, mathematicians would call G the *stabilizer* of p^*.
12 But if you are interested in the other cases it works just the same!

9.3.3 Classifying the Representations

We list all the physically relevant representations of $\mathcal{P}^*$ in Section 9.6 and proceed to a first investigation of the physical properties of the corresponding particles in Section 9.7.[13] This is made easy by an important feature of the method outlined in Section 9.3.1: it is not so much the point p^* which matters as its orbit under the action of $SL(2,\mathbb{C})$. We may describe the same representation using any other point of the orbit. Thus to classify all the representations we really have to look at the orbits in $\mathbb{R}^{1,3}$ under the action of $SL(2,\mathbb{C})$, and we may pick for p^* *any convenient point in the orbit.* The physically relevant orbits are $X_0 \setminus \{0\}$ and

$$X_m := \{p \in \mathbb{R}^{1,3}; p^2 = m^2c^2, p^0 \geq 0\}$$

for $m > 0$. We find two series of particles:

9.3.4 Massive Particles

The most important representations arise when for some $m > 0$ (which is argued to be the mass of the particle),[14]

$$p^* = (mc,0,0,0) = mce_0 \in X_m. \tag{9.13}$$

Then an element A of $SL(2,\mathbb{C})$ belongs to G_{p^*} if and only if $\kappa(A)(e_0) = e_0$, and as observed right after Exercise 8.4.4 this means that $A \in SU(2)$.

$$\text{For } m > 0 \text{ the little group } G_{p^*} \text{ is } SU(2). \tag{9.14}$$

A unitary representation of the little group $SU(2)$ is one of the representations π_j of Definition 8.2.3. The associated representation of $\mathcal{P}^*$ corresponds to a particle of mass m and spin $j/2$. Within this theory:[15]

A massive particle is characterized by its mass and its spin.

9.3.5 Massless Particles

The second most important case is[16]

$$p^* = (1,0,0,1) \in X_0 \setminus \{0\}, \tag{9.15}$$

13 If you are very impatient to learn about the particles you are made of, you may skip Sections 9.6 and 9.8!

14 The choice of p^* is just an element of X_m as simple as possible.

15 This means that as far as the action of $\mathcal{P}^*$ is concerned. There are other characteristics of the particle which have nothing to do with that, such as the electrical charge. Please also note that at this stage, the theory does not include parity.

16 There is no magic here either, p^* is just a simple choice of an element of X_0. We are focusing on the representations of $\mathcal{P}^*$ corresponding to actual particles. As we will argue later, the mass m of the particle satisfies $(p^*)^2 = m^2c^2 \geq 0$. However, given any element p^* of $\mathbb{R}^{1,3}$, even with $(p^*)^2 < 0$ and any unitary representation of its little group, one may also construct a unitary representation of $\mathcal{P}^*$. The hypothetical particles corresponding to this case would have imaginary mass, and could travel only at speeds greater than the speed of light. These are the mythical tachyons.

and the associated representations of $\mathcal{P}^*$ correspond to massless particles. The little group G_{p^*} is then more complicated. It turns out to consist of the matrices of the form

$$A = \begin{pmatrix} a & b \\ 0 & a^* \end{pmatrix}, \tag{9.16}$$

where $a, b \in \mathbb{C}$ and $aa^* = 1$. There is one important family $\hat{\pi}_j$ for $j \in \mathbb{Z}$ of one-dimensional representations of G_{p^*}.[17] Within this family:[18]

Massless particles are characterized by an integer $j \in \mathbb{Z}$. The number $j/2$ is called the helicity *of the particle.* (9.17)

9.3.6 Massless Particles and Parity

The properties of massless particles are elusive. In a sense that will be made mathematically precise, we will prove that the mirror image of a massless particle of helicity $j/2 \neq 0$ is a different particle, of helicity $-j/2$.

9.4 Elementary Construction of Induced Representations

Throughout the chapter we consider a number $m \geq 0$. For $m > 0$ we define p^* by (9.13) and for $m = 0$ we define p^* by (9.15).[19] For each $p \in X_m$ we fix once and for all $D_p \in SL(2,\mathbb{C})$ such that[20]

$$D_p(p^*)(= \kappa(D_p)(p^*)) = p. \tag{9.18}$$

When $m > 0$ there is a canonical choice for such a map, a unique positive definite Hermitian matrix, as provided by Lemma 8.4.6. We shall see in Proposition 9.6.7 that for $m = 0$ there is also a rather natural choice for D_p giving rise to appealing explicit formulas.[21] More importantly, our constructions using such a choice of D_p *will in some sense not depend on it.*

Lemma 9.4.1 *For each p and $A \in SL(2,\mathbb{C})$, it holds that*

$$D_p^{-1} A D_{A^{-1}(p)} \in G_{p^*}.$$

Proof

$$D_p^{-1} A D_{A^{-1}(p)}(p^*) = D_p^{-1} A(A^{-1}(p)) = D_p^{-1}(p) = p^*. \qquad \square$$

Theorem 9.4.2 *Consider a unitary representation V of the little group G_{p^*} in a finite-dimensional Hilbert space $\mathcal{V}$, and the Hilbert space $L^2 = L^2(X_m, \mathcal{V}, d\lambda_m)$ of $\mathcal{V}$-valued*

17 No particle corresponding to any of the other representations has been observed.
18 Again, parity is not considered here. When it is, matters become rather more complicated.
19 Again, the choices hold *throughout the entire chapter*.
20 It goes without saying that we assume that D_p depends regularly enough on p so that all functions we write are Lebesgue measurable.
21 In the case $m = 0$, D_p is very different from a pure boost. This is why we moved away from the notation of Definition 4.3.1.

functions on X_m provided with the norm induced by $\mathrm{d}\lambda_m$. Then the operators $\tilde{U}(a,A)$ on L^2 given for $(a,A) \in \mathcal{P}^*$ and $\varphi \in L^2$ by the formula

$$\tilde{U}(a,A)(\varphi)(p) = \exp(\mathrm{i}(a,p)/\hbar)V(D_p^{-1}AD_{A^{-1}(p)})[\varphi(A^{-1}(p))] \tag{9.7}$$

form a unitary representation of $\mathcal{P}^*$.

To gain a first understanding we observe that this is the same formula as (4.61) when $\mathcal{V} = \mathbb{C}$ and V is trivial, and that the formula makes sense according to Lemma 9.4.1: The operator $V(D_p^{-1}AD_{A^{-1}(p)})$ on $\mathcal{V}$ is well defined and can be applied to the element $\varphi(A^{-1}(p))$ of this space. The problem with the unappealing formula (9.7) is that the product $D_p^{-1}AD_{A^{-1}(p)}$ belongs to the little group but not the individual factors, so that it is not possible to expand $V(D_p^{-1}AD_{A^{-1}(p)})$. Not surprisingly this formula is not convenient for practical computations. In the next section we will discover much more pleasant formulas.

Proof of Theorem 9.4.2 That the maps $\tilde{U}(a,A)$ are unitary follows from the fact that the exponential has modulus 1, that V is unitary, and that $\mathrm{d}\lambda_m$ is invariant under Lorentz transformations.

To prove the representation property, we compute $\tilde{U}(a,A)\tilde{U}(c,C)(\psi)$ for a given function ψ. For this we apply $\tilde{U}(a,A)$ given by the formula (9.7) to the function

$$\varphi(p) := \tilde{U}(c,C)(\psi)(p) = \exp(\mathrm{i}(c,p)/\hbar)V(D_p^{-1}CD_{C^{-1}(p)})[\psi(C^{-1}(p))],$$

and we obtain

$$\exp(\mathrm{i}(a,p)/\hbar)\exp(\mathrm{i}(c,A^{-1}(p))/\hbar)W[\psi(C^{-1}(A^{-1}(p)))], \tag{9.19}$$

where

$$\begin{aligned} W &= V(D_p^{-1}AD_{A^{-1}(p)})V(D_{A^{-1}(p)}^{-1}CD_{C^{-1}(A^{-1}(p))}) \\ &= V(D_p^{-1}AD_{A^{-1}(p)}D_{A^{-1}(p)}^{-1}CD_{C^{-1}(A^{-1}(p))}) \\ &= V(D_p^{-1}ACD_{(AC)^{-1}(p)}). \end{aligned} \tag{9.20}$$

Moreover, $(c,A^{-1}(p)) = (c,\kappa(A)^{-1}(p)) = (\kappa(A)(c),p) = (A(c),p)$. The proof is then completed as in Theorem 4.8.2. □

In (9.20) we use the formula $V(A_1A_2) = V(A_1)V(A_2)$ for operators A_1, A_2 which depend on p, and the proof does not work if we only assume that V is a projective representation. This is why here it is important to use this theorem for $\mathcal{P}^*$ rather than $\mathcal{P}$: For $m > 0$ the little group is $SU(2)$ rather than $SO(3)$. According to Theorem 8.6.1, true representations of $SU(2)$ suffice, but this is not the case for $SO(3)$. However, the theorem is true for $\mathcal{P}$ rather than $\mathcal{P}^*$ when V is a true representation of the little group (which is then defined as a subgroup of $SO^{\uparrow}(1,3)$).

The representation constructed by Theorem 9.4.2 seems to depend on the arbitrary choice of the elements D_p. The next exercise shows that it is a very simple matter to prove that different choices of D_p give unitarily equivalent representations.[22]

[22] This also follows from the abstract presentation of induced representations given in Section 9.8.

Exercise 9.4.3 (a) Prove that $D_p^{-1}D'_p \in SU(2)$ whenever D'_p denotes another choice of D_p satisfying (9.18).
(b) Prove that the map $W(\varphi)$ given by $W(\varphi)(p) = V(D_p^{-1}D'_p)(\varphi(p))$ is unitary and an intertwining map between the two corresponding representations.

Thus, up to unitary equivalence, the representation constructed by Theorem 9.4.2 depends only on the representation V of the little group G_{p^*}, and the following definition makes sense.

Definition 9.4.4 The representation $\tilde{U}$ of $\mathcal{P}^*$ is called the representation *induced* by the unitary representation V of the little group G_{p^*}.

The choice of D_p in Theorem 9.4.2 is unimportant, as different choices produce equivalent representations. Next we show that the choice of p^* is also unimportant; any other choice produces an equivalent representation. Any other possible such choice $\tilde{p} \in X_m$ is of the form $C(p^*)$ for $C \in SL(2,\mathbb{C})$.

Proposition 9.4.5 *In Theorem 9.4.2 we obtain the same representation if we replace p^* by another point $C(p^*)$ (where $C \in SL(2,\mathbb{C})$) and the representation V of the little group of p^* by the representation $V' : A \mapsto V(C^{-1}AC)$ of the little group of $C(p^*)$.*

Proof Let $\tilde{p} = C(p^*)$. Then if A belongs to $G_{\tilde{p}}$, i.e. $A(\tilde{p}) = \tilde{p}$, $C^{-1}AC$ belongs to G_{p^*}, i.e. $C^{-1}AC(p^*) = p^*$. Further, if D_p satisfies $D_p(p^*) = p$ then $D'_p := D_pC^{-1}$ satisfies $D'_p(\tilde{p}) = p$. Moreover

$$C^{-1}((D'_p)^{-1}AD'_{A^{-1}(p)})C = D_p^{-1}AD_{A^{-1}(p)},$$

so that

$$V(D_p^{-1}AD_{A^{-1}(p)}) = V'((D'_p)^{-1}AD'_{A^{-1}(p)})$$

where V' is the representation $A \mapsto V(C^{-1}AC)$ of $G_{\tilde{p}}$. □

Proposition 9.4.6 *The representation induced by V is irreducible if and only if V is irreducible.*

The proof of this important result is postponed until Section 9.8.

Exercise 9.4.7 Suppose you want to study by the same method the (unphysical) representations of $\mathcal{P}^*$ arising from the point $p^* = (0, mc, 0, 0)$. You would need an invariant measure on the orbit of this point. Describe this orbit and find a formula for the invariant measure. (Warning: The unitary representations of the little group of p^* are not simple and you are not asked to study them.)

9.5 Variegated Formulas

Changing notation, given a unitary representation V' of the little group, we have defined a representation of $\mathcal{P}^*$ by the formula (9.7):

$$\tilde{U}(a, A)(\varphi)(p) := \exp(\mathrm{i}(a, p)/\hbar)V'(D_p^{-1}AD_{A^{-1}(p)})(\varphi(A^{-1}(p))).$$

Nobody relishes the term $V'(D_p^{-1}AD_{A^{-1}(p)})$ in this formula. In this section, we will learn how to express the corresponding representation in a much more pleasant way. This method will allow us to create a variety of formulas, but it will be easy to interpret their meaning. With this knowledge, the content of the concrete formulas often introduced without justification in the literature will become transparent. Better still, in the sequel we will be able to write our own concrete formulas for the most important representations of physics.

The problem with the term $V'(D_p^{-1}AD_{A^{-1}(p)})$ is that V' is defined on this product but not on the individual factors. One might look then for a representation V which extends V' to $SL(2,\mathbb{C})$. For this one may need to go a larger space $\mathcal{H}_0$ than $\mathcal{V}$, which we will also assume to be finite-dimensional.[23] We do not worry yet how this will be done, as it will be rather obvious how to proceed on the examples we need. So, suppose that $\mathcal{V}$ is a subspace of a finite-dimensional space $\mathcal{H}_0$ on which there exists a representation[24] V of $SL(2,\mathbb{C})$ with the following property

$$u \in \mathcal{V}, A \in G_{p^*} \Rightarrow V'(A)(u) = V(A)(u). \tag{9.21}$$

Then

$$\begin{aligned} V'(D_p^{-1}AD_{A^{-1}(p)})[\varphi(A^{-1}(p))] &= V(D_p^{-1}AD_{A^{-1}(p)})[\varphi(A^{-1}(p))] \\ &= V(D_p^{-1})V(A)V(D_{A^{-1}(p)})[\varphi(A^{-1}(p))], \end{aligned} \tag{9.22}$$

and we may rewrite (9.7) as

$$V(D_p)\tilde{U}(a,A)(\varphi)(p) = \exp(\mathrm{i}(a,p)/\hbar)V(A)V(D_{A^{-1}(p)})[\varphi(A^{-1}(p))]. \tag{9.23}$$

For a function[25] $\varphi\colon X_m \to \mathcal{H}_0$ and $(a,A) \in \mathcal{P}$ let us define[26]

$$U(a,A)(\varphi)(p) = \exp(\mathrm{i}(a,p)/\hbar)V(A)[\varphi(A^{-1}(p))]. \tag{9.24}$$

Let us further define

$$W(\varphi)(p) = V(D_p)(\varphi(p)). \tag{9.25}$$

Then (9.23) reads as

$$W\tilde{U}(a,A)(\varphi) = U(a,A)W(\varphi). \tag{9.26}$$

Our plan is to show that (9.26) means that W transports the representation $\tilde{U}(a,A)$ to a unitary representation given by the much nicer formula (9.24). We can guess right away that this representation will live on the space $\mathcal{H}$ of functions $\psi\colon X_m \to \mathcal{H}_0$ such that

$$\forall p \in X_m\ ,\ \psi(p) \in \mathcal{V}_p := V(D_p)(\mathcal{V}), \tag{9.27}$$

23 Even if V' extends to a representation of $SL(2,\mathbb{C})$ on $\mathcal{V}$, one may rather like to use another extension using a larger space $\mathcal{H}_0$ in order to get a cleaner formula for the representation.

24 As usual for representations of $SL(2,\mathbb{C})$, this representation is *not* assumed to be unitary.

25 Of course we always only consider "reasonable" functions, e.g. Borel-measurable.

26 Note that it is U and not $\tilde{U}$ which occurs here. We have reserved the simpler notation for this formula which we will use much more than (9.7).

for which

$$\|\psi\|^2 = \int \mathrm{d}\lambda_m(p)\|\psi(p)\|_p^2 < \infty, \tag{9.28}$$

where

$$\|u\|_p = \|V(D_p)^{-1}(u)\|, \tag{9.29}$$

and provided with the norm (9.28). We introduce first some important terminology.

Definition 9.5.1 Consider a representation V of $SL(2,\mathbb{C})$ on a space $\mathcal{H}_0$ and a subspace $\mathcal{V}$ of $\mathcal{H}_0$. Consider a subgroup G of $SL(2,\mathbb{C})$ such that $V(A)(\mathcal{V}) \subset \mathcal{V}$ for $A \in G$. The *restriction of V to $\mathcal{V}$ and G* is the representation of G on $\mathcal{V}$ obtained in the obvious manner by considering $V(A)(u)$ only for $A \in G$ and $u \in \mathcal{V}$.

This should not be confused with the restriction of V to G, which is a representation on $\mathcal{H}_0$. Even if the restriction of V to $\mathcal{V}$ and G is a unitary representation of G, the restriction of V to G need not be unitary (but then $\mathcal{V}$ is an invariant subspace of this restriction and *on that subspace V* is unitary).

One should also be aware that the restriction of an irreducible representation to a subgroup is often not irreducible. For example the representation κ of $SL(2,\mathbb{C})$ on $\mathbb{R}^{1,3}$ is irreducible, but its restriction to $SU(2)$ is not irreducible, since it consists of rotations, which leave e_0 invariant.

To simplify terminology, we will describe the situation where $V(A)(\mathcal{V}) \subset \mathcal{V}$ for each $A \in G$ and where the restriction of V to G and $\mathcal{V}$ is unitary simply by saying that *"the restriction of V to $\mathcal{V}$ and G is unitary"*, keeping implicit the fact that this representation is well defined. Note that then $V(A)(\mathcal{V}) = \mathcal{V}$ for each $A \in G$.

Let us recapitulate. We start with a unitary representation V' of the little group G_{p^*} in the space $\mathcal{V}$. We find a representation V of $SL(2,\mathbb{C})$ on a larger finite-dimensional space $\mathcal{H}_0$ with the property (9.21). In the words of Definition 9.5.1, the restriction of V to $\mathcal{V}$ and G_{p^*} is unitary (since this restriction equals V'). Turning things around, we now start from a representation of $SL(2,\mathbb{C})$ into a finite-dimensional space $\mathcal{H}_0$ and a subspace $\mathcal{V}$ of $\mathcal{H}_0$.

Lemma 9.5.2 *Let us assume that the restriction of V to $\mathcal{V}$ and G_{p^*} is unitary. Given $p \in X_m$ the vector space*

$$\mathcal{V}_p = V(A)(\mathcal{V}) \tag{9.30}$$

is independent of the choice of $A \in SL(2,\mathbb{C})$ satisfying $A(p^*) = p$. For $u \in \mathcal{V}_p$ the norm

$$\|u\|_p = \|V(A^{-1})(u)\| \tag{9.31}$$

is independent of $A \in SL(2,\mathbb{C})$ satisfying $A(p^*) = p$. Furthermore,

$$p \in X_m\ ,\ u \in \mathcal{V}_p\ ,\ A \in SL(2,\mathbb{C}) \Rightarrow V(A)(u) \in \mathcal{V}_{A(p)}\ ;\ \|V(A)(u)\|_{A(p)} = \|u\|_p. \tag{9.32}$$

In particular the space (9.27) and the norm (9.29) are independent of the choice of D_p such that $D_p(p^*) = p$.

Proof Consider $A_1, A_2 \in SL(2,\mathbb{C})$ with $A_1(p^*) = A_2(p^*)$. Then $A := A_2^{-1}A_1$ satisfies $A(p^*) = p^*$ so that $A \in G_{p^*}$. Thus by assumption $V(A)(\mathcal{V}) = \mathcal{V}$. Moreover $A_1 = A_2A$, so that $V(A_1)(\mathcal{V}) = V(A_2)(V(A)(\mathcal{V})) = V(A_2)(\mathcal{V})$. We have shown that $V(B)(\mathcal{V})$ is independent of $B \in SL(2,\mathbb{C})$ satisfying $B(p^*) = p$. As a consequence

$$v \in \mathcal{V}\,,\, C \in SU(2) \Rightarrow V(C)(v) \in \mathcal{V}_{C(p^*)}. \tag{9.33}$$

Since $A_1 = A_2A$ we have $V(A_1)^{-1} = V(A)^{-1}V(A_2)^{-1}$ and thus for $u \in V(A_1)(\mathcal{V}) = V(A_2)(\mathcal{V})$ it holds that

$$\|V(A_1)^{-1}(u)\| = \|V(A)^{-1}(V(A_2)^{-1}(u))\| = \|V(A_2)^{-1}(u)\|,$$

using in the last equality that $V(A)$ is an isometry on $\mathcal{V}$ by hypothesis. We have thus shown that $\|V(A)^{-1}(u)\|$ is independent of $A \in SL(2,\mathbb{C})$ satisfying $A(p^*) = p$. It remains to prove (9.32). Consider $u \in \mathcal{V}_p$, so that $u = V(B)(v)$ where $v \in \mathcal{V}$ and $B(p^*) = p$. Then $C := AB$ satisfies $C(p^*) = A(p)$ and $V(A)(u) = V(C)(v)$. Thus by (9.33) we have $V(A)(u) \in \mathcal{V}_{C(p^*)} = \mathcal{V}_{A(p)}$. Since $C(p^*) = A(p)$ by (9.31) we have $\|V(A)(u)\|_{A(p)} = \|V(C)^{-1}V(A)(u)\| = \|v\|$. Since $B(p^*) = p$ we have similarly $\|u\|_p = \|V(B)^{-1}(u)\| = \|v\|$. □

Theorem 9.5.3 *Consider a representation V of $SL(2,\mathbb{C})$ into a finite-dimensional space $\mathcal{H}_0$ and a subspace $\mathcal{V}$ of $\mathcal{H}_0$. Assume that the restriction of V to $\mathcal{V}$ and G_{p^*} is unitary, and consider the spaces $\mathcal{V}_p$ and the norms $\|\cdot\|_p$ as defined in Lemma 9.5.2. Consider the space $\mathcal{H}$ of functions $\varphi\colon X_m \to \mathcal{H}_0$ such that $\varphi(p) \in \mathcal{V}_p$ for each p, provided with the norm (9.28),*

$$\|\varphi\|^2 = \int d\lambda_m(p)\|\varphi(p)\|_p^2. \tag{9.28}$$

Then the formula (9.24)

$$U(a,A)(\varphi)(p) = \exp(\mathrm{i}(a,p)/\hbar)V(A)[\varphi(A^{-1}(p))] \tag{9.24}$$

defines a unitary representation of $\mathcal{P}^*$ on $\mathcal{H}$. This representation is unitarily equivalent to the representation induced by the restriction of V to $\mathcal{V} = \mathcal{V}_{p^*}$ and G_{p^*}.

We will sometimes use the straightforward fact that (9.24) defines a representation of $\mathcal{P}$ itself (rather than $\mathcal{P}^*$) when V is a representation of $SO^{\uparrow}(1,3)$ (rather than $SL(2,\mathbb{C})$). Let us note the following, which will be of constant use.

Corollary 9.5.4 *Up to unitary equivalence the representation U of Theorem 9.5.3 is completely determined by the restriction*[27] *of V to $\mathcal{V}_{p^*}$ and G_{p^*}.*

Proof of Theorem 9.5.3 We give a self-contained proof, although one may also deduce some of the statements from Theorem 9.4.2. Let us first compute $U(a,A)U(b,B)(\psi)$ for a given function ψ. For this we apply (9.24) to the function

$$\varphi(p) := U(b,B)(\psi)(p) = \exp(\mathrm{i}(b,p)/\hbar)V(B)[\psi(B^{-1}(p))],$$

27 This restriction is of course unitary, as assumed in the theorem.

we obtain

$$\exp(\mathrm{i}(a,p)/\hbar)\exp(\mathrm{i}(b,A^{-1}(p))/\hbar)V(A)V(B)\psi(B^{-1}(A^{-1}(p))), \tag{9.34}$$

and we conclude as in the case of Theorem 4.8.2.

To prove that the representation is unitary we use (9.32) for $u = \varphi(A^{-1}(p))$ and $A^{-1}(p)$ instead of p to obtain

$$\|U(a,A)(\varphi)(p)\|_p = \|V(A)(\varphi(A^{-1}(p)))\|_p = \|\varphi(A^{-1}(p))\|_{A^{-1}(p)},$$

and unitarity follows from the Lorentz invariance of $\mathrm{d}\lambda_m$. The last statement is proved by the formula (9.26). □

The next exercise provides an alternative formula for the norm (9.28) when $m > 0$.

Exercise 9.5.5 Consider $m > 0$ and as usual $p^* = (mc,0,0,0)$. The group G_{p^*} is then the compact group $SU(2)$. As does every compact group it has a *Haar probability*, that is a probability measure ν which is invariant by left translations: For each continuous function f on $SU(2)$ and any $C \in SU(2)$ one has $\int \mathrm{d}\nu(B)f(CB) = \int \mathrm{d}\nu(B)f(B)$.
(a) Prove that the measure $\mathrm{d}\mu$ on $SL(2,\mathbb{C})$ given for each continuous function f with compact support by $\int \mathrm{d}\mu(A)f(A) = \iint \mathrm{d}\lambda_m(p)\mathrm{d}\nu(B)f(D_pB)$ is left-invariant, i.e. $\int \mathrm{d}\mu(A)f(CA) = \int \mathrm{d}\mu(A)f(A)$ for each $C \in SL(2,\mathbb{C})$.
(b) In the setting of Theorem 9.5.3 prove that for a function $\varphi \in \mathcal{H}$ we have

$$\|\varphi\|^2 = \int \mathrm{d}\mu(A)\|V(A)^{-1}\varphi(A(p^*))\|^2.$$

(c) What happens in the case $m = 0$?

Conceptually the statement of Theorem 9.5.3 is very instructive, but in concrete situations, one may like to describe the constraint (9.27) on ψ by equations that ψ should satisfy, and to get explicit formulas for the norm $\|u\|_p$. We give one example of such an explicit formula now in a fundamental situation. We will give many more examples later (the most elaborate of which are given in Section G.2).

We consider the situation where $m > 0$ and V is the defining representation of $SL(2,\mathbb{C})$, where the elements of $SL(2,\mathbb{C})$ act as matrices on $\mathcal{H}_0 = \mathcal{V} = \mathbb{C}^2$, so that the space $\mathcal{H}$ of Theorem 9.5.3 is the space of all measurable functions from X_m to $\mathcal{H}_0$. Since $m > 0$ the little group is $SU(2)$, see (9.14), and the restriction of V to the little group $SU(2)$ is the action of $SU(2)$ on $\mathbb{C}^2$ by matrix multiplication. As we have seen in Sections 8.2 and 8.3 this is the simplest non-trivial unitary representation of $SU(2)$. It will become apparent later that the representation of $\mathcal{P}^*$ produced then by Theorem 9.5.3 corresponds to particles of spin 1/2, the most important particles of the universe.

We recall the matrices $M(x)$ of (8.19) and the parity operator $P\colon (p^0,\boldsymbol{p}) \mapsto (p^0,-\boldsymbol{p})$.

Proposition 9.5.6 *Let us assume that $m > 0$, and that V is the defining representation of $SL(2,\mathbb{C})$, where the elements of $SL(2,\mathbb{C})$ act as matrices on $\mathcal{H}_0 = \mathcal{V} = \mathbb{C}^2$. Then*

$$\|u\|_p^2 = \frac{1}{mc}(M(P(p))(u),u). \tag{9.35}$$

Proof Consider $A \in SL(2,\mathbb{C})$ with $A(p^*) = p$. Then

$$\|u\|_p^2 = \|A^{-1}(u)\|^2 = (A^{\dagger -1}A^{-1}(u),u) = ((AA^\dagger)^{-1}(u),u),$$

and the result follows from (9.37).[28] □

Lemma 9.5.7 *Recalling that $p^* = (mc,0,0,0)$, if $A(p^*) = p$ then*

$$AA^\dagger = \frac{1}{mc}M(p) \tag{9.36}$$

and

$$(AA^\dagger)^{-1} = \frac{1}{mc}M(P(p)). \tag{9.37}$$

Proof We recall that $A(p^*)$ stands for $\kappa(A)(p^*)$ and that $AM(x)A^\dagger = M(\kappa(A)(x))$. Using this formula for $x = p^*$ yields $mcAA^\dagger = M(p)$. Recalling (8.38), and since $\det M(p) = p^2 = m^2c^2$ we obtain $(AA^\dagger)^{-1} = mcM(p)^{-1} = M(P(p))/mc$. □

Let us then summarize the situation:

- A representation of $\mathcal{P}^*$ (as constructed by our methods[29]) is characterized by a unitary representation of the little group G_{p^*}.
- Given such a representation of G_{p^*} the corresponding induced representation of $\mathcal{P}^*$ is given by Theorem 9.4.2.
- Unfortunately, bitter experience teaches us that computations involving the complicated formula (9.7) are hard.
- Given a unitary representation $\bar{V}$ of G_{p^*} on $\mathcal{V}$, we simply find a (possibly non-unitary) representation V of $SL(2,\mathbb{C})$, in a possibly larger space $\mathcal{H}_0 \supset \mathcal{V}$, such that its restriction to G_{p^*} and $\mathcal{V}$ coincides with $\bar{V}$, and we rather describe the representation induced by $\bar{V}$ using V and Theorem 9.5.3.
- In this manner the same representation of $\mathcal{P}^*$ can be described in many different ways, but we do not get confused because we know that all that matters is the restriction of V' to $\mathcal{V}$ and G_{p^*}.

These methods will be illustrated in Section 9.6.

Exercise 9.5.8 Consider the case $m > 0$ and $\mathcal{H}_0 = \mathcal{V} = \mathbb{C}^2$, and denote by U_R the representation constructed in Theorem 9.5.3 with $V(A) = A$, and U_L the representation constructed with $V(A) = A^{\dagger -1}$. Since $A = A^{\dagger -1}$ for $A \in SU(2)$ these representations are unitarily equivalent. The purpose of this exercise is to check this directly. We assume that D_p is Hermitian, so that $\kappa(D_p)$ is a pure boost. Prove that the map W from $\mathcal{H}$ to $\mathcal{H}$ given by $W(\varphi)(p) = D_p^{-2}(\varphi(p))$ satisfies $\|W(\varphi)(p)\|_{L,p} = \|\varphi(p)\|_{R,p}$ and check that $U_L W = WU_R$.

[28] More generally when $V(A)^\dagger = V(A^\dagger)$ one has $\|u\|_p^2 = (V((D_pD_p^\dagger)^{-1})(u),u)$. This holds true for all the representations of $SL(2,\mathbb{C})$ constructed from tensor products.

[29] That is, we did not prove that different representations could not be found using other methods.

9.6 Fundamental Representations

Given $m \geq 0$ and the point p^* of X_m, considering an irreducible unitary representation V of the little group G_{p^*}, Theorem 9.4.2 produces an irreducible representation of $\mathcal{P}^*$. In this section we name the most important representations of $\mathcal{P}^*$ obtained in this manner, and we show how they can be described within the more friendly setting of Theorem 9.5.3.

9.6.1 Massive Particles

Let us start with the case $m > 0$. Then the little group G_{p^*} of p^* is $SU(2)$, see (9.14). We have constructed the fundamental family π_j of unitary representations of this group in Section 8.2. We will however use the equivalent form (8.14) of these representations using tensor products, which will be recalled in a few lines. The importance of the following representations cannot be overstated.

Definition 9.6.1 For $m > 0$, $j \geq 0$, the representation $\pi_{m,j}$ is the unitary representation of $\mathcal{P}^*$ induced by the representation π_j of the little group $SU(2)$.

It should be stressed that this representation depends on m (as does the point p^*, see (9.13)) despite the fact that the little group does not depend on $m > 0$.

The representation $\pi_{m,0}$ is the representation of Section 4.8. A concrete realization of $\pi_{m,1}$ is given in Proposition 9.5.6, and we can give a concrete realization of $\pi_{m,j}$ for $j > 1$ by a rather straightforward generalization of that construction. We consider the space $\mathcal{V} = \mathcal{H}_0 = \mathcal{S}_j$ of symmetric tensors $x = (x_{n_1,n_2,\ldots,n_j})$ with $n_k \in \{1,2\}$, and the representation V of $SL(2,\mathbb{C})$ on $\mathcal{S}_j$ given by

$$V(A)(x) = \Big(\sum_{k_1,\ldots,k_j} A_{n_1,k_1} \cdots A_{n_j,k_j} x_{k_1,\ldots,k_j}\Big). \tag{9.38}$$

It is proved in Proposition 8.3.1 that the restriction of V to $\mathcal{V}$ and the little group $SU(2)$ is unitarily equivalent to π_j, so that the representation constructed by Theorem 9.5.3 for this choice of V is equivalent to $\pi_{m,j}$. We then have $\mathcal{V}_p = \mathcal{V}$ for each p. According to Lemma 9.5.7 and the footnote on page 222 the norm $\|u\|_p$ is given by $\|u\|_p^2 = (V(M(P(p))/mc)(u), u)$. One may also use the formula of Exercise 9.5.5 for the norm.

9.6.2 Massless Particles

In the remainder of this section we study the case of massless particles, $m = 0$. This case is far more elusive than the case $m > 0$ due to the more complicated nature of the little group. This material will be instrumental in understanding what photons are in Section 9.13, but this in itself is a side story to our main line of investigation, and it may be omitted at first reading.[30] When we work specifically in the case $m = 0$ we will denote by G_0 the

30 Generally speaking, all the considerations concerning massless particles constitute a side story and can be omitted at first reading. The reason massless particles are treated in great detail is that the author wanted to truly understand this situation, but could not easily find a satisfactory treatment in the literature.

little group G_{p^*}, and the first step is to compute this group.[31] We recall that when $m = 0$, $p^* = (1,0,0,1)$ and then the matrix $M(p^*)$ of (8.19) is given by

$$M(p^*) = \begin{pmatrix} 2 & 0 \\ 0 & 0 \end{pmatrix}. \tag{9.39}$$

The condition that $A \in G_0$, i.e. $\kappa(A)(p^*) = p^*$, that is $AM(p^*)A^\dagger = M(p^*)$, yields in a straightforward manner that A has to be of the type

$$A = \begin{pmatrix} a & b \\ 0 & a^* \end{pmatrix}, \tag{9.40}$$

where $a, b \in \mathbb{C}$ and $aa^* = 1$.

Definition 9.6.2 For $j \in \mathbb{Z}$ and $A \in G_0$ as in (9.40) we define $\hat{\pi}_j(A) = a^{-j} \in \mathbb{C}$.

The reason for the minus sign will become apparent later. Please keep in mind that $\hat{\pi}_j(A)$ is a *number*. However, we can think of $\hat{\pi}_j$ as a one-dimensional representation, by the formula $\hat{\pi}_j(A)(x) = \hat{\pi}_j(A)x$.

We recall that for a matrix A, we denote by A^* the conjugate matrix, where every entry has been replaced by its complex conjugate. The following is trivial but useful.

Lemma 9.6.3 *The map $A \to A^*$ is a group automorphism of G_0 and $\hat{\pi}_j(A^*) = \hat{\pi}_{-j}(A)$.*

Definition 9.6.4 Assume $m = 0$. For $j \in \mathbb{Z}$, $j \neq 0$, the representation $\pi_{0,j}$ is the unitary representation of $\mathcal{P}^*$ induced by the representation $\hat{\pi}_j$ of the little group G_0.

The following exercise provides a closer look at the little group G_0, and brings forward the fact that this group is itself a semidirect product.

Exercise 9.6.5 (a) Consider the group $\mathbb{U}$ of complex numbers of modulus 1. Prove that the operation on $\mathbb{C} \times \mathbb{U}$ defined by

$$(c,a)(c',a') = (c + a^2c', aa')$$

is a group operation. This group is denoted by G.
(b) Prove that G can be naturally identified as a double cover of the group of transformations of $\mathbb{R}^2$ generated by translations and rotations. Hint: Consider the operation of G on $\mathbb{C}$ given by $(c,a) \cdot z = c + a^2 z$. Prove that it is an action, i.e. $(c',a') \cdot ((c,a) \cdot z) = ((c',a')(c,a)) \cdot z$.
(c) Prove that the map that sends the matrix (9.40) to the element (ab, a) of G is a group isomorphism from G_0 to G.
(d) Given $j \in \mathbb{Z}$ and $\alpha \in \mathbb{C}$, prove that the formula

$$U(c,a)(f)(w) = a^j \exp(\mathrm{i}\,\mathrm{Im}(\alpha c^* w)) f(a^{-2}w)$$

[31] The reader should observe that there are only two possible values for the little group: G_0 and $SU(2)$.

(where $\operatorname{Im} z$ is the imaginary part of the complex number z) defines a unitary representation of G in the set of square-integrable functions on $\mathbb{C}$. Do you see any connection between this formula and the formulas defining representations of $\mathcal{P}^*$? Why is this the case? Hint: Study Section A.5.

One may construct unitary representations of $\mathcal{P}^*$ using the unitary representations of G_0 exhibited in the previous exercise, but they do not correspond to any known particle, and for physics the important case is that of Definition 9.6.4.

We now explore concrete realizations of the representation $\pi_{0,j}$. These realizations are neither very important nor very enlightening, but they do make the point that we are dealing with very non-trivial structures.

Proposition 9.6.6 *For*

$$A = \begin{pmatrix} a & b \\ c & d \end{pmatrix} \in SL(2,\mathbb{C}) \tag{9.41}$$

and $p \in X_0$ let us define $\xi(A,p) = d(p^0 + p^3) - b(p^1 + \mathrm{i}p^2)$. Then given A, λ_0-a.e. we have $\xi(A,p) \neq 0$. The formula

$$U(a,A)(\varphi)(p) = \exp(\mathrm{i}(a,p)/\hbar)\Big(\frac{\xi(A,p)}{|\xi(A,p)|}\Big)^j \varphi(A^{-1}(p)) \tag{9.42}$$

defines a unitary representation of $\mathcal{P}^*$ in $L^2 := L^2(X_0, \mathrm{d}\lambda_0)$. This representation is unitarily equivalent to $\pi_{0,j}$.

One striking feature of the formula is that the crucial factor $(\xi(A,p)/|\xi(A,p)|)$ does not change if we replace p by λp for some $\lambda > 0$.

We are going to show that this formula is simply the formula (9.7) when V is the representation $\hat{\pi}_j$ of G_0 and when we use an adequate choice of D_p.

Proposition 9.6.7 *For $p \in X_0$ we have $p^0 + p^3 \geq 0$. When $p^0 + p^3 \neq 0$ the matrix*

$$D_p = \frac{1}{\sqrt{2(p^0 + p^3)}} \begin{pmatrix} p^0 + p^3 & 0 \\ p^1 + \mathrm{i}p^2 & 2 \end{pmatrix} \in SL(2,\mathbb{C}) \tag{9.43}$$

satisfies $D_p(p^*) = p$, where $p^* = (1,0,0,1)$.

Proof The first claim follows from the fact that $(p^1)^2 + (p^2)^2 = (p^0)^2 - (p^3)^2$ and $p^0 \geq 0$. When A is as in (9.41) it is straightforward to use the definition $M(A(p^*)) = AM(p^*)A^\dagger$ of $A(p)(= \kappa(A)(p))$ to check that the relation $p := A(p^*)$ amounts to

$$2|a|^2 = p^0 + p^3 \; ; \; 2a^*c = p^1 + \mathrm{i}p^2 \; ; \; 2|c|^2 = p^0 - p^3, \tag{9.44}$$

from which one readily discovers the choice $a = \sqrt{(p^0 + p^3)/2}$ and $c = (p^1 + \mathrm{i}p^2)/\sqrt{2(p^0 + p^3)}$, the last relation of (9.44) being then a consequence of the relation $(p^1)^2 + (p^2)^2 = (p^0)^2 - (p^3)^2$. The natural choice $b = 0$ together with the requirement $D_p \in SL(2,\mathbb{C})$ then leads to the formula (9.43). □

Proof of Proposition 9.6.6 We make the choice (9.43) when $p^0 + p^3 \neq 0$ (and any arbitrary choice on the negligible set where $p^3 = -p^0$), and we carry out the computation of the expression (9.7).

The basic observation is that for a matrix $\tilde{A}$ of the type (9.40), if we know that $a = \lambda\alpha$ for some $\alpha \in \mathbb{C}$ and some $\lambda > 0$ then since $|a| = 1$ we have $a = \alpha/|\alpha|$ and therefore $\hat{\pi}_j(\tilde{A}) = (\alpha/|\alpha|)^{-j}$. Recalling that $p^0 + p^3 \geq 0$, D_p^{-1} is a lower-triangular matrix with diagonal coefficients in $\mathbb{R}^+$. When we consider a matrix $B = (B_{i,j})$ with $\tilde{A} := D_p^{-1}B \in G_0$ it follows that $\tilde{A}_{1,1} = \lambda B_{1,1}$ with $\lambda > 0$. Consequently $\hat{\pi}_j(D_p^{-1}B) = \tilde{A}_{1,1}^{-j} = (B_{1,1}/|B_{1,1}|)^{-j}$. Consider now $A \in SL(2,\mathbb{C})$. Then $\hat{\pi}_j(D_p^{-1}AD_{A^{-1}(p)}) = (B_{1,1}/|B_{1,1}|)^{-j}$ where $B = AD_{A^{-1}(p)}$. From (9.43) the first column of D_p is the same as λ times the first column of $M(p)$ for a certain $\lambda > 0$. Thus the first column of $D_{A^{-1}(p)}$ is the same as λ times the first column of $M(A^{-1}(p))$ for a certain $\lambda > 0$. Consequently $B_{1,1}/|B_{1,1}| = C_{1,1}/|C_{1,1}|$ where $C = AM(A^{-1}(p))$. But then $M(A^{-1}(p)) = A^{-1}M(p)(A^{-1})^\dagger$ so that $C = M(p)(A^{-1})^\dagger$. A straightforward calculation yields $C_{1,1} = \xi(A,p)^*$. Finally $(\xi(A,p)^*/|\xi(A,p)|)^{-j} = (\xi(A,p)/|\xi(A,p)|)^j$. □

This near-miraculous calculation does not shed much light on why the formula (9.42) defines a representation. The following exercise will provide a direct proof of this fact.

Exercise 9.6.8 (a) Prove that the formula

$$U(a,A)(\varphi)(p) = \exp(\mathrm{i}(a,p)/\hbar)g(A,p)\varphi(A^{-1}(p))$$

defines a representation of $\mathcal{P}^*$ if for each $A, B \in SL(2,\mathbb{C})$ and each $p \in X_0$ we have

$$g(B,p)g(A,B^{-1}(p)) = g(BA,p). \tag{9.45}$$

In the sequel, please do not worry about the rare cases where the definitions make no sense.

(b) For $z \in \mathbb{C}$ and $A \in SL(2,\mathbb{C})$ as in (9.41) define $A \cdot z = (c + dz)/(a + bz)$. Prove that this defines an action of $SL(2,\mathbb{C})$ on $\mathbb{C}$, i.e. $A \cdot (B \cdot z) = (AB) \cdot z$.

(c) For $A \in SL(2,\mathbb{C})$ and $z \in \mathbb{C}$ define $f(A,z) = d - bz$. Prove that

$$f(B,z)f(A,B^{-1} \cdot z) = f(BA,z). \tag{9.46}$$

(d) What is the relation with Proposition 9.6.6? Hint: Find a suitable identification of the rays of X_0 with $\mathbb{C}$, which transforms the natural action of $SL(2,\mathbb{C})$ on these rays into the previous one.

Exercise 9.6.9 Think of $v \in \mathbb{C}^2$ as a column matrix. We denote by v_1, v_2 the components of $v \in \mathbb{C}^2$. We recall the action of matrix multiplication Av for $A \in SL(2,\mathbb{C})$ and $v \in \mathbb{C}^2$.

(a) Prove that for $v \in \mathbb{C}^2$ there exists a unique $p(v) \in X_0$ such that $M(p(v)) = vv^\dagger$.

(b) Consider $j \in \mathbb{Z}$ and the space $\mathcal{H}_j \subset L^2(\mathbb{C}^2)$ consisting of functions f such that for any complex number θ of modulus 1 we have $f(\theta v) = \theta^j f(v)$. Prove that the formula

$$V(a,A)(f)(v) = \exp(\mathrm{i}(a,p(v))/\hbar)f(A^{-1}v) \tag{9.47}$$

defines a unitary representation of $\mathcal{P}^*$ in $\mathcal{H}_j$.
(c) For $p \in X_0$ define (whenever possible) $w(p) \in \mathbb{C}^2$ by $\sqrt{p^0+p^3}\,w(p) = \begin{pmatrix} p^0+p^3 \\ p^1+\mathrm{i}p^2 \end{pmatrix}$. Prove that $p(w(p)) = p$. Prove that if $vv^\dagger = M(p)$ then $v = (v_1/|v_1|)w(p)$.
(d) Prove that $Bw(p) = (Bw(p)_1/|Bw(p)_1|)w(Bp)$ for $B \in SL(2,\mathbb{C})$.
(e) Prove that the representations (9.42) and (9.47) are unitarily equivalent. Hint: Use the map $T : \mathcal{H}_j \to L^2(X_0, \mathrm{d}\lambda_0)$ given by $T(f)(p) = f(w(p))$.

In the remainder of this section we study the description of the representation $\pi_{0,-j}$ of $\mathcal{P}^*$ within the framework of Theorem 9.5.3, using this theorem for $m = 0$. We assume $j \geq 1$. We consider again the space $\mathcal{H}_0 = \mathcal{S}_j$ of symmetric tensors and the representation V of (9.38). In the important case $j = 1$ we then have $\mathcal{H}_0 = \mathbb{C}^2$ and V is the action on $\mathbb{C}^2$ by matrix multiplication, $V(A)(x) = A(x)$. Consider the tensor $g \in \mathcal{S}_j$ given by $g_{n_1,\ldots,n_j} = 0$ unless all indices are equal to 1, in which case $g_{n_1,\ldots,n_j} = 1$. Thus, when $j = 1$, g is the vector with $g_1 = 1$ and $g_2 = 0$. We denote by $\mathcal{V}$ the linear span of g, which is then a space of dimension 1. For $A \in G_0$ given by (9.40) we have $V(A)(g) = a^j g = (a^*)^{-j} g = \hat{\pi}_{-j}(A)g$. Thus $\mathcal{V}$ is invariant under $V(A)$, and the restriction of V to G_0 and $\mathcal{V}$ is unitarily equivalent to the representation $\hat{\pi}_{-j}$ of G_0. Consequently, the representation of $\mathcal{P}^*$ constructed by Theorem 9.5.3 is unitarily equivalent to the representation $\pi_{0,-j}$.

Let us now compute the norm $\|u\|_p$ of an element u of $\mathcal{V}_p$. Such an element is of the type

$$u = V(A)(\lambda g) = (\lambda A_{n_1,1}A_{n_2,1}\cdots A_{n_j,1}),$$

and $p = A(p^*) = \kappa(A)(p^*)$ i.e. $AM(p^*)A^\dagger = M(p)$. Using (9.39), it is straightforward to see that if A is as in (9.41) then $p^0 = |a|^2 + |c|^2$. Thus

$$\|u\|_p^2 = \|V(A)^{-1}(u)\|^2 = \|\lambda g\|^2 = |\lambda|^2$$

whereas since $|A_{1,1}|^2 + |A_{2,1}|^2 = |a|^2 + |c|^2$,

$$\|u\|^2 = |\lambda|^2 \sum_{n_1,\ldots,n_j} |A_{n_1,1}|^2 \cdots |A_{n_j,1}|^2 = |\lambda|^2(|a|^2+|c|^2)^j = (p^0)^j|\lambda|^2.$$

Consequently $\|u\|_p^2 = \|u\|^2/(p^0)^j$. We have proved the following.

Proposition 9.6.10 *For $j \geq 1$ consider the space $\mathcal{H}$ of functions $\varphi : X_0 \to \mathcal{S}_j$ which satisfy $\varphi(p) \in \mathcal{V}_p$ for each p, provided with the norm*

$$\|\varphi\|^2 = \int \mathrm{d}\lambda_0(p)\frac{\|\varphi(p)\|^2}{(p^0)^j}. \tag{9.48}$$

Then the formula

$$U(a,A)(\varphi)(p) = \exp(\mathrm{i}(a,p)/\hbar)V(A)[\varphi(A^{-1}(p))] \tag{9.49}$$

defines a unitary representation of $\mathcal{P}^*$ on $\mathcal{H}$, which is unitarily equivalent to the representation $\pi_{0,-j}$.

Exercise 9.6.11 Find the corresponding statement for the representation $\pi_{0,j}$.

When $j = 1$ our next result describes the space $\mathcal{V}_p$ by an explicit formula.

Lemma 9.6.12 *When $j = 1$ we have*

$$\mathcal{V}_p = \{u \in \mathcal{H}_0 = \mathbb{C}^2 \,;\, M(Pp)(u) = 0\}. \tag{9.50}$$

Proof We recall that when $j = 1$, we have $\mathcal{H}_0 = \mathbb{C}^2$ and $g \in \mathbb{C}^2$ is the vector whose components are $g_1 = 1$ and $g_2 = 0$. Its linear span $\mathcal{V}$ is the set of vectors which have a second component equal to zero. Thus (9.50) holds for $p = p^*$ since, as in (9.39), we have

$$M(Pp^*) = \begin{pmatrix} 0 & 0 \\ 0 & 2 \end{pmatrix}.$$

For the general case, we observe that since $PA(p^*) = P\kappa(A)(p^*) = \kappa(A^{\dagger -1})(Pp^*)$ by (8.36), it holds that

$$M(PA(p^*)) = A^{\dagger -1} M(Pp^*) A^{-1}. \tag{9.51}$$

If now $v \in \mathcal{V}_p$ we have $v = D_p(u)$ for a certain $u \in \mathcal{V}$. Applying the previous relation to v and taking $A = D_p$ shows that $M(Pp)(v) = 0$ for $v \in \mathcal{V}_p$. Since $\mathcal{V}_p$ has dimension 1, this proves (9.50). □

Therefore the space $\mathcal{H}$ has a very clean description as the set of functions φ from X_0 to $\mathbb{C}^2$ which satisfy the equation $M(Pp)(\varphi(p)) = 0$. Let us again state this result.

Proposition 9.6.13 *Consider the space $\mathcal{H}$ of functions $\varphi\colon X_0 \to \mathbb{C}^2$ which satisfy the equation $M(Pp)(\varphi(p)) = 0$, provided with the norm (9.48). Then the formula*

$$U(a,A)(\varphi)(p) = \exp(\mathrm{i}(a,p)/\hbar) A[\varphi(A^{-1}(p))] \tag{9.52}$$

defines a unitary representation of $\mathcal{P}^*$ on $\mathcal{H}$, which is unitarily equivalent to the representation $\pi_{0,-1}$.

Recalling that $M(p) = p^\mu \sigma_\mu$ we can consider the matrices $\bar{\sigma}_\mu$ such that $M(Pp) = p^\mu \bar{\sigma}_\mu$. That is $\bar{\sigma}_0 = \sigma_0$ and $\bar{\sigma}_i = -\sigma_i$ for $1 \le i \le 3$.[32] The equation $M(Pp)(\varphi(p)) = 0$ is then written as $p^\mu \bar{\sigma}_\mu v(p) = 0$ in physics books.

Exercise 9.6.14 Find the corresponding statement for the representation $\pi_{0,1}$.

Exercise 9.6.15 In the case of a general value of $j \ge 1$ prove that $u \in \mathcal{V}_p$ if and only if $(D_p D_p^\dagger M(Pp) \otimes I \otimes \cdots \otimes I)(u) = 0$.

9.7 Particles, Spin, Representations

The present section is a continuation of Section 4.9, which the reader may like to review now. What are the properties of a particle to which corresponds the representation $\pi_{m,j}$ or

[32] So that, indeed, $\bar{\sigma}_\mu = \sigma^\mu$. The point of the new notation is that the formalism forbids the expression $p^\mu \sigma^\mu$.

$\pi_{0,j}$? Let us first consider the case of $\pi_{m,j}$ where $m > 0$. As on page 127 we may argue that *the particle is of mass m* and here we investigate its spin. We certainly expect from Section 9.1 that

The representation $\pi_{m,j}$ of Definition 9.6.1 describes a particle of spin $j/2$. (9.53)

In Definition 8.7.1 we defined what is a particle of spin $j/2$ in Non-relativistic Quantum Mechanics. This theory is however not a part of Quantum Field Theory, just like, say, Newton's theory of gravitation is not a part of General Relativity. So we cannot check that we are here in the setting of Definition 8.7.1. The best we can do is to decide that we will *define* the spin of a particle described by the representation $\pi_{m,j}$ as being $j/2$ and to convince ourselves that this is reasonably consistent with Definition 8.7.1. For this let us denote by $D_p \in SL(2,\mathbb{C})$ the unique positive definite Hermitian matrix with $D_p(p^*) = (\kappa(D_p)(p^*)) = p$ as provided by Lemma 8.4.6.

Lemma 9.7.1 *We have $D_p^{-1} A D_{A^{-1}(p)} = A$ for each $A \in SU(2)$ and each $p \in X_m$.*

Proof The matrix $C := A^{-1} D_p A = A^\dagger D_p A$ is positive Hermitian. Since $A \in SU(2)$ it holds that $A(p^*) = \kappa(A)(p^*) = p^*$ so that $C(p^*) = A^{-1} D_p A(p^*) = A^{-1}(p)$. Thus C equals $D_{A^{-1}(p)}$, the unique positive Hermitian matrix with this property. □

Consequently, for $A \in SU(2)$, (9.7) yields

$$\tilde{U}(0, A)(\varphi)(p) = V(A)(\varphi(A^{-1}(p))). \tag{9.54}$$

To understand what this means we compare it with (8.34), which we write

$$(U \otimes \pi_j)(A)(\varphi)(\boldsymbol{p}) = \pi_j(A)(\varphi(A^{-1}(\boldsymbol{p}))), \tag{9.55}$$

since $A^{-1}(\boldsymbol{p})$ is the notation we now use for $\kappa(A^{-1})(\boldsymbol{p})$. When $V = \pi_j$, the overwhelming analogy between (9.54) and (9.55) supports (9.53).[33]

The heuristic approach of Section 9.1 effortlessly discovers the method of induced representations (9.6). That is, *assuming that the particle is characterized by its spin and its mass,* it discovers how the Poincaré group acts on it. On the other hand, Wigner *proved* that all the projective representations of $\mathcal{P}^*$ are obtained as induced representations. This goes much further: The way the Poincaré group acts on a particle of a given positive mass depends only on its spin. It *proves* that there is no other possible characteristic of the particle involved there.

Next we turn to the case of the representation $\pi_{0,j}$ of Definition 9.6.4, acting as in this definition on the space $L^2(X_0, \lambda_0)$. As on page 127 we argue that a particle described by this representation must be massless.

What property of this particle is reflected by the representation $\hat{\pi}_j$ of the little group? In imprecise but picturesque terms, we will show that

[33] None of the arguments above pretends to be a rigorous deduction of anything.

a rotation of angle θ around the direction of motion $\boldsymbol{p}$ *multiplies the state by a phase* $\exp(\mathrm{i}j\theta/2)$. (9.56)

Keeping in mind the result of Exercise 8.8.2 we describe the situation by saying: "the particle has helicity $j/2$".[34] We will think of a particle with helicity $j/2$ as having spin $|j|/2$. Thus a massless particle of given spin $j/2$ ($j > 0$) comes in two versions, with helicity $\pm j/2$. In contrast, a massive particle is determined by its mass and its spin.

The similarity of terminology should not hide that the massive and massless cases are very different.

Exercise 9.7.2 What differences do you see?

To prove (9.56), let us first consider the ideal case where $\varphi(p) = 0$ unless $p = \lambda p^* = (\lambda, 0, 0, \lambda)$ for some $\lambda > 0$. The direction of motion is then the z-axis. The element $A = \exp(-\mathrm{i}\theta\sigma_3/2) = \begin{pmatrix} \exp(-\mathrm{i}\theta/2) & 0 \\ 0 & \exp(\mathrm{i}\theta/2) \end{pmatrix}$ is such that $\kappa(A)$ corresponds to a rotation of angle θ around the z-axis. The quantity $\xi(A, p)$ of Proposition 9.6.6 equals $2\lambda \exp(\mathrm{i}\theta/2)$ and the quantity $\xi(A, p)/|\xi(A, p)|$ then equals $\exp(\mathrm{i}\theta/2)$. Thus (9.42) takes the form $U(0, A)(\varphi)(p) = \exp(\mathrm{i}j\theta/2)\varphi(p)$ which proves the claim (9.56).

To prove (9.56) in the general case, the basic observation is that if R is a rotation of angle θ around the direction of a vector $\boldsymbol{u}$, and S is another rotation, then SRS^{-1} is a rotation of angle θ around the direction of $S(\boldsymbol{u})$. Considering a rotation $\kappa(U_p)$ which sends $(p^0, 0, 0, |\boldsymbol{p}|)$ to $p = (p^0, \boldsymbol{p})$, and

$$A = U_p \exp(-\mathrm{i}\theta\sigma_3/2) U_p^{-1}, \tag{9.57}$$

then $\kappa(A)$ is a rotation of angle θ around the direction of $\boldsymbol{p}$, and

$$U(0, A)(\varphi) = U(0, U_p) U(0, \exp(-\mathrm{i}\theta\sigma_3/2)) U(0, U_p^{-1})(\varphi). \tag{9.58}$$

In the ideal case where $\varphi(p')$ is $\neq 0$ only when $\boldsymbol{p}'$ is in the direction of $\boldsymbol{p}$, then (9.42) shows that $\psi := U(0, U_p^{-1})(\varphi)$ is such that $\psi(p') \neq 0$ only for $\boldsymbol{p}'$ in the direction of e_3. We have then shown that $U(0, \exp(-\mathrm{i}\theta\sigma_3/2))(\psi) = \exp(\mathrm{i}j\theta/2)\psi$ and (9.58) implies as desired that $U(0, A)(\varphi) = \exp(\mathrm{i}j\theta/2)\varphi$.

Exercise 9.7.3 The purpose of the present challenging exercise is to try to look at the representation $\pi_{0,j}$ using physicists' tools. We denote by $R(\theta, \boldsymbol{u})$ the rotation of angle θ and axis $\boldsymbol{u}$, where $\boldsymbol{u}$ is a unit vector. Considering a representation W of $SO(3)$ (which models the action of rotations) one defines a self-adjoint operator $J_{\boldsymbol{u}}$ by

$$J_{\boldsymbol{u}} = \lim_{\theta \to 0} \frac{\hbar}{\mathrm{i}\theta}(W(R(\theta, \boldsymbol{u})) - 1). \tag{9.59}$$

It is called the *angular momentum with respect to axis* $\boldsymbol{u}$.

[34] It is the desire to have $\pi_{0,j}$ describe particles of helicity $j/2$ (rather than $-j/2$) which dictates the choice $\hat{\pi}_j(A) = a^{-j}$ as opposed to the seemingly more natural choice $\hat{\pi}_j(A) = a^j$.

(a) When we consider instead a representation of $SU(2)$, convince yourself after studying the first three sections of Appendix D that instead of (9.59) one should use the formula

$$J_{\boldsymbol{u}} = \lim_{\theta\to 0} \frac{\hbar}{\mathrm{i}\theta}(W(\exp(-\mathrm{i}\theta \boldsymbol{u}\cdot\boldsymbol{\sigma}/2)) - 1), \tag{9.60}$$

where the notations are those of Appendix D.

We first assume $j = 1$ and consider the realization of $\pi_{0,-1}$ described in Proposition 9.6.13. In this case the representation W modeling the action of rotations is given by $W(A) = U(0, A)$, where $U(a, A)$ is defined in (9.52).

(b) In the ideal case where φ is non-zero only for a given value of $p = (p^0, \boldsymbol{p})$ and where the unit vector $\boldsymbol{u}$ is $\boldsymbol{p}/|\boldsymbol{p}|$ show that $J_{\boldsymbol{u}}(\varphi)$ is non-zero only at p and that

$$J_{\boldsymbol{u}}(\varphi)(p) = -\frac{\hbar}{2}\boldsymbol{u}\cdot\boldsymbol{\sigma}(\varphi(p)). \tag{9.61}$$

(c) Use the condition $M(Pp)(\varphi(p)) = 0$ to show that the previous relation implies

$$J_{\boldsymbol{u}}(\varphi)(p) = -\frac{\hbar}{2}\varphi(p),$$

and interpret this relation as "the spin in the direction of motion is $-1/2$".

(d) Generalize the previous considerations to the case of an arbitrary value of j.

Physicists find the considerations of the previous two pages as self-evident, and write the following argument: "Since $-\hbar\boldsymbol{u}\cdot\boldsymbol{\sigma}/2$ measures the angular momentum in the direction of the unit vector $\boldsymbol{u}$, the operator $-\hbar\boldsymbol{p}\cdot\boldsymbol{\sigma}/(2|\boldsymbol{p}|)$ measures the angular momentum in the direction of motion, but according the equation $M(Pp)(\varphi(p)) = 0$, this operator is simply multiplication by $-\hbar/2$".[35]

Let us then summarize the situation:

- For $m > 0$, the representation $\pi_{m,j}$ corresponds to a particle of mass m and spin $j/2$.
- The representation $\pi_{0,j}$ corresponds to a massless particle of spin $|j|/2$ and helicity $j/2$.

The mathematical subtleties of the treatment of massless particles are largely irrelevant in physics.[36] It is possible in physics to assume that every particle has a tiny mass. The results of every experiment conceivable today would be the same if photons had a rest mass of 10^{-1000} kg, and articles finding experimental upper bounds for this rest mass do get published.[37] Still, as the mathematics are interesting, starting with Section 9.12 we will indulge for a few pages in thinking about massless particles.

[35] I do not see how to give a precise meaning to this without entering the considerations of the previous exercise. If you think I am nit-picking ask yourself what is the state space on which these operators act.

[36] As of today, the photon is the only confirmed massless particle. The hypothetical graviton would also be massless, but with spin 2. Neutrinos were long thought to be be massless, but now are believed to have a very small positive mass.

[37] Try the search for the words "photon mass limit"! The upper bound 10^{-16}e.V. seems very solid.

9.8 Abstract Presentation of Induced Representations

In the present section we give a more abstract, but more intrinsic description of the previous representations. Using it certain results become far simpler. We will witness this when proving Proposition 9.4.6, again in Section 9.12, and crucially in Section 10.4. The method is a special case of a very general method in representation theory, the so-called "method of induced representations". In the setting of representation of a semidirect product $N \rtimes H$ (with N abelian) the key idea of the method can be implemented in a simple way.[38]

We carry out the method only in the case of the physically relevant representations of the Poincaré group.[39] We start with a fixed point $p^* \in X_m$ ($m \geq 0$), with little group $G_{p^*} = \{A \in SL(2,\mathbb{C}); A(p^*) = p^*\}$ for the action of $SL(2,\mathbb{C})$. For each $p \in X_m$ we fix an element $D_p \in SL(2,\mathbb{C})$ with $D_p(p^*) = p$. Consider a unitary representation V of G_{p^*} in a finite-dimensional space $\mathcal{V}$, and the induced representation $\tilde{U}(a,A)$ of Theorem 9.4.2. It acts on the space $L^2 = L^2(X_m, \mathcal{V}, \mathrm{d}\lambda_m)$. The magic idea (which will transport the representation to something much nicer) is to associate to each function $\varphi \in L^2$ a function $T(\varphi)\colon SL(2,\mathbb{C}) \to \mathcal{V}$ by the formula

$$T(\varphi)(A) = V(A^{-1}D_{A(p^*)})\varphi(A(p^*)). \tag{9.62}$$

This makes sense because $A^{-1}D_{A(p^*)}(p^*) = A^{-1}A(p^*) = p^*$ so that $A^{-1}D_{A(p^*)} \in G_{p^*}$. Recalling the formula (9.7), let $\psi := \tilde{U}(a,A)(\varphi)$ so that

$$\psi(p) = \tilde{U}(a,A)(\varphi)(p) = \exp(\mathrm{i}(a,p)/\hbar)V(D_p^{-1}AD_{A^{-1}(p)})[\varphi(A^{-1}(p))].$$

For $B \in SL(2,\mathbb{C})$ we compute

$$\begin{aligned} T(\tilde{U}(a,A)(\varphi))(B) &= T(\psi)(B) = V(B^{-1}D_{B(p^*)})[\psi(B(p^*))] \\ &= \exp(\mathrm{i}(a,B(p^*))/\hbar)V(B^{-1}D_{B(p^*)})V(D_{B(p^*)}^{-1}AD_{A^{-1}B(p^*)})[\varphi(A^{-1}B(p^*))] \\ &= \exp(\mathrm{i}(a,B(p^*))/\hbar)V(B^{-1}AD_{A^{-1}B(p^*)})[\varphi(A^{-1}B(p^*))] \\ &= \exp(\mathrm{i}(a,B(p^*))/\hbar)T(\varphi)(A^{-1}B). \end{aligned} \tag{9.63}$$

For a function $f\colon SL(2,\mathbb{C}) \to \mathcal{V}$ let us then define

$$\pi(a,A)(f)(B) = \exp(\mathrm{i}(a,B(p^*))/\hbar)f(A^{-1}B), \tag{9.64}$$

so that (9.63) means

$$T\tilde{U}(a,A) = \pi(a,A)T. \tag{9.65}$$

Thus, if we transport the representation $\tilde{U}(a,A)$ by T it is given by the much simpler formula (9.64). The function $f = T(\varphi)$ of (9.62) satisfies

$$A \in SL(2,\mathbb{C}), C \in G_{p^*} \Rightarrow f(AC) = V(C^{-1})f(A), \tag{9.66}$$

[38] It was a character-forming experience to figure this out.

[39] If it helps you to think in more abstract and general terms, a general scheme to construct unitary representations of semidirect products is presented in the form of a sequence of exercises at the end of the section.

since $C(p^*) = p^*$ and $V(C^{-1}A^{-1}D_{A(p^*)}) = V(C^{-1})V(A^{-1}D_{A(p^*)})$. Conversely, if the function f satisfies (9.66), then $f = T(\varphi)$ where the function φ is given by $\varphi(p) = f(D_p)$. Indeed,

$$T(\varphi)(A) = V(A^{-1}D_{A(p^*)})f(D_{A(p^*)}) = f(A),$$

using (9.66) for $C = D_{A(p^*)}^{-1}A$.

Theorem 9.8.1 *Given a unitary representation V of the little group G_{p^*} on a space $\mathcal{V}$, consider the space $\mathcal{F}$ of functions $f : SL(2,\mathbb{C}) \to \mathcal{V}$ which satisfy the condition (9.66), provided with the norm*

$$\|f\|^2 = \int d\lambda_m(p)\|f(D_p)\|^2. \tag{9.67}$$

The formula (9.64) defines a unitary representation of $\mathcal{P}^*$ on $\mathcal{F}$. This representation is unitarily equivalent to the representation induced by V.

Proof We have shown that T is unitary from L^2 to $\mathcal{F}$, so that (9.65) proves the theorem. □

Exercise 9.8.2 Prove directly that the formula (9.64) defines a unitary representation on $\mathcal{F}$. Prove that the norm (9.67) is independent of the choice of $D_p \in SL(2,\mathbb{C})$ with $D_p(p^*) = p$.

Exercise 9.8.3 Following the method of Exercise 9.5.5 prove that when $m > 0$ there is a left-invariant measure $d\mu$ on $SL(2,\mathbb{C})$ such that for $f \in \mathcal{F}$ we have $\|f\|^2 = \int d\mu(A)\|f(A)\|^2$.

Exercise 9.8.4 Find a proof of Proposition 9.4.5 using the presentation of the present section.

We are now ready to prove that the method of induced representation constructs irreducible representations.

Proof of Proposition 9.4.6 It suffices to prove the corresponding result for the representations of Theorem 9.8.1. The "only if" part is obvious, since if $\mathcal{W}$ is a subspace of $\mathcal{V}$ which is invariant under V, the space of $\mathcal{W}$-valued functions is invariant under each $\pi(a, A)$.

The non-trivial proof of the converse may be skipped at first reading. We denote by g a non-zero element of $\mathcal{F}$ and by $(\cdot,\cdot)$ the inner product in $\mathcal{V}$. Consider an element $f \in \mathcal{F}$ and assume that f is orthogonal to all the elements $\pi(a, A)(g)$. We have to prove that $f = 0$ in the space of functions provided with the norm (9.67), that is we have to prove that $f(D_p) = 0$ $d\lambda_m$-a.e. The inner product in $\mathcal{F}$ is given by $(f, g) = \int d\lambda_m(p)\langle f(D_p), g(D_p)\rangle$ so that our hypothesis is that for each value of a, A the integral $\int d\lambda_m(p)\langle f(D_p), \exp(i(a, D_p(p^*))/\hbar)g(A^{-1}D_p)\rangle$ is zero. Since $D_p(p^*) = p$ we prove as in Proposition 4.8.4 that for each A in $SL(2,\mathbb{C})$ the function $p \mapsto \langle f(D_p), g(A^{-1}D_p)\rangle$ has to be zero $d\lambda_m$-a.e. Changing A into A^{-1}, we have proved that for each $A \in SL(2,\mathbb{C})$ the function $p \mapsto \langle f(D_p), g(AD_p)\rangle$ has to be zero $d\lambda_m$-a.e.

We now use that on the locally compact group $SL(2,\mathbb{C})$ there is a right-invariant measure $d\mu$.[40] That is, $d\mu$ is invariant by the transformations $A \mapsto AB$ for $B \in SL(2,\mathbb{C})$. In particular if a function $A \mapsto h(A)$ is zero $d\mu$-a.e. then for each $B \in SL(2,\mathbb{C})$ the function $A \mapsto h(AB)$ is zero $d\mu$-a.e. (and conversely). Using Fubini's theorem for the measure $d\lambda_m \otimes d\mu$, for $d\lambda_m$-almost each p the function $A \mapsto (f(D_p), g(AD_p))$ has to be zero $d\mu$-a.e. Let us fix now such a value of p with the goal of showing that $f(D_p) = 0$ (which will conclude the proof). By right-invariance of μ the function $A \mapsto (f(D_p), g(A))$ is zero $d\mu$-a.e. and hence, again by right-invariance, for every $C \in G_{p^*}$ the function $A \mapsto (f(D_p), g(AC)) = (f(D_p), V(C^{-1})g(A))$ is zero $d\mu$-a.e.

Denoting by $d\gamma$ the Haar measure of G_{p^*} and using Fubini's theorem again, for $d\mu$-almost all values of A the function $C \mapsto (f(D_p), V(C^{-1})g(A))$ is zero $d\gamma$-a.e., and hence everywhere as this function is continuous. In particular this occurs for a value A for which $g(A) \neq 0$. But since V is irreducible, this proves as desired that $f(D_p) = 0$. □

In the following exercises, we sketch a general method to construct unitary representations of a semidirect product $N \rtimes H$ where N is commutative and N and H are countable.[41]

Exercise 9.8.5 Consider a countable group G and a subgroup H. The relation $A\mathcal{R}B$ iff $B^{-1}A \in H$ is an equivalence relation on G. The quotient of G by this equivalence relation is denoted by G/H. Describe a natural action of G on G/H, and consider a positive measure λ on G/H which is invariant under this action. Let us assume that we are given a unitary representation V of H in a Hilbert space $\mathcal{V}$. If a function f from G to $\mathcal{V}$ satisfies

$$A \in G, C \in H \Rightarrow f(AC) = V(C^{-1})f(A), \tag{9.68}$$

prove that it makes sense to define

$$\|f\|^2 = \int_{G/H} d\lambda(p) \|f(D_p)\|^2$$

where $D_p \in G$ is such that its class in G/H is p. Consider the Hilbert space $\mathcal{H}$ of these functions for which $\|f\| < \infty$. Prove that we may define a unitary representation π of G in $\mathcal{H}$ by the formula $\pi(A)(f)(B) = f(A^{-1}B)$.

Exercise 9.8.6 This exercise continues the previous one. We consider a countable abelian group N and a semidirect product $N \rtimes H$. The law of this group is given by $(a,A)(b,B) = (a + \kappa(A)(b), AB)$ where $\kappa(AB) = \kappa(A)\kappa(B)$ and $\kappa(A)(a+b) = \kappa(A)(a) + \kappa(A)(b)$. A *character* of N is a unitary representation of N on $\mathbb{C}$, and $\hat{N}$ denotes the set of these representations, called the *dual* of N. Prove that we can define an action $\hat{\kappa}$ of H on $\hat{N}$ by $\hat{\kappa}(A)(w)(x) = w(\kappa(A^{-1})(x))$ for $w \in \hat{N}$ and $x \in H$. Let us then fix $w \in \hat{N}$, and define the little group H_w of w as the set of $A \in H$ for which

[40] On a locally compact group there exist both right-invariant and left-invariant measures, which may be different, although it can be shown that this does not happen in the case of $SL(2,\mathbb{C})$. A left-invariant measure is constructed in Exercise 9.5.5.

[41] If you know about locally compact groups, you may assume that these groups are locally compact, adding the proper continuity assumptions whenever required.

$\hat{\kappa}(A)(w) = w$. Explain why the orbit of w under the action of H identifies with the quotient H/H_w. Consider a unitary representation V of H_w in a Hilbert space $\mathcal{V}$ and the space $\mathcal{H}$ of functions from H to $\mathcal{V}$ defined in the previous exercise. Prove that we define a unitary representation π of $N \rtimes H$ by the formula

$$\pi(a,A)f(B) = \hat{\kappa}(B)(w)(a)f(A^{-1}B).$$

Relate this to formula (9.64).

9.9 Particles and Parity

Now, what about parity? We can act on a particle by translations, rotations, even (in principle) Lorentz transformations. There is however no machine that takes a particle and produces a mirror image of it. Still, many situations are mirror images of each other, and to account for this we should try to build models which involve a parity operator. The first step in this direction is to enlarge the group $\mathcal{P}^*$ by "adding a parity element". This is done simply by considering the group $\mathcal{P}^{*+} = \mathbb{R}^{1,3} \rtimes SL^+(2,\mathbb{C})$ where the group $SL^+(2,\mathbb{C})$ is as in Definition 8.9.3. That is, one adds a new element P' to $SL(2,\mathbb{C})$, so the group $SL^+(2,\mathbb{C})$ consists of the elements $P', A, P'A$ where $A \in SL(2,\mathbb{C})$ with the multiplication rule (8.41): $P'I = P'$, $P'AP' = A^{\dagger-1}$, and κ is extended to $SL^+(2,\mathbb{C})$ by (8.42), i.e. $\kappa(P'A) = P\kappa(A)$, where P is the parity operator $(p^0, \boldsymbol{p}) \to (p^0, -\boldsymbol{p})$. The "parity element" of $\mathcal{P}^{*+}$ is then $(0, P')$.

The obvious question is then how to construct meaningful representations of $\mathcal{P}^{*+}$.

The mass shell X_m is still an orbit under the action of $\mathcal{P}^{*+}$ and Theorem 9.4.2 extends in a straightforward manner to the setting where $SL(2,\mathbb{C})$ is replaced by $SL^+(2,\mathbb{C})$ and $\mathcal{P}^*$ is replaced by $\mathcal{P}^{*+}$. This extension will be called "Extended Theorem 9.4.2".

Let us first say a few words about the really easy case $m > 0$ and $p^* = (mc,0,0,0)$. Then the little group of $SL^+(2,\mathbb{C})$ consists of the group $SU^+(2)$ generated by $SU(2)$ and P', and the element P' commutes with every element of $SU^+(2)$.

Let us compare the irreducible representations of $SU(2)$ and $SU^+(2)$. Consider first an irreducible representation V of $SU^+(2)$. Since P' commutes with every element of $SU^+(2)$ then $V(P')$ commutes with every operator $V(A)$. Since V is irreducible, $V(P')$ is a multiple of the identity 1 by Schur's lemma (Lemma 4.5.7). Since P'^2 is the identity, there are two cases: Either $V(P')$ is the identity or it is minus the identity. Conversely an irreducible representation of $SU(2)$ can be extended in exactly two ways to $SU^+(2)$, by setting $V(P') = 1$ or by setting $V(P') = -1$.

As a consequence of this and of the extended Theorem 9.4.2 the representation $\pi_{m,j}$ of Definition 9.6.1 can naturally be extended to $\mathcal{P}^{*+}$ in two different ways, one differing from the other in the fact that for one of them the action of parity involves an extra minus sign. In physics these two extensions are distinguished by saying that the particle has *intrinsic parity* ± 1.

The study of massive particles and parity continues in the next two sections, while the last two sections deal with massless particles.

9.10 Dirac Equation

The reader should review Section 8.10 and in particular the matrix $\gamma(x)$ of (8.46) and the γ matrices γ_ν of (8.49), so that $\gamma(x) = x^\nu \gamma_\nu$. Our goal is to apply the extended Theorem 9.5.3 to the case where $\mathcal{H}_0 = \mathbb{C}^4$, where $m > 0$ and where $V = S$ is the representation of $SL^+(2,\mathbb{C})$ given by (8.44) and (8.45):

$$S(A) := \begin{pmatrix} A^{\dagger-1} & 0 \\ 0 & A \end{pmatrix} \; ; \; S(P') = \begin{pmatrix} 0 & I \\ I & 0 \end{pmatrix}.$$

There are several reasons to do this. First S is in a sense the simplest non-trivial representation of $SL^+(2,\mathbb{C})$. Second this will lead us to the Dirac equation, which played a fundamental historical role in quantum mechanics.

The little group is the group $SU^+(2)$ generated by $SU(2)$ and parity P'. The restriction of S to $SU^+(2)$ is unitary, but it is not irreducible. This is because P' commutes with every element of $SU^+(2)$ so that the eigenspaces of $S(P') = \gamma_0$ are invariant under this restriction. Since $\gamma_0^2 = 1$ the eigenvalues are ± 1 and the corresponding eigenspaces are

$$\mathcal{V} = \{u \in \mathbb{C}^4 \; ; \; \gamma_0 u = u\} \; ; \; \mathcal{V}' = \{u \in \mathbb{C}^4 \; ; \; \gamma_0 u = -u\},$$

each of which has dimension two, and we are going to apply the extended Theorem 9.5.3 to the space $\mathcal{V}$. Given $p \in X_m$, and $D_p \in SL(2,\mathbb{C})$ such that $p = D_p(p^*)$, the space $\mathcal{V}_p = S(D_p)(\mathcal{V})$ is independent of D_p and is two-dimensional. The extended Theorem 9.5.3 constructs a representation on the space $\mathcal{H}$ of functions $\varphi\colon X_m \to \mathbb{C}^4$ such that $\varphi(p) \in \mathcal{V}_p$ for each p.

Lemma 9.10.1 *For each p it holds that*

$$\mathcal{V}_p = \{u \; ; \; \gamma(p)u = mcu\}.$$

Proof First we observe that $\gamma(p^*) = mc\gamma_0$ so that the elements u of $\mathcal{V}$ are characterized by the equation $\gamma(p^*)u = mcu$. The relation (8.47) implies here

$$\gamma(p)S(D_p) = S(D_p)\gamma(p^*), \tag{9.69}$$

which shows that the elements u of $\mathcal{V}_p = S(D_p)\mathcal{V}$ are characterized by the equation $\gamma(p)u = mcu$. □

Let us define the Dirac operator $\widehat{D}$ on the space of functions $\varphi\colon X_m \to \mathbb{C}^4$ by

$$\widehat{D}(\varphi)(p) = \gamma(p)\varphi(p). \tag{9.70}$$

Therefore the space $\mathcal{H}$ is simply the space of functions which satisfy the equation $\widehat{D}\varphi = mc\varphi$, i.e. $\gamma(p)\varphi(p) = mc\varphi(p)$ for each p. This equation is known as the *Dirac equation*, and will be discussed in the next section.

Exercise 9.10.2 The extended Theorem 9.5.3 defines the representation $U(a, A)$ of $\mathcal{P}^{*+}$ by the formula (9.73). Prove that the operator $\hat{D}$ commutes with the operators $U(a, A)$.

Let us also observe for further use that the identity $\gamma(x)^2 = x^2 1$ implies

$$\widehat{D}^2(\varphi)(p) = p^2 \varphi(p). \tag{9.71}$$

We provide now a simple alternative expression for the norm (9.31). Given a Dirac spinor, that is, an element u of $\mathbb{C}^4$ seen as a column vector, let us denote by $u^\dagger$ its transpose-conjugate, which is a row vector. Given two Dirac spinors u, v we consider the quantity

$$u^\dagger \gamma_0 v,$$

which is a number. It is linear in v and anti-linear in u, but one need not have $u^\dagger \gamma_0 u \geq 0$. The key fact is that this sesqui-linear form is invariant under each map $S(A)$ as shown by the next lemma.

Lemma 9.10.3 *For each $A \in SL^+(2,\mathbb{C})$ we have*

$$(S(A)u)^\dagger \gamma_0 S(A) v = u^\dagger \gamma_0 v.$$

Proof Since $(S(A)u)\dagger = u^\dagger S(A)^\dagger$ it suffices to show that whenever $A \in SL^+(2,\mathbb{C})$ we have

$$S(A)^\dagger \gamma_0 S(A) = \gamma_0. \tag{9.72}$$

Since $\gamma_0 = S(P')$, and since S is a representation, the above is equivalent to $S(A)^\dagger S(P'A) = S(P')$, and one simply checks all the cases. For example if $A = P'$ this holds since $S(P')^\dagger = S(P')$, while for $A \in SL(2,\mathbb{C})$ we have $S(A)^\dagger S(P'A) = S(A^\dagger) S(P'A) = S(A^\dagger P'A)$ and $A^\dagger P'A = P'$. □

The nice thing is that for $u \in \mathcal{V}$ we have $u^\dagger \gamma_0 u = u^\dagger u = \|u\|^2$. Consequently, for $u \in \mathcal{V}_p$, since $S(D_p)^{-1} u \in \mathcal{V}$, we have

$$\|u\|_p^2 = \|S(D_p)^{-1} u\|^2 = \left(S(D_p)^{-1} u\right)^\dagger \gamma_0 S(D_p)^{-1} u = u^\dagger \gamma_0 u,$$

where we use Lemma 9.10.3 for $A = D_p^{-1}$ in the last equality. This provides a simple expression for $\|u\|^2$. In summary:

Theorem 9.10.4 *Consider the space $\mathcal{H}_{\mathrm{Dirac}}$ of functions $\varphi\colon X_m \to \mathbb{C}^4$ which satisfy the Dirac equation $\widehat{D}\varphi = mc\varphi$, provided with the inner product*

$$(\varphi, \psi) = \int \mathrm{d}\lambda_m(p) \varphi^\dagger(p) \gamma_0 \psi(p).$$

Then the formula

$$U(a, A)(\varphi)(p) = \exp(\mathrm{i}(a, p)/\hbar) S(A) [\varphi(A^{-1}(p))] \tag{9.73}$$

defines a unitary representation of $\mathcal{P}^{*+}$ on $\mathcal{H}_{\mathrm{Dirac}}$.

Exercise 9.10.5 State a similar result for the space of functions that satisfy the equation $\widehat{D}(\varphi) = -mc\varphi$.

9.11 History of the Dirac Equation

Schrödinger's equation (2.80) is not Lorentz invariant.[42] While looking for a version of this equation which would be consistent with Special Relativity, Paul Adrien Maurice Dirac was led to an equation of the type

$$i\hbar a^{\mu}\partial_{\mu}f = mcf, \tag{9.74}$$

where the differential operator ∂_{μ} is defined by

$$\partial_{\mu}f := \frac{\partial f}{\partial x^{\mu}}, \tag{9.75}$$

and where the quantities a^{μ} should satisfy the commutation relations[43]

$$a^{\mu}a^{\nu} + a^{\nu}a^{\mu} = 2\eta^{\mu\nu}. \tag{9.76}$$

The relations (9.76) cannot be satisfied by complex numbers, so that in (9.74) it is not possible to use for f an ordinary wave function, that is, a map from Minkowski space $\mathbb{R}^{1,3}$ to $\mathbb{C}$. Analysis of these commutation relations led him to discover the matrices γ_{μ}, and to the fundamental idea that f should rather be a "multicomponent" wave function, a map φ from $\mathbb{R}^{1,3}$ to $\mathbb{C}^4$. Recalling that $\gamma^{\mu} = \eta^{\mu\nu}\gamma_{\nu}$, the equation (9.74) then takes the form

$$i\hbar\gamma^{\mu}\partial_{\mu}f = mcf, \tag{9.77}$$

that we shall call the *Dirac equation.*[44] The reader is certainly not expected to understand the physical content of this equation, or even its relationship with the Schrödinger equation just by looking at the formula (9.77). Doing so requires very significant work.

The standard way to study the Dirac equation goes roughly as follows. Having guessed this equation following Dirac's intuition, it should be "Lorentz invariant": All legitimate (inertial) observers should agree that the *same* equation describes the evolution of "wave functions". Different observers however use different coordinate systems, so will use different "wave functions" to describe the same particle. That is, there must exist a transformation rule of the wave function when one changes coordinate systems (i.e. one performs a Lorentz transformation A) that preserves the form of the Dirac equation. To discover this transformation rule one analyzes the case where "A is infinitesimally close to the identity", or in other words one uses Lie algebras. In this manner one discovers formula (9.78). The Dirac equation then reveals plenty of symmetries, including parity.

Reversing this historical approach we have shown that the simple desire to have a parity operator more or less automatically leads us to invent the Dirac matrices, and to discover the Dirac equation (as in the previous section).

[42] Please review Section 6.1 if you forgot what this means.

[43] One of the ideas of Dirac is that the equation (9.74) "should be the square root of the Klein-Gordon equation (6.3)". The relations (9.76) are just the relations required for this. They imply that the square of the operator $i\hbar a^{\mu}\partial_{\mu}$ is $-\hbar^2\partial_{\mu}\partial^{\mu}$.

[44] The relationship through the Fourier transform between this formula and what we called the Dirac equation in the previous section will be explained shortly.

The Dirac equation, seen as a (kind of) relativistic wave equation of a single particle, is a very interesting topic in the sense that it mixes triumphs and disasters. Among its triumphs is a better prediction of the energy levels of the hydrogen atom than the Schrödinger equation. In the end the Dirac equation is however *not* part of a consistent theory, since there seems to be no way to avoid the disaster of "negative energy states".[45] We shall not study this aspect of Dirac's equation.[46] The topic is treated in countless physics textbooks, and sketched in Folland's book [31]. The books explain in particular why the Schrödinger equation is the "non-relativistic limiting case" of the Dirac equation, which is certainly not obvious from (9.77). The ideas behind the Dirac equation are still very much relevant to Quantum Field Theory, but there this equation is no longer interpreted as a kind of wave equation.

Let us now prove the Lorentz invariance of the Dirac equation. This property is just another way to express the result of Exercise 9.10.2 (as can be shown using the Fourier transform), but we give a direct proof. Now that we are somewhat familiar with representations of the Poincaré group, it is easy to guess what the transformation rule for the wave function should be. For a function $f : \mathbb{R}^{1,3} \to \mathbb{C}^4$ and $A \in SL^+(2,\mathbb{C})$ let us define the function $T_A(f)$ by

$$T_A(f)(x) = S(A)f(\kappa(A^{-1})(x)). \tag{9.78}$$

Here for clarity we write $\kappa(A)(x)$ rather than the shorthand $A(x)$. Then (9.78) defines a representation of $SL^+(2,\mathbb{C})$ in the space of functions on $\mathbb{R}^{1,3}$ valued in the Dirac spinors. This is checked as in the proof of Theorem 9.5.3.

Theorem 9.11.1 *The Dirac operator $D := \gamma^\mu \partial_\mu$ satisfies*

$$D(T_A(f)) = T_A(D(f)). \tag{9.79}$$

Here of course $\gamma^\mu = \eta^{\mu\nu}\gamma_\nu$, and we assume that f is smooth.

Proof Since the matrix $S(A)$ does not depend on x it commutes with the differentiation ∂_μ, so that

$$\gamma^\mu \partial_\mu \big(S(A)f(\kappa(A^{-1})(x))\big) = \gamma^\mu S(A)\partial_\mu \big(f(\kappa(A^{-1})(x))\big). \tag{9.80}$$

Now, since $\kappa(A^{-1})(x)^\nu = (\kappa(A^{-1}))^\nu{}_\mu x^\mu$, by the chain rule we get

$$\partial_\mu \big(f(\kappa(A^{-1})(x))\big) = \kappa(A^{-1})^\nu{}_\mu (\partial_\nu f)(\kappa(A^{-1})(x)),$$

while by (8.51) (used for A^{-1} instead of A, and since $S(A)^{-1} = S(A^{-1})$) it holds that $\gamma_\mu S(A) = S(A)\kappa(A)^\nu{}_\mu \gamma_\nu = S(A)\gamma_\nu \kappa(A)^\nu{}_\mu$ which by raising and lowering the appropriate indices we rewrite as

$$\gamma^\mu S(A) = S(A)\gamma^\rho \kappa(A^{-1})_\rho{}^\mu.$$

[45] Sadly enough, some of the negative legacies of this equation, such as the brilliant but nonsensical idea of the "Dirac sea" to cope with this problem still found their way in textbooks published not so long ago.

[46] The study is part of the *history* of Quantum Field Theory but not of this theory itself.

Exercise 9.11.2 Justify this statement in detail.

Thus the quantity (9.80) is exactly

$$S(A)\gamma^\rho \kappa(A^{-1})_\rho^{\ \mu} \kappa(A^{-1})^\nu_{\ \mu} (\partial_\nu f)(\kappa(A^{-1})(x)),$$

and since $\kappa(A^{-1})_\rho^{\ \mu} \kappa(A^{-1})^\nu_{\ \mu} = \delta^\nu_\rho$ by (4.18) this is $T_A(D(f))$. □

As a consequence, we obtain the fundamental fact that the Dirac equation (9.77) is Lorentz invariant: Its form does not change under a Lorentz transformation, *provided* after such a transformation A we use it for the function $T_A(f)$ of (9.78) rather than for f.

In the next exercise we relate our two versions of the Dirac equation. We recall the Fourier transform of a function $\mathbb{R}^{1,3} \to \mathbb{C}$ as defined in (4.58):

$$\hat{f}(p) = \int \mathrm{d}^4x \exp(\mathrm{i}(x,p)/\hbar) f(x) = \int \mathrm{d}^4x \exp(\mathrm{i}(x^0 p^0 - \boldsymbol{x} \cdot \boldsymbol{p})/\hbar) f(x). \quad (4.58)$$

The Fourier transform of a function $\mathbb{R}^{1,3} \to \mathbb{C}^4$ is taken component by component.

We recall the operator $\widehat{D}$ of (9.70), given by $\widehat{D}(\varphi)(p) = \gamma(p)\varphi(p)$.

Exercise 9.11.3 (a) Prove that for a function $f : \mathbb{R}^{1,3} \to \mathbb{C}$ one has $\widehat{\partial_\mu(f)}(p) = -\mathrm{i}p_\mu \hat{f}(p)/\hbar$, so that $\mathrm{i}\hbar\widehat{\partial_\mu(f)} = p_\mu \hat{f}(p)$. Compare with (1.23).

(b) Use the commutation relations between the Dirac matrices to compute the square of the Dirac operator D. Prove that if a function f satisfies the Dirac equation each of its components satisfies the Klein-Gordon equation.

(c) Prove that for a function $f : \mathbb{R}^{1,3} \to \mathbb{C}^4$ one has $\widehat{D}(\hat{f}) = \mathrm{i}\hbar\widehat{D(f)}$.

(d) Prove that a function $f : \mathbb{R}^{1,3} \to \mathbb{R}^4$ which satisfies the Dirac equation is determined by the restriction of its Fourier transform $\varphi = \hat{f}$ to X_m, and that this restriction satisfies the Dirac equation $\widehat{D}(\varphi) = mc\varphi$. (This is formal: We pretend that φ is well defined and is a function.)

(e) Recalling the formula (9.78), prove that for a function f on position space, $U(0,A)(\hat{f}) = \widehat{T_A(f)}$.

(f) Relate (9.79) and Exercise 9.10.2.

9.12 Parity and Massless Particles

Our true goal is to understand the representation of $\mathcal{P}^{*+}$ corresponding to the photon, the most common particle of the universe. This is the object of the next section. In order to really understand why the representation corresponding to the photon is what it is, we must look first into the relationship between parity and massless particles. This is considerably more subtle than in the massive case and is the object of the present section. This material, as well as the material of the next section, is not required to follow the main story.

At the root of what we discuss here is the following striking fact: If we look at the image in a mirror of say, a cube, we see a cube. But if we look at the image of a left hand, we see a right hand, or maybe more to the point, if we look at the image of a corkscrew, we do not see a standard corkscrew, but the version for left-handed people.[47] One of the main conclusions of this section is that if we look at the mirror image of the representation $\pi_{0,j}$ we see the representation $\pi_{0,-j}$, which is *not* equivalent to $\pi_{0,j}$ unless $j = 0$. As a striking consequence, for $j \neq 0$, the representation $\pi_{0,j}$ does not extend to $\mathcal{P}^{*+}$.

Going back to mathematics, we start with the following algebraic identity in $\mathcal{P}^{*+}$, which has a fundamental bearing on the structure of the representations of $\mathcal{P}^{*+}$. If $A \in SL(2,\mathbb{C})$ then

$$(0,P')(a,A)(0,P') = (Pa, P'AP') = (Pa, A^{\dagger -1}). \tag{9.81}$$

As a first consequence, we get the following.

Lemma 9.12.1 *The map $(a,A) \mapsto (Pa, A^{\dagger -1})$ from $\mathcal{P}^*$ to $\mathcal{P}^*$ is a group homomorphism. Consequently if π is a representation of $\mathcal{P}^*$ the map $(a,A) \mapsto \tilde{\pi}(a,A) := \pi(Pa, A^{\dagger -1})$ is also a representation of $\mathcal{P}^*$.*

As we try later to explain in words, we may think of the representation $\tilde{\pi}$ as "the mirror image" of the representation π. A key result of this section is as follows.

Proposition 9.12.2 *The representation $\tilde{\pi}_{0,j}$ of $\mathcal{P}^*$ given by*

$$\tilde{\pi}_{0,j}(a,A) = \pi_{0,j}(Pa, A^{\dagger -1}), \tag{9.82}$$

whenever $a \in \mathbb{R}^{1,3}$ and $A \in SL(2,\mathbb{C})$ is unitarily equivalent to $\pi_{0,-j}$.

We will deduce this later from a general result, Proposition 9.12.8, but it also has a simple direct proof using the concrete realizations of $\pi_{0,j}$ which we have obtained, as is shown in the following exercises.

Exercise 9.12.3 In the case $j = -1$ find an alternative proof of Proposition 9.12.2 using Proposition 9.6.13 and Exercise 9.6.14.

Exercise 9.12.4 Find a direct proof of Proposition 9.12.2 using the representation (9.47). Hint: Recalling the matrix J of Lemma 8.1.4 use the map $T(f)(v) = f(Jv^*)$ from $\mathcal{H}_j$ to $\mathcal{H}_{-j}$.

Lemma 9.12.5 *The representations $\pi_{0,j}$ and $\pi_{0,-j}$ are not unitarily equivalent.*

This is intuitively obvious from our study of helicity, since the rotations of space do not act in the same manner in these representations. Generally speaking, it is true that if we consider unitary representations V and V' of the little group, the corresponding induced representations are unitarily equivalent if and only if V and V' are unitarily equivalent, and in the present case $\hat{\pi}_j$ and $\hat{\pi}_{-j}$ are not unitarily equivalent. We give a self-contained proof of Lemma 9.12.5 in the present case.[48]

[47] That is a way to say: in practice, left-handed persons do not get the version of scissors and corkscrews adapted to them, but have to cope with the standard version. Intolerable, and at times, lethal discrimination. It does cause deaths in the construction/forest industry.

[48] As usual, some of the measure-theoretic details are not covered properly.

Proof In the setting of Theorem 9.4.2 these two representations both live in the space $L^2(X_0, d\lambda_0)$. Assume that there is a unitary map T such that

$$T\pi_{0,j}(a,A) = \pi_{0,-j}(a,A)T \tag{9.83}$$

for each $(a,A) \in \mathcal{P}^*$. Taking A the identity, for each $a \in \mathbb{R}^{1,3}$ and each $\varphi \in L^2$ the image by T of the function $p \mapsto \exp(\mathrm{i}(a,p)/\hbar)\varphi(p)$ is the function $p \mapsto \exp(\mathrm{i}(a,p)/\hbar)T(\varphi)(p)$. Taking suitable averages over a, for any test function ξ the image by T of the function $p \mapsto \xi(p)\varphi(p)$ is the function $\xi(p)T(\varphi)(p)$, i.e. $T(\xi\varphi)(p) = \xi(p)T(\varphi)(p)$. Next we prove the existence of a function $\beta(p)$ such that

$$T(\psi)(p) = \beta(p)\psi(p). \tag{9.84}$$

Consider a test function ξ with compact support and a test function φ which is equal to 1 on the support of ξ. Then $\xi = \xi\varphi$ so that for each p we have $T(\xi)(p) = T(\xi\varphi)(p) = \xi(p)T(\varphi)(p)$. When $\xi(p) \neq 0$ this shows that $T(\varphi)(p)$ does not depend on φ. It is then a function $\beta(p)$ and we have shown that $T(\xi)(p) = \beta(p)\xi(p)$. This holds for any test function and (9.84) follows by continuity.

Going back to the formula (9.7) it should be obvious that a map of the type (9.84) cannot satisfy (9.83) for all elements $(0,A)$. □

Lemma 9.12.6 *A representation π of $\mathcal{P}^*$ can be extended to $\mathcal{P}^{*+}$ if and only if π and $\tilde{\pi}$ are unitarily equivalent.*

Proof Denoting by π' an extension of π to $\mathcal{P}^{*+}$, (9.81) implies

$$\pi(Pa, A^{\dagger-1}) = U\pi(a,A)U^{-1}$$

where $U = U^{-1} = \pi'(0,P')$ is unitary. Thus if π can be extended to $\mathcal{P}^{*+}$ then π and $\tilde{\pi}$ are unitarily equivalent. Conversely if U is a unitary operator such that $\pi(Pa, A^{\dagger-1}) = U\pi(a,A)U^{-1}$ then by checking cases one sees that it is possible to extend π to $\mathcal{P}^{*+}$ by setting $\pi(0,P') = U$. □

Since $\pi_{0,j}$ and $\tilde{\pi}_{0,j} = \pi_{0,-j}$ are not unitarily equivalent, this has the following striking consequence, showing that the situation is completely different from the massive case.

Theorem 9.12.7 *The representation $\pi_{0,j}$ of Definition 9.6.4 cannot be extended to $\mathcal{P}^{*+}$.*

We will now prove the following very clean generalization of Proposition 9.12.2.

Proposition 9.12.8 *Consider a unitary representation V of the little group G_0 and the corresponding induced representation π of $\mathcal{P}^*$. Then the representation $\tilde{\pi}$ defined in Lemma 9.12.1 is equivalent to the representation induced by the unitary representation $V': A \mapsto V(A^*)$ of G_0.*

To prepare for the proof we recall the matrix $J = \begin{pmatrix} 0 & 1 \\ -1 & 0 \end{pmatrix}$ of (8.1).

Lemma 9.12.9 *(a) If* $p = (p^0, p^1, p^2, p^3)$ *we have*

$$J(p) = (p^0, -p^1, p^2, -p^3). \tag{9.85}$$

(b) If $A(p^*) = (p^0, p^1, p^2, p^3)$ *then* $A^*(p^*) = (p^0, p^1, -p^2, p^3)$.
(c) Furthermore for $A \in SL(2, \mathbb{C})$ *we have*

$$JA^*(p^*) = P'A(p^*). \tag{9.86}$$

Proof To prove (a) we compute

$$J\begin{pmatrix} a & b \\ c & d \end{pmatrix} J^\dagger = \begin{pmatrix} d & -c \\ -b & a \end{pmatrix},$$

we use that $M(J(p)) = JM(p)J^\dagger$ and the definition of $M(p)$. To prove (b), since $M(p^*)$ is real, taking the complex conjugate of the equality $AM(p^*)A^\dagger = M(A(p^*))$ yields $M(A^*(p^*)) = M(A(p^*))^*$ and the result. Furthermore (c) follows by combining (a) and (b). □

Proof of Proposition 9.12.8 Consider the space $\mathcal{F}'$ of functions $f : SL(2, \mathbb{C}) \to \mathcal{V}$ which satisfy the condition (9.66) for V':

$$A \in SL(2, \mathbb{C}),\ C \in G_0 \Rightarrow f(AC) = V'(C^{-1})f(A) = V(C^*)^{-1}f(A),$$

provided with the norm (9.67)

$$\|f\|^2 = \int \mathrm{d}\lambda_m(p)\|f(D_p)\|^2.$$

Then by Theorem 9.8.1 the representation π' of $\mathcal{P}^*$ on $\mathcal{F}'$ given by the formula

$$\pi'(a, A)(f)(B) = \exp(\mathrm{i}(a, B(p^*))/\hbar)f(A^{-1}B) \tag{9.87}$$

is unitarily equivalent to the representation induced by V'. Consider the space $\mathcal{F}$ of functions $f : SL(2, \mathbb{C}) \to \mathcal{V}$ which satisfy the condition (9.66) for V

$$A \in SL(2, \mathbb{C}),\ C \in G_0 \Rightarrow f(AC) = V(C^{-1})f(A),$$

also provided with the norm (9.67). According to Theorem 9.8.1, the representation π of $\mathcal{P}^*$ on $\mathcal{F}$ given by the right-hand side of (9.87) is equivalent to the representation induced by V.

For $f \in \mathcal{F}'$ we define the function $T(f)$ by

$$T(f)(A) = f(JA^*). \tag{9.88}$$

We will show that this map is unitary and witnesses the equivalence of the previous two representations, finishing the proof. It is obvious that $T(f) \in \mathcal{F}$. We prove first that for $(a, A) \in \mathcal{P}^*$ and $f \in \mathcal{F}'$ it holds that

$$T\pi'(a, A)(f) = \pi(Pa, A^{\dagger -1})T(f). \tag{9.89}$$

We compute

$$T[\pi'(a, A)(f)](B) = \pi'(a, A)(f)(JB^*) = \exp(\mathrm{i}(a, JB^*(p^*))/\hbar)f(A^{-1}JB^*). \tag{9.90}$$

From (9.86) we have $(a, JB^*(p^*)) = (Pa, B(p^*))$, and $A^{-1}J = J(A^\dagger)^*$, so that $A^{-1}JB^* = J(A^\dagger B)^*$ and $f(A^{-1}JB^*) = f(J(A^\dagger B)^*) = T(f)(A^\dagger B)$. Thus the right-hand side of (9.90) is indeed $\pi(Pa, A^{\dagger -1})T(f)(B)$.

To prove that T is unitary we observe from (9.86) that $JD_p^*(p^*) = P'D_p(p^*) = Pp = D_{Pp}(p^*)$. Now, since $\|f(A)\|$ depends only on $A(p^*)$ it holds that $\|T(f)(D_p)\| = \|f(JD_p^*)\| = \|f(D_{Pp})\|$, and thus

$$\int \mathrm{d}\lambda_0(p)\|T(f)(D_p)\|^2 = \int \mathrm{d}\lambda_0(p)\|f(D_{Pp})\|^2 = \int \mathrm{d}\lambda_0(p)\|f(D_p)\|^2,$$

making the change of variable $p \to Pp$ in the last equality. □

To end this section we discuss some considerations often read in physics textbooks about "mirror images" and "parity images". A reformulation of Proposition 9.12.2 is as follows.

Theorem 9.12.10 *The parity image of a massless particle of helicity $j/2$ is a massless particle of the same spin but opposite helicity $-j/2$.*

To make sense of these words, one has to define "how the group $\mathcal{P}^*$ operates on the parity image of a particle". We first apply the parity operator to go back from the parity image of the particle to the particle itself, we act on the particle by an element of $\mathcal{P}^*$ and then we apply the parity operator again to go back to the parity image of the particle. If we were working with $\mathcal{P}$ itself rather than $\mathcal{P}^*$ this would mean that the action of $(a, A) \in \mathcal{P}$ on the parity image is given by $\pi_{0,j}(Pa, PAP)$, but since we are working with $\mathcal{P}^*$, where the parity operator is P' we have to replace PAP by $P'AP' = A^{\dagger -1}$. Therefore Theorem 9.12.10 indeed follows from Proposition 9.12.2.

Exercise 9.12.11 We have defined the "parity image" of a particle using the parity transformation $(x^0, \boldsymbol{x}) \to (x^0, -\boldsymbol{x})$. Replacing the symmetry through the origin $\boldsymbol{x} \to -\boldsymbol{x}$ by a symmetry through a given plane, define the "mirror image" of a particle. Show that it is the same as the "parity image". Hint: A symmetry through the origin is the composition of a symmetry through a plane and a rotation of angle π.

The physicists describe Theorem 9.12.10 with the words "the mirror image of a massless particle is a massless particle of the same spin but with opposite helicity". Hence this mirror image (just like in the case of a corkscrew) is a different object. This suggests the following convenient terminology.

Definition 9.12.12 We say that a particle is its own mirror image if it is described by a representation π of $\mathcal{P}^{*+}$.

One should be cautious that this does *not* mean that $\pi(0, P')$ will act as the identity on the state space.[49] The terminology makes sense according to Lemma 9.12.6. If a particle is described by a representation π of $\mathcal{P}^*$ which is equivalent to its "mirror image" $\tilde{\pi}$, the representation of Lemma 9.12.1, then π can be extended to $\mathcal{P}^{*+}$.

49 As is made particularly clear by the next section.

9.13 Photons

In this section we investigate (at the level of representations) photons, ubiquitous massless "spin 1" particles, which are their own mirror image.

Starting with a representation π of $\mathcal{P}^*$, one may construct a representation π^+ of $\mathcal{P}^{*+}$ as follows.

Proposition 9.13.1 *Given a unitary representation π of $\mathcal{P}^*$ on a Hilbert space $\mathcal{H}$ there is a unique unitary representation π^+ of $\mathcal{P}^{*+}$ on $\mathcal{H} \times \mathcal{H}$ such that*

$$\pi^+(a,A)(x,y) = (\pi(a,A)(x), \tilde{\pi}(a,A)(y)) = (\pi(a,A)(x), \pi(Pa, A^{\dagger-1})(y)) \tag{9.91}$$

and

$$\pi^+(0,P')(x,y) = (y,x). \tag{9.92}$$

It is not difficult to invent this construction, as the basic idea already occurs in Exercise 8.9.4. The proof is straightforward.

Exercise 9.13.2 Check the proof in all its details. Hint: This is really simple, the key identity to check is $\pi^+(0,P')\pi^+(a,A)\pi^+(0,P') = \pi^+(Pa, A^{\dagger-1})$.

Definition 9.13.3 We denote by $\pi^+_{0,2}$ the representation of $\mathcal{P}^{*+}$ constructed by Proposition 9.13.1 in the case $\pi = \pi_{0,2}$.

This representation models the *photon*. The state space of the photon is a space of functions on X_0 valued in a space of dimension two. Consequently the state space of a photon is of the type $\mathcal{H} \times \mathcal{H}$, and a photon can naturally be seen as a mixture of two components, belonging to the spaces $\mathcal{H} \times \{0\}$ and $\{0\} \times \mathcal{H}$ and corresponding to the representations $\pi_{0,2}$ and $\pi_{0,-2}$. As both of these represent particles of spin 1 (but with opposite helicity) it is quite natural to say "the photon is of spin 1". Photons for which one of these two components is zero are called in physics photons of left or right circular polarization.

More generally one could consider the representation $\pi^+_{0,j}$, which would describe a massless particle "of spin $j/2$" which is its own mirror image. The corresponding particles for $j \neq 2$ apparently do not occur in Nature.[50] Massless particles of spin 1/2 (neutrinos and anti-neutrinos, in a theory where they are massless) are not their own mirror image, but are described by the representations $\pi_{0,\pm 1}$, which cannot be extended to include parity.

Our next goal is to introduce a natural representation of $\mathcal{P}^{*+}$ and to show that it is equivalent to $\pi^+_{0,2}$. Let us define a sesqui-linear form on $\mathbb{C}^4$ by the formula

$$(x,y)_\eta := -(x^\mu)^* y^\nu \eta_{\mu\nu} = -x^{0*}y^0 + \sum_{i\le 3} x^{i*}y^i, \tag{9.93}$$

where the subscript η in the expression $(x,y)_\eta$ prevents confusion with the Lorentz bilinear form. (So, this sesqui-linear form is the natural extension to $\mathbb{C}^4$ of the *opposite* of the Lorentz

[50] Gravitons, massless particles of spin 2, have been theorized but not observed yet.

bilinear form.[51]) We may consider each transformation in $O(1,3)$ as operating on $\mathbb{C}^4$ (by the same matrix) and such a transformation A satisfies

$$\forall x, y \in \mathbb{C}^4 \ , \ (Ax, Ay)_\eta = (x, y)_\eta, \tag{9.94}$$

because this is true if $x, y \in \mathbb{R}^4$ since $A \in O(1,3)$, and because both the left and the right-hand sides are linear in y and anti-linear in x. Given $p \in \mathbb{R}^{1,3}$, $p \neq 0$ we define

$$\mathcal{V}_p = \{x \in \mathbb{C}^4 \ ; \ (x, p)_\eta = 0\}. \tag{9.95}$$

Thus, $p \in \mathcal{V}_p$ if and only if $p^2 = 0$.

Lemma 9.13.4 *If $p^2 \geq 0$, for $x \in \mathcal{V}_p$ we have $(x,x)_\eta \geq 0$. If $p^2 > 0$ it holds that $(x,x)_\eta > 0$ for $x \in \mathcal{V}_p$ and $x \neq 0$. If $p^2 = 0$, for $x \in \mathcal{V}_p$ we have $(x,x)_\eta = 0$ if and only if $x = \alpha p$ for some $\alpha \in \mathbb{C}$.*

Proof This is rather obvious by Lorentz invariance (9.94). When $p^2 > 0$ we have $p = A(p^*)$ where $p^* = (a,0,0,0)$ and $A \in O(1,3)$. If $x \in \mathcal{V}_p$ we have $(x,p)_\eta = (x, A(p^*))_\eta = (A^{-1}(x), p^*)_\eta$ so $A^{-1}(x) \in \mathcal{V}_{p^*}$. Since $(x,x)_\eta = (A^{-1}(x), A^{-1}(x))_\eta$ it suffices to prove the result for $p = p^*$, which is straightforward. We proceed similarly when $p^2 = 0$ but now with $p^* = (a,0,0,a)$. □

Consider $m \geq 0$ and the space $\mathcal{H}_m$ of functions φ from X_m to $\mathbb{C}^4$ such that

$$\forall p \in X_m \ , \ \varphi(p) \in \mathcal{V}_p.$$

As a consequence of Lemma 9.13.4, the inner product

$$(\varphi, \psi)_\eta = \int \mathrm{d}\lambda_m(p) (\varphi(p), \psi(p))_\eta \tag{9.96}$$

on $\mathcal{H}_m$ is positive, and is also positive-definite if $m > 0$.

Exercise 9.13.5 Assume $m > 0$. Prove that the formula

$$V(a,A)(\varphi)(p) = \exp(\mathrm{i}(a,p)/\hbar) A(\varphi(A^{-1}(p))) \tag{9.97}$$

defines a unitary representation of $\mathbb{R}^{1,3} \rtimes O(1,3)$ on $\mathcal{H}_m$. Use (a variation of) Theorem 9.4.2 to describe it. Hint: Observe that $A(\mathcal{V}_p) = \mathcal{V}_{A(p)}$ for $A \in O(1,3)$.

From now on $m = 0$, and $\mathcal{H} := \mathcal{H}_0$ denotes the space of functions from X_0 to $\mathbb{C}^4$ such that for each p, $\varphi(p) \in \mathcal{V}_p$. For φ in $\mathcal{H}$ and $(a,A) \in \mathcal{P}^{*+}$ we define $V(a,A)(\varphi)$ by the formula (9.97). There, for $A \in SL(2,\mathbb{C})$ and $x \in \mathbb{C}^4$ we denote by $A(x)$ the element $\kappa(A)(x)$ where $\kappa(A) \in O^+(1,3)$ is seen as a 4×4 matrix.[52] More specifically, we now view the map (8.19) as being defined on $\mathbb{C}^4$:

$$x \in \mathbb{C}^4 \leftrightarrow M(x) := x^\mu \sigma_\mu = \begin{pmatrix} x^0 + x^3 & x^1 - \mathrm{i}x^2 \\ x^1 + \mathrm{i}x^2 & x^0 - x^3 \end{pmatrix}, \tag{9.98}$$

[51] If you remember the convention of (4.16) to raise and lower indices you may write this as $(x,y)_\eta = -(x^\mu)^* y_\mu = -x^*_\mu y^\mu$.

[52] We could as well here define a representation of $\mathbb{R}^{1,3} \rtimes O^+(1,3)$ by the same formula, but we use $\mathcal{P}^{*+} = \mathbb{R}^{1,3} \rtimes SL^+(2,\mathbb{C})$ to make the comparison easier with $\pi^+_{0,2}$.

and for $A \in SL(2,\mathbb{C})$, $A(x) = \kappa(A)(x)$ is given by $M(\kappa(A)(x)) = AM(x)A^\dagger$, so that

$$M(A(x)) = AM(x)A^\dagger. \tag{9.99}$$

Then $V(a,A)(\varphi) \in \mathcal{H}$ because $A(\mathcal{V}_p) \subset \mathcal{V}_{Ap}$. Furthermore, it is straightforward to check the formula

$$V(a,A)V(b,B) = V((a,A)(b,B)). \tag{9.100}$$

According to Lemma 9.13.4, $(\varphi,\varphi)_\eta = 0$ if and only if $\varphi(p)$ is proportional to p for λ_0-almost each p. Thus $\mathcal{H}$ is not a Hilbert space, but the set

$$\mathcal{H}_{\text{null}} = \{\varphi \in \mathcal{H}\ ;\ (\varphi,\varphi)_\eta = 0\}$$

is a subspace of $\mathcal{H}$ and the quotient $\mathcal{H}_{\text{phys}} = \mathcal{H}/\mathcal{H}_{\text{null}}$ is a Hilbert space. For φ in $\mathcal{H}$ let us denote by $[\varphi]$ its image in $\mathcal{H}_{\text{phys}}$.

Proposition 9.13.6 *The formula*

$$U(a,A)([\varphi]) = [V(a,A)(\varphi)] \tag{9.101}$$

defines a unitary representation of $\mathcal{P}^{*+}$ on $\mathcal{H}_{\text{phys}}$. This representation is unitarily equivalent to the representation $\pi^+_{0,2}$ constructed at the beginning of the present section.

Formula (9.101) makes sense because (trivially) $V(a,A)(\mathcal{H}_{\text{null}}) = \mathcal{H}_{\text{null}}$, and defines a representation because of (9.100), and this representation is obviously unitary. The appeal of Proposition 9.13.6 is the simple formula defining the representation.

Let us now explain why this representation is natural from the physical point of view, using the language of Section E.2. Consider a four-momentum p and $b \in \mathbb{C}^4$ which satisfies $(b^\mu)^* p_\mu = 0$, or simply $b^\mu p_\mu = 0$ since the components of p are real numbers. Thinking like physicists, the ideal case where $\varphi \in \mathcal{H}$ is of the type $b\delta_p$ corresponds to a state $|p,b\rangle$. Furthermore, if b and b' differ only by a multiple of p, $|p,b\rangle$ and $|p,b'\rangle$ represent the same state. The action of an element $A \in SL^+(2,\mathbb{C})$ on the state $|p,b\rangle$ is as simple as it can be: it sends this state to $|Ap, Ab\rangle$. The point of this formulation is that the data of p and b with $(p,b)_\eta = 0$ is closely connected to the data which describes classically a plane wave of the electromagnetic field, as explained in Section E.2.

We turn to the proof of Proposition 9.13.6. Our first task will be to study the action on $\mathcal{V}_{p^*}$ of some elements of the little group $G_0^+ = \{A \in SL^+(2,\mathbb{C}); A(p^*) = p^*\}$. Since $p^* = (1,0,0,1)$, $\mathcal{V}_{p^*}$ contains the vectors

$$x_L = (0,1,\mathrm{i},0)\ ;\ x_R = (0,1,-\mathrm{i},0). \tag{9.102}$$

We recall the representation $\hat{\pi}_j$ of G_0 constructed in Definition 9.6.2 and the matrix J of (8.1).[53]

[53] Let us recall that $\hat{\pi}_j(A) \in \mathbb{C}$ for $A \in G_0$.

Lemma 9.13.7 *(a) If $A \in G_0$ we have*

$$A(x_L) - \hat{\pi}_{-2}(A)x_L \in \mathbb{C}p^* \; ; \; A(x_R) - \hat{\pi}_2(A)x_R \in \mathbb{C}p^*. \tag{9.103}$$

(b) The element $Q := P'J$ belongs to G_0^+. We have $Q(x_L) = x_R$ and $Q(x_R) = x_L$.

Proof We observe the relations

$$M(x_L) = \begin{pmatrix} 0 & 2 \\ 0 & 0 \end{pmatrix} \; ; \; M(x_R) = \begin{pmatrix} 0 & 0 \\ 2 & 0 \end{pmatrix}, \tag{9.104}$$

and we recall the relation (9.99): $M(A(x)) = AM(x)A^\dagger$. Consider $A = \begin{pmatrix} a & b \\ 0 & a^* \end{pmatrix} \in G_0$, so that $\hat{\pi}_{-2}(A) = a^2$ by Definition 9.6.2 and we have

$$A\begin{pmatrix} 0 & y_1 \\ y_2 & 0 \end{pmatrix} A^\dagger = \begin{pmatrix} c & a^2 y_1 \\ a^{*2} y_2 & 0 \end{pmatrix} \tag{9.105}$$

for a certain $c \in \mathbb{C}$. We then compute $M(A(x_L)) - M(\hat{\pi}_{-2}(A)x_L)$ to find that it is of the type $\begin{pmatrix} c & 0 \\ 0 & 0 \end{pmatrix}$ and hence of the type $M(z)$ for $z \in \mathbb{C}p^*$. Since M is a linear isomorphism this proves the first part of (9.103).
(b) It follows from (9.86) that $J(p^*) = P(p^*)$, $J(x_L) = -x_R$ and $J(x_R) = -x_L$. Furthermore $P(x_L) = -x_L$ and $P(x_R) = -x_R$. □

Let us define the following subspaces of $\mathcal{V}_{p^*}$:

$$\mathcal{L} = \mathrm{span}(x_L, p^*) \; ; \; \mathcal{R} = \mathrm{span}(x_R, p^*).$$

Since the spaces $\mathcal{L}$ and $\mathcal{R}$ are invariant under the action of G_0, for each $A \in SL(2,\mathbb{C})$ the spaces $A(\mathcal{L})$ and $A(\mathcal{R})$ depend only on $A(p^*)$, see the proof of Lemma 9.5.2. Thus we may define $\mathcal{L}_p := A(\mathcal{L})$ whenever $A(p^*) = p$, and $\mathcal{R}_p$ similarly. Then $A(\mathcal{L}_p) = \mathcal{L}_{Ap}$ and $A(\mathcal{R}_p) = \mathcal{R}_{Ap}$ for $A \in SL(2,\mathbb{C})$. Furthermore $\mathcal{L}_p$ and $\mathcal{R}_p$ span $\mathcal{V}_p$ and have the property that $(x,y)_\eta = 0$ for $x \in \mathcal{L}_p$ and $y \in \mathcal{R}_p$, because this is true for $p = p^*$, and the structure of $\mathcal{V}_p$ is simply obtained by transporting the structure on $\mathcal{V}_{p^*}$.

Lemma 9.13.8 *For each $p \in X_0$ we have $P'(\mathcal{L}_p) = \mathcal{R}_{Pp}$ and $P'(\mathcal{R}_p) = \mathcal{L}_{Pp}$.*

Proof In the relation $A(\mathcal{L}_p) = \mathcal{L}_{Ap}$ let us replace A by $P'AP'$ and p by Pp. Keeping in mind that $P'AP'(Pp) = P(Ap)$ we obtain the relation $A(\mathcal{S}_p) = \mathcal{S}_{Ap}$ where $\mathcal{S}_p := P'(\mathcal{L}_{Pp})$. We want to prove that $\mathcal{S}_p = \mathcal{R}_p$ for all p. Since $A(\mathcal{R}_p) = \mathcal{R}_{Ap}$ it suffices to prove this for $p = p^*$. Since $J(p^*) = P(p^*)$ we have $J(\mathcal{L}_{p^*}) = \mathcal{L}_{J(p^*)} = \mathcal{L}_{P(p^*)}$. Thus $\mathcal{S}_{p^*} = P'(\mathcal{L}_{P(p^*)}) = P'(J(\mathcal{L}_{p^*})) = Q(\mathcal{L}_{p^*})$. Since $Q(p^*) = p^*$ and $Q(x_L) = x_R$ by (b) of Lemma 9.13.7 the proof is complete. □

Proof of Proposition 9.13.6 The properties of $\mathcal{L}_p$ and $\mathcal{R}_p$ described above should make it obvious that have a natural orthogonal decomposition $\mathcal{H}_{\mathrm{phys}} = \mathcal{H}_R \oplus \mathcal{H}_L$ where $\mathcal{H}_L$ consists of the classes $[\varphi]$ of functions where $\varphi(p) \in \mathcal{L}_p$ for each p and similarly for $\mathcal{H}_R$. It follows from Lemma 9.13.8 that $U(0,P')(\mathcal{H}_R) = \mathcal{H}_L$, so that the map $T\colon (x,y) \mapsto x + U(0,P')(y)$ is unitary from $\mathcal{H}_R \oplus \mathcal{H}_R$ to $\mathcal{H}_{\mathrm{phys}}$. Denoting by π the restriction of U

to $\mathcal{P}^*$ and $\mathcal{H}_R$, we claim that the map T witnesses that U is unitarily equivalent to the representation π^+ constructed from π in Proposition 9.13.1. It is straightforward to check this claim. For example to prove that for $A \in SL(2,\mathbb{C})$ we have $T(\pi^+(a,A)(x,y)) = U(a,A)T(x,y)$ all it takes is that $U(0,P')U(a,A)U(0,P') = U(Pa,A^{\dagger-1})$. To conclude the proof of the proposition, we now prove that π is unitarily equivalent to $\pi_{0,2}$. Given a function $\varphi\colon X_0 \to \mathbb{C}$ we consider the function $\tilde{\varphi} \in \mathcal{H}$ given by $\tilde{\varphi}(p) = \varphi(p)D_p(x_R)$ so that the class $W(\varphi) := [\tilde{\varphi}]$ of $\tilde{\varphi}$ in $\mathcal{H}_{\text{phys}}$ belong to $\mathcal{H}_R$. Our goal is to prove that

$$U(a,A)(W(\varphi)) = W(\pi_{0,2}(a,A)(\varphi)). \tag{9.106}$$

Since it is straightforward to check that W is proportional to an isometry the claim the π and $\pi_{0,2}$ are unitarily equivalent follows. Recalling the formula (9.7) defining $\pi_{0,2}$ the righthand side of (9.106) is the class $[\theta]$ in $\mathcal{H}_{\text{phys}}$ of the function

$$\theta(p) = \exp(\mathrm{i}(a,p)/\hbar)\hat{\pi}_2(D_p^{-1}AD_{A^{-1}(p)})\varphi(A^{-1}(p))D_p(x_R). \tag{9.107}$$

Now, using the second part of (9.103) for the element $D_p^{-1}AD_{A^{-1}(p)} \in G_0$ we obtain that

$$\hat{\pi}_2(D_p^{-1}AD_{A^{-1}(p)})x_R - D_p^{-1}AD_{A^{-1}(p)}(x_R) \in \mathbb{C}p^*,$$

and thus, recalling that $\pi_2(D_p^{-1}AD_{A^{-1}(p)}) \in \mathbb{C}$,

$$\hat{\pi}_2(D_p^{-1}AD_{A^{-1}(p)})D_p(x_R) - AD_{A^{-1}(p)}(x_R) \in \mathbb{C}p.$$

Let us now define ψ by

$$\psi(p) = \exp(\mathrm{i}(a,p)/\hbar)\varphi(A^{-1}(p))AD_{A^{-1}(p)}(x_R).$$

Then $\psi(p) - \theta(p) \in \mathbb{C}p$ so that $\psi - \theta \in \mathcal{H}_{\text{null}}$ and $[\psi] = [\theta]$. Recalling the formula (9.97) we see that $\psi = V(a,A)(\tilde{\varphi})$. Since $U(a,A)(W(\varphi)) = U(a,A)([\tilde{\varphi}]) = [V(a,A)(\tilde{\varphi})]$ by definition of U we have proved (9.106).

Key ideas to remember:

- To consider only true representations one replaces the Poincaré group by $\mathcal{P}^* = \mathbb{R}^{1,3} \rtimes SL(2,\mathbb{C})$.
- Given $p^* \in \mathbb{R}^{1,3}$ its little group is $G_{p^*} = \{A \in SL(2,\mathbb{C})\ ,\ k(A)(p^*) = p^*\}$. When $p^* = (mc,0,0,0)$ its little group is $SU(2)$.
- An irreducible finite-dimensional unitary representation of the little group induces an irreducible unitary representation of $\mathcal{P}^*$.
- The physically relevant representations of $\mathcal{P}^*$ come in two families. (i) The representations $\pi_{m,j}$ ($j \in \mathbb{N}$) that correspond to massive particles of mass m and spin $j/2$. (ii) The representations $\pi_{0,j}$($j \in \mathbb{Z}$) which correspond to massless particles of helicity $j/2$.
- Parity is brought into the picture by studying representations of $\mathcal{P}^{*+} = \mathbb{R}^{1,3} \rtimes SL^+(2,\mathbb{C})$.
- The mirror image of a massless particle is a massless particle of opposite helicity.
- Photons are massless particles "of spin 1" which are their own mirror images.

10

Basic Free Fields

The free field discussed in Chapter 5 is *scalar* in the sense that it has a single component. This free field is a mathematically canonical object, but Nature uses it sparingly. Rather, Nature[1] uses multicomponent fields (the most apparent of which is the electromagnetic field) so these are of crucial importance. Studying them is the object of this chapter. Each component of such a field is a function (or rather, a distribution) with values in the space of operators on a certain Fock space.[2] If our models are to describe the bewildering complexity of the subatomic world, we must expect to meet a whole menagerie of fields.[3] This is confusing at first, and any study of the topic must strive to bring some organization to it. Two main roads can be followed for this purpose.

In the majority of physics textbooks, including some very recent ones such as that of Schwartz [80], the free quantum fields are introduced through the following procedure: One systematically looks for Lagrangians with suitable invariance properties. Each of these Lagrangians gives rise to a classical field, and one quantizes this field in a (more or less) canonical manner to arrive at a corresponding quantum field. The procedure however faces serious obstacles because certain quantum objects simply *do not have a classical equivalent*. (This is typically the case for anything with spin 1/2.) To pretend otherwise, one must bend the rules of mathematical sanity to a point where the usefulness of the whole approach becomes questionable. This will become particularly apparent in Sections 10.15 and 10.19, where we attempt to present the Dirac field as a quantization of a classical field.[4]

As mathematicians we are more comfortable with the far more solid approach put forward by Steven Weinberg [87]: The natural condition under which such fields should transform under the action of the Poincaré group almost determines their entire structure. Let us insist on the miracle which gives its appeal to the present chapter. A few very natural and weak-looking constraints considerably limit the possible quantum fields, and while exploring how to satisfy these constraints, we will be naturally led to the discovery of the main free quantum fields relevant to the physical world.

1 Or, more accurately, our description of Nature.
2 It will not always be boson Fock space.
3 For this intrinsic reason, whatever our efforts of exposition may be, the chapter is demanding.
4 Classical Mechanics is a wonderful theory. But why do so many people accept so many contortions to pretend that it explains certain quantum phenomena, which are remarkable precisely because they have no classical equivalent?

The starting point of the construction of a free quantum field is an irreducible unitary representation of the Poincaré group, or in other words, a particle.[5] The situation is described by saying that *particles of this type are the quanta of the free field.* We will then face a fundamental fact: When the quanta of the field have spin 1/2, it is not possible to use boson Fock space to describe multiparticle states. Another fundamental construction, fermion Fock space, is required. Fermion Fock space is based on anti-symmetry rather than symmetry, and commutators give way to anti-commutators.

Quantum fields will appear later as natural building blocks from which Hamiltonians describing interactions between different particles can be constructed. This will be explained at the end of Section 13.3. To study interactions between real-world particles, one must use these quantum fields. But the real world is a complicated matter, and in the book we concentrate on toy models, for which the Hamiltonian describing interactions uses only the free scalar field of Chapter 5 (and not the free fields constructed in this chapter). Thus the material of the present chapter, however beautiful, is not a prerequisite for the sequel.

The reader should have Chapters 5 and 9 fresh in her mind before continuing.

10.1 Charged Particles and Anti-particles

The present section is a kind of prelude to the heart of the matter. Its purpose is to demonstrate a fascinating aspect of Quantum Field Theory: The existence of anti-particles is almost unavoidable. We will reach this conclusion by heuristic arguments analyzing electrical charge, but the rest of the chapter is entirely rigorous. This analysis is based on a fundamental experimental fact:

- Particles have a property called electrical charge, which, just as their rest mass or their spin, is an intrinsic property of the particle. The electrical charge can be positive or negative. A particle whose electrical charge is zero is called *neutral.* A particle whose electrical charge is not zero is called *charged.*
- In any particle interaction, the sum of the electrical charges of the incoming particles equals the sum of the electrical charges of the outgoing particles. Electrical charge is conserved.[6]

In the framework of Quantum Mechanics, electrical charge is measured by a self-adjoint operator Q. This operator must satisfy remarkable commutation relations with the creation and annihilation operators of a particle of charge q:

$$[Q,a^{\dagger}(\boldsymbol{p})] = qa^{\dagger}(\boldsymbol{p})\ ;\ [Q,a(\boldsymbol{p})] = -qa(\boldsymbol{p}). \tag{10.1}$$

To explain these relations (which may not be intuitive at first) we appeal to their physical meaning. Consider a state $|\varphi\rangle$ of well-defined charge μ so that it is an eigenstate of Q,

5 In principle it could happen that two different particles correspond to the same representation of the Poincaré group, for example if the particles have the same mass, the same spin, and different electrical charge. But in Nature this occurs only for the pairs particle–anti-particle.

6 In a nutshell, the content of this section is that conservation of electrical charge "implies" the existence of anti-particles.

$Q|\varphi\rangle = \mu|\varphi\rangle$. Then $a^\dagger(\boldsymbol{p})Q|\varphi\rangle = \mu a^\dagger(\boldsymbol{p})|\varphi\rangle$. On the other hand the creation operator $a^\dagger(\boldsymbol{p})$ creates a particle of well-defined electrical charge q, so that the charge of $a^\dagger(\boldsymbol{p})|\varphi\rangle$ is $\mu + q$, i.e. $a^\dagger(\boldsymbol{p})|\varphi\rangle$ is an eigenstate of Q of eigenvalue $q + \mu$, that is $Qa^\dagger(\boldsymbol{p})|\varphi\rangle = (q+\mu)a^\dagger(\boldsymbol{p})|\varphi\rangle$. Therefore $[Q,a^\dagger(\boldsymbol{p})]|\varphi\rangle = qa^\dagger(\boldsymbol{p})|\varphi\rangle$. This should hold for sufficiently many $|\varphi\rangle$ that indeed $[Q,a^\dagger(\boldsymbol{p})] = qa^\dagger(\boldsymbol{p})$. Taking adjoints, since Q is self-adjoint, we also have $[Q,a(\boldsymbol{p})] = -qa(\boldsymbol{p})$.

On the other hand, since electrical charge is conserved in every interaction the operator Q has to commute with the Hamiltonian governing the interaction. As we will see, the most important such Hamiltonians are constructed from the quantum field, and it does not seem possible to achieve this unless the quantum fields themselves have simple commutation relations with Q. But the formula (5.30)

$$\varphi(x) = \int \frac{\mathrm{d}^3\boldsymbol{p}}{(2\pi\hbar)^3\sqrt{2c\omega_{\boldsymbol{p}}}}(e^{-\mathrm{i}(x,p)/\hbar}a(\boldsymbol{p}) + e^{\mathrm{i}(x,p)/\hbar}a^\dagger(\boldsymbol{p}))$$

shows that there is no simple commutation relation between $\varphi(x)$ and Q. This is a compelling reason why the formula (5.30) must be modified.[7] A brilliant solution to this problem is, for each particle, say of mass m and charge q, to consider *another particle*, called the *anti-particle* of the original particle, with the same mass m and opposite charge $-q$. We then consider the field

$$\varphi(x) = \int \frac{\mathrm{d}^3\boldsymbol{p}}{(2\pi\hbar)^3\sqrt{2c\omega_{\boldsymbol{p}}}}(\exp(-\mathrm{i}(x,p)/\hbar)a(\boldsymbol{p}) + \exp(\mathrm{i}(x,p)/\hbar)b^\dagger(\boldsymbol{p})) \tag{10.2}$$

and its conjugate

$$\varphi^\dagger(x) = \int \frac{\mathrm{d}^3\boldsymbol{p}}{(2\pi\hbar)^3\sqrt{2c\omega_{\boldsymbol{p}}}}(\exp(-\mathrm{i}(x,p)/\hbar)b(\boldsymbol{p}) + \exp(\mathrm{i}(x,p)/\hbar)a^\dagger(\boldsymbol{p})). \tag{10.3}$$

Here, as in much of the chapter, we lighten the presentation by pretending that distributions are functions and we let the reader state the results in terms of distributions if she likes. In these formulas, $b(\boldsymbol{p})$ and $b^\dagger(\boldsymbol{p})$ are respectively annihilation and creation operators for the anti-particle. That is, we consider a copy $\mathcal{B}'$ of the boson Fock space $\mathcal{B}$ of the particle, and the basic state space is now $\mathcal{B}\otimes\mathcal{B}'$. There $a(\boldsymbol{p})$ acts only on $\mathcal{B}$. More precisely, $a(\boldsymbol{p}) = a'(\boldsymbol{p})\otimes 1$ where $a'(\boldsymbol{p})$ is an operator on $\mathcal{B}$, and similarly $b(\boldsymbol{p}) = 1\otimes b'(\boldsymbol{p})$. Moreover the "operators" $a(\boldsymbol{p}), a^\dagger(\boldsymbol{p}), b(\boldsymbol{p})$ and $b^\dagger(\boldsymbol{p})$ satisfy the commutation relations

$$[a(\boldsymbol{p}),a^\dagger(\boldsymbol{p}')] = (2\pi\hbar)^3\delta^{(3)}(\boldsymbol{p}-\boldsymbol{p}')1 \; ; \; [b(\boldsymbol{p}),b^\dagger(\boldsymbol{p}')] = (2\pi\hbar)^3\delta^{(3)}(\boldsymbol{p}-\boldsymbol{p}')1, \tag{10.4}$$

with all other commutators zero. Let us stress once and for all that in this construction the operators $a(\boldsymbol{p}), a^\dagger(\boldsymbol{p})$ commute with the operators $b(\boldsymbol{p}'), b^\dagger(\boldsymbol{p}')$ "because they live on different spaces".

[7] The preceding is a very rough version of the argument of Weinberg [87, page 199], to which we refer for more details.

This construction behaves very well with respect to the charge operator Q. Since the anti-particle has charge $-q$ it satisfies $[Q,b^\dagger(p)] = -qb^\dagger(p)$ and therefore[8] $[Q,\varphi(x)] = -q\varphi(x)$.

At this stage the decision as to which is the particle and which is the anti-particle is pretty arbitrary. So it is really by convention that we call "the field" the expression (10.2) which destroys a particle and creates an anti-particle, and the "conjugate field" the expression (10.3) which does the opposite.[9]

The following fundamental fact is proved just as Theorem 5.1.5. The notion of microcausality is explained on page 136 and is detailed further at the beginning of Section 10.7.

Theorem 10.1.1 (Microcausality) *Consider φ and $\varphi^\dagger$ as in (10.2) and (10.3). If x and y are causally separated, i.e. $(x-y)^2 < 0$ then*

$$[\varphi(x),\varphi(y)] = [\varphi^\dagger(x),\varphi^\dagger(y)] = [\varphi(x),\varphi^\dagger(y)] = 0. \tag{10.5}$$

As we explained, charged particles "must" have an anti-particle with opposite charge.[10] It turns out that neutral particles also have anti-particles. This anti-particle may be different from the particle itself. In this case the corresponding quantum field is still given by (5.30). It may also happen that the neutral particle is *its own anti-particle*. In this case the operators a and b in the formulas (10.2) and (10.3) are identical, as in the case of the massive scalar field of Chapter 5. Particles which are their own anti-particles are rare in physics. Nonetheless, we shall also cover this case in detail, if only because the extra effort required is minimal.

10.2 Lorentz Covariant Families of Fields

We start now the real study of families of quantum fields.

Our basic setting consists of a unitary representation $U(c,C)$ of $\mathcal{P}^*$ on a Hilbert space $\mathcal{F}$ and its extension $U_{\mathcal{B}}(c,C)$ on the corresponding boson Fock space $\mathcal{B}$. (Our universe still contains only bosons at this stage.) We consider a family $(\psi_k)_{1\le k\le N}$ of quantum fields. These are operator-valued distributions as in Section 5.1, that is, for $f \in \mathcal{S}^4$, $\psi_k(f)$ is an

[8] To obtain this key relation, it is crucial that the anti-particle has charge $-q$.

[9] In many textbooks, this quantum field is often obtained as "quantization of the complex Klein-Gordon field" (which does not seem to occur in Nature more often than its real counterpart). Whichever way the argument is wrapped, its depth is exactly the same as that of the following argument: Given two real numbers a and b, the complex numbers $a+ib$ and $a-ib$ are complex conjugates. A complex Klein-Gordon field is simply a pair of real Klein-Gordon fields, independent of each other. Quantizing them gives the two fields

$$\varphi_i(x) = \int \frac{\mathrm{d}^3 p}{(2\pi\hbar)^3\sqrt{2c\omega_p}}(\exp(-\mathrm{i}(x,p)/\hbar)a_i(p) + \exp(\mathrm{i}(x,p)/\hbar)a_i^\dagger(p))$$

for $i = 1,2$, and the fields

$$\varphi(x) = \frac{1}{\sqrt{2}}(\varphi_1(x) + \mathrm{i}\varphi_2(x))\ ;\ \varphi^\dagger(x) = \frac{1}{\sqrt{2}}(\varphi_1(x) - \mathrm{i}\varphi_2(x)),$$

are the fields we are looking for. The purpose of the $\sqrt{2}$ factor is to ensure that the operators $a(p) = (a_1(p) + ia_2(p))/\sqrt{2}$ and $b(p) = (a_1(p) - ia_2(p))/\sqrt{2}$ satisfy the commutation relations (10.4).

[10] This is not a mathematical theorem, but Nature has agreed with our argument and made things that way.

operator on $\mathcal{B}$ which is defined at least on the space $\mathcal{B}_0$ of (3.26). We recall the definition (5.2) of $V(c,C)(f)$ for $f \in \mathcal{S}^4$ i.e. $V(c,C)(f)(x) = f(C^{-1}(x-c))$.

Definition 10.2.1 The family $(\psi_k)_{k\leq N}$ is *Lorentz covariant* if there exists an representation S of $SL(2,\mathbb{C})$ on $\mathbb{C}^N$ such that the following occurs: For each $c \in \mathbb{R}^{1,3}$, $C \in SL(2,\mathbb{C})$, $k \leq N$, $f \in \mathcal{S}^4_{\mathbb{R}}$ we have

$$U_{\mathcal{B}}(c,C) \circ \psi_k(f) \circ U_{\mathcal{B}}(c,C)^{-1} = \sum_{\ell\leq N} S(C^{-1})_{k,\ell}\psi_\ell(V(c,C)(f)). \tag{10.6}$$

Here we have denoted the composition of operators on $\mathcal{B}$ by $\circ$ for legibility. When $N=1$ the Lorentz covariance condition coincides with (5.5), and the reader should review the paragraph following this equation, as the motivation behind (10.6) is the same. Equation (10.6) looks scary at first, but this formula describes *the simplest possible situation*. Acting on the fields by a Poincaré transformation as in the left-hand side simply amounts to combining them linearly.

The strange term $S(C^{-1})$ in the right-hand side has to do with ever-confusing matters of linear algebra. In abstract terms, coordinates on $\mathbb{C}^N$ are linear functionals, which belong to the dual of $\mathbb{C}^N$ and the way an operator on $\mathbb{C}^N$ acts on a space and its dual are transpose of each other. In more concrete terms, if (e_ℓ) form a basis, an operator of matrix $(V_{k,\ell})$ satisfies $V(e_k) = \sum_\ell V_{\ell,k}e_\ell$. In (10.6) the indices k and ℓ are exchanged compared to this formula. Transposition of matrices reverses the order of multiplication, and this is compensated by the other reversing of factors due to the C^{-1}. Nonetheless, we recommend that the reader carry out the following exercise in complete detail.

Exercise 10.2.2 Consider elements e_k of a vector space $\mathcal{G}$, an N-dimensional representation S of $SL(2,\mathbb{C})$ and linear maps $W(C)$ on $\mathcal{G}$ that satisfy

$$W(C)(e_k) = \sum_{\ell\leq N} S(C^{-1})_{k,\ell}e_\ell. \tag{10.7}$$

Prove that these maps form a representation of $SL(2,\mathbb{C})$ in the linear span of the vectors $(e_k)_{k\leq N}$, i.e. prove that $W(D)W(C) = W(DC)$.

The reader may also wonder why rather than (10.6) we do not use the simpler formula

$$U_{\mathcal{B}}(c,C) \circ \psi_k(f) \circ U_{\mathcal{B}}(c,C)^{-1} = \sum_{\ell\leq N} S(C)_{\ell,k}\psi_\ell(V(c,C)(f)), \tag{10.8}$$

which amounts to changing the representation S into the representation $C \mapsto S(C^{-1})^T$. It will only become apparent later that (10.6) is the natural formulation, as it is the representation S itself which will come up in the arguments.

Pretending as usual that distributions are functions we write (10.6) as

$$U_{\mathcal{B}}(c,C) \circ \psi_k(x) \circ U_{\mathcal{B}}(c,C)^{-1} = \sum_{\ell\leq N} S(C^{-1})_{k,\ell}\psi_\ell(c + C(x)). \tag{10.9}$$

It is not required in Definition 10.2.1 that the representation $U(c,C)$ be irreducible because this definition is designed to accommodate models which would describe families of particles. It could also happen that the representation S involves block matrices, so that it is simply a direct sum of smaller representations. We however concentrate on the simplest situation, where S is irreducible. It is then appropriate to think of the ψ_k as being the components of a single N-component quantum field.

10.3 Road Map I

Our goal is to investigate how one might construct Lorentz covariant families of quantum fields that satisfy a suitable version of the microcausality condition (5.10),[11] and as far as possible to find all such families of fields, at least within a certain setting. The microcausality condition is essential, as it will put considerable constraints on the construction.

We make *from the start* a number of hypotheses on our quantum fields. In order to obtain fields that satisfy the microcausality condition we copy the scheme on page 134. We assume that the fields have both an annihilation part and a creation part. We assume that both the annihilation and the creation part of the fields are Lorentz covariant. It is for the microcausality condition that both parts are needed together.

We first discuss the annihilation part ψ_k^+ of the fields.[12]

At the root of the construction is a Hilbert space $\mathcal{F}$ provided with an irreducible representation $U(c,C)$ of $\mathcal{P}^*$. According to Wigner's idea, such a representation describes an elementary particle. *Particles of this type are called the quanta of the field.*[13] In other words, for a given representation $U(c,C)$ of $\mathcal{P}^*$ and a quantum field built upon this representation, the quanta of the field are the particles described by the very same representation.[14] Whenever we talk of "the particle" we mean the particle type described by this representation. The representation associated to the particle does not entirely describe it as there are other characteristics of a particle than its mass and its spin, and first of all its electrical charge. The charge of the particle is not involved in the present construction. Simply, the anti-particle has to be of opposite charge. (So, in particular, when the particle is identical to its anti-particle it is neutral, its electrical charge is zero.)

The quantum fields ψ_k^+ are valued in the space of operators on the boson Fock space $\mathcal{B}$ of $\mathcal{F}$. More precisely we assume *from the start* that $\psi_k^+(x)$ is of the type $A(f_k(x))$ for a certain element $f_k(x)$ of $\mathcal{F}$, where $A(f_k(x))$ denotes the corresponding annihilation operator in $\mathcal{B}$. We express this by saying that "the annihilation part of the field annihilates particles", being understood that it acts only on the particles which are the quanta of this field.

In Section 10.4 we translate the Lorentz covariance condition (10.6) in terms of the properties of the family of functions (f_k), and we soon reach a rather detailed understanding of the possible forms of the annihilation fields.

[11] The reader might like to review this notion now, as it will play an important part in the present chapter.

[12] Denoting the *annihilation* part of the field by ψ^+ is common practice. However this notation creates a psychological conflict, since the sign $+$ and positivity have beneficial connotations, whereas annihilation is typically considered detrimental.

[13] Poetic expressions such as "a particle is an excitation of a quantum field" found in physics textbooks have no deeper meaning.

[14] There is a kind of duality between quantum fields and particles, although many different quantum fields may correspond to the same quanta. As will gradually become clear, it is the quantum fields that are the fundamental objects.

We will then have to perform the same analysis for the creation part of the fields, replacing the particle by its anti-particle (which may be the same as the particle or not).[15] We again assume from the start that the creation part of the field $\psi_k^-(x)$ is of the type $B^\dagger(f_k'(x))$ where now $B^\dagger$ is the creation operator on the boson Fock space of the anti-particle. In words, we say that the creation part of the field creates anti-particles. The analysis is not entirely identical to the analysis of the annihilation part of the field because the map $B^\dagger$ is linear in its argument and the map A is anti-linear in its argument.

Having determined the general form of both the annihilation and the creation parts of the field, we investigate in Section 10.7 how we might achieve microcausality. Our general considerations end there, and from Section 10.9 we investigate concrete examples, discovering in turn the most important fields used by Nature.

10.4 Form of the Annihilation Part of the Fields

Having made these precise (and restrictive) assumptions as to the nature of the space $\mathcal{F}$ and the form of the fields ψ_k^+ we now turn to the determination of the structure of these fields. We advise reviewing the argument on page 134, since the arguments of the present section are in the same spirit (although technically more complicated).

In order to describe the irreducible representation $U(c,C)$ of $\mathcal{P}^*$ at the root of the construction, we use the setting provided by Theorem 9.8.1, which we briefly review. We start with a mass $m \geq 0$. We consider the action of $SL(2,\mathbb{C})$ on X_m, and the little group G_{p^*} of the specific point p^* given by $p^* = (mc,0,0,0)$ if $m > 0$ and $p^* = (1,0,0,1)$ if $m = 0$. We consider a finite-dimensional Hilbert space $\mathcal{V}$, and an irreducible unitary representation V of G_{p^*} in $\mathcal{V}$. The state space $\mathcal{F}$ is then the set of functions $f : SL(2,\mathbb{C}) \to \mathcal{V}$ that satisfy the condition

$$A \in SL(2,\mathbb{C}), B \in G_{p^*} \Rightarrow f(AB) = V(B)^{-1} f(A), \tag{10.10}$$

provided with the norm $\|f\|^2 = \int \mathrm{d}\lambda_m(p)\|f(D_p)\|^2$, where as usual $D_p \in SL(2,\mathbb{C})$ is such that $D_p(p^*) = p$, and the operators $U(a,A)$ are given for $(a,A) \in \mathcal{P}^*$ by

$$U(a,A)(f)(B) = \exp(\mathrm{i}(a, B(p^*))/\hbar) f(A^{-1}B). \tag{10.11}$$

If we were physicists, we would immediately write down explicit formulas for the representations V of (10.10) and use their specific properties. These specific properties are however not helpful. Furthermore, our more conceptual approach is valid both for massive and massless particles. Matters are transparent. The action of the group $\mathcal{P}^*$ on the state space $\mathcal{F}$ is completely determined by the irreducible representation V of the little group G_{p^*}, and we will prove that condition (10.6) is equivalent to a precise relationship between this representation and the representation S. Only later we will specialize our results to the cases of physical interest: The case where $m > 0$, $G_{p^*} = SU(2)$ and V is the representation π_j

[15] When the anti-particle is different from the particle, the field $\psi_k(x) = \psi_k^+(x) + \psi_k^-(x)$ operates on the tensor product of the boson Fock space of the particle and the boson Fock space of the anti-particle. To formulate the notion of Lorentz covariant family of fields we need to extend in the obvious manner Definition 10.2.1. This is not really needed since we require from the start that both the creation and the annihilation part of the field are Lorentz covariant.

($j \geq 0$) of Definition 8.2.3 and also the case where $m = 0$, and where V is the representation $\hat{\pi}_j$ ($j \in \mathbb{Z}$) of Definition 9.6.2 of the little group G_0.

We denote by $(e_k)_{k \leq N}$ the canonical basis of $\mathbb{C}^N$. We recall that for an element f of the state space $\mathcal{F}$, $A(f)$ denotes the corresponding annihilation operator on the boson Fock space $\mathcal{B}$.

Proposition 10.4.1 *The annihilation fields*

$$\psi_k^+(x) = A(f_k(x)) \tag{10.12}$$

for $k \leq N$ satisfy the condition (10.6) of Lorentz covariance if and only if the functions $f_k(x) \in \mathcal{F}$ are of the type

$$f_k(x)(B) = \exp(\mathrm{i}(x, B(p^*))/\hbar)\Pi S(B)^\dagger(e_k) \tag{10.13}$$

where Π is a linear map $\mathbb{C}^N \to \mathcal{V}$ which satisfies the condition

$$\forall C \in G_{p^*},\ V(C)^{-1}\Pi = \Pi S(C)^\dagger. \tag{10.14}$$

$$\begin{array}{ccc} \mathbb{C}^N & \xrightarrow{\ \Pi\ } & \mathcal{V} \\ \big\downarrow{\scriptstyle S(C)^\dagger} & & \big\downarrow{\scriptstyle V(C)^{-1}} \\ \mathbb{C}^N & \xrightarrow{\ \Pi\ } & \mathcal{V} \end{array}$$

In order for equation (10.13) to make sense, the right-hand side, seen as a function of B, must belong to $\mathcal{F}$, i.e. must satisfy (10.10). Condition (10.14) is exactly what is required for this.

The meaning of the formula (10.13) is not transparent at this stage, but will gradually become clear as we learn to write it in different ways.

Before we start the preparations for the proof of Proposition 10.4.1 we interpret the condition (10.14). Taking adjoints and using that $V(C)$ is unitary this condition is equivalent to saying that $W := \Pi^\dagger : \mathcal{V} \to \mathbb{C}^N$ satisfies

$$\forall C \in G_{p^*},\ WV(C) = S(C)W, \tag{10.15}$$

and we now focus on the meaning of this condition.

Proposition 10.4.2 *Unless $W = 0$, condition (10.15) occurs if and only if the following occur: The range $\mathcal{G}$ of W is invariant under the restriction of the representation S on $\mathbb{C}^N$ to the little group G_{p^*}, W is one-to-one from $\mathcal{V}$ to $\mathcal{G}$ and*

$$\forall C \in G_{p^*};\ V(C) = W^{-1}S(C)W. \tag{10.16}$$

Furthermore the map W is determined by (10.16) up to a multiplicative constant.

Consequently, in order to determine the possible forms of the annihilation part of a Lorentz covariant field, it is sufficient to determine the representations S and the subspaces $\mathcal{G}$ of $\mathbb{C}^N$ which are invariant under the restriction of S to G_{p^*}, and are such that the restriction of S to G_{p^*} and $\mathcal{G}$ is equivalent to V. We will turn to this clean mathematical problem later. Its analysis will lead us to discover in a completely natural manner the many fields considered in physics.

Proof Condition (10.15) implies that the kernel of W is invariant under the action of all the maps $V(C)$. Since we assume that V is irreducible, this kernel is either $\{0\}$ or $\mathcal{V}$. Thus when W is not identically zero this kernel is $\{0\}$ and W is one-to-one on its image, and (10.15) therefore implies $V(C) = W^{-1}S(C)W$. Conversely it is obvious that (10.16) implies (10.15).

Finally, consider a map $W' : \mathcal{V} \to \mathcal{G}$ which satisfies $V(C) = W'^{-1}S(C)W'$ for $C \in G_{p^*}$. Then $V(C) = W^{-1}W'V(C)W'^{-1}W$. Since we assume that V is irreducible, $W^{-1}W'$ is a multiple of the identity by Schur's lemma. □

We now start the preparations for the proof of Proposition 10.4.1.

Lemma 10.4.3 *The fields (10.12) satisfy the Lorentz covariance condition (10.6) if and only if the following equality holds for each values of $k \le N, x, p, c \in \mathbb{R}^{1,3}, C, B \in SL(2,\mathbb{C})$:*

$$\exp(\mathrm{i}(c, B(p^*))/\hbar) f_k(x)(C^{-1}B) = \sum_{\ell \le N} S(C^{-1})^*_{k,\ell} f_\ell(c + C(x))(B). \tag{10.17}$$

Proof Use of Proposition 3.5.2 shows that

$$U_{\mathcal{B}}(c,C) \circ \psi_k^+(x) \circ U_{\mathcal{B}}(c,C)^{-1} = A(U(c,C) f_k(x)).$$

On the other hand, since A is anti-linear, we get

$$\sum_{\ell \le N} S(C^{-1})_{k,\ell} \psi_\ell^+(c + C(x)) = A\Big(\sum_{\ell \le N} S(C^{-1})^*_{k,\ell} f_\ell(c + C(x))\Big).$$

Consequently the Lorentz covariance condition is satisfied if and only if we always have $U(c,C) f_k(x) = \sum_{\ell \le N} S(C^{-1})^*_{k,\ell} f_\ell(c + C(x))$, which is the formula (10.17) using the definition of $U(c,C)$, $(U(c,C) f_k(x))(B) = \exp(\mathrm{i}(c, B(p^*))/\hbar) f_k(x)(C^{-1}B)$. □

Lemma 10.4.4 *If the fields (10.12) satisfy the Lorentz covariance condition (10.6) then the functions f_k are of the type (10.13) and the condition (10.14) is satisfied.*

Proof We proved in the previous lemma that under the present assumptions (10.17) holds. In this relation we take C equal to the identity I and $x = 0$ to obtain

$$f_k(c)(B) = \exp(\mathrm{i}(c, B(p^*))/\hbar) G_k(B), \tag{10.18}$$

where we have set $G_k(B) := f_k(0)(B)$. We then define $\Pi : \mathbb{C}^N \to \mathcal{V}$ by $\Pi(e_k) = G_k(I)$. Using now (10.17) with $c = x = 0$, $B = I$ and changing C into C^{-1} gives

$$\begin{aligned} G_k(C) &= \sum_{\ell \le N} S(C)^*_{k,\ell} G_\ell(I) = \sum_{\ell \le N} S(C)^*_{k,\ell} \Pi(e_\ell) \\ &= \Pi\Big(\sum_{\ell \le N} S(C)^*_{k,\ell} e_\ell\Big) = \Pi S(C)^\dagger(e_k). \end{aligned}$$

Finally (10.14) follows from (10.10). □

Proof of Proposition 10.4.1 It remains only to prove that under (10.15) the fields corresponding to the functions (10.13) are Lorentz covariant. For this one simply checks that the identity (10.17) holds for the functions (10.13), which is straightforward, using only that S is a representation. □

It could help the reader to think of the previous construction in an equivalent way, but where now the representation U of $\mathcal{P}^*$ is constructed through Theorem 9.5.3.

Exercise 10.4.5 In this (challenging) exercise we consider a representation S of $SL(2,\mathbb{C})$ on $\mathbb{C}^N$. We consider a subspace $\mathcal{G}$ of $\mathbb{C}^N$, which is invariant under $S(C)$ for each $C \in G_{p^*}$ and we assume that the restriction of S to $\mathcal{G}$ and G_{p^*} is unitary. We are therefore in the setting of Theorem 9.5.3. The formula

$$U(a,A)(\varphi)(p) = \exp(\mathrm{i}(a,p)/\hbar)S(A)[\varphi(A^{-1}(p))] \tag{10.19}$$

defines a unitary representation of $\mathcal{P}^*$ in the space $\mathcal{H}$ of functions $\varphi\colon X_m \to \mathbb{C}^N$ such that $\varphi(p) \in \mathcal{G}_p = S(D_p)(\mathcal{G})$ provided with the norm

$$\|\varphi\|^2 = \int \mathrm{d}\lambda_m(p)\|S(D_p)^{-1}(\varphi(p))\|^2. \tag{10.20}$$

We denote by W the injection $\mathcal{G} \to \mathbb{C}^N$ and by Π the adjoint of W, so that $\Pi\colon \mathbb{C}^N \to \mathcal{G}$.
(a) Using that $S(C)$ is unitary on $\mathcal{G}$ for $C \in G_{p^*}$ prove that

$$S(C^{-1})\Pi = \Pi S(C)^{\dagger}. \tag{10.21}$$

(b) Show that the map $\Pi_p\colon \mathbb{C}^N \to \mathcal{G}_p$ given by $\Pi_p = S(D_p)\Pi S(D_p)^{\dagger}$ is independent of the element $D_p \in SL(2,\mathbb{C})$ with $D_p(p^*) = p$. Show that for $C \in SL(2,\mathbb{C})$ we have $S(C)\Pi_p S(C)^{\dagger} = \Pi_{C(p)}$.

For a function $\varphi\colon X_m \to \mathbb{C}^N$ we then define $\Pi(\varphi)$ by $\Pi(\varphi)(p) = \Pi_p(\varphi(p))$ so that $\Pi(\varphi) \in \mathcal{H}$. (At least when φ is not "too large" so that the integral corresponding to (10.20) exists.)
(c) Let $(g_k)_{k\le N}$ be the canonical basis of $\mathbb{C}^N$. For a function $f \in \mathcal{S}^4_{\mathbb{R}}$, consider its Fourier transform $\hat{f}$, seen as a function on X_m. Then $\hat{f} g_k$ is a function $X_m \to \mathbb{C}^N$ and the function $\Pi(\hat{f} g_k)$ belongs to $\mathcal{H}$. Prove that the family of quantum fields defined by

$$\psi_k(f) = A(\Pi(\hat{f} g_k)),$$

(where A in the right-hand side corresponds to the boson Fock space of $\mathcal{H}$) satisfies (10.8). Hint: Recall that an operator B on $\mathbb{C}^N$ satisfies $B(g_k) = \sum_{\ell\le N} g_\ell B_{\ell,k}$. Prove that the functions $\varphi_k(f) := \Pi(\hat{f} g_k)$ satisfy

$$U(c,C)(\varphi_k(f)) = \sum_{\ell\le N} S(C^{-1})^*_{k,\ell}\varphi_\ell(V(c,C)f) \tag{10.22}$$

and prove then that this relation implies (10.8).

10.5 Explicit Formulas

Let us pause now and learn how to write the quantum field given by (10.12) and (10.13) the way a physicist would do. The first step is to replace the space $\mathcal{F}$ of functions from $SL(2,\mathbb{C})$ to $\mathcal{V}$ by the more usual space $\mathcal{H} = L^2(X_m, \mathcal{V}, d\lambda_m)$, using the map $T : \mathcal{F} \to \mathcal{H}$ given by $T(f)(p) = f(D_p)$. The operators $\psi_k^+(x)$ will then operate on the boson Fock space of $\mathcal{H} = L^2(X_m, \mathcal{V}, d\lambda_m)$. The state space and its boson Fock space are not the same as in the previous section. Still we lighten notation by denoting again $A(f)$ the annihilation operator on the boson Fock space corresponding to the element f of the state space, as the notation is unambiguous.

Using that $D_p(p^*) = p$, formula (10.12) then becomes

$$\psi_k^+(x) = A(F_k(x)) \tag{10.23}$$

where the functions $F_k(x)$ of $p \in X_m$ are of the type

$$F_k(x)(p) = \exp(\mathrm{i}(x,p)/\hbar)\Pi S(D_p)^\dagger(e_k) = \exp(\mathrm{i}(x,p)/\hbar)\sum_{\ell \le N} S(D_p)^*_{k,\ell}\Pi(e_\ell). \tag{10.24}$$

Throughout the rest of this section

$$n \textit{ is the dimension of } \mathcal{V} \textit{ and } n \le N,$$

and $(f_s)_{s \le n}$ is an orthonormal basis[16] of $\mathcal{V}$. Let us then define an operator-valued distribution $a(\boldsymbol{p},s)$ just as in the case of the scalar free field, by the formula, holding for any test function ξ,

$$\int \frac{d^3\boldsymbol{p}}{\sqrt{2\omega_{\boldsymbol{p}}(2\pi\hbar)^3}} a(\boldsymbol{p},s)\xi(\boldsymbol{p}) = A(\xi_s) \tag{10.25}$$

where $\xi_s \in L^2(X_m, \mathcal{V}, d\lambda_m)$ is given by $\xi_s(p) = \xi(\boldsymbol{p})^* f_s$ for $p = (\omega_{\boldsymbol{p}}, \boldsymbol{p})$.[17] The commutation relations between these operators are given by

$$[a(\boldsymbol{p},s), a^\dagger(\boldsymbol{p}',s')] = \delta_s^{s'}(2\pi\hbar)^3\delta^{(3)}(\boldsymbol{p}-\boldsymbol{p}')1, \tag{10.26}$$

with all other commutators being zero.

For now it serves no purpose to specify which basis of $\mathcal{V}$ we use, but in practice one should strive to choose a basis of $\mathcal{V}$ which makes physical sense. For example, in the case where $m > 0$, the irreducible representation V is unitarily equivalent to one of the π_j, (with $n = j+1$) and there is an orthonormal basis of $\mathcal{V}$ which corresponds to particles having "given spin in the z direction", and such a choice is very natural for a physicist. Then each operator $a(\boldsymbol{p},s)$ can be interpreted as "destroying a particle of momentum $\boldsymbol{p}$ and given spin in the z direction".

In our next result we keep the notation of the previous section and in particular W is the map of Proposition 10.4.2.

16 Thus, s denotes an integer in the rest of this chapter.

17 The occurrence of ξ^* rather than ξ in the definition of ξ_s is motivated by the fact that A is anti-linear, not linear.

Proposition 10.5.1 *Consider $u(\boldsymbol{p},s) = (u(\boldsymbol{p},s)_k)_{k\le N} \in \mathbb{C}^N$ given by the formula*

$$u(\boldsymbol{p},s) = S(D_p)W(f_s). \tag{10.27}$$

Then the annihilation part of the quantum field ψ_k is given by the formula

$$\psi_k^+(x) = \sum_{s\le n}\int \frac{\mathrm{d}^3\boldsymbol{p}}{\sqrt{2\omega_{\boldsymbol{p}}(2\pi\hbar)^3}} \exp(-\mathrm{i}(x,p)/\hbar)u(\boldsymbol{p},s)_k a(\boldsymbol{p},s). \tag{10.28}$$

To understand this formula we simply use that for $\xi \in \mathcal{S}^4$ we have $\psi_k^+(\xi) = \int \mathrm{d}^4x\psi_k^+(x)$ $\xi(x)$ and we formally compute this integral using the expression (10.28). We find

$$\widehat{\xi}(p)\sum_{s\le n}\int \frac{\mathrm{d}^3\boldsymbol{p}}{\sqrt{2\omega_{\boldsymbol{p}}(2\pi\hbar)^3}} u(\boldsymbol{p},s)_k a(\boldsymbol{p},s),$$

and using the definition (10.25) this is $A(\Xi)$ where $\Xi(p) = \widehat{\xi}(p)^*\sum_{s\le n} u(\boldsymbol{p},s)_k^* f_s'$. The formula (10.27) is remarkably simple. We should think of the map W as a way to identify $\mathcal{V}$ and the range $\mathcal{G}$ of W. This identification provides $\mathcal{G}$ with a natural inner product,[18] and for this inner product the vectors $W(f_s)$ are *simply an orthonormal basis of $\mathcal{G}$.*

Proof Let us decompose $\Pi(e_\ell)$ in the basis (f_s), i.e. $\Pi(e_\ell) = \sum_{s\le n} f_{s,\ell} f_s$, so that

$$W(f_s)_\ell = (e_\ell, W(f_s)) = (\Pi(e_\ell), f_s) = f_{s,\ell}^*$$

and

$$\sum_{\ell\le N} S(D_p)_{k,\ell}^*\Pi(e_\ell) = \sum_{s\le n}\Big(\sum_{\ell\le N} S(D_p)_{k,\ell}^* f_{s,\ell}\Big) f_s.$$

We then rewrite (10.24) as

$$F_k(x)(p) = \exp(\mathrm{i}(x,p)/\hbar)\sum_{s\le n} u(\boldsymbol{p},s)_k^* f_s \tag{10.29}$$

where

$$u(\boldsymbol{p},s)_k := \sum_{\ell\le N} S(D_p)_{k,\ell} f_{s,\ell}^* = \sum_{\ell\le N} S(D_p)_{k,\ell} W(f_s)_\ell, \tag{10.30}$$

i.e. $u(\boldsymbol{p},s) = (u(\boldsymbol{p},s)_k)_{k\le N}$ satisfies (10.27). Then (10.28) follows from (10.25) and (10.29). □

We may rewrite (10.28) in a compact way as

$$\psi^+(x) = \sum_{s\le n}\int \frac{\mathrm{d}^3\boldsymbol{p}}{\sqrt{2\omega_{\boldsymbol{p}}(2\pi\hbar)^3}} \exp(-\mathrm{i}(x,p)/\hbar)u(\boldsymbol{p},s)a(\boldsymbol{p},s). \tag{10.31}$$

Since W is determined only up to a multiplicative constant this is also the case for the field (10.31). One may then decide at a later stage on a reasonable condition to impose to

[18] This inner product has the property that when $\mathcal{G}$ is provided with it, the restriction of S to $\mathcal{G}$ and G_{p^*} is unitary and irreducible. It is shown in Proposition D.6.5 that this inner product is unique up to a multiplicative constant.

normalize the field, i.e. to fix this constant. One may also choose this multiplicative constant in order to achieve a given dimensionality of the field. Such dimensionality considerations motivated the factor $\sqrt{c}$ in the denominator of the formula (5.30). For similar reasons we may get other factors, e.g. in the case of the Dirac field to be investigated in Section 10.19.

Let us now summarize the method to construct annihilation fields that satisfy the Lorentz covariance condition (10.6).

- We start with a given mass $m \geq 0$, and a specific point p^* of X_m. Let G_{p^*} be its little group with respect to the natural action of $SL(2,\mathbb{C})$ on X_m. We are given an irreducible representation V of G_{p^*} in a finite-dimensional Hilbert space $\mathcal{V}$. We consider the unitary representation $U(c,C)$ of $\mathcal{P}^*$ on the space $\mathcal{H} = L^2(X_m, \mathcal{V}, \mathrm{d}\lambda_m)$ provided by Theorem 9.4.2. This representation determines "the particle", the type of the quantum of the field.
- We consider an irreducible[19] representation S of $SL(2,\mathbb{C})$ on $\mathbb{C}^N$. Its restriction to the little group G_{p^*} need not be irreducible. We look for an invariant subspace $\mathcal{G} \subset \mathbb{C}^N$ on which this restriction is irreducible and for a one-to-one linear map W from $\mathcal{V}$ to $\mathcal{G}$ which satisfies $V(C) = W^{-1}S(C)W$ for each $C \in G_{p^*}$.[20]
- The annihilation field is then given by the formula (10.31) for $u(\boldsymbol{p},s) = S(D_p)W(f_s)$ as prescribed by (10.27).

Exercise 10.5.2 Prove that every component of the quantum field constructed above satisfies the Klein-Gordon equation in the sense of distributions. Hint: Copy the argument of Proposition 6.1.3.

Exercise 10.5.3 Try to rewrite the arguments leading to (10.27) the way a physicist might write them. That is, first study the transformation properties of the operators $a(\boldsymbol{p},s)$ in the spirit of (5.43), and then look for annihilation fields of the type

$$\psi_k^+(x) = \sum_{s \leq n} \int \frac{\mathrm{d}^3\boldsymbol{p}}{\sqrt{2\omega_{\boldsymbol{p}}(2\pi\hbar)^3}} u(\boldsymbol{p},x,s)_k a(\boldsymbol{p},s).$$

Express the transformation properties of the functions $u(\boldsymbol{p},x,s)_k$ implied by (10.6), and reach your own version of (10.27). Warning: This is a small research project. Look into Duncan's textbook [24] if this is too hard.

10.6 Creation Part of the Fields

We now need to consider the creation part $\psi_k^-(x)$ of the field. We will assume that it is of the type $B^\dagger(f_k(x))$ where $B^\dagger$ is the creation operator on the boson Fock space of the anti-particle. We assume from the start that the anti-particle has the same mass as the particle. Since the little group depends only on whether the mass is > 0 or 0, it is the same for the anti-particle and the particle. The anti-particle is described again by a representation given

[19] As we already mentioned this condition is not essential and we impose it for simplicity.
[20] Since V is irreducible, such a W can exist only if the restriction of S to G_{p^*} and $\mathcal{G}$ is irreducible!

by Theorem 9.4.2, and we denote by $\mathcal{V}'$ the corresponding finite-dimensional space and V' the corresponding representation of the little group. (This will also cover the case of a neutral particle which is its own anti-particle.) We assume from the start that $\mathcal{V}'$ has the same dimension n as $\mathcal{V}$. It simplifies notation and does not decrease generality to then take $\mathcal{V}' = \mathcal{V}$. In the case $m > 0$ there is a unique choice for V, namely the representation π_j of Definition 8.2.3 for $j + 1 = n$. Then $V' = V$. However, in the case of massless particles there are several possible choices for V: if $n = 1$, all the representations $\widehat{\pi}_j$ of Definition 9.6.2 are of dimension one. In that case one may have $V' \neq V$.

A difference between creation and annihilation is that the map $B^\dagger$ is linear in its argument rather than anti-linear, so that in (10.17) the numbers $S(C^{-1})^*_{k,\ell}$ have to be replaced by their complex conjugates. This brings the need for the following.

Definition 10.6.1 Given a representation S of a group G on $\mathbb{C}^N$, we denote by S^* the representation of G on $\mathbb{C}^N$ such that for $C \in G$ the matrix of $S^*(C)$ is the conjugate $S(C)^*$ of the matrix of $S(C)$, i.e. each entry in the matrix is replaced by its complex conjugate.

We next summarize the situation in the case of the creation part of the fields. The proofs are identical to the case of the annihilation part of the fields.

Proposition 10.6.2 *The creation fields satisfying the Lorentz covariance condition (10.6) are obtained as follows. There is a one-to-one linear map W' from $\mathcal{V}$ to a subspace $\mathcal{G}'$ of $\mathbb{C}^N$ which is invariant under the restriction of the representation S^* to G_{p^*}, and such that*

$$\forall C \in G_{p^*}\ ;\ V'(C) = W'^{-1} S^*(C) W', \tag{10.32}$$

and the creation part of the quantum fields are given by

$$\psi_k^-(x) = \sum_{s \le n} \int \frac{\mathrm{d}^3 \boldsymbol{p}}{\sqrt{2\omega_{\boldsymbol{p}}(2\pi\hbar)^3}} \exp(\mathrm{i}(x,p)/\hbar) v(\boldsymbol{p},s)_k b^\dagger(\boldsymbol{p},s) \tag{10.33}$$

where

$$v(\boldsymbol{p},s)_k := \sum_{\ell \le N} S(D_p)_{k,\ell} W'(f_s)^*_\ell. \tag{10.34}$$

Here $(f_s)_{s \le n}$ is an orthogonal basis for $\mathcal{V}$ and $b^\dagger(\boldsymbol{p},s)$ is defined similarly to $a(\boldsymbol{p},s)$. In a compact form we may write (10.33) as

$$\psi^-(x) = \sum_{s \le n} \int \frac{\mathrm{d}^3 \boldsymbol{p}}{\sqrt{2\omega_{\boldsymbol{p}}(2\pi\hbar)^3}} \exp(\mathrm{i}(x,p)/\hbar) v(\boldsymbol{p},s) b^\dagger(\boldsymbol{p},s), \tag{10.35}$$

and (10.34) as

$$v(\boldsymbol{p},s) = S(D_p) W'(f_s)^*. \tag{10.36}$$

Here $W'(f_s)^*$ denotes the *column* vector with components $W'(f_s)^*_\ell$.

Exercise 10.6.3 The purpose of the present (very challenging) exercise is to prove that when $m > 0$ (and provided one has chosen an appropriate basis) there is a very tight relation between the creation and annihilation parts of the fields. Since $m > 0$ the only possibility is $V = V' = \pi_n$ for some $n \geq 0$. We choose in $\mathcal{V}$ the special basis of Proposition 8.2.8 (which is the numbered from 0 to $j = n - 1$ rather than from 1 to $n = j + 1$ as was done in the present section). You are going to prove that for some number λ, and all $0 \leq s \leq j$ it holds that

$$v(\boldsymbol{p}, s) = \lambda(-1)^s u(\boldsymbol{p}, j - s). \tag{10.37}$$

(a) Study the results of Appendix D to convince yourself that since we assume that S is irreducible there is at most one single subspace $\mathcal{G}$ of $\mathbb{C}^N$ such that the restriction of S to $\mathcal{G}$ and $SU(2)$ is equivalent to π_n.
(b) Recall the matrix J of Lemma 8.1.4. Thinking of (10.32) as an identity between matrices, take the complex conjugate of this equality to obtain the relation

$$\forall C \in G = SU(2)\,,\; V(C) = V(J)Z^{-1}S(C)ZV(J^{-1}), \tag{10.38}$$

where $Z := W'^*$ (in the sense that the matrices of Z and W' are conjugate).
(c) Deduce from Proposition 10.4.2 that for some $\lambda \in \mathbb{C}$ we have $ZV(J^{-1}) = \lambda W$, and deduce from (10.36) that $v(\boldsymbol{p}, s) = \lambda S(D_p)WV(J)(f_s)$. Conclude using (8.12).

10.7 Microcausality

The massive scalar free field φ which we constructed in Chapter 5 is self-adjoint, so that the values $\varphi(x)$ correspond to observables. We used this fact to argue that preservation of causality implies the microcausality condition (5.10). We cannot use this argument for quantum fields that are not self-adjoint, since they do not correspond to observables. The necessity of microcausality is then less obvious, but let us try to explain it in a few words. As will be explained at the beginning of Chapter 13 the main purpose of quantum fields is to construct interaction Hamiltonian densities out of products of such fields. As will be detailed in Section 13.5, in order to obtain a sensible theory these Hamiltonian densities must commute at causally separated points. One method to obtain this property is to impose the following condition, which will again be called microcausality.[21] For any $k, k' \leq N$,

$$(x - y)^2 < 0 \Rightarrow [\psi_k(x), \psi_{k'}(y)] = 0\,;\; [\psi_k^\dagger(x), \psi_{k'}^\dagger(y)] = 0\,;\; [\psi_k(x), \psi_{k'}^\dagger(y)] = 0. \tag{10.39}$$

Quantum fields which do not satisfy this condition (or the appropriate substitutes which we will meet later) *are simply not physically interesting*.

Microcausality is known to hold in great generality for massive particles of integer spin and S irreducible (where we previously explained that both the creation and the annihilation parts of the fields ψ_k are unique up to a multiplicative constant). Unfortunately the proof

21 The second condition follows automatically from the first one.

is difficult and beyond the scope of this book.[22] Thus we will give a separate argument in each of the examples we shall present. A part of the computations is common to all these examples. It is presented here. The reader eager for concrete examples of free fields should move now to the next section, coming back to the present results only when required.

The following identities are a straightforward consequence of the commutation relations (10.26) (although it takes a bit of patience to get the signs right). We formulate a first version in the case where the particle is its own anti-particle, where the fields ψ^+ and ψ^- are given by (10.28) and (10.33) for $b = a$. Let us observe that even in this case, there is no obvious reason why the fields ψ_k should be self-adjoint.

Theorem 10.7.1 *When the particle is its own anti-particle the quantum fields $\psi_k(x) = \psi_k^+(x) + \psi_k^-(x)$ satisfy*

$$[\psi_k(x), \psi_{k'}(y)] = (G_{k,k'}(y-x) - G_{k',k}(x-y))1 \tag{10.40}$$

where for $z \in \mathbb{R}^{1,3}$ we define

$$G_{k,k'}(z) = \int \frac{\mathrm{d}^3\boldsymbol{p}}{2\omega_{\boldsymbol{p}}(2\pi\hbar)^3} \exp(\mathrm{i}(z,p)/\hbar) \sum_{s\le n} u(\boldsymbol{p},s)_k v(\boldsymbol{p},s)_{k'}. \tag{10.41}$$

Moreover,

$$[\psi_k(x), \psi_{k'}^\dagger(y)] = (G'_{k,k'}(y-x) - G''_{k,k'}(x-y))1, \tag{10.42}$$

where

$$G'_{k,k'}(z) = \int \frac{\mathrm{d}^3\boldsymbol{p}}{2\omega_{\boldsymbol{p}}(2\pi\hbar)^3} \exp(\mathrm{i}(z,p)/\hbar) \sum_{s\le n} u(\boldsymbol{p},s)_k u(\boldsymbol{p},s)^*_{k'}, \tag{10.43}$$

and

$$G''_{k,k'}(z) = \int \frac{\mathrm{d}^3\boldsymbol{p}}{2\omega_{\boldsymbol{p}}(2\pi\hbar)^3} \exp(\mathrm{i}(z,p)/\hbar) \sum_{s\le n} v(\boldsymbol{p},s)_k v(\boldsymbol{p},s)^*_{k'}. \tag{10.44}$$

Exercise 10.7.2 Prove the reassuring fact that the quantities

$$\sum_{s\le n} u(\boldsymbol{p},s)_k v(\boldsymbol{p},s)_{k'}\ ;\ \sum_{s\le n} u(\boldsymbol{p},s)_k u(\boldsymbol{p},s)^*_{k'}\ ;\ \sum_{s\le n} v(\boldsymbol{p},s)_k v(\boldsymbol{p},s)^*_{k'}$$

are independent of the choice of the element $D_p \in SL(2,\mathbb{C})$ such that $D_p(p^*) = p$ and of the choice of the orthogonal basis in $\mathcal{V}$.

Corollary 10.7.3 *Assume that $G_{k,k'}(z) = G_{k',k}(z)$ and that this function* is even *in the domain $z^2 < 0$. Assume also that $G'_{k,k'}(z) = G''_{k,k'}(z)$ and is* even *in the domain $z^2 < 0$. Then the quantum fields (10.28) satisfy the microcausality condition (10.39).*

22 Among the many obstacles which a mathematician faces while trying to read a textbook of physics, one is particularly unexpected: Arguments presented as a "proof" but which do not even approach the main difficulty, as in pages 182–183 in Duncan's textbook [24] concerning the present point.

Generally speaking parity arguments about the functions $G_{k,k'}, G'_{k,k'}, G''_{k,k'}$ such as in the previous corollary will be of constant use. To prove a parity property of a function of this type we will use the following.

Lemma 10.7.4 *If the function $f(\boldsymbol{p})$ is a homogeneous polynomial in the components of $\boldsymbol{p}$ then on the domain $z^2 < 0$ the function*

$$\int \frac{\mathrm{d}^3 \boldsymbol{p}}{2\omega_{\boldsymbol{p}}(2\pi\hbar)^3} \exp(\mathrm{i}(z,p)/\hbar) f(\boldsymbol{p})$$

is even if f is of even degree and is odd if f is of odd degree.

Proof The function $I(z) = \int \mathrm{d}^3 \boldsymbol{p} \exp(\mathrm{i}(z,p)/\hbar)/(2\omega_{\boldsymbol{p}}(2\pi\hbar)^3)$ of (5.13) is even on this domain, so that its partial derivatives of even order are even and its partial derivatives of odd order are odd. □

We must face the fact that we must compute the three different functions of $\boldsymbol{p}$ occurring in the statement of Exercise 10.7.2. This is not fun.

When the particle is different from its anti-particle we have the following version of Theorem 10.7.1, which is obtained again by a straightforward computation.

Theorem 10.7.5 *When the particle is distinct from its anti-particle the quantum field $\psi_k(x) = \psi_k^+(x) + \psi_k^-(x)$ satisfies $[\psi_k(x), \psi_{k'}(y)] = 0$ and $[\psi_k^\dagger(x), \psi_{k'}^\dagger(y)] = 0$ while the formulas (10.42) to (10.44) remain true.*

Therefore in that case to check microcausality it is sufficient to prove that for $k, k' < N$,

$$(x-y)^2 < 0 \Rightarrow [\psi_k(x), \psi_{k'}^\dagger(y)] = 0. \tag{10.45}$$

Corollary 10.7.6 *Assume that the particle is distinct from its anti-particle. Assume that $G'_{k,k'}(z) = G''_{k,k'}(z)$ and is* even *in the domain $z^2 < 0$. Then the quantum fields (10.28) satisfy the microcausality condition (10.39).*

The program of the rest of the chapter can be roughly described as follows. We will try to describe the quantum fields with a given type of quanta that satisfy the microcausality condition, or try to show that such fields do not exist. (This is a rough description indeed, and the precise path we shall follow is laid out very carefully in the next section.) In such a construction there are always two versions, one where the particle is its own anti-particle and one where it differs from its anti-particle. Assume that we have constructed a quantum field for which the particle is its own anti-particle and ψ^- is given by (10.35) where $b^\dagger = a^\dagger$, and that we have proved through Corollary 10.7.3 that this field satisfies the microcausality condition. Consider now "the same" quantum field where the particle differs from its anti-particle. This new quantum field also satisfies the microcausality condition because there are fewer conditions in Corollary 10.7.6 than in 10.7.3. So, when trying to construct quantum fields that satisfy the micro-causality condition, it makes sense to attempt first the most difficult construction, where the particle is its own anti-particle. On the other hand, if you try to show that such a field does not exist, then it is better to prove that this is the case even when the anti-particle differs from the particle, because then the other version of field, where the particle is its own anti-particle will not satisfy the microcausality condition either.

10.8 Road Map II

In the rest of the chapter we will mostly work out a large number of examples, trying to construct quantum fields, and discovering in a natural manner the most important of them.

We gave a step-by-step description of how to construct Lorentz covariant quantum fields. In this description we assumed that we had already decided what are $\mathcal{V}$ and the representation V, or in other words that we look for a quantum field with a specific type of quanta. (Let us recall the terminology here. The pair $(\mathcal{V}, V)$ determines a representation of $\mathcal{P}^*$, and such a representation is associated with a particle. Particles of this type are called the quanta of the field.)

In practice, however, we will proceed differently. We will *start* with the representation S, *then* look for an invariant subspace of the restriction of S to G_{p^*}, and use it to find suitable $\mathcal{V}$ and V, determining in this manner what the quanta of the field should be. It seems therefore necessary to restate the method.

- We decide the mass $m \geq 0$ of the quanta. This determines p^* and its little group G_{p^*}.
- We consider an irreducible representation S of $SL(2,\mathbb{C})$ on $\mathbb{C}^N$. We look for a subspace $\mathcal{G}$ of $\mathbb{C}^N$ such that the restriction of S to $\mathcal{G}$ and G_{p^*} is irreducible.
- We look for a Hilbert space $\mathcal{V}$ of the same dimension as $\mathcal{G}$ and a one-to-one map W from $\mathcal{V}$ to $\mathcal{G}$ such that the representation V of G_{p^*} on $\mathcal{V}$ given by $V(C) = W^{-1}S(C)W$ for each $C \in G_{p^*}$ is unitary and irreducible. The map W is completely determined up to a multiplicative constant.
- Theorem 9.4.2 then provides the representation $U(c,C)$ of $\mathcal{P}^*$ on the space $\mathcal{H} = L^2(X_m, \mathcal{V}, \mathrm{d}\lambda_m)$. This representation determines the quanta of the field.
- The annihilation part of the quantum fields are then given by the formula (10.31) where $u(\boldsymbol{p},s) = S(D_p)W(f_s)$ as prescribed by (10.27).
- We look for a subspace $\mathcal{G}'$ of $\mathbb{C}^N$ with $\dim\mathcal{G} = \dim\mathcal{G}'$ such that the restriction of S^* to $\mathcal{G}'$ and G_{p^*} is irreducible, and for a one-to-one map W' from $\mathcal{V}$ to $\mathcal{G}'$ such that the representation V' of G_{p^*} on $\mathcal{V}$ given by $V'(C) = W'^{-1}S^*(C)W'$ is unitary and irreducible.
- Theorem 9.4.2 *used for the representation* V' then provides a representation $U'(c,C)$ of $\mathcal{P}^*$ on the space $\mathcal{H} = L^2(X_m, \mathcal{V}, \mathrm{d}\lambda_m)$. *This representation must correspond to the anti-particle of the quanta of the field.*[23]
- The creation part of the quantum fields are then given by the formula (10.33) where $v(\boldsymbol{p},s) = S(D_p)W'(f_s)^*$ as prescribed by (10.36).
- Finally we check that the microcausality condition holds, for otherwise the field is not physically interesting.

In all our constructions we will follow this exact pattern, trying first to construct quantum fields for which the quanta of the fields are their own anti-particle, since as explained above this is the most difficult construction to perform. We will investigate the case where S is

[23] In the case of massless Weyl spinors, this will impose a strong requirement: The anti-particle must be of opposite helicity from the particle.

the trivial representation ($N = 1$), the case where S is the $(1,1)$ representation in disguise[24] ($N = 4$) and the case where S is the $(1,0)$ representation ($N = 2$).

Mathematically, the most interesting cases are probably those of the massive and massless Weyl spinors, and the reader may like to concentrate on these at first reading.

10.9 The Simplest Case ($N = 1$)

The first example is almost trivial. It is the case where $N = 1$ and where S is the trivial representation, $S(C) \equiv 1$. Then $n = 1$, and the indices s disappear. The only possible choices are $\mathcal{G} = \mathcal{G}' = \mathcal{V} = \mathbb{C}$. Then V and V' are trivial.

How do we choose the map W? As always the maps W and W' are determined up to a multiplicative constant. Denoting by e_1 the basis vector of $\mathbb{C}$ and f_1 the basis vector of $\mathcal{V}$, we must have $W(f_1) = \lambda e_1$, and similarly $W'(f_1) = \tau e_1$. Thus, assuming that the particle is its own anti-particle,

$$\psi^+(x) = \lambda \int \frac{\mathrm{d}^3 \boldsymbol{p}}{\sqrt{2\omega_{\boldsymbol{p}}(2\pi\hbar)^3}} \exp(-\mathrm{i}(x,p)/\hbar) a(\boldsymbol{p}) \tag{10.46}$$

and

$$\psi^-(x) = \tau \int \frac{\mathrm{d}^3 \boldsymbol{p}}{\sqrt{2\omega_{\boldsymbol{p}}(2\pi\hbar)^3}} \exp(\mathrm{i}(x,p)/\hbar) a^\dagger(\boldsymbol{p}). \tag{10.47}$$

Thus we exactly recover the formula (5.9), which is not surprising as we derived that formula by a special case of the present arguments. Recalling the function I of (5.13) and the functions of (10.43) and (10.44) we find

$$G'_{1,1}(z) = |\lambda|^2 I(z)\,;\; G''_{1,1}(z) = |\tau|^2 I(z).$$

Consequently since I is an even function we obtain from (10.42) that $[\psi_1(x), \psi_1^\dagger(y)] = 0$ for $(x-y)^2 < 0$ if and only if $|\lambda| = |\tau|$. In that case we can write $\lambda = \alpha \exp(\mathrm{i}\theta)$, $\tau = \alpha \exp(-\mathrm{i}\theta)$ where $\alpha \in \mathbb{C}$, $\theta \in \mathbb{R}$. The parameter θ can be eliminated by changing $a(\boldsymbol{p})$ into $\exp(\mathrm{i}\theta) a(\boldsymbol{p})$. The multiplicative factor α should always be real if we want ψ^+ to be the adjoint of ψ^-. Modulo the multiplicative constant α we just find the field (5.30).

In this case nothing fruitful came out of the parameters λ and τ. This is a general fact and we shall no longer consider them. Since W and W' are determined up to a multiplicative constant, we will make a convenient choice for these constants without further comment.

10.10 A Very Simple Case ($N = 4$)

We next turn to another very simple situation. In general we work with $SL(2,\mathbb{C})$ rather than $SO^\uparrow(1,3)$ so as to deal only with true representations; but in the present section and the next we do not need $SL(2,\mathbb{C})$ and we can work using only $SO^\uparrow(1,3)$. We consider the simplest non-trivial representation S of $SO^\uparrow(1,3)$, the "defining representation" as a

24 See Definition 8.3.2 for the (j,ℓ) representation of $SL(2,\mathbb{C})$.

group of matrices operating on $\mathbb{C}^4$. Let us note right away that in this case the entries of the matrices are real, so that $S = S^*$.[25]

Our first problem in studying this situation is a conflict of notation.[26] In the general setting of a representation $S(C)$ on $\mathbb{C}^N$ we label the coordinates from 1 to N and we denote a matrix $S(C)$ by $S(C)_{k,\ell}$ where k indicates the row and ℓ indicates the column. On the other hand to study $SO^{\uparrow}(1,3)$ we number the coordinates from 0 to 3 and denote matrices differently: For a matrix $C \in SO^{\uparrow}(1,3)$ we denote its elements by $C^{\mu}_{\ \nu}$ where μ indicates the row and ν indicates the column, the notations we will use in this section. To be consistent with the fact that the coordinates of a point x in $\mathbb{R}^{1,3}$ are denoted by x^{ν} we will give upper indices to the field components. To construct the field:

- We choose $m > 0$, so $p^* = (mc,0,0,0)$ and the little group G_{p^*} is the orthogonal group $SO(3)$.
- The restriction of S to $SO(3)$ has two invariant subspaces, namely the spans of e_0 and of e_1, e_2, e_3 respectively. We choose for $\mathcal{G}$ the span of e_0.[27]
- Thus here $n = 1$ and the representation of $SO(3)$ on $\mathcal{G}$ is the trivial one. We must have $\mathcal{G}' = \mathcal{G}$ as this is the only one-dimensional space invariant under the action of $SO(3)$. Denoting by f_1 the basis vector of $\mathcal{V} = \mathbb{C}$ we simply choose $W(f_1) = W'(f_1) = e_0$. (We may also, as we will often do, simply identify $\mathcal{V}$ with the span of e_0.) The indices s disappear and there is only one creation operator $a^{\dagger}(\boldsymbol{p})$, without indices s and one annihilation operator $a(\boldsymbol{p})$.

In the present setting D_p denotes the unique pure boost which sends p^* to p. According to the first part of (4.25) and since $D_p = B_{p/mc}$ one has

$$(D_p)^{\mu}_{\ 0} = \frac{p^{\mu}}{mc}. \tag{10.48}$$

Thus $u(\boldsymbol{p})^{\mu} = p^{\mu}/mc = v(\boldsymbol{p})^{\mu}$ and then

$$\psi^{-\mu}(x) := (\psi^{-})^{\mu} = \frac{1}{mc}\int \frac{\mathrm{d}^3\boldsymbol{p}}{\sqrt{2\omega_{\boldsymbol{p}}(2\pi\hbar)^3}} \exp(\mathrm{i}(x,p)/\hbar)p^{\mu}a^{\dagger}(\boldsymbol{p}), \tag{10.49}$$

and a similar formula for $\psi^{+\mu}$. In particular $\psi^{-\mu}$ is proportional to $\partial^{\mu}\varphi^{-}(x) = \partial\varphi^{-}(x)/\partial x_{\mu}$ where $\varphi^{-}(x)$ is the creation part of the field (5.30), and thus we are not doing anything really new here.

10.11 The Massive Vector Field ($N = 4$)

As in the previous section we consider the case where S is the defining representation of $SO^{\uparrow}(1,3)$ on $\mathbb{C}^4$, but where now $\mathcal{G}$ is the span of e_1, e_2, e_3. The restriction of S to $\mathcal{G}$ and $G_{p^*} = SO(3)$ is the representation of $SO(3)$ by matrices acting on $\mathbb{C}^3$ in the natural manner.

[25] An alternative route here is to stick with $SL(2,\mathbb{C})$ and to use the representation κ of this group. We will eventually see that κ is equivalent to the representation $(1,1)$ of $SL(2,\mathbb{C})$ so that we are studying this representation in disguise.

[26] The reader may wonder why we choose as a first example a representation S with this extra difficulty of notation. This is simply because this is the simplest choice before one has learned what fermions are, as we will soon do.

[27] In the next section we will consider the case where $\mathcal{G}$ is the span of e_1, e_2, e_3.

We simplify our mental picture by taking for $\mathcal{V} = \mathcal{G}$ the span of e_1, e_2, e_3 and W the identity.[28] To describe the corresponding representation of $\mathcal{P}$ (since here there is no need to consider $\mathcal{P}^*$), we consider the space $L^2 = L^2(X_m, \mathbb{C}^3, \mathrm{d}\lambda_m)$. For $p \in X_m$ we denote by D_p the unique pure boost which sends p^* to p. Then the unitary representation of $\mathcal{P}$ is given by the following formula

$$U(a,A)(\varphi)(p) = \exp(\mathrm{i}(a,p)/\hbar)(D_p^{-1} A D_{A^{-1}(p)})(\varphi(A^{-1}(p))). \tag{10.50}$$

We know from Section 9.7 that it describes a spin-one massive particle. We perform the construction in the case where it is its own anti-particle.

Let us turn to the creation part of the quantum fields. The only invariant subspace for $S^* = S$ of $\mathbb{C}^4$ of dimension 3 is $\mathcal{G}' = \mathcal{G}$. Choosing as usual the multiplicative constant in W' to our liking, the only choice for W' is then the identity.

The last task is to choose a basis of $\mathcal{V}$ to get explicit formulas. A physicist would probably like to choose a basis that describes the spin in the z direction, but as mathematicians we will accept the simpler basis e_1, e_2, e_3. The formulas for the quantum field $\psi^\mu(x) = \psi^{-\mu}(x) + \psi^{+\mu}(x) := (\psi^-)^\mu + (\psi^+)^\mu$ then take the very simple form

$$\psi^{+\mu}(x) = \sum_{1\le s\le 3} \int \frac{\mathrm{d}^3\boldsymbol{p}}{\sqrt{2c\omega_{\boldsymbol{p}}}(2\pi\hbar)^3} \exp(-\mathrm{i}(x,p)/\hbar)(D_p)^\mu_s a(\boldsymbol{p},s) \tag{10.51}$$

$$\psi^{-\mu}(x) = \sum_{1\le s\le 3} \int \frac{\mathrm{d}^3\boldsymbol{p}}{\sqrt{2c\omega_{\boldsymbol{p}}}(2\pi\hbar)^3} \exp(\mathrm{i}(x,p)/\hbar)(D_p)^\mu_s a^\dagger(\boldsymbol{p},s). \tag{10.52}$$

As in the case of the free field, a factor $1/\sqrt{c}$ has been inserted to give the fields the dimensionality consistent with the forthcoming Lagrangian density (10.57).

Theorem 10.11.1 *The quantum fields $\psi^\mu(x)$ are self-adjoint and satisfy the micro-causality condition*

$$(x-y)^2 < 0 \Rightarrow [\psi^\mu(x), \psi^\nu(y)] = 0. \tag{10.53}$$

Proof Defining $T^{\mu\nu}(p) = \sum_{1\le s\le 3}(D_p)^\mu_s (D_p)^\nu_s$ and

$$G^{\mu\nu}(z) = \int \frac{\mathrm{d}^3\boldsymbol{p}}{2\omega_{\boldsymbol{p}}(2\pi\hbar)^3} \exp(\mathrm{i}(z,p)/\hbar) T^{\mu\nu}(p), \tag{10.54}$$

we use (10.40) or proceed directly from the relations (10.27) to obtain

$$c[\psi^\mu(x), \psi^\nu(y)] = (G^{\mu\nu}(x-y) - G^{\nu\mu}(y-x))1. \tag{10.55}$$

Since the matrix of D_p is symmetric, as in (4.8) the condition that D_p is a Lorentz transformation implies

$$\sum_{0\le s,s'\le 3} \eta^{ss'} (D_p)^\mu_s (D_p)^\nu_{s'} = \eta^{\mu\nu} \tag{10.56}$$

28 Since by the last assertion of Proposition 10.4.2 this is basically the only choice.

and thus, using (10.56) in the second inequality and again (10.48) in the last equality,

$$T^{\mu\nu}(p) = \sum_{1\le s\le 3} (D_p)^\mu_s (D_p)^\nu_s = -\eta^{\mu\nu} + (D_p)^\mu_0 (D_p)^\nu_0 = -\eta^{\mu\nu} + \frac{p^\mu p^\nu}{m^2c^2}.$$

It then follows from Lemma 10.7.4 that $G^{\mu\nu}(z) = G^{\nu\mu}(-z)$, and the conclusion follows from Corollary 10.7.3. □

10.12 The Classical Massive Vector Field

The previous quantum field can be viewed as the quantization of a classical four-dimensional field (A^μ), "the massive vector field". We investigate the massive vector field in the present section which can be skipped at first reading. To follow the forthcoming arguments, the reader should review Sections 6.4 and 6.5. We will not attempt to quantize the massive vector field and to show that the result of this quantization is the field of the present section, as this is a significant undertaking. In this section we use the convention (4.16) for raising and lowering indices in Lorentz space.[29] We first give the Lagrangian density of the massive vector field by fiat (postponing a brief discussion to the end of this section):

$$\mathcal{L} = -\frac{\hbar^2 c^2}{4} F_{\nu\mu} F^{\nu\mu} + \frac{m^2 c^4}{2} A_\nu A^\nu, \tag{10.57}$$

where

$$F_{\nu\mu} = \partial_\nu A_\mu - \partial_\mu A_\nu. \tag{10.58}$$

Here $m \ge 0$ is a parameter. When $m > 0$, it is the last term in (10.57) "which makes the field massive".[30] The Euler-Lagrange equations then take the form

$$\frac{\partial \mathcal{L}}{\partial A_\mu} = \partial_\nu \frac{\partial \mathcal{L}}{\partial(\partial_\nu A_\mu)}, \tag{10.59}$$

where this confusing but standard notation pretends that $\partial_\nu A_\mu$ are independent variables. To make the Euler-Lagrange equations explicit one has to learn how to get the signs right. For example, writing $A_\mu A^\mu = A_0^2 - \sum_{1\le\mu\le 3}(A_\mu)^2$ one obtains

$$\frac{\partial \mathcal{L}}{\partial A_\mu} = m^2 c^4 A^\mu, \tag{10.60}$$

and again with some patience[31]

$$\frac{\partial \mathcal{L}}{\partial(\partial_\nu A_\mu)} = -c^2\hbar^2 F^{\nu\mu}. \tag{10.61}$$

29 So that $F^{\mu,\nu} = F_{\mu,\nu}$ if $\mu = \nu = 0$ or if $\mu,\nu > 0$ and $F^{\mu,\nu} = -F_{\mu,\nu}$ otherwise. We also use the convention that repeated indices in the same term, one of which is a subscript and the other one is a superscript, are summed.

30 Thus the name "massive vector field" is reserved for the case $m > 0$. When $m = 0$, the Lagrangian density becomes that of four-potential of the electromagnetic field, see Section E.2, which now is a good time to read.

31 The reason the factor 1/4 of (10.57) disappears is that the term $\partial_\nu A_\mu$ occurs four times, in the terms $F_{\nu,\mu}, F^{\nu,\mu}, F_{\mu,\nu}$ and $F^{\mu,\nu}$. As for understanding the signs, I see no other way than writing things down in complete detail, in each of the situations which determine the sign of $F^{\mu,\nu}$ compared to $F_{\mu,\nu}$.

This gives the Euler-Lagrange equations $m^2c^2A^\mu + \hbar^2\partial_\nu F^{\nu\mu} = 0$, or, equivalently

$$m^2c^2A_\mu + \hbar^2\partial^\nu F_{\nu\mu} = 0. \tag{10.62}$$

The analysis here differs when $m = 0$ or $m > 0$. When $m = 0$ the above equation reduces to $\partial^\nu F_{\nu\mu} = 0$, which, as you will learn in Section E.2 are the famous Maxwell equations,[32] so we assume $m \neq 0$. Since $F_{\nu\mu} = -F_{\mu\nu}$ we have $\partial^\mu\partial^\nu F_{\nu\mu} = 0$, so that the previous relation implies $m^2c^2\partial^\mu A_\mu = 0$ and thus

$$\partial^\mu A_\mu = 0. \tag{10.63}$$

This implies in turn that $\partial^\nu F_{\nu\mu} = \partial^\nu(\partial_\nu A_\mu - \partial_\mu A_\nu) = \partial^\nu\partial_\nu A_\mu$, and therefore (10.62) becomes

$$m^2c^2A_\mu + \hbar^2\partial^\nu\partial_\nu A_\mu = 0. \tag{10.64}$$

Thus each of the components A_μ satisfies the Klein-Gordon equation, a condition which is certainly expected if we are to succeed in quantizing the field as is shown by Exercise 10.5.2. Still, at this stage it is not obvious at all yet that the massive vector field has any relationship with the quantum field of the previous section. Some relation is brought to light by the following exercise.

Exercise 10.12.1 Prove that in the sense of distributions the quantum field (10.52), (10.51) satisfies the equations of motion corresponding to (10.63) and (10.64).

In order to compute the Hamiltonian density corresponding to the Lagrangian density (10.57), we would like to use the formula (6.46). For this one introduces the variables $\pi_\nu = \partial\mathcal{L}/\partial(\partial_t A_\nu)$ which are conjugate to the variables A_ν in the sense of Lagrangian Mechanics, see (6.33) and (6.45). One tries as in (6.35) to express the variables $\partial_t A_\nu$ using the variables π_ν. An ominous sign of trouble[33] is that $\pi_0 = \partial\mathcal{L}/\partial(\partial_t A_0) = 0$ according to (10.61), and the procedure fails. This is a reflection of the fact that the variables A_ν are not independent (they are related by (10.63)). As we show in Section H.1 it is nonetheless possible to identify A_1, A_2, A_3 "as a proper set of independent variables". This suggests that the variables ψ_1, ψ_2, ψ_3 (where ψ^μ are the components of the quantum field of Section 10.11) and their "conjugate variables" (in the sense of (6.45)) satisfy remarkable commutations relations. This is the content of the next result. This result is made rigorous by methods similar to those of Proposition 6.9.1. The proof offers no new difficulty and is left to the reader.

Theorem 10.12.2 (Equal time commutation relations) *The fields (10.52) and (10.51) satisfy the relation*

$$[\psi_k(ct,\boldsymbol{x}), \pi_k(ct,\boldsymbol{y})] = i\hbar\delta^{(3)}(\boldsymbol{x}-\boldsymbol{y})1$$

where $\psi_k = -\psi^k$ *and* $\pi_k = c\hbar^2(\partial_0\psi_k - \partial_k\psi_0)$. *All the other equal time commutators are zero, e.g.* $[\psi_k(ct,\boldsymbol{x}), \psi_{k'}(ct,\boldsymbol{y})] = 0$, $[\psi_k(ct,\boldsymbol{x}), \pi_{k'}(ct,\boldsymbol{y})] = 0$ *if* $k \neq k'$.

32 When the electromagnetic fields is expressed in terms of a four-potential.

33 In this situation the Legendre transform is singular, and the passage from the Lagrangian to the Hamiltonian formulation is not readily available. A deeper theory, the Dirac–Bergmann constraint analysis is needed here.

Let now briefly discuss the formula (10.57). It is an instance of a quite amazing fact: The correct Lagrangian is often the simplest which makes sense. To follow our discussion, the reader should study Section E.2. To produce equations of motion the Lagrangian should contain terms such as $\partial_0 A_\nu$, which occur naturally as one of the components of the tensor field $G_{\nu\mu} := \partial_\nu A_\mu$ (here tensor field is in the sense explained in Section E.2). The use of the tensor field $F_{\nu\mu}$ is however more natural because in the case of the electromagnetic field $m = 0$ it creates equations of motions that do not involve the (non-physically measurable) four potential A but only the measurable components of the electromagnetic field. To produce a Lorentz invariant term the natural method is then to perform a contraction $F_{\nu\mu}F^{\nu\mu}$ (you can learn more about this general procedure in Section D.12). The term $A^\nu A_\nu$ is also completely natural. In principle one could also have terms such as $A^\nu A^\mu F_{\nu\mu}$, but these would lead to non-linear equations of motion, and again this is not the most natural choice.

10.13 Massive Weyl Spinors, First Attempt ($N = 2$)

In this section we fix $m > 0$, and we investigate the case where S is the defining representation of $SL(2,\mathbb{C})$, $S(C) = C$, so that $N = 2$. Our goal is to construct a quantum field satisfying the microcausality condition based on this representation (sometimes called the massive Weyl spinor).[34] Our first attempt will fail, not because we are clumsy, but because it is doomed from the start, due to a deep fact, called the "spin-statistic theorem". Informally stated this theorem[35] asserts that when the spin of the quanta of the field is not an integer (here it will be 1/2) we must use a new quantization method which we develop in the next section.

The representation S^* where $S(C)$ is replaced by its conjugate will simply be written as $S^*(C) = C^*$. Let us then apply the theory of Sections 10.4 and 10.6 to build a Lorentz covariant quantum field. We investigate the less constraining case where the particle is different from its anti-particle. The restrictions of S and S^* to $SU(2) = G_{p^*}$ are irreducible, so the spaces $\mathcal{G}$ and $\mathcal{G}'$ of Propositions 10.4.2 and 10.6.2 have to be equal to $\mathbb{C}^2$. There is no loss of generality to choose $\mathcal{V} = \mathcal{V}' = \mathbb{C}^2$ and W the identity (so that the particle has spin 1/2). To perform the computations we choose as a basis of $\mathcal{V}$ the vectors

$$f_1 = \begin{pmatrix} 1 \\ 0 \end{pmatrix} \; ; \; f_2 = \begin{pmatrix} 0 \\ 1 \end{pmatrix},$$

so that the components $f_{s,\ell}$ of these vectors are given by $f_{s,\ell} = \delta_s^\ell$. We then have to find a map W' which satisfies (10.32). Recalling the matrix J of Lemma 8.1.4,

$$J = \begin{pmatrix} 0 & 1 \\ -1 & 0 \end{pmatrix} \tag{10.65}$$

34 One would expect that a situation related to such a basic representation should be fundamental, and you may wonder why you have not heard of it before. The corresponding particles do not occur in Nature, but the corresponding structure is indeed fundamental, as it can be thought of as a theory of electrons that does not yet incorporate parity.

35 In axiomatic Quantum Field Theory, this is a legitimate mathematical theorem.

for $C \in SU(2)$ we have $J^{-1}CJ = J^{-1}S(C)J = S^*(C) = C^*$. Thus $W' = J$ satisfies (10.32). (The most general choice is $W' = \lambda J$, and this does not change our conclusion.) Thus

$$f_1' := W'(f_1) = J(f_1) = -f_2 \; ; \; f_2' := W'(f_2) = J(f_2) = f_1,$$

so that the components of these vectors are given by[36]

$$f_{1,1}' = 0 \; ; \; f_{1,2}' = -1 \; ; \; f_{2,1}' = 1 \; ; \; f_{2,2}' = 0.$$

Let us then rewrite the formulas for the quantum field:

$$\psi^+(x) = \sum_{s=1,2} \int \frac{\mathrm{d}^3\boldsymbol{p}}{\sqrt{2\omega_{\boldsymbol{p}}(2\pi\hbar)^3}} \exp(-\mathrm{i}(x,p)/\hbar) u(\boldsymbol{p},s) a(\boldsymbol{p},s). \tag{10.66}$$

$$\psi^-(x) = \sum_{s=1,2} \int \frac{\mathrm{d}^3\boldsymbol{p}}{\sqrt{2\omega_{\boldsymbol{p}}(2\pi\hbar)^3}} \exp(\mathrm{i}(x,p)/\hbar) v(\boldsymbol{p},s) b^\dagger(\boldsymbol{p},s), \tag{10.67}$$

where

$$u(\boldsymbol{p},s) = D_p(f_s) \; ; \; v(\boldsymbol{p},s) = D_p(J(f_s)) = D_p(f_s'). \tag{10.68}$$

Lemma 10.13.1 *In the domain $z^2 < 0$ we have $G'_{k,k'}(z) = G''_{k,k'}(z)$ and this function is* odd.

Recalling from (10.42) that

$$[\psi_k(x), \psi_{k'}^\dagger(y)] = (G'_{k,k'}(y-x) - G''_{k,k'}(x-y))1 = 2G'_{k,k'}(x-y)1,$$

there is no reason why this quantity should be zero. Our attempt to construct a quantum field which satisfies the microcausality condition has failed.

Despite this negative result the lemma is important. It will imply (a substitute of) microcausality after we learn a new method of quantization in the next section. For further reference, we perform a part of the computation separately.

Lemma 10.13.2 *The quantity*

$$T'_{k,k'}(\boldsymbol{p}) := \sum_{s\le 2} u(\boldsymbol{p},s)_k u(\boldsymbol{p},s)^*_{k'}$$

is the coefficient $B_{k,k'}$ of the matrix $B := M(p)/mc = M(p/(mc))$. Moreover

$$T''_{k,k'}(\boldsymbol{p}) := \sum_{s\le 2} v(\boldsymbol{p},s)_k v(\boldsymbol{p},s)^*_{k'} = T'_{k,k'}(\boldsymbol{p}). \tag{10.69}$$

Proof To lighten notation we set $D_{k,\ell} = (D_p)_{k,\ell}$, so that $u(\boldsymbol{p},s)_k = \sum_{\ell\le 2} D_{k,\ell} f_{s,\ell}$, etc. We compute

$$T'_{k,k'}(\boldsymbol{p}) = \sum_{s\le 2} \sum_{\ell,\ell'\le 2} D_{k,\ell} f_{s,\ell} D^*_{k',\ell'} f_{s,\ell'} = \sum_{\ell,\ell'\le 2} D_{k,\ell} D^*_{k',\ell'} a_{\ell,\ell'},$$

36 This argument is a special case of the general argument leading to (10.37).

where $a_{\ell,\ell'} = \sum_{s\le 2} f_{s,\ell} f_{s,\ell'}$, so that $a_{1,2} = a_{2,1} = 0$ and $a_{1,1} = a_{2,2} = 1$. Thus $T'_{k,k'}(\boldsymbol{p}) = \sum_{\ell\le 2} D_{k,\ell} D^*_{k',\ell}$. Since the matrix D_p is Hermitian, $D^*_{k',\ell} = D_{\ell,k'}$ so that $T'_{k,k'}(\boldsymbol{p})$ is the coefficient $B_{k,k'}$ of the matrix $B = D_p^2 = D_p D_p^\dagger$, which equals $M(p)/mc$ by (9.36). We then obtain (10.69) by the same computation. □

Proof of Lemma 10.13.1 It follows in particular from Lemma 10.13.2 that $T'_{k,k'}(\boldsymbol{p})$ is a homogeneous first-degree polynomial in the quantities p^μ, and Lemma 10.7.4 implies that $G'_{k,k'}(z) = G''_{k,k'}(z)$ is an odd function of z in the domain $z^2 < 0$. □

10.14 Fermion Fock Space

The most important particles (such as electrons and protons) are fermions, which behave quite differently from bosons. The boson Fock space of Section 3.4 (which the reader should review now) is not appropriate to describe multiparticle systems consisting of fermions. We need another construction, fermion Fock space, which we describe in this section. We recall that S_n denotes the group of permutations of $\{1,\dots,n\}$.

Principle 8 *Consider the state space $\mathcal{H}$ of a fermion, provided with a basis $(e_i)_{i\ge 1}$. Then the state space which represents a system of n such identical fermions is*

$$\mathcal{H}_{n,a} = \Big\{ \sum_{i_1,\dots,i_n} \alpha_{i_1,i_2,\dots,i_n} e_{i_1} \otimes \cdots \otimes e_{i_n} \; ; \; \forall \sigma \in \mathsf{S}_n \, , \, \alpha_{i_1,\dots,i_n} = \operatorname{sign}\sigma \, \alpha_{i_{\sigma(1)},\dots,i_{\sigma(n)}} \Big\}, \tag{10.70}$$

where $\operatorname{sign}\sigma \in \{-1,1\}$ *is the signature of the permutation* σ.

The index a stands for "anti-symmetric". The norm there is the obvious quadratic norm. We observe that

$$\mathcal{H} = \mathcal{H}_{1,a}.$$

Let us pause for a moment. We model here the two-particle states by $\mathcal{H}_{2,a} \subset \mathcal{H}\otimes\mathcal{H}$. This innocent-looking hypothesis has extremely deep consequences. For example $x \otimes x \notin \mathcal{H}_{2,a}$. Two identical fermions cannot be in the same state. The model obeys *Pauli's exclusion principle.*[37]

For $n = 0$ we define $\mathcal{H}_{0,a} = \mathbb{C}$, and we denote by $e_\emptyset$ its basis element (i.e. 1). It represents the vacuum, the state where no particle is present. We define $\mathcal{F}_0 = \bigoplus_{n\ge 0} \mathcal{H}_{n,a}$ the algebraic sum of the spaces $\mathcal{H}_{n,a}$, so that any element α of $\mathcal{F}_0$ is a sequence $\alpha = (\alpha(n))_{n\ge 0}$ with $\alpha(n) \in \mathcal{H}_{n,a}$ and $\alpha(n) = 0$ for n large enough. We define the inner product as in the case of the boson Fock space. The fermion Fock space $\mathcal{F}$ is then the completion of $\mathcal{F}_0$ for the norm (3.27) but we will never need elements of $\mathcal{F}$ which are not in $\mathcal{F}_0$. We will somewhat abuse notation by considering each $\mathcal{H}_{n,a}$, and in particular $\mathcal{H} = \mathcal{H}_{1,a}$ as a subspace of $\mathcal{F}_0$.

37 The consequences of Pauli's exclusion principle are hard to overstate. Following this principle, the electrons organize into orbitals in atoms, and this is what makes chemistry (and us) possible.

The fermion Fock space has properties very similar to the boson Fock space, with the all-important difference that commutators are replaced by *anti-commutators*. The anti-commutator of two operators A, B is defined as

$$\{A,B\} = AB + BA, \tag{10.71}$$

so that in particular $\{A,B\} = \{B,A\}$.

Theorem 10.14.1 (The creation and annihilation operators) *For ξ in $\mathcal{H}$ we can*[38] *define operators $A(\xi)$ and $A^\dagger(\xi)$ on $\mathcal{F}_0$ with the following properties.*

$$\textit{For } n \geq 1\,,\ A(\xi) \textit{ maps } \mathcal{H}_{n,a} \textit{ into } \mathcal{H}_{n-1,a}. \tag{10.72}$$

$$\textit{For } n \geq 0\,,\ A^\dagger(\xi) \textit{ maps } \mathcal{H}_{n,a} \textit{ into } \mathcal{H}_{n+1,a}. \tag{10.73}$$

$$A(\xi)(e_\emptyset) = 0\,;\ A^\dagger(\xi)(e_\emptyset) = \xi \in \mathcal{H} = \mathcal{H}_{1,a} \subset \mathcal{F}_0. \tag{10.74}$$

$$\forall \xi \in \mathcal{H}\,,\ \forall \alpha,\beta \in \mathcal{F}_0\,,\ (A^\dagger(\xi)(\alpha),\beta) = (\alpha, A(\xi)(\beta)). \tag{10.75}$$

$$\forall \xi,\eta \in \mathcal{H}\,,\ \{A(\xi), A^\dagger(\eta)\} = (\xi,\eta)1. \tag{10.76}$$

$$\forall \xi,\eta \in \mathcal{H}\,,\ \{A(\xi), A(\eta)\} = \{A^\dagger(\xi), A^\dagger(\eta)\} = 0. \tag{10.77}$$

$$\textit{The map } \xi \mapsto A^\dagger(\xi) \textit{ is linear. The map } \xi \mapsto A(\xi) \textit{ is anti-linear}. \tag{10.78}$$

The situation here is much nicer than in the boson case: The operators $A^\dagger(\xi)$ and $A(\xi)$ are defined everywhere and are adjoint of each other. Indeed, taking $\eta = \xi$ in (10.76) implies that

$$A(\xi)A^\dagger(\xi) + A^\dagger(\xi)A(\xi) = (\xi,\xi)1$$

whenever $\xi \in \mathcal{H}$. Consequently, for any $x \in \mathcal{F}_0$ we have

$$(x, A(\xi)A^\dagger(\xi)(x)) + (x, A^\dagger(\xi)A(\xi)(x)) = (\xi,\xi)\|x\|^2.$$

It then follows from (10.75) that for $x \in \mathcal{F}_0$

$$\|A(\xi)(x)\|^2 = (A(\xi)(x), A(\xi)(x)) = (x, A^\dagger(\xi)A(\xi)(x))$$

and similarly for $A^\dagger$. Therefore we have

$$\|A(\xi)(x)\|^2 + \|A^\dagger(\xi)(x)\|^2 = (\xi,\xi)\|x\|^2$$

and thus the operators $A(\xi)$ and $A^\dagger(\xi)$ are both of norm $\leq \|\xi\|$.

To describe the construction we will think of an element of $\mathcal{H}_{n,a}$ as a tensor $(\alpha_{i_1,\ldots,i_n})_{i_1,\ldots,i_n}$ which is antisymmetric in its indices, and of $\xi \in \mathcal{H} = \mathcal{H}_{1,a}$ as a sequence $(\xi_i)_{i\geq 1}$. The fact that $\alpha_{i_1,\ldots,i_n} = 0$ whenever the indices are not all different is of constant use.

[38] I am trying to save some space here, but the situation is the same as for the boson Fock space: There is a unique interesting version of these operators, which is the one we will always use.

For $\xi, \eta \in \mathcal{H}$ and $\alpha \in \mathcal{H}_{n,a}$ we define $A(\xi)(\alpha) = 0$ if $n = 0$ and otherwise

$$A(\xi)(\alpha)_{i_1,\ldots,i_{n-1}} = \sqrt{n} \sum_{i \geq 1} \xi_i^* \alpha_{i,i_1,\ldots,i_{n-1}}, \tag{10.79}$$

and we define

$$A^\dagger(\eta)(\alpha)_{i_1,\ldots,i_{n+1}} = \frac{1}{\sqrt{n+1}} \sum_{\ell \leq n+1} (-1)^{\ell+1} \eta_{i_\ell} \alpha_{i_1,\ldots,\widehat{i_\ell},\ldots,i_{n+1}}, \tag{10.80}$$

where the notation $\widehat{i_\ell}$ means that this term has been removed.

The tensor α is anti-symmetric in its indices, so that this is also the case of the left-hand side of (10.79). To see that the left-hand side of (10.80) is also anti-symmetric in its indices, notice first that it suffices to prove that it reverses its sign each time we exchange two consecutive indices. Look at what happens when you exchange i_1 and i_2: The first two terms of the summation get exchanged and their signs reversed, while all the other terms change sign.

The proof of Theorem 10.14.1 is simple if we calculate in a suitable basis, which by itself is very interesting. Given a set $I = \{j_1, \ldots, j_n\}$ of integers, we consider the element[39] $|I\rangle = (\alpha_{i_1,\ldots,i_n})$ of $\mathcal{H}_{n,a}$ such that $\alpha_{i_1,\ldots,i_n} = 0$ when the indices $i_1, \ldots, i_n$ are not a permutation of the elements of I, whereas when these indices are a permutation of the elements of I, i.e. $i_\ell = j_{\sigma(\ell)}$ for a certain $\sigma \in \mathsf{S}_n$,

$$\alpha_{i_1,\ldots,i_n} = \alpha_{j_{\sigma(1)},\ldots,j_{\sigma(n)}} = \frac{1}{\sqrt{n!}} \operatorname{sign} \sigma .$$

It is basically obvious that the elements $|I\rangle$ form an orthonormal basis of $\mathcal{H}_{n,a}$. The physical interpretation is transparent: The particle comes in different states $1, 2, \ldots$, and $|I\rangle$ represents an n-particle state, with one particle in each of the states $j_1, \ldots, j_n$. Let us say that $k \in I$ *has rank* m if there are exactly $m - 1$ elements of I which are $< k$.

Lemma 10.14.2 *Let us set $a(k) = A(e_k)$ and $a^\dagger(k) = A^\dagger(e_k)$. Then*
(a) If $k \in I$ then $a^\dagger(k)|I\rangle = 0$.
(b) If $k \notin I$ then $a^\dagger(k)|I\rangle = (-1)^{m+1}|I \cup \{k\}\rangle$ where m is the rank of k in $I \cup \{k\}$.
(c) If $k \notin I$ then $a(k)|I\rangle = 0$.
(d) If $k \in I$ then $a(k)|I\rangle = (-1)^{m+1}|I \setminus \{k\}\rangle$ where m is the rank of k in I.

Proof During the proof we use the notation $|I\rangle = (\alpha_{i_1,\ldots,i_n})$. Using the formula (10.80) for $\eta_i = \delta_i^k$ shows that $a^\dagger(k)(\alpha)_{i_1,\ldots,i_{n+1}}$ is 0 unless for some $\ell \leq n+1$ one has $i_\ell = k$, and then

$$a^\dagger(k)(\alpha)_{i_1,\ldots,i_{n+1}} = \frac{1}{\sqrt{n+1}} (-1)^{\ell+1} \alpha_{i_1,\ldots,\widehat{i_\ell},\ldots,i_{n+1}} .$$

To prove (a) observe that since $i_1, \ldots, i_{n+1}$ are distinct, $k = i_\ell$ does not belong to the list $i_1, \ldots, \widehat{i_\ell}, \ldots, i_{n+1}$ so this list cannot be a permutation of the elements of I when $k \in I$.

[39] One may also use the notation $|n_1, n_2, \ldots\rangle$ rather than $|I\rangle$ where $n_i \in \{0,1\}$ is 1 if and only if $i \in I$. This parallels the notation in the boson case.

When $k \notin I$ the elements $i_1,\dots,i_\ell = k,\dots,i_{n+1}$ are obtained as a permutation σ' of the elements of $I \cup \{k\}$ if and only if the elements $i_1,\dots,i_{\ell-1},i_{\ell+1},\dots,i_{n+1}$ are obtained as a permutation σ of the elements of I. Furthermore $\operatorname{sign}\sigma' = (-1)^{\ell+m}\operatorname{sign}\sigma$. To see this, we recall that $\operatorname{sign}\sigma$ equals $(-1)^s$ where s is the number of transpositions one has to apply to the elements of I to bring them in the order $i_1,\dots,i_{\ell-1},i_{\ell+1},\dots,i_{n+1}$. To bring the elements $I \cup \{k\}$ in the order $i_1,\dots,i_{n+1}$ one first brings k, which is in the mth position in $I \cup \{k\}$, into the ℓth position, for an extra factor $(-1)^{|\ell-m|} = (-1)^{\ell+m}$. This shows that $\operatorname{sign}\sigma' = (-1)^{\ell+m}\operatorname{sign}\sigma$ and implies (b).

It follows from (10.79) that

$$a(k)(\alpha)_{i_1,\dots,i_{n-1}} = \sqrt{n}\,\alpha_{k,i_1,\dots,i_{n-1}}.$$

This makes (c) obvious. The elements $i_1,\dots,i_{n-1}$ are obtained by a permutation σ' of the elements of $I \setminus \{k\}$ if and only if the elements $k,i_1,\dots,i_{n-1}$ are obtained by a permutation σ from the elements of I, and $\operatorname{sign}\sigma' = (-1)^{m+1}\operatorname{sign}\sigma$. This implies (d). □

To prove the commutation relations of Theorem 10.14.1 it suffices to prove the following

$$\{a(k),a(k')\} = 0\;;\;\{a^\dagger(k),a^\dagger(k')\} = 0\;;\;\{a(k),a^\dagger(k')\} = \delta^k_{k'}1,$$

which are proved in a tedious but straightforward fashion by computing the value of these operators when applied to a state $|I\rangle$ using Lemma 10.14.2. For example to prove the third relation when $k = k'$ we observe that if $k \in I$ then $a^\dagger(k)a(k)|I\rangle = |I\rangle$ and $a(k)a^\dagger(k)|I\rangle = 0$ whereas if $k \notin I$, $a^\dagger(k)a(k)|I\rangle = 0$ and $a(k)a^\dagger(k)|I\rangle = |I\rangle$, so that in any case $(a^\dagger(k)a(k)+a(k)a^\dagger(k))|I\rangle = |I\rangle$. The first two relations are proved similarly.

Given $\eta_1,\dots,\eta_n \in \mathcal{H}$ we define

$$\varphi_n(\eta_1,\dots,\eta_n) = \frac{1}{\sqrt{n!}}\sum_{\sigma\in S_n}\operatorname{sign}\sigma\;\eta_{\sigma(1)}\otimes\cdots\otimes\eta_{\sigma(n)}. \tag{10.81}$$

As in the case of bosons, the following is proved by reducing to the case of basis vectors and using (10.79) and (10.80).

Proposition 10.14.3 *For $\eta,\xi,\eta_1,\dots,\eta_n \in \mathcal{H}$ one has*

$$A^\dagger(\eta)(\varphi_n(\eta_1,\dots,\eta_n)) = \varphi_{n+1}(\eta,\eta_1,\dots,\eta_n),$$

$$A(\xi)(\varphi_n(\eta_1,\dots,\eta_n)) = \sum_{\ell\le n}(-1)^{\ell+1}(\xi,\eta_\ell)\varphi_{n-1}(\eta_1,\dots,\widehat{\eta_\ell},\dots,\eta_n).$$

In particular, repeated applications of n creation operators $A^\dagger(\eta_k)$ to the vacuum (the state $e_\emptyset$) results in a state consisting of n particles in the corresponding states.

Every unitary operator U on $\mathcal{H}$ has a canonical extension to a unitary operator $U_{\mathcal{F}}$ on $\mathcal{F}$. Any time-evolution taking place in $\mathcal{H}$ can be extended in a canonical way to the whole of $\mathcal{F}$, describing the evolution of non-interacting particles. The argument is entirely similar to the boson case. In particular, to a Hamiltonian H on $\mathcal{H}$ corresponds a Hamiltonian $H_{\mathcal{F}}$ on $\mathcal{F}$. As follows from (3.5) it satisfies the relation

$$H_{\mathcal{F}}\Big(\sum_{\sigma\in S_n} \operatorname{sign}\sigma\, x_{\sigma(1)}\otimes\cdots\otimes x_{\sigma(n)}\Big)$$
$$= \sum_{k\le n,\sigma\in S_n} \operatorname{sign}\sigma\, x_{\sigma(1)}\otimes\cdots\otimes x_{\sigma(k-1)}\otimes H(x_{\sigma(k)})\otimes x_{\sigma(k+1)}\otimes\cdots\otimes x_{\sigma(n)}. \tag{10.82}$$

This means that the energy of a non-interacting multiparticle system is the sum of the energies of the individual particles.

The following is proved as in the boson case.

Proposition 10.14.4 *For any vector ξ we have*

$$U_{\mathcal{F}}A(\xi)U_{\mathcal{F}}^{-1} = A(U(\xi))\;;\; U_{\mathcal{F}}A^{\dagger}(\xi)U_{\mathcal{F}}^{-1} = A^{\dagger}(U(\xi)). \tag{10.83}$$

10.15 Massive Weyl Spinors, Second Attempt

The constructions of Sections 10.4 and 10.6 can be repeated verbatim when one replaces boson Fock space by fermion Fock space, so we may use the results of this analysis to keep investigating Lorentz covariant fields. The only difference is that commutators are replaced by anti-commutators. In particular we have now

$$\{a(\boldsymbol{p},s),a^{\dagger}(\boldsymbol{p}',s')\} = \delta_s^{s'}(2\pi\hbar)^3\delta^{(3)}(\boldsymbol{p}-\boldsymbol{p}')1, \tag{10.84}$$

with all other anti-commutators being zero, and we obtain the following mirror image of Theorem 10.7.1.

Theorem 10.15.1 *The quantum field $\psi_k(x) = \psi_k^+(x)+\psi_k^-(x)$ constructed in Sections 10.4 and 10.6 but with fermion Fock space rather than boson Fock space satisfies*

$$\{\psi_k(x),\psi_{k'}(y)\} = (G_{k,k'}(y-x)+G_{k',k}(x-y))1 \tag{10.85}$$

where $G_{k,k'}(z)$ is given by (10.41). Moreover,

$$\{\psi_k(x),\psi_{k'}^{\dagger}(y)\} = (G'_{k,k'}(y-x)+G''_{k,k'}(x-y))1, \tag{10.86}$$

where $G'_{k,k'}(z)$ is given by (10.43) and $G''_{k,k'}(z)$ is given by (10.44).

We now want to show that the construction of Section 10.13 produces a physically relevant field if one uses fermion Fock spaces. We consider the most difficult case, where the particle is its own anti-particle. In the next statement, we use the notation of Section 10.13, and in particular the vectors $f_1, f_2, f_1' = -f_2, f_2' = f_1$.

Theorem 10.15.2 *The quantum field given by the formulas*

$$\psi^+(x) = \sum_{s=1,2}\int \frac{\mathrm{d}^3\boldsymbol{p}}{\sqrt{2\omega_{\boldsymbol{p}}(2\pi\hbar)^3}} \exp(-\mathrm{i}(x,p)/\hbar)u(\boldsymbol{p},s)a(\boldsymbol{p},s), \tag{10.87}$$

$$\psi^-(x) = \sum_{s=1,2}\int \frac{\mathrm{d}^3\boldsymbol{p}}{\sqrt{2\omega_{\boldsymbol{p}}(2\pi\hbar)^3}} \exp(\mathrm{i}(x,p)/\hbar)v(\boldsymbol{p},s)a^{\dagger}(\boldsymbol{p},s), \tag{10.88}$$

where

$$u(\boldsymbol{p},s) = D_p(f_s)\ ;\ v(\boldsymbol{p},s) = D_p(J(f_s)) = D_p(f'_s) \tag{10.89}$$

satisfies

$$\{\psi_k(x),\psi_{k'}(y)\} = 0\ ;\ \{\psi_k^\dagger(x),\psi_{k'}^\dagger(y)\} = 0\ ;\ \{\psi_k(x),\psi_{k'}^\dagger(y)\} = 0 \tag{10.90}$$

whenever $(x-y)^2 < 0$.

The formulas defining the field look the same as the formulas[40] (10.66) to (10.68), but it is understood here that quantization is done using fermion rather than boson Fock space.

Condition (10.90) is *different* from the microcausality condition (10.39). As we explained at the beginning of Section 10.7, one use of the microcausality condition is to ensure that Hamiltonians constructed out of products of quantum fields commute at causally separated points. Condition (10.90) ensures that Hamiltonians constructed out of products of an *even* number of such fields commute at causally separated points.[41] This is why the fields satisfying (10.90) are physically relevant.

Proof Due to the change of sign in (10.85) compared to (10.40), Lemma 10.13.1 implies the last part of (10.90), and we turn to the proof of the first part of this statement that $\{\psi_k(x),\psi_{k'}(y)\} = 0$ (the second being similar). This requires again a few lines of computation. It holds that

$$\begin{aligned} T_{k,k'} := &\sum_{s\le 2} u(\boldsymbol{p},s)_k v(\boldsymbol{p},s)_{k'} \\ = &\sum_{s\le 2}\sum_{\ell,\ell'\le 2} D_{k,\ell} f_{s,\ell} D_{k',\ell'} f'_{s,\ell'} \\ = &\sum_{\ell,\ell'\le 2} D_{k,\ell} D_{k',\ell'} a_{\ell,\ell'}, \end{aligned}$$

where $a_{\ell,\ell'} = \sum_{s\le 2} f_{s,\ell} f'_{s,\ell'} = \sum_{s\le 2}\delta_s^\ell f'_{s,\ell'} = f'_{\ell,\ell'}$ so that $a_{1,1} = a_{2,2} = 0$ while $a_{1,2} = -a_{2,1} = -1$. Consequently, $T_{1,1} = T_{2,2} = 0$, while $T_{1,2} = -T_{2,1} = -1$, because the 2×2 matrix $(D_{k,\ell})$ is of determinant 1. The functions $G_{k,k}$ of (10.41) then satisfy $G_{1,1} = G_{2,2} = 0$ and $G_{1,2} = -G_{2,1} = I$ where I is the function (5.13). It then follows from (10.85) that $\{\psi_1(x),\psi_1(y)\} = \{\psi_2(x),\psi_2(y)\} = 0$ whereas

$$\{\psi_1(x),\psi_2(y)\} = (I(y-x) - I(x-y))1,$$

Since the function $I(z)$ is even on the domain $z^2 < 0$ this quantity is zero. □

We now investigate the "equal-time anti-commutation relations". In order to prepare for the results of Section 10.19, we assume that the particle is different from its anti-particle. In that case (10.67) is replaced by

40 In (10.88) we now have the operator $a^\dagger$ rather than $b^\dagger$, since we assume that the particle is its own anti-particle.

41 It is very simple to check this, assuming of course that all fields commute at a given point.

$$\psi^-(x) = \sum_{s=1,2} \int \frac{\mathrm{d}^3 \boldsymbol{p}}{\sqrt{2\omega_{\boldsymbol{p}}(2\pi\hbar)^3}} \exp(\mathrm{i}(x,p)/\hbar) v(\boldsymbol{p},s) b^\dagger(\boldsymbol{p},s), \tag{10.91}$$

where b denotes a creation operator for the anti-particle. Proceeding as in the proof of Proposition 6.9.1 it is possible to give a precise mathematical meaning to the following result, but we content ourselves with a formal proof.

Theorem 10.15.3 (Equal-time anti-commutation relations[42]) *We have*

$$\{\psi_k(ct,\boldsymbol{x}), \psi_{k'}(ct,\boldsymbol{y})\} = 0\,;\ \{\psi_k^\dagger(ct,\boldsymbol{x}), \psi_{k'}^\dagger(ct,\boldsymbol{y})\} = 0,$$

$$\{\psi_k(ct,\boldsymbol{x}), \psi_{k'}^\dagger(ct,\boldsymbol{y})\} = \frac{1}{mc}\delta_k^{k'}\delta^{(3)}(\boldsymbol{x}-\boldsymbol{y})1. \tag{10.92}$$

The factor $1/(mc)$ in the right-hand side of the last equation looks strange, but we must remember that the quantum field is defined only up to a multiplicative constant. One may then decide that getting the constant 1 in the right-hand side of (10.92) would be a good method to normalize this field. This can be done by introducing a factor $\sqrt{mc}$ in formulas such as (10.91).

Proof As in Theorem 10.15.1 we obtain (after a change of variable $\boldsymbol{p} \to -\boldsymbol{p}$ in one of the integrals)

$$\{\psi_k(ct,\boldsymbol{x}), \psi_{k'}^\dagger(ct,\boldsymbol{y})\} = 1 \int \frac{\mathrm{d}^3 \boldsymbol{p}}{2\omega_{\boldsymbol{p}}(2\pi\hbar)^3} \exp(\mathrm{i}(\boldsymbol{x}-\boldsymbol{y})\cdot \boldsymbol{p}/\hbar) C_{k,k'}(\boldsymbol{p}),$$

where

$$C_{k,k'}(\boldsymbol{p}) = \sum_{s\le 2} u(\boldsymbol{p},s)_k u(\boldsymbol{p},s)_{k'}^* + \sum_{s\le 2} v(-\boldsymbol{p},s)_k v(-\boldsymbol{p},s)_{k'}^*.$$

Now, as follows from Lemma 10.13.2, we have $C_{k,k'}(\boldsymbol{p}) = 0$ if $k \neq k'$ while $C_{k,k}(\boldsymbol{p}) = 2p^0/mc$, and since $p^0 = \omega_{\boldsymbol{p}}$ the result follows from (a version of) (1.19). The other case is similar. □

10.16 Equation of Motion for the Massive Weyl Spinor

In the present section we investigate whether the previous quantum field can be seen as a quantization of a classical field. This section may be skipped at first reading. Let us accept that when this is the case the quantum and the classical field satisfy the same equations of motion (as we have seen in the case e.g. in the situation of Exercise 10.12.1). Our strategy will then be to find an "equation of motion" for the quantum field, and then to look for a classical field satisfying this equation of motion. We first describe such an equation of motion.

42 Just as in the case of equal-time commutation relations, some authors use these relations as a road to discover Weyl spinors, but this road is fraught with dangers as we discuss below.

We consider the field $\psi^\dagger$ given by $(\psi^\dagger)_k = (\psi_k)^\dagger$. Thus $\psi^\dagger$ is the *column* vector the kth row of which is the adjoint of the kth row of ψ. Hence

$$\psi^\dagger = \psi^{\dagger +} + \psi^{\dagger -} \tag{10.93}$$

where

$$\psi^{\dagger +} = \sum_{s=1,2} \int \frac{\mathrm{d}^3 \boldsymbol{p}}{\sqrt{2\omega_{\boldsymbol{p}}(2\pi\hbar)^3}} \exp(-\mathrm{i}(x,p)/\hbar) v(\boldsymbol{p},s)^* a(\boldsymbol{p},s) \tag{10.94}$$

and

$$\psi^{\dagger -} = \sum_{s=1,2} \int \frac{\mathrm{d}^3 \boldsymbol{p}}{\sqrt{2\omega_{\boldsymbol{p}}(2\pi\hbar)^3}} \exp(\mathrm{i}(x,p)/\hbar) u(\boldsymbol{p},s)^* a^\dagger(\boldsymbol{p},s), \tag{10.95}$$

where $v(\boldsymbol{p},s)^*$ and $u(\boldsymbol{p},s)^*$ are the column vectors of $\mathbb{C}^2$ whose components are the complex conjugates of the components of $v(\boldsymbol{p},s)$ and $u(\boldsymbol{p},s)$ defined in (10.89). We recall the Pauli matrices σ_μ given by (2.3), and the matrix J of (10.65).

Proposition 10.16.1 *In the sense of distributions, the quantum field of Theorem 10.15.2 satisfies the equations of motion*

$$\mathrm{i}\hbar\rho^\mu \partial_\mu \psi = mc\psi^\dagger, \tag{10.96}$$

where $\rho^\mu = J\sigma_\mu$.

The key point is the following simple fact.

Lemma 10.16.2 *For* $p \in X_m$ *we have*

$$JM(P(p))u(\boldsymbol{p},s) = mcv(\boldsymbol{p},s)^* \; ; \; JM(P(p))v(\boldsymbol{p},s) = -mcu(\boldsymbol{p},s)^*. \tag{10.97}$$

Proof It is obvious from (4.28) that

$$D_{P(p)} = D_p^{-1}. \tag{10.98}$$

Using (9.36) and (10.98) we have $M(P(p)) = mcD_p^{-2}$, so that, using that D_p is Hermitian in the second equality and Lemma 8.1.4 in the third we get

$$M(P(p))D_p = mcD_p^{-1} = mcD_p^{\dagger -1} = mcJ^{-1}D_p^* J, \tag{10.99}$$

and multiplying on the left by J we obtain

$$JM(P(p))D_p = mcD_p^* J \tag{10.100}$$

so that

$$JM(P(p))D_p(f_s) = mcD_p^* J(f_s) = mc(D_p J(f_s))^*$$

because the components of $J(f_s)$ are real. Since by (10.68) we have $u(\boldsymbol{p},s) = D_p(f_s)$ and $v(\boldsymbol{p},s) = D_p J(f_s)$ this proves the first part of (10.97). Multiplying (10.100) on the right by J, and since $J^2 = -I$ one gets

$$JM(P(p))D_p J = -mcD_p^*$$

and this similarly implies the second part of (10.97). □

Proof of Proposition 10.16.1 We only perform a formal calculation, as there is no difficulty to turn it into a rigorous proof. From (10.87) we obtain (using that $(x,p) = x^\mu p_\mu$)

$$\mathrm{i}\hbar\rho^\mu\partial_\mu\psi^+ = \sum_{s=1,2}\int \frac{\mathrm{d}^3\boldsymbol{p}}{\sqrt{2\omega_{\boldsymbol{p}}(2\pi\hbar)^3}}\exp(-\mathrm{i}(x,p)/\hbar)p_\mu\rho^\mu u(\boldsymbol{p},s)a(\boldsymbol{p},s). \tag{10.101}$$

Now,

$$JM(P(p)) = J\sum_{0\le\mu\le 3} p_\mu\sigma_\mu = p_\mu\rho^\mu,$$

and (10.97) implies $p_\mu\rho^\mu u(\boldsymbol{p},s) = mcv(\boldsymbol{p},s)^*$. Consequently (10.101) and (10.94) imply $\mathrm{i}\hbar\rho^\mu\partial_\mu\psi^+ = mc\psi^{\dagger+}$. The case of ψ^- is similar. □

To make the quantum field of Theorem 10.15.2 appear as the quantization of a classical field, the natural path is to find a Lagrangian yielding the equation of motion (10.96), and then to quantize the corresponding Hamiltonian. It does not seem possible to construct such a Lagrangian. Several standard textbooks try to find the quantum fields of interest by investigating the classical fields arising from natural Lagrangians. Since the method utterly fails (as it does not lead to the discovery of the present field) they argue that in the case of spin 1/2, where anti-commutators replace commutators, one should "replace real numbers by Grassmann numbers" i.e. assume that the components ψ_k of ψ take as values anti-commuting objects which satisfy $xy = -yx$, and write a "Lagrangian" built from such objects.[43] Discovering quantum fields through their invariance properties as done in this chapter avoids this perilous gymnastics.

10.17 Massless Weyl Spinors

We consider the case where $m = 0$ (so the little group G_0 is given by (9.40)) and where again the representation S of $SL(2,\mathbb{C})$ is the defining representation on $\mathbb{C}^2$. To construct the annihilation part of the fields we have to look for subspaces of $\mathbb{C}^2$ which are invariant under all the operators $S(C)$ for C in the little group. The only such subspace is the span $\mathcal{G}$ of the first basis vector, and the representation of the little group G_0 thus obtained is the representation $\hat{\pi}_{-1}$ of Definition 9.6.2, $\hat{\pi}_{-1}(A) = a$ for A as in (9.40). Thus the quanta of the field are massless particles of helicity $-1/2$. Since the spin is $1/2$, quantization will use

43 We let the reader make up her mind as to whether the consideration of Lagrangians built out of Grassmann numbers really promotes our understanding, or whether it just represents a desperate last-ditch attempt to force a genuine quantum question into the framework of Classical Mechanics. Why this should be an attractive program is anyway a mystery to this author.

fermion Fock spaces. To construct the creation part of the fields, we then have to look for subspaces of $\mathbb{C}^2$ which are invariant under all the operators $S^*(C)$ for C in the little group. There is a single one-dimensional space with this property, the same space $\mathcal{G}$ as above. The restriction of S^* to this space and G_0 is unitarily equivalent to $\hat{\pi}_1$, not to $\hat{\pi}_{-1}$. That means that the anti-particle of the quantum of the field is a massless particle with helicity 1/2. (So in particular we cannot construct the desired quantum field when the particle is its own anti-particle.) The creation part of the quantum fields are then given by

$$\psi_k^-(x) = \int \frac{\mathrm{d}^3\boldsymbol{p}}{\sqrt{2\omega_{\boldsymbol{p}}(2\pi\hbar)^3}} \exp(\mathrm{i}(x,p)/\hbar)(D_p)_{k,1}a^\dagger(\boldsymbol{p}) \tag{10.102}$$

and the annihilation parts of the quantum fields are given by

$$\psi_k^+(x) = \int \frac{\mathrm{d}^3\boldsymbol{p}}{\sqrt{2\omega_{\boldsymbol{p}}(2\pi\hbar)^3}} \exp(-\mathrm{i}(x,p)/\hbar)(D_p)_{k,1}b(\boldsymbol{p}), \tag{10.103}$$

where $b(\boldsymbol{p})$ is the operator which annihilates an anti-particle of momentum $\boldsymbol{p}$.

Theorem 10.17.1 *The quantum field $\psi(x) = \psi^+(x) + \psi^-(x)$ satisfies*

$$\{\psi(x), \psi(y)\} = 0\ ;\ \{\psi^\dagger(x), \psi^\dagger(y)\} = 0\ ;\ \{\psi(x), \psi^\dagger(y)\} = 0$$

whenever $(x-y)^2 < 0$.

As in the case of the massive Weyl spinors, it suffices to prove here that $T_{k,k'} = (D_p)_{k,1}(D_p)^*_{k',1}$ is a first degree polynomial in p^0, p^1, p^2, p^3. Recalling (9.39), this is because $2T_{k,k'}$ is the coefficient $B_{k,k'}$ of the matrix

$$B = D_p M(p^*) D_p^\dagger = M(\kappa(D_p)(p^*)) = M(p).$$

Massless Weyl spinors model neutrinos, at least in a theory where these are of mass zero, which is now known to be only approximately true.

10.18 Parity

We would like to discover physically relevant quantum fields "when there exists a parity operator". One way to reach this goal is simply to replace everywhere in the previous analysis the group $SL(2,\mathbb{C})$ by the group $SL^+(2,\mathbb{C})$ of Section 8.9, the group generated by $SL(2,\mathbb{C})$ and a new element P' which represents parity. Lorentz covariance, as defined in Definition 10.2.1, has a "version with parity", simply by replacing everywhere $SL(2,\mathbb{C})$ by $SL^+(2,\mathbb{C})$, so that $U(c,C)$ has to be a representation of $\mathcal{P}^{*+} = \mathbb{R}^{1,3} \rtimes SL^+(2,\mathbb{C})$. All our considerations carry over to that setting.

Assuming for the moment $m > 0$ the little group is now the group $SU^+(2)$ generated by $SU(2)$ and P', and in Section 9.9 we have discussed how to obtain representations of $\mathcal{P}^{*+}$ through the extended Theorem 9.4.2. Condition (10.16) must now be satisfied for the little group $SU^+(2)$ and not only for $SU(2)$.

We will describe two models for fermions constructed by this procedure. As they are fermions, quantization is based on fermion Fock space. In the next section we examine

"Dirac fermions", which are different from their anti-particle. This is a fundamental structure, as it is in particular the successful model for electrons.

10.19 Dirac Field

In this section and the next $D_p \in SL(2,\mathbb{C})$ such that $D_p(p^*) = p$ is positive Hermitian as provided by Lemma 8.4.6, and by (4.28) it satisfies

$$D_{P(p)} = D_p^{-1} = D_p^{\dagger -1}. \tag{10.98}$$

Let us recall the representation S of $SL^+(2,\mathbb{C})$ constructed in Section 8.10 (which the reader might want to review at this stage). Thus, for $A \in SL(2,\mathbb{C})$, $S(A)$ is the 4×4 matrix

$$S(A) := \begin{pmatrix} A^{\dagger -1} & 0 \\ 0 & A \end{pmatrix}, \tag{10.104}$$

where, as everywhere in the present section 4×4 matrices are decomposed in blocks of size 2×2. Recalling the matrices γ_μ given in (8.49) and in particular

$$\gamma_0 = \begin{pmatrix} 0 & I \\ I & 0 \end{pmatrix}, \tag{10.105}$$

where I is the 2×2 identity matrix, then $S(P') = \gamma_0$.

There are *two* distinguished subspaces of $\mathbb{C}^4$ which are invariant under the restriction of S to the little group $SU^+(2)$, i.e. which are invariant under both the action of the restriction of S to $SU(2)$ and under γ_0. Writing $x \in \mathbb{C}^4$ as a column vector, these are

$$\mathcal{G} = \{x \; ; \; \gamma_0 x = x\} \; ; \; \mathcal{G}' = \{x \; ; \; \gamma_0 x = -x\}. \tag{10.106}$$

Thus $x = (x_k)_{1 \le k \le 4}$ belongs to $\mathcal{G}$ if and only if $x_1 = x_3$ and $x_2 = x_4$. We needed such an invariant space to construct the annihilation part of the fields, and another such invariant space to construct the creation part of the fields. Thus we have several choices. We may use the *same* space for both the annihilation and the creation part, a choice we examine in Section 10.21. Or we may use one of these spaces for the annihilation part of the fields and the *other one* for the creation part. This is what we do in the present section. Since then $S(P')$ is the identity on $\mathcal{G}$ and minus the identity on $\mathcal{G}'$, this shows that the fermion and the anti-fermion are of opposite intrinsic parities (which one is the fermion and which one is the anti-fermion is a matter of convention).

We consider the vectors

$$f_1 = \begin{pmatrix} 1 \\ 0 \\ 1 \\ 0 \end{pmatrix} \; ; \; f_2 = \begin{pmatrix} 0 \\ 1 \\ 0 \\ 1 \end{pmatrix} \; ; \; f_1' = \begin{pmatrix} 1 \\ 0 \\ -1 \\ 0 \end{pmatrix} \; ; \; f_2' = \begin{pmatrix} 0 \\ 1 \\ 0 \\ -1 \end{pmatrix}$$

so that f_1, f_2 form a basis of $\mathcal{G}$ and f_1', f_2' form a basis of $\mathcal{G}'$. (These bases are orthogonal but not orthonormal. This is simply a matter of normalization: We take for $\mathcal{V}$ the space $\mathcal{G}$ equipped with a norm which is $1/\sqrt{2}$ the norm of $\mathcal{G}$.) Let us now consider $u(\boldsymbol{p},s)_k$ given by (10.30) and $v(\boldsymbol{p},s)_k$ given by (10.34). Thus, denoting by $u(\boldsymbol{p},s)$ the column vector with components $u(\boldsymbol{p},s)_k$ and similarly for $v(\boldsymbol{p},s)$ we have

$$u(\boldsymbol{p},s) = S(D_p)f_s \; ; \; v(\boldsymbol{p},s) = S(D_p)f'_s. \tag{10.107}$$

The components of the Dirac field are then defined by $\psi_k(x) = \psi_k^+(x) + \psi_k^-(x)$ where

$$\psi_k^+(x) = \sum_{s\le 2}\int \frac{\mathrm{d}^3\boldsymbol{p}}{(2\pi\hbar)^3}\frac{c\sqrt{m}}{\sqrt{2\omega_{\boldsymbol{p}}}}\exp(-\mathrm{i}(x,p)/\hbar)u(\boldsymbol{p},s)_k a(\boldsymbol{p},s), \tag{10.108}$$

$$\psi_k^-(x) = \sum_{s\le 2}\int \frac{\mathrm{d}^3\boldsymbol{p}}{(2\pi\hbar)^3}\frac{c\sqrt{m}}{\sqrt{2\omega_{\boldsymbol{p}}}}\exp(\mathrm{i}(x,p)/\hbar)v(\boldsymbol{p},s)_k b^\dagger(\boldsymbol{p},s). \tag{10.109}$$

The Dirac field is intrinsically defined only up to a multiplicative constant, and the factor $c\sqrt{m}$ in the previous formulas is unimportant and is motivated by the desire to get the correct dimension, as we will see later. We may rewrite the formulas (10.108) and (10.109) in a compact way as

$$\psi^+(x) = \sum_{s\le 2}\int \frac{\mathrm{d}^3\boldsymbol{p}}{(2\pi\hbar)^3}\frac{c\sqrt{m}}{\sqrt{2\omega_{\boldsymbol{p}}}}\exp(-\mathrm{i}(x,p)/\hbar)u(\boldsymbol{p},s)a(\boldsymbol{p},s), \tag{10.110}$$

$$\psi^-(x) = \sum_{s\le 2}\int \frac{\mathrm{d}^3\boldsymbol{p}}{(2\pi\hbar)^3}\frac{c\sqrt{m}}{\sqrt{2\omega_{\boldsymbol{p}}}}\exp(\mathrm{i}(x,p)/\hbar)v(\boldsymbol{p},s)b^\dagger(\boldsymbol{p},s). \tag{10.111}$$

Exercise 10.19.1 Write the components of the column vectors $u(\boldsymbol{p},s)$ and $v(\boldsymbol{p},s)$ using Corollary G.1.3 to enjoy explicit formulas like a physicist.

The following exercise is a bit challenging.

Exercise 10.19.2 Check directly that the Lorentz covariance condition corresponding to (10.9) holds for $c = 0, C = P'$. Hint: It helps to use (10.98).

We now use the notation of (10.30), denoting by $u(\boldsymbol{p},s)^\dagger$ the row vector which is the conjugate-transpose of the column vector $u(\boldsymbol{p},s)$. The following computation prepares the proof of microcausality.

Lemma 10.19.3 *We have*

$$\sum_{s\le 2} u(\boldsymbol{p},s)u(\boldsymbol{p},s)^\dagger = \frac{1}{mc}\gamma(p)\gamma_0 + \gamma_0 \tag{10.112}$$

$$\sum_{s\le 2} v(\boldsymbol{p},s)v(\boldsymbol{p},s)^\dagger = \frac{1}{mc}\gamma(p)\gamma_0 - \gamma_0. \tag{10.113}$$

Proof We first prove by direct computation that

$$\sum_{s\le 2} f_s f_s^\dagger = \begin{pmatrix} I & I \\ I & I \end{pmatrix} \; ; \; \sum_{s\le 2} f'_s f_s'^\dagger = \begin{pmatrix} I & -I \\ -I & I \end{pmatrix}.$$

We then have in succession

$$\sum_{s\le 2} u(\boldsymbol{p},s)u(\boldsymbol{p},s)^\dagger = S(D_p)\begin{pmatrix} I & I \\ I & I \end{pmatrix} S(D_p)^\dagger$$

$$= \begin{pmatrix} (D_p)^{\dagger -1} & 0 \\ 0 & D_p \end{pmatrix}\begin{pmatrix} I & I \\ I & I \end{pmatrix}\begin{pmatrix} D_p^{-1} & 0 \\ 0 & D_p^\dagger \end{pmatrix}$$

$$= \begin{pmatrix} (D_pD_p^\dagger)^{-1} & I \\ I & D_pD_p^\dagger \end{pmatrix}. \tag{10.114}$$

In Lemma 9.5.7 we proved that if $A(p^*) = p$ then $AA^\dagger = M(p)/(mc)$ and $(AA^\dagger)^{-1} = M(P(p))/(mc)$. Thus (10.114) proves (10.112) and the case of (10.113) is similar. □

Theorem 10.19.4 *For* $(x-y)^2 < 0$ *we have*

$$\{\psi_k(x), \psi_{k'}(y)\} = 0\ ;\ \{\psi_k^\dagger(x), \psi_{k'}^\dagger(y)\} = 0\ ;\ \{\psi_k(x), \psi_{k'}^\dagger(y)\} = 0.$$

Proof The first two equalities are basically trivial because the anti-particle is different from the particle (just as in the first statement of Theorem 10.7.5). The last equality is proved from (10.86), using Lemma 10.19.3. The reader should observe the beauty of this structure: The first term on the right of (10.112) produces an odd contribution to $G_{k,k'}(z)$, while the second term produces an even contribution, but this second term has a minus sign in (10.113). □

We have also the following more precise version, which is proved exactly as Theorem 10.15.3.

Theorem 10.19.5 (Equal-time anti-commutation relations) *It holds that*

$$\{\psi_k(ct,\boldsymbol{x}), \psi_{k'}(ct,\boldsymbol{y})\} = 0\ ;\ \{\psi_k(ct,\boldsymbol{x}), \psi_{k'}^\dagger(ct,\boldsymbol{y})\} = c\delta_k^{k'}\delta^{(3)}(\boldsymbol{x}-\boldsymbol{y})1. \tag{10.115}$$

The last relation of (10.115) is a way to normalize the Dirac field (fixing the multiplicative constant on which it depends).

Our last result is that the Dirac field satisfies the Dirac equation in the sense of distributions. Several textbooks and in particular that of Greiner and Reinhardt [35] recover the formulas (10.109) and (10.112) from the assumption that the quantum field satisfies the Dirac equation as well as the anti-commutation relations of Theorem 10.19.5. We denote by $\psi(x)$ the column vector whose rows are the fields $\psi_k(x)$. We should not be confused between the components $\psi_1, \psi_2, \psi_3, \psi_4$ of ψ and the components $x^\mu, 0 \le \mu \le 3$ of x. The following should be compared to Proposition 6.1.3.

Theorem 10.19.6 *The quantum field* ψ *satisfies the Dirac equation (in the sense of distributions)*

$$i\hbar\gamma^\mu\partial_\mu\psi = mc\psi. \tag{10.116}$$

Proof One has to prove that both the creation and the annihilation parts satisfy this relation. We treat only the case of the annihilation part, at the level of a formal calculation, as it is clear how this can be turned into a rigorous argument. Recalling (10.107) and that $\gamma(p) = p_\mu\gamma^\mu$, let us define

$$U(\boldsymbol{p},s) := p_\mu\gamma^\mu u(\boldsymbol{p},s) = \gamma(p)S(D_p)f_s.$$

Using that $\partial_\mu = \partial/\partial x^\mu$ and that $(x,p) = x^\mu p_\mu$ we obtain

$$\mathrm{i}\hbar\partial_\mu \exp(-\mathrm{i}(x,p)/\hbar) = p_\mu \exp(-\mathrm{i}(x,p)/\hbar)$$

and hence, recalling (10.108),

$$\mathrm{i}\hbar\gamma^\mu\partial_\mu\psi^+(x) = \sum_{s\le 2}\int \frac{\mathrm{d}^3\boldsymbol{p}}{(2\pi\hbar)^3}\frac{c\sqrt{m}}{\sqrt{2\omega_{\boldsymbol{p}}}}\exp(-\mathrm{i}(x,p)/\hbar)U(\boldsymbol{p},s)a(\boldsymbol{p},s). \tag{10.117}$$

Now (8.47) means with our current notation that $\gamma(p)S(A) = S(A)\gamma(A^{-1}(p))$. For $A = D_p$, we have $\gamma(A^{-1}(p)) = \gamma(p^*) = mc\gamma_0$ and thus $\gamma(p)S(D_p) = mcS(D_p)\gamma_0$. Therefore since $\gamma_0 f_s = f_s$ we get

$$U(\boldsymbol{p},s) = mcS(D_p)\gamma_0 f_s = mcS(D_p)f_s = mcu(\boldsymbol{p},s).$$

Consequently $\mathrm{i}\hbar\gamma^\mu\partial_\mu\psi^+(x) = mc\psi^+(x)$. □

Applying the Dirac operator $\mathrm{i}\hbar\gamma^\mu\partial_\mu$ to (10.116) (and using that the square of the Dirac operator is given by (P.20)) we obtain $\hbar^2\partial_\mu\partial^\mu\psi + m^2c^2\psi = 0$, showing as expected that each component of ψ satisfies the Klein-Gordon equation.

10.20 Dirac Field and Classical Mechanics

In this section we explain as best we can some material which is ubiquitous in elementary textbooks.[44] It attempts to relate the Dirac quantum field to considerations of Classical Mechanics, and to make this quantum field appear as the quantization of the classical[45] field given by the Dirac equation (10.116). The reader must be warned that all this has to be taken with a grain of salt, since the project is unsound from the beginning:[46] Anything involving half-integer spins is genuine quantum behavior with no real classical equivalent.[47]

The classical field to start with is a quadruple of *complex-valued* functions ψ^μ over $\mathbb{R}^{1,3}$ that satisfy the Dirac equation

$$\mathrm{i}\hbar\gamma^\mu\partial_\mu\psi = mc\psi.$$

This "classical Dirac field" is determined by eight real-valued functions over $\mathbb{R}^{1,3}$. It is not observed in Nature.

44 In my opinion this material is of questionable value and very confusing, sometimes apparently even for the people who write about it. I cover it only to attempt to clear some of the confusion. Possibly this material is of historical importance, but the danger for the innocent reader is that it is presented without any warning about its potentially dubious quality.

45 The word "classical" simply stresses that this is not a quantum field, but just a collection of functions over $\mathbb{R}^{1,3}$.

46 Here again I would like to thank G. Folland for making this clear to me.

47 It seems natural that the discoverers of radically new theories such as Quantum Mechanics stretch the rules of sanity to relate their discoveries to the established theories. But I cannot explain why this contrived material finds its way into textbooks several generations later.

To write simple formulas, given a Dirac spinor, that is a column vector $\xi \in \mathbb{C}^4$ we define its *Dirac adjoint* $\bar{\xi}$ as the row vector $\xi^\dagger \gamma_0$ where $\xi^\dagger$ denotes as usual the conjugate-transpose of ξ. Thus $\bar{\xi}\xi$ is a number. Given the classical field $\psi = (\psi^\mu)$ we then define its Dirac adjoint $\bar{\psi}$ by $\bar{\psi}(x) = \bar{\xi}$ for $\xi = \psi(x)$.

First, ψ satisfies the Dirac equation if and only if $\bar{\psi}$ satisfies the adjoint Dirac equation

$$-i\hbar\partial_\mu \bar{\psi}\gamma^\mu = mc\bar{\psi}. \tag{10.118}$$

This is proved by taking the adjoint of the Dirac equation i.e.

$$-i\hbar\partial_\mu \psi^\dagger (\gamma^\mu)^\dagger = mc\psi^\dagger,$$

by multiplying to the right by γ^0 and by observing the relations

$$(\gamma^\mu)^\dagger = \gamma_0 \gamma^\mu \gamma_0, \tag{10.119}$$

which are apparent from the formulas (8.49) since the Pauli matrices σ_i are self-adjoint.

The goal now is to write a Lagrangian density that will produce the correct equation of motion, that is such that the corresponding Euler-Lagrange equations coincide with the Dirac equation. Let us recall first that the Lagrangian has to be *real-valued*[48] and is used to produce equations for real-valued functions, in the sense that unknown complex-valued functions have to be considered as pairs of real-valued functions. We consider the Lagrangian density

$$\mathcal{L} = \frac{i\hbar}{2}(\bar{\psi}\gamma^\mu \partial_\mu \psi - \partial_\mu \bar{\psi}\gamma^\mu \psi) - mc\bar{\psi}\psi. \tag{10.120}$$

In this very compact notation four complex-valued functions, i.e. eight real-valued functions over $\mathbb{R}^{1,3}$ give the components of ψ. Using (10.119) we get

$$(\bar{\psi}\gamma^\mu \partial_\mu \psi)^\dagger = \partial_\mu \psi^\dagger (\gamma^\mu)^\dagger \gamma_0 \psi = \partial_\mu \bar{\psi}\gamma^\mu \psi,$$

so that the two numbers $\bar{\psi}\gamma^\mu \partial_\mu \psi$ and $\partial_\mu \bar{\psi}\gamma^\mu \psi$ are complex conjugate and the Lagrangian is real as it should be.

Exercise 10.20.1 Perform dimensional analysis, and explain the normalization of the Dirac field. Hint: Review Exercise 3.8.3.

It is no fun to write the Euler-Lagrange equations, but matters simplify greatly if we observe that these equations are just the same as those obtained by pretending that the unknowns are the four components of ψ, and the four components of $\bar{\psi}$, these being independent of each other.[49] We will use even more compact notation to describe the Euler-Lagrange equations, e.g. by writing $\partial\mathcal{L}/\partial\bar{\psi}$ for the vector with four components corresponding to the partial derivatives of $\mathcal{L}$ with respect to each of the components of $\bar{\psi}$. Thus

48 A basic reason why a Lagrangian should be real is that a complex Lagrangian would typically give too many constraints when writing the Euler-Lagrange equations, as both the real and the complex part of the Lagrangian would give such an equation.

49 This is basically a change of variables. The idea is that rather than using independent real variables x, y, we may use independent variables $x + iy$ and $x - iy$.

$$\frac{\partial \mathcal{L}}{\partial \bar{\psi}} = \frac{i\hbar}{2}\gamma^\mu \partial_\mu \psi - mc\psi \; ; \; \frac{\partial \mathcal{L}}{\partial(\partial_\mu \bar{\psi})} = -\frac{i\hbar}{2}\gamma^\mu \psi$$

and the corresponding Euler-Lagrange equation $\partial\mathcal{L}/\partial\bar{\psi} = \partial_\mu(\partial\mathcal{L}/\partial(\partial_\mu\bar{\psi}))$ simply says that ψ must satisfy the Dirac equation. Similarly one also finds that $\bar{\psi}$ satisfies the adjoint Dirac equation (10.118).

The Lagrangian density (10.120) gives rise to conjugate momenta

$$\pi = \frac{\partial \mathcal{L}}{\partial(\partial\psi/\partial t)} = \frac{\partial \mathcal{L}}{c\partial(\partial_0\psi)} = \frac{i\hbar}{2c}\bar{\psi}\gamma^0 = \frac{i\hbar}{2c}\psi^\dagger, \tag{10.121}$$

$$\bar{\pi} = \frac{\partial \mathcal{L}}{c\partial(\partial_0\bar{\psi})} = -\frac{i\hbar}{2c}\gamma^0\psi. \tag{10.122}$$

Suppose now that we want to argue that "the Dirac field of Section 10.19 is a quantization of the classical Dirac field satisfying the canonical anti-commutation relations". Defining $\pi_k = (i\hbar/2c)\psi_k^\dagger$, the second part of (10.115) can be then reformulated as

$$\{\psi_k(t,\boldsymbol{x}), \pi_{k'}(t,\boldsymbol{y})\} = \frac{i\hbar}{2}\delta_k^{k'}\delta^{(3)}(\boldsymbol{x}-\boldsymbol{y})1. \tag{10.123}$$

"Canonical anti-commutation relations" would require not having this factor 1/2 on the right-hand side of (10.123). This is not a real problem since anyway we are not doing "canonical quantization" which would require commutators rather than anti-commutators. Unfortunately several textbooks deal with this factor 1/2 using in cold blood the following fallacious argument: They consider the "Lagrangian density"

$$\mathcal{L} = \bar{\psi}(i\hbar\gamma^\mu\partial_\mu\psi - mc\psi), \tag{10.124}$$

which gives the same equation of motion as (10.120). For this density (which is not real) the conjugate momentum is $\pi = i\hbar\psi^\dagger/c$, which conveniently removes the factor 1/2 in (10.123) allowing the author to call this "canonical commutation relations". Even if one does not object to complex Lagrangians when they provide the correct equation of motion, why then is the choice of (10.124) more canonical than the choice $\mathcal{L} = (i\hbar\partial_\mu\bar{\psi}\gamma^\mu + mc\bar{\psi})\psi$ which also gives the same equation of motion but gives the very different conjugate momentum $\pi = 0$?[50]

Let us try now to compute the Hamiltonian corresponding to the Lagrangian (10.120). It is not possible to compute the "velocity" $\partial_0\psi$ as a function of the "momentum" π (since the expression of π involves only ψ and no derivative).[51] Still one may hope (just as in the formula (H.2)) that the quantity

$$\mathcal{H} := c\pi\,\partial_0\psi + c\partial_0\bar{\psi}\bar{\pi} - \mathcal{L} \tag{10.125}$$

will represent a Hamiltonian density (observe that the order of the terms is dictated by the way the matrices multiply). Unfortunately it can be shown that the resulting Hamiltonian

50 I am just here making the point that what is printed in basic physics textbooks need not make sense, yet another obstacle to self-education. The whole approach is flawed anyway because one needs the theory of constrained systems for a proper analysis, see e.g. Dirac's textbook [22].

51 We refer the reader to Section H.1 for an example of a similar situation.

obtained by integrating this density is not bounded below. This is a very serious problem if it is supposed to represent an energy. The root of this problem is that the Dirac equation deals with fermions, and that those have no equivalent in Classical Mechanics, so here (again!) we are trying to force a true quantum situation into the framework of Classical Mechanics.[52] Nonetheless it takes more than that to stop us, so we compute the quantity $\int \mathrm{d}^3\boldsymbol{x}\mathcal{H}$ where $\mathcal{H}$ is the quantity (10.125), and when we substitute in $\mathcal{H}$ for π and $\bar{\pi}$ the values (10.121) and (10.122) and for ψ and $\bar{\psi}$ the values $\psi(x)$ and $\bar{\psi}(x)$ given by (10.108) and (10.109). The computation is simplified by the fact that ψ satisfies the Dirac equation and $\bar{\psi}$ satisfies the adjoint Dirac equation, so that after the substitution the value of $\mathcal{L}$ is zero and the value of $\mathcal{H}$ reduces to $\mathrm{i}\hbar(\psi^\dagger(x)\partial_0\psi(x) - (\partial_0\psi^\dagger(x))\psi(x))/2$. Our candidate for the Hamiltonian is then

$$\frac{\mathrm{i}\hbar}{2}\int \mathrm{d}^3\boldsymbol{x}(\psi^\dagger(x)\partial_0\psi(x) - \partial_0(\psi^\dagger(x))\psi(x)), \tag{10.126}$$

where $x = (ct, \boldsymbol{x})$ for a certain t, and should be independent of this value of t.

Our next goal is to actually compute the expression (10.126) when we replace $\psi(x)$ by the Dirac field (as we did for the basic free field in Section 6.11), and for this we need the following preparatory lemma.

Lemma 10.20.2 *We have*

$$u(\boldsymbol{p},s)^\dagger u(\boldsymbol{p},s') = v(\boldsymbol{p},s)^\dagger v(\boldsymbol{p},s') = \frac{2p_0}{mc}\delta_s^{s'}. \tag{10.127}$$

Proof Using (9.36) and (9.37) we obtain

$$S(D_p)^\dagger S(D_p) = \begin{pmatrix} D_p^{-1} & 0 \\ 0 & D_p^\dagger \end{pmatrix}\begin{pmatrix} D_p^{\dagger -1} & 0 \\ 0 & D_p \end{pmatrix} = \frac{1}{mc}\begin{pmatrix} M(P(p)) & 0 \\ 0 & M(p) \end{pmatrix}$$

and the result follows easily since $u(\boldsymbol{p},s)^\dagger u(\boldsymbol{p},s') = f_s^\dagger S(D_p)^\dagger S(D_p) f_{s'}$ (etc.). □

Proposition 10.20.3 *At the level of formal computation we have*

$$\frac{\mathrm{i}\hbar}{2}\int \mathrm{d}^3\boldsymbol{x}(\psi^\dagger(x)\partial_0\psi(x) - \partial_0(\psi^\dagger(x))\psi(x))$$
$$= \int \frac{\mathrm{d}^3\boldsymbol{p}}{(2\pi\hbar)^3} c\omega_{\boldsymbol{p}} \sum_{s=1,2}(a^\dagger(\boldsymbol{p},s)a(\boldsymbol{p},s) - b(\boldsymbol{p},s)b^\dagger(\boldsymbol{p},s)). \tag{10.128}$$

Proof Expressions such as $b(\boldsymbol{p},s)b^\dagger(\boldsymbol{p},s)$ cannot be defined meaningfully, but let us perform this computation seen as encoding a correct computation in the case where the universe has been put in a box. We have

$$\int \mathrm{d}^3\boldsymbol{x}\psi^\dagger(x)\partial_0\psi(x) = \int \mathrm{d}^3\boldsymbol{x}(\psi^{+\dagger}(x) + \psi^{-\dagger}(x))\partial_0(\psi^+(x) + \psi^-(x)), \tag{10.129}$$

[52] The problem of the unbounded negative energy of the Hamiltonian could be solved by writing a Hamiltonian made up of anti-commuting Grassmann quantities, but this is hardly satisfactory either.

and we compute each of the corresponding four terms.

$$\psi^{-\dagger}(x) = \sum_{s\le 2}\int \frac{\mathrm{d}^3\boldsymbol{p}}{(2\pi\hbar)^3}\frac{c\sqrt{m}}{\sqrt{2\omega_{\boldsymbol{p}}}}\exp(\mathrm{i}(x,p)/\hbar)u(\boldsymbol{p},s)^\dagger a^\dagger(\boldsymbol{p},s), \tag{10.130}$$

$$\partial_0\psi^-(x) = -\frac{\mathrm{i}}{\hbar}\sum_{s\le 2}\int \frac{\mathrm{d}^3\boldsymbol{p}}{(2\pi\hbar)^3}\frac{c\sqrt{m\omega_{\boldsymbol{p}}}}{\sqrt{2}}\exp(-\mathrm{i}(x,p)/\hbar)u(\boldsymbol{p},s)a(\boldsymbol{p},s), \tag{10.131}$$

and using that $\int \mathrm{d}^3\boldsymbol{x}\exp(\mathrm{i}(x,p-p')/\hbar) = (2\pi\hbar)^3\delta^{(3)}(\boldsymbol{p}-\boldsymbol{p}')$ and Lemma 10.20.2 we then find

$$\int \mathrm{d}^3\boldsymbol{x}\,\psi^{-\dagger}(x)\partial_0\psi^-(x) = -\frac{\mathrm{i}c}{\hbar}\int\frac{\mathrm{d}^3\boldsymbol{p}}{(2\pi\hbar)^3}\,\omega_{\boldsymbol{p}}\sum_{s=1,2} a^\dagger(\boldsymbol{p},s)a(\boldsymbol{p},s),$$

and in a isomorphic computation

$$\int \mathrm{d}^3\boldsymbol{x}\,\psi^{+\dagger}(x)\partial_0\psi^+(x) = \frac{\mathrm{i}c}{\hbar}\int\frac{\mathrm{d}^3\boldsymbol{p}}{(2\pi\hbar)^3}\,\omega_{\boldsymbol{p}}\sum_{s=1,2} b(\boldsymbol{p},s)b^\dagger(\boldsymbol{p},s).$$

The other terms are computed in a similar fashion. □

We now reach the climax, but you may need to review Section 6.11 and make sure you understand the formula (5.36) to appreciate it. Despite the fact that the term $b(\boldsymbol{p},s)b^\dagger(\boldsymbol{p},s)$ makes no sense, we use the anti-commutation relation

$$b^\dagger(\boldsymbol{p},s)b(\boldsymbol{p}',s) + b(\boldsymbol{p}',s)b^\dagger(\boldsymbol{p},s) = (2\pi\hbar)^3\delta^{(3)}(\boldsymbol{p}-\boldsymbol{p}')1$$

for $\boldsymbol{p} = \boldsymbol{p}'$ to argue that modulo an irrelevant infinite constant the Hamiltonian is

$$\int\frac{\mathrm{d}^3\boldsymbol{p}}{(2\pi\hbar)^3}\,c\omega_{\boldsymbol{p}}\sum_{s=1,2}(a^\dagger(\boldsymbol{p},s)a(\boldsymbol{p},s) + b^\dagger(\boldsymbol{p},s)b(\boldsymbol{p},s)).$$

We expect the Hamiltonian to represent the energy of the system. This is indeed what we obtain here, the sum of the energies of the particles plus the sum of the energy of the anti-particles. More precisely the term

$$\int\frac{\mathrm{d}^3\boldsymbol{p}}{(2\pi\hbar)^3}\,c\omega_{\boldsymbol{p}}\sum_{s=1,2} a^\dagger(\boldsymbol{p},s)a(\boldsymbol{p},s)$$

expresses the extension to the fermion Fock space of the Hamiltonian "multiplication by the function $c\omega_{\boldsymbol{p}}$" represents the sum of the energy of the particles.[53] and the other term represents the sum of the energy of the anti-particles.

[53] You should convince yourself that the sum over s in the previous expression compared to (5.36) occurs because there are two possible values for the spin.

10.21 Majorana Field

In Section 10.19 (of which we keep the notation) we used the space $\mathcal{G}$ of (10.106) for the particle and the space $\mathcal{G}'$ for the anti-particle. As explained just below (10.106), we may also choose the same space $\mathcal{G}$ or $\mathcal{G}'$ for both the particle and the anti-particle. The anti-particle may then be the same as the particle, or may be different. The resulting quantum field describes fermions (spin 1/2). These, called "Majorana fermions", are *not the same* as the usual fermions (which are called Dirac fermions in this context). The mathematical structure corresponding to Majorana fermions is just as natural as the structure corresponding to Dirac fermions. Dirac fermions are ubiquitous. Majorana fermions have not yet been observed.[54] Experimentalists are currently trying to determine whether neutrinos (which are now known to have a tiny mass) are Dirac or Majorana fermions.

For the mathematical construction of the Majorana field, we assume that the particle is its own anti-particle. There are two versions of the construction, one with intrinsic parity 1 using $\mathcal{G}$ and one with intrinsic parity -1 using $\mathcal{G}'$. Let us give the formulas, first in the case where the intrinsic parity of the fermion is 1. We consider the vectors

$$f_1 = \begin{pmatrix}1\\0\\1\\0\end{pmatrix} ; \; f_2 = \begin{pmatrix}0\\1\\0\\1\end{pmatrix} ; \; f_1' = -f_2 = \begin{pmatrix}0\\-1\\0\\-1\end{pmatrix} ; \; f_2' = f_1 = \begin{pmatrix}1\\0\\1\\0\end{pmatrix} \tag{10.132}$$

so that f_1 and f_2 form a basis of $\mathcal{G} = \mathcal{V}$. The quantum field is given by the formulas (10.66) and (10.67) (with $b^\dagger(s,\boldsymbol{p})$ replaced by $a^\dagger(s,\boldsymbol{p})$ because the anti-particle now coincides with the particle) and where now

$$u(\boldsymbol{p},s) = S(D_p)(f_s) \, ; \; v(\boldsymbol{p},s) = S(D_p)(f_s'), \tag{10.133}$$

and where S is the same representation of $SL^+(2,\mathbb{C})$ on $\mathbb{C}^4$ as in Section 10.19. (When using these formulas we keep the normalization of the fields (10.66) and (10.67), which may not be what the physicists use.) Modifying the formulas (10.132) by changing the signs of the components in the last two lines, one obtains another version of the construction where the Majorana fermion is of intrinsic parity -1.

Exercise 10.21.1 Prove that these fields satisfy the microcausality anti-commutation relations of Theorem 10.19.4.

10.22 Lack of a Suitable Field for Photons

In Section 9.13 we constructed a representation $\pi^+_{0,2}$ of $\mathcal{P}^{*+}$ corresponding to photons. Does there exist a four-component quantum field based on this representation and satisfying

[54] Nature sometimes simply does not make use of possible mathematical structures, however beautiful.

"Lorentz covariance with parity", as discussed in Section 10.18? Such a field could be a suitable model for a quantized version of the electromagnetic field, whose quanta are photons.[55] The natural choice for S is the extension of κ to $SL^+(2,\mathbb{C})$ defined by (8.42). In the present section we point out that *no suitable field constructed within the previous framework exists*. This is one reason for which the quantization of the electromagnetic field is not as simple as one wishes.[56]

The reason for this failure is not deep. Let us denote by G_0^+ the little group of $SL^+(2,\mathbb{C})$ corresponding to the point $(1,0,0,1)$. The representation $\pi_{0,2}^+$ is induced (in a sense which is a rather direct generalization of the way we used this terminology in Theorem 9.4.2) by a certain representation V^+ of G_0^+ on $\mathbb{C}^2$ which is described below. Reproducing the analysis of Proposition 10.4.2, to be able to construct the annihilation part of the fields one has to find a subspace $\mathcal{G}$ of $\mathbb{C}^4$ so that the restriction of S to $\mathcal{G}$ and G_0^+ is equivalent to the representation V^+. It is an exercise to show that no such subspace exists when S is as previously described.

Exercise 10.22.1 (a) Prove that the element $Q := P'J \in SL^+(2,\mathbb{C})$ belongs to G_0^+, where J is the matrix (8.1). Prove that $Q^2 = -I$.
(b) Prove that each element of G_0^+ is of the type A or AQ for $A \in G_0$.
(c) Prove that for each $A \in SL^+(2,\mathbb{C})$ we have $QA^* = AQ$.
(d) Prove that given a unitary representation V of G_0 in a finite-dimensional space $\mathcal{V}$ one may define a unitary representation V^+ of G_0^+ on $\mathcal{V} \times \mathcal{V}$ by the formula

$$A \in SL^+(2,\mathbb{C}) \Rightarrow V^+(A)(x,y) = (V(A)x, V(A^*)y)\,;\; V^+(Q)(x,y) = (y, V(-I)x).$$

It can be shown that if π is the representation of $\mathcal{P}^*$ induced by the unitary representation V then π^+ (defined in Proposition 9.13.1) is induced by the unitary representation V^+.

Exercise 10.22.2 Prove it. If you succeed, you really have mastered this material. Hint: It helps a lot to use the appropriate version of Theorem 9.8.1 to describe the induced representations, but this is not so easy to do.

In particular when $\mathcal{G} = \mathbb{C}$ and $V = \hat{\pi}_2$ the representation induced by V^+ is $\pi_{0,2}^+$.

Exercise 10.22.3 (a) Give an example of a representation S of $SL^+(2,\mathbb{C})$ in $\mathbb{C}^8$ such that there is a subspace $\mathcal{G}$ such that the restriction of S to $\mathcal{G}$ and G_0^+ is the preceding representation V^+ (for $V = \hat{\pi}_2$).
(b) Prove that when S is the extension of κ to $SL^+(2,\mathbb{C})$ (as defined by (8.42)) no such subspace exists.

Key ideas to remember:

- Conservation of charge almost implies the existence of anti-particles.
- The fruitful multicomponent quantum fields are Lorentz covariant and satisfy the microcausality condition.

[55] It is actually the four-dimensional potential vector of the electromagnetic field which gets quantized.
[56] A particularly intriguing approach to such a quantization is detailed in Appendix I.

- To each such quantum field is associated an elementary particle called the quantum of the field.
- When the quanta of the fields have a non-integer spin the construction of the quantum field must use fermion Fock space rather than boson Fock space.
- The anti-particle of a massless particle has opposite helicity.
- The Dirac field is associated to particles of spin 1/2 in a theory that includes parity.

Part III

Interactions

From now on we use a system of units in which $\hbar = 1$ and $c = 1$, although at times for clarity we will write their symbols, as in the formula $E = mc^2$, which looks better than $E = m$ unless you are a professional physicist.

We should also stress that in Part III we are really doing physics. This is the domain of approximations and of hand-waving arguments. Therefore any statement called "Theorem" has to be taken with a grain of salt: It is not intended to be understood as a totally sound mathematical result, and the "proofs" are not ε-δ proofs in the mathematical sense. The few occasions where we make a real attempt at mathematical rigor (such as Theorem 12.4.1) will be apparent enough.

11

Perturbation Theory

Assume that we are given a quantum system governed by a Hamiltonian which we understand well, and that we make a small change to this Hamiltonian (a perturbation). How do we relate the behavior of the system with and without the perturbation? This is a fundamentally important question because we can often think of the system of real interest as a small perturbation of a much simpler system.[1] In our approach to Quantum Field Theory, an interacting system will be viewed as a perturbation of a simpler non-interacting system.

Thus, assuming we know well the "unperturbed" Hamiltonian H_0 we want to study its perturbation $H_0 + gH_I$ where g is small. The notation H_I reflects the way we will mostly use this: H_I represents "the interaction". The quantity g that governs the intensity of the perturbation (or the strength of the interaction when H_I represents an interaction) is called the *coupling constant*. The coupling constant is denoted by the letter g throughout this work.

The basic idea will then be to study whatever quantity we are interested in for the full Hamiltonian $H_0 + gH_I$ through an expansion in powers of g. As g is supposed to be small, terms containing high powers of g should be small too.

Definition 11.1 Working at the kth order in perturbation theory means throwing out of the computations any term containing a power g^n with $n > k$.

This will typically be done without any knowledge of the actual size of the terms that are thrown out. Specifically, we may find ourselves approximating $g + 10^{10000}g^2$ by g. This is horrendous for a mathematician but it typically works wonderfully in practice.[2]

11.1 Time-independent Perturbation Theory

The "time-independent Schrödinger equation" is simply the eigenvalue equation for the Hamiltonian. Time-independent perturbation theory studies how the eigenvalues and the eigenvectors of the Hamiltonian change under a small change of this Hamiltonian. A typical application of this result would be the study of how much the energy levels of the electron of a hydrogen atom are modified by a small magnetic field.

[1] Very few problems in physics can be solved exactly. They are called exactly solved models. For the others one has to make approximations.

[2] Typically but not always. In the study of strong interactions the coupling constant is of order unity and perturbation theory is of limited use.

Matters need not be simple. Recalling the Pauli matrices (2.3), the vector $\binom{1}{1}$ is an eigenvector of $\sigma_0 = I$ but is not close to an eigenvector of $\sigma_0 + g\sigma_3$ however small g is. Here we are in a special situation: The eigenspace of σ_0 we start from has dimension > 1. This situation is important in physics, but we shall not study it.

Let us assume instead that all goes well: The unperturbed Hamiltonian H_0 has an eigenvector v_0 with eigenvalue λ_0 while the perturbed Hamiltonian $H_0 + gH_I$ has an eigenvector v_g with eigenvalue λ_g, both depending smoothly on g, in such a way that we have an expansion

$$\lambda_g = \lambda_0 + a_1 g + a_2 g^2 + O(g^3)\ ;\ v_g = v_0 + g w_1 + g^2 w_2 + O(g^3), \tag{11.1}$$

for numbers a_1, a_2 and vectors w_1, w_2. There are of course theorems to the effect that these expansions hold, but our approach in this chapter is heuristic. Assuming that the expansions hold, our arguments assume only that H_0 and H_I are symmetric. We assume moreover that v_g has been normalized so that[3]

$$(v_0, v_g) = 1. \tag{11.2}$$

Taking $g = 0$ we obtain in particular $(v_0, v_0) = 1$.

We wish to compute the coefficients of the expansions (11.1).[4] In such an expansion it is the first non-zero coefficient that has the greatest importance. The "generic case" where the first-order coefficient is not zero has countless applications in Quantum Mechanics. Here, however we have in mind a specific application in Section 11.5 to a situation where $a_1 = 0$[5] and for this reason we focus on the computation of a_2. First we observe that

$$1 = (v_0, v_g) = 1 + g(v_0, w_1) + g^2(v_0, w_2) + O(g^3),$$

so that

$$(v_0, w_1) = (v_0, w_2) = 0. \tag{11.3}$$

Let us now write

$$(H_0 + gH_I)(v_g) = \lambda_g v_g = (\lambda_0 + a_1 g + a_2 g^2)(v_0 + g w_1 + g^2 w_2) + O(g^3)$$

and also

$$(H_0 + gH_I)(v_g) = (H_0 + gH_I)(v_0 + g w_1 + g^2 w_2) + O(g^3).$$

Equating the coefficients of g and g^2 respectively yields

$$a_1 v_0 + \lambda_0 w_1 = H_I v_0 + H_0 w_1 \tag{11.4}$$

and

$$\lambda_0 w_2 + a_1 w_1 + a_2 v_0 = H_0 w_2 + H_I w_1. \tag{11.5}$$

[3] This leads to simpler calculations than the more natural choice $(v_g, v_g) = 1$.

[4] In the absence of concrete error bounds, this does not tell us for which values of g the expansions (11.1) are useful. In physics however, the procedure is to use the approximation and see if the corresponding prediction matches experiments. Rigorous bounds on the error terms are anyway typically far too pessimistic to be of use.

[5] Our goal is to motivate the fundamental fact that "interactions change the observed mass of the particle".

Taking the inner product of (11.4) with v_0, and since $(v_0, v_0) = 1$ and $(v_0, H_0 w_1) = (H_0 v_0, w_1) = \lambda_0 (v_0, w_1) = 0$ we obtain

$$a_1 = (v_0, H_I v_0). \tag{11.6}$$

Taking the inner product of (11.5) with v_0 we obtain similarly

$$a_2 = (v_0, H_I w_1). \tag{11.7}$$

We observe that the projection P on the orthogonal complement of $\mathbb{C}v_0$ is given by $Pu = u - v_0(v_0, u)$. Now, using (11.4) in the first equality below and (11.6) in the second one,

$$(H_0 - \lambda_0) w_1 = -(H_I - a_1) v_0 = -(H_I v_0 - (v_0, H_I v_0) v_0) = -P H_I v_0. \tag{11.8}$$

Since w_1 belongs to the orthogonal complement of v_0, on which $H_0 - \lambda_0$ is (hopefully...) invertible, we should be able to compute w_1 from (11.8) and then a_2 from (11.7). It is instructive to write the resulting formulas the way physicists do. Assume for this that H_0 has an orthonormal basis $(v_n)_{n \geq 0}$ of eigenvectors, with corresponding eigenvalues λ_n, and assume $\lambda_n \neq \lambda_0$ for $n \geq 1$. Observe that

$$(v_n, H_0 w_1) = (H_0 v_n, w_1) = \lambda_n (v_n, w_1),$$

and

$$(v_n, P H_I v_0) = (P v_n, H_I v_0) = (v_n, H_I v_0),$$

so that taking the inner product of (11.8) with v_n we obtain

$$(v_n, w_1) = \frac{(v_n, H_I v_0)}{\lambda_0 - \lambda_n}. \tag{11.9}$$

We note for further use that this means

$$w_1 = \sum_{n \geq 1} v_n \frac{(v_n, H_I v_0)}{\lambda_0 - \lambda_n} \tag{11.10}$$

and thus

$$\|g w_1\|^2 = \sum_{n \geq 1} \frac{|(v_n, g H_I v_0)|^2}{(\lambda_0 - \lambda_n)^2}. \tag{11.11}$$

Moreover, since

$$(v_0, H_I w_1) = (H_I v_0, w_1) = \sum_{n \geq 1} (H_I v_0, v_n)(w_1, v_n)$$

and since $(w_1, v_n) = (v_n, w_1)^*$ we obtain from (11.7) and (11.9) that

$$a_2 = \sum_{n \geq 1} \frac{|(v_n, H_I v_0)|^2}{\lambda_0 - \lambda_n}. \tag{11.12}$$

One remarkable aspect of this formula is how it depends on the whole structure of H_0, and in particular on the eigenvalues which may be quite different from λ_0. One may picture the eigenvalues larger than λ_0 as pushing a_2 down and those smaller than λ_0 as pushing it up. In particular if λ_0 is the smallest eigenvalue, then $a_2 \le 0$. This is to be expected: The minimum of $(x, H_0 x)$ over the set of unit vectors is obtained at $x = v_0$ and when $a_1 = 0$, we have $(v_0, H_I v_0) = 0$ and the small perturbation gH_I should decrease this minimum.[6]

By definition "the shift at the first order of perturbation theory of the eigenvalue λ_0" is defined as the quantity $a_1 g$ of (11.1). When it is zero "the shift at the second order of perturbation theory of the eigenvalue" is defined as $a_2 g^2$. According to (11.12) we may state:

Proposition 11.1.1 *When the shift of an eigenvalue is zero at the first order of perturbation theory, this shift is given at the second order of perturbation theory by*[7]

$$\sum_{n \ge 1} \frac{|(v_n, gH_I v_0)|^2}{\lambda_0 - \lambda_n}. \tag{11.13}$$

The next exercise provides a simple application of the present theory. A far more complex application will be given in Section 11.5.

Exercise 11.1.2 In this exercise we use the notation of Section 2.18. We take $H_0 = a^\dagger a$ and $H_I = \gamma a^\dagger + \gamma^* a$ where $\gamma \in \mathbb{C}$. The smallest eigenvalue of H_0 is zero.

(a) Writing $H_0 + gH_I = (a^\dagger + g\gamma^* 1)(a + g\gamma 1) - g^2|\gamma|^2 1$, prove that the smallest eigenvalue of $H_0 + gH_I$ is $-g^2|\gamma|^2$. Hint: Prove that for an operator B, all the eigenvalues of $B^\dagger B$ are ≥ 0.

(b) Prove that in this case the shift of the smallest eigenvalue of H is zero at the first-order perturbation theory, and that at second-order perturbation theory (11.13) provides the exact value of the shift.

We end this section with a side story. We examine the specific problem of finding an upper bound for the smallest eigenvalue of $H_0 + gH_I$. A general method to bound from above the ground state energy (= smallest eigenvalue) of a self-adjoint operator U is to observe that for any non-zero vector z (assuming for simplicity that U has a basis of eigenvectors), U has an eigenvalue $\le (z, Uz)/(z, z)$.[8] Of course one has to make a smart choice for z to get a useful bound.

To use this method in the present setting, consider a self-adjoint operator H_0, its smallest eigenvalue λ_0 and a unit vector v_0 with $H_0 v_0 = \lambda_0 v_0$. To lighten notation, without loss of generality we may assume that $\lambda_0 = 0$ by replacing H_0 by $H_0 - \lambda_0 1$. Consider a perturbation gH_I of this operator, and assume that

$$(v_0, H_I v_0) = 0, \tag{11.14}$$

[6] Think of what happens when making the small perturbation $f + gh$ of the function $f(x) = x^2$, where $h(0) = 0$. Unless $h'(0) = 0$, the minimum decreases.

[7] We focus on this case rather than on the generic case where the shift at first order is not zero not because it is more important but because it is needed for the examples we have in mind in Section 11.5.

[8] If U does not have a basis of eigenvectors, one uses spectral theory to conclude that the infimum of the spectrum of U is $\le (z, Uz)/(z, z)$.

so that we are "in the situation of second-order perturbation theory" (as opposed to first-order perturbation theory).

Let us then consider a vector z with $(v_0, z) = 0$, so that $(H_0 v_0, z) = (v_0, H_0 z) = 0$. Then, using these relations we obtain

$$(v_0 + z, (H_0 + gH_I)(v_0 + z)) = (z, (H_0 + gH_I)z) + 2\mathrm{Re}\,(v_0, gH_I z), \tag{11.15}$$

so that a bound from above of the ground state energy of $H_0 + gH_I$ is given by the quantity

$$\frac{1}{\|v_0 + z\|^2}\big((z, (H_0 + gH_I)z) + 2\mathrm{Re}\,(v_0, gH_I z)\big). \tag{11.16}$$

Any choice of z provides an upper bound for the ground state.

Exercise 11.1.3 Convince yourself that the perturbation method at the second order amounts to choosing z in (11.16) such as to minimize the quantity

$$(z, H_0 z) + 2\mathrm{Re}\,(v_0, gH_I z), \tag{11.17}$$

to pretend that $\|v_0 + z\| = 1$ and then to neglect the term $(z, H_I z)$. This makes clear that when $\|z\|$ is not small the method lacks credibility.

11.2 Time-dependent Perturbation Theory and the Interaction Picture

Now, in the same setting, we want to compare the time-evolutions under the unperturbed Hamiltonian H_0 and the full Hamiltonian $H_0 + gH_I$. It will be useful to consider more generally a perturbation of H_0 of the type $H_0 + g(t)H_I$. This allows us in particular to consider the case of perturbations that oscillate with a given frequency, and will also allow us to turn on and off the perturbation in a smooth way. Still more generally we consider a perturbation of H_0 of the type $H_0 + H_1(t)$, where, as the notation indicates, the operator $H_1(t)$ now depends on the time t. A solid mathematical framework would assume for example that H_0 is self-adjoint, and that $H_1(t)$ is self-adjoint and bounded. Then the operator $H_0 + H_1(t)$ is self-adjoint with the same domain as H_0. Unfortunately, the cases we will really need will be quite more pathological, so the forthcoming manipulations are pretty formal.

First we have to explain what is the time-evolution of a system under the time-varying Hamiltonian $H(t) = H_0 + H_1(t)$. It is given[9] by the unitary operator $U(t)$ such that $U'(t) := \mathrm{d}U(t)/\mathrm{d}t = -\mathrm{i}H(t)U(t)$. If we want to prove rigorous theorems the domain of validity of this equation is an issue, but we are not concerned with such issues here. (In the good case where $H_1(t)$ is bounded this equation holds in the domain of H_0.)

At this stage, the reader might want to brush up on the Schrödinger and the Heisenberg pictures of Section 2.16.

The basic idea is that we know the time-evolution[10] $U_0(t) = \exp(-\mathrm{i}tH_0)$ and that we want to study the unknown time-evolution $U(t)$. Presumably these time-evolutions will not be much different from each other, so that we will study their "difference":

[9] Keeping in mind that our system of units is now such that $\hbar = 1$.

[10] The important case in practice is that not only we know it, but also that it is pretty simple, much more so than the evolution under $H_0 + H_1(t)$.

$$V(t) := U_0(t)^{-1}U(t) = U_0(-t)U(t) \tag{11.18}$$

from which we can reconstruct

$$U(t) = U_0(t)V(t) \tag{11.19}$$

since we assume that $U_0(t)$ is known. We observe that by Lemma 2.14.6 we have $U_0'(t) = -\mathrm{i}H_0U_0(t) = -\mathrm{i}U_0(t)H_0$. Then,

$$\begin{aligned} V'(t) &= -U_0'(-t)U(t) + U_0(-t)U'(t) \\ &= \mathrm{i}U_0(-t)H_0U(t) - \mathrm{i}U_0(-t)H(t)U(t) \\ &= -\mathrm{i}U_0(-t)H_1(t)U(t) \\ &= -\mathrm{i}U_0(-t)H_1(t)U_0(t)V(t), \end{aligned}$$

and finally

$$V'(t) = -\mathrm{i}\tilde{H}_1(t)V(t), \tag{11.20}$$

where

$$\tilde{H}_1(t) := U_0(-t)H_1(t)U_0(t). \tag{11.21}$$

The situation where we time-evolve the operators A according to

$$A(t) = U_0(-t)AU_0(t) \tag{11.22}$$

and the states according to

$$x(t) = V(t)(x) = U_0(-t)U(t)x, \tag{11.23}$$

is called the *interaction picture*.[11] It is intermediate between the Schrödinger picture (where $A(t) = A$ and $x(t) = U(t)x$) and the Heisenberg picture (where $A(t) = U(-t)AU(t)$ and $x(t) = x$) and it is consistent with these, since at time t the average value of $A(t)$ for a system in state $x(t)$ is given by

$$(x(t), A(t)x(t)) = (U_0(-t)U(t)x, U_0(-t)AU(t)x) = (U(t)x, AU(t)x),$$

see (2.84). One should be cautious here that the operators $V(t)$ do *not* form a one-parameter group. The evolution of a state between times s and t is given by $V(t)V(s)^{-1}$ which need not be the same as $V(t-s)$.

Exercise 11.2.1 This exercise presents another way to reach the interaction picture, closer to what is found in traditional physics texts. We want to solve the Schrödinger equation

$$\mathrm{i}\frac{\partial}{\partial t}|\varphi(t)\rangle = H(t)|\varphi(t)\rangle$$

11 This is one more of the contributions of Paul Dirac.

where $H(t) = H_0 + H_1(t)$. The idea is to "factor out" the time-evolution due to H_0 by defining $|\psi(t)\rangle = \exp(\mathrm{i}tH_0)|\varphi(t)\rangle = U_0(-t)|\varphi(t)\rangle$. Prove that $|\psi(t)\rangle$ satisfies the equation

$$\mathrm{i}\frac{\partial}{\partial t}|\psi(t)\rangle = \tilde{H}_1(t)|\psi(t)\rangle$$

where $\tilde{H}_1(t)$ is given by (11.21).

How does one solve (11.20)? You may think at first that

$$V(t) \stackrel{?}{=} \exp\Big(-\mathrm{i}\int_0^t \mathrm{d}\theta\, \tilde{H}_1(\theta)\Big), \tag{11.24}$$

but this is completely wrong because the operators $\tilde{H}_1(\theta)$ do not commute with each other. Rather, we integrate (11.20) to get

$$V(t) = 1 - \mathrm{i}\int_0^t \mathrm{d}\theta\, \tilde{H}_1(\theta)V(\theta). \tag{11.25}$$

We then get a zeroth-order approximation $V(t) \simeq 1$. Setting $V_0(t) \equiv 1$ and replacing $V(\theta)$ by this zeroth-order approximation in the integral yields a first-order approximation

$$V_1(t) = 1 - \mathrm{i}\int_0^t \mathrm{d}\theta\, \tilde{H}_1(\theta). \tag{11.26}$$

Of course the expression "first-order approximation" is somewhat optimistic as we have no bound for the error term. Using in turn the first-order approximation in the integral rather than the exact value of $V(t)$ we get the second-order approximation

$$V_2(t) = 1 - \mathrm{i}\int_0^t \mathrm{d}\theta\, \tilde{H}_1(\theta) + (-\mathrm{i})^2\int_0^t \mathrm{d}\theta_1\, \tilde{H}_1(\theta_1)\int_0^{\theta_1} \mathrm{d}\theta_2\, \tilde{H}_1(\theta_1). \tag{11.27}$$

Here the intermediate variables θ_1, θ_2 are labeled[12] so that $\theta_2 < \theta_1$.

We can continue this process and obtain the kth order approximation

$$V_k(t) = 1 + \sum_{n\le k} W_n(t), \tag{11.28}$$

where

$$W_n(t) = (-\mathrm{i})^n \int_0^t \mathrm{d}\theta_1 \int_0^{\theta_1} \mathrm{d}\theta_2 \ldots \int_0^{\theta_{n-1}} \mathrm{d}\theta_n\, \tilde{H}_1(\theta_1)\cdots\tilde{H}_1(\theta_n). \tag{11.29}$$

We can then write in a purely formal way

$$V(t) = 1 + \sum_{n\ge 1} W_n(t), \tag{11.30}$$

[12] When using time-ordering I try to consistently number the points in decreasing order.

a series that is called the *Dyson series*. (In particular we make no claim here as to the convergence of that series.) Recalling (11.19) we obtain from (11.30)

$$U(t) = U_0(t) + \sum_{n\geq 1} U_n(t), \tag{11.31}$$

where $U_n(t) = U_0(t)W_n(t)$, so that since

$$U_0(t)\tilde{H}_1(\theta) = U_0(t-\theta)H_1(\theta)U_0(\theta),$$

we obtain

$$U_1(t) = -\mathrm{i}\int_0^t \mathrm{d}\theta\, U_0(t-\theta)H_1(\theta)U_0(\theta), \tag{11.32}$$

and similarly

$$U_2(t) = (-\mathrm{i})^2 \iint_{0\leq\theta_2\leq\theta_1\leq t} \mathrm{d}\theta_1\mathrm{d}\theta_2 U_0(t-\theta_1)H_1(\theta_1)U_0(\theta_1-\theta_2)H_1(\theta_2)U_0(\theta_2), \tag{11.33}$$

etc. It helps the intuition to form the following picture: For (11.32) the system evolves freely (i.e. according to U_0) until time θ, where it is hit by the interaction, and then freely again from time θ to time t. For (11.33), the system is hit twice by the interaction at times $0 < \theta_2 < \theta_1 < t$ and in between evolves freely.

It is unpleasant to deal with the complicated domain of integration in (11.29). A simple and very important trick simplifies this domain. To understand this trick in the simplest case we observe the equality

$$\iint_{0\leq\theta_2\leq\theta_1\leq t} \mathrm{d}\theta_1\mathrm{d}\theta_2 f(\theta_1,\theta_2) = \frac{1}{2}\iint_{0\leq\theta_1,\theta_2\leq t} \mathrm{d}\theta_1\mathrm{d}\theta_2 g(\theta_1,\theta_2)$$

where $g(\theta_1,\theta_2) = f(\theta_1,\theta_2)$ if $\theta_2 < \theta_1$ and $g(\theta_1,\theta_2) = f(\theta_2,\theta_1)$ if $\theta_1 < \theta_2$.

Definition 11.2.2 The time-ordered[13] product $\mathcal{T}\tilde{H}_1(\theta_1)\cdots\tilde{H}_1(\theta_n)$ is the product $\tilde{H}_1(\theta_{\sigma(1)})\cdots\tilde{H}_1(\theta_{\sigma(n)})$ where the permutation σ of $\{1,\ldots,n\}$ satisfies $\theta_{\sigma(n)} \leq \ldots \leq \theta_{\sigma(1)}$.[14]

An equivalent expression for the time-ordered product (still not caring about lower-dimensional sets) is given by

$$\mathcal{T}\tilde{H}_1(\theta_1)\cdots\tilde{H}_1(\theta_n) = \sum_{\sigma\in S_n} 1_{\{\theta_{\sigma(n)}\leq\ldots\leq\theta_{\sigma(1)}\}}\tilde{H}_1(\theta_{\sigma(1)})\cdots\tilde{H}_1(\theta_{\sigma(n)}).$$

In (11.29) the domain of integration is the set $\{0\leq\theta_n\leq\theta_{n-1}\leq\ldots\leq\theta_1\leq t\}$ so that we can rewrite this equality as

$$W_n(t) = \frac{(-\mathrm{i})^n}{n!}\int_0^t \mathrm{d}\theta_1\ldots\int_0^t \mathrm{d}\theta_n \mathcal{T}\tilde{H}_1(\theta_1)\cdots\tilde{H}_1(\theta_n). \tag{11.34}$$

[13] This definition is appropriate for bosons, the only case we will consider from now on.
[14] This permutation is not unique if the numbers θ_j are not distinct, but we do not care about what happens on lower-dimensional sets.

This is because (neglecting lower-dimensional sets) the domain of integration can be split into $n!$ sets on each of which the permutation giving the time-ordering is constant, and each of these domains has the same contribution. This multiple counting is compensated by the $n!$ in the denominator. You can see from this formula that *when the operators* $\tilde{H}_1(\theta)$ *commute with each other* (and assuming one can sum the Dyson series), the formula (11.24) is correct.

Suppose now that $H_1(t) \equiv g(t)H_I$ for some function $g(t)$. Then setting $H_I(t) = U_0(-t)H_I U_0(t)$, (11.34) takes the form

$$W_n(t) = \frac{(-\mathrm{i})^n}{n!} \int_0^t \mathrm{d}\theta_1 \ldots \int_0^t \mathrm{d}\theta_n g(\theta_1) \cdots g(\theta_n) \mathcal{T} H_I(\theta_1) \cdots H_I(\theta_n). \tag{11.35}$$

When $g(t) = g$ is constant this becomes

$$W_n(t) = \frac{(-\mathrm{i}g)^n}{n!} \int_0^t \mathrm{d}\theta_1 \ldots \int_0^t \mathrm{d}\theta_n \mathcal{T} H_I(\theta_1) \cdots H_I(\theta_n). \tag{11.36}$$

Let us then summarize the situation for further use. Recalling (11.18) and (11.30) we have

$$V(t) = U_0(-t)U(t) = 1 + \sum_{n \geq 1} \frac{(-\mathrm{i}g)^n}{n!} \int_0^t \mathrm{d}\theta_1 \ldots \int_0^t \mathrm{d}\theta_n \mathcal{T} H_I(\theta_1) \cdots H_I(\theta_n). \tag{11.37}$$

Even though we have no real data about the size of the multiple integral, the presence of the small coefficient g^n creates a near-irresistible temptation to feel that the terms of the series are getting small and that one should get useful results by considering only the first few terms of this series. We will yield to that temptation in the next chapter. The resulting computations are the cornerstone of Quantum Field Theory.

Exercise 11.2.3 Generalize (11.37) as follows: If $s < t$ then

$$V(t)V(s)^{-1} = 1 + \sum_{n \geq 1} \frac{(-\mathrm{i}g)^n}{n!} \int_s^t \mathrm{d}\theta_1 \ldots \int_s^t \mathrm{d}\theta_n \mathcal{T} H_I(\theta_1) \cdots H_I(\theta_n). \tag{11.38}$$

Hint: $V(t)V(s)^{-1} = U_0(-s)V(t-s)U_0(s)$.

11.3 Transition Rates

Let us now assume that we know an eigenbasis $(e_n)_{n \geq 0}$ of H_0 so that $H_0 e_n = E_n e_n$ for the corresponding eigenvalues E_n. Then the time-evolution of e_n under $U_0(t)$ is given by $U_0(t)e_n = \exp(-\mathrm{i}tE_n)e_n$: In the Schrödinger picture the state e_n evolves only by a phase.[15] Thus if the system is in state e_n at time zero it is in state $\exp(-\mathrm{i}tE_n)e_n$ at time t. According to Section 2.9, if $m \neq n$, the probability that at time t we measure the state e_m is $|(e_m, \exp(-\mathrm{i}tE_n)e_n)|^2 = |(e_m, e_n)|^2 = 0$. When there is an interaction this is no longer true. The interaction produces transitions between the states e_k and for small t we are going to compute "to the lowest order" in t the probability that they have already occurred at time t.

15 Let us recall that a phase is simply a complex number of modulus 1.

We consider first the case where $H_1(t) \equiv gH_I$ so that the full Hamiltonian is $H_0 + gH_I$. Using the first-order approximation $U(t) \simeq U_0(t) + U_1(t)$ and (11.32) to compute $U_1(t)$, we obtain

$$(e_m, U(t)e_n) \simeq -\mathrm{i}g \int_0^t \mathrm{d}\theta (e_m, U_0(t-\theta)H_I U_0(\theta)e_n). \tag{11.39}$$

Since

$$\begin{aligned}(e_m, U_0(t-\theta)H_I U_0(\theta)e_n) &= (U_0(t-\theta)^\dagger e_m, H_I U_0(\theta)e_n)\\ &= \exp(-\mathrm{i}(t-\theta)E_m - \mathrm{i}\theta E_n)(e_m, H_I e_n)\\ &= \exp(-\mathrm{i}tE_m)\exp(\mathrm{i}\theta(E_m - E_n))(e_m, H_I e_n)\end{aligned}$$

we obtain from (11.39)

$$c_{n,m}(t) := |(e_m, U(t)e_n)|^2 \simeq a_{n,m} \frac{|\exp(\mathrm{i}t(E_m - E_n)) - 1|^2}{(E_m - E_n)^2}, \tag{11.40}$$

where

$$a_{n,m} := g^2 |(e_m, H_I e_n)|^2. \tag{11.41}$$

According to Section 2.9, $c_{n,m}(t)$ is the probability that when we measure at time t a system that was in state e_n at time zero, we find it in state e_m. We may say then in flowery language that this is the probability that the transition from state e_n to state e_m has already taken place at time t.

A quantity such as $(e_m, H_I e_n)$ will be called a *matrix element* of H_I. These matrix elements determine H_I. Let us consider the function

$$f(\alpha, t) := \frac{|\exp(\mathrm{i}\alpha t) - 1|^2}{\alpha^2} = \frac{\sin^2(t\alpha/2)}{(\alpha/2)^2},$$

so that (11.40) takes the form

$$c_{n,m}(t) \simeq a_{n,m} f(E_m - E_n, t). \tag{11.42}$$

Let us stress that the computation is heuristic: We do not know how small t should be for the previous approximations to be valid.[16]

Now, using here the standard integral (N.6), it holds that

$$\int_{-\infty}^{\infty} \mathrm{d}\alpha \, f(\alpha, 1) = 2\int_{-\infty}^{\infty} \mathrm{d}x \frac{\sin^2 x}{x^2} = 2\pi.$$

[16] If we wanted to know it, among the many problems one would have to solve is the following one. The probability (11.40) is proportional to the quantity $|(e_m, H_I e_n)|^2$, and captures only the possibility of a "direct transition from e_n to e_m". It could very well happen that this probability is very small for these particular values of m and n, and that a direct transition from e_n to e_m is much less likely than a transition through a certain intermediate state e_k. This may be the case if both $|(e_k, H_I e_n)|^2$ and $|(e_m, H_I e_k)|^2$ are very much larger than $|(e_m, H_I e_n)|^2$. Such issues concern mathematicians much more than physicists. A sensible physicist will have no qualms about the method as long as it is not contradicted by experiments.

Thus, as $t \to \infty$, for any function $\xi \in \mathcal{S}$ we have

$$\lim_{t\to\infty} \frac{1}{2\pi t}\int_{-\infty}^{\infty} \mathrm{d}\alpha\, f(\alpha,t)\xi(\alpha) = \lim_{t\to\infty} \frac{1}{2\pi}\int_{-\infty}^{\infty} \mathrm{d}\beta \frac{f(\beta/t,t)}{t^2}\xi(\beta/t)$$
$$= \lim_{t\to\infty} \frac{1}{2\pi}\int_{-\infty}^{\infty} \mathrm{d}\beta\, f(\beta,1)\xi(\beta/t) = \xi(0), \qquad (11.43)$$

as is seen by making the change of variable $\alpha = \beta/t$ and since $f(\beta/t,t)/t^2 = f(\beta,1)$. You may wonder what the condition $t \to \infty$ means in physics. It simply means that t is much larger than the timescale at which the process you are studying takes place. We will make much use of this fact.

It will greatly help to follow our forthcoming heuristic arguments to picture the function $\alpha \mapsto f(\alpha,t)$ as taking its significant values for $|\alpha|$ of order $1/t$, while having an integral $2\pi t$. We may then write

$$f(\alpha,t) \simeq 2\pi t\delta(\alpha). \qquad (11.44)$$

Despite the fact that classical textbooks such as that of Messiah [54] perform a thorough and correct analysis of the present situation, more recent textbooks unfortunately use the approximation (11.44) into (11.42). This simply makes no sense when E_m and E_n are fixed numbers because the delta function is not defined point-wise, but only when integrated against a test function.

On the other hand there is a natural situation where such approximations make sense. To describe it, let us fix a given e_n. Let us then consider the total transition probability from e_n to *any other* state. This is the norm of the projection of $U(t)(e_n)$ on the span of the e_m for $m \neq n$. The square of this norm is given by

$$\sum_{m\neq n} |(e_m, U(t)(e_n))|^2 = \sum_{m\neq n} c_{n,m}(t),$$

and at the first order (and if all goes well) using (11.42) this equals

$$\sum_{m\neq n} a_{n,m} f(E_n - E_m, t). \qquad (11.45)$$

In many situations of physical interest we will be able to simplify this expression. For this let us assume that the energy levels[17] E_m are labeled as an increasing sequence. Let us also make further regularity assumptions, which should hold at least in an interval around the energy E_n of our given state:

- The energy levels E_m are *extremely* close to each other.
- They are also very evenly spaced.[18] That is, there is an extremely small scale ε such that the difference $E_{m+1} - E_m$ is about $\varepsilon/\rho(E_m)$ for a smooth function ρ. Equivalently when x is close to E_n the number of energy levels contained in a small interval of length $\alpha \gg \varepsilon$ around x is about $\alpha\rho(x)/\varepsilon$.

[17] This expression is used here as a shorthand for "the eigenvalues of the Hamiltonian H_0".
[18] Such as the case considered in the next section of the energy levels of a particle contained in a macroscopic box.

- The dependence on m of the numbers $a_{n,m}$ is nice. Since by (11.41) we have $\sum_m a_{n,m} = g^2\|H_I e_n\|^2$, then, at least for E_m close to E_n, $a_{n,m}$ should be of order ε. It is then reasonable to assume that $a_{n,m} \simeq \varepsilon h_n(E_m)$ for a nice function h_n.[19]

Under these conditions,

$$a_{n,m} f(E_n - E_m, t) \simeq \varepsilon h_n(E_m) f(E_n - E_m, t) \simeq \int_{E_m}^{E_{m+1}} \mathrm{d}x h_n(x)\rho(x) f(E_n - x, t), \tag{11.46}$$

where the error is of order ε^2. By summation over m, with an error of order ε the transition probability is about

$$\int \mathrm{d}x h_n(x)\rho(x) f(E_n - x, t).$$

At that level it makes sense to perform the approximation (11.44), and to write

$$\int \mathrm{d}x h_n(x)\rho(x) f(E_n - x, t) \simeq 2\pi t h_n(E_n)\rho(E_n). \tag{11.47}$$

Thus, within these approximations, the probability that the transition has taken place before time t is proportional to t. The coefficient of t is called *the rate of transition per unit of time*. Here, it is simply $2\pi h_n(E_n)\rho(E_n)$.

Let us stress the conditions under which these approximations are valid. For (11.47) to be valid, t has to be large enough so that both functions h_n and ρ vary little on an interval of length about $1/t$. For (11.46) to be valid, $1/t$ has to be much larger than ε, so that t cannot be too large either. When ε is extremely small, there is plenty of room to satisfy both conditions.

Another reason for which t should not be too large is that the right-hand side of (11.47) is supposed to be a probability so it has to be ≤ 1 and our approximations cannot be correct for t too large. Thus we expect the approximation to be valid in a certain interval of values of t and to be unreliable outside this interval. As the boundaries of the interval are ill defined, a mathematician might feel uncomfortable but there are many situations of interest where the whole approach works beautifully.

11.4 A Side Story: Oscillating Interactions

Although we will not need it for our main line of work, let us consider now the case where the time-dependence of the interaction is given by $H_1(t) = g(t)H_I$ where for some $\omega > 0$ we have $g(t) = 2\cos\omega t = \exp \mathrm{i}\omega t + \exp(-\mathrm{i}\omega t)$.[20] This case is important in physics, but physicists often write formulas that cannot be taken at face value. In that situation (11.39) becomes

[19] There is no reason whatsoever that $a_{n,m}$ should have a simple expression in terms of E_m, but expressing the dependence on m of $a_{n,m}$ in that form simplifies our exposition.

[20] This normalization is chosen to yield the same result as the case $g(t) = \exp \mathrm{i}\omega t$ usually considered in the literature. The case of periodic $g(t)$ is the second most important after the constant case.

$$(e_m, U(t)e_n) \simeq -\mathrm{i}\int_0^t \mathrm{d}\theta\, g(\theta)(e_m, U_0(t-\theta)H_I U_0(\theta)e_n)$$

and (11.40) becomes

$$c_{n,m}(t) \simeq a_{n,m}|A+B|^2, \tag{11.48}$$

where

$$A = \frac{\exp(\mathrm{i}t(E_n - E_m + \omega)) - 1}{E_n - E_m + \omega}\ ;\ B = \frac{\exp(\mathrm{i}t(E_n - E_m - \omega)) - 1}{E_n - E_m - \omega}.$$

The situation of typical interest is completely different from the situation of (11.45) we considered earlier. Here we are interested in transitions between two specific energy levels $E_n \neq E_m$. For example, we are looking for transitions between the energy levels of an electron in a given atom. In that case the term $|A+B|$ can be large only if one of the terms $|A|$ or $|B|$ is large, and this can happen only if the corresponding denominator is small. If, say, $E_m > E_n$, then

$$|E_n - E_m - \omega| = E_m - E_n + \omega \geq |E_n - E_m|$$

and only the denominator of the term A can be small (when $\omega \simeq E_m - E_n$). Thus only the term A can be large, and then $|A+B|^2 \simeq |A|^2$, so that (11.48) becomes

$$c_{n,m}(t) \simeq a_{n,m} f(E_n - E_m + \omega, t). \tag{11.49}$$

Despite what Weinberg himself writes on page 186 of his textbook [88], one again has to be cautious here to use the approximation (11.44), since $\delta(\alpha)$ is a distribution and makes no sense for a given value of α.

Writing explicitly the dependence of the left-hand side on ω, this approximation means

$$c_{n,m}(t,\omega) \simeq 2\pi a_{n,m} t\delta(E_n - E_m + \omega), \tag{11.50}$$

in the sense that we expect that when integrating both sides against a test function we get near equality.[21] The reason why the approximation is correct from a physical point of view is that it is never true in physics that $g(t)$ is exactly of the type $g_\omega(t) := 2\cos\omega t$, which would represent a "truly monochromatic perturbation".[22] There is always some "spectral width" which means that the perturbation is a mixture of perturbations $g_\omega(t)$ for ω varying in a small interval, and it is averaging over these values which validates the approximation by the delta function. With the same approximations, physicists write the rate of transition $\Gamma(n \to m)$ per unit of time between energy levels E_n and E_m as

$$\Gamma(n \to m) = 2\pi a_{n,m}\delta(E_n - E_m + \omega). \tag{11.51}$$

This is known as *Fermi's golden rule*.

[21] You are certainly aware that here the near equality does not hold uniformly over all test functions. Rather, given a fixed test function, as our approximations get better, the integrals of both sides of (11.50) against this test function get close to each other.

[22] This physicists's terminology simply means that there is a single frequency, here ω, in the perturbation.

Exercise 11.4.1 Assume now that the energy levels (E_n) are very close to each other. Using the same approximations and notation as before, prove that the rate of transition from energy level n per unit of time is $2\pi h_n(E_n+\omega)\rho(E_n+\omega)$.

11.5 Interaction of a Particle with a Field: A Toy Model

This section attempts to give a flavor (of a small part) of Section XXI.13 of Messiah [54]. The model, a very simplified version of the interaction of electrons and photons, is arguably of marginal interest[23] and will never be used in the sequel, but it allows us to make a truly fundamental point: The interaction of a particle with a field *changes* the apparent mass of the particle. The model will also let us use the methods of Sections 11.1 and 11.3 in a reasonably simple but non-trivial situation, to compute how the interaction with "photons" shifts the energy levels of an "electron", and at what rate this interaction induces transitions between these energy levels. At the same time, we will be able to test the limits of these methods.

What we would really like to understand is how an electron interacts with photons, the quanta of the electromagnetic field. The electron might be free, or, say, part of a hydrogen atom. As this is a very difficult problem, we will replace both the electron and the electromagnetic field by simpler objects. We replace the electron by a non-relativistic particle; this particle cannot be destroyed. In the model there exists one (and only one) such particle. We put the spatial universe in a large box $B=[-L/2,L/2]^3$, and the state space of the particle is the space L^2_B of square-integrable functions on B, which we met before (3.63). The Hamiltonian is

$$H=\frac{\mathbf{P}^2}{2M}+V, \tag{11.52}$$

where $\mathbf{P}$ is the momentum operator $-\mathrm{i}\partial/\partial\boldsymbol{x}$ and M the mass of the particle. (In full, $H(f)=(-1/2M)\sum_{i\le 3}\partial^2 f/\partial x_i^2+Vf$.) When the potential V is zero the "electron" is free. We assume that there exists an orthonormal basis $(\xi_n)_{n\ge 0}$ of eigenvectors of H, with $H\xi_n=E_n\xi_n$, and we assume $E_0\le E_1\le\ldots$. When the potential is not zero, it is typically the first eigenvectors, representing bound states of the particle, which are of interest (so, think of these as modeling the various energy states of an electron in a hydrogen atom). As we sometimes use Dirac's notation in this section we also denote the function ξ_n by $|n\rangle$.

Besides the particle, our model includes identical "corpuscles". These can be created or destroyed. Their common rest mass $\mu>0$ is different from the rest mass M of the particle. Even though the corpuscle is a simplified version of the photon, we give it the positive rest mass μ because the case of massless particles is technically genuinely more difficult. One single corpuscle is modeled by *another copy* of L^2_B, which we denote by L^2_C. As we learned in Section 3.4, the corresponding boson Fock space $\mathcal{B}$ is then the appropriate state space to describe a varying number of corpuscles.

23 Still, it *does* carry a flavor of the real thing.

To pursue the description of the model we need an appropriate basis for the boson Fock space $\mathcal{B}$, and we construct it now. We recall the set $\mathcal{K}$ of elements $\boldsymbol{k} \in \mathbb{R}^3$ such that all the coordinates of $\boldsymbol{k}$ are of the type $2\pi n/L$ for $n \in \mathbb{Z}$. For $\boldsymbol{k} \in \mathcal{K}$ the functions

$$f_{\boldsymbol{k}}(\boldsymbol{x}) = \frac{1}{L^{3/2}} \exp(\mathrm{i}\boldsymbol{x} \cdot \boldsymbol{k}) \tag{3.63}$$

form an orthogonal basis of L^2_C, and $f_{\boldsymbol{k}}$ is an eigenvector of eigenvalue k_j for the momentum operator $-\mathrm{i}\partial/\partial x_j$. Each of the functions $f_{\boldsymbol{k}}$ describes a certain state of a single corpuscle, and quite naturally we will describe corpuscles in state $f_{\boldsymbol{k}}$ as "corpuscles of momentum $\boldsymbol{k}$". Since the functions $f_{\boldsymbol{k}}$ form an orthonormal basis of L^2_C, as explained in (3.10), we obtain from these functions an orthonormal basis of the corresponding boson Fock space $\mathcal{B}$. We denote the elements of this basis by $|(n_{\boldsymbol{\ell}})\rangle$. In this notation $(n_{\boldsymbol{\ell}})_{\boldsymbol{\ell}\in\mathcal{K}}$ is a sequence of integers indexed by $\mathcal{K}$, all of which but a finite number are zero. The number $n_{\boldsymbol{\ell}}$ should be interpreted as the number of corpuscles in the state described by $f_{\boldsymbol{\ell}}$ i.e. of corpuscles of momentum $\boldsymbol{\ell}$. Let us call $\omega_{\boldsymbol{\ell}}$ the energy of a corpuscle of momentum $\boldsymbol{\ell}$. (Although the precise value of $\omega_{\boldsymbol{k}}$ does not really matter, we may as well assume for consistency that as usual $\omega_{\boldsymbol{k}} := \sqrt{\mu^2 + \boldsymbol{k}^2}$.) Then, when these corpuscles do not interact with each other, according to (3.43), the Hamiltonian H_c governing the evolution of the corpuscles is given by

$$H_c|n_{\boldsymbol{\ell}}\rangle = E_{(n_{\boldsymbol{\ell}})}|n_{\boldsymbol{\ell}}\rangle, \tag{11.53}$$

where

$$E_{(n_{\boldsymbol{\ell}})} := \sum_{\boldsymbol{\ell}\in\mathcal{K}} n_{\boldsymbol{\ell}}\omega_{\boldsymbol{\ell}}. \tag{11.54}$$

In words, the energy of the system where several corpuscles are present is simply the sum of the energies of the individual corpuscles.

The system where both the particle and the corpuscles are present is modeled by the state space $\mathcal{H} = L^2_B \otimes \mathcal{B}$. The states $|n,(n_{\boldsymbol{\ell}})\rangle := |n\rangle \otimes |(n_{\boldsymbol{\ell}})\rangle$ form an orthonormal basis of this space.[24] On this space consider the Hamiltonian H_0 such that

$$H_0|n,(n_{\boldsymbol{\ell}})\rangle = (E_n + E_{(n_{\boldsymbol{\ell}})})|n,(n_{\boldsymbol{\ell}})\rangle = \Big(E_n + \sum_{\boldsymbol{\ell}\in\mathcal{K}} n_{\boldsymbol{\ell}}\omega_{\boldsymbol{\ell}}\Big)|n,(n_{\boldsymbol{\ell}})\rangle, \tag{11.55}$$

which is the Hamiltonian on $L^2_B \otimes \mathcal{B}$ describing the system where the particle and the corpuscles do not interact, the particle evolving according to the Hamiltonian (11.52) and the corpuscles according to the Hamiltonian (11.53). The ground state, that is the state with the lowest possible energy, is the state where $n = 0$ and $n_{\boldsymbol{\ell}} = 0$ for each $\boldsymbol{\ell}$.

To model the interaction of the particle with the corpuscles we will use a quantum field whose quanta are the corpuscles. Since we have put space in a box, such a field is given by the formula (5.31). Here however we can expect trouble with the infinite summation $\sum_{\boldsymbol{k}\in\mathcal{K}}$ and we will have to somehow suppress the terms with large values of $\boldsymbol{k}^2$. In any case it does not make too much sense to consider corpuscles of high energy ($\gg M = Mc^2$) since a

24 Make sure you understand that n means that the particle has energy E_n whereas $n_{\boldsymbol{\ell}}$ denotes the number of corpuscles.

realistic model would allow creation of new particles at these high energies. To cut off the large values we consider a function $0 \le \theta \le 1$ on $\mathbb{R}$, which we assume to satisfy

$$k^2 \le M^2 \Rightarrow \theta(k^2) = \frac{1}{\sqrt{2\omega_k}}. \tag{11.56}$$

This simply means that the cutoff acts only on large values of k^2. We also assume that θ decreases fast enough at infinity to make everything we need converge (for example by taking $\theta(k^2)$ equal to zero for k^2 large enough). Instead of (5.31) let us then rather use as quantum field the following function on $\mathbb{R}^3$, with values in the operators on $\mathcal{B}$:

$$\varphi(x) = \sum_{k \in \mathcal{K}} \theta(k^2)(f_k(x)a_k + f_k(x)^* a_k^\dagger), \tag{11.57}$$

where

$$a_k = A(f_k)\ ;\ a_k^\dagger = A^\dagger(f_k), \tag{11.58}$$

and where $A^\dagger(\gamma)$ and $A(\gamma)$ denote respectively for $\gamma \in L^2_C$ the creation and annihilation operators on $\mathcal{B}$ constructed in Definition 3.4.1. Thus a_k and $a_k^\dagger$ are operators on $\mathcal{B}$, whose action on the basis elements $|(n_\ell)\rangle$ is given by (3.36) and (3.37).

To construct the interaction we first observe that $\mathcal{H} = L^2_B \otimes \mathcal{B}$ identifies with the space $L^2(B,\mathcal{B})$ of square-integrable functions on B valued in $\mathcal{B}$ as follows: The function corresponding to an element $f \otimes b$ is the function $x \mapsto f(x)b$. So, an element of $L^2_B \otimes \mathcal{B}$ can be considered a function $\psi : B \to \mathcal{B}$, and for each $x \in B$, $\psi(x) \in \mathcal{B}$. Thus given two such functions ψ_1 and ψ_2 their inner product is given by the formula

$$(\psi_1, \psi_2) = \int_B \mathrm{d}^3 x (\psi_1(x), \psi_2(x)), \tag{11.59}$$

where the inner product in the integral is the inner product in $\mathcal{B}$.

Recalling the formula (11.57) we consider the *interaction Hamiltonian* H_I such that for $\psi \in L^2(B,\mathcal{B})$, $H_I(\psi) \in L^2(B,\mathcal{B})$ is given by

$$H_I(\psi)(x) = \varphi(x)(\psi(x)). \tag{11.60}$$

The formula makes sense because $\varphi(x)$ is an operator on $\mathcal{B}$, and here we simply compute the value of this operator on the element $\psi(x) \in \mathcal{B}$. The full Hamiltonian will then be $H_0 + gH_I$ for a coupling constant g describing the strength of the interaction between the particle and the corpuscles.

At first this formula seems very strange, but it makes sense under closer examination. At some imprecise level, the quantum field $\varphi(x)$ creates and destroys corpuscles located at x, while the wave function $\psi(x)$ is related to the probability of the particle being located at x. Thus (11.60) looks like it has to do with corpuscles interacting with the particle by being created or destroyed precisely where the particle is located.

Our next goal is to study "the transition rates of the electron between the different energy levels due to the interaction with photons". As we saw in Section 11.3, these rates are governed by the matrix elements

$$\langle m, (m_\ell)| H_I | n, (n_\ell)\rangle \tag{11.61}$$

and we first compute these. Replacing φ by its value (11.57) in (11.60) yields

$$H_I = \sum_{k \in \mathcal{K}} \theta(k^2)(H_{I,k} + H_{I,k}^\dagger), \tag{11.62}$$

where for any function $\psi : B \to \mathcal{B}$ one has

$$H_{I,k}(\psi)(x) = f_k(x) a_k(\psi(x)) \; ; \; H_{I,k}^\dagger(\psi)(x) = f_k(x)^* a_k^\dagger(\psi(x)). \tag{11.63}$$

The element $|n,(n_\ell)\rangle$ of $\mathcal{H} = L_B^2 \otimes \mathcal{B}$ corresponds to the specific function $\psi : B \to \mathcal{B}$ given by $\psi(x) = \xi_n(x)|(n_\ell)\rangle$ so that $H_{I,k}|n,(n_\ell)\rangle$ corresponds to the function $\psi_2(x) = \xi_n(x) f_k(x) a_k |(n_\ell)\rangle$. Using then (11.59) for $\psi_1(x) = \xi_m(x)|(m_\ell)\rangle$ yields the matrix element

$$\langle m,(m_\ell)|H_{I,k}|n,(n_\ell)\rangle = \langle (m_\ell)|a_k|(n_\ell)\rangle \int_B \mathrm{d}^3 x \, \xi_m(x)^* \xi_n(x) f_k(x). \tag{11.64}$$

Let us now define the quantities

$$C(m,n,k) := \int_B \mathrm{d}^3 x \, \xi_m(x)^* \xi_n(x) f_k(x). \tag{11.65}$$

Lemma 11.5.1 *The matrix element $\langle m,(m_\ell)|H_I|n,(n_\ell)\rangle$ is zero unless one of the following happens:*

$$\exists k \,,\, \forall \ell \in \mathcal{K} \,,\, \ell \neq k \Rightarrow m_\ell = n_\ell \,,\, m_k = n_k - 1 \tag{11.66}$$

$$\exists k \,,\, \forall \ell \in \mathcal{K} \,,\, \ell \neq k \Rightarrow m_\ell = n_\ell \,,\, m_k = n_k + 1. \tag{11.67}$$

In the first case we have

$$\langle m,(m_\ell)|H_I|n,(n_\ell)\rangle = \theta(k^2)\sqrt{n_k}\, C(m,n,k), \tag{11.68}$$

and in the second case it holds that

$$\langle m,(m_\ell)|H_I|n,(n_\ell)\rangle = \theta(k^2)\sqrt{n_k + 1}\, C(m,n,-k). \tag{11.69}$$

Observe that in the case (11.66) a corpuscle of momentum k is destroyed while in the case (11.67) such a corpuscle is created.

Proof We use (11.64). We compute $\langle (m_\ell)|a_k|(n_\ell)\rangle$ using (3.37). It is zero unless (11.66) occurs, and then it is $\sqrt{n_k}$. We proceed in the same way for $H_{I,k}^\dagger$. □

To better understand the physics of the situation, we look at the case of the free particle, where the potential V is zero. In that case we may use the functions f_n (rather than the functions ξ_n) as a basis of eigenvectors of H, so that $C(m,n,k)$ has to be replaced by

$$C(m,n,k) := \int_B \mathrm{d}^3 x \, f_m(x)^* f_n(x) f_k(x). \tag{11.70}$$

From (3.63) it holds that $f_k f_n = L^{-3/2} f_{k+n}$. Thus the integral in (11.70) is zero unless $m = k + n$, and then it is $L^{-3/2}$. According to the previous lemma, there are exactly two cases where the matrix element $\langle m,(m_\ell)|H_I|n,(n_\ell)\rangle$ is not zero (and where therefore the interaction creates transitions from the state $|n,(n_\ell)\rangle$ to the state $|m,(m_\ell)\rangle$). The first case is given by (11.67) where $m = n + k$. The particle gains momentum k, and a corpuscle of momentum $\boldsymbol{k}$ is being destroyed. Similarly, the situation of (11.68) corresponds to the case where a corpuscle of momentum $\boldsymbol{k}$ is being created and the momentum of the particle decreases by $\boldsymbol{k}$. This again makes sense according to the next exercise.

Exercise 11.5.2 In the previous situation write down a "total momentum operator" that measures the momentum of the particle and the corpuscles together. Show that this operator commutes with the Hamiltonian, and explain why momentum is conserved.

Going back to the case where there is a potential V, let us illustrate how to compute the transition rate from the state where the particle is at energy level m and where no corpuscle is present to the state where the particle is at energy state n and a single corpuscle of momentum $\boldsymbol{k}$ is present, that is, from the state $|m,(0)\rangle$ to the state $|n,(\delta_{\boldsymbol{\ell}}^{\boldsymbol{k}})\rangle$.[25] The probability $c(m,n,\boldsymbol{k},t)$ that this transition has taken place before time t is given at the first order by the formula (11.42) which here is

$$c(m,n,\boldsymbol{k},t) = g^2\theta(\boldsymbol{k}^2)^2|C(m,n,\boldsymbol{k})|^2 f(E_m - E_n - \omega_{\boldsymbol{k}},t). \tag{11.71}$$

The probability that before time t the particle has emitted a corpuscle of any momentum is, at the first order, the sum of these probabilities over the various values of $\boldsymbol{k}$.

To sum the quantities (11.71) over $\boldsymbol{k}$ we recall that each $\boldsymbol{k} \in \mathcal{K}$ corresponds to a little box of volume $(2\pi/L)^3$ so (since L is the size of the box representing the entire universe) we may replace the sum by the integral

$$g^2 \int \frac{\mathrm{d}^3\boldsymbol{p}}{(2\pi)^3}\theta(\boldsymbol{p}^2)^2|D(m,n,\boldsymbol{p})|^2 f(E_m - E_n - \omega_{\boldsymbol{p}},t),$$

where

$$D(m,n,\boldsymbol{p}) := \int_B \mathrm{d}^3\boldsymbol{x}\xi_m(\boldsymbol{x})^*\xi_n(\boldsymbol{x})\exp(\mathrm{i}\boldsymbol{p}\cdot\boldsymbol{k}).$$

At that level we may make the approximation (11.44) to obtain the transition rate per unit of time

$$g^2 \int \frac{\mathrm{d}^3\boldsymbol{p}}{(2\pi)^3}\theta(\boldsymbol{p}^2)^2|D(m,n,\boldsymbol{p})|^2(2\pi)\delta(E_m - E_n - \omega_{\boldsymbol{p}}).$$

In any reasonable situation we have $E_m - E_n < M\,(= Mc^2)$ so that when $\omega_{\boldsymbol{p}} = E_m - E_n$ we have $\boldsymbol{p}^2 \leq M^2$ and thus $\theta(\boldsymbol{p}^2)^2 = 1/2\omega_{\boldsymbol{p}}$. Therefore we have reached the following:[26]

Proposition 11.5.3 *The transition rate per unit of time from the state $|m,(0)\rangle$ to a state of the form $|n,(\delta_{\boldsymbol{\ell}}^{\boldsymbol{k}})\rangle$ where n is given and $\boldsymbol{k}$ is arbitrary is*

$$g^2 \int \frac{\mathrm{d}^3\boldsymbol{p}}{(2\pi)^2 2\omega_{\boldsymbol{p}}}|D(m,n,\boldsymbol{p})|^2\delta(E_m - E_n - \omega_{\boldsymbol{p}}). \tag{11.72}$$

The right-hand side of (11.72) may be computed, and this is the purpose of the next exercise.

[25] That is, the particle drops from energy level E_m to energy level E_n by emitting a corpuscle of momentum $\boldsymbol{k}$.
[26] Although the force of habit leads us to use words such as "proved", let us repeat that we are doing physics in the whole of Part III, which contains very few rigorous results in the sense a mathematician would understand it.

Exercise 11.5.4 (a) Show that for a well-behaved function ξ on $\mathbb{R}^3$ one has

$$\int \mathrm{d}^3 \boldsymbol{p}\, \xi(\boldsymbol{p}) = 4\pi \iint_{r\geq 0} \mathrm{d}r \mathrm{d}\rho_r(\boldsymbol{p}) r^2 \xi(\boldsymbol{p})$$

where $\mathrm{d}\rho_r$ is the uniform probability on the sphere of radius r.
(b) Conclude that the quantity (11.72) equals $\int_0^\infty \mathrm{d}r\, \eta(r)\delta(h(r))$, where $h(r) = E_m - E_n - \sqrt{r^2+\mu^2}$ and

$$\eta(r) = \frac{g^2}{2\pi\sqrt{\mu^2+r^2}} \int \mathrm{d}\rho_r(\boldsymbol{p}) |D(m,n,\boldsymbol{p})|^2.$$

(c) Conclude that the quantity (11.72) equals

$$\frac{g^2\sqrt{(E_m-E_n)^2-\mu^2}}{2\pi} \int \mathrm{d}\rho(\boldsymbol{p}) |D(m,n,\boldsymbol{p})|^2$$

where $\mathrm{d}\rho$ is the uniform probability measure on the sphere of radius

$$\sqrt{(E_m-E_n)^2-\mu^2}.$$

Exercise 11.5.5 When the potential is zero, show that the rate of transition per unit time from a state of momentum $\boldsymbol{m}$ (with no corpuscles present) to any other state is given (when $|\boldsymbol{m}|$ is not too large) by

$$g^2 \int \frac{\mathrm{d}^3\boldsymbol{p}}{(2\pi)^2 2\omega_{\boldsymbol{p}}} \delta\Big(\frac{\boldsymbol{m}^2}{2M} - \frac{(\boldsymbol{m}-\boldsymbol{p})^2}{2M} - \omega_{\boldsymbol{p}}\Big).$$

In Section 11.1 we studied how the eigenvalues of an operator are modified by a perturbation. Our next goal is to use these results to compute the shift Δ_n in the eigenvalue E_n corresponding to the eigenvector $|n,(0)\rangle$ due the interaction Hamiltonian. At the first order of perturbation theory this shift Δ_n is ga_1 where $a_1 = \langle n,(0)|H_I|n,(0)\rangle$ is given by the formula (11.6). This is zero according to Lemma 11.5.1, so that we will compute the shift Δ_n "at the level of the second-order perturbation theory", i.e. by the formula (11.13) (assuming that we can use this formula here). Here $\lambda_0 = E_n$ and the summation over $n \geq 1$ in (11.13) becomes the summation over all $m \geq 0$ and (m_ℓ) for $(m,(m_\ell)) \neq (n,(0))$, so that

$$\Delta_n = g^2 \sum_{(m,(m_\ell))\neq(n,(0))} \frac{|\langle m,(m_\ell)|H_I|n,(0)\rangle|^2}{E_n - (E_m + \sum_{\ell\in\mathcal{K}} m_\ell \omega_\ell)}.$$

Use of the formula (11.69) then gives

$$\Delta_n = g^2 \sum_{m,\boldsymbol{k}} \frac{\theta(\boldsymbol{k}^2)^2 |C(m,n,-\boldsymbol{k})|^2}{E_n - E_m - \omega_{\boldsymbol{k}}} = g^2 \sum_{m,\boldsymbol{k}} \frac{\theta(\boldsymbol{k}^2)^2 |C(m,n,\boldsymbol{k})|^2}{E_n - E_m - \omega_{\boldsymbol{k}}}, \tag{11.73}$$

where the summation does not include the case $m = n, \boldsymbol{k} = \boldsymbol{0}$. We know how to deal with the summation over $\boldsymbol{k}$ by replacing it by an integral, but it is a harder issue to handle the summation over m. A first problem is that the quantities $E_n - E_m - \omega_{\boldsymbol{k}}$ may be of different signs, so there might be cancellation between the different terms. We put this aside by considering only the case $n = 0$, so that then $E_n - E_m - \omega_{\boldsymbol{k}} < 0$. One then has to understand which values of m really contribute to the summation, an issue which is clarified by the next lemma.

Lemma 11.5.6 *Given any values of n and* $\boldsymbol{k}$ *we have*

$$\sum_m |C(m,n,\boldsymbol{k})|^2 = \frac{1}{L^3} \tag{11.74}$$

and

$$\sum_m \left(E_m - E_n - \frac{\boldsymbol{k}^2}{2M}\right)^2 |C(m,n,\boldsymbol{k})|^2 \le \frac{1}{L^3 M^2}\|\boldsymbol{k}\cdot\mathbf{P}\xi_n\|^2. \tag{11.75}$$

We will use this by arguing that in our non-relativistic situation, the right-hand side of (11.75) is very much smaller than the right-hand side of (11.74) so that most of the contribution to the sum in (11.74) is from terms for which $E_m \simeq E_n + \boldsymbol{k}^2/2M$.

Proof Consider the function

$$\eta(\boldsymbol{x}) := \xi_n(\boldsymbol{x}) f_{\boldsymbol{k}}(\boldsymbol{x}) = L^{-3/2}\xi_n(\boldsymbol{x})\exp(\mathrm{i}\boldsymbol{k}\cdot\boldsymbol{x}),$$

and note that

$$(\xi_m,\eta) = C(m,n,\boldsymbol{k})\ ;\ \|\eta\|^2 = \frac{1}{L^3}, \tag{11.76}$$

which already implies (11.74) since (ξ_m) is an orthonormal basis. Next, using Leibniz's rule,

$$\boldsymbol{P}^2\eta = \boldsymbol{P}^2(\xi_n f_{\boldsymbol{k}}) = (\boldsymbol{P}^2\xi_n)f_{\boldsymbol{k}} + 2f_{\boldsymbol{k}}\boldsymbol{k}\cdot\boldsymbol{P}\xi_n + \boldsymbol{k}^2 f_{\boldsymbol{k}}\xi_n.$$

Moreover since the operator V is just multiplication by the function $V(\boldsymbol{x})$ we have $V\eta = f_{\boldsymbol{k}}(V\xi_n)$. Consequently the Hamiltonian H of (11.52) satisfies

$$H\eta = f_{\boldsymbol{k}}\Big(H\xi_n + \frac{\boldsymbol{k}^2}{2M}\xi_n\Big) + f_{\boldsymbol{k}}\frac{\boldsymbol{k}\cdot\mathbf{P}\xi_n}{M} = a\eta + f_{\boldsymbol{k}}\frac{\boldsymbol{k}\cdot\mathbf{P}\xi_n}{M}$$

where $a := E_n + \boldsymbol{k}^2/2M$. Therefore, if $\psi := f_{\boldsymbol{k}}\boldsymbol{k}\cdot\mathbf{P}\xi_n/M$ then

$$E_m(\xi_m,\eta) = (H\xi_m,\eta) = (\xi_m,H\eta) = a(\xi_m,\eta) + (\xi_m,\psi),$$

and thus $(E_m - a)(\xi_m,\eta) = (\xi_m,\psi)$. Since ξ_m form an orthonormal basis and since $(\xi_m,\eta) = C(n,m,\boldsymbol{k})$, taking as m varies the sum of the squares of the moduli in this equality yields

$$\sum_m (E_m - a)^2 |C(n,m,\boldsymbol{k})|^2 \le \|\psi\|^2 = \frac{1}{L^3 M^2}\|\boldsymbol{k}\cdot\mathbf{P}\xi_n\|^2,$$

where we use in the last equality that $|f_{\boldsymbol{k}}(\boldsymbol{x})| = L^{-3/2}$. □

To use this result, let us bound the right-hand side of (11.75) when $n = 0$. We first use the bound $\|\boldsymbol{k}\cdot\mathbf{P}\xi_0\|^2 \le \boldsymbol{k}^2\|\mathbf{P}\xi_0\|^2$. Since in Classical Mechanics the momentum is the product of the mass and the speed, the quantity $\|\mathbf{P}\xi_0\|/M$ should be of the same order as the velocity the particle in state ξ_0. It is believable that the kinetic energy of the particle in state ξ_0 should be of the same order as $|E_0|$ (despite the fact that E_0 is the sum of the kinetic and potential energies, which are of opposite signs). Thus the square of the speed of the particle should be of order $|E_0|/M$. Since we are in the non-relativistic case $|E_0| = \alpha M$ with α much smaller than 1.[27] Therefore the right-hand side of (11.75) should be of order $\alpha \boldsymbol{k}^2/L^3$.

Let us now set $n = 0$ and estimate Δ_0. Comparing (11.75) with (11.74) means that the contribution to this sum comes mostly from terms such that $E_m - E_n \simeq \boldsymbol{k}^2/2M$, so in (11.73) we simply replace $E_m - E_n$ by $\boldsymbol{k}^2/2M$, and with an error which becomes small with α we may write, using (11.74) in the equality,

$$\Delta_0 \simeq -g^2 \sum_{m,\boldsymbol{k}} \frac{\theta(\boldsymbol{k}^2)^2 |C(m,n,\boldsymbol{k})|^2}{\omega_{\boldsymbol{k}} + \boldsymbol{k}^2/2M} = -\frac{g^2}{L^3} \sum_{\boldsymbol{k}} \frac{\theta(\boldsymbol{k}^2)^2}{\omega_{\boldsymbol{k}} + \boldsymbol{k}^2/2M}. \tag{11.77}$$

As usual we can replace the sum by the integral

$$\Delta_0 \simeq -g^2 \int \frac{\mathrm{d}^3 \boldsymbol{p}}{(2\pi)^3} \frac{\theta(\boldsymbol{p}^2)^2}{\omega_{\boldsymbol{p}} + \boldsymbol{p}^2/2M}. \tag{11.78}$$

Exercise 11.5.7 If you enjoy ε-δ proofs, write a rigorous version of the previous argument, letting $\alpha \to 0$ and $L \to \infty$.

Quite interestingly the downward shift in E_0 is independent of the potential V (at this level of accuracy). A fishy feature though is that the quantity (11.78) depends on how we cut off the large values of $\boldsymbol{p}$. This unfortunately is a recurring feature of quantum field theory, and it is a highly difficult problem to avoid.

When the mass μ of the corpuscles is less than the mass M of the particle, the integral in (11.78) is at least of order M. To see this we estimate the contribution of the region $M/2 \le |\boldsymbol{p}| \le M$ to this integral. The volume of integration is of order M^3. Recalling (11.56) the integrand is $1/(2\omega_{\boldsymbol{p}}(\omega_{\boldsymbol{p}} + \boldsymbol{p}^2/(2M)))$ and is of order $1/M^2$ in that region. So, the downward shift of the ground state is predicted by perturbation theory to be at least of order $g^2 M$.

Now let us evaluate the credibility of perturbation theory in the present calculation. With the notation of Section 11.1, we used the eigenvector $v_0 + g w_1 + \cdots$ which is supposed to be a small perturbation of v_0. According to (11.11) the square of the norm of the vector $g w_1$ is given here by

$$g^2 \sum_{m,\boldsymbol{k}} \frac{\theta(\boldsymbol{k}^2)^2 |C(m,n,\boldsymbol{k})|^2}{(E_0 - E_m - \omega_{\boldsymbol{k}})^2} \tag{11.79}$$

where the summation does not include the case $m = 0$ and $\boldsymbol{k} = \boldsymbol{0}$. Just as before we approximate this quantity by

$$g^2 \int \frac{\mathrm{d}^3 \boldsymbol{p}}{(2\pi)^3} \frac{\theta(\boldsymbol{p}^2)^2}{(\omega_{\boldsymbol{p}} + \boldsymbol{p}^2/2M)^2}.$$

[27] In non-relativistic Quantum Mechanics as here, the energy of a particle does not include the energy $M(= Mc^2)$ of its rest mass, in contrast with the relativistic setting of (4.30).

In the limit of zero-mass corpuscles we have $\omega_p = |p|$, and recalling (11.56) this is at least

$$g^2 \int_{|p|\le M} \frac{\mathrm{d}^3 p}{(2\pi)^3} \frac{1}{2|p|(|p| + p^2/2M)^2}.$$

This infinite quantity is hardly the square of the norm of the supposedly small vector $g w_1$, and consequently the credibility of perturbation theory in computing this precise downward shift is minimal.[28] Perturbation theory may well break down. This will not prevent us from ruthlessly using similar methods.

Exercise 11.5.8 Convince yourself that crude choices of z in (11.16) would actually *prove*[29] that in our toy situation the downward shift Δ_0 of the ground state is indeed of order $g^2 M$ at least for small g. (Thus, perturbation theory fares well after all!)

Coming back to the real world of photons and electrons, the shifts of the energy levels of an electron due to its interactions with the electromagnetic field cannot be experimentally measured, for the simple reason that it is impossible to "turn off" this interaction. Denoting by $E'_0, \dots, E'_n, \dots$ the actual energy levels of the electron in a potential (the prime example being the potential created by a single proton) spectroscopy measures the differences $E'_m - E'_n$. Naive computation of the energy levels E'_n (for example by solving Schrödinger's equation) does not account for the measured values. This difference persists even when one uses improvements of Schrödinger's equation such as Dirac's equation, because these equations do not reflect the interaction of the electrons with photons. It takes a surprising amount of effort to really explain the measured values.[30] The previous methods are powerless for this, but nonetheless overall point in the correct direction.

A subtle point is that the mass of the electron, as it is measured in the lab, need not be the same as the number M occurring in the Hamiltonian H_0. We will discuss later in more detail this mysterious fact. It is, however, essential to understand that in Quantum Field Theory we may write a Hamiltonian with some parameters which we call mass, charge, etc., but that may not be what we measure in the lab. And only what we measure in the lab is "real". In the present situation one may argue (as Folland does [31]) that since the interaction with the field lowers the ground state energy, and since the only energy a particle at rest may have is the energy of its rest mass, the interaction with the field lowers the apparent rest mass of the particle.[31] It would be more satisfactory to give a precise definition of the apparent mass of the particle, based on how one might actually measure it in the lab, and to show that it is different from the parameter M of the Hamiltonian. Such a definition would presumably be obtained by studying the relation between the energy of the particle and its momentum. It is not obvious how to do this, in particular because the energy and momentum operators have no common eigenvectors.

28 It can be shown in the present situation that nonetheless perturbation theory still provides the correct order of magnitude for this downward shift.

29 I admit I did not really check this, leaving you a chance to get a refund for this book.

30 A famous example of this is the Lamb shift.

31 As this is not a relativistic theory, one may also say "mass" rather than "rest mass".

Key ideas to remember:

- A complex system is often studied as a small perturbation of a simpler system.
- The interaction picture allows thinking of the time-evolution of the complicated system as a perturbation of the time-evolution of the simpler system. This time-evolution is expressed through the Dyson series.
- Perturbations of a system induce transitions between the energy levels.

12

Scattering, the Scattering Matrix and Cross-Sections

We remind the reader that from Chapter 11 on we have chosen a system of units such that $\hbar = 1$ and $c = 1$.

The majority of our data concerning the particle world comes from scattering experiments, and the theoretical analysis of these is of fundamental importance. In a typical scattering experiment, two particles start far away from each other. They then get close to each other, interact and the products of the interaction, which may consist of several particles, again get far from each other. The important point is that at the beginning and at the end of the experiment the particles do not interact with each other. In a similar but simpler experiment a particle is thrown from far away at a scattering potential,[1] and after interacting with the potential moves again to infinity. How do we describe the situation mathematically? The analysis has two parts. In the first part, the properties of the scattering are encoded in an object called the S-matrix. The computation of the S-matrix is a main objective of Quantum Field Theory. The second part of the analysis is to relate the S-matrix to quantities that can actually be measured in our laboratory, the so-called *cross-sections*.

In the present chapter we explain heuristically, through the analysis of situations of increasing complexity, what the S-matrix is, but we do not try yet to compute it. We then turn to the relation between the S-matrix and cross-sections. This non-trivial task is not central to Quantum Field Theory. The fine points of this analysis are unimportant as long as one gets the correct formula, which one may discover with a few lines of heuristics as in Folland's book [31, page 137]. The author, finding this simplified analysis unfulfilling,[2] tried to learn more in standard textbooks such as that of Sakurai and Napolitano [72]. This attempt ran into the usual deep differences of language and perspective from a mathematical point of view.[3] It seems therefore useful to provide a detailed analysis in mathematical language on how the S-matrix relates to cross-sections. We obtain rather precise statements, requiring no higher mathematics than taking limits over two different parameters in the correct order. Understanding the details of this analysis is *certainly not* required for the sequel, so you may skip the proof of Theorem 12.4.1 and the entire Section 12.6 if these arguments do not thrill you.

1 The precise setting will be explained soon and the meaning of these words will become transparent.
2 One of the steps of the analysis is to make an average over the whole universe, not the most natural method to study the interaction taking place in your lab.
3 Interestingly, the older book of J. Taylor [82], a model of writing excellence, is far more accessible.

12.1 Heuristics in a Simple Case of Classical Mechanics

Let us think of the motion of a classical particle feeling only the influence of a potential of limited range (that is, the potential is identically zero outside a bounded region). Its equation of motion is then given by (6.26), that is $m\ddot{x} = -\nabla_x V$, but we make no attempt to solve this equation. The particle comes from far away. Before it feels the influence of the potential, it moves in a straight line like a "free particle" (that is, a particle which does not feel any external influence). It then enters the region where the potential acts and its trajectory bends. The particle may get captured by the potential, or it may become free again and move far away. When it gets far away from the potential it behaves like a free particle again. The problem is to describe the way the action of the potential has modified the trajectory of the particle, and more precisely to describe it in a way which will easily generalize to far more complex situations. This requires some effort, although the basic idea is obvious: Knowing how the trajectory looked in the far past, we have to describe how it looks in the far future. However the simplest descriptions one might consider, say by giving the angle by which the trajectory bends,[4] do not easily generalize. The reasons behind the approach we outline in the present section will be understood only at the beginning of the next section, when we move to Quantum Mechanics.

Let us denote by $\boldsymbol{x}_t$ and $\boldsymbol{v}_t$ the position and velocity of the particle at time t, respectively. A free particle has a motion of the type $\boldsymbol{x}_t = \boldsymbol{x} + t\boldsymbol{v}$, $\boldsymbol{v}_t = \boldsymbol{v}$ so that in the far past our scattered particle has a motion of this type, while in the far future it has a motion of the same type $\boldsymbol{x}' + t\boldsymbol{v}'$, where $\boldsymbol{x}'$ and $\boldsymbol{v}'$ are entirely determined by $\boldsymbol{x}$ and $\boldsymbol{v}$.[5] The pair $(\boldsymbol{x}, \boldsymbol{v})$ entirely describes the motion of the particle in the far past and the pair $(\boldsymbol{x}', \boldsymbol{v}')$ entirely describes this motion in the far future. The map

$$S : (\boldsymbol{x}, \boldsymbol{v}) \to (\boldsymbol{x}', \boldsymbol{v}') \tag{12.1}$$

describes how the trajectory in the far future is determined by the trajectory in the far past. It encodes a lot of the properties of the scattering action. It need not be defined everywhere, since some particles may get trapped by the potential.

Consider now the case of a potential with an infinite range, but which becomes weaker fast enough far from the origin. Then we should expect that the particle[6] has asymptotes as $t \to \pm\infty$. That is, if $\boldsymbol{v}_t$ denotes the velocity of the particle, the limits

$$\boldsymbol{v} = \lim_{t \to -\infty} \boldsymbol{v}_t$$

and

$$\boldsymbol{x} = \lim_{t \to -\infty} \boldsymbol{x}_t - t\boldsymbol{v}$$

[4] Conservation of energy implies that the kinetic energy of the particle is the same before and after the interaction, so that the speed, i.e. the modulus of the velocity is unchanged by the interaction, but the direction of the velocity is typically changed.

[5] We think of course of the potential as given and fixed.

[6] It is always assumed in such a statement that the particle feels the influence of the potential. A particle which does not feel this influence is referred to as a *free particle*.

should exist,[7] so that for large negative t one has $\boldsymbol{x}_t \simeq t\boldsymbol{v} + \boldsymbol{x}$, and again the motion of the particle as it gets in the interaction (i.e. for large negative t) is entirely determined by $\boldsymbol{x}$ and $\boldsymbol{v}$. Moreover it is likely that the pair $(\boldsymbol{x}, \boldsymbol{v})$ entirely determines the trajectory of the particle. When the particle escapes the potential we define $\boldsymbol{v}' = \lim_{t\to\infty} \boldsymbol{v}_t$, $\boldsymbol{x}' = \lim_{t\to\infty} \boldsymbol{x}_t - t\boldsymbol{v}'$, and we define again S by (12.1), and again this operator describes how the way the trajectory looks in the far future is determined by the way it looked in the far past.

To generalize this simple idea it helps to state it in a more abstract way. We denote by $\mathcal{P}$ the space of pairs $(\boldsymbol{x}, \boldsymbol{v})$ where $\boldsymbol{x}$ is a position and $\boldsymbol{v}$ is a velocity. One may like to think of $\mathcal{P}$ as the "state space" describing a moving particle at a given time.[8] We define an operator Ω_+ from $\mathcal{P}$ to $\mathcal{P}$ as follows. Starting with the free particle of position $\boldsymbol{x}$ and velocity $\boldsymbol{v}$ at time zero, we run time backward until a large negative time $-t_0$ (so that the position of the particle is then $\boldsymbol{x} - t_0\boldsymbol{v}$ and its velocity is still $\boldsymbol{v}$). We then turn the potential on and run time forward t_0, reaching time zero again. The position and velocity of the particle at this time zero then define a point of $\mathcal{P}$, which depends both on $(\boldsymbol{x}, \boldsymbol{v})$ and on t_0.[9] Hopefully the limit of this point as $t_0 \to \infty$ exists and we call it $\Omega_+(\boldsymbol{x}, \boldsymbol{v})$.

One also uses the notation

$$\Omega_+(\boldsymbol{x}, \boldsymbol{v}) = (\boldsymbol{x}, \boldsymbol{v})_{\text{in}}.$$

Thus (as you will read many times in the sequel), $(\boldsymbol{x}, \boldsymbol{v})_{\text{in}}$ represents the state at time $t = 0$ of the particle whose motion in the far past resembled the motion of a free particle of position and velocity $(\boldsymbol{x}, \boldsymbol{v})$ at time 0.

One defines similarly an operator Ω_- by starting with the free particle of position $\boldsymbol{x}$ and velocity $\boldsymbol{v}$ at time zero, running time forward until a large positive time t_0, turning the potential on, running time backward until we reach time zero again, and taking the limit as $t_0 \to \infty$.[10] One uses also the notation $(\boldsymbol{x}, \boldsymbol{v})_{\text{out}} = \Omega_-(\boldsymbol{x}, \boldsymbol{v})$. The scattering operator S, on the domain where this makes sense, is then the transformation $(\boldsymbol{x}, \boldsymbol{v}) \to (\boldsymbol{x}', \boldsymbol{v}')$, where $(\boldsymbol{x}, \boldsymbol{v})$ and $(\boldsymbol{x}', \boldsymbol{v}')$ are related by $(\boldsymbol{x}, \boldsymbol{v})_{\text{in}} = (\boldsymbol{x}', \boldsymbol{v}')_{\text{out}}$, i.e. $\Omega_+(\boldsymbol{x}, \boldsymbol{v}) = \Omega_-(\boldsymbol{x}', \boldsymbol{v}')$.[11] Thus one may reformulate the definition of S as

$$S = \Omega_-^{-1}\Omega_+.$$

12.2 Non-relativistic Quantum Scattering by a Potential

We consider the analog of the previous section in non-relativistic Quantum Mechanics: The scattering of a particle of mass m by a potential V. When no potential is present we call

[7] These limits exist if the potential vanishes fast enough at infinity. In the case of the electrostatic potential, the limit $\boldsymbol{v}$ exists, but the limit $\boldsymbol{x}$ is infinite. Many potentials of interest for particle interactions decrease faster than the electrostatic potential so the present heuristic is not misleading.

[8] It would be more appropriate to use momentum rather than velocity and to call $\mathcal{P}$ the phase space, but this could be confusing for the reader having no exposure to Classical Mechanics.

[9] Thus, equivalently, this point is obtained by running time of a length t_0, starting at position $\boldsymbol{x} - t_0\boldsymbol{v}$ with velocity $\boldsymbol{v}$.

[10] Please observe that Ω_+ corresponds to $t \to -\infty$. This is the standard notation and I do not want to go against it.

[11] So, the motion of a particle such that $(\boldsymbol{x}, \boldsymbol{v})_{\text{in}} = (\boldsymbol{x}', \boldsymbol{v}')_{\text{out}}$ in the far past resembles the motion of a free particle which at time zero has position $\boldsymbol{x}$ and velocity $\boldsymbol{v}$, whereas in the far future it resembles the motion of a free particle which at time zero has position $\boldsymbol{x}'$ and velocity $\boldsymbol{v}'$.

the particle a free particle. In position state space $\mathcal{H} = L^2(\mathbb{R}^3)$ the Hamiltonian of the free particle is given by

$$H_0 = \frac{\boldsymbol{P}^2}{2m} := -\frac{1}{2m}\sum_{i\le 3}\frac{\partial^2}{\partial x_i^2}. \tag{12.2}$$

Before we bring in the potential we will explain a phenomenon known as "the spreading of the wave function". This phenomenon has no classical equivalent. Heuristically, if we start at time zero with the free particle in a given state, and let it evolve, as time passes its localization becomes more and more uncertain. Intuitively the reason for this phenomenon is that at time zero the momentum of the particle is not precisely known. The spreading of the wave function makes it very difficult to give a meaning to the quantum equivalent of classical statements such as "consider a particle coming from infinity...", because if we start with a well-localized particle very far away, by the time it reaches the location of the scattering center, its localization will be so spread out that the likelihood that it gets close to the scattering center will be very small, and most likely it will not be scattered at all. To bypass the difficulty, we control the localization of the particle when it reaches the region of the scattering center. That is, the starting point of our scattering experiment is a particle in a given state which we should think of as being localized in the same region of space as the scattering center. We run time backward for a large duration T. The particle we obtain is the starting point of our scattering experiment. Its location is not welldefined, but it will be when time runs forward by the same duration T. This is exactly what motivates the construction of the operator Ω_+.

Let us now be more precise. An element of the position state space $\mathcal{H}$ is a function φ on $\mathbb{R}^3$ and the standard terminology is to call it the *wave function*, or sometimes the *wave packet.*[12] One thinks of the wave function as depending on t through time-evolution $U_0(t) = \exp(-itH_0)$, so that we consider $\varphi(\boldsymbol{x},t) := U_0(t)(\varphi)(\boldsymbol{x})$. The "spreading of the wave function" is the fact that as $t \to \infty$, the function $\boldsymbol{x} \mapsto \varphi(\boldsymbol{x},t)$ has a tendency to spread out, to be less and less localized. In the next lemma we compute exactly what happens for some specific wave functions. We recall the notation $\boldsymbol{x}^2 = \sum_{i\le 3} x_i^2$.

Lemma 12.2.1 *Consider $\boldsymbol{a} \in \mathbb{R}^3$ and $\alpha > 0$. Define*

$$\varphi_{\alpha,\boldsymbol{a}}(\boldsymbol{x}) := \exp(-(\boldsymbol{x}-\boldsymbol{a})^2/2\alpha) \tag{12.3}$$

then

$$U_0(t)(\varphi_{\alpha,\boldsymbol{a}})(\boldsymbol{x}) = \Big(\frac{m\alpha}{\mathrm{i}t + m\alpha}\Big)^{3/2}\exp\Big(-\frac{(\boldsymbol{x}-\boldsymbol{a})^2}{2(\alpha+\mathrm{i}t/m)}\Big), \tag{12.4}$$

where the power $3/2$ of the complex number is determined by continuity starting with $t = 0$. Consequently, if $\boldsymbol{a} \in \mathbb{R}^3$, we have

$$|U_0(t)(\varphi_{\alpha,\boldsymbol{a}})(\boldsymbol{x})| \le (1 + t^2/(m\alpha)^2)^{-3/4}. \tag{12.5}$$

[12] The reason for the terminology "packet" is to distinguish this situation from a "plane wave" $\exp(\mathrm{i}\boldsymbol{x}\cdot\boldsymbol{p})$ that is not an element of $\mathcal{H}$, but what physicists call "a non-renormalizable state" and which, as they say, is "unphysical" but nonetheless ubiquitous in their considerations.

Thus, for large t, $U_0(t)(\varphi_{\alpha,\boldsymbol{a}})$ has a small supremum norm, and represents a state without precise localization.

Proof It suffices to show that the function $\varphi(\boldsymbol{x},t)$ given by the right-hand side of (12.4) satisfies the Schrödinger equation $\mathrm{i}\partial\varphi/\partial t = H_0\varphi$, which is really straightforward. □

Corollary 12.2.2 *For each $\boldsymbol{a} \in \mathbb{R}^3$, each $\alpha > 0$ and each $\psi \in \mathcal{H}$ we have*

$$\lim_{t\to\infty}(\psi, U_0(t)(\varphi_{\alpha,\boldsymbol{a}})) = 0. \tag{12.6}$$

Proof If $\varphi \in \mathcal{H}$ we have

$$(\psi,\varphi) = \int \mathrm{d}^3\boldsymbol{x}\,\psi(\boldsymbol{x})^*\varphi(\boldsymbol{x}) \le \sup_{\boldsymbol{x}} |\varphi(\boldsymbol{x})| \int \mathrm{d}^3\boldsymbol{x}\,|\psi(\boldsymbol{x})|.$$

Using this inequality for $\psi = U_0(t)(\varphi_{\alpha,\boldsymbol{a}})$ in (12.5) proves (12.6) when $\int \mathrm{d}^3\boldsymbol{x}|\psi(\boldsymbol{x})| < \infty$. The subset of elements of $\mathcal{H}$ with this property is dense in $\mathcal{H}$, and the set of $\psi \in \mathcal{H}$ for which (12.5) holds is closed. □

Exercise 12.2.3 (a) Using the Fourier transform and the fact that in momentum state space, $U_0(t)$ is simply multiplication by $\exp(\mathrm{i}t\boldsymbol{p}^2/2m)$, explain how one may discover the formula (12.4). (Warning: One must be fluent with Gaussian integrals.)
(b) More generally prove the formula

$$U_0(t)(\varphi)(\boldsymbol{x}) = \int \mathrm{d}^3\boldsymbol{y}\,K(t,\boldsymbol{x},\boldsymbol{y})\varphi(\boldsymbol{y})$$

for a certain kernel $K(t,\boldsymbol{x},\boldsymbol{y})$. Hint: This is rather challenging but can be found in many textbooks, and in particular in Hall's book [39, Section 4.2].

Lemma 12.2.1 is significant because there are many functions of the type $\varphi_{\alpha,\boldsymbol{a}}$, as the following shows.

Lemma 12.2.4 $\mathcal{H} = L^2 := L^2(\mathbb{R}^3, \mathrm{d}^3\boldsymbol{x})$ *is the closed linear span of the functions of the type*[13] $\varphi_{\alpha,\boldsymbol{a}}$ *for* $\alpha > 0$ *and* $\boldsymbol{a} \in \mathbb{R}^3$.

Proof Let us stay informal. Since for $\boldsymbol{a} \in \mathbb{R}^3$ the functions $\varphi_{\alpha,\boldsymbol{a}}(\boldsymbol{x})$ are well peaked around $\boldsymbol{a}$ for α small, every test function is well approximated in L^2 by a mixture of such functions, and the space of test functions is dense in L^2. Consequently every state function can be well approximated in L^2 by a linear combination of the functions $\varphi_{\alpha,\boldsymbol{a}}$. □

Exercise 12.2.5 (a) Prove that for any wave function φ and any $\varepsilon > 0$ for t large enough the function $U_0(t)(\varphi)$ is within ε distance (for the L^2 norm) of a function which is uniformly bounded by ε.

(b) Construct a uniformly bounded[14] normalized wave function, and a sequence (t_n) of positive times such that $U_0(t_n)(\varphi)(0) \ge 2^n$. Hint: Look for φ of the type $\varphi = \sum_{n\ge1} U_0(-t_n)(\varphi_n)$ where $\varphi_n(0)$ is very large while the L^2 norm of φ_n is very small.

[13] As is easily seen these are the wave functions of particles of average momentum zero. This does not mean however that all particles have average momentum zero because the momentum operator is not continuous.

[14] That is $|\varphi(\boldsymbol{x})| \le C$ for all $\boldsymbol{x}$ and a number C independent of $\boldsymbol{x}$.

Consider now a potential V, that is a function on $\mathbb{R}^3$. Denoting also by V the operator "multiplication by the function $V(\boldsymbol{x})$", under suitable regularity conditions on the potential V, the formula

$$H = H_0 + V \tag{12.7}$$

defines a self-adjoint operator,[15] which is the Hamiltonian of the particle submitted to the potential V. We may then consider the unitary operator $U(t) = \exp(-itH)$. We would like to describe the way the interaction with the potential modifies the "trajectory" of a particle, a phenomenon which we will describe as *scattering by a potential*. The key for doing this is a suitable definition of the operators Ω_+ and Ω_-.

Proposition 12.2.6 *Assume that*

$$\|V\|^2 := \int \mathrm{d}^3\boldsymbol{x}\, V(\boldsymbol{x})^2 < \infty. \tag{12.8}$$

Then for each φ in the state space $\mathcal{H}$ the limits[16]

$$\Omega_+(\varphi) = \lim_{t\to-\infty} U(t)^{-1}U_0(t)(\varphi) \tag{12.9}$$

$$\Omega_-(\varphi) = \lim_{t\to\infty} U(t)^{-1}U_0(t)(\varphi) \tag{12.10}$$

exist.

The hypothesis on the potential in Proposition 12.2.6 is rather restrictive, making the result of limited use. Our intention is simply to give a flavor of the methods through the study of as simple as possible an example of a really non-trivial situation. We refer the reader to Reed and Simon's book [69] for far more general conditions on the potential V which imply similar results.

We may as in the previous section interpret $\Omega_+(\varphi)$ as the state at time zero of the motion (with the influence of the potential V) which asymptotically at time $t \to -\infty$ resembles the motion of the free particle which at time zero is in state φ. In Dirac's notation people often write

$$|\xi\rangle_{\text{in}} := \Omega_+|\xi\rangle \; ; \; |\xi\rangle_{\text{out}} := \Omega_-|\xi\rangle. \tag{12.11}$$

Exercise 12.2.7 Prove (heuristically) that $|\xi\rangle_{\text{in}}$ is the state which in the interaction picture evolves at $t = -\infty$ into $|\xi\rangle$. Hint: Use (11.23).

The use of the operator $U(t)^{-1}U_0(t)$ certainly reminds us of the considerations of Section 11.2, but we will study the connection with these only later.

15 More accurately, it defines an essentially self-adjoint operator, which can then be extended into a self-adjoint operator. Results in this direction may be found in Reed and Simon's book [69]. A typical condition on V is $V \in L^2 + L^\infty$, V is the sum of a bounded function and a square-integrable function. Sharper results can be found in Kato's paper [46].

16 Again here: Ω_+ corresponds to $t \to -\infty$.

Proof of Proposition 12.2.6 The set of φ for which the right-hand side of (12.9) converges is a linear space, and it is closed for the distance induced by the L^2 norm, as one sees using Cauchy's convergence criterion and the fact that $U(t)$ and $U_0(t)$ are norm-preserving. Therefore using Lemma 12.2.4 it suffices to prove convergence when $\varphi = \varphi_{\alpha,\boldsymbol{a}}$. Let us then define

$$\begin{aligned}\psi(t) &:= \frac{\mathrm{d}}{\mathrm{d}t} U(t)^{-1} U_0(t)(\varphi_{\alpha,\boldsymbol{a}}) \\ &= \mathrm{i}U(t)^{-1}(H - H_0)U_0(t)(\varphi_{\alpha,\boldsymbol{a}}) \\ &= \mathrm{i}U(t)^{-1} V U_0(t)(\varphi_{\alpha,\boldsymbol{a}}),\end{aligned} \tag{12.12}$$

where we have used that $\mathrm{d}U(t)/\mathrm{d}t = -\mathrm{i}HU(t) = -\mathrm{i}U(t)H$ and that U is unitary, $U(t)^{-1} = U(t)^\dagger = U(-t)$, and similar relations for U_0. Consequently, for $u < 0$,

$$U(u)^{-1}U_0(u)(\varphi_{\alpha,\boldsymbol{a}}) = \varphi_{\alpha,\boldsymbol{a}} - \int_u^0 \mathrm{d}t\,\psi(t). \tag{12.13}$$

Moreover, since $U(t)$ is unitary,

$$\|\psi(t)\| = \|VU_0(t)(\varphi_{\alpha,\boldsymbol{a}})\| \le \|V\|(1 + t^2/(m\alpha)^2)^{-3/4},$$

where we use (12.5) together with the simple fact that for $\varphi \in \mathcal{H}$

$$\|V\varphi\| \le \|V\| \sup_{\boldsymbol{x}\in\mathbb{R}^3} |\varphi(\boldsymbol{x})|.$$

Consequently $\int \mathrm{d}t\|\psi(t)\| < \infty$ and this implies the result in the case of Ω_+. The case of Ω_- is similar. □

In the rest of this section, limits of operators will be for the strong topology. That is, if $W(t)$ and W are operators we say that $W(t) \to W$ (say, as $t \to \infty$) if $W(t)(\varphi) \to W(\varphi)$ for each $\varphi \in \mathcal{H}$. Thus (12.9) means that $U(t)^{-1}U_0(t) \to \Omega_+$ as $t \to -\infty$.

Let us note right away the following important property.

Lemma 12.2.8 *We have*

$$H\Omega_+ = \Omega_+ H_0\ ;\ H\Omega_- = \Omega_- H_0. \tag{12.14}$$

Proof We prove only the first part since the second one is similar. For any number τ we have

$$\begin{aligned}\Omega_+ = \lim_{t\to-\infty} U(t)^{-1}U_0(t) &= \lim_{t\to-\infty} U(t+\tau)^{-1}U_0(t+\tau) \\ &= \lim_{t\to-\infty} U(\tau)^{-1}U(t)^{-1}U_0(t)U_0(\tau) \\ &= U(\tau)^{-1}\Omega_+ U_0(\tau),\end{aligned} \tag{12.15}$$

so that $U(\tau)\Omega_+ = \Omega_+ U_0(\tau)$ and taking the derivative at $\tau = 0$ of the equality yields the results. □

The intuitive content of this result is that Ω_+ transforms states of a given energy for H_0 into states of the same energy for H.

Both Ω_+ and Ω_- are isometries on their respective ranges $\mathcal{R}_+$ and $\mathcal{R}_-$. How do we describe these spaces? Let us start with a simple observation.

Lemma 12.2.9 *Any eigenvector ψ of H is orthogonal to $\mathcal{R}_+$ and $\mathcal{R}_-$.*

This result connects to the discussion on page 62: Typically the eigenvectors correspond to negative eigenvalues and represent "bound states". A particle in a bound state never gets away from the potential. The lemma holds because the motion in the far past of a particle in a bound state does not look like the motion of free particle (which would be far away from the potential).

Proof Recalling the definition (12.3), and according to Lemma 12.2.4, it suffices to show that $(\psi, \Omega_+(\varphi_{\alpha,\boldsymbol{a}})) = 0$. Now, calling E the eigenvalue corresponding to ψ for any t we have

$$(\psi, U(t)^{-1}U_0(t)(\varphi_{\alpha,\boldsymbol{a}})) = (U(t)(\psi), U_0(t)(\varphi_{\alpha,\boldsymbol{a}})) = \exp(\mathrm{i}Et)(\psi, U_0(t)(\varphi_{\alpha,\boldsymbol{a}})).$$

As a consequence of (12.6) the last inner product goes to 0 as $|t| \to \infty$. The same argument works in the case of $\mathcal{R}_-$. □

Let us stress that in this lemma we assume ψ to be an element of $\mathcal{H}$, so we *cannot* use it when ψ is an "improper state".

It is not true in general that $\mathcal{R}_+ = \mathcal{R}_-$. Points in $\mathcal{R}_+ \setminus \mathcal{R}_-$ are the quantum equivalent of a particle that, coming from infinity, gets trapped by the potential. Its motion in the far future does not resemble the motion of a free particle.

The important and difficult theorem is as follows. Let us denote by $\mathcal{B}$ the closed linear span of the eigenvectors of H and by $\mathcal{B}^\perp$ its the orthogonal complement. Then if the potential V is nice enough we have

$$\mathcal{R}_+ = \mathcal{R}_- = \mathcal{B}^\perp. \tag{12.16}$$

This condition is called *asymptotic completeness*. Here nice enough means that V decreases fast enough at infinity and is not too singular at the origin. Proving asymptotic completeness is a very difficult task which we do not attempt.[17] Assuming these conditions, we define next the scattering operator which, in a sense, describes how the scattering changes the trajectory of the particle.

Definition 12.2.10 Assuming asymptotic completeness, the *scattering operator* S is defined by

$$S = \Omega_-^{-1}\Omega_+. \tag{12.17}$$

It is unitary and satisfies

$$H_0 S = S H_0. \tag{12.18}$$

[17] One further reason for not giving any proof is that there are very few rigorous results in Quantum Field Theory where one succeeds to prove a suitable version of asymptotic completeness.

The definition makes sense because Ω_+ is unitary between $\mathcal{H}$ and $\mathcal{R}_+$ and Ω_-^{-1} is unitary from $\mathcal{R}_- = \mathcal{R}_+$ to $\mathcal{H}$. The last assertion follows from (12.14), since

$$SH_0 = \Omega_-^{-1}\Omega_+ H_0 = \Omega_-^{-1} H\Omega_+ = H_0\Omega_-^{-1}\Omega_+ = H_0 S.$$

With the notation (12.11) one writes

$$\langle\eta|S|\xi\rangle = \langle\eta|\Omega_-^\dagger\Omega_+|\xi\rangle = {}_{\text{out}}\langle\eta|\xi\rangle_{\text{in}}.$$

Here of course ${}_{\text{out}}\langle\eta|$ is the vector $|\eta\rangle_{\text{out}}$ seen as a linear functional on $\mathcal{H}$, or in other words, ${}_{\text{out}}\langle\eta|\xi\rangle_{\text{in}}$ is the inner product of the vectors $|\eta\rangle_{\text{out}}$ and $|\xi\rangle_{\text{in}}$.

We have $\Omega_- = \lim_{t\to\infty} U(t)^{-1}U_0(t)$ so that $\Omega_-^{-1} = \lim_{t\to\infty} U_0(t)^{-1}U(t) = \lim_{t\to\infty} U_0(-t)U(t)$. Since $\Omega_+ = \lim_{t\to-\infty} U(t)^{-1}U_0(t) = \lim_{t\to-\infty} U(-t)U_0(t)$ we deduce from (12.17) that

$$S = \lim_{t_1\to\infty, t_2\to-\infty} U_0(-t_1)U(t_1-t_2)U_0(t_2), \tag{12.19}$$

a formula which will be fundamental later in computing S.

12.3 The Scattering Matrix in Non-relativistic Quantum Scattering

Having defined the scattering *operator* in the previous section, *in the same setting* we define the scattering *matrix*, or S-matrix. We use the Fourier transform to view the scattering operator as an operator from momentum state space to itself, which we still call S.[18] Heuristically the S-matrix is simply the matrix of the operator S using the customary continuous basis of improper states $|\boldsymbol{p}\rangle$ for the momentum state space. In case you are allergic to this formalism, let us attempt a rigorous definition. Consider two Schwartz functions $\xi,\eta \in \mathcal{S}^3_{\mathbb{R}}$, the space of real-valued Schwartz functions on $\mathbb{R}^3$. (We consider first only real-valued functions to avoid problems with anti-linearity.) We define the quantity[19]

$$\Theta(\xi,\eta) := (\xi, S\eta) = (\Omega_-\xi, \Omega_+\eta). \tag{12.20}$$

The formula (12.20) defines a bilinear functional on $\mathcal{S}^3_{\mathbb{R}}$, which is a distribution in one of the variables ξ or η when the other variable is fixed. A result of Laurent Schwartz ("Schwartz Kernel Theorem" [33]) then asserts that there exists a tempered distribution Φ on $\mathbb{R}^6$ such that for any choice of ξ and η one has

$$\Theta(\xi,\eta) = \Phi(\xi\otimes\eta),$$

where $\xi\otimes\eta(x,y) := \xi(x)\eta(y)$. *This distribution Φ is called the S-matrix.*

We pretend as usual that Φ is a function rather than a distribution

$$(\xi, S\eta) = \Phi(\xi\otimes\eta) = \iint \frac{\mathrm{d}^3\boldsymbol{p}_1}{(2\pi)^3}\frac{\mathrm{d}^3\boldsymbol{p}_2}{(2\pi)^3}\xi(\boldsymbol{p}_1)\eta(\boldsymbol{p}_2)\Phi(\boldsymbol{p}_1,\boldsymbol{p}_2), \tag{12.21}$$

[18] If you think as a physicist, for whom the state space is independent of any representation you do not need this step.

[19] This definition does not require the asymptotic completeness condition, but to do correct physics S has to be unitary anyway.

and then we define the S-matrix as the "function $\Phi(\boldsymbol{p}_1, \boldsymbol{p}_2)$". Then, allowing now ξ and η to be complex valued,

$$(\xi, S\eta) = \Phi(\xi \otimes \eta) = \iint \frac{\mathrm{d}^3\boldsymbol{p}_1}{(2\pi)^3} \frac{\mathrm{d}^3\boldsymbol{p}_2}{(2\pi)^3} \xi(\boldsymbol{p}_1)^* \eta(\boldsymbol{p}_2) \Phi(\boldsymbol{p}_1, \boldsymbol{p}_2). \tag{12.22}$$

Observe here that the S-matrix entirely determines the scattering operator. Looking again at the same question from the point of view of a physicist, we witness yet another triumph of Dirac's formalism. The scattering operator acts on the state space, which we think of as independent of any kind of representation. Considering two states $|\alpha\rangle$ and $|\beta\rangle$, we use twice the completeness relation

$$1 = \int \frac{\mathrm{d}^3\boldsymbol{p}}{(2\pi)^3} |\boldsymbol{p}\rangle\langle\boldsymbol{p}|$$

to write

$$\langle\alpha|S|\beta\rangle = \iint \frac{\mathrm{d}^3\boldsymbol{p}_1}{(2\pi)^3} \frac{\mathrm{d}^3\boldsymbol{p}_2}{(2\pi)^3} \langle\alpha|\boldsymbol{p}_1\rangle\langle\boldsymbol{p}_1|S|\boldsymbol{p}_2\rangle\langle\boldsymbol{p}_2|\beta\rangle. \tag{12.23}$$

When a mathematician uses the function η in momentum state space to describe a state $|\beta\rangle$ this means that $\langle\boldsymbol{p}|\beta\rangle = \eta(\boldsymbol{p})$, and $\langle\beta|\boldsymbol{p}\rangle = \eta(\boldsymbol{p})^*$. Consequently (12.23) has exactly the same meaning as (12.21), except that the S-matrix is now written as $\langle\boldsymbol{p}_1|S|\boldsymbol{p}_2\rangle$ rather than $\Phi(\boldsymbol{p}_1, \boldsymbol{p}_2)$.

In the sequel we shall use physics notation $\langle\boldsymbol{p}_1|S|\boldsymbol{p}_2\rangle$, but we keep in mind that despite this notation, this *quantity is a distribution* and makes sense only when integrated against a test function.

Continuing to describe the situation the way a physicist might do, we denote by $|\boldsymbol{p}\rangle$ the (improper, non-normalizable and unphysical) state $(2\pi)^3\delta^{(3)}_{\boldsymbol{p}}$ (in momentum state space) of wave function $\exp(\mathrm{i}\boldsymbol{p}\cdot\boldsymbol{x})$ (in position state space), and we "define"

$$|\boldsymbol{p}\rangle_{\mathrm{in}} = \Omega_+|\boldsymbol{p}\rangle \tag{12.24}$$

(the precise meaning of which will be explained below). Since $|\boldsymbol{p}\rangle$ is an eigenvector of H_0, (12.14) shows that $|\boldsymbol{p}\rangle_{\mathrm{in}}$ is an eigenvector for H. The physicists say that

$$|\boldsymbol{p}\rangle_{\mathrm{in}} \text{ is the state of } \mathcal{H} \text{ which as time } t \to -\infty \text{ approaches the state } |\boldsymbol{p}\rangle. \tag{12.25}$$

They are well aware that they are walking on eggshells while making such a statement because it is false as it stands: Since $U(t)|\boldsymbol{p}\rangle_{\mathrm{in}}$ is proportional to $|\boldsymbol{p}\rangle_{\mathrm{in}}$ because $|\boldsymbol{p}\rangle_{\mathrm{in}}$ is an eigenvector of H, it never gets close to $|\boldsymbol{p}\rangle$ whatever the value of t.

The statement (12.25) is true only in the distributional sense, as is explained in more detail below. In a similar manner, one defines $|\boldsymbol{p}\rangle_{\mathrm{out}} = \Omega_-|\boldsymbol{p}\rangle$, the state "which approaches $|\boldsymbol{p}\rangle$ as $t \to \infty$". Using the second equality in (12.20) one has

$$\langle\boldsymbol{p}_1|S|\boldsymbol{p}_2\rangle = {}_{\mathrm{out}}\langle\boldsymbol{p}_1|\boldsymbol{p}_2\rangle_{\mathrm{in}}, \tag{12.26}$$

a formula which is often given as a definition of the S-matrix.

Let us now explain more precisely the meaning of (12.24). The three-dimensional version of the formula (2.37) i.e.

$$|\varphi\rangle = \int \frac{\mathrm{d}^3\boldsymbol{p}}{(2\pi)^3}|\boldsymbol{p}\rangle\langle\boldsymbol{p}|\varphi\rangle, \tag{12.27}$$

allows one to view the "states" $|\boldsymbol{p}\rangle$ as a continuous basis of the state space. These states are eigenvectors of H_0 but not of H. On the other hand, the states $|\boldsymbol{p}\rangle_{\mathrm{in}}$ also form a continuous basis (but not of the whole of $\mathcal{H}$), which is far more appropriate to the study of H because they are eigenvectors of H, and whenever one has (12.27) application of (12.24) also yields

$$\Omega_+|\varphi\rangle = \int \frac{\mathrm{d}^3\boldsymbol{p}}{(2\pi)^3}|\boldsymbol{p}\rangle_{\mathrm{in}}\langle\boldsymbol{p}|\varphi\rangle, \tag{12.28}$$

which actually *defines* the states $|\boldsymbol{p}\rangle_{\mathrm{in}}$ in a kind of distributional sense: These states $|\boldsymbol{p}\rangle_{\mathrm{in}}$ make mathematical sense *only* when integrated against a function as in (12.28). They satisfy the normalization condition

$${}_{\mathrm{in}}\langle\boldsymbol{p}_1|\boldsymbol{p}_2\rangle_{\mathrm{in}} = (2\pi)^3\delta^{(3)}(\boldsymbol{p}_1 - \boldsymbol{p}_2), \tag{12.29}$$

which is a simple consequence of the similar relation for the states $|\boldsymbol{p}\rangle$ and of the fact that Ω_+ is an isometry. Also in consequence of this fact, the states $|\boldsymbol{p}\rangle_{\mathrm{in}}$ form a "continuous basis" of $\mathcal{R}_+$. Here we should stress how tricky it can be to use "improper states". The improper states $|\boldsymbol{p}\rangle_{\mathrm{in}}$ are eigenvalues of H and belong to $\Omega_+\mathcal{H} = \mathcal{R}_+$ despite the result of Lemma 12.2.9 (but, as we pointed out, this lemma holds only for true states).

The continuous basis $|\boldsymbol{p}\rangle_{\mathrm{in}}$ is appropriate for the study of what happens in the far past, because we know how to compute there the evolution of a mixture of such states. Indeed, when $t \ll 0$, by the very definition of Ω_+ one has

$$U(t)\int \frac{\mathrm{d}^3\boldsymbol{p}}{(2\pi)^3}|\boldsymbol{p}\rangle_{\mathrm{in}}\langle\boldsymbol{p}|\varphi\rangle \simeq U_0(t)\int \frac{\mathrm{d}^3\boldsymbol{p}}{(2\pi)^3}|\boldsymbol{p}\rangle\langle\boldsymbol{p}|\varphi\rangle = U_0(t)|\varphi\rangle.$$

The states $|\boldsymbol{p}\rangle_{\mathrm{out}}$ are another continuous basis, this time adapted to what happens in the far future. In much more general situations it will still be true that the S-matrix *relates a continuous basis adapted to the past to a continuous basis adapted to the future.* More about the states $|\boldsymbol{p}\rangle_{\mathrm{in}}$ may be found in Appendix J.

What can we say a priori about the S-matrix? A first observation is that when there is no interaction,

$$\langle\boldsymbol{p}_1|S|\boldsymbol{p}_2\rangle = (2\pi)^3\delta^{(3)}(\boldsymbol{p}_1 - \boldsymbol{p}_2), \tag{12.30}$$

so that when there is interaction it is natural to consider the difference

$$\langle\boldsymbol{p}_1|S|\boldsymbol{p}_2\rangle - (2\pi)^3\delta^{(3)}(\boldsymbol{p}_1 - \boldsymbol{p}_2). \tag{12.31}$$

Also, a consequence of (12.18) is that the distribution (12.31) is supported on the set $\{\boldsymbol{p}_1^2 = \boldsymbol{p}_2^2\}$. The physicist's proof of this is to use $H_0|\boldsymbol{p}\rangle = (\boldsymbol{p}^2/2m)|\boldsymbol{p}\rangle$ to write

$$\frac{\boldsymbol{p}_1^2}{2m}\langle\boldsymbol{p}_1|S|\boldsymbol{p}_2\rangle = \langle\boldsymbol{p}_1|H_0S|\boldsymbol{p}_2\rangle = \langle\boldsymbol{p}_1|SH_0|\boldsymbol{p}_2\rangle = \frac{\boldsymbol{p}_2^2}{2m}\langle\boldsymbol{p}_1|S|\boldsymbol{p}_2\rangle$$

so that $\langle \boldsymbol{p}_1|S|\boldsymbol{p}_2\rangle = 0$ unless $\boldsymbol{p}_1^2 = \boldsymbol{p}_2^2$. What is less obvious is that in many circumstances we will have a relation

$$\langle \boldsymbol{p}_1|S|\boldsymbol{p}_2\rangle = (2\pi)^3\delta^{(3)}(\boldsymbol{p}_1 - \boldsymbol{p}_2) + \mathrm{i}\delta(\boldsymbol{p}_1^2 - \boldsymbol{p}_2^2)\Xi(\boldsymbol{p}_1, \boldsymbol{p}_2), \tag{12.32}$$

where Ξ is a continuous (or even smooth) function of $\boldsymbol{p}_1$ and $\boldsymbol{p}_2$. The factor i is conventional.[20] To understand this formula it might help you to read Appendix K. To clarify the meaning of the distribution $\delta(\boldsymbol{p}_1^2 - \boldsymbol{p}_2^2)\Xi(\boldsymbol{p}_1, \boldsymbol{p}_2)$, it can be computed as follows: Given a test function $\eta(\boldsymbol{p}_1, \boldsymbol{p}_2)$, then

$$\begin{aligned}\iint \frac{\mathrm{d}^3\boldsymbol{p}_1}{(2\pi)^3}\frac{\mathrm{d}^3\boldsymbol{p}_2}{(2\pi)^3}\delta(\boldsymbol{p}_1^2 - \boldsymbol{p}_2^2)\Xi(\boldsymbol{p}_1, \boldsymbol{p}_2)\eta(\boldsymbol{p}_1, \boldsymbol{p}_2)\\ = \lim_{\varepsilon\to 0}\frac{1}{2\varepsilon}\iint_{\{|\boldsymbol{p}_1^2 - \boldsymbol{p}_2^2|\le\varepsilon\}}\frac{\mathrm{d}^3\boldsymbol{p}_1}{(2\pi)^3}\frac{\mathrm{d}^3\boldsymbol{p}_2}{(2\pi)^3}\Xi(\boldsymbol{p}_1, \boldsymbol{p}_2)\eta(\boldsymbol{p}_1, \boldsymbol{p}_2).\end{aligned} \tag{12.33}$$

In somewhat imprecise terms, this distribution is carried by the manifold $\boldsymbol{p}_1^2 = \boldsymbol{p}_2^2$ and is simply given by integration against a smooth function on this manifold.

As this book is not a treatise of Quantum Mechanics we will not provide conditions for the formula (12.32) to hold. Rather, we behave as physicists, assuming in our analysis that the simplest possible structure holds until proven wrong.[21] In the case which really concerns us, the S-matrix in Quantum Field Theory, we will eventually be able to find concrete formulas (order by order in perturbation theory) which will vindicate our regularity assumptions.

12.4 The Scattering Matrix and Cross-Sections, I

The purpose of a physical theory is to predict the behavior of the physical world. The purpose of Quantum Field Theory is specifically to predict how particles interact. In the next chapter we will learn how to use Quantum Field Theory to compute the S-matrix. To check the validity of our theory, we perform scattering experiments. How then do we relate the S-matrix to the results of these experiments?

Our goal in the present section is to relate, in the simpler framework of Quantum Mechanics, and in a simple case, the abstract S-matrix with quantities which can be measured in an actual scattering experiment, although our experiment will be very idealized. In this experiment a device[22] accelerates and focuses a beam of particles that is sent in the direction of the z-axis from the negative to the positive direction toward a scattering potential located in a very small region around the origin.[23] Detectors placed above the origin measure the momentum of the scattered particles. Knowing the S-matrix, we want to calculate the rate at which the detector gets triggered, so that we can compare it with experimental results to validate our theory.

20 The standard way to write this formula is different from ours. It includes numerical factors in the last term. These factors are not introduced here as they can be motivated only with foresight.
21 In contrast with negative-minded mathematicians always wondering what could go wrong.
22 It will be referred to in this section as "the device".
23 Think of the potential as being generated by a single particle, and of the range of this potential as being of the order of the size of an atom, extremely smaller than the distance between the origin and the detectors.

Before we get into that, let us describe a related entirely classical experiment. We are shooting bullets in the direction of the z-axis toward a target located near the origin, but whose exact position and shape is unknown. In our model of this situation the bullets are punctual and the target is a compact subset of $\mathbb{R}^3$. Let us denote by D the projection of the target on the (x,y) plane, and let us describe the trajectory of the bullet by its intersection $u = (u_1, u_2)$ with the (x,y) plane. Thus the bullet will hit the target if and only if $u = (u_1, u_2) \in D$. It is easier to hit the target if the area A of D is large. This area is called the *cross-section* of the target. In order to try to hit the target, a sensible strategy is to choose (u_1, u_2) uniformly at random within a distance R of the origin, where R is large enough that every point of D is within distance R of the origin. The probability of hitting the target is then $A/(\pi R^2)$. This becomes vanishingly small as R becomes large, and to compensate for this it is better to shoot about πR^2 bullets[24] (one per unit of surface in the disc $\|u\| \le R$) and then the mean number of bullets hitting the target will be A.

Let us now modify the previous experiment. Instead of a target, there is a scattering center near the origin, but whose exact position and properties are unknown. We keep throwing bullets at this scattering center, in the direction of the z-axis. The trajectory of the bullets (before they approach the scattering center) is again described by a point (u_1, u_2) of the (x,y) plane. If this point (u_1, u_2) is far from the origin, the influence of the scattering center will not be much felt. So if we want to study how often the bullets are scattered in a given direction,[25] it still makes sense to choose (u_1, u_2) uniformly at random within distance R of the origin and to shoot about πR^2 bullets. Then we expect that the mean number of bullets scattered in a given direction will have a finite limit as $R \to \infty$. The setting is described in Figure 12.1.

Going back to the our Quantum Mechanical scattering experiment, the scattering matrix S encodes in a rather indirect way the information about the scattering operator, that is how the scattering potential alters the trajectory of particles. This information is described *in the sense of distributions* by the collection of numbers $\langle p_1|S|p_2\rangle$. These quantities have to be integrated against test functions. This is not a matter of hair-splitting search for rigor. Notwithstanding the heuristic arguments which may be found in some textbooks, it *makes no sense* to assume that the incoming particles are in an "improper state" of given momentum because such particles have no localization whatsoever and thus cannot interact with the scattering potential. To perform a sensible analysis we have to assume that the incoming particles are described by true elements of the state space, by *wave packets* as the physicists say, and then only take a limit as the momenta of the incoming particles become more localized. This is what makes the analysis non-trivial.

We have to model both the detector and the beam of particles, keeping in mind that the art of building models is to capture the essential features while at the same time keeping matters simple. We must also be aware that when relating a mathematical model with the physical world we may have to make statements which make physical sense but are hard to precisely formulate mathematically.

24 The random choices of the corresponding u's being of course independent for different particles.
25 Or more accurately in a given set of directions.

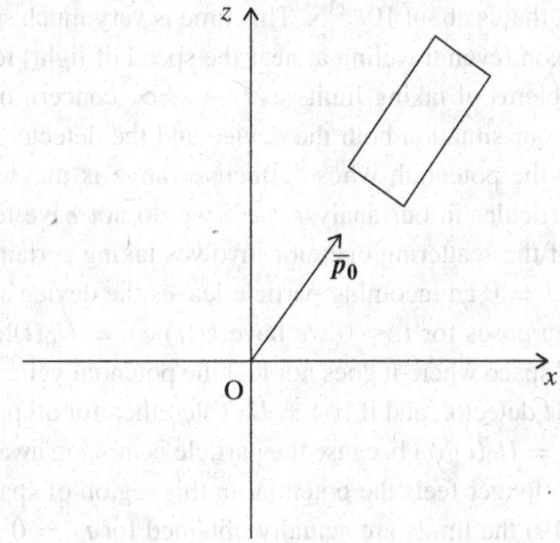

Figure 12.1 The detector is placed so that a particle of momentum $\boldsymbol{p}_0$ leaving from near the origin will enter it.

Let us first model the detector. Given a value of $\boldsymbol{p}_0$, we try to model a detector, which, at the time a particle enters it "measures if its momentum is equal to $\boldsymbol{p}_0$". Let $|\psi\rangle$ be the state of the particle as it enters the detector[26] (and assume that it is normalized, $\langle\psi|\psi\rangle = 1$), so that the probability density of the momentum of this particle is the function $\boldsymbol{p} \mapsto |\langle\boldsymbol{p}|\psi\rangle|^2$.

An *ideal* detector could measure for any set A the probability that the momentum belongs to A, that is the integral over A of the function $|\langle\boldsymbol{p}|\psi\rangle|^2$. We must account, (and this is absolutely essential) for the fact that *a real-life detector has a finite accuracy*. It is then natural to assume that the probability that the detector answers "yes" to the question "is the momentum of the particle equal to $\boldsymbol{p}_0$?" is given by an expression

$$\int \frac{\mathrm{d}^3\boldsymbol{p}}{(2\pi)^3}\theta(\boldsymbol{p})|\langle\boldsymbol{p}|\psi\rangle|^2, \tag{12.34}$$

where $0 \leq \theta \leq 1$ is a smooth function which is equal to 1 in a small neighborhood of $\boldsymbol{p}_0$ and to 0 outside a somewhat larger neighborhood of $\boldsymbol{p}_0$. In the limit case where the particle has exactly momentum $\boldsymbol{p}$, the content of (12.34) is that the detector is triggered with probability $\theta(\boldsymbol{p})$. This equals 1 if $\boldsymbol{p}$ is close to $\boldsymbol{p}_0$, equals 0 if $\boldsymbol{p}$ is far from $\boldsymbol{p}_0$, and is in between in the intermediate region. The function θ models the fact that the detector has a finite accuracy and plays a crucial role in the analysis.

Let us stress again that we are doing physics in this section, not mathematics. In physics it is fundamental to keep in mind the scale at which we are studying Nature and to make suitable approximations. For example in nuclear physics a typical "strong interaction" lasts not much longer than the time it takes for light, traveling at $3 \cdot 10^8$ m/s, to cross a nucleon

[26] Here is the first occurrence of a statement which is hard to make mathematically precise. Let us note that when the particle reaches the detector it is a free particle and, as we will make precise, "its momentum does not change", so precision is not essential here.

of 10^{-15} m of diameter, that is about 10^{-23} s. This time is very much smaller than the time it will take for the nucleon (even traveling at near the speed of light) to reach the detector. For this reason the problems of taking limits as $t \to \pm\infty$ concern only mathematicians but not physicists.[27] In our situation both the device and the detector are situated at least centimeters away from the potential, whose effective range is maybe of order 10^{-10} m if not much less. In particular in our analysis here we do not have to be concerned that the definition (12.19) of the scattering operator involves taking certain limits. Let us then assume that at the time $t = 0$ an incoming particle leaves the device and its state is $|\varphi\rangle$.[28] Then for all practical purposes for $t \leq 0$ we have $U(t)|\varphi\rangle = U_0(t)|\varphi\rangle$ because then the particle is in a region of space where it does not feel the potential yet.[29] Then if the particle takes time T to reach the detector, and if $|\psi\rangle = U(T)|\varphi\rangle$ then for all practical purposes for $t \geq T$ we have $U(t)|\psi\rangle = U_0(t)|\psi\rangle$ because the particle is moving away from the potential past the detector and no longer feels the potential in this region of space. Then we should understand that in (12.19) the limits are actually obtained for $t_2 = 0$ and $t_1 = T$ so that, using again that $U(T)|\varphi\rangle = |\psi\rangle$,

$$S|\varphi\rangle = U_0(-T)U(T)|\varphi\rangle = U_0(-T)|\psi\rangle.$$

The particle is in state $|\psi\rangle = U(T)|\varphi\rangle$ as it enters the detector, so the probability that it triggers the detector is given by

$$\Phi(|\varphi\rangle) := \int \frac{d^3\boldsymbol{p}}{(2\pi)^3}\theta(\boldsymbol{p})|\langle\boldsymbol{p}|\psi\rangle|^2 = \int \frac{d^3\boldsymbol{p}}{(2\pi)^3}\theta(\boldsymbol{p})|\langle\boldsymbol{p}|S|\varphi\rangle|^2, \tag{12.35}$$

because

$$|\langle\boldsymbol{p}|\psi\rangle| = |\langle\boldsymbol{p}|U_0(T)S|\varphi\rangle| = |\langle\boldsymbol{p}|S|\varphi\rangle|,$$

since in momentum state space $U_0(T)$ is simply multiplication by $\exp(-iT\boldsymbol{p}^2/2m)$.

A quantity such as (12.35) is a probability. To measure it[30] we must perform the scattering experiment many times. To do this we have to know how to reliably produce an incoming particle which is exactly in state $|\varphi\rangle$ at the time it leaves the device. It is rather questionable whether this can be done, but, remarkably, our analysis will not depend on this.

Next we have to model the fact that the incoming particle belongs to a beam perpendicular to the (x, y) plane which is *extremely wide* at the scale of the scattering potential.[31] For this we consider lateral shifts of the position of this incoming particle in directions contained in the (x, y) plane. These shifts are parameterized by a point $u = (u_1, u_2) \in \mathbb{R}^2$, and the fact that the incoming beam is cylindrical centered on the z-axis and of radius R is reflected

[27] So, however horrifying it may be at first for a mathematician, physicists are perfectly correct to blankly take $t = \infty$ at times.

[28] Again it might be difficult to precisely formulate mathematically or even physically what this exactly means, but the incoming particles do have to enter the picture at some stage, and we have to know something about them when they do, so what else may we assume?

[29] This is an idealization, which ignores what happened to the particle while it was accelerated by the device. This however is another story, which does not concern the scattering experiment. For the purposes of the scattering experiment, the particle just comes from far and happens to be in state $|\varphi\rangle$ at time $t = 0$.

[30] We will not be able to measure (12.35) itself, but only a related quantity, as is explained later.

[31] For example the radius of the beam of the Large Hadron Collider is of the order 10 microns $= 10^{-5}$ m in the collision areas, much larger than the order of the size of a nucleon, 10^{-15} m. The value of R is still very small at our scale, but for the purpose of the present analysis it is ∞.

by the fact that $\|u\| \le R$. Thus one may think of u as a random variable which models the position of the incoming particle within the beam.[32] Given $u \in \mathbb{R}^2$ let $\boldsymbol{u} = (u_1, u_2, 0)$. Given a state $|\varphi\rangle$ let us then consider the state $|\varphi\rangle_u$ after the shift by $\boldsymbol{u}$. It is defined by

$$\langle \boldsymbol{x}|\varphi\rangle_u = \langle \boldsymbol{x} - \boldsymbol{u}|\varphi\rangle, \tag{12.36}$$

or, equivalently

$$\langle \boldsymbol{p}|\varphi\rangle_u = \exp(-\mathrm{i}\boldsymbol{u}\cdot\boldsymbol{p})\langle \boldsymbol{p}|\varphi\rangle. \tag{12.37}$$

We state informally the main result of this section, using as is traditional for this result the notion of solid angle. Solid angle $\mathrm{d}\Omega$ is just the natural surface measure on the unit sphere $\mathbb{S}^2$ of $\mathbb{R}^3$, so that $\Omega(\mathbb{S}^2) = 4\pi$.[33]

Theorem 12.4.1 *Let us assume that the scattering operator S satisfies the condition (12.32). That is*

$$\langle \boldsymbol{p}_1|S|\boldsymbol{p}_2\rangle = (2\pi)^3\delta^{(3)}(\boldsymbol{p}_1 - \boldsymbol{p}_2) + \mathrm{i}\delta(\boldsymbol{p}_1^2 - \boldsymbol{p}_2^2)\Xi(\boldsymbol{p}_1, \boldsymbol{p}_2), \tag{12.32}$$

for a continuous function Ξ. *Let us recall the function* Φ *of (12.35) and make the further assumption that the function* θ *occurring there is equal to zero when the momentum* $\boldsymbol{p}$ *is in a neighborhood of the direction of the z-axis.*[34] *Consider a given* $\tilde{\boldsymbol{p}}$ *of the type* $(0,0,\tilde{p})$ *with* $\tilde{p} > 0$. *Then, given* $\varepsilon > 0$ *there exists a neighborhood V of* $\tilde{\boldsymbol{p}}$ *such that if* φ *satisfies the condition*

$$\boldsymbol{p} \notin V \Rightarrow \langle \boldsymbol{p}|\varphi\rangle = 0 \tag{12.38}$$

then

$$\left| \lim_{R\to\infty} \int_{\|u\|\le R} \mathrm{d}^2u\, \Phi(|\varphi\rangle_u) - \frac{1}{(8\pi^2)^2}\int_{\mathbb{S}^2} \mathrm{d}\Omega(\boldsymbol{v})\theta(\tilde{p}\boldsymbol{v})|\Xi(\tilde{p}\boldsymbol{v}, \tilde{\boldsymbol{p}})|^2 \right| \le \varepsilon. \tag{12.39}$$

In words, we compute the limit of the quantity $\int_{\|u\|\le R} \mathrm{d}^2u\, \Phi(|\varphi\rangle_u)$ as first $R \to \infty$[35] and then "the momenta of the incoming particles getting close to $\tilde{\boldsymbol{p}}$". This later property is required in a strong sense: $\langle \boldsymbol{p}|\varphi\rangle$ has to be zero outside a small neighborhood of $\tilde{\boldsymbol{p}}$.[36] This order of limits is certainly required by Heisenberg's uncertainty principle: If we know that the first two components of the momentum of the incoming particle are nearly zero, we cannot say much about the first two coordinates of the position of the particle.

The use of the uniform measure on the disc $\|u\| \le R$ in the quantity $\int_{\|u\|\le R} \mathrm{d}^2u\, \Phi(|\varphi\rangle_u)$ reflects the fact that the lateral position of the incoming particles is uniformly distributed in

32 These choices being independent for different particles.

33 Thus the uniform probability on $\mathbb{S}^2$ is $\mathrm{d}\Omega/(4\pi)$.

34 The function θ models the accuracy of the detector. The hypothesis that this function is zero when the direction of $\boldsymbol{p}$ is in a neighborhood of the direction of the z-axis physically means that within the accuracy of the detector, the direction of $\tilde{\boldsymbol{p}}$ is different from the direction of the z-axis.

35 Please keep in mind that this simply means that R is much larger than the "size of the scattering center", which is certainly satisfied for $R = 10^{-5}$m.

36 As the momenta of the incoming particles concentrate around $\tilde{\boldsymbol{p}}$, "the incoming wave packets become close to plane waves". But as we pointed out, it makes no sense to assume that the incoming particles are plane waves.

that disc, but (just as in the classical case discussed earlier) this quantity does not contain the factor $1/(\pi R^2)$ which would express that we average over the lateral position of the particle. This would not be the proper normalization: One single particle somewhere in a wide beam is likely to miss the scattering potential and to produce no interaction. Rather our normalization expresses that our wide beam contains on average one single particle per unit area. Thus the quantity

$$\frac{1}{(8\pi^2)^2}\int_{\mathbb{S}^2} \mathrm{d}\Omega(\boldsymbol{v})\theta(\tilde{p}\boldsymbol{v})|\Xi(\tilde{p}\boldsymbol{v},\tilde{\boldsymbol{p}})|^2 \tag{12.40}$$

represents (in the iterated limit discussed above) the expected number of particles detected if one sends an average of πR^2 particles in the beam.[37]

Finally, the hypothesis that $\theta(\boldsymbol{p})$ is zero for $\boldsymbol{p}$ close to the z-axis implies that this result is of use only to study scattering in directions which are *different* from the direction of the incoming particles.

Exercise 12.4.2 Prove that the conclusion of Theorem 12.4.1 cannot hold unless $\theta(\boldsymbol{p})|\langle\boldsymbol{p}|\varphi\rangle|^2$ is zero almost everywhere. This justifies the need of a strong hypothesis as (12.38).

A fundamental point of the theorem is that the right-hand side does not depend on the precise "shape" of $|\varphi\rangle$, but only on the fact that the corresponding momentum concentrates around $\tilde{\boldsymbol{p}}$. Equally fundamental is the following:

The probability that the detector gets triggered is entirely determined by the scattering matrix. (12.41)

How does this single-particle experiment result relate to what happens when a beam of particles is thrown at the potential? Consider a beam of particles of radius R, and assume that during any second there is an average number ρ of particles per unit area which passes into a horizontal plane. The different particles inside the beam do not interact with each other and each performs the scattering experiment independently. Let us make the (highly unrealistic) assumption that we can pretend that each of these particles performs the same single-particle experiment, with the *same* value of $|\varphi\rangle$ and with u uniformly distributed over a circle of radius R. The particle beam is very wide compared to the range of the potential, and the incoming particles having traveled together a long time in the particle accelerator should have a rather well-defined momentum. Then it is reasonable to assume that each incoming particle will trigger the detector with probability about $1/(\pi R^2)$ times the right-hand side of (12.39). Since there is a total of $\pi R^2\rho$ particles per second, the average number of times $\langle n\rangle$ the detector is triggered each second will be ρ times the right-hand side of (12.39) i.e.

$$\langle n\rangle = \rho\frac{1}{(8\pi^2)^2}\int_{\mathbb{S}^2} \mathrm{d}\Omega(\boldsymbol{v})\theta(\tilde{p}\boldsymbol{v})|\Xi(\tilde{p}\boldsymbol{v},\tilde{\boldsymbol{p}})|^2. \tag{12.42}$$

[37] Again, the reason that the number of particles detected stays bounded as $R\to\infty$ is that most of the particles pass too far from the scattering center to be scattered toward the detector.

What then of our highly unrealistic assumption that each particle corresponds to the same $|\varphi\rangle$? It is here that the keypoint that the right-hand side does not depend on $|\varphi\rangle$ matters. We will have to consider different types of $|\varphi\rangle$ and we simply cross our fingers[38] and hope that the averaging phenomenon described by Theorem 12.4.1 takes place for each of these types and that it remains true that the average number of times the detector is triggered per second is given by (12.42).

To bring forward the physical content of this fact we observe that $\langle n\rangle$ is dimensionless since it represents an average number. Since ρ is the number of particles per unit area it has the dimension of the inverse of an area, the right-hand side of (12.39) has the dimension of an area. It is called the *cross-section* that the detector is triggered. The reason for this name is that the mean number of particles which trigger the detector is the same as the number of (classical[39]) particles which would hit an area of this size placed perpendicularly to the beam.

To interpret the relation (12.42) let us consider a unit vector $\boldsymbol{v}$ with a direction different from the direction of $\tilde{\boldsymbol{p}}$. Let us consider a small neighborhood U of $\boldsymbol{v}$ on $\mathbb{S}^2$ such that the function $\boldsymbol{w} \mapsto \Xi(\tilde{p}\boldsymbol{w}, \tilde{\boldsymbol{p}})$ is almost constant on U. Pretending now that we can build detectors of arbitrary accuracy, consider θ such that $\theta(\tilde{p}\boldsymbol{w}) = 1$ if $\boldsymbol{w} \in U$ and $\theta(\tilde{p}\boldsymbol{w}) = 0$ if $\boldsymbol{w} \notin U$. Then the average number of times the detector is triggered per second is nearly given by $\rho\mathrm{d}\Omega|\Xi(\tilde{p}\boldsymbol{v}, \tilde{\boldsymbol{p}})/(8\pi^2)|^2$ where $\mathrm{d}\Omega$ is the solid angle corresponding to U. The coefficient $|\Xi(\tilde{p}\boldsymbol{v}, \tilde{\boldsymbol{p}})/(8\pi^2)|^2$ of $\rho\mathrm{d}\Omega$ is called in physics the *differential cross-section*. Thus when the direction of the unit vector $\boldsymbol{v}$ is different from the direction of $\tilde{\boldsymbol{p}}$ we have reached a remarkably simple formula. When the length of the incoming momentum is $\tilde{p}$, the differential cross-section in the direction of $\boldsymbol{v}$ is given by

$$\left|\frac{\Xi(\tilde{p}\boldsymbol{v}, \tilde{\boldsymbol{p}})}{8\pi^2}\right|^2. \tag{12.43}$$

The numerical constants here are sensitive to the way we have normalized the measure in momentum space. The coefficient $8\pi^2$ can be removed by redefining the function Ξ of (12.32). This is the standard way this result is presented in the literature, refer to J. Taylor's book [82]. In the case where the situation is rotationally invariant, the differential cross-section depends only of course on $\tilde{p}$, the momentum of the particles being scattered, and on the angle of scattering.

Some of the hypotheses of the theorem are perfectly sensible, but the fact that we throw the particles at a single scattering center is totally unrealistic: Rather, in an actual experiment the beam is directed either at a whole piece of material (such as a very thin gold sheet in Rutherford's historical experiment) containing many such centers, or at another beam of particles coming in the opposite direction. Fortunately however using this more realistic hypothesis does not really change the analysis so we do not give the details. Still it is another matter to determine if the theorem can be applied for a given experiment.

38 Or try to work much harder.
39 I.e. non-quantum.

Proof of Theorem 12.4.1 We stay informal but the argument can be made rigorous. The core of the proof consists in establishing that provided

$$\theta(\boldsymbol{p}) \neq 0 \Rightarrow \langle \boldsymbol{p}|\varphi\rangle = 0 \tag{12.44}$$

we have the identity

$$\lim_{R\to\infty}\int_{\|u\|\le R} \mathrm{d}^2u\, \Phi(|\varphi\rangle_u) = \int \frac{\mathrm{d}^3\boldsymbol{p}'}{(2\pi)^3} W(\boldsymbol{p}')|\langle \boldsymbol{p}'|\varphi\rangle|^2, \tag{12.45}$$

where W is a certain explicit continuous function. Let us first complete the proof assuming this has been proved. Since W is continuous, given $\varepsilon > 0$ there exists a neighborhood V of $\tilde{\boldsymbol{p}}$ such that $\boldsymbol{p} \in V \Rightarrow |W(\boldsymbol{p}) - W(\tilde{\boldsymbol{p}})| \le \varepsilon$. By decreasing V if necessary we may assume that $\boldsymbol{p} \in V \Rightarrow \theta(\boldsymbol{p}) = 0$ because by assumption the set $\{\boldsymbol{p};\theta(\boldsymbol{p}) = 0\}$ is a neighborhood of $\tilde{\boldsymbol{p}}$.

Let us now assume that $\langle \boldsymbol{p}|\varphi\rangle = 0$ for $\boldsymbol{p} \notin V$. Since $\int \frac{\mathrm{d}^3\boldsymbol{p}'}{(2\pi)^3}|\langle \boldsymbol{p}'|\varphi\rangle|^2 = 1$, we then get

$$\left|\int \frac{\mathrm{d}^3\boldsymbol{p}'}{(2\pi)^3} W(\boldsymbol{p}')|\langle \boldsymbol{p}'|\varphi\rangle|^2 - W(\tilde{\boldsymbol{p}})\right| \le \varepsilon. \tag{12.46}$$

Furthermore, (12.44) holds by construction of V, so that (12.45) holds, and we have shown that

$$\left|\lim_{R\to\infty}\int_{\|u\|\le R} \mathrm{d}^2u\, \Phi(|\varphi\rangle_u) - W(\tilde{\boldsymbol{p}})\right| \le \varepsilon.$$

To prove Theorem 12.4.1 it thus suffices to prove the identity (12.45) as well as the fact that the actual value of $W(\tilde{\boldsymbol{p}})$ is given by (12.40). This is obtained through calculation. Unfortunately, this computation has a technical aspect: Having reached in a rather straightforward manner the multiple integral (12.52) containing many delta functions, we have to use these delta functions to reduce the number of variables we integrate. This is *all there is* in the argument. If this does not thrill you, please skip this part as well as Section 12.6. The only thing which really matters for the sequel is (12.41): The probability that the detector is triggered depends only on the S-matrix.

We start the argument with (12.35) used for $|\varphi\rangle_u$. Exchanging the order of integration we obtain the identity

$$\int_{\|u\|\le R} \mathrm{d}^2u\, \Phi(|\varphi\rangle_u) = \int \frac{\mathrm{d}^3\boldsymbol{p}}{(2\pi)^3}\theta(\boldsymbol{p}) f_R(\boldsymbol{p}), \tag{12.47}$$

where

$$f_R(\boldsymbol{p}) := \int_{\|u\|\le R} \mathrm{d}^2u |\langle \boldsymbol{p}|S|\varphi\rangle_u|^2,$$

and we proceed to compute f_R. First, using (12.37) in the second equality,

$$\langle \boldsymbol{p}|S|\varphi\rangle_u = \int \frac{\mathrm{d}^3\boldsymbol{p}_1}{(2\pi)^3}\langle \boldsymbol{p}|S|\boldsymbol{p}_1\rangle\langle \boldsymbol{p}_1|\varphi\rangle_u = \int \frac{\mathrm{d}^3\boldsymbol{p}_1}{(2\pi)^3}\langle \boldsymbol{p}|S|\boldsymbol{p}_1\rangle\langle \boldsymbol{p}_1|\varphi\rangle \exp(-\mathrm{i}\boldsymbol{p}_1\cdot \boldsymbol{u}),$$

so that

$$|\langle \boldsymbol{p}|S|\varphi\rangle_u|^2 = \langle \boldsymbol{p}|S|\varphi\rangle_u \langle \boldsymbol{p}|S|\varphi\rangle_u^*$$
$$= \iint \frac{\mathrm{d}^3\boldsymbol{p}_1}{(2\pi)^3}\frac{\mathrm{d}^3\boldsymbol{p}_2}{(2\pi)^3}\langle \boldsymbol{p}|S|\boldsymbol{p}_1\rangle\langle \boldsymbol{p}|S|\boldsymbol{p}_2\rangle^*\langle \boldsymbol{p}_1|\varphi\rangle\langle \boldsymbol{p}_2|\varphi\rangle^*\exp(\mathrm{i}(\boldsymbol{p}_2-\boldsymbol{p}_1)\cdot\boldsymbol{u}), \quad (12.48)$$

and thus

$$f_R(\boldsymbol{p}) = \int_{\|u\|\le R}\mathrm{d}^2u\iint \frac{\mathrm{d}^3\boldsymbol{p}_1}{(2\pi)^3}\frac{\mathrm{d}^3\boldsymbol{p}_2}{(2\pi)^3}\langle \boldsymbol{p}|S|\boldsymbol{p}_1\rangle\langle \boldsymbol{p}|S|\boldsymbol{p}_2\rangle^*\langle \boldsymbol{p}_1|\varphi\rangle\langle \boldsymbol{p}_2|\varphi\rangle^*\exp(\mathrm{i}(\boldsymbol{p}_2-\boldsymbol{p}_1)\cdot\boldsymbol{u}). \quad (12.49)$$

To pursue the computation we will use that $\langle \boldsymbol{p}|S|\boldsymbol{p}_1\rangle = (2\pi)^3\delta^{(3)}(\boldsymbol{p}-\boldsymbol{p}_1) + \mathrm{i}\delta(\boldsymbol{p}^2-\boldsymbol{p}_1^2)\,\Xi(\boldsymbol{p},\boldsymbol{p}_1)$ from (12.32), and similarly for $\langle \boldsymbol{p}|S|\boldsymbol{p}_2\rangle$. This produces four terms. The first term is obtained by replacing $\langle \boldsymbol{p}|S|\boldsymbol{p}_1\rangle$ by $(2\pi)^3\delta^{(3)}(\boldsymbol{p}-\boldsymbol{p}_1)$ and $\langle \boldsymbol{p}|S|\boldsymbol{p}_2\rangle$ by $(2\pi)^3\delta^{(3)}(\boldsymbol{p}-\boldsymbol{p}_2)$ in (12.49). The δ functions let us compute the integrals in $\boldsymbol{p}_1$ and $\boldsymbol{p}_2$ to obtain

$$|\langle \boldsymbol{p}|\varphi\rangle|^2\int_{\|u\|\le R}\mathrm{d}^2u,$$

which does not contribute to (12.47) because we assume through (12.44) that $\theta(\boldsymbol{p})|\langle \boldsymbol{p}|\varphi\rangle|^2 = 0$. A second term is obtained by replacing $\langle \boldsymbol{p}|S|\boldsymbol{p}_1\rangle$ by $(2\pi)^3\delta^{(3)}(\boldsymbol{p}-\boldsymbol{p}_1)$ and $\langle \boldsymbol{p}|S|\boldsymbol{p}_2\rangle$ by $\mathrm{i}\delta(\boldsymbol{p}^2-\boldsymbol{p}_2^2)\Xi(\boldsymbol{p},\boldsymbol{p}_2)$. We can then use the delta function to perform the integral in $\boldsymbol{p}_1$, yielding

$$-\mathrm{i}\int_{\|u\|\le R}\mathrm{d}^2u\int\frac{\mathrm{d}^3\boldsymbol{p}_2}{(2\pi)^3}\delta(\boldsymbol{p}^2-\boldsymbol{p}_2^2)\Xi(\boldsymbol{p},\boldsymbol{p}_2)^*\langle \boldsymbol{p}|\varphi\rangle\langle \boldsymbol{p}_2|\varphi\rangle^*\exp(\mathrm{i}(\boldsymbol{p}_2-\boldsymbol{p})\cdot\boldsymbol{u})$$

and this term again does not contribute to (12.47) because the product $\theta(\boldsymbol{p})\langle \boldsymbol{p}|\varphi\rangle$ is identically zero by (12.38). A third similar term does not contribute either and consequently we have shown that (12.47) holds with now

$$f_R(\boldsymbol{p}) \quad (12.50)$$
$$= \iint\frac{\mathrm{d}^3\boldsymbol{p}_1}{(2\pi)^3}\frac{\mathrm{d}^3\boldsymbol{p}_2}{(2\pi)^3}\delta(\boldsymbol{p}^2-\boldsymbol{p}_1^2)\delta(\boldsymbol{p}^2-\boldsymbol{p}_2^2)\Xi(\boldsymbol{p},\boldsymbol{p}_1)\Xi(\boldsymbol{p},\boldsymbol{p}_2)^*\langle \boldsymbol{p}_1|\varphi\rangle\langle \boldsymbol{p}_2|\varphi\rangle^* A_R(\boldsymbol{p}_1,\boldsymbol{p}_2),$$

where $A_R(\boldsymbol{p}_1,\boldsymbol{p}_2) := \int_{\|u\|\le R}\mathrm{d}^2u\exp(\mathrm{i}(\boldsymbol{p}_2-\boldsymbol{p}_1)\cdot\boldsymbol{u})$. Now, denoting p_{11}, p_{12}, p_{13} the components of $\boldsymbol{p}_1$ (etc.),

$$\lim_{R\to\infty}A_R(\boldsymbol{p}_1,\boldsymbol{p}_2) = \lim_{R\to\infty}\int_{\|u\|\le R}\mathrm{d}^2u\exp(\mathrm{i}(\boldsymbol{p}_2-\boldsymbol{p}_1)\cdot\boldsymbol{u}) = (2\pi)^2\delta(p_{21}-p_{11})\delta(p_{22}-p_{12}),$$

so that we have reached the identity

$$\lim_{R\to\infty}\int_{\|u\|\le R}\mathrm{d}^2u\,\Phi(|\varphi\rangle_u) = \int\frac{\mathrm{d}^3\boldsymbol{p}}{(2\pi)^3}\theta(\boldsymbol{p})f(\boldsymbol{p}), \quad (12.51)$$

where

$$f(\boldsymbol{p}) = \iint\frac{\mathrm{d}^3\boldsymbol{p}_1}{(2\pi)^3}\frac{\mathrm{d}^3\boldsymbol{p}_2}{(2\pi)^3}\Xi(\boldsymbol{p},\boldsymbol{p}_1)\Xi(\boldsymbol{p},\boldsymbol{p}_2)^*\langle \boldsymbol{p}_1|\varphi\rangle\langle \boldsymbol{p}_2|\varphi\rangle^*\delta, \quad (12.52)$$

and where

$$\delta := \delta(\boldsymbol{p}^2 - \boldsymbol{p}_1^2)\delta(\boldsymbol{p}^2 - \boldsymbol{p}_2^2)(2\pi)^2\delta(p_{21} - p_{11})\delta(p_{22} - p_{12}).$$

This expression contains four delta functions, three of which involve $\boldsymbol{p}_2$. These three delta functions will let us eliminate the three components of $\boldsymbol{p}_2$ and perform the integral in $\boldsymbol{p}_2$ in (12.52) at $\boldsymbol{p}$ and $\boldsymbol{p}_1$ fixed. We start this procedure now. Integrating in p_{21} and p_{22} simply means that we fix $p_{21} = p_{11}$ and $p_{22} = p_{12}$. To integrate in p_{23} we have to perform an integral of the type

$$\int \frac{\mathrm{d}p_{23}}{2\pi}\delta(p_{23}^2 - a)w(p_{23}), \tag{12.53}$$

for a certain function w, where $a = \boldsymbol{p}^2 - p_{11}^2 - p_{12}^2$. We may assume that $\boldsymbol{p}^2 = \boldsymbol{p}_1^2$, for otherwise the contribution is zero due to the factor $\delta(\boldsymbol{p}^2 - \boldsymbol{p}_1^2)$ in δ. Then $a = \boldsymbol{p}_1^2 - p_{11}^2 - p_{12}^2 = p_{13}^2$, so that the integral (12.53) is

$$\int \frac{\mathrm{d}p_{23}}{2\pi}\delta(p_{23}^2 - p_{13}^2)w(p_{23}).$$

We proceed as in (4.40) to find that this is

$$\frac{1}{2\pi}\frac{1}{2p_{13}}w(p_{13}),$$

where we have taken into account the fact that the function $w(p_{23})$ is not zero only when $\boldsymbol{p}_2 \simeq \tilde{\boldsymbol{p}}$ (because of the term $\langle \boldsymbol{p}_2|\varphi\rangle$ in (12.52)) and in particular only when $p_{23} > 0$, so that $w(-p_{13}) = 0$. In words, to perform the integration in $\boldsymbol{p}_2$, we replace the factor δ by a factor $\delta(\boldsymbol{p}^2 - \boldsymbol{p}_1^2)/(4\pi p_{13})$ and we set $\boldsymbol{p}_2 = \boldsymbol{p}_1$ in the integrand. That is, we have

$$f(\boldsymbol{p}) = \frac{1}{2\pi}\int \frac{\mathrm{d}^3\boldsymbol{p}_1}{(2\pi)^3}\delta(\boldsymbol{p}^2 - \boldsymbol{p}_1^2)\frac{1}{2p_{13}}|\Xi(\boldsymbol{p},\boldsymbol{p}_1)\langle \boldsymbol{p}_1|\varphi\rangle|^2,$$

and therefore

$$\int \frac{\mathrm{d}^3\boldsymbol{p}}{(2\pi)^3}\theta(\boldsymbol{p})f(\boldsymbol{p}) = \int \frac{\mathrm{d}^3\boldsymbol{p}_1}{(2\pi)^3}W(\boldsymbol{p}_1)|\langle \boldsymbol{p}_1|\varphi\rangle|^2, \tag{12.54}$$

where the continuous function W is given by

$$W(\boldsymbol{p}_1) = \frac{1}{2\pi}\frac{1}{2p_{13}}\int \frac{\mathrm{d}^3\boldsymbol{p}}{(2\pi)^3}\theta(\boldsymbol{p})\delta(\boldsymbol{p}^2 - \boldsymbol{p}_1^2)|\Xi(\boldsymbol{p},\boldsymbol{p}_1)|^2.$$

We have proved (12.45), and it remains only to compute $W(\tilde{\boldsymbol{p}})$. For this we move to polar coordinates. We make the change of variable $\boldsymbol{p} = r\boldsymbol{v}$ where $r \geq 0$ and $\boldsymbol{v} \in \mathbb{S}^2$, the unit sphere of $\mathbb{R}^3$, so that $\boldsymbol{p}^2 = r^2$. The normalization of the solid angle is such that $\mathrm{d}^3\boldsymbol{p} = r^2\mathrm{d}r\mathrm{d}\Omega(\boldsymbol{v})$, so that

$$W(\tilde{\boldsymbol{p}}) = \frac{1}{(2\pi)^4}\frac{1}{2\tilde{p}}\int_{r\geq 0}\mathrm{d}r r^2\delta(r^2 - \tilde{p}^2)\int_{\mathbb{S}^2}\mathrm{d}\Omega(\boldsymbol{v})\theta(\tilde{p}\boldsymbol{v})|\Xi(\tilde{p}\boldsymbol{v},\tilde{\boldsymbol{p}})|^2.$$

Performing the integral in r as we have done multiple times yields the formula (12.39). □

12.5 Scattering Matrix in Quantum Field Theory

Let us now move back to Quantum Field Theory. The setting is somewhat different, particles do not interact with a scattering potential, but with each other. The definition of the S-matrix is far more difficult. We do not even begin to address the main difficulty in the present section: The interaction of a particle with itself in a sense *changes the nature of single-particle states,* a question we will examine in Chapter 14. We do not examine either the problem of constructing Hamiltonians governing the scattering process.

Our focus instead is finding the proper tool, the S-matrix, to describe the results of scattering experiments. Let us examine again the physics we are trying to model. We may not be able to probe exactly what happens during the interaction, but a long time before it, and a long time after it, our system consists of well-separated particles, and as such they do not interact with each other. Here "long time" means at the time-scale of the interactions, and again we will never have to worry about taking limits.

The state space $\mathcal{H}_{\text{part}}$ describing a collection of non-interacting particles is our familiar boson Fock space, or a tensor product of such spaces if there are several particle types. For example, if we have two particles of types 1 and 2, then $\mathcal{H}_{\text{part}} = \mathcal{H}_1 \otimes \mathcal{H}_2$ where $\mathcal{H}_1$ and $\mathcal{H}_2$ are the boson Fock spaces of Chapter 3 modeling indistinguishable particles of the same type. One would naively think at first that the same state space would describe interacting particles, but there is no reason to believe that, while interacting, particles can still be seen as distinct individuals, so the state space $\mathcal{H}$ should be more complicated than that. Here we hit a difficulty in generalizing (12.9) and (12.10): $U_0(t)$ operates on $\mathcal{H}_{\text{part}}$ whereas $U(t)$ operates on the whole state space $\mathcal{H}$.

At the heuristic level though the meaning is clear. The only states with which we can experiment look like collections of free particles long before and long after the experiment. Given states $|\psi\rangle, |\varphi\rangle \in \mathcal{H}_{\text{part}}$ which describe collections of non-interacting particles, $\langle\psi|S|\varphi\rangle$ is the amplitude that a state which "looks like $|\varphi\rangle$" before the interaction "looks like $|\psi\rangle$" after the interaction. Here "looks like" means being composed of the same particle types with the same distribution of momenta.

Trying to be more precise, we assume that there is a unitary operator Ω_+ from $\mathcal{H}_{\text{part}}$ to $\mathcal{H}$, which expresses that $|\varphi\rangle_{\text{in}} := \Omega_+|\varphi\rangle$ is the element of the state space which in the far past looks like the collection $|\varphi\rangle$ of free particles.[40] The hypothesis that Ω_+ is unitary and hence onto contains the strong assumption that each state is of the type $|\varphi\rangle_{\text{in}}$, that is, it looks in the far past like a collection of free particles. We assume that $\Omega_+ H_0 = H\Omega_+$ (as in (12.14)). Similarly there is a unitary operator Ω_- which describes what happens in the far future, with $H\Omega_- = \Omega_- H_0$ and we define $|\varphi\rangle_{\text{out}} = \Omega_-|\varphi\rangle$. The S-matrix is defined for $|\varphi\rangle$ and $|\psi\rangle$ in $\mathcal{H}_{\text{part}}$ by

$$\langle\psi|S|\varphi\rangle = {}_{\text{out}}\langle\psi|\varphi\rangle_{\text{in}}. \tag{12.55}$$

In other words $S = \Omega_-^{-1}\Omega_+$ is a unitary operator on $\mathcal{H}_{\text{part}}$ which will completely describe the effect of the interactions (as we have already claimed and will explain further).

40 In the sense that as $t \to -\infty$ the collection $U(t)|\varphi\rangle_{\text{in}}$ of interacting particles looks more and more like the collection $U_0(t)|\varphi\rangle$ of free particles.

You might object that the present paragraph just repeats in jargon what was said informally in the previous one without explaining what is really meant by "look in the far past like..." There are at least two good reasons why most textbooks stop at the present level of hand-waving:[41] First, it is really difficult to go beyond it. Second, these considerations might be important to construct a sound general theory, but they are completely irrelevant to what we will do. We will work in perturbation theory, or, equivalently, in the interaction picture. There we will implicitly pretend that $\mathcal{H}_{\rm part} \subset \mathcal{H}$. We will also implicitly pretend that $U(t)$ operates on $\mathcal{H}_{\rm part}$.[42] Most of the previous difficulties then just disappear, and we will never be concerned about elements of $\mathcal{H} \setminus \mathcal{H}_{\rm part}$.

How then do we describe the S-matrix, and what general form might it have? To describe the family of numbers $\langle\psi|S|\varphi\rangle$ for $|\psi\rangle, |\varphi\rangle \in \mathcal{H}_{\rm part}$, we first need to describe the elements of $\mathcal{H}_{\rm part}$. This is best done using a "continuous basis" of improper states. A natural continuous basis to describe one-particle states is *denoted* by $a_i^\dagger(\boldsymbol{p})|0\rangle$. Here $\boldsymbol{p} \in \mathbb{R}^3$ and the index i accounts for the particle type, its spin, etc., but to keep matters simple we consider only the case of spinless particles. The element $a_i^\dagger(\boldsymbol{p})|0\rangle$ represents a particle of type i and given momentum $\boldsymbol{p}$. This quantity is an improper state which makes mathematical sense only when integrated against a test function. To describe incoming two-particle states, we will similarly use a continuous basis denoted $a_{i_1}^\dagger(\boldsymbol{p}_1)a_{i_2}^\dagger(\boldsymbol{p}_2)|0\rangle$,[43] which similarly represents the two-particle state where a particle of type i_1 and momentum $\boldsymbol{p}_1$ and a particle of type i_2 and momentum $\boldsymbol{p}_2$ are present.

Thus, if e.g. $|\varphi\rangle = |a_1^\dagger(\boldsymbol{p}_1)a_2^\dagger(\boldsymbol{p}_2)|0\rangle$ and $|\psi\rangle = a_5^\dagger(\boldsymbol{p}_5)a_4^\dagger(\boldsymbol{p}_4)a_3^\dagger(\boldsymbol{p}_3)|0\rangle$ the quantity $\langle\psi|S|\varphi\rangle$ equals

$$\langle 0|a_3(\boldsymbol{p}_3)a_4(\boldsymbol{p}_4)a_5(\boldsymbol{p}_5)|S|a_1^\dagger(\boldsymbol{p}_1)a_2^\dagger(\boldsymbol{p}_2)|0\rangle. \tag{12.56}$$

This quantity is the amplitude[44] that starting with a state which, in the far past, looks like it is made up of the two particles of types 1 and 2 and momenta $\boldsymbol{p}_1$ and $\boldsymbol{p}_2$ we end up in the far future with a state which looks like it is made of the three particles of types 3, 4, 5 and respective momenta $\boldsymbol{p}_3, \boldsymbol{p}_4, \boldsymbol{p}_5$. This statement is true only in a distributional sense, and the quantity (12.56) itself makes sense only as a distribution.[45]

The S-matrix is then the collection of all quantities of similar type as (12.56) for various choices of incoming and outgoing particles. Let us investigate the consequences of conser-

[41] You are probably also wondering how much credit should be given to claims which involve that much hand-waving. You are right, but for the time being we will stop there and delay further examination of these matters until Chapter 13.

[42] As we will show in Section 13.4 these assumptions are mathematically untenable, but it takes more than that to stop us, doesn't it?

[43] In physics textbooks one often finds the following notation: If α is a label that describes particle types and their momenta, there corresponds to α an element $|\alpha\rangle$ of $\mathcal{H}_{\rm part}$, and $|\alpha\rangle_{\rm in}$ and $|\alpha\rangle_{\rm out}$ are the states "made of the same particles as α in the far past and the far future" respectively. This makes sense only in the distributional sense.

[44] Let us recall that inner products in the state space are often called amplitudes (see page 46) and that their squares represent probabilities.

[45] To avoid the minor complications created by the fact that identical particles cannot be distinguished from each other we consider only the case of particles of different types $i_1 = 1, i_2 = 2$.

vation of four-momentum on the S-matrix. Denoting by m_i the mass of the particle of type i the initial four-momentum is

$$p_{\rm in} = \left(\sqrt{\boldsymbol{p}_1^2 + m_1^2} + \sqrt{\boldsymbol{p}_2^2 + m_2^2}, \boldsymbol{p}_1 + \boldsymbol{p}_2\right),$$

and the final momentum $p_{\rm out}$ is defined similarly,

$$p_{\rm out} = \left(\sqrt{\boldsymbol{p}_3^2 + m_3^2} + \sqrt{\boldsymbol{p}_4^2 + m_4^2} + \sqrt{\boldsymbol{p}_5^2 + m_5^2}, \boldsymbol{p}_3 + \boldsymbol{p}_4 + \boldsymbol{p}_5\right).$$

Conservation of four-momentum[46] implies that the quantity (12.56) will be zero unless $p_{\rm in} = p_{\rm out}$. In other words, the distribution (12.56) is supported by the set $\{p_{\rm in} = p_{\rm out}\}$, so that we should be able to write[47]

$$\langle 0|a_3(\boldsymbol{p}_3)a_4(\boldsymbol{p}_4)a_5(\boldsymbol{p}_5)|S|a_1^\dagger(\boldsymbol{p}_1)a_2^\dagger(\boldsymbol{p}_2)|0\rangle = \mathrm{i}(2\pi)^4\delta^{(4)}(p_{\rm in} - p_{\rm out})M((\boldsymbol{p}_i)_{i\le 5}) \quad (12.57)$$

for a certain quantity M. The factor i is conventional. In the good cases it will happen that M is a *continuous* function of $\boldsymbol{p}_1, \ldots, \boldsymbol{p}_5$. Our choice of incoming particles and outgoing particles of a different nature was not accidental, but helps to make this hypothesis plausible. If we had, say, two outgoing particles of types 1 and 2, we would have to take into account the possibility that these are the same as the incoming particles, with no interaction having taken place, and in M there would be terms with delta functions to express this.[48]

12.6 Scattering Matrix and Cross-Sections, II

In this section we show how to relate the quantity M of (12.57) with cross-sections that can be measured (at least in principle) in the spirit of (12.43). Our sample result assumes that M is a *continuous function*. The same type of arguments can be used in more complex situations but we simply try to show that besides technicalities, everything works as in Section 12.4. One simply has to display even more agility when computing integrals with many delta functions in the integrand. If you are already convinced of the importance of the S-matrix it is harmless to skip this section.

We perform the following experiment. A beam of particles of type 1 intersects a beam of particles of type 2. Some particles of type 1 interact with a particle of type 2. We are interested only in the case where this results in the production of a triple of particles of types 3, 4, 5 respectively. Given the function M we try to compute how often this happens.

The first step is to precisely understand how we work with elements of $\mathcal{H}_{\rm part}$. For this we assume that our operators $a_i(\boldsymbol{p})$ are normalized as usual,

$$a_i(\boldsymbol{q})a_i^\dagger(\boldsymbol{p}) = a_i^\dagger(\boldsymbol{p})a_i(\boldsymbol{q}) + (2\pi)^3\delta^{(3)}(\boldsymbol{q} - \boldsymbol{p})1, \quad (12.58)$$

with all the other commutators being zero. An element of $\mathcal{H}_{\rm part}$ consisting of three particles of types 3, 4, 5 respectively is then of the form

46 Conservation of four-momentum is a fundamental physical principle. When we will actually compute the S-matrix, the delta function enforcing this conservation will automatically appear in our calculations.

47 Mathematically this argument is wishful thinking. For example, the derivative δ' of the delta function "is zero outside the origin", and there is no way to write $\delta' = \delta M$ for a meaningful M. However, in the cases where we will be able to compute the S-matrix, order by order in perturbation theory, it will often have the form (12.57), where M is continuous.

48 Just as the first term in the right-hand side of (12.32) takes into account the possibility that no interaction took place between the particle and the potential.

$$|\psi\rangle = \iiint \prod_{i=3,4,5} \frac{\mathrm{d}^3 \boldsymbol{p}_i}{(2\pi)^3} \psi(\boldsymbol{p}_3, \boldsymbol{p}_4, \boldsymbol{p}_5) a_3^\dagger(\boldsymbol{p}_3) a_4^\dagger(\boldsymbol{p}_4) a_5^\dagger(\boldsymbol{p}_5)|0\rangle, \tag{12.59}$$

where $\psi \in L^2(\mathbb{R}^9, \prod_{i=3,4,5} \mathrm{d}^3\boldsymbol{p}_i/(2\pi)^3)$. The normalization (12.58) ensures that $\langle\psi|\psi\rangle = \|\psi\|_2^2$, the square of the L^2 norm of ψ.

We next have to model our detector. This amazing top-of-the line detector will not be triggered unless it detects exactly three particles, one of each of the types 3, 4, 5. We try to build it so that it gets triggered exactly when these three particles are of given momenta, but however much money we spend on it, it cannot do this exactly. Rather, the probability that it will be triggered by a state of the type (12.59) is of the type

$$\iiint \prod_{i=3,4,5} \frac{\mathrm{d}^3 \boldsymbol{p}_i}{(2\pi)^3} \theta(\boldsymbol{p}_3, \boldsymbol{p}_4, \boldsymbol{p}_5) |\psi(\boldsymbol{p}_3, \boldsymbol{p}_4, \boldsymbol{p}_5)|^2, \tag{12.60}$$

for a certain smooth function θ.

We then have to model incoming pairs of particles. As these come from different beams they are not entangled,[49] so that it suffices to consider elements of $\mathcal{H}_{\mathrm{part}}$ which are of the type

$$|\varphi_1\varphi_2\rangle = \iint \prod_{i=1,2} \frac{\mathrm{d}^3 \boldsymbol{p}_i}{(2\pi)^3} \varphi_1(\boldsymbol{p}_1)\varphi_2(\boldsymbol{p}_2) a_1^\dagger(\boldsymbol{p}_1) a_2^\dagger(\boldsymbol{p}_2)|0\rangle \tag{12.61}$$

where $\varphi_1, \varphi_2 \in L^2(\mathbb{R}^3, \mathrm{d}^3\boldsymbol{p}/(2\pi)^3)$ are of norm 1.

To lighten the exposition we use creative notation: $\boldsymbol{p}_{\mathrm{out}} = (\boldsymbol{p}_3, \boldsymbol{p}_4, \boldsymbol{p}_5)$ and

$$\mathrm{d}^{\mathrm{out}}\boldsymbol{p}_{\mathrm{out}} = \prod_{i=3,4,5} \frac{\mathrm{d}^3 \boldsymbol{p}_i}{(2\pi)^3}\,;\ \langle\mathrm{out}| = \langle 0|a_3(\boldsymbol{p}_3)a_4(\boldsymbol{p}_4)a_5(\boldsymbol{p}_5)|. \tag{12.62}$$

Then, following the analysis of (12.35) the probability that the detector gets triggered when the incoming pair of particles is described by $|\varphi_1\varphi_2\rangle$ is given by

$$\int \mathrm{d}^{\mathrm{out}}\boldsymbol{p}_{\mathrm{out}}\theta(\boldsymbol{p}_{\mathrm{out}})|\langle\mathrm{out}|S|\varphi_1\varphi_2\rangle|^2. \tag{12.63}$$

The whole analysis depends on the cancellations produced by modeling the fact that the incoming particles of types 1 and 2 belong to a wide beam. We lighten the exposition by pretending rather that a particle of type 2 is just sitting at the origin (and plays the role of the scattering center of Section 12.4) modeling that the particle of type 1 belongs to a wide beam by random lateral displacements of these in the (x, y) plane. We know the actions of translations on one-particle state are expressed by the representation of the Poincaré group: A translation by a four-vector w amounts to multiplication of the state by $\exp(\mathrm{i}(w, p))$. For $u = (u_1, u_2) \in \mathbb{R}^2$ we consider the four-vector $y_u = (0, u_1, u_2, 0)$ and the vector $\boldsymbol{u} = (u_1, u_2, 0)$. Since the time-component of y_u is zero, $(y_u, p) = y_u^0 p^0 - \boldsymbol{u}\cdot\boldsymbol{p} = -\boldsymbol{u}\cdot\boldsymbol{p}$, and lateral displacement of u amounts to multiplication of the state by $\exp(\mathrm{i}(y_u, p)) = \exp(-\mathrm{i}\boldsymbol{u}\cdot\boldsymbol{p})$, and replaces the state $|\varphi_1\varphi_2\rangle$ of (12.61) by the state

$$|\varphi_1\varphi_2\rangle_u := \iint \prod_{i=1,2} \frac{\mathrm{d}^3 \boldsymbol{p}_i}{(2\pi)^3} \exp(-\mathrm{i}\boldsymbol{u}\cdot\boldsymbol{p}_1)\varphi_1(\boldsymbol{p}_1)\varphi_2(\boldsymbol{p}_2) a_1^\dagger(\boldsymbol{p}_1) a_2^\dagger(\boldsymbol{p}_2)|0\rangle. \tag{12.64}$$

49 In the precise Quantum Mechanical sense.

To lighten notation further, let us set

$$M(\boldsymbol{p}_1, \boldsymbol{p}_2, \boldsymbol{p}_{\text{out}}) := M(\boldsymbol{p}_1, \boldsymbol{p}_2, \boldsymbol{p}_3, \boldsymbol{p}_4, \boldsymbol{p}_5)$$

and note that then (12.57) takes the form

$$\langle \text{out}|S a_1^\dagger(\boldsymbol{p}_1) a_2^\dagger(\boldsymbol{p}_2)|0\rangle = \mathrm{i}(2\pi)^4 \delta^{(4)}(p_{\text{in}} - p_{\text{out}}) M(\boldsymbol{p}_1, \boldsymbol{p}_2, \boldsymbol{p}_{\text{out}}). \tag{12.65}$$

The next theorem is the main result of the present section. It corresponds to (12.39): *We compute how the probability that our detector is triggered depends on the function M.*[50] To simplify notation, we use the relativistic relation $\boldsymbol{p}/p^0 = \boldsymbol{v}$ between the velocity $\boldsymbol{v}$ of a relativistic particle and its four-momentum $(p^0, \boldsymbol{p})$, a relation which is proved in (N.3), so that e.g. $\tilde{v}_{13} = \tilde{p}_{13}/\tilde{p}_{10}$ for $\tilde{p}_1 = (\tilde{p}_{10}, \tilde{p}_{11}, \tilde{p}_{12}, \tilde{p}_{13})$.

Theorem 12.6.1 *Assume that (12.65) holds, where the function M is continuous. As $R \to \infty$ and then the supports of φ_1 and φ_2 concentrate around values $\tilde{\boldsymbol{p}}_1$ and $\tilde{\boldsymbol{p}}_2$, the quantity*

$$\int_{\|u\| \le R} \mathrm{d}^2 u \int \mathrm{d}^{\text{out}} \boldsymbol{p}_{\text{out}} \theta(\boldsymbol{p}_{\text{out}}) |\langle \text{out}|S|\varphi_1\varphi_2\rangle_u|^2 \tag{12.66}$$

converges to

$$\int \mathrm{d}^{\text{out}} \boldsymbol{p}_{\text{out}} \theta(\boldsymbol{p}_{\text{out}}) \frac{1}{|\tilde{v}_{13} - \tilde{v}_{23}|} (2\pi)^4 \delta^{(4)}(\tilde{p}_{\text{in}} - p_{\text{out}}) |M(\tilde{\boldsymbol{p}}_1, \tilde{\boldsymbol{p}}_2, \boldsymbol{p}_{\text{out}})|^2. \tag{12.67}$$

The ε-δ meaning of the expression "the support of φ_1 concentrates around the value $\tilde{\boldsymbol{p}}_1$" is that given a neighborhood V of $\tilde{\boldsymbol{p}}_1$ and $\varepsilon > 0$, we may require that the integral of $|\langle \boldsymbol{p}|\varphi_2\rangle|^2$ on the complement of V be $\le \varepsilon$. This condition is quite weaker than the corresponding condition in Theorem 12.4.1, where we required that $\langle \boldsymbol{p}|\varphi_2\rangle$ be zero outside V. This is made possible by the fact that in the right-hand side of (12.65) we do not have the troublesome first term on the right-hand side of (12.32).

To prove Theorem 12.6.1 we proceed as in Theorem 12.4.1. We will show that the quantity

$$\mathcal{L}(\varphi_1, \varphi_2) := \lim_{R \to \infty} \int_{\|u\| \le R} \mathrm{d}^2 u \int \mathrm{d}^{\text{out}} \boldsymbol{p}_{\text{out}} \theta(\boldsymbol{p}_{\text{out}}) |\langle \text{out}|S|\varphi_1\varphi_2\rangle_u|^2 \tag{12.68}$$

is of the form

$$\mathcal{L}(\varphi_1, \varphi_2) = \iint \frac{\mathrm{d}^3 \boldsymbol{p}_1}{(2\pi)^3} \frac{\mathrm{d}^3 \boldsymbol{p}_2}{(2\pi)^3} W(\boldsymbol{p}_1, \boldsymbol{p}_2) |\varphi_1(\boldsymbol{p}_1)|^2 |\varphi_2(\boldsymbol{p}_2)|^2 \tag{12.69}$$

where the continuous function $W(\boldsymbol{p}_1, \boldsymbol{p}_2)$ is given by

$$W(\boldsymbol{p}_1, \boldsymbol{p}_2) = \int \mathrm{d}^{\text{out}} \boldsymbol{p}_{\text{out}} \theta(\boldsymbol{p}_{\text{out}}) \frac{1}{|\tilde{v}_{13} - \tilde{v}_{23}|} (2\pi)^4 \delta^{(4)}(\tilde{p}_{\text{in}} - p_{\text{out}}) |M(\boldsymbol{p}_1, \boldsymbol{p}_2, \boldsymbol{p}_{\text{out}})|^2. \tag{12.70}$$

[50] There are special features in our sample result. When some of the outgoing particles are of the same type as the incoming particles, it is always necessary to assume that the direction of scattering of these outgoing particles is *different* from the direction of the incoming particles, for otherwise other terms of the S-matrix would be involved.

When the supports of φ_1 and φ_2 concentrate around $\tilde{\boldsymbol{p}}_1$ and $\tilde{\boldsymbol{p}}_2$ respectively, then $\mathcal{L}(\varphi_1, \varphi_2)$ approaches $W(\tilde{\boldsymbol{p}}_1, \tilde{\boldsymbol{p}}_2)$, proving Theorem 12.6.1.

Lemma 12.6.2 *We have*

$$\lim_{R\to\infty} \int_{\|u\|\le R} \mathrm{d}^2 u |\langle \mathrm{out}|S|\varphi_1\varphi_2\rangle_u|^2 = \iint \frac{\mathrm{d}^3\boldsymbol{p}_1}{(2\pi)^3}\frac{\mathrm{d}^3\boldsymbol{p}_2}{(2\pi)^3} Z(\boldsymbol{p}_1, \boldsymbol{p}_2) \tag{12.71}$$

where

$$Z(\boldsymbol{p}_1, \boldsymbol{p}_2) = \frac{1}{|\tilde{v}_{13} - \tilde{v}_{23}|}(2\pi)^4 \delta^{(4)}(\tilde{p}_{\mathrm{in}} - p_{\mathrm{out}})|M(\boldsymbol{p}_1, \boldsymbol{p}_2, \boldsymbol{p}_{\mathrm{out}})|^2 |\varphi_1(\boldsymbol{p}_1)|^2 |\varphi_2(\boldsymbol{p}_2)|^2.$$

To prove (12.69) we substitute (12.71) into (12.68). To prove the lemma we compute first

$$\lim_{R\to\infty} \int_{\|u\|\le R} \mathrm{d}^2 u |\langle \mathrm{out}|S|\varphi_1\varphi_2\rangle_u|^2 = \lim_{R\to\infty} \int_{\|u\|\le R} \mathrm{d}^2 u \langle \mathrm{out}|S|\varphi_1\varphi_2\rangle_u \langle \mathrm{out}|S|\varphi_1\varphi_2\rangle_u^*. \tag{12.72}$$

For this we observe that

$$\langle \mathrm{out}|S|\varphi_1\varphi_2\rangle_u = \iint \frac{\mathrm{d}^3\boldsymbol{p}_1}{(2\pi)^3}\frac{\mathrm{d}^3\boldsymbol{p}_2}{(2\pi)^3} \exp(-\mathrm{i}\boldsymbol{u}\cdot\boldsymbol{p}_1)\varphi_1(\boldsymbol{p}_1)\varphi_2(\boldsymbol{p}_2)\langle \mathrm{out}|S|a_1(\boldsymbol{p}_1)^\dagger a_2(\boldsymbol{p}_2)^\dagger|0\rangle. \tag{12.73}$$

We substitute in (12.72), and write the product of the two double integrals as a quadruple integral. We use (12.65) and the relation

$$\lim_{R\to\infty} \int_{\|u\|\le R} \mathrm{d}^2 u \exp(-\mathrm{i}\boldsymbol{u}\cdot(\boldsymbol{p}_1 - \boldsymbol{p}'_1)) = (2\pi)^2 \delta(p_{11} - p'_{11})\delta(p_{12} - p'_{12}),$$

where $\boldsymbol{p}_1 = (p_{11}, p_{12}, p_{13})$ and similarly for $\boldsymbol{p}'_1$. Using (12.65) we then obtain that the quantity (12.72) is

$$\iiiint \frac{\mathrm{d}^3\boldsymbol{p}_1}{(2\pi)^3}\frac{\mathrm{d}^3\boldsymbol{p}_2}{(2\pi)^3}\frac{\mathrm{d}^3\boldsymbol{p}'_1}{(2\pi)^3}\frac{\mathrm{d}^3\boldsymbol{p}'_2}{(2\pi)^3} A\delta, \tag{12.74}$$

where

$$A = M(\boldsymbol{p}'_1, \boldsymbol{p}'_2, \boldsymbol{p}_{\mathrm{out}})^* M(\boldsymbol{p}_1, \boldsymbol{p}_2, \boldsymbol{p}_{\mathrm{out}})\varphi_1(\boldsymbol{p}'_1)^*\varphi_2(\boldsymbol{p}'_2)^*\varphi_1(\boldsymbol{p}_1)\varphi_2(\boldsymbol{p}_2)$$

and

$$\delta = (2\pi)^{10}\delta^{(4)}(p_{\mathrm{in}} - p_{\mathrm{out}})\delta^{(4)}(p'_{\mathrm{in}} - p_{\mathrm{out}})\delta(p_{11} - p'_{11})\delta(p_{12} - p'_{12}). \tag{12.75}$$

Lemma 12.6.3 *With the previous notation,*

$$\iint \frac{\mathrm{d}^3\boldsymbol{p}'_1}{(2\pi)^3}\frac{\mathrm{d}^3\boldsymbol{p}'_2}{(2\pi)^3} A\delta = \frac{1}{|v_{13} - v_{23}|}(2\pi)^4\delta^{(4)}(p_{\mathrm{in}} - p_{\mathrm{out}})|M(\boldsymbol{p}_1, \boldsymbol{p}_2, \boldsymbol{p}_{\mathrm{out}})|^2|\varphi_1(\boldsymbol{p}_1)|^2|\varphi_2(\boldsymbol{p}_2)|^2. \tag{12.76}$$

Substitution in (12.74) proves (12.71) and finishes the proof of Theorem 12.6.1.

Proof This is the part which is a bit technical and demanding. The main idea is that the vectors $\boldsymbol{p}'_1$ and $\boldsymbol{p}'_2$ have six components between the two of them, and in the term $(2\pi)^6\delta^{(4)}(p'_{\rm in}-p_{\rm out})\delta(p_{11}-p'_{11})\delta(p_{12}-p'_{12})$ there are six delta functions which determine these components (counting a four-dimensional delta function as four one-dimensional delta functions). This allows performing the integration in these variables $\boldsymbol{p}'_1$ and $\boldsymbol{p}'_2$.

As a first step we observe that due to the factor $(2\pi)^4\delta^{(4)}(p_{\rm in}-p_{\rm out})$ in (12.75) we may assume $p_{\rm in}=p_{\rm out}$ so that $(2\pi)^4\delta^{(4)}(p'_{\rm in}-p_{\rm out})=(2\pi)^4\delta^{(4)}(p'_{\rm in}-p_{\rm in})$. This quantity is the product of the four one-dimensional delta functions

$$\delta_{(\nu)}:=2\pi\delta(p'_{1\nu}+p'_{2\nu}-p_{1\nu}-p_{2\nu}), \tag{12.77}$$

for $\nu=0,1,2,3$. In the left-hand side of (12.76) let us first perform the integration in p'_{11} and p'_{21}. The term $2\pi\delta(p_{11}-p'_{11})$ in (12.75) means that integrating in p'_{11} simply amounts to fixing $p'_{11}=p_{11}$. The quantity $\delta_{(1)}$ of (12.77) then becomes $2\pi\delta(p'_{21}-p_{21})$, so that integrating in p'_{21} simply amounts to fixing $p'_{21}=p_{21}$. The same pattern occurs when integrating in p'_{12} and p'_{22}, and after these four integrations the left-hand side of (12.76) is of the type

$$\iint {\rm d}p'_{13}{\rm d}p'_{23}2\pi\delta(p'_{13}+p'_{23}-p_{13}-p_{23})\delta_{(0)}B, \tag{12.78}$$

where B is a function of p'_{13} and p'_{23}. Taking the integral in p'_{13} amounts to fixing $p'_{13}=p_{13}+p_{23}-p'_{23}$. It remains then to take the integral in p'_{23}. We observe that

$$p'_{10}=\sqrt{m_1^2+p_{11}^2+p_{12}^2+(p'_{13})^2}=\sqrt{m_1^2+p_{11}^2+p_{12}^2+(p_{13}+p_{23}-p'_{23})^2}$$

$$p'_{20}=\sqrt{m_2^2+p_{21}^2+p_{22}^2+(p'_{23})^2}$$

so that $p'_{10}+p'_{20}=\xi(p'_{23})$ where

$$\xi(x)=\sqrt{m_1^2+p_{11}^2+p_{12}^2+(p_{13}+p_{23}-x)^2}+\sqrt{m_2^2+p_{21}^2+p_{22}^2+x^2}.$$

In a similar manner we have $p_{10}+p_{20}=\xi(p_{23})$. The quantity $\delta_{(0)}$ is thus $2\pi\delta(\rho(p'_{23}))$ where $\rho(x)=\xi(x)-\xi(p_{23})$. Thus $\rho^{-1}(0)=\pm p_{13}$. Moreover the derivative ρ' of ρ satisfies

$$|\rho'(p_{23})|=\left|\frac{p_{13}}{\sqrt{m_1^2+\boldsymbol{p}_1^2}}-\frac{p_{23}}{\sqrt{m_2^2+\boldsymbol{p}_2^2}}\right|, \tag{12.79}$$

which, using the relation $\boldsymbol{p}/p^0=\boldsymbol{v}$ we may write as

$$|\rho'(p_{23})|=|v_{13}-v_{23}|,$$

and using (4.40) the integral (12.78) becomes

$$\frac{1}{|v_{13}-v_{23}|}B(p_{13},p_{23}),$$

since as in Section 12.4 we only have to consider the contribution of p_{23} but not of $-p_{23}$.

In summary, to compute an integral like (12.78) we fix $p'_{13} = p_{13}$ and $p'_{23} = p_{23}$ and we replace the delta functions by the factor $1/|v_{13} - v_{23}|$. That is, we have shown that computing the left-hand side of (12.76) amounts to fixing $\boldsymbol{p}'_1 = \boldsymbol{p}_1, \boldsymbol{p}'_2 = \boldsymbol{p}_2$ and replacing the factor $(2\pi)^6\delta^{(4)}(p'_{\text{in}} - p_{\text{out}})\delta(p_{11} - p'_{11})\delta(p_{12} - p'_{12})$ by $1/|v_{12} - v_{23}|$, which is exactly the meaning of the formula (12.76). □

Key ideas to remember:

- The S-matrix encodes the effect of interactions.
- In Quantum Field Theory the S-matrix relates how a collection of far-away particles looked before the interaction to the way it looks after the interaction.
- The S-matrix entirely determines the results of scattering experiments.

13

The Scattering Matrix in Perturbation Theory

In this chapter we learn how to compute the S-matrix in perturbation theory. Similar computations (in a far more elaborate form than is presented here) form the bulk of many Quantum Field Theory textbooks. This chapter marks a transition, as we are now gradually going to drift further and further from respectable mathematics. Do not worry then if you find that the arguments are getting fuzzier and the computations more formal than ever.[1] These methods are justified by their tremendous practical success.

In the first five sections we examine the overall method and its mathematical sanity. The consideration of actual models starts in Section 13.6. We consider spinless massive particles with really simple interaction Hamiltonians. This situation is far simpler than the models that actually describe Nature, but it has already many features of these. Up to Section 13.16 we gradually build the tools to systematically express scattering amplitudes (at a given order of perturbation theory) as the sum of the values of certain diagrams, the celebrated Feynman diagrams. The values of these diagrams are expressed as integrals, which often diverge when the diagrams contain loops. This is the main technical difficulty of Quantum Field Theory. From Section 13.17 to Section 13.20 we study in considerable detail one of the simplest cases of such divergences, certain diagrams with one loop in ϕ^4 theory. In the last four sections we attack the far more difficult case of two loops, which already touches the central issues of renormalization.

13.1 The Scattering Matrix and the Dyson Series

Let us go back to the situation of Section 11.2 (which the reader should review now), of an unperturbed Hamiltonian H_0 and of a Hamiltonian H defined by the formula

$$H = H_0 + gH_I, \tag{13.1}$$

where g is the coupling constant controlling the strength of the interaction and H_I is an "interaction term". Crossing our fingers, let us consider the "scattering operator"

$$S = \lim_{t_1 \to \infty, t_2 \to -\infty} U_0(-t_1)U(t_1 - t_2)U_0(t_2). \tag{13.2}$$

[1] Think positively: this prepares you for the next chapter!

Let us recall from (11.37) the formal expansion

$$U_0(t_2-t_1)U(t_1-t_2) = 1 + \sum_{n\geq 1} \frac{(-ig)^n}{n!} \int_0^{t_1-t_2} \mathrm{d}\theta_1 \ldots \int_0^{t_1-t_2} \mathrm{d}\theta_n \mathcal{T} H_I(\theta_1)\cdots H_I(\theta_n), \tag{13.3}$$

where $H_I(t) = \exp(\mathrm{i}tH_0)H_I\exp(-\mathrm{i}tH_0) = U_0(-t)H_IU_0(t)$ is the evolution in the interaction picture of the operator H_I and $\mathcal{T}$ is time-ordering. We observe the identity

$$U_0(-t_2)(\mathcal{T}H_I(\theta_1)\cdots H_I(\theta_n))U_0(t_2) = \mathcal{T}H_I(\theta_1+t_2)\cdots H_I(\theta_n+t_2).$$

Multiplying (13.3) to the left by $U_0(-t_2)$ and to the right by $U_0(t_2)$, making changes of variables to obtain integrals from t_2 to t_1 and letting $t_1 \to \infty$ and $t_2 \to -\infty$ we obtain from (13.2) the fundamental formula:

$$S = 1 + \sum_{n\geq 1} S_n \; ; \; S_n = \frac{(-ig)^n}{n!} \int \mathrm{d}t_1 \ldots \int \mathrm{d}t_n \mathcal{T} H_I(t_1)\cdots H_I(t_n), \tag{13.4}$$

which will be the basis of every computation. This formula applies to each operator S of the type (13.2), and in particular to the operator S of (12.19) (scattering by a potential), as we use in the next section. The series $S = 1 + \sum_{n\geq 1} S_n$ is called the *Dyson series*.

We will further pretend that we are also permitted to use the Dyson series (13.4) for the S-matrix in the setting of Section 12.5.[2]

Let us repeat that writing (13.4) involves no claim about the convergence of the series, which we consider as a formal series in powers of g. Let us also recall the following.

Definition 13.1.1 Performing computations *at the* k*th-order in perturbation theory* means replacing S by $1 + \sum_{1\leq n\leq k} S_n$, and more generally ignoring any term with a coefficient g^{k+1} in the computations.

This has strictly nothing to do with the convergence of the series and is being done without any estimate of the size of the terms being thrown away. Furthermore, as we will soon realize, the terms thrown away have a tendency to be infinite. Even when these "infinite" terms are properly "normalized", i.e. replaced by a proper suitable finite quantity as we will learn to do, in the cases of real interest it does not seem that the Dyson series is convergent.[3] The use of the method is justified by the following empirical fact: Using the very first few terms of the Dyson series yields predictions that wonderfully agree with experiments.[4]

[2] This means that in particular we pretend that both time-evolutions $U(t)$ and $U_0(t)$ operate on the state space.

[3] It seems that the most optimistic hypothesis one can make for the theories of real interest is that the Dyson series is an "asymptotic series". A real-valued function $f(g)$ defined for $g > 0$ has the series $\sum_n a_n g^n$ as an asymptotic series if given any integer $N > 0$ then $|f(g) - \sum_{0\leq n\leq N} a_n g^n| < g^N$ for g small enough (depending on N). However, g does not tend to 0. Its value is provided by Nature. The knowledge of an asymptotic series for the function f is only of theoretical interest. For the actual value of g, it does not say that any of the partial sums of the series is a good approximation of $f(g)$.

[4] An interesting point here is that up to now using more terms in the Dyson series has led to better agreement with experiments, but if the series is indeed divergent, at some point using more terms will produce worse predictions!

13.2 Prologue: The Born Approximation in Scattering by a Potential

Before we start computing the S-matrix in Quantum Field Theory, let us give a much simpler example of use of the formula (13.4). For this we go back to the setting and the notation of Section 12.2, the Hamiltonian $H_0 = \boldsymbol{P}^2/2m := -(1/2m)\sum_{i\le 3} \partial^2/\partial x_i^2$, and the Hamiltonian H_I "multiplication by the function $V(\boldsymbol{x})$", and we compute "at the first order in perturbation theory" the corresponding S-matrix. In other words we use the approximation $S \simeq 1 - \mathrm{i}gA$ where

$$A = \int_{\mathbb{R}} \mathrm{d}t\, H_I(t). \tag{13.5}$$

The computation of A is best done in a basis where H_0 is diagonal, "the continuous basis of states $|\boldsymbol{p}\rangle$ of given momentum" because $H_0|\boldsymbol{p}\rangle = (\boldsymbol{p}^2/2m)|\boldsymbol{p}\rangle$. We will write things as a physicist would write them, using both states $|\boldsymbol{p}\rangle$ of given momentum and states $|\boldsymbol{x}\rangle$ of given position, but it is not difficult here to write the same calculation in a more formal way in momentum space. Thus

$$\begin{aligned}\langle \boldsymbol{p}_1|A|\boldsymbol{p}_2\rangle &= \int_{\mathbb{R}} \mathrm{d}t\,\langle \boldsymbol{p}_1|\exp(\mathrm{i}tH_0)V\exp(-\mathrm{i}tH_0)|\boldsymbol{p}_2\rangle \\ &= \int_{\mathbb{R}} \mathrm{d}t\,\exp(\mathrm{i}t(\boldsymbol{p}_1^2 - \boldsymbol{p}_2^2)/2m)\langle \boldsymbol{p}_1|V|\boldsymbol{p}_2\rangle \\ &= 2\pi\delta((\boldsymbol{p}_1^2 - \boldsymbol{p}_2^2)/2m)\langle \boldsymbol{p}_1|V|\boldsymbol{p}_2\rangle. \end{aligned} \tag{13.6}$$

Now, $\langle \boldsymbol{x}|V|\boldsymbol{y}\rangle = V(\boldsymbol{y})\langle \boldsymbol{x}|\boldsymbol{y}\rangle = V(\boldsymbol{y})\delta^{(3)}(\boldsymbol{x}-\boldsymbol{y})$, and thus

$$\begin{aligned}\langle \boldsymbol{p}_1|V|\boldsymbol{p}_2\rangle &= \iint \mathrm{d}^3\boldsymbol{x}\,\mathrm{d}^3\boldsymbol{y}\,\langle \boldsymbol{p}_1|\boldsymbol{x}\rangle\langle \boldsymbol{x}|V|\boldsymbol{y}\rangle\langle \boldsymbol{y}|\boldsymbol{p}_2\rangle \\ &= \iint \mathrm{d}^3\boldsymbol{x}\,\mathrm{d}^3\boldsymbol{y}\,\exp(\mathrm{i}\boldsymbol{p}_2\cdot\boldsymbol{y} - \mathrm{i}\boldsymbol{p}_1\cdot\boldsymbol{x})\delta^{(3)}(\boldsymbol{y}-\boldsymbol{x})V(\boldsymbol{x}) \\ &= \int \mathrm{d}^3\boldsymbol{x}\,\exp(-\mathrm{i}(\boldsymbol{p}_1 - \boldsymbol{p}_2)\cdot\boldsymbol{x})V(\boldsymbol{x}) \\ &= \widehat{V}(\boldsymbol{p}_1 - \boldsymbol{p}_2), \end{aligned} \tag{13.7}$$

where $\widehat{V}$ is the Fourier transform of V. Thus, *at the first order of perturbation theory*,

$$\langle \boldsymbol{p}_1|S|\boldsymbol{p}_2\rangle = (2\pi)^3\delta^{(3)}(\boldsymbol{p}_1 - \boldsymbol{p}_2) - 2\pi\mathrm{i}g\delta((\boldsymbol{p}_1^2 - \boldsymbol{p}_2^2)/2m)\widehat{V}(\boldsymbol{p}_1 - \boldsymbol{p}_2). \tag{13.8}$$

This is called the *Born approximation*. It will help us to recognize interactions involving the potential V. It is not an easy matter to develop intuition about the meaning of this formula, because information is encoded in an indirect way. Again, $|\boldsymbol{p}\rangle$ represents a particle with given momenta, so it is not localized in space and is unlikely to feel the influence of the scattering center. However, this formula leads to a differential cross-section which can be experimentally tested.

Exercise 13.2.1 What is the differential cross-section predicted by the Born approximation?

13.3 Interaction Terms in Hamiltonians

We now go back to Quantum Field Theory. We have already built good models for systems consisting of non-interacting particles of a given type. For example, in the case of a single type of massive, spinless particle the state space is the boson Fock space $\mathcal{B}$ constructed on the space $L^2 = L^2(X_m, \mathrm{d}\lambda_m)$, with Hamiltonian H_0 given by (5.36).[5] In order to bring in some physics, besides the Hamiltonian H_0 describing the free particles, we consider a term H_I representing the interaction between these particles.

A fruitful method to construct such interaction terms is to use a *Hamiltonian density* $\mathcal{H}(x)$, an operator-valued distribution on $\mathbb{R}^{1,3}$. Denoting by $U_0(t)$ the time-evolution of the free particles under the Hamiltonian H_0, the Hamiltonian density is chosen as to have the property

$$U_0(-t) \circ \mathcal{H}(0,\boldsymbol{x}) \circ U_0(t) = \mathcal{H}(t,\boldsymbol{x}), \tag{13.9}$$

The interaction Hamiltonian is then defined by the formula

$$H_I = \int \mathrm{d}^3\boldsymbol{x}\,\mathcal{H}(0,\boldsymbol{x}). \tag{13.10}$$

When writing this formula we assume that it makes sense to consider $\mathcal{H}(0,\boldsymbol{x})$ so that $\mathcal{H}(x)$ is indeed a special type of operator-valued distribution. We also assume that the integral makes sense.[6] In the interaction picture, the Hamiltonian $H_I(t)$ is given by the formula (11.21) i.e. $H_I(t) := U_0(-t)H_I U_0(t)$. Combining (13.9) and (13.10) we then obtain

$$H_I(t) = \int \mathrm{d}^3\boldsymbol{x}\,\mathcal{H}(t,\boldsymbol{x}), \tag{13.11}$$

and combining with (13.4) suggests the very nice formula

$$S_n = \frac{(-\mathrm{i}g)^n}{n!} \int \mathrm{d}^4x_1 \ldots \int \mathrm{d}^4x_n \mathcal{T}\mathcal{H}(x_1)\cdots\mathcal{H}(x_n). \tag{13.12}$$

Presumably this means that given $x_1, \ldots, x_n$ the time-ordering of the operators $\mathcal{H}(x_k)$ is performed accordingly to the time-components of these points. This does not always make sense. In general the meaning of $\mathcal{T}\mathcal{H}(x_1)\mathcal{H}(x_2)$ when $\mathcal{H}(x)$ is a distribution is not so clear. Here, however, we will consider only the case where $\mathcal{H}(x)$ is a product of free fields. These are well defined at a given value of t. The time-ordered product $\mathcal{T}\mathcal{H}(x_1)\mathcal{H}(x_2)$ is then well defined unless x_1 and x_2 have the same time-component.[7] But since free fields commute at equal times as shown by the equal-time commutation relations (6.63), the time-ordering is well defined in all cases.[8]

5 When we consider several particle types, we simply take a tensor product of such spaces as will be detailed later.

6 Generally speaking, and for reasons that will become apparent soon, we will no longer worry about such mundane matters, or about exchanging orders of integrations, etc.

7 And assuming that the product of the distributions $\mathcal{H}(t_1,\boldsymbol{x}_1)$ and $\mathcal{H}(t_2,\boldsymbol{x}_2)$ makes sense.

8 However there are situations where this is not the case, when one considers "derivative couplings", i.e. if $\mathcal{H}(x)$ is a product of terms some of which are of the type $\partial_t\varphi(x)$, and this creates an additional source of complications, see (6.62). We will not meet such situations.

Exercise 13.3.1 Use the expansion $S = \sum_{n\geq 0} S_n$ to compute formally $S^\dagger S$ as a formal series in g. Prove that the coefficients of g, g^2 and g^3 are zero. The last case is harder. A proof that the general coefficient is zero is even harder.

13.4 Prickliness of the Interaction Picture

When one modifies the Hamiltonian H_0 by introducing an interaction term H_I, it is not obvious what a natural choice of state space is, and how one defines the full Hamiltonian as a self-adjoint operator on the state space. Unfortunately, it is not only un-obvious how to achieve these minimal goals, it is just *mathematically impossible* to achieve them, at least while preserving all the desirable properties one would expect from such a construction. As everything in the present section is at the level of hand-waving, our description of the basic obstacle will be at the same level. Together with the free Hamiltonian H_0 is a free field φ^{free} (as defined in Chapter 5), which, as shown in Section 6.9, has a well defined value φ_t^{free} at a given value of t, i.e. $\varphi_t^{\text{free}}(\boldsymbol{x}) := \varphi^{\text{free}}(t,\boldsymbol{x})$. It satisfies the evolution equation (2.85)

$$\dot{\varphi}_t^{\text{free}} := \frac{\partial}{\partial t}\varphi_t^{\text{free}} = \mathrm{i}[H_0, \varphi_t^{\text{free}}]. \tag{13.13}$$

On the other hand, there should exist an "interaction quantum field φ_t", which will simply be the Heisenberg evolution of the field $\varphi_0 = \varphi_0^{\text{free}}$ according to the *full* Hamiltonian $H = H_0 + gH_I$, reflecting the fact that at time $t = 0$, the Schrödinger, Heisenberg and interaction picture all coincide. The interaction quantum field satisfies the corresponding evolution equation

$$\dot{\varphi}_t = \mathrm{i}[H, \varphi_t]. \tag{13.14}$$

In particular (13.14) implies that

$$\dot{\varphi}_0 = \mathrm{i}[H,\varphi_0] = \mathrm{i}[H,\varphi_0^{\text{free}}] = \mathrm{i}g[H_I,\varphi_0^{\text{free}}] + \mathrm{i}[H_0,\varphi_0^{\text{free}}] = \mathrm{i}g[H_I,\varphi_0^{\text{free}}] + \dot{\varphi}_0^{\text{free}}.$$

In the usual cases we have

$$[H_I, \varphi_0^{\text{free}}] = 0 \tag{13.15}$$

and then $\dot{\varphi}_0 = \dot{\varphi}_0^{\text{free}}$. In Appendix M we state a theorem of Haag that shows that this condition, together with the condition $\varphi_0 = \varphi_0^{\text{free}}$ implies that φ is itself (provided it behaves nicely) a free field of mass m. This in turn (heuristically) means that the interactions are trivial. This is certainly not the case in general. Something went terribly wrong.

To get a better feeling of what went wrong, we can look at a very simple case. We proved in Proposition 6.11.3 that a Hamiltonian density for the free field of mass m is given by the formula

$$:\frac{1}{2}\Big(\dot{\varphi}^{\text{free}}(x)^2 + \sum_{1\leq\nu\leq 3}(\partial_\nu\varphi^{\text{free}}(x))^2\Big) + \frac{1}{2}m^2\varphi^{\text{free}}(x)^2: , \tag{13.16}$$

where we recall that the funny notation : : means normal ordering, writing all operators $a^\dagger$ to the left of the operators a. Let us choose

$$\mathcal{H}(x) = \frac{1}{2} :\varphi^{\text{free}}(x)^2: \tag{13.17}$$

as the Hamiltonian density for the interaction term H_I. If we add g times the quantity (13.17) to the quantity (13.16), this amounts to replacing in (13.16) the quantity m^2 by m^2+g. This means that the Hamiltonian $H_0 + gH_I$ is the Hamiltonian of a free field of mass m' with $m'^2 = m^2 + g$. This Hamiltonian is perfectly well defined (and we studied it in Section 6.11). Next, let us check that (13.15) is satisfied. We have $2H_I = \int \mathrm{d}^3\boldsymbol{y} :\varphi_0^{\text{free}}(\boldsymbol{y})^2:$. Going back to (5.30), let us write

$$\varphi^+(\boldsymbol{y}) = \int \frac{\mathrm{d}^3\boldsymbol{p}}{(2\pi)^3\sqrt{2\omega_{\boldsymbol{p}}}} \exp(\mathrm{i}\boldsymbol{y}\cdot\boldsymbol{p})a(\boldsymbol{p})\,;\ \varphi^-(\boldsymbol{y}) = \int \frac{\mathrm{d}^3\boldsymbol{p}}{(2\pi)^3\sqrt{2\omega_{\boldsymbol{p}}}} \exp(\mathrm{i}\boldsymbol{y}\cdot\boldsymbol{p})a^\dagger(\boldsymbol{p})$$

the annihilation and creation parts of $\varphi_0^{\text{free}}(\boldsymbol{y})$, so that $\varphi_0^{\text{free}}(\boldsymbol{y}) = \varphi^+(\boldsymbol{y}) + \varphi^-(\boldsymbol{y})$. Proceeding rather formally, we have $:\varphi_0^{\text{free}}(\boldsymbol{y})^2: = \varphi^-(\boldsymbol{y})^2 + 2\varphi^-(\boldsymbol{y})\varphi^+(\boldsymbol{y}) + \varphi^+(\boldsymbol{y})^2$ so that $:\varphi_0^{\text{free}}(\boldsymbol{y})^2: - \varphi_0^{\text{free}}(\boldsymbol{y})^2 = [\varphi^-(\boldsymbol{y}),\varphi^+(\boldsymbol{y})]$ and according to (3.34) such a commutator is a multiple of the identity. Consequently to compute $[H_I,\varphi_0^{\text{free}}]$, in the expression $H_I = (1/2)\int \mathrm{d}^3\boldsymbol{y}\ :\varphi_0^{\text{free}}(\boldsymbol{y})^2:$ we may replace $:\varphi_0^{\text{free}}(\boldsymbol{y})^2:$ by $\varphi_0^{\text{free}}(\boldsymbol{y})^2$. It then follows from the equal-time commutation relation (6.63) and the identity $[AB,C] = A[B,C]+[A,C]B$ that $[H_I,\varphi_0^{\text{free}}] = 0$.

However, as shown by Haag's theorem, Theorem M.2.1, it cannot be true that the free field φ_t of mass m' and the free field φ_t^{free} of mass m satisfy both $\varphi_0 = \varphi_0^{\text{free}}$ and $\dot{\varphi}_0 = \dot{\varphi}_0^{\text{free}}$.[9]

Even though H and H_0 both make perfect sense, *it is the (implicit) assumption that they operate on the same state space that does not.*

Disregarding these problems we shall pretend that the interaction picture exists and proceed accordingly. If a mathematician really corners us into explaining what is the meaning of what we are doing, we will argue that these obstacles are an artifact of continuous models, and that what we really mean is that we have put the universe in a box, and that we have used an ultraviolet cutoff[10] to remove the remaining diverging series. (The difficult part would then be to prove that we obtain results independent of this cutoff.) But when no mathematician is in sight we will keep as always making formal computations in the continuous model, keeping our eyes open for the infinite series that might occur then in the form of divergent integrals. Putting the universe in a box also takes care of another problem which will become glaring when we write actual formulas for the Hamiltonian density $\mathcal{H}$. These formulas involve products of quantities that are distributions, and, even though one can formally compute with them, what they really mean is far from obvious.

9 This statement actually follows from Corollary M.3.6, a special case of Haag's theorem, which is much easier than the general theorem, and which we prove in complete detail.

10 Let us recall that using an ultraviolet cutoff to make sense of a diverging series simply amounts to discarding all the terms of the series beyond a certain rank.

13.5 Admissible Hamiltonian Densities

Having brilliantly succeeded in not confronting the question of really defining either the Hamiltonian or the state space, we keep pretending that the state space is just the same as if we had only the free field. In the present section we examine two general issues, first how to find Hamiltonian densities that satisfy (13.9) and then whether the resulting interacting theory is Lorentz invariant. We expect that the result of our scattering experiments does not depend on the inertial observers. A natural way to ensure this is to require that if $U(c,C)$ denotes the action of $\mathcal{P}^*$ on $\mathcal{H}_{\rm part}$ then[11] whenever $|{\rm in}\rangle$ and $|{\rm out}\rangle$ belong to $\mathcal{H}_{\rm part}$,

$$\langle {\rm out} U(c,C)^\dagger|S|U(c,C){\rm in}\rangle = \langle {\rm out}|S|{\rm in}\rangle,$$

or, equivalently, that S *commutes with* $U(c,C)$.

Certainly (13.9) is too weak to give us a chance to obtain Lorentz invariance. Rather (in line with (5.6)) we will assume that[12]

$$U(c,C)\circ\mathcal{H}(x)\circ U(c,C)^{-1} = \mathcal{H}(c+C(x)). \tag{13.18}$$

This is more general than (13.9) because the time-evolution $U_0(t)$ is just translation by $-t$ along the time coordinate, i.e. $U(b,1)^{-1}$ for $b=(t,0,0,0)$.

We will consider only cases where (13.18) is trivially satisfied, because the Hamiltonian density $\mathcal{H}(x)$ will be a product of free scalar fields that satisfy this condition.[13] More generally, the construction of densities transforming as (13.18) is based on families of fields that transform according to (10.6). Such densities are of fundamental importance in the study of the real world, and provide the true motivation for all the work of Chapter 10. An example of such a construction is provided in the next exercise.

Exercise 13.5.1 (a) Consider two families of fields $(\varphi_k)_{k\le N}, (\varphi'_k)_{k\le N}$ and assume that for a representation T of $SL(2,\mathbb{C})$ in $\mathbb{C}^N$ we have

$$U(c,C)\circ\varphi_k(x)\circ U(c,C)^{-1} = \sum_{\ell\le N} T(C^{-1})_{k,\ell}\varphi_\ell(c+C(x)), \tag{13.19}$$

as well as the same condition for φ'_k. Prove that the Hamiltonian density

$$\mathcal{H}(x) = \sum_{k,k'} \alpha_{k,k'}\varphi_k(x)\varphi'_{k'}(x)$$

satisfies (13.18) as soon as the following identity holds true

$$\alpha_{\ell,\ell'} = \sum_{k,k'\le N} \alpha_{k,k'} T_{k,\ell}(C^{-1})T_{k',\ell'}(C^{-1}), \tag{13.20}$$

i.e. as soon as $(\alpha_{k,k'})$ is invariant under the representation $T\otimes T$ on $\mathbb{C}^N\otimes\mathbb{C}^N$.
(b) When T is the defining representation of $SL(2,\mathbb{C})$ on $\mathbb{C}^2$ (multiplying a matrix by a column vector), prove that (13.20) is satisfied by the choice $\alpha_{2,2}=\alpha_{1,1}=0$, $\alpha_{1,2}=1=-\alpha_{2,1}$.

[11] There also could be in principle a phase factor.
[12] A physicist may express this relation by saying that $\mathcal{H}$ *transforms as a scalar*.
[13] Furthermore, as is shown in Section 6.9, these fields are really defined at a given value of t as required in (13.10).

In the rest of this section we provide a simple criterion for our theory to be Lorentz invariant. It shows in particular that the theories we consider later in the chapter are Lorentz invariant.

Theorem 13.5.2 *Assume that the Hamiltonian density $\mathcal{H}$ satisfies (13.18), is smooth enough (as will be discussed below) and satisfies the condition*

$$(x-y)^2 < 0 \Rightarrow [\mathcal{H}(x), \mathcal{H}(y)] = 0. \tag{13.21}$$

Then the S-matrix is Lorentz invariant within perturbation theory.

The expression "within perturbation theory" is related to Definition 13.1.1. It means that *each term* of the formal series given by Dyson's formula is Lorentz invariant, in the sense that it commutes with the operators $U(c,C)$. This is about the best we can do since we do not have a real formula for the S-matrix. Condition 13.21 is the usual causality condition: Events happening at two points that are causally separated cannot influence each other. So, we have to prove that

$$\int \mathrm{d}^4x_1 \ldots \int \mathrm{d}^4x_n U(c,C) \circ \mathcal{T}\mathcal{H}(x_1)\cdots\mathcal{H}(x_n) \circ U(c,C)^{-1}$$
$$= \int \mathrm{d}^4x_1 \ldots \int \mathrm{d}^4x_n \mathcal{T}\mathcal{H}(x_1)\cdots\mathcal{H}(x_n). \tag{13.22}$$

If we did not have the time-ordering operator $\mathcal{T}$ in (13.12) this would be an obvious consequence of (13.18) by Lorentz invariance of the measure d^4x, even without requiring (13.21).

The words "smooth enough" in the statement of the theorem indicate a technical difficulty, as there can be pathological behavior of the distributions involved here. But since we are in hand-waving mode, we pretend that $\mathcal{H}(x)$ is a function of x, assuming moreover that $\mathcal{H}(x)$ and $\mathcal{H}(y)$ commute if x and y have the same time-component, so that the time-ordering is well defined. A hand-waving argument meets our purpose here since the point of Theorem 13.5.2 is simply to reassure us that we are doing something sensible. To "prove" the theorem it is sufficient to establish the identity

$$U(c,C) \circ \mathcal{T}\mathcal{H}(x_1)\cdots\mathcal{H}(x_n) \circ U(c,C)^{-1} = \mathcal{T}\mathcal{H}(c+C(x_1))\cdots\mathcal{H}(c+C(x_n)). \tag{13.23}$$

There is no difficulty with the term c so we lighten notation by taking $c = 0$ and assuming that the product $\mathcal{H}(x_1)\cdots\mathcal{H}(x_n)$ is already time-ordered. Then

$$U(0,C) \circ \mathcal{H}(x_1)\cdots\mathcal{H}(x_n) \circ U(0,C)^{-1} = \mathcal{H}(C(x_1))\cdots\mathcal{H}(C(x_n)). \tag{13.24}$$

The problem is that the right-hand side is not time-ordered. To analyze its time-ordering, let us denote by t_i the time-component of x_i and by t'_i the time-component of $x'_i := C(x_i)$.

Lemma 13.5.3 *If a pair (i,j) is such that $t_i \geq t_j$ but $t'_i < t'_j$ then the operators $\mathcal{H}(x'_i)$ and $\mathcal{H}(x'_j)$ commute.*

Proof By (13.21) it suffices to prove that $(x'_i - x'_j)^2 = (x_i - x_j)^2 < 0$. But if we had $(x'_i - x'_j)^2 = (x_i - x_j)^2 > 0$ it would hold that $t'_i \geq t'_j$ by Lemma 4.1.3. □

Let us now sketch the proof that

$$\mathcal{H}(C(x_1))\cdots\mathcal{H}(C(x_n)) = T\mathcal{H}(C(x_1))\cdots\mathcal{H}(C(x_n)), \tag{13.25}$$

thereby completing the proof of (13.23) and of Theorem 13.5.2. The proof of (13.25) can be described by one sentence: To go from the natural ordering to the time-ordering, we have to exchange the order of certain operators, but these commute. To make the argument more precise, consider a permutation σ of $\{1,\dots,n\}$ such that $t'_{\sigma(1)} \geq t'_{\sigma(2)} \geq \dots$, so that the time-ordered product of $\mathcal{H}(x'_1)\cdots\mathcal{H}(x'_n)$ is $\mathcal{H}(x'_{\sigma(1)})\cdots\mathcal{H}(x'_{\sigma(n)})$. To transform the product $\mathcal{H}(x'_1)\cdots\mathcal{H}(x'_n)$ into the time-ordered product we first bring the term $\mathcal{H}(x'_{\sigma(1)})$ from the position $\sigma(1)$ in the product to the first position. This is possible because it commutes with all the terms $\mathcal{H}(x'_i)$ for $i < \sigma(1)$ by the previous argument, and then we proceed recursively in this manner.

13.6 Simple Models for Interacting Particles

There are two different obstacles when trying to compute S-matrices in the interaction picture.

- Some integrals diverge and it is difficult to assign a meaning to them.
- There is a lot of bookkeeping to be done to keep track of complications such as spin.

A physicist has no choice but to deal with Nature's complications. In the present setting however it pays to separate real difficulty (the divergent integrals) from accessory complications. For this reason we will concentrate on models that are as free as possible of accessory complications. These models have no claim whatsoever to have any realistic physical content.[14] Nonetheless they will teach us how to deal with the greatest technical difficulty of Quantum Field Theory, the divergent integrals. Once this is mastered, it still requires stamina and creativity to learn how to proceed with the realistic models, but the main difficulty is arguably behind us.

We will consider only spinless particles which are their own anti-particles.[15] In the basic model called ϕ^3 theory (or, equivalently, called the ϕ^3 model) one considers a single such particle type, with a (tentative) interacting Hamiltonian density[16]

$$\mathcal{H}(x) = \frac{\varphi(x)^3}{3!}. \tag{13.26}$$

Here, the factor 3! is conventional (and allows for somewhat simpler numerical factors in the computations), and $\varphi(x)$ is the free field associated to the particle.[17] Thus (13.21) is a consequence of microcausality, Theorem 5.1.5 (if one pretends that microcausality means that $[\varphi(x),\varphi(y)] = 0$ when $(x-y)^2 < 0$).

[14] Worse, some of these models are non-physical, in the sense that the Hamiltonian is not bounded below. This however is irrelevant at the level of perturbation theory, which does not "see" this.

[15] Thus such particles must be neutral, i.e. they cannot carry an electrical charge.

[16] I make the convention that the coupling constant g is not part of the interacting density.

[17] In this chapter we no longer consider the "true interaction quantum field" of Section 13.4, and all the fields are free fields.

You may wonder of course why we choose the formula (13.26). Generally speaking the fruitful method to construct admissible Hamiltonian densities is as products of free fields. Formula (13.26) is about the simplest you can write, as it uses one single free field of the simplest type. The exponent 3 is the smallest which gives a non-trivial theory. (In particular we have seen in the Section 13.4 that a term $\varphi(x)^2$ simply amounts to changing the mass of the particle.) What is remarkable however (and will become clear only gradually) is that this theory already presents the very basic difficulties of Quantum Field Theory, while being free of the auxiliary complications that clutter realistic models. It is thus far more than a toy example, and we will later spend considerable time studying it. In the present chapter, after a little detour, we will however focus on the similarly defined ϕ^4 theory.

The goal is then to use the formula (13.10) to define the interaction Hamiltonian H_I (and then to study the full Hamiltonian $H_0 + gH_I$ for the interacting particles). If we were doing mathematics, we would immediately ask two questions: First, since φ is a distribution, what is the meaning of $\varphi(x)^3$? Second, are we in the setting previously described, that is, can we make sense of the Hamiltonian H_I given by (13.10)? But we are physicists, so we proceed immediately to the important matter, the computation of S-matrix elements. A short answer to the preceding questions is that at least it is possible to make some sense out of $:\varphi(x)^3:$, which we will later choose as an interacting Hamiltonian. On the other hand, it is harder to make sense of H_I, see Exercise 13.14.2.[18]

Before we start studying the ϕ^3 theory, we first study a related model, for which some marginal aspects of the early computations are simpler than in the ϕ^3 model. In this model, we consider *three different types* of spinless particles which are their own anti-particles, which with a frightful lack of imagination we will call the a-particles, b-particles and c-particles. We will use the indices a, b, c to make clear to which particle a given quantity pertains.[19]

First, let us consider the model where a, b and c-particles are present, but where there is no interaction between particles. Denoting by $\mathcal{B}_a$ the boson Fock space of particle a (etc.), the state space is the tensor product $\mathcal{H}_{\text{part}} := \mathcal{B}_a \otimes \mathcal{B}_b \otimes \mathcal{B}_c$ and the Hamiltonian is

$$H_0 = H_a \otimes 1 \otimes 1 + 1 \otimes H_b \otimes 1 + 1 \otimes 1 \otimes H_c. \tag{13.27}$$

The interaction between particles will be given by the Hamiltonian density

$$\mathcal{H}(x) = \varphi_a(x)\varphi_b(x)\varphi_c(x). \tag{13.28}$$

When we write such a formula, we identify an operator Φ on $\mathcal{B}_a$ with the operator $\Phi \otimes 1 \otimes 1$ on the space $\mathcal{H}_{\text{part}} = \mathcal{B}_a \otimes \mathcal{B}_b \otimes \mathcal{B}_c$, so that φ_a, φ_b and φ_c are distributions valued in this

[18] Let us openly admit that our argument that time-ordering does not create problems is flawed. One may argue in fact that the dreadful divergences we are going to face are a consequence of an improper definition of time-ordering of distributions. Working much harder with distributions, these divergences can be removed. This is the Epstein-Glaser approach to renormalization [28]. We will not explore this direction.

[19] If you had already some exposure to Feynman diagrams, the point of considering these different particles types is that each line pertains to a particular particle type. This makes it easier to distinguish lines from each other, and marginally simplifies counting arguments.

space of operators on $\mathcal{H}_{\rm part}$, which furthermore have the nice property of being defined at a given value of t.[20]

We pretend that the interaction picture makes sense. As we discussed in Section 12.5 the state space $\mathcal{H}$ appropriate to describe the interacting particles should be different from $\mathcal{H}_{\rm part}$. We assume that $\mathcal{H}_{\rm part}$ is a subspace of $\mathcal{H}$ (as there seems little choice on how to proceed otherwise).[21]

The relevant part of the S-matrix will be between two elements of $\mathcal{H}_{\rm part}$. We will represent the elements of $\mathcal{H}_{\rm part}$ as we did on page 344. That is, denoting by $a^\dagger$, $b^\dagger$ and $c^\dagger$ the creation operators relative to the a, b and c-particles respectively, elements of $\mathcal{H}_{\rm part}$ are described by quantities such as $a^\dagger(\boldsymbol{p}_1)b^\dagger(\boldsymbol{p}_2)c^\dagger(\boldsymbol{p}_3)|0\rangle$. Denoting generically such an element by $|\text{in}\rangle$, and the corresponding bra vector by $\langle\text{out}|$, we will compute matrix elements $\langle\text{out}|S|\text{in}\rangle$ "in perturbation theory" in the sense of Definition 13.1.1, that is when we make a computation at order n we replace S by $1 + S_1 + \cdots + S_n$ where S_n is given by the concrete formula (13.12). To learn how to compute S_n there are two steps:

- Learn how to compute $\langle\text{out}|\mathcal{T}\mathcal{H}(x_1)\cdots\mathcal{H}(x_n)|\text{in}\rangle$.
- Learn how to integrate in $x_1,\ldots,x_n$.

A first observation is that since we are working in a tensor product $\mathcal{B}_a \otimes \mathcal{B}_b \otimes \mathcal{B}_c$ the contributions of the various types of particles separate. That is, if $|\text{in}\rangle = |\text{in}_a\rangle \otimes |\text{in}_b\rangle \otimes |\text{in}_c\rangle$ and similarly for $|\text{out}\rangle$ we have

$$\langle\text{out}|\mathcal{T}\mathcal{H}(x_1)\cdots\mathcal{H}(x_n)|\text{in}\rangle = \langle\text{out}_a|\mathcal{T}\varphi_a(x_1)\cdots\varphi_a(x_n)|\text{in}_a\rangle \\ \times \langle\text{out}_b|\mathcal{T}\varphi_b(x_1)\cdots\varphi_b(x_n)|\text{in}_b\rangle \times \langle\text{out}_c|\mathcal{T}\varphi_c(x_1)\cdots\varphi_c(x_n)|\text{in}_c\rangle. \tag{13.29}$$

As a specific example of this formula,

$$\langle 0|a(\boldsymbol{p}_4)b(\boldsymbol{p}_3)\varphi_a(x_1)\varphi_b(x_1)a^\dagger(\boldsymbol{p}_1)b^\dagger(\boldsymbol{p}_2)|0\rangle \\ = \langle 0|a(\boldsymbol{p}_4)\varphi_a(x_1)a^\dagger(\boldsymbol{p}_1)|0\rangle \times \langle 0|b(\boldsymbol{p}_3)\varphi_b(x_1)b^\dagger(\boldsymbol{p}_2)|0\rangle. \tag{13.30}$$

13.7 A Computation at the First Order

In the setting of the previous section, at the first order in perturbation theory, we compute the amplitude

$$\langle 0|b(\boldsymbol{p}_2)c(\boldsymbol{p}_3)Sa^\dagger(\boldsymbol{p}_1)|0\rangle \tag{13.31}$$

that starting with a single a-particle of momentum $\boldsymbol{p}_1$ we end up with a b-particle of momentum $\boldsymbol{p}_2$ and a c-particle of momentum $\boldsymbol{p}_3$. We first observe that the term of

20 In (13.28) we are writing a quantity which is in fact a product of distributions, without examining what it might mean. But let us proceed!

21 We keep in mind however that the space $\mathcal{H}_{\rm part}$ describes particles as we see them after and before the interactions. The masses of the particles described by this space are the masses *as we measure them*. It will turn out that these measured masses may not be the "masses" occurring as parameters in the explicit expression of H_0. Do not worry yet if this sounds mysterious. This is related to the difficult issue of "mass renormalization" which we will explain gradually.

order zero i.e. $\langle 0|b(\boldsymbol{p}_2)c(\boldsymbol{p}_3)a^\dagger(\boldsymbol{p}_1)|0\rangle$ is zero. To compute the term of order one, that is $\langle 0|b(\boldsymbol{p}_2)c(\boldsymbol{p}_3)S_1a^\dagger(\boldsymbol{p}_1)|0\rangle$, we start by computing

$$\langle 0|b(\boldsymbol{p}_2)c(\boldsymbol{p}_3)\mathcal{H}(x)a^\dagger(\boldsymbol{p}_1)|0\rangle = \langle 0|b(\boldsymbol{p}_2)c(\boldsymbol{p}_3)\varphi_a(x)\varphi_b(x)\varphi_c(x)a^\dagger(\boldsymbol{p}_1)|0\rangle. \quad (13.32)$$

Using (13.29) this is

$$\langle 0|b(\boldsymbol{p}_2)\varphi_b(x)|0\rangle\langle 0|c(\boldsymbol{p}_3)\varphi_c(x)|0\rangle\langle 0|\varphi_a(x)a^\dagger(\boldsymbol{p}_1)|0\rangle. \quad (13.33)$$

The first task is then to compute the three terms above. It helps here to think in abstract terms, and for this we go back to the setting of Theorem 3.4.2. Given the Hilbert space $\mathcal{H}$, for elements ξ,η of $\mathcal{H}$ we consider the operators $A(\xi)$ and $A^\dagger(\eta)$. We recall the absolutely central formula (3.33):

$$\forall \xi,\eta \in \mathcal{H}\,,\ [A(\xi),A^\dagger(\eta)] = (\xi,\eta)\mathbf{1}. \quad (3.33)$$

as well as (3.34), which are the cornerstones of all the calculations.

Lemma 13.7.1 *Consider elements $\xi_1,\xi_2,\eta_1,\eta_2$ and for $j=1,2$ let $B_j = A(\xi_j) + A^\dagger(\eta_j)$. Then $\langle 0|B_j|0\rangle = 0$. Moreover, recalling the inner product $(\cdot,\cdot)$ in $\mathcal{H}$ we have*

$$\langle 0|B_1B_2|0\rangle = (\xi_1,\eta_2). \quad (13.34)$$

Proof Let us first note the crucial fact that

$$A(\xi)|0\rangle = 0\ ;\ \langle 0|A^\dagger(\eta) = 0. \quad (13.35)$$

The first relation is simply the fact (3.31) that $A(\xi)(e_\emptyset) = 0$ written in Dirac's notation, and the second fact is a consequence of the first, see Exercise 2.5.20. In mathematical notation, $(e_\emptyset, A^\dagger(\eta)(x)) = (A(\eta)(e_\emptyset),x) = 0$. This proves that $\langle 0|B_j|0\rangle = 0$.

Four terms occur in the expression $\langle 0|B_1B_2|0\rangle$, but because of (13.35) only the term $\langle 0|A(\xi_1)A^\dagger(\eta_2)|0\rangle$ contribute. By (3.33) it is equal to $\langle 0|A^\dagger(\eta_2)A(\xi_1)|0\rangle + (\xi_1,\eta_2)\langle 0|0\rangle = (\xi_1,\eta_2)$. □

To compute the terms in (13.33) we go back to the case of the particle of mass m, so that $\omega_{\boldsymbol{p}}^2 = m^2c^2 + \boldsymbol{p}^2$. The following formulas will be of constant use.

Lemma 13.7.2 *It holds that $\langle 0|\varphi(x)|0\rangle = 0$. Moreover*[22]

$$\langle 0|a(\boldsymbol{p})a^\dagger(\boldsymbol{p}')|0\rangle = (2\pi)^3\delta^{(3)}(\boldsymbol{p}-\boldsymbol{p}')\,, \quad (13.36)$$

$$\langle 0|a(\boldsymbol{p})\varphi(x)|0\rangle = \frac{1}{\sqrt{2\omega_{\boldsymbol{p}}}}\exp(\mathrm{i}(x,p))\ ;\ \langle 0|\varphi(x)a^\dagger(\boldsymbol{p})|0\rangle = \frac{1}{\sqrt{2\omega_{\boldsymbol{p}}}}\exp(-\mathrm{i}(x,p)). \quad (13.37)$$

Proof Even though we write a quantity such as $\langle 0|a(\boldsymbol{p})\varphi(x)|0\rangle$ it is a distribution, and it makes sense only when integrated against a function $\xi(\boldsymbol{p})f(x)$. We recall that $J\xi$ is defined

[22] One would expect to see also the formula for $\langle 0|\varphi(x_1)\varphi(x_2)|0\rangle$, but the useful formula is for $\langle 0|\mathcal{T}\varphi(x_1)\varphi(x_2)|0\rangle$ and we will come to this later.

as $J\xi(p) = \sqrt{2\omega_{\boldsymbol{p}}}\xi(\boldsymbol{p})$. Then, using (5.28) in the first line, the definition of φ in the second line and (13.34) in the third line,

$$\begin{aligned}\iint \frac{\mathrm{d}^3 \boldsymbol{p}}{(2\pi)^3}\mathrm{d}^4 x\xi(\boldsymbol{p})f(x)\langle 0|a(\boldsymbol{p})\varphi(x)|0\rangle &= \langle 0|A(J\xi)\varphi(f)|0\rangle \\ &= \langle 0|A(J\xi)(A(\widehat{f^*}) + A^\dagger(\hat{f}))|0\rangle \\ &= (J\xi, \hat{f}) = \int \mathrm{d}\lambda_m(p)J\xi(p)^*\hat{f}(p) \\ &= \int \frac{\mathrm{d}^3 \boldsymbol{p}}{(2\pi)^3\sqrt{2\omega_{\boldsymbol{p}}}}\xi(\boldsymbol{p})^*\hat{f}(p) \\ &= \iint \frac{\mathrm{d}^3 \boldsymbol{p}}{(2\pi)^3}\mathrm{d}^4 x\xi(\boldsymbol{p})f(x)\frac{\exp(\mathrm{i}(x,p))}{\sqrt{2\omega_{\boldsymbol{p}}}},\end{aligned} \tag{13.38}$$

which proves the first equality in (13.37). The rest is similar. □

Thus, the quantity (13.33) equals[23]

$$\frac{1}{\sqrt{8\,\omega^b_{\boldsymbol{p}_2}\omega^c_{\boldsymbol{p}_3}\omega^a_{\boldsymbol{p}_1}}}\exp(\mathrm{i}(x, p_2 + p_3 - p_1)).$$

Here, of course, $(\omega^a_{\boldsymbol{p}_1})^2 = m_a^2 + \boldsymbol{p}_1^2$, and we lighten the notation by keeping implicit the fact that $p_1 = (\omega^a_{\boldsymbol{p}_1}, \boldsymbol{p}_1)$, etc.

The next step is to perform the integration in x, but this is trivial: As a consequence of (1.19) we have

$$\int \mathrm{d}^4 x \exp(\mathrm{i}(x, p_2 + p_3 - p_1)) = (2\pi)^4\delta^{(4)}(p_2 + p_3 - p_1).$$

We have obtained our first non-trivial amplitude:

$$\langle 0|b(\boldsymbol{p}_2)c(\boldsymbol{p}_3)S_1 a^\dagger(\boldsymbol{p}_1)|0\rangle = (-\mathrm{i}g)\frac{1}{\sqrt{8\,\omega^b_{\boldsymbol{p}_2}\omega^c_{\boldsymbol{p}_3}\omega^a_{\boldsymbol{p}_1}}}(2\pi)^4\delta^{(4)}(p_2 + p_3 - p_1). \tag{13.39}$$

Let us discuss the physical meaning of this result. A striking feature of (13.39) is that the numerical factor of the delta function in the right-hand side does not depend on the directions of the momenta $\boldsymbol{p}_2$ and $\boldsymbol{p}_3$. This is a feature of our simple model and certainly not a general fact.

The delta function simply enforces conservation of energy-momentum,[24] $p_1 = p_2 + p_3$. Conservation of four-momentum is a fundamental physical principle. An important point is that this conservation appears here as a consequence of the rules by which the amplitudes are computed. The relation $p_1 = p_2 + p_3$ can happen only if $m_a \geq m_b + m_c$. This is obvious if one chooses a reference frame in which $p_1 = (m_a, 0, 0, 0)$ because then

[23] The ugly factor in front of the exponential is the price to pay for having chosen a non-Lorentz invariant normalization of the creation and annihilation operators. We will soon adopt a more natural normalization to remove such factors.

[24] Thus, there is both conservation of momentum, $\boldsymbol{p}_1 = \boldsymbol{p}_2 + \boldsymbol{p}_3$ and conservation of energy, $E_1 = E_2 + E_3$.

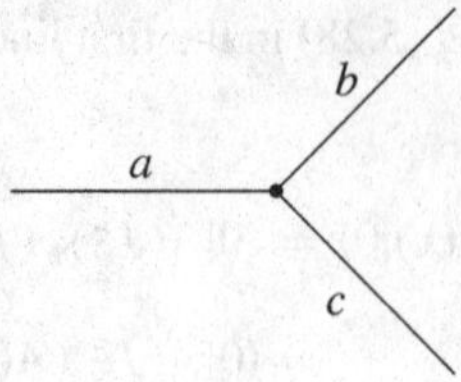

Figure 13.1 An a-particle decomposing into a b- and a c-particle.

$m_a = p_1^0 = p_2^0 + p_3^0 \geq m_b + m_c$.[25] The amplitude $\langle 0|b(\boldsymbol{p}_2)c(\boldsymbol{p}_3)S_1 a^\dagger(\boldsymbol{p}_1)|0\rangle$ is not really related to an interaction but to a single a-particle spontaneously disintegrating into a b- and a c-particle, so that amplitudes such as (13.39) are related to decay rates rather than to scattering amplitudes. To study particles that might decay we would need to modify our setting, as it does not make sense to consider a collection of particles, which "in the far past" contained an unstable particle (which must have decayed meanwhile). We do not pursue this subject here, and we assume that each of the masses m_a, m_b, m_c is less than the sum of the two others. Then the quantity (13.39) is identically zero (as we could have guessed beforehand from the conservation of four-momentum).

While learning how to make computations we will learn at the same time how to encode them into *diagrams*. The present computation is encoded in the diagram of Figure 13.1.

Here the three lines on the diagram are each of a different nature, as is indicated by the little letters a, b and c, reflecting the fact that such a line brings information about a particle of the same type. The most important idea about this diagram is that each edge corresponds to a term in the product (13.33).

Let us agree on some general conventions and terminology. First one must specify how one reads diagrams. It is convenient to think of a diagram as representing (part of) what happens during an interaction. There is before (the incoming particles) and after (the outgoing particles). As we write from left to right, we also read the diagram from left to right. On the left is what we have at the beginning of the interaction and on the right is what we have at the end.[26] The diagram of Figure 13.1 has one *initial vertex* on the left, corresponding to the incoming particle, and two *final vertices* to the right corresponding to the two outgoing particles. The diagram has also one *internal vertex*, representing "where the interaction takes place". Internal vertices are represented with a $\bullet$. A vertex that is either initial or final is called an *external vertex*. The edges of the diagrams are also called lines. There is exactly one line ending at each external vertex. Such a line is called an *external line*. The external vertices are not represented in the diagram, they are just sitting at the free end of the external lines. It is important to think of each external line (or, equivalently each external

[25] Generally speaking interactions of particles are simpler to study if one uses a reference frame following the center of mass of the system.

[26] In order to match the fact in the S-matrix, $\langle \text{out}|S|\text{in}\rangle$, the situation before the interaction is described at the *right*, it would have been more logical to orient diagrams from the right to the left, but almost nobody in the literature uses this convention, with the notable exception of Coleman's book [13].

vertex) as uniquely identified by the incoming or outgoing particle it represents.[27] In the present model, this line is of the same type as the particle the external vertex represents. For legibility we will often label each external vertex/line by the momentum of the particle it represents (although there was no point doing so in the case of Figure 13.1). A general fact is that in this model three lines come out of each internal vertex, one for each type of particle.

Each diagram will have a value (which we will learn how to calculate). The value of the diagram is a distribution.[28] In the present case, the value of the diagram of Figure 13.1 is the right-hand side of (13.39).

13.8 Wick's Theorem

The following extension of Lemma 13.7.1 is a basic tool.

Lemma 13.8.1 (Wick's theorem[29]) *In the setting of Lemma 13.7.1, consider operators $B_i = A(\xi_i) + A^\dagger(\eta_i)$ for $i \leq k$. Then*

$$\langle 0|B_1 \cdots B_k|0\rangle = 0 \tag{13.40}$$

when k is odd. When $k = 2$, $\langle 0|B_1 B_2|0\rangle = (\xi_1, \eta_2)$. When k is even, $k = 2n$, then

$$\langle 0|B_1 \cdots B_k|0\rangle = \sum \prod_{\{\ell,\ell'\}\in\mathcal{P}} (\xi_\ell, \eta_{\ell'}) = \sum \prod_{(\ell,\ell')\in\mathcal{P}} \langle 0|B_\ell B_{\ell'}|0\rangle, \tag{13.41}$$

where the sum is over all partitions $\mathcal{P}$ of the set $\{1,\ldots,k = 2n\}$ into n subsets $\{\ell,\ell'\}$ of 2 elements with $\ell < \ell'$.

Proof It goes by induction over k and relies on the identities (13.35). These imply the result for $k = 1$ and for $k = 2$ it is the object of Lemma 13.7.1. This case $k = 2$ proves the second equality in (13.41).[30]

For the general induction step, the main argument is as follows. Since the term $A(\xi_k)|0\rangle$ does not contribute we may assume $\xi_k = 0$ and $B_k = A^\dagger(\eta_k)$. The strategy is then to move this term from the right to the extreme left, where $\langle 0|B_k$ will not contribute. The first step is to write, using (3.33), and (3.34),

$$[B_{k-1}, A^\dagger(\eta_k)] = [A(\xi_{k-1}) + A^\dagger(\eta_{k-1}), A^\dagger(\eta_k)] = [A(\xi_{k-1}), A^\dagger(\eta_k)] = (\xi_{k-1}, \eta_k)1$$

and thus $B_{k-1}A^\dagger(\eta_k) = A^\dagger(\eta_k)B_{k-1} + (\xi_{k-1}, \eta_k)1$, so that

$$\langle 0|B_1 \cdots B_k|0\rangle = \langle 0|B_1 \cdots B_{k-2}A^\dagger(\eta_k)B_{k-1}|0\rangle + (\xi_{k-1}, \eta_k)\langle 0|B_1 \cdots B_{k-2}|0\rangle. \tag{13.42}$$

[27] The point of this statement will become clearer when we start "exchanging" identical lines. A given external line cannot be exchanged with any another line.

[28] In our definition the overall delta function is part of the value of the diagram. This is not what is usually done. The one single advantage of this convention is to be able to say that "the scattering amplitude is the sum of the values of the Feynman diagrams".

[29] Calling this lemma Wick's theorem is convenient but somewhat inaccurate. What is usually called Wick's theorem is a more general statement about products of polynomials in creation and annihilation operators of which the previous lemma is an immediate consequence, but which we shall not use.

[30] The reason for stating this second equality is that the expression $\langle 0|B_{\ell_j} B_{\ell'_j}|0\rangle$ will often be a convenient way to write the quantity $(\xi_{\ell_j}, \eta_{\ell'_j})$.

Using the induction hypothesis, we are now able to compute the last term. When k is odd this term is zero. When k is even it is exactly the contribution to (13.41) of the partitions of the set $\{1,\dots,k=2n\}$ into n subsets of two elements such that $\{k-1,k\}$ is one element of this partition. In the second step we write, using (3.33),

$$\begin{aligned}\langle 0|B_1\cdots B_{k-2}A^\dagger(\eta_k)B_{k-1}|0\rangle &= \langle 0|B_1\cdots B_{k-3}A^\dagger(\eta_k)B_{k-2}B_{k-1}|0\rangle\\ &\quad+(\xi_{k-2},\eta_k)\langle 0|B_1\cdots B_{k-3}B_{k-1}|0\rangle.\end{aligned}$$

Using the induction hypothesis, the last term is zero when k is odd. When k is even it is the contribution to (13.41) of the partitions of the set $\{1,\dots,k=2n\}$ into n subsets of two elements such that $\{k-2,k\}$ is one element of this partition. Iterating this procedure completes the induction step and the proof. □

In order to deal with time-ordering we will need the following variation on Lemma 13.8.1, which uses the same notations and hypotheses.

Proposition 13.8.2 *Consider operators $B_i = A(\xi_i)+A^\dagger(\eta_i)$ for $i\le k$. Assume that to B_i is associated a time t_i. Assume that B_i and B_j commute if $t_i=t_j$. We may then define the time-ordered product $\mathcal{T}B_1\cdots B_k$ as $B_{\sigma(1)}\cdots B_{\sigma(k)}$ where σ is a permutation of the indices such that $t_{\sigma(1)}\ge t_{\sigma(2)}\dots\ge t_{\sigma(k)}$. Then*

$$\langle 0|\mathcal{T}B_1\cdots B_k|0\rangle = 0 \tag{13.43}$$

when k is odd. When k is even, $k=2n$, then

$$\langle 0|\mathcal{T}B_1\cdots B_k|0\rangle = \sum\prod_{\{\ell,\ell'\}\in\mathcal{P}}\langle 0|\mathcal{T}B_\ell B_{\ell'}|0\rangle, \tag{13.44}$$

where the sum is over all partitions $\mathcal{P}$ of the set $\{1,\dots,k=2n\}$ into n subsets $\{\ell,\ell'\}$ of two elements with $\ell<\ell'$.

Proof This should be an obvious consequence of Lemma 13.8.1 applied to the operators $B'_j := B_{\sigma(j)}$. □

When the indices ℓ and ℓ' belong to the same element of the partition of $\{1,\dots,k\}$ in two-elements sets, we will say that *we contract* the operators B_ℓ and $B_{\ell'}$, and that the *contraction* of these operators[31] is the number $\langle 0|\mathcal{T}B_\ell B_{\ell'}|0\rangle$. Throughout the chapter we will say that a certain quantity *is a contraction* if it is the contraction a certain operator. Let us then recapitulate how we compute $\langle 0|\mathcal{T}B_1\cdots B_k|0\rangle$:

- We partition the set of indices into two-elements sets.
- We contract together the operators with indices in the same element of the partition, and we take the product of the corresponding contractions.
- We sum the previous quantities over all possible partitions.

[31] One may also consider the version of the same principle without time-ordering as in Lemma 13.8.1. The contraction of the operators is then defined as the number $\langle 0|B_\ell B_{\ell'}|0\rangle$. It is for this version that the condition $\ell<\ell'$ matters.

We will use this method formally, allowing the terms B_ℓ to be of types such as $\varphi(x)$, $a(\boldsymbol{p})$ or $a^\dagger(\boldsymbol{p})$ (as was already the case e.g. in Lemma 13.7.2).

The practical problem we will face is that there are many partitions of the set of indices in two-element sets. The solution to the problem is a fundamental advance brought to us by Richard Feynman: Keep track of the many partitions using diagrams.

13.9 Interlude: Summing the Dyson Series

In this section we use the Dyson series to compute the time-evolution of a simple perturbation of the harmonic oscillator. This computation is absolutely atypical, as we will be able to sum the whole Dyson series rather than just using the first few terms of it; but it illustrates well some useful principles in a very simple case, and in particular the use of Wick's theorem. This section can be skipped without harm. We use the notation of Section 2.18. We advise the reader to review Section 11.2. We are interested in computing the time-evolution of the time-dependent Hamiltonian

$$H(t) = H_0 + \gamma(t)a^\dagger + \gamma(t)^* a = H_0 + H_1(t), \tag{13.45}$$

where $H_0 = \omega(a^\dagger a + (1/2)1)$ is the Hamiltonian of a harmonic oscillator, and where $\gamma(t)$ is a nice function of t. This Hamiltonian is seen as a small perturbation of the Hamiltonian H_0. According to (11.21), in the interaction picture the interaction part $\tilde{H}_1(t) = U_0(-t)H_1(t)U_0(t)$ of the Hamiltonian equals

$$\tilde{H}_1(t) = \alpha(t)a^\dagger + \alpha(t)^* a, \tag{13.46}$$

where we have used the formulas (2.105), $U_0(-t)aU_0(t) = \exp(-i\omega t)a$ and the similar formula (2.104) and where

$$\alpha(t) = \gamma(t)\exp(i\omega t). \tag{13.47}$$

In the present case, the time-evolution equation (11.20) can be solved easily because, while the operators $\tilde{H}_1(t)$ do not commute with each other for different values of t, they have the remarkable property that the commutator of any two such operators is a multiple of the identity. In that case, based on an appropriate version of Proposition C.1.4 one can directly show that a small modification of the formula (11.24) works. The present example is "too simple" because it is completely solvable. Let us pretend however that we did not notice this, and let us set the goal of calculating the probability that starting at time zero in the ground state we end up at time t with k quanta of oscillation. Denoting by $U(t)$ the time-evolution of the system under the Hamiltonian (13.45) this probability is $|\langle 0|a^k U(t)|0\rangle|^2 = |\langle 0|a^k V(t)|0\rangle|^2$, where $V(t) = U_0(-t)U(t)$, and where we use in the equality that $a^\dagger|0\rangle$ is an eigenvector of H_0 so that $U_0(t)a^{\dagger k}|0\rangle = ca^{\dagger k}|0\rangle$ where $|c| = 1$. We will use Dyson's series (11.30) to compute the amplitude $\langle 0|a^k V(t)|0\rangle$ through the formula $V(t) = 1+\sum_{n\geq 1} W_n(t)$ where $W_n(t)$ is given by (11.34). Thus

$$\langle 0|a^k W_n(t)|0\rangle = \frac{(-\mathrm{i})^n}{n!}\int\cdots\int_{0\leq\theta_1,\theta_2,\ldots,\theta_n\leq t} \mathrm{d}\theta_1\ldots\mathrm{d}\theta_n \langle 0|a^k T\tilde{H}_1(\theta_n)\cdots\tilde{H}_1(\theta_1)|0\rangle. \tag{13.48}$$

Assigning time $+\infty$ to a, to compute

$$\langle 0|a^k \mathcal{T}\tilde{H}_1(\theta_n)\cdots\tilde{H}_1(\theta_1)|0\rangle = \langle 0|\mathcal{T}a^k\tilde{H}_1(\theta_n)\cdots\tilde{H}_1(\theta_1)|0\rangle, \tag{13.49}$$

we are exactly in the setting of Proposition 13.8.2. Indeed, the state space of the harmonic oscillator is the boson Fock space of a space of dimension 1, and if the basis vector of this space is denoted by e_1 we then have $a = A(e_1)$ and $a^\dagger = A^\dagger(e_1)$, so that each operator a and $\tilde{H}_1(\theta_i)$ is of the form considered in that proposition. We will compute the quantity (13.49) using the formula 13.44. The only non-zero terms in the sum (13.44) occur when the k powers of a are contracted with terms $\tilde{H}_1(\theta_{b_1}),\ldots,\tilde{H}_1(\theta_{b_k})$ for distinct integers $1 \le b_1 < \ldots < b_k \le n$, and where the remaining $\tilde{H}(\theta_\ell)$ are contracted in pairs. This can happen only if $n \ge k$ and $n-k$ is even, so that $n = k+2m$. The total number of such contractions is then

$$n(n-1)\ldots(n-k+1)\frac{(2m)!}{m!\,2^m} = \frac{n!}{m!\,2^m}. \tag{13.50}$$

It will turn out that the contribution of each such contraction to the integral (13.49) is the same. We fix the contraction and we compute this contribution now. We denote by $(\ell_j,\ell'_j)_{j\le m}$ the pairs of indices such that $\tilde{H}_1(\theta_{\ell_j})$ and $\tilde{H}_1(\theta_{\ell'_j})$ are contracted together. Observing that $\langle 0|a\tilde{H}_1(\theta)|0\rangle = \alpha(\theta)$ and $\langle 0|\tilde{H}_1(\theta)\tilde{H}_1(\theta')|0\rangle = \alpha(\theta)^*\alpha(\theta')$ the contribution of the given contraction to the quantity (13.49) is

$$\prod_{i\le k}\alpha(\theta_{b_i})\prod_{j\le m}c(\ell_j,\ell'_j), \tag{13.51}$$

where

$$c(\ell,\ell') = \begin{cases}\alpha(\theta_\ell)^*\alpha(\theta_{\ell'}) \text{ if } \theta_\ell > \theta_{\ell'}\\ \alpha(\theta'_\ell)^*\alpha(\theta_\ell) \text{ if } \theta_{\ell'} > \theta_\ell.\end{cases} \tag{13.52}$$

It is then easy to perform the integration (13.48), since the various factors in (13.51) depend on different variables. This integral is $A^k B^m$, where

$$A = \int_0^t \alpha(\theta)\mathrm{d}\theta\ ;\ B = 2\iint_{0\le\theta_1\le\theta_2\le t}\mathrm{d}\theta_1\mathrm{d}\theta_2\alpha(\theta_2)^*\alpha(\theta_1).$$

Taking into account the number of contractions, the quantity (13.48) is therefore

$$\langle 0|a^k W_n(t)|0\rangle = \langle 0|a^k W_{k+2m}(t)|0\rangle = (-\mathrm{i})^n A^k\frac{B^m}{m!\,2^m}.$$

Summation of the Dyson series amounts to summing over $m \ge 0$, giving

$$\langle 0|a^k V(t)|0\rangle = (-\mathrm{i}A)^k\exp(-B/2). \tag{13.53}$$

This provides a complete description of $V(t)|0\rangle$:

$$V(t)|0\rangle = \exp(-B/2)\sum_{k\ge 0}(-\mathrm{i}A)^k(a^\dagger)^k|0\rangle.$$

Exercise 13.9.1 (a) Perform a little calculus to prove that $\mathrm{Re}B = |A|^2$.
(b) Using the first part of (2.88) to recover a crucial factor, check on the formula (13.53) that the sum over k of the probabilities that at time t the system has k quanta of energy is indeed equal to 1.

It is possible to show that $U(t)|0\rangle$ is a "coherent state" in the sense of Section C.2.

13.10 The Feynman Propagator

We go back to the study of the S-matrix. To compute a term such as $\langle 0|a(\boldsymbol{p})\mathcal{T}\varphi(x_1)\varphi(x_2) a^\dagger(\boldsymbol{p}')|0\rangle$ using Proposition 13.8.2, we simply assign time $+\infty$ to each term $a(\boldsymbol{p})$ and time $-\infty$ to each term $a^\dagger(\boldsymbol{p}')$, in line with the intuition that the creation operators act much before the interaction itself and the annihilation operators act much later than the interaction.[32] We then write

$$\langle 0|a(\boldsymbol{p})\mathcal{T}\varphi(x_1)\varphi(x_2)a^\dagger(\boldsymbol{p}')|0\rangle = \langle 0|\mathcal{T}a(\boldsymbol{p})\varphi(x_1)\varphi(x_2)a^\dagger(\boldsymbol{p}')|0\rangle.$$

Applying Proposition 13.8.2 to this quantity brings forward contractions such as $\langle 0|\mathcal{T}a(\boldsymbol{p})\varphi(x_1)|0\rangle = \langle 0|a(\boldsymbol{p})\varphi(x_1)|0\rangle$ which we have already computed. It also brings forward the new quantity

$$\langle 0|\mathcal{T}\varphi(x_1)\varphi(x_2)|0\rangle. \tag{13.54}$$

The goal of the present section is to investigate this quantity. To compute it we need the following fundamental object.

Definition 13.10.1 The Feynman propagator is the tempered distribution given for $x = (t, \boldsymbol{x})$ by

$$\Delta_F(x) = \Delta_F(t, \boldsymbol{x}) = \mathrm{i}\int \frac{\mathrm{d}^3\boldsymbol{p}}{(2\pi)^3 2\omega_{\boldsymbol{p}}} \exp(-\mathrm{i}|t|\omega_{\boldsymbol{p}} + \mathrm{i}\boldsymbol{p}\cdot\boldsymbol{x}). \tag{13.55}$$

The Feynman propagator depends on the mass of the particle, so we will put indices a, b, c when needed. The factor i in this definition is conventional. Its purpose is to make the Feynman propagator a fundamental solution of the Klein-Gordon equation.[33]

The Feynman propagator is closely related to the distribution Δ_0 of Definition 5.1.6 and Lemma 5.1.7. In the present mood of not distinguishing distributions and functions we pretend that Δ_0 is given by the formula (5.11):

$$\Delta_0(x) = \int \mathrm{d}\lambda_m(p)\exp(\mathrm{i}(x,p)) = \int \frac{\mathrm{d}^3\boldsymbol{p}}{(2\pi)^3 2\omega_{\boldsymbol{p}}} \exp(\mathrm{i}(x^0\omega_{\boldsymbol{p}} - \boldsymbol{x}\cdot\boldsymbol{p})). \tag{13.56}$$

Thus it holds that

$$t = x^0 > 0 \Rightarrow \Delta_F(x) = \mathrm{i}\Delta_0(-x)\ ;\ t = x^0 < 0 \Rightarrow \Delta_F(x) = \mathrm{i}\Delta_0(Px), \tag{13.57}$$

[32] Please try to remember this convention, how to assign time to the terms $a(\boldsymbol{p})$ and $a^\dagger(\boldsymbol{p})$. It will be used again.
[33] For the relationship between the Feynman propagator and the Klein-Gordon equation, and the way physics textbooks treat this question, please see Appendix N and in particular Theorem N.2.1.

where $Px = (x^0, -\boldsymbol{x})$. The factors i in these formulas simply occur because of the factor i in (13.55) which is not present in the definition of Δ_0.[34]

Thus the tempered distribution Δ_F is actually a function outside the light cone, because this is the case for Δ_0. We observe also that $\Delta_F(\pm t, \pm \boldsymbol{x}) = \Delta_F(t, \boldsymbol{x})$, as is shown by the change of variables $\boldsymbol{p} \to -\boldsymbol{p}$.

Lemma 13.10.2 *It holds that* $\langle 0|T\varphi(x_1)\varphi(x_2)|0\rangle = -\mathrm{i}\Delta_F(x_2 - x_1)$.

Proof Writing symbolically the definition (5.1) of the free field as $\varphi(x) = A(f_x) + A^\dagger(f_x)$ where $f_x(p) = \exp(\mathrm{i}(x,p))$ we have

$$\langle 0|\varphi(x_2)\varphi(x_1)|0\rangle = (f_{x_2}, f_{x_1}) = \int \frac{\mathrm{d}^3\boldsymbol{p}}{(2\pi)^3 2\omega_{\boldsymbol{p}}} \exp(\mathrm{i}(x_1 - x_2, p))$$

$$= \int \frac{\mathrm{d}^3\boldsymbol{p}}{(2\pi)^3 2\omega_{\boldsymbol{p}}} \exp(-\mathrm{i}\omega_{\boldsymbol{p}}(t_2 - t_1) + \mathrm{i}\boldsymbol{p}\cdot(\boldsymbol{x}_2 - \boldsymbol{x}_1)), \tag{13.58}$$

which is indeed $-\mathrm{i}\Delta_F(x_2 - x_1)$ when $t_2 > t_1$. The other case holds since $\Delta_F(x_1 - x_2) = \Delta_F(x_2 - x_1)$. The argument can be made rigorous by integrating against test functions. □

Exercise 13.10.3 Deduce informally the result of the previous lemma from (5.30) using only the commutation relations between the $a(\boldsymbol{p})$ and the $a^\dagger(\boldsymbol{p}')$.

Our computations will require integration over the variables x_i. Certain such integrations can be done very easily using (1.19) if we express the Feynman propagator as an inverse Fourier transform, which is our next task.

Lemma 13.10.4 *As tempered distributions, we have*

$$\Delta_F(x) = \lim_{\varepsilon\to 0^+} \int \frac{\mathrm{d}^4 p}{(2\pi)^4} \frac{\exp(-\mathrm{i}(x,p))}{-p^2 + m^2 - \mathrm{i}\varepsilon}. \tag{13.59}$$

We may also express this in the more compact form

$$\widehat{\Delta}_F(p) = \lim_{\varepsilon\to 0^+} \frac{1}{-p^2 + m^2 - \mathrm{i}\varepsilon}, \tag{13.60}$$

where $\widehat{\Delta}_F$ denotes the Fourier transform as a distribution; but (13.59) is the form under which the formula is proved and used. It is obvious from this formula that the Feynman propagator is *Lorentz invariant.*

Proof Let us write again $x = (t, \boldsymbol{x})$. The simplest proof of (13.59) consists in computing the integral

$$\int \frac{\mathrm{d}p^0}{2\pi} \frac{\exp(-\mathrm{i}tp^0)}{-(p^0)^2 + \boldsymbol{p}^2 + m^2 - \mathrm{i}\varepsilon}$$

using a contour integral (these are reviewed in Appendix N) and the holomorphic function

$$f(z) = \frac{\exp(-\mathrm{i}tz)}{-z^2 + \boldsymbol{p}^2 + m^2 - \mathrm{i}\varepsilon}. \tag{13.61}$$

[34] We could have defined Δ_0 with a similar conventional factor i, but this would have looked completely artificial at that time.

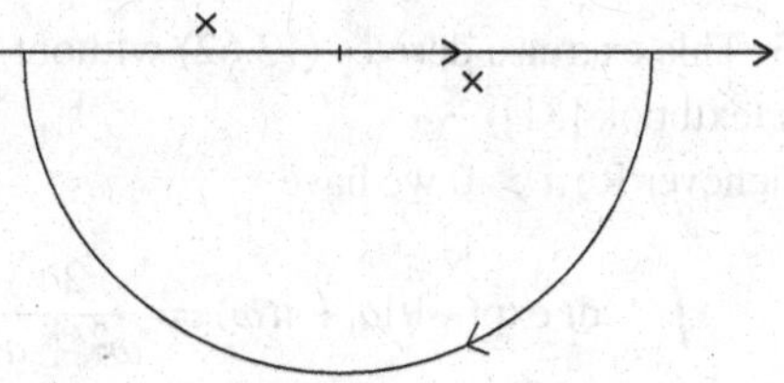

Figure 13.2 The position of the poles is indicated by the little crosses.

Let us denote by $\omega_{p,\varepsilon}$ the point such that $\omega_{p,\varepsilon}^2 = p^2+m^2-\mathrm{i}\varepsilon$ and $\mathrm{Re}\,\omega_{p,\varepsilon} > 0$, so at the first order in ε we have $\omega_{p,\varepsilon} \simeq \sqrt{p^2+m^2} - \mathrm{i}\varepsilon/(2\sqrt{p^2+m^2})$. The poles of the function (13.61) are the points $\pm\omega_{p,\varepsilon}$ and the residues of f at these points are $-\exp(\mp\mathrm{i}t\omega_{p,\varepsilon})/\pm 2\omega_{p,\varepsilon}$. Assuming e.g. $t > 0$ (so that $\mathrm{Re}\,(-\mathrm{i}tz) < 0$ for $\mathrm{Im}\,z < 0$) the contour C of Figure 13.2 contains only the pole at $\omega_{p,\varepsilon}$. Cauchy's theorem (N.1) yields

$$\int \frac{\mathrm{d}p^0}{2\pi} \frac{\exp(-\mathrm{i}tp^0)}{-(p^0)^2 + p^2 + m^2 - \mathrm{i}\varepsilon} = -\mathrm{i}\Big(-\frac{\exp(-\mathrm{i}|t|\omega_{p,\varepsilon})}{2\omega_{p,\varepsilon}}\Big) = \frac{\mathrm{i}\exp(-\mathrm{i}|t|\omega_{p,\varepsilon})}{2\omega_{p,\varepsilon}}, \quad (13.62)$$

where the first minus sign occurs because C is oriented clockwise. A similar argument proves that this still holds for $t < 0$. If we pretend that distributions are functions, the proof is finished by using this formula in the right-hand side of (13.59) and letting $\varepsilon \to 0$. If we are self-conscious about behaving this way,[35] we consider a test function $\xi(x)$ and we compute the integral

$$\int \frac{\mathrm{d}^4 p}{(2\pi)^4} \frac{1}{-p^2 + m^2 - \mathrm{i}\varepsilon} \int \mathrm{d}^4 x \exp(-\mathrm{i}(x,p))\xi(x).$$

Integrating first in $\mathrm{d}^3\boldsymbol{x}$, this becomes

$$\int \mathrm{d}t \frac{\mathrm{d}^4 p}{(2\pi)^4} \frac{\exp(-\mathrm{i}tp^0)}{-p^2 + m^2 - \mathrm{i}\varepsilon} \eta(t,\boldsymbol{p}) \quad (13.63)$$

where

$$\eta(t,\boldsymbol{p}) = \int \mathrm{d}^3\boldsymbol{x} \exp(\mathrm{i}\boldsymbol{x}\cdot\boldsymbol{p}))\xi(t,\boldsymbol{x}).$$

In (13.63) we may now apply (13.62) at given t and $\boldsymbol{p}$, and use dominated convergence to take the limit as $\varepsilon \to 0$, obtaining

$$\mathrm{i}\int \mathrm{d}t \frac{\mathrm{d}^3\boldsymbol{p}}{(2\pi)^3} \frac{\exp(-\mathrm{i}|t|\omega_{\boldsymbol{p}})}{2\omega_{\boldsymbol{p}}} \eta(t,\boldsymbol{p}) = \mathrm{i}\int \frac{\mathrm{d}^3\boldsymbol{p}}{(2\pi)^3 2\omega_{\boldsymbol{p}}} \int \mathrm{d}^4 x \exp(-\mathrm{i}|t|\omega_{\boldsymbol{p}} + \mathrm{i}\boldsymbol{x}\cdot\boldsymbol{p})\xi(x),$$

which is $\int \mathrm{d}^4 x \Delta_F(x)\xi(x)$ by the very meaning of (13.55). □

35 It is then urgent to relax, as it would be overwhelming to keep trying to be rigorous.

Exercise 13.10.5 This exercise derives (13.62) without using contour integrals (following Folland's textbook [31]).
(a) Prove that whenever $\mathrm{Re}\, a > 0$ we have

$$\int_{-\infty}^{\infty} \mathrm{d}t \exp(-|t|a + \mathrm{i}t\omega) = \frac{2a}{\omega^2 + a^2}.$$

(b) Using the inverse Fourier transform, prove that

$$\frac{1}{2a}\exp(-|t|a) = \int \frac{\mathrm{d}\omega}{2\pi} \frac{\exp(-\mathrm{i}t\omega)}{\omega^2 + a^2}. \tag{13.64}$$

(c) Deduce (13.62) using a suitable value of a.

Exercise 13.10.6 (a) Can you use (13.59) to define the square of the Feynman propagator? Why?
(b) Can you define the square of the distribution Δ_0 of Definition 5.1.6?

One of the benefits of denoting distributions as functions is that it prevents us from facing the unpleasant fact that propagators are distributions and that we are not permitted to multiply them.[36] For two distributions $\Delta(x)$ and $\Delta'(y)$ there is no problem defining a product $\Delta(x)\Delta'(y)$ but it is another matter, say, to define the square of a distribution. For example, the square of the delta function does not make sense. As we have seen in Exercise 13.10.6, even the square of the Feynman propagator is actually ill defined. *Ill-defined products of propagators will haunt us in the form of divergent integrals.*[37] For the time being however *we have found a method to pretend that we can multiply propagators.* The formula (13.59) gives us the value of the propagator as the limit as $\varepsilon \to 0$ of a perfectly well-defined quantity depending on $\varepsilon > 0$. To find a candidate for the product of propagators we exchange the operations of taking products and limits: We take the product of the corresponding well-defined quantities at given $\varepsilon > 0$ and then we take the limit as $\varepsilon \to 0$.[38]

The proof of the following is nearly identical to the proof of Lemma 13.10.4. It will not be used until Chapter 16 (so that you can ignore it until then).

Lemma 13.10.7 *As tempered distributions, we have*

$$\Delta_F(x) = \lim_{\varepsilon \to 0^+} \int \frac{\mathrm{d}^4 p}{(2\pi)^4} \frac{\exp(-\mathrm{i}(x,p))}{-(1+\mathrm{i}\varepsilon)(p^0)^2 + \boldsymbol{p}^2 + m^2}. \tag{13.65}$$

In a more compact form

$$\widehat{\Delta}_F(p) = \lim_{\varepsilon \to 0^+} \frac{1}{-(1+\mathrm{i}\varepsilon)(p^0)^2 + \boldsymbol{p}^2 + m^2}. \tag{13.66}$$

36 This statement is intended as a joke.

37 As usual we will pretend that we have put the universe in a box or somehow regularized the situation so that our manipulations make sense.

38 It is just too bad that after taking this limit we obtain ill-defined integrals.

13.11 Redefining the Incoming and Outgoing States

The cumbersome factor $1/\sqrt{8\,\omega^b_{p_2}\omega^c_{p_3}\omega^a_{p_1}}$ in (13.39) is a serious nuisance for the clarity of the exposition. The presence of this term stems from our choice of normalization of the operators $a(\boldsymbol{p})$, as was explained in Section 5.4. This choice might be reasonable for other purposes but it is simply absurd in the present setting, where it is Lorentz invariance that matters. We will get rid of this factor by setting

$$a(p) := \sqrt{2\omega_{\boldsymbol{p}}}a(\boldsymbol{p})\,,\; a^\dagger(p) := \sqrt{2\omega_{\boldsymbol{p}}}a^\dagger(\boldsymbol{p}),$$

and likewise for the other particles. Here $\omega^2_{\boldsymbol{p}} = m_a^2 + \boldsymbol{p}^2$, where m_a is the mass of the a-particle, and where it is understood that $p \in X_{m_a}$. With this convention (13.39) takes the more elegant form

$$\langle 0|b(p_2)c(p_3)S_1a^\dagger(p_1)|0\rangle = (-\mathrm{i}g)(2\pi)^4\delta^{(4)}(p_2+p_3-p_1).$$

The relation (13.37) takes the form

$$\langle 0|a(p)\varphi(x)|0\rangle = \exp(\mathrm{i}(x,p))\;;\;\langle 0|\varphi(x)a^\dagger(p)|0\rangle = \exp(-\mathrm{i}(x,p)). \tag{13.67}$$

As it is the relations (13.37) which created the factors $1/\sqrt{2\omega_{\boldsymbol{p}}}$ these will never occur anymore.

13.12 A Computation at Order Two with Trees

The goal of the present section is to compute in our model with interaction density (13.28), at the second order of perturbation theory, the matrix elements[39]

$$\langle 0|a(p_3)b(p_4)Sa^\dagger(p_1)b^\dagger(p_2)|0\rangle = \gamma\langle 0|a(\boldsymbol{p}_3)b(\boldsymbol{p}_4)Sa^\dagger(\boldsymbol{p}_1)b^\dagger(\boldsymbol{p}_2)|0\rangle, \tag{13.68}$$

where $\gamma = \sqrt{16\,\omega^a_{\boldsymbol{p}_3}\omega^a_{\boldsymbol{p}_1}\omega^b_{\boldsymbol{p}_4}\omega^b_{\boldsymbol{p}_2}}$. This is the amplitude that starting with an a-particle of momentum $\boldsymbol{p}_1$ and a b-particle of momentum $\boldsymbol{p}_2$ we end up with two particles of the same types but now momenta $\boldsymbol{p}_3$ and $\boldsymbol{p}_4$. The same type of computations as we present here for realistic models yields a number of remarkable results, as is explained in detail e.g. in Peskin and Schroeder's book [64].

As we are interested only in the case where scattering really occurs we assume $\boldsymbol{p}_3 \neq \boldsymbol{p}_1$ and $\boldsymbol{p}_4 \neq \boldsymbol{p}_2$.[40] The reader may think at first that quantities such as (13.68) must be in some sense trivial because the a and b-particles do not directly interact with each other, but our model will illustrate the fundamental fact that they may indirectly interact with each other through their joint interaction with the c-particle.[41] (Quantum Electrodynamics, the most important Quantum Field Theory, teaches us that electrons interact with each other through their interaction with the photons of the electromagnetic field.)

[39] We do not pursue the example (13.31) as the next non-zero-order terms are really tricky.

[40] Since momentum is conserved in interactions i.e. $\boldsymbol{p}_1 + \boldsymbol{p}_2 = \boldsymbol{p}_3 + \boldsymbol{p}_4$, the conditions $\boldsymbol{p}_3 \neq \boldsymbol{p}_1$ and $\boldsymbol{p}_4 \neq \boldsymbol{p}_2$ are equivalent.

[41] The previous expressions do not have a mathematically precise meaning. They intuitively convey what happens at the level of diagrams, and their meaning will become obvious as we learn to work with these. For example, the diagram (a) in Figure 13.3 makes it impossible to resist saying that a and b-particle interact through a c-particle. Furthermore, all the relevant diagrams involve at least a c-particle, which leads us to say that the a and b-particles do not interact "directly" with each other.

Let us first detail a rather trivial fact, namely that

$$\langle 0|a(p_3)b(p_4)S_1 a^\dagger(p_1)b^\dagger(p_2)|0\rangle = 0. \tag{13.69}$$

This is because

$$\langle 0|a(p_3)b(p_4)\varphi_a(x)\varphi_b(x)\varphi_c(x)a^\dagger(p_1)b^\dagger(p_2)|0\rangle = 0$$

since using (13.29) to write this as a product of three terms corresponding to the different types of particles brings forward the factor $\langle 0|\varphi_c(x)|0\rangle = 0$.

Therefore we need only to compute

$$\langle 0|a(p_3)b(p_4)S_2 a^\dagger(p_1)b^\dagger(p_2)|0\rangle. \tag{13.70}$$

Keeping (13.12) in mind, we start by the computation of

$$\langle 0|a(p_3)b(p_4)\mathcal{T}\mathcal{H}(x_1)\mathcal{H}(x_2)a^\dagger(p_1)b^\dagger(p_2)|0\rangle, \tag{13.71}$$

or, equivalently, of

$$\langle 0|a(p_3)b(p_4)\mathcal{T}\varphi_a(x_1)\varphi_b(x_1)\varphi_c(x_1)\varphi_a(x_2)\varphi_b(x_2)\varphi_c(x_2)a^\dagger(p_1)b^\dagger(p_2)|0\rangle. \tag{13.72}$$

We will do this computation in great detail, as its understanding is the key to take the mystery out of diagrams. According to (13.29) the quantity (13.72) is the product of the following quantities:

$$\langle 0|a(p_3)\mathcal{T}\varphi_a(x_1)\varphi_a(x_2)a^\dagger(p_1)|0\rangle \tag{13.73}$$

$$\langle 0|b(p_4)\mathcal{T}\varphi_b(x_1)\varphi_b(x_2)b^\dagger(p_2)|0\rangle, \tag{13.74}$$

$$\mathcal{C} := \langle 0|\mathcal{T}\varphi_c(x_1)\varphi_c(x_2)|0\rangle. \tag{13.75}$$

Let us then look at a typical expression such as

$$\langle 0|a(p_3)\mathcal{T}\varphi_a(x_1)\varphi_a(x_2)a^\dagger(p_1)|0\rangle = \langle 0|\mathcal{T}B_1B_2B_3B_4|0\rangle, \tag{13.76}$$

where

$$B_1 = a(p_3)\ ;\ B_2 = \varphi_a(x_1)\ ;\ B_3 = \varphi_a(x_2)\ ;\ B_4 = a^\dagger(p_1).$$

To apply Proposition 13.8.2 we need to group the elements 1, 2, 3, 4 in pairs. There are three ways to do this,[42] but since $p_3 \neq p_1$ the contraction of $a(p_3)$ and $a^\dagger(p_1)$ is zero, as follows from (13.36). So we obtain only the following two terms:

$$\langle 0|\varphi_a(x_1)a^\dagger(p_1)|0\rangle\langle 0|a(p_3)\varphi_a(x_2)|0\rangle\ ;\ \langle 0|a(p_3)\varphi_a(x_1)|0\rangle\langle 0|\varphi_a(x_2)a^\dagger(p_1)|0\rangle, \tag{13.77}$$

which we may compute using (13.67). For example, the term to the right equals

$$\exp(\mathrm{i}(x_1,p_3) - \mathrm{i}(x_2,p_1)).$$

42 To describe in words how to proceed, we might say that there are three possibilities to contract the term B_1 with another term, and that once this is done, the other two elements must be contracted together.

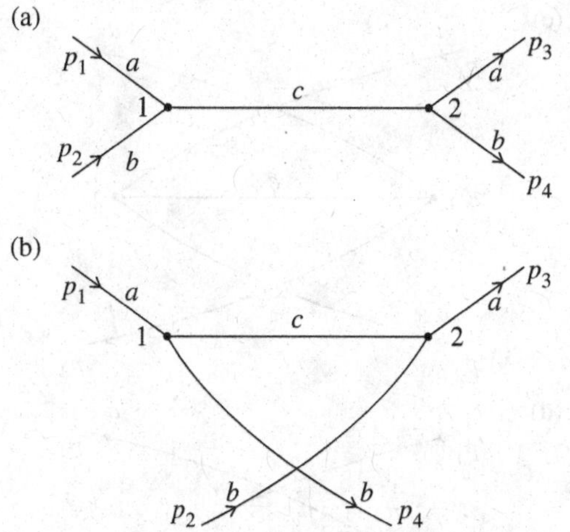

Figure 13.3 Two of the four possible contraction diagrams. There is no vertex at the crossing of the a or b-lines.

In an entirely similar manner the quantity (13.74) is the sum of the following two terms:

$$\langle 0|\varphi_b(x_1)b^\dagger(p_2)|0\rangle\langle 0|b(p_4)\varphi_b(x_2)|0\rangle\ ;\ \langle 0|b(p_4)\varphi_b(x_1)|0\rangle\langle 0|\varphi_b(x_2)b^\dagger(p_2)|0\rangle.$$

Therefore, recalling the term $\mathcal{C}$ of (13.75) the quantity (13.72) is the sum of the following four terms:

$$\mathcal{C}\langle 0|\varphi_a(x_1)a^\dagger(p_1)|0\rangle\langle 0|a(p_3)\varphi_a(x_2)|0\rangle\langle 0|\varphi_b(x_1)b^\dagger(p_2)|0\rangle\langle 0|b(p_4)\varphi_b(x_2)|0\rangle \quad (13.78)$$

$$\mathcal{C}\langle 0|\varphi_a(x_1)a^\dagger(p_1)|0\rangle\langle 0|a(p_3)\varphi_a(x_2)|0\rangle\langle 0|\varphi_b(x_2)b^\dagger(p_2)|0\rangle\langle 0|b(p_4)\varphi_b(x_1)|0\rangle \quad (13.79)$$

$$\mathcal{C}\langle 0|\varphi_a(x_2)a^\dagger(p_1)|0\rangle\langle 0|a(p_3)\varphi_a(x_1)|0\rangle\langle 0|\varphi_b(x_1)b^\dagger(p_2)|0\rangle\langle 0|b(p_4)\varphi_b(x_2)|0\rangle \quad (13.80)$$

$$\mathcal{C}\langle 0|\varphi_a(x_2)a^\dagger(p_1)|0\rangle\langle 0|a(p_3)\varphi_a(x_1)|0\rangle\langle 0|\varphi_b(x_2)b^\dagger(p_2)|0\rangle\langle 0|b(p_4)\varphi_b(x_1)|0\rangle. \quad (13.81)$$

Each of the four terms is a product of contractions. We will describe this product by a diagram. We need four different diagrams. These diagrams will be called *contraction diagrams* because each edge represents a certain term which is the contraction of two operators.[43] Each of these diagrams has two initial vertices, two internal vertices, and two final vertices. The terms (13.78) and (13.79) are represented by the diagrams of Figure 13.3, and the terms (13.80) and (13.81) are represented by the diagrams of Figure 13.4. These diagrams are drawn in a way that the external lines, which represent incoming and outgoing particles are in the same position in each of these diagrams.

The internal vertices are labeled 1 and 2, as they are respectively associated with the variables x_1 and x_2. There is now an *internal line* in these diagrams, i.e. a line between two internal vertices.

[43] Equivalently, a diagram tells you which contraction we performed when applying Wick's theorem.

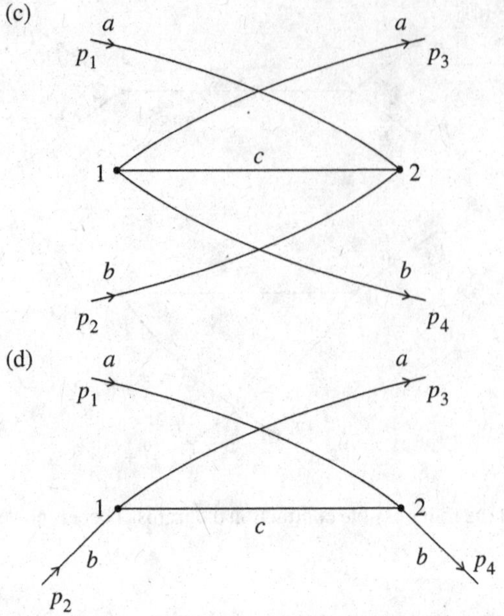

Figure 13.4 The other two possible contraction diagrams. Three of the crossing are not vertices.

Each line in a diagram represents a specific contraction. The internal line, which is labeled c (so we call it a c-line) represents the contraction $\mathcal{C} = \langle 0|\varphi_c(x_1)\varphi_c(x_2)|0\rangle$. It is the same in all diagrams because this term occurs in all four products we consider. Writing "i.v." and "e.v." for internal vertex and external vertex, respectively, the other four contractions occurring in the term (13.78) correspond to lines of the diagram (a) of Figure 13.3 as follows:

$\langle 0\|\varphi_a(x_1)a^\dagger(p_1)\|0\rangle$	a-line between the e.v. labeled p_1 and the i.v. labeled 1
$\langle 0\|a(p_3)\varphi_a(x_2)\|0\rangle$	a-line between the i.v. labeled 2 and the e.v. labeled p_3
$\langle 0\|\varphi_b(x_1)b^\dagger(p_2)\|0\rangle$	b-line between the e.v. labeled p_2 and the i.v. labeled 1
$\langle 0\|b(p_4)\varphi_b(x_2)\|0\rangle$	b-line between the i.v. labeled 2 and the e.v. labeled p_4.

There is no possible ambiguity about the order of the terms in the contraction, which is the natural time-ordering with incoming particles at time $-\infty$ and out-going particles at time $+\infty$.

We now define the *value* of each of our diagrams.

- We make the product of the contractions corresponding to the edges of the diagram as described above.
- We integrate in x_1 and x_2.
- We multiply by the factor $(-ig)^2/2!$.

Recalling (13.12), our desired amplitude is *simply the sum of the values of the four contraction diagrams.*

Let us then compute the values of these four diagrams. Using the formula (13.37) and Lemma 13.10.2 it is straightforward that the product of the contractions described by the diagram (a) in Figure 13.3 equals

$$-\mathrm{i}\Delta_F^c(x_2 - x_1)\exp\bigl(\mathrm{i}(x_2, p_3 + p_4) - \mathrm{i}(x_1, p_1 + p_2)\bigr). \tag{13.82}$$

Here the superscript c refers to the fact that this is the propagator corresponding to the c-particle. It is understood here that $p_1 = (\omega^a_{\boldsymbol{p}_1}, \boldsymbol{p}_1)$, $p_2 = (\omega^b_{\boldsymbol{p}_2}, \boldsymbol{p}_2)$, and so on. Using (13.59) the term (13.82) equals

$$-\mathrm{i}\lim_{\varepsilon\to 0^+}\int \frac{\mathrm{d}^4 p}{(2\pi)^4}\frac{\exp\bigl(\mathrm{i}(x_2, -p + p_3 + p_4) + \mathrm{i}(x_1, p - p_1 - p_2)\bigr)}{-p^2 + m_c^2 - \mathrm{i}\varepsilon}. \tag{13.83}$$

Proceeding in a rather formal manner, we now see that the integrations in x_1 and x_2 become trivial (at least if we assume that it is legal to exchange the order of integration with the limit $\varepsilon \to 0^+$). The result of these integrations is

$$\begin{aligned}
&-\mathrm{i}\lim_{\varepsilon\to 0^+}\int \frac{\mathrm{d}^4 p}{(2\pi)^4}\frac{(2\pi)^8\delta^{(4)}(-p + p_3 + p_4)\delta^{(4)}(p - p_1 - p_2)}{-p^2 + m_c^2 - \mathrm{i}\varepsilon}\\
&= -\mathrm{i}(2\pi)^4\lim_{\varepsilon\to 0^+}\frac{\delta^{(4)}(p_3 + p_4 - p_1 - p_2)}{m_c^2 - (p_1 + p_2)^2 - \mathrm{i}\varepsilon}\\
&= -\mathrm{i}(2\pi)^4\frac{\delta^{(4)}(p_3 + p_4 - p_1 - p_2)}{m_c^2 - (p_1 + p_2)^2},
\end{aligned} \tag{13.84}$$

assuming in the third line that $m_c^2 - (p_1 + p_2)^2 \neq 0$.[44] The reason why this computation is formal is that first, even though we write the quantity (13.82) as a function, it is really a tempered distribution, and the integral of this quantity over x_1 and x_2 is not defined. What is defined on the other hand is the integral

$$\iint \mathrm{d}^4x_1\mathrm{d}^4x_2\xi(x_1, x_2)\Bigl(-\mathrm{i}\Delta_F^c(x_2 - x_1)\exp\bigl(\mathrm{i}(x_2, p_3 + p_4) - \mathrm{i}(x_1, p_1 + p_2)\bigr)\Bigr),$$

whenever $\xi \in \mathcal{S}^8$. The meaning of the previous computation is that, as ξ converges to the identity function the previous integral converges to the quantity (13.84). The task of proving this is left to the mathematically inclined reader.

By the same methods, the product of the contractions encoded by diagram (b) of Figure 13.3 equals

$$-\mathrm{i}\Delta_F^c(x_2 - x_1)\exp\bigl(\mathrm{i}(x_2, p_3 - p_2) + \mathrm{i}(x_1, p_4 - p_1)\bigr), \tag{13.85}$$

and the integral over x_1 and x_2 is found to equal

$$-\mathrm{i}(2\pi)^4\frac{\delta^{(4)}(p_4 + p_3 - p_1 - p_2)}{m_c^2 - (p_4 - p_1)^2}. \tag{13.86}$$

Now we turn to the diagrams of Figure 13.4. The pleasant fact is that these diagrams differ from the corresponding diagrams of Figure 13.3 only by the fact that the vertices

44 We will not mention anymore such restrictions in the sequel.

labeled 1 and 2 have been exchanged. Thus their value is the same. We will investigate this important phenomenon in the next section.

Not forgetting the factor $(-ig)^2/2 = -g^2/2$ we have found the value of the quantity (13.70):

$$ig^2(2\pi)^4\delta^{(4)}(p_3 + p_4 - p_1 - p_2)\Big(\frac{1}{m_c^2 - (p_1 + p_2)^2} + \frac{1}{m_c^2 - (p_4 - p_1)^2}\Big). \quad (13.87)$$

There are two rather different terms in this formula, reflecting the fact that the two diagrams of Figure 13.3 are "non-isomorphic" ways to contribute to the scattering amplitude. One then says that there are different *channels*. At some intuitive level these two contributions correspond to two physically different processes. In the case of diagram (a), the incoming a and b-particles transform into a c-particle,[45] which transforms back into an a and a b-particle. In the case of diagram (b) the incoming a and b-particles transform in b and a-particles respectively while "exchanging a c-particle". It however makes little sense to consider these processes separately, since the corresponding amplitudes have to be added.

The four diagrams considered in this section are *trees* i.e. contain no loop.[46] Typically trees do not create problems. Unfortunately, as we will experience soon, diagrams that contain loops are another matter.

In the rest of the section we try to understand some of the physical content of the previous formulas. We assume $m_a = m_b$, and we investigate the non-relativistic limit, and the limit $m_c \to 0$ (hoping to get some glimpse about Quantum Electrodynamics, with the c-particle playing the role of the massless photon). Recalling that the speed of light is 1 in our system of units, for a particle of rest mass m_a and four-momentum p, in the non-relativistic limit $\boldsymbol{p}^2 \ll m_a^2$ we have

$$p^0 = \sqrt{m_a^2 + \boldsymbol{p}^2} \simeq m_a + \frac{\boldsymbol{p}^2}{2m_a}.$$

In that case $(p_4^0 - p_1^0)^2 \simeq (\boldsymbol{p}_4^2 - \boldsymbol{p}_1^2)^2/(2m_a)^2$, and since $\sqrt{\boldsymbol{p}^2}$ is just the length $\|\boldsymbol{p}\|$ of $\boldsymbol{p}$, we have

$$|\boldsymbol{p}_4^2 - \boldsymbol{p}_1^2| = |\|\boldsymbol{p}_4\|^2 - \|\boldsymbol{p}_1\|^2| = (\|\boldsymbol{p}_4\| - \|\boldsymbol{p}_1\|)(\|\boldsymbol{p}_4\| + \|\boldsymbol{p}_1\|) \ll 2m_a\|\boldsymbol{p}_4 - \boldsymbol{p}_1\|,$$

so that $(p_4^0 - p_1^0)^2 \ll (\boldsymbol{p}_4 - \boldsymbol{p}_1)^2$ and therefore

$$-(p_4 - p_1)^2 \simeq (\boldsymbol{p}_4 - \boldsymbol{p}_1)^2.$$

Exercise 13.12.1 Prove that if $p_1 + p_2 = p_3 + p_4$, in the reference frame with origin the center of mass of the incoming particles, the previous relation is exact.

On the other hand, $(p_1 + p_2)^2 \simeq (p_1^0 + p_2^0)^2 \simeq 4m_a^2$. We then see that (recalling that we assume $m_b = m_a$) in the limit $m_c \to 0$ the term (13.84) is very much smaller than the term (13.86) and can be neglected.

[45] Such a particle is called "a virtual particle" in physics. It exists as a line on a diagram, but its claim to existence beyond this are tenuous. It cannot be detected in any conceivable experiment, for the good reason that "it is not on the shell", it does not satisfy the relation $p^2 = m^2$.

[46] A loop is a closed path that uses any given edge at most once.

Within this approximation we have found that

$$\langle 0|a(p_3)b(p_4)S_2 a^\dagger(p_1)b^\dagger(p_2)|0\rangle \simeq ig^2(2\pi)^4\frac{\delta^{(4)}(p_3+p_4-p_1-p_2)}{(p_4-p_1)^2}. \tag{13.88}$$

We would like to compare this formula with (13.8). The formula (13.8) refers to the different situation of a particle scattered by a potential. But suppose that we study the case of the scattering of two (possibly different) particles that attract (or repel) each other with a potential depending only on their mutual positions. Shifting to hand-waiving mode, we may argue that if we study these in the reference frame centered at their common center of mass, each individual particle will behave as if it is attracted by this common center of mass. Then (13.88) shows that the potential V such that

$$\widehat{V}(\boldsymbol{p}) = -\frac{1}{\boldsymbol{p}^2} \tag{13.89}$$

is involved here. It is a simple computation (using spherical coordinates) to check that the electrostatic potential

$$V(\boldsymbol{x}) = -\frac{1}{4\pi|\boldsymbol{x}|} \tag{13.90}$$

has precisely the property (13.89).[47] At this point we refer the reader to Appendix O for a related and more physical model bringing forward the so-called Yukawa potential $V(\boldsymbol{x}) = -\exp(-m_c|\boldsymbol{x}|)/4\pi|\boldsymbol{x}|$.

13.13 Feynman Diagrams and Symmetry Factors

Having seen simple examples we should be ready to discuss some general features of the computation of n-order terms in $\langle \mathrm{out}|S|\mathrm{in}\rangle$ in the present model, that is of terms

$$\frac{(-ig)^n}{n!}\int\dots\int d^4x_1\dots d^4x_n\langle \mathrm{out}|T\varphi_a(x_1)\varphi_b(x_1)\varphi_c(x_1)\dots\varphi_a(x_n)\varphi_b(x_n)\varphi_c(x_n)|\mathrm{in}\rangle. \tag{13.91}$$

The brackets $\langle\dots\rangle$ are computed through Wick's theorem (here in the form of Proposition 13.8.2). Each term obtained through this proposition is described by a contraction diagram. The value of this diagram is defined by integrating the corresponding term in the variables $x_1,\dots,x_n$ and multiplying by $(-ig)^n/n!$. Therefore we reach the following statement:

The value of (13.91) is the sum of the values of all the contraction diagrams with n internal vertices and the appropriate initial and final vertices (corresponding to the given incoming and outgoing particles). (13.92)

All the contraction diagrams we consider in this sum have the "same" external vertices and have internal vertices labeled 1 to n. We say that two contraction diagrams are the same

[47] The electrostatic potential describes the interaction of electrons through photons.

if one of the diagrams has a line of a certain type between two given vertices if and only if the other diagram has a line of the same type between the corresponding vertices.[48] When drawing diagrams one has to be cautious that the same diagram (in the previous sense) may be drawn in different-looking ways. Referring to two different-looking pictures, we will then say "that they represent the same diagram drawn differently". A very simple example of this situation occurs in diagram (a) of Figure 13.6. If one exchanges the labels 1 and 2, one obtains the same diagram, although it is drawn differently. The sum in (13.92) is taken over all possible *different* diagrams.

To prevent uninteresting situations, where not all the particles actually interact,[49] we typically assume that there does not exist a subset of incoming particles and a subset of outgoing particles whose sums of four-momenta are equal. In a contraction diagram, each external vertex is then linked by an external line to an internal vertex, and three lines (one of each type a, b, c) come out of each internal vertex. It may be worth stressing this important feature, which arises from the form of the interaction Hamiltonian $\mathcal{H}(x) = \varphi_a(x)\varphi_b(x)\varphi_c(x)$. Each of the three terms creates a line of the corresponding nature out of an internal vertex.

Let us then state the rules to compute the value of a contraction diagram. We state these rules formally, not worrying yet whether the corresponding operations are really well defined.

- For each external line linking a final vertex with an internal vertex labeled j (corresponding to the variable x_j), write a factor $\exp(\mathrm{i}(x_j, p))$, where p is the four-momentum flowing on this external line from the internal vertex to the final vertex. If instead the external line links an initial vertex with an internal vertex labeled j write a factor $\exp(-\mathrm{i}(x_j, p))$, where p is the four-momentum flowing from the external vertex to the internal vertex.[50]
- For each internal line with endpoints labeled i and j write a factor equal to $-\mathrm{i}\Delta_F^d(x_j - x_i)$, where again d is the type of the particle.
- For each internal vertex write a factor $-ig$.
- Multiply all these factors and integrate over all x_i's.[51]
- Divide by $n!$.

To explain why these rules are correct, we observe that according to (13.37) the quantity assigned in the first item above to an external line is just the value of the contraction corresponding to this line, and that by Lemma 13.10.2 the quantity assigned to an internal line is just the value of the contraction corresponding to this line. The factor $-ig$ assigned to each internal vertex and the factor $1/n!$ of the last item combine to give the factor $(-ig)^n/n!$ in front of (13.12).

48 That definition may have to be adjusted for different models.
49 Or where the particles can be split into two groups that interact separately. In such cases one has more terms to consider and this makes the exposition heavier.
50 You have certainly noticed that the reason why the factors are different in the case of a line joining an internal vertex to a final or initial vertex is that the corresponding external lines are oriented differently. An external line is oriented from an initial vertex to an internal vertex, but from an internal vertex to a final vertex.
51 As we are going to see soon, this integration raises problems.

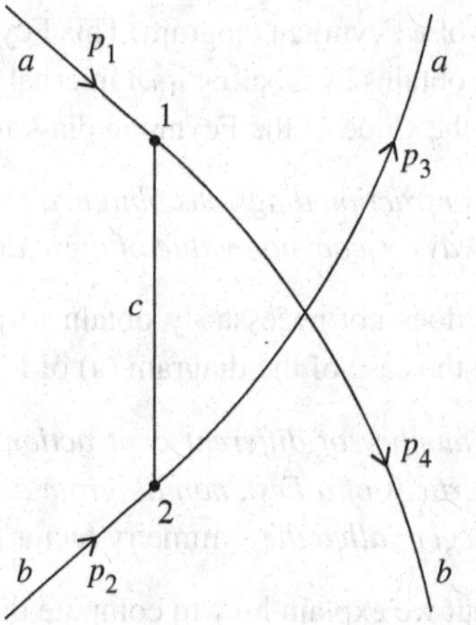

Figure 13.5 Another way to draw diagram (b) of Figure 13.3. The last crossing is not a vertex.

There are many factors i. Generally speaking when computing diagrams it is tricky to get the powers of i and π right. It is customary for authors to apologize for the unavoidable mistakes they might have made for such factors, and they often add that they made a real effort to get them right. The situation is different here. As these factors bear on the practical use of the computations but not on the theoretical understanding which concerns us here, *I made no special effort to get them right!*

At last we are ready to introduce *Feynman diagrams*. Two contraction diagrams that differ only in the numbering of their internal vertices have the same value, because when one integrates in the x_i one does not care how the variables that are integrated are named. For example, if one exchanges the labels 1 and 2 in diagram (a) in Figure 13.3 one obtains the diagram (c) in Figure 13.4. In order not to do the same computation twice it is then natural to *identify* these two diagrams. The result of this identification is called a *Feynman diagram*. So (in this model), a Feynman diagram is a contraction diagram where the labeling of the internal vertices has been erased. Whenever convenient in our pictures of Feynman diagrams the internal vertices might be labeled, but this labeling is *not part* of the diagram, and any other labeling would represent the same Feynman diagram (but probably another contraction diagram). After this identification, the four diagrams of Figures 13.3 and 13.4 reduce to the two diagrams shown in Figure 13.3.[52] The diagram (b) is often drawn in textbooks as in Figure 13.5.

The goal now is to *define* the value of Feynman diagrams in such a manner that we may replace (13.92) with the following:

The value of (13.91) is the sum of the values of all the Feynman *diagrams with n internal vertices and the appropriate initial and final vertices.* (13.93)

[52] Again, when these are seen as Feynman diagrams, one must forget about the labeling of the internal vertices.

This determines the value of a Feynman diagram. This Feynman diagram arises from all the contraction diagrams one obtains by labeling its n internal vertices any of the $n!$ possible ways. It is then obvious that the value of the Feynman diagram should be given by

Number of different contraction diagrams obtained by labeling the internal vertices all possible ways $\times$ common value of these diagrams. (13.94)

The difficulty is that one does not necessarily obtain $n!$ possible different contraction diagrams this way (as e.g. in the case of the diagram (a) of Figure 13.6).

Lemma 13.13.1 *The number of different contraction diagrams one obtains when labeling the internal vertices of a Feynman diagram all possible ways is of the form $n!/S$ where S is an integer called the* symmetry factor *of the diagram.*

We will prove this later, but we explain how to compute the value of a general Feynman diagram in order for (13.93) to hold. The rules to compute this value are known as *Feynman rules in position space,*[53] and you are surely aware that they are fundamental. Again we consider only the case where all particles are scattered.

- Number the internal vertices from 1 to n any way you like.
- For each external line linking a final vertex with an internal vertex labeled j (corresponding to the variable x_j), write a factor $\exp(\mathrm{i}(x_j, p))$, where p is the four-momentum flowing from the internal vertex to the final vertex. If instead the external line links an initial vertex with an internal vertex labeled j write a factor $\exp(-\mathrm{i}(x_j, p))$, where p is the four-momentum flowing from the external vertex to the internal vertex.
- For each internal line with endpoints labeled i and j write a factor equal to $-\mathrm{i}\Delta_F^d(x_j - x_i)$, where again d is the type of the particle.
- For each internal vertex write a factor $-ig$.
- Multiply all these factors and integrate over all x_i's.
- *Divide by the symmetry factor S of the diagram.*

Only the final step differs from the rules computing the value of a contraction diagram. As some textbooks put it, "the symmetry factor of the Feynman diagram prevents over-counting".

Let us go back to the discussion of Lemma 13.13.1, and define the symmetry factor of a contraction diagram as being the symmetry factor of the corresponding Feynman diagram.

Lemma 13.13.2 *The symmetry factor S of a contraction diagram is the number of ways one may relabel the internal vertices without changing the diagram.*

The next exercise is highly recommended.

[53] The exact form of the rules depends on the model under study. The name "position space" is motivated by the fact "that the interactions occur at the positions x_i".

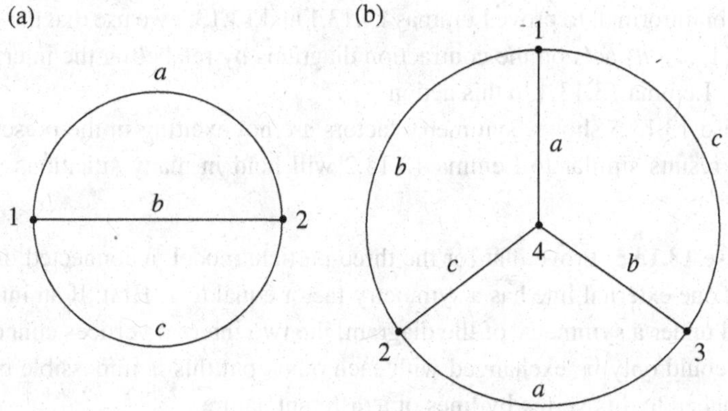

Figure 13.6 Two vacuum bubbles with $S = 2$ for (a) and $S = 4$ for (b).

Exercise 13.13.3 (a) Show that the symmetry factor S of a contraction diagram is the number of permutations σ of the set of internal vertices that have the following properties. First the internal vertex i is connected to the external vertex v by a line of a certain type if and only if $\sigma(i)$ is connected to v by a line of the same type. Second, two internal vertices i and j are connected by a line of a certain type if and only if $\sigma(i)$ and $\sigma(j)$ are connected by a line of the same type.
(b) A permutation as in (a) is called a symmetry of the diagram. Prove that if an internal vertex is connected to an external vertex by an external line, it is invariant under any symmetry of the diagram.

Consider for example the contraction diagrams of Figure 13.6. (Such diagrams without external legs are called *vacuum bubbles*. They are further discussed at the end of this section.) It should be obvious that the diagram (a) has a symmetry factor 2. For diagram (b) the symmetry factor is 4: There are four different ways to label the vertices which just represent the same diagram drawn a different way.[54]

The proof of Lemmas 13.13.1 and 13.13.2 is very simple, and it might help to state an abstract version of the argument.

Lemma 13.13.4 *If a finite group G operates on a set, given any point a of orbit $O = \{g \cdot a; g \in G\}$ and its little group $V = \{g \in G; g \cdot a = a\}$, one has* card O card V = card G.

Proof Consider the map $T : g \mapsto g \cdot a$ from G to O. Then $g \cdot (h \cdot a) = h \cdot a$ if and only if $(h^{-1}gh) \cdot a = a$ i.e. $g \in hVh^{-1}$. Thus card $T^{-1}(h \cdot a) =$ card V. Summation of this equality over the different values of $h \cdot a \in O$ yields the result. □

[54] One may exchange simultaneously vertices 1 with 2 and 3 with 4; vertices 1 with 3 and 2 with 4, and vertices 1 with 4 and 2 with 3.

Staying a bit informal, to prove Lemmas 13.13.1 and 13.13.2 we use that the permutation group G of $\{1,\ldots,n\}$ acts on the contraction diagrams by relabeling the internal vertices, and we apply Lemma 13.13.4 to this action.

As Exercise 13.13.5 shows, symmetry factors are not exciting in the present situation. Nonetheless results similar to Lemma 13.13.2 will hold in many situations with similar proofs.

Exercise 13.13.5 Prove that for the three-particle model, a connected diagram with at least one external line has a symmetry factor equal to 1. Hint: If an internal vertex is fixed under a symmetry of the diagram, the two internal vertices connected to this vertex could only be exchanged with each other, but this is impossible because they are attached to this vertex by lines of a different nature.

Soon we will learn to deal with models where there is a single type of line. In this case, the possibility of exchanging lines that cannot be distinguished will contribute to the symmetry factor of a diagram.

When actually computing amplitudes, Feynman diagrams are certainly the way to go. For general theoretical arguments, however, symmetry factors are a nuisance. This nuisance is entirely our own doing, since it does not exist if one sticks with contraction diagrams. We will however follow the crowd and cope with this nuisance.

Coming back to vacuum bubbles as in Figure 13.6, it is obvious that they spell trouble. For example, the value of the bubble diagram (a) of Figure 13.6 is given by

$$(-ig)^2 \iint \mathrm{d}^4x_1\mathrm{d}^4x_2\Delta_F^a(x_2-x_1)\Delta_F^b(x_2-x_1)\Delta_F^c(x_2-x_1).$$

It cannot be finite because the integrand depends only on $x_2 - x_1$.[55] These infinities are a consequence of our continuous model.[56] Physicists think of vacuum bubbles as representing "vacuum fluctuations" which occur with a constant space-time density. There is no reason to believe that they should interfere with our scattering experiment. *We will ignore diagrams containing vacuum bubbles to compute scattering amplitudes.* If you find that this rule is not well grounded, you cannot be blamed, but further support will be obtained in Section 14.5.

13.14 The ϕ^4 Model

From this point on we consider only models with a single type of particle. The model we introduce in this section will be of constant use. It does not pretend to be realistic. It however presents many of the difficulties inherent to Quantum Field Theory in a form which is at the same time rather generic and free of accessory complications. In this model there is a single massive spinless particle which is its own anti-particle. The (tentative) interaction Hamiltonian density is given by the formula

55 Besides, as we will learn soon, this integrand is an ill-defined product of three distributions.
56 They disappear if we put the universe (both spatial and temporal) in a box.

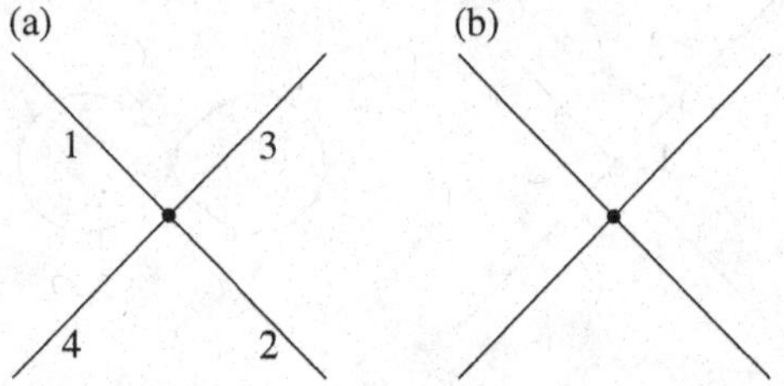

Figure 13.7 Scattering two particles at order one; (a) contraction diagram and (b) Feynman diagram.

$$\mathcal{H}(x) = \frac{\varphi(x)^4}{4!}. \tag{13.95}$$

This model is called the ϕ^4 model.[57] We will learn first how to deal with the minor complications this model suffers compared with the previous model, but our main objective is the computation at the second order of perturbation theory of the scattering amplitude

$$\langle 0|a(p_3)a(p_4)Sa^\dagger(p_1)a^\dagger(p_2)|0\rangle. \tag{13.96}$$

We assume that the momenta p_3, p_4, p_1, p_2 are all different from each other. Let us start by the computation of

$$\langle 0|a(p_3)a(p_4)\varphi(x)^4 a^\dagger(p_1)a^\dagger(p_2)|0\rangle. \tag{13.97}$$

When applying Wick's theorem to compute this quantity, one may not contract $a(p_3)$ with either of $a^\dagger(p_1)$ or $a^\dagger(p_2)$, etc. Indeed if we contract $a(p_3)$ with $a^\dagger(p_1)$ we get a factor $(2\pi)^4\delta^{(4)}(p_3 - p_1)$ which is zero as we assume that the incoming momenta are different from the outgoing momenta. In order to keep proper track of the contractions, it greatly helps to distinguish between the four different copies of $\varphi(x)$ which occur in (13.97). For this we *number* them from one to four. A contraction is then completely determined by knowing with which of these copies each of the terms $a(p_3)$, $a(p_4)$ etc. is contracted. We symbolize such a contraction by a contraction diagram such as diagram (a) in Figure 13.7. Each line out of a vertex, instead of carrying the information as to which particle type it refers, now carries the information of which copy of $\varphi(x)$ it reaches in the internal vertex. The diagram (a) of Figure 13.7 represents one of the 4! such diagrams (recalling that the external lines are uniquely identified). Each of the contraction diagrams gives the same contribution

$$\exp(\mathrm{i}(x, p_3 + p_4 - p_1 - p_2)), \tag{13.98}$$

and the denominator 4! in (13.95) is included exactly to cancel the number of these diagrams. Quite naturally we will identify contraction diagrams which differ only by the numbering of the lines. For this we simply remove this numbering, obtaining in the present case the *Feynman diagram* labeled (b) in Figure 13.7, which now represents 4! contraction

[57] Please do not embarrass me by asking me what is the exact meaning of the fourth-power of the distribution $\varphi(x)$. As we will see soon, our contractions will create only sensible terms. More precisely, we will keep only the sensible terms, throw the other terms out and wish for the best.

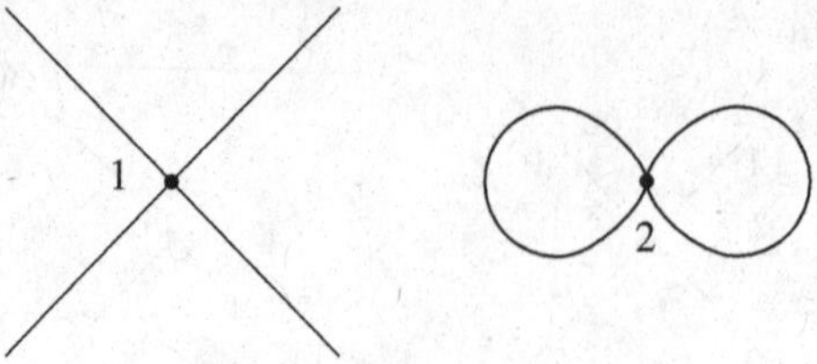

Figure 13.8 Lines from one vertex to itself.

diagrams. Identification of contraction diagrams which differ only by the numbering of their lines will give rise to the same nuisance as identification of contraction diagrams which differ only by the labeling of their internal vertices: Different ways to number these lines may not give different diagrams, but just the same diagram drawn differently. Examples to come in the next section will make this matter clearer.

From the contribution (13.98) we reach the result

$$\langle 0|a(p_3)a(p_4)S_1a^\dagger(p_1)a^\dagger(p_2)|0\rangle = -ig(2\pi)^4\delta^{(4)}(p_3+p_4-p_1-p_2). \tag{13.99}$$

Let us now turn to the computation of

$$\langle 0|a(p_3)a(p_4)\varphi(x_1)^4\varphi(x_2)^4a^\dagger(p_1)a^\dagger(p_2)|0\rangle. \tag{13.100}$$

The first observation is that when using Wick's theorem it is possible to contract the four terms with a's with terms $\varphi(x_1)$ and to contract some of the terms $\varphi(x_2)$ with other such terms, as in the diagram of Figure 13.8 (where the irrelevant numbering of the lines has not been indicated).

A line from a vertex labeled x to itself corresponds to an infinite term $(f_x, f_x) = \langle 0|\varphi(x)\varphi(x)|0\rangle$, see (13.58). We will simply get rid of these lines with the following.

Exclusion Rule: *Lines between an internal vertex and itself are forbidden.*

The Exclusion Rule is *enforced from that point on,* for every type of diagram we will ever consider. The following exercise shows that the Exclusion Rule is most reasonable.[58]

Exercise 13.14.1 (a) Convince yourself that after replacing $\varphi(x)^4$ by $:\varphi(x)^4:$ the terms of the S matrix are still computed as the sums of the values of certain diagrams (but that there are fewer of them).

(b) Convince yourself that enforcing the exclusion rule amounts to replacing the Hamiltonian density of the interacting term by its normal-ordered version $:\varphi(x)^4:/4!$.

As a consequence of the previous exercise, as we enforce the exclusion rule, it is more appropriate from now on to write the Hamiltonian density of the interaction term for the Hamiltonian as[59]

[58] Please keep in mind that the goal is to build a fruitful model, so that the notion of "reasonable" is rather elastic. If one does not like this rule, there are other methods to take care of the lines from one vertex to itself, but these are hardly satisfactory for a mathematician.

[59] If you have difficulties solving Exercise 13.14.1, please do not be discouraged by this detail. Just think that the Hamiltonian density is $\varphi(x)^4/4!$ and that we enforce the exclusion rule.

$$\mathcal{H}(x) = :\frac{\varphi(x)^4}{4!}: \tag{13.101}$$

The formula (13.99) remains true for this new Hamiltonian.

Exercise 13.14.2 (a) Let us examine the meaning of (13.101). The term S_1 of the expansion $S = 1 + S_1 + \cdots$ of the S matrix is $S_1 = (-ig)\int d^4x \,:\varphi^4(x):\, /4!$, which should be an operator defined on a certain domain of the Fock space. Can you give a sensible definition of $S_1|0\rangle$?
(b) Can you show that the operator S_1 is defined on many elements of the Fock space? Hint: This is harder.
(c) The interaction Hamiltonian is supposed to be defined by $H_I(t) = \int d^3\boldsymbol{x}\mathcal{H}(t,\boldsymbol{x})$. Can you make sense of $H_I(0)|0\rangle$?

Exercise 13.14.3 At least at the formal level, find an argument that the Hamiltonian (13.101) satisfies the condition (13.21).

13.15 A Closer Look at Symmetry Factors

We now compute the quantity (13.100) using contraction diagrams and applying the exclusion rule. One typical such diagram is drawn as diagram (a) in Figure 13.9.

The lines out of each internal vertex labeled i are numbered 1 to 4 to tell us which copy of $\varphi(x_i)$ has been used in that vertex. Each internal line carries two numbers between 1 to 4, one at each end, whereas each external line carries one single such number.[60] The diagrams fall into six different types. Three of these types are represented in Figure 13.10. The other three types of diagrams are obtained by exchanging x_1 and x_2, or equivalently by exchanging the labels 1 and 2 on the vertices.

In these pictures the numbering of the lines out of the internal vertices has not been indicated for clarity. There are $(4!)^2/2$ possible ways of numbering lines for each of these diagrams. To see this we observe there are $(4 \times 3)^2$ ways to number the external lines, and given these there are only two ways (rather than four) to number the internal lines to get different diagrams. For example, diagram (a) of Figure 13.9, the internal line which has the

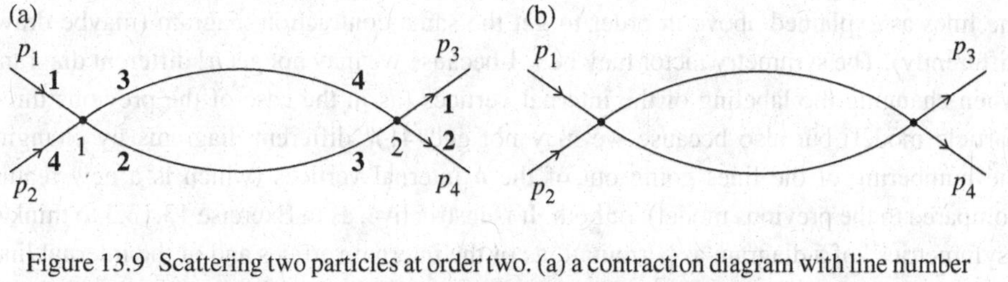

Figure 13.9 Scattering two particles at order two. (a) a contraction diagram with line number boldface and vertex numbers and (b) the corresponding Feynman diagram.

[60] In the next few pages, when we talk of "numbering the lines" we mean determining for each line what are these numbers.

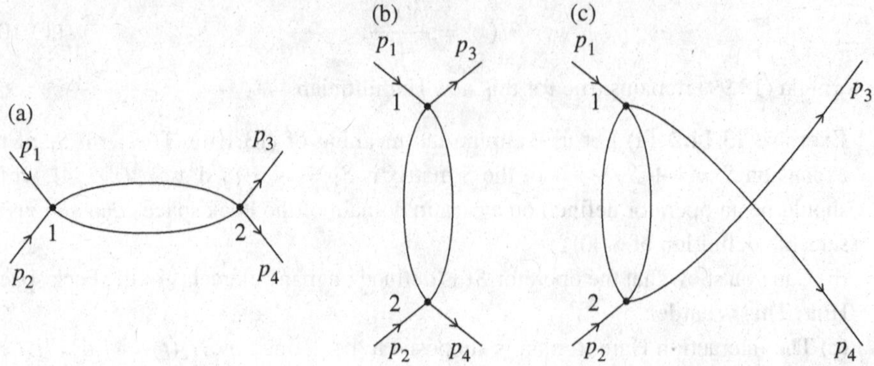

Figure 13.10 The three different channels. The last crossing on the right is not a vertex.

label 3 on the side of the vertex numbered 1 may have label either 4 or 3 on the side of the vertex numbered 2. (To put things differently, if you exchange the labels 2 and 3 on the internal lines getting out of the vertex numbered 1 and you also exchange the labels 3 and 4 on the internal lines getting out of the internal vertex numbered 2, you get the same diagram, drawn differently.) The number $(4!)^2/2$ fails by a factor 2 to compensate the $1/(4!)^2$. One then says that *each of the diagrams of Figure 13.10 has a symmetry factor* 2. Equivalently, one may number the lines of the diagram in two different ways while getting the same contraction diagram. These two different numberings differ just by exchanging the two internal lines between the internal vertices.

Exercise 13.15.1 Write the products of contractions corresponding to the three diagrams of Figure 13.10.

Let us then summarize some general features of contraction diagrams. In a contraction diagram, the internal vertices are labeled from 1 to n, and each line out of an internal vertex is numbered from 1 to 4 (so that an internal line receives two such numbers). A Feynman diagram is obtained from a contraction diagram by removing both the labeling of the internal vertices and the numbering of the lines out of the internal vertices. The symmetry factor of the diagram is the number of ways one may label the internal vertices and number the lines as explained above in order to get the same contraction diagram (maybe drawn differently). The symmetry factor may be >1 because we may not get $n!$ different diagrams when changing the labeling of the internal vertices (as in the case of the previous three-particle model) but also because we may not get $(4!)^n$ different diagrams by changing the numbering of the lines going out of the n internal vertices (which is a new feature compared to the previous model), or both. It is instructive, as in Exercise 13.13.3 to think of "symmetries" of a diagram as permutations of the internal vertices and of the internal lines that do not change the diagram, and of the symmetry factor of a diagram as the number of such symmetries.

Exercise 13.15.2 State the results corresponding to Lemmas 13.13.1 and 13.13.2.

We illustrate symmetry factors on several more examples.

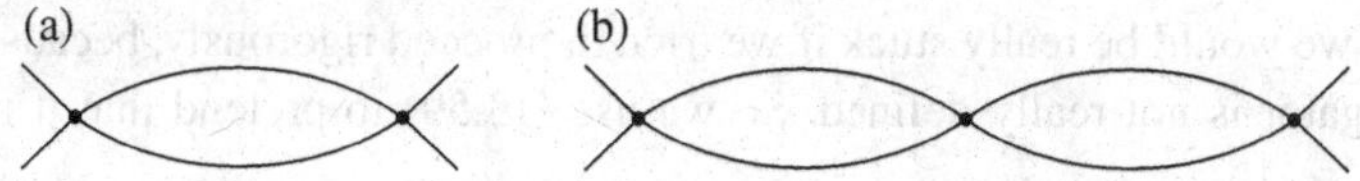

Figure 13.11 Symmetry factor of 2 in (a) and 4 in (b).

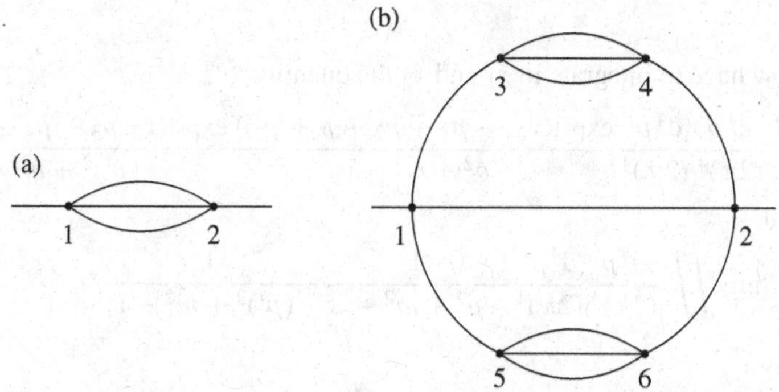

Figure 13.12 Symmetry factors of 3! in (a) and $2 \times (3!)^2$ in (b).

In the diagrams of Figure 13.11, the symmetry factors are respectively 2 and 4, which come from the possibilities of permuting the internal lines without changing the diagrams. We have not yet seen diagram (b) of this figure. It will occur when we make computations of the same scattering amplitude at order three.[61]

In diagram (a) of Figure 13.12 the symmetry factor is 3!, for all the possible permutations of lines between vertices 1 and 2. In diagram (b), the symmetry factor is $2 \times (3!)^2$, where the factor 2 arises from exchanging the vertices 3 and 4 with the vertices 5 and 6, whereas the factors 3! are created by the permutations of the three lines between vertices 3 and 4 and the three lines between vertices 5 and 6.

Exercise 13.15.3 Convince yourself that generally speaking the symmetry factor S of a diagram is of the type $S_v S_l$ where S_v is the number of permutations of the internal vertices which do not change the diagram, while S_l is the number of permutations of internal lines which do not change the diagram. In the case of diagram (b) of Figure 13.12 we have $S_v = 2$ and $S_l = (3!)^2$.

13.16 A Computation at Order Two with One Loop

Going back to the computation of the term (13.100), we compute the value of each of the diagrams of Figure 13.10. For diagram (a) we will have to integrate the function

$$(-ig)^2(-i\Delta_F(x_2 - x_1))^2 \exp(i(x_2, p_3 + p_4) - i(x_1, p_1 + p_2)). \tag{13.102}$$

[61] Generally speaking, any diagram with n internal vertices in which four lines stem from each internal vertex appears in the computation of the scattering amplitude at order n. The reader may like to attempt drawing all diagrams with four internal vertices to get a feeling for the magnitude of this task.

At this point we would be really stuck if we tried to proceed rigorously, because the square of the propagator is not really defined. So we use (13.59) to pretend that it makes sense to write

$$\Delta_F(x_2 - x_1)^2 = \lim_{\varepsilon \to 0^+} \iint \frac{\mathrm{d}^4 p}{(2\pi)^4} \frac{\mathrm{d}^4 p'}{(2\pi)^4} \frac{\exp(-\mathrm{i}(x_2 - x_1, p))}{-p^2 + m^2 - \mathrm{i}\varepsilon} \frac{\exp(-\mathrm{i}(x_2 - x_1, p'))}{-(p')^2 + m^2 - \mathrm{i}\varepsilon}, \tag{13.103}$$

so that we now have to integrate in x_1 and x_2 the quantity

$$g^2 \lim_{\varepsilon \to 0^+} \iint \frac{\mathrm{d}^4 p}{(2\pi)^4} \frac{\mathrm{d}^4 p'}{(2\pi)^4} \frac{\exp(\mathrm{i}(x_1, -p_1 - p_2 + p + p'))}{-p^2 + m^2 - \mathrm{i}\varepsilon} \frac{\exp(\mathrm{i}(x_2, p_3 + p_4 - p - p'))}{-(p')^2 + m^2 - \mathrm{i}\varepsilon},$$

and we obtain

$$g^2 \lim_{\varepsilon \to 0^+} \iint \frac{\mathrm{d}^4 p}{(2\pi)^4} \frac{\mathrm{d}^4 p'}{(2\pi)^4} \frac{1}{-p^2 + m^2 - \mathrm{i}\varepsilon} \frac{1}{-(p')^2 + m^2 - \mathrm{i}\varepsilon} \delta, \tag{13.104}$$

where

$$\delta := (2\pi)^8 \delta^{(4)}(p_3 + p_4 - p - p')\delta^{(4)}(p + p' - p_1 - p_2). \tag{13.105}$$

Here we see the difference with the tree case of the previous section. As a consequence of the fact that the diagram contains a loop, there are not enough delta functions to render the integrals over p and p' trivial. Still, the delta function allows to reduce the number of variables, since it forces the relation $p' = p_1 + p_2 - p$. There remains an integral in p. We have to worry about the convergence of this integral, and also about the existence of the limit as $\varepsilon \to 0^+$.

Before we pursue this, we stress what we have been doing. When trying to compute the value of diagrams containing loops, we are faced with ill-defined products of propagators. In order to proceed, we resort to the rather desperate device of writing these products as limits of multiple integrals as in (13.59).[62] Of course this is not rigorous, but we hope that we will eventually be able to extract numerical values out of our manipulations.

It should be apparent to write the expression (13.104) directly from diagram (a) in Figure 13.10. The method is encapsulated in a set of rules called *Feynman rules in momentum space*.[63] This is very useful when one must compute many such diagrams. (These rules are obtained using mathematically questionable manipulations such as in (13.103).)

- Orient each internal line in an arbitrary direction. To each internal line corresponds a variable p integrated over momentum space, and the corresponding factor $-\mathrm{i}/(-p^2 + m^2 - \mathrm{i}\varepsilon)$.
- To each internal vertex assign a factor $(-ig)(2\pi)^4\delta^{(4)}(\Sigma)$, where the argument Σ is an algebraic sum of the four-momenta corresponding to the lines going into that vertex,

62 In particular, we make the choice to use the same value of ε in each of the limits 13.59 for the different propagators, because in retrospect this is a successful strategy.

63 These rules depend on the model. Here they are stated for the ϕ^4 model, and similar rules hold for the ϕ^3 model which we study later. The name "rules in momentum space" is motivated by the fact that the main parameter is momentum p.

with a minus sign when the line is oriented toward the vertex and a plus sign when the line is oriented away from the vertex.[64]

- Multiply all these factors. Use the delta functions to eliminate as many of the p variables as possible and integrate over the remaining ones.
- Divide by the symmetry factor of the graph if needed.

These rules are fundamental and will be of constant use, so it essential that you understand them fully. Please note that we attach to an internal line with corresponding variable p the factor $-\mathrm{i}/(-p^2+m^2-\mathrm{i}\varepsilon)$ because we have been using the formula (13.60). If instead we had been using the formula (13.66), as we will do in Chapter 16, we would attach a factor $-\mathrm{i}/(-(1+\mathrm{i}\varepsilon)(p^0)^2+\boldsymbol{p}^2+m^2)$.

We now comment in more detail on these rules. Another way to express in words the content of the second bullet is to say that "we assign a factor ig at each internal vertex and we enforce the conservation of four-momentum at this vertex". We will at times use this convenient language. Let us now explain in more detail this second bullet. To compute the value of the diagram we number the internal vertices from 1 to n and we attach a variable x_i to the vertex labeled i, and the term $(2\pi)^4\delta^{(4)}(\Sigma)$ corresponding to the vertex labeled i is obtained as the integral $\int \mathrm{d}^4x_i \exp(\mathrm{i}(x_i,\Sigma))$. Here Σ is an algebraic sum of all the four-momenta on lines adjacent to the internal vertex labeled i. How do we get the signs right in that sum? For an external line, this is ensured by our conventions. For an internal line oriented from the vertex numbered i to the vertex numbered j, we express the propagator $\Delta_F(x_j-x_i)$ as

$$\Delta_F(x_j-x_i)=\lim_{\varepsilon\to 0^+}\int\frac{\mathrm{d}^4p}{(2\pi)^4}\frac{\exp(-\mathrm{i}(x_j-x_i,p))}{-p^2+m^2-\mathrm{i}\varepsilon},$$

whereas when the internal line is oriented from the vertex numbered j to the vertex numbered i we make the change of variable $p\to -p$ in the previous formula.

To compute the value of a term such as (13.96) at order k of perturbation theory, one "simply"[65] adds the values of all the Feynman diagrams with at most k internal vertices.

Let us now explain how conservation of four-momentum during an interaction is built in the Feynman rules. The quantity to integrate contains a factor $\delta(\Sigma_v)$ for each internal vertex v, where Σ_v is an algebraic sum of the four-momenta corresponding to the lines going into v, with a minus sign when the line is oriented toward the vertex and a plus sign when the line is oriented away from the vertex. Thus the product of the $\delta(\Sigma_v)$ can be non-zero only if each Σ_v is zero. Physically, four-momentum is conserved at every vertex, and the quantity of four-momentum exiting the diagram equals the quantity of four-momentum entering it. More formally the sum of the Σ_v over the internal vertices v equals $p_{\mathrm{out}}-p_{\mathrm{in}}$ where p_{out} is the sum of the four-momenta exiting through the final vertices and p_{in} the

[64] This point is not totally obvious and will be elaborated soon, but the basic reason the internal lines may be oriented any way we wish is simply that changing the orientation of such a line simply amounts to changing the corresponding momentum variable from p to $-p$.

[65] The number of diagrams to consider explodes when one goes to higher orders of perturbation theory. Some computations are currently undertaken which involve hundreds of thousands of diagrams. I cannot conceive which strategy one must adopt in order not to forget any diagram and not to make computational mistakes.

sum of the four-momenta entering through the initial vertices. This is pretty obvious since all the other terms cancel out, as they occur twice with different signs. Thus the value of the Feynman diagram will be zero unless $p_{\rm out} = p_{\rm in}$, which means that four-momentum is conserved in the interaction. This explains why when we compute the value of a Feynman diagram this value always contains a factor $(2\pi)^4\delta^{(4)}(p_{\rm out} - p_{\rm in})$. We will show later, in Theorem 15.7.1, that this is a general fact.

Returning to the quantity (13.104), we can use the delta function (13.105) to perform the integral in p', getting the result

$$g^2(2\pi)^4\delta^{(4)}(p_3 + p_4 - p_1 - p_2)$$
$$\times \lim_{\varepsilon\to 0^+} \int \frac{\mathrm{d}^4 p}{(2\pi)^4} \frac{1}{-p^2 + m^2 - \mathrm{i}\varepsilon} \frac{1}{-(p_1 + p_2 - p)^2 + m^2 - \mathrm{i}\varepsilon}. \tag{13.106}$$

Looking at the second line of this expression, there are two issues.

- As $\varepsilon \to 0$, the denominators get small for $p^2 \simeq m^2$ and $(p_1 + p_2 - p)^2 \simeq m^2$, and that may make the integral diverge.
- The integrand decays as $(p^2)^{-2}$ which is not fast enough to make the integral converge.

The integral is not finite because we forgot (again) to introduce an ultraviolet cutoff. (This divergence is the price to pay for considering an ill-defined product of distributions.) As it turns out, the first issue, while not trivial, will not be a major obstacle, so we ignore it for now. The second issue is far more serious, and we deal with it in the next four sections.

13.17 One Loop: A Simple Case of Renormalization

Generally speaking, renormalization is a procedure by which a finite value is assigned to scattering amplitudes that involve diagrams whose value is given by divergent integrals.[66] In this section we start the task of renormalizing the scattering amplitude (13.96) ϕ^4 theory, computed at order two in perturbation theory. The divergent diagrams have one single loop.

To tame a divergent integral such as (13.106) we will introduce a cutoff. Before we decide on it, we should know what we are going to make of our cutoff-dependent integral. So, we assume now that we have decided on a cutoff with parameter θ, which makes the integral finite, and we assume that moreover the limit as $\varepsilon \to 0^+$ exists. This limit then depends on θ and $p_1 + p_2$. We also have to consider the other relevant diagrams, which also involve divergent integrals, and we compute the contribution of each of them by the same method. In this manner we obtain the relation

$$\langle 0|a(p_3)a(p_4)S_2 a^\dagger(p_1)a^\dagger(p_2)|0\rangle = g^2(2\pi)^4\delta^{(4)}(p_3 + p_4 - p_1 - p_2)M, \tag{13.107}$$

where $M = M(p_1, p_2, p_3, p_4, \theta)$ is a quantity that depends on the cutoff θ and on the momenta of the incoming and outgoing particles. If we take (13.99) into account, we obtain that at order two in perturbation theory we have

66 That is, the objective is not necessarily to assign a value to each diagram, but only to certain sums of these diagrams.

$$\langle 0|a(p_3)a(p_4)Sa^\dagger(p_1)a^\dagger(p_2)|0\rangle = (2\pi)^4\delta^{(4)}(p_3+p_4-p_1-p_2)(-\mathrm{i}g+g^2M). \tag{13.108}$$

For readability let us rewrite this amplitude as

$$A(g+\mathrm{i}g^2M), \tag{13.109}$$

where

$$A := -\mathrm{i}(2\pi)^4\delta^{(4)}(p_3+p_4-p_1-p_2) \tag{13.110}$$

is a quantity completely determined by the values of p_1,p_2,p_3,p_4. (This quantity will be used until the end of this chapter, even though at times the letter A will also be used for other purposes.)

To interpret the preceding result, we have to remember a basic fact, which we already stressed: Writing an equation with a parameter does not mean that we can measure this parameter in the lab. The only "real" quantities are those which can be measured. The parameter g of our model quantifies the strength of the interaction between particles. But how do we *measure* the strength of this interaction? A sensible experiment is to try to scatter two particles and measure cross-sections. These cross-sections are likely to depend on p_1,p_2,p_3,p_4. A sensible choice (among others) is to perform this experiment at small velocity, $p_1,p_2,p_3,p_4 \simeq p^* := (m,\mathbf{0})$. Measuring cross-sections allows us to measure the modulus of the scattering amplitude, but let us pretend that we can measure the amplitude itself. Let us *define* the number g_{lab} by

At small velocity we measure the scattering amplitude Ag_{lab}.

Thus the quantity g_{lab} is determined by Nature and has nothing to do with our model. The subscript lab insists that it is a quantity we can measure in our lab. How does g_{lab} relate to the parameter g of our model? This model predicts (at order two of perturbation theory) a scattering amplitude at small velocity given by the formula (13.109), that is $A(g+\mathrm{i}g^2M^0)$ where

$$M^0 := M(p^*,p^*,p^*,p^*,\theta). \tag{13.111}$$

The definition of g_{lab} then implies that

$$g_{\mathrm{lab}} \simeq g+\mathrm{i}g^2M^0, \tag{13.112}$$

where the $\simeq$ is a reminder that the right-hand side is not an exact value, but has been computed at order two in perturbation theory. The quantity on the left-hand side is fixed by Nature. On the right-hand side the quantity M^0 depends on our cutoff. We then cannot escape the conclusion that, in order for our model to make sense, we must allow the coupling constant g to *depend on the cutoff*.

In order to get a physical prediction from our model, we must express the predicted amplitude (13.109) as a function of the known quantity g_{lab}, not of the unknown parameter g. The normal mathematical procedure for that would be to solve the equation (13.112) in g and plug this value of g in (13.109). This is however not what we are going to do, because it does not work. In particular if we took the equation (13.112) at face value, we would be let to believe that $g \to 0$ as the cutoff gets removed whereas it will turn out that $g \to \infty$.

Instead of thinking as mathematicians, we are going to learn to "work in perturbation theory" the way the physicists do, so please take a deep breath before you continue reading. We have obtained (13.112) within order two perturbation theory, that is we have ignored any term containing a higher power of g, so that we should rewrite (13.112) taking into account these terms as

$$g_{\mathrm{lab}} = g + \mathrm{i}g^2 M^0 + O(g^3), \tag{13.113}$$

where the symbol $O(g^3)$ takes into account "the contribution of all the terms containing a coefficient g^n for some $n \geq 3$". Generally speaking, $O(g^3)$ will denote any quantity which is a sum of terms containing a coefficient g^n for some $n \geq 3$,[67] *irrelevant* of the size of the term that multiplies g^n. The same symbol may denote different quantities at different occurrences. We have now to express our predicted amplitude $A(g + \mathrm{i}g^2 M + O(g^3))$ using g_{lab} rather than g. First, using (13.113), we obtain

$$g + \mathrm{i}g^2 M + O(g^3) = g_{\mathrm{lab}} + \mathrm{i}g^2(M - M^0) + O(g^3). \tag{13.114}$$

Next, squaring the relation (13.113) we obtain

$$g_{\mathrm{lab}}^2 = g^2 + O(g^3), \tag{13.115}$$

and replacing g^2 by $g_{\mathrm{lab}}^2 + O(g^3)$ in the right-hand side of (13.114) this gives

$$g + \mathrm{i}g^2 M + O(g^3) = g_{\mathrm{lab}} + \mathrm{i}g_{\mathrm{lab}}^2(M - M^0) + O(g^3). \tag{13.116}$$

The occurrence of the quantity $M - M^0$ is great news, because this difference of large quantities could remain finite as we remove the cutoff. In fact, our main finiteness problem with M came from an integrand which decayed as $(p^2)^{-2}$. The difference $M - M^0$ involves a difference of two such integrands, which has a faster rate of decay. *The divergence is less severe.* Assume now that indeed the difference $M - M^0$ remains finite as the cut-off is removed. Then according to the formula (13.116) the scattering amplitude (13.109) becomes

$$A(g_{\mathrm{lab}} + \mathrm{i}g_{\mathrm{lab}}^2(M - M^0)) + O(g^3). \tag{13.117}$$

Assuming that the terms $O(g^3)$ are indeed small, this formula contains a *prediction*[68] on how the cross-section depends on the four-momenta of the incoming and outgoing particles, and this prediction can be tested against experiment.

The symbol O has not been used the way a mathematician would use it.[69] The standard mathematical meaning of $O(g^3)$ is to denote a quantity $B(g)$ depending on g with the property that $|B(g)| \leq |g|^3 C$ where C does not depend on g, so that such a term $O(g^3)$ does

[67] This standard physics notation unfortunately conflicts with the usual meaning of this symbol in mathematics. This is explained a bit later.
[68] Once you have measured g_{lab} of course.
[69] The way the symbol O is used in physics textbooks can be an endless source of wonder for a mathematician. A particularly striking instance is the way these books "justify" path integrals.

indeed become small at a controlled speed as $g \to 0$. Here, by contrast, we agglomerate in $O(g^3)$ *any* term of the type g^3C *however large is the quantity* C. The intended mathematical meaning of a relation such as (13.115) is at the very least obscure. Furthermore, one really sees no reason why the term $O(g^3)$ should be small in (13.117), in particular since as we will see later $g \to \infty$ as the cutoff gets removed.

Let us summarize. On the positive side, we have obtained a physical prediction about scattering amplitudes. On the negative side, calling a spade a spade, *the mathematical value of our procedure is strictly nil.*[70] Let us elaborate on this point. In the next section we will use perfectly rigorous mathematics to prove that as the cutoff gets removed, that quantity $M - M^0$ has a finite limit. Later, in Part IV, we will again use perfectly sound mathematics to assign a finite value to each Feynman diagram, a procedure known as renormalization. What is well outside mathematics is the argument that this procedure should produce the correct prediction for the experimentally measured values.

Many predictions of Quantum Field Theory were originally obtained by methods that are technically more elaborate but are in essence similar to the way we obtained (13.117), with equally nil mathematical value.[71] The astounding fact is that these predictions are experimentally verified, sometimes with fantastic accuracy. One should of course wonder how and why such a miracle takes place.[72]

In the next section we develop the tools to prove that the difference $M - M_0$ in (13.117) remains finite as the cutoff gets removed. In Section 13.20 we explain the traditional description of the procedure leading to (13.117) using "counter-terms". In the last four sections we extend the procedure of the present section to the far more challenging situation of two loops.

13.18 Wick Rotation and Feynman Parameters

The program of this section is to show how to make sense of the difference $M - M^0$ in (13.117) as the cutoff gets removed. The quantity M is the sum of different terms, arising from different diagrams. We focus on the term arising from the integral (13.106). (At this stage we worry only about the integral itself. We will examine later what to do about the various powers of i.)

Let us stress that, as mathematicians, we want to understand how things work and why what we are doing makes sense. In the present case we will find explicit formulas, but actual computations will not be our emphasis, as these are available in any physics textbook.

Up to this point our only serious encounter with infinities was the Casimir effect of Chapter 7, and we got away with flying colors, a clean argument showing that the final

70 This is typically not stated clearly in physics textbooks. Rather the whole issue is often downplayed to a quite remarkable extent, see e.g. Lancanster and Blumdell's book [50]. Whereas the procedure predicts reality very accurately, it does not achieve anything to pretend that our manipulations are innocuous.

71 That is, one finds ways to eliminate certain infinite terms (in the form of divergent integrals which cancel each other) from expressions involving quantities that can actually be measured, while ignoring whole series of even larger terms.

72 I wish I could explain it to you. I cannot, but I am unsure as to whether anybody really could. A more recent method, "the renormalization group approach", is probably less unsatisfactory, although my understanding is that it does not really solve the problem either. We refer to Lancanster and Blumdell's book [50] for this topic.

answer was independent of the regularization procedure. It would certainly be desirable to achieve the same result in all cases where regularization is used. Unfortunately, this is a difficult program. For example, a physically satisfying cutoff would preserve Lorentz invariance. But such a cutoff does not make the integral absolutely convergent, basically because the volume of the set of p with p^2 bounded is infinite. And if one works with integrals that are not absolutely convergent, it gets much harder to rigorously justify anything. This program is also of limited interest. Once one has found a numerical value by a regularization method, if this value matches experimental results, the proof that another regularization method would yield the same value is unlikely to attract much attention. Several physics textbooks claim that the results obtained while renormalizing diagrams are independent of the renormalization method, but they give neither arguments nor references to original papers.[73]

In any case, since the method we will eventually develop for renormalization does not use a cutoff at all, our overwhelming desire when considering cutoffs will be to avoid technicalities, at the expense of being somewhat brutish. To make the integral (13.106) converge, we simply introduce a factor $\theta(\boldsymbol{p})$ in the integrand, where $0 \le \theta(\boldsymbol{p}) \le 1$ is a function that takes the value 1 unless $\boldsymbol{p}$ is very large and that goes to 0 fast enough at infinity.[74]

We will call the function θ a *cutoff*. The physical results are obtained when we *remove the cutoff*. That is, we assume that for some number $R > 0$

$$\boldsymbol{p}^2 \le R^2 \Rightarrow \theta(\boldsymbol{p}) = 1$$

and we show that the limits of our θ-dependent quantities exist as $R \to \infty$. Let us simplify notation by setting $w = p_1 + p_2$, so our goal is to study the quantity

$$U(w,\theta) := \lim_{\varepsilon \to 0^+} \int \frac{\theta(\boldsymbol{p})\mathrm{d}^4 p}{(2\pi)^4} \frac{1}{-p^2 + m^2 - i\varepsilon} \frac{1}{-(w-p)^2 + m^2 - i\varepsilon}, \tag{13.118}$$

The integral is now well-defined and absolutely convergent. Our goal is to prove the following:

- As $\varepsilon \to 0^+$ (and provided the cutoff is "far enough away") the limit exists in (13.118).
- Given w, w', as the cutoff gets removed, the quantity $U(w,\theta) - U(w',\theta)$ has a finite limit.

Even though we cannot make much sense of the quantity (13.106), the *difference* of two such quantities for different values of $w = p_1 + p_2$ is well defined. For as simple as possible an example of a similar situation, for $a > 0$ the integrals

$$I(a) = \int_{-\infty}^{\infty} \mathrm{d}x \frac{1}{a + |x|}$$

do not make sense. However, the quantity $I(a) - I(b)$ is well defined.

[73] A common method in physics books to "show that renormalization results are independent of the renormalization method" consists of showing that two different specific methods yield the same results.

[74] As already pointed out, such a non-Lorentz invariant cutoff is unphysical.

The reader might now like to review contour integrals in Appendix N, as we shall use this to perform the integrals in p^0.

Before we start describing the physicists' magical methods, it is a good idea to convince yourself that even the existence (not to even mention the computation) of the limit in (13.118) is not trivial. As $\varepsilon \to 0^+$, singularities develop at the points where either $-m^2 + p^2 = 0$ or $-m^2 + (w - p)^2 = 0$, or, worse of all, at the points where *both* these quantities are zero. The most obvious approach to proving the limit in (13.118) would be to compute the integral in p^0 at given $\boldsymbol{p}$ and then to use dominated convergence. The purpose of the next exercise is to show that this does not work.[75]

Exercise 13.18.1 Consider the function $f_{\boldsymbol{p},\varepsilon}(z)$ of one complex variable given by

$$f_{\boldsymbol{p},\varepsilon}(z) = \frac{1}{-z^2 + \boldsymbol{p}^2 + m^2 - i\varepsilon} \frac{1}{-(z - w^0)^2 + (\boldsymbol{w} - \boldsymbol{p})^2 + m^2 - i\varepsilon}. \tag{13.119}$$

(a) Prove that for $(\sqrt{\boldsymbol{p}^2 + m^2} - w^0)^2 \neq (\boldsymbol{w} - \boldsymbol{p})^2 + m^2$ and $(w_0 + \sqrt{(\boldsymbol{w} - \boldsymbol{p})^2 + m^2})^2 \neq \boldsymbol{p}^2 + m^2$ the limit

$$A(\boldsymbol{p}) = \lim_{\varepsilon \to 0^+} \int \frac{\mathrm{d}p^0}{2\pi} f_{\boldsymbol{p},\varepsilon}(p^0)$$

exists whereas this limit does not exist if $(\sqrt{\boldsymbol{p}^2 + m^2} - w^0)^2 = (\boldsymbol{w} - \boldsymbol{p})^2 + m^2$ or $(w_0 + \sqrt{(\boldsymbol{w} - \boldsymbol{p})^2 + m^2})^2 = \boldsymbol{p}^2 + m^2$.

(b) Prove however that for R large enough one has

$$\int_{\boldsymbol{p}^2 \leq R^2} \frac{\mathrm{d}^3 \boldsymbol{p}}{(2\pi)^3} |A(\boldsymbol{p})| = \infty. \tag{13.120}$$

Hint: For (a) compute the limit using contour integrals. For (b) show that for α large, the volume of the set of $\boldsymbol{p}$ for which $|A(\boldsymbol{p})| \geq \alpha$ is of order $1/\alpha$. Think before you compute.

An obvious reason why it is tricky to compute a limit such as $\lim_{\varepsilon \to 0^+} \int \mathrm{d}x f_{\boldsymbol{p},\varepsilon}(x)$ where $f_{\boldsymbol{p},\varepsilon}$ is given in (13.119) is that the poles of this function get close to the contour of integration (here the real line) as $\varepsilon \to 0^+$. If we replace the contour of integration by another contour which stays away from these poles as $\varepsilon \to 0^+$, then it is obvious that there will be no problem in taking the limit $\varepsilon \to 0^+$. We should try to choose the new contour as simply as we can. A prime candidate as a simple contour is the imaginary axis. The technique to do this is known as Wick rotation and is useful enough to be stated in its own right.

Proposition 13.18.2 (Wick rotation) *If the function $f(z) = f(x+iy)$ is holomorphic in an open domain containing the quadrants $x, y \geq 0$ and $x, y \leq 0$ and decays fast enough at infinity in these quadrants ($|f(z)| \leq |z|^{-1-\alpha}$ for some $\alpha > 0$ and all $|z|$ large enough suffices), we have*

[75] There are further cancellations which are not immediately apparent.

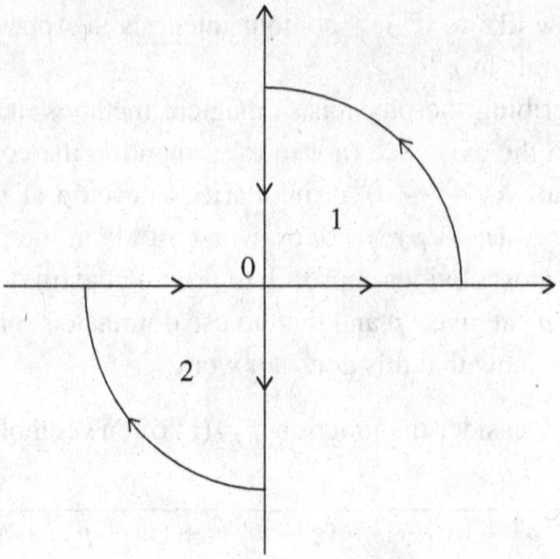

Figure 13.13 Wick rotation.

$$\int \mathrm{d}x f(x) = \mathrm{i} \int \mathrm{d}x f(\mathrm{i}x). \tag{13.121}$$

Proof We write that the integral of f on the contours 1 and 2 of Figure 13.13 is zero and we let the radius of these contours go to infinity. □

At this point we should stress that the formula $\hat{\Delta}_F(p) = \lim_{\varepsilon\to 0} 1/(-p^2 + m^2 - \mathrm{i}\varepsilon)$ is terribly misleading. It basically looks like this quantity is a real number, since the imaginary part has this small coefficient ε. This however is not true at all. Indeed, using the formula (13.121) for the function $f(z) = 1/(-z^2 + \boldsymbol{p}^2 + m^2 - \mathrm{i}\varepsilon)$ shows that the quantity

$$\lim_{\varepsilon\to 0^+} \int \frac{\mathrm{d}p^0}{-p^2 + m^2 - \mathrm{i}\varepsilon} = \mathrm{i} \int \frac{\mathrm{d}p^0}{(p^0)^2 + \boldsymbol{p}^2 + m^2}$$

is purely imaginary.

Recalling (13.119), one may not use Wick rotation to study the integral $\int \mathrm{d}x f_{\boldsymbol{p},\varepsilon}(x)$ unless $|w^0| < \sqrt{(\boldsymbol{p} - \boldsymbol{w})^2 + m^2}$ because then otherwise $f_{\boldsymbol{p},\varepsilon}$ has poles in the wrong place. Richard Feynman has however found a magical way to transform the integral in (13.118) to bypass this difficulty. It relies on the following identity. If two complex numbers A, B are such that $uA + (1-u)B \neq 0$ for $0 \le u \le 1$ then

$$\frac{1}{AB} = \int_0^1 \frac{\mathrm{d}u}{(Au + B(1-u))^2}. \tag{13.122}$$

This follows from the fact that the integrand is the derivative of $1/((B-A)(Au+B(1-u)))$. The number u is called a *Feynman parameter*.

We use (13.122) for

$$A = -(w-p)^2 + m^2 - \mathrm{i}\varepsilon \; ; \; B = -p^2 + m^2 - \mathrm{i}\varepsilon$$

so that we have

$$Au + B(1-u) = -(p-uw)^2 - u(1-u)w^2 + m^2 - \mathrm{i}\varepsilon.$$

Making the change of variables $p \to p + uw$ (13.118) becomes

$$U(w,\theta) = \lim_{\varepsilon\to 0^+} \int_0^1 \mathrm{d}u \int \frac{\theta(\boldsymbol{p}+u\boldsymbol{w})\mathrm{d}^4 p}{(2\pi)^4} \frac{1}{(-p^2 - u(1-u)w^2 + m^2 - \mathrm{i}\varepsilon)^2}. \quad (13.123)$$

Here we see the magic of the transformation. We have an extra parameter u, but given u the integral in p is *much simpler*. In particular we may now use Wick rotation (13.121) on p^0. Using the notation

$$\|p\|^2 := (p^0)^2 + \boldsymbol{p}^2$$

for the Euclidean norm of a four-vector, after Wick rotation (13.123) becomes

$$U(w,\theta) = \mathrm{i} \lim_{\varepsilon\to 0^+} \int_0^1 \mathrm{d}u \int \frac{\theta(\boldsymbol{p}+u\boldsymbol{w})\mathrm{d}^4 p}{(2\pi)^4} \frac{1}{(\|p\|^2 - u(1-u)w^2 + m^2 - \mathrm{i}\varepsilon)^2}. \quad (13.124)$$

When $\|p\|$ is large (say $\|p\|^2 \ge w^2$) the denominator cannot be small and there are no problems with the limit $\varepsilon \to 0^+$. The unfortunate fact is that it can be shown that[76] $w^2 > 4m^2$. There are then values of u for which $-u(1-u)w^2 + m^2 < 0$ and singularities occur for certain values of p as $\varepsilon \to 0^+$. We should stay cool at this stage.

Exercise 13.18.3 (a) Prove that for $\varepsilon > 0$ one has

$$\int_{\infty}^{\infty} \frac{\mathrm{d}x}{(x - \mathrm{i}\varepsilon)^2} = 0.$$

(b) Does it make sense to set $\varepsilon = 0$ in the previous identity?
(c) Does it make sense to replace ε by 0 in the integral in (13.124)?

The next goal is to show that when "the cutoff is far enough away", the limit $\varepsilon \to 0^+$ exists in (13.124). For this we will use the following explicit formula.[77]

Lemma 13.18.4 *Consider a complex number C, of the type $C = v - \mathrm{i}\varepsilon$ where $v \in \mathbb{R}$ and $\varepsilon > 0$. Then*

$$\int_{\|p\|\le R} \frac{\mathrm{d}^4 p}{(2\pi)^4} \frac{1}{(\|p\|^2 + C)^2} = \frac{1}{16\pi^2}\Big(\log(R^2 + C) - \log C - \frac{R^2}{R^2 + C}\Big). \quad (13.125)$$

In this lemma we use the function "log", the logarithm of a complex number. This function logarithm is defined for non-zero complex numbers but it is "multivalued". At present we need only the quantity $\log(v - \mathrm{i}\varepsilon)$ for $v \in \mathbb{R}$. For v large, there is no ambiguity

[76] Indeed, recalling that $w = p_1 + p_2$, to compute the Lorentz invariant quantity w^2 we may as well assume that we are in the "center of mass" reference frame, $\boldsymbol{p}_1 + \boldsymbol{p}_2 = 0$, in which case it is apparent that $w^2 = (p_1^0 + p_2^0)^2 \ge (2m)^2 = 4m^2$, with strict inequality unless the situation is trivial.

[77] Do not think however that the specific computation here has any importance whatsoever in the existence of the limit. Later we will obtain such limits in a general setting just by moving to polar coordinates and integrating by parts.

about what this means, and for other values of v we simply "follow the branch" as we decrease v. In particular for $v < 0$, we have $\lim_{\varepsilon \to 0^+} \log(v - i\varepsilon) = \log|v| - i\pi$.

Proof As a first step we use polar coordinates to obtain

$$\int_{\|p\| \le R} \frac{d^4 p}{(\|p\|^2 + C)^2} = 2\pi^2 \int_0^R \frac{dr r^3}{(r^2 + C)^2}, \tag{13.126}$$

where the factor $2\pi^2$ is just the surface area of the sphere in $\mathbb{R}^4$. We then write

$$\frac{r^3}{(r^2 + C)^2} = \frac{r}{r^2 + C} - \frac{rC}{(r^2 + C)^2},$$

and we integrate to obtain the result. □

Observe that when $v < 0$ it is not possible to pass to the limit $\varepsilon \to 0^+$ inside the integral (13.125) because the resulting integral[78] is $+\infty$.

Given a number B and $w \in \mathbb{R}^{1,3}$ we define the following number:

$$C(B, w) = \sup\{(\boldsymbol{p} + u\boldsymbol{w})^2 \; ; \; \boldsymbol{p}^2 < B \; ; \; 0 < u < 1\}.$$

Let us also consider the following function

$$W(w) = \frac{i}{16\pi^2} \int_0^1 du \log(m^2 - u(1-u)w^2). \tag{13.127}$$

It is understood here that for $m^2 - u(1-u)w^2 < 0$ the value of the logarithm is obtained "by following the correct branch", i.e.

$$\log(m^2 - u(1-u)w^2) = \log|m^2 - u(1-u)w^2| - i\pi.$$

Proposition 13.18.5 *(a) Given w and a cutoff θ such that*

$$\boldsymbol{p}^2 \le C(w^2, w) \Rightarrow \theta(\boldsymbol{p}) = 1, \tag{13.128}$$

the limit $U(w,\theta)$ exists in (13.118) as $\varepsilon \to 0^+$.

(b) Given w and w', as the cutoff gets removed (in the sense defined just above (13.118)), the difference $U(w,\theta) - U(w',\theta)$ has the limit $W(w) - W(w')$.

Thus, the precise meaning of (b) is that given w, w' and $\varepsilon_1 > 0$ there exists $R > 0$ such that $|U(w,\theta) - U(w',\theta) - (W(w) - W(w'))| < \varepsilon_1$ whenever the cutoff θ satisfies $\theta(\boldsymbol{p}) = 1$ for $\boldsymbol{p}^2 < R^2$.

Condition (13.128) is very mild and gets satisfied when the cutoff is removed.

In broader terms, as the cutoff gets removed we have $U(w,\theta) \simeq U(0,\theta) + W(w) - W(0)$ and also $U(0,\theta) \to \infty$. In flowery language we could say that the whole family of diverging integrals depends only on a single infinite parameter. This parameter cancels out of each expression that has a physical content.

[78] In a number of respected textbooks, the present calculation is precisely handled by first setting $\varepsilon = 0$ inside the integral (13.124) and then using the formal relation (13.125) when $C < 0$. It is all right in the end but neither of these procedures makes mathematical sense. Having been trained from kindergarten to avoid risky manipulations I cannot explain how physicists succeed in routinely using (and teaching!) such methods without falling into utter nonsense.

Proof of (a) of Proposition 13.18.5 Let us fix a number $R > 0$ with $R^2 \geq w^2$. Assuming

$$p^2 \leq C(R^2, w) \Rightarrow \theta(p) = 1, \tag{13.129}$$

for $\|p\| \leq R$ we have $(p+uw)^2 \leq C(R^2,w)$ and $\theta(p+uw) = 1$ so the integral in (13.124) is the sum of the following two pieces:

$$\int_0^1 \mathrm{d}u \int_{\|p\|\leq R} \frac{\mathrm{d}^4 p}{(2\pi)^4} \frac{1}{(\|p\|^2 - u(1-u)w^2 + m^2 - \mathrm{i}\varepsilon)^2}, \tag{13.130}$$

$$\int_0^1 \mathrm{d}u \int_{\|p\|\geq R} \frac{\theta(p+uw)\mathrm{d}^4 p}{(2\pi)^4} \frac{1}{(\|p\|^2 - u(1-u)w^2 + m^2 - \mathrm{i}\varepsilon)^2}. \tag{13.131}$$

Lemma 13.18.4 shows that as $\varepsilon \to 0^+$ the first piece has the limit

$$U_1(w,R) := \frac{1}{16\pi^2} \int_0^1 \mathrm{d}u \Big(\log(R^2 - u(1-u)w^2 + m^2) - \log(-u(1-u)w^2 + m^2) + \frac{R^2}{R^2 - u(1-u)w^2 + m^2} \Big). \tag{13.132}$$

Since we assume that $R^2 \geq w^2$, on the domain of integration of (13.131) we have $\|p\|^2 - u(1-u)w^2 + m^2 \geq R^2 - w^2 + m^2 \geq m^2$, so that as $\varepsilon \to 0^+$ the quantity (13.131) has the limit

$$U_2(w,R,\theta) := \int_0^1 \mathrm{d}u \int_{\|p\|\geq R} \frac{\theta(p+uw)\mathrm{d}^4 p}{(2\pi)^4} \frac{1}{(\|p\|^2 - u(1-u)w^2 + m^2)^2}. \tag{13.133}$$

This proves the existence of the limit in (13.118) and the formula

$$U(w,\theta) = \mathrm{i}U_1(w,R) + \mathrm{i}U_2(w,R,\theta). \tag{13.134}$$

Taking $R^2 = w^2$, we have proved (a) since in that case (13.129) reduces to (13.128). □

The main idea to Prove Part (b) is to use (13.134) together with the fact that for another number w' it follows from (13.132) that

$$\lim_{R\to\infty} \mathrm{i}(U_1(w,R) - U_1(w',R)) = W(w) - W(w'). \tag{13.135}$$

One then shows that the terms $U_2(w,R,\theta)$ have a vanishing importance as $R \to \infty$. As this technical argument is not essential to follow the main story, it is better left as an exercise.

Exercise 13.18.6 Complete the proof of part (b) of Proposition 13.18.5.

13.19 Explicit Formulas

We have completed the proof that (13.117) does lead to a physical prediction (assuming that the term $O(g^3)$ is small). It is another matter to check that the prediction makes real sense, e.g. is Lorentz invariant, but this is going to be obvious soon. Our next task is to be

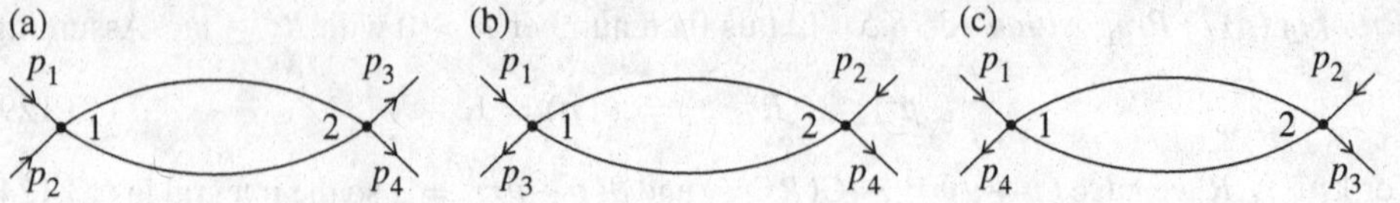

Figure 13.14 Another way to present the three channels of Figure 13.10.

more explicit about the value of M, to recover the physicist's formulas. We recall that we considered a (suitable) cutoff θ and we defined

$$U(w,\theta) = \lim_{\varepsilon \to 0^+} \int \frac{\theta(\boldsymbol{p})\mathrm{d}^4 p}{(2\pi)^4} \frac{1}{-p^2 + m^2 - \mathrm{i}\varepsilon} \frac{1}{-(w-p)^2 + m^2 - \mathrm{i}\varepsilon}. \tag{13.136}$$

This is the integral occurring in diagram (a) of Figure 13.10. Recalling the quantity A of (13.110) the value of the diagram itself is

$$(-\mathrm{i}g)^2(-\mathrm{i})^2(2\pi)^4\delta^{(4)}(p_1 + p_2 - p_3 - p_4)U(w,\theta) = A\mathrm{i}g^2 U(w,\theta).$$

If we draw the diagrams of Figure 13.10 as in Figure 13.14 it is plain that they "are the same" but that $w = p_1 + p_2$ has to be replaced respectively by $p_1 - p_3$ and $p_1 - p_4$.

Thus we have

$$M_2 := M\big(p_1, p_2, p_3, p_4, \theta\big) = \frac{1}{2}\big(U(p_1 + p_2, \theta) + U(p_1 - p_3, \theta) + U(p_1 - p_4, \theta)\big). \tag{13.137}$$

Here we introduce the subscript 2 to the quantity M to indicate that we have taken into account all the contributions of the diagrams with two vertices. The factor $1/2$ occurs because the symmetry factor of the diagrams of Figure 13.14 is 2. As the function $W(w)$ of (13.127) depends only on w^2, there exists a function V such that $W(w) = V(w^2)$. The traditional notation when studying scattering two particles into two particles is to use the variables[79]

$$s := (p_1 + p_2)^2 \ ; \ t := (p_1 - p_3)^2 \ ; \ u := (p_1 - p_4)^2. \tag{13.138}$$

In the specific situation

$$p_1 = p_2 = p_3 = p_4 = p^* = (m,0,0,0), \tag{13.139}$$

we have $s = 4m^2, t = u = 0$. As the cutoff is removed, using (b) of Proposition 13.18.5 our prediction for the scattering amplitude becomes

$$A\Big(g_{\mathrm{lab}} + \frac{1}{2}\mathrm{i}g_{\mathrm{lab}}^2\big(V(s) + V(t) + V(u) - V(4m^2) - 2V(0)\big)\Big). \tag{13.140}$$

Our prediction 13.117 has used the special choice (13.139) and the corresponding value M^0 given by (13.111). Surely the choice (13.139) is reasonable, but other choices are possible. We argue now that all these choices *lead to the same prediction of the scattering amplitude.*

[79] s, t, u are called the Mandelstam variables in physics.

Rather than the specific point (13.139) let us use another arbitrary point, and denote by a "prime" the corresponding quantities, so that (13.113) is replaced by

$$g'_{\text{lab}} = g + \mathrm{i}g^2 M'^0 + O(g^3), \tag{13.141}$$

and the prediction (13.117) for the scattering amplitude is replaced by

$$A(g'_{\text{lab}} + \mathrm{i}g'^2_{\text{lab}}(M - M'^0) + O(g^3)). \tag{13.142}$$

Squaring the relation (13.141) we obtain that $g'^2_{\text{lab}} = g^2$ at order $O(g^3)$. Comparing (13.113) with (13.141) we obtain at this order

$$g'_{\text{lab}} = g_{\text{lab}} + \mathrm{i}g^2(M'^0 - M^0).$$

Plugging this into (13.142) and using that $g'^2_{\text{lab}} = g^2 = g^2_{\text{lab}}$ at order $O(g^3)$ recovers (13.117), showing that the two predictions are identical at the order of error involved.[80]

The incoming and outgoing particles satisfy $p_1^2 = p_2^2 = p_3^2 = p_4^2 = m^2$. This is however never used, and it is fruitful in the previous considerations to think of p_1, p_2, p_3, p_4 as any points of $\mathbb{R}^{1,3}$. Having argued that the specific point used to perform the renormalization does not matter, rather than (13.139) it is tempting to use the (unphysical but more convenient) choice $p_1 = p_2 = p_3 = p_4 = 0$, to which correspond the more symmetrical values $s = t = u = 0$. This idea will be much used later.

13.20 Counter-terms, I

The gist of our method toward renormalization has been to write the predicted scattering amplitude as a function of the parameter g_{lab} in (13.117). As we will experience in the next section, this method becomes cumbersome at higher orders of perturbation theory, and physicists have developed another way to perform this procedure. This is the so-called counter-terms method. It is more efficient, but it is also very mysterious-looking at the beginning. In the present section we provide a first introduction to this method.

Let us look back at the computations of Section 13.17. Our main parameter g has to depend on the cutoff, and cannot be physically measured. It therefore seems appropriate to make a change of variables, and to take g_{lab} as the main parameter. We then write

$$g = g_{\text{lab}} + D, \tag{13.143}$$

where D depends on the cutoff. Let use this notation to redo the computations of Section 13.17, to introduce in its simplest version a scheme of computation which will be used again and again. We now write (13.109) as

$$A(g_{\text{lab}} + D + \mathrm{i}(g_{\text{lab}} + D)^2 M) + O((g_{\text{lab}} + D)^3), \tag{13.144}$$

whereas (13.113) becomes

$$g_{\text{lab}} = g_{\text{lab}} + D + \mathrm{i}(g_{\text{lab}} + D)^2 M^0 + O((g_{\text{lab}} + D)^3),$$

[80] The whole argument is based on the *fairy tale* that it makes sense to compute quantities at order $O(g^3)$. Unfortunately, if one rejects the fairy tale, there is little to say about the whole matter.

i.e.

$$D = -\mathrm{i}(g_{\mathrm{lab}} + D)^2 M^0 + O((g_{\mathrm{lab}} + D)^3). \tag{13.145}$$

The fruitful way to use this relation is not in the direction of standard mathematical thinking. We do not make estimates on the size of the terms, or such sensible approaches. Rather, we pretend that D is *a formal series in powers of* g_{lab}, with cutoff-dependent coefficients, $D = \sum_{n\geq 1} D_n g_{\mathrm{lab}}^n$.[81] We substitute this expression in (13.145). Identifying the coefficients of g_{lab} in the right and left-hand side gives $D_1 = 0$, and then $D_2 = -\mathrm{i}M^0$ so that $D = -\mathrm{i}g_{\mathrm{lab}}^2 M^0 + O(g_{\mathrm{lab}}^3)$.

We should stress how far we are from rigorous arguments here.[82] The relation $D = -\mathrm{i}g_{\mathrm{lab}}^2 M^0 + O(g_{\mathrm{lab}}^3)$ leads us to believe that as the cutoff gets removed we have $D \to \infty$, so that $g = g_{\mathrm{lab}} + D \to \infty$ and

(13.144) is an expansion in powers of a quantity that goes to infinity!

Going back to (13.144), the prediction for the scattering amplitude is

$$A(g_{\mathrm{lab}} + \mathrm{i}g_{\mathrm{lab}}^2(M - M^0)) + O(g_{\mathrm{lab}}^3). \tag{13.146}$$

In parallel with our decomposition $g = g_{\mathrm{lab}} + D$ it will be fruitful to write the interaction term in the Hamiltonian as

$$\frac{g}{4!} :\varphi^4: = \frac{g_{\mathrm{lab}}}{4!} :\varphi^4: + \frac{D}{4!} :\varphi^4: , \tag{13.147}$$

and to think of the first term on the right as the correct interaction term (which produces the measured value g_{lab} at small velocity) and of the other term as a *counter-term*. The standard picture (to which we will gradually lead the reader) is that

Counter-terms can be adjusted to cancel the divergences of the diagrams.

Later we will learn which version of Feynman diagrams should be used for Hamiltonians containing counter-terms as in (13.147). At the present stage it should be rather obvious that the term $D :\varphi^4: /4!$ simply creates a term AD in the amplitude (see (13.99)), which therefore becomes

$$A(g_{\mathrm{lab}} + \mathrm{i}g_{\mathrm{lab}}^2 M + D).$$

Since $D = -\mathrm{i}g_{\mathrm{lab}}^2 M^0 + O(g_{\mathrm{lab}}^3)$, this recovers the formula (13.146). We see here the effect of the counter-term. It replaces the divergent term M by the convergent term $M - M^0$.

It is a general feature of the method of counter-terms that the coefficient D has an expansion in powers of g_{lab}, which is computed term by term. The next term of this expansion will be computed in Section 13.24, and the general method is explained in Chapter 17.

13.21 Two Loops: Toward the Central Issues

The renormalization of diagrams containing one single loop already yields spectacular results in Quantum Electrodynamics (as can be found in any book on this topic). The main

[81] Doing so, we pretend of course that g_{lab} is a variable, not a number.
[82] It is really unfortunate that certain physics textbooks make nonsensical claims about the legitimacy of this approach, see e.g. Schwartz's book [80].

textbooks consider the study of diagrams containing two loops as an advanced topic. The purpose of this section and the next two is to extend the analysis of Section 13.17 to the computation of scattering at order three in perturbation theory, still in ϕ^4 theory. This involves many different two-loops diagrams. The possibility of renormalization in this case already hinges on very non-trivial issues. These issues will be addressed in a systematic manner in Chapters 15 and 16.

The difference between order two and order three of perturbation theory is that we now have to include in our computation the contribution $\langle 0|a(p_3)a(p_4)S_3a^\dagger(p_1)a^\dagger(p_2)|0\rangle$. Recalling the quantity $A = -\mathrm{i}(2\pi)^4\delta^{(4)}(p_3 + p_4 - p_1 - p_2)$ of (13.110) let us denote this contribution by

$$-Ag^3M_3, \tag{13.148}$$

where $M_3 = M_3(p_1, p_2, p_3, p_4, \theta)$ depends on the cutoff and the momenta of the incoming and outgoing particles. Thus the relation (13.148) *defines* M_3. The minus sign is conventional, and the reason behind it will appear when we start making actual computations.

At order $O(g^4)$ the scattering amplitude is then

$$A(g + \mathrm{i}g^2M_2 - g^3M_3) + O(g^4), \tag{13.149}$$

where M_2 is the very same as the quantity M in (13.109). In particular

$$g_{\mathrm{lab}} = g + \mathrm{i}g^2M_2^0 - g^3M_3^0 + O(g^4). \tag{13.150}$$

We have already explained that the choice of the non-measurable quantity g as parameter is not clever, so that we use again the change of variable $g = g_{\mathrm{lab}} + D$, where D depends on the cutoff, and all our computations are at the order g_{lab}^3. Then (13.150) implies the equality

$$g_{\mathrm{lab}} = g_{\mathrm{lab}} + D + \mathrm{i}(g_{\mathrm{lab}} + D)^2M_2^0 - (g_{\mathrm{lab}} + D)^3M_3^0 + O((g_{\mathrm{lab}} + D)^4),$$

i.e.

$$D = -\mathrm{i}(g_{\mathrm{lab}} + D)^2M_2^0 + (g_{\mathrm{lab}} + D)^3M_3^0 + O((g_{\mathrm{lab}} + D)^4). \tag{13.151}$$

To use this relation, we proceed as in Section 13.20: We try to find[83] D as a power series in g_{lab}, with cut-off dependent coefficients: $D = D_1g_{\mathrm{lab}} + D_2g_{\mathrm{lab}}^2 + D_3g_{\mathrm{lab}}^3 + O(g_{\mathrm{lab}}^4)$. We substitute this relation into (13.151) and we identify in turn the coefficients of g_{lab}^k for $k = 1, 2, 3$ in the left-hand and the right-hand sides. Since there is no term in g_{lab} in the right-hand side, we obtain $D_1 = 0$. Thus the term in g_{lab}^2 in the right-hand side is $-\mathrm{i}g_{\mathrm{lab}}^2M_2^0$ so that $D_2 = -\mathrm{i}M_2^0$. Then the term in g_{lab}^3 in the right hand side is $g_{\mathrm{lab}}^3(M_3^0 - 2(M_2^0)^2)$, and we have shown that

$$D = -\mathrm{i}g_{\mathrm{lab}}^2M_2^0 + g_{\mathrm{lab}}^3(M_3^0 - 2(M_2^0)^2) + O(g_{\mathrm{lab}}^4).$$

Going back to (13.149), replacing there g by its value $g_{\mathrm{lab}} + D = g_{\mathrm{lab}} - \mathrm{i}g_{\mathrm{lab}}^2M_2^0 + g_{\mathrm{lab}}^3(M_3^0 - 2(M_2^0)^2) + O(g_{\mathrm{lab}}^4)$ and collecting again all terms that contain a term g_{lab}^n for $n \leq 3$, a straightforward computation shows that the prediction for the scattering amplitude is now

$$A\Big(g_{\mathrm{lab}} + \mathrm{i}g_{\mathrm{lab}}^2(M_2 - M_2^0) + g_{\mathrm{lab}}^3\big(2M_2^0(M_2 - M_2^0) - M_3 + M_3^0\big)\Big) + O(g_{\mathrm{lab}}^4). \tag{13.152}$$

83 Once again, this is probably not the first thing which a mathematician would try, but this is what works.

In Section 13.18 we proved that as $\varepsilon \to 0^+$ and then the cutoff is removed, the quantity $M_2 - M_2^0$ has a finite limit, giving us a physical prediction. In order to get a physical prediction at the present third order of perturbation theory, it must be true that

$$\text{As } \varepsilon \to 0^+ \text{ and then the cutoff gets removed, the quantity}$$
$$2M_2^0(M_2 - M_2^0) - M_3 + M_3^0 \text{ has a finite limit.} \tag{13.153}$$

This fact involves a *truly remarkable* "cancellation of infinities" between the various terms. We will prove this in the next three sections. There is no doubt that this is a significant undertaking. Our motivation is to drive home the point that the possibility of renormalization is a non-trivial and near-miraculous fact. Our general study of renormalization will however follow another path, so that the material of Sections 13.23 and 13.24 is not an absolute prerequisite for the sequel.

13.22 Analysis of Diagrams

We turn to the computation of $\langle 0|a(p_3)a(p_4)S_3a^\dagger(p_1)a^\dagger(p_2)|0\rangle$. This involves all diagrams with three internal vertices and four external lines, and containing no vacuum bubble. Four lines come out of each internal vertex. The first task is to understand the structure of these diagrams. For each diagram we may consider a triplet of integers describing how many external lines are attached to a given internal vertex. Within permutation of the vertices the possible triplets are $(0,0,4)$, $(0,1,3)$, $(0,2,2)$ and $(1,1,2)$. The first type is ruled out because in such a diagram the first two vertices form a vacuum bubble, and we have ruled out these. Let us recall that lines from a vertex to itself are forbidden. For the type $(0,1,3)$, let us call a,b,c the internal vertices to which are attached zero, one or three external lines. Then b is touched by three internal lines and c is touched by one internal line. Thus one of the four internal lines going out of a must attach to c and three of the four internal lines going out of a must attach to b. This gives the shape (1) of Figure 13.15. A similar argument shows that the type $(0,2,2)$ gives the shape (2) of Figure 13.15. To analyze the type $(1,1,2)$, there are two internal lines out of the vertex touched by two external lines. When these two internal lines are attached one to each of the other two vertices, one obtains the shape (3) in Figure 13.15. It is not possible that these two internal lines attach to the same other vertex, because then there would be an internal line from the third vertex to itself, which is forbidden. In conclusion we obtain only the three shapes in Figure 13.15.

We may then write

$$\langle 0|a(p_3)a(p_4)S_3a^\dagger(p_1)a^\dagger(p_2)|0\rangle = \mathrm{I} + \mathrm{II},$$

where I is the contribution of the four diagrams[84] of shape 1 and II is the contribution of the diagrams of shapes 2 and 3. Even though this is absolutely not obvious yet, the proper way to compute the scattering amplitude is to consider only the term II.[85] We will accept

[84] The blob that is on the incoming edge in the picture might sit on any of the external edges.

[85] The term I will be shown later to contribute to "mass renormalization". We study mass renormalization starting with Section 14.10.

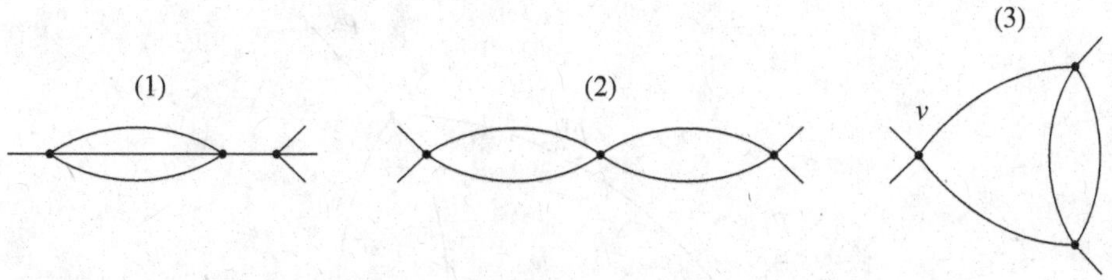

Figure 13.15 The three shapes of diagrams.

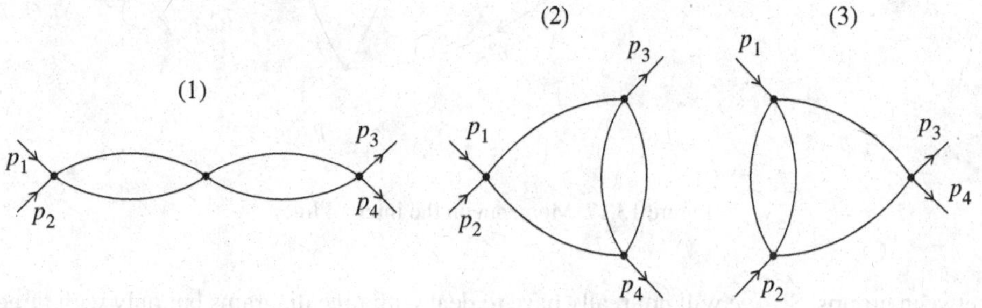

Figure 13.16 The three diagrams in the s group.

this now without further comments, so we have to take into account only the diagrams of shape 2 and 3. To obtain a diagram from its shape described in Figure 13.15, we have to assign the in-coming and out-going momenta to the external legs. We find that there are three diagrams of shape 2, which correspond to the diagrams of Figure 13.14, except that the loop is repeated. Each of these diagrams has a symmetry factor 4 (since in each loop the edges can be exchanged). On the other hand, there are $6 = \binom{4}{2}$ diagrams of shape 3, because such a diagram is determined by which two external momenta appear "at the vertex of the ice cone", indicated by the letter v on the figure. Each of these diagrams has a symmetry factor 2.

Therefore we have nine diagrams to contend with. Fortunately we can bring some organization to this collection of nine diagrams: It naturally breaks down in three collections of three diagrams each, which we will call respectively the s, t and u groups. The diagrams in the s group are represented in Figure 13.16.

Observe that the momenta on the external legs to the left of these diagrams are the same as the momenta on the external legs to the left of diagram (a) of Figure 13.14. The name "s group" is then motivated by the notation (13.138). The other two collections of three diagrams are defined in the obvious manner: The momenta on the external legs to the left of the diagrams are as in the second (respectively third) diagrams of Figure 13.14. They will be called respectively the t and u groups. The motivation for defining these groups may not be obvious now by will appear when we make the computation. In brief, the contributions of the diagrams within one given group "fit nicely together" and the computations are similar

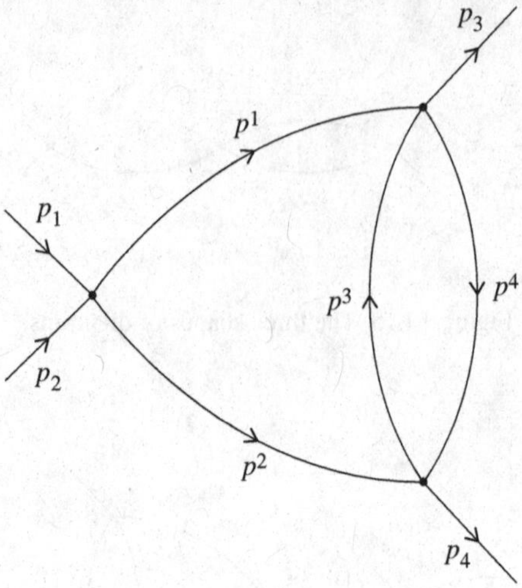

Figure 13.17 Momenta on the internal lines.

between groups. So, we will not really have to deal with nine diagrams but only with three, and the task will not be as difficult as we might have feared.

We start by computing the value of diagram (2) of Figure 13.16. For this we use the Feynman rules in momentum space, and we do not yet use a cutoff for clarity. These rules assign a four-momentum p to each internal line. We call these p^1, p^2, p^3, p^4, with upper indices to distinguish them from the momenta of the particles, and they are as in Figure 13.17.

To lighten notation we set

$$G_m(k) := \frac{1}{-k^2 + m^2 - \mathrm{i}\varepsilon},$$

the value of ε being implicit. Denoting by S the symmetry factor of the diagram (here $S = 2$), the value assigned by the Feynman rules in momentum space on page 390 to the diagram is then

$$\frac{1}{S}(-\mathrm{i}g)^3(-\mathrm{i})^4 \lim_{\varepsilon \to 0^+} \iiiint \prod_{1 \le i \le 4} \frac{\mathrm{d}^4 p^i}{(2\pi)^4} \prod_{1 \le i \le 4} G_m(p^i)\delta. \tag{13.154}$$

Here the factor $(-\mathrm{i}g)^3$ occurs because we assign a factor $-\mathrm{i}g$ to each of the three internal vertices. The factor $(-\mathrm{i})^4$ occurs because we assign a factor $-\mathrm{i}G_m(p)$ to an internal line with flow p. The factor $1/S$ occurs because in the end we divide by the symmetry factor S. Finally, δ is the product of the delta functions enforcing conservation of four-momentum at each internal vertex:

$$\delta := (2\pi)^{12}\delta^{(4)}(p_1 + p_2 - p^1 - p^2)\delta^{(4)}(p^2 - p^3 + p^4 - p_4)\delta^{(4)}(p^1 + p^3 - p^4 - p_3). \tag{13.155}$$

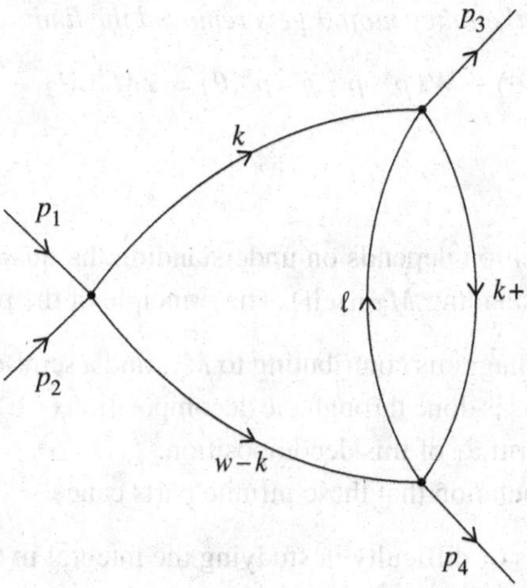

Figure 13.18 Relabeled momenta on the internal lines.

Writing as usual $w = p_1 + p_2$, the term $(2\pi)^4\delta^{(4)}(w - p^1 - p^2)$ forces the relation $p^2 = w - p^1$, and the term $(2\pi)^4\delta^{(4)}(p^1 + p^3 - p^4 - p_3)$ forces the relation $p^4 = p^1 + p^3 - p_3$. With these relations the term $(2\pi)^4\delta^{(4)}(p^2 - p^3 + p^4 - p_4)$ becomes $(2\pi)^4\delta^{(4)}(p_1 + p_2 - p_3 - p_4)$. We lighten notation by letting $k := p^1$ and $\ell := p^3$. The corresponding flows are then described in Figure 13.18.

Recalling the quantity $A = -\mathrm{i}(2\pi)^4\delta^{(4)}(p_3 + p_4 - p_1 - p_2)$ of (13.110), the expression (13.154) becomes

$$-Ag^3 \times \frac{1}{S}\lim_{\varepsilon\to 0^+}\iint \frac{\mathrm{d}^4k}{(2\pi)^4}\frac{\mathrm{d}^4\ell}{(2\pi)^4}G_m(k)G_m(w-k)G_m(\ell)G_m(k+\ell-p_3). \quad (13.156)$$

We then understand the reason for the sign convention in (13.148): The contribution of the diagram to the quantity M_3 is given by the simple expression

$$\frac{1}{S}\lim_{\varepsilon\to 0^+}\iint \frac{\mathrm{d}^4k}{(2\pi)^4}\frac{\mathrm{d}^4\ell}{(2\pi)^4}G_m(k)G_m(w-k)G_m(\ell)G_m(k+\ell-p_3). \quad (13.157)$$

13.23 Cancellation of Infinities

The goal of the present section is to explain how to prove (13.153). More precisely, keeping the notation $p^* = (m,0,0,0)$, our goal is the following.

Proposition 13.23.1 *Consider the quantity*

$$W(p_1,p_2,p_3,p_4,\theta) = 2M_2^0M_2 - M_3.$$

Then as $\varepsilon \to 0^+$ and then the cutoff θ gets removed the limit of

$$W(p_1,p_2,p_3,p_4,\theta) - W(p^*,p^*,p^*,p^*,\theta) = 2M_2^0(M_2 - M_2^0) - M_3 + M_3^0 \tag{13.158}$$

exists and is finite.[86]

The proof of this proposition depends on understanding the quantity $M_3 - M_3^0$ (which might be easier than understanding M_3 itself). The principle of the proof is simple.

- For each of the nine diagrams contributing to M_3, find a semi-explicit way to express "its infinite part". This is done through the decomposition (13.171); the "infinite part" is contained in the term J_1 of this decomposition.
- Check through computation that these infinite parts cancel.

There are several sources of difficulty in studying the integral in (13.157). The problem of the vanishingly small denominators as $\varepsilon \to 0^+$ is far less trivial than in Section 13.17 but again the main difficulty is the slow decrease of the integrand at infinity. To separate these two difficulties and bypass the problem of small denominators let us examine first what happens when we replace G_m in (13.157) by the *Euclidean propagator* E given by

$$E(k) := \frac{1}{\|k\|^2 + m^2}. \tag{13.159}$$

Replacing G_m by E, the integral in (13.157) is then replaced by

$$\iint \frac{\mathrm{d}^4k}{(2\pi)^4}\frac{\mathrm{d}^4\ell}{(2\pi)^4} E(k)E(w-k)E(\ell)E(k+\ell-p_3). \tag{13.160}$$

This integral cannot converge because the denominator does not grow fast enough at infinity. More precisely, there are only four powers of either $\|k\|^2$ or $\|\ell\|^2$ in the denominator, so that this denominator is bounded above by $C(1+\|k\|^2+\|\ell\|^2)^4$, whereas, setting $x = (k,\ell) \in (\mathbb{R}^{1,3})^2$, the integral

$$\iint \frac{\mathrm{d}^4k}{(2\pi)^4}\frac{\mathrm{d}^4\ell}{(2\pi)^4}\frac{1}{(1+\|k\|^2+\|\ell\|^2)^4} = \int \frac{\mathrm{d}^8x}{(2\pi)^8}\frac{1}{(1+\|x\|^2)^4}$$

is "logarithmically" divergent.[87] This type of divergence of (13.160) is called an *overall divergence*. We already faced such a problem in Section 13.17. There is however a further problem: In (13.160) the integral in ℓ itself is divergent, because there are basically only two powers of $\|\ell\|^2$ in the denominator. The divergence of the integral in ℓ will be referred to as a *sub-divergence*.[88] There is no free lunch. The overall divergence and the sub-divergence have to be addressed separately. We first remove the sub-divergence in the ℓ integral. The strategy for this is to replace the term $E(\ell)E(k+\ell-p_3)$ by the term

86 As we will see later, this is true only in the sense of distributions, when the previous quantity is integrated against a test function of $p_1,\ldots,p_4$.

87 This argument is a special case of the technique of *power counting* to be developed later.

88 More generally when there is a divergence "inside" another divergence one talks of "nested divergences".

$E(\ell)(E(k+\ell-p_3)-E(\ell))$. This improves the integrability in ℓ, since by considering the difference $E(k+\ell-p_3)-E(\ell)$ "we gain a power of ℓ in the denominator". The function

$$F(k,\ell,w,p_3) := E(k)E(w-k)E(\ell)(E(k+\ell-p_3)-E(\ell)) \tag{13.161}$$

is now integrable in ℓ and

$$\int \frac{\mathrm{d}^4\ell}{(2\pi)^4} F(k,\ell,w,p_3) = E(k)E(w-k)I(k-p_3)$$

where

$$I(k-p_3) := \int \frac{\mathrm{d}^4\ell}{(2\pi)^4} E(\ell)(E(\ell+k-p_3)-E(\ell)).$$

We however did nothing to address the problem of the overall divergence, so let us now look into that. It is not difficult to get convinced that $I(k-p_3)$ depends smoothly on the parameter $k-p_3$, and grows logarithmically with this parameter. Actually this integral can be explicitly computed by a formula similar to (13.127), and these claims are obvious from the resulting formula. In loose words the function $E(k)E(w-k)I(k-p_3)$ behaves like $\|k\|^{-4}\log\|k\|$ for large k. The function

$$F(k,\ell,w,p_3) - F(k,\ell,2p^*,p^*) \tag{13.162}$$

will then have a faster rate of decrease at infinity and will be integrable.[89] Thus, writing now the cutoff the integral

$$\iint \frac{\mathrm{d}^4k}{(2\pi)^4}\frac{\mathrm{d}^4\ell}{(2\pi)^4}\theta(\boldsymbol{k})\theta(\boldsymbol{\ell})\big(F(k,\ell,w,p_3)-F(k,\ell,2p^*,p^*)\big) \tag{13.163}$$

has a limit as the cutoff is removed.

All this is very good, but let us go back to what we really try to study, the integral (13.160), which, writing now the cutoff is

$$H(w,p_3,\theta) := \iint \frac{\mathrm{d}^4k}{(2\pi)^4}\frac{\mathrm{d}^4\ell}{(2\pi)^4}\theta(\boldsymbol{k})\theta(\boldsymbol{\ell})E(k)E(w-k)E(\ell)E(k+\ell-p_3). \tag{13.164}$$

The trick is to write this as $H(w,p_3,\theta) = H_1(w,p_3,\theta) + H_2(w,p_3,\theta)$ where

$$H_1(w,p_3,\theta) := \iint \frac{\mathrm{d}^4k}{(2\pi)^4}\frac{\mathrm{d}^4\ell}{(2\pi)^4}\theta(\boldsymbol{k})\theta(\boldsymbol{\ell})E(k)E(w-k)E(\ell)^2 \tag{13.165}$$

$$H_2(w,p_3,\theta) = \iint \frac{\mathrm{d}^4k}{(2\pi)^4}\frac{\mathrm{d}^4\ell}{(2\pi)^4}\theta(\boldsymbol{k})\theta(\boldsymbol{\ell})F(k,\ell,w,p_3). \tag{13.166}$$

Two very remarkable things happen then.

- The integral (13.165) is much simpler than the integral (13.164) because the integrations in $\boldsymbol{k}$ and $\boldsymbol{\ell}$ separate.

[89] One may subtract $F(k,\ell,w^*,p_3^*)$ for any fixed value of w^*, p_3^*. It would be less natural here to choose $w^* = p_3^* = 0$ but there are good reasons for which this second choice is better in the long run, as will be apparent later.

- As the cutoff gets removed, the quantity $H_2(w,p_3,\theta) - H_2(2p^*,p^*,\theta)$ has a finite limit, see (13.163).

As a consequence of the second point above, when trying to prove the existence of the limit in (13.158) *we may replace* $H(w,p_3,\theta)$ *by the much simpler quantity* $H_1(w,p_3,\theta)$. This allows computations to proceed.

We follow the same strategy when dealing with the true propagator rather than the Euclidean propagator. That is, we consider the function

$$F_m(k,\ell,w,p_3,\varepsilon) = G_m(k)G_m(w-k)G_m(\ell)(G_m(k+\ell-p_3) - G_m(\ell)), \tag{13.167}$$

and (writing now the cutoff) the integral

$$\iint \frac{\mathrm{d}^4k}{(2\pi)^4}\frac{\mathrm{d}^4\ell}{(2\pi)^4}\theta(\boldsymbol{k})\theta(\boldsymbol{\ell})\big(F_m(k,\ell,w,p_3,\varepsilon) - F_m(k,\ell,2p^*,p^*,\varepsilon)\big). \tag{13.168}$$

We may then hope that

As $\varepsilon \to 0^+$ *this integral has a limit, and that as the cutoff gets removed this limit has itself a finite limit.* (13.169)

The problem of the limit as $\varepsilon \to 0^+$ makes this a far less trivial statement than in the previous case of the Euclidean propagator. To prove it we basically have to show that when we control a certain expression involving Euclidean propagators, we also control a similar expression involving the true propagators. Later, in Theorem 18.2.1 we will prove a general theorem to that effect. This theorem holds only "in distribution", that is only after integration against a test function of the momenta of the out-going and in-coming particles. This is not surprising since we expect that the scattering amplitudes are only defined in the sense of distributions even though we denote them as functions. In the same manner, Proposition 13.23.1 is true only "in distribution". Proposition 13.23.1 does not immediately follow from Theorem 18.2.1, which uses slightly different hypotheses, but since our purpose in the present section is simply to illustrate the near-miraculous "cancellation of infinities" we will not write a specialized argument to prove it, and we will simply accept that (13.169) is in some sense true (as well as corresponding results for the other diagrams).

Going back to (13.157), let

$$J := \lim_{\varepsilon\to 0^+}\iint \frac{\mathrm{d}^4k}{(2\pi)^4}\frac{\mathrm{d}^4\ell}{(2\pi)^4}\theta(\boldsymbol{k})\theta(\boldsymbol{\ell})G_m(k)G_m(w-k)G_m(\ell)G_m(k+\ell-p_3). \tag{13.170}$$

Following the same idea as in the decomposition of the function $H(w,p_3,t)$ of (13.164) we write

$$J = J_1 + J_2 \tag{13.171}$$

where

$$J_1 = \lim_{\varepsilon\to 0^+}\iint \frac{\mathrm{d}^4k}{(2\pi)^4}\frac{\mathrm{d}^4\ell}{(2\pi)^4}\theta(\boldsymbol{k})\theta(\boldsymbol{\ell})G_m(k)G_m(w-k)G_m(\ell)^2, \tag{13.172}$$

$$J_2 = \lim_{\varepsilon\to 0^+}\iint \frac{\mathrm{d}^4k}{(2\pi)^4}\frac{\mathrm{d}^4\ell}{(2\pi)^4}\theta(\boldsymbol{k})\theta(\boldsymbol{\ell})F_m(k,\ell,w,p_3,\varepsilon). \tag{13.173}$$

The quantity J_1 is much simpler than J because the integrations in k and ℓ separate in (13.172). Indeed, recalling the notation (13.136), the quantity J_1 equals

$$U(w,\theta)U(0,\theta).$$

On the other hand, (13.169) asserts that as the cutoff gets removed, the quantity J_2 contributes only a finite value to the quantity (13.153), so that it is simply irrelevant in our proof "that the infinities cancel out".

That is the key step: We can basically express any contribution we wish using the functions $U(\cdot,\theta)$, so that we can witness the cancellation of infinities through calculation. Since the graph of Figure 13.18 has a symmetry factor 2, its contribution (not taking into account an irrelevant finite contribution) to $M_3 - M_3^0$ is then

$$\frac{1}{2}(U(w,\theta) - U(2p^*,\theta))U(0,\theta). \tag{13.174}$$

Exactly by the same method, one sees that the third graph of Figure 13.16 provides the same contribution (again within a finite quantity).

Finally there is a contribution from the first graph of Figure 13.16. The computation of the value of this graph is much easier than for the second graph of the same figure. We let the reader check that (13.170) is replaced by

$$\lim_{\varepsilon\to 0^+} \iint \frac{\mathrm{d}^4k}{(2\pi)^4}\frac{\mathrm{d}^4\ell}{(2\pi)^4}\theta(k)\theta(\ell)G_m(k)G_m(w-k)G_m(\ell)G_m(w-\ell)$$
$$= \Big(\lim_{\varepsilon\to 0^+}\int \frac{\mathrm{d}^4k}{(2\pi)^4}\theta(k)G_m(k)G_m(w-k)\Big)^2 = U(w,\theta)^2.$$

Since the graph has a symmetry factor 4 its contribution[90] to $M_3 - M_3^0$ is then

$$\frac{1}{4}(U(w,\theta)^2 - U(2p^*,\theta)^2).$$

The contribution of all three graphs of the s group to $M_3 - M_3^0$ is the sum of this quantity and of twice (13.174), that is

$$\frac{1}{2}(U(w,\theta) - U(2p^*,\theta))\Big(\frac{1}{2}U(w,\theta) + \frac{1}{2}U(2p^*,\theta) + 2U(0,\theta)\Big). \tag{13.175}$$

On the other hand, we recall (13.137):

$$M_2 = \frac{1}{2}(U(w,\theta) + U(p_1 - p_3,\theta) + U(p_1 - p_4,\theta)),$$

and its consequence

$$M_2^0 = \frac{1}{2}(U(2p^*,\theta) + 2U(0,\theta)). \tag{13.176}$$

90 I am very much grateful to Jean Bernard Zuber for having provided the most useful help with symmetry factors in this computation.

The quantity $2M_2^0(M_2 - M_2^0)$ is thus the sum of the following three terms

$$\frac{1}{2}(U(w,\theta) - U(2p^*,\theta))(U(2p^*,\theta) + 2U(0,\theta)) \tag{13.177}$$

$$\frac{1}{2}(U(p_1 - p_3,\theta) - U(0,\theta))(U(2p^*,\theta) + 2U(0,\theta)) \tag{13.178}$$

$$\frac{1}{2}(U(p_1 - p_4,\theta) - U(0,\theta))(U(2p^*,\theta) + 2U(0,\theta)). \tag{13.179}$$

The term (13.177) exactly "cancels the infinite part of (13.175)" because the difference between these two quantities has a finite limit as the cutoff gets removed, since this is the case for the differences $U(v,\theta) - U(v',\theta)$ according to Proposition 13.18.5.

In other words, the term (13.177) cancels the infinite part of the contribution of the three graphs of the s group of Figure 13.16 to the quantity $M_3 - M_3^0$. To complete the proof of Proposition 13.23.1 and of (13.153) it then suffices to proceed in an entirely similar manner with the t and the u groups of diagrams to see that the infinite parts of these groups are canceled respectively by the terms (13.178) and (13.179).

Exercise 13.23.2 Check this in complete detail.

13.24 Counter-terms, II

Let us now present the computation leading to (13.152) in a manner that is closer to the physicists' standard presentation.[91] We write the interacting Hamiltonian as in (13.147)

$$\frac{g_{\mathrm{lab}}}{4!} :\varphi^4: + \frac{D}{4!} :\varphi^4:, \tag{13.180}$$

where D depends on the cutoff. From now on g_{lab} is our main parameter,[92] so *we change notation* and write g rather than g_{lab}. The former meaning of $g = g_{\mathrm{lab}} + D$ will no longer be used.

Our goal is to compute

$$\langle 0|a(p_3)a(p_4)Sa^\dagger(p_1)a^\dagger(p_2)|0\rangle \tag{13.181}$$

in perturbation theory, thinking now of the interaction Hamiltonian as made up of two terms as in (13.147). The computation of the S-matrix now uses Feynman diagrams with two types of internal vertices. The internal vertices corresponding to the term $g :\varphi^4: /4!$ are represented with a • and are called ordinary vertices. The internal vertices corresponding to the counter-term $D :\varphi^4: /4!$ are represented with a □ and are called *counter-term vertices*. The Feynman rules for these diagrams are nearly identical to the Feynman rules for ordinary diagrams, except that counter-term vertices are given the coefficient $-iD$, as opposed to the coefficient $-ig$ of the ordinary vertices. We let the reader convince herself of that.

91 Please be ready for a certain lack of mathematical crispness. Also, please keep in mind that however hazy this section sounds, its actual content is exactly the same as that of the previous section.

92 This is physically sensible since g_{lab} is the parameter that can be measured, while we have no way to measure $g = g_{\mathrm{lab}} + D$.

It is fruitful to think of D not as a single quantity, but as a sum of terms filling different purposes. There is one term in the sum for each order in g, starting at order two, so that

$$D = g^2 D_2 + g^3 D_3 + O(g^4). \tag{13.182}$$

A nice feature is that there is no term in g. Thus each counter-term vertex brings a factor g^2 and counter-term vertices contribute *twice as much* as ordinary vertices to reach a given order in g and the value of a diagram that contains a ordinary vertices and b counter-term vertices is $O(g^{a+2b})$. Hence, to compute amplitudes at order k (that is, throwing away all terms that contain a power of g higher than k) we need to consider only diagrams for which $0 \leq a+2b \leq k$. For $k = 2$ we have either $b = 0, a \leq 2$ (ordinary diagrams with at most two internal vertices) or $b = 1, a = 0$ (the diagram with a single internal counter-term vertex). For $k = 3$ either $b = 0$ (ordinary diagrams with at most three internal vertices) or else $b = 1$ and $a \leq 1$ (diagrams containing one counter-term vertex and at most one ordinary vertex).

We recall yet again the quantity $A = -\mathrm{i}(2\pi)^4\delta^{(4)}(p_3 + p_4 - p_1 - p_2)$ of (13.110). The counter-term method consists of adjusting in succession the value of D_k order by order to enforce the condition

At the given value $p_1 = p_2 = p_3 = p_4 = p^*$

the scattering amplitude is Ag. (13.183)

This simply says that at small velocity the model predicts the scattering amplitude which is actually measured in the lab. Two lines below (13.146) we have already determined the value of D_2 which enforces this condition at order two but the argument starts afresh here.

We first compute the scattering amplitude at order two. For this we take into account four ordinary diagrams: the one-vertex diagram, which gives a contribution Ag and the three diagrams of Figure 13.10 which together give a contribution of the type $A(\mathrm{i}g^2 M_2)$ where M_2 is a function of p_1, p_2, p_3, p_4 and the cutoff θ. We also need to consider the diagram with a single counter-term vertex, which is shown on diagram (a) of Figure 13.19. It should be obvious that the value of this diagram is AD. This gives us a scattering amplitude of

$$A(g + \mathrm{i}g^2 M_2 + D) + O(g^3). \tag{13.184}$$

Recalling (13.182), we find the value of D_2 by enforcing (13.183) within error $O(g^3)$, that is $Ag = A(g + \mathrm{i}g^2 M_2^0 + g^2 D_2) + O(g^3)$, where M_2^0 is the value of M_2 at the point $p_1 = p_2 = p_3 = p_4 = p^*$. This provides the condition $D_2 = -\mathrm{i}M_2^0$. That is exactly what we did in (13.146). The scattering amplitude then takes the form $A(g + \mathrm{i}g^2(M_2 - M_2^0)) + O(g^3)$.

To compute the scattering amplitude at order three we need to take into account other contributions:

- The contribution $Ag^3 D_3$ of the term $AD = Ag^2 D_2 + Ag^3 D_3 + O(g^4)$ in (13.184).
- The contribution $Ag^3(-M_3)$ of the nine ordinary diagrams with three internal vertices which we considered in the previous section.
- The contribution of the six diagrams which each have exactly one ordinary and one counter-term internal vertex.

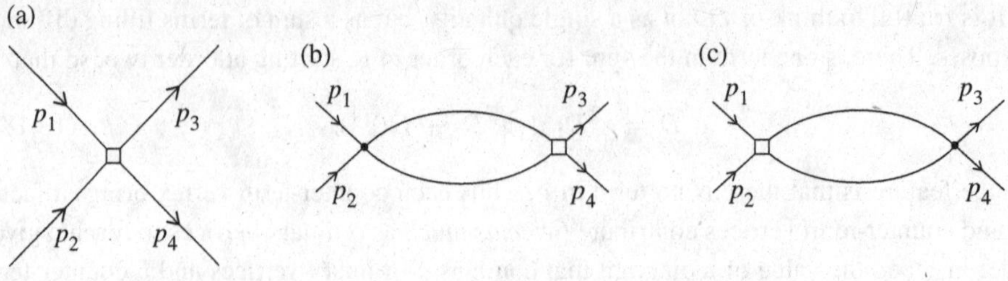

Figure 13.19 Some diagrams with counter-terms vertices.

Two of these six diagrams are shown as (b) and (c) in Figure 13.19. To each of the two diagrams (b) and (c) corresponds two other diagrams as in Figure 13.14. Let us compute the value of diagram (b) in Figure 13.19. It has a symmetry factor 2 and it should be quite obvious that this value is

$$\frac{1}{2}(-\mathrm{i}g)(-\mathrm{i}D)(-\mathrm{i})^2(2\pi)^4\delta(p_1+p_2-p_3-p_4)\lim_{\varepsilon\to 0^+}\int\frac{\mathrm{d}^4k}{(2\pi)^4}G_m(k)G_m(w-k)$$

$$=\frac{1}{2}A(\mathrm{i}gD)U(w,\theta)/2, \tag{13.185}$$

where $w=p_1+p_2$ and $U(w,\theta)$ is defined in (13.136). The diagram (c) in Figure 13.19 has the same value. Since $D=-\mathrm{i}g^2M_2^0+O(g^3)$, the total contribution of these two diagrams is

$$g^3AM_2^0U(w,\theta)+O(g^4). \tag{13.186}$$

The total contribution of the six diagrams with two internal vertices, exactly one of which is a counter-term vertex is therefore, using (13.137) in the last equality,

$$Ag^3M_2^0\big(U(p_1+p_2,\theta)+U(p_1-p_3,\theta)+U(p_1-p_4,\theta)\big)+O(g^4)=2Ag^3M_2^0M_2+O(g^4).$$

Therefore at order three the contribution of all diagrams to the scattering amplitude is

$$A\Big(g+\mathrm{i}g^2(M_2-M_2^0)+g^3\big(2M_2^0M_2-M_3+D_3\big)\Big)+O(g^4). \tag{13.187}$$

In words, the effect of the counter-term g^2D_2 is to replace M_2 by $M_2-M_2^0$ and to replace $-M_3$ by $-M_3+2M_2M_2^0$.

We now enforce (13.183) at order three. This determines the choice $D_3=M_3^0-2(M_2^0)^2$. Then the predicted scattering amplitude (13.187) coincides with (13.152). This is hardly a surprise since we just performed in this other guise the very same computation leading to (13.152).

Having recovered the equation (13.152) for the predicted scattering amplitude in the language of counter-terms, our next task is to recover the "cancellation of infinities" ensuring that this prediction remains finite as the cutoff gets removed. For this we will *entirely forget about* (13.152) and we think entirely in terms of diagrams. We will show that in the words of the physicists "the counter-terms cancel the divergences" at order three.[93] The diagram (a)

[93] We are not concerned here with the easier problem of what happens as $\varepsilon\to 0$.

of Figure 13.19 contributes as $AD = -Ag^2 M_2^0 + O(g^3)$. The term $-\mathrm{i}Ag^2 M_2^0$ cancels the divergences of the diagrams of Figure 13.14 with two internal vertices. This was explained in Sections 13.17 and 13.18 and we do not come back to this.

Let us now explain what happens at order three. We keep the cutoffs implicit for clarity. When we add the contributions of the type $g^3 C$ (that is, the contribution containing exactly a factor g^3) of the six diagrams with exactly one ordinary and one counter-term vertex with the similar contributions of the nine ordinary diagrams with three internal vertices we obtain the term $Ag^3(2M_2^0 M_2 - M_3)$. The quantity $2M_2^0 M_2 - M_3$ is of the type

$$\lim_{\varepsilon \to 0^+} \iint \frac{\mathrm{d}^4 k}{(2\pi)^4} \frac{\mathrm{d}^4 \ell}{(2\pi)^4} T(k, \ell, \varepsilon, p_1, p_2, p_3, p_4), \tag{13.188}$$

where the integral diverges. On the other hand, the contribution $Ag^3 D_3$ of the diagram (a) of Figure 13.19 is $-Ag^3(2(M_2^0)^2 - M_3^0)$. Adding these contributions together replaces the integral in (13.188) by

$$\iint \frac{\mathrm{d}^4 k}{(2\pi)^4} \frac{\mathrm{d}^4 \ell}{(2\pi)^4} (T(k, \ell, \varepsilon, p_1, p_2, p_3, p_4) - T(k, \ell, \varepsilon, p^*, p^*, p^*, p^*)). \tag{13.189}$$

As we will prove below, the integral in (13.189) is "convergent".[94] This fact can be stated in flowery language by saying that "although the integral (13.188) is divergent, it suffices to remove an overall divergence (that is, the integral of $-T(k, \ell, p^*, p^*, p^*, p^*)$, which independent of the external momenta) to make it convergent".

The physicists describe the situation as follows:

- The contributions (13.186) of the six diagrams with exactly one ordinary and one counter-term vertex "cancel the sub-divergences" of the nine ordinary diagrams.
- The remaining overall divergence is canceled by the contribution $Ag^3 D_3$ of the diagram (a) of Figure 13.19.

The first bullet *does not* mean that the integrals in k and ℓ separately converge in (13.188) (this does not seem to be the case).[95] Rather, it means that, as stated in the second bullet the integral (13.188) can "be made convergent by subtracting an overall divergence" (independent of the external momenta).

Let us now stress the "miracle" happening here. The term D_2 is determined by looking only at diagrams with two internal vertices. It determines in turn the contributions at order three of the diagrams with exactly one ordinary and one counter-term vertex. This affects the diagrams with three internal vertices just in the correct manner to make the procedure work. This is *extremely remarkable.*

Let us now proceed to the proof of the convergence of (13.189). The whole procedure is very similar to the decomposition 13.171, but is expressed in a different language. Again, we do not have to deal with the nine ordinary diagrams at the same time, we can group them

94 We are here interested only in showing that the counter-terms solve the problem of the integrands which do not decrease fast enough at infinity, so when we say that an integral is "convergent" we mean that it would be if we replace the propagators by their Euclidean version.

95 But the books I have seen certainly do not make this clear!

into three groups of three diagrams, as in Section 13.22. We will prove that the contributions (13.186) of the two diagrams (b) and (c) of Figure 13.19 "cancel the sub-divergences" of the three diagrams in the s group.

As follows from (13.186) and (13.176), at order three the total contribution of the two diagrams (b) and (c) of Figure 13.19 is

$$Ag^3 M_2^0 U(w,\theta) = Ag^3 \frac{1}{2} U(w,\theta)(U(2p^*,\theta) + 2U(0,\theta)) + O(g^4). \tag{13.190}$$

We think of this quantity as the sum of *three*[96] terms: Twice the term $Ag^3U(w,\theta)U(0,\theta)/2$, and one time the term $Ag^3U(w,\theta)U(2q^*,\theta)/2$. We will prove that subtracting the appropriate term of the previous type from the value of a diagram of the s group "cancels the sub-divergences of this diagram", in the sense that it suffices then to "remove an overall divergence" to make the integral converge. This completes the argument (as one proceeds similarly for the t and the u groups).

We start this program with diagram (2) of Figure 13.16 (which is also the diagram of Figure 13.18). Taking into account the symmetry factor 2 of this diagram, subtracting the term $Ag^3U(w,\theta)U(0,\theta)/2$ from the value of this diagram replaces the integral in (13.157) (keeping the cutoff and the dependence in ε implicit) by the integral

$$\iint \frac{\mathrm{d}^4k}{(2\pi)^4} \frac{\mathrm{d}^4\ell}{(2\pi)^4} R(k,\ell,w,p_3), \tag{13.191}$$

where

$$R(k,\ell,w,p_3) := G_m(k)G_m(w-k)\big(G_m(\ell)G_m(k+\ell-p_3) - G_m(\ell)G_m(-\ell)\big),$$

which is the function of (13.167), and the corresponding integral of $R(k,\ell,w,p_3) - R(k,\ell,2p^*,p^*)$ is "convergent", completing the argument for this diagram (the overall divergence removed being the integral of $R(k,\ell,2p^*,p^*)$).

We proceed in the same way for diagram (3) of Figure 13.16 (using up the other term $Ag^3U(w,\theta)U(0,\theta)/2$)!) and we consider finally the case of diagram (1) of Figure 13.16. Taking into account the symmetry factor 4, subtraction of the term $Ag^3U(w,\theta)U(2p^*,\theta)/2$ confronts us with the integral

$$\iint \frac{\mathrm{d}^4k}{(2\pi)^4} \frac{\mathrm{d}^4\ell}{(2\pi)^4} S(k,\ell,w)$$

where

$$S(k,\ell,w) = G_m(k)G_m(w-k)G_m(\ell)G_m(w-\ell) - 2G_m(k)G_m(w-k)G_m(\ell)G_m(2p^*-\ell),$$

and replacing $S(k,\ell,w)$ by $S(k,\ell,w) - S(k,\ell,2p^*)$, (subtracting an overall divergence!) the integral becomes the perfectly well-defined quantity[97]

[96] One for each of the diagrams in the s group!
[97] Well-defined at least if we replace the propagators by their Euclidean version.

$$\Big(\int \frac{d^4k}{(2\pi)^4}\big(G_m(k)G_m(w-k) - G_m(k)G_m(2p^* - k)\big)\Big)^2.$$

This completes the argument for this diagram.

Thus we have provided an alternative approach to Proposition 13.23.1 (which is actually just another way to write the previous argument). One expresses this result by saying that "the counter-terms have made all diagrams with at most three internal vertices convergent".[98]

It is by no means obvious how to continue the procedure to higher orders of perturbation theory. Furthermore, matters will then become distinctly more complicated, because at higher order it does not suffice to renormalize the coupling constant. One also has to renormalize the mass and the field itself. The next chapter will present (non-rigorous) arguments trying to justify these facts. In the final chapters, we will investigate systematic methods to perform renormalization. These methods can be seen as a considerable extension of the ideas of Section 13.23.

Key ideas to remember:

- The S-matrix is studied through the Dyson series, order by order in perturbation theory.
- Models for interaction are constructed using Hamiltonian densities.
- The computation of the S-matrix proceeds despite the fact that the interaction picture is not mathematically sound.
- Feynman diagrams are a picturesque way to encode the terms arising from the application of Wick's theorem.
- When using Feynman diagrams one must watch for treacherous symmetry factors.
- The value of a Feynman diagram is given by an integral.
- Wick rotations are an effective method to tame singularities in the denominators of the integrands.
- Tree computations are fine, but loops in diagrams give rise to divergent integrals, reflecting the fact that we make ill-defined products of distributions.
- In ϕ^4 theory, by expressing the results of the computation as a function of the strength of the interaction as it is measured in the lab (as opposed to the corresponding parameter in the Hamiltonian) one can however make a sensible prediction of certain scattering amplitudes, at least at the one and two loops levels.
- These predictions involve a remarkable "cancellation of infinities" which the physicists express by introducing counter-terms.

[98] This might sound foggy now, but the whole procedure should become clearer when it is made systematic in Chapter 16.

14

Interacting Quantum Fields

The mathematical approach to Quantum Field Theory consists of setting a number of axioms and developing consequences of these axioms. In Appendix M we describe the most popular axiomatic setting, the Wightman axioms, and we strive to give an idea of some of the underlying mathematics. Unfortunately it is not known as of today whether in four space-time dimensions any non-trivial theory (in the sense that it is not equivalent to a theory without interactions) satisfies these axioms.[1] Many books on this topic, such as that of Araki [2], Bogoliubov [9], Derezinski and Gérard [19], Jost [45], Streater and Wightman [79], and Rivasseau [70] should be fairly accessible to a reader having reached this point and read Appendix M. We shall not pursue this direction in this chapter.

The physicists' approach is more pragmatic. Without worrying about axioms, they construct all kinds of arguments (sometimes very subtle). These types of considerations go under the general name of "theoretical physics", not to be confused with the often fully rigorous "mathematical physics". Overall it is not an easy task for a mathematician to understand what they are doing. Among the various obstacles is the fact that many conditions are implicit or not precisely stated. This is to be expected, for otherwise it would probably be possible to build a rigorous theory. The present introduction to Quantum Field Theory, which is first and foremost a physical theory, must include a sample of these types of arguments. The current chapter presents some of them.[2]

It is appropriate first to renew our warnings that we are exploring here a very different domain from mathematics. The previous sections saw plenty of hand-waving, but here lack of rigor takes place at a different scale. Even if some of the vocabulary is familiar, it might not have the expected meaning. A "theorem" need not be always true, but rather is expected to hold under normal circumstances.[3] As mathematical sanity has been left behind, we will use the word "theorem" the way physicists use it, even when it concerns an ill-defined statement with a purely formal supporting argument. In order however to help the reader distinguish between what is a theorem in the sense of mathematics and what is a theorem in

[1] Another somewhat different rigorous approach, the so-called algebraic approach, suffers from basically the same problems.
[2] Which of course we have tried to write in a near-mathematical language.
[3] This is in particular the case of the Gell-Mann–Low theorem.

the sense of physics, the latter do not have "proofs" but have only "supporting arguments".[4] It is not only that these "theorems" have not received a rigorous proof yet. Rather, it is that *the whole story* is a kind of fairy tale, as we make-believe that certain structures exist really when this is far from being clear. Authors rarely make the uncertain status of their assertions explicit. This can be very confusing for a mathematician who wanders there for the first time.

The reader must also keep in mind that there does not exist a whole coherent mathematical model. Therefore, the best one can do is to tell some little stories, each relating to a certain aspect of things. It is not necessarily easy to make these little stories fully compatible with each other. In other words, the global picture is far fuzzier than our individual little stories. The reader should be reassured that this is the theory's problem, not hers.

14.1 Interacting Quantum Fields and Particles

As the name should indicate, quantum fields (and not particles) are arguably the central concept of Quantum Field Theory. In this entire chapter we consider the simplest possible case, when the quantum of the field is a neutral spinless particle which is its own anti-particle.

Let us describe in words some of the properties we expect from our quantum field. (A more precise description is the object of the Wightman axioms of Appendix M.) The state space $\mathcal{H}$ encodes a description of all the objects of our theory, and in particular all multiparticle systems. It contains a distinguished element, the *vacuum*, which we denote $|\Omega\rangle$. (In the present chapter, we use Dirac's notation, so our notation does not coincide with that of the more mathematical Appendix M.) On the state space is given a unitary representation $U(c,C)$ of the group $\mathcal{P}^* = \mathbb{R}^{1,3} \rtimes SL(2,\mathbb{C})$. The vacuum is an eigenvector for each operator $U(c,C)$,[5] and it shares this property only with the states of the type $\alpha|\Omega\rangle$ for $\alpha \in \mathbb{C}$.[6] The operators $U(c,1)$ form a commutative group of unitary operators. For such a group, we will accept that there is suitable generalization of (2.72) which for $c = (c_0, c_1, c_2, c_3)$ allows us to write

$$U(c,1) = \exp(\mathrm{i}(c_0 P_0 - c_1 P_1 - c_2 P_2 - c_3 P_3)) := \exp(\mathrm{i} c^\mu P_\mu),$$

for certain self-adjoint operators P_μ, which are called the momentum operators. To form a mental picture of the situation, one may consider a positive measure $\mathrm{d}\lambda$ on $\mathbb{R}^{1,3}$ and the unitary operator $U(c,1)$ on $L^2 = L^2(\mathbb{R}^{1,3}, \mathrm{d}\lambda)$ given by multiplication by $\exp(\mathrm{i} c^\mu p_\mu)$. Then P_μ is simply the operator "multiplication by p_μ" on L^2 given for $f \in L^2$ by $P_\mu(f)(p) = p_\mu f(p)$. The self-adjoint operator $P^\mu P_\mu = P_0^2 - \sum_{1\le j\le 3} P_j^2$ should be seen as corre-

4 Despite the fact that we have often used the word "proof" in Part III, this does *not* imply that all the results we have obtained in the previous chapters of this part are rigorous. Still, the line has to be drawn somewhere, and it is not appropriate to use the word "proof" when the result under consideration may fail.

5 Since the operators $U(c,C)$ are unitary, this is equivalent to saying that $U(c,C)|\Omega\rangle = \alpha|\Omega\rangle$ where α is of modulus 1, and is equivalent to saying that $U(c,C)|\Omega\rangle$ represents the same state as $|\Omega\rangle$. If you think about the physical content of this statement, it should appear very natural. If I look at the vacuum, and then move any way I want, it will appear just the same.

6 Which means that any other state than the vacuum will not be invariant under all the transformations $U(c,C)$.

sponding to the observable "square of the mass" and to make the theory physical it should have only non-negative eigenvalues.[7]

The quantum field $\varphi(x)$ is an (unbounded) operator on $\mathcal{H}$. Actually we cannot really think of φ as being defined at a given point. It has to be "smeared" as the physicists say. In other words, it is really an operator-valued distribution. The quantum fields behave as expected with respect to the action of the Poincaré group. In an informal way relativistic invariance (5.6) becomes

$$\varphi(c + Cx) = U(c,C) \circ \varphi(x) \circ U(c,C)^{-1}. \tag{14.1}$$

The previous setting says nothing about particles. The assumption that the model describes particles is separate from the assumptions involving the existence of fields. To describe particles we specifically assume that the state space $\mathcal{H}$ contains a subspace $\mathcal{H}_1$ that describes single-particle states (recalling that we consider only the case of a single particle type). The space $\mathcal{H}_1$ is invariant under all operators $U(c,C)$. The restriction of these operators to the space $\mathcal{H}_1$ is then a unitary representation of the Poincaré group, and it is natural to assume that this representation is irreducible, reflecting the fact that the particle is elementary.

Calling μ the mass of the particle[8] we assume that $\mathcal{H}_1$ consists of eigenvectors of the operator $P^\nu P_\nu$, with eigenvalue μ^2. We always assume $\mu > 0$ because the case of massless particles is far more difficult.

14.2 Road Map I

The overall goal of the chapter is to provide support for some of the trickiest points in the entire theory, the issues of mass and field renormalization. One of the central ideas is the use of *diagrammatics*: Quantities of interests are expressed as the sum of the values of (lots of) diagrams. This is just what we have been doing all along, but now there is an implicit change of perspective. We keep saving face by stating that our expansions make sense only as formal series in the coupling constant, but somewhere deep inside we start to believe that there is more to it, and that these expansions should be trusted more than what is strictly permitted.

A central result in the whole approach is Proposition 14.5.4 which lets us compute the Green functions[9]

$$G(y_1,\dots,y_k) := \langle\Omega|T\varphi(y_1)\dots\varphi(y_k)|\Omega\rangle \tag{14.2}$$

as a sum of the values of certain diagrams. The reader is not expected to understand immediately why these objects are important, this will appear only gradually.

The purpose of Section 14.3 is primarily to support Proposition 14.5.4, although this is done through results arguably of independent interest. Supporting is not proving, and to be

[7] More formally, this operator should have a positive spectrum. In our mental picture of the operator $U(c,1)$ just described, this operator $P^\mu P_\mu$ is just "multiplication by p^2" and the condition that it has a non-negative spectrum means exactly that $d\lambda$ lives on the set $\{p^2 \geq 0\}$.

[8] The change of notation from m to μ prepares us to accept the fact that the mass μ of the particle will not be the parameter m of the free field.

[9] "Green functions" occur in many areas of physics. Typically they encode the response of the system to a small perturbation, but we will not use or develop this point of view.

on the safe side in Section 14.4 we present another line of approach, also of independent interest, supporting again Proposition 14.5.4. The only use of the material of Sections 14.3 and 14.4 in the rest of the chapter is through Proposition 14.5.4, so these sections should be thought of as independent of the main story.

14.3 The Gell-Mann–Low Formula and Theorem

Trying to learn the Gell-Mann–Low theorem proved to be a real obstacle race. This theorem belongs to a circle of ideas, and different authors use different ingredients (none of which is rigorous). They also call different formulas the "Gell-Mann–Low formula".[10] We will try to clear the confusion and tell several aspects of the story in the present section and the next, while later in the chapter we explain how this result will provide us guidance on choosing renormalization parameters.

The object of interest is a true interaction field φ. Please make sure you notice the change of notation, now φ is the interacting field, not the free field. The time-evolution of φ is governed by a Hamiltonian H. A key hypothesis is that we assume that the "vacuum" of the previous section is the unique ground state of H. It will be denoted by $|\Omega\rangle$, and is sometimes called the physical ground state. Assuming also (in a way which will be made precise later) that φ is a perturbation of a free field φ_0, the purpose of the section is twofold:

- Relate the physical ground state $|\Omega\rangle$ with the ground state $|0\rangle$ of the non-interacting system.[11]
- Then use this relationship to express the Green functions $\langle\Omega|T\varphi(y_1)\ldots\varphi(y_k)|\Omega\rangle$ of (14.2) using only quantities related to the simpler non-interacting system.[12]

As we use the interaction picture, we recommend that the reader reviews now Section 11.2, and also Sections 13.3 and 13.4.

We have a free field φ_0 (as in Section 5.1) with associated mass m, whose time-evolution is governed by a Hamiltonian H_0. Writing $U_0(t) = \exp(-itH_0)$, according to Theorem 6.11.1, it holds that

$$\varphi_0(t,\boldsymbol{x}) = U_0(t)^{-1} \circ \varphi_0(0,\boldsymbol{x}) \circ U_0(t). \tag{14.3}$$

We assume that the Hamiltonian H of the true interacting field φ is of the type $H = H_0 + gH_I$, where H_I is an interacting Hamiltonian[13] H_I. We assume that H has a unique ground state $|\Omega\rangle$. We denote by U the unitary evolution $U(t) = \exp(-itH)$, so that as a consequence of (14.1) the true interacting field φ satisfies

$$\varphi(t,\boldsymbol{x}) = U(t)^{-1} \circ \varphi(0,\boldsymbol{x}) \circ U(t). \tag{14.4}$$

10 Furthermore, the "imaginary times" considered in Peskin and Schroeder's book [64] do not really contribute to clarify the picture, but fortunately we will avoid such extravagance.

11 You are certainly aware that we make plenty of implicit assumptions, like e.g. the fact that the interacting quantum field and the non-interaction quantum field correspond to the same state space.

12 As we will see later, the Green functions are intrinsically related to scattering amplitudes. For now we take their importance for granted.

13 Although the argument is quite general, we may think of H_I as being given by a density $\mathcal{H}(x) =: \varphi_0^k(x)/k!:$.

We assume, as is typical in the interaction picture, that $\varphi(0,\boldsymbol{x}) = \varphi_0(0,\boldsymbol{x})$.[14] Then, it follows from (14.3) and (14.4) that we may describe φ in the interaction picture by

$$\varphi(t,\boldsymbol{x}) = U(t)^{-1} \circ \varphi_0(0,\boldsymbol{x}) \circ U(t) = V(t)^{-1} \circ \varphi_0(t,\boldsymbol{x}) \circ V(t), \tag{14.5}$$

with our usual notation[15]

$$V(t) = U_0(-t)U(t). \tag{14.6}$$

Consider states $|\xi\rangle, |\psi\rangle$ and points $y_1,\ldots,y_k \in \mathbb{R}^{1,3}$. Let $t_1,\ldots,t_k$ be the time-components of $y_1,\ldots,y_k$. Consider two times t_- and t_+ such that $t_- \le t_1,\ldots,t_k \le t_+$. The following purely algebraic formula is called the Gell-Mann–Low formula in Duncan's textbook [24]. It is a first step in our program of expressing the Green functions using the free field only. We recall the notation $H_I(t) = U_0(t)^{-1} H_I U_0(t)$.

Proposition 14.3.1 (The Gell-Mann–Low formula) *It holds that*

$$\langle\xi|V(t_+)\mathcal{T}\varphi(y_1)\ldots\varphi(y_k)V(t_-)^{-1}|\psi\rangle = \sum_{n\ge 0}\frac{(-\mathrm{i}g)^n}{n!}\int_{t_-}^{t_+}\mathrm{d}\theta_1\ldots\int_{t_-}^{t_+}\mathrm{d}\theta_n\mathcal{K}_n, \tag{14.7}$$

where

$$\mathcal{K}_n = \langle\xi|\mathcal{T}\varphi_0(y_1)\ldots\varphi_0(y_k)H_I(\theta_1)\cdots H_I(\theta_n)|\psi\rangle. \tag{14.8}$$

The special case $k = 0, t_+ = \infty, t_- = -\infty$ is Dyson's formula (13.4) for the S-matrix. The meaning of (14.7) is the same as that of Dyson's formula: In perturbation theory, one approximates the left-hand side by the first few terms of the series in the right-hand side (by definition of perturbation theory). The point of the formula is that the left-hand side, an expression involving the complicated full interaction field, is expressed in the right-hand side using only the simpler free field φ_0.

Proof To start the derivation we define

$$V(t,s) = V(t)V(s)^{-1}. \tag{14.9}$$

We may assume without loss of generality that the time-components $t_1,\ldots,t_k$ of $y_1,\ldots,y_k$ satisfy

$$t_0 := t_+ \ge t_1 \ge t_2 \ldots \ge t_k \ge t_{k+1} := t_-.$$

[14] As we explained in Section 13.4, the interaction picture is not mathematically consistent, and the assumption that $\varphi(0,\boldsymbol{x}) = \varphi_0(0,\boldsymbol{x})$ is probably untenable. One may try to salvage the situation as follows. For the diagrammatic computations that follow, all that matters is that $\varphi_0(t,\boldsymbol{x})$ is given by the formula (5.30) where the operators $a(\boldsymbol{p})$ and $a^\dagger(\boldsymbol{p}')$ satisfy the usual commutation relations. We may assume as a basic hypothesis that the field φ satisfies the Klein-Gordon equation and the canonical commutation relations. We have then argued that $\varphi(0,\boldsymbol{x})$ is given by the formula (6.72) where the operators $a(\boldsymbol{p})$ and $a^\dagger(\boldsymbol{p}')$ satisfy the usual commutation relations. One may then *define* $\varphi_0(t,\boldsymbol{x}) := U_0(t)^{-1} \circ \varphi(0,\boldsymbol{x}) \circ U_0(t)$ where $U_0(t)$ is the time-evolution under the Hamiltonian (5.36). The field $\varphi_0(t,\boldsymbol{x})$ is then given by the formula (5.30) for the same operators $a(\boldsymbol{p})$ and $a^\dagger(\boldsymbol{p})$ and everything works out.

[15] Using the interaction picture precisely means that we are considering this operator.

We observe that (14.5) means $\varphi(y_\ell) = V(t_\ell)^{-1}\varphi_0(y_\ell)V(t_\ell)$, so that the left-hand side of (14.7) is

$$\langle\xi|V(t_+,t_1)\varphi_0(y_1)V(t_1,t_2)\dots\varphi_0(y_k)V(t_k,t_-)|\psi\rangle. \tag{14.10}$$

The plan is now to manipulate the right-hand side of (14.7) and to show that it equals (14.10). Let us define

$$W_n := \int_{t_-}^{t_+} \mathrm{d}\theta_1 \dots \int_{t_-}^{t_+} \mathrm{d}\theta_n \mathcal{K}_n, \tag{14.11}$$

where $\mathcal{K}_n$ is as in (14.8). Thus the right-hand side of (14.7) is $\sum_{n\geq 0}(-ig)^n W_n/n!$. We will decompose the domain of integration in the integral W_n into different pieces, according to exactly which the points $\theta_1,\dots,\theta_n$ fall into each of the intervals $(t_\ell,t_{\ell-1})$, for $1\leq \ell\leq k+1$ (with $t_0 = t_-$ and $t_{k+1} = t_+$). Consider a partition $J_1,\dots,J_{k+1}$ of $\{1,\dots,n\}$ into $k+1$ sets, and denote $D(J_1,\dots,J_{k+1})$ the set of all $(\theta_1,\dots,\theta_n)$ with $t_- \leq \theta_1,\dots,\theta_n \leq t_+$ such that $\theta_j \in (t_\ell,t_{\ell-1})$ whenever $j \in J_\ell$ (and not caring about lower-dimensional sets where one θ_j may equal one t_ℓ). Let us denote

$$W_{J_1,\dots,J_{k+1}} = \int \dots \int_{D(J_1,\dots,J_{k+1})} \mathrm{d}\theta_1 \dots \mathrm{d}\theta_n \mathcal{K}_n,$$

so that

$$W_n = \sum_{J_1,\dots,J_{k+1}} W_{J_1,\dots,J_{k+1}}, \tag{14.12}$$

for a sum over all possible partitions $J_1,\dots,J_{k+1}$ of $\{1,\dots,n\}$. Now, if $(\theta_1,\dots,\theta_n) \in D(J_1,\dots,J_{k+1})$ we know how to perform the time-ordering

$$\mathcal{T}\varphi_0(y_1)\dots\varphi_0(y_k)H_I(\theta_1)\cdots H_I(\theta_n).$$

It is simply

$$\mathcal{U}_1\varphi_0(y_1)\mathcal{U}_2\dots\varphi_0(y_k)\mathcal{U}_{k+1}$$

where $\mathcal{U}_\ell$ is the time-ordered product $\mathcal{T}\prod_{j\in J_\ell} H_I(\theta_j)$. Hence

$$\mathcal{K}_n = \langle\xi|\mathcal{U}_1\varphi_0(y_1)\mathcal{U}_2\dots\varphi_0(y_k)\mathcal{U}_{k+1}|\psi\rangle.$$

Consequently

$$W_{J_1,\dots,J_{k+1}} = \langle\xi|V_{1,m_1}\varphi_0(y_1)V_{2,m_2}\dots\varphi_0(y_k)V_{k+1,m_{k+1}}|\psi\rangle$$

where $m_\ell = \operatorname{card} J_\ell$ and

$$V_{\ell,m} = \int_{t_\ell}^{t_{\ell-1}} \mathrm{d}\theta_1 \dots \int_{t_\ell}^{t_{\ell-1}} \mathrm{d}\theta_m \mathcal{T} H_I(\theta_1)\cdots H_I(\theta_m).$$

Thus $W_{J_1,\dots,J_{k+1}}$ depends only on $m_1,\dots,m_{k+1}$, $W_{J_1,\dots,J_{k+1}} = W_{m_1,\dots,m_{k+1}}$, and (14.12) implies

$$W_n = \sum_{J_1,\dots,J_{k+1}} W_{J_1,\dots,J_{k+1}} = \sum_{m_1,\dots,m_{k+1}} \frac{n!}{m_1!\cdots m_{k+1}!} W_{m_1,\dots,m_{k+1}},$$

and finally

$$\frac{W_n}{n!} = \sum_{m_1,\dots,m_{k+1}} \frac{1}{m_1!\cdots m_{k+1}!} W_{m_1,\dots,m_{k+1}}$$

$$= \sum \langle \xi | \frac{V_{1,m_1}}{m_1!} \varphi_0(y_1) \frac{V_{2,m_2}}{m_2!} \dots \varphi_0(y_k) \frac{V_{k+1,m_{k+1}}}{m_{k+1}!} |\psi\rangle, \tag{14.13}$$

where the summation is over all ways to decompose n as a sum $m_1 + \cdots + m_{k+1}$. Next, we note that if for $\ell = 1,\dots,k+1$ we consider generic formal series $A_{\ell,m}$ then

$$\prod_{\ell \le k+1} \sum_m A_{\ell,m} = \sum_n \sum_{m_1+\cdots+m_{k+1}=n} A_{1,m_1} \dots A_{k+1,m_{k+1}},$$

where the second summation is over all ways to decompose n as a sum $m_1 + \cdots + m_{k+1}$. Multiplying equality (14.13) by $(-ig)^n$, summing over $n \ge 0$ and using the previous identity we then obtain

$$\sum_{n\ge 0} \frac{(-ig)^n}{n!} W_n = \langle \xi | V_1 \varphi_0(y_1) V_2 \dots \varphi_0(y_k) V_{k+1} |\psi\rangle, \tag{14.14}$$

where

$$V_\ell = \sum_{m\ge 0} \frac{(-ig)^m}{m!} \int_{t_\ell}^{t_{\ell-1}} \mathrm{d}\theta_1 \dots \int_{t_\ell}^{t_{\ell-1}} \mathrm{d}\theta_m \mathcal{T} H_I(\theta_1)\cdots H_I(\theta_m).$$

Finally, we recall from (11.38) that for $s \le t$ we have

$$V(t,s) = \sum_{m\ge 0} \frac{(-ig)^m}{m!} \int_s^t \mathrm{d}\theta_1 \dots \int_s^t \mathrm{d}\theta_m \mathcal{T} H_I(\theta_1)\cdots H_I(\theta_m).$$

This proves that $V_\ell = V(t_{\ell-1}, t_\ell)$. We have achieved our desired goal to show that the right-hand side of (14.14) coincides with (14.10), i.e. with the left-hand side of (14.7). □

Let us specialize the Gell-Mann–Low formula to the case $|\xi\rangle = |\psi\rangle = |0\rangle$, the ground state of the non-interacting system. Since $H_0|0\rangle = 0$, we have $U_0(t)|0\rangle = |0\rangle$ so that $V(t)^{-1}|0\rangle = U(t)^{-1}|0\rangle$. Looking back at (14.7) we observe that if we can relate $U(t)^{-1}|0\rangle$ with $|\Omega\rangle$ (the ground state of the interacting system) then we can express the Green functions using quantities related to the non-interacting model.

Let us first observe that since we assume that $|\Omega\rangle$ is an eigenstate of every operator $U(c,C)$ it is in particular an eigenstate of $U(t)$, and we have

$$U(t)|\Omega\rangle = \exp(-itE)|\Omega\rangle \tag{14.15}$$

where E is the energy of the ground state $|\Omega\rangle$, $H|\Omega\rangle = E|\Omega\rangle$.[16] Since for a complex number α of modulus 1, $\alpha|\Omega\rangle$ represents the same state as $|\Omega\rangle$, (14.15) means that the true vacuum is the same at all times. Consider now a state $|\psi\rangle$ and assume that it is not

16 Some authors do not bother to mention this phase factor (e.g. Schwartz [80, page 80]), but it cannot be 1 unless one has arranged that $E = 0$.

a mixture of $|\Omega\rangle$ and another state, i.e. assume that $\langle\psi|\Omega\rangle = 0$. Then, as we time-evolve $|\psi\rangle$, i.e. we consider $U(t)|\psi\rangle$ we expect that this state "spreads out and goes to infinity" as $t \to \pm\infty$, just as what happens in standard Quantum Mechanics for the spreading of the wave function that we met in Section 12.2. More specifically, if we fix any state $|\xi\rangle$ we expect that the probability that for $t \to \pm\infty$ the state $U(t)|\psi\rangle$ is measured as being $|\xi\rangle$ vanishes, i.e. we expect that $\lim_{t\to\pm\infty}\langle\psi|U(t)^{-1}|\xi\rangle = 0$. In the next section we will show that this physically obvious fact is also mathematically reasonable.

To continue our analysis, we *assume* something a bit weaker, namely that for any state $|\psi\rangle$ one has the following

$$\langle\psi|\Omega\rangle = 0 \Rightarrow \lim_{t\to\pm\infty}\langle\psi|U(t)^{-1}|0\rangle = 0. \tag{14.16}$$

Lemma 14.3.2 *Under (14.16) there exists a phase factor $r(t)$, i.e. a complex number with $|r(t)| = 1$ with the following property: Given any state $|\psi\rangle$, we have*

$$\langle\psi|U(t)^{-1}|0\rangle = r(t)\langle\psi|\Omega\rangle\langle\Omega|0\rangle + \varepsilon(t), \tag{14.17}$$

where $\lim_{t\to\pm\infty}\varepsilon(t) = 0$.

In particular this shows that when (14.16) holds, this implication can be reversed if $\langle\Omega|0\rangle \neq 0$.

Proof Assuming without loss of generality that $\langle\Omega|\Omega\rangle = 1$, consider the state $|\xi\rangle$ given by $\langle\xi| = \langle\psi| - \langle\psi|\Omega\rangle\langle\Omega|$, so that, using (14.15),

$$\langle\psi|U(t)^{-1}|0\rangle = \langle\xi|U(t)^{-1}|0\rangle + \exp(itE)\langle\psi|\Omega\rangle\langle\Omega|0\rangle.$$

Since $\langle\xi|\Omega\rangle = 0$ the result follows using (14.16) for ξ. □

We are now ready to argue that, *assuming* (14.16), one has the following, which is also sometimes called the "Gell-Mann–Low theorem". The word "theorem" is certainly not to be taken in the mathematical sense, in particular since the assumption (14.16) appears nowhere in the statement of the theorem. But we will wave our hands later to argue that "(14.16) should be typically true".

Theorem 14.3.3 (The Gell-Mann–Low Theorem) *It holds that*

$$\langle\Omega|\mathcal{T}\varphi(y_1)\ldots\varphi(y_k)|\Omega\rangle = \frac{A}{B} \tag{14.18}$$

where

$$A = \sum_{n\geq 0}\frac{(-ig)^n}{n!}\int_{-\infty}^{\infty} d\theta_1 \ldots \int_{-\infty}^{\infty} d\theta_n \langle 0|\mathcal{T}\varphi_0(y_1)\ldots\varphi_0(y_k)H_I(\theta_1)\cdots H_I(\theta_n)|0\rangle \tag{14.19}$$

and

$$B = \sum_{n\geq 0}\frac{(-ig)^n}{n!}\int_{-\infty}^{\infty} d\theta_1 \ldots \int_{-\infty}^{\infty} d\theta_n \langle 0|\mathcal{T}H_I(\theta_1)\cdots H_I(\theta_n)|0\rangle. \tag{14.20}$$

Supporting argument. Assuming (14.16), and using (14.17) for $\langle\psi| = \langle 0|V(t_+)\mathcal{T}\varphi(y_1)\ldots\varphi(y_k)|$, we can then believe that for given t_+, when $-t_-$ becomes large the left-hand side of (14.7) when $|\xi\rangle = |\eta\rangle = |0\rangle$ is about

$$r(t_-)\langle\Omega|0\rangle\langle 0|V(t_+)\mathcal{T}\varphi(y_1)\ldots\varphi(y_k)|\Omega\rangle. \tag{14.21}$$

Repeating the argument (using now the assumption (14.16) for $t\to\infty$) we can further believe that for t_+ large this is about

$$r(t_-)r(t_+)\langle\Omega|0\rangle\langle 0|\Omega\rangle\langle\Omega|\mathcal{T}\varphi(y_1)\ldots\varphi(y_k)|\Omega\rangle. \tag{14.22}$$

In summary, the quantity (14.22) is about the right-hand side of (14.7). This formula is not useful yet because the first four terms are not known. To get rid of them we consider the same expression, but with $k=0$. Taking the ratio of these expressions eliminates the unknown quantity $r(t_-)r(t_+)\langle\Omega|0\rangle\langle 0|\Omega\rangle$, and taking the limit as $-t_-\to\infty$, $t_+\to\infty$ we obtain the expression (14.18) for the Green functions.

For this procedure to make any sense at all it is absolutely necessary that $\langle\Omega|0\rangle \neq 0$. As to why this should be true, I repeat the argument put forward in Peskin and Schroeder's textbook [64]: The Hamiltonian H is supposed to be a small perturbation of H_0 and this would not be the case if its ground state $|\Omega\rangle$ was too different from the ground state $|0\rangle$ of H_0.

Let us now analyze (14.16) more precisely, with the goal of arguing that it should typically be true. Changing t into $-t$ we want that for a state $|\psi\rangle$

$$\langle\psi|\Omega\rangle = 0 \Rightarrow \lim_{t\to\infty}\langle\psi|U(t)|0\rangle = 0. \tag{14.23}$$

Denoting by (x,y) the generic point of $\mathbb{R}^2$ let us recall the "multiplication by x" operator of Exercise 2.5.19 on the space $L^2(\mathbb{R}^2, d\mu)$, where μ is a certain positive measure μ on $\mathbb{R}^2$.[17] As stated without proof just before Exercise 2.5.19, spectral theory asserts that a general self-adjoint operator is (essentially) not more complicated than this "multiplication by x" operator. So to assess the credibility of (14.23) we might as well assume that the state space is this L^2, and that H is of this type. In this case the time-evolution $U(t)$ is simply multiplication by $\exp(-itx)$.What are then the reasonable hypotheses one should make about μ? First, without loss of generality we may assume that the physical vacuum $|\Omega\rangle$ has zero energy, $H|\Omega\rangle = 0$, by adding a constant to H if necessary. Also, a standard hypothesis is that the ground state for H is unique (all eigenstates of zero energy are proportional to the physical vacuum). According to Exercise 2.5.18 the hypothesis that the multiplication by x operator has zero as an eigenvalue translates into

$$\mu(\{0\}\times\mathbb{R}) = \mu(\{(0,y)\,;\, y\in\mathbb{R}\}) > 0. \tag{14.24}$$

[17] It might be difficult to fully appreciate the material that follows unless one has some idea about what a positive measure is, other than a measure defined by a density, or concentrated at a few points. I did learn this in my third year of college, but I am unsure that this is taught properly at every institution. If you do not know this, please jump to a few lines before (14.32).

Each element of $L^2(\mathbb{R}^2, \mathrm{d}\mu)$ that is not zero only for $x = 0$ provides an eigenvector with eigenvalue zero. The hypothesis that the eigenspace corresponding to the eigenvalue zero has dimension 1 implies that for some y_0 we have[18]

$$\mu(\{(0,y_0)\}) = \mu(\{(0,y)\,;\; y \in \mathbb{R}\}) > 0. \tag{14.25}$$

Thus the ground state $|\Omega\rangle$ corresponds to the function χ of $L^2(\mathbb{R}^2, \mathrm{d}\mu)$ given by $\chi(x,y) = 1/\sqrt{\mu(\{(0,y_0)\})}$ if $(x,y) = (0,y_0)$ and by $\chi(x,y) = 0$ otherwise. Hence, for a function $g \in L^2$,

$$\int \mathrm{d}\mu g\chi = \mu(\{0,y_0\})\chi(0,y_0)g(0,y_0) = \sqrt{\mu(\{0,y_0\})}g(0,y_0).$$

If f is an element of $L^2(\mathbb{R}^2, \mathrm{d}\mu)$ which we denote $|\xi\rangle$ in Dirac's notation, then $\langle\xi|\Omega\rangle$ is the inner product (f,χ),

$$\langle\xi|\Omega\rangle = \int \mathrm{d}\mu\, f^*\chi = \sqrt{\mu(\{(0,y_0)\})}\,f^*(0,y_0). \tag{14.26}$$

Let us now assume that

$$|\Omega\rangle \text{ is the only eigenstate.} \tag{14.27}$$

As the next exercise shows, this is the case for free particles, so that the assumption is not unreasonable. Let us stress that here we talk about any eigenstate at all, independently of the eigenvalue, and we talk about true states, elements of $\mathcal{H}$. The "improper states of given momentum" of physics do not qualify as true states!

> **Exercise 14.3.4** Prove that the Hamiltonian (5.36) on the boson Fock space has only multiples of the vacuum as eigenvectors.

According to Exercise 2.5.18 this condition translates into

$$\forall x \neq 0\,;\; \mu(\{(x,y)\,;\; y \in \mathbb{R}\}) = 0. \tag{14.28}$$

Now, it holds that

$$\langle\xi|U(t)|0\rangle = \int \mathrm{d}\mu(x,y) f(x,y)^* h(x,y)\exp(-itx),$$

where f and h are the elements of the state space $L^2(\mathbb{R}^2, \mathrm{d}\mu)$ which we have denoted $|\xi\rangle$ and $|0\rangle$ in Dirac's notation. Assuming that $\langle\xi|\Omega\rangle = 0$, by (14.26) we have $f(0,y_0) = 0$. From (14.25) we get

$$\int_{\{x=0\}} \mathrm{d}\mu(x,y) f(x,y)^* h(x,y)\exp(-itx) = \mu(\{(0,y_0)\}) f(0,y_0)^* h(0,y_0) = 0$$

[18] This is because the only spaces $L^2(\nu)$ that are one-dimensional are those for which ν is concentrated at one point.

so that

$$\langle \xi | U(t) | 0 \rangle = \int_{\{x \neq 0\}} \mathrm{d}\mu(x,y) f(x,y)^* h(x,y) \exp(-itx). \tag{14.29}$$

As the integral is now restricted to $x \neq 0$ the term $\exp(-itx)$ provides oscillatory behavior at $t \to \infty$. Several textbooks at this point appeal to the "Riemann–Lebesgue Lemma". This lemma states that for a Lebesgue integrable function g

$$\lim_{t \to \infty} \int \mathrm{d}x g(x) \exp(-itx) = 0. \tag{14.30}$$

It is proved (once one knows something about the Lebesgue integral) by approximating g by a continuous function with compact support. From a purely mathematical point of view it is not quite true that the hypothesis (14.28) implies that the term in (14.29) goes to zero.[19] In order to appeal to the Riemann–Lebesgue lemma one certainly has to assume more than (14.28),[20] but assuming this extra regularity is again completely reasonable by analogy with the case of free particles.

In conclusion, we have waved our hands in believing that (14.16) should "typically" be true and that it is safe to use the Gell-Mann–Low "theorem".

14.4 Adiabatic Switching of the Interaction

In this section we support the plausibility of the Gell-Mann–Low theorem (14.18) from a related but different direction. Our previous analysis relied on arguing that the quantity

$$\int_{\{x \neq 0\}} \mathrm{d}\mu(x,y) f(x,y)^* h(x,y) \exp(-itx) \tag{14.31}$$

of (14.29) goes to zero as $t \to \infty$. A particularly effective way to achieve this is to insert a factor $\exp(-\varepsilon t|x|)$ in the integrand.[21] In this section we investigate the related idea of *adiabatic switching* of the interaction. The word "adiabatic" comes from Thermodynamics, where it refers to processes without exchange of heat between the system and the outside world. The important case is when these processes are "infinitely slow" and here the word "adiabatic" refers to this. The interaction is turned on "infinitely slowly" by considering for $\varepsilon > 0$ the Hamiltonian

$$H_\varepsilon(t) = H_0 + g \exp(-\varepsilon|t|) H_I, \tag{14.32}$$

and letting $\varepsilon \to 0$. To clarify the following discussion let us call *adiabatic evolution*[22] of a state $|\alpha\rangle$ the following procedure (which assumes that all goes well). We start with the state

[19] The point here is that there exist positive measures ν on $\mathbb{R}$ of finite mass such that $\nu(\{x\}) = 0$ for *each* x, but for which $\int \mathrm{d}\nu(x) \exp(-itx) \not\to 0$ as $t \to \infty$.

[20] Specifically one should assume that image of μ by the projection on the x-axis has a density with respect to Lebesgue's measure.

[21] Since $\exp(-itx)\exp(-\varepsilon tx) = \exp(-itx(1 - i\varepsilon)) = \exp(-i\hat{t}x)$ where $\hat{t} = t(1 - i\varepsilon)$, this is what motivates the "imaginary times" $t(1 - i\varepsilon)$ in Peskin and Schroeder's book [64], the idea being to let $t \to \infty$ and then $\varepsilon \to 0$. Unfortunately, even at the informal level, the details of recovering (14.18) with this approach lack appeal.

[22] I am making up this non-standard terminology for clarity

$|\alpha\rangle$ at time $t = -\infty$ and we turn on the interaction very slowly.[23] The state we obtain at time $t = 0$ is the adiabatic evolution of the original state $|\alpha\rangle$ (by definition of adiabatic evolution). The basic expectation in the approach is that if all goes well, the adiabatic evolution of the ground state of $H_0 = H_\varepsilon(-\infty)$ is the ground state of $H_\varepsilon(0) = H$, the full Hamiltonian.

In the first part of this section we concentrate on this phenomenon, and at the end of the section we explain why our results provide a set of conditions under which (14.18) holds.

The adiabatic evolution of the ground state of H_0 was originally studied by Gell-Mann and Low in the theory of multiparticle systems.[24] The arguments supporting the main result, Theorem 14.4.1, look closer to respectable mathematics than the previous considerations, but the conclusion of that theorem is somewhat different from what we need. It simply says that the adiabatic evolution of an eigenvector of the non-interacting system (and in particular the ground state of this system) is an eigenvector of the interacting system. It does *not* say that the adiabatic evolution of the ground state of the non-interacting system is the ground state of the interacting system. There are natural models (in the theory of multiparticle systems) where this is not the case, refer to Fetter and Walecka's book [29]. Nonetheless we may expect that "usually" this will be the case, and this belief provides further support for (14.18). A reasonable (and very common) hypothesis one might make about Hamiltonians in Quantum Field Theory is the existence of a *mass gap*. Assuming that we have added a constant to the Hamiltonian to ensure that $|\Omega\rangle$ has energy zero, this means that anything that is orthogonal to the vacuum $|\Omega\rangle$ will have energy $\geq \mu_0 > 0$. One can express this by requiring that

$$\langle\Omega|\xi\rangle = 0 \Rightarrow \langle\xi|H|\xi\rangle \geq \mu_0\langle\xi|\xi\rangle. \tag{14.33}$$

Intuitively this means that μ_0 "is the smallest mass in the universe".[25] One might believe that if for each $0 \leq \lambda \leq 1$ there is a "mass gap" for the interaction with Hamiltonian $H_0 + \lambda g H_I$, then there are far better chances that the adiabatic evolution of the ground state of H_0 is the true ground state of the interacting system, for the simple reason that under this hypothesis the ground state of the interacting system is "far more distinguished". The mass gap hypothesis looks reasonable when studying models involving a single massive particle.

We now start the detailed study of the adiabatic evolution of the ground state of H_0. Let us explain the setting. We start with a Hamiltonian H_0 and a true eigenstate $|\eta\rangle$,

$$H_0|\eta\rangle = E_0|\eta\rangle.$$

We consider a perturbation of this Hamiltonian,

$$H = H_0 + gH_I,$$

[23] That is, as will be made precise later, taking the limit of the previous procedure as $\varepsilon \to 0$.

[24] Consequently, Theorem 14.4.1 is also called "the Gell-Mann–Low theorem", just as Theorem 14.3.3, greatly contributing to the linguistic confusion.

[25] This is a bit informal. An eigenvalue of the Hamiltonian is an energy, and technically μ_0 is an energy. In our unit system where the speed of light is $c = 1$, there is equivalence between energy and mass.

and for $\varepsilon > 0$ we consider the time-dependent Hamiltonian (14.32). Generally speaking the time-evolution $U(t)$ of a time-dependent Hamiltonian $H(t)$ satisfies[26] $\mathrm{i}\mathrm{d}U(t)/\mathrm{d}t = H(t)U(t)$.[27] As usual we denote by $U_0(t)$ the time-evolution under the Hamiltonian H_0, and we denote by $\bar{U}_\varepsilon(t)$ the time-evolution under the Hamiltonian $H_\varepsilon(t)$ of (14.32). We consider the operators $V_\varepsilon(t) := U_0(-t)\bar{U}_\varepsilon(t)$, which satisfy the equation

$$\begin{aligned}\mathrm{i}\frac{\partial}{\partial t}V_\varepsilon(t) &= -U_0(-t)H_0\bar{U}_\varepsilon(t) + U_0(-t)(H_0 + g\exp(-\varepsilon|t|)H_I)\bar{U}_\varepsilon(t)\\ &= g\exp(-\varepsilon|t|)H_I(t)V_\varepsilon(t),\end{aligned} \tag{14.34}$$

where $H_I(t) = U_0(-t)H_IU_0(t)$, a relation that is itself equivalent to

$$\mathrm{i}\frac{\mathrm{d}}{\mathrm{d}t}H_I(t) = -[H_0, H_I(t)]. \tag{14.35}$$

Because of the exponential decay in (14.34), we believe that the limits $V_\varepsilon(\pm\infty) := \lim_{t\to\pm\infty} V_\varepsilon(t)$ exist. For the time being we consider only $t \le 0$ and we define $V_\varepsilon(t, t_0) := V_\varepsilon(t)V_\varepsilon(t_0)^{-1}$. Since $V_\varepsilon(0)$ is the identity, we have $V_\varepsilon(0, t_0) = V_\varepsilon(t_0)^{-1}$. Furthermore (14.34) implies

$$\mathrm{i}\frac{\partial}{\partial t}V_\varepsilon(t, t_0) = g\exp(\varepsilon t)H_I(t)V_\varepsilon(t, t_0). \tag{14.36}$$

These formulas contain a dependence on g which is kept implicit, but which motivates the use of the partial derivatives below. We define

$$|\psi_\varepsilon\rangle = V_\varepsilon(0, -\infty)|\eta\rangle = V_\varepsilon(-\infty)^{-1}|\eta\rangle. \tag{14.37}$$

Theorem 14.4.1 *Assume the following limits exist:*

$$|\psi\rangle = \lim_{\varepsilon\to 0}\frac{|\psi_\varepsilon\rangle}{\langle\eta|\psi_\varepsilon\rangle} \tag{14.38}$$

$$\lim_{\varepsilon\to 0}\mathrm{i}\varepsilon g\frac{\partial}{\partial g}\log\langle\eta|\psi_\varepsilon\rangle, \tag{14.39}$$

$$0 = \lim_{\varepsilon\to 0}\varepsilon\frac{\partial}{\partial g}\frac{|\psi_\varepsilon\rangle}{\langle\eta|\psi_\varepsilon\rangle}. \tag{14.40}$$

Then typically[28] *$|\psi\rangle$ is an eigenvector of H of eigenvalue*

$$E = E_0 + \lim_{\varepsilon\to 0}\mathrm{i}\varepsilon g\frac{\partial}{\partial g}\log\langle\eta|\psi_\varepsilon\rangle. \tag{14.41}$$

The actual computation (14.41) of the eigenvalue is remarkable. You would not see conditions (14.39) and (14.40) in physics papers since no physicist is going to worry about these limits which have no particular reason *not* to exist. Rigorous mathematical results exist

[26] Recalling that we take $\hbar = 1$ in this chapter.
[27] It is by no means obvious when this actually exists but we stay here at the heuristic level.
[28] The purpose of the word "typically" is to remind you that this is not a mathematical theorem, and that it is not guaranteed that its conclusions hold with absolute certainty.

in the direction of Theorem 14.4.1 as in Nenciu and Rasche's paper [60], but unfortunately they require hypotheses that are not satisfied in the cases of interest here.

To understand better Theorem 14.4.1, it helps to investigate the trivial case where $H_0 = 0$ (or more generally where H_0 is a multiple of the identity). Then $H_I(t) = H_I$ and the equation (14.36) has the solution

$$V_\varepsilon(t,t_0) = \exp\Big(\frac{gH_I}{\mathrm{i}\varepsilon}(\exp(\varepsilon t) - \exp(\varepsilon t_0))\Big).$$

Then if $|\eta\rangle$ is an eigenvector of H_I, $H_I|\eta\rangle = \alpha|\eta\rangle$,

$$|\psi_\varepsilon\rangle = V_\varepsilon(0,-\infty)|\eta\rangle = \exp\Big(\frac{g\alpha}{\mathrm{i}\varepsilon}\Big)|\eta\rangle.$$

Thus $|\psi_\varepsilon\rangle$ will not have a limit as $\varepsilon \to 0$ because of the oscillating factor $\exp(g\alpha/\mathrm{i}\varepsilon)$.[29] On the other hand $|\psi_\varepsilon\rangle/\langle\eta|\psi_\varepsilon\rangle = |\eta\rangle/\langle\eta|\eta\rangle$ is independent of ε and g, so that (14.40) holds and furthermore

$$\lim_{\varepsilon\to 0} \mathrm{i}\varepsilon g\frac{\partial}{\partial g}\log\langle\eta|\psi_\varepsilon\rangle = g\alpha.$$

Based on this example we may believe that condition (14.40) is reasonable, and that the partial derivative in g has a tendency to stay bounded as $\varepsilon \to 0$.

Another equally trivial example is of interest: In dimension two, consider the case where H_0 is diagonal with diagonal elements 1 and -1, and denote by $|\pm 1\rangle$ the corresponding eigenvectors. Take $H_I = -2H_0$ and $g = 1$. Then $|\eta\rangle = |-1\rangle$ is the ground state of H_0. Obviously in the present case the operators $V_\varepsilon(t,t_0)$ are diagonal, so that the vector $|\psi\rangle$ of (14.38) is $|\eta\rangle$ itself, but this is *not* the ground state of $H = -H_0$. This is related to the fact that a singular situation occurs for $H_0 + 1/2H_I$, where the ground state is not unique. This does not contradict the theorem, which does not claim that $|\psi\rangle$ should be the ground state, but it dashes our hopes that $|\psi\rangle$ should be the ground state unless pathology occurs. As the type of phenomenon occurring in this example is however ruled out if one assumes that there is a mass gap for any strength of the interaction, we may still hope that everything will be fine in the cases we need.

Lemma 14.4.2 *We have*

$$\begin{aligned}[H_0, V_\varepsilon(t,t_0)] &= -g\exp(\varepsilon t)H_I(t)V_\varepsilon(t,t_0)\\ &\quad + g\exp(\varepsilon t_0)V_\varepsilon(t,t_0)H_I(t_0) + \mathrm{i}\varepsilon g\frac{\partial}{\partial g}V_\varepsilon(t,t_0). \end{aligned}\tag{14.42}$$

Supporting argument. It holds that

$$\begin{aligned}\mathrm{i}\frac{\partial}{\partial t}[H_0, V_\varepsilon(t,t_0)] &= [H_0, \mathrm{i}\frac{\partial}{\partial t}V_\varepsilon(t,t_0)]\\ &= [H_0, g\exp(\varepsilon t)H_I(t)V_\varepsilon(t,t_0)],\end{aligned}\tag{14.43}$$

[29] The purpose of the denominator $\langle\eta|\psi_\varepsilon\rangle$ in (14.38) is to cancel this factor.

using (14.36) in the last line. Next, we use the formula $[A, BC] = [A, B]C + B[A, C]$ to obtain

$$\mathrm{i}\frac{\partial}{\partial t}[H_0, V_\varepsilon(t, t_0)] = g\exp(\varepsilon t)\Big([H_0, H_I(t)]V_\varepsilon(t, t_0) + H_I(t)[H_0, V_\varepsilon(t, t_0)]\Big).$$

Recalling that $[H_0, H_I(t)] = -\mathrm{i}\mathrm{d}H_I(t)/\mathrm{d}t$ from (14.35), we have shown that the operator-valued function $S(t) := [H_0, V_\varepsilon(t, t_0)]$ satisfies the differential equation

$$DS(t) = -\mathrm{i}g\exp(\varepsilon t)\frac{\mathrm{d}}{\mathrm{d}t}(H_I(t))V_\varepsilon(t, t_0), \tag{14.44}$$

where the differential operator D is given by

$$DS(t) = \mathrm{i}\frac{\partial}{\partial t}S(t) - g\exp(\varepsilon t)H_I(t)S(t). \tag{14.45}$$

Let us denote by $W(t)$ the right-hand side of (14.42). Then, since $V_\varepsilon(t_0, t_0) = 1$ we have $W(t_0) = S(t_0) = 0$. We will then show that $W(t)$ satisfies the differential equation (14.44). Thus $G(t) := S(t) - W(t)$ satisfies $G(t_0) = 0$ and $DG(t) = 0$ and hence $G(t) = 0$ for all t, concluding the proof.

Let us denote by $W_1(t)$, $W_2(t)$, $W_3(t)$ the three terms in the right-hand side of (14.42). Then

$$\begin{aligned} DW_1(t) = -\mathrm{i}g\frac{\mathrm{d}}{\mathrm{d}t}(\exp(\varepsilon t)H_I(t))V_\varepsilon(t, t_0) \\ -\mathrm{i}g\exp(\varepsilon t)H_I(t)\frac{\mathrm{d}}{\mathrm{d}t}V_\varepsilon(t, t_0) - g\exp(\varepsilon t)H_I(t)W_1(t), \end{aligned}$$

and using (14.36) the last two terms cancel out so that

$$DW_1(t) = -\mathrm{i}g\varepsilon\exp(\varepsilon t)H_I(t)V_\varepsilon(t, t_0) - \mathrm{i}g\exp(\varepsilon t)\frac{\mathrm{d}}{\mathrm{d}t}(H_I(t))V_\varepsilon(t, t_0). \tag{14.46}$$

Next, using (14.36) again, we find $D(W_2(t)) = 0$. Finally, using (14.36) in the second equality,

$$\begin{aligned} \mathrm{i}\frac{\partial}{\partial t}W_3(t) &= \mathrm{i}\varepsilon g\frac{\partial}{\partial g}(\mathrm{i}\frac{\partial}{\partial t}V_\varepsilon(t, t_0)) \\ &= \mathrm{i}\varepsilon g\frac{\partial}{\partial g}g\exp(\varepsilon t)H_I(t)V_\varepsilon(t, t_0) \\ &= \mathrm{i}\varepsilon g\exp(\varepsilon t)H_I(t)V_\varepsilon(t, t_0) + \mathrm{i}\varepsilon g^2\exp(\varepsilon t)H_I(t)\frac{\partial}{\partial g}V_\varepsilon(t, t_0) \\ &= \mathrm{i}\varepsilon g\exp(\varepsilon t)H_I(t)V_\varepsilon(t, t_0) + g\exp(\varepsilon t)H_I(t)W_3(t), \end{aligned} \tag{14.47}$$

and thus

$$DW_3(t) = \mathrm{i}\varepsilon g\exp(\varepsilon t)H_I(t)V_\varepsilon(t, t_0). \tag{14.48}$$

Combining the previous relations we obtain that indeed

$$DW(t) = \sum_{i\le 3} DW_i(t) = -\mathrm{i}g\exp(\varepsilon t)\frac{\mathrm{d}}{\mathrm{d}t}(H_I(t))V_\varepsilon(t, t_0) = DS(t).$$

The preceding argument has no claim to constitute a proof as a mathematician would understand it. As soon as we consider the derivative of an operator-valued function, there are issues as to what this really means and where it is defined. One would need in particular to investigate under what conditions one can actually prove that $DA(t) = 0$ and $A(t_0) = 0$ imply $A(t) = 0$.

Supporting argument for Theorem 14.4.1 We use (14.42) for $t = 0$ and $t_0 \to -\infty$ and apply the resulting equality to $|\eta\rangle$ to obtain, since $|\psi_\varepsilon\rangle = V_\varepsilon(0, -\infty)|\eta\rangle$,

$$[H_0, V_\varepsilon(0, -\infty)]|\eta\rangle = -gH_I|\psi_\varepsilon\rangle + \mathrm{i}\varepsilon g\frac{\partial}{\partial g}|\psi_\varepsilon\rangle. \tag{14.49}$$

Since $H_0|\eta\rangle = E_0|\eta\rangle$ the left-hand side of (14.49) is $(H_0 - E_0)|\psi_\varepsilon\rangle$ and therefore

$$H|\psi_\varepsilon\rangle = E_0|\psi_\varepsilon\rangle + \mathrm{i}\varepsilon g\frac{\partial}{\partial g}|\psi_\varepsilon\rangle. \tag{14.50}$$

Using the relations

$$\frac{\partial}{\partial g}\frac{|\psi_\varepsilon\rangle}{\langle\eta|\psi_\varepsilon\rangle} = \frac{1}{\langle\eta|\psi_\varepsilon\rangle}\frac{\partial}{\partial g}|\psi_\varepsilon\rangle - \frac{|\psi_\varepsilon\rangle}{\langle\eta|\psi_\varepsilon\rangle^2}\frac{\partial}{\partial g}\langle\eta|\psi_\varepsilon\rangle\ ;\ \frac{\partial}{\partial g}\log\langle\eta|\psi_\varepsilon\rangle = \frac{1}{\langle\eta|\psi_\varepsilon\rangle}\frac{\partial}{\partial g}\langle\eta|\psi_\varepsilon\rangle,$$

we obtain the identity

$$\frac{1}{\langle\eta|\psi_\varepsilon\rangle}\frac{\partial}{\partial g}|\psi_\varepsilon\rangle = \frac{\partial}{\partial g}\frac{|\psi_\varepsilon\rangle}{\langle\eta|\psi_\varepsilon\rangle} + \frac{|\psi_\varepsilon\rangle}{\langle\eta|\psi_\varepsilon\rangle}\frac{\partial}{\partial g}\log\langle\eta|\psi_\varepsilon\rangle,$$

which we use to transform (14.50) into

$$\Big(H - E_0 - \mathrm{i}\varepsilon g\frac{\partial}{\partial g}\log\langle\eta|\psi_\varepsilon\rangle\Big)\frac{|\psi_\varepsilon\rangle}{\langle\eta|\psi_\varepsilon\rangle} = \mathrm{i}\varepsilon g\frac{\partial}{\partial g}\frac{|\psi_\varepsilon\rangle}{\langle\eta|\psi_\varepsilon\rangle},$$

and letting $\varepsilon \to 0$, using (14.38) to (14.40) concludes the argument.

We end this section by showing how Theorem 14.4.1 provides support for (14.18). Consider $\varepsilon > 0$ and define $g_\varepsilon(t) = g\exp(-\varepsilon|t|)$. Consider the time-evolution $\bar{U}_\varepsilon(t)$ of the system under the Hamiltonian $H_\varepsilon(t) := H_0 + g_\varepsilon(t)H_I$ of (14.32). Recalling (14.4), let us define the quantum field φ^ε by

$$\varphi^\varepsilon(t, \boldsymbol{x}) = \bar{U}_\varepsilon(t)^{-1} \circ \varphi_0(0, \boldsymbol{x}) \circ \bar{U}_\varepsilon(t). \tag{14.51}$$

One can then copy the proof of (14.7) (in the specific case $|\xi\rangle = |\psi\rangle = |0\rangle$ and $t_- = -\infty$, $t_+ = \infty$) to obtain (please compare (11.35) and (11.36) to understand the mechanism)

$$\langle 0|V_\varepsilon(\infty)\mathcal{T}\varphi^\varepsilon(y_1)\ldots\varphi^\varepsilon(y_k)V_\varepsilon(-\infty)^{-1}|0\rangle = \sum_{n\geq 0}\frac{(-\mathrm{i})^n}{n!}\int_{-\infty}^{\infty}\mathrm{d}\theta_1\ldots\int_{-\infty}^{\infty}\mathrm{d}\theta_n\mathcal{K}_n, \tag{14.52}$$

where

$$\mathcal{K}_n = g_\varepsilon(\theta_1)\ldots g_\varepsilon(\theta_n)\langle 0|\mathcal{T}\varphi_0(y_1)\ldots\varphi_0(y_k)H_I(\theta_1)\cdots H_I(\theta_n)|0\rangle.$$

We set $|\eta\rangle = |0\rangle$ and we use the notation (14.37): $|\psi_\varepsilon\rangle = V_\varepsilon(-\infty)^{-1}|0\rangle$. Since $|0\rangle$ is an eigenvector of H_0, if all goes well we can use Theorem 14.4.1. Then the limit

$$|\psi\rangle = \lim_{\varepsilon\to 0} \frac{|\psi_\varepsilon\rangle}{\langle 0|\psi_\varepsilon\rangle} \tag{14.53}$$

exists, and is an eigenvector of $H = H_\varepsilon(0)$. This eigenvector is likely (especially under the mass gap hypothesis) to be a ground state, so that it is proportional to $|\Omega\rangle : |\psi\rangle = \lambda|\Omega\rangle$.[30] We may then reformulate (14.53) as saying that as $\varepsilon \to 0$,

$$V_\varepsilon(-\infty)^{-1}|0\rangle = |\psi_\varepsilon\rangle \simeq \lambda\langle 0|\psi_\varepsilon\rangle|\Omega\rangle.$$

Using the same procedure for $t \to \infty$, as $\varepsilon \to 0$ we should have $\langle 0|V_\varepsilon(\infty) \simeq \lambda'\langle 0|\psi'_\varepsilon\rangle\langle\Omega|$. We may now believe that as $\varepsilon \to 0$ the left-hand side of (14.52) behaves as

$$\lambda\lambda'\langle 0|\psi_\varepsilon\rangle\langle 0|\psi'_\varepsilon\rangle\langle\Omega|\mathcal{T}\varphi(y_1)\ldots\varphi(y_k)|\Omega\rangle.$$

Using the same result for $k=0$ and taking the ratio cancels the unknown factor $\lambda\lambda'\langle 0|\psi_\varepsilon\rangle\langle 0|\psi'_\varepsilon\rangle$ and provides for us an expression of the Green functions. As $\varepsilon \to 0$ the terms $g_\varepsilon(\theta_i)$ in the right-hand side of (14.52) have limit g and we have recovered the formula (14.18).

14.5 Diagrammatic Interpretation of the Gell-Mann–Low Theorem

Although some arguments are quite general, from this point on we assume that in the interaction picture the interacting Hamiltonian is given by a density $\mathcal{H}(x)$ as in (13.10). Then (14.19) becomes

$$A = \sum_{n\geq 0} \frac{(-\mathrm{i}g)^n}{n!} \int \mathrm{d}^4x_1 \ldots \int \mathrm{d}^4x_n \langle 0|\mathcal{T}\varphi_0(y_1)\ldots\varphi_0(y_k)\mathcal{H}(x_1)\ldots\mathcal{H}(x_n)|0\rangle. \tag{14.54}$$

We will further assume that the Hamiltonian density is one of our favorite types $:\varphi_0(x)^3:/3!$ or the similar case $:\varphi_0(x)^4:/4!$.[31] These two cases differ only by the fact that in the first case, in Feynman diagrams (and in the other types of diagrams we will consider), three lines come out of each internal vertex while in the second case four lines come out of such vertices. The mass associated to the free field φ_0 will be called m. *All pictures will pertain to the case of ϕ^3 theory.*

From now on many expressions we write will be considered as formal series in g. The reader who has never thought about formal series should greatly benefit from the following exercise.

Exercise 14.5.1 Formal series in g are expressions of the type $\sum_{n\geq 0} a_n g^n$. Provided with the natural addition and multiplication, the set of formal series in g form a commutative ring.

(a) Prove that if $a_0 \neq 0$ then the element $\sum_{n\geq 0} a_n g^n$ has an inverse.

[30] Applying $\langle 0|$ to the previous equality implies $1 = \lambda\langle 0|\Omega\rangle$, so that here again the approach can make sense only if $\langle 0|\Omega\rangle \neq 0$.

[31] Let us recall that the corresponding models are called respectively ϕ^3 and ϕ^4 theory.

(b) Let us now think of g as a formal series $g = g' + \sum_{n\geq 1} a_n(g')^n$ in the new quantity g'. Prove that each formal series in g becomes a formal series in g', and that conversely g' may be expressed as a formal series in g.

In this section we interpret the formula (14.18) in terms of Feynman diagrams. The right-hand side of (14.18) is considered as a formal series in g (see Exercise 14.5.1)[32] so the result is a statement about the coefficients of this formal series. All the infinite sums in this section are similarly considered as formal series in g. We will use Wick's theorem to compute

$$\langle 0|T\varphi_0(y_1)\ldots\varphi_0(y_k)\mathcal{H}(x_1)\ldots\mathcal{H}(x_n)|0\rangle,$$

just as in the case of the S-matrix, and to understand why Proposition 14.5.3 is true, the reader should be familiar with the contents of Section 13.8 and Sections 13.12–13.16. Each term of the series above (which is a function of $y_1,\ldots,y_k$) can be represented as a "sum of the values of certain Feynman diagrams" of a slightly new type (but the name "Feynman diagram" will be strictly reserved for the usual Feynman diagrams). In these diagrams, each term $\varphi_0(y_j)$ is represented by an external vertex (adjacent to a single line), and the external vertices no longer represent incoming or outgoing particles. There will also be internal vertices corresponding to the terms $(\mathcal{H}(x_i))_{i\leq n}$.

The value of a diagram depends both on its "shape" and on the quantities $y_1,\ldots,y_k$. In the long run it clarifies matters to adopt a notation that brings forward these two sources of dependence, in particular because our arguments will involve many summations over various diagrams. Thus the value of the diagram is going to be expressed as a function $I_D(y_1,\ldots,y_k)$ of $y_1,\ldots,y_k$ where D is an object that describes the "shape" of the diagram. This object is itself a diagram, and we next describe it.

We introduce a class of diagrams which we will call *diagrams with k external vertices labeled* 1 *to k*, although for simplicity most of the time we will simply say "diagrams with k external vertices". These diagrams have $k \geq 0$ external vertices, which are labeled 1 to k. They have also $n \geq 0$ internal vertices. The idea is that the external vertex labeled j corresponds to the term $\varphi_0(y_j)$. In such a diagram, each external vertex is adjacent to a single line, and sits at one end of this line. A line adjacent to an external vertex is called an external line. An external line may connect two external vertices (in which case the whole diagram is not connected), or it may connect an external vertex and an internal vertex. The internal vertices are adjacent to three lines for the density $:\varphi_0(x)^3:/3!$ and four lines for the density $:\varphi_0(x)^4:/4!$. No line joins a vertex to itself. The diagrams need not be connected, even if there is no line joining two external vertices.

Given a diagram D with k external vertices, we define a function $I_D(y_1,\ldots,y_k)$ where $y_j \in \mathbb{R}^{1,3}$. (When $k = 0$ there are no variables and I_D is just a number.) The idea of the notation is that $I_D(y_1,\ldots,y_k)$ represents the value of the diagram of shape D when the value y_j is assigned to the external vertex labeled j. Recalling that $\Delta_{F,m}$ denotes the Feynman propagator corresponding to the mass m of the free field this function is computed by the following rules:

[32] The denominator is not a problem there, since the quantity $1/B$ is a formal series in g, as is seen for the identity $(1-C)^{-1} = \sum_{n\geq 0} C^n$ used for $C = \sum_{k\geq 1} a_k g^k$.

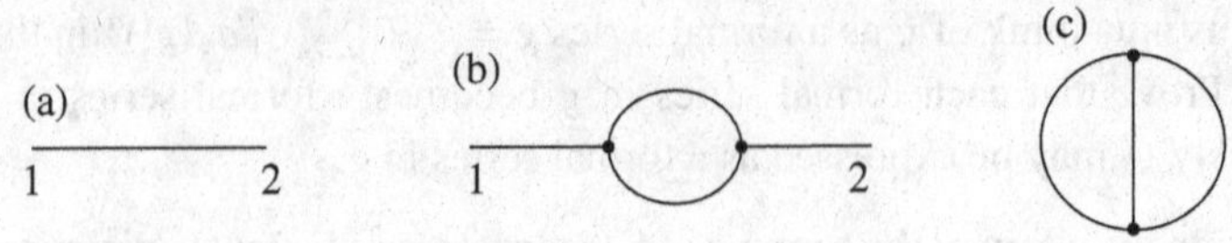

Figure 14.1 (a) The simplest diagram D_0, with $k = 2$ and $n = 0$: two external vertices labeled 1 and 2 joined by a line. (b) A more complicated diagram, with $k = 2$ and $n = 2$. (c) A vacuum bubble with $k = 0$ and $n = 2$.

- For $j \leq k$ attach the number y_j to the external vertex labeled j.
- Label the n internal vertices from 1 to n any way we want.
- To the internal vertex labeled by i attach a variable x_i, and a factor $-ig$.
- To each external line between the external vertex labeled y_j and the internal vertex labeled i attach the factor $-\mathrm{i}\Delta_{F,m}(x_i - y_j)$.
- To each internal line between the internal vertices labeled i and j attach the factor $-\mathrm{i}\Delta_{F,m}(x_j - x_i)$.[33]
- To each external line between the external vertices y_j and $y_{j'}$ attach the factor $-\mathrm{i}\Delta_{F,m}(y_{j'} - y_j)$.
- Take the product of all the previous factors and integrate over the variables x_i.
- Divide by the symmetry factor S.[34]

Here, just as in the case of Feynman diagrams, where we think of the external lines as being completely identified by the particles they represent, (so that these lines cannot be exchanged when computing symmetry factors) we think of the external vertices as completely identified by their label. For example, the symmetry factor of the diagram (a) of Figure 14.1 is 1 and the symmetry factor of the diagram (b) is 2 (corresponding to the exchange of the two internal lines) and the symmetry factor of the vacuum bubble (c) of this figure is $2 \times 3!$, where the factor 2 corresponds to the exchange of the two internal vertices and the factor 3! to the permutations of the internal lines.

Lemma 14.5.2 *The functions I_D are translation invariant, $I_D(y_1, \ldots, y_k) = I_D(y_1 + y, \ldots, y_k + y)$.*

Proof This is obvious by the change of variables $x_i \to x_i + y$. □

For the diagram D_0 labeled (a) in Figure 14.1, we have $I_{D_0}(y_1, y_2) = -\mathrm{i}\Delta_{F,m}(y_2 - y_1) = \langle 0|\mathcal{T}\varphi_0(y_1)\varphi_0(y_2)|0\rangle$. The following should be obvious from the work of Chapter 13.

Proposition 14.5.3 *Recalling the quantity A of (14.54) we have the identity*

$$A = \sum_D I_D(y_1, \ldots, y_k), \tag{14.55}$$

where the sum is over all the diagrams with k external vertices.

[33] One has to be more cautious in theories where the propagator is not an even function, as is often the case in "real" theories such as Quantum Electrodynamics (QED).

[34] Symmetry factors were studied in detail in Section 13.13 in the case of usual Feynman diagrams, and are defined similarly here, although we will not perform a detailed analysis.

We will call a *vacuum bubble* a subset of *internal* vertices that is connected by internal lines and such that there is no line between a vertex of this subset and a vertex (external or internal) which does not belong to this subset. The diagram (c) of Figure 14.1 is such a vacuum bubble.

Proposition 14.5.4 *We have the identity*

$$G(y_1,\dots,y_k) = \langle\Omega|T\varphi(y_1)\dots\varphi(y_k)|\Omega\rangle = \sum_D I_D(y_1,\dots,y_k) \tag{14.56}$$

where the sum is over all diagrams with k external vertices containing no vacuum bubble.

According to the case $k=0$ of (14.55), the value of the quantity B of (14.20) is given by

$$B = \sum_D I_D, \tag{14.57}$$

where the sum is over all diagrams with no external vertex, i.e. diagrams that are unions of vacuum bubbles. In words, the meaning of Proposition 14.5.4 is that the vacuum bubbles cancel between the numerator and the denominator of the right-hand side of (14.18).

Proof of Proposition 14.5.4 Let us denote by C the right-hand side of (14.56). We will prove that the quantity A of (14.54) equals BC. The result then follows from (14.18).

Given a diagram D with k external vertices we will construct a pair (D_1, D_2) of diagrams in such a way that

- D_1 has k external vertices and contains no vacuum bubble.
- D_2 has no external vertex so is a union of vacuum bubbles.
- The map $D \mapsto (D_1, D_2)$ is one-to-one between the set of diagrams with k external vertices and the set of pairs (D_1, D_2) as above.
- Given $y_1,\dots,y_k$ we have

$$I_D(y_1,\dots,y_k) = I_{D_1}(y_1,\dots,y_k)I_{D_2}. \tag{14.58}$$

The required identity $A = BC$ then follows by summation of the equality (14.58) over all possible choices of D and using (14.55) and (14.57).

To proceed to this construction we consider a diagram D with k external vertices. Denoting by D_2 the union of the vacuum bubbles of D, we decompose D in a unique way as the union of two Feynman diagrams D_1 and D_2 such that

- D_1 and D_2 are disjoint.
- D_1 has k external vertices and contains no vacuum bubble.
- D_2 is a union of vacuum bubbles and therefore has no external vertex.

It should be obvious that the map $D \mapsto (D_1, D_2)$ is one-to-one from the set of diagrams with k external vertices to the set of pairs (D_1, D_2) as above.

To complete the proof, it suffices to prove the relation (14.58). To compute the value of $I_D(y_1,\dots,y_k)$ we use the rules stated above. That is, we construct a certain function of the variables attached to the internal vertices, we integrate this function and divide by the symmetry factor. It should be obvious from the rules that the function attached to D is just the product of the functions attached to D_1 and D_2, and that these depend on disjoint sets of variables. The issue then is to prove that the symmetry factor S of D is the product of the symmetry factors S_1 and S_2 of D_1 and D_2. For this it is better to stay at the intuitive level than to give a formal proof. The symmetry factor of a diagram with k external vertices is simply (as shown by the appropriate version of Lemma 13.13.2) the cardinality of the group of symmetries of this diagram. Symmetries fix external vertices and external lines, and exchange internal vertices and internal lines without changing the diagram. The vertices belonging to D_1 are exactly those vertices of D that are connected by a path in D to an external vertex. This makes it obvious that a symmetry can exchange only the vertices of D_1 with vertices of D_1 and the vertices of D_2 with vertices of D_2. That is, the group of symmetries of D is naturally the product of the group of symmetries of D_1 and the group of symmetries of D_2, so that its cardinality is the product of the cardinalities of these two groups. □

14.6 Road Map II

Now that we have established our main tool, Proposition 14.5.4, the main story can start. A central part of this story is the fundamental matter of *mass renormalization*, and we first try to describe it at a high level.

Let us review first some conclusions of Section 13.17. To make sense of the divergent integrals, we introduce a cutoff θ. Our interaction Hamiltonian contains a parameter g, the coupling constant, which models the strength of the interaction. However we realized that the parameter g is not the coupling constant g_{lab} which is measured in the lab. Furthermore to make sense of the model we must assume that the parameter g is cutoff-dependent.

A somewhat similar story occurs for the other parameter of the model, m, which everything in our analysis made us believe should represent the mass of the particle we study, as we measure it in our lab. This however *turns out to be wrong*. This phenomenon is called mass renormalization. A shocking conclusion is that while taming divergent integrals, the parameter m of our model will also have to be made cutoff-dependent. However, the story of diverging integrals and the story of mass renormalization are intertwined but different. As we will explain, mass renormalization occurs even in the absence of divergent integrals.

To make sure that we do not confuse the parameter m of our model and the mass μ we measure in our lab, we call μ, the mass we measure, the *physical mass*. The expression "physical mass" *does not refer to any new concept*: it is the mass as it appears to be in everyday physics.

A major problem is then to find how the physical mass μ fits in the model (which involves parameters m and g). There is a two-pronged attack to the problem:

- In Section 14.8 we perform the key analysis through the study of the most important Green function, the two-point function $\langle\Omega|T\varphi(y_1)\varphi(y_2)|\Omega\rangle$. We relate *through physical arguments* the location of the singularities of (the Fourier transform of) this function with the physical mass. This is the Källén–Lehmann representation.

- In Section 14.9 we compute the *same* two-point Green function, using *an entirely different approach*, diagrammatics, from the parameters of the model.
- In Section 14.10 we *compare* both expressions to relate the parameters of the model and the physical mass of the particle it studies.

Before performing diagrammatic computations, we must however take care of a technical matter. It is not the functions $I_D(y_1,\dots,y_n)$ of Proposition 14.5.4 which are convenient to work with, but their Fourier transforms, which we learn how to compute in Section 14.7. At the same time we discover an important link between the S-matrix and the Green functions, revealing their fundamental importance.

We are then ready, in Section 14.9, to diagrammatically compute the value of the two-point Green function, or, more precisely of its Fourier transform.

The last step of the program is performed in Section 14.10: The comparison of the two very different-looking expressions for the two-point Green function teaches us how to relate the parameters m and g of the model with the physical mass μ of the particle the model studies. The reconciliation of these two very different expressions is however a rather delicate balancing act, and the tight-rope walking to that effect continues in Section 14.11.

Conquering mass renormalization is however not the end of the story, and even darker mysteries are revealed in Section 14.12.

14.7 Green Functions and S-matrix

In this section we prove a Fourier transform identity of interest for general Green functions. This identity will be used in a crucial way starting with Section 14.9. It also brings forward the link between Green functions and the S-matrix.

In Section 14.5 we introduced the concept of diagrams with k external vertices labeled 1 to k. For such a diagram D we defined a function $I_D(y_1,\dots,y_k)$, using the version of the "Feynman rules in position state space" stated on page 437. We assume from now on that D is *connected*. In particular for $k \geq 3$ there do not exist lines joining two external vertices, because then the diagram consisting of these vertices and the line connecting them is not connected to the rest. From the "Feynman rules in position state space", we will deduce "Feynman rules in momentum state space"[35] by using formula (13.59) for each propagator.

In order to state our result properly, we introduce a new type of diagram: *diagrams with k external lines labeled* $1,2,\dots,k$. For simplicity, we will often simply call these diagrams "diagrams with k external lines". You may worry about meeting a third type of diagram. However you will soon realize that these diagrams are in fact a rather straightforward generalization of the usual Feynman diagrams. (Furthermore, you will also realize the diagrams with k external lines are a subclass of diagrams with k external vertices.) Diagrams with k external lines have internal vertices. They have internal lines, connecting two internal vertices (but a line never connects a vertex to itself). They have external lines,

[35] As on page 390.

each connected to one single internal vertex. As the name indicates, there are k such external lines, which are labeled 1 to k. There is an external vertex at the end of each external line. Each internal vertex is adjacent to three lines for the density $:\varphi_0(x)^3: /3!$ and four lines for the density $:\varphi_0(x)^4: /4!$

To each diagram D with k external lines labeled $1, \ldots, k$ and each $\varepsilon > 0$ we associate a function $J_D(p_1, \ldots, p_k, \varepsilon)$ in the variables $p_1, \ldots, p_k \in \mathbb{R}^{1,3}$ as follows:

- For each $j \le k$, we assign to the external line labeled j a four-momentum p_j *flowing from the internal vertex to the external vertex.*[36]
- We orient each internal line any way we like and assign a four-momentum p flowing in that direction. We then assign to the line the factor $-\mathrm{i}/(-p^2 + m^2 - \mathrm{i}\varepsilon)$.
- To each internal vertex we assign the factor $-\mathrm{i}g(2\pi)^4\delta^{(4)}(\Sigma)$ where Σ is the algebraic sum of all the four-momenta flowing on the lines adjacent to this vertex.
- We take the product of the corresponding factors.
- We integrate this product over all momenta corresponding to the internal lines.
- We divide by the symmetry factor S.

The integral over the internal momenta might not (and usually will not) be convergent, but implicitly we are using cutoffs to make sure this does not happen. Let us note the role of the external lines: no factor is attached to these. As in the case of Feynman diagrams, we expect that at least in the sense of distributions we may make sense of

$$J_D(p_1, \ldots, p_k) := \lim_{\varepsilon \to 0^+} J_D(p_1, \ldots, p_k, \varepsilon).$$

The rules by which we compute the function $J_D(p_1, \ldots, p_k)$ are very similar to the rules by which we compute the value of a Feynman diagram, so it is worth understanding clearly the differences:

- We do not assume that $p_i^2 = m^2$.
- We do not distinguish between initial and final vertices (so we have only external vertices), and that each external line is conventionally oriented from the internal vertex to the corresponding external vertex.

This should make it clear that the value of a given Feynman diagram will always be obtained as the value of a certain function J_D. As we are going to see now, the functions J_D are naturally related to Green functions. Thus, through the functions J_D, Green functions are related to Feynman diagrams and to scattering amplitudes.

A diagram D with k external lines labeled 1 to k can be considered as a diagram with k external vertices labeled 1 to k simply by attaching to the external vertex at the end of an external line the label of this line.[37] For such a diagram D we now relate the functions I_D and J_D.

[36] This is conventional.

[37] Conversely a connected diagram with k external vertices has k external lines since no two external vertices may be joined with each other by a line.

Lemma 14.7.1 *The Fourier transform of the function $I_D(y_1,\ldots,y_k)$ is given by*

$$\hat{I}_D(p_1,\ldots,p_k) = \lim_{\varepsilon\to 0^+} J_D(p_1,\ldots,p_k,\varepsilon)\prod_{j\le k}\frac{-\mathrm{i}}{-p_j^2+m^2-\mathrm{i}\varepsilon}. \tag{14.59}$$

This is of course in the sense of distributions. In some imprecise way the contribution of the external lines to $\hat{I}_D(p_1,\ldots,p_k)$ is represented in the previous formula by the factors $-\mathrm{i}/(-p_j^2+m^2-\mathrm{i}\varepsilon)$ (as the external lines do not contribute to the value of $J_D(p_1,\ldots,p_k)$).

Proof We express the function $I_D(y_1,\ldots,y_k)$ using the "Feynman rules in position state space" stated on page 438. It gives a value of the type

$$I_D(y_1,\ldots,y_k) = \frac{1}{S}\frac{(-\mathrm{i}g)^n}{n!}\int \mathrm{d}x_1^4\ldots\int \mathrm{d}x_n^4\Pi$$

where S is the symmetry factor and Π is a product of quantities of the type $-\mathrm{i}\Delta_{F,m}(z)$, one for each line of the diagram. If the line is an internal line between vertices labeled x_i and x_j, then $z = x_j - x_i$. If the line is the external line between the internal vertex labeled x_i and the external vertex labeled y_j then $z = y_j - x_i$. For each such quantity we use the formula (13.59),

$$\Delta_{F,m}(z) = \lim_{\varepsilon\to 0^+}\int\frac{\mathrm{d}^4 p}{(2\pi)^4}\frac{\exp(-\mathrm{i}(z,p))}{-p^2+m^2-\mathrm{i}\varepsilon}, \tag{13.59}$$

pretending as in (13.103) that we can use it within ill-defined products of propagators. This introduces a variable of integration p for each line of the diagram. We call p_j the variable corresponding to the external line labeled j, adjacent to the external vertex labeled j. Thus if the other end of this external line is the internal vertex numbered i and associated to the variable x_i, the propagator corresponding to this external line is

$$\Delta_{F,m}(y_j - x_i) = \lim_{\varepsilon\to 0^+}\int\frac{\mathrm{d}^4 p_j}{(2\pi)^4}\frac{\exp(-\mathrm{i}(y_j - x_i, p_j))}{-p_j^2+m^2-\mathrm{i}\varepsilon}.$$

Using the formula (1.19) to perform the integration in the variables x_i we then obtain the expression

$$I_D(y_1,\ldots,y_k) = \lim_{\varepsilon\to 0^+}\int\frac{\mathrm{d}^4 p_1}{(2\pi)^4}\cdots\int\frac{\mathrm{d}^4 p_k}{(2\pi)^4}J(p_1,\ldots,p_k,\varepsilon)\prod_{j\le k}\frac{-\mathrm{i}\exp(-\mathrm{i}(y_j,p_j))}{-p_j^2+m^2-\mathrm{i}\varepsilon}.$$

Taking the Fourier transform of both sides (and of course assuming that there is no problem with the limit $\varepsilon\to 0^+$) implies (14.59). □

The formula (14.59) brings forward the fact that the function $\hat{I}_D(p_1,\ldots,p_k)$ has singularities when $p_j^2 = m^2$. When computing the S matrix, we will need to use this formula when the p_j represent four-momenta of incoming or outgoing particles of mass m, which precisely satisfy the relation $p_j^2 = m^2$. Still, one may hope that even then it makes some sense to rewrite (14.59) as

$$J_D(p_1,\ldots,p_k) = (\mathrm{i})^k \lim \prod_{j\le k}(-q_j^2+m^2)\hat{I}_D(q_1,\ldots,q_k), \tag{14.60}$$

where the function in the limit in the right-hand side is well-defined when $q_j^2 \neq m^2$ for each j and the limit is taken as $q_j \to p_j, q_j^2 \neq m^2$ for each j.

Let us now turn to the relation between the S-matrix and Green functions, which we will state on a (rather generic) example. Consider four-momenta $p_1, \ldots, p_k$.[38] Our goal is to establish the following relation between the scattering amplitude $\langle 0|a(p_3)\cdots a(p_k)Sa^\dagger(p_1) a^\dagger(p_2)|0\rangle$ and the Green functions.

The Tentative Lehmann–Symanzik–Zimmermann (LSZ) reduction formula. *Assume that no strict subset of outgoing particles has the same total four-momenta as a subset of the incoming particles.*[39]*. Then we have the formula*

$$\langle 0|a(p_3)\ldots a(p_k)Sa^\dagger(p_1)a^\dagger(p_2)|0\rangle = (\mathrm{i})^k \lim \prod_{j\leq k}(-q_j^2+m^2)\hat{G}(-q_1,-q_2,q_3,\ldots,q_k), \tag{14.61}$$

where we expect the right-hand side to be defined when each q_j^2 is close to m^2 but not equal to it, and where the limit is taken as $q_j \to p_j$.

In this manner we have computed the scattering amplitude from the Green function. We have used the word "tentative" in this statement because we will further analyze the Green functions, and discover that, for a rather subtle reason, the formula (14.61) has to be modified.[40] Let us also mention that(14.61) extends to the case of several incoming particles.[41]

In order to establish (14.61) we first consider the scattering amplitude in the left-hand side. In Chapter 13 we have seen that this scattering amplitude is the sum of the values of certain Feynman diagrams which contain no vacuum bubbles.[42] (More accurately, and this is important for the sequel, *we hand-waved ourselves into believing this*.) As we noticed already, the values of Feynman diagrams are closely related to the quantities $J_D(p_1,\ldots,p_k)$ and our next goal is to rephrase the value of a scattering amplitude in terms of these quantities.

Tentative Lemma. *Assume that no strict subset of outgoing particles has the same total four-momenta as a subset of the incoming particles. Then we have the formula*

$$\langle 0|a(p_k)\cdots a(p_3)Sa^\dagger(p_1)a^\dagger(p_2)|0\rangle = \sum_D J_D(-p_1,-p_2,p_3,\ldots,p_k), \tag{14.62}$$

where the sum is over all connected diagrams D with k external lines.

[38] The four-momenta here are those of the incoming and outgoing particles. After we understand the difference between the physical mass μ and the parameter m, it will become obvious that these four-momenta should be computed using the physical mass μ, not the parameter m.

[39] When this condition is not satisfied, there is still a formula, but it is quite more complicated.

[40] In plain language, (14.61) is wrong. Unlike mathematics, physics progresses toward the truth through a sequence of approximations.

[41] It is this more general version which should properly be called the LSZ formula.

[42] Let us recall that on page 386 we decided to exclude diagrams containing vacuum bubbles.

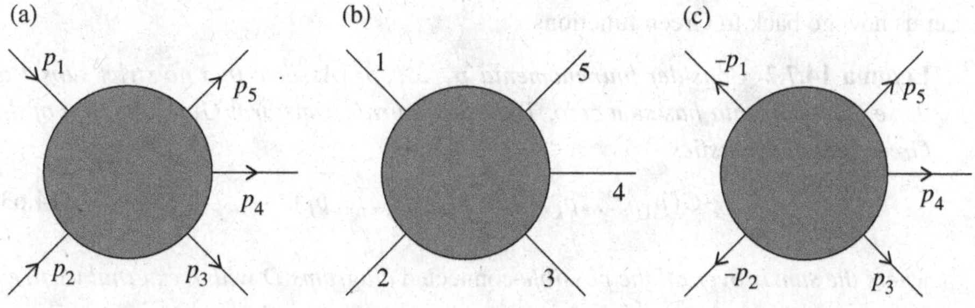

Figure 14.2 (a) A Feynman diagram. (b) The corresponding diagram D with five external lines. (c) The assignment of four-momenta to these external lines for the computation of $J_D(-p_1, -p_2, p_3, p_4, p_5)$.

We call this result "tentative" because unfortunately the formula (14.62) is wrong, as we will see later. The problem is not with the derivation below, but stems from the fact that our formula for expressing scattering amplitudes as a sum of values of Feynman diagrams is itself incorrect. We will discover the correct formula later, in Section 14.12.

Proof We (wrongly) convinced ourselves that the scattering amplitude to the left is the sum of the values of the Feynman diagrams with the appropriate external lines. We will show first that the Feynman diagrams that are not connected have value zero. Indeed, as the Feynman diagrams do not contain vacuum bubbles, each connected part must have some external line attached to it (for otherwise this part would be a vacuum bubble). Let us denote by p_{in} the sum of the four-momenta of the incoming particles on these external lines entering this connected part (respectively p_{out} the sum of the four-momenta of the particles on the external lines leaving the connected part). Unless the connected part is the whole diagram, by our hypothesis we have $p_{\text{in}} \neq p_{\text{out}}$. As explained on page 391, the value of the Feynman diagram consisting of this connected part is then zero. The value of the whole Feynman diagram is then zero because it is the product of the values of the connected parts.

Next we express the value of a connected Feynman diagram with k external lines as follows. To this diagram we associate a diagram D with k external lines labeled 1 to k as follows: We assign the label j to the external line corresponding to the four-momentum p_j.[43] Now, comparing the Feynman rules in momentum space on page 390 and the rules on page 442 makes it obvious that the value of the Feynman diagram is simply $J_D(-p_1, -p_2, p_3, \ldots, p_k)$, see Figure 14.2. The reason for the minus sign in front of p_1 and p_2 is that when computing J_D we orient all external lines from the internal vertex toward the corresponding external vertex. In the case where the external line is attached to an initial vertex, this is the opposite orientation to what was chosen for the computation of the value of the Feynman diagram. □

43 We also forget the information about the four-momenta flowing through the external lines, as this information is not part of a diagram with k external lines.

Let us now go back to Green functions.

Lemma 14.7.2 *Consider four-momenta $p'_1, \ldots, p'_k$. Assume that no strict subset of these four-momenta has sum zero. Then the Fourier transform $\widehat{G}(p'_1, \ldots, p'_k)$ of the Green function satisfies*

$$\widehat{G}(p'_1, \ldots, p'_k) = \sum_D \hat{I}_D(p'_1, \ldots, p'_k), \tag{14.63}$$

where the sum is over all the possible connected *diagrams D with k external vertices.*

Proof Taking the Fourier transform of (14.56) we obtain a sum as in (14.63) but with the sum over all possible D diagrams with k external vertices. It suffices to prove that the disconnected diagrams do not contribute. According to Lemma 14.5.2, the function $I_D(y_1, \ldots, y_k)$ is translation invariant, $I_D(y_1, \ldots, y_k) = I_D(y_1 + y, \ldots, y_k + y)$, so that its Fourier transform contains a factor $\delta^{(4)}(p'_1 + \cdots + p'_k)$ (see Lemma 14.9.2). When the diagram is not connected, the corresponding function $I_D(y_1, \ldots, y_k)$ is a product of the functions corresponding to the connected parts (which depend on disjoint sets of variables), and its Fourier transform is the product of the corresponding Fourier transforms. These contain delta functions whose arguments are sums of the values of certain p'_j and these sums are not zero according to our hypothesis. □

Supporting argument for the LSZ formula (14.61) Consider four-momenta $p_1, \ldots, p_k$ as in (14.62), such that no strict subset of the outgoing four-momenta $p_3, \ldots, p_k$ has a sum that is equal to either p_1 or p_2, and set $p'_1 = -p_1, p'_2 = -p_2, p'_j = p_j$ for $3 \le j \le k$. Then it is easy to see that no strict subset of $p'_1, \ldots, p'_k$ has sum zero and therefore (14.63) implies

$$\widehat{G}(-p_1, -p_2, p_3, \ldots, p_k) = \sum_D \hat{I}_D(-p_1, -p_2, p_3, \ldots, p_k), \tag{14.64}$$

where the sum is over all the possible connected diagrams D with k external vertices.

In a connected diagram with $k \ge 3$ external vertices labeled $1, \ldots, k$, each external vertex connects to an internal vertex.[44] Thus such a diagram is obtained from a connected diagram D with k external lines as in (14.59), by labeling j the external vertex at the end of the external line labeled j. That is, combining (14.60) with (14.64) and comparing with (14.62) we reach (14.61), where we expect the right-hand side to be defined when each $q_j^2 \neq m^2$ and where the limit is taken as $q_j \to p_j$.

We have justified the LSZ formula as an identity between two expressions which are sums of diagrams, because it is the simplest way to bring it about. This aspect is however fallacious.[45] Our shaky derivation of the formula for the S-matrix was based on frantic hand-waving. The correct analysis first establishes the LSZ formula, and then deduces from it the correct expression of the S-matrix in terms of Feynman diagrams.[46] We will come back to that in Section 14.12.

[44] This is wrong for $k = 2$, as the two external vertices may be connected by an external line.

[45] I was confused here for a long time by the presentation in Folland's textbook [31] which does not make this point clear.

[46] This aspect is clearly put forward in the textbook of Peskin and Schroeder [64].

14.8 The Dressed Propagator in the Källén–Lehmann Representation

The goal of this section is to relate the simplest Green function, the two-point function $\langle\Omega|\mathcal{T}\varphi(y_1)\varphi(y_2)|\Omega\rangle$, to certain physical features of the system, in particular the physical mass.

It follows from (14.1) and the invariance of the vacuum under the action of the Poincaré group that the two-point Green function depends only on the difference $y_1 - y_2$. It is conventional to define the following version of this function.

Definition 14.8.1 *The dressed propagator* Δ is defined by

$$-\mathrm{i}\Delta(y_1 - y_2) = \langle\Omega|\mathcal{T}\varphi(y_1)\varphi(y_2)|\Omega\rangle. \tag{14.65}$$

One should hope that this makes sense at least as a tempered distribution. We should not confuse the dressed propagator Δ with the Feynman propagator Δ_F, which, according to Lemma 13.10.2 satisfies

$$-\mathrm{i}\Delta_F(y_1 - y_2) = \langle 0|\mathcal{T}\varphi_0(y_1)\varphi_0(y_2)|0\rangle.$$

This motivates the factor $-\mathrm{i}$ in the previous definition. As will become apparent soon the dressed propagator is far more complicated than the Feynman propagator.

Closely related to the dressed propagator (14.65) is the "function" W given by

$$W(y_1 - y_2) = \langle\Omega|\varphi(y_1)\varphi(y_2)|\Omega\rangle. \tag{14.66}$$

The similarities and differences with (14.65) should be observed now.

In axiomatic Quantum Field Theory (a taste of which is given in Appendix M), it is part of the axioms that W has to be a tempered distribution. Just by using the fact that it is Lorentz invariant, we show in this appendix that one obtains the following. Recalling the measure $\mathrm{d}\lambda_m$ of (4.36), for some constant c and some non-decreasing function ρ on $\mathbb{R}^+$ such that $\rho(s) \le s^k$ for s large and a certain k one has

$$W(x) = c + \int_0^\infty \mathrm{d}\rho(s) \int \mathrm{d}\lambda_s(p) \exp(-\mathrm{i}(x, p)), \tag{14.67}$$

see (M.54). Here (assuming without loss of generality that ρ is right-continuous) $\mathrm{d}\rho$ means that we integrate with respect to the unique measure $\mathrm{d}\nu$ on $\mathbb{R}$ such that $\rho(s) = \nu((0, s])$.[47] A physicist would write $\rho'(s)\mathrm{d}s$ instead of $\mathrm{d}\rho(s)$, but $\rho'(s)$ need not be a real function, it can be a distribution.[48] The growth condition on ρ is a consequence of the technical condition that W is a tempered distribution.

In this section we do not work in the setting of axiomatic Quantum Field Theory. We do *not* make the assumptions of Appendix M, which should be considered as an independent story. Rather, we derive (14.67) and obtain information on the form of ρ from the physics of the system, in particular bringing the physical mass into the picture.

Let us derive the formula (14.67) as a physicist might do. The reader can look into Srednicki's book [77, Chapter 13], or Peskin and Schroeder's book [64, Section 7.1] for

47 Please note for further use that for any s we have $\nu(\{s\}) = \rho(s) - \lim_{s' \to s, s' < s} \rho(s')$.
48 These technicalities are unimportant here.

examples of this approach. We try to use a more mathematical language, but the arguments are definitely not rigorous. There are a large number of reasonable but unproven assumptions. The first step is to write

$$\langle\Omega|\varphi(y_1)\varphi(y_2)|\Omega\rangle = \sum_\alpha \langle\Omega|\varphi(y_1)|\alpha\rangle\langle\alpha|\varphi(y_2)|\Omega\rangle. \tag{14.68}$$

Here α denotes the label of a "continuous basis" of states of given total four-momenta. By definition "continuous basis" means that it satisfies the completeness relation $1 = \sum_\alpha |\alpha\rangle\langle\alpha|$, of which (14.68) is a consequence. The sum $\sum_\alpha$ is highly symbolic, as a continuous basis is not countable. A physicist would say that such a basis exists because the momentum operators commute between themselves. A mathematician would describe the same result by saying that the unitary operators $U(c,1)$ commute between themselves and thus have a common spectral resolution.

Saying that the states $|\alpha\rangle$ have a definite four-momentum means that for certain $p_\alpha \in \mathbb{R}^{1,3}$ we have

$$U(c,1)|\alpha\rangle = \exp(\mathrm{i}(c, p_\alpha))|\alpha\rangle. \tag{14.69}$$

(Compare with (4.66).) As a special case of (14.1) we get

$$\varphi(y) = U(y,1) \circ \varphi(0) \circ U(y,1)^{-1}. \tag{14.70}$$

Combining with (14.69), and since by assumption $U(y,1)^{-1}|\Omega\rangle = |\Omega\rangle$[49] we obtain

$$\langle\alpha|\varphi(y_2)|\Omega\rangle = \langle\alpha|U(y_2,1)\varphi(0)|\Omega\rangle = \exp(\mathrm{i}(y_2, p_\alpha))\langle\alpha|\varphi(0)|\Omega\rangle. \tag{14.71}$$

One might object that since φ is a distribution, $\varphi(0)$ makes no sense and neither does (14.70). However, based on the case of the free field, one may believe that the quantity $\langle\alpha|\varphi(0)|\Omega\rangle$ is a well-defined number and that (14.71) does hold. We will also assume that $\varphi(x)$ is *self-adjoint* so that (as is easily seen by going back to the mathematical notation)

$$\langle\Omega|\varphi(0)|\alpha\rangle = \langle\alpha|\varphi(0)|\Omega\rangle^* \tag{14.72}$$

and (14.68) becomes

$$\langle\Omega|\varphi(y_1)\varphi(y_2)|\Omega\rangle = \sum_\alpha \exp(\mathrm{i}(y_2 - y_1, p_\alpha))|\langle\alpha|\varphi(0)|\Omega\rangle|^2. \tag{14.73}$$

Before we analyze this expression, let us bring forward the main tool, Lorentz invariance.

Lemma 14.8.2 *For $C \in SO^\uparrow(1,3)$ we have*

$$\langle\alpha|\varphi(0)|\Omega\rangle = \langle\alpha|U(0,C)\varphi(0)|\Omega\rangle = \langle C\alpha|\varphi(0)|\Omega\rangle, \tag{14.74}$$

where $|C\alpha\rangle = U(0,C)^{-1}|\alpha\rangle$, see Exercise 2.5.20.

Proof Indeed (14.1) implies that for $C \in SO^\uparrow(1,3)$,

$$U(0,C) \circ \varphi(0) \circ U(0,C)^{-1} = \varphi(0),$$

and the invariance of Ω under $U(0,C)^{-1}$ yields (14.74). □

49 This simply asserts that the vacuum has zero four-momentum.

Let us now investigate what kind of states $|\alpha\rangle$ form our (continuous) basis of states of given four-momentum. Certainly this basis should include the vacuum $|\Omega\rangle$. According to (14.72), the number $A := \langle\Omega|\varphi(0)|\Omega\rangle$ is real, so the term for $\alpha = \Omega$ in the sum in (14.73) is $\exp(\mathrm{i}(y_2 - y_1, p_\alpha))A^2 = \exp(\mathrm{i}(y_2 - y_1, p_\alpha))|A|^2$. To proceed further we make some more assumptions on the content of the theory.

We assume first that our theory describes a single type of particle of *physical mass* μ. In accordance with Section 14.1 the state space contains a subspace $\mathcal{H}_1$ (with $\mathcal{H}_1 \simeq L^2(X_\mu, \mathrm{d}\lambda_\mu)$) describing these particles. This subspace $\mathcal{H}_1$ contains (improper) states $|p\rangle$ for $p \in X_\mu$ (i.e. $p^2 = \mu^2$) which describe a single occurrence of this particle when it has momentum p. Then the operator $\int \mathrm{d}\lambda_\mu(p)|p\rangle\langle p|$ is valued in $\mathcal{H}_1$. According to (4.70), this operator is the identity on $\mathcal{H}_1$, and since it is Hermitian, it coincides with the orthogonal projection $1_{\mathcal{H}_1}$ on $\mathcal{H}_1$:

$$1_{\mathcal{H}_1} = \int \mathrm{d}\lambda_\mu(p)|p\rangle\langle p|.$$

The contribution of the basis elements $|p\rangle$ to (14.73) should then be (reproducing the computation that leads to (14.73))

$$\langle\Omega|\varphi(y_1)1_{\mathcal{H}_1}\varphi(y_2)|\Omega\rangle = \int \mathrm{d}\lambda_\mu(p)\exp(\mathrm{i}(y_2 - y_1, p))|\langle p|\varphi(0)|\Omega\rangle|^2. \tag{14.75}$$

As a consequence of (14.74) and since $SO^\uparrow(1,3)$ acts transitively[50] on X_μ, the quantity $|\langle p|\varphi(0)|\Omega\rangle|$ does not depend on $p \in X_\mu$. According to (14.75) the contribution of the basis elements $|p\rangle$ to (14.73) is then

$$Z\int \mathrm{d}\lambda_\mu(p)\exp(\mathrm{i}(y_2 - y_1, p)), \tag{14.76}$$

where

$$Z := |\langle p|\varphi(0)|\Omega\rangle|^2 \tag{14.77}$$

is a positive number. Furthermore, we assume, as is reasonable, that there exists no object in the theory which is lighter than a single particle. It is reasonable to propose that the states contributing to the sum (14.73), besides the vacuum and the single-particle states, will consist of multiple particles. Let us say that a system consisting of several particles is in a *bound state*[51] if the energy of the total system is less than the sum of the energy of the individual particles.[52] We assume from now on that such bound states do not exist.[53] We show then that all multiparticle states will have a four-momentum p_α with $p_\alpha^2 \geq 4\mu^2$. To see this we compute the four-momentum in a frame of reference where the momentum is zero, so that $p_\alpha = (p_\alpha^0, 0, 0, 0)$ and then $p_\alpha^2 = (p_\alpha^0)^2$. But since we assume that the state is not a bound state, p_α^0, the energy of the multiparticle state, is at the least the sum of the energy of the individual particles, each of which is at least μ and thus $p_\alpha^2 \geq 4\mu^2$.

50 That is, given any two points of X_μ, one may find an element of $SO^\uparrow(1,3)$ which sends the first point to the second one.

51 This is very close to the concept of bound state of a particle interacting with a potential considered on page 329.

52 Thus, one has to spend energy to separate these particles from each other.

53 However the existence of such bound states does not fundamentally change the following analysis.

The next stage is to argue that the multiparticle states should contribute to the sum (14.73) as a (continuous) sum of terms of the type $\int d\lambda_s(p)\exp(i(y_2-y_1,p))$ where $s \geq 2\mu$. For this we assume that our continuous basis has been chosen with the property that if $|\alpha\rangle$ belongs to the basis and $C \in SO^{\uparrow}(1,3)$ then $U(0,C)|\alpha\rangle$ also belongs to the basis.[54] We then group together the multiparticle states $|\alpha\rangle$ which are related by a transformation $U(0,C)$, or in more mathematical terms we consider the equivalence relation $\mathcal{R}$ given by $|\alpha\rangle\mathcal{R}|\beta\rangle$ if and only if there exists $C \in SO^{\uparrow}(1,3)$ such that $|\alpha\rangle = U(0,C)|\beta\rangle$. As Peskin and Schroeder [64] put it, "the states organize themselves into hyperboloids". For linguistic simplicity, let us call an "hyperboloid"(i.e. an equivalence class of the relation $\mathcal{R}$) of states $|\alpha\rangle$ a *class*. As a consequence of (14.74) the value of $|\langle\alpha|\varphi(0)|\Omega\rangle|^2$ is constant over a given class.

The contribution to the sum (14.73) of the terms for which $|\alpha\rangle$ is in a given class should be of the type $\int d\mu(\alpha)\exp(i(y_2-y_1,p_\alpha))$ where μ is a measure on that class which is invariant under the transformations $U(0,C)$, because this is the only Lorentz invariant method to "sum" such contributions. Therefore it should be a multiple of $\int d\lambda_s(p)\exp(i(y_2-y_1,p))$, where s^2 is the common value of the p_α^2 for $|\alpha\rangle$ in this class, because the image of μ under the map $\alpha \mapsto p_\alpha$ is an invariant measure on X_s and hence is proportional to λ_s. Summing over the classes[55] produces a continuous sum of such contributions. The mathematical way to express this continuous sum is that there is a non-decreasing function ρ for which

$$\langle\Omega|\varphi(y_1)\varphi(y_2)|\Omega\rangle = A^2 + \int_0^\infty d\rho(s)\int d\lambda_s(p)\exp(i(y_2-y_1,p)). \tag{14.78}$$

Furthermore, we have assumed that in our continuous basis, the only states α for which $0 < p_\alpha^2 < 4\mu^2$ are the one-particle states $|p\rangle$ for which $p_\alpha^2 = \mu^2$. This translates into the fact that ρ is zero on the interval $(0,\mu)$, has a jump of exactly Z at μ and then is constant again in the interval $(\mu,2\mu)$. The jump of ρ at μ captures the contribution of the single-particle states, and the increments of ρ beyond the value 2μ capture the contributions of the multiparticle states. Another property which we use later is that we expect ρ to be strictly increasing and continuous on the interval $(2\mu,\infty)$. This is because given p with $p^2 \geq 4\mu^2$ there exist two-particle systems of total four-momentum p. This is seen by computing the total four-momentum $(2\sqrt{\mu^2+\boldsymbol{v}^2},\mathbf{0})$ in the center of mass referential, where $\boldsymbol{v}$ and $-\boldsymbol{v}$ are the velocities of the two particles. In some sense, this parameter $\boldsymbol{v}$ spreads out the contributions of the two-particle states over the whole range $p^2 \geq 4\mu^2$. The same spreading should occur for multiparticle systems and we therefore expect that ρ will be continuous and strictly increasing on the range $(2\mu,\infty)$.

Taking into account (14.66), (14.78) coincides with the formula (14.67) but now with the extra information $c = A^2$ and precise information on the structure of ρ (although we have not shown that $\rho(s) \leq s^k$ for large s).

We now show that the formula (14.78) actually gives an explicit expression of the dressed propagator as a mixture of the Feynman propagators $\Delta_{F,s}$ of a free field of mass s as s varies.

[54] This is not unreasonable since the continuous basis of $\mathcal{H}_1$, namely the set of single-particle states $|p\rangle$, has exactly this property, see Proposition 4.10.1.

[55] There may be many classes with a given value of s^2.

Proposition 14.8.3 (Källén–Lehmann representation) *It holds that*

$$\Delta(x) = \mathrm{i}A^2 + Z\Delta_{F,\mu}(x) + \int_{2\mu}^{\infty} \mathrm{d}\rho(s)\Delta_{F,s}(x). \tag{14.79}$$

The fundamental point here bears repetition: μ is the *physical mass of the particle*, the mass you measure in your lab by whatever experiment you design for this purpose. Again, this number *need not* be the value of any parameter called "mass" in the Hamiltonian.[56] As we will see in Section 14.10, in our class of models μ and m *will be* different. This is the phenomenon of mass renormalization.

Proof Assume first that the product $\varphi(y_1)\varphi(y_2)$ is already time-ordered. Then by (14.65)

$$\Delta(y_1 - y_2) = \mathrm{i}\langle\Omega|\varphi(y_1)\varphi(y_2)|\Omega\rangle.$$

The time-component of $y_1 - y_2$ is > 0 and using the first part of (13.57) as well as (13.56), the Feynman propagator $\Delta_{F,s}$ of the free field of mass s satisfies

$$\Delta_{F,s}(y_1 - y_2) = \mathrm{i}\int \mathrm{d}\lambda_s(p)\exp(\mathrm{i}(y_2 - y_1, p)),$$

and then (14.78) yields

$$\Delta(y_1 - y_2) = \mathrm{i}A^2 + \int_0^{\infty} \mathrm{d}\rho(s)\Delta_{F,s}(y_1 - y_2). \tag{14.80}$$

When $\varphi(y_2)\varphi(y_1)$ is time-ordered, one has to replace $y_1 - y_2$ by $y_2 - y_1$ in the right-hand side, but this is the same formula as (14.80) because $\Delta_{F,s}(y) = \Delta_{F,s}(-y)$. Therefore (14.80) holds for any values of y_1 and y_2. This implies (14.79) where Z equals the jump of ρ at μ. □

Taking the Fourier transform of (14.79) and using (13.60) we obtain

$$\hat{\Delta}(p) = -\mathrm{i}A^2(2\pi)^4\delta^{(4)}(p) + \lim_{\varepsilon\to 0^+}\left(\frac{Z}{-p^2 + \mu^2 - \mathrm{i}\varepsilon} + \int_{2\mu}^{\infty} \mathrm{d}\rho(s)\frac{1}{-p^2 + s^2 - \mathrm{i}\varepsilon}\right). \tag{14.81}$$

A very important feature of this formula is that it exhibits μ^2 as the smallest value of $p^2 > 0$ for which $\hat{\Delta}(p)$ has a singularity. To make this more formal we will compute the imaginary part $\mathrm{Im}\,\hat{\Delta}$ of $\hat{\Delta}$. This distribution is defined by

$$\int \frac{\mathrm{d}^4 p}{(2\pi)^4}\xi(p)\mathrm{Im}\,\hat{\Delta}(p) := \mathrm{Im}\left(\int \frac{\mathrm{d}^4 p}{(2\pi)^4}\xi(p)\hat{\Delta}(p)\right)$$

for each *real-valued* test function ξ. To compute $\mathrm{Im}\,\hat{\Delta}$ we write $\mathrm{Im}\,(1/(x - \mathrm{i}\varepsilon)) = \varepsilon/(x^2 + \varepsilon^2)$, and the change of variables $x \to \varepsilon x$ shows that in the sense of distributions[57]

$$\lim_{\varepsilon\to 0^+} \mathrm{Im}\,(1/(x - \mathrm{i}\varepsilon)) = \pi\delta(x). \tag{14.82}$$

56 As physicists put it, "the self-interaction of the particles changes the apparent mass".
57 For more details, the reader is referred to Appendix P at this point.

Assuming reasonably that in (14.81) the number Z is real, (14.81) yields the formula

$$\mathrm{Im}\,\hat{\Delta}(p) = -A^2(2\pi)^4\delta^{(4)}(p) + \pi Z\delta(p^2-\mu^2) + \pi\int_{s\geq 2\mu} \mathrm{d}\rho(s)\delta(p^2-s^2). \quad (14.83)$$

We can informally state the following conclusion, which will be crucially used in the sequel. *Assuming that our theory describes a particle of physical mass μ, then*

$$\mathrm{Im}\,\hat{\Delta}(p) = \begin{cases} 0 & \text{if } 0 < p^2 < \mu^2 \text{ or } \mu^2 < p^2 < 4\mu^2 \\ \neq 0 & \text{if } p^2 = \mu^2 \text{ or } p^2 > 4\mu^2. \end{cases} \quad (14.84)$$

It is the last term in (14.83), together with the fact that $\mathrm{d}\rho(s) > 0$ for $s > 2\mu$ which implies that $\mathrm{Im}\,\hat{\Delta}(p) \neq 0$ for $p^2 > 4\mu^2$. Furthermore, we expect that $\mathrm{Im}\,\hat{\Delta}(p)$ depends on p^2 only, has a delta function at $p^2 = \mu^2$ and is strictly increasing and continuous on the interval $p^2 \geq 4\mu^2$.

The next lemma sheds some light on the value of Z. It shows that if we normalize φ in a natural way, then $0 \leq Z \leq 1$.[58]

Lemma 14.8.4 *If the quantum field φ satisfies the equal-time commutation relations (6.62), then $\mathrm{d}\rho$ is a probability measure, i.e. $\int_0^\infty \mathrm{d}\rho(s) = 1$.*

So if φ satisfies the equal-time commutation relations, then $1 = Z + \int_{2\mu}^\infty \mathrm{d}\rho(s)$ and thus $0 \leq Z \leq 1$.

Proof From (14.78) we obtain

$$\langle\Omega|[\varphi(y_1),\varphi(y_2)]|\Omega\rangle = \int_0^\infty \mathrm{d}\rho(s)\int \mathrm{d}\lambda_s(p)(\exp(\mathrm{i}(y_2-y_1,p)) - \exp(\mathrm{i}(y_1-y_2,p))),$$

so that differentiating in the time-component of y_2, and setting $\pi = \partial\varphi/\partial t$,

$$\langle\Omega|[\varphi(y_1),\pi(y_2)]|\Omega\rangle = \mathrm{i}\int_0^\infty \mathrm{d}\rho(s)\int \mathrm{d}\lambda_s(p)p^0(\exp(\mathrm{i}(y_2-y_1,p)) + \exp(\mathrm{i}(y_1-y_2,p))). \quad (14.85)$$

Let us then assume the canonical commutation relations, $[\varphi(t,\boldsymbol{y}_1),\pi(t,\boldsymbol{y}_2)] = \mathrm{i}\delta^{(3)}(\boldsymbol{y}_1 - \boldsymbol{y}_2)1$. When $y_1 = (t,\boldsymbol{y}_1)$ and $y_2 = (t,\boldsymbol{y}_2)$ we have $(y_2-y_1,p) = -(\boldsymbol{y}_2-\boldsymbol{y}_1)\cdot\boldsymbol{p}$ and (14.85) yields

$$\delta^{(3)}(\boldsymbol{y}_1-\boldsymbol{y}_2) = \int_0^\infty \mathrm{d}\rho(s)\int \mathrm{d}\lambda_s(p)p^0\big(\exp(\mathrm{i}(\boldsymbol{y}_1-\boldsymbol{y}_2)\cdot\boldsymbol{p}) + \exp(\mathrm{i}(\boldsymbol{y}_2-\boldsymbol{y}_1)\cdot\boldsymbol{p})\big).$$

Since $\int \mathrm{d}\lambda_s(p)\ldots = \int \mathrm{d}^3\boldsymbol{p}/((2\pi)^3 2p^0)\ldots$, the equality $(2\pi)^3\delta^{(3)}(\boldsymbol{x}) = \int \mathrm{d}\boldsymbol{p}^3 \exp(\mathrm{i}\boldsymbol{x}\cdot\boldsymbol{p})$ shows that the right-hand side is just $\delta^{(3)}(\boldsymbol{y}_1-\boldsymbol{y}_2)\int_0^\infty \mathrm{d}\rho(s)$, and this proves the result.[59] □

[58] We will however learn in Section 14.12 that this is not the correct way to normalize φ.

[59] This irresistibly convincing argument is probably one of the lowest points of this book. Rigorous results, bearing on models which are much tamer than the physicist's objects, predict only that $\rho(s)$ does not grow faster than a certain power of s as $s \to \infty$ while here we have *proved* that $\rho \leq 1$. The problem is probably that the assertion that the field satisfies the canonical commutation relations is not tenable.

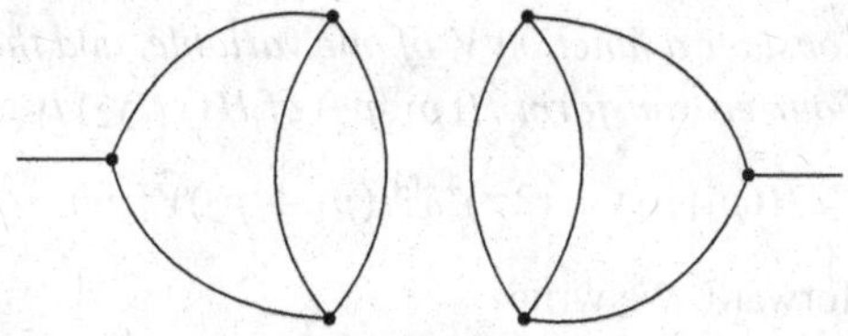

Figure 14.3 A disconnected diagram without bubbles.

14.9 Diagrammatic Computation of the Dressed Propagator

In this section we compute the dressed propagator through the parameters of the model. Let us recall that the quantity $\langle\Omega|\varphi(y)|\Omega\rangle$, being Lorentz invariant, is a constant, which we called A. We recall that in Section 14.5, we defined diagrams with k external lines and for such a diagram D the function $I_D(y_1,\ldots,y_k)$.

Proposition 14.9.1 *We have*

$$-\mathrm{i}\Delta(y_1-y_2)-A^2=\sum_D I_D(y_1,y_2),$$

where the sum is over all connected *diagrams with two external vertices labeled 1 and 2.*

The sum includes the diagram D_0 of Figure 14.1, and $I_{D_0}(y_1,y_2)=\langle 0|\mathcal{T}\varphi(y_1\varphi(y_2)|0\rangle=-\mathrm{i}\Delta_{F,m}(y_1-y_2)$.

Supporting argument for Proposition 14.9.1 According to Definition 14.8.1 and Proposition 14.5.4, $-\mathrm{i}\Delta(y_1-y_2)=\sum_D I_D(y_1,y_2)$ where the sum is over all diagrams with two external vertices which contain no vacuum bubble. Therefore it suffices to prove that

$$A^2=\sum_D I_D(y_1,y_2),$$

where the sum is over all *disconnected* diagrams with two external vertices which contain no vacuum bubble. Figure 14.3 shows an example of such a diagram. As in the proof of Proposition 14.5.4, this sum is shown to be $C(y_1)C(y_2)$ where $C(y)=\sum_D I(y)$, for a sum over the connected diagrams with one single external vertex labeled y. (These diagrams are called *tadpoles*.) But according to Proposition 14.5.4, $C(y)=A$.

Our goal is to compute the Fourier transform $\hat{\Delta}(p)$ of the dressed propagator from Proposition 14.9.1, and to compare it with the expression (14.81). The key consideration is the location of the singularities of $\hat{\Delta}(p)$, and for this the value of A is irrelevant, *so we lighten presentation by pretending that* $A=0$.[60] But we need some preparations before we can start computing $\widehat{\Delta}(p)$.

[60] Only minor changes are required to take A into account, but one may fear that the integrals involved in its computation are divergent.

Lemma 14.9.2 *Consider a function V of one variable, and the function $H(y_1, y_2) = V(y_1 - y_2)$. The Fourier transform $\widehat{H}(p_1, p_2)$ of $H(y_1, y_2)$ is given by the formula*

$$\widehat{H}(p_1, p_2) = (2\pi)^4 \delta^{(4)}(p_1 + p_2)\widehat{V}(p_1). \tag{14.86}$$

Proof This is straightforward. We write

$$\begin{aligned}\hat{H}(p_1, p_2) &= \iint \mathrm{d}^4 y_1 \mathrm{d}^4 y_2 V(y_1 - y_2) \exp(\mathrm{i}(y_1, p_1) + \mathrm{i}(y_2, p_2)) \\ &= \int \mathrm{d}^4 y_3 V(y_3) \exp(\mathrm{i}(y_3, p_1)) \int \mathrm{d}^4 y_2 \exp(\mathrm{i}(y_2, p_1 + p_2)),\end{aligned} \tag{14.87}$$

using the change of variables $y_1 = y_2 + y_3$. □

Given a diagram D with two external lines and given $\varepsilon > 0$, in Section 14.7 we have constructed the distribution $J_D(p_1, p_2, \varepsilon)$. This distribution is supported by the set $p_1 + p_2 = 0$. This is because when $J_D(p_1, p_2, \varepsilon) \neq 0$ all the delta functions enforcing conservation of momentum at each vertex of the graph are non-zero. Thus four-momentum is conserved at each vertex and no net momentum can enter or leave the graph. We have in fact the following more precise result.

Lemma 14.9.3 *We have*

$$J_D(p_1, p_2, \varepsilon) = (2\pi)^4 \delta^{(4)}(p_1 + p_2) K_D(p_1, \varepsilon) \tag{14.88}$$

for a nice function $K_D(p_1, \varepsilon)$.

Here again we assume that we have made a cutoff so that all integrals are convergent. At the intuitive level the result is obvious,[61] and we are really making our lives miserable by using distributions here. When $p_1 + p_2 = 0$ the constraints that the algebraic sum of the momenta entering each internal vertex are zero are compatible, and just amount to choosing certain four-momenta flows on the internal lines with respect to which we integrate, expressing the other four-momenta as functions of these. As we will use this later, it is worth stating the rules to compute the function $K_D(p, \varepsilon)$:

- We assign to the external line labeled 1 a four-momentum p flowing from the internal vertex to the external vertex and to the external line labeled 2 a four-momentum $-p$ flowing from the internal vertex to the external vertex.
- We orient each internal line the way we like and we assign a four-momentum q flowing in that direction. We then assign to the line the factor $-\mathrm{i}/(-q^2 + m^2 - \mathrm{i}\varepsilon)$.
- To each internal vertex we assign the factor $-\mathrm{i}g(2\pi)^4 \delta^{(4)}(\Sigma)$ where Σ is the algebraic sum of all the four-momenta flowing on the lines adjacent to this vertex.
- We make the product of the corresponding factors.
- We integrate over all momenta corresponding to the internal lines.
- We divide by the symmetry factor S.

[61] A complete proof of a more general statement will be given later in Theorem 15.7.2.

We recall that the dressed propagator (14.65) is a function of a single variable, and we denote by $\hat{\Delta}(p)$ its Fourier transform.

Proposition 14.9.4

$$-\mathrm{i}\hat{\Delta}(p) = \lim_{\varepsilon\to 0^+}\Big(\frac{-\mathrm{i}}{-p^2+m^2-\mathrm{i}\varepsilon}+\big(\frac{-\mathrm{i}}{-p^2+m^2-\mathrm{i}\varepsilon}\big)^2\sum_D K_D(p,\varepsilon)\Big), \tag{14.89}$$

where the sum is over all connected diagrams D with two external lines.

Supporting argument. We apply Lemma 14.9.2 to the function $G(y_1,y_2) = -\mathrm{i}\Delta(y_1-y_2) = \langle\Omega|\mathcal{T}\varphi(y_1\varphi(y_2)|\Omega\rangle$ to obtain

$$\widehat{G}(p_1,p_2) = (2\pi)^4\delta^{(4)}(p_1+p_2)\big(-\mathrm{i}\hat{\Delta}(p_1)\big). \tag{14.90}$$

On the other hand, according to Proposition 14.9.1, and recalling that we assume $A = 0$, $\widehat{G}(p_1,p_2) = \sum_D \hat{I}_D(p_1,p_2)$, where the sum is over all possible connected diagrams with two external vertices.

One of these diagrams is very special, the diagram D_0 of Figure 14.1, for which $I_{D_0}(y_1,y_2) = V(y_1-y_2)$ where $V(y) = -\mathrm{i}\Delta_{F,m}(y)$ and $-\mathrm{i}\widehat{V}(p) = \lim_{\varepsilon\to 0^+}(-\mathrm{i}/(-p^2+m^2-\mathrm{i}\varepsilon))$, so that

$$\hat{I}_{D_0}(p_1,p_2) = (2\pi)^4\delta^{(4)}(p_1+p_2)\lim_{\varepsilon\to 0^+}\frac{-\mathrm{i}}{-p_1^2+m^2-\mathrm{i}\varepsilon}. \tag{14.91}$$

Any other connected diagram with two external vertices labeled y_1 and y_2 has at least one internal vertex and two external lines. Thinking of this diagram as a diagram D with two external lines we may appeal to the case $k = 2$ of (14.59) to obtain

$$\hat{I}_D(p_1,p_2) = \lim_{\varepsilon\to 0^+} J_D(p_1,p_2,\varepsilon)\prod_{j=1,2}\frac{-\mathrm{i}}{-p_j^2+m^2-\mathrm{i}\varepsilon} \tag{14.92}$$

and combining with (14.88), we obtain

$$\hat{I}_D(p_1,p_2) = \lim_{\varepsilon\to 0^+}(2\pi)^4\delta^{(4)}(p_1+p_2)\Big(\frac{-\mathrm{i}}{-p_1^2+m^2-\mathrm{i}\varepsilon}\Big)^2 K_D(p_1,\varepsilon). \tag{14.93}$$

Therefore by summation of these relations we have obtained the expression

$$\widehat{G}(p_1,p_2) = (2\pi)^4\delta^{(4)}(p_1+p_2)$$
$$\times\lim_{\varepsilon\to 0^+}\Big(\frac{-\mathrm{i}}{-p_1^2+m^2-\mathrm{i}\varepsilon}+\big(\frac{-\mathrm{i}}{-p_1^2+m^2-\mathrm{i}\varepsilon}\big)^2\sum_D K_D(p_1,\varepsilon)\Big),$$

where the sum is over all connected diagrams with two external lines. Comparing with (14.90) supports (14.89).

Our next goal is to study connected diagrams with two external lines.

Definition 14.9.5 A diagram is *1-particle irreducible* (1-PI) if it cannot be split into two disjoint parts by removing an internal line.

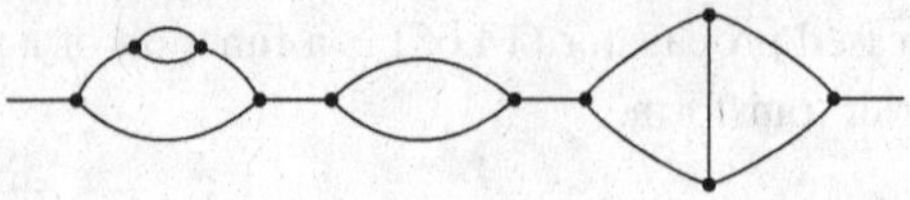

Figure 14.4 A sausage with three links.

The obvious but important fact is the "sausage theorem": A connected diagram with two external lines has the following structure: one external line, followed by a 1-PI blob, which is linked by a single internal line to another 1-PI blob, etc., ending with the second external line, see Figure 14.4.

Each 1-PI blob has two distinguished vertices, and if we attach external lines to these vertices we obtain a diagram with two external lines. These diagrams will be called the components of D.

Lemma 14.9.6 *Consider a diagram D with two external lines, and its 1-PI components $D_1, \ldots, D_k$. Then*

$$K_D(p,\varepsilon) = \left(\frac{-\mathrm{i}}{-p^2+m^2-\mathrm{i}\varepsilon}\right)^{k-1} K_{D_1}(p,\varepsilon)\ldots K_{D_k}(p,\varepsilon). \tag{14.94}$$

Proof If the momentum through one external line is p, conservation of momentum ensures that the momentum through each of the internal lines joining two consecutive blobs is also p. The result should be obvious from the rules of computation of $K_D(p,\varepsilon)$. □

Let us define the function $\Sigma(p,\varepsilon)$ by:

$$-\mathrm{i}\Sigma(p,\varepsilon) \text{ is the sum of the functions } K_D(p,\varepsilon) \text{ corresponding to all the 1-PI diagrams } D \text{ with two external lines.} \tag{14.95}$$

This quantity has to be understood as a formal series in g, where the coefficients of the powers of g are functions of p. The factor $-\mathrm{i}$ is conventional. Let us try to motivate this factor. When computing the value of a diagram, there is a factor i for each vertex, and a factor i in the numerator of each propagator. The computation of the value of the diagram involves the computation of a certain integral (as will be detailed later in Section 15.7). This computation will create a factor i "for each loop variable occurring in this integral" when performing a Wick rotation. If there are L internal lines and V internal vertices, we will prove later in Section 15.7 that there are $L-V+1$ such loop variables, so that we get as grand total the odd number $2L+1$ of factors i. Therefore "in $\Sigma(p,\varepsilon)$ there are $2L$ factors i and we may expect that $\Sigma(p,\varepsilon)$ should have a tendency to be real" (although it will still often have an imaginary part).

Theorem 14.9.7 *As a formal series in the coupling constant we have*

$$\hat{\Delta}(p) = \lim_{\varepsilon\to 0^+} \frac{1}{-p^2+m^2-\mathrm{i}\varepsilon+\Sigma(p,\varepsilon)}. \tag{14.96}$$

Proof We compute the sum $\sum_D K_D(p,\varepsilon)$ in (14.89) by summation of the formula (14.94) over all possible values of k and of $D_1,\dots,D_k$. Writing

$$B(p,\varepsilon) = \frac{-\mathrm{i}}{-p^2+m^2-\mathrm{i}\varepsilon},$$

and summing (14.94) over all possible 1-PI components[62] yields

$$\sum_D K_D(p,\varepsilon) = \sum_{k\geq 1} B^{k-1}(p,\varepsilon)(-\mathrm{i}\Sigma(p,\varepsilon))^k,$$

and thus

$$B(p,\varepsilon)+B(p,\varepsilon)^2\sum_D K_D(p,\varepsilon) = \sum_{k\geq 0} B^{k+1}(p,\varepsilon)(-\mathrm{i}\Sigma(p,\varepsilon))^k = \frac{B(p,\varepsilon)}{1+\mathrm{i}B(p,\varepsilon)\Sigma(p,\varepsilon)},$$

which is the desired result. □

The equality (14.96) is between formal series in g.[63] Still, in the next section we will basically pretend that it is an equality between functions, and that the values of p where the denominator vanishes correspond to singularities of $\hat{\Delta}(p)$, a huge leap of faith indeed.

14.10 Mass Renormalization

The main topic of the present section is the phenomenon of mass renormalization, which is created by the self-interaction of the particle. *It has nothing to do with diverging integrals.* To make this clear in this section and the next we work in the so-called ϕ_2^3 theory, where the interaction term is $g :\varphi_0^3: /3!$ and where space-time has dimension two. As we will prove in the next chapter, in this theory all the Feynman diagrams are convergent.[64] Besides g, the theory has one other parameter, the mass m of the free field.

We then may compute in perturbation theory the values of all S-matrix elements, and we have (order by order in perturbation theory) a complete theory which models the interactions between different particles of the same type.

To use this theory, the question is then to decide which are the appropriate values of the parameters m and g. Obviously we can choose these parameters only by requiring that certain predictions of the theory fit the measurements in our lab. We will examine this question in full in Section 14.13, but here we examine how to fix the parameters[65] m and g so that the physical mass μ predicted by the theory coincides with the physical mass we actually measure. The key point we make in this section is that the physical mass *μ does not coincide with the parameter m*. It is this phenomenon which is called "mass renormalization".

[62] Such a component can be any 1-PI diagram with two external legs.
[63] Formal summation of a geometric series may yield non-sensical results as exemplified by the well-known identity $-1 = \sum_{n\geq 0} 2^n$.
[64] The problem of possible singularities as $\varepsilon \to 0^+$ remains, but we ignore it.
[65] These parameters are called the *bare parameters* in physics. They cannot be directly measured.

Within our theory[66] we can compute $\hat{\Delta}(p)$ and its imaginary part via (14.96). The plan is to compare the result of this computation with what we learned on general grounds in Section 14.8 about $\hat{\Delta}(p)$ for a theory modeling a particle of physical mass μ, and in particular (14.84):

$$\operatorname{Im}\hat{\Delta}(p) = \begin{cases} 0 & \text{if } 0 < p^2 < \mu^2 \text{ or } \mu^2 < p^2 < 4\mu^2 \\ \neq 0 & \text{if } p^2 = \mu^2 \text{ or } p^2 > 4\mu^2, \end{cases}$$

with furthermore $\operatorname{Im}\hat{\Delta}$ having a delta function at $p^2 = \mu^2$.

How can (14.96) produce an imaginary part? To investigate this we make a series of reasonable assumptions (which become all the more reasonable after one learns to compute the value of diagrams). We pretend that $\lim_{\varepsilon\to 0^+}\Sigma(p,\varepsilon)$ exists. By Lorentz invariance this limit should be a function $\Gamma(p^2)$ of p^2. First it may happen that $\Gamma(p^2)$ has a non-zero imaginary part. This will be examined in the next section. Another way for $\hat{\Delta}$ to get an imaginary part is from (14.82), when $-p^2 + m^2 + \Gamma(p^2) = 0$ and we examine it now. Let us consider the smallest root θ of the equation

$$-\theta + m^2 + \Gamma(\theta) = 0. \tag{14.97}$$

For p^2 close to θ we have $\Gamma(p^2) \simeq \Gamma(\theta) + \Gamma'(\theta)(p^2 - \theta)$. Crossing our fingers we may expect that for some number $A(p^2)$ one has

$$\Sigma(p,\varepsilon) \simeq \Sigma(p,0) + A(p^2)\varepsilon = \Gamma(p^2) + A(p^2)\varepsilon. \tag{14.98}$$

Then for p^2 close to θ the formula (14.96) becomes

$$\hat{\Delta}(p) = \lim_{\varepsilon\to 0^+} \frac{1}{(-p^2+\theta)(1-\Gamma'(\theta)) + A(\theta)\varepsilon - \mathrm{i}\varepsilon}.$$

We can repeat the argument of (14.82) to obtain that if α is a complex number with $\operatorname{Re}\alpha > 0$ then $\lim_{\varepsilon\to 0^+} 1/(x - \mathrm{i}\alpha\varepsilon) = \pi\delta(x)$. Using this for $x = (-p^2+\theta)(1-\Gamma'(\theta))$ and $\alpha = 1 + \mathrm{i}A(\theta)$, and assuming $\operatorname{Im} A(\theta) < 1$ so that $\operatorname{Re}\alpha > 0$, this gives

$$\operatorname{Im}\hat{\Delta}(p) = \pi\delta((-p^2+\theta)(1-\Gamma'(\theta))) = \frac{\pi}{|1-\Gamma'(\theta)|}\delta(p^2-\theta),$$

where we use (1.15) in the second equality. This should account for the delta function part of $\operatorname{Im}\hat{\Delta}$ at $p^2 = \mu^2$ in (14.84) i.e. the term $\pi Z\delta(p^2-\mu^2)$ in (14.83). This motivates a fundamental fact:

$$\textit{If our theory describes a particle of physical mass } \mu \textit{ then } \theta = \mu^2. \tag{14.99}$$

This also motivates the relation

$$Z = \frac{1}{|1-\Gamma'(\mu^2)|}. \tag{14.100}$$

In particular, the physical mass μ satisfies the equation

$$-\mu^2 + m^2 + \Gamma(\mu^2) = 0. \tag{14.101}$$

[66] And in contrast with (14.79) which is based on very general arguments.

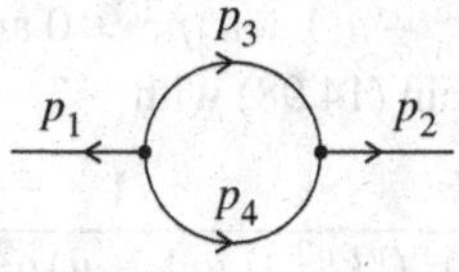

Figure 14.5 The one-loop diagram.

Unless $\Gamma(m^2) = 0$ (which has no special reason to hold) then $\mu \neq m$. The parameter m of the model does not represent the physical mass of the particle under study.

Again, everything is meant order by order in perturbation theory. For example, if $\Gamma(p^2) = g^2\Gamma_2(p^2) + O(g^4)$ then (14.97) gives $\theta = m^2 + \Gamma(\theta) = m^2 + g^2\Gamma_2(\theta) + O(g^4)$. Plugging back the expression $\theta = m^2 + g^2\Gamma_2(\theta) + O(g^4)$ into $\Gamma_2(\theta)$ gives $\Gamma_2(\theta) = \Gamma_2(m^2) + O(g^2)$ and hence $\theta = m^2 + g^2\Gamma_2(m^2) + O(g^4)$.

Exercise 14.10.1 Check that if $\Gamma(p^2) = g^2\Gamma_2(p^2) + g^4\Gamma_4(p^2) + O(g^6)$ then $\theta = m^2 + g^2\Gamma_2(m^2) + g^4(\Gamma_2(m^2)\Gamma_2'(m^2) + \Gamma_4(m^2)) + O(g^6)$.

Let us now check by a concrete computation that all the optimistic hypotheses we made are sensible. We consider the one-loop diagram D of Figure 14.5 and we compute $J_D(p_1, p_2, \varepsilon)$ according to the rules on page 442:

$$J_D(p_1, p_2, \varepsilon) = (-\mathrm{i}g)^2(-\mathrm{i})^2 \iint \frac{\mathrm{d}^2 p_3}{(2\pi)^2}\frac{\mathrm{d}^2 p_4}{(2\pi)^2}\delta\frac{1}{(-p_3^2 + m^2 - \mathrm{i}\varepsilon)}\frac{1}{(-p_4^2 + m^2 - \mathrm{i}\varepsilon)}$$

where $\delta = (2\pi)^4\delta^{(2)}(p_1 + p_3 + p_4)\delta^{(2)}(p_2 - p_3 - p_4)$, and, as predicted by Lemma 14.9.3

$$J_D(p_1, p_2, \varepsilon) = (2\pi)^2\delta^{(2)}(p_1 + p_2)K_D(p_1, \varepsilon)$$

for the nice function

$$K_D(p, \varepsilon) = g^2 \int \frac{\mathrm{d}^2 k}{(2\pi)^2}\frac{1}{(-k^2 + m^2 - \mathrm{i}\varepsilon)}\frac{1}{(-(p+k)^2 + m^2 - \mathrm{i}\varepsilon)}.$$

Using Feynman parameters and Wick rotation, that is, copying the arguments which lead to (13.124) we obtain

$$K_D(p, \varepsilon) = \mathrm{i}g^2 \int_0^1 \mathrm{d}u \int \frac{\mathrm{d}^2 k}{(2\pi)^2}\frac{1}{(\|k\|^2 - u(1-u)p^2 + m^2 - \mathrm{i}\varepsilon)^2}. \tag{14.102}$$

Exercise 14.10.2 Prove that in ϕ_d^3 theory, a diagram with two external lines has an even number of internal vertices.

According to (14.95) and the previous exercise, we obtain $\Sigma(p, \varepsilon) = \mathrm{i}K_D(p, \varepsilon) + O(g^4)$, because all the other diagrams in the sum have at least four internal vertices and contribute as $O(g^4)$. Thus

$$\Sigma(p, \varepsilon) = -g^2 \int_0^1 \mathrm{d}u \int \frac{\mathrm{d}^2 k}{(2\pi)^2}\frac{1}{(\|k\|^2 - u(1-u)p^2 + m^2 - \mathrm{i}\varepsilon)^2} + O(g^4). \tag{14.103}$$

For $p^2 < 4m^2$ we have $\inf_{0\le u\le 1} m^2 - u(1-u)p^2 > 0$ and we can expand the integrand as $(a - \mathrm{i}\varepsilon)^{-2} \simeq a^{-2} + 2\mathrm{i}\varepsilon a^{-3}$ to obtain (14.98) with

$$\Gamma(p^2) = -g^2 \int_0^1 \mathrm{d}u \int \frac{\mathrm{d}^2k}{(2\pi)^2} \frac{1}{(\|k\|^2 - u(1-u)p^2 + m^2)^2} + O(g^4), \tag{14.104}$$

$$A(p^2) = -\mathrm{i}g^2 \int_0^1 \mathrm{d}u \int \frac{\mathrm{d}^2k}{(2\pi)^2} \frac{2}{(\|k\|^2 - u(1-u)p^2 + m^2)^3} + O(g^4). \tag{14.105}$$

so that $\Gamma(p^2) < 0$ and $\operatorname{Im} A(p^2) < 0$. Since $\Gamma(p^2) < 0$ at order g^2 and $-\mu^2 + m^2 + \Gamma(\mu^2) = 0$ by (14.97), we should have $\mu < m$: Mass renormalization *decreases* the mass, at least at order g^2.

In a more realistic theory such as ϕ_4^4 theory some diagrams will diverge. We introduce a cutoff to make sure that everything is welldefined. We argued in Section 13.17 that it is not possible to directly measure the "bare" coupling constant g, and to get a sensible model we must make g depend on the cutoff.[67] It is not possible to measure the bare mass m either. It turns out that as the cutoff gets removed, in order that the physical mass remains constant, the bare mass must go to infinity, just like the bare coupling constant (everything here makes sense order by order in perturbation theory).

For further use let us end this section by one more observation, going back to ϕ_2^3 theory. We expect that for $p^2 \neq \mu^2$ but close to μ^2 we will have $-p^2 + m^2 + \Gamma(p^2) \neq 0$ so that $\hat{\Delta}(p^2)$ is well defined and equals $1/(-p^2 + m^2 + \Gamma(p^2))$. By L'Hospital rule we get

$$\lim_{p^2 \to \mu^2} (p^2 - \mu^2)\hat{\Delta}(p) = \frac{1}{-1 + \Gamma'(\mu^2)}. \tag{14.106}$$

14.11 Difficult Reconciliation

Let us go back to the problem of the reconciliation of (14.96) with (14.83). We still assume that we are in the ϕ_2^3 theory in two space-time dimensions. All physical quantities are given by formal series in g, with well-defined coefficients. This is also the case of the quantity θ given by the equation (14.97).

We recall that we assume that $\Gamma(p^2) := \lim_{\varepsilon\to 0^+} \Sigma(p,\varepsilon)$ exists. Then (assuming that the denominator is not zero)

$$\operatorname{Im}\hat{\Delta}(p) = \frac{-\operatorname{Im}\Gamma(p^2)}{(-p^2 + m^2 + \operatorname{Re}(\Gamma(p^2)))^2 + (\operatorname{Im}\Gamma(p^2))^2}, \tag{14.107}$$

Thus $\operatorname{Im}\hat{\Delta}(p) \neq 0$ if $\operatorname{Im}\Gamma(p^2) \neq 0$. But what are the values of p for which $\operatorname{Im}\Gamma(p^2) \neq 0$?

At a given order n of perturbation theory,

$$\operatorname{Im}\Gamma(p^2) = \sum_{2\le k\le n} g^k f_k(p^2) + O(g^{n+1}) \tag{14.108}$$

67 The term "bare" is physics terminology. It refers to a parameter of the model which cannot be directly measured.

for certain functions f_k. It is a simple exercise to show that, with the exception of maybe countably many values of g, the support of a function of the type $\sum_{0\le k\le n} g^k f_k$ is the union of the supports of the functions f_k.

$$\textit{The support of } \operatorname{Im}\Gamma(p^2) \textit{ is generically independent of } g. \tag{14.109}$$

On the other hand, according to (14.84) we rather expect the support of $\operatorname{Im}\Gamma(p^2)$ to be $[4\mu^2,\infty)$ where μ is the physical mass. Since the physical mass μ satisfies the equation (14.101) which depends on g, we do not expect μ^2 to be independent of g. This seems to contradict (14.109).

Before we pursue this matter, let us work a bit harder to actually compute exactly the union of the supports of the functions f_k. As we have seen from (14.104) the one-loop diagram does not produce an imaginary part in $\Gamma(p^2)$ when $p^2 < 4m^2$. Once one has learned how to compute the values of more complicated diagrams one may show that each of the other diagrams contributing to $\Gamma(p^2)$ can produce an imaginary part only for $p^2 > 4m^2$. Therefore the support of $\operatorname{Im}\Gamma(p^2)$ is contained in $[4m^2,\infty)$. One may also show from (14.103) that $g^2 f_2(p^2)$ is not zero for $p^2 > 4m^2$.[68] Consequently the union of the supports of the functions f_k is the interval $[4m^2,\infty)$ and we have improved (14.109) into the following.

$$\textit{The support of } \operatorname{Im}\Gamma(p^2) \textit{ is generically } [4m^2,\infty). \tag{14.110}$$

In conclusion:

- According to (14.107), at any order of expansion in powers of g, generically $\operatorname{Im}\hat{\Delta}(p) \neq 0$ for $4m^2 < p^2 < \infty$ but $\operatorname{Im}\hat{\Delta}(p) = 0$ for $m^2 < p^2 < 4m^2$.
- According to (14.84) and (14.99), $\operatorname{Im}\hat{\Delta}(p) \neq 0$ for $4\mu^2 < p^2 < \infty$ but $\operatorname{Im}\hat{\Delta}(p) = 0$ for $\mu^2 < p^2 < 4\mu^2$.
- Typically $m \neq \mu$.

This discrepancy stems from the fact that we have done wrong physics in the previous argument. It is the observable physical mass μ which should be taken as a parameter for the model, not the un-observable m. Similarly it is a certain "observable version" ρ of the interaction strength which should be taken as a parameter, not the un-observable g. Such a change of parameterization is called a *finite renormalization*. Then at given μ, m and g are determined by ρ (as formal power series).[69] When $\rho = 0$ there is no interaction and we have $m = \mu$, so we expect the series for m to start as $m = \mu + \cdots$. We also expect the series for g to start as $g = \rho + \cdots$ because for ρ small, ρ and g should be nearly equal. In summary we expect that $m = \mu + O(\rho)$ and $g = \rho + O(\rho^2)$. The functions f_k of (14.108) also depend on m^2, so we denote them by $f_k(p^2,m^2)$. To obtain the expansion of $\operatorname{Im}\hat{\Delta}(p)$ in powers of ρ we simply substitute in (14.108) the expression of g and m as power series in ρ to obtain an expression

[68] This non-trivial fact will look less mysterious after you study Chapter 18. You should even be able to give a complete proof, using the fact proved there that the integral in k can be explicitly performed.

[69] One of the conditions determining m and g is the equality $\mu^2 = \theta$.

$$\operatorname{Im}\Gamma(p^2) = \sum_{2\le k\le n} \rho^k h_k(p^2) + O(\rho^{n+1}). \tag{14.111}$$

Since $m^2 = \mu^2 + O(\rho)$ the first term $\rho^2 h_2(p^2)$ is $\rho^2 f_2(p^2,\mu^2)$ which we have shown to be non-zero exactly for $p^2 > 4\mu^2$. Furthermore, for any k the function $h_k(p^2)$ is zero for $p^2 < 4\mu^2$ because all the functions $m \mapsto f_\ell(p^2,m^2)$ are identically zero for $p^2 < 4m^2$ and because $\mu^2 < m^2$. Thus the union of the supports of the functions h_k is $[4\mu^2,\infty)$. Let us stress what happened:

After the finite renormalization it appears from (14.111) that $\operatorname{Im}\hat{\Delta}(p) \neq 0$ *for* $p^2 > 4\mu^2$ *and* $\operatorname{Im}\hat{\Delta}(p) = 0$ *for* $\mu^2 < p^2 < 4\mu^2$.

This is now consistent with (14.84).[70] However hard it is to accept the first time one sees it, *the finite renormalization changes the apparent support of* $\operatorname{Im}\hat{\Delta}(p)$.

The previous argument is not fallacious. The situation may seem strange to a mathematician because she is not used to working in perturbation theory. In perturbation theory, we have only access to a finite number of terms of the series (14.111), from which we cannot reconstruct the expression (14.108).[71]

The previous arguments showed how tricky perturbation theory can be. Still we trusted it in Section 14.10 to detect the shift in the smallest value of p^2 for which there is a singularity of $\hat{\Delta}(p)$. This provides the correct answer, but the mathematical value of this procedure is infinitesimal.

14.12 Field Renormalization

The matter of field renormalization is related to the factor Z in (14.77), and it is a real challenge to clearly understand it.[72] The present section will probably look to the reader as the most puzzling of the entire book, so a few words of orientation might help. As for the practical matter of assigning a finite value to each diagram, we will describe in Chapter 17 an algorithm to that effect. The matter of "field normalization" is perfectly clear at the level of this algorithm: it requires adding certain "counter-terms" to cancel divergences. The considerations of the present section attempt to somehow justify this mysterious procedure. One of the reasons for which the present section is puzzling is that (at least in the eyes of this author) the attempt is only partially successful.

There are ominous signs of trouble in our picture. On the one hand, the relation (14.61) seems to indicate that the functions $\hat{G}(-q_1,-q_2,q_3,\ldots q_k)$ blow up at $q_j^2 \to m^2$ (so that the left-hand side is not zero). On the other hand, combining (14.90) and (14.106) show that the singularities of $\hat{G}(p_1,p_2)$ occur not for $p_j^2 = m^2$ but for $p_j^2 = \mu^2$. Why should the Green function of two variables behave in a fundamentally different way than the Green function of three variables? Further, by comparing (14.106) with (14.100) we see that the quantity Z should play a role. This problem will be solved in this section.

70 It took me a looong time to figure this out, and I thank Marc Srednicki for putting me on the right track.
71 Obviously the situation is not fully satisfactory, and yet we cannot really argue that it is non-sensical.
72 The difficulty might be measured by the fact that despite us having brilliantly proved that $0 \le Z \le 1$, when computing Z in perturbation theory it has an irresistible tendency to be infinite.

One may picture the quantum field φ as creating both single-particle states and more complicated states when applied to the vacuum. Arguing at a (truly!) heuristic level, it is not really the whole field φ which matters when studying the S-matrix, but "only that part of φ which creates single particles".[73] This argument looks flimsy at first, but it is not. Within axiomatic Quantum Field Theory a rigorous theory of scattering has been developed, the Haag-Ruelle scattering theory.[74] This theory proceeds to a rigorous construction of the out and in states, and the first part of the construction is precisely to isolate "the part of the quantum field which creates single particles". It is *this* part of the field which should be properly normalized, *not* the whole field. (For example, in Lemma 14.8.4 we normalized the whole field by imposing the equal-time commutation relations.) If one does not work within the framework of axiomatic Quantum Field Theory, but heuristically as in Section 14.8 "the part of the quantum field which creates single particles" is identified by the equation (14.77). To ensure that this part of the field is properly normalized, one has to "renormalize the field by multiplying it by a factor $Z^{-1/2}$". The LSZ formula (14.61) then has to be modified in two different ways:

- One has to insert a correction factor $Z^{-k/2}$ in the right-hand side.
- The term $\prod_{j\le k}(-q_j^2+m^2)$ has to be replaced by $\prod_{j\le k}(-q_j^2+\mu^2)$.

The first modification accounts for field renormalization, and the second one for mass renormalization. Thus, the formula (14.61) has to be replaced by

$$\langle 0|a(p_3)\dots a(p_k)Sa^\dagger(p_1)a^\dagger(p_2)|0\rangle$$
$$= Z^{-k/2}(\mathrm{i})^k \lim \prod_{j\le k}(-q_j^2+\mu^2)\hat{G}(-q_1,-q_2,q_3,\dots,q_k). \tag{14.112}$$

Let us face it: This is not an easy result. The main difficulty is that it no longer works to be cavalier about the in and the out states: One has to find a way to express these using the field φ, and it is at that stage that the mysterious factor Z come out, through (14.77). In the framework of axiomatic Quantum Field Theory, to construct the in and out states and to prove the LSZ formula at the level of detail of this book would require several chapters. As for heuristic approaches, we recommend Coleman's book [13] or Ticciati's book [85].[75] Different approaches may be found in Peskin and Schroeder's book [64] or, for the truly daring, in Weinberg' book [87, Section 10.2]. We do not enter these arguments here. We will have a very limited use of the LSZ formula, in Chapter 17. Furthermore, as we explain below, the way the LSZ formula is actually used requires a considerable act of faith[76] (with no rigorous support), so that it is hardly fundamental for us to try to carefully justify the formula itself.

Let us go back to the problem of computing the scattering amplitude using Feynman diagrams. We established our tentative version (14.61) of the LSZ formula from our naive

[73] This is to be taken as an image and not as a technical statement.
[74] The very accessible textbook of Jost [45] covers this topic well. One may also look in Duncan's textbook [24, Chapter 9].
[75] It is again very difficult to learn this result on your own because several of the main physics textbooks simply sweep the difficulties under the rug. But Coleman's textbook [13] is remarkable in truly confronting difficulties.
[76] I am much grateful again here to J. B. Zuber for helping me understand this point, as this inglorious aspect of QFT is certainly not put forward clearly in any of the main books.

hand-waving computation of this amplitude using Feynman diagrams. The fact that we have to modify (14.61) indicates that this naive computation is wrong. The plan now is to travel backward along the same road we traveled to prove (14.61). We start from (14.112) and our goal is to deduce from this formula the correct way to compute scattering amplitudes using Feynman diagrams.

To perform the main step of the computation we compute the Fourier transform of a Green function as a sum over diagrams, and we reorganize this sum. Since none of this really holds at the mathematical level, let us simplify matters by pretending that we are dealing with functions rather than distributions and that we may just take $\varepsilon = 0$ (so that dependence on ε of all our quantities is removed). Thus we simply write (14.59) as

$$\hat{I}_D(p_1,\ldots,p_k) = J_D(p_1,\ldots,p_k)\prod_{j\le k}\frac{-\mathrm{i}}{-p_j^2+m^2}. \tag{14.113}$$

Definition 14.12.1 A diagram with k external lines is said to be *amputated* if one cannot disconnect an external line from all the other external lines by removing one single internal line.

We also define an *amputated Feynman diagram* exactly the same way. For example, the diagram (1) of Figure 13.15 is not amputated (and the present section will motivate our decision not to include it in the computation of the scattering amplitude).

Proposition 14.12.2 *The following formula holds, where $\widehat{G}(p_1,\ldots,p_k)$ is the Fourier transform of the Green function:*

$$\widehat{G}(p_1,\ldots,p_k) = \prod_{j\le k}(-\mathrm{i}\hat{\Delta}(p_j))\sum_{D'} J_{D'}(p_1,\ldots,p_k), \tag{14.114}$$

where the sum of over all amputated *diagrams D' with k external lines.*

Proof Taking the Fourier transform of (14.56) and combining with (14.113) we obtain

$$\widehat{G}(p_1,\ldots,p_k) = \prod_{j\le k}\frac{-\mathrm{i}}{-p_j^2+m^2}\sum_{D} J_{D}(p_1,\ldots,p_k), \tag{14.115}$$

where the sum is over all diagrams D with k external vertices labeled $1,\ldots,k$ and containing no vacuum bubbles.

The reader is referred to Figure 14.6 to follow the diagram surgery which we perform. We consider a diagram D with external lines labeled $1,\ldots,k$. Given an external line labeled j, there are two cases to consider. First it may happen that one cannot disconnect this external line from the rest of the diagram by removing a single internal line. In that case we say that the external line is *thin*. The second case is when this can be done. We then say that the external line is *fat*. When the external line labeled j is fat, starting from it there is a last[77] internal line such that removing this internal line disconnects this external line from the rest of the diagram. (This line is called a in the example of Figure 14.6.) Let us call this part

[77] In the sense that it is the furthest from the external line as measured by the graph distance, the length of the shortest path between vertices.

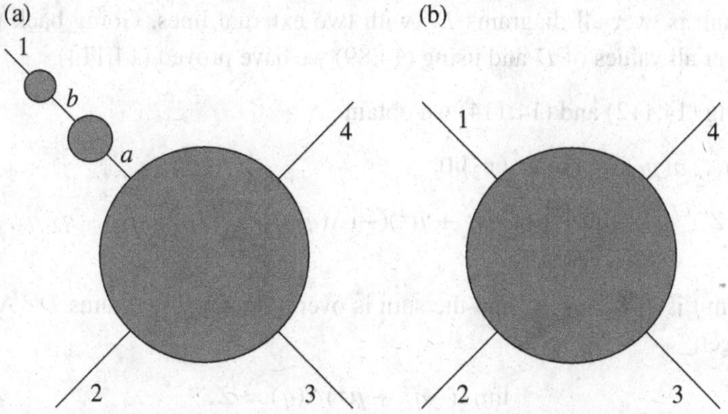

Figure 14.6 (a) The fat line corresponding to the external line labeled 1 consists of this external lines, the internal lines labeled a and b and the two blobs in between. The lines labeled 2 to 4 are thin. The big blob is shaded to convey the idea that it is solid: You cannot separate an external line from the rest of the blob by cutting a single internal line. (b) The corresponding amputated diagram.

of the diagram, including the external line labeled j and the last internal line *the fat line labeled* j. Seen by itself, the fat line labeled j is just a diagram $D(j)$ with two external lines. One of these external lines is an internal line of D (this is again the line called a in the example of the picture). To the diagram D we can associate a diagram D' by "amputating the fat lines". Each fat line is removed and replaced by an ordinary external line (with the same label). It should be obvious that D' is amputated.

Looking[78] at the rules used to compute $J_D(p_1, \ldots, p_k)$ one obtains

$$J_D(p_1, \ldots, p_k) = J_{D'}(p_1, \ldots, p_k) \prod_{j \le k} A(j), \tag{14.116}$$

where the quantity $A(j)$ equals 1 if the external line labeled j is thin, and where

$$A(j) = \frac{-\mathrm{i}}{-p_j^2 + m^2} K_{D(j)}(p_j), \tag{14.117}$$

otherwise, where $D(j)$ is the diagram with two external lines corresponding to the fat line j, and where the function K_D occurs in Lemma 14.9.3. The factor $-\mathrm{i}/(-p_j^2 + m^2)$ occurs because when computing either $J_{D'}(p_1, \ldots, p_k)$ or $K_{D(j)}(p_j)$ the external lines of these diagrams do not contribute, but here we have added an internal line connecting D_j with the rest of the diagram. (This is again the internal line labeled a in Figure 14.6.) Thus, if we sum the quantities (14.116) over all diagrams D for which D' has a given value, we obtain the quantity

$$J_{D'}(p_1, \ldots, p_k) \prod_{j \le k} \Big(1 + \frac{-\mathrm{i}}{-p_j^2 + m^2} \sum_{D''} K_{D''}(p_j)\Big),$$

78 You may have to think a few minutes here to convince yourself that this is the case.

where the sum is over all diagrams D'' with two external lines. Going back to (14.115), summing over all values of D and using (14.89) we have proved (14.114). □

Combining (14.112) and (14.114) we obtain

$$\langle 0|a(p_3)\ldots a(p_k)Sa^\dagger(p_1)a^\dagger(p_2)|0\rangle \tag{14.118}$$
$$= Z^{-k/2}(\mathrm{i})^k \lim \prod_{j\le k}\Big((-q_j^2+\mu^2)(-\mathrm{i}\hat{\Delta}(q_j))\Big)\sum_{D'} J_{D'}(-q_1,-q_2,q_3,\ldots,q_k),$$

where the limit is as $q_j \to p_j$ and the sum is over amputated diagrams D'. According to (14.81) we get

$$\lim_{q^2\to\mu^2}(-q^2+\mu^2)\hat{\Delta}(q) = Z, \tag{14.119}$$

and (14.118) becomes

$$\langle 0|a(p_3)\ldots a(p_k)Sa^\dagger(p_1)a^\dagger(p_2)|0\rangle = Z^{k/2}\sum_{D'} J_{D'}(-p_1,-p_2,p_3,\ldots,p_k), \tag{14.120}$$

which is our corrected version of (14.62).

If we recall how we relate the value of a Feynman diagram with a certain function J_D (as is explained a few lines above Figure 14.2 and as was used in deriving (14.62)), the following is now transparent: To compute the scattering amplitude $\langle 0|a(p_3)\ldots a(p_k)Sa^\dagger(p_1)a^\dagger(p_2)|0\rangle$,

We multiply by the factor $Z^{k/2}$ the sum of the values of corresponding amputated *Feynman diagrams.* (14.121)

Thus our naive method has to be modified in two ways. First, we consider only amputated diagrams.[79] Second we multiply the result by a factor $Z^{k/2}$.

How, then, are we supposed to use this formula? Let us recall here that we have been working in perturbation theory, using the "bare mass" m in the interaction Hamiltonian, and using a cutoff to make all integrals convergent. For the model to make sense, the bare mass and the interaction strength have to be cutoff-dependent. How then do we extract information as the cutoff gets removed? At the end of Chapter 13 we showed how to assign values to one-loop and two-loop diagrams as the cutoff gets removed, by introducing "counter-terms". There was however very little theoretical support that the values thus assigned contain any information whatsoever about the true behavior of the model, see page 395. In Part IV, we will generalize these procedures, producing a "renormalized theory", which assigns a "renormalized value" to each diagram. We also generalize the leap of faith[80] made in Chapter 13 that the renormalized values of the diagrams actually contain relevant information. The basic act of faith is the following:

79 Note however that we need to consider diagrams which are not necessarily 1-PI. Removing an internal line may split the diagram into two parts, but at least two external lines are attached to each of the parts.

80 If I understand correctly, the theoretical support for this fundamental point is *not stronger* than, say, the theoretical support we have to believe that (13.117) is a prediction of reality. Furthermore this whole procedure is kept largely *implicit* in the major textbooks, probably because in the minds of the authors, it goes without saying that this is the correct approach. They are certainly supported by experimental evidence. The resulting formulas are very predictive.

All the previous considerations are assumed to be relevant when the values of the diagrams are replaced by their renormalized values. (14.122)

In other words, to compute the scattering amplitudes in the renormalized theory we will mimic the previous procedure, and we take as an act of faith that this works.[81] Let us explain how this is done. The renormalized theory depends on two parameters m and g and will allow us to assign a well-defined value[82] to each Feynman diagram.[83] As we assign a (renormalized) value to each diagram, we may compute the function $-\mathrm{i}\Gamma_{\mathrm{renorm}}(p^2)$, the sum of the (renormalized) values of all the 1-PI diagrams with two external lines, and the function

$$\Phi(p^2) = \frac{1}{-p^2 + m^2 + \Gamma_{\mathrm{renorm}}(p^2)}. \tag{14.123}$$

Here again Γ_{renorm} is a formal series in g, and all computations are performed at a given order of perturbation theory, that is at order k one keeps only the terms with powers of g that are $\leq k$. The function $\Phi(p^2)$ of (14.123) has a singularity at the smallest value $p^2 = \mu^2$, and μ is the physical mass of the particle. There is no reason why one should have $\mu = m$. One may then consider[84]

$$z := \lim_{p^2 \to \mu^2} (-p^2 + \mu^2)\Phi(p^2). \tag{14.124}$$

Let us then accept without further justification[85] that the correct way to compute the scattering matrix is the formula (14.121) where the diagrams are now renormalized and where Z is replaced by z. That is:

The scattering amplitude is $z^{k/2}$ times the sum of the values of the corresponding renormalized amputated Feynman diagrams. (14.125)

Exercise 14.12.3 Prove that if $\Gamma_{\mathrm{renorm}}(m^2) = \Gamma'_{\mathrm{renorm}}(m^2) = 0$ then $m = \mu$ and $z = 1$.

14.13 Putting It All Together

In this section we give more details on how to make predictions about scattering amplitudes in ϕ_2^3 theory. The situation is made simpler by the fact that no renormalization is required (so that in particular $\Gamma_{\mathrm{renorm}} = \Gamma$), but the procedure would be exactly the same if the theory had to be renormalized. One would use the renormalized values of the diagrams rather than their actual values.

[81] Again, this is justified by the tremendous agreement with the experiments.
[82] At least in the sense of distributions. The present discussion is imprecise and only provided for global orientation.
[83] The way we will proceed to renormalization does not use cutoffs at all. Diverging integrals are made convergent not by introducing cutoffs, but by modifying the integrand.
[84] Again, this definition is motivated by (14.119), but there is *no direct relationship between Z and z*. We are simply reproducing, within the setting of the normalized theory, the (presumably correct) method (14.121) to compute scattering amplitudes, and we cross our fingers that we obtain the correct result.
[85] To the best of my understanding such a justification *does not* exist, at least in the sense a mathematician would understand it.

The model depends on two un-observable parameters g and m. We will show how to determine (order by order) these parameters as a function of two observable parameters, and how we can then make physical predictions. One of the parameters is the physical mass μ. The other is a scattering amplitude ρ that we define below.

We will study the scattering amplitude of two particles of momenta[86] p_1, p_2 into two particles of momenta p_3, p_4, which we may measure in the lab. Let us lighten notation by writing $\delta := (2\pi)^2\delta^{(2)}(p_1+p_2-p_3-p_4)$. Among many other possible sensible choices let us define ρ such that $\delta\rho$ is the amplitude at small velocity for the incoming and the outgoing particles,

$$p_1 = p_2 = p_3 = p_4 = p^* := (\mu, 0). \tag{14.126}$$

This is our second parameter.

The goal is now to relate μ, ρ, m and g. Let us consider the function

$$F(p^2, m) := \int_0^1 \mathrm{d}u \int \frac{\mathrm{d}^2k}{(2\pi)^2} \frac{1}{(\|k\|^2 - u(1-u)p^2 + m^2)^2}, \tag{14.127}$$

so that (14.104) becomes $\Gamma(p^2) = -g^2F(p^2,m) + O(g^4)$. The relation (14.101) $-\mu^2 + m^2 + \Gamma(\mu^2) = 0$ becomes $m^2 = \mu^2 - g^2F(\mu^2,m) + O(g^4)$. Since $m = \mu + O(g^2)$ we have $m^2 = \mu^2 - g^2F(\mu^2,\mu) + O(g^4)$ and (14.123) becomes

$$\Phi(p^2) = \frac{1}{-p^2 + \mu^2 - g^2F(\mu^2,\mu) + g^2F(p^2,\mu)} + O(g^4),$$

so that using L'Hospital rule (14.124) gives

$$\begin{aligned} z &= \lim_{a\to 0} \frac{a}{a + g^2(F(\mu^2 - a, \mu) - F(\mu^2,\mu))} + O(g^4) \\ &= \frac{1}{1 - g^2G(\mu)} + O(g^4) = 1 + g^2G(\mu) + O(g^4), \end{aligned} \tag{14.128}$$

where $G(\mu) := \mathrm{d}F(x,\mu)/\mathrm{d}x\big|_{x=\mu^2}$. We recall that there are three different channels, see Figure 13.10.[87] We set $s = (p_1+p_2)^2, t = (p_1-p_3)^2, u = (p_1-p_4)^2$. At a given order g^{2n}, the scattering amplitude is given by z^2 times[88] the sum of the values of the amputated diagrams with at most $2n$ internal vertices. Assuming that we are in the generic situation where $s,t,u \neq m^2$, the sum of the diagrams with two internal vertices is

$$\delta(-\mathrm{i}g)^2(-\mathrm{i}) \sum_{\theta=s,t,u} \frac{1}{-\theta + m^2} =: g^2\delta U(s,t,u,m). \tag{14.129}$$

It is more complicated to compute the sum $V(s,t,u,m)$ of all amputated diagrams with four internal vertices, as these are of several types, some of which are illustrated in Figure 14.7.

[86] These are no longer four-momenta, since we now have only two dimensions for space.
[87] Don't be confused though that this figure pertain to ϕ^4 theory while here we work in ϕ^3 theory and there is only one internal line, and the momentum flowing through this line is completely determined!
[88] The exponent of z is half the number of particles, and there are four particles here.

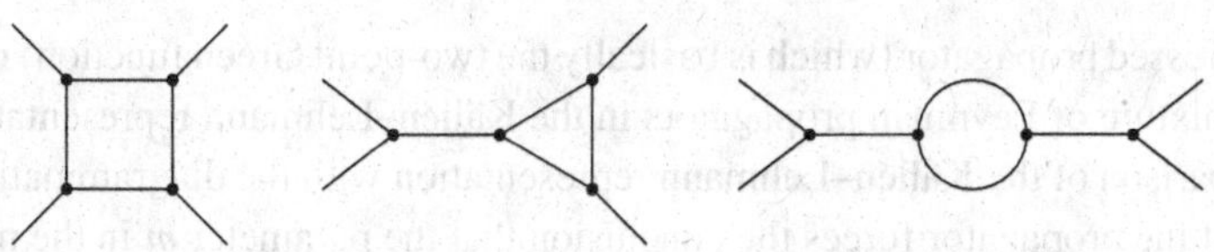

Figure 14.7 Several types of amputated diagrams with four internal vertices and four external vertices.

This is nonetheless possible with a bit of patience. Taking into account that $m^2 = \mu^2 - g^2 F(\mu^2,\mu) + O(g^4)$ we can eliminate m in the expressions $U(s,t,u,m)$ and $V(s,t,u,m)$ at order $O(g^4)$ and find for the scattering amplitude an expression of the type

$$\delta(g^2 A(s,t,u) + g^4 B(s,t,u)) + O(g^6), \tag{14.130}$$

where $A(s,t,u)$ and $B(s,t,u)$ depend on μ but do not depend on m or g. We observe that both mass renormalization and (14.128) *do* contribute to the term $B(s,t,u)$. Defining $v := (2p^*)^2$, and specializing to the case (14.126), (14.130) gives

$$\delta(g^2 A(v,0,0) + g^4 B(v,0,0)) + O(g^6) = \delta\rho. \tag{14.131}$$

We solve this equation, yielding the value of g^2 as a function of the observable parameters μ and ρ:

$$g^2 = \rho \frac{1}{A(v,0,0)} - \rho^2 \frac{B(v,0,0)}{A(v,0,0)^3} + O(\rho^3).$$

We then plug back in (14.130) to get our prediction for the scattering amplitude:

$$\delta\left(\rho \frac{A(s,t,u)}{A(v,0,0)} + \rho^2 \frac{A(v,0,0)B(s,t,u) - A(s,t,u)B(v,0,0)}{A(v,0,0)^3}\right) + O(\rho^3).$$

14.14 Conclusions

Mathematicians and physicists are unlikely to agree about how convincing the arguments presented in this chapter are. As a mathematician I feel that whereas our theoretical work describes convincing little pieces of structure, in the end, however, this information justifies the correct computational procedure (14.125) only in a rather general sense. I regret that the status of these arguments is not explained in a clearer way in textbooks of physics, and I found the authoritative way these arguments are presented very confusing, although again this is a cultural problem. In particular the very important point (14.122) is certainly not put forward clearly, and I might never have figured it out myself had not J. B. Zuber explained it to me.

Key ideas to remember:

- The Green functions $\langle\Omega|T\varphi(y_1)\ldots\varphi(y_k)|\Omega\rangle$ encompass many properties of an interacting quantum field.
- As a consequence of the Gell-Mann–Low "theorem", the Green function is the sum of the values of the diagrams with k external vertices containing no vacuum bubble.

- The dressed propagator (which is basically the two-point Green function) is expressed as a mixture of Feynman propagators in the Källén–Lehmann representation.
- Comparison of the Källén–Lehmann representation with the diagrammatic computation of the propagator forces the conclusion that the parameter m in the model is not the same as the physical mass μ of the particle.
- Our naive formula to compute scattering amplitudes was wrong. One has to account for field renormalization through the LSZ formula.
- The LSZ formula is (magically) valid for the "renormalized theory" where the diagrams have been made finite through a certain "renormalization procedure".

Part IV

Renormalization

From now on integrals occurring in the computation of Feynman diagrams will be called *diagram integrals.*

As we have now experienced many times, the study of diagram integrals runs into difficulties of different natures. On the one hand, integrals may diverge because the integrands do not decrease fast enough at infinity. On the other hand, singularities arise as $\varepsilon \to 0^+$. The first problem is by far the most difficult. Renormalization is the art to attack it. Several lifetimes would be required to learn all aspects of this topic, so obviously choices have to be made. We will study in depth the first truly successful approach, the so-called Bogoliubov-Parasiuk-Hepp-Zimmermann (BPHZ) method. Mathematically speaking, its appeal is that the underlying mathematics are both elementary and extremely ingenious.* Physically speaking, its appeal is that it connects very well with the physicist's methods (as is explained in Chapter 17). The main idea is that the diverging integrals are tamed by replacing the integrand F by a regularized version $\mathcal{R}F$ with better properties. The description of the procedure and the analysis of its consequences are the object of Chapter 16, whereas the proof that the procedure actually succeeds is the object of Chapters 18 and 19.

If the goal is simply to define renormalized amplitudes, there exist somewhat shorter methods; refer to Bergère's work [7]. However, these methods use the specific form of the propagator, a parametric representation of this propagator, and the computations are not intuitive. As a mathematician, I am far more comfortable with the methods presented here which rely on far more general principles.†

Among later and more advanced approaches to renormalization many follow a different route. They concentrate their efforts on the "Euclidean case"(where one replaces the Fourier transform $1/(-p^2+m^2-\mathrm{i}\varepsilon)$ of the Feynman propagator by its "Euclidean version" $1/(\|p\|^2+m^2)$). They then prove that if one can build a Euclidean theory with sufficiently

* Furthermore, this approach does not seem to have been presented in book form, and the original papers are not easy to follow.

† And my goal is not brevity but conceptual clarity.

nice properties, one can deduce a satisfactory theory for the true propagator. This is a difficult theorem, which we will avoid entirely by not following this path.‡ There is no doubt however that these methods deepen the understanding of renormalization; refer to Rivasseau's book [70].

‡ Another problem with this approach is that building a Euclidean theory which satisfies the hypothesis of this theorem is not a goal within easy reach. This problem is serious enough that it has not been solved for either of the models we consider here, ϕ_4^4 and ϕ_6^3.

15

Prologue: Power Counting

In the present chapter, we perform preliminary work in different directions, so the chapter consists of several parts which are independent of each other to a large extent. The first direction of investigation addresses the problem of the integrands that do not decrease fast enough at infinity, trying to reach a precise understanding of the problem before attempting to solve it. This topic occupies Sections 15.1 and 15.2. One removes the secondary problem of the singularities as $\varepsilon \to 0^+$ by replacing the Fourier transform $1/(-p^2+m^2-\mathrm{i}\varepsilon)$ of the Feynman propagator by its "Euclidean version" $1/(\|p\|^2+m^2)$.[1] Diagram integrals are then of the form $\int \mathrm{d}^n x\, P(x)/Q(x)$, where P and Q are polynomials (with $P=1$ and Q of a very special form). We study in a rather general setting the absolute convergence of such integrals. The basic idea is that "the degree of the denominator has to be sufficiently larger than the degree of the numerator". Estimating these degrees is called "power counting" in physics, and the central result concerning such integrals is Weinberg's power counting theorem, Theorem 15.2.1.

In the second part of the chapter, starting with Section 15.3 we then apply these general results to actual diagram integrals (still in the Euclidean setting). Recalling that the Feynman rules in momentum space impose at each vertex a delta function that enforces the conservation of four-momentum at this vertex, we study the space $\ker \mathcal{L}$ of four-momenta which satisfy this condition. We compute its dimension. We also develop basic concepts of graph theory which play a fundamental role in the sequel. Finally in the last section, Section 15.7, we learn how to parameterize diagram integrals.

15.1 What Is Power Counting?

The simplest form of power counting is as follows.

Lemma 15.1.1 *Consider two polynomials P and Q in the variable x with $P \neq 0$ (and of course $Q \neq 0$). If the integral*

$$\int_0^\infty \mathrm{d}x \left|\frac{P(x)}{Q(x)}\right|$$

converges, then $\deg P - \deg Q < -1$.

[1] At first sight the Euclidean propagator looks much smaller than Feynman propagator. As we will gradually understand, matters are more subtle however, because there is a lot of cancellation in integrals involving the Feynman propagator. This cancellation is brought forward by Wick rotations.

Proof For large x, we have $P(x)/Q(x) \sim cx^{\deg P - \deg Q}$ where $c \neq 0$ and the integral $\int_1^\infty \mathrm{d}x\, x^k$ converges only for $k < -1$. □

A complex-valued function P on $\mathbb{R}^n$ is called a *polynomial* if it is of the type

$$P(x) = \sum_{\nu_1,\ldots,\nu_n \geq 0} \alpha_{\nu_1,\ldots,\nu_n} x_1^{\nu_1} \ldots x_n^{\nu_n},$$

where $x_1,\ldots,x_n$ are the coordinates of x, where $\nu_1,\ldots,\nu_n$ are integers and where only finitely many coefficients $\alpha_{\nu_1,\ldots,\nu_n}$ are not zero. A term in the summation is called a *monomial*, and the *degree* of this monomial is $\nu_1 + \cdots + \nu_n$. The *degree* of the polynomial is the degree of its monomial of largest degree with a non-zero coefficient. A polynomial is called *homogeneous of degree* k if all the monomials that occur in it are of degree k. A polynomial decomposes in a unique manner as a sum of homogeneous polynomials of different degrees.

If A is a linear operator from $\mathbb{R}^m \to \mathbb{R}^n$ and if P is a polynomial on $\mathbb{R}^n$ then it is rather obvious that $Q(y) := P(A(y))$ is a polynomial on $\mathbb{R}^m$ with $\deg Q \leq \deg P$. In particular we have the following.

Exercise 15.1.2 Prove that a change of basis in $\mathbb{R}^n$ transforms a polynomial into a polynomial of the same degree.

Throughout this chapter we denote by $\|x\|$ the Euclidean norm of $x \in \mathbb{R}^n$. Since $|x_i| \leq \|x\|$, a polynomial of degree k satisfies an estimate

$$|P(x)| \leq C(\|x\|^k + 1) \tag{15.1}$$

for a constant C independent of x. It is however false that $|P(x)|$ looks like $\|x\|^k$ whenever $\|x\|$ is large, as is shown by the example $P(x) = x_1^k$.

In the present section we need some basic concepts of measure theory, and first of all the concept of a set of measure zero. This will be used only for the n-dimensional volume measure on $\mathbb{R}^n$, or for the uniform measure on a sphere. The reader who has not studied measure theory should not be afraid of this, as nothing of any depth takes place here. A set of measure zero is in a precise technical sense a "small set". A single point is of measure zero. One basic property of sets of measure zero is that a countable union of such sets is still of measure zero, and in particular a countable set is of measure zero. Say we have a certain property that depends on a point $y \in \mathbb{R}^k$. It will happen that this property is not true for all values of y, but that there are only a few exceptional values of y for which it fails. Saying that this set of exceptional values is of measure zero is just a technical way to quantify that these values are exceptional.

A property of a point x in a measure space is said to be true for almost all x if the set of points where it fails is of measure zero. In particular a function is said to be zero almost everywhere (a.e.) if the set of points where it is not zero is of measure zero. If a function $f \geq 0$, which may take infinite values, is of integral zero, it is zero a.e. If its integral is finite, it is finite a.e.

Many arguments rely on the Fubini–Tonelli theorem, one of the basic theorems of measure theory. Consider a function f on a product space of two measure spaces. Then if $|f|$ is integrable for the product measure, Fubini's theorem states that

$$\iint \mathrm{d}\mu(x)\mathrm{d}\nu(y)f(x,y) = \int \mathrm{d}\mu(x)\Big(\int \mathrm{d}\nu(y)f(x,y)\Big), \tag{15.2}$$

whereas Tonelli's theorem states that is also the case if $f \geq 0$. We will use this only when $\mathrm{d}\nu$ is the usual measure on some $\mathbb{R}^m$ and when μ is either the usual measure on some $\mathbb{R}^k$ or on a sphere. When $f \geq 0$, among the most notable consequences of this formula:

- If the left-hand side of (15.2) is finite, then the integral $\int \mathrm{d}\nu(y)f(x,y)$ is finite for almost all x.
- The left-hand side is zero if and only if the integral $\int \mathrm{d}\nu(y)f(x,y)$ is zero for almost all x.

Lemma 15.1.3 *If a polynomial P on $\mathbb{R}^n$ is not identically zero then the set of its zeros is of measure zero.*

If you are not familiar with the notion of set of measure zero, the intuitive meaning that "the set of zeros of a non-zero polynomial is a very small set" suffices perfectly.

Proof The proof goes by induction on n. For $n = 1$ the result follows from the fact that a non-zero polynomial has a finite number of zeros. For the induction step from $n-1$ to n we group the terms in P to write

$$P(x) = \sum_{\nu \geq 0} w_\nu(x_2,\ldots,x_n)x_1^\nu,$$

where w_ν is a polynomial. Since P is not zero, then there exists ν such that w_ν is not zero. By the induction hypothesis, the set where $w_\nu(x_2,\ldots,x_n)$ is zero is of measure zero. When $w_\nu(x_2,\ldots,x_n) \neq 0$, there are only finitely many values of x_1 for which $P(x_1,x_2,\ldots,x_n) = 0$. This makes the set of zeros of P appear as the union of following two sets: the set of points $(x_1,x_2,\ldots x_n)$ where $w_\nu(x_2,\ldots,x_n) = 0$ and a set that has a finite intersection with every set of the type $\mathbb{R} \times (x_2,\ldots,x_n)$. Both sets are of measure zero, as one sees applying (15.2) to the indicator function of the second set.[2] □

Corollary 15.1.4 *If a homogeneous polynomial P on $\mathbb{R}^n$ is not identically zero, the set of its zeros on the sphere S of radius 1 is of measure zero for the surface measure μ on S.*

Proof Since P is homogeneous, for $x \neq 0$, $P(x) = 0$ if and only if $P(x/\|x\|) = 0$. The result then follows from Lemma 15.1.3, computing the measure of the set of zeros of P in polar coordinates and using (15.2). □

[2] Again don't worry about the details if you have not learned measure theory, the argument should make it clear that the set of zeros of P is really small.

The following extends Lemma 15.1.1.

Lemma 15.1.5 *Consider two polynomials P and Q on $\mathbb{R}^n$ with $P \neq 0$ (and of course $Q \neq 0$). If the integral*

$$\int \mathrm{d}^n x \left| \frac{P(x)}{Q(x)} \right| \tag{15.3}$$

is convergent, then $\deg P - \deg Q < -n$.

The basic reason why this is true is that the integral $\int_{\|x\|\geq 1} \mathrm{d}^n x \|x\|^k$ is finite only for $k < -n$. Furthermore, $|P(x)|$ is about as large as $\|x\|^{\deg P}$ (and hence $|P(x)/Q(x)|$ is about as large as $\|x\|^{\deg P - \deg Q}$) in sufficiently many directions for the conclusion to be correct. Writing this argument in a precise way requires a little bit of work. We will use polar coordinates. We denote by S the sphere of radius 1 of $\mathbb{R}^n$ and $\mathrm{d}\mu$ its surface measure. Given $x \in S$ we denote $\deg_t P(tx)$ the degree of the polynomial $t \mapsto P(tx)$.

Lemma 15.1.6 *For all $x \in S$ outside a set of μ-measure zero one has* $\deg_t P(tx) = \deg P$.

Proof Let us write $P = P_0 + P_1$ where $P_0 \neq 0$ is homogeneous of degree $\deg P$ whereas $\deg P_1 < \deg P$. According to Corollary 15.1.4 we have $P_0(x) \neq 0$ outside a set of μ-measure zero. Moreover since $\deg_t P_1(tx) \leq \deg P_1$ and $P_0(tx) = t^{\deg P} P_0(x)$ it holds that $\deg_t P(tx) = \deg P$ whenever $P_0(x) \neq 0$. □

Corollary 15.1.7 *For all $x \in S$ outside a set of μ-measure zero, for large t it holds that*

$$\left| \frac{P(tx)}{Q(tx)} \right| \simeq C_x t^{\deg P - \deg Q}, \tag{15.4}$$

for a number $C_x > 0$ independent of t.

Proof For $x \in S$ outside a set of μ-measure zero one has $\deg_t P(tx) = \deg P$ and $\deg_t Q(tx) = \deg Q$. □

Proof of Lemma 15.1.5 Passing to polar coordinates and using (15.2), the convergence of the integral (15.3) implies the convergence of the integral

$$\int_S \mathrm{d}\mu(x) \left(\int_0^\infty \mathrm{d}t\, t^{n-1} \left| \frac{P(tx)}{Q(tx)} \right| \right).$$

Then outside a set of measure zero the inner integral is finite and (15.4) holds, so that $\deg P - \deg Q < -n$. □

Now, we would like to go into the reverse direction of Lemma 15.1.5, to find conditions on the degree of P and Q to ensure the convergence of the integral (15.3). An instructive example is given for $n = 2$ by the case $P(x) = 1$ and $Q(x) = x_1^m + 1$ for some $m \geq 1$. It shows that a condition of degree does not suffice, as even when m is large there is a divergence when we try to integrate in x_2, i.e. in the direction of the second vector basis. Such a problematic direction need not be parallel to a space generated by coordinate vectors as shown by the example $Q(x) = (x_1 - x_2)^m + 1$. In the rest of this section we put forward a stronger necessary condition than that of Lemma 15.1.5. This condition will also be shown to be sufficient for the special type of polynomials that are relevant to us. To understand the idea of this condition, consider $k \le n$, so we may think of $x \in \mathbb{R}^n$ as a pair (y,z) with $y \in \mathbb{R}^k$ and $z \in \mathbb{R}^{n-k}$. Using (15.2) the integral (15.3) is then

$$\int \mathrm{d}^k y \left(\int \mathrm{d}^{n-k} z \left| \frac{P(y,z)}{Q(y,z)} \right| \right). \tag{15.5}$$

As explained above, when this integral is finite, Tonelli's theorem implies that outside a set of measure zero the integral

$$\int \mathrm{d}^{n-k} z \left| \frac{P(y,z)}{Q(y,z)} \right| \tag{15.6}$$

is finite. Denoting by p and q the degrees of the polynomials $z \mapsto P(y,z)$ and $z \mapsto Q(y,z)$, Lemma 15.1.5 shows that the convergence of the integral (15.5) imposes the condition $p - q < -(n-k)$. This condition does not follow from the condition $\deg P - \deg Q < -n$.

We have obtained a non-trivial condition by decomposing $\mathbb{R}^n = \mathbb{R}^k \oplus \mathbb{R}^{n-k}$ but we will also obtain non-trivial conditions by decomposing $\mathbb{R}^n = E^\perp \oplus E$ for any subspace E. We now explore this idea, in sufficient generality for further use.

Definition 15.1.8 Consider a polynomial P on $\mathbb{R}^n$ and a subspace E of $\mathbb{R}^n$, $E \neq \{0\}$. Then given a point f of $\mathbb{R}^n$ and a basis $f_1, \ldots, f_k$ of E, the function $P_{E,f}(u_1, \ldots, u_k) := P(f + \sum_{j \le k} u_j f_j)$ is a polynomial in the variables $u_1, \ldots, u_k$, whose degree is independent of the basis $f_1, \ldots, f_k$. The maximum degree of this polynomial as f varies is denoted by $\deg_E P$.

That the degree of $P_{E,f}$ is independent of the choice of the basis follows from Exercise 15.1.2. One should think of $\deg_E P$ as "the degree of P in the direction of E". This interpretation will become transparent after we examine the case where E is given by setting some of the coordinates equal to 0.

Exercise 15.1.9 Consider the case $n = 2$ and $P(x) = x_1^2$. Show that when E is of dimension 1, then $\deg_E P = 2$ unless E is the subspace of equation $x_1 = 0$. Assuming now $P(x) = (x_1 - x_2)^2$ show that when E is of dimension 1 then $\deg_E P = 2$ unless E is the subspace of equation $x_1 = x_2$.

As we are going to show, for the typical value of f, the polynomial $P_{E,f}$ is of degree $\deg_E P$. It is only for a few exceptional values of f that $P_{E,f}$ can be of degree $< \deg_E P$. Our goal in this section is the following.

Proposition 15.1.10 *If the integral (15.3) is convergent then for each subspace E of $\mathbb{R}^n$, $E \neq \{0\}$, we have*

$$\deg_E P - \deg_E Q < -\dim E. \tag{15.7}$$

Consider k,m with $n = k+m$, $y \in \mathbb{R}^k$, $z \in \mathbb{R}^m$ and $x = (y,z) \in \mathbb{R}^{k+m} = \mathbb{R}^n$. Thus a polynomial $P(x)$ is a finite sum

$$P(x) = \sum a_{\nu_1,\ldots,\nu_k,\mu_1,\ldots,\mu_m} y_1^{\nu_1} \ldots y_k^{\nu_k} z_1^{\mu_1} \ldots z_m^{\mu_m}, \tag{15.8}$$

over integers $\nu_1,\ldots,\nu_k,\mu_1,\ldots,\mu_m \geq 0$. We defined the degree $\deg P$ of P as the largest value of $\nu_1 + \cdots + \nu_k + \mu_1 + \cdots + \mu_m$ for which the corresponding coefficient is not zero. In a similar manner *we define the degree* $\deg_z P$ *in the variable* z *as the largest value of* $\mu_1 + \cdots + \mu_m$ *for which there exists* $\nu_1,\ldots,\nu_k$ *such that* $a_{\nu_1,\ldots,\nu_k,\mu_1,\ldots,\mu_m} \neq 0$. We may group the terms in (15.8) as

$$P(x) = \sum_{\mu_1,\ldots,\mu_m} R_{\mu_1,\ldots,\mu_m}(y) z_1^{\mu_1} \ldots z_m^{\mu_m}, \tag{15.9}$$

where $R_{\mu_1,\ldots,\mu_m}(y)$ are polynomials in y, and $\deg_z P$ is the largest value of $\mu_1 + \cdots + \mu_m$ for which the polynomial $R_{\mu_1,\ldots,\mu_m}(y)$ is not identically zero.

Definition 15.1.11 Consider a polynomial $P(x)$ as in (15.8), where $x = (y,z)$. Then, given $y \in \mathbb{R}^k$ we define the polynomial P_y in z by $P_y(z) = P(y,z)$.

The tricky part of the notation is that here we think of y as a fixed point in $\mathbb{R}^k$ whereas when we write $P(x)$ for $x = (y,z)$ we think of it as a variable. Again, to obtain the polynomial P_y we replace each variable quantity y_i by its numerical value, and obtain a polynomial in z only. For example, if $P(y,z) = yz$ then $P_2(z) = 2z$ but $P_0(z) = 0$. On the other hand, for this polynomial, $\deg_z P = 1$.

Lemma 15.1.12 *We have* $\deg P_y \leq \deg_z P$ *and the set of* y *for which* $\deg P_y < \deg_z P$ *is of measure zero. Finally, if* $E = \{0\} \times \mathbb{R}^m = \{(y,z) \in \mathbb{R}^k \times \mathbb{R}^m ; y = 0\}$ *then* $\deg_E P = \deg_z P$.

Proof It should be obvious that $\deg P_y \leq \deg_z P$. Recalling (15.9), by definition of $\deg_z P$ we may find $\mu_1,\ldots,\mu_m$ with $\mu_1 + \cdots + \mu_m = \deg_z P$ and the polynomial $R_{\mu_1,\ldots,\mu_m}$ is not the zero polynomial. Then outside a set of measure zero $R_{\mu_1,\ldots,\mu_m}(y) \neq 0$ and then P_y is of degree $\mu_1 + \cdots + \mu_m = \deg_z P$. To prove the last assertion, we use the canonical basis $(e_{k+1},\ldots,e_n)$ of E to compute $\deg_E P$, so that if $f = (f_\ell)_{\ell \leq n}$ and $f' = (f_\ell)_{1 \leq \ell \leq k}$ then

$$\begin{aligned} P_{E,f}(u) &= P(f_1,\ldots,f_k,f_{k+1}+u_1,\ldots,f_n+u_m) \\ &= \sum_{\mu_1,\ldots,\mu_m} R_{\mu_1,\ldots,\mu_m}(f')(f_{k+1}+u_1)^{\mu_1} \ldots (f_n+u_m)^{\mu_m}. \end{aligned} \tag{15.10}$$

The maximum possible degree in u as f varies is the largest possible sum $\mu_1 + \cdots + \mu_n$ for which $R_{\mu_1,\ldots,\mu_n}$ is not the zero polynomial. That is $\deg_z P$. □

We are now ready to prove Proposition 15.1.10 in the special case where E is spanned by a set of basis vectors.

Lemma 15.1.13 *Assume that* $E = \{0\} \times \mathbb{R}^{n-k}$. *Assume that the integral (15.5) is convergent. Then*

$$\deg_E P - \deg_E Q < -(n-k) = -\dim E.$$

Proof By Tonelli's theorem the integral (15.6) is convergent outside a set of y of measure zero. We can find a value of y for which the integral (15.6) is finite, and at the same time the polynomial $z \mapsto P(y,z)$ is of degree $\deg_E P$ whereas the polynomial $z \mapsto Q(y,z)$ is of degree $\deg_E Q$, because the set of points that fail these three conditions is of measure zero. The result then follows from Lemma 15.1.5. □

The following relation is almost obvious from the definitions, but it is important because it allows us to reduce to the case where E is as above.

Lemma 15.1.14 *Consider an invertible linear operator A on $\mathbb{R}^n$, and for a polynomial P on $\mathbb{R}^n$ consider the polynomial $P \circ A$ given by $P \circ A(x) = P(A(x))$. Then for a subspace E of $\mathbb{R}^n$ we have*

$$\deg_{A(E)} P = \deg_E(P \circ A). \tag{15.11}$$

Proof of Proposition 15.1.10 Consider a linear transformation A such that $A(E)$ is generated by basis vectors, so that it is of the type $E = \{0\} \times \mathbb{R}^{n-k}$. Consider the polynomials $P' = P \circ A^{-1}$ and $Q' = Q \circ A^{-1}$ so that the integral $\int \mathrm{d}^n x |P'(x)/Q'(x)|$ is convergent (as is seen by change of variables) and therefore that integral $\int \mathrm{d}^k y \int \mathrm{d}^{n-k} z |P'(y,z)/Q'(y,z)|$ is convergent. It follows by Lemma 15.1.13 that $\deg_{A(E)} P' - \deg_{A(E)} Q' < -\dim A(E) = -\dim E$. But $\deg_{A(E)} P' - \deg_{A(E)} Q' = \deg_E P - \deg_E Q$ by (15.11). □

Let us examine in detail a non-trivial example, which we urge the reader to fully understand before proceeding further. The features of this example are quite representative of the difficulties of the general situation to be studied in the next section. We take $n = 2$, and $Q(x_1,x_2) = (1+x_1^2)(1+x_2^2)(1+(x_1-x_2)^2)$ so that $\deg Q = 6$. Which are the polynomials P that satisfy (15.7)? Using this condition for $E = \mathbb{R}^2$ means $\deg P \le 3$. Furthermore when $\dim E = 1$ condition (15.7) means $\deg_E P - \deg_E Q < -1$ i.e. $\deg_E P \le \deg_E Q - 2$. This is a non-trivial condition when E is given by the equation $x_1 = 0$ or $x_2 = 0$ or $x_1 = x_2$, since in each of these cases $\deg_E Q = 4 < \deg Q = 6$. So that we must then also have $\deg_E P \le 2$ for each of the three preceding choices of E. The first two cases imply that P may only contain monomials $x_1^a x_2^b$ with $a,b \le 2$ and since $\deg P \le 3$ it also holds that $a+b \le 3$. The case where E is given by the equation $x_1 = x_2$ imposes a further condition: that the terms of degree 3 cancel out when $x_2 = x_1$. This means that these terms of degree 3 are of the type $(x_2 - x_1)R$ where R is of degree 2. But since P contains only monomials $x_1^a x_2^b$ for $a,b \le 2$, R has to be of the type $a x_1 x_2$ for $a \in \mathbb{C}$. In summary the conditions (15.7) are satisfied exactly when $P(x_1,x_2) = P_1(x_1,x_2) + a(x_1 - x_2)x_1x_2$ where P_1 is of degree 2 and $a \in \mathbb{C}$.

It turns out that for these P the integral (15.3) is convergent. The reader can convince herself that this is not a completely trivial statement.

We end this section by gathering tools for later purposes.

Lemma 15.1.15 *Consider a polynomial P on $\mathbb{R}^n$ and its decomposition $P = \sum_\ell P_\ell$ as a sum of homogeneous polynomials of different degrees. Then for each subspace E of $\mathbb{R}^n$ and each ℓ we have*

$$\deg_E P_\ell \le \deg_E P. \tag{15.12}$$

Proof We use (15.11) as well as the fact that $P \circ A = \sum_\ell P_\ell \circ A$ is the decomposition of $P \circ A$ into a sum of homogeneous polynomials to reduce to the situation where E is of the type $E = \{0\} \times \mathbb{R}^{n-k}$. The result is then obvious by Lemma 15.1.12. □

Lemma 15.1.16 *Consider $x \in \mathbb{R}^{k+m}$ as a pair $(y,z) \in \mathbb{R}^k \times \mathbb{R}^m$. Consider a polynomial*

$$P(x) = \sum_{\nu_1,\ldots,\nu_k \geq 0} y_1^{\nu_1} \ldots y_k^{\nu_k} S_{\nu_1,\ldots,\nu_k}(z), \tag{15.13}$$

where each $S_{\nu_1,\ldots,\nu_k}$ is a polynomial in z. Recall that given $y \in \mathbb{R}^k$, P_y denotes the polynomial in z given by $P_y(z) = P(y,z)$. Consider a subspace E' of $\mathbb{R}^m$ and the subspace $E = \{0\} \times E'$ of $\mathbb{R}^{k+m}$. Then it holds that

$$\sup_y \deg_{E'} P_y = \sup_{\nu_1,\ldots,\nu_k} \deg_{E'} S_{\nu_1,\ldots,\nu_k} = \deg_E P. \tag{15.14}$$

To understand this statement, observe first that P_y and $S_{\nu_1,\ldots,\nu_k}$ are polynomials in $z \in \mathbb{R}^m$ so that it makes sense to consider $\deg_{E'}$ for these polynomials where E' is a subspace of $\mathbb{R}^m$. On the other hand, P is a polynomial on $\mathbb{R}^{k+m}$ and it makes sense to consider $\deg_E P$ since E is a subspace of $\mathbb{R}^{k+m}$.

The first equality in (15.14) has a remarkable consequence. Let us think of P_y as a polynomial in z that depends also on the parameter y. If it is true that whatever the choice of the parameter y it holds that $\deg_{E'} P_y \leq r$ then it is already the case that each of the polynomials $S_{\nu_1,\ldots,\nu_k}$ satisfies $\deg_{E'} S_{\nu_1,\ldots,\nu_k} \leq r$ (the converse being obvious).

Proof By linear change of variables we may assume that E' is the space given by $z_1 = \ldots = z_{m'} = 0$ for some $m' \leq m$, so that $E = \{0\} \times \mathbb{R}^{m-m'}$. Then according to Lemma 15.1.12, $\deg_E P$ is simply the degree of P in the variables $z_{m'+1}, \ldots, z_m$. Since $\deg_{E'} S_{\nu_1,\ldots,\nu_k}$ is the degree of $S_{\nu_1,\ldots,\nu_k}$ in the same variables, this should make it obvious that $\sup_{\nu_1,\ldots,\nu_k} \deg_{E'} S_{\nu_1,\ldots,\nu_k} \leq \deg_E P$. It should also be obvious that for each y we have $\deg_{E'} P_y \leq \sup_{\nu_1,\ldots,\nu_k} \deg_{E'} S_{\nu_1,\ldots,\nu_k}$. Finally, Lemma 15.1.12 asserts that simply by avoiding a set of measure zero in $\mathbb{R}^{k+m'}$ we may choose $y \in \mathbb{R}^k$ and $z_1, \ldots, z_{m'}$ such that fixing the values of these variables the resulting polynomial in the variables $z_{m'+1}, \ldots, z_m$ is of degree $\deg_E P$. This choice of $z_1, \ldots, z_{m'}$ witnesses then that $\deg_{E'} P_y = \deg_E P$. □

15.2 Weinberg's Power Counting Theorem

Our goal in this section is to study a "Euclidean version" of diagram integrals. Matters are greatly clarified by adopting a sufficiently general setting as we do now. Consider some integer n' and for $1 \leq i \leq s$ consider linear maps $T_i : \mathbb{R}^n \to \mathbb{R}^{n'}$ and $w_i \in \mathbb{R}^{n'}$. Consider a polynomial $P(x)$ on $\mathbb{R}^n$. We are interested in the convergence of the integral

$$I(w_1, \ldots, w_s) := \int \mathrm{d}^n x \frac{|P(x)|}{\prod_{i \leq s} (\|T_i(x) + w_i\|^2 + 1)}. \tag{15.15}$$

Let us consider the polynomials

$$Q(x) := \prod_{i \le s} (\|T_i(x)\|^2 + 1) \; ; \; Q_w(x) := \prod_{i \le s} (\|T_i(x) + w_i\|^2 + 1). \tag{15.16}$$

Theorem 15.2.1 (Weinberg's power counting theorem[3]) *Recalling (15.16), given any values of $w_1, \dots, w_s$ the integral (15.15) is absolutely convergent if and only if for each subspace $S \neq \{0\}$ of $\mathbb{R}^n$ we have*

$$\deg_S P - \deg_S Q < -\dim S. \tag{15.17}$$

We start the proof by a simple observation.

Lemma 15.2.2 *We have*

$$\deg_S Q = \deg_S Q_w. \tag{15.18}$$

Proof Given $f \in \mathbb{R}^n$ and a basis $f_1, \dots, f_k$ of S, the polynomial $\|T_i(f + \sum_{j \le k} u_j f_j) + w_i\|^2 + 1$ in the variables u_j is of degree 0 if $T_i(f_j) = 0$ for each $j \le k$ and of degree 2 otherwise, independently of the value of w_i. Thus both $\deg_S Q$ and $\deg_S Q_w$ equal twice the number of indices i for which $T_i(f_j) \neq 0$ for some $j \le k$, or equivalently for which T_i is not identically zero on S.[4] □

The necessity of the condition (15.17) in Weinberg's theorem then follows from Proposition 15.1.10 since $\deg_S P - \deg_S Q = \deg_S P - \deg_S Q_w$.

We now start the preparations for the hard part of the proof of Theorem 15.2.1, that (15.17) suffices for the convergence of the integral.

Lemma 15.2.3 *If $\|w_i\| \le B$ for $i \le s$ then*

$$\frac{1}{\prod_{i \le s} (\|T_i(x) + w_i\|^2 + 1)} \le \frac{C(B,s)}{\prod_{i \le s} (\|T_i(x)\|^2 + 1)}, \tag{15.19}$$

where $C(B,s)$ depends on B and s only.

Proof Using the inequality $\|x - y\|^2 \le 2(\|x\|^2 + \|y\|^2)$ in the second line we obtain

$$\begin{aligned} \|T_i(x)\|^2 + 1 &= \|T_i(x) + w_i - w_i\|^2 + 1 \\ &\le 2(\|T_i(x) + w_i\|^2 + B^2) + 1 \\ &\le (2 + 2B^2)(\|T_i(x) + w_i\|^2 + 1). \end{aligned} \tag{15.20}$$

□

The proof of Theorem 15.2.1 proceeds by induction over n. Without loss of generality we assume that $T_i \neq 0$ for $i \le s$. According to (15.12) we may and do *assume that P is homogeneous*. Given $j \le s$, we consider the set

$$D_j := \{x \in \mathbb{R}^n \; ; \; \forall i \le s \, , \; \|T_j(x)\| \le \|T_i(x)\|\}. \tag{15.21}$$

[3] This theorem is proved in Weinberg's paper [86] but the idea of power counting occurs in a much earlier paper of F. Dyson, refer to [26, page 1746].

[4] I will help you later to make this important argument absolutely clear in your mind.

Since the sets D_j for $j \le s$ cover $\mathbb{R}^n$, using Lemma 15.2.3, it suffices to prove that for any j,

$$\int_{D_j} \mathrm{d}^n x \frac{|P(x)|}{\prod_{i\le s}(\|T_i(x)\|^2+1)} < \infty. \tag{15.22}$$

We fix j until the end of the proof. Making a suitable linear change of variables we may assume that $\|T_j(x)\|^2 = \sum_{\ell\le k} x_\ell^2$, where $k > 0$ is the rank of T_j. Let $m := n - k < n$ and write $x \in \mathbb{R}^n$ as $(y,z) \in \mathbb{R}^k \times \mathbb{R}^m$, so we have $\|T_j(x)\| = \|y\|$. We have to show that

$$\iint_{D_j} \mathrm{d}^k y \mathrm{d}^m z \frac{|P(y,z)|}{\prod_{i\le s}(\|T_i(y,z)\|^2+1)} < \infty. \tag{15.23}$$

We will split the integral over y in the domains $\|y\| \le 1$ and $\|y\| \ge 1$. It follows from the next result that the integral on the first domain is finite.

Lemma 15.2.4 *If* $\|y\| \le 1$ *then*

$$\int \mathrm{d}^m z \frac{|P(y,z)|}{\prod_{i\le s}(\|T_i(y,z)\|^2+1)} \le C, \tag{15.24}$$

where C *is independent of* y.

It turns out that the integral is finite for any value of y, but the uniform bound holds only for $\|y\|$ bounded.

Proof We write $P(x) = \sum_{\nu_1,\ldots,\nu_k\ge 0} y_1^{\nu_1}\cdots y_k^{\nu_k} S_{\nu_1,\ldots,\nu_k}(z)$ as in (15.13). Since $\|y\| \le 1$ we have $|y_j| \le 1$ for $j \le k$. Since $T_i(y,z) = T_i(y,0) + T_i(0,z)$ it suffices from Lemma 15.2.3 to prove that for each $\nu_1,\ldots,\nu_k$

$$\int \mathrm{d}^m z \frac{|S_{\nu_1,\ldots,\nu_k}(z)|}{\prod_{i\le s}(\|T_i(0,z)\|^2+1)} \le C. \tag{15.25}$$

Since $m < n$ we may appeal to the induction hypothesis. We have to show that the hypotheses of Weinberg's theorem are satisfied here. Consider a subspace $E' \subset \mathbb{R}^m$, and the space $E = \{0\} \times E' \subset \mathbb{R}^n$. Then $\dim E = \dim E'$. The hypothesis (15.17) implies $\deg_E P - \deg_E Q < -\dim E$. First, the last equality in (15.14) implies $\deg_{E'} S_{\nu_1,\ldots,\nu_k} \le \deg_E P$. Second, denoting by $Q'(z)$ the denominator in (15.25) we have $\deg_{E'} Q' = \deg_E Q = 2a$ where a is the number of $1 \le i \le s$ such that T_i is not identically zero on E. This implies that $\deg_{E'} S_{\nu_1,\ldots,\nu_k} - \deg_{E'} Q' \le \deg_E P - \deg_E Q < -\dim E = -\dim E'$ as required. □

Recalling that $\|T_j(x)\| = \|y\|$, we have $x = (y,z) \in D_j$ if and only if $z \in D(y) := \{z; \forall i \le s, \|T_i(y,z)\| \ge \|y\|\}$. Consequently to prove (15.23) and conclude the proof of Weinberg's theorem it suffices to show that

$$\int_{\{\|y\|\ge 1\}} \mathrm{d}^k y \int_{D(y)} \mathrm{d}^m z \frac{|P(y,z)|}{\prod_{i\le s}(\|T_i(y,z)\|^2+1)} < \infty. \tag{15.26}$$

To compute the first integral we move to polar coordinates. Let $t = \|y\|$ and $y' = y/t$. In the inner integral, we make the change of variables $z = tz'$. Thus, if $\mathrm{d}\mu$ is the surface measure of the unit sphere of $\mathbb{R}^k$, the previous integral is proportional to

$$\int_{t\geq 1} \mathrm{d}t\, t^{n-1} \int \mathrm{d}\mu(y') \int_{D(y')} \mathrm{d}^m z' \frac{|P(ty', tz')|}{\prod_{i\leq s}(t^2\|T_i(y',z')\|^2 + 1)}, \tag{15.27}$$

where $D(y') = \{z' \ ; \ \forall i \leq s\,,\ \|T_i(y',z')\| \geq 1\}$. We use that for a number $B \geq 1$ and $t \geq 1$ one has $t^2 B + 1 \geq t^2(B+1)/2$ to obtain

$$\prod_{i\leq s}(t^2\|T_i(y',z')\|^2 + 1) \geq \left(\frac{t^2}{2}\right)^s \prod_{i\leq s}(\|T_i(y',z')\|^2 + 1).$$

Also, since we assume that P is homogeneous, we have $|P(ty',tz')| = t^{\deg P}|P(y',z')|$. Combining these estimates we bound the integral (15.27) by

$$\int_1^\infty \mathrm{d}t\, t^{n-1+\deg P - 2s} C_1 \tag{15.28}$$

where

$$C_1 = 2^s \int \mathrm{d}\mu(y') \int \mathrm{d}^m z' \frac{|P(y',z')|}{\prod_{i\leq s}(\|T_i(y',z')\|^2 + 1)} \tag{15.29}$$

is finite by Lemma 15.2.4. Furthermore the integral (15.28) is finite because

$$n - 1 + \deg P - 2s = n + \deg P - \deg Q - 1 \leq -2. \qquad \square$$

15.3 The Fundamental Space ker $\mathcal{L}$

Given a Feynman diagram, the graph of interest to us is the graph consisting of its *internal* vertices and its *internal* edges. We denote by $\mathcal{E}$ the set of internal edges and by $\mathcal{V}$ the set of internal vertices. This notation is used throughout the rest of the book and must be remembered. The graph thus associated to a Feynman diagram is not naturally oriented, but we turn it into an oriented graph by picking an arbitrary orientation for each edge. Given an edge e we denote by e_o and e_t its originating and terminating vertices, which we always assume to be different. There can be many edges with the same originating and terminating vertices. A vertex v is *adjacent* to an edge e if either $v = e_o$ or $v = e_t$.

The edges of a Feynman diagram carry four-momentum. The four-momenta carried by the edges will be described by a point x of $(\mathbb{R}^{1,3})^{\mathcal{E}}$, so x_e represents the four-momentum carried by the edge e. A fundamental property of the Feynman rules in momentum space stated on page 390 is that they enforce conservation of momentum at each internal vertex: The algebraic sum of the four-momenta (including four-momenta flowing on the external lines) *entering*[5] each internal vertex is zero.

[5] Choice of this orientation is conventional and unimportant: If four-momentum p enters a vertex, four-momentum $-p$ leaves the same vertex.

To make this formal we denote by $\mathcal{L}$ the map from $(\mathbb{R}^{1,3})^{\mathcal{E}}$ to $(\mathbb{R}^{1,3})^{\mathcal{V}}$ which calculates at each vertex the amount of four-momentum that enters this vertex through the adjacent internal edges, by the formula

$$\mathcal{L}(x)_v = \sum_{e_t=v} x_e - \sum_{e_o=v} x_e. \tag{15.30}$$

Going back to Feynman diagrams, let us denote by w_v the amount of four-momentum *leaving* the vertex v through the external lines adjacent to v. (Thus $w_v = 0$ where there are no such external lines.) Conservation of four-momentum at vertex v means that $\mathcal{L}(x)_v - w_v = 0$.

According to the Feynman rules in momentum space, the value of a Feynman diagram is given (forgetting some numerical coefficients[6]) by

$$\lim_{\varepsilon\to 0^+} \int \cdots \int \prod_{e\in\mathcal{E}} \frac{\mathrm{d}^4 x_e}{(2\pi)^4} f(x_e) \prod_{v\in\mathcal{V}} (2\pi)^4 \delta^{(4)}(\mathcal{L}(x)_v - w_v), \tag{15.31}$$

where $f(x_e) = -\mathrm{i}/(-x_e^2 + m^2 - \mathrm{i}\varepsilon)$. Here, the dependence of the propagators on ε is kept implicit. The term $\delta^{(4)}(\mathcal{L}(x)_v - w_v)$ is simply the term $\delta^{(4)}(\Sigma)$ where Σ is the algebraic sum of the four-momenta entering the vertex v: Four-momentum $\mathcal{L}(x)_v$ enters through the internal lines and four-momentum $-w_v$ enters through the external lines.

Looking at (15.31) it is not difficult to *guess* the following fundamental fact (which we will prove in detail later):

$$\textit{Diagram integrals are integrals over}\ \ker\mathcal{L}. \tag{15.32}$$

The space $\ker\mathcal{L}$ is a finite-dimensional vector space. As such it has a translation-invariant measure, which is entirely determined up to a multiplicative constant.[7] The precise meaning of (15.32) is that diagram integrals are integrals over $\ker\mathcal{L}$ for such a translation-invariant measure.

This brings forward the fundamental nature of $\ker\mathcal{L}$. We will later prove the following.

Theorem 15.3.1 *The range of $\mathcal{L}$ is the space*

$$\mathcal{N} = \Big\{ w = (w_v)_{v\in\mathcal{V}} \in (\mathbb{R}^{1,3})^{\mathcal{V}};\ \sum_{v\in\mathcal{V}} w_v = 0 \Big\}, \tag{15.33}$$

and

$$\dim\ker\mathcal{L} = 4(\operatorname{card}\mathcal{E} - \operatorname{card}\mathcal{V} + 1). \tag{15.34}$$

15.4 Power Counting in Feynman Diagrams

Leaving aside for the moment the proof of (15.32) and (15.34), let us denote by $\mathrm{d}\mu_{\mathcal{L}}$ a suitable normalization of the translation-invariant measure on $\ker\mathcal{L}$. We will determine

6 That is, not writing some factors that are finite numbers.
7 In the case of the finite-dimensional space $\mathbb{R}^n$, the usual volume measure is such a translation-invariant measure.

this normalization later, but its precise definition is irrelevant to the purpose of the present section, which is to investigate through Weinberg's theorem the convergence of the integral

$$\int \mathrm{d}\mu_{\mathcal{L}}(x) \frac{1}{\prod_{e \in \mathcal{E}} (\|x_e + y_e\|^2 + 1)}, \tag{15.35}$$

where y_e are any points of $\mathbb{R}^{1,3}$. As we will see later in (15.46) such integrals are exactly what we need to compute diagram integrals, except that here we have replaced the propagators by their Euclidean version.

We will also introduce notions which will be essential in the next chapter in the study of renormalization.

Definition 15.4.1 Given a subset $\mathcal{E}'$ of $\mathcal{E}$, consider the subset $\mathcal{V}'$ of $\mathcal{V}$ consisting of the vertices adjacent to $\mathcal{E}'$. The graph with set of edges $\mathcal{E}'$ and set of vertices $\mathcal{V}'$ is called a *subdiagram* of the original Feynman diagram. The edges of $\mathcal{E}'$ are oriented the same way as the edges of $\mathcal{E}$.

As it is always understood in all our considerations that we start from a given Feynman diagram, we will simply say "subdiagram" rather than "subdiagram of the original Feynman diagram". There is a largest subdiagram, with set of edges $\mathcal{E}$ and set of vertices $\mathcal{V}$. It is simply the original Feynman diagram to which one has removed external vertices and external edges. This largest subdiagram will nearly always be denoted by γ.[8]

Definition 15.4.2 Given a subset $\mathcal{V}'$ of $\mathcal{V}$, consider the subset $\mathcal{E}'$ of $\mathcal{E}$ consisting of all edges between two vertices in $\mathcal{V}'$. The graph with set of vertices $\mathcal{V}'$ and set of edges $\mathcal{E}'$ is called a *subgraph* of the original Feynman diagram.

A subgraph with no isolated vertices is a subdiagram (namely the subdiagram corresponding to its set of edges), but a subdiagram need not be a subgraph. For a subdiagram, there may exist a pair of vertices v_1, v_2 in $\mathcal{V}'$ and one edge e in $\mathcal{E}$ adjacent to both vertices that does not belong to $\mathcal{E}'$.

Given a subdiagram with set $\mathcal{E}'$ of edges and set $\mathcal{V}'$ of vertices we may define a map $\mathcal{L}' : (\mathbb{R}^{1,3})^{\mathcal{E}'} \to (\mathbb{R}^{1,3})^{\mathcal{V}'}$ as in (15.30). These maps will be of constant use. The following fact is obvious from the definition of $\mathcal{L}$, but it will also be of constant use.

Lemma 15.4.3 *Consider a subdiagram with set of edges $\mathcal{E}'$ and set of vertices $\mathcal{V}'$. Consider $x = (x_e)_{e \in \mathcal{E}} \in (\mathbb{R}^{1,3})^{\mathcal{E}}$ and $y = (y_e)_{e \in \mathcal{E}'} \in (\mathbb{R}^{1,3})^{\mathcal{E}'}$. Assume that $x_e = 0$ if $e \notin \mathcal{E}'$ and that $y_e = x_e$ if $e \in \mathcal{E}'$. Then for $v \in \mathcal{V}'$ one has $\mathcal{L}(x)_v = \mathcal{L}'(y)_v$.*

Definition 15.4.4 The superficial degree of divergence of a subdiagram with set of edges $\mathcal{E}'$ is given by

$$d' := \dim \ker \mathcal{L}' - 2 \operatorname{card} \mathcal{E}'. \tag{15.36}$$

The reason for that name will become gradually clear. Our next goal is to prove the following consequence of Theorem 15.2.1.

8 Please keep in mind that a subdiagram has no external lines, and that a Feynman diagram is not a subdiagram of itself!

Theorem 15.4.5 *The integral (15.35) is convergent if and only if the superficial degree of divergence of each subdiagram is* < 0.

Proof Assume first that the superficial degree of divergence of each subdiagram is <0. Consider a subspace S of $\ker \mathcal{L}$. Denote by $\mathcal{E}' \subset \mathcal{E}$ the set of edges such that $x \mapsto x_e$ is *not* identically zero on S. Then if $Q(x) := \prod_{e\in\mathcal{E}}(\|x_e + y_e\|^2 + 1)$ we have $\deg_S Q = 2\,\mathrm{card}\,\mathcal{E}'$. Consider the subdiagram with set of edges $\mathcal{E}'$. Consider the map $\Pi_{\mathcal{E}'} : \mathbb{R}^{\mathcal{E}} \to \mathbb{R}^{\mathcal{E}'}$ given by $\Pi_{\mathcal{E}'}((x_e)_{e\in\mathcal{E}}) = (x_e)_{e\in\mathcal{E}'}$. Then, by definition of $\mathcal{E}'$ the map $\Pi_{\mathcal{E}'}$ is one to one on S, and by Lemma 15.4.3, we have $\Pi_{\mathcal{E}'}(S) \subset \ker \mathcal{L}'$ so that $\dim S = \dim \Pi_{\mathcal{E}'}(S) \le \dim\ker\mathcal{L}'$ and thus $\dim S - \deg_S Q \le \dim\ker\mathcal{L}' - 2\,\mathrm{card}\,\mathcal{E}'$. By hypothesis we have $\dim\ker\mathcal{L}' - 2\,\mathrm{card}\,\mathcal{E}' < 0$. Thus we have proved (15.17) when $P = 1$ and by Theorem 15.2.1 the integral converges.[9]

Conversely assume that the integral converges. Consider a subdiagram with set of edges $\mathcal{E}'$. Consider the subspace S of $\ker\mathcal{L} \subset (\mathbb{R}^{1,3})^{\mathcal{E}}$ defined by the equations $x_e \equiv 0$ for $e \notin \mathcal{E}'$, where we recall the notation $x = (x_e)_{e\in\mathcal{E}}$. For $y \in \ker\mathcal{L}' \subset (\mathbb{R}^{1,3})^{\mathcal{E}'}$ consider $x \in (\mathbb{R}^{1,3})^{\mathcal{E}}$ defined by $x_e = y_e$ if $e \in \mathcal{E}'$ and $x_e = 0$ otherwise. Then by Lemma 15.4.3, the map $y \mapsto x$ is one-to-one and onto from $\ker\mathcal{L}'$ to S so that $\dim S = \dim\ker\mathcal{L}'$ and since $\deg_S Q = 2\,\mathrm{card}\,\mathcal{E}'$ we have $\dim\ker\mathcal{L}' - 2\,\mathrm{card}\,\mathcal{E}' < 0$ by the necessary condition of Theorem 15.2.1. That is, we have proved that the superficial degree of divergence of the subdiagram with set of edges $\mathcal{E}'$ is < 0. □

In the rest of the section we analyze the superficial degree of divergence of connected subdiagrams in the case of ϕ^k_d theory. This theory is a natural generalization of the ϕ^4_4 theory of (13.95). It has an interaction term $:\varphi^k/k!:$ and it is assumed that *space-time has dimension* d, i.e. $\mathbb{R}^{1,3}$ is replaced by $\mathbb{R}^{1,d-1}$. This might feel weird at first.[10] The purpose of allowing d space-time dimensions is to let us consider the so-called ϕ^3_6 model, with interaction $:\varphi^3/3!:$ in six space-time dimensions, which is a particularly instructive toy model.

Definition 15.4.6 A subdiagram is *connected* if it is connected as a graph, that is if one cannot split its set of vertices in two non-empty parts such that there are no edges linking a vertex of the first part and a vertex of the second part.

Definition 15.4.7 Two subdiagrams are *disjoint* if they have no *vertex* in common.

Observe that disjoint subdiagrams have no edges in common, but that subdiagrams with no edges in common need not be disjoint. It should be obvious how to decompose a subdiagram as a union of disjoint connected subdiagrams, which are called its connected components. The following should also be obvious.

Lemma 15.4.8 *The superficial degree of divergence of a subdiagram is the sum of the superficial degrees of divergence of its connected components.*

We go back to the analysis of the superficial degree of divergence in ϕ^k_d theory. We consider a Feynman diagram, and a subdiagram with set of edges $\mathcal{E}'$ and set of vertices $\mathcal{V}'$.

[9] If it is disturbs you that we apply Theorem 15.2.1 to $\ker\mathcal{L}$ while this theorem is stated for $\mathbb{R}^n$, just fix a basis of $\ker\mathcal{L}$ to identify this space with $\mathbb{R}^n$.

[10] Let us hurry to say that this has nothing to do with the extra dimensions considered by string theory.

Let us consider the set $\mathcal{E}_1$ of edges of the Feynman diagram (including the external edges) such that exactly one of their end vertices belongs to $\mathcal{V}'$. Let us consider the set $\mathcal{E}_2$ of edges of the Feynman diagram such that both the end vertices belong to $\mathcal{V}'$. Thus $\mathcal{E}' \subset \mathcal{E}_2$. Now, since k lines go out of each internal vertex,

$$k \operatorname{card} \mathcal{V}' = \operatorname{card} \mathcal{E}_1 + 2 \operatorname{card} \mathcal{E}_2,$$

which we rewrite as

$$2 \operatorname{card} \mathcal{E}' = k \operatorname{card} \mathcal{V}' - b, \tag{15.37}$$

where

$$b := \operatorname{card} \mathcal{E}_1 + 2 \operatorname{card} (\mathcal{E}_2 \setminus \mathcal{E}'). \tag{15.38}$$

In picturesque terms, b counts the number of edges of the Feynman diagram, including the external edges, which do not belong to $\mathcal{E}'$ but which spring out of one of the vertices of $\mathcal{V}'$.[11]

As will be apparent when we prove it, the proper version of (15.34) is that

$$\dim \ker \mathcal{L}' = d(\operatorname{card} \mathcal{E}' - \operatorname{card} \mathcal{V}' + 1), \tag{15.39}$$

where d is the dimension of space-time. Combining this relation with (15.37), the superficial degree of divergence $d' = d(\operatorname{card} \mathcal{E}' - \operatorname{card} \mathcal{V}' + 1) - 2 \operatorname{card} \mathcal{E}'$ of the subdiagram satisfies

$$d' = (d-2) \operatorname{card} \mathcal{E}' - d \operatorname{card} \mathcal{V}' + d = (d - 2 - 2d/k) \operatorname{card} \mathcal{E}' + d(1 - b/k). \tag{15.40}$$

When the quantity $d - 2 - 2d/k$ is zero, something remarkable happens: The superficial degree of divergence of a subdiagram depends only on b. This is certainly not obvious now, but it turns out that the case $d - 2 - 2d/k = 0$ is the critical case for renormalization: In this case renormalization is possible but hard (as we will experience).[12] We concentrate our efforts on this case $d - 2 - 2d/k = 0$. For our real world $d = 4$ this happens for $k = 4$, but there are other integer solutions of this equation, and first of all the solution $d = 6$ and $k = 3$, the case of ϕ_6^3 theory.[13] The appeal of this case is that the smaller value of k makes matters somehow simpler, so this case is an excellent toy model to understand the case $k = d = 4$.[14]

Recalling that when $k = 4$ then b is even by (15.37), we state for further use:

Lemma 15.4.9 *If $d = k = 4$ then b is even and*

$$d' = 4(1 - b/4), \tag{15.41}$$

so that in particular $d' = 0$ if $b = 4$ and $d' = 2$ if $b = 2$.

If $d = 6$ and $k = 3$ then $d' = 6(1 - b/3)$ so that $d' = 0$ if $b = 3$ and $d' = 2$ if $b = 2$.

[11] If two vertices of $\mathcal{V}'$ are linked by an internal edge that does not belong to $\mathcal{E}'$ this edge contributes twice to b, one time for each of its endpoints.

[12] In the case where $d - 2 - 2d/k < 0$, e.g. in ϕ_4^3 theory, there are only finitely many diagrams with $d' \geq 0$ and renormalization is easier. In the case where $d - 2 - 2d/k > 0$ the theory is "non-renormalizable".

[13] When working in ϕ_6^3 some minor changes have to be made, e.g. $(2\pi)^4\delta^{(4)}$ should be replaced by $(2\pi)^6\delta^{(6)}$, etc. This is left to the reader.

[14] We will not consider the case $d = 3$, $k = 6$ which is more complicated than the case $d = k = 4$ of main interest.

As a consequence of Theorem 15.4.5 and Lemma 15.4.9 we get the following.

Proposition 15.4.10 *In ϕ_6^3 theory, the Euclidean version of a diagram integral is convergent if for each set of internal vertices containing at least two such vertices there are at least four edges out of these vertices that lead to vertices (possibly external) not in this set. In ϕ_4^4 theory this is the case if for each set of internal vertices containing at least two such vertices there are at least five edges out of these vertices that lead to vertices not in this set.*

This proposition unfortunately does not cover the case of many diagrams.

From now on "Feynman diagram" will always have to be understood as "*connected* Feynman diagram". Also, since from now on we almost always work in ϕ_6^3 theory, where the number of space-time dimensions is fixed to six, we change notation: We use the letter d (rather than d') to denote the superficial degree of divergence of a subdiagram.

Proposition 15.4.11 *In ϕ_6^3 theory consider a Feynman diagram with at least two external lines. Consider a subdiagram α that cannot be disconnected from the rest of the Feynman diagram by removing a single internal edge of the Feynman diagram. Then:*

(a) The number c of edges (internal or external) of the Feynman diagram with exactly one endpoint in α satisfies $c \geq 2$.

(b) If $d(\alpha) \geq 0$ then α is a subgraph, c equal the number b of edges springing out of a vertex of α and $b = 2$ or $b = 3$.

Proof (a) We cannot have $c = 0$ since the Feynman diagram is connected. Assume for contradiction that $c = 1$, and denote by e the unique edge of the Feynman diagram out of a vertex of α. First, e cannot be an internal edge because removing it would disconnect α from the rest of the Feynman diagram, contrary to our hypothesis. Assuming now that e is an external edge, let us prove then that the Feynman diagram consists of just α and this external edge. For otherwise there are internal vertices that do not belong to α. But there is no internal edge out of α to connect any of these vertices to α, so that the Feynman diagram cannot be connected–a contradiction. But the Feynman diagram cannot consist of α and the external edge e because we assume that it has two external edges.
(b) According to Lemma 15.4.9 we have $d(\alpha) = 6(1 - b/3)$ so that if $d(\alpha) \geq 0$ we have $b \leq 3$, that is there are at most three edges of the Feynman diagram springing out of vertices of α which are not edges of α. If α is not a subgraph, there is an edge between two vertices of α which is not an edge of α. This edge already accounts for two of the previous $b \leq 3$ edges, so there is at most one other edge of the Feynman diagram with one endpoint in α. This is impossible by (a). Thus α is a subgraph so that $b = c$ and then $2 \leq b \leq 3$. □

Exercise 15.4.12 Let us say that a subgraph of a Feynman diagram is a *biped* if two edges of the Feynman diagram (which may be external edges) have exactly one endpoint in this subgraph, and that it is a *quadruped* if there are exactly four such edges. In ϕ_4^4 theory prove that if a subdiagram α of a connected diagram is different from the diagram itself and has a superficial degree of divergence $d(\alpha) \geq 0$, then either it is a subgraph, or else it is obtained from a subgraph by deleting a single edge. More precisely, either $d(\alpha) = 2$ and α is a biped, or $d(\alpha) = 0$ and either α is a quadruped or is obtained from a biped by removing a single edge.

15.5 Proof of Theorem 15.3.1

Let us go back to the proof of Theorem 15.3.1. The reason for the factor 4 in (15.34) is rather obvious: Since each of the four components of four-momentum gets conserved, we have "four copies of the same situation", a situation in which four-momentum is replaced by a scalar that is conserved at each vertex. A typical such scalar quantity is electrical current. This motivates us to start thinking of a graph as an electrical network, where the current is conserved but is allowed to enter or leave the graph at each vertex. This simple idea will be fundamental for renormalization.

We start with some definitions. Let us consider an oriented graph, with set of vertices $\mathcal{V}$ and set of edges $\mathcal{E}$. We say that two vertices v_1 and v_2 are connected by an edge e if either $v_1 = e_o$ and $v_2 = e_t$ or $v_1 = e_t$ and $v_2 = e_0$. A *path* is a sequence of vertices $v_1, \ldots, v_k$, all different, such that any two consecutive vertices are connected by an edge.[15] We then say that the path connects v_1 and v_k.

Exercise 15.5.1 We have defined a connected graph as a graph such that the set of vertices cannot be split into two subsets such that there is no edge linking a vertex of the first set and a vertex of the second set. Prove that a graph is connected if and only if any two different vertices can be connected by a path.

Observe that in a connected graph, each vertex is adjacent to at least one edge.

Let us now think of a graph as a part of an electrical network. The electrical network consists of the graph, and of "feeding edges", one of which is attached to each vertex of the graph, allowing electrical current to leave or enter the graph through this vertex. To each edge $e \in \mathcal{E}$ we attach a real number x_e which we think of as describing the intensity of the (time-independent) "electrical current" flowing through the oriented edge e. Thus the current through each edge is described by $x = (x_e)_{e\in\mathcal{E}} \in \mathbb{R}^{\mathcal{E}}$. The current is conserved in the network, in the sense that the algebraic sum of the current leaving each vertex through both the edges of the graph and the feeding edge is zero. Consequently, the amount of current leaving the graph through the feeding edge attached to the vertex v is given by

$$\mathcal{L}_0(x)_v = \sum_{e_t=v} x_e - \sum_{e_o=v} x_e. \tag{15.42}$$

Formally this is identical to the formula (15.30). The difference is that here $x_e \in \mathbb{R}$ whereas in (15.30) $x_e \in \mathbb{R}^{1,3}$. It is important to understand that "$\ker \mathcal{L}$ consists of four copies of $\ker \mathcal{L}_0$". To make this explicit, write $x_e \in \mathbb{R}^{1,3}$ as $x_e = (x_e^\nu)_{0\le\nu\le3}$. Then for $x = (x_e) \in (\mathbb{R}^{1,3})^{\mathcal{E}}$ and $0 \le \nu \le 3$ we can define $x^\nu \in \mathbb{R}^{\mathcal{E}}$ by $x^\nu = (x_e^\nu)_{e\in\mathcal{E}}$. Then it should be obvious that $x \in \ker \mathcal{L}$ if and only if $x^\nu \in \ker \mathcal{L}_0$ for $0 \le \nu \le 3$.

[15] Observe that we *do not* require that the orientation of the edges be consistent along the path.

Exercise 15.5.2 Consider the subspace A of $\mathbb{R}^n$ given by the equation $\sum_{i\le n} x_i = 0$. Prove that this space is spanned by the vectors of A that have exactly two non-zero components.

Theorem 15.3.1 is a consequence of the following.

Theorem 15.5.3 *The range of the map* $\mathcal{L}_0 \colon \mathbb{R}^{\mathcal{E}} \to \mathbb{R}^{\mathcal{V}}$ *is the space*

$$\mathcal{N}_0 := \Big\{ w = (w_v)_{v\in\mathcal{V}} \in \mathbb{R}^{\mathcal{V}} \,;\, \sum_v w_v = 0 \Big\}, \tag{15.43}$$

and

$$\dim\ker \mathcal{L}_0 = \operatorname{card}\mathcal{E} - \operatorname{card}\mathcal{V} + 1. \tag{15.44}$$

Proof Since $\mathcal{L}_0(x)_v$ is the quantity of current leaving the network at vertex v, the fact that the range of $\mathcal{L}_0$ is contained in $\mathcal{N}_0$ simply expresses conservation of current: The total quantity of current leaving the network has to be zero. For a formal proof we use (15.42): In the quantity $\sum_v \mathcal{L}_0(x)_v$ each term x_e occurs once with positive sign for $v = e_o$ and once with negative sign for $v = e_t$ so that $\sum_v \mathcal{L}_0(x)_v = 0$.

To prove that the range of $\mathcal{L}_0$ is $\mathcal{N}_0$ we consider two distinct vertices v_a and v_b. Since we assume that the graph is connected according to Exercise 15.5.1 we can find a path between v_a and v_b. That is we find vertices $v_a = v_1, \ldots, v_k = v_b$ and edges $e_1, \ldots, e_{k-1}$ such that e_ℓ connects v_ℓ to $v_{\ell+1}$. We construct an element $x \in \mathbb{R}^{\mathcal{E}}$ as follows. First, $x_e = 0$ if e is not an edge of the path. Next $x_e = 1$ if $e = e_\ell$ is oriented from v_ℓ to $v_{\ell+1}$ (that is $e_o = v_\ell$ and $e_t = v_{\ell+1}$). Finally $x_e = -1$ if $e = e_\ell$ is oriented from $v_{\ell+1}$ to v_ℓ. It is straightforward to see that $\mathcal{L}_0(x)_v$ is zero unless $v = v_a$ or $v = v_b$ and that $\mathcal{L}_0(x)_{v_a} + \mathcal{L}_0(x)_{v_b} = 0$. According to Exercise 15.5.2 the elements of this type span $\mathcal{N}_0$, and this concludes the proof of (15.43).

Linear algebra tells us that $\dim \mathbb{R}^{\mathcal{E}} = \dim\ker \mathcal{L}_0 + \dim \operatorname{range} \mathcal{L}_0$ i.e. $\operatorname{card}\mathcal{E} = \dim\ker \mathcal{L}_0 + \operatorname{card}\mathcal{V} - 1$ and this proves (15.44). □

15.6 A Side Story: Loops

A *loop* is a sequence of vertices $v_1, \ldots, v_{k-1}, v_k = v_1$, such that $v_1, \ldots, v_{k-1}$ are all different, and a sequence of edges $e_1, \ldots, e_{k-1}$, which are all different,[16] such that for $1 \le \ell \le k-1$ the edge e_ℓ connects v_ℓ and $v_{\ell+1}$.

Consider a loop L defined by its vertices $v_1, \ldots, v_k = v_1$ and the connecting edges $e_1, \ldots, e_{k-1}$, so that for $1 \le \ell \le k-1$ the endpoints of e_ℓ are v_ℓ and $v_{\ell+1}$, but e_ℓ may be oriented from v_ℓ to $v_{\ell+1}$ or the other way around. To this loop we associate the point $\tau(L) \in \mathbb{R}^{\mathcal{E}}$ as follows. First $\tau(L)_e = 0$ unless e is one of the edges e_ℓ. If $e = e_\ell$ then $\tau(L)_e = 1$ if $e_o = v_\ell$ and $e_t = v_{\ell+1}$, that is if e_ℓ is oriented from v_ℓ to $v_{\ell+1}$, whereas $\tau(L)_e = -1$ if $e_o = v_{\ell+1}$ and $e_t = v_\ell$, that is if e_ℓ is oriented from $v_{\ell+1}$ to v_ℓ, and the loop crosses this edge in the opposite direction of its orientation.

[16] In particular when $k = 2$, to go back from v_2 to v_1 you have to use a different edge than the one you used to go from v_1 to v_2.

Definition 15.6.1 The *number of independent loops* of a graph is the largest integer n such that there exist loops $L_1, \dots, L_n$ for which $\tau(L_1), \dots, \tau(L_n)$ are linearly independent.

As an example, consider the diagram with two vertices and three edges e_1, e_2, e_3 directed from the first vertex to the second one. Then there exist independent loops L_1 and L_2 with $\tau(L_1) = (1, -1, 0)$, $\tau(L_2) = (0, 1, -1)$, whereas $\tau(L_1) + \tau(L_2) = (1, 0, -1) := \tau(L_3)$ for another loop L_3, but the loops L_1, L_2, L_3 are not independent. In this example, there are two independent loops.

We will use several times the following general principle, where we recall the map $\mathcal{L}_0$ of (15.42).

Lemma 15.6.2 *Consider a Feynman diagram with set $\mathcal{E}$ of internal edges, and $e^* \in \mathcal{E}$. Assume that removing e^* disconnects the Feynman diagram. Then for $x \in \ker \mathcal{L}_0$ we have $x_{e^*} = 0$.*

Proof The argument is a refinement of the proof of Theorem 15.5.3. Obviously removing e^* decomposes the Feynman diagram in two components. Denote by $\mathcal{V}_1$ and $\mathcal{V}_2$ the sets of vertices and by $\mathcal{E}_1$ and $\mathcal{E}_2$ the set of edges of the connected components remaining after if e^* has been removed. Consider the quantity $\sum_{v \in \mathcal{V}_1} \mathcal{L}_0(x)_v$. In this sum the values x_e for each $e \in \mathcal{E}_2$ never contribute because the edges of $\mathcal{E}_2$ are not adjacent to the vertices of $\mathcal{V}_1$. The contributions of x_e for $e \in \mathcal{E}_1$ cancel out because they occur two times with opposite signs, corresponding to the two endpoints of e. But the term x_{e^*} occurs only once with a $\pm$ sign. Thus $x_{e^*} = \pm \sum_{v \in \mathcal{V}_1} \mathcal{L}_0(x)_v$. Now, if $x \in \ker \mathcal{L}_0$ we have $\mathcal{L}_0(x)_v = 0$ for each v, so that $x_{e^*} = 0$. □

We note that there is another version of this lemma, with $\mathcal{L}$ instead of $\mathcal{L}_0$.

Theorem 15.6.3 *(a) For each loop L we have $\tau(L) \in \ker \mathcal{L}_0$.*
(b) The elements of the type $\tau(L)$ where L is a loop span $\ker \mathcal{L}_0$.

In words: The number of independent loops of a graph is $\operatorname{card} \mathcal{E} - \operatorname{card} \mathcal{V} + 1$ and the independent loops span $\ker \mathcal{L}_0$.

Proof It should be obvious that (a) holds true. We prove (b) by induction over $\operatorname{card} \mathcal{E}$. The argument is important and will be used again. If there is only one edge, there are two vertices and no loop and the result is proved since $\dim \ker \mathcal{L}_0 = 0$. Assume then that there are at least two edges, and select one such edge e^*. Consider $x \in \ker \mathcal{L}_0$. We have to show that x is a combination of elements $\tau(L)$. Consider first the case where the graph obtained after removing e^* is not connected. It follows from Lemma 15.6.2 that $x_{e^*} = 0$, and the conclusion follows by applying the induction hypothesis to the two components. Consider next the case where the graph obtained upon removing e^* is still connected. Thus there is a path from the vertex e^*_o to the vertex e^*_t which does not use the edge e^*. Completing this path by e^* gives a loop L going through the edge e^* and, as we have seen, $\tau(L) \in \ker \mathcal{L}_0$. Since $x \in \ker \mathcal{L}_0$, for $t \in \mathbb{R}$ we have $x' = x - t\tau(L) \in \ker \mathcal{L}_0$. Since $\tau(L)_{e^*} \neq 0$ we may find t for which $x'_{e^*} = 0$. We then apply the induction hypothesis to the graph from which

e^* has been removed to obtain that x' is a linear combination of elements of the type $\tau(L')$ where L' is a loop of that graph. Then $x = x' + t\tau(L)$ is a linear combination of elements of the type $\tau(L')$. □

The following is an immediate consequence of (b) of the previous result.

Corollary 15.6.4 *Given $x \in \ker \mathcal{L}_0$ any edge e such that $x_e \neq 0$ belongs to a loop.*

15.7 Parameterization of Diagram Integrals

When we first met diagram integrals, we simply said that "we use the delta functions to eliminate as many momenta x_e as possible and we integrate over the remaining ones". This works fine in practice for small diagrams, but for theoretical considerations it pays to be more systematic.

In the present section we make precise the idea put forward in (15.32), that diagram integrals are integrals over $\ker \mathcal{L}$ (where $\mathcal{L}$ is given by (15.30)). For this we will have a closer look at the integrals

$$\int \cdots \int \prod_{e\in\mathcal{E}} \frac{\mathrm{d}^4 x_e}{(2\pi)^4} F(x) \prod_{v\in\mathcal{V}} (2\pi)^4 \delta^{(4)}(\mathcal{L}(x)_v - w_v), \tag{15.45}$$

where F is a (sufficiently integrable[17]) function on $(\mathbb{R}^{1,3})^{\mathcal{E}}$ and $w = (w_v)_{v\in\mathcal{V}} \in (\mathbb{R}^{1,3})^{\mathcal{V}}$.

The following simple fact will play a fundamental part in our later constructions. We denote by $\mathcal{Q}$ the orthogonal complement $(\ker \mathcal{L})^\perp$ of $\ker \mathcal{L}$ (for the Euclidean structure on $(\mathbb{R}^4)^{\mathcal{E}}$).

Theorem 15.7.1 *There exists a linear map $\mathcal{Y}$ from $(\mathbb{R}^{1,3})^{\mathcal{V}}$ to $\mathcal{Q} = (\ker \mathcal{L})^\perp$ and a normalization $\mu_{\mathcal{L}}$ of the translation-invariant measure on* $\ker \mathcal{L}$ *such that the integral (15.45) equals*

$$(2\pi)^4 \delta^{(4)}\Big(\sum_{v\in\mathcal{V}} w_v\Big) \int \mathrm{d}\mu_{\mathcal{L}}(x) F(x + \mathcal{Y}(w)). \tag{15.46}$$

Furthermore for $e \in \mathcal{E}$ we have $\mathcal{Y}(w)_e = \sum_{v\in\mathcal{V}} y_{e,v} w_v$ for real numbers $y_{e,v}$.

Before we start the proof, we state the consequence we will actually use.

Theorem 15.7.2 *For a certain normalization $\mu_{\mathcal{L}}$ of the translation-invariant measure on* $\ker \mathcal{L}$ *the integral*

$$\int \cdots \int \prod_{e\in\mathcal{E}} \frac{\mathrm{d}^4 x_e}{(2\pi)^4} F(x) \prod_{v\in\mathcal{V}} (2\pi)^4 \delta^{(4)}(\mathcal{L}(x)_v - w_v),$$

[17] The gist of the renormalization method that we will use in the next chapter is to modify the integrand in a way that the integral becomes well-defined, so that this assumption is actually harmless.

equals

$$(2\pi)^4\delta^{(4)}\Big(\sum_{j\le r} p_j\Big)\int \mathrm{d}\mu_{\mathcal{L}}(x)F(x+\mathcal{X}(p)), \tag{15.47}$$

where $\mathcal{X}(p)\in\mathcal{Q}\subset(\mathbb{R}^{1,3})^{\mathcal{E}}$ is a linear function of the external momenta $p=(p_j)_{j\le r}$, which is of the type $\mathcal{X}(p)_e=\sum_{j\le r} b_{e,j}p_j$ for real numbers $b_{e,j}$.

Proof We use Theorem 15.7.1 in the case where w_v is the sum of the momenta entering the internal vertex v through external lines. □

Proof of Theorem 15.7.1 It is usually fine to work with distributions at an intuitive level, but in the present case I felt the need to go beyond that and to give a real argument. Please accept the result if this proof puts you off. Let us set $m=\operatorname{card}\mathcal{V}$. Given the function F, the quantity (15.45) has to be understood as a distribution $\Phi(w)$ such that for a test function ξ on $(\mathbb{R}^{1,3})^{\mathcal{V}}$ we have

$$\int \frac{\mathrm{d}^{4m}w}{(2\pi)^{4m}}\xi(w)\Phi(w)=\int\cdots\int\prod_{e\in\mathcal{E}}\frac{\mathrm{d}^4x_e}{(2\pi)^4}F(x)\xi(\mathcal{L}(x)). \tag{15.48}$$

The integral on the right is for the translation-invariant measure $\prod_{e\in\mathcal{E}}(\mathrm{d}^4x_e/(2\pi)^4)$ on $(\mathbb{R}^{1,3})^{\mathcal{E}}=\ker\mathcal{L}\oplus\mathcal{Q}$. This measure is the image of $\mathrm{d}\mu'_{\mathcal{L}}\otimes\mathrm{d}\mu_{\mathcal{Q}}$ under the map $(z,q)\mapsto z+q$ for certain translation-invariant measures $\mathrm{d}\mu'_{\mathcal{L}}$ and $\mathrm{d}\mu_{\mathcal{Q}}$ on $\ker\mathcal{L}$ and $\mathcal{Q}$. Therefore

$$\int\cdots\int\prod_{e\in\mathcal{E}}\frac{\mathrm{d}^4x_e}{(2\pi)^4}F(x)\xi(\mathcal{L}(x))=\iint \mathrm{d}\mu'_{\mathcal{L}}(z)\mathrm{d}\mu_{\mathcal{Q}}(q)F(z+q)\xi(\mathcal{L}(z+q)).$$

Using Fubini's theorem, and since $z\in\ker\mathcal{L}$ so that $\mathcal{L}(z)=0$, we obtain

$$\int\cdots\int\prod_{e\in\mathcal{E}}\frac{\mathrm{d}^4x_e}{(2\pi)^4}F(x)\xi(\mathcal{L}(x))=\int \mathrm{d}\mu_{\mathcal{Q}}(q)G(q)\xi(\mathcal{L}(q)), \tag{15.49}$$

where

$$G(q)=\int \mathrm{d}\mu'_{\mathcal{L}}(z)F(z+q). \tag{15.50}$$

Since $\mathcal{Q}$ is the orthogonal complement of $\ker\mathcal{L}$, the restriction of $\mathcal{L}$ to $\mathcal{Q}$ is one-to-one from $\mathcal{Q}$ to the range $\mathcal{N}$ of $\mathcal{L}$, where $\mathcal{N}$ is given by (15.33). This map therefore has an inverse $\mathcal{W}:\mathcal{N}\to\mathcal{Q}$ and $\mathcal{Q}=\mathcal{W}(\mathcal{N})$. There is a translation-invariant measure $\mathrm{d}\mu_{\mathcal{N}}$ on $\mathcal{N}$ whose image under $\mathcal{W}$ is $\mathrm{d}\mu_{\mathcal{Q}}$. Making the change of variable $q=\mathcal{W}(u)$ and combining (15.48) to (15.50) yields

$$\int \frac{\mathrm{d}^{4m}w}{(2\pi)^{4m}}\xi(w)\Phi(w)=\int \mathrm{d}\mu_{\mathcal{N}}(u)G(\mathcal{W}(u))\xi(u). \tag{15.51}$$

On the other hand, it should be clear from the definition (15.33) of $\mathcal{N}$ that for all test functions η one has

$$\int \frac{\mathrm{d}^{4m}w}{(2\pi)^{4m}}(2\pi)^4\delta^{(4)}\Big(\sum_{v\in\mathcal{V}} w_v\Big)\eta(w)=\int \mathrm{d}\mu'_{\mathcal{N}}(u)\eta(u), \tag{15.52}$$

where $\mathrm{d}\mu'_{\mathcal{N}}$ is a translation-invariant measure on $\mathcal{N}$, which is therefore proportional to $\mathrm{d}\mu_{\mathcal{N}}$, i.e. of the type $C\mathrm{d}\mu_{\mathcal{N}}$. Thus we may rewrite (15.52) as

$$\int \mathrm{d}\mu_{\mathcal{N}}(u)\eta(u) = \frac{1}{C}\int \frac{\mathrm{d}^{4m}w}{(2\pi)^{4m}}(2\pi)^4\delta^{(4)}\Big(\sum_{v\in\mathcal{V}} w_v\Big)\eta(w).$$

Applying this identity to the function $\eta(w) = G(\mathcal{W}(P(w)))\xi(w)$ where P is a linear projection of $(\mathbb{R}^{(1,3)})^{\mathcal{V}}$ onto $\mathcal{N}$ (so that $P(u) = u$ for $u \in \mathcal{N}$), we obtain the identity

$$\int \mathrm{d}\mu_{\mathcal{N}}(u)G(\mathcal{W}(u))\xi(u) = \frac{1}{C}\int \frac{\mathrm{d}^{4m}w}{(2\pi)^{4m}}(2\pi)^4\delta^{(4)}\Big(\sum_{v\in\mathcal{V}} w_v\Big)G(\mathcal{W}(P(w)))\xi(w).$$

Therefore the left-hand side of (15.51) equals the right-hand side of the above identity. This is true for each test function ξ, and therefore, as distributions, we have

$$\Phi(w) = \frac{1}{C}(2\pi)^4\delta^{(4)}\Big(\sum_{v\in\mathcal{V}} w_v\Big)G(\mathcal{W}(P(w))).$$

Recalling (15.50), this proves (15.46) with $\mathcal{Y}(w) := \mathcal{W}(P(w)) \in \mathcal{Q}$ and $\mathrm{d}\mu_{\mathcal{L}} = C^{-1}\mathrm{d}\mu'_{\mathcal{L}}$. It remains only to prove that $\mathcal{Y}(w)_e = \sum_{v\in\mathcal{V}} y_{e,v}w_v$ for real numbers $y_{e,v}$, but we let the reader convince herself of this.[18] □

Exercise 15.7.3 Prove (15.52) in complete detail.

15.8 Parameterization of Diagram Integrals by Loops

In this section we describe the traditional parameterization of diagram integrals by loops, which we will systematically use in our examples. One advantage of this parameterization is that built into it is the important information that "$\ker\mathcal{L}$ consists of four independent copies of $\ker\mathcal{L}_0$" (where $\mathcal{L}_0$ is defined in (15.42)), so it will be convenient to also use the parameterization for arguments related to Lorentz invariance, which precisely rely on this feature of $\ker\mathcal{L}$. This section is a continuation of Section 15.6. Given a loop L we have constructed an element $\tau(L) = (\tau(L)_e)_{e\in\mathcal{E}} \in \ker\mathcal{L}_0 \subset \mathbb{R}^{\mathcal{E}}$. Given $k \in \mathbb{R}^{1,3}$, the element $k\tau(L) := (k\tau(L)_e)_{e\in\mathcal{E}} \in \ker\mathcal{L} \subset (\mathbb{R}^{1,3})^{\mathcal{E}}$. If there exist n independent loops $L_1, \dots, L_n$, given $k = (k_i)_{i\le n}$ where $k_i \in \mathbb{R}^{1,3}$ the elements

$$x(k) := \sum_{i\le n} k_i\tau(L_i) \tag{15.53}$$

belong to $\ker\mathcal{L}$. If n is the largest possible number of independent loops every element of $\ker\mathcal{L}$ is of this type since the independent loops span $\ker\mathcal{L}$.[19] One unimportant but amusing feature here is that the components of $x(k)$ are of the form $\sum_{i\le n}\eta_i k_i$ where $\eta_i \in \{-1,0,1\}$. The integral in (15.47) takes the form

[18] The reason being "that we do four copies of the same thing for the indices $\nu = 0, 1, 2, 3$".
[19] This is for ϕ^4_4 theory, the case of ϕ^3_6 theory requires obvious modifications.

$$\int \mathrm{d}\mu_{\mathcal{L}}(x)F(x+\chi(p)) = C\int\cdots\int\prod_{i\leq n}\frac{\mathrm{d}^4k_i}{(2\pi)^4}F(x(k)+\chi(p)) \tag{15.54}$$

for a certain constant C. Another amusing but unimportant feature is that if we construct the independent loops as in the proof of Theorem 15.6.3, the constant C in (15.54) equals 1. As we will not use this fact, we leave it as a challenge to the reader.

Key ideas to remember:

- Computation of a diagram integral reduces to the computation of an integral of a rational function over a linear space.
- The absolute convergence of this integral is decided by looking at the degree of this rational function on the linear space and its subspaces.
- It is rare that the integral converges.
- The ϕ_4^4 and ϕ_6^3 theories are critical cases to understand renormalization.

16

The Bogoliubov–Parasiuk–Hepp–Zimmermann Scheme

The Bogoliubov–Parasiuk–Hepp–Zimmermann (BPHZ) scheme, which we study here in Zimmermann's version, allows us to consistently assign a definite value to a diagram integral with any number of loops. It is mathematically very clean. The divergent integrals are tamed not by introducing cutoffs, but by modifying the integrand itself, very much in the spirit of Section 13.21: One introduces "compensating terms" to remove the divergences.

While using no mathematics beyond calculus, the BPHZ scheme is an admirable achievement. Its relevance to the practice of Quantum Field Theory is however marginal. For real-world calculations what matters are diagram integrals with a few loops,[1] often only one or two loops, and physicists have long known how to renormalize these. It is however one thing to renormalize a diagram with a few loops, and quite another to conceive a scheme that will work in the complicated configurations possible in large diagrams. Still, no major physics textbook comes even close to explaining in detail the main features of the BPHZ scheme.[2] Even more conceptually oriented textbooks such as that of Duncan [24] refer the reader to the original papers. Worse, many specialized physics textbooks dealing with renormalization adopt the same attitude, see e.g. [1, 15].[3]

Mathematicians have developed renormalization theory along different lines, see e.g. [25].[4] There are several good reasons however to focus here on the BPHZ scheme: This scheme is historically the first, it is the most elementary, and it connects clearly to the counter-term method used in practice in physics. Reading the original papers is quite challenging, and specialized books such as those of Manoukian [52] and Rivasseau [70] require a serious investment, so the beauties of the method are not easily accessible. To remedy this, in the present chapter we describe the method in complete detail, not attempting yet to prove that it succeeds. In our discussion preceding Lemma 15.4.9, in our class of ϕ_d^k theories we identified the important cases of the ϕ_4^4 and ϕ_6^3 theories, and we will cover both cases. In our formulas we assume that we are in four space-time dimensions, e.g. by writing the

[1] One of the triumphs of Quantum Electrodynamics is the computation of the "anomalous magnetic moment of the electron". This has been computed up to four loop diagrams. The properly cyclopean undertaking of computing all diagrams up to five loops is underway.

[2] Here is a quote from Zee's book [93, page 179]: "If you demand rigor, you should consult the many field theory tomes on renormalization theory, and I do mean tomes, in which such arcane topics as overlapping divergences and Zimmerman's forest formula are discussed in exhaustive and exhausting detail".

[3] A partial reason for this is that the conceptual understanding of renormalization has been deepened by the ideas of the "renormalization group", a more advanced topic which we cannot cover here.

[4] Their work seems largely ignored by mainstream physics.

delta functions as $\delta^{(4)}$. The reader will have no difficulty making the necessary changes in the case of ϕ_6^3 theory (replacing everywhere four-vectors by "six-vectors", since there are six space-time dimensions in ϕ_6^3 theory). The reason for considering ϕ_6^3 theory is that it is marginally simpler than ϕ_4^4 theory, and all of our figures and examples pertain to this case.[5]

The relationship between the BPHZ method and the traditional counter-term method of physics is the object of a separate chapter, Chapter 17. A completely rigorous self-contained proof of the validity of the BPHZ scheme for both ϕ_4^4 and ϕ_6^3 theories (which are cases where the difficulty of renormalization is rather generic) will be given in Chapters 18 and 19.

16.1 Overall Approach

Our goal is to assign a value to each Feynman diagram. Let us recall that in Theorem 15.7.2 we have reduced the computation of the value of a Feynman diagram to the computation of an integral of the type

$$I(p,\varepsilon) = \int \mathrm{d}\mu_{\mathcal{L}}(x) F(x + \mathcal{X}(p)) \tag{16.1}$$

where (keeping the dependence in ε implicit) $F(x)$ is the product $F(x) = \prod_{e\in\mathcal{E}} f(x_e) = \prod_{e\in\mathcal{E}}(-\mathrm{i}/(-x_e^2 + m^2 - \mathrm{i}\varepsilon))$, where $\mathrm{d}\mu_{\mathcal{L}}$ is a certain normalization of the volume measure on $\ker \mathcal{L}$ and where $\mathcal{X}(p) \in (\mathbb{R}^{1,3})^{\mathcal{E}}$ is a linear function of the external momenta of a special type: $\mathcal{X}(p)_e = \sum_{j\le r} b_{e,j} p_j$. For reasons which will soon become clear, we will call throughout the chapter the computation of an integral such as (16.1) *integrating over the internal degrees of freedom.*

Throughout the chapter (for reasons that will gradually become clear) we use the form (13.66) of the Fourier transform of the propagator. To lighten notation, for a four-vector k we write

$$W_\varepsilon(k) = -(1+\mathrm{i}\varepsilon)(k^0)^2 + \boldsymbol{k}^2, \tag{16.2}$$

so that now[6]

$$f(x_e) = \frac{-\mathrm{i}}{W_\varepsilon(x_e) + m^2}, \tag{16.3}$$

and the function F on $(\mathbb{R}^{1,3})^{\mathcal{E}}$ is given by

$$F(x) = \prod_{e\in\mathcal{E}} f(x_e) = \prod_{e\in\mathcal{E}} \frac{-\mathrm{i}}{(W_\varepsilon(x_e) + m^2)}. \tag{16.4}$$

- The problem: the integral (16.1) diverges.
- The solution: replace the integrand F by a better-behaved regularized version $\mathcal{R}F$.

Much of the present chapter is devoted to building the tools required to construct $\mathcal{R}F$. In Chapters 18 and 19 we will prove the following.

[5] The ϕ_6^3 theory is non-physical, as it can be shown that the Hamiltonian is not bounded below, but this is irrelevant for our purposes.

[6] As explained on page 391.

Theorem 16.1.1 *The integral* $\int d\mu_{\mathcal{L}}(x)\mathcal{R}F(x+\mathcal{X}(p))$ *is absolutely convergent. As* $\varepsilon \to 0$ *its value converges in distribution to a well-defined Lorentz invariant distribution in the external momenta p of the Feynman diagram.*[7]

Thus replacing F by $\mathcal{R}F$ has *removed the dreaded divergences*, and we can now (at least in the sense of distributions) assign a value to each Feynman diagram.

To state this result in plain terms, using the notation of (15.45) and denoting by S the symmetry factor we now compute the value of Feynman diagrams as[8]

$$\lim_{\varepsilon\to 0^+}\frac{1}{S}\int\cdots\int\prod_{e\in\mathcal{E}}\frac{d^4x_e}{(2\pi)^4}\mathcal{R}F(x)\prod_{v\in\mathcal{V}}(2\pi)^4\delta^{(4)}(\mathcal{L}(x)_v - w_v).$$

Theorem 16.1.1 is far from trivial, so that the reader should not expect to have much intuition about why it should be true even after studying the present chapter. The physical content of this result is explored in Section 16.8.

A clean aspect of Zimmermann's version of the BPHZ method[9] is that the essential structure is not so much the specific form (16.3) of the Fourier transform of the propagator as the structure of the integrand $F(x)=\prod_{e\in\mathcal{E}} f(x_e)$ as a product, and for such a function F we quite generally define its regularized version $\mathcal{R}F$.[10] This regularized version is obtained by adding compensating terms designed to cancel the subdivergences. These compensating terms are defined through a kind of Taylor operation which is described in the next section. The function $\mathcal{R}F$ is given by an explicit formula, Zimmermann's forest formula, as a (rather large) sum of F and compensating terms. The construction is not really complicated. The difficult part is to show that it succeeds, but we will worry about this in later chapters.

16.2 Simple Examples

Before we go into general considerations, let us treat two simple examples by hand. We start with a diagram of the simplest type, one single loop, labeled (a) in Figure 16.1. Here we have to integrate over $q \in \mathbb{R}^{1,3}$ and p_1, p_2 are parameters. Using (15.54) with $C=1$ the value of this diagram is given by the integral

$$\int\frac{d^4q}{(2\pi)^4}f(q+(p_1+p_2)/2)f(q-(p_1+p_2)/2)f(q-(p_2-p_1)/2),$$

7 I try not to be pedantic, but in view of Theorem 15.7.2 one should really prove that the quantity $\delta(\sum_{j\le r}p_j)I(p,\varepsilon)$ converges in distributions. This time I am not cheating, the proof is the same.

8 Fine points about distributions are swept under the rug.

9 This approach is complicated by a number of technical considerations related to the need of specifying a "routing"of a flow through a network. The need for these considerations is however an artifact of the historical path through which the ideas developed, and we bypass it to a very large extent. This is just a matter of presentation, we have no real improvement to offer to Zimmermann's magnificent proof.

10 The shortest approach to the BPHZ theorem probably uses, as in Bergère's paper [7] and subsequent work, a representation of the propagator as an integral over a parameter, introducing then a space of parameters in which one performs the required constructions.

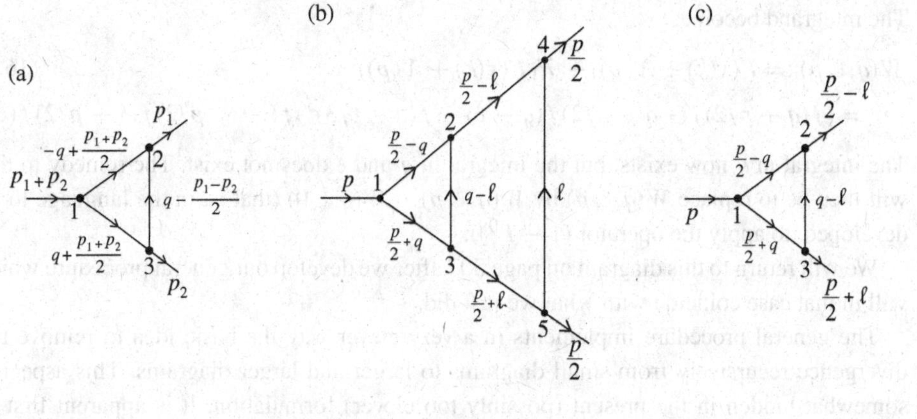

Figure 16.1 Simple diagrams.

which unfortunately diverges. To make it converge a natural procedure is to subtract a term $f(q)^3$ from the integrand.[11] For reasons which will become clear in the next section, a better choice is to subtract a term $f(q-(p_2-p_1)/6)^3$, or, in other words, to add a *compensating term* $-f(q-(p_2-p_1)/6)^3$.

Let us then look at diagram (b) of Figure 16.1, which has two loops. The value of this diagram is given by (using notation which looks strange now but will be explained fully later)

$$\iint \frac{\mathrm{d}^4 q}{(2\pi)^4}\frac{\mathrm{d}^4 \ell}{(2\pi)^4} F(x(k)+\mathcal{X}(p)), \tag{16.5}$$

where

$$F(x(k)+\mathcal{X}(p)) := f(q+p/2)f(-q+p/2)f(q-\ell)f(-\ell+p/2)f(\ell+p/2)f(\ell). \tag{16.6}$$

The first problem is that the integral in q does not exist at a fixed value of ℓ. (This phenomenon is called a subdivergence.) This problem does not occur because of the whole diagram, but only from the part of the diagram (c) of Figure 16.1. This part of the diagram is identical to the diagram (a) of the figure for the choice $p=p_1+p_2$ and $\ell=(p_2-p_1)/2$. This means that we have *already solved* the problem of this divergence. It is removed by adding to the integrand $F(x(k)+\mathcal{X}(p))$ the compensating term (using again notation which will be explained later)

$$\mathcal{F}_1 F(x(k)+\mathcal{X}(p)) := -f(q-\ell/3)^3 f(-\ell+p/2)f(\ell+p/2)f(\ell). \tag{16.7}$$

[11] It is not entirely obvious that after this subtraction the integral converges, but this will be taken care of in full detail later. In the examples analyzed here, we stay at the heuristic level: Our compensating terms remove the obvious power-counting obstacles to convergence.

The integrand becomes

$$\begin{aligned} W(q,\ell,p) &:= F(x(k)+\mathcal{X}(p)) + \mathcal{F}_1 F(x(k)+\mathcal{X}(p)) \qquad (16.8)\\ &= (f(q+p/2)f(-q+p/2)f(q-\ell) - f(q-\ell/3)^3)f(-\ell+p/2)f(\ell+p/2)f(\ell). \end{aligned}$$

The integral in q now exists, but the integral in q and ℓ does not exist. The remedy to this will then be to replace $W(q,\ell,p)$ by $W(q,\ell,p) - W(q,\ell,0)$ (that is, in the language to be developed, to apply the operator $(1-T^0)$).

We will return to this diagram on page 507 after we develop our general procedure which will in that case coincide with what we just did.

The general procedure implements in a very clever way the basic idea to remove the divergence recursively from small diagrams to larger and larger diagrams. This aspect is somewhat hidden in the present (possibly too clever) formulation. It is apparent first in Lemma 16.5.3. We will control the divergence of a diagram in two steps:

- In the first step we remove the divergences which already occur in smaller diagrams (adding the compensating term (16.8) in the previous example).
- In the second step we remove the part of the divergence which is left in the whole diagram (replacing $W(q,\ell,p)$ by $W(q,\ell,p) - W(q,\ell,0)$ in the previous example).

16.3 Canonical Flow and the Taylor Operation

Historically, the fundamental idea is as follows. Assume that you assign a momentum to each internal line of a Feynman diagram. Consider a function of these momenta (such as the integrand relevant to the computation of the value of this diagram). Then this function can be seen in a canonical way as depending on the momentum flow through the external lines and on other "internal degrees of freedom". We may then compute Taylor polynomials of the function seen as a function of this flow through the external lines, and subtract these from the function to improve its integrability.

This idea is at the root of the entire process. In the present section we formulate it in a more abstract manner.[12] Let us consider generally an oriented graph with set of edges $\mathcal{E}$ and set of vertices $\mathcal{V}$ and recall the map $\mathcal{L}: (\mathbb{R}^{1,3})^{\mathcal{E}} \to (\mathbb{R}^{1,3})^{\mathcal{V}}$ of (15.30) given by $\mathcal{L}(x)_v = \sum_{e_t=v} x_e - \sum_{e_o=v} x_e$. We decompose

$$(\mathbb{R}^{1,3})^{\mathcal{E}} = \ker \mathcal{L} \oplus \mathcal{Q}, \qquad (16.9)$$

where $\mathcal{Q} = (\ker \mathcal{L})^{\perp}$. Here the orthogonality is for the ordinary Euclidean dot product of $(\mathbb{R}^{1,3})^{\mathcal{E}}$.[13] We then define two orthogonal projections $\mathcal{I}$ and $\mathcal{T}$ by

$$x = \mathcal{I}(x) + \mathcal{T}(x)\,;\ \mathcal{I}(x) \in \ker \mathcal{L}\,,\ \mathcal{T}(x) \in \mathcal{Q} = \ker \mathcal{I}. \qquad (16.10)$$

We will use countless times the fact that $x = \mathcal{I}(x)$ if and only if $\mathcal{L}(x) = 0$.

[12] This very simple shift of perspective produces considerable technical simplification.

[13] If you wonder what the Euclidean dot product is doing here, you may use instead the dot product given $(x,y) = \sum_{e\in\mathcal{E}}(x_e, y_e)$ where (x_e, y_e) is the Lorentz dot product. This gives the same result, a fact which is due to the special structure of "having four independent copies of $\mathbb{R}^{\mathcal{E}}$".

The idea is that $\mathcal{I}(x)$ represents the "internal degrees of freedom of x" whereas $\mathcal{T}(x)$ represents the "part of x which is determined by how much momentum leaves each vertex". Let us explain this. Since $\ker \mathcal{L} \cap \mathcal{Q} = \{0\}$, $\mathcal{L}$ is one-to-one from $\mathcal{Q}$ onto its range

$$\mathcal{N} = \Big\{ w \in (\mathbb{R}^{1,3})^{\mathcal{V}} \,;\, \sum_{v \in \mathcal{V}} w_v = 0 \Big\},$$

and there is a linear map $\mathcal{W}$ from $\mathcal{N}$ to $\mathcal{Q}$ which is the inverse of $\mathcal{L}$. Thus, given a point $w \in \mathcal{N} \subset (\mathbb{R}^{1,3})^{\mathcal{V}}$, which describes how much four-momentum leaves the network at each node there is a unique $y = \mathcal{W}(w) \in \mathcal{Q} \subset (\mathbb{R}^{1,3})^{\mathcal{E}}$ such that $\mathcal{L}(y) = w$. This point y represents a flow through the network: y_e tells you how much momentum flows through the edge e. We will call y the *canonical flow* of w through the network.[14] (We already used this map in the proof of Theorem 15.7.2, at the end of the previous chapter.)

Since $y = \mathcal{W}(\mathcal{L}(y))$ for $y \in \mathcal{Q}$ and since $\mathcal{T}(x) \in \mathcal{Q}$ we have $\mathcal{T}(x) = \mathcal{W}(\mathcal{L}\mathcal{T}(x))$. Since $\mathcal{L}(\mathcal{T}(x)) = \mathcal{L}(x)$ we obtain $\mathcal{T}(x) = \mathcal{W}(\mathcal{L}(x))$. That is, $\mathcal{T}(x)$ is the canonical flow corresponding to $\mathcal{L}(x)$, which is determined by $\mathcal{L}(x)$, that is by how much momentum enters each vertex. The remainder $x - \mathcal{T}(x) = \mathcal{I}(x)$ is in a sense not influenced by how much momentum enters each vertex, so we might say that it represents the internal degrees of freedom of x, with the word "internal" referring to the fact that nothing external to the network (such as a four-momentum leaving through the external lines) is involved.

Let us go back to Feynman diagrams, where $\mathcal{E}$ is the set of *internal* edges of the diagram, and where four-momentum is allowed to leave or enter only through the external edges. In Figure 16.2 we have represented the canonical flow for a number of simple Feynman diagrams in ϕ^3 theory. The external edges and how much flow enters through them are also represented in these diagrams.

The map $x \mapsto (\mathcal{I}(x), \mathcal{T}(x))$ provides a natural identification between $(\mathbb{R}^{1,3})^{\mathcal{E}}$ and $(\ker \mathcal{L}) \times \mathcal{Q}$. The inverse map is given by $(z,q) \mapsto z + q$. In this manner a function F on $(\mathbb{R}^{1,3})^{\mathcal{E}}$ may be considered as a function of $z \in \ker \mathcal{L}$ and $q \in \mathcal{Q}$. This idea is fundamental for the entire construction. In particular it will allow us as we explain next to consider derivatives with respect to q.

Let us recall that given a function $H(q)$ on $\mathbb{R}^n$, its dth-order Taylor polynomial at *zero* is the polynomial in q given by

$$H(0) + \sum_{1 \le k \le d} \frac{1}{k!} \sum_{i_1,\ldots,i_k \le n} q_{i_1} \cdots q_{i_k} \frac{\partial^k H}{\partial q_{i_1} \ldots \partial q_{i_k}}\Big|_{q=0}. \tag{16.11}$$

More generally given a function H on a finite-dimensional space we may define its dth-order polynomial by picking a basis and using the previous definition (16.11). It is straightforward that the resulting polynomial does not depend on the basis we used for the computation.

Consider then again a function F on $(\mathbb{R}^{1,3})^{\mathcal{E}}$. Given $z \in \ker \mathcal{L}$ we may consider the function $q \mapsto F(z+q)$ defined on $\mathcal{Q}$, and as we just explained we may compute the Taylor polynomial of order d at $q = 0$ of this function. This Taylor polynomial is a function

[14] This idea will be used only for illustration, and is no longer part of the proof. It is exactly Zimmermann's routing procedure.

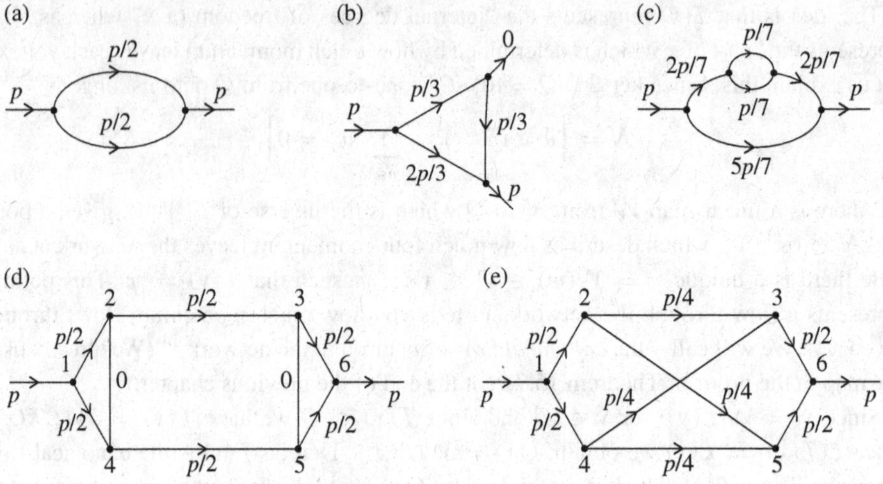

Figure 16.2 Canonical flows.

$G_d(z,q)$. We define the function $T^d F$ on $(\mathbb{R}^{1,3})^{\mathcal{E}}$ by $T^d F(x) = G_d(\mathcal{I}(x), \mathcal{T}(x))$. The case $d = 0$ is of particular interest: Then $G_0(z,q) = F(z)$ and thus

$$T^0 F(x) = F(\mathcal{I}(x)). \tag{16.12}$$

To describe the procedure in words:

Definition 16.3.1 Given a function F on $(\mathbb{R}^{1,3})^{\mathcal{E}}$ and an integer $d \geq 0$ the function $T^d F$ on $(\mathbb{R}^{1,3})^{\mathcal{E}}$ is the order-d Taylor polynomial of F at $q = 0$ when F is considered as a function of $z \in \ker \mathcal{L}$ and $q \in \mathcal{Q}$, z being fixed. If $d < 0$ we *define* $T^d F = 0$.

As an illustration, consider the case of the "eye" graph consisting of two vertices numbered 1 and 2 and two edges joining them oriented from vertex 1 to vertex 2. These edges carry four-momenta k_1 and k_2 and for $x = (k_1, k_2)$ we have $\mathcal{L}(x) = (-k_1 - k_2, k_1 + k_2)$ so that $\mathcal{I}(x) = ((k_1 - k_2)/2, (-k_1 + k_2)/2)$ and $\mathcal{T}(x) = ((k_1 + k_2)/2, (k_1 + k_2)/2)$. Therefore

$$T^0 F(k_1, k_2) = F((k_1 - k_2)/2, (-k_1 + k_2)/2). \tag{16.13}$$

Exercise 16.3.2 Consider the graph consisting of three vertices labeled 1, 2, 3 and three edges oriented from vertex 1 to vertex 2, from vertex 2 to vertex 3, from vertex 3 to vertex 1. These carry momenta $k_{1,2}, k_{2,3}, k_{3,1}$. For $x = (k_{1,2}, k_{2,3}, k_{3,1})$ show that $\mathcal{L}(x) = (k_{3,1} - k_{1,2}, k_{1,2} - k_{2,3}, k_{2,3} - k_{3,1})$ and that $\mathcal{I}(x) = (\bar{k}, \bar{k}, \bar{k})$ where $3\bar{k} = k_{1,2} + k_{2,3} + k_{3,1}$.

Exercise 16.3.3 Given a Lorentz transformation A for $x = (x_e)_{e \in \mathcal{E}} \in (\mathbb{R}^{1,3})^{\mathcal{E}}$ write $Ax = (Ax_e)_{e \in \mathcal{E}}$. Given a function F on $(\mathbb{R}^{1,3})^{\mathcal{E}}$ denote by F_A the function given by $F_A(x) = F(Ax)$. Prove that $T^d(F_A) = (T^d F)_A$.

16.4 Subdiagrams

Let us review some definitions now. Consider a given Feynman diagram with a set of internal edges $\mathcal{E}$ and a set of internal vertices $\mathcal{V}$. Given a subset $\mathcal{E}'$ of $\mathcal{E}$, we define a subdiagram (of the original Feynman diagram) as the graph with a set of edges $\mathcal{E}'$, and a set of vertices $\mathcal{V}'$, the set of vertices adjacent to at least one edge of $\mathcal{E}'$. When furthermore every edge of $\mathcal{E}$ with endpoints in $\mathcal{V}'$ belongs to $\mathcal{E}'$, we say that the subdiagram is a subgraph.

Subdiagrams will be denoted by lowercase Greek letters. A subdiagram α has a set of edges $\mathcal{E}_\alpha$ and a set of vertices $\mathcal{V}_\alpha$. The largest subdiagram (with $\mathcal{E}_\gamma = \mathcal{E}$ and hence $\mathcal{V}_\gamma = \mathcal{V}$) will almost always be denoted by γ. Thus γ is simply obtained from the Feynman diagram by removing the external lines.

Definition 16.4.1 For two subdiagrams α, β we say that $\beta \subset \alpha$ if every edge of β is an edge of α. Then every vertex of β is a vertex of α.

Note that however even if every vertex of β is a vertex of α this need not imply that $\beta \subset \alpha$.

We recall Definition 15.4.6 of connected subdiagrams and Definition 15.4.7 of disjoint subdiagrams. We also recall from Definition 14.9.5 that a Feynman diagram is called 1-particle irreducible or 1-PI if it cannot be split into two disjoint diagrams by removing one internal line, a concept related to the following definition.

Definition 16.4.2 A connected subdiagram is *proper* if it cannot be split into two disjoint graphs by removing a single edge.[15]

Figure 16.3 illustrates some improper subdiagrams. The following simple lemma explains to a large extent the relevance of proper subdiagrams.

Lemma 16.4.3 *A connected subdiagram is proper if and only if each of its edges belongs to a loop of this subdiagram.*

Proof It should be obvious that if each edge belongs to a loop the subdiagram is proper. The converse follows by an argument familiar since the proof of Theorem 15.5.3: If, upon removing an edge the subdiagram is still connected, this edge belongs to a loop of the subdiagram. □

Recalling Definition 14.9.5, the following result is obvious enough to deserve no proof.

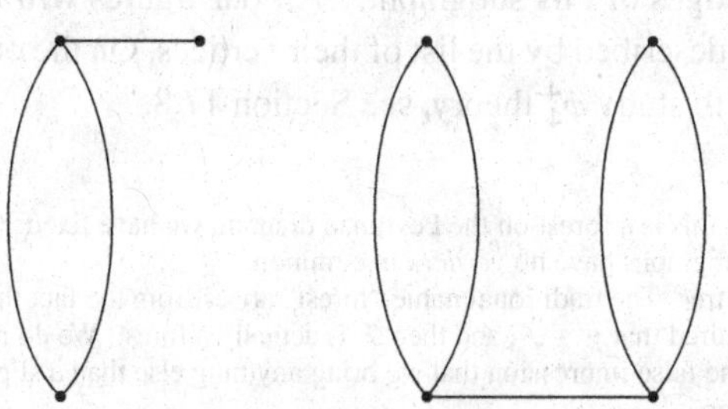

Figure 16.3 Some improper diagrams.

[15] In particular, there is no dangling edge that has a vertex not belonging to any other edge.

Lemma 16.4.4 *A Feynman diagram is 1-PI if and only if the subdiagram γ consisting of its internal vertices and its internal edges is proper.*

16.5 Forests

If a Feynman diagram is not 1-PI, one can split it into two disconnected components by removing an internal line; this internal line is not part of any loop and the integral in (16.1) basically breaks into the product of two smaller integrals. We will detail more this process later, but this argument brings forward the fact that the critical case is the case where the Feynman diagram is 1-PI. We now fix such a diagram, and we denote by γ the proper subdiagram formed by its internal edges and its internal vertices. We recall from (15.36) that the superficial degree of divergence $d(\alpha)$ of a subdiagram α is defined as

$$d(\alpha) := \dim \ker \mathcal{L}_\alpha - 2\,\mathrm{card}\,(\mathcal{E}_\alpha). \tag{16.14}$$

We recall also that in Lemma 15.4.9 we have computed the superficial degree of divergence as a function of the number b of edges which stem out of a vertex of α, but are not edges of α: $d(\alpha) = 4 - b$ in ϕ_4^4 theory and $d(\alpha) = 6 - 2b$ in ϕ_6^3 theory.

Definition 16.5.1 A *forest* $\mathcal{F}$[16] is a collection of proper connected subdiagrams (of the original Feynman diagram) such that $\gamma \in \mathcal{F}$ which has the following property: Any two subdiagrams α, β belonging to $\mathcal{F}$ are either disjoint[17] or one contains the other.

Following the common terminology on ordered sets, it would perhaps be more appropriate to call $\mathcal{F}$ a tree rather than a forest.[18]

Definition 16.5.2 A *renormalization set* is a proper connected subdiagram α with a superficial degree of divergence $d(\alpha) \geq 0$.

According to Theorem 15.4.5, it is not surprising that these diagrams are troublesome.

It will turn out as we will soon show that only the forests such that all the subdiagrams they contain are renormalization sets (except maybe γ itself) will actually contribute to renormalization. It is however technically convenient to consider all possible forests. In ϕ_6^3 theory, renormalization sets are subgraphs, as shown in Proposition 15.4.11.[19] Furthermore these are exactly the subgraphs such that at most three edges stem out of the vertices of the subgraph but are not edges of this subgraph.[20] All our figures will illustrate this situation. The subgraphs will be described by the list of their vertices. On the other hand, subdiagrams seem really necessary to study ϕ_4^4 theory, see Section 17.8.

[16] It is understood here that this is a forest on the Feynman diagram we have fixed.

[17] Recalling that disjoint subgraphs have no *vertices* in common.

[18] Thus γ is the root of the tree. The traditional name "forest" arises from the fact that in the traditional presentation it is not required that $\gamma \in \mathcal{F}$, and then $\mathcal{F}$ is actually a forest. We do not change the name because we do not want to give the false impression that we bring anything else than a slight variation of the usual presentation of these ideas.

[19] As we will explain in the next section, Feynman diagrams with zero or one external lines will be renormalized to zero, so that the only case of interest here is when the diagram has at least two external lines.

[20] Since the whole diagram is proper, there are either two or three edges stemming out of the vertices of the subgraph which are not edges of the subgraph.

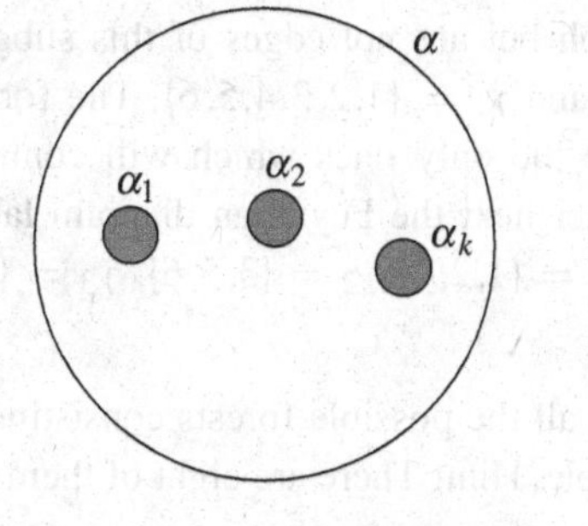

Figure 16.4 Maximal subdiagrams.

The following lemma is almost trivial but is fundamental.

Lemma 16.5.3 *Consider a forest $\mathcal{F}$ and $\alpha \in \mathcal{F}$. Then there exists a (possibly empty) collection of subdiagrams $\alpha_1, \ldots, \alpha_{\tilde{n}}$ of $\mathcal{F}$ strictly contained in α such that every element of $\mathcal{F}$ strictly contained in α is contained in one of the elements α_j.*

Definition 16.5.4 We will call $\alpha_1, \ldots, \alpha_{\tilde{n}}$ *the maximal subdiagrams* of α.[21]

The situation is described in Figure 16.4. To prove that all elements of a forest have a certain property, one may then proceed inductively, from the "small elements" to the "larger ones". The lemma also brings forward the recursive nature of the notion of forest. For each $i \le \tilde{n}$, $\mathcal{F}_i := \{\alpha \in \mathcal{F}; \alpha \subset \alpha_i\}$ is a forest of α_i and these forests entirely determine $\mathcal{F}$.

Proof of Lemma 16.5.3 If α does not strictly contain an element of $\mathcal{F}$ there is nothing to prove. Otherwise consider β strictly contained in α. The family of elements τ of $\mathcal{F}$ containing β but strictly contained in α is totally ordered by the definition of a forest, because any two elements of this family have a non-empty intersection. Therefore this family has a largest element. Thus every β strictly contained in α is contained in an element that is maximal with respect to this property. These maximal elements $\alpha_1, \ldots, \alpha_{\tilde{n}}$ are disjoint by the definition of a forest. □

The following even more obvious result should be noted.

Lemma 16.5.5 *If one of the vertices v adjacent to an edge of a subdiagram β of a forest $\mathcal{F}$ is adjacent to an edge of $\alpha \in \mathcal{F}$ then $\alpha \subset \beta$ or $\beta \subset \alpha$.*

Proof By definition of a subdiagram v is a vertex of β and a vertex of α. By definition of a forest if two different subdiagrams of a forest have a vertex in common one contains the other. □

Let us now give some examples. Consider the Feynman diagram labeled (e) in Figure 16.2 (and forget about the canonical flow represented on this figure). To find the renormalization sets we look for subgraphs such that at most three edges stem out of

21 These depend also on the forest $\mathcal{F}$, which is understood as being given.

the vertices of the subgraph but are not edges of this subgraph. We find[22] $\gamma_1 = \{1,2,3,4,5\}$, $\gamma_2 = \{2,3,4,5,6\}$ and $\gamma = \{1,2,3,4,5,6\}$. The forests consisting only of renormalization sets (which are the only ones which will contribute to renormalization) are $\{\gamma\},\{\gamma_1,\gamma\},\{\gamma_2,\gamma\}$. Consider next the Feynman diagram labeled (d) in Figure 16.2. The renormalization sets are $\gamma_1 = \{1,2,4\}, \gamma_2 = \{3,5,6\}$, $\gamma_3 = \{1,2,3,4,5\}$, $\gamma_4 = \{2,3,4,5,6\}$ and γ.

Exercise 16.5.6 List all the possible forests consisting only of renormalization sets in the previous example. Hint: There are eight of them.

16.6 Renormalizing the Integrand: The Forest Formula

Given a 1-PI Feynman diagram, with integrand[23] F, we will define a "renormalized integrand" $\mathcal{R}F$ for which the integral will be convergent. We assume that the diagram has at least two external legs. Diagrams with no external leg (vacuum bubbles) or one external leg (tadpoles) are assigned the value zero in the renormalized theory. The Feynman diagram is considered to be fixed throughout the whole discussion.

We consider the subdiagram γ consisting of the internal vertices and the internal edges of the Feynman diagram. Let $\mathcal{E} = \mathcal{E}_\gamma$ be the set of edges of γ. The integrand F of (16.1) is given by (16.4). For the time being however, it only matters that $F(x) = \prod_{e\in\mathcal{E}} f(x_e)$, and the specific choice of f is irrelevant. We show how to regularize a function of this type. The main construction will associate to F and each forest $\mathcal{F}$[24] a function $\mathcal{F}F$. The regularization $\mathcal{R}F$ is the function on $(\mathbb{R}^{1,3})^{\mathcal{E}}$ given by the formula

$$\mathcal{R}F = (1 - T^{d(\gamma)}) \sum_{\mathcal{F}\text{ forest}} \mathcal{F}F, \tag{16.15}$$

where the sum is over all possible forests $\mathcal{F}$, and where $d(\gamma)$ is given by (16.14). At some intuitive level one may think of the quantity $\sum_{\mathcal{F}} \mathcal{F}F$ as the sum of F and of terms canceling all the subdivergences of F. The operator $(1 - T^{d(\gamma)})$ then cancels the overall divergence, recalling that we have defined $T^{d(\gamma)} = 0$ if $d(\gamma) < 0$.[25] You should keep this interpretation in mind as you read the proofs, but we will be able to justify it in full generality only much later.[26] On the other hand, on simple concrete cases such as those we treat below, this interpretation is easy to justify, and there is probably no better way to get a feeling for the method than working out these examples and exercises in complete detail with pencil and paper.

Let us perform the main construction: For each forest $\mathcal{F}$ we construct the function $\mathcal{F}F$ by the following procedure. For a subdiagram α which belongs to $\mathcal{F}$ we define recursively two functions F_α and F'_α on $(\mathbb{R}^{1,3})^{\mathcal{E}_\alpha}$, where $\mathcal{E}_\alpha$ is the set of edges of α.[27] First consider the case where α contains no other element of $\mathcal{F}$. Then for $x \in (\mathbb{R}^{1,3})^{\mathcal{E}_\alpha}$ we set

22 We recall that in our examples we describe subgraphs by the list of their vertices.
23 The integrand depends on the parameter $\varepsilon > 0$ but the dependence is kept implicit.
24 Here "forest" means on the Feynman diagram we have fixed in this section.
25 This corresponds to the two steps described at the end of Section 16.2.
26 However this picture is not literally true in some subtle cases, see Exercise 16.6.3.
27 These functions do not depend only on α but also on which subdiagrams of α belong to $\mathcal{F}$. This stays implicit in the notation.

$$F_\alpha(x) = \prod_{e \in \mathcal{E}_\alpha} f(x_e). \tag{16.16}$$

Recalling Definition 16.3.1 we then set

$$F'_\alpha := -T^{d(\alpha)} F_\alpha, \tag{16.17}$$

and we recall that this means $F'_\alpha = 0$ when $d(\alpha) < 0$. Next we define recursively F_α and F'_α for each diagram $\alpha \in \mathcal{F}$. Assume, following Lemma 16.5.3, that α contains maximal disjoint elements $\alpha_1, \ldots, \alpha_{\tilde{n}}$ of $\mathcal{F}$, for which the functions F'_{α_i} have already been defined. For each $i \le \tilde{n}$ consider the natural projection U_i from $(\mathbb{R}^{1,3})^{\mathcal{E}_\alpha}$ onto $(\mathbb{R}^{1,3})^{\mathcal{E}_{\alpha_i}}$ (so that $U_i(x)_e = x_e$ for $e \in \mathcal{E}_{\alpha_i}$). We then define

$$F_\alpha(x) = \prod_{i \le \tilde{n}} F'_{\alpha_i}(U_i(x)) \prod f(x_e), \tag{16.18}$$

where the last product is over the edges e of α which are not edges of any of the α_i. We then define F'_α by the formula (16.17). We have now defined F_α and F'_α for each $\alpha \in \mathcal{F}$, and in particular for $\alpha = \gamma$. We then define

$$\mathcal{F}F := F_\gamma, \tag{16.19}$$

It should be obvious that when $\mathcal{F} = \{\gamma\}$ we have $\mathcal{F}F = F$.

Lemma 16.6.1 *If for some $\beta \in \mathcal{F}, \beta \subset \alpha, \beta \ne \alpha$ one has $d(\beta) < 0$ then $F_\alpha = 0$.*

Proof If $F_\alpha \not\equiv 0$ it follows from (16.17) and (16.18) that for each maximal subdiagram α_i one has $F_{\alpha_i} \not\equiv 0$ and $d(\alpha_i) \ge 0$, and the result is then obvious by recursion. □

As a consequence, $\mathcal{F}F \ne 0$ only for the forests $\mathcal{F}$ such that all the subdiagrams different from γ they contain are renormalization sets.

Let us now turn to examples to illustrate the previous procedure. In most of these examples we parameterize $\ker \mathcal{L}$ as is traditional, using the formula (15.53) $x = x(k) = \sum_{i \le n} k_i \tau(L_i)$ where k_i are six-vectors,[28] and L_i are independent loops.[29] However we will lighten notation by using names such as q and ℓ for k_1 and k_2.

Let us start with the one-loop diagram (a) of Figure 16.1. The only renormalization set is γ and the formula (16.15) boils down to $\mathcal{R}F = (1 - T^0)F$. To compute $T^0 F$ we use (16.12) and the result of Exercise 16.3.2. One has to be careful about the orientation of the edges, i.e. $k_{1,2} = -q + (p_1 + p_2)/2, k_{2,3} = -q + (p_2 - p_1)/2, k_{3,1} = -q - (p_1 + p_2)/2$ so that $\bar{k} := (k_{1,2} + k_{2,3} + k_{3,1})/3 = -q + (p_2 - p_1)/6$. Since $f(\ell) = f(-\ell)$ we obtain then that $-T^0 F$ is just the compensating term $-f(q - (p_2 - p_1)/6)^3$ we used in Section 16.2.

Our next example is the diagram (b) of Figure 16.1, a typical Feynman diagram.[30] We will show that our general procedure, when specialized to that diagram, performs exactly

28 Since in our examples we work in ϕ^3_6 theory.

29 In one of our exercises we will use a different parameterization.

30 The flow through the external lines in the diagram is not the most general possible, but suffices to illustrate our purpose.

the manipulations of Section 16.2. The quantity $F(x(k)+\mathcal{X}(p))$ is given by the formula (16.6). When f is the propagator (16.3), even the integral over q of (16.6) diverges.

There are two renormalization sets, namely $\{1,2,3\}$ and γ. One of these sets contains the other, a situation which is called *nested divergences*. There are two forests consisting only of renormalization sets, namely $\{\gamma\}$ and $\mathcal{F}_1=\{\{1,2,3\},\gamma\}$. Therefore

$$\sum_{\mathcal{F}\text{ forest}}\mathcal{F}F(x(k)+\mathcal{X}(p))=F(x(k)+\mathcal{X}(p))+\mathcal{F}_1F(x(k)+\mathcal{X}(p)). \tag{16.20}$$

Furthermore, using the computation we did for the one-loop diagram, we obtain that $\mathcal{F}_1F(x(k)+\mathcal{X}(p))=-f(q-\ell/3)^3f(p/2-\ell)f(p/2+\ell)f(\ell)$ as in (16.7). Thus the quantity (16.20) is exactly the quantity (16.8). At this stage we have removed the divergence in q,[31] but the integral in q and ℓ is still divergent. To tame it we apply the operator $1-T^0$, as we did in Section 16.2.

It is far less obvious how to handle the diagram of Figure 16.5. This is because the two diverging subdiagrams of vertices $\{1,2,3\}$ and $\{2,3,4\}$ are neither disjoint nor nested. This situation is known as *overlapping divergences*. The historical approach to renormalization was to "recursively remove the divergences from the small diagrams to the large diagrams" and the existence of overlapping divergences is a major obstacle to that program. It took considerable efforts to pass it,[32] although now that the solution has been found, the difficulty is no longer apparent.

In the next two exercises f is the propagator (16.3).

Exercise 16.6.2 Consider the diagram of Figure 16.5.

(a) Prove that if $\mathcal{F}_1$ is the forest consisting of γ and of the diagram $\alpha=\{1,2,3\}$ then

$$\mathcal{F}_1F=-f(q-\ell/3)^3f(p/2+\ell)f(p/2-\ell).$$

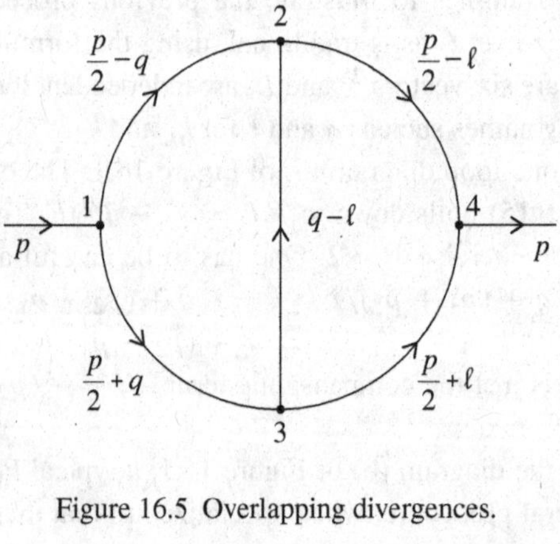

Figure 16.5 Overlapping divergences.

[31] We are not claiming here and at other places that the integral in q is convergent, as there are some other details that will be taken care of later. We are claiming that we have lowered the degree in q in such a way that crude power-counting is now consistent with the convergence of this integral.

[32] A brief entertaining account of the history of renormalization is given in Folland's book [31, Section 7.12].

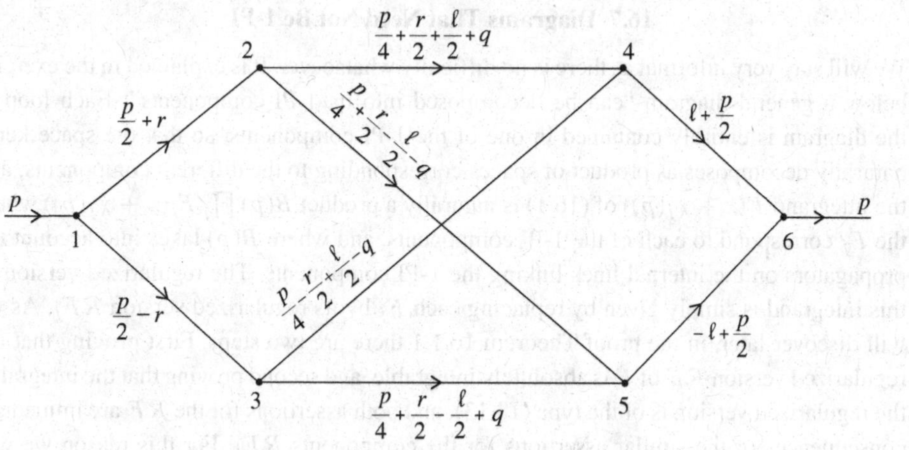

Figure 16.6 r, ℓ, q denote four-momenta.

(b) Considering now the forest $\mathcal{F}_2$ consisting of γ and of $\beta = \{2,3,4\}$. Prove that the quantity

$$F + \mathcal{F}_1 F + \mathcal{F}_2 F$$

no longer exhibits subdivergences in the diagrams $\{1,2,3\}$ and $\{2,3,4\}$, in the sense that the integrals in each of q and ℓ "converge" when the other variable is fixed (see footnote 30).

Exercise 16.6.3 Consider the diagram of Figure 16.6 where $\ker \mathcal{L}$ has been parameterized in a way that should help you compute the canonical flows.[33]
(a) If the forest $\mathcal{F}_1$ consists of γ and $\alpha = \{1,2,3,4,5\}$ show that

$$\mathcal{F}_1 F = -f(r)f(-r)f(r/2+q)^2 f(-r/2-q)^2 f(\ell+p/2)f(-\ell+p/2).$$

(b) If the forest $\mathcal{F}_2$ consists of γ and $\beta = \{2,3,4,5,6\}$ prove that the quantity

$$F + \mathcal{F}_1 F + \mathcal{F}_2 F$$

still has divergences in the diagram $\{1,2,3,4,5\}$ (in the sense that the integral in ℓ and q does not converge), due to the term $\mathcal{F}_1 F$.[34]

Exercise 16.6.4 Try to convince yourself of the following important fact. Once one has understood that the good way to improve the integrability of a function F is to add the compensating term $-T^d(\gamma)F$, the most natural way to apply this method recursively to improve the subdivergences is to consider the forest formula. (But this heuristic argument is powerless to explain why the forest formula will actually remove all divergences.)

33 Note however that this is not a standard parameterization using loops!
34 This example does not compromise the validity of the BPHZ scheme since there are other compensating terms. It shows however how subtle matters are.

16.7 Diagrams That Need Not Be 1-PI

We will stay very informal as there is no difficulty whatsoever. It is explained in the exercise below a general diagram "can be decomposed into its 1-PI components". Each loop in the diagram is entirely contained in one of the 1-PI components, so that the space $\ker \mathcal{L}$ naturally decomposes as product of spaces corresponding to the different components. and the integrand $F(x + \chi_j(p))$ of (16.1) is naturally a product $B(p) \prod_j F_j(x + \chi_j(p))$ where the F_j correspond to each of the 1-PI components, and where $B(p)$ takes into account the propagators on the internal lines linking the 1-PI components. The regularized version of this integrand is simply given by replacing each F_j by its regularized version $\mathcal{R}F_j$. As we will discover later, in the proof Theorem 16.1.1 there are two steps. First proving that the regularized version $\mathcal{R}F$ of F is absolutely integrable, and second proving that the integral of the regularized version is of the type (18.13), and both assertions for the $\mathcal{R}F$ are immediate consequences of the similar assertions for the components $\mathcal{R}F_j$. For this reason we will care only about 1-PI diagrams in the sequel.

Exercise 16.7.1 Given a Feynman diagram consider the collection of internal edges such that removing one of these edges disconnects the diagram. Prove that each connected component of the remaining diagram is 1-PI. Prove that the diagram obtained by contracting each of these components to a single vertex is a tree (i.e. does not contain any loop).

16.8 Interpretation

We have completed our description of the construction of $\mathcal{R}F$. According to Theorem 16.1.1, given two parameters m and g the BPHZ method assigns values to all Feynman diagrams. Thus it assigns values to all scattering amplitudes, in the form of a formal series in the coupling constant g. We have not proved that the corresponding series converges. This convergence is in fact very unlikely because of the combinatorial explosion of the number of diagrams involved in computations at high orders of perturbation theory. Nonetheless we may hope that computing the first few terms of the expansion will provide useful numerical values, and we have actually built a physical theory.

One should then perform consistency checks that this theory makes physical sense. It is not even apparent at this stage that our theory is Lorentz invariant,[35] because the expression (16.2) is not Lorentz invariant. Fortunately, Lorentz invariance is recovered when one takes the limit $\varepsilon \to 0^+$ as will be shown in Theorem 18.2.1. Lorentz invariance is not sufficient for the theory to make physical sense. For example, the S-matrix should be unitary. This imposes very non-trivial relations. Indeed, writing as is customary $S = 1 + \mathrm{i}\mathcal{M}$ the relation $S^\dagger S = 1$ implies $\mathrm{i}(\mathcal{M} - \mathcal{M}^\dagger) + \mathcal{M}^\dagger \mathcal{M} = 0$, and we should check this relation on matrix elements. It does not seem that this consistency property has been directly checked for the BPHZ method, but it is known through other approaches, see [28].

[35] In the sense that it predicts scattering amplitudes that transform appropriately under Lorentz transformations.

Let us modestly admit that after following such a convoluted path as we did, any argument that our theory describes reality is rather flimsy. However, the predictions of the theory can be tested against experiments.[36] Our ϕ_6^3 theory depends on two parameters, m and g, and models a certain self-interacting massive particle. As argued in Chapter 14, the parameter m *is not* the physical mass μ of the particle. In a similar manner the parameter g controls the strength of the interaction but does not really describe a particular scattering amplitude. These parameters μ and g have to be determined as described in Section 14.13 (which the reader should review now) to fit the actual physical mass and certain scattering amplitudes. The procedure is identical to the procedure described in Section 14.13, except that we now assign to each diagram the values provided by the BPHZ method. Let us briefly describe again how to proceed. First, as in Section 14.13 we have a relation of the type $m^2 = \mu^2 - Fg^2 + O(g^4)$ for a certain number F. We can use this relation to eliminate m in favor of the measurable quantity μ in any expression we wish.

Let us study the scattering amplitude of two particles of four-momenta p_1, p_2 into two particles of four-momenta p_3, p_4. At order $O(g^6)$, with the notation of Section 14.13 the scattering amplitude calculated by the BPHZ method is of the form

$$(g^2 A(s,t,u) + g^4 B(s,t,u))\delta + O(g^6), \tag{16.21}$$

where $\delta = (2\pi)^6 \delta^{(6)}(p_1 + p_2 - p_3 - p_4)$. The quantities $A(s,t,u)$ and $B(s,t,u)$ depend on the parameter m, which may be eliminated in favor of m.[37] To describe the strength of the interaction, we may then choose (among other reasonable choices) as a measurable quantity the number $\rho_{\rm lab}$ such that at a small velocity $p_1 = p_2 = p_3 = p_4 = p^* := (\mu, 0,0,0,0,0)$ the scattering amplitude is given by $\rho_{\rm lab}\delta$, so that, setting $v = 4(p^*)^2 = 4\mu^2$ we obtain the relation

$$\rho_{\rm lab} = g^2 A(v,0,0) + g^4 B(v,0,0) + O(g^6)$$

which we invert to find a relation $g^2 = C\rho_{\rm lab} + D\rho_{\rm lab}^2 + O(\rho_{\rm lab}^3)$. We may use this relation to express our scattering predictions as functions of the measurable quantity $\rho_{\rm lab}$.

In ϕ_4^4 theory the situation is pretty similar. We have to relate the parameters m and g of the theory to quantities measured in the lab. For mass renormalization, this is the same procedure as in ϕ_3^6 theory. At order $O(g^4)$ the scattering amplitudes predicted by the BPHZ method are of the type $(-ig + g^2 A(s,t,u) + g^3 B(s,t,u))\delta$, where $\delta = (2\pi)^4 \delta(p_1 + p_2 - p_3 - p_4)$. A reasonable choice of a measurable quantity is the quantity $g_{\rm lab}$ such that at a small velocity the scattering amplitude is $\mathrm{i} g_{\rm lab}\delta$. We then have the relation $-\mathrm{i}g_{\rm lab} = \mathrm{i}g + g^2 A(4\mu^2, 0, 0) + g^3 B(4\mu^2, 0, 0)$ which we invert to obtain a relation of the type $g = g_{\rm lab} + g_{\rm lab}^2 C + g_{\rm lab}^3 D + O(g_{\rm lab}^4)$ and express the scattering amplitudes as functions of the measurable quantity $g_{\rm lab}$.

[36] In the present case they cannot since this theory is a toy theory. In the somewhat similar case of Quantum Electrodynamics, the theory of interactions of electrons and photons, the predictions obtained by similar methods are in excellent agreement with measurements.

[37] Please note that this influences the value of B!

16.9 Specificity of the Parameterization

The values assigned by the BPHZ method to a diagram have special properties, which we put forward and explain now. A diagram with two external lines will be called a *biped.*

Proposition 16.9.1 *In ϕ_6^3 theory the value assigned by the BPHZ method to a diagram with three external legs, more than one internal vertex, and flows p_1, p_2, p_3 through these legs is of the type $(2\pi)^6\delta^{(6)}(p_1 + p_2 + p_3)W(p_1, p_2, p_3)$ where the function W satisfies $W(0,0,0) = 0$. The value assigned by this method to a biped and flows p_1, p_2 through these legs is of the type $(2\pi)^6\delta^{(6)}(p_1 + p_2)G(p_1^2)$ where the function G satisfies $G(0) = G'(0) = 0$.*

In particular the function $\Gamma(p^2) = \Sigma(p,0)$ (where as defined in (14.95), $-\mathrm{i}\Sigma(p,\varepsilon)$ is the sum of the functions G over all bipeds) is such that $\Gamma(0) = \Gamma'(0) = 0$. One may wonder why such a condition should hold. The short answer is that this condition is not a law of Nature, but a consequence of the way the theory is parameterized. This issue will be discussed in detail in the next chapter. We will show that in some sense there is a unique way to parameterize the theory in a way that this condition holds, and this parameterization is just built into the BPHZ construction.

As we have not yet proved Theorem 16.1.1, we cannot really prove Proposition 16.9.1 but we can explain why it is true (pretending that in Theorem 16.1.1 the limit exists as a function). The value of a diagram (not taking into account the delta function factor) is given by the limit as $\varepsilon \to 0$ of the integral $\int \mathrm{d}\mu_{\mathcal{L}}(x)\mathcal{R}F(x + \mathcal{X}(p))$, where $\mathcal{R}F = (1 - T^{d(\gamma)})\sum_{\mathcal{F}\text{ forest}} \mathcal{F}F$. For the present purpose the important part of the formula is the factor $(1 - T^{d(\gamma)})$, together with the fact that $\mathcal{X}(p) \in \mathcal{Q}$. For diagrams with three external legs we have $d(\gamma) = 0$ and the operator $1 - T^0$ ensures that integrand $\mathcal{R}F(x + \mathcal{X}(p))$ is zero for $p = 0$, so this is also the case for the limit of the integral as $\varepsilon \to 0$. This applies to all diagrams with three external legs except of course the diagram with a single internal vertex, which does not need to be renormalized. For diagrams with two external legs, $d(\gamma) = 2$ and the operator $1 - T^2$ ensures that all the partial derivatives of order ≤ 2 of the integrand $\mathcal{R}F(x + \mathcal{X}(p))$ in the components of p are zero at $p = 0$. We will eventually show[38] that in the limit $\varepsilon \to 0$ the value of the integral $\int \mathrm{d}\mu_{\mathcal{L}}(x)\mathcal{R}F(x + \mathcal{X}(p))$ depends only on p^2 where p is the four-momentum entering the biped, so it is a function $G(p^2)$ with $G(0) = G'(0) = 0$, and this concludes our supporting argument for Proposition 16.9.1.

The BPHZ method is of little practical value. The integrands produced by this method are very complicated. They are linear combinations of many simpler terms. In practice it is convenient to be able to assign a value to each individual simpler term. The method currently the most used for this purpose is called *dimensional regularization.*[39] The principle is as follows: Facing a divergent integral in, say, four dimensions, we try to formally guess what should be the value of the integral if it were computed in d dimensions where d is a complex number. The integral is then expressed as a function of d with a singularity at $d = 4$. Removing this singularity produces an actual value for $d = 4$. Clever implementations of

38 By proving Lorentz invariance, see Theorem 18.2.1.

39 It is applied after Wick rotation, so that the diverging integrands are invariant under rotations of $\mathbb{R}^{1,3}$.

the method have all kinds of desirable properties, and in particular respect the symmetries of the theory. This procedure is described in every recent textbook. As I find it very hard to become wiser by looking at the resulting formulas, I see no need to write any of them here.

Key ideas to remember:

- The BPHZ method tames divergent integrals without using cutoffs, by replacing the integrand by a better-behaved one.
- The better-behaved integrand is given by the explicit forest formula.
- The regularized values of diagram integrals thus constructed allow physical predictions.

17

Counter-terms

The BPHZ method, as we presented it, is the crowning achievement of a long line of research. The historical approach to taming divergences, the counter-term method, is markedly different. The counter-term method is taught in nearly all physics textbooks. In most textbooks only the case of diagrams with one (and rarely two) loops is ever explained and the general procedure is sketched in the foggiest terms. In this chapter we describe the counter-term method and how it relates to the conceptually much clearer BPHZ method. The counter-term method was historically developed to "remove the divergences" in theories such as the ϕ_6^3 theory. Besides removing divergences, the counter-term method "re-parameterizes the theory". The removal of the divergences and the re-parameterization of the theory are intertwined but independent stories. Matters will be far clearer if we separate these stories. Up to Section 17.4 we will focus on the aspect "re-parameterizing the theory" in the simpler setting of ϕ_2^3 theory, where there are no divergences. Please understand however that much of what we will do will be motivated later in the chapter when we will reinterpret the BPHZ method through the counter-term method starting with Section 17.5. The point of these early sections, (besides providing an introduction to the counter-term method in a very simple setting) is to explain some properties of the values attached to certain diagrams by the BPHZ method (see (17.7)). These properties are by no means a law of nature, but a consequence of the way the theory is parameterized. The main story starts in Section 17.7, where we start to use the counter-term method in ϕ_6^3 theory for its true purpose: canceling the divergences. We will heuristically deduce the possibility of canceling these divergences with counter-terms from the fact that the BPHZ method works (which we will prove in the last two chapters).

In order to avoid disappointment, it might be useful to set realistic expectations about what you may or may not learn in this chapter. You should understand (at a high level) the counter-term method by which physicists compute scattering amplitudes, and why it yields finite, meaningful results. You will see a very striking use of the correcting term $z^{k/2}$ in the formula (14.125). You might understand better the meaning of the expression "field renormalization", first at the level of combinatorics, then more deeply in Section 17.9. But your understanding may not progress on some of the key issues. The whole approach to computing scattering amplitudes can be seen as an algorithm to compute these amplitudes, validated by its tremendous agreement with experimental data. The somewhat stretched efforts we made in Chapter 14 to justify parts of this algorithm may however overall not become much clearer after you finish the present chapter. This author is not solely responsible for this unpleasant state of affairs. The efforts of Chapter 14 are indeed stretched, and gaping holes remain in any theoretical attempt to justify the success of the procedure. The specific place where these efforts utterly fail is in providing any reason whatsoever that the renormalized value of the diagrams should contain information about the physical world.[1]

17.1 What Is the Counter-term Method?

The purpose of a physical theory such as the ϕ_2^3 theory is to provide a framework in which one can predict the results of experiments. This theory $T(m,g)$ depends on the two parameters m and g. It allows one to compute the value of scattering amplitudes as the sum of the values of the corresponding Feynman diagrams. This sum is defined order by order in g, or, if one prefers, as a formal series in g.

Suppose now that for some reason we are not satisfied with our theory $T(m,g)$. Rather than starting from scratch elsewhere we try to salvage what we already have and to *deform* our theory into a new one. We want to keep the same setting: The value of a scattering amplitude will be the sum of the values of certain Feynman diagrams. We will extend our class of Feynman diagrams by allowing a new type, Δ, of vertices. These are always adjacent to three lines and are often called *counter-term vertices*.[2] A Feynman diagram containing only ordinary vertices will be called an ordinary Feynman diagram.

Given a parameter D the Feynman rules for the Feynman diagrams in the deformed theory are as follows:[3]

- Assign to each ordinary vertex a factor $-ig$.
- Assign to each counter-term vertex Δ a factor $-iD$.
- Assign to each internal line with momentum p the factor $-i/(-p^2 + m^2 - i\varepsilon)$.
- Take the product of the previous quantities.
- Enforce conservation of momentum at each vertex by multiplying the previous product by a product of suitable delta functions, one for each vertex.
- Integrate the product over the internal degrees of freedom of the diagram.[4]
- Take the limit $\varepsilon \to 0^+$.
- Divide if necessary by the corresponding symmetry factor.

In ϕ_2^3 theory, all diagrams are convergent, so there is no problem when integrating over the internal degrees of freedom. There are however still potential problems as $\varepsilon \to 0^+$: singularities may develop. These problems are a nuisance rather than the main issue, so for clarity we will *pretend that they do not exist*.

[1] The author certainly stands by his qualification that the mathematical value of that step is strictly nil, irrespective of the fact that this is downplayed to a properly remarkable extent in the main textbooks.
[2] In short we will say that counter-term vertices have three legs.
[3] You may review for comparison the Feynman rules on momentum state space on page 390 at this point.
[4] If you forgot what this means go to page 497.

In this manner we obtain a theory $T(m,g,D)$. The coefficient D itself will be a formal series in g, with no constant term. In the theory $T(m,g,D)$ we are able to compute scattering amplitudes at any order in perturbation theory i.e. as formal series in g. When $D = 0$ any Feynman diagram containing a counter-term vertex has value zero, so $T(m,g,0)$ is just our original theory $T(m,g)$. It is for $D \neq 0$ that we get a deformation of our original theory.

17.2 A Very Simple Case: Coupling Constant Renormalization

In the theory $T(m,g)$ the sum of the values of the diagrams with three external lines and flows p_1, p_2, p_3 through these lines is of the type[5] $(-\mathrm{i})(2\pi)^2\delta^{(2)}(p_1+p_2+p_3)W(p_1,p_2,p_3,m,g)$ for a certain formal series $W(p_1,p_2,p_3,m,g) = g + \sum_{k\geq 3} g^k W_k(p_1,p_2,p_3,m)$.[6] We will now consider a deformed theory $T(m,g,D)$ where the counter-term is a formal series in g,[7]

$$D = \sum_{k\geq 3} g^k D_k. \tag{17.1}$$

Thus the deformed theory depends on m, g, and $\bar{D} := (D_k)_{k\geq 3}$, and we denote this deformed theory by $T(m,g,\bar{D})$. In the deformed theory the sum of the values of the diagrams (including of course the diagrams with counter-term vertices) with three external lines and flows p_1, p_2, p_3 through these lines is of the type $(-\mathrm{i})(2\pi)^2\delta^{(2)}(p_1+p_2+p_3)W(p_1,p_2,p_3,m,g,\bar{D})$ for a certain formal series $W(p_1,p_2,p_3,m,g,\bar{D}) = g + \sum_{k\geq 3} g^k W_k(p_1,p_2,p_3,m,\bar{D})$. Suppose now that we fancy deforming our theory (i.e. determining $\bar{D}$) to enforce the conditions

$$\forall k \geq 3\ ;\ W_k(0,0,0,m,\bar{D}) = 0. \tag{17.2}$$

Why you would ever want to do such a thing will only be apparent later. (In short it is because a similar condition holds in the BPHZ scheme.)

The goal of the present section is to show how we can deform the theory to enforce (17.2). As we will show in detail, the deformed theory is in fact the same theory as the original theory (i.e. it predicts the same scattering amplitudes) but it is "parameterized in a different way". The implication of the result is that the condition (17.2) has nothing to do with a deep characteristic of the model, but rather can be obtained "for free" using a suitable parameterization of this model. This situation is a kind of toy situation for the case of real interest to us, the BPHZ scheme, for which the corresponding result is described in Section 17.4.

In order to achieve (17.2) we will construct the coefficients D_k recursively. To start the construction we compute $a_3 := W_3(0,0,0,m,(0,0,\ldots)) = W_3(0,0,0,m)$.

[5] Please do not forget that we work in two space-time dimensions here so that our $\delta^{(2)}$ functions came with factors $(2\pi)^2$.

[6] There is no term in g^2 because in ϕ_2^3 theory a diagram with three external lines has either a single vertex or at least three vertices.

[7] There is no term in g^2 for the reason explained in the previous footnote.

We then consider a parameter D_3 and we compute the sum of the values of the diagrams with three external legs and flows p_1, p_2, p_3 through these legs in the theory $T(m, g, (D_3, 0, \ldots))$. First there is the contribution

$$(-i)(2\pi)^2 \delta^{(2)}(p_1 + p_2 + p_3) W(p_1, p_2, p_3, m, g)$$

of the diagrams which do not contain a vertex $\triangle$. The diagram with a single internal vertex $\triangle$ brings a contribution

$$(-i)(2\pi)^2 \delta^{(2)}(p_1 + p_2 + p_3) g^3 D_3 + O(g^4). \tag{17.3}$$

Then there is the contribution of all the other diagrams. These contain at least a vertex $\triangle$ and contribute at a higher order in g. Identification of the coefficients of g^3 yields

$$W_3(0, 0, 0, m, (D_3, 0, \ldots)) = W_3(0, 0, 0, m) + D_3 = a_3 + D_3.$$

Thus if we fix the value $D_3 = -a_3$ we ensure that $W_3(0, 0, 0, m, (D_3, 0, \ldots)) = 0$. We are on our way!

We then compute the value of $a_4 := W_4(0, 0, 0, m, (D_3, 0, \ldots))$. We observe the crucial relations

$$W_3(0, 0, 0, m, (D_3, D_4, 0, \ldots)) = W_3(0, 0, 0, m, (D_3, 0, \ldots)) = 0$$

$$W_4(0, 0, 0, m, (D_3, D_4, 0, \ldots)) = a_4 + D_4.$$

The first relation is obvious: The new term $g^4 D_4$ in D does not contribute at order three. The second relation is because the only *new* contribution at order four brought by the new term $g^4 D_4$ in D is again provided by the diagram with the single vertex $\triangle$. We then choose D_4 appropriately (i.e. $D_4 = -a_4$) to ensure that

$$W_4(0, 0, 0, m, (D_3, D_4, 0, \ldots)) = 0.$$

We then continue in this manner to recursively construct the terms D_k.

The interesting part of the story is that (as we detail below) the deformed theory is simply the theory $T(m, g')$ where $g' = g + D = g + \sum_{k \geq 3} g^k D_k$, in the sense that both theories predict the same scattering amplitudes. In other words, the deformation does not change the theory, it only changes the parameters we use to describe it. In more detail, the theory predicts scattering amplitudes through the formula (14.125). When we compute a scattering amplitude in the theory $T(m, g')$, it is a formal series in g', with coefficients depending on m (and on the momenta of the scattered particles). When substituting in this series the value $g' := g + \sum_{k \geq 3} g^k D_k = g + D$ one obtains a formal series in g *which is the same formal series as provided by the deformed theory* $T(m, g, \bar{D})$. To prove this it suffices to prove the following claim: The sum of the values of the Feynman diagrams with given external lines and given flows through these lines in the theory $T(m, g')$ (seen as a formal series in g through the relation $g' = g + D$) is the same as the sum of the values of the Feynman diagrams in the deformed theory with the same external lines and the same flows through these lines. To see this, let us say that two diagrams are equivalent if we can pass from one to the other by replacing certain ordinary vertices by counter-term

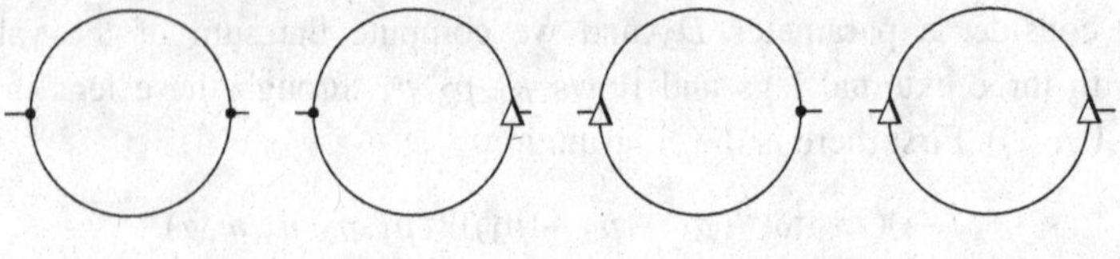

Figure 17.1 Four equivalent diagrams.

vertices Δ and certain counter-term vertices by ordinary vertices. So, for diagrams with n vertices their equivalence class contains 2^n diagrams (see Figure 17.1). To prove our claim we sum the values of the diagrams within a single equivalence class. We remember that an ordinary vertex contributes a factor $-\mathrm{i}g$ and a counter-term vertex a factor $-\mathrm{i}D$. We then use the identity $\sum_{I\subset\{1,\ldots,n\}}(-\mathrm{i}g)^{\operatorname{card} I}(-\mathrm{i}D)^{n-\operatorname{card} I} = (-\mathrm{i}(g+D))^n$ to prove our claim. This argument however ignores the issue of symmetry factors. Issues with symmetry factors are our own doing to a large extent, because we insisted on working with Feynman diagrams rather than with contraction diagrams. This is particularly clear in the present case. If, rather than using Feynman diagrams we keep the numbering of the vertices as part of the diagram (and in the last step of the computation of the value of a diagram we divide by $n!$ rather than by the symmetry factor) the issue disappears. We will let the reader convince herself of this. Unfortunately arguments using contraction diagrams get bogged down in notational problems and in order not to obscure the simple and beautiful ideas of the present chapter the issues of symmetry factors will almost be swept under the rug.

17.3 Mass and Field Renormalization: Diagrammatics

In this section we learn how to deform our basic ϕ_2^3 theory a different way than in Section 17.1, by adding another type of counter-term vertex $\otimes$ with two legs and special properties. (An example of such a diagram is the one labeled (a) in Figure 17.2.) As we will argue toward the end of the section, this deformation amounts in a certain way to "renormalizing the field" (although our deformed theory is defined only at the level of diagrams). The goal is similar to that of Section 17.2. Our deformed theory will be in fact the same theory as the original theory (but parameterized in a different way) and will satisfy a new condition of the same nature as (17.2). This condition is described in Proposition 17.3.1. The conclusion is the same as in the previous section: The conditions of Proposition 17.3.1 do not express a deep characteristic of the model, but "came for free" using a suitable parameterization of this model.

From now on a Feynman diagram with two external lines is called a *biped*. We recall the expression (14.96) of the Fourier transform of the dressed propagator:

$$\hat{\Delta}(p) = \lim_{\varepsilon\to 0^+} \frac{1}{-p^2+m^2-\mathrm{i}\varepsilon+\Sigma(p,\varepsilon)}, \tag{14.96}$$

where we recall now how the quantity $-\mathrm{i}\Sigma(p,\varepsilon)$ is defined (in (14.95)). For each biped W with flows p_1, p_2 through the external lines we compute the quantity $J_W(p_1,p_2,\varepsilon)$ by the rules on page 442. These rules are exactly the rules by which one computes the value of a Feynman diagram, except that we do not take the limit $\varepsilon \to 0^+$. It then happens that $J_W(p_1, p_2, \varepsilon)$ is of the form $(2\pi)^2\delta^{(2)}(p_1 + p_2)K_W(p_1, \varepsilon)$ for a certain quantity $K_W(p_1, \varepsilon)$, see (14.88). The quantity $-\mathrm{i}\Sigma(p, \varepsilon)$ is the sum of the quantities $K_W(p, \varepsilon)$ over all possible bipeds.

Assuming that there is no problem with the limit $\varepsilon \to 0^+$ let us define *the reduced value* of the biped W with flow p through its external lines[8] as being the quantity $K_W(p, 0)$. The quantity $-\mathrm{i}\Gamma(p^2) := -\mathrm{i}\Sigma(p, 0)$ is then the sum of the reduced values $K_W(p, 0)$ of all 1-PI bipeds with a flow of p through the external legs. It is a formal series in g, and the first non-zero term is of the type ag^2.

Our new fancy is to deform the theory (using a new method of deformation) in order to obtain in the deformed theory the conditions $\Gamma(m^2) = \Gamma'(m^2) = 0$. These conditions are called *on-shell renormalization conditions.* As we remember from Section 14.10, the point of the condition $\Gamma(m^2) = 0$ is that then the parameter m represents the physical mass of the particle. The point of the condition $\Gamma'(m^2) = 0$ is that the corrective term z of (14.124) occurring in the LSZ formula (14.125) equals 1,[9] see Exercise 14.12.3.

We deform the theory by introducing counter-term vertices $\otimes$. These vertices have two legs. The really new feature is that the factor assigned to each counter-term vertex $\otimes$ depends on the momentum flow $p \in \mathbb{R}^{1,1}$ through its legs. This factor involves two parameters $B = \sum_{\ell \geq 2} g^\ell B_\ell$ and $C = \sum_{\ell \geq 2} g^\ell C_\ell$ which are again formal series in g. The Feynman rules in the deformed theory are as follows.

- Assign to each ordinary vertex a factor $-\mathrm{i}g$.
- Assign to each counter-term vertex $\otimes$ a factor $-\mathrm{i}(p^2 B + C)$ where p is the four-momentum flowing through the legs of the vertex.[10]
- Assign to each internal line with momentum p the factor $-\mathrm{i}/(-p^2 + m^2 - \mathrm{i}\varepsilon)$.
- Take the product of the previous quantities.
- Enforce conservation of momentum at each vertex.
- Integrate over the internal degrees of freedom of the diagram and take the limit $\varepsilon \to 0^+$.
- Divide if necessary by the corresponding symmetry factor.

The next goal is to define in the deformed theory the quantities $\Gamma_d(p^2)$ which correspond to the quantities $\Gamma(p^2)$ in the original theory. For this we follow exactly the same procedure as in the original theory, except that now we have more diagrams to consider. Let us repeat the procedure for clarity. For a biped W we compute the quantity $J_{W,d}(p_1, p_2, \varepsilon)$ by the Feynman rules above, except that we do not take the limit $\varepsilon \to 0^+$. The function $J_{W,d}(p_1, p_2, \varepsilon)$ turns out to be of the type $(2\pi)^2\delta^{(2)}(p_1+p_2)K_W(p_1, \varepsilon)$. Assuming again that there are no problems in the limit $\varepsilon \to 0^+$, in the deformed theory we define the reduced value of a biped with flow p through the external lines as being $K_{W,d}(p, 0)$. We denote by

8 This means that flow p enters through one of the external lines and then exits through the other external line. It does not matter through which of the external lines the flow p enters since $K_W(p, 0) = K_W(-p, 0)$.

9 We could just as well enforce other conditions, such as the condition $\Gamma(0) = \Gamma'(0) = 0$, which occurs in the BPHZ scheme.

10 The factor p^2 is a new feature here, related to the fact that this has to do with "field renormalization". The value of p is defined only up to a sign, as it depends on how one orients the legs of the vertex, but the value of p^2 is well defined.

$\Gamma_d(p^2)$ the quantity $\Gamma(p^2)$ computed in the deformed theory, that is, the quantity $-\mathrm{i}\Gamma_d(p^2)$ is the sum of the reduced values $K_{W,d}(p,0)$ over all bipeds W including those which contain $\otimes$ vertices.

Let us now turn to the problem of enforcing the condition $\Gamma_d(m^2) = \Gamma'_d(m^2) = 0$ in the deformed theory. This is done order by order in g just as in the preceding section. There the key of the construction was (17.3). Here the key of the construction is that (according to the Feynman rules) the biped with a single vertex $\otimes$ has a reduced value $-\mathrm{i}(p^2B + C)$, so that in the deformed theory it contributes to $\Gamma_d(p^2)$ as $p^2B + C$ (this is the key point). In particular the contribution at order $k+1$ of this quantity is $g^{k+1}(p^2B_{k+1} + C_{k+1})$. To describe the procedure by which the coefficients B_ℓ, C_ℓ are recursively determined, let us denote by Γ_1 the value of the function Γ_d when it is computed in the original theory, and for $k \geq 2$ by Γ_k its value when it is computed in the deformed theory where $B = \sum_{2\leq\ell\leq k} g^\ell B_\ell$ and $C = \sum_{2\leq\ell\leq k} g^\ell C_\ell$. We then recursively enforce for $k \geq 1$ the conditions $\Gamma_k(m^2) = O(g^{k+1})$ and $\Gamma'_k(m^2) = O(g^{k+1})$. These automatically hold for $k = 1$. Assuming that for some $k \geq 1$ we have already achieved that $\Gamma_k(m^2) = O(g^{k+1}) = dg^{k+1} + O(g^{k+2})$ and $\Gamma'_k(m^2) = O(g^{k+1}) = bg^{k+1} + O(g^{k+2})$, we then simply determine B_{k+1} and C_{k+1} by the equations $m^2B_{k+1} + C_{k+1} + d = 0$ and $B_{k+1} + b = 0$ to obtain that $\Gamma_{k+1}(m^2) = O(g^{k+2})$ and $\Gamma'_{k+1}(m^2) = O(g^{k+2})$. Let us note for further use that since we have arranged our factors i so that $\Sigma(m^2,0)$ is real, this is also the case for B_{k+1} and C_{k+1}. We have then proved the following:

Proposition 17.3.1 *We may deform the* ϕ^3_2 *theory in such a way that* $\Gamma_d(m^2) = \Gamma'_d(m^2) = 0$.

The coefficients B, C (as formal series in g) have been entirely determined by the previous procedure. We thus have built a deformed theory $T_d(m,g)$, depending on two parameters m and g.

Assuming once for all that we may take $\varepsilon = 0$, the function

$$\widehat{\Delta}_d(p) := \frac{1}{-p^2 + m^2 + \Gamma_d(p^2)} \tag{17.4}$$

plays in the deformed theory the role of the Fourier transform of the dressed propagator (see (14.89)).[11] As a consequence of the conditions $\Gamma_d(m^2) = \Gamma'_d(m^2) = 0$, when computing scattering amplitudes in the deformed theory by the prescription (14.125), the correction factor z equals 1.

We now turn to the really fascinating fact: The proof that "the deformed theory $T_d(m,g)$ coincides with the original theory $T(m',g')$ for certain values g' and m'", where m' and g' are formal series in g. This is why we say that the deformed theory is a re-parameterization of the original theory. In some sense, there is a correspondence between the parameters (m,g) and (m',g') such that $T_d(m,g) = T(m',g')$. We will make these ideas precise at

[11] Since the deformed theory is defined only at the level of diagrams, we can only define this quantity as a sum of values of certain diagrams.

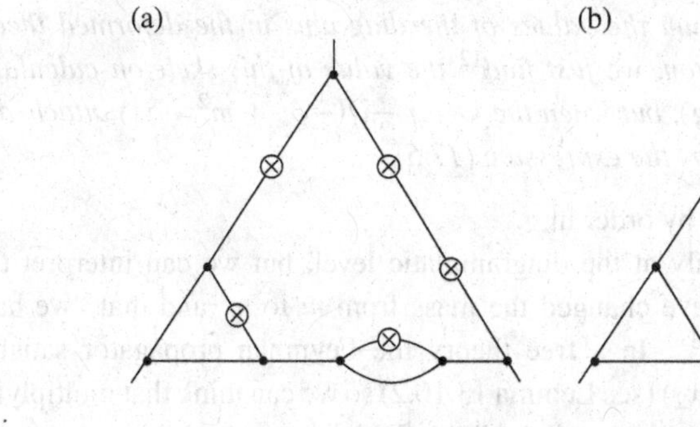

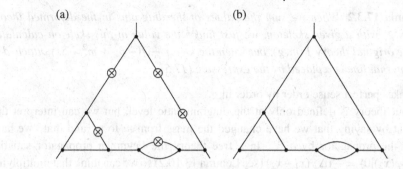

Figure 17.2 A diagram and its skeleton.

the end of the section. Before we can come to this, we turn to the computation of the scattering amplitudes in the deformed theory. For this computation we use the prescription (14.125). We have to sum over amputated diagrams with given external lines and given flows through these lines. (We simplify the exposition by saying from now on "given external lines" and keeping implicit that the flows through these lines are also given.) We recall that an amputated diagram is such that an external line cannot be disconnected from the other external lines by removing a single internal line. In an amputated diagram an internal vertex attached to an external line must be an ordinary vertex (because the vertex $\otimes$ has only two edges).

Consider an amputated Feynman diagram F in the deformed theory. It has ordinary vertices and vertices $\otimes$. We define the *skeleton* of F as follows. The skeleton of F is an ordinary diagram, with the same external lines as F. Its vertices are the ordinary vertices of F. Each path of F crossing only counter-term vertices is replaced by a single edge, see Figure 17.2. Thus, to obtain the skeleton of a diagram, one simply "removes the counter-term vertices $\otimes$". Obviously this skeleton is an amputated diagram.

Consider two ordinary vertices and a path between them consisting only of edges and vertices $\otimes$, such that there is a flow $p \in \mathbb{R}^{1,1}$ through these edges. If there are k edges and $k-1$ vertices, this path contributes as

$$\Big(\frac{-\mathrm{i}}{-p^2+m^2-\mathrm{i}\varepsilon}\Big)^k(-\mathrm{i}(p^2B+C))^{k-1}.$$

We sum this geometric series $\sum_{k\geq 1}\alpha^k\beta^{k-1} = \alpha/(1-\alpha\beta)$ to obtain

$$\frac{-\mathrm{i}}{-p^2(1-B)+m^2+C-\mathrm{i}\varepsilon} = a^{-1}\frac{-\mathrm{i}}{-p^2+m'^2-\mathrm{i}\varepsilon'}, \tag{17.5}$$

where $a := 1-B$, $am'^2 := m^2+C$ and $a\varepsilon' = \varepsilon$. (Recall that B and C are formal series in g with real coefficients, and that $B \neq 1$ since it starts with a g^2 term.) We have proved the following.

Lemma 17.3.2 *When we sum the values of the diagrams in the deformed theory $T_d(m,g)$ with a given skeleton, we just find*[12] *the value of this skeleton calculated in the original theory $T(m,g)$, but when the factor $-\mathrm{i}/(-p^2+m^2-\mathrm{i}\varepsilon)$ attached to an internal line is replaced by the expression (17.5).*

This makes perfect sense order by order in g.

Here, our theory is defined only at the diagrammatic level, but we can interpret this replacement by saying that we have changed the mass from m to m' and that "we have multiplied the propagator by a^{-1}". In a free theory the Feynman propagator satisfies $\langle 0|T\varphi(x_1)\varphi(x_2)|0\rangle = -\mathrm{i}\Delta_F(x_1-x_2)$ (see Lemma 13.10.2) so we can think that multiplying the propagator by a^{-1} amounts to "renormalizing" the field φ by multiplying it by $a^{-1/2}$, hence the name "field renormalization."

Our next two lemmas are specific to ϕ^3 theory.

Lemma 17.3.3 *For a diagram in ϕ^3 theory the numbers I of internal lines, V of internal vertices and E of external vertices satisfy $3V-2I=E$.*

Proof A total $3V$ lines spread out of the internal vertices, $3V-E$ of which are internal lines. Each of these lines touches two vertices. Thus $3V-E=2I$. □

Lemma 17.3.4 *The sum of the values of the amputated diagrams with n given external lines in the deformed theory $T_d(m,g)$ equals $a^{n/2}$ times the sum of the amputated diagrams with the same external lines in the theory $T(m',g')$ where $g'=a^{-3/2}g$.*

Proof It suffices to show that the sum of the values in the deformed theory $T_d(m,g)$ of the diagrams with a given skeleton and n external lines equals $a^{n/2}$ times the value of this skeleton in the theory $T(m',g')$. We compute the sum of these values using Lemma 17.3.2: It is the value of this skeleton computed in the original theory $T(m,g)$ but when the factor $-\mathrm{i}/(-p^2+m^2-\mathrm{i}\varepsilon)$ attached to an internal line is replaced by the expression (17.5). Comparing this value with the value of the skeleton computed in the theory $T(m',g')$ brings a factor a^{-1} for each internal line and a factor $a^{3/2}$ for each internal vertex (because to each such internal vertex is affected the coefficient $-\mathrm{i}g=a^{3/2}(-\mathrm{i}g')$ in the theory $T(m,g)$ but the coefficient $-\mathrm{i}g'$ in the theory $T(m',g')$). Using Lemma 17.3.3 this is a total factor of $a^{3V/2-I}=a^{n/2}$. □

The following has exactly the same proof, because an amputated diagram in the deformed theory is 1-PI if and only if its skeleton in 1-PI.[13]

Lemma 17.3.5 *The sum of the values of all the amputated 1-PI diagrams with n given external lines in the deformed theory equals $a^{n/2}$ times the sum of the values of all 1-PI diagrams with the same external lines in the theory $T(m',g')$ where $g'=a^{-3/2}g$.*

Where things really get beautiful is when one computes $\widehat{\Delta}_d$ as a function of quantities of the theory $T(m',g')$.

[12] Putting again the issue of symmetry factors under the rug. The reader may like to investigate in the case of the "eye diagram" why these come out right.

[13] It should be obvious that in the deformed theory a 1-PI diagram is always amputated.

Proposition 17.3.6 *Denote by $\tilde{\Gamma}(p^2)$ the function corresponding to Γ in the theory $T(m',g')$. Then*

$$\widehat{\Delta}_d(p) = a^{-1}\frac{1}{-p^2 + m'^2 + \tilde{\Gamma}(p^2)}. \tag{17.6}$$

Proof To understand this proof you have to be familiar with the proof of Theorem 14.9.7. The argument is exactly in the spirit of Lemma 17.3.4. The starting point is that $-\mathrm{i}\widehat{\Delta}_d(p)$ is the sum of the reduced values of all bipeds with flow p through their external lines in the deformed theory. Such a biped has the following structure. Starting from the left external vertex there is a path containing only counter-term vertices ⊗ until we meet the first ordinary vertex which is the beginning of a 1-PI block (there may be zero counter-term vertices on this path, in which case the path consists of a single edge). We are not concerned with the structure of the 1-PI block, which may or may not contain counter-term vertices. The 1-PI block ends with an ordinary vertex, and from there is path containing only counter-term ⊗ vertices, until we meet an ordinary vertex which is the beginning of the second 1-PI block. This path may contain zero counter-term vertices, in which case it consists of a single edge. And so on. According to the case $n = 2$ of Lemma 17.3.5 the sum of the reduced values over the 1-PI blocks is a times the sum of the reduced values of these same blocks in the theory $T(m',g')$. That is, this sum is $-\mathrm{i}a\tilde{\Gamma}(p^2)$ where $\tilde{\Gamma}$ is the function Γ of the theory $T(m',g')$. On the other hand summing over the paths crossing only vertices ⊗ produces the quantity U of (17.5). Consequently the sum of the reduced values of all biped with exactly k 1-PI blocks is $U^{k+1}(-\mathrm{i}a\tilde{\Gamma}(p^2))^k$, for a total contribution

$$-\mathrm{i}\widehat{\Delta}_d(p) = \sum_{k\geq 0} U^{k+1}(-\mathrm{i}a\tilde{\Gamma}(p^2))^k.$$

Summing the geometric series gives the required expression (17.6). □

Lemma 17.3.7 *The physical mass of the theory $T(m',g')$ is m. The quantity z of (14.124) relative to this theory equals a.*

Proof Comparing the expressions (17.4) and (17.6) we obtain

$$\frac{1}{-p^2 + m'^2 + \tilde{\Gamma}(p^2)} = a\frac{1}{-p^2 + m^2 + \Gamma_d(p^2)},$$

where we recall that $\Gamma_d(m^2) = \Gamma_d'(m^2) = 0$. We recall that the physical mass of the theory is given by the smallest root of the equation (14.101). Here $-m^2 + m'^2 + \tilde{\Gamma}(m^2) = 0$, and m appears to be the physical mass of the theory $T(m',g')$.[14] We also obtain, using L'Hospital rule,

$$z := \lim_{p^2\to m^2}\frac{-p^2 + m^2}{-p^2 + m'^2 + \tilde{\Gamma}(p^2)} = a \lim_{p^2\to m^2}\frac{-p^2 + m^2}{-p^2 + m^2 + \Gamma_d(p^2)} = a.$$ □

[14] We do not provide an argument showing that m is the *smallest* root of the equation (14.101). This is *physics*, and we assume that everything goes well unless proved otherwise.

Proposition 17.3.8 *The scattering amplitudes predicted by the deformed theory $T_d(m,g)$ are the same as those predicted by the theory $T(m',g')$.*

Please make sure you understand this proof, which sheds light on the role of mysterious factor z of the LSZ formula.

Proof The scattering amplitudes are computed by the formula (14.125). For the deformed theory $z = 1$ by construction, and for the theory $T(m',g')$ we have $z = a$. The result then follows from Lemma 17.3.5. □

Let us recall the setting of Section 14.13. The particle has a physical mass μ. Writing $\delta := (2\pi)^2\delta^{(2)}(p_1+p_2-p_3-p_4)$, we have defined ρ such that at small velocity for the incoming and the outgoing particles, the scattering amplitude of two particles into two particles is $\rho\delta$ (which for brevity we will simply call the scattering amplitude at small velocity). Thus, μ and ρ are measured in the lab, and we want to predict scattering amplitudes. We have two different theories, $T_d(m,g)$ and $T(m,g)$ and we are going to argue at a high level that they give the same predictions.

Whichever theory we use we have to adjust the parameters in a way that within this theory the physical mass is μ and the scattering amplitude at small velocity is $\delta\rho$. So, if we use the theory $T_d(m,g)$, where m is the physical mass, we take $m = \mu$ and g is a function $g(\rho)$. In Lemma 17.3.7 and Proposition 17.3.8 we have found values m' and g' such that the physical mass of the theory $T(m',g')$ is μ and the scattering amplitudes computed by the theory $T(m',g')$ are those computed by the theory $T_d(\mu,g(\rho))$. In particular the theory $T(m',g')$ predicts the correct physical mass and the correct scattering amplitude at small velocity, so that if we use the theory $T(m,g)$ to make predictions we should precisely use these parameters. But then the theory $T(m',g')$ gives the same prediction for all scattering amplitudes as does the theory $T_d(\mu,g(\rho))$.[15]

17.4 The BPHZ Renormalization Prescription

Still working with the ϕ_2^3 theory $T(m,g)$, and keeping the notation of the previous sections, let us now fancy deforming our theory in order to achieve in the deformed theory the conditions

$$W(0,0,0) = g\ ;\ \Gamma(0) = \Gamma'(0) = 0. \tag{17.7}$$

We will leave as an exercise to the reader to convince herself that this can be achieved by the previous methods, introducing now both $\triangle$ and $\otimes$ vertices. Moreover it can be shown that the resulting theory is simply a theory $T(m',g')$ for certain values of m' and g' which are formal series in g.

The point of this remark is simple: The conditions (17.7) are not mysterious and can be obtained by a suitable parameterization of the theory. The reason for the title of the section

15 The previous argument is rather sketchy, since we should really prove that the two theories provide the same predictions "order by order in perturbation theory", but this is better left to the interested reader.

is Proposition 16.9.1: The conditions (17.7) are satisfied when one computes the values of diagrams following the BPZH method.[16]

17.5 Cancelling Divergences with Counter-terms

Our task up to this point has been easy because we have been dealing with ϕ_2^3 theory where there are no divergences. The main story starts now: using counter-terms to *cancel the divergences*, in a way that we are going to make precise. The principle of the method is as follows:

- We introduce a cutoff to make all integrals convergent.
- We fix $\varepsilon > 0$.
- We deform the theory by allowing new vertices in our diagrams. The coefficients attached to these vertices depend on the cutoff and on $\varepsilon > 0$.
- We arrange that when computing the scattering amplitudes in the deformed theory these amplitudes have a finite limit as the cutoff gets removed and ε remains constant.
- Thus at given $\varepsilon > 0$ we attach a finite value to each scattering amplitude.
- Finally we send ε to zero.

What happens in the second-to-last bullet is expressed by the words "the counter-terms cancel the divergences". This occurs order by order in perturbation theory: Cancellation occurs between the terms containing the same power of the coupling constant g.

The important part of the procedure is the "cancellation of infinities". This occurs at a given $\varepsilon > 0$. Thus throughout all the subsequent considerations we think of ε as being fixed once and for all. In particular this *changes the meaning* of the expression "value of a Feynman diagram" as the value is now computed *at the given value of* $\varepsilon > 0$. This will however remain implicit as we are not going to indicate the dependence on ε in the notation. It is just understood once and for all that ε is sent to zero at the end of the procedure (and that we obtain a finite limit this way).

The details of the procedure depend on the model, and for now we consider the simplest case of ϕ_6^3 theory. It is simpler because in that case, as proved in Proposition 15.4.11, a proper subdiagram with a superficial degree of divergence $d \geq 0$ is a subgraph. The case of ϕ_4^4 theory, which requires a significant twist, is discussed in Section 17.8.

The Feynman diagrams in the deformed theory involve two types of counter-term vertices, denoted respectively $\otimes$ and Δ, which are exactly the vertices we met earlier in this chapter. The first one has two legs and the second one has three legs.[17] A Feynman diagram with only ordinary vertices will be called an ordinary Feynman diagram. The Feynman rules for the new types of vertices depend on three cutoff-dependent parameters B, C, D, which are formal series in g.[18] Each of the terms B, C, D has a physical

[16] The value g in the left part of (17.7) comes from the contribution of the triped with a single internal vertex.

[17] As will become gradually clear the reason why exactly these two counter-terms are needed has to do with the last statement of Lemma 15.4.9. In ϕ_6^3 theory, the subdiagrams with external degree of divergence ≥ 0 have $b = 3$ or $b = 2$.

[18] These also depend on $\varepsilon > 0$, which we think of as being fixed through the whole procedure.

interpretation.[19] Namely, as in Section 13.20 the term D occurs because the coefficient g in the Hamiltonian is not what is measured in the lab. In a similar manner the term C occurs because of mass renormalization. It can be argued that the term B occurs because of the "field renormalization" discussed in Section 14.12, as will be investigated in Section 17.9. Field renormalization also contributes to the other two terms.

The Feynman rules for the Feynman diagrams in the deformed theory are as follows:

- Assign to each ordinary vertex a factor $-ig$.
- Assign to each vertex $\otimes$ a factor $-\mathrm{i}(p^2 B + C)$ where $p \in \mathbb{R}^{1,5}$ is the momentum flowing through the legs of the vertex.
- Assign to each vertex Δ a factor $-\mathrm{i}D$.
- Assign to each internal line with momentum p the factor $-\mathrm{i}/(-p^2 + m^2 - \mathrm{i}\varepsilon)$.
- Take the product of the previous quantities.
- Enforce conservation of momentum at each vertex.
- Integrate over the internal degrees of freedom of the diagram.
- Divide if necessary by the corresponding symmetry factor.

An important idea is that requiring the counter-terms to cancel the divergences at best "determines the counter-terms B, C, D *only* up to a finite part". One has to fix some rule to determine this finite part. Such a rule is called a *renormalization prescription.* Although we will not prove it in general, different renormalization prescriptions however amount simply to parameterizing the theory in different ways, as we explored in special cases in the beginning of the chapter.[20]

When using the counter-term method in relation with the BPHZ method, it is natural to use the renormalization prescription (17.7) since we already argued that conditions (17.7) hold if we use only ordinary diagrams, but compute their values according to the BPHZ method. To describe again this prescription, we recall that a Feynman diagram with two external lines is called a biped. A Feynman diagram with three external lines will be called a *triped.* In the deformed theory the sum of the values of the tripeds with given six-momenta flows p_1, p_2, p_3 through their external lines is of the form

$$(-\mathrm{i})(2\pi)^6\delta^{(6)}(p_1 + p_2 + p_3)W(p_1, p_2, p_3),$$

and the sum of the values of the bipeds with six-momentum flow $p_1, p_2 \in \mathbb{R}^{1,5}$ through the external lines is of the form

$$(2\pi)^6\delta^{(6)}(p_1 + p_2)\Gamma(p_1^2).$$

Again, there is nothing special about the renormalization condition (17.7) (except that it is naturally associated to the BPHZ method). Other renormalization conditions are typically used by physicists, and lead "to the same theory, but parameterized in a different way".

19 This interpretation is not needed if we think that we are following an algorithmic procedure.

20 In these sections, we were dealing with ϕ_2^3 theory, which has no divergence, and we were dealing only with the "finite part" of the counter-terms and different renormalization prescriptions.

17.6 Determining the Counter-terms from BPHZ

That it is actually possible to cancel the divergences as outlined in the previous section is a highly non-trivial fact. We will deduce it from Theorem 16.1.1 (the proof of which will be completed in the next two chapters). The argument will be heuristic. It is unlikely that there exists a clean mathematical argument, for the simple reason that it does not seem that the counter-term method can be made rigorous unless one moves to the Euclidean setting, replacing propagators by their Euclidean versions.[21] Even in this setting, it is a non-trivial task to construct cutoffs with the required invariance properties.

One difficulty is that to have clean arguments (and as will become apparent when we go into the details), we need that the cutoff we use to make our integrals absolutely convergent is Lorentz invariant.[22] Such a cutoff unfortunately does not exist.[23] As we try to clarify matters rather than obscuring them by getting bogged down in details, we take drastic simplifying steps:

- We *pretend* that we can use a Lorentz invariant cutoff.
- We *pretend* that in the BPHZ method we use the standard Lorentz invariant propagators with Fourier transform (13.60) instead of the regularized propagators with Fourier transform (16.3), that is, we *pretend*[24] that in the formulas of Section 16.6 we have

$$f(k) = \frac{-\mathrm{i}}{-k^2 + m^2 - \mathrm{i}\varepsilon}.$$

In contrast with the rest of Part IV we do not give complete mathematical proofs in the rest of this chapter, which may rather feel like an excerpt from a physics textbook.[25] This might make it difficult to penetrate for a mathematically inclined reader, especially if she had no previous exposure to physics' style, but a reader not motivated by physics' methods should simply skip this material.

The principle of the approach is very simple. The BPHZ method replaces certain integrands F by their regularization $\mathcal{R}F$. The forest formula presents $\mathcal{R}F$ as a sum of many terms. These are not individually integrable, it is only their sum that is integrable. In the presence of a cutoff however, one may integrate these terms individually. This means that the value attributed to a Feynman diagram by the BPHZ method appears as a sum of many terms. The basic idea is that (at a given cutoff and a given value of ε)

Each of these terms can be interpreted as the value of a certain Feynman diagram in the split *deformed theory.* (17.8)

The *split* deformed theory is a variation of the deformed theory where each of the counter-term vertices has been split into a whole sequence of counter-term vertices. It will appear

[21] Almost every textbook of physics discussing renormalization does this, usually through the following single-sentence magic incantation: "using a Wick rotation, it suffices to consider the Euclidean setting".

[22] In the sense that the value assigned to an integral remains the same if the same Lorentz transformation is applied to each of the four-momenta involved in this integral.

[23] As we pointed out in Section 13.18, this is because the volume of the set where p^2 is bounded is infinite.

[24] Please do not be confused here. The *actual* BPHZ scheme uses the formula (16.3) to which we go back in later chapters.

[25] In particular we provide crisp examples of the powerful "proof by examples" and "let us give the main idea" methods.

in a completely natural manner. The important property at this stage is that the value of a diagram in the deformed theory appears naturally as the sum of the values of a collection of diagrams in the split deformed theory.[26]

Furthermore it will be true that summing the values of the BPHZ terms contributing to a given scattering amplitude exactly amounts to summing all the values of the Feynman diagrams contributing to the same amplitude in the deformed theory.[27] When removing the cutoff, the (cutoff-dependent) sum of the Feynman diagrams in the deformed theory corresponding to a given amplitude[28] has a finite limit which is just the value of the amplitude given by the BPHZ method. This cancellation is expressed by saying that the counter-terms cancel the divergences order by order in perturbation theory: Cancellation occurs between the terms containing the same power of the coupling constant g.

Let us now go into the details. We must first introduce some notation. We start with a 1-PI Feynman diagram, with at least two external lines.[29] In the BPHZ method we consider its subdiagrams, which are denoted by lowercase Greek letters. The important subdiagrams are the renormalization sets. Since we are working in ϕ_6^3 theory we may appeal to Proposition 15.4.11: such a subdiagram α is a subgraph, that is its set of vertices is a subset of the set of internal vertices of the Feynman diagram, and its edges consist of all internal edges between its vertices. Furthermore, there are $b \in \{2,3\}$ edges (internal or external) of the Feynman diagram out of the vertices of α which are not edges of α (and $d(\alpha) = 6 - 2b$).

The subgraph α is certainly not a Feynman diagram because it does not have external lines. We can however turn it into a Feynman diagram by adding the edges out of α as external lines. It will be convenient not to distinguish between α and the corresponding Feynman diagram. In particular when $b = 2$, we will say that α is a biped and when $b = 3$ we will say that α is a triped. Conversely Feynman diagrams which are either bipeds or tripeds will be denoted by lowercase Greek letters.

Recalling that the value of a Feynman diagram is now calculated at a given $\varepsilon > 0$, given a diagram with a set $\mathcal{V}$ of n internal vertices and a set $\mathcal{E}$ of internal edges, the value of this diagram is given by the Feynman rules as

$$\frac{1}{S}(-ig)^n \int \cdots \int \prod_{e \in \mathcal{E}} \frac{d^6 x_e}{(2\pi)^6} F(x) \prod_{v \in \mathcal{V}} (2\pi)^6 \delta^{(6)}(\mathcal{L}(x)_v - w_v), \tag{17.9}$$

where S is the symmetry factor. Here F is the function on $(\mathbb{R}^{1,5})^{\mathcal{E}}$ given by

$$F(x) = \prod_{e \in \mathcal{E}} f(x_e)\ ;\ f(k) = \frac{-i}{-k^2 + m^2 - i\varepsilon}, \tag{17.10}$$

and w_v is the sum of the six-momenta leaving the internal vertex v through the external lines. On the other hand, the renormalized value attached to the diagram (again at a given $\varepsilon > 0$) by the BPHZ procedure[30] is simply the quantity (17.9) where F has been replaced by

26 This sum is infinite, but makes sense as a formal series in g.

27 And also to summing all the values of the Feynman diagrams contributing to the same amplitude in the split deformed theory.

28 A very satisfactory aspect of the counter-term method is that one just has to compute a sum of diagrams, business as usual.

29 The case of diagrams with a single external line will be considered later.

30 In the remainder of this chapter, "the renormalized value" of a diagram is the value attached to this diagram at a given $\varepsilon > 0$ by the BPHZ method.

its regularized version $\mathcal{R}F$ given by (16.15). That is, denoting by γ the largest subdiagram of the Feynman diagram, we have

$$\mathcal{R}F = \sum_{\mathcal{F}}(-T^{d(\gamma)}\mathcal{F}F + \mathcal{F}F),$$

where the sum is over all forests on γ. Thus, the renormalized value of the diagram appears as the sum $\sum_{\mathcal{F}}(\mathrm{I}_{\gamma,\mathcal{F}} + \mathrm{II}_{\gamma,\mathcal{F}})$ where $\mathrm{I}_{\gamma,\mathcal{F}}$ is obtained by the expression (17.9) when we replace F by $-T^{d(\gamma)}\mathcal{F}F$ and $\mathrm{II}_{\gamma,\mathcal{F}}$ is obtained by the expression (17.9) when we replace F by $\mathcal{F}F$. For further use let us write the formula

$$\mathrm{I}_{\gamma,\mathcal{F}} = \frac{1}{S}(-\mathrm{i}g)^n \int\cdots\int \prod_{e\in\mathcal{E}} \frac{\mathrm{d}^6 x_e}{(2\pi)^6}(-T^{d(\gamma)}\mathcal{F}F)(x) \prod_{v\in\mathcal{V}}(2\pi)^6\delta^{(6)}(\mathcal{L}(x)_v - w_v). \quad (17.11)$$

Please observe that the dependence on the original Feynman diagram is indicated as a dependence on γ.

The whole approach relies on the following lemma.

Lemma 17.6.1 *(a) If γ is a triped then we have*

$$\mathrm{I}_{\gamma,\mathcal{F}} = -\mathrm{i}(2\pi)^6\delta^{(6)}(\Sigma)D_{\gamma,\mathcal{F}} \quad (17.12)$$

where Σ is the sum of the six-momenta leaving the diagram through the external edges, and where the number $D_{\gamma,\mathcal{F}}$ depends only on $\gamma,\mathcal{F}$ and ε (but not on the momenta through the external lines).

(b) Assume that γ is a biped, and let p_1 and p_2 be the six-momenta leaving the diagram through the external edges. Then

$$\mathrm{I}_{\gamma,\mathcal{F}} = -\mathrm{i}(2\pi)^6\delta^{(6)}(\Sigma)(p_1^2 B_{\gamma,\mathcal{F}} + C_{\gamma,\mathcal{F}}), \quad (17.13)$$

where $\Sigma = p_1 + p_2$, and where the numbers $B_{\gamma,\mathcal{F}}$ and $C_{\gamma,\mathcal{F}}$ depend only on $\gamma,\mathcal{F}$ and ε (but not on the momenta p_1 and p_2).[31]

(c) If γ is neither a biped nor a triped, then $\mathrm{I}_{\gamma,\mathcal{F}}$ is zero.

Proof Let us recall from Section 16.3 that for a function H on $(\mathbb{R}^{1,5})^{\mathcal{E}}$ we consider it as a function $H(z+q)$ for $z \in \ker\mathcal{L}$, $q \in \mathcal{Q} = \ker\mathcal{L}^{\perp}$ and that $T^d H$ denotes the Taylor polynomial of order d at $q = 0$ of the function $q \mapsto H(z+q)$. In particular $(T^0 H)(z+q) = H(z)$.

To prove (a) we note that $d(\gamma) = 0$ and we apply Theorem 15.7.2 (to the function $T^0\mathcal{F}F$ rather than F) to obtain

$$\mathrm{I}_{\gamma,\mathcal{F}} = \frac{1}{S}(-\mathrm{i}g)^n(2\pi)^6\delta^{(6)}(\Sigma)\int \mathrm{d}\mu_{\mathcal{L}}(x)(-T^0\mathcal{F}F)(x + \mathcal{X}(p)),$$

where $\mathcal{X}(p) \in \mathcal{Q} \subset (\mathbb{R}^{1,5})^{\mathcal{E}}$ is the function of the six-momenta leaving the diagram through the external lines (which are symbolically denoted by p). The integral is independent of p because $T^0\mathcal{F}F(x + \mathcal{X}(p)) = T^0\mathcal{F}F(x)$.

31 Observe that the expression (17.13) is the same if we replace p_1 by the momentum p_2 leaving through the other external edge. Thus we may write the right-hand side of (17.13) as $-\mathrm{i}(2\pi)^6\delta^{(6)}(\Sigma)(p^2 B_{\gamma,\mathcal{F}} + C_{\gamma,\mathcal{F}})$ where "p is the six-momentum flowing through the external edges".

(b) When γ is a biped we have $d(\gamma) = 2$, and as before we have now

$$\mathrm{I}_{\gamma,\mathcal{F}} = \frac{1}{S}(-\mathrm{i}g)^n(2\pi)^6\delta^{(6)}(\Sigma)\int \mathrm{d}\mu_{\mathcal{L}}(x)(-T^2\mathcal{F}F)(x+\mathcal{X}(p_1)), \tag{17.14}$$

where $T^2\mathcal{F}F(x+\mathcal{X}(p_1))$ is a second-degree polynomial in the components of p_1. It is of course Lorentz invariant,[32] see Exercise 17.6.2. Now comes the key point: Since we pretend that our implicit cutoff is Lorentz invariant the integral in (17.14) is a second-degree Lorentz invariant polynomial in the components of p_1, so that it is of the form $Bp_1^2 + C$ where B and C are independent of p_1.

(c) This is obvious since then $d(\gamma) < 0$ and we defined $T^{d(\gamma)}\mathcal{F}F = 0$. □

Exercise 17.6.2 Get into more detail as to why the quantity (17.14) is Lorentz invariant.

Now comes the basic observation. When we compute amplitudes, we have to take sums over diagrams with given external lines. When we sum the quantities (17.12) over given external lines, we obtain the quantity $-\mathrm{i}(2\pi)^6\delta^{(6)}(\Sigma)D$ where

$$D = \sum_{\alpha} D_{\alpha,\mathcal{F}}. \tag{17.15}$$

Here the sum is over all 1-PI tripeds[33] α and all forests $\mathcal{F}$ on α. The sum makes sense because the coefficient $D_{\alpha,\mathcal{F}}$ contains a factor g^n when α has n internal vertices. In this manner D appears as a formal series in g. Furthermore we can give a meaning to the expression "order by order in g".

In words, with this choice of D, *the sum of quantities (17.12) over all possible diagrams with the same external lines is exactly the value* $-\mathrm{i}(2\pi)^6\delta^{(6)}(\Sigma)D$ *of a diagram with the same external lines and the unique vertex* Δ *in the deformed theory.*

Similarly to (17.15) we will set $B = \sum_\beta B_{\beta,\mathcal{F}}$ and $C = \sum_\beta C_{\beta,\mathcal{F}}$ with now a sum over all 1-PI bipeds β and all forests $\mathcal{F}$ on β.

Since the value of D in (17.15) appears naturally as a sum, we may think of "splitting the corresponding counter-term vertex into pieces", each piece corresponding to one of the terms of this sum. That is, formally, we think of the counter-term vertex as a sum $\Delta = \sum \Delta_{\alpha,\mathcal{F}}$, with the understanding that in the Feynman rules we now attach a factor $-\mathrm{i}D_{\alpha,\mathcal{F}}$ to the vertex $\Delta_{\alpha,\mathcal{F}}$. Splitting in the same manner the counter-term vertex $\otimes$ in counter-term vertices $\otimes_{\beta,\mathcal{F}}$, where β ranges over all bipeds and $\mathcal{F}$ over all forests on β, we obtain the *split deformed theory*. More formally, in the split deformed theory, besides the ordinary vertices, there are counter-term vertices $\Delta_{\alpha,\mathcal{F}}$ and $\otimes_{\beta,\mathcal{F}}$, with the obvious generalization of the Feynman rules.

As illustrated by Figure 17.3 the value of any diagram in the deformed theory is canonically a sum of values of diagrams in the split deformed theory.

The point of introducing the split deformed theory is that we can now go back to (17.12) and state that the term $\mathrm{I}_{\gamma,\mathcal{F}}$ is exactly the value in the split deformed theory of a diagram

[32] Here Lorentz invariance refers to the adaptation of this concept to $\mathbb{R}^{1,5}$.
[33] The diagram consisting of one single ordinary vertex is not considered to be a triped.

$$\cdots = \sum_{\substack{\alpha \text{ triped, } \mathcal{F} \text{ forest} \\ \beta \text{ biped, } \mathcal{F}' \text{ forest}}} \cdots$$

$\beta, \mathcal{F}'$

$\alpha, \mathcal{F}$

Figure 17.3 Splitting the vertices $\otimes$ and $\triangle$.

with the same external lines and single vertex $\triangle_{\gamma, \mathcal{F}}$. We have accomplished our goal (17.8) in the case of the term $I_{\gamma, \mathcal{F}}$ where γ is a triped, and similarly also when γ is a biped.

Another benefit of introducing the split deformed method is that it clarifies one aspect of the counter-term method which is rather hard to grasp in physics books. The counter-term is the sum of different terms each of which fills a specific purpose (which we can describe here as removing the overall divergence of a certain diagram).

In the next section we will achieve our goal (17.8) also for the terms $II_{\gamma, \mathcal{F}}$. This will prove (assuming that the BPHZ method works) that the previous choices for the counter-terms succeed in removing all divergences. However before that the reader should think a bit by herself and try to solve the next two exercises. Their purpose is to show that in two simple cases a straightforward use of the counter-term method enforcing the renormalization prescription (17.7) yields exactly the previous values of B, C, D at the lowest order in g.

Exercise 17.6.3 The simplest triped with a divergence is the triangle diagram α_0 (three internal vertices and three internal edges). What is the value of this triped? What is the value of D which achieves the first condition in (17.7) at order three? How does it compare with the value (17.15)?

Exercise 17.6.4 Consider the simplest biped, the "eye diagram" β_0 with six-momentum flow p through the external legs. What is the value of this diagram? How do you choose B and C to enforce the second part of (17.7) at order g^2?

17.7 From BPHZ to the Counter-term Method

In this section we complete the "proof" of the following.[34] Assuming that the BPHZ method works, the method of the previous section determine counter-terms that cancel the divergences (in the sense explained in Section 17.5) in such a manner that the resulting deformed theory predicts exactly the same scattering amplitudes as the BPHZ method and that the condition (17.7) holds. This is the content of Theorem 17.7.1.[35] It is in fact true that if we tried to determine the counter-terms recursively to cancel the divergences while implementing the renormalization conditions (17.7) we would be led exactly to the same choice of counter-terms, but we will not show this.

[34] Modulo the fact that we take for granted that symmetry factors work out.
[35] One more time, I am shamefully cheating here, since computing amplitudes requires considering also some 1-PI diagrams, which I omit, but doing this requires only minor complications which I want to avoid.

Theorem 17.7.1 *The sum of the renormalized values of the ordinary 1-PI Feynman diagrams with given external lines, and given momenta through these lines, is the sum of the values of the 1-PI Feynman diagrams for the deformed theory with the same external lines and momenta through these lines.*

The principle of the proof is to show that the renormalized value of a given Feynman diagram is the sum of the values of certain diagrams of the split deformed theory with the same external lines and the same momenta through these external lines. It will be basically obvious from the construction that as we sum over the ordinary 1-PI Feynman diagrams with given external lines and given momenta through these lines, we obtain exactly all the Feynman diagrams in the deformed theory with the same external lines once and only once, thereby proving the theorem.[36]

So, let us now fix a 1-PI Feynman diagram with n internal vertices and given momenta through the external lines. As usual γ is the largest subdiagram γ, $\mathcal{V}$ is the set of internal vertices and $\mathcal{E}$ is the set of internal edges. As we have recalled in the previous section the renormalized value of the Feynman diagram is given as a sum $\sum_{\mathcal{F}}(\mathrm{I}_{\gamma,\mathcal{F}} + \mathrm{II}_{\gamma,\mathcal{F}})$ over all forests. The content of Lemma 17.6.1 is that, unless $\mathrm{I}_{\gamma,\mathcal{F}}$ is zero, this term equals the value of a certain diagram with one single counter-term vertex in the split deformed theory.

So we have only to consider the term $\mathrm{II}_{\gamma,\mathcal{F}}$. Our goal is to show that this term is the value of a certain diagram in the split deformed theory (completing our goal (17.8)). When the forest $\mathcal{F}$ consists of the single subdiagram γ, then $\mathcal{F}F = F$, and then $\mathrm{II}_{\gamma,\mathcal{F}}$ is just the value of the diagram, seen as an ordinary diagram in the deformed theory.

Thus we may assume that $\mathcal{F}$ does not consist only of the subdiagram γ. Then, as in Lemma 16.5.3, consider the maximal subdiagrams $\alpha_1, \ldots, \alpha_{\tilde{n}}$ of $\mathcal{F}\backslash\{\gamma\}$. As we may assume that $\mathcal{F}F \neq 0$, these are all renormalization sets: they are all bipeds or tripeds. For each such biped or triped α_i we define

$$\mathcal{F}_i = \{\xi \in \mathcal{F}\ ;\ \xi \subset \alpha_i\},$$

which is a forest on α_i since $\alpha_i \in \mathcal{F}$. We then define a diagram in the split deformed theory by *contracting* each triped α_i into the vertex $\triangle_{\alpha_i,\mathcal{F}_i}$ and each biped α_i into the vertex $\otimes_{\alpha_i,\mathcal{F}_i}$. This diagram will be called the *contracted diagram*, see Figure 17.4.

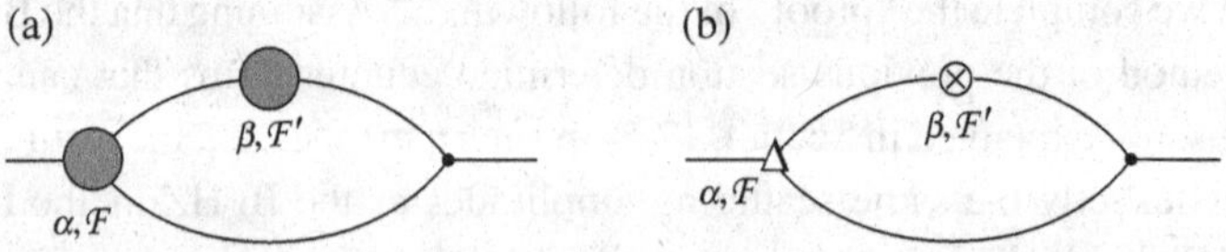

Figure 17.4 Contracting a triped and a biped. In (a) the shaded areas represent whole subdiagrams, which in (b) are replaced by a single counter-term vertex in the split deformed theory.

36 As we have already waived our hands in Proposition 16.9.1 into believing that the conditions (17.7) hold for the BPHZ method, we will not come back to that point. The drastic simplifying steps of the previous section actually help toward the arguments of Proposition 16.9.1.

Proposition 17.7.2 *Modulo a multiplicative factor depending on the symmetry factor, the quantity* $\mathrm{II}_{\gamma,\mathcal{F}}$ *equals the value of the contracted diagram.*

Proof By definition (and forgetting in this proof all the symmetry factors) we have

$$\mathrm{II}_{\gamma,\mathcal{F}} = (-\mathrm{i}g)^n \int \cdots \int \prod_{e\in\mathcal{E}} \frac{\mathrm{d}^6 x_e}{(2\pi)^6} \mathcal{F}F(x) \prod_{v\in\mathcal{V}} (2\pi)^6 \delta^{(6)}(\mathcal{L}(x)_v - w_v), \tag{17.16}$$

where as before w_e is the sum of the six-momenta leaving e through the external lines of γ. Combining (16.17), (16.18) and (16.19) we obtain

$$\mathcal{F}F(x) = \prod_{i\le\tilde{n}} (-T^{d(\alpha_i)} F_{\alpha_i})(U_i(x)) \prod f(x_e), \tag{17.17}$$

where the last product is over the edges e of γ which are not edges of any of the α_i.

To compute the integral in (17.16), we may first, for each $i \le \tilde{n}$, integrate in the variables x_e for $e \in \mathcal{E}_{\alpha_i}$. We claim that if we perform these integrations, we obtain the value of the contracted diagram in the split deformed theory, proving the required identity. To see how this goes, denoting by n_i the number of vertices of the diagram α_i, let us consider the quantity

$$\mathcal{I}_i := (-\mathrm{i}g)^{n_i} \int \cdots \int \prod_{e\in\mathcal{E}_{\alpha_i}} \frac{\mathrm{d}^6 x_e}{(2\pi)^6} (-T^{d(\alpha_i)} F_{\alpha_i})(U_i(x)) \prod_{v\in\mathcal{V}_{\alpha_i}} (2\pi)^6 \delta^{(6)}(\mathcal{L}(x)_v - w_v). \tag{17.18}$$

We note that for $v \in \mathcal{V}_i$ the quantity $\mathcal{L}_v(x)$ depends only on that x_e where the edge e has an endpoint in α_i. Thus the quantity $\mathcal{I}_i$ depends only on the x_e for which the edge e has an endpoint in α_i but is not an edge of α_i.

Let us denote by $n^* := n - \sum_{i\le\tilde{n}} n_i$ the number of vertices of γ which are not vertices of any α_i, by $\tilde{\mathcal{E}}$ the set of edges of γ which are not edges of any α_i and by $\tilde{\mathcal{V}}$ the set of vertices of γ which are not vertices of any α_i. Thus, if we integrate in (17.16) in the variables x_e for $e \in \mathcal{E}_i$ and each $i \le \tilde{n}$ we obtain the identity

$$\mathrm{II}_{\gamma,\mathcal{F}} = (-\mathrm{i}g)^{n^*} \int \cdots \int \prod_{e\in\tilde{\mathcal{E}}} \frac{\mathrm{d}^6 x_e}{(2\pi)^6} \prod_{i\le\tilde{n}} \mathcal{I}_i \prod_{e\in\tilde{\mathcal{E}}} f(x_e) \prod_{v\in\tilde{\mathcal{V}}} (2\pi)^6 \delta^{(6)}(\mathcal{L}(x)_v - w_v), \tag{17.19}$$

and we have to show that this is precisely the value of the contracted diagram. We prove first that within a symmetry factor

$$\mathcal{I}_i = (2\pi)^6 \delta^{(6)}(\Sigma_i) G_i, \tag{17.20}$$

where Σ_i is the algebraic sum of the six-momenta leaving the subdiagram α_i and $G_i = -\mathrm{i}D_{\alpha_i,\mathcal{F}_i}$ for tripeds and $G_i = -\mathrm{i}(p^2 B_{\alpha_i,\mathcal{F}_i} + C_{\alpha_i,\mathcal{F}_i})$ for bipeds, where p is the flow out of the biped. Writing $y = U_i(x) = (x_e)_{e\in\mathcal{E}_i}$ and denoting by $\mathcal{L}_i$ the map $\mathcal{L}$ corresponding to the diagram α_i, for $v \in \mathcal{V}_{\alpha_i}$ we claim that we have the identity

$$\mathcal{L}(x)_v - w_v = \mathcal{L}_i(y)_v - w_v^i, \tag{17.21}$$

where w_v^i denotes the sum of the six-momenta leaving v through the external lines of α_i (that is all the edges of γ which are not edges of α_i but touch α_i). To see this, assume for example that there is exactly one edge e of γ which is adjacent to v but is not an edge of α_i and that this edge is oriented toward v. Thus six-momentum x_e enters v through e, and since e is not an internal edge of α_i we have $\mathcal{L}(x)_v = \mathcal{L}_i(y)_v + x_e$. Furthermore, e is an external edge of α_i and six-momentum $-x_e$ leaves α_i through the edge e, so that $w_v^i = w_v - x_e$, i.e. $w_v = w_v^i + x_e$, completing the proof of (17.21). This relation implies that the quantity $\mathcal{I}_i$ of (17.18) then equals

$$(-\mathrm{i}g)^{n_i} \int \cdots \int \prod_{e\in\mathcal{E}_{\alpha_i}} \frac{\mathrm{d}^6 x_e}{(2\pi)^6} (-T^{d(\alpha_i)} F_{\alpha_i})(y) \prod_{v\in\mathcal{V}_{\alpha_i}} (2\pi)^6 \delta^{(6)}(\mathcal{L}_i(y)_v - w_v^i). \qquad (17.22)$$

The formula (17.11) shows that up to the symmetry factor, this is just the quantity $\mathrm{I}_{\alpha_i,\mathcal{F}_i}$ so that the equality (17.20) follows from the identities (17.12) and (17.13) of Lemma 17.6.1.

Finally it remains to convince ourselves that, taking (17.20) into account, the right-hand side of (17.19) is the value of the contracted diagram in the split deformed theory. This is because the term G_i in (17.20) is exactly the factor associated to the corresponding counter-term vertex by the Feynman rules, and the delta function in (17.20) exactly enforces conservation of flow at the counter-term vertices of the contracted diagram. □

In this manner we have shown that the renormalized value of a 1-PI Feynman diagram with n internal vertices is in a natural manner the sum of the values of certain diagrams in the split deformed theory. These diagrams occur as contractions of the original diagram, and their counter-term vertices carry the information of which subdiagram was contracted, showing that they can arise from one and only one 1-PI diagram.

The issue of the symmetry factors is non-trivial (but we ignore it).[37] It is best explained on a simple example, in Figure 17.5. Starting with diagram (a) (with symmetry factor 8) if one considers the subgraphs {3,4} and {5,6} and contracts each of them, one gets diagram (b), with symmetry factor 2. Here each ⊗ vertex has a weight corresponding to the contracted diagram, which carries a symmetry factor 2 and the symmetry factors match because $8 = 2 \times (2)^2$. If on the other hand one contracts only the subgraph {3,4} one gets diagram (c), which has a symmetry factor of 2. Here the ⊗ vertex has a symmetry factor 2, so the symmetry factor of this graph is 2×2, which is only half of the symmetry factor of

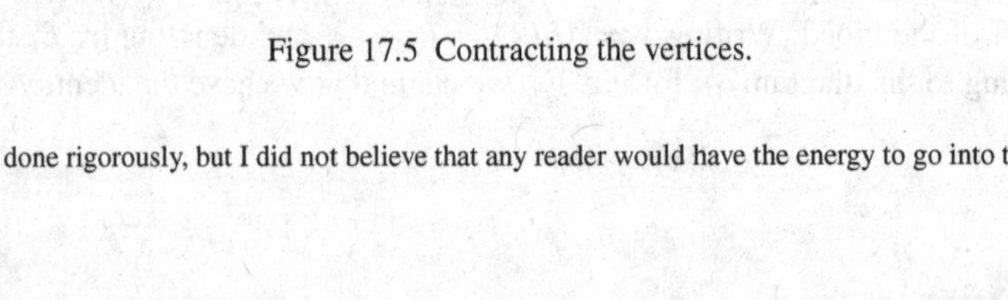

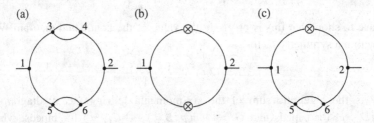

Figure 17.5 Contracting the vertices.

[37] It can be done rigorously, but I did not believe that any reader would have the energy to go into this at this stage.

the graph (a). The missing factor 2 is provided by the fact that if in the graph (a) one now contracts only the subgraph $\{5,6\}$ one gets again the graph (c), drawn differently.

We have not addressed the secondary issue of tadpoles (i.e. diagrams with a single external leg). The value of each tadpole is set to zero in the BPHZ method. To make matters consistent, the value of tadpoles has to be set equal to zero also in the deformed theory. There are different levels of sophistication to do this. Our crude method just declares by fiat that this is the case. A more elaborate method is to introduce yet another type of counter-term vertex with a single leg. The coefficient of this vertex is adjusted order by order in g to ensure that the value of each tadpole is zero. This formalism wonderfully matches what we did for the vertices $\otimes$ and Δ, but it does not really prove anything either.

17.8 What Happened to Subdiagrams?

In this section we perform the work of Section 17.7 in the case of ϕ_4^4 theory. It is stated everywhere in physics books that "counter-terms cancel the divergences in sub*graphs*". Since the BPHZ method involves sub*diagrams*, there is a little riddle there.[38] Where did the subdiagrams go? In words, as will become clearer with the proof, the answer to that riddle is that the contributions of the various subdiagrams can be grouped together according to the value of the smallest subgraphs which contain them.

The central fact on which the analysis relies is that (as a consequence of (15.41)) as we have seen in Exercise 15.4.12 that given a subdiagram α with superficial degree of divergence $d = d(\alpha) \geq 0$, either $d = 2$ and then α is a biped, or else $d = 0$ and either α is a quadruped or else α is obtained from a biped $\bar{\alpha}$ just by removing one single edge of $\bar{\alpha}$.

A broad overview of the counter-term method in ϕ_4^4 theory is that it involves two counter-term vertices, our familiar vertex $\otimes$, and a four-legged vertex $\square$ which replaces the three-legged vertex Δ, with Feynman rules similar to those of Section 17.5.

To understand the difference between the ϕ_4^4 theory and the ϕ_6^3 theory, we examine the proof of Theorem 17.7.1 and Proposition 17.7.2. We fix a 1-PI Feynman diagram with n internal vertices and largest subdiagram γ and we examine how to take care of the term $\mathrm{II}_{\gamma,\mathcal{F}}$ when $\mathcal{F}$ does not consist only of the subdiagram γ. Then, as in Lemma 16.5.3, consider the maximal subdiagrams $\alpha_1, \ldots, \alpha_{\tilde{n}}$ of $\mathcal{F} \setminus \{\gamma\}$. For each α_i we define

$$\mathcal{F}_i = \{\xi \in \mathcal{F}\,;\ \xi \subset \alpha_i\},$$

which is a forest on α_i since $\alpha_i \in \mathcal{F}$. We also consider the smallest subgraph $\bar{\alpha}_i$ containing α_i, and we note that these subgraphs have disjoint sets of vertices.[39] As we may assume that $\mathcal{F}F \neq 0$, all the α_i are renormalization sets. Thus for each i there are three cases:

- $\alpha_i = \bar{\alpha}_i$ is a quadruped, and $d(\alpha_i) = 0$.
- $\alpha_i = \bar{\alpha}_i$ is a biped and $d(\alpha_i) = 2$.
- $\bar{\alpha}_i$ is a biped, α_i is obtained from $\bar{\alpha}_i$ by removing a single edge e_i of $\bar{\alpha}_i$ and $d(\alpha_i) = 0$.

[38] I could not find this clarified in any of the textbooks I used. Even ambitious works such as that of Duncan [24] entirely ignore it.

[39] It might well happen that $\tilde{n} = 1$ and that $\bar{\alpha}_1 = \gamma$.

We will contract each *subgraph* $\bar{\alpha}_i$ to a vertex in the split deformed theory. You will not be surprised that in the first case $\bar{\alpha}_i = \alpha_i$ is contracted to a vertex $\square_{\alpha_i,\mathcal{F}_i}$, and in the second case it is contracted to a vertex $\otimes_{\alpha_i,\mathcal{F}_i}$, but the third case creates a new situation where $\bar{\alpha}_i$ is contracted to new type of vertex $\tilde{\otimes}_{\bar{\alpha}_i,\mathcal{F}_i}$. These new vertices $\tilde{\otimes}_{\beta,\mathcal{F}}$ are defined when β is a biped and $\mathcal{F}$ is a forest, not on β itself, but on a subdiagram α obtained from β by removing a single edge e. As in the proof of Proposition 17.7.2 for each $i \le \tilde{n}$ we will integrate in the variables x_e for $e \in \mathcal{E}_{\bar{\alpha}_i}$. We concentrate on the case where $\bar{\alpha}_i \ne \alpha_i$ which is the new feature. Recalling (17.17) the dependence of $\mathcal{F}F$ in the variables x_e, $e \in \bar{\alpha}_i$ is through

$$G_i(x) := (-T^{d(\alpha_i)}F_{\alpha_i})(U_i(x))f(x_{e_i}) = -T^0 F_{\alpha_i}(U_i(x))f(x_{e_i})$$

where e_i is the edge of $\bar{\alpha}_i$ which does not belong to α_i, and where the second equality uses that $d(\alpha_i) = 0$. Denoting by n_i the number of vertices of $\bar{\alpha}_i$ we will prove the following:

Lemma 17.8.1 *Within a symmetry factor, the quantity*

$$(-\mathrm{i}g)^{n_i} \int \cdots \int \prod_{e \in \mathcal{E}_{\bar{\alpha}_i}} \frac{\mathrm{d}^4 x_e}{(2\pi)^4} G_i(x) \prod_{v \in \mathcal{V}_{\alpha_i}} (2\pi)^4 \delta^{(4)}(\mathcal{L}(x)_v - w_v) \tag{17.23}$$

equals

$$-\mathrm{i}(2\pi)^4 \delta^{(4)}(\Sigma_i) \tilde{C}_{\bar{\alpha}_i,\mathcal{F}_i}, \tag{17.24}$$

where $\tilde{C}_{\bar{\alpha}_i,\mathcal{F}_i}$ is a number (depending on $\bar{\alpha}_i$ and on $\mathcal{F}_i$ but independent of the flow of four-momentum through $\bar{\alpha}_i$).

To prove this lemma we first reproduce the argument which took us from (17.18) to (17.22), and then we use Theorem 15.7.2 to reduce the proof to the following.

Lemma 17.8.2 *Consider a Feynman diagram β with exactly two external edges and a subdiagram α obtained from β by removing a single internal edge e. Let $\mathcal{E}$ be the set of internal edges of β and $\mathcal{E}' = \mathcal{E} \setminus \{e\}$ the set of edges of α. Consider the projection U from $(\mathbb{R}^{1,3})^{\mathcal{E}}$ to $(\mathbb{R}^{1,3})^{\mathcal{E}'}$ and a function H on $(\mathbb{R}^{1,3})^{\mathcal{E}'}$. Consider the function $G(x) = T^0 H(U(x)) f(x_e)$. Then if $\chi(p) \in \mathcal{Q} = \ker \mathcal{L}^{\perp}$ the integral*

$$\int \mathrm{d}\mu_{\mathcal{L}}(x) G(x + \chi(p)) \tag{17.25}$$

is independent of p.

We first explain why this lemma is true in the simple case of the biped β of Figure 17.6. Let us denote by e_0, e_1, e_2 the three internal edges (starting from the top), and consider the subdiagram α with edges e_1 and e_2 with flows $\ell - k/2 + p/3$ and $-\ell - k/2 + p/3$. This subdiagram is the "eye" considered in formula (16.13). According to this formula, we have $T^0 H(\ell - k/2 + p/3, -\ell - k/2 + p/3) = H(\ell, -\ell)$. Then

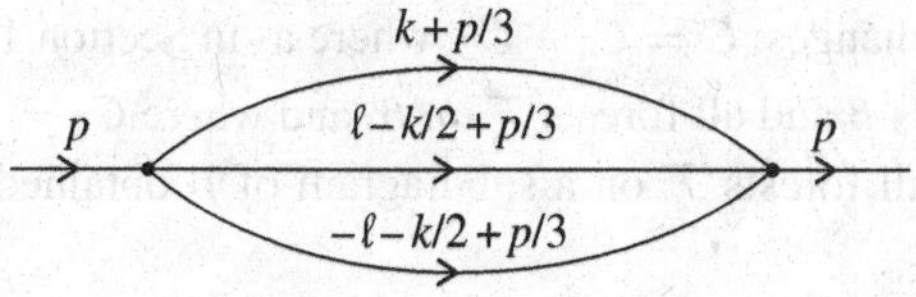

Figure 17.6 A simple biped.

$$\begin{aligned}\int \mathrm{d}\mu_{\mathcal{L}}(z)G(z+\mathcal{X}(p)) &= \iint \frac{\mathrm{d}^4k}{(2\pi)^4}\frac{\mathrm{d}^4\ell}{(2\pi)^4} f(k+p/3)H(\ell,-\ell) \\ &= W\int \frac{\mathrm{d}^4k}{(2\pi)^4} f(k+p/3),\end{aligned}$$

for a certain quantity W which does not depend on p. As desired (at least if we keep pretending that our implicit cutoff has magic properties which it does not really have) this quantity is *independent of* p as seen by the change of variable $k \to k - p/3$.[40]

Proof of Lemma 17.8.2 Let us consider the map $V : (\mathbb{R}^{1,3})^{\mathcal{E}'} \theta o (\mathbb{R}^{1,3})^{\mathcal{E}}$ such that $V(y)_e = 0$ and $U(V(y)) = y$. Consider the map $\mathcal{L}$ of (15.30) and the corresponding map $\mathcal{L}'$ relative to α. It should be obvious that $V(\ker \mathcal{L}') \subset \ker \mathcal{L}$. Let us consider an orthogonal decomposition $\ker \mathcal{L} = E \oplus V(\ker \mathcal{L}')$. For $z \in \ker \mathcal{L}$ we can then decompose $z = z^1 + z^2$ where $z^1 \in E$ and $z^2 \in V(\ker \mathcal{L}')$. We claim that

$$(T^0 H)(U(z+\mathcal{X}(p))) = (T^0 H)(U(z^2)). \tag{17.26}$$

According to (16.12), with obvious notation, it suffices to prove that $\mathcal{I}'(U(z+\mathcal{X}(p))) = \mathcal{I}'(U(z^2))$, i.e. $\mathcal{I}'(U(z^1+\mathcal{X}(p))) = 0$, which by definition of $\mathcal{I}'$ means that $U(z^1+\mathcal{X}(p))$ is orthogonal to $\ker \mathcal{L}'$. Now, if $x \in (\mathbb{R}^{1,3})^{\mathcal{E}}$ and $y \in (\mathbb{R}^{1,3})^{\mathcal{E}'}$ it is obvious that the inner product of $U(x)$ and y is the same as the inner product of x and $V(y)$. Thus it suffices to show that $z^1+\mathcal{X}(p)$ is orthogonal to $V(\ker \mathcal{L}')$. For z^1 this is by construction, and for $\mathcal{X}(p)$ this is because $\mathcal{X}(p)$ is already orthogonal to the larger space $\ker \mathcal{L}$.

We have proved (17.26), and consequently we have obtained the formula

$$\int \mathrm{d}\mu_{\mathcal{L}}(z)G(z+\mathcal{X}(p)) = \int \mathrm{d}\mu_{\mathcal{L}}(z)f(z^1_e+\mathcal{X}(p)_e)W(z^2) \tag{17.27}$$

for a certain function W of z^2, where we have also used the fact that since $z^2 \in V(\ker \mathcal{L}')$ we have $z^2_e = 0$, so that $(z+\chi(p))_e = z^1_e + \chi(p)_e$. To the decomposition $\ker \mathcal{L} = E \oplus V(\ker \mathcal{L}')$ corresponds a decomposition $\mathrm{d}\mu_{\mathcal{L}} = \mathrm{d}\mu_1 \otimes \mathrm{d}\mu_2$, and using Fubini's theorem (17.27) implies $\int \mathrm{d}\mu_{\mathcal{L}}(z)G(z+\mathcal{X}(p)) = \int \mathrm{d}\mu_1(z_1)f(z^1_e+\mathcal{X}(p)_e)\int \mathrm{d}\mu_2(z^2)W(z^2)$. The first factor on the right is $\int \mathrm{d}\mu'(u)f(u+\mathcal{X}(p)_e)$ where μ' is the image of μ_1 under the map $z^1 \mapsto z^1_e$. Since $\mathrm{d}\mu'$ is obviously translation invariant, a change of variable shows as desired that the previous quantity is independent of p. □

The Feynman rules on page 526 for the deformed theory are modified as follows (besides the obvious replacement of $\triangle$ by $\square$): To each $\otimes$ vertex we still assign a factor $-\mathrm{i}(p^2 B + C)$,

[40] Going to the Euclidean setting, one may use dimensional regularization to make things work, see e.g. [15, Chapter 4].

but the definition of C changes: $C = C_1 + C_2$, where as in Section 17.6, $C_1 = \sum_{\beta,\mathcal{F}} C_{\beta,\mathcal{F}}$ for a sum over all bipeds β and all forests $\mathcal{F}$ on β, and where $C_2 = \sum_{\beta,\mathcal{F}} \tilde{C}_{\beta,\mathcal{F}}$, for a sum over all bipeds β and all forests $\mathcal{F}$ on a subdiagram of β obtained by removing a single edge of β.

17.9 Field Renormalization, II

In Section 17.3, just before Lemma 17.3.3, we have argued that deforming a theory by introducing the appropriate new types of vertices amounts to "renormalizing the field" (even though there is no field in the deformed theory which is defined only at the level of diagrams). To complete our mental picture, it would be nice to argue in a more direct way that field renormalization (in the sense of making the change of variable $\varphi \to \sqrt{Z}\varphi$ as we do below) amounts to introducing these new vertices. This, unfortunately, is difficult to do in our approach. The standard treatment of this question which is followed in all the main textbooks is through the technique of *path integrals*, also called *Feynman's integrals*. (A particularly clear presentation is given in Srednicki's book [77].) Unfortunately, path integrals are mathematically ill defined,[41] so we cannot follow this route. The goal of the present section is to explain in broad terms what happens within our formalism when we "renormalize the field". The reader needs to be fluent with Sections 6.4 and 6.5. In this section, we truly graduate as physicists: we use a system of units where $\hbar = 1$ and $c = 1$.

At the classical level, the Lagrangian density of ϕ^3 theory is given by

$$\mathcal{L} = \frac{1}{2}\partial^\mu\varphi\partial_\mu\varphi - \frac{1}{2}m^2\varphi^2 - \frac{g}{3!}\varphi^3.$$

After renormalization $\varphi \to Z^{1/2}\varphi$ this gives the Lagrangian density

$$\mathcal{L} = \frac{Z}{2}\partial^\mu\varphi\partial_\mu\varphi - \frac{Z}{2}m^2\varphi^2 - \frac{gZ^{3/2}}{3!}\varphi^3.$$

so that, writing $\dot\varphi = \partial\varphi/\partial x^0$,[42]

$$\pi := \frac{\partial\mathcal{L}}{\partial\dot\varphi} = Z\dot\varphi.$$

Thus $\dot\varphi = \pi/Z$. The Hamiltonian density is then given by

$$\mathcal{H} = \pi\dot\varphi - \mathcal{L} = \frac{1}{2Z}\pi^2 + \frac{Z}{2}\sum_{1\le i\le 3}(\partial_i\varphi)^2 + \frac{Z}{2}m^2\varphi^2 + \frac{gZ^{3/2}}{3!}\varphi^3$$

which leads us to write $\mathcal{H} = \mathcal{H}_0 + \mathcal{H}_{\text{int}}$ where

$$\mathcal{H}_0 = \frac{1}{2}\pi^2 + \frac{1}{2}\sum_{1\le i\le 3}(\partial_i\varphi)^2 + \frac{1}{2}m^2\varphi^2, \tag{17.28}$$

$$\mathcal{H}_{\text{int}} = (Z-1)\Big(-\frac{1}{2Z}\pi^2 + \frac{1}{2}\sum_{1\le i\le 3}(\partial_i\varphi)^2\Big) + \frac{Z-1}{2}m^2\varphi^2 + \frac{gZ^{3/2}}{3!}\varphi^3. \tag{17.29}$$

41 Which is why we did not use them.
42 Since $c = 1$ is our system of units!

Crossing our fingers, we can hope that after quantization, the proper interaction Hamiltonian density should be given by the formula (17.29), where now $\pi = \dot{\varphi}$ (according to (17.28)).

It is absolutely not obvious that such an interaction Hamiltonian gives rise to a Lorentz invariant theory,[43] and that the scattering amplitudes in this theory should be computed by the diagrams in the deformed theory which we previously described. In fact one could easily form the wrong impression that the theory with interacting Hamiltonian arising from the Hamiltonian density (17.29) cannot be Lorentz invariant, because the term $\pi = \dot{\varphi} = \partial_0\varphi$ does not play the same role as the terms $\partial_i\varphi$.[44] Matters are tricky however. When trying to extend the method of Feynman diagrams to this theory, applying Wick's theorem we meet the contractions $\langle 0|\mathcal{T}\dot{\varphi}(x)\dot{\varphi}(y)|0\rangle$. To calculate these we need the following:

Lemma 17.9.1 *We have*

$$\mathcal{T}\dot{\varphi}(x)\dot{\varphi}(y) = \frac{\partial^2}{\partial x^0 \partial y^0}\mathcal{T}\varphi(x)\varphi(y) - i\delta^{(4)}(x-y)1.$$

Proof Consider the formula

$$\mathcal{T}\varphi_1(x)\varphi_2(y) = \theta(x^0 - y^0)\varphi_1(x)\varphi_2(y) + \theta(y^0 - x^0)\varphi_2(y)\varphi_1(x)$$

where for a real number u, $\theta(u) = 1$ if $u \geq 0$ and $\theta(u) = 0$ if $u < 0$.[45] This function has the property that its derivative is the delta function. Thus

$$\begin{aligned}\frac{\partial}{\partial x^0}\mathcal{T}\varphi_1(x)\varphi_2(y) &= \theta(x^0 - y^0)\frac{\partial}{\partial x^0}\varphi_1(x)\varphi_2(y) + \theta(y^0 - x^0)\varphi_2(y)\frac{\partial}{\partial x^0}\varphi_1(x)\\ &\quad + \delta(x^0 - y^0)(\varphi_1(x)\varphi_2(y) - \varphi_2(y)\varphi_1(x))\\ &= \mathcal{T}\frac{\partial}{\partial x^0}\varphi_1(x)\varphi_2(y) + \delta(x^0 - y^0)[\varphi_1(x), \varphi_2(y)]. \qquad (17.30)\end{aligned}$$

Using this for $\varphi_2 = \dot{\varphi}$ and $\varphi_1 = \varphi$ gives

$$\begin{aligned}\mathcal{T}(\dot{\varphi}(x)\dot{\varphi}(y)) &= \frac{\partial}{\partial x^0}\mathcal{T}\varphi(x)\dot{\varphi}(y) - \delta(x^0 - y^0)[\varphi(x), \dot{\varphi}(y)]\\ &= \frac{\partial}{\partial x^0}\mathcal{T}\varphi(x)\dot{\varphi}(y) - i\delta^{(4)}(x-y)1, \qquad (17.31)\end{aligned}$$

where we have used (6.62) in the second line. We use now (17.30) for $\varphi_1 = \varphi_2 = \varphi$. Since $\varphi(x)$ and $\varphi(y)$ commute at equal times by (6.63), the last term is zero and we obtain

$$\mathcal{T}\varphi(x)\dot{\varphi}(y) = \frac{\partial}{\partial y^0}\mathcal{T}\varphi(x)\varphi(y).$$

Combining with (17.31) yields the result. □

[43] Checking that a theory is Lorentz invariant is a kind of sanity check: A theory that is not Lorentz invariant cannot be correct.

[44] I am very grateful to Mark Srednicki for having pointed out to me Preskill's notes [67, pages 33–36], which explains the way out of this puzzle.

[45] This function is known as the Heaviside function.

As a consequence, recalling the Feynman propagator Δ_F we have the formula

$$\langle 0|\mathcal{T}\dot{\varphi}(x)\dot{\varphi}(y)|0\rangle = -\mathrm{i}\frac{\partial^2}{\partial x^0 \partial y^0}\Delta_F(x-y) - \mathrm{i}\delta^{(4)}(x-y),$$

and one can prove that the last term in this formula re-establishes Lorentz invariance. This proof is a definitely non-trivial task, as one may group the terms in seemingly miraculous ways.[46] As the purpose of the present approach is mainly to perform a sanity check, we will not enter this proof. The mechanism of the computation is already apparent in the following case, which we leave as a challenge to the reader.

Exercise 17.9.2 Consider a parameter $g \neq -1$ and assume as in Section 13.3 that the interacting Hamiltonian is given by the formula $H_I = \int \mathrm{d}^3\boldsymbol{x}\mathcal{H}_{\mathrm{int}}(0,\boldsymbol{x})$ where

$$\mathcal{H}_{\mathrm{int}} = g\Big(-\frac{1}{2(1+g)}(\partial_0\varphi)^2 + \frac{1}{2}\sum_{1\leq i\leq 3}(\partial_i\varphi)^2\Big). \tag{17.32}$$

Consider the theory where the time-evolution in the interaction picture of the field φ is given by H_I. Check the formula (13.9) to ensure that this makes sense. Recall the dressed propagator Δ given by (14.65). Prove that (as a formal series in g) the Fourier transform of this dressed propagator is given by the formula

$$\widehat{\Delta}(p) = \lim_{\varepsilon\to 0+}\frac{1}{-p^2(1+g)+m^2-\mathrm{i}\varepsilon}.$$

Hint: Use the appropriate version of Proposition 14.5.3 to compute the order-two Green function. The value of the diagrams is computed using Wick's theorem, and it is convenient to develop a version of "Feynman's rules in momentum space". The relation $g = \sum_{k\geq 1}(g/(g+1))^k$ is crucial.

Even after one has shown that the Hamiltonian density (17.29) leads to a Lorentz invariant theory, it is still absolutely not obvious that this theory predicts the same scattering amplitudes as the deformed theory obtained by adding new vertices Δ and $\otimes$ as in Sections 17.2 and 17.3. To see this, one has to perform a "re-summation" as on page 521 on both the deformed theory and the theory with Hamiltonian density (17.29).

In conclusion, matters are consistent. Field renormalization (in the sense of the present section, making the change of variable $\varphi \to \sqrt{Z}\varphi$) does yield a theory equivalent to the theory of Sections 17.2 and 17.3, but it does not do it in a clean way. This is the price to pay for using Hamiltonian formalism, which breaks Lorentz invariance.[47]

Key ideas to remember:

- The introduction of new types of vertices in the counter-term method increases the number of diagrams to be considered when computing scattering amplitudes.

[46] One may wonder if there exists a mathematical structure explaining this magic.

[47] In contrast, the path integral formalism respects this invariance at every step. Path integrals are however not mathematically defined, which is why we have not used them at all.

- The same theory may be parameterized in many different ways.
- The parameterization is fixed by a set of rules called a renormalization prescription.
- The physicists use counter-terms to cancel the divergences. When the suitable renormalization prescription is used to fix the finite part of the counter-terms this gives the same result as the BPHZ method.

18

Controlling Singularities

The task ahead of us is to prove Theorem 16.1.1. There are two steps:

- Prove that replacing the integrand F by its regularization $\mathcal{R}F$ actually removes all divergences.
- Having proved the above, prove that there is a well-defined limit as $\varepsilon \to 0^+$.

The first item above is the most difficult, and is the object of the final Chapter 19. In the present chapter we perform the comparatively easier task of proving the existence of the limit as $\varepsilon \to 0^+$.

18.1 Basic Principle

We first bring forward some general ideas. Ultimately all results in the present chapter rely on the following simple principle.

Lemma 18.1.1 *Consider a function $\varphi(x)$ for $-1 \leq x \leq 1$ and an integer $s \geq 1$. Assume that φ is $s+1$ times differentiable. Then the limit*

$$\lim_{\varepsilon \to 0^+} \int_{-1}^{1} \mathrm{d}x \frac{\varphi(x)}{(x - \mathrm{i}\varepsilon)^s} \tag{18.1}$$

exists, and is bounded by a constant times $\sup\{|\varphi^{(k)}(x)| \; ; \; k \leq s \; , \; |x| \leq 1\}$.

Proof The proof relies on integration by parts. Each integration by parts reduces the singularity. After performing s such integrations by parts one gets a term $\int_{-1}^{1} \mathrm{d}x \varphi^{(s)}(x) \log(x - \mathrm{i}\varepsilon)$ for which the singularity is mild enough that the existence of the limit is obvious since the integral $\int_{-1}^{1} \mathrm{d}x |\log x|$ is finite. □

Exercise 18.1.2 Find a more general version of this principle where φ depends also on ε.

Exercise 18.1.3 Show that the limit

$$\lim_{\varepsilon \to 0^+} \int_{-1}^{1} \mathrm{d}x \frac{1}{x^2 - \mathrm{i}\varepsilon}$$

does not exist. Hint: Set $x = \sqrt{\varepsilon} t$.

Exercise 18.1.4 For which values of the integer s does the limit

$$\lim_{\varepsilon\to 0^+}\int_{\{\|y\|\le 1\}} \mathrm{d}^4 y \frac{1}{(\|y\|^2 - \mathrm{i}\varepsilon)^s}$$

exist, where $\|y\|^2 = \sum_{i\le 4} y_i^2$? Hint: Set $y = \sqrt{\varepsilon}x$ and use polar coordinates.

Exercise 18.1.5 Prove a multidimensional version of (18.1). Find conditions under which the limit

$$\lim_{\varepsilon\to 0^+}\iint_{\{|x|,|y|\le 1\}} \mathrm{d}x\mathrm{d}y \frac{\varphi(x,y)}{(x-\mathrm{i}\varepsilon)^s (y-\mathrm{i}\varepsilon)^{s'}}$$

exists.

Before we start to prove specific results, it is worth spending a page or so to explain the difficulties our program runs into, and why the "obvious approach" of combining Lemma 18.1.1 with a change of variables does not work in the situations of interest to us.

Let us consider the following general situation. Given two nice smooth real-valued functions U, V on the set $C := \{y \in \mathbb{R}^n \; ; \; \forall i \le n, |y_i| \le 1\}$, we want to prove the existence of the limit

$$\lim_{\varepsilon\to 0^+}\int_C \mathrm{d}^n y \frac{V(y)}{(U(y) - \mathrm{i}\varepsilon)^s}. \tag{18.2}$$

As $\varepsilon \to 0$ the denominator develops singularities on the set $W := \{y \in C; U(y) = 0\}$. Let us assume that this set is a nice smooth surface. The easy case is where

$$\forall y \in W\,,\ \nabla U(y) \neq 0, \tag{18.3}$$

where ∇U denotes the gradient of U. In that case we can cover the set W by small boxes with disjoint interiors, in such a manner that on each of them we can make a change of variables so that $U(y)$ is one of the new coordinates. Integrating in the other coordinates, we are reduced to the situation of Lemma 18.1.1, and in this case the limit (18.2) exists as soon as sufficiently many derivatives of the function V exist.

As Exercise 18.1.3 shows, things however may go wrong when (18.3) fails. As we explain now, (18.3) fails in the situation we would really like to control. To see this let us try to prove a rigorous version of (13.169). Certainly this will lead us to consider the quantity

$$A_\varepsilon(k) = \int \frac{\mathrm{d}^4\ell}{(2\pi)^4}\theta(\boldsymbol{\ell})\frac{1}{(-\ell^2 + m^2 - \mathrm{i}\varepsilon)}\frac{1}{(-(k+\ell)^2 + m^2 - \mathrm{i}\varepsilon)}, \tag{18.4}$$

where $\theta(\boldsymbol{\ell})$ is a cutoff making the integral meaningful as in Section 13.18. We would like to show that this quantity has a limit as $\varepsilon \to 0^+$. We follow the physicists in transforming this quantity using Feynman parameters as in (13.123) and performing a Wick rotation,

$$A_\varepsilon(k) = \mathrm{i}\int_0^1 \mathrm{d}u \int \frac{\mathrm{d}^4\ell}{(2\pi)^4}\theta(\boldsymbol{\ell} + u\boldsymbol{k})\frac{1}{(\|\ell\|^2 - u(1-u)k^2 + m^2 - \mathrm{i}\varepsilon)^2}. \tag{18.5}$$

This certainly reminds us of (18.2), with the function $U(\ell, u) = \|\ell\|^2 - u(1-u)k^2 + m^2$. The really unfortunate fact is that in the case where $k^2 = 4m^2$, at the point $\ell = 0$ and $u = 1/2$

both the function U and its gradient are zero, and (18.3) fails.[1] For this reason one may not use the principles above, and deciding whether the limit of the quantity (18.5) exists would require more effort.

Exercise 18.1.6 For $k^2 = 4m^2$ decide if the previous integral has a limit as $\varepsilon \to 0^+$.

We will be able to bypass this problem by further smoothing in the external momenta. This will be demonstrated in the next section, but here we provide the main tool, Lemma 18.1.7. Let us recall that a quadratic form $Q(y)$ of $y \in \mathbb{R}^n$ is an expression of the type $\sum_{i,j\le n} a_{i,j} y_i y_j$, where $a_{i,j}$ are real numbers. This form is *positive semidefinite* if $Q(y) \ge 0$ for all $y \in \mathbb{R}^n$. It is *positive definite* if $Q(y) > 0$ unless $y = 0$. A fundamental result of elementary linear algebra is that any positive semidefinite quadratic form may be reduced to the form $\sum_{i\le n} \alpha_i y_i^2$ for numbers $\alpha_i \ge 0$ by a change of orthogonal basis.

Lemma 18.1.7 *Consider an infinitely differentiable function φ with compact support on $\mathbb{R}^n$. Then for any positive semidefinite quadratic form Q and any integer $s \ge 1$ the limit*

$$\lim_{\varepsilon\to 0+} \int \mathrm{d}^n y \frac{\varphi(y)}{(1 - \mathrm{i}\varepsilon - Q(y))^s} \tag{18.6}$$

exists. This limit is uniform over the choice of Q and of φ, as long as the support of φ stays in a given compact set and the partial derivatives of φ of order $\le s$ remain bounded.

A remarkable part of the lemma is that the limit is uniform in Q. No hypothesis is made concerning the "size" of Q. The lemma will follow from a similar result which assumes that "Q is bounded below", which we state now.

Lemma 18.1.8 *Consider an infinitely differentiable function φ on $\mathbb{R}^n$ with compact support, an integer $s \ge 1$ and a number $\alpha > 0$. Consider numbers $\alpha_i \ge \alpha$ for $i \le n$. Then the limit*

$$\lim_{\varepsilon\to 0+} \int \mathrm{d}^n y \frac{\varphi(y)}{\left(1 - \mathrm{i}\varepsilon - \sum_{i\le n} \alpha_i y_i^2\right)^s} \tag{18.7}$$

exists uniformly over φ as above and on the numbers $\alpha_i \ge \alpha$.[2]

Proof of Lemma 18.1.7 Making a change of orthonormal basis, we may then assume that $Q(y) = \sum_{i\le n} \alpha_i y_i^2$ where $\alpha_i \ge 0$ and the sequence (α_i) is non-increasing, $\alpha_1 \ge \alpha_2 \ge \ldots \ge \alpha_n$. Considering a number $B > 0$, we will prove the uniformity of the limit (18.7) over the functions φ with support contained in $[-B, B]^n$ such that their derivatives of order $\le s$ remain bounded, and over the numbers α_i. Set

$$\alpha = \frac{1}{2nB^2}.$$

[1] The case $k^2 = 4m^2$ has no special physical importance, and is just brought forward to show the limit of the "obvious" approach.
[2] But we see no reason why that limit would be uniform over $\alpha > 0$.

If all the coefficients α_i are $\geq \alpha$, we are in the situation of Lemma 18.1.8 and the proof is finished. Otherwise there exists an integer $r \leq n$ with

$$r \leq i \leq n \Rightarrow \alpha_i \leq \alpha.$$

Letting $y' = (y_1, \ldots, y_{r-1})$ it then suffices for fixed $y_r, \ldots, y_n$ to prove the existence (and uniformity) of the limit

$$\lim_{\varepsilon \to 0^+} \int \mathrm{d}^{r-1}y' \frac{\varphi(y)}{(1 - \mathrm{i}\varepsilon - \sum_{i \leq n} \alpha_i y_i^2)^s}. \tag{18.8}$$

We may assume here that for $r \leq i \leq n$ we have $|y_i| \leq B$ for otherwise $\varphi(y) = 0$. Then

$$\beta := \sum_{r \leq i \leq n} \alpha_i y_i^2 \leq \alpha n B^2 \leq 1/2,$$

and the quantity (18.8) equals

$$\lim_{\varepsilon \to 0^+} \int \mathrm{d}^{r-1}y' \frac{\varphi(y)}{(1 - \beta - \mathrm{i}\varepsilon - \sum_{i \leq r-1} \alpha_i y_i^2)^s}$$

$$= \frac{1}{(1-\beta)^s} \lim_{\varepsilon \to 0^+} \int \mathrm{d}^{r-1}y' \frac{\varphi(y)}{(1 - \mathrm{i}\varepsilon - \sum_{i \leq r-1} \alpha'_i y_i^2)^s}, \tag{18.9}$$

where $\alpha'_i = \alpha_i/(1-\beta) \geq \alpha$. The existence and uniformity of the limit then follows from Lemma 18.1.8 used with $r - 1$ rather than n. □

Proof of Lemma 18.1.8 It suffices to prove the existence and uniformity over φ (under the same conditions as in Lemma 18.1.7) and over the numbers $\alpha_i \geq \alpha$ of the limit

$$\lim_{\varepsilon \to 0^+} \int_D \mathrm{d}^n y \frac{\varphi(y)}{\left(1 - \mathrm{i}\varepsilon - \sum_{i \leq n} \alpha_i y_i^2\right)^s}, \tag{18.10}$$

where

$$D = \Big\{ y \in \mathbb{R}^n \; ; \; 1/2 \leq \sum_{i \leq n} \alpha_i y_i^2 \leq 2 \Big\},$$

because outside D the denominator never gets small as $\varepsilon \to 0$ (and because φ has compact support). Let us then define the linear operator A on $\mathbb{R}^n$ by $A(z)_i = z_i/\sqrt{\alpha_i}$. Making the change of variables $y = A(z)$ the integral in (18.10) is

$$\frac{1}{\sqrt{\prod_{i \leq n} \alpha_i}} \int_{\{1/2 \leq \sum_{i \leq n} z_i^2 \leq 2\}} \mathrm{d}^n z \frac{\varphi(A(z))}{\left(1 - \mathrm{i}\varepsilon - \sum_{i \leq n} z_i^2\right)^s}. \tag{18.11}$$

The important fact is that the size of the partial derivatives of the function $\psi(z) := \varphi(A(z))$ is controlled by the size of the partial derivatives of φ because $\alpha_i \geq \alpha$. In particular, it does not matter that certain α_i can be large. (This only improves the situation.) To finish the proof, it suffices to transform the integral (18.11) in polar coordinates (r, ξ), where ξ belongs to the unit sphere of $\mathbb{R}^n$, to set $x = 1 - r^2$ and to apply Lemma 18.1.1 to the integral in x (now from -1 to $1/2$) at given ξ. □

18.2 Zimmermann's Theorem

This theorem relates convergence of integrals in the Euclidean and Minkowski version of the propagators, and provides a final solution to the problem of singularities as $\varepsilon \to 0^+$.[3] We will state this theorem in a form tailored to prove that as $\varepsilon \to 0^+$ the integral

$$\int \mathrm{d}\mu_{\mathcal{L}}(x)\mathcal{R}F(x + \mathcal{X}(p)) \tag{18.12}$$

has a limit (although the actual application of our result to that case will only be done in Section 19.1).

Consider an integer $n \geq 1$ and for $j \leq n$ consider a four-momentum k_j. The idea here is that we have parameterized ker $\mathcal{L}$ using independent loops in the traditional way of (15.54). To lighten notation, integration with respect to all the k_j will be simply denoted by $\mathrm{d}^{4n}k$. We consider external momenta p_j, $j \leq r$, and in line with our synthetic notation we write $p = (p_j)_{j \leq r}$. We consider another integer s (which represents the number of propagators) and real numbers $a_{i,j}$, $1 \leq i \leq s$, $1 \leq j \leq n$ and $b_{i,j}$, $1 \leq i \leq s$, $1 \leq j \leq r$. As we will show in complete detail later, the integral (18.12) has the form

$$I(p,\varepsilon) = \int \mathrm{d}^{4n}k \frac{P(k,p,\varepsilon)}{\prod_{i \leq s}(W_\varepsilon(\sum_{j \leq n} a_{i,j}k_j + \sum_{j \leq r} b_{i,j}p_j) + m^2)}, \tag{18.13}$$

where $P(k,p,\varepsilon)$ is a polynomial, and where as usual $W_\varepsilon(s) = -(1+\mathrm{i}\varepsilon)(s^0)^2 + s^2$ for $s \in \mathbb{R}^{1,3}$. We recall that $\|\cdot\|$ denotes the Euclidean norm of a four-vector.

Theorem 18.2.1 (Zimmermann's theorem) *Assume that for each p and each $\varepsilon > 0$ the integral*

$$\int \mathrm{d}^{4n}k \frac{|P(k,p,\varepsilon)|}{\prod_{i \leq s}(\|\sum_{j \leq n} a_{i,j}k_j\|^2 + 1)} \tag{18.14}$$

is convergent. Then for each $\varepsilon > 0$ and each p the integral (18.13) is absolutely convergent. As $\varepsilon \to 0^+$, the quantity (18.13) converges to a distribution $I(p) = I(p_1, \ldots, p_r)$, in the sense that for each test function $\varphi \in \mathcal{S}^{4r}$ with compact support *we have $\lim_{\varepsilon \to 0^+} \int \mathrm{d}^{4r}pI(p,\varepsilon)\varphi(p) = \int \mathrm{d}^{4r}pI(p)\varphi(p)$.[4] Furthermore when $P(k,p,0)$ is Lorentz invariant, in the sense that for each Lorentz transformation A the value of $P(k,p,0)$ does not change if one replaces p_j and k_j by $A(p_j)$ and $A(k_j)$, then $I(p)$ is Lorentz invariant.*

In the case of the integral (18.12) the convergence of the integral (18.14) will be checked in the next chapter using Weinberg's power-counting theorem. It will clarify matters to explain right now how this is done.

Proposition 18.2.2 *Let us denote by $Q(k,p,\varepsilon)$ the denominator of the integrand in (18.13). Consider elements $(k_\ell)_{0 \leq \ell \leq N}$ of $(\mathbb{R}^{1,3})^n$ and assume that the $(k_\ell)_{1 \leq \ell \leq N}$*

[3] Provided that we have already solved the problem of the diverging integrals, as will become clear below.
[4] Using similar methods but more work one may remove the condition that φ has compact support. Here is not the place for these technicalities.

are independent. Consider variables $u_1, \ldots, u_N$ and the element $k(u) = k_0 + \sum_{1 \le \ell \le N} u_\ell k_\ell$. Then $P(k(u), p, \varepsilon)$ and $Q(k(u), p, \varepsilon)$ are polynomials in u. The degree of such a polynomial is denoted by $\deg_u$. Assume that for all the previous choices (with all values of p and all $\varepsilon > 0$) we have

$$\deg_u P(k(u), p, \varepsilon) - \deg_u Q(k(u), p, \varepsilon) < -N. \tag{18.15}$$

Then for each p and each $\varepsilon > 0$ the integral (18.14) is absolutely convergent.

Proof Let us denote by $Q'(k)$ the denominator of the integrand (18.14). The proof relies on the fact that $\deg_u Q'(k(u)) = \deg_u Q(k(u), p, \varepsilon)$. This holds because this degree is $2N_1$ where N_1 is the number of factors that actually depend on u i.e. for which $\sum_{j \le n} a_{i,j} k(u)_j$ is not independent of u.

Consider a subspace S of $(\mathbb{R}^{1,3})^n$. Let $N = \dim S$ and consider a basis $(k(\ell))_{1 \le \ell \le N}$ of S. Then, using (18.15) in the first inequality and the definition of $\deg_S Q'$ in the second one, we have

$$\deg_u P(k(u), p, \varepsilon) \le -\dim S + \deg_u Q'(k(u)) \le -\dim S + \deg_S Q'(k).$$

Taking the supremum over all possible choices of the $(k_\ell)_{0 \le \ell \le N}$ yields[5] $\deg_S P(k, p, \varepsilon) - \deg_S Q'(k) < -\dim S$, so that Weinberg's power-counting theorem implies the result. □

Remark 18.2.3 Given a rational function $R = P/Q$ where P and Q are polynomials in variables $u, p, \varepsilon, \ldots$ we will define

$$\deg_u R = \deg_u P - \deg_u Q$$

so that denoting by R the integrand of (18.13), the condition (18.15) can simply be written as

$$\deg_u R(k(u), p, \varepsilon) < -N.$$

During the proof of Theorem 18.2.1 we will furthermore assume that the matrix $(a_{i,j})$ is of rank n. This is not a restriction since in the integrand of (18.13) we can multiply both numerator and denominator by a quantity $\prod_{i \le n}(W_\varepsilon(\sum_{j \le n} a'_{i,j} k_j) + m^2)$ where the matrix $(a'_{i,j})$ is of rank n.

Exercise 18.2.4 Use Weinberg's power-counting theorem to prove that if a polynomial $P(k, p, \varepsilon)$ is such that the integral (18.14) is convergent for each p and each $\varepsilon > 0$, then $P(k, p, \varepsilon) = \sum_M P_M(k) M(p, \varepsilon)$ where $M(p, \varepsilon)$ are (different) monomials in p and ε, and each P_M is a polynomial in k such that the integral

$$\int \mathrm{d}^{4n} k \frac{|P_M(k)|}{\prod_{i \le s}(\| \sum_{j \le n} a_{i,j} k_j \|^2 + 1)} \tag{18.16}$$

is convergent.

[5] We denote by $\deg_S P(k, p, \varepsilon)$ the quantity $\deg_S P'$ where P' is the polynomial $k \mapsto P(k, p, \varepsilon)$.

The absolute convergence of the integrals (18.13) follows from (18.14) and the following elementary lemma.

Lemma 18.2.5 *Consider real numbers $c, d \geq 0$ and $0 < \alpha \leq 1/2$. Consider a complex number z. Assume that either $\mathrm{Re}\, z \geq 0$ or that $|\mathrm{Im}\, z| \geq \alpha |z|$. Then*

$$\frac{\alpha}{4}(c + d|z|) \leq |c + zd| \leq (c + d|z|). \tag{18.17}$$

In words: When z is fixed and is not a negative real number, for $c, d \geq 0$, $c + zd$ is about of the same size as $c + d|z|$.

Proof The second inequality is obvious so we prove the first one. We assume first that $\mathrm{Re}\, z \geq 0$. Then $|c+dz| \geq \mathrm{Re}(c+dz) = c + d\mathrm{Re}\, z$. Assume first that $\mathrm{Re}\, z = |\mathrm{Re}\, z| \geq |z|/2$. Then $c + d\mathrm{Re}\, z \geq c + d|z|/2$ and the proof is finished. Otherwise we have $|\mathrm{Im} z| \geq |z|/2$, so that

$$|c + dz| \geq |\mathrm{Im}(c + dz)| \geq d|\mathrm{Im}\, z| \geq d|z|/2, \tag{18.18}$$

and since $|c + dz| \geq c$ we have $|c + dz| \geq c/2 + d|z|/4$ and the proof is finished.

Assume now that $|\mathrm{Im}\, z| \geq \alpha|z|$. Then as in (18.18) we have $|c + dz| \geq \alpha d|z|$ so that the proof is finished if $\alpha d|z| \geq \alpha(c + d|z|)/4$ and thus we may assume that $\alpha d|z| \leq \alpha(c + d|z|)/4$. Then $3d|z| \leq c$ so that $|c + zd| \geq c - |z|d \geq c/3 + |z|d \geq \alpha(c + d|z|)/4$ and the proof is finished again. □

Consider a four-vector k. Using (18.17) for $c = \boldsymbol{k}^2 + m^2$, $d = (k^0)^2$ and $z = -(1 + i\varepsilon)$ (so that $|\mathrm{Im}\, z| \geq \varepsilon|z|/2$) we obtain

$$|W_\varepsilon(k) + m^2| \geq \frac{\varepsilon}{16}(\|k\|^2 + m^2). \tag{18.19}$$

Using this inequality together with (15.19) proves that the absolute convergence of the integrals (18.13) follows from the convergence of (18.14).

We now start the main argument of Theorem 18.2.1. The overall strategy is to transform the integral (18.13) until we are in a position to apply Lemma 18.1.8. The transformations follow the methods which the physicists invented to actually compute diagram integrals.

To perform the first transformation we need a general version of the Feynman parameters of (13.122).

Lemma 18.2.6 *Consider complex numbers $A_1, \ldots, A_s$ with $\mathrm{Im}\, A_i < 0$. Then*

$$\frac{1}{A_1 \cdots A_s} = (s-1)! \int_{\mathcal{D}} \mathrm{d}\mu(u) \frac{1}{\left(\sum_{i \leq s} u_i A_i\right)^s}, \tag{18.20}$$

where $\mathcal{D}$ is the set

$$\Big\{u = (u_i)_{1 \leq i \leq s}\ ;\ u_i \geq 0\,,\ \sum_{i \leq s} u_i = 1\Big\},$$

and $\mathrm{d}\mu$ is the surface measure on $\mathcal{D}$, such that $\mathrm{d}\mu(u) = \delta(1 - \sum_{i \leq s} u_i)\mathrm{d}^s u$.

The u_i are called *Feynman parameters*.

Proof We refer to Folland's textbook [31, page 201] for a plain argument and we give a fancier proof, based on the fact that for a function F on $\mathbb{R}^s$, sufficiently integrable, we have

$$\int_{(\mathbb{R}^+)^s} \mathrm{d}^s x F(x) = \int_{\mathcal{D}} \mathrm{d}\mu(u) \int_0^\infty \mathrm{d}t\, t^{s-1} F(tu). \tag{18.21}$$

This kind of polar decomposition is obtained by making the change of variables

$$x_1 = tu_1, \ldots, x_{s-1} = tu_{s-1}, x_s = t(1 - u_1 - \ldots - u_{s-1}).$$

Replacing the numbers A_i by the numbers $\mathrm{i}A_i$ we may rather assume that $\mathrm{Re}\, A_i > 0$ for each i. We then note for a complex number A with $\mathrm{Re}\, A > 0$ the relation $\int_0^\infty \mathrm{d}t \exp(-tA) = 1/A$ and its consequence

$$\int_0^\infty t^{s-1} \mathrm{d}t \exp(-tA) = (s-1)!\, / A^s$$

through differentiation in A, and we apply (18.21) to the function $F(x) = \exp(-\sum_{i \le s} x_i A_i)$. □

In the remainder of the proof we denote $\mathcal{D}' \subset \mathcal{D}$ the set:

$$\mathcal{D}' = \Big\{u = (u_i)_{1 \le i \le s}\ ;\ \forall i \le s\,,\ u_i > 0\,,\ \sum_{i \le s} u_i = 1\Big\},$$

so that in (18.20) we can take the integral over $\mathcal{D}'$ rather than over $\mathcal{D}$. Let us lighten notation by writing until further notice

$$\forall i \le s\ ;\ q_i = \sum_{j \le r} b_{i,j} p_j. \tag{18.22}$$

We now transform the product in the denominator in (18.13) into the sth power of a single term using Feynman parameters. Since $\sum_{i \le s} u_i m^2 = m^2$ for $u \in \mathcal{D}'$, this single term is

$$\sum_{i \le s} u_i W_\varepsilon\Big(\sum_{j \le n} a_{i,j} k_j + q_i\Big) + m^2, \tag{18.23}$$

and

$$I(p,\varepsilon) = (s-1)! \int_{\mathcal{D}} \mathrm{d}\mu(u) I(p,\varepsilon,u), \tag{18.24}$$

where

$$I(p,\varepsilon,u) := \int \mathrm{d}^{4n}k \frac{P(k,p,\varepsilon)}{\Big(\sum_{i \le s} u_i W_\varepsilon\big(\sum_{j \le n} a_{i,j} k_j + q_i\big) + m^2\Big)^s}. \tag{18.25}$$

We define

$$|I|(p,\varepsilon,u) := \int \mathrm{d}^{4n}k \frac{|P(k,p,\varepsilon)|}{\Big|\sum_{i \le s} u_i W_\varepsilon\big(\sum_{j \le n} a_{i,j} k_j + q_i\big) + m^2\Big|^s}. \tag{18.26}$$

Lemma 18.2.7 *Assuming the hypothesis of Theorem 18.2.1, for each $\varepsilon > 0$ and p we have*

$$\int_{\mathcal{D}'} \mathrm{d}\mu(u)|I(p,\varepsilon,u)| \le \int_{\mathcal{D}'} \mathrm{d}\mu(u)|I|(p,\varepsilon,u) < \infty. \tag{18.27}$$

Proof The first inequality is obvious. For the second one we use Lemma 18.2.5 just as in (18.19) to obtain

$$\Big|\sum_{i\le s} u_i W_\varepsilon\Big(\sum_{j\le n} a_{i,j}k_j + q_i\Big) + m^2\Big|^s \ge \Big(\frac{\varepsilon}{16}\Big)^s \Big(\sum_{i\le s} u_i \Big\|\sum_{j\le n} a_{i,j}k_j + q_i\Big\|^2 + m^2\Big)^s,$$

so that

$$|I|(p,\varepsilon,u) \le \Big(\frac{16}{\varepsilon}\Big)^s \int \mathrm{d}^{4n}k \frac{|P(k,p,\varepsilon)|}{\Big(\sum_{i\le s} u_i \|\sum_{j\le n} a_{i,j}k_j + q_i\|^2 + m^2\Big)^s}. \tag{18.28}$$

To prove that $\int \mathrm{d}\mu(u)|I|(p,\varepsilon,u)$ is finite we use (18.20) again to obtain

$$\int \mathrm{d}\mu(u)|I|(p,\varepsilon,u) \le \frac{1}{(s-1)!}\Big(\frac{16}{\varepsilon}\Big)^s \int \mathrm{d}^{4n}k \frac{|P(k,p,\varepsilon)|}{\prod_{i\le s}(\|\sum_{j\le n} a_{i,j}k_j + q_i\|^2 + m^2)}$$

which is finite using (15.19) and the absolute convergence of the integral (18.14). □

The study of the integral $I(p,\varepsilon,u)$ of (18.25) is at the center of the proof. Brutish estimates will later prove the absolute convergence of this integral for $u \in \mathcal{D}'$.[6] The miracle of the Feynman parameters is there. The delicate issue of the convergence of the integral (18.14) translates into the convergence of the integral (18.27), but the individual integrals $I(p,\varepsilon,u)$ are robustly convergent. These integrals can basically be explicitly calculated, and the proof of Theorem 18.2.1 will reduce to the use of Lemma 18.1.8 at a given value of u.

We start the program of "calculating" the integrals $I(p,\varepsilon,u)$, which will end when we reach the expression of Proposition 18.2.12.

In the expression (18.25) quadratic forms in the variables k_j occur. We first perform some standard algebra to transform these quadratic forms. This algebra is a generalization of the method of "completing the square" when studying a second-degree polynomial in one variable.

Lemma 18.2.8 *Consider numbers $u_i > 0$ for $i \le s$. Consider a vector $y = (y_i)_{i\le s}$, and the quadratic function*

$$F(x,y,u) = \sum_{i\le s} u_i \Big(\sum_{j\le n} a_{i,j}x_j + y_i\Big)^2 \tag{18.29}$$

[6] This convergence is rather obvious. Since $\int \mathrm{d}\mu(u)|I|(p,\varepsilon,u)$ is finite, then $|I|(p,\varepsilon,u)$ must be finite for at least one value of u. Denoting by $J_u(k,p,\varepsilon)$ the integrand in (18.28) for any other value u' one has $J_{u'}(k,p,\varepsilon) \le C J_u(k,p,\varepsilon)$ where C depends only on u and thus $|I|(p,\varepsilon,u')$ is also finite. This indirect argument is however confusing.

of $x = (x_j)_{j\le n}$. Given y and u the system of n equations $\partial F/\partial x_j = 0$ i.e.

$$\forall j \le n\ ;\ \sum_{i\le s} u_i a_{i,j}\Big(\sum_{j'\le n} a_{i,j'}x_{j'} + y_i\Big) = 0 \tag{18.30}$$

has a unique solution. Let us denote this solution by $B(y,u) = (B_j(y,u))_{j\le n}$, which is therefore linear in y. It holds that

$$F(x,y,u) = \sum_{i\le s} u_i\Big(\sum_{j\le n} a_{i,j}(x_j - B_j(y,u))\Big)^2 + F(B(y,u),y,u). \tag{18.31}$$

Furthermore

$$F(B(y,u),y,u) = \sum_{i,i'\le s} F_{i,i'}(u) y_i y_{i'} \tag{18.32}$$

for certain numbers $F_{i,i'}(u)$, and

$$0 \le F(B(y,u),y,u) \le \sum_{i\le s} u_i y_i^2. \tag{18.33}$$

Proof The essential fact is that the matrix $(a_{i,j})_{i\le s, j\le n}$ is of rank n. As a consequence the quadratic form

$$\sum_{i\le s} u_i\Big(\sum_{j\le n} a_{i,j}x_j\Big)^2$$

is positive definite (i.e. it is positive semidefinite and cannot be zero unless x itself is zero). Thus the matrix $b_{j,j'} = \sum_{i\le s} u_i a_{i,j} a_{i,j'}$ is positive definite and hence invertible. As (18.30) means

$$\forall j \le n\ ;\ \sum_{j'\le n} b_{j,j'}x_{j'} + \sum_{i\le s} u_i a_{i,j} y_i = 0,$$

these equations have a unique solution, and (18.31) follows by algebra.

Since $B(y,u)$ is linear in y, (18.32) is an obvious consequence of (18.29). The second statement is a consequence of the fact that from (18.31) it holds that

$$F(B(y,u),y,u) = \inf_x F(x,y,u) \le F(0,y,u).$$

□

Exercise 18.2.9 Complete the algebra in the previous proof.

In the lemma we assume that all the u_i are > 0. Matters might blow up at the boundary of $\mathcal{D}$. The exquisite feature of the proof is that we do not need a detailed understanding of what happens. Everything is taken care of by the convergence of the integral (18.27). In the following exercises you are asked to show that matters are not as bad as they could be, but these difficult facts are not required for our argument.

Exercise 18.2.10 Prove the remarkable fact that there exists a number S *independent of* $u \in \mathcal{D}'$ such that for each j one has

$$|B_j(y,u)| \le S \sum_{i \le s} |y_i|. \tag{18.34}$$

This is very challenging.

We may also observe that, according to (18.33), the eigenvalues of the matrix $(F_{i,i'}(u))$ (and hence the numbers $F_{i,i'}(u)$ themselves) remain bounded uniformly over $\mathcal{D}$. The following exercise shows that we have even more, but this will not be needed in the rest of the argument. The quadratic form $\sum_{i,i' \le s} F_{i,i'}(u) y_i y_{i'}$ will be used through Lemma 18.1.7 which does not require this good behavior.

Exercise 18.2.11 In this exercise we prove that the functions $F_{i,i'}(u)$ are restrictions to $\mathcal{D}'$ of *continuous* functions[7] on $\mathcal{D}$. This requires minimal fluency in topology in metric spaces. We fix $y \in \mathbb{R}^s$.
(a) Prove that given $u \in \mathcal{D}$ there exists at least one x_0 such that $F(x_0, y, u) = G(y,u) := \inf_x F(x,y,u)$.
(b) Prove that if $u_n \in \mathcal{D}'$ satisfies $u_n \to u$ then $\limsup_{n\to\infty} G(y,u_n) \le G(y,u)$.
(c) Consider u_n as above and $x_n = B(y,u_n)$. Using that (x_n) has a converging subsequence by (18.34), prove that $\liminf_{n\to\infty} G(y,u_n) \ge G(y,u)$.
(d) Conclude that $u \mapsto G(y,u)$ extends to a continuous function on $\mathcal{D}$.
(e) Prove that $F_{i,i'}(u)$ is the restriction to $\mathcal{D}'$ of a continuous function on $\mathcal{D}$. Hint: You may need an interpolation principle that you will meet on page 553.

To continue the analysis we introduce synthetic notation. Given the four-vectors $(q_i)_{i \le s}$ of (18.22) and $0 \le \nu \le 3$ we define $q^\nu = (q_i^\nu)_{i \le s} \in \mathbb{R}^s$. For $j \le n$ we denote $B_j(q,u)$ the four-vector $(B_j(q^\nu,u))_{0 \le \nu \le 3}$, and we write $B(q,u) = (B_j(q,u))_{j \le n}$. Given four-vectors $(k_j)_{j \le n}$ and $0 \le \nu \le 3$ we define $k^\nu = (k_j^\nu)_{j \le n}$. From the definitions (16.2) of W_ε and (18.29) of F we observe that

$$\sum_{i \le s} u_i W_\varepsilon\Big(\sum_{j \le n} a_{i,j} k_j + q_i\Big) = -(1+\mathrm{i}\varepsilon) F(k^0, q^0, u) + \sum_{1 \le \nu \le 3} F(k^\nu, q^\nu, u).$$

In the right-hand side we apply (18.31) to each of the four terms to get

$$\sum_{i \le s} u_i W_\varepsilon\big(\sum_{j \le n} a_{i,j} k_j + q_i\big) = \sum_{i \le s} u_i W_\varepsilon\big(\sum_{j \le n} a_{i,j}(k_j - B_j(q,u))\big) + \tilde{C}(q,u,\varepsilon), \tag{18.35}$$

where

$$\tilde{C}(q,u,\varepsilon) := -(1+\mathrm{i}\varepsilon) F(B(q^0,u), q^0, u) + \sum_{1 \le \nu \le 3} F(B(q^\nu,u), q^\nu, u). \tag{18.36}$$

[7] In his original paper [94], Zimmerman seems to consider this property as evident. We respectfully disagree, and we do not see how this can be proved without knowing the result of the previous exercise.

Recalling again that $q = q(p)$ by (18.22) we now define

$$C(p,u,\varepsilon) = \tilde{C}(q(p),u,\varepsilon).$$

Let us observe for further use that it follows from (18.32) and (18.33) that

$$C(p,u,\varepsilon) = -(1+\mathrm{i}\varepsilon)Q_u(p^0) + \sum_{1\le\nu\le 3} Q_u(p^\nu), \tag{18.37}$$

where Q_u is a positive semidefinite quadratic form on $\mathbb{R}^n$ (depending on u).

We now perform the change of variables $k \to k + B(q(p),u)$ in (18.25) to obtain

$$I(p,\varepsilon,u) = \int \mathrm{d}^{4n}k \frac{T_u(k,p,\varepsilon)}{\left(\sum_{i\le s} u_i W_\varepsilon\left(\sum_{j\le n} a_{i,j}k_j\right) + C(p,u,\varepsilon) + m^2\right)^s}, \tag{18.38}$$

where

$$T_u(k,p,\varepsilon) := P(k + B(q(p),u),p,\varepsilon) \tag{18.39}$$

is a polynomial in k, p, ε depending on u.

Proposition 18.2.12 *The quantity $I(p,\varepsilon,u)$ is of the form*

$$I(p,\varepsilon,u) = (-1-\mathrm{i}\varepsilon)^{-a}\frac{R(p,\varepsilon,u)}{(C(p,u,\varepsilon)+m^2)^{s-2n}}, \tag{18.40}$$

where $2a$ is an integer ≥ 0 depending only on P, and where $R(p,\varepsilon,u)$ is a polynomial in p and ε which depends[8] *on u. Furthermore when $P(k,p,0)$ is Lorentz invariant, for each u the polynomial $p \mapsto R(p,0,u)$ is Lorentz invariant.*

We delay the proof of this key result and continue the analysis.

Lemma 18.2.13 (Interpolation lemma) *Consider positive integers n and N. Then given integers $\mu_1,\ldots,\mu_n \ge 0$ with $\sum_{\ell\le n}\mu_\ell \le N$ there exists a finite subset F of $\mathbb{R}^n$ and complex numbers $(\alpha_t)_{t\in F}$ such that for any polynomial*

$$P(x) = \sum_{\nu_1,\ldots,\nu_n} a_{\nu_1,\ldots,\nu_n} x_1^{\nu_1}\cdots x_n^{\nu_n}$$

on $\mathbb{R}^n$, of degree $\le N$ we have the identity

$$a_{\mu_1,\ldots,\mu_n} = \sum_{t\in F}\alpha_t P(t).$$

In words, the coefficients of the polynomial may be computed as certain finite linear combinations of the values of the polynomial at certain points. When $n = 1$ there is a classical formula, Bernstein's formula, which does just this, and it can be generalized to several dimensions. On the other hand, an explicit formula is irrelevant for our purposes, and the result itself requires only elementary linear algebra. To follow the proof make sure the following elementary facts are clear in your mind.

[8] This dependence on u need not be polynomial.

Exercise 18.2.14 (a) A subset F of a finite-dimensional space $\mathcal{Q}$ spans $\mathcal{Q}$ if and only if a linear form on $\mathcal{Q}$ which is zero at each point of F is identically zero.
(b) A subset F of the dual $\mathcal{Q}'$ of $\mathcal{Q}$ spans $\mathcal{Q}'$ if and only if the only point of $\mathcal{Q}$ where all the elements of F are zero is 0. Hint: Use (a) for $\mathcal{Q}'$.

Proof of Lemma 18.2.13 The space $\mathcal{Q}$ of polynomials of degree $\leq N$ is a finite-dimensional linear space. A polynomial that takes the value zero at every point has all its coefficients zero. Consequently, the linear forms "evaluation at a given point" are such that the intersection of their kernels reduces to zero. Thus by (b) of the previous exercise they span the dual of $\mathcal{Q}$. In particular the linear form on $\mathcal{Q}$ which associates to a polynomial the coefficient of the monomial of a given degree in that polynomial is a linear combination of linear forms of the type "evaluation at a given point". □

We write the polynomial $R(p,\varepsilon,u)$ in (18.40) as

$$R(p,\varepsilon,u) = \sum_{\eta,\gamma} a_{\eta,\gamma}(u)\varepsilon^{\eta} M_{\gamma}(p), \tag{18.41}$$

where η is an integer, where $\gamma = (\gamma_j^{\nu})_{j\leq r,\nu=0,1,2,3}$ for integers $\gamma_j^{\nu} \geq 0$, where $a_{\eta,\gamma}(u) \in \mathbb{C}$ and where $M_{\gamma}(p)$ is the monomial $\prod_{\nu=0,1,2,3,\, j\leq r}(p_j^{\nu})^{\gamma_j^{\nu}}$.

Lemma 18.2.15 *For each η and γ we have*

$$\int_{\mathcal{D}'} \mathrm{d}\mu(u)|a_{\eta,\gamma}(u)| < \infty. \tag{18.42}$$

Proof Given p and ε it follows from (18.33) that the quantity $C(p,\varepsilon,u)$ stays uniformly bounded as u varies, so that from (18.27) and (18.40) for each p and each $\varepsilon > 0$ it holds that $\int_{\mathcal{D}} \mathrm{d}\mu(u)|R(p,\varepsilon,u)| < \infty$. Using Lemma 18.2.13 for the variables ε and p_j^{ν} rather than x_i we can write $a_{\eta,\gamma}(u)$ as a linear combination $\sum_{(p,\varepsilon)\in F} \alpha_{(p,\varepsilon)} R(p,\varepsilon,u)$. This implies (18.42). □

We also need two simple facts.

Lemma 18.2.16 *Consider a positive semidefinite quadratic form Q on $\mathbb{R}^r$. Then the quantity $-Q(p^0) + \sum_{1\leq\nu\leq 3} Q(p^{\nu})$ is Lorentz invariant.*

Here of course $p^{\nu} = (p_j^{\nu})_{j\leq r}$.

Proof This is obvious when Q is diagonal, i.e. $Q(y) = \sum_{j\leq r} a_j y_j^2$ and one reduces to that case by change of basis. □

Exercise 18.2.17 Carry out the details of the previous proof.

Lemma 18.2.18 *Uniformly in $\alpha, a \in \mathbb{R}$ we have*

$$\lim_{\varepsilon\to 0} \frac{a(1+\mathrm{i}\varepsilon)+\alpha}{a+(1-\mathrm{i}\varepsilon)\alpha} = 1.$$

Proof

$$\left|\frac{a(1+\mathrm{i}\varepsilon)+\alpha}{a+(1-\mathrm{i}\varepsilon)\alpha}-1\right| = \left|\frac{\varepsilon(a+\alpha)}{a+(1-\mathrm{i}\varepsilon)\alpha}\right| \leq \varepsilon. \qquad \square$$

Proof of Theorem 18.2.1 assuming (18.40) Consider an infinitely differentiable function φ on $\mathbb{R}^r$, with compact support. From (18.24) it holds that

$$\int \mathrm{d}^{4r} p\varphi(p)I(p,\varepsilon) = (s-1)! \int_{\mathcal{D}'} \mathrm{d}\mu(u) \int \mathrm{d}^{4r} p\varphi(p)I(p,\varepsilon,u), \tag{18.43}$$

whereas (18.40) and (18.41) imply

$$\int \mathrm{d}^{4r} p\varphi(p)I(p,\varepsilon,u) = (-1-\mathrm{i}\varepsilon)^{-a} \sum_{\eta,\gamma} \varepsilon^{\eta} a_{\eta,\gamma}(u) \int \mathrm{d}^{4r} p \frac{M_\gamma(p)\varphi(p)}{(C(p,u,\varepsilon)+m^2)^{s-2n}}. \tag{18.44}$$

To prove the convergence of the integral (18.43) as $\varepsilon \to 0^+$, in view of (18.42) it suffices to prove the uniform convergence over $u \in \mathcal{D}'$ as $\varepsilon \to 0^+$ of the integrals

$$\int \mathrm{d}^{4r} p \frac{M_\gamma(p)\varphi(p)}{(C(p,u,\varepsilon)+m^2)^{s-2n}}. \tag{18.45}$$

Since φ has a compact support, we do not have to worry about the convergence of the integral itself. Recalling (18.37) and using Lemma 18.2.18 for $a = -Q_u(p^0)$ and $\alpha := m^2 + \sum_{1\leq\nu\leq3} Q_u(p^\nu)$ implies that as far as taking the limit as $\varepsilon \to 0^+$ goes, in (18.45) we may replace $C(p,u,\varepsilon)+m^2 = \alpha(1+\mathrm{i}\varepsilon)+a$ by

$$a+(1-\mathrm{i}\varepsilon)\alpha = -Q_u(p^0)+(1-\mathrm{i}\varepsilon)(m^2+\sum_{1\leq\nu\leq3} Q_u(p^\nu)).$$

That is, it suffices to prove the uniform convergence over u as $\varepsilon \to 0^+$ of the integrals

$$\int \mathrm{d}^{4r} p \frac{M_\gamma(p)\varphi(p)}{(-Q_u(p^0)+(1-\mathrm{i}\varepsilon)\alpha)^{s-2n}} = \frac{1}{\alpha^{s-2n}} \int \mathrm{d}^{4r} p \frac{M_\gamma(p)\varphi(p)}{(1-\mathrm{i}\varepsilon-Q'_u(p^0))^{s-2n}},$$

where $Q'_u(p^0) = Q_u(p^0)/\alpha$. This follows from Lemma 18.1.7 used at given values of p^ν for $\nu \geq 1$ and completes the proof of convergence as $\varepsilon \to 0^+$ of the integrals (18.43).

It remains to prove Lorentz invariance. First we note that for $\alpha > 0$ we have $\lim_{\varepsilon\to0^+} f(\alpha\varepsilon) = \lim_{\varepsilon\to0^+} f(\varepsilon)$. Using this for $\alpha = m^2+\sum_{1\leq\nu\leq3} Q_u(p^\nu)$, and since we take limits in integrals in p^0 at given values of p^ν, $1 \leq \nu \leq 3$, in (18.45) we may also replace

$$-Q_u(p^0)+(1-\mathrm{i}\varepsilon)(m^2+\sum_{1\leq\nu\leq3} Q_u(p^\nu)) = m^2-\mathrm{i}\varepsilon\alpha-Q_u(p^0)+\sum_{1\leq\nu\leq3} Q_u(p^\nu)$$

by

$$m^2-\mathrm{i}\varepsilon-Q_u(p^0)+\sum_{1\leq\nu\leq3} Q_u(p^\nu).$$

Assuming now that $P(k,p,0)$ is Lorentz invariant, let us prove the Lorentz invariance of the limiting distribution. In (18.44) only the terms for which $\eta = 0$ contribute. Since $\sum_\gamma a_{0,\gamma}(u)M_\gamma(p) = R(p,0,u)$ we have in fact

$$\lim_{\varepsilon\to 0+}\int d^{4r}p\varphi(p)I(p,\varepsilon,u)$$
$$= \lim_{\varepsilon\to 0+}\int d^{4r}p\frac{R(p,0,u)\varphi(p)}{\left(m^2-\mathrm{i}\varepsilon-Q_u(p^0)+\sum_{1\le\nu\le 3}Q_u(p^\nu)\right)^{s-2n}}. \qquad (18.46)$$

The right-hand side is Lorentz invariant, in the sense that it does not change if we replace $\varphi(p)=\varphi(p_1,\ldots,p_r)$ by the function $\varphi(A(p_1),\ldots,A(p_r))$ for a Lorentz transformation A. This follows from the change of variable $p'_j=A(p_j)$, using that $R(p,0,u)$ is Lorentz invariant and Lemma 18.2.16. □

18.3 Proof of Proposition 18.2.12

To prove Theorem 18.2.1 it only remains to prove Proposition 18.2.12, a task which we start now. Our main tool is the following elementary but fundamental fact about analytic functions: A non-zero analytic function defined in a connected open set of the complex plane has only isolated zeroes. Consequently if two such functions defined in the same connected open set coincide, say, on a line segment, they are identical. Suppose then that we have an integral $\mathcal{I}(z)$ depending on a complex parameter z, and which defines an analytic function in a connected domain containing, say, $\mathbb{R}^+$. If for $z\in\mathbb{R}^+$ we succeed in computing this integral by a change of variable, proving that then $\mathcal{I}(z)=\mathcal{J}(z)$ for another analytic function $\mathcal{J}(z)$ defined in the same connected domain as $\mathcal{I}$, the equality $\mathcal{I}(z)=\mathcal{J}(z)$ will hold not only for $z\in\mathbb{R}^+$ but for all values of z in the connected domain. It will suffice to apply this principle twice to obtain Proposition 18.2.12.

Keeping the value of u implicit consider the quadratic form on $\mathbb{R}^n$ given by

$$V(x):=\sum_{i\le s}u_i\Big(\sum_{j\le n}a_{i,j}x_j\Big)^2\ge 0,$$

and note that

$$\sum_{i\le s}u_iW_\varepsilon\Big(\sum_{j\le n}a_{i,j}k_j\Big)=-(1+\mathrm{i}\varepsilon)V(k^0)+\sum_{1\le\nu\le 3}V(k^\nu), \qquad (18.47)$$

so that (18.38) becomes

$$I(p,\varepsilon,u)=\int d^{4n}k\frac{T_u(k,p,\varepsilon)}{\left(-(1+\mathrm{i}\varepsilon)V(k^0)+\sum_{1\le\nu\le 3}V(k^\nu)+C(p,u,\varepsilon)+m^2\right)^s}. \qquad (18.48)$$

Consider the subset $\mathcal{U}$ of the complex plane defined as

$$\mathcal{U}=\{-c+(1+\mathrm{i}\varepsilon)d\ ;\ c,d\in\mathbb{R}\ ,\ c,d\ge 0\}, \qquad (18.49)$$

so that for $z\notin\mathcal{U}$ the denominator in the integral

$$\theta(z):=\int d^{4n}k\frac{T_u(k,p,\varepsilon)}{\left(-(1+\mathrm{i}\varepsilon)V(k^0)+\sum_{1\le\nu\le 3}V(k^\nu)+z\right)^s} \qquad (18.50)$$

never vanishes. Next we prove that for $z \notin \mathcal{U}$ the integral (18.50) is absolutely convergent. First, from the convergence of (18.14) it follows by power-counting that $\deg_k P(k,p,\varepsilon) < 2s - 4n$ because the degree of the denominator is $\leq 2s$. Consequently by (15.1) it holds that

$$|T_u(k,p,\varepsilon)| = |P(k + B(q(p),u), p,\varepsilon)| \leq C_1(\|k\|^{2s-4n-1} + 1), \tag{18.51}$$

where $\|k\|^2 = \sum_{j\leq n} \|k_j\|^2$. Here and below $C_1, \ldots$ are constants depending only on u, ε, p but certainly not on k. Next, by (18.17) it holds that

$$\Big| -(1+\mathrm{i}\varepsilon)V(k^0) + \sum_{1\leq\nu\leq 3} V(k^\nu)\Big| \geq \frac{1}{C_2}\sum_{0\leq\nu\leq 3} V(k^\nu). \tag{18.52}$$

Since the matrix $(a_{i,j})$ is of rank n, when all u_i are >0 the quadratic form $V(x)$ is positive definite, so that $V(x) \geq \|x\|^2/C_3$. Combining with (18.52) yields

$$\Big| -(1+\mathrm{i}\varepsilon)V(k^0) + \sum_{1\leq\nu\leq 3} V(k^\nu)\Big| \geq \frac{1}{C_2C_3}\|k\|^2. \tag{18.53}$$

Comparing with (18.51)(and using that the denominator never vanishes for $z \notin \mathcal{U}$) should make the required convergence of the integral obvious.

Exercise 18.3.1 Give a direct proof of the convergence of the integral (18.26).

The polynomial $T_u(k,p,\varepsilon)$ is a sum of monomials. Making the change of variable $k^\nu \to -k^\nu$ for a certain $0 \leq \nu \leq 3$ shows that such a monomial contributes to the integral (18.50) only if its total degree in the variables k_j^ν is *even* (since the denominator is invariant by the change of variable). We assume from now on that this property holds for each monomial in T.

It is rather obvious from the previous estimates that we are permitted to differentiate in z inside the integral (18.50). The formula (18.50) thus defines an analytic function $\theta(z)$ in the complement of $\mathcal{U}$. This complement contains the positive real line. For $z \in \mathbb{R}^+$ the change of variable $k \to k\sqrt{z}$ shows that

$$\theta(z) = z^{2n-s}\int \mathrm{d}^{4n}k \frac{T_u(k\sqrt{z},p,\varepsilon)}{(-(1+\mathrm{i}\varepsilon)V(k^0) + \sum_{1\leq\nu\leq 3} V(k^\nu) + 1)^s}. \tag{18.54}$$

The right-hand side defines another analytic function of z in the complement of $\mathcal{U}$, so that these two analytic functions coincide in the entire complement of $\mathcal{U}$. Now, recalling the quantity $C(p,u,\varepsilon)$ of (18.36), we claim that $z = m^2 + C(p,u,\varepsilon) \notin \mathcal{U}$. Otherwise, arguing by contradiction, we have $m^2 + C(p,u,\varepsilon) = -c + d(1+\mathrm{i}\varepsilon)$ where $c,d \geq 0$. Thus $C(p,u,\varepsilon) = -m^2 - c + d(1+\mathrm{i}\varepsilon)$. On the other hand, by (18.36) we have $C(p,u,\varepsilon) = c' - d'(1+\mathrm{i}\varepsilon)$ where $c', d' \geq 0$. Comparing the imaginary parts of these two expressions, we see that $d = d' = 0$, so that $-m^2 - c = c'$ and $-m^2 = c + c'$ which is impossible since $c, c' \geq 0$ and $-m^2 < 0$. We have proved that $z = m^2 + C(p,u,\varepsilon) \notin \mathcal{U}$, so that one may use (18.54) for this value of z to obtain that for any p, ε, u one has

$$I(p,\varepsilon,u) = \frac{1}{(C(q,u,\varepsilon) + m^2)^{s-2n}}\int \mathrm{d}^{4n}k \frac{T_u(k\sqrt{C(p,u,\varepsilon) + m^2}, p,\varepsilon)}{(-(1+\mathrm{i}\varepsilon)V(k^0) + \sum_{1\leq\nu\leq 3} V(k^\nu) + 1)^s}.$$

We have made good progress toward (18.40), and we now study the integral above, which is absolutely convergent according to the estimates (18.51) and (18.53). This integral is $\sigma(-1-\mathrm{i}\varepsilon)$ where the function of $\sigma(z)$ is given by

$$\sigma(z)=\int \mathrm{d}^{4n}k\,\frac{T_u(k\sqrt{C(q,u,\varepsilon)+m^2},p,\varepsilon)}{\left(zV(k^0)+\sum_{1\le\nu\le3}V(k^\nu)+1\right)^s}.$$

The denominator never vanishes if z belongs to the complement of $\mathbb{R}^-=\{x\in\mathbb{R},x<0\}$, and one can check as before that then the integral is absolutely convergent. For $z\in\mathbb{R}^+$ we may calculate $\sigma(z)$ by the change of variable $k^0\to k^0z^{-1/2}$, to obtain

$$\sigma(z)=z^{-n/2}\int \mathrm{d}^{4n}k\,\frac{T_u(k_z\sqrt{C(q,u,\varepsilon)+m^2},p,\varepsilon)}{\left(\sum_{0\le\nu\le3}V(k^\nu)+1\right)^s},$$

where the notation k_z means that the component k^0 has been changed to $k^0z^{-1/2}$. Defining $z^{-1/2}$ in the obvious manner in the complement of $\mathbb{R}^-$, the preceding formula[9] defines an analytic function of z there, and this function coincides with $\sigma(z)$. In particular,

$$\sigma(-1-\mathrm{i}\varepsilon)=(-1-\mathrm{i}\varepsilon)^{-n/2}\int \mathrm{d}^{4n}k\,\frac{T_u(k(\varepsilon)\sqrt{C(q,u,\varepsilon)+m^2},p,\varepsilon)}{\left(\sum_{0\le\nu\le3}V(k^\nu)+1\right)^s},$$

where the notation $k(\varepsilon)$ means that the component k^0 has been changed to $k^0(-1-\mathrm{i}\varepsilon)^{-1/2}$. Let us call b the largest possible even degree of a monomial in $T_u(k,p,\varepsilon)$ in the variables k_j^0 so that b is even as noted. Consider the quantity

$$\begin{aligned}R(p,\varepsilon,u)&:=(-1-\mathrm{i}\varepsilon)^{(b+n)/2}\sigma(-1-\mathrm{i}\varepsilon)\\&=(-1-\mathrm{i}\varepsilon)^{b/2}\int \mathrm{d}^{4n}k\,\frac{T_u(k(\varepsilon)\sqrt{C(p,u,\varepsilon)+m^2},p,\varepsilon)}{\left(\sum_{0\le\nu\le3}V(k^\nu)+1\right)^s}.\end{aligned}$$

In every monomial, the factor $\sqrt{C(p,u,\varepsilon)+m^2}$ has been raised to an even power, so it contributes as a polynomial in p and ε. The quantity $R(p,\varepsilon,u)$ is a polynomial in ε because the term $(-1-\mathrm{i}\varepsilon)^{b/2}$ removes all the negative powers of $(-1-\mathrm{i}\varepsilon)$ created by the terms $k_j^0(-1-\mathrm{i}\varepsilon)^{-1/2}$ (because in each monomial there is an even number $d\le b$ of such terms). Thus we have proved (18.40) with $a=(b+n)/2\ge 0$.[10] Letting $\varepsilon\to 0^+$, we observe that $(-1-\mathrm{i}\varepsilon)^{-1/2}\to\mathrm{i}$ and we obtain

$$R(p,0,u)=(-1)^{b/2}\int \mathrm{d}^{4n}k\,\frac{T_u(k(0)\sqrt{C(p,u,0)+m^2},p,0)}{\left(\sum_{0\le\nu\le3}V(k^\nu)+1\right)^s},\tag{18.55}$$

where the notation $k(0)$ means that the component k^0 has been changed to $\mathrm{i}k^0$.

9 What we are doing here is basically a Wick rotation on each of the components k_j^0. It is not so simple to do in a rigorous manner, and the present approach solves this technical difficulty. To relate our method to Wick rotation, observe that $(-1-\mathrm{i}\varepsilon)^{-1/2}\sim\mathrm{i}$.

10 One may pursue the computation of the right-hand side integral by diagonalizing the quadratic form $V(x)$ in an orthonormal basis, but the information gained this way is irrelevant to our argument.

It remains only to show that $R(p,0,u)$ is Lorentz invariant. Let us use synthetic notation. For $p = (p_j)_{j\le r}$ and a Lorentz transformation A we write $A(p) = (A(p_j))_{j\le r}$, etc. Recalling that $q = q(p)$ by (18.22), by linearity we have $q(A(p)) = A(q(p))$. Writing what this means it is straightforward that $B(A(q),u) = A(B(q,u))$.

Exercise 18.3.2 Prove it.

Going back to (18.39) and using that $P(k,p,0)$ is Lorentz invariant by hypothesis we obtain

$$T_u(A(k),A(p),0) = T_u(k,p,0). \tag{18.56}$$

From (18.37) and Lemma 18.2.16, it holds that

$$C(A(p),u,0) = C(p,u,0). \tag{18.57}$$

Thus, using (18.57) in the first line and (18.56) in the second line,

$$\begin{aligned} R(A(p),0,u) &= (-1)^{b/2}\int \mathrm{d}^{4n}k\frac{T_u(k(0)\sqrt{C(p,u,0)+m^2},A(p),0)}{\left(\sum_{0\le\nu\le3}V(k^\nu)+1\right)^s} \\ &= (-1)^{b/2}\int \mathrm{d}^{4n}k\frac{T_u(A^{-1}(k(0))\sqrt{C(p,u,0)+m^2},p,0)}{\left(\sum_{0\le\nu\le3}V(k^\nu)+1\right)^s}. \end{aligned}$$

To lighten notation, let us think of p and u as fixed once and for all, and let us consider the polynomial in k given by

$$H(k) := T_u(k\sqrt{C(p,u,0)+m^2},p,0). \tag{18.58}$$

Changing A into A^{-1} we conclude the proof by showing that for a Lorentz transformation A we have

$$\int \mathrm{d}^{4n}k\frac{H(A(k(0)))}{\left(\sum_{0\le\nu\le3}V(k^\nu)+1\right)^s} = \int \mathrm{d}^{4n}k\frac{H(k(0))}{\left(\sum_{0\le\nu\le3}V(k^\nu)+1\right)^s}. \tag{18.59}$$

Let us first consider the case where A is a rotation, so that $A(k)^0 = k^0$. It should be obvious that then $A(k(0)) = A(k)(0)$, and the equality (18.59) follows by making the change of variables $k_j \to A^{-1}(k_j)$ in the integral in the left of (18.59), and using that both the integration measure and the denominator are invariant under this transformation (as is apparent following the argument of Lemma 18.2.16).

Since we have proved on page 111 that a Lorentz transformation is the product of a pure boost and a rotation, it suffices to prove (18.59) when A is a pure boost. According to (4.23) it suffices to consider the case where A is of the type (recalling (4.21) and denoting by x an element of $\mathbb{R}^{1,3}$)

$$A(x) = (\cosh z\, x^0 + \sinh z\, x^3, x^1, x^2, \sinh z\, x^0 + \cosh z\, x^3),$$

and where $z \in \mathbb{R}$. This formula also makes sense when the x^ν are complex numbers. Denoting by $x(0)$ the element of $\mathbb{C}^4$ obtained by changing x^0 into $\mathrm{i}x^0$ and leaving the other components unchanged, it is straightforward that $A(x(0)) = U_z(x)(0)$ where

$$U_z(x) = (\cosh z\, x^0 - \mathrm{i}\sinh z\, x^3, x^1, x^2, \mathrm{i}\sinh z\, x^0 + \cosh z\, x^3). \tag{18.60}$$

Thus all we have to prove is the equality

$$\int \mathrm{d}^{4n}k \frac{H(U_z(k)(0))}{\left(\sum_{0\le\nu\le 3} V(k^\nu)+1\right)^s} = \int \mathrm{d}^{4n}k \frac{H(k(0))}{\left(\sum_{0\le\nu\le 3} V(k^\nu)+1\right)^s}. \tag{18.61}$$

The left-hand side defines an analytical function of $z \in \mathbb{C}$. When $z \in \mathrm{i}\mathbb{R}$, $z = \mathrm{i}z'$, $z' \in \mathbb{R}$, the equalities $\cosh z = \cos z'$ and $-\mathrm{i}\sinh z = \sin z'$ together with (18.60) prove that U_z defines a rotation of $\mathbb{R}^4$, in which case we have already proved that (18.61) holds. Therefore it holds for all values of z, completing the proof of Proposition 18.2.12.

18.4 A Side Story: Feynman Diagrams and Wick Rotations

Euclidean propagators are not only useful in theoretical considerations but also in practice. The value of Feynman diagrams is often computed through the following strategy.

- One performs a reverse Wick rotation on the external momenta, that is the components p_j^0 are replaced by $-\mathrm{i}p_j^0$.
- One performs a Wick rotation on the variables k_j^0 and one computes the value of the resulting diagram for the new external momenta.

As shown below, the two Wick rotations basically "amount to replacing the true propagator by the Euclidean propagator". We cannot really expect to prove easily a theorem in full generality to the effect that the above procedure works, since we know how to define the value of a Feynman diagram only in the sense of distributions as in Zimmermann's theorem. Rather we will prove that the procedure does work when the energy component of the external momenta is small enough (and unphysical). We then simply cross our fingers that the actual formulas obtained in this manner extend to all values through analytic continuation without too many singularities. In the next statement, for $r = (r^\nu)_{0\le\nu\le 3} \in \mathbb{C}\times\mathbb{R}^3$ we write

$$Q(r) = \sum_{0\le\nu\le 3} (r^\nu)^2, \tag{18.62}$$

so that $Q(r) = \|r\|^2$ when $r^0 \in \mathbb{R}$.

Theorem 18.4.1 *Consider a polynomial $P(k)$.*[11] *Assume that the matrix $(a_{i,j})$ is of rank n, and that the integral*

$$\int \mathrm{d}^{4n}k \frac{|P(k)|}{\prod_{i\le s}(\|\sum_{j\le n} a_{i,j}k_j\|^2+1)} \tag{18.63}$$

is convergent. Then the formula

$$J(q_1,\ldots,q_s) = \lim_{\varepsilon\to 0^+} \int \mathrm{d}^{4n}k \frac{P(k)}{\prod_{i\le s}\left(W_\varepsilon(\sum_{j\le n} a_{i,j}k_j + q_i)+m^2\right)} \tag{18.64}$$

[11] We could also consider a polynomial $P(k,p,\varepsilon)$ with the same argument, but this makes the notation more complicated, and we do not want to obscure what is a very simple observation.

defines a function of $q_1, \ldots, q_s \in \mathbb{R}^{1,3}$ in the domain $\{\forall i \le s, |q_i^0| < m\}$. Consider the polynomial $\tilde{P}(k)$ obtained from $P(k)$ by replacing each k_j^0 by $\mathrm{i}k_j^0$. Consider four-vectors $(r_i)_{i \le n}$ where now r_i^0 is allowed to take complex values i.e. $r_i \in \mathbb{C} \times \mathbb{R}^3$, with $|\mathrm{Im}\, r_i^0| < m$. Then, recalling the notation (18.62), the integral

$$I(r_1, \ldots, r_s) = \int \mathrm{d}^{4n}k \frac{\tilde{P}(k)}{\prod_{i \le s}(Q(\sum_{j \le n} a_{i,j}k_j + r_i) + m^2)} \tag{18.65}$$

converges. Furthermore,

$$J(q_1, \ldots, q_s) = \mathrm{i}^n I(R(q_1), \ldots, R(q_s)) \tag{18.66}$$

where $R(q_i) = (-\mathrm{i}q_i^0, q_i^1, q_i^2, q_i^3)$.

In words, the true value of the $J(q_1, \ldots, q_s)$ may be obtained by making a reverse Wick rotation of the external momentum q_i^0, and a Wick rotation of the variables k_j^0. This "replaces the Minkowski by the Euclidean propagators".

Proof We first prove the existence of the integral (18.65). Fixing $i \le s$ let $\alpha = \sum_{j \le n} a_{i,j}k_j^0$ and $r_i^0 = \beta + \mathrm{i}\gamma$ ($= \mathrm{Re}\, r_i^0 + \mathrm{i}\mathrm{Im}\, r_i^0$) where $\beta, \gamma \in \mathbb{R}$, so that

$$\Big(\sum_{j \le n} a_{i,j}k_j^0 + r_i^0\Big)^2 = (\alpha + \beta + \mathrm{i}\gamma)^2 = (\alpha + \beta)^2 + 2\mathrm{i}(\alpha + \beta)\gamma - \gamma^2,$$

and thus

$$\mathrm{Re}\Big(\sum_{j \le n} a_{i,j}k_i^0 + r_i^0\Big)^2 = \Big(\sum_{j \le n} a_{i,j}k_j^0 + \mathrm{Re}\, r_i^0\Big)^2 - |\mathrm{Im}\, r_i^0|^2.$$

If we denote by $\mathrm{Real}\, r_i$ the element of $\mathbb{R}^4$ obtained from r_i by replacing r_i^0 by its real part, it follows that

$$\mathrm{Re}\, Q\Big(\sum_{j \le n} a_{i,j}k_j + r_i\Big) = \Big\|\sum_{j \le n} a_{i,j}k_j + \mathrm{Real}\, r_i\Big\|^2 - |\mathrm{Im}\, r_i^0|^2.$$

For the complex number $z = Q(\sum_{j \le n} a_{i,j}k_j + r_i) + m^2$ we have $|z| \ge \mathrm{Re}\, z$ and thus

$$\Big|Q\Big(\sum_{j \le n} a_{i,j}k_j + r_i\Big) + m^2\Big| \ge \Big\|\sum_{j \le n} a_{i,j}k_j + \mathrm{Real}\, r_i\Big\|^2 + m^2 - |\mathrm{Im}\, r_i^0|^2. \tag{18.67}$$

On the other hand, by the triangle inequality we have $\|x\| \le \|x + y\| + \|y\|$ so that $\|x\|^2 \le 2(\|x + y\|^2 + \|y\|^2)$, and we obtain, using also that $|\mathrm{Im}\, r_i^0| < m$,

$$\begin{aligned}\Big\|\sum_{j \le n} a_{i,j}k_j\Big\|^2 + 1 &\le 2\Big(\Big\|\sum_{j \le n} a_{i,j}k_j + \mathrm{Real}\, r_i\Big\|^2 + \big\|\mathrm{Real}\, r_i\big\|^2 + 1\Big) \\ &\le C\Big(\Big\|\sum_{j \le n} a_{i,j}k_j + \mathrm{Real}\, r_i\Big\|^2 + m^2 - |\mathrm{Im}\, r_i^0|^2\Big),\end{aligned} \tag{18.68}$$

where

$$C = 2 + 2 \max_{i \le n} \frac{\|\mathrm{Real}\, r_i\|^2 + 1}{m^2 - |\mathrm{Im}\, r_i^0|^2}.$$

Combining with (18.67) yields $\| \sum_{j \le n} a_{i,j} k_j \|^2 + 1 \le C |Q(\sum_{j \le n} a_{i,j} k_j + r_i) + m^2|$. The convergence of the integral (18.65) is then a consequence of the convergence of the integral

$$\int \mathrm{d}^{4n} k \frac{|\tilde{P}(k)|}{\prod_{i \le s} (\| \sum_{j \le n} a_{i,j} k_j \|^2 + 1)}, \tag{18.69}$$

which itself follows from the convergence of the integral (18.63) through the power-counting theorem, since obviously $\deg_E P = \deg_E \tilde{P}$ for any subspace E.

Next we consider the function

$$\theta(z) = \int \mathrm{d}^{4n} k \frac{P(k)}{\prod_{i \le s} \left(z \left(\sum_{j \le n} a_{i,j} k_j^0 + q_j^0 \right)^2 + \left(\sum_{j \le n} a_{i,j} \boldsymbol{k}_j + \boldsymbol{q}_i \right)^2 + m^2 \right)}. \tag{18.70}$$

Let us prove that it is well defined for $z \notin \mathbb{R}^-$. For this, we note first that according to Lemma 18.2.5 for numbers $c, d > 0$ we have $|c + zd| \ge \alpha(z)(c + d|z|)$ where $\alpha(z) = 1/4$ if $\mathrm{Re}\, z > 0$ and $\alpha(z) = |\mathrm{Im}\, z|/4|z|$ if $\mathrm{Im}\, z \ne 0$. We then appeal to the argument of (15.19) to obtain that the convergence of the integral in (18.70) follows from the convergence of the integral (18.63). The formula (18.70) then defines an analytic function of z in the complement of $\mathbb{R}^-$.

The integral in the right-hand side of (18.64) equals $\theta(-1 - \mathrm{i}\varepsilon)$. For $z \in \mathbb{R}^+$ the change of variable $k^0 \to k^0 z^{-1/2}$ shows that then $\theta(z)$ is given by the formula

$$z^{-n/2} \int \mathrm{d}^{4n} k \frac{P(k_z)}{\prod_{i \le s} \left(\left(\sum_{j \le n} a_{i,j} k_j^0 + z^{1/2} q_j^0 \right)^2 + \left(\sum_{j \le n} a_{i,j} \boldsymbol{k}_j + \boldsymbol{q}_i \right)^2 + m^2 \right)},$$

where the notation k_z means that the components k_i^0 have been replaced by $k_i^0 z^{-1/2}$. As we have seen this integral converges whenever $|\mathrm{Im}\, z^{1/2} q_i^0| < m$ for all $i \le s$. On this domain it obviously defines an analytic function of z. This function must coincide with the function (18.70). Taking $z = -(1 + \mathrm{i}\varepsilon)$ and letting $\varepsilon \to 0^+$ (so that $\sqrt{z} \to -\mathrm{i}$) we obtain the existence of the limit in (18.64) and the equality (18.66). □

Key idea to remember: The problem of the singularities created by the small denominators has received a complete solution.

19

Proof of Convergence of the BPHZ Scheme

In this chapter we prove Theorem 16.1.1 through Zimmerman's original approach [95]. The delicate part of the proof is to show that the regularized integrands satisfy the hypotheses of Theorem 18.2.1, i.e. that the hypotheses of Weinberg's counting theorem are satisfied. This requires grouping the compensating terms in a magnificently clever way, and will be done in Section 19.3. The rest is simply elaborate bookkeeping.

The goal of this chapter is to prove that given a 1-PI Feynman diagram and a subspace $S \subset \ker \mathcal{L}$ with $S \neq \{0\}$ we have the following.

Theorem 19.0.1 *Consider* $f_0 \in (\mathbb{R}^{1,3})^{\mathcal{E}}$ *and a basis* $(f_j)_{j \le N}$ *of* S. *Given* $u = (u_j)_{j \le N}$ *consider the point* $x(u) = f_0 + \sum_{j \le N} u_j f_j$. *Then, whatever the value of* f_0 *we have*[1]

$$\deg_u \mathcal{R}F(x(u)) < -\dim S. \tag{19.1}$$

19.1 Proof of Theorem 16.1.1

In this section we use Theorem 19.0.1 to complete the proof of Theorem 16.1.1. We will complete this proof only in the case of 1-PI diagrams. The necessary modifications for the general case have been sketched in Section 16.7, and completing the details offers no difficulty and is left to the interested reader. Our main effort goes toward proving the following.

Proposition 19.1.1 *When we parameterize* $\ker \mathcal{L}$ *by* $x = x(k) = \sum_{j \le n} k_j \tau(L_j)$ *as in Section 15.8 the integral* $\int \mathrm{d}\mu_{\mathcal{L}}(x) \mathcal{R}F(x + \mathcal{X}(p))$ *takes the form (18.13).*

Proof of Theorem 16.1.1 According to Proposition 19.1.1 the integral

$$\int \cdots \int \prod_{i \le n} \frac{\mathrm{d}^4 k_i}{(2\pi)^4} \mathcal{R}F(x(k) + \chi(p))$$

falls within the framework of Theorem 18.2.1, so the desired convergence will follow from that theorem, provided we can prove that the integral (18.14) is absolutely convergent. For

[1] The degree of a rational function is the degree of its numerator minus the degree of its denominator. You will soon know everything about degrees.

this we will appeal to Proposition 18.2.2. Consider $k(u) = k_0 + \sum_{1\le\ell\le N} u_\ell k_\ell$ as in this proposition. Then $x(u) := x(k(u)) = f_0 + \sum_{1\le\ell\le N} u_\ell f_\ell$ where $f_\ell = x(k_\ell)$ and the vectors $(f_\ell)_{1\le\ell\le N}$ are independent. Denoting by S the span of these vectors, $\dim S = N$ and the vectors $(f_\ell)_{1\le\ell\le N}$ form a basis of S. The condition (18.15) then follows from (19.1) (see Remark 18.2.3). □

We turn to the proof of Proposition 19.1.1, which occupies the rest of this section. We need first to make explicit the fact that "we deal with four copies of the same situation". An element $x \in (\mathbb{R}^{1,3})^{\mathcal{E}}$ is a family $x = (x_e)_{e\in\mathcal{E}}$ where $x_e = (x_e^0, x_e^1, x_e^2, x_e^3) \in \mathbb{R}^{1,3}$. Thus for $\nu = 0,1,2,3$ we may define $x^\nu = (x_e^\nu)_{e\in\mathcal{E}} \in \mathbb{R}^{\mathcal{E}}$.

Definition 19.1.2 We say that a linear map J from $(\mathbb{R}^{1,3})^{\mathcal{E}}$ to itself is a *tensor map* if there is a linear map J^0 from $\mathbb{R}^{\mathcal{E}}$ to itself such that for any $x \in (\mathbb{R}^{1,3})^{\mathcal{E}}$ and any $0 \le \nu \le 3$ we have $J(x)^\nu = J^0(x^\nu)$.[2]

Lemma 19.1.3 *The map $\mathcal{I}(x)$ of (16.10) is a tensor map.*

Proof The basic fact is that $x \in \ker\mathcal{L}$ if and only if $x^\nu \in \ker\mathcal{L}_0$ for each $0 \le \nu \le 3$. Let us then denote by $\mathcal{I}^0$ the orthogonal projection of $\mathbb{R}^{\mathcal{E}}$ on $\ker\mathcal{L}^0$, so that for $y \in \mathbb{R}^{\mathcal{E}}$ the point $y - \mathcal{I}^0(y)$ is orthogonal to $\ker\mathcal{L}_0$. Then given $x \in (\mathbb{R}^{1,3})^{\mathcal{E}}$ the point z given by $z^\nu = \mathcal{I}^0(x^\nu)$ for $0 \le \nu \le 3$ is such that $z \in \ker\mathcal{L}$ and $x - z$ is orthogonal to $\ker\mathcal{L}$. Thus $z = \mathcal{I}(x)$. □

Given a forest $\mathcal{F}$, we recall the functions F_α and F'_α on $(\mathbb{R}^{1,3})^{\mathcal{E}_\alpha}$ defined recursively on page 506.

Lemma 19.1.4 *The functions $F_\alpha(x)$ and $F'_\alpha(x)$ are of the type $P'(x,\varepsilon)/Q'(x,\varepsilon)$ where P' and Q' are polynomials in x and ε, Q' being of the type*

$$Q'(x,\varepsilon) = \prod_{i\le s}(W_\varepsilon(J_i(x)_{e_i}) + m^2), \tag{19.2}$$

where $e_i \in \mathcal{E}$,[3] and where J_i is a tensor linear map from $(\mathbb{R}^{1,3})^{\mathcal{E}_\alpha}$ into itself.[4] In particular the function $\mathcal{R}F(x)$ is of this type (for $\alpha = \gamma$).

Proof The sum of two rational functions of the above type is of the same type, as is seen by reducing to the same denominator. Thus it suffices to recursively prove the claim for the functions F_α and F'_α. Everything is straightforward, except to check that when a function F on $(\mathbb{R}^{1,3})^{\mathcal{E}}$ is of the previous type, this is also that case for T^dF. Let us repeat how this function is defined. Given $z \in \ker\mathcal{L}$ we consider the function $q \mapsto F(z+q)$ defined on $\mathcal{Q}$, and we compute the Taylor polynomial of order d at $q = 0$ of this function. This Taylor polynomial is a function $G_d(z,q)$. It is a sum of terms of the type $P(q)D^\beta F(z)$ where $P(q)$ is a polynomial in q and where $D^\beta F(z)$ is a partial derivative of the function $q \mapsto F(z+q)$ at $q = 0$. The function T^dF on $(\mathbb{R}^{1,3})^{\mathcal{E}}$ is defined by $T^dF(x) = G_d(\mathcal{I}(x), \mathcal{T}(x))$. It is a

[2] The reason for the name is that we may identify $(\mathbb{R}^{1,3})^{\mathcal{E}}$ with $\mathbb{R}^{1,3} \otimes \mathbb{R}^{\mathcal{E}}$ and that a tensor map J is then a map of the type $\mathrm{Id} \otimes J^0$.

[3] The point here is that the same edge e may occur several times as e_i for different values of i.

[4] So that $J_i(x)$ is an element $(J_i(x)_e)_{e\in\mathcal{E}_\alpha}$ of $(\mathbb{R}^{1,3})^{\mathcal{E}_\alpha}$.

sum of terms of the type $P(\mathcal{I}(x))D^\beta F(\mathcal{I}(x))$. The result then follows from the following three facts. First, taking a partial derivative a rational function with a denominator of the type (19.2) does not create new types of denominators. Second, $\mathcal{I}$ is a tensor map by Lemma 19.1.3. Third, the composition of two tensor maps is a tensor map (so that replacing x by $\mathcal{I}(x)$ in an expression of the type (19.2) one obtains an expression of the same type.) □

Lemma 19.1.5 *If J is a tensor map when we parameterize* $\ker\mathcal{L}$ *by* $x = x(k) = \sum_{j\le n} k_j\tau(L_j)$ *for independent loops* L_j *as in (15.54), then* $J(x(k)) = \sum_{j\le n} k_j J^0(\tau(L_j))$.

Proof The definition of the expression $k_j\tau(L_j), x(k)$ is such that $x(k)^\nu = \sum_{j\le n} k_j^\nu \tau(L_j)$, so that $J(x(k))^\nu = J^0(x(k)^\nu) = \sum_{j\le n} k_j^\nu J^0(\tau(L_j))$. □

Proof of Proposition 19.1.1 Use of (15.54) shows that

$$\int \mathrm{d}\mu_{\mathcal{L}}(x)\mathcal{R}F(x+\chi(p)) = C\int\cdots\int\prod_{i\le n}\frac{\mathrm{d}^4k_i}{(2\pi)^4}\mathcal{R}F(x(k)+\chi(p)).$$

The issue is to show that the denominator of the function $\mathcal{R}F(x(k)+\mathcal{X}(p))$ has the correct form. This follows from the fact that the function $\mathcal{R}F(x)$ is of the type described in Lemma 19.1.4. Concerning the dependence on k, since J_i is a tensor map, by Lemma 19.1.5 we have $J_i(x(k))_{e_i} = \sum_{j\le n} k_j a_{i,j}$ where $a_{i,j} = J_i^0(\tau(L_j))_{e_i}$. A similar argument works for the dependence in p, using the special form of $\mathcal{X}(p)$ described in Theorem 15.7.2. □

19.2 Simple Facts

The material and the notation of Chapter 16 should be fresh in the reader's mind.

We fix a 1-PI Feynman diagram throughout the entire chapter. A subdiagram α has set of edges $\mathcal{E}_\alpha$ and set of vertices $\mathcal{V}_\alpha$. The largest subdiagram is denoted γ, and is obtained by removing the external lines and the external vertices of the Feynman diagram. Let us recall yet again the fundamental tools of our construction. We denote by $\mathcal{L}_\alpha$ the map $(\mathbb{R}^{1,3})^{\mathcal{E}_\alpha} \to (\mathbb{R}^{1,3})^{\mathcal{V}_\alpha}$ as constructed in (15.30) and $\mathcal{Q}_\alpha = (\ker\mathcal{L}_\alpha)^\perp$. For $x \in (\mathbb{R}^{1,3})^{\mathcal{E}_\alpha}$ we define as before the orthogonal projections $\mathcal{I}_\alpha(x)$ and $\mathcal{T}_\alpha(x)$ of x on $\ker\mathcal{L}_\alpha$ and $\mathcal{Q}_\alpha$ respectively, so that

$$x = \mathcal{I}_\alpha(x) + \mathcal{T}_\alpha(x). \tag{19.3}$$

The restriction of $\mathcal{L}_\alpha$ to $\mathcal{Q}_\alpha$ is one-to-one on its image $\mathcal{N}_\alpha$, and there is a linear map $\mathcal{W}_\alpha$ from $\mathcal{N}_\alpha$ to $\mathcal{Q}_\alpha$ such that

$$\mathcal{T}_\alpha(x) = \mathcal{W}_\alpha\mathcal{L}_\alpha(x). \tag{19.4}$$

Given two subdiagrams $\beta \subset \alpha$ we denote by $U_{\beta,\alpha}$ the map from $(\mathbb{R}^{1,3})^{\mathcal{E}_\alpha} \to (\mathbb{R}^{1,3})^{\mathcal{E}_\beta}$ which forgets the values of w_e for $e \notin \mathcal{E}_\beta$,

$$e \in \mathcal{E}_\beta \Rightarrow U_{\beta,\alpha}(w)_e = w_e.$$

The following trivial fact will be used many times. It is an immediate consequence of the definitions.

Lemma 19.2.1 *Consider two subdiagrams $\beta \subset \alpha$ and $w \in (\mathbb{R}^{1,3})^{\mathcal{E}_\alpha}$. Then for each vertex v of β the quantity*

$$\mathcal{L}_\alpha(w)_v - \mathcal{L}_\beta U_{\beta,\alpha}(w)_v$$

is a linear combination of the quantities w_e where e is an edge of α adjacent to v and e is not an edge of β. In particular if all these w_e are zero we have

$$\mathcal{L}_\alpha(w)_v = \mathcal{L}_\beta U_{\beta,\alpha}(w)_v. \tag{19.5}$$

Observe that (19.5) is just a reformulation of Lemma 15.4.3.

Of crucial importance is the "hereditary property of canonical flows", to which we turn now. Let us say that $y \in (\mathbb{R}^{1,3})^{\mathcal{E}_\alpha}$ is a *canonical flow for α* if it belongs to $\mathcal{Q}_\alpha$. The hereditary property states that y then induces a canonical flow in any subdiagram, i.e. for $\beta \subset \alpha$, $U_{\beta,\alpha}(y)$ is a canonical flow for β.

Lemma 19.2.2 *Consider two subdiagrams $\beta \subset \alpha$. Then*

$$U_{\beta,\alpha}(\mathcal{Q}_\alpha) \subset \mathcal{Q}_\beta.$$

Proof It suffices to show that if $y \in \mathcal{Q}_\alpha = (\ker \mathcal{L}_\alpha)^\perp$ then $U_{\beta,\alpha}(y) \in (\ker \mathcal{L}_\beta)^\perp = \mathcal{Q}_\beta$. To see this consider $z \in \ker \mathcal{L}_\beta \subset (\mathbb{R}^{1,3})^{\mathcal{E}_\beta}$. Consider the element $\bar{z} \in (\mathbb{R}^{1,3})^{\mathcal{E}_\alpha}$ given by $\bar{z}_e = z_e$ if $e \in \mathcal{E}_\beta$ and $z_e = 0$ otherwise. Then from Lemma 19.2.1 we have $\bar{z} \in \ker \mathcal{L}_\alpha$, so that $\bar{z} \perp y$ since $y \in (\ker \mathcal{L}_\alpha)^\perp$. Now the dot product of $\bar{z}$ and y equals the dot product of z and $U_{\beta,\alpha}(y)$ because the components of $\bar{z}$ corresponding to the edges of $\mathcal{E}_\alpha \setminus \mathcal{E}_\beta$ are zero. Hence $z \perp U_{\beta,\alpha}(y)$. □

Since $\mathcal{Q}_\beta = \ker \mathcal{I}_\beta$, one may reformulate the previous result as

$$\mathcal{I}_\beta U_{\beta,\alpha}(\mathcal{Q}_\alpha) = \{0\}. \tag{19.6}$$

The following formula is not intuitive, but it is important. It will allow us to calculate. It will be used many times.

Lemma 19.2.3 *Consider two subdiagrams $\beta \subset \alpha$. Then*

$$\mathcal{I}_\beta U_{\beta,\alpha} \mathcal{I}_\alpha = \mathcal{I}_\beta U_{\beta,\alpha}. \tag{19.7}$$

Proof Since $\mathcal{T}_\alpha(z) \in \mathcal{Q}_\alpha$, applying $\mathcal{I}_\beta U_{\beta,\alpha}$ to the equality $z = \mathcal{I}_\alpha(z) + \mathcal{T}_\alpha(z)$ (19.7) follows from (19.6). □

We will be very much concerned by the following problem. Consider two subdiagrams $\beta \subset \alpha$ and a function F on $(\mathbb{R}^{1,3})^{\mathcal{E}_\beta}$. Then $\tilde{F} := F U_{\beta,\alpha}$ is a function on $(\mathbb{R}^{1,3})^{\mathcal{E}_\alpha}$. How do we relate F, seen as a function on $(\ker \mathcal{L}_\beta) \times \mathcal{Q}_\beta$, with $\tilde{F}$, seen as a function on $(\ker \mathcal{L}_\alpha) \times \mathcal{Q}_\alpha$? The following is an important step in this program.

Lemma 19.2.4 *Consider two subdiagrams $\beta \subset \alpha$ and $w \in \ker \mathcal{L}_\alpha \subset (\mathbb{R}^{1,3})^{\mathcal{E}_\alpha}$. Then for $q \in \mathcal{Q}_\alpha$ we have*

$$U_{\beta,\alpha}(w + q) = \mathcal{I}_\beta U_{\beta,\alpha}(w) + c(w) + b(q), \tag{19.8}$$

where $b(q) \in \mathcal{Q}_\beta$ is a linear function of $q \in \mathcal{Q}_\alpha$ and where $c(w) \in \mathcal{Q}_\beta$ is a linear function of the variables w_e where $e \in \mathcal{E}_\alpha \setminus \mathcal{E}_\beta$ is an edge with at least a vertex in β.

Proof Let $z := U_{\beta,\alpha}(w+q)$, so that $\mathcal{I}_\beta(z) = \mathcal{I}_\beta U_{\beta,\alpha}(w)$ by (19.6). Then, recalling (19.3) and (19.4),

$$z = \mathcal{I}_\beta(z) + \mathcal{T}_\beta(z) = \mathcal{I}_\beta U_{\beta,\alpha}(w) + \mathcal{W}_\beta \mathcal{L}_\beta(z). \tag{19.9}$$

Now,

$$\mathcal{L}_\beta(z) = \mathcal{L}_\beta U_{\beta,\alpha}(w) + \mathcal{L}_\beta U_{\beta,\alpha}(q). \tag{19.10}$$

Since $\mathcal{L}_\alpha(w) = 0$, by Lemma 19.2.1, the first term $\mathcal{L}_\beta U_{\beta,\alpha}(w)$ on the right is a linear function $c'(w)$ of the quantities w_e where e is an edge adjacent to a vertex of β but is not an edge of β. The second term $\mathcal{L}_\beta U_{\beta,\alpha}(q)$ on the right is a linear function $b'(q)$ of q. Thus $\mathcal{W}_\beta \mathcal{L}_\beta(z) = \mathcal{W}_\beta c'(w) + \mathcal{W}_\beta b'(q)$. Combining with (19.9) this is the required result with $c(w) = \mathcal{W}_\beta c'(w)$ and $b(q) = \mathcal{W}_\beta b'(q)$. □

Let us note also the following.

Lemma 19.2.5 *Consider $x \in (\mathbb{R}^{1,3})^{\mathcal{E}}$ with $\mathcal{L}(x) = 0$ and $x \neq 0$. Consider the subdiagram τ consisting of all the edges e with $x_e \neq 0$. Then the connected components of this subdiagram are proper subdiagrams.*

Proof Let $x' = U_{\tau,\gamma}(x)$ so that $\mathcal{L}_\tau(x') = 0$ by (19.5). We apply now Corollary 15.6.4 to the diagram τ. Each edge e of τ is such that $x'(e) = x(e) \neq 0$, so that it belongs to a loop of τ. The conclusion follows by applying Lemma 16.4.3 to the connected components of τ. □

19.3 Grouping the Terms

The convergence of the renormalized integrand will be proved using Theorem 15.2.1. Throughout this section we fix once and for all a linear subspace S of $\ker \mathcal{L} \subset (\mathbb{R}^{1,3})^{\mathcal{E}}$, with $S \neq \{0\}$. Thus

$$\forall x \in S\,,\ \mathcal{L}(x) = 0. \tag{19.11}$$

Our ultimate goal is to prove Theorem 19.0.1. Depending on S, we will group the terms in the expression $\mathcal{R}F = (1 - T^{d(\gamma)}) \sum_{\mathcal{F}\ \mathrm{forest}}$ to bring forward cancellation between these terms. This is the marvelously clever part of Zimmermann's proof. The main grouping of the terms is obtained in the equation (19.19), $\mathcal{R}F = \sum_{\mathcal{G}\ \mathrm{complete}} \mathcal{R}_{\mathcal{G}} F$. The magic of that formula is that the quantities $\mathcal{R}_{\mathcal{G}} F$ may be recursively computed and that at certain stages the recursion takes the form (19.23), $H'_\alpha(x) = (1 - T^{d(\alpha)}) H_\alpha(x)$, where the crucial factor $1 - T^{d(\alpha)}$ produces cancellation.

For a subdiagram α we recall the maps $\mathcal{L}_\alpha$, $\mathcal{I}_\alpha$ and $\mathcal{T}_\alpha$ of Section 19.2. For $x \in (\mathbb{R}^{1,3})^{\mathcal{E}}$ and a subdiagram α we define

$$M_\alpha(x) := \mathcal{I}_\alpha U_{\alpha,\gamma}(x) \in \ker \mathcal{L}_\alpha \subset (\mathbb{R}^{1,3})^{\mathcal{E}_\alpha}. \tag{19.12}$$

An important idea is that "when working with α one does not see S but only $M_\alpha(S)$", so these spaces are essential. As a consequence of (19.7) we note that if $\beta \subset \alpha$ then

$$\mathcal{I}_\beta U_{\beta,\alpha} M_\alpha = M_\beta. \tag{19.13}$$

We will also use many times that (as follows from (19.12)),

$$z \in M_\alpha(S) \Rightarrow \mathcal{L}_\alpha(z) = 0.$$

Definition 19.3.1 An edge e of a subdiagram α is called *α-passive* if the function $z \mapsto z_e$ is identically zero on $M_\alpha(S)$. Otherwise the edge is called *α-active*.

The following simple fact is fundamental.

Lemma 19.3.2 *Consider two diagrams $\beta \subset \alpha$. Assume that all edges of α which are not edges of β but are adjacent to a vertex of β are α-passive. Then for $x \in S$ one has*

$$M_\beta(x) = U_{\beta,\alpha} M_\alpha(x). \tag{19.14}$$

In particular an edge of β is β-active if and only if it is α-active.

Proof Consider $x \in S$ and $z = M_\alpha(x)$, so that $\mathcal{L}_\alpha(z) = 0$. Since we assume that all edges of α that are not edges of β but are adjacent to a vertex of β are α-passive, for such an edge e we have $z_e = 0$ and thus by (19.5) we get $\mathcal{L}_\beta U_{\beta,\alpha}(z) = \mathcal{L}_\alpha(z) = 0$. Thus $\mathcal{I}_\beta U_{\beta,\alpha}(z) = U_{\beta,\alpha}(z)$ and therefore by (19.13) it holds that

$$M_\beta(x) = \mathcal{I}_\beta U_{\beta,\alpha} M_\alpha(x) = \mathcal{I}_\beta U_{\beta,\alpha}(z) = U_{\beta,\alpha}(z) = U_{\beta,\alpha} M_\alpha(x). \qquad \square$$

Let us recall that a forest $\mathcal{F}$ is a collection of proper connected subdiagrams (of the given Feynman diagram) containing γ, and such that if $\alpha, \beta \in \mathcal{F}$ and $\alpha \cap \beta \neq \emptyset$, then either $\alpha \subset \beta$ or $\beta \subset \alpha$.

Definition 19.3.3 Given a subdiagram α belonging to a forest $\mathcal{F}$, we denote by $\mathcal{E}(\mathcal{F},\alpha)$ the set of edges of α which are not edges of any $\beta \in \mathcal{F}$ which is strictly contained in α.

Equivalently, in the formulation of Lemma 16.5.3, the edges in $\mathcal{E}(\mathcal{F},\alpha)$ are the edges of α which are not edges of one of the α_i.

Definition 19.3.4 A forest $\mathcal{G}$ is called *complete* if for each $\alpha \in \mathcal{G}$ either all the edges of $\mathcal{E}(\mathcal{G},\alpha)$ are α-active or all these edges are α-passive.

Our program is as follows:

- Given a forest $\mathcal{F}$, we will construct in a natural manner a complete forest $\bar{\mathcal{F}}$ containing $\mathcal{F}$.
- We will group together the forests $\mathcal{F}$ with a given value of $\bar{\mathcal{F}}$ and this will bring cancellation forwards.

Let us try to provide some intuition for these claims, which are certainly not obvious, by examining the forest $\{\gamma\}$.[5] This forest has no reason to be complete, since some edges will be γ-active and some will be γ-passive. Let us call τ the subdiagram generated by the γ-active edges. This subdiagram need not be connected, so we decompose it into its connected components $\tau_1, \ldots, \tau_{\tilde{n}}$. Let us show that the forest $\mathcal{G} = \{\gamma, \tau_1, \ldots, \tau_{\tilde{n}}\}$ is complete. First, it is obvious that each edge of $\mathcal{E}(\mathcal{G}, \gamma)$ is γ-passive by construction. Next, consider a subdiagram τ_i. Each edge adjacent to a vertex of τ_i but not in τ_i is γ-passive since τ_i is a connected component of τ. Then by Lemma 19.3.2 each edge of τ_i is τ_i-active because it is γ-active, completing the proof that $\mathcal{G}$ is complete. For a subset I of $\{1, \ldots, \tilde{n}\}$ consider the forest $\mathcal{F}_I$ which consists of γ and of the diagrams τ_i for $i \in I$. It is not obvious now, but we will prove that a forest $\mathcal{F}$ is such that $\mathcal{G} = \bar{\mathcal{F}}$ (the natural complete forest containing $\mathcal{F}$ as we will construct below), if and only if there is a subset I of $\{1, \ldots, \tilde{n}\}$ such that $\mathcal{F} = \mathcal{F}_I$. Thus summing over the forests $\mathcal{F}$ such that $\bar{\mathcal{F}} = \mathcal{G}$ amounts to summing over the forests $\mathcal{F}_I$. Let us then compute[6] $\mathcal{F}_I F$ as in (16.18). For $i \leq \tilde{n}$ and $x \in (\mathbb{R}^{1,3})^{\mathcal{E}_{\tau_i}}$ let $F_i(x) = \prod_{e \in \mathcal{E}_{\tau_i}} f(x_e)$. For $x \in (\mathbb{R}^{1,3})^{\mathcal{E}_\gamma}$ let $U_i(x) = (x_e)_{e \in \mathcal{E}_{\tau_i}}$, so that $F_i(U_i(x)) = \prod_{e \in \mathcal{E}_{\tau_i}} f(x_e)$. It should then be obvious from (16.18) that for $x \in (\mathbb{R}^{1,3})^{\mathcal{E}_\gamma}$ we have

$$\mathcal{F}_I F(x) = \prod_{i \in I} (-T^{d(\tau_i)} F_i(U_i(x))) \prod f(x_e)$$

where the last product is over the edges of γ which are not edges of any τ_i for $i \in I$. Consequently,

$$\mathcal{F}_I F(x) = \prod_{i \in I} (-T^{d(\tau_i)} F_i(U_i(x))) \prod_{i \notin I} F_i(U_i(x)) \prod f(x_e)$$

where the last product is now over the edges of γ which are not edges of any τ_i, $i \leq \tilde{n}$. When we sum these quantities over I we obtain the identity

$$\sum_{I \subset \{1, \ldots, \tilde{n}\}} \mathcal{F}_I F = \prod_{i \leq \tilde{n}} (1 - T^{d(\tau_i)}) F_i(U_i(x)) \prod f(x_e). \tag{19.15}$$

and each of the terms $1 - T^{d(\tau_i)}$ creates cancellation. This will be explained in detail right after the statement of Lemma 19.4.1.

Given a forest $\mathcal{F}$ and $\alpha \in \mathcal{F}$, for every α in $\mathcal{F}$ we perform the following construction. (We will refer to this construction as "the basic construction relative to α".) We remove from α all the edges in $\mathcal{E}(\mathcal{F}, \alpha)$ which are α-passive, if such edges exist. (It will turn out that if no such edge exists, the construction does not do anything.) We consider the subdiagram τ generated by the remaining edges. Equivalently, τ is the subdiagram whose edges are edges of α which are either α-active, or are edges of one of the maximal subdiagrams $(\alpha_i)_{i \leq \tilde{n}}$ of α. This subdiagram τ need not be connected, so we decompose it in its connected components $\tau_1, \ldots, \tau_r$. The key property of this construction is as follows.

[5] All the ideas are already pretty apparent on that case, and the general case uses more technique but no more ideas.

[6] Of course $\mathcal{F}_I F$ denotes the function constructed in (16.19) when $\mathcal{F} = \mathcal{F}_I$.

Lemma 19.3.5 *(a) Each maximal subdiagram α_i is contained in one of the τ_j.*[7] *Consequently, each $\beta \in \mathcal{F}$ which is strictly contained in α is contained in one of the τ_j.*
(b) If $\beta \in \mathcal{F}$ contains one of the τ_j with $\tau_j \notin \mathcal{F}$ then β contains α.
(c) All the τ_j are proper subdiagrams.

Proof (a) holds because we have not removed any edge of any α_i (since by definition these edges do not belong to $\mathcal{E}(\mathcal{F},\alpha)$) and because α_i is connected by the definition of a forest.
(b) Such a β intersects α, so since $\mathcal{F}$ is a forest, if we assume that β does not contain α, then β is strictly contained in α, so it is contained in one of the α_i and by (a) β is contained in a certain $\tau_{j'}$. Since $\tau_j \subset \beta$, we must have $j = j'$ and $\beta = \tau_j$. This is impossible since we assume that $\tau_j \notin \mathcal{F}$.
(c) We will use Lemma 16.4.3 and prove that each edge e of τ_j is part of a loop contained in τ_j. Assume first that e is an edge of a certain α_i. Then by (a) $\alpha_i \subset \tau_j$. Since α_i is a proper subdiagram there exists a loop of α_i using that edge, and this loop is contained in τ_j.

On the other hand, by the definition of τ, if the edge e of τ is not an edge of any α_i it must be α-active. Thus there exists $z \in M_\alpha(S)$ with $z_e \neq 0$. Since $\mathcal{L}_\alpha(z) = 0$, by Lemma 19.2.5 the edge e is part of a loop contained in the set of edges where $z \neq 0$, and thus in τ. Since the loop passes through e which is an edge of the connected subdiagram τ_j, it is contained in τ_j. □

One may think of the previous construction as creating diagrams "between α and its maximum subdiagrams". One should not be confused by this sentence. Some of the τ_j may very well not contain any $\beta \in \mathcal{F}$. Let us also note that when all the edges in $\mathcal{E}(\mathcal{F},\alpha)$ are α-active, then in the previous construction we have $\tau = \alpha$. In that case, the construction does not create any new subdiagram, and the τ_j are simply the α_i. Of course Lemma 19.3.5 remains (trivially) true in that case.

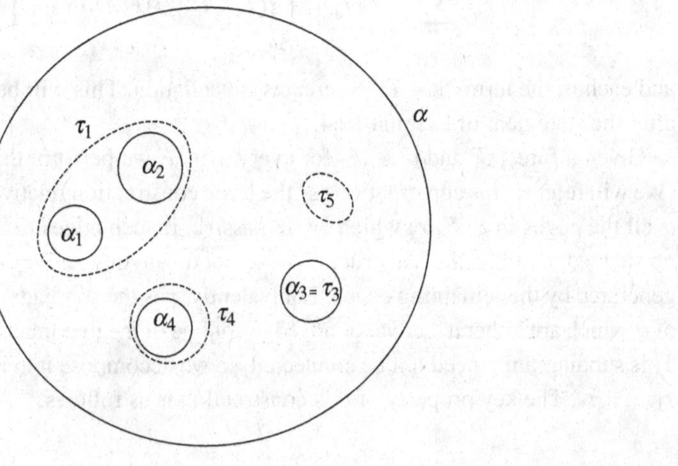

Figure 19.1 The main construction.

[7] But as shown in Figure 19.1 some of the τ_j may not contain any α_i, or may contain several of them.

Definition 19.3.6 Consider a forest $\mathcal{F}$ and a subdiagram $\beta \neq \gamma$. We do not require that $\beta \in \mathcal{F}$. We denote by $\beta^+_{\mathcal{F}}$ the smallest element of $\mathcal{F}$ which *strictly* contains β.

Thus β is strictly contained in $\beta^+_{\mathcal{F}}$. Until further notice we lighten notation by writing β^+ for $\beta^+_{\mathcal{F}}$.

Definition 19.3.7 Given a forest $\mathcal{F}$ we denote by $\bar{\mathcal{F}}$ the union of $\mathcal{F}$ and of all the sets τ_j generated performing the basic construction relative to all elements α of $\mathcal{F}$.

Lemma 19.3.8 *(a) If $\beta \in \bar{\mathcal{F}} \setminus \mathcal{F}$ then β is constructed by applying the basic construction to β^+.*
(b) Consider $\beta \in \bar{\mathcal{F}} \setminus \mathcal{F}$ and $\alpha \in \mathcal{F}$. If $\alpha \cap \beta \neq \emptyset$ and α is strictly contained in β^+ then $\alpha \subset \beta$.

Proof (a) By definition of $\bar{\mathcal{F}}$, β is obtained by applying the basic construction to a certain element α of $\mathcal{F}$, which by Lemma 19.3.5 (b) is the smallest element β^+ of $\mathcal{F}$ containing β.
(b) According to (a) β is obtained by applying the basic construction to β^+. The connected components τ_j of the diagram τ of this basic construction are disjoint and β is one of them. We apply Lemma 19.3.5 (a) to the pair β^+, α[8] to obtain that α is contained in one of the τ_j. Since $\alpha \cap \beta \neq \emptyset$ then $\beta = \tau_j$ and $\alpha \subset \beta$ since $\alpha \subset \tau_j$. □

Lemma 19.3.9 *$\bar{\mathcal{F}}$ is a forest.*

Proof Consider $\beta, \beta' \in \bar{\mathcal{F}}$, and assume that $\beta \cap \beta' \neq \emptyset$. We want to show that one of these sets contains the other. We simply have to patiently check all the cases. We already know that this is the case if they both belong to $\mathcal{F}$ so assume that $\beta \notin \mathcal{F}$. Thus β is obtained by applying the basic construction to $\beta^+ \in \mathcal{F}$. Assume first that $\beta' \in \mathcal{F}$. Thus $\beta' \cap \beta^+ \neq \emptyset$ and one of these sets contains the other since $\mathcal{F}$ is a forest. If $\beta^+ \subset \beta'$ then $\beta \subset \beta^+ \subset \beta'$. If β' is strictly contained in β^+, then $\beta' \subset \beta$ by Lemma 19.3.8 (b). Assume now that $\beta' \in \bar{\mathcal{F}} \setminus \mathcal{F}$, so that β' is obtained by applying the basic construction to β'^+. Again since $\mathcal{F}$ is a forest, one of the sets β^+, β'^+ contains the other. If $\beta^+ = \beta'^+$ it should be obvious that $\beta = \beta'$ (since they are the same connected component of the diagram τ of the basic construction). If β^+ is strictly contained in β'^+, by Lemma 19.3.8 (b) applied to $\alpha = \beta^+ \in \mathcal{F}$ we obtain that $\beta^+ \subset \beta'$, so that then $\beta \subset \beta^+ \subset \beta'$. Similarly if β'^+ is strictly contained in β^+ then $\beta' \subset \beta$. □

There is no reason why the τ_j should be renormalization sets, so $\bar{\mathcal{F}}$ need not consist of renormalization sets. This is why it is convenient to allow a forest to contain proper connected subdiagrams which are not renormalization sets. Only later we will prove that $\bar{\mathcal{F}}$ is the smallest complete forest containing $\mathcal{F}$.

Exercise 19.3.10 Consider $\alpha \in \mathcal{F}$ and its maximum subdiagrams $(\alpha_i)_{i \leq \tilde{n}}$. Prove that if all the edges in $\mathcal{E}(\mathcal{F},\alpha)$ are α-active, then the elements of $\bar{\mathcal{F}}$ which meet α either contain α or are contained in one of the α_i.

[8] That is the α of Lemma 19.3.5 is β^+ and the β of the lemma is α.

Lemma 19.3.11 *Consider $\beta \in \bar{\mathcal{F}} \setminus \mathcal{F}$. Then β^+ is the smallest element of $\bar{\mathcal{F}}$ which contains strictly β.*

The point here is that β^+ is defined as the smallest element of $\mathcal{F}$ containing β. The lemma shows that it is also the smallest element of the larger forest $\bar{\mathcal{F}}$ containing β. Recalling Definition 19.3.6 one may reformulate Lemma 19.3.11 as follows: If $\beta \in \bar{\mathcal{F}} \setminus \mathcal{F}$ then $\beta^+_{\bar{\mathcal{F}}} = \beta^+_{\mathcal{F}}$.

Proof Consider the smallest $\alpha \in \bar{\mathcal{F}}$ such that β is strictly contained in α, so that $\alpha \subset \beta^+$ since β is strictly contained in β^+ and $\beta^+ \in \bar{\mathcal{F}}$. We have to prove that $\alpha = \beta^+$. For this it suffices to prove that $\alpha \in \mathcal{F}$ since then $\alpha = \beta^+$ by definition of $\beta^+ \in \bar{\mathcal{F}}$. Assume for contradiction that $\alpha \in \bar{\mathcal{F}} \setminus \mathcal{F}$, so that by Lemma 19.3.8 (a) α is obtained by applying the basic construction to α^+. Since $\alpha \subset \beta^+ \in \mathcal{F}$ and since $\alpha \notin \mathcal{F}$, α is strictly contained in β^+ so that $\alpha^+ \subset \beta^+$. Assume first that α^+ is strictly contained in β^+. Since $\beta \subset \alpha \subset \alpha^+$ we have $\alpha^+ \cap \beta \neq \emptyset$ and we may apply Lemma 19.3.8 (b) to the pair $\alpha^+ \in \mathcal{F}$ and $\beta \in \bar{\mathcal{F}} \setminus \mathcal{F}$. We obtain that $\alpha \subset \beta$, which contradicts the fact that β is strictly contained in α. Assume finally that $\alpha^+ = \beta^+$. Then $\alpha = \beta$ because they are the same connected component of the diagram τ of the basic construction, and this contradicts again the fact that β is strictly contained in α. □

Proposition 19.3.12 *The forest $\bar{\mathcal{F}}$ is complete. More precisely,*
(a) Given $\alpha \in \mathcal{F}$ then all the edges in $\mathcal{E}(\bar{\mathcal{F}},\alpha)$ are α-passive.
(b) Given $\beta \in \bar{\mathcal{F}} \setminus \mathcal{F}$ all the edges in $\mathcal{E}(\bar{\mathcal{F}},\beta)$ are β-active.

Proof When $\alpha \in \mathcal{F}$ by construction $\mathcal{E}(\bar{\mathcal{F}},\alpha)$ consists only of α-passive edges. This proves (a). To prove (b) consider $\beta \in \bar{\mathcal{F}} \setminus \mathcal{F}$. By construction of $\bar{\mathcal{F}}$, such a β has been obtained as follows: There is α (=β^+) in $\mathcal{F}$ such that β is one of the connected components τ_j of the subdiagram τ of α obtained by removing from α the α-passive edges in $\mathcal{E}(\mathcal{F},\alpha)$. Consider an edge e of $\mathcal{E}(\bar{\mathcal{F}},\beta)$. We have to prove that e is β-active. Consider the maximum subdiagrams α_i of α. Next, we prove that e is not an edge of an α_i. By Lemma 19.3.5 (a) each of the α_i is contained in one of the τ_j. These do not have vertices in common. So, either α_i is contained in β or disjoint from β. Thus if e is an edge of an α_i, then $\alpha_i \subset \beta$, but since $\alpha_i \in \mathcal{F} \subset \bar{\mathcal{F}}$ this contradicts the assumption that $e \in \mathcal{E}(\bar{\mathcal{F}},\beta)$. We have proved that e is not an edge of an α_i, so that $e \in \mathcal{E}(\mathcal{F},\alpha)$. Since e belongs to τ, it is α-active by construction. Since β is a connected component of τ, any edge of $\mathcal{E}_\alpha \setminus \mathcal{E}_\beta$ adjacent to a vertex of β is not an edge of τ, so that by construction it is α-passive. Then Lemma 19.3.2 proves that e is β-active. □

Definition 19.3.13 Consider a forest $\mathcal{G}$ and a subdiagram $\alpha \in \mathcal{G}$. If all edges of $\mathcal{E}(\mathcal{G},\alpha)$ are α-active we say that α is *active*. If all edges of $\mathcal{E}(\mathcal{G},\alpha)$ are α-passive we say that α is *passive*.

By definition of a complete forest (Definition 19.3.4) each diagram in a complete forest is either active or passive.

We recall Definition 19.3.6.

Definition 19.3.14 Given a complete forest $\mathcal{G}$ we define the forest $\widehat{\mathcal{G}} \subset \mathcal{G}$ as consisting of γ and the subdiagrams of $\alpha \in \mathcal{G}$ which satisfy *either* of the following properties:

$$\alpha \text{ is passive} \tag{19.16}$$

or else

$$\alpha^+_{\mathcal{G}} \text{ is active.} \tag{19.17}$$

In some sense the important part of $\mathcal{G}$ is $\mathcal{G} \setminus \widehat{\mathcal{G}}$, which consists of the subdiagrams $\alpha \in \mathcal{G}$ for which α is active *and* $\alpha^+_{\mathcal{G}}$ is passive.

Proposition 19.3.15 *Given a complete forest $\mathcal{G}$, a forest $\mathcal{F}$ satisfies $\bar{\mathcal{F}} = \mathcal{G}$ if and only if $\widehat{\mathcal{G}} \subset \mathcal{F} \subset \mathcal{G}$.*

For clarity we will break the proof in three lemmas. Throughout the proof for a diagram $\alpha \in \mathcal{G}, (\alpha \neq \gamma)$ we use the notation $\alpha^+ = \alpha^+_{\mathcal{G}}$. (Thus our previous convention that $\alpha^+ = \alpha^+_{\mathcal{F}}$ is replaced from this point on by the convention that $\alpha^+ = \alpha^+_{\mathcal{G}}$.) Until the end of the proof *the notions of active and passive diagrams refer to the forest $\mathcal{G}$.*[9]

Lemma 19.3.16 *If $\bar{\mathcal{F}} = \mathcal{G}$ then $\widehat{\mathcal{G}} \subset \mathcal{F} \subset \mathcal{G}$.*

Proof First we have $\mathcal{F} \subset \bar{\mathcal{F}} = \mathcal{G}$. To prove that $\widehat{\mathcal{G}} \subset \mathcal{F}$ we consider $\alpha \in \bar{\mathcal{F}} \setminus \mathcal{F}$, with the goal of proving that $\alpha \notin \widehat{\mathcal{G}}$. Consider $\beta = \alpha^+_{\mathcal{F}}$. Since $\bar{\mathcal{F}} = \mathcal{G}$, it follows from Lemma 19.3.11 that $\beta = \alpha^+ (= \alpha^+_{\mathcal{G}})$. Moreover, by Proposition 19.3.12, α is active and $\beta = \alpha^+$ is passive. Thus as desired $\alpha \notin \widehat{\mathcal{G}}$. □

Lemma 19.3.17 *If $\widehat{\mathcal{G}} \subset \mathcal{F} \subset \mathcal{G}$ then $\mathcal{G} \subset \bar{\mathcal{F}}$.*

Proof Considering $\alpha \in \mathcal{G} \setminus \widehat{\mathcal{G}}$ we have to show that $\alpha \in \bar{\mathcal{F}}$. We may assume that $\alpha \notin \mathcal{F}$. By definition of $\widehat{\mathcal{G}}$, α^+ is passive and α is active. In particular $\alpha^+ \in \widehat{\mathcal{G}} \subset \mathcal{F}$. We first prove the following statement.

An edge e of α^+ which is adjacent to a vertex of α, but is not an edge of α is α^+-passive. (19.18)

We prove that $e \in \mathcal{E}(\mathcal{G},\alpha^+)$. Arguing by contradiction, otherwise e is an edge of a subdiagram β of $\mathcal{G}$ which is strictly contained in α^+. Since e is an edge of β, β has a vertex in common with α, and since $\mathcal{G}$ is a forest we have $\beta \subset \alpha$ by definition of α^+, so that e is an edge of α, in contradiction with our hypothesis. We have proved that $e \in \mathcal{E}(\mathcal{G},\alpha^+)$, so that e is α^+-passive since α^+ is passive.

Combining (19.3) with Lemma 19.3.2 proves that an edge of α is α-active if and only if it is α^+-active.

Consider the diagram τ constructed when we apply the basic construction to $\alpha^+ \in \mathcal{F}$. We now prove that an edge e of α is an edge of τ. If $e \in \mathcal{E}(\mathcal{G},\alpha)$ then e is α-active because α is active. According to (19.3) and Lemma 19.3.2 then e is α^+ active, and therefore is an

[9] On the other hand, given a diagram $\alpha \in \mathcal{F} \subset \mathcal{G}$ and an edge e of α, the notion that e is α active does not depend on whether we think of α as an element of $\mathcal{F}$ or of $\mathcal{G}$.

edge of τ by construction of τ. If $e \notin \mathcal{E}(\mathcal{G},\alpha)$ then e belongs to a subdiagram $\beta \in \mathcal{G}$ of α, $\beta \neq \alpha$, and we may assume that β is maximal with this property. Since $\beta \subset \alpha$ we have $\beta \subset \beta^+ \subset \alpha$ and then $\beta^+ = \alpha$ for otherwise β would not be maximal with respect to the property that $\beta \subset \alpha$ and $\beta \neq \alpha$. Then $\beta^+ = \alpha$ is active so that $\beta \in \widehat{\mathcal{G}} \subset \mathcal{F}$. Since $\beta \in \mathcal{F}$ is strictly contained in α, by construction of τ we have $\beta \subset \tau$ and again e is an edge of τ.

We have proved that an edge of α is an edge of τ so that $\alpha \subset \tau$. Since α is connected, it is then contained in a connected component τ_j of τ, and it follows from (19.3) that $\tau_j = \alpha$ and thus that $\alpha \in \bar{\mathcal{F}}$. □

Lemma 19.3.18 *If $\widehat{\mathcal{G}} \subset \mathcal{F} \subset \mathcal{G}$ then $\bar{\mathcal{F}} \subset \mathcal{G}$.*

Proof Let us consider $\alpha \in \mathcal{F} \subset \mathcal{G}$. We want to prove that the subdiagrams we add to $\mathcal{F}$ while performing the basic construction relative to α belong to $\mathcal{G}$. Consider the maximal subdiagrams $\alpha_1, \ldots, \alpha_{\bar{n}} \in \mathcal{G}$ of α as in Lemma 16.5.3. Let us stress that *we apply this lemma to $\mathcal{G}$, not to $\mathcal{F}$.*

Since $\mathcal{G}$ is complete, α is either active or passive.[10]

Assume first that α is active. By definition of $\widehat{\mathcal{G}}$, and since $\alpha_i^+ = \alpha$ is active, each α_i belongs to $\widehat{\mathcal{G}} \subset \mathcal{F}$ and thus $\mathcal{E}(\mathcal{G},\alpha) = \mathcal{E}(\mathcal{F},\alpha)$. Thus all edges of $\mathcal{E}(\mathcal{F},\alpha)$ are α-active and we add no element to $\mathcal{F}$ while applying the basic construction to α (as observed just before Definition 19.3.6).

Thus it is enough to consider the case where α is passive. Consider the usual diagram τ whose edges are either α-active, or are edges of some $\beta \in \mathcal{F}, \beta \subset \alpha, \beta \neq \alpha$. To construct this diagram we remove all the passive edges of $\mathcal{E}(\mathcal{F},\alpha)$, and in particular all the edges of $\mathcal{E}(\mathcal{G},\alpha)$, so that τ is contained in the union of the α_i.

Since each α_i is connected, a connected component τ_j of τ is contained in one of the α_i. We prove that then $\tau_j = \alpha_i \in \mathcal{G}$, concluding the proof of the lemma. Since α_i is connected it suffices to prove that $\alpha_i \subset \tau$. Assume first that α_i is passive. Then $\alpha_i \in \widehat{\mathcal{G}} \subset \mathcal{F}$ so that $\alpha_i \subset \tau$ by Lemma 19.3.5 (a). Assume next that α_i is active, and consider an edge e of α_i. If e is α_i-active, since α is passive then e is α-active by Lemma 19.3.2, so that e belongs to τ. If e is α_i-passive, since α_i is active, e is an edge of a subdiagram $\beta \in \mathcal{G}$ with $\beta \subset \alpha_i$ and $\beta \neq \alpha_i$. If β is maximal with respect to this property then $\beta^+ = \alpha_i$. Since $\beta^+ = \alpha_i$ is active then $\beta \in \widehat{\mathcal{G}} \subset \mathcal{F}$ and $\beta \subset \tau$ by Lemma 19.3.5 (a) and again e is an edge of τ. □

We have proved Proposition 19.3.15. It provides us with a method for regrouping the various forests $\mathcal{F}$ according to the value of $\bar{\mathcal{F}}$. This is done by writing (16.15) as

$$\mathcal{R}F = \sum_{\mathcal{G}\text{ complete}} \mathcal{R}_{\mathcal{G}}F, \tag{19.19}$$

where the sum is over all complete forests $\mathcal{G}$, and where

$$\mathcal{R}_{\mathcal{G}}F = (1 - T^{d(\gamma)}) \sum_{\widehat{\mathcal{G}}\subset\mathcal{F}\subset\mathcal{G}} \mathcal{F}F. \tag{19.20}$$

We show in the next section that cancellation occurs between the different terms of $\mathcal{R}_{\mathcal{G}}$.

[10] Recalling that these notions are with respect to $\mathcal{G}$ throughout the proof.

19.4 Bringing Forward Cancellation

In this section we prove a magical formula to compute $\mathcal{R}_{\mathcal{G}}F$, a formula that will bring forward the steps at which cancellation occurs. A key point here is that any set $\mathcal{F}$ of diagrams containing γ and such that $\widehat{\mathcal{G}} \subset \mathcal{F} \subset \mathcal{G}$ is a forest.

To describe the procedure, given a complete forest $\mathcal{G}$, for $\alpha \in \mathcal{G}$ we define recursively functions H'_α and H_α on $(\mathbb{R}^{1,3})^{\mathcal{E}_\alpha}$ as follows.[11]

Consider first the case where α contains no other diagram of $\mathcal{G}$. We set

$$H_\alpha(x) = \prod_{e \in \mathcal{E}_\alpha} f(x_e). \tag{19.21}$$

When $\alpha \in \widehat{\mathcal{G}}$ we define

$$H'_\alpha(x) = -T^{d(\alpha)} H_\alpha(x), \tag{19.22}$$

and when $\alpha \in \mathcal{G} \setminus \widehat{\mathcal{G}}$ or $\alpha = \gamma$ we set

$$H'_\alpha(x) = (1 - T^{d(\alpha)}) H_\alpha(x), \tag{19.23}$$

with the usual understanding that if $d(\alpha) < 0$ the term $T^{d(\alpha)}$ is 0.

Consider now the case where α contains other diagrams of $\mathcal{G}$. We consider the maximal elements $\alpha_1, \ldots, \alpha_{\tilde{n}}$ of α as in Lemma 16.5.3 and we set

$$H_\alpha(x) = \prod_{i \le \tilde{n}} H'_{\alpha_i}(U_i(x)) \prod f(x_e), \tag{19.24}$$

where the last product is over the edges in $\mathcal{E}(\mathcal{G},\alpha)$ and where $U_i = U_{\alpha_i,\gamma}$. We then define H'_α by (19.22) and (19.23). Let us stress right away the main features of this construction. The procedures (19.22) and (19.23) are very different. In the procedure (19.22) we basically replace a function by a certain Taylor polynomial. This does nothing to reduce the degree of the function. The crucial procedure is (19.23), where cancellation occurs. This procedure is applied exactly when α is active and α^+ is passive (or when $\alpha = \gamma$).

Here is the magic formula.

Lemma 19.4.1 *We have $\mathcal{R}_{\mathcal{G}}F = H'_\gamma$.*

The idea is very simple, and it is best to explain it in the simplest non-trivial case, where $\widehat{\mathcal{G}} = \{\gamma\}$, and where $\mathcal{G} \setminus \widehat{\mathcal{G}}$ consists of disjoint diagrams $\alpha_1, \ldots, \alpha_{\tilde{n}}$, which are the maximal subdiagrams of γ. (We recall at this point the notations used in the analysis of the completion of the forest $\{\gamma\}$ performed on page 569.) Using (19.22) and (19.23) we may rewrite the identity (19.15) as

$$\sum_{I \subset \{1,\ldots,\tilde{n}\}} \mathcal{F}_I F(x) = H_\gamma(x).$$

[11] To avoid using complicated notation, we do not indicate in the notation that this construction *depends* on the complete forest $\mathcal{G}$, which is fixed throughout the construction.

Now, it follows from Proposition 19.3.15 that the forests $\mathcal{F}_I$ are exactly the forests $\mathcal{F}$ such that $\bar{\mathcal{F}} = \mathcal{G}$, so that recalling (16.15) and applying the operator $(1 - T^{d(\alpha)})$ to this previous identity yields $\mathcal{R}_{\mathcal{G}} F = H'_\gamma$.

Proof of Lemma 19.4.1 The subsets $\mathcal{J}$ of $\mathcal{G} \setminus \widehat{\mathcal{G}}$ are in one-to-one correspondence with the forests $\mathcal{F}$ with $\widehat{\mathcal{G}} \subset \mathcal{F} \subset \mathcal{G}$, the correspondence being given by the map $\mathcal{J} \mapsto \mathcal{F}_{\mathcal{J}} := \mathcal{J} \cup \widehat{\mathcal{G}}$.

Given a set $\mathcal{J} \subset \mathcal{G} \setminus \widehat{\mathcal{G}}$ we define recursively functions $F'_{\alpha,\mathcal{J}}$ and $F_{\alpha,\mathcal{J}}$ on $(\mathbb{R}^{1,3})^{\mathcal{E}_\alpha}$ for $\alpha \in \mathcal{G}$ as follows. If α contains no other element of $\mathcal{G}$ we define $F_{\alpha,\mathcal{J}}(x) = \prod_{e \in \mathcal{E}_\alpha} f(x_e)$. If $\alpha \neq \gamma$ we then define $F'_{\alpha,\mathcal{J}}(x) = -T^{d(\alpha)} F_{\alpha,\mathcal{J}}(x)$ if $\alpha \in \mathcal{F}_{\mathcal{J}}$ and $F'_{\alpha,\mathcal{J}}(x) = F_{\alpha,\mathcal{J}}(x)$ if $\alpha \notin \mathcal{F}_{\mathcal{J}}$. If α contains other elements of $\mathcal{G}$ we denote by $\alpha_1, \ldots, \alpha_{\tilde{n}}$ these maximal elements and we define

$$F_{\alpha,\mathcal{J}}(x) = \prod_{i \le \tilde{n}} F'_{\alpha_i,\mathcal{J}}(U_i(x)) \prod f(x_e), \tag{19.25}$$

where the last product is over the edges in $\mathcal{E}(\mathcal{G},\alpha)$. If $\alpha \neq \gamma$ we then define $F'_{\alpha,\mathcal{J}}$ as previously: $F'_{\alpha,\mathcal{J}}(x) = -T^{d(\alpha)} F_{\alpha,\mathcal{J}}(x)$ if $\alpha \in \mathcal{F}_{\mathcal{J}}$ and $F'_{\alpha,\mathcal{J}}(x) = F_{\alpha,\mathcal{J}}(x)$ if $\alpha \notin \mathcal{F}_{\mathcal{J}}$. Finally, if $\alpha = \gamma$ we define $F'_{\gamma,\mathcal{J}} = (1 - T^{d(\gamma)}) F_{\gamma,\mathcal{J}}$. It should then be intuitively obvious that for $\alpha \in \mathcal{F}_{\mathcal{J}}$ we have $F_{\alpha,\mathcal{J}} = F_\alpha$, where the notation F_α refers to the construction described on page 506 relative to the forest $\mathcal{F}_{\mathcal{J}}$. The reason for this is that in our construction of $F_{\alpha,\mathcal{J}}$ we "do nothing" for the elements $\alpha \notin \mathcal{F}_{\mathcal{J}}$, so that the construction should be the same as if we replace $\mathcal{G}$ by $\mathcal{F}_{\mathcal{J}}$, in which case this construction is exactly the construction of F_α.[12] Consequently, by definition (16.19) of $\mathcal{F}_{\mathcal{J}} F$ we have $\mathcal{F}_{\mathcal{J}} F = F_{\gamma,\mathcal{J}}$ and

$$F'_{\gamma,\mathcal{J}} = (1 - T^{d(\gamma)}) F_{\gamma,\mathcal{J}} = (1 - T^{d(\gamma)}) \mathcal{F}_{\mathcal{J}} F.$$

By summation of this formula over all sets $\mathcal{J} \subset \mathcal{G} \setminus \widehat{\mathcal{G}}$ and comparing with (19.20) we have proved the formula

$$\mathcal{R}_{\mathcal{G}} F = \sum_{\mathcal{J} \subset \mathcal{G} \setminus \widehat{\mathcal{G}}} F'_{\gamma,\mathcal{J}}. \tag{19.26}$$

To prove Lemma 19.4.1 we need only to prove that the right-hand side is H'_γ. For this we have to find an appropriate statement which we can prove by recursion. We are going to prove that for each $\alpha \in \mathcal{G}$ we have

$$H'_\alpha = \sum_{\mathcal{J} \in \mathcal{Z}_\alpha} F'_{\alpha,\mathcal{J}}, \tag{19.27}$$

where

$$\mathcal{Z}_\alpha = \{\mathcal{J} \subset \mathcal{G} \setminus \widehat{\mathcal{G}}\, ;\, \beta \in \mathcal{J} \Rightarrow \beta \subset \alpha\}.$$

For $\alpha = \gamma$ the right-hand sides of (19.26) and (19.27) are the same and this proves the required equality $\mathcal{R}_{\mathcal{G}} F = H'_\gamma$.

We will prove (19.27) recursively. To explain the key features of the argument we first consider separately the case where $\alpha \in \mathcal{G}$ is such that it does not contain any other element

[12] It is of course not hard to give a formal proof of this fact.

of $\mathcal{G}$, so that $\{\alpha\} = \{\beta \in \mathcal{G}; \beta \subset \alpha\}$. If $\alpha \in \widehat{\mathcal{G}}$ we have $\mathcal{Z}_\alpha = \{\emptyset\}$, and $F_{\alpha,\emptyset}(x) = \prod_{e\in\mathcal{E}_\alpha} f(x_e) = H_\alpha(x)$. Since $\alpha \in \widehat{\mathcal{G}} \subset \mathcal{F}_\emptyset$ we have $F'_{\alpha,\emptyset} = -T^{d(\alpha)}F_{\alpha,\emptyset}$ and this implies $F'_{\alpha,\emptyset} = H'_\alpha$ as desired. If $\alpha \in \widehat{\mathcal{G}} \setminus \mathcal{G}$ then $\mathcal{Z}_\alpha = \{\emptyset, \{\alpha\}\}$, and $F_{\alpha,\emptyset}(x) = F_{\alpha,\{\alpha\}}(x) = \prod_{e\in\mathcal{E}_\alpha} f(x_e) = H_\alpha(x)$. Now, according to the definitions, since $\alpha \notin \mathcal{F}_\emptyset$ we have $F'_{\alpha,\emptyset} = F_{\alpha,\emptyset} = H_\alpha$ and since $\alpha \in \mathcal{F}_{\{\alpha\}}$ we have $F'_{\alpha,\{\alpha\}} = -T^{d(\alpha)}F_{\alpha,\{\alpha\}} = -T^{d(\alpha)}H_\alpha$. Thus

$$\sum_{\mathcal{J}\in\mathcal{Z}_\alpha} F'_{\alpha,\mathcal{J}} = F'_{\alpha,\emptyset} + F'_{\alpha,\{\alpha\}} = (1 - T^{d(\alpha)})H_\alpha = H'_\alpha$$

as desired.

Suppose now that α contains other elements of $\mathcal{G}$ and let $\alpha_1, \ldots, \alpha_{\tilde{n}}$ be the maximal elements as in Lemma 16.5.3, so that by the recursion hypothesis, for each $i \le \tilde{n}$ we have $H'_{\alpha_i} = \sum_{\mathcal{J}\in\mathcal{Z}_{\alpha_i}} F'_{\alpha_i,\mathcal{J}}$.

Consider first the simpler case where $\alpha \in \widehat{\mathcal{G}}$. For $\mathcal{J} \subset \mathcal{G}\setminus\widehat{\mathcal{G}}$ define $\mathcal{J}_i = \{\beta \in \mathcal{J}; \beta \subset \alpha_i\}$ so that if $\mathcal{J} \in \mathcal{Z}_\alpha$ then $\mathcal{J} = \bigcup_{i\le\tilde{n}} \mathcal{J}_i$, and this defines a one-to-one correspondence between $\{\mathcal{J}; \mathcal{J} \in \mathcal{Z}_\alpha\}$ and the set of sequences $(\mathcal{J}_i)_{i\le\tilde{n}}$ such $\mathcal{J}_i \in \mathcal{Z}_{\alpha_i}$. It then follows from (19.25) that

$$F_{\alpha,\mathcal{J}}(x) = \prod_{i\le\tilde{n}} F'_{\alpha_i,\mathcal{J}}(U_i(x)) \prod f(x_e).$$

Now, since the construction of $F_{\alpha,\mathcal{J}}$ depends only on those $\beta \in \mathcal{J}$ such that $\beta \subset \alpha$ we have $F'_{\alpha_i,\mathcal{J}} = F'_{\alpha_i,\mathcal{J}_i}$ so that

$$F_{\alpha,\mathcal{J}}(x) = \prod_{i\le\tilde{n}} F'_{\alpha_i,\mathcal{J}_i}(U_i(x)) \prod f(x_e),$$

and hence

$$\begin{aligned}\sum_{\mathcal{J}\in\mathcal{Z}_\alpha} F_{\alpha,\mathcal{J}}(x) &= \sum_{\mathcal{J}\in\mathcal{Z}_\alpha}\prod_{i\le\tilde{n}} F'_{\alpha_i,\mathcal{J}_i}(U_i(x)) \prod f(x_e) \\ &= \prod_{i\le\tilde{n}}\sum_{\mathcal{J}_i\in\mathcal{Z}_{\alpha_i}} F'_{\alpha_i,\mathcal{J}_i}(U_i(x)) \prod f(x_e) \\ &= \prod_{i\le\tilde{n}} H'_{\alpha_i}(U_i(x)) \prod f(x_e) = H_\alpha(x), \end{aligned} \tag{19.28}$$

where we use the recursion hypothesis in one before last equality. This implies (19.27) from our definitions.

Consider next the case $\alpha \in \mathcal{G} \setminus \widehat{\mathcal{G}}$. The class of sets $\mathcal{J} \in \mathcal{Z}_\alpha$ splits into two parts. First the class $\mathcal{Z}_\alpha^-$ of sets $\mathcal{J}$ for which $\alpha \notin \mathcal{J}$. For these sets as above $\mathcal{J} = \bigcup_{i\le\tilde{n}} \mathcal{J}_i$. Second, the class $\mathcal{Z}_\alpha^+$ of sets $\mathcal{J}$ for which $\alpha \in \mathcal{J}$. For these sets $\mathcal{J} = \{\alpha\} \cup \bigcup_{i\le\tilde{n}} \mathcal{J}_i$. Just as in (19.28) we prove that

$$\sum_{\mathcal{J}\in\mathcal{Z}_\alpha^-} F_{\alpha,\mathcal{J}} = \sum_{\mathcal{J}\in\mathcal{Z}_\alpha^+} F_{\alpha,\mathcal{J}} = H_\alpha. \tag{19.29}$$

Then, for $\mathcal{J} \in \mathcal{Z}_\alpha^-$ we have $\alpha \notin \mathcal{J}$ so that $F'_{\alpha,\mathcal{J}} = F_{\alpha,\mathcal{J}}$. And for $\mathcal{J} \in \mathcal{Z}_\alpha^+$ we have $\alpha \in \mathcal{J}$ so that $F'_{\alpha,\mathcal{J}} = -T^{d(\alpha)} F_{\alpha,\mathcal{J}}$. Consequently (19.29) implies

$$\sum_{\mathcal{J} \in \mathcal{Z}_\alpha^-} F'_{\alpha,\mathcal{J}} = H_\alpha \; ; \; \sum_{\mathcal{J} \in \mathcal{Z}_\alpha^+} F'_{\alpha,\mathcal{J}} = -T^{d(\alpha)} H_\alpha$$

and thus

$$\sum_{\mathcal{J} \in \mathcal{Z}_\alpha} F'_{\alpha,\mathcal{J}} = \sum_{\mathcal{J} \in \mathcal{Z}_\alpha^-} F'_{\alpha,\mathcal{J}} + \sum_{\mathcal{J} \in \mathcal{Z}_\alpha^+} F'_{\alpha,\mathcal{J}} = (1 - T^{d(\alpha)}) H_\alpha = Z'_\alpha.$$

□

19.5 Regular Rational Functions

We have made significant progress toward the proof of (19.1). Indeed, if we combine (19.19) and Lemma 19.4.1, we see that it suffices to prove that for any complete forest $\mathcal{G}$, we have

$$\deg_u H'_\gamma(x(u)) < -\dim S. \tag{19.30}$$

We will think now as if S and the complete forest $\mathcal{G}$ are fixed once and for all. A basic idea is to recursively control the degree of the functions H_α of Section 19.4 with respect to certain variables. To lighten notation we write

$$x_\alpha(u) = M_\alpha(x(u)) = \mathcal{I}_\alpha U_{\alpha,\gamma}(x(u)) \in \ker \mathcal{L}_\alpha \subset (\mathbb{R}^{1,3})^{\mathcal{E}_\alpha}.$$

We recall the functions H_α and H'_α constructed in (19.24) and (19.23), and that our goal is to control $\deg_u H'_\gamma(x_\gamma(u))$. Our induction hypotheses will involve control of the functions *$H_\alpha(x_\alpha(u)+q)$, which are seen as functions of $u \in \mathbb{R}^N$ and $q \in \mathcal{Q}_\alpha$.*[13] The control of these functions will be either from their degree in u or their degree in u and q.[14]

Let us first recall generalities about the degree of rational functions. We consider variables $u = (u_i)_{i \le N}, q = (q_j)_{j \le N'}$. We recall that a function $P(u,q)$ is called a polynomial in u and q if it is of the type

$$P(u,q) = \sum_{\nu_1,\ldots,\nu_N;\mu_1,\ldots\mu_{N'}} c_{\nu_1,\ldots,\nu_N;\mu_1,\ldots\mu_{N'}} u_1^{\nu_1} \cdots u_N^{\nu_N} q_1^{\mu_1} \cdots q_{N'}^{\mu_{N'}}, \tag{19.31}$$

where the ν_i and the μ_j are integers ≥ 0. The degree $\deg_{u,q} P(u,q)$ is the largest value of $\nu_1 + \cdots + \nu_N + \mu_1 + \cdots + \mu_{N'}$ for which the corresponding coefficient $c_{\nu_1,\ldots,\nu_N;\mu_1,\ldots\mu_{N'}}$ is not zero. On the other hand, $\deg_u P(u,q)$ is the degree of P when considering q as a constant, i.e. it is the largest value of $\nu_1 + \cdots + \nu_N$ for which a coefficient $c_{\nu_1,\ldots,\nu_N;\mu_1,\ldots\mu_{N'}}$ is not zero. The quantity $\deg_q P(u,q)$ is defined similarly. Obviously

$$\deg_u P(u,q) \le \deg_{u,q} P(u,q) \le \deg_u P(u,q) + \deg_q P(u,q). \tag{19.32}$$

13 The reader may at first wonder why instead we do not rather control the functions $H'_\alpha(x_\alpha(u)+q)$, since after all our goal is to control the function $H'_\gamma(x(u)+q)$. The reason for this is simple: the functions H_α depend only on which subgraphs of α belong to $\mathcal{G}$. On the other hand there are two different procedures to go from H_α to H'_α which are reflected by (19.22) and (19.23). Deciding which of these procedures applies requires knowing whether α^+ is active or passive.

14 The art of recursion is to find the proper recursion hypothesis!

For a rational function $K(u,q) = P(u,q)/Q(u,q)$ we define

$$\deg_{u,q} K(u,q) = \deg_{u,q} P(u,q) - \deg_{u,q} Q(u,q),$$

and $\deg_u K(u,q)$ similarly.[15] Controlling the degree of a polynomial is easy through (19.32). It is another matter to control the degree of rational functions, because the denominators have a sneaky tendency to reduce their degree when we do not watch them. For example it is *not* true in general that $\deg_u K(u,q) \le \deg_{u,q} K(u,q)$, e.g. $K(u,q) = 1/q$. It is not true either that $\deg_u K(u,q) \le \deg_u K(u,0)$, e.g. $K(u,q) = 1/(1+uq)$.

The good news is that the rational functions we care about have denominators of a very special form. Just as in Lemma 19.1.4 one sees that the denominators of the rational functions $H_\alpha(x)$ are always of the type $\prod_{i\le s}(W_\varepsilon(J_i(x)_{e_i}) + m^2)$ where J_i is a map from $(\mathbb{R}^{1,3})^{\mathcal{E}_\alpha}$ to itself.[16] Recalling that $x_\alpha(u) = M_a(x(u))$ and that $x(u) = f_0 + \sum_{j\le N} u_j f_j$, $J_i(x(u)+q)_{e_i}$ is of the form $\sum_{j\le N} u_j a_{i,j} + J_i(q)_{e_i} + a_i$ where $a_{i,j}, a_i \in \mathbb{R}^{1,3}$. Consequently, the denominator of a function $H_\alpha(x_\alpha(u)+q)$ is a product $\prod_{i\le s}(W_\varepsilon(\sum_{j\le N} u_j a_{i,j} + b_i(q)) + m^2)$, where $a_{i,j} \in \mathbb{R}^{1,3}$ and where $b_i(q) \in \mathbb{R}^{1,3}$ is the sum of a constant term and a linear function of q (we will call such a function *a first-degree polynomial* in q).

We now formalize this result by defining convenient special classes of rational functions, and by showing that for these functions, a number of pathologies (such as the ones described above) never occur. Rather than thinking of q as a point of $\mathbb{R}^{N'}$ it is more convenient to think of q as a point in a finite-dimensional space $\mathcal{Q}$, and this is what we do from now on. We also think of ε as fixed once and for all.

Definition 19.5.1 A *semiregular function* $K(u,q)$, $u = (u_i)_{i\le N} \in \mathbb{R}^N$, $q \in \mathcal{Q}$ is a rational function of the form

$$K(u,q) = \frac{P(u,q)}{Q(u,q)}, \tag{19.33}$$

where P, Q are polynomials in the components of u and q and Q is of the type

$$Q(u,q) = \prod_{i\le N_1} \Big(W_\varepsilon\Big(\sum_{j\le N} u_j a_{i,j} + b_i(q)\Big) + m^2\Big). \tag{19.34}$$

Here, as in (16.2), $W_\varepsilon(k) = -(1+\mathrm{i}\varepsilon)(k^0)^2 + \boldsymbol{k}^2$, $b_i(q) \in \mathbb{R}^{1,3}$ is a first-degree polynomial in q, and $a_{i,j} \in \mathbb{R}^{1,3}$. Moreover, when each term in the product (19.34) depends on u, that is when

$$\forall i \le N_1 \;;\; \exists j \le N \;,\; a_{i,j} \ne 0, \tag{19.35}$$

we will say that the function is *regular*.

We have shown the following.

Lemma 19.5.2 *The functions $H_\alpha(x_\alpha(u)+q)$ are semiregular.*

15 There are many ways to write the same function K as a ratio of two polynomials, but if $K = P'/Q'$ then $PQ' = P'Q$ so that $\deg P - \deg Q = \deg P' - \deg Q'$ and the previous definition makes sense.
16 We do not use here that J_i is a tensor map.

These functions will often be regular, and regular functions have nice properties which we explore first.

Lemma 19.5.3 *(a) A partial derivative of a regular function with respect to a component of q or u is also a regular function.*
(b) A product of regular functions is a regular function.
(c) For a regular function K one has

$$\deg_u K(u,0) \le \deg_u K(u,q) \le \deg_{u,q} K(u,q). \tag{19.36}$$

Proof (a) and (b) are easily checked. For (c) since each term in the product (19.34) actually depends on u, we have

$$\deg_u Q(u,0) = \deg_u Q(u,q) = \deg_{u,q} Q(u,q) = 2N_1, \tag{19.37}$$

from which the result follows since $\deg_u P(u,0) \le \deg_u P(u,q) \le \deg_{u,q} P(u,q)$. □

Lemma 19.5.4 *For a regular function K we have*

$$\deg_{u,q} T_q^d K \le \deg_{u,q} K. \tag{19.38}$$

Here and everywhere $T_q^d K$ denotes the Taylor polynomial at zero in q of order d of the function K.[17]

Proof $T_q^d K$ is a sum of terms of the type $K_1(q)K_2(u,0)$ where $K_1(q)$ is polynomial of degree $k \le d$ in q and where K_2 is obtained from K by taking k derivatives with respect to certain components of q. Thus

$$\deg_{u,q} K_1(q)K_2(u,0) = k + \deg_u K_2(u,0) \le k + \deg_{u,q} K_2,$$

where we used Lemma 19.5.3 (a) and (c) in the last inequality. Now $\deg_{u,q} K_2 \le \deg_{u,q} K - k$ because taking a derivative with respect to a component of q of a rational function $R(u,q)$ lowers $\deg_{u,q} R$ by at least 1, and this proves (19.38). □

Proposition 19.5.5 *If K is a regular function then for an integer $d \ge 0$ the function $(1 - T_q^d)K := K - T_q^d K$ is regular and we have*

$$\deg_u (1 - T_q^d)K \le \deg_{u,q} K - d - 1. \tag{19.39}$$

This result is the crux of the BPHZ method, as it achieves a reduction of degree. We start our preparation for the proof with the following.

Lemma 19.5.6 *(a) Consider a rational function $R = P/Q$ where P, Q are polynomials in u and q, with Q never zero. Then if $T_q^d R = 0$ we have $T_q^d P = 0$.*
(b) If a polynomial P satisfies $T_q^d P = 0$ then $\deg_u P \le \deg_{u,q} P - d - 1$.

[17] The operator T_q^d should not be confused with the operator T^d of Definition 16.3.1. We will study the relationship between these operators in the next section.

Proof (a) Stating that for a function K of u,q we have $T_q^d K = 0$ means that if we compute any partial derivative of order $\le d$ of K with respect to the components of q and set $q = 0$ we obtain zero. Writing $P = RQ$ and computing the partial derivatives of P using Leibniz rule yields the result.
(b) Writing P as in (19.31), the condition $T_q^d P = 0$ means that $c_{\nu_1,\ldots,\nu_N;\mu_1,\ldots\mu_{N'}} = 0$ unless $\mu_1 + \cdots + \mu_{N'} \ge d+1$ and then the result is obvious by definition of $\deg_u$ and $\deg_{u,q}$. □

Lemma 19.5.7 *A regular function K with $T_q^d K = 0$ satisfies $\deg_u K \le \deg_{u,q} K - d - 1$.*

Proof Since K is a regular function it is of the type P/Q where Q is as in (19.34). Combining (a) and (b) of Lemma 19.5.6 we have $\deg_u P \le \deg_{u,q} P - d - 1$. According to (19.37) we have $\deg_u Q = \deg_{u,q} Q$ and then $\deg_u K = \deg_u P - \deg_u Q \le \deg_{u,q} P - d - 1 - \deg_{u,q} Q = \deg_{u,q} K - d - 1$. □

Proof of Proposition 19.5.5 It follows from Lemma 19.5.12 that the function $K' := (1 - T_q^d)K$ is regular and from Lemma 19.5.4 that $\deg_{u,q} K' \le \deg_{u,q} K$. The Taylor polynomial of degree d of a polynomial of degree d is itself. So $T_q^d(T_q^d K) = T_q^d K$ and then $T_q^d K' = 0$ so that by Lemma 19.5.7 we obtain $\deg_u K' \le \deg_{u,q} K' - d - 1 \le \deg_{u,q} K - d - 1$. □

Lemma 19.5.8 *(a) Taking a derivative with respect to a component of q of a semiregular function yields a semiregular function.*
(b) If K is semiregular then $\deg_u K(u,0) \le \deg_u K(u,q)$.

Proof (a) is obvious. To prove (b) we observe that (obviously) $K(u,q)$ is of the type $R(u,q)/Q'(q)$ where $R(u,q)$ is a regular rational function and $Q'(q)$ is a polynomial. Then $\deg_u K(u,0) = \deg_u R(u,0) \le \deg_u R(u,q) = \deg_u K(u,q)$ using (19.36) in the inequality. □

Lemma 19.5.9 *For a semiregular function $K(u,q)$ we have*

$$\deg_u T_q^d K \le \deg_u K \tag{19.40}$$

and

$$\deg_{u,q} T_q^d K \le \deg_u K + d. \tag{19.41}$$

Proof Similar to Lemma 19.5.4 using now Lemma 19.5.8. □

To understand the difference between (19.38) and (19.41) one may consider the case where $R(u,q)$ does not depend on u.

Definition 19.5.10 Consider variables u and q as in Definition 19.5.1. A function $K(u,q)$ is called *fully regular* if it is semiregular and $Q(u,q)$ does not depend on q.

Lemma 19.5.11 *A fully regular function is regular.*

Proof When $Q(u,q)$ does not depend on q, each factor of $Q(u,q)$ is either constant or actually depends on u, so we can always write $Q(u,q) = C\prod_{i\le N_2}(W_\varepsilon(\sum_{j\le N} u_j a_{i,j} + b_i) + m^2)$ where C is a constant and where $\forall i \le N_2$; $\exists j \le N$, $a_{i,j} \ne 0$. □

Lemma 19.5.12 *For a semiregular function K, the function $T_q^d K$ is fully regular.*

Proof Setting $q = 0$ in the derivatives removes the dependence on q in the denominators. □

Definition 19.5.13 If $\mathcal{Q}$ and $\mathcal{Q}'$ are finite dimensional spaces we say that a function $q(u,q')\colon \mathbb{R}^N \times \mathcal{Q}' \to \mathbb{R}^N \times \mathcal{Q}$ is a first-degree polynomial if it is the sum of a linear function in u and q' and of a constant term. A first-degree polynomial $q(q')\colon \mathcal{Q} \to \mathcal{Q}'$ is defined similarly.

Lemma 19.5.14 *When the function K is fully regular and we make a change of variable $q = q(u,q')$ where $q(u,q')$ is a first-degree polynomial, then the function $(u,q') \mapsto K(u,q(u,q'))$ is regular*[18] *and*

$$\deg_{u,q'} K(u,q(u,q')) \leq \deg_{u,q} K(u,q). \tag{19.42}$$

We will need to make the change of variable $q \to q(u,q')$ and we need to keep controlling the degree after we make this change of variable. It is this lemma that motivated the idea of a fully regular function.

Proof When $q(u,q')$ is a first-degree polynomial, for any polynomial $P(u,q)$, one has

$$\deg_{u,q'} P(u,q(u,q')) \leq \deg_{u,q} P(u,q). \tag{19.43}$$

The potential problem would be that the change of variable lowers the degree of the denominator, but since the function $K(u,q)$ is fully regular, the substitution $q \to q(u,q')$ does not change the denominator of this function since this denominator does not depend on q. □

Lemma 19.5.15 *When the function K is regular and we make a change of variable $q = q(q')$ where $q(q')$ is a first-degree polynomial, then the function $K(u,q(q'))$ is regular and*

$$\deg_{u,q'} K(u,q(q')) \leq \deg_{u,q} K(u,q)\,; \tag{19.44}$$

$$\deg_u K(u,q(q')) \leq \deg_u K(u,q). \tag{19.45}$$

We will need to make the change of variable $q \to q(q')$ and we need to keep controlling the degree after we make this change of variable. It is this lemma that motivated the idea of a regular function.

Proof When the function is regular as in (19.33) the substitution $q \to q(q')$ does not change the form $Q(u,q)$ of its denominator, and it does not change its degree in u, because the degree in u of $Q(u,q)$ is simply twice the number of factors, see (19.37). This substitution does not change either the degree in u and q of the denominator, again by (19.37). Using (19.43) for the numerator yields (19.44). In the case of (19.45) one proceeds similarly using that $\deg_u P(u,q(q')) \leq \deg_u P(u,q)$ instead of (19.43). □

[18] It is true that this function is fully regular, but we will only use that it is regular, so we state the lemma this way.

19.6 Controlling the Degree

We are ready to prove (19.30), by recursively controlling the functions H'_α. A first idea is that, when working with the diagram α "one does not see S and one sees only $M_\alpha(S)$". Thus the dimension $m(\alpha)$ of this space is likely to be very important.

Definition 19.6.1 Recalling the map $M_\alpha = \mathcal{I}_\alpha U_{\alpha,\gamma}$ of (19.12) we define

$$m(\alpha) := \dim M_\alpha(S). \tag{19.46}$$

In particular $\dim S = m(\gamma)$ so that our goal is to prove that

$$\deg_u H'_\gamma(x(u)) < -m(\gamma). \tag{19.47}$$

For our recursion arguments we need to relate $m(\alpha)$ to the same quantity calculated for the maximum subdiagrams of α, and doing this is our first task.

Lemma 19.6.2 *Consider α and its maximal subdiagrams $\alpha_1, \ldots, \alpha_{\tilde{n}}$ with respect to $\mathcal{G}$. Then if α is passive we have*

$$m(\alpha) \leq \sum_{i \leq \tilde{n}} m(\alpha_i), \tag{19.48}$$

whereas if α is active we have

$$m(\alpha) \leq \sum_{i \leq \tilde{n}} m(\alpha_i) + \dim \ker \mathcal{L}_\alpha - \sum_{i \leq \tilde{n}} \dim \ker \mathcal{L}_{\alpha_i}. \tag{19.49}$$

The underlying idea is rather simple. When α is passive, in some sense $M_\alpha(S)$ cannot be larger than the sum of the $M_{\alpha_i}(S)$ because $M_\alpha(S)$ does not use the edges in $\mathcal{E}(\alpha, \mathcal{G})$. When α is active, whatever constraints were limiting the size of $m(\alpha_i)$ do not go away. Viewing these constraints as lower bounds on $\dim \ker \mathcal{L}_{\alpha_i} - m(\alpha_i)$ we should have

$$\dim \ker \mathcal{L}_\alpha - m(\alpha) \geq \sum_{i \leq \tilde{n}} (\dim \ker \mathcal{L}_{\alpha_i} - m(\alpha_i)),$$

which is just (19.49).

Proof For each $i \leq \tilde{n}$ define $V_i := \mathcal{I}_{\alpha_i} U_{\alpha_i,\alpha}$. It follows from (19.7) that

$$\forall x \in (\mathbb{R}^{1,3})^{\mathcal{E}} \; ; \; V_i(M_\alpha(x)) = M_{\alpha_i}(x). \tag{19.50}$$

Consequently $V_i(M_\alpha(S)) = M_{\alpha_i}(S)$ and thus

$$\dim V_i(M_\alpha(S)) = m(\alpha_i). \tag{19.51}$$

Elementary linear algebra teaches us that given a linear operator W between two finite-dimensional spaces E and F one has $\dim E \leq \dim \ker W + \dim F$. Using this for $E = M_\alpha(S)$ and the operator $W(x) = (V_i(x))_{i \leq \tilde{n}}$ yields

$$m(\alpha) = \dim M_\alpha(S) \leq \dim \Big(M_\alpha(S) \cap \bigcap_{i \leq \tilde{n}} \ker V_i \Big) + \sum_{i \leq \tilde{n}} m(\alpha_i). \tag{19.52}$$

The task now is to control the dimension of $T := M_\alpha(S) \cap \bigcap_{i\le\bar{n}} \ker V_i$. Consider $x \in S$ with $z = M_\alpha(x) \in T$. Note that by (19.50) we have $M_{\alpha_i}(x) = 0$.

Assume first that α is passive. By (19.14) we have $U_{\alpha_i,\alpha}M_\alpha(x) = M_{\alpha_i}(x) = 0$ so that $z_e = 0$ for $e \in \mathcal{E}_{\alpha_i}$. Since α is passive we have $z_e = 0$ for $e \in \mathcal{E}(\alpha,\mathcal{F})$. Thus $z = 0$. Therefore $T = \{0\}$ and (19.52) implies (19.48).

Assume next that α is active. Let θ_i be the map from $(\mathbb{R}^{1,3})^{\mathcal{E}_{\alpha_i}}$ to $(\mathbb{R}^{1,3})^{\mathcal{E}_\alpha}$ given by $\theta_i(s)_e = s_e$ for $e \in \mathcal{E}_{\alpha_i}$ and $\theta_i(s)_e = 0$ otherwise. Then by (19.5)

$$\theta_i(\ker \mathcal{L}_{\alpha_i}) \subset \ker \mathcal{L}_\alpha. \tag{19.53}$$

Since $\mathcal{I}_{\alpha_i}U_{\alpha_i,\alpha}(z) = V_i(z) = 0$ then $U_{\alpha_i,\alpha}(z) \in \ker \mathcal{I}_{\alpha_i} = (\ker \mathcal{L}_{\alpha_i})^\perp$, so that $z \perp \theta_i(\ker \mathcal{L}_{\alpha_i})$. Consequently $T \subset \ker \mathcal{L}_\alpha$ and at the same time $T \perp \theta_i(\ker \mathcal{L}_{\alpha_i})$. These later spaces are orthogonal to each other because the diagrams α_i have no edge in common since they have no vertex in common. Since they are subspaces of $\ker \mathcal{L}_\alpha$, we get

$$\dim T \le \dim\ker \mathcal{L}_\alpha - \sum_{i\le n} \dim\ker \mathcal{L}_{\alpha_i},$$

and this concludes the proof. □

We recall the notation $x_\alpha(u) = M_\alpha(x(u)) \in \ker \mathcal{L}_\alpha$. We also recall *that the function $H_\alpha(x_\alpha + q)$ are always semiregular*. We state our main induction hypotheses.

Proposition 19.6.3 *(a) If α is active the function $H_\alpha(x_\alpha(u)+q)$ is regular and satisfies*

$$\deg_{u,q} H_\alpha(x_\alpha(u)+q) \le d(\alpha) - m(\alpha). \tag{19.54}$$

(b) If α is passive the function $H_\alpha(x_\alpha(u)+q)$ (is semiregular and) satisfies

$$\deg_u H_\alpha(x_\alpha(u)+q) \le -m(\alpha). \tag{19.55}$$

Moreover, if $m(\alpha) > 0$ it holds that

$$\deg_u H_\alpha(x_\alpha(u)+q) < -m(\alpha). \tag{19.56}$$

A first understanding of the content of Proposition 19.6.3 is reached by considering the case where α contains no other subdiagrams of $\mathcal{G}$. Then

$$H_\alpha(x_\alpha(u)+q) = \prod_{e\in\mathcal{E}_\alpha} f((x_\alpha(u)+q)_e). \tag{19.57}$$

Since α contains no other subdiagram of $\mathcal{G}$, when α is passive every edge is α-passive, which means that $x_\alpha(u)$ does not depend on u, so that the right-hand side does not depend on u and thus its degree in u is $0 = m(\alpha)$, and (19.55) holds. When α is active, each term in the right-hand side of (19.57) contributes -2 to the degree, so that

$$\deg_u H_\alpha(x_\alpha(u)+q) = -2\,\mathrm{card}\,\mathcal{E}_\alpha = d(\alpha) - \dim(\ker \mathcal{L}(\alpha)) \le d(\alpha) - m(\alpha),$$

using that $M_\alpha(S) \subset \ker \mathcal{L}(\alpha)$ and hence $m(\alpha) \le \dim(\ker \mathcal{L}(\alpha))$. Furthermore the function is regular because it is of the type $1/Q(u,q)$ where Q is as in (19.34).

Let us also note that the information we carry is not the same when α is active or when α is passive. When α is active, the function $H_\alpha(x_\alpha(u)+q)$ is regular and we control its degree in u and q. When α is passive this function may not be regular (but it is semiregular), and we control only its degree in u.

In general there may be many diagrams α which are active but such that α^+ is passive. It is at these stages of the construction that degree reduction takes place, always through (19.23). At the other stages we do not lose or gain anything about degrees, although we have to account precisely for this.[19] This uninspiring bookkeeping is not very intuitive, but all it takes is a little care and patience.[20]

Our arguments are recursive, as they must be. We consider $\alpha \in \mathcal{G}$ and its maximal subdiagrams $\alpha_1, \ldots, \alpha_{\tilde{n}}$, and we assume that these subdiagrams satisfy the conditions of Proposition 19.6.3 , a fact we call "the recursion hypothesis". That is, if α_i is active, the function $H_{\alpha_i}(x_{\alpha_i}(u)+q)$ is regular and satisfies (19.54), whereas if α_i is passive this function satisfies (19.56). The goal is to prove that α satisfies (19.54) and (19.56) and that the function $H_\alpha(x_\alpha(u)+q)$ has suitable regularity properties. We lighten notation by defining $U_i = U_{\alpha_i,\alpha}$.

The first step of the proof of Proposition 19.6.3 is to write that according to (19.24) we have

$$H_\alpha(x_\alpha(u)+q) = \prod_{i \le \tilde{n}} H'_{\alpha_i}(U_i(x_\alpha(u)+q)) \prod_{e \in \mathcal{E}(\alpha,\mathcal{G})} f((x_\alpha(u)+q)_e). \tag{19.58}$$

It bears repeating that we consider each term in the right-hand side as a function of u and q. Before we turn to the details, we may try to give the overall idea of why the induction is going to work.

- When α is active, each term in the second product in (19.58) contributes with a degree -2. We may then expect to control the degree $\deg_{u,q}$ of each term in the first product through the recursion hypothesis (19.54). We will then prove (19.54) using (19.49). The quantity (19.58) is a regular function because the second product in (19.24) is a regular function and each term in the first product is fully regular. This is because $\alpha_i \in \widehat{\mathcal{G}}$ since $\alpha_i^+ = \alpha$ is active, so that we compute H'_{α_i} as in (19.22). by applying an operator T_q^d, and we then appeal to Lemma 19.5.12.
- When α is passive, the terms in the second product in (19.58) do not depend on u. The key point that for each $i \le \tilde{n}$ we have $\deg_u H'_{\alpha_i}(U_i(x_\alpha(u)+q)) \le -m(\alpha_i)$. There are two different reasons for which this is the case, depending on whether α_i is active or passive. If α_i is passive, the induction hypothesis (19.55) asserts that $\deg_u H_{\alpha_i} \le -m(\alpha_i)$, and then $H'_{\alpha_i} = -T^{d(\alpha_i)} H_{\alpha_i}$ and T^d preserves $\deg_u$ by (19.40). If α_i is active, the recursion hypothesis (19.54) asserts that $\deg_u H_{\alpha_i} \le d(\alpha_i) - m(\alpha_i)$. But in that case $H'_{\alpha_i} = (1 - T^{d(\alpha_i)})H_{\alpha_i}$, and this degree-reducing operation reduces the degree $d(\alpha_i) - m(\alpha_i)$ to the degree $-m(\alpha_i) - 1$. This is better than the desired degree $-m(\alpha_i)$, which explains why one can even get (19.56). We will then conclude using (19.48).

[19] Something technically important happens at these stages though: we replace our functions by much more regular ones.

[20] After 40+ years of practice, I am still amazed at the role of preconception in discovering proofs. Probably if you really believe my word that the proof will work, you may figure out the details yourself. On the other hand, the surest way to miss a simple argument is to get convinced for wrong reasons that it does not exist.

Let us now turn to the details. We must relate the functions $H'_{\alpha_i}(U_i(x_\alpha(u)+q))$ on $\mathbb{R}^N \times \mathcal{Q}_\alpha$ to the functions $H_{\alpha_i}(x_{\alpha_i}(u)+q_i)$ on $\mathbb{R}^N \times \mathcal{Q}_{\alpha_i}$ of the recursion hypothesis.[21] We have to work twice, once to account for the fact that the argument of H'_{α_i} is $U_i(x_\alpha(u)+q)$ rather than $x_{\alpha_i}(u)+q_i$ and once to relate the functions H_{α_i} and H'_{α_i}.

We first study the argument $U_i(x_\alpha(u)+q)$.

Lemma 19.6.4 *For each α we have*

$$U_i(x_\alpha(u)+q) = x_{\alpha_i}(u)+q_i(u,q), \tag{19.59}$$

where the function $q_i(u,q) \in \mathcal{Q}_{\alpha_i}$ is a first-degree polynomial. Furthermore if α is passive then

$$U_i(x_\alpha(u)+q) = x_{\alpha_i}(u)+q_i(q), \tag{19.60}$$

where $q_i(q) \in \mathcal{Q}_{\alpha_i}$ is a first-degree polynomial in q.

Proof We apply (19.8) to the case $\beta = \alpha_i$ and to $w := x_\alpha(u) = \mathcal{I}_\alpha U_{\alpha,\gamma}(x(u))$. This yields

$$U_i(x_\alpha(u)+q) = \mathcal{I}_{\alpha_i} U_i \mathcal{I}_\alpha U_{\alpha,\gamma}(x(u)) + b_i(q) + c_i(x_\alpha(u)).$$

Now, by (19.7) it holds that

$$\mathcal{I}_{\alpha_i} U_i \mathcal{I}_\alpha U_{\alpha,\gamma}(x(u)) = \mathcal{I}_{\alpha_i} U_{\alpha_i,\gamma}(x(u)) = x_{\alpha_i}(u).$$

It should be obvious that the quantity $c_i(x_\alpha(u))$ is of the first degree in u, so that the function $q_i(u,q) := b_i(q)+c_i(x_\alpha(u))$ is of the first degree in q and u. When α is passive, for all the edges $e \in \mathcal{E}(\mathcal{G},\alpha)$ the quantity $x_\alpha(u)_e$ is independent of u. Since $c_i(w)$ depends only on the values w_e for $e \in \mathcal{E}(\mathcal{G},\alpha)$, in this case $c_i(x_\alpha(u))$ does not depend on u, and is a constant. Then $q_i(u,q)$ does not depend on u and thus is of the first degree in q. □

Putting things another way, if we know the function $K(u,q_i) := H'_{\alpha_i}(x_{\alpha_i}(u)+q_i)$, the function $K'(u,q) := H'_{\alpha_i}(U_i(x_\alpha(u)+q))$ is given by the formula $K'(u,q) = K(u,q_i(u,q))$. This result motivated Lemmas 19.5.15 and 19.5.14.

Next we must relate H_{α_i} to H'_{α_i}. This is the purpose of the next four lemmas.

Lemma 19.6.5 *For $i \leq \tilde{n}$ the function $T^{d(\alpha_i)}_{q_i}[H_{\alpha_i}(x_{\alpha_i}(u)+q_i)]$ is fully regular and its degree in u and q_i is $\leq d(\alpha_i)-m(\alpha_i)$.*

Here $T^{d(\alpha_i)}_{q_i}[H_{\alpha_i}(x_{\alpha_i}(u)+q_i)]$ denotes the Taylor polynomial in the variable q_i at $q_i = 0$ of the function $(u,q_i) \mapsto H_{\alpha_i}(x_{\alpha_i}(u)+q_i)$.

Proof Assume first that α_i is active, so that by the induction hypothesis the function $H_{\alpha_i}(x_{\alpha_i}(u)+q_i)$ is regular and satisfies $\deg_{u,q_i} H_{\alpha_i}(x_{\alpha_i}(u)+q_i) \leq d(\alpha_i)-m(\alpha_i)$. According to Lemma 19.5.12 the function $T^{d(\alpha_i)}_q[H_{\alpha_i}(x_{\alpha_i}(u)+q_i)]$ is fully regular and since in particular it is regular, according to (19.38) its degree in u and q_i is $\leq d(\alpha_i)-m(\alpha_i)$.

[21] In order to distinguish the variable $q \in \mathcal{Q}_\alpha$ from a variable in $\mathcal{Q}_{\alpha_i}$ the latter is denoted by q_i.

Assume next that α_i is a passive. By the induction hypothesis the rational function $H_{\alpha_i}(x_{\alpha_i}(u)+q_i)$ satisfies $\deg_u H_{\alpha_i}(x_{\alpha_i}(u)+q_i) \le -m(\alpha_i)$. Applying (19.41) and Lemma 19.5.12 we obtain that the function $T_q^{d(\alpha_i)}[H_{\alpha_i}(x_{\alpha_i}(u)+q_i)]$ is again fully regular and that its degree in u and q_i is $\le d(\alpha_i) - m(\alpha_i)$. □

Lemma 19.6.6 *For $i \le \tilde{n}$ the function $(T^{d(\alpha_i)} H_{\alpha_i})(x_{\alpha_i}(u)+q_i)$ is fully regular and*

$$\deg_{u,q}(T^{d(\alpha_i)} H_{\alpha_i})(x_{\alpha_i}(u)+q_i) \le d(\alpha_i) - m(\alpha_i). \tag{19.61}$$

Let us first explain the notation. For any subdiagram α and any function H on $(\mathbb{R}^{1,3})^{\mathcal{E}_\alpha}$ we have defined in Section 16.3 the function $T^d H$ on $(\mathbb{R}^{1,3})^{\mathcal{E}_\alpha}$. Thinking of H_{α_i} as a function on $(\mathbb{R}^{1,3})^{\mathcal{E}_{\alpha_i}}$ we apply this operator $T^{d(\alpha_i)}$ to obtain a new function $(T^{d(\alpha_i)} H_{\alpha_i})$ which we then compute at the point $x_{\alpha_i}(u)+q_i$.

Proof Let us recall that the function $T^d H(x+q)$ is the Taylor polynomial of degree d of the function $q \mapsto H(x+q)$ at $q=0$. In particular $(T^{d(\alpha_i)} H_{\alpha_i})(x_{\alpha_i}(u)+q_i)$ is the Taylor polynomial of degree $d(\alpha_i)$ of the function $q_i \mapsto H_{\alpha_i}(x_{\alpha_i}(u)+q_i)$ at $q_i = 0$, i.e.

$$(T^{d(\alpha_i)} H_{\alpha_i})(x_{\alpha_i}(u)+q_i) = T_q^{d(\alpha_i)}[H_{\alpha_i}(x_{\alpha_i}(u)+q_i)].$$

Then Lemma 19.6.5 implies (19.61). □

Lemma 19.6.7 *For $i \le \tilde{n}$ the function $(T^{d(\alpha_i)} H_{\alpha_i})(U_i(x_\alpha(u)+q))$ is fully regular and*

$$\deg_{u,q}(T^{d(\alpha_i)} H_{\alpha_i})(U_i(x_\alpha(u)+q)) \le d(\alpha_i) - m(\alpha_i). \tag{19.62}$$

Proof This follows from Lemma 19.6.6, (19.59) and Lemma 19.5.14. □

Lemma 19.6.8 *Assume that α is passive.*

(a) If α_i is passive the function

$$(T^{d(\alpha_i)} H_{\alpha_i})(U_i(x_\alpha(u)+q))$$

is regular[22] *and satisfies*

$$\deg_u(T^{d(\alpha_i)} H_{\alpha_i})(U_i(x_\alpha(u)+q)) \le -m(\alpha_i). \tag{19.63}$$

If moreover $m(\alpha_i) > 0$ one has

$$\deg_u(T^{d(\alpha_i)} H_{\alpha_i})(U_i(x_\alpha(u)+q)) < -m(\alpha_i). \tag{19.64}$$

(b) If α_i is active the function

$$((1-T^{d(\alpha_i)})H_{\alpha_i})(U_i(x_\alpha(u)+q))$$

is regular and satisfies

$$\deg_u((1-T^{d(\alpha_i)})H_{\alpha_i})(U_i(x_\alpha(u)+q)) < -m(\alpha_i). \tag{19.65}$$

The condition that α is passive is critical and will be used through (19.60).

22 This function is actually fully regular, although we will not use this fact.

Proof As in Lemma 19.6.6 it holds that

$$(T^{d(\alpha_i)} H_{\alpha_i})(x_{\alpha_i}(u) + q_i) = T_q^{d(\alpha_i)}[H_{\alpha_i}(x_{\alpha_i}(u) + q_i)]. \tag{19.66}$$

(a) Assume that α_i is passive. Using the induction hypothesis and Lemma 19.5.12, we obtain that the function (19.66) is fully regular and by (19.40) its degree in u is $\leq \deg_u H_{\alpha_i}(x_{\alpha_i}(u)) + q_i)$. The conclusion then follows from (19.60) and (19.45).
(b) Assume now that α_i is active. The recursion hypothesis (19.54) implies that the function $H_{\alpha_i}(x_{\alpha_i}(u)+q)$ is regular and that $\deg_{u,q} H_{\alpha_i}(x_{\alpha_i}(u)+q) \leq d(\alpha_i)-m(\alpha_i)$. Thus Proposition 19.5.5 implies that the function $(1 - T_q^{d(\alpha_i)})[H_{\alpha_i}(x_{\alpha_i}(u) + q_i)]$ is regular and satisfies

$$\deg_u(1 - T_q^{d(\alpha_i)})[H_{\alpha_i}(x_{\alpha_i}(u) + q_i)] < \deg_{u,q} H_{\alpha_i}(x_{\alpha_i}(u) + q) - d(\alpha_i) \leq -m(\alpha_i). \tag{19.67}$$

Consequently, the function

$$((1 - T^{d(\alpha_i)})H_{\alpha_i})(x_{\alpha_i}(u) + q_i) = (1 - T_q^{d(\alpha_i)})[H_{\alpha_i}(x_{\alpha_i}(u) + q_i)]$$

is regular and satisfies

$$\deg_u((1 - T^{d(\alpha_i)})H_{\alpha_i})(x_{\alpha_i}(u) + q_i) < -m(\alpha_i). \tag{19.68}$$

The conclusion then follows from (19.60) and (19.45). □

Proof of Proposition 19.6.3 We first recall (19.58):

$$H_\alpha(x_\alpha(u) + q) = \prod_{i \leq \tilde{n}} H'_{\alpha_i}(U_i(x_\alpha(u) + q)) \prod_{e \in \mathcal{E}(\alpha,\mathcal{G})} f((x_\alpha(u) + q)_e). \tag{19.58}$$

Assume first that α is active, so then each α_i belongs to $\widehat{\mathcal{G}}$ and

$$H'_{\alpha_i} = -T^{d(\alpha_i)} H_{\alpha_i}. \tag{19.69}$$

According to Lemma 19.6.6 each of the functions $H'_{\alpha_i}(U_i(x_\alpha(u)+q))$ is regular, so that the function (19.58) is product of regular functions and hence it is regular. Furthermore each of the terms in the last product of (19.58) has degree -2 in u because the edge e is α-active. Thus

$$\deg_{u,q} H_\alpha(x_\alpha(u) + q) \leq \sum_{i \leq \tilde{n}} \deg_{u,q} H'_{\alpha_i}(U_i(x_\alpha(u) + q)) - 2\operatorname{card} \mathcal{E}(\mathcal{G},\alpha), \tag{19.70}$$

while (19.61) implies

$$\deg_{u,q} H'_{\alpha_i}(U_i(x_\alpha(u) + q)) \leq d(\alpha_i) - m(\alpha_i).$$

Now, using that

$$\operatorname{card} \mathcal{E}_\alpha = \operatorname{card} \mathcal{E}(\mathcal{G},\alpha) + \sum_{i \leq \tilde{n}} \operatorname{card} \mathcal{E}_{\alpha_i}$$

in the third line and (19.49) in the fourth line,

$$
\begin{aligned}
&\sum_{i\le\tilde{n}}(d(\alpha_i)-m(\alpha_i))-2\mathrm{card}\,\mathcal{E}(\mathcal{G},\alpha)\\
&=\sum_{i\le\tilde{n}}(\dim\ker\mathcal{L}_{\alpha_i}-2\mathrm{card}\,\mathcal{E}_{\alpha_i}-m(\alpha_i))-2\mathrm{card}\,\mathcal{E}(\mathcal{G},\alpha)\\
&=\sum_{i\le\tilde{n}}(\dim\ker\mathcal{L}_{\alpha_i}-m(\alpha_i))-2\mathrm{card}\,\mathcal{E}_{\alpha}\\
&\le\dim\ker\mathcal{L}_{\alpha}-m(\alpha)-2\mathrm{card}\,\mathcal{E}_{\alpha}\\
&=d(\alpha)-m(\alpha),
\end{aligned}
$$

and this completes the proof of (19.54) when α is active.

Assume now that α is passive. Then by (19.58),

$$H_\alpha(x_\alpha(u)+q)=\frac{1}{Q(q)}\prod_{i\le\tilde{n}}H'_{\alpha_i}(U_i(x_\alpha(u)+q)),\tag{19.71}$$

where Q is a polynomial in q. Next we prove that the function $H'_{\alpha_i}(U_i(x_\alpha(u)+q))$ is regular and satisfies

$$\deg_u H'_{\alpha_i}(U_i(x_\alpha(u)+q))\le -m(\alpha_i),\tag{19.72}$$

and that furthermore if $m(\alpha_i)>0$ we also have

$$\deg_u H'_{\alpha_i}(U_i(x_\alpha(u)+q))< -m(\alpha_i).\tag{19.73}$$

The argument depends on whether α_i is active or passive. If α_i is active then $\alpha_i\in\mathcal{G}\setminus\widehat{\mathcal{G}}$, so $H'_{\alpha_i}=(1-T^{d(\alpha_i)})H_{\alpha_i}$ and (19.73) follows from (19.65) while H'_{α_i} is regular as stated in Lemma 19.6.8. If α_i is passive, then $\alpha_i\in\widehat{\mathcal{G}}$, so that (19.69) holds, and (19.72) and (19.73) follow respectively from (19.63) and (19.64), while H'_{α_i} is again regular by Lemma 19.6.8.

It then follows that the right-hand side of (19.71) is a semiregular function of degree in $u\le-\sum_{i\le\tilde{n}}m(\alpha_i)\le-m(\alpha)$ by (19.48). Furthermore if $m(\alpha)>0$ then for at least one i we have $m(\alpha_i)>0$ and using (19.73) this finishes the recursion. □

Proof of (19.47) We recall that $H'_\gamma=(1-T^{d(\gamma)})H_\gamma$. We observe that $m(\gamma)>0$ because we assume $S\ne\{0\}$. When γ is passive, (19.47) is a consequence of (19.56) and of the fact that according to (19.40) the degree in u can only decrease under $T^{d(\gamma)}$ because the function H_γ is semiregular. When γ is active, it is a consequence of (19.54) used for $\alpha=\gamma$, of (19.39), and of the fact that $(T^{d(\gamma)}H_\gamma)(x(u)+q)=T_q^{d(\gamma)}[H_\gamma(x(u)+q)]$. □

Key idea to remember: The cancellation of degree central to the BPHZ method is brought forward by suitably grouping the terms of the forest formula and appealing to Weinberg's power-counting theorem.

Part V

Complements

The appendices that follow are of different natures. Some are small stories, which my curiosity led me to understand, and there is no better way to understand something than trying to explain it to others. Others deal with important matters. Mathematicians certainly cannot ignore the issues raised in Appendix A. The most important one is Appendix C, where the Stone–von Neumann theorem is established. This theorem is a fundamental sanity check for Quantum Mechanics. Appendix D is also important. While this material is widely available, we strove to present the absolute minimum required to understand spin. Appendix M should help the reader to taste the flavor of Haag's theorem without plunging into the specialized literature.

Appendix A

Complements on Representations

The present appendix consists of six sections. The first three sections are related, but are to a large extent independent of the last three. Section A.1 can be read at the same time as Chapter 2. Section A.2 addresses some fine points of topology. These are reserved for mathematically inclined, but the main lemma of this section is basic for Section A.3, where we prove that in rather general circumstances finite-dimensional projective representations arise from true representations. Section A.5 motivates the use of induced representations in Chapter 9, whereas Section A.6 justifies the way physicists use angular momentum and can be read at any time.

A.1 Projective Unitary Representations of $\mathbb{R}$

In this section we present a mathematical proof of Theorem 2.13.2.[1] We restate the theorem for the convenience of the reader.

Theorem 2.13.2 *A strongly continuous projective unitary representation of $\mathbb{R}$ arises from a true representation. That is, for such a projective representation $U(u)$ we have $U(u) = \lambda(u)V(u)$ where $V(u)$ is a true representation and $|\lambda(u)| = 1$.*

Our proof does not assume differentiability and is valid in infinite dimensions. After this proof we briefly discuss the relationship between true and projective representations.

Differentiability issues are irrelevant for physics. Apparently all functions there are differentiable (even sometimes when they are not really defined).[2] This is even more so in Quantum Field Theory, where the mathematical issues are far more challenging than nit-picking about differentiability. On the other hand, for a mathematician assuming differentiability is a big issue. This makes proving a theorem easier, but makes its statement weaker. For example, what are the maps λ from $\mathbb{R}$ to the complex numbers which satisfy the relation

$$\lambda(a+b) = \lambda(a)\lambda(b)\ ? \tag{A.1}$$

Assuming that λ is differentiable, it satisfies the relation $\lambda'(a) = \lambda'(0)\lambda(a)$. Since $\lambda(0) = \lambda(0)^2$ we get $\lambda(0) = 0$ or $\lambda(0) = 1$, and thus $\lambda(a) \equiv 0$ or $\lambda(a) = \exp(a\lambda'(0))$. This conclusion is still valid if one assumes only that the function λ is continuous, but the proof is quite harder.[3] A mathematician will feel that the "correct" result assumes only continuity of λ.

We turn to the proof of Theorem 2.13.2. Consider a projective unitary representation U of $\mathbb{R}$. Since $\mathbb{R}$ is commutative we write (2.47) as

$$U(a+b) = r(a,b)U(a)U(b). \tag{A.2}$$

1 In particular, we use only the hypotheses stated in the theorem!
2 This is intended as a joke. Please refrain from suing me.
3 On the other hand this conclusion does not hold if one does not assume some kind of regularity for λ.

Since $a+b+c=(a+b)+c=a+(b+c)$, using (A.2) implies

$$r(a,b)r(a+b,c)=r(a,b+c)r(b,c), \tag{A.3}$$

and also, assuming $U(0)=1$ without loss of generality,

$$r(0,a)=r(a,0)=1. \tag{A.4}$$

Lemma A.1.1 *Consider a continuous function r from a neighborhood of the origin in $\mathbb{R}^2$ into the complex numbers of modulus 1 which satisfies (A.3) and (A.4). Then there is an $\varepsilon>0$ and a continuous function λ from $\mathbb{R}$ into the complex numbers of modulus 1 which satisfies*

$$r(a,b)=\lambda(a+b)/(\lambda(a)\lambda(b)) \tag{A.5}$$

whenever $|a|$ and $|b|<\varepsilon$.

To understand that this lemma cannot be completely trivial, imagine that I secretly choose a continuous but otherwise very nasty function λ with $\lambda(0)=1$ and that I form the function $r(a,b)$ of (A.5). Then I give you this function r, and from it you will have to reconstruct my nasty function λ. This corresponds to the situation when starting with a true representation $V(a)$ one constructs the projective representation $U(a)=\lambda(a)V(a)$ for a continuous but otherwise nasty function λ. For a physicist such an idea is absurd, but nonetheless, while, as will be shown later, all continuous finite-dimensional representations of $\mathbb{R}$ are differentiable, this is certainly not true for projective representations.

Mathematician's proof of Theorem 2.13.2 Let us first assume that the function r of (A.2) is continuous, so that we may apply Lemma A.1.1 to find $\varepsilon>0$ and λ which satisfy (A.5). We then define $W(a)=\lambda(a)^{-1}U(a)$. This is a projective unitary representation of $\mathbb{R}$ and it satisfies $W(a+b)=W(a)W(b)$ for $|a|,|b|<\varepsilon$. Consequently $W(2^{-n}a)^{2^n}$ does not depend on n for n large enough (depending on a). We may then set $V(a):=\lim_{n\to\infty}W(2^{-n}a)^{2^n}$. It is straightforward that this defines a true unitary representation of $\mathbb{R}$. Finally we prove that the projective representation U arises from V. To see this, fix $a\in R$, and consider an integer n. Since U is a projective representation we have $U(a)=\mu U(2^{-n}a)^{2^n}$ where $|\mu|=1$. By definition of W we have $U(2^{-n}a)^{2^n}=\mu' W(2^{-n}a)^{2^n}$ where $|\mu'|=1$. If n is large enough we have $W(2^{-n}a)^{2^n}=V(a)$, so finally as desired we have $U(a)=\lambda V(a)$ where $|\lambda|=1$.

To complete the argument, we need to prove the continuity of r. For this we simply observe that for any vector x, the vectors $U(a+b)(x)$ and $U(a)U(b)(x)$ are continuous functions of (a,b) (which requires a little proof for the second case), and the continuity of the function r at a given point (a,b) is then obtained by considering a vector x for which $U(a+b)(x)\neq 0$. □

A trained topologist might have noticed that we used a notion of continuity for projective representations which is not the natural one, a matter which is clarified in the next section.

Proof of Lemma A.1.1 The proof contains no physical idea but is exquisite. Using continuity we find $\varepsilon_0>0$ for which $|r(a,b)-1|<1$ for $|a|,|b|<\varepsilon_0$. For these values of a,b we then define $f(a,b)=\mathrm{i}^{-1}\log r(a,b)$, i.e. $f(a,b)$ is the number $-\pi/3\leq\theta\leq\pi/3$ for which $\exp\mathrm{i}\theta=r(a,b)$. This function is continuous, real-valued (since $r(a,b)$ is of modulus 1) and for $|a|,|b|,|c|\leq\varepsilon_0/2$ it satisfies

$$f(a,b)+f(a+b,c)=f(a,b+c)+f(b,c) \tag{A.6}$$

and also

$$f(0,a)=f(a,0)=0. \tag{A.7}$$

Our goal is to find a real-valued function θ such that

$$f(a,b)=\theta(a+b)-\theta(a)-\theta(b). \tag{A.8}$$

The function $\lambda(a) = \exp(\mathrm{i}\theta(a))$ then satisfies (A.5) and the proof is finished. The proof would be easy if we knew that f is differentiable (as we will show later). We will reduce that case by a clever argument.

Let us then consider a real-valued infinitely differentiable function ψ with support contained in the interval $]-\varepsilon_0,\varepsilon_0[$ and of integral 1. For $|a| < \varepsilon_0$ let us define[4]

$$\varphi(a) = \int f(a,t)\psi(t)\mathrm{d}t,$$

so that $\varphi(0) = 0$, and for $|a|,|b| \le \varepsilon_0/2$ let us define

$$g(a,b) = f(a,b) - \varphi(a) - \varphi(b) + \varphi(a+b). \tag{A.9}$$

Our goal is to find a function γ such that

$$g(a,b) = \gamma(a+b) - \gamma(a) - \gamma(b). \tag{A.10}$$

The function g satisfies the functional equation (A.6), and moreover, since $\int \psi(t)\mathrm{d}t = 1$, and using the functional equation (A.6) for f in the second line,

$$\begin{aligned} g(a,b) &= \int (f(a,b) - f(a,t) - f(b,t) + f(a+b,t))\psi(t)\mathrm{d}t \\ &= \int (f(a,b+t) - f(a,t))\psi(t)\mathrm{d}t. \end{aligned} \tag{A.11}$$

Since

$$\int f(a,b+t)\psi(t)\mathrm{d}t = \int f(a,u)\psi(u-b)\mathrm{d}u$$

this proves that the function $b \mapsto g(a,b)$ is differentiable. (This magic and crucial argument is apparently due to Masatake Kuranishi [49].) Let $g'(a,b)$ be its differential at b. Differentiating (A.10) in b at $b=0$ gives $g'(a,0) = \gamma'(a) - \gamma'(0)$, and there is no loss of generality to assume that $\gamma'(0) = 0$ by replacing $\gamma(x)$ by $\gamma(x) - x\gamma'(0)$, so that $\gamma'(a) = g'(s,0)$. This tells us how to find γ! So, we now define

$$\gamma(a) = \int_0^a g'(s,0)\mathrm{d}s.$$

Writing (A.6) as

$$g(a,b) + g(a+b,c) = g(a,b+c) + g(b,c)$$

and differentiating in c at $c = 0$ then yields

$$g'(a+b,0) = g'(a,b) + g'(b,0). \tag{A.12}$$

Then, using (A.12) in the third line,

$$\begin{aligned} \gamma(a+b) - \gamma(a) &= \int_a^{a+b} g'(s,0)\mathrm{d}s \\ &= \int_0^b g'(a+s,0)\mathrm{d}s \\ &= \int_0^b g'(s,0)\mathrm{d}s + \int_0^b g'(a,s)\mathrm{d}s \\ &= \gamma(b) + g(a,b), \end{aligned} \tag{A.13}$$

[4] The majority of the appendices are purely mathematical, and from this point on we use the standard mathematics convention to put the d in an integrand at the end.

where we have used that $g(a,0) = 0$. We have proved (A.10). Therefore we have proved (A.8) for $\theta(a) = \gamma(a) - \varphi(a)$. Observing that θ is real-valued the proof is finished. □

Despite the ease with which we proved Lemma A.1.1, we must stress that this is due to the fact that we considered the group $\mathbb{R}$. To understand this, let us consider the functional equation (A.6) in the case where $\mathbb{R}$ is replaced by $\mathbb{R}^2$, i.e. $a = (a_1, a_2) \in \mathbb{R}^2$. It is then straightforward that this equation is satisfied for

$$f(a,b) = f((a_1,a_2),(b_1,b_2)) = a_2 b_1.$$

However $f(a,b) \neq f(b,a)$, so that it is impossible to find a function $\gamma(a)$ for which

$$f(a,b) = a_2 b_1 = \gamma(a+b) - \gamma(a) - \gamma(b). \tag{A.14}$$

As a striking consequence of this fact, let us recall the operators U and V given by (2.53) and (2.54).

Proposition A.1.2 *For $(s_1,s_2) \in \mathbb{R}^2$ the formula $W(s_1,s_2) = U(s_1)V(s_2)$ defines a projective unitary representation of $\mathbb{R}^2$ which does not arise from a true representation.*

Proof We recall (2.74):

$$V(s)U(t) = \exp(its/\hbar)U(t)V(s). \tag{2.74}$$

Then

$$\begin{aligned} W(s_1,s_2)W(t_1,t_2) &= U(s_1)V(s_2)U(t_1)V(t_2) \\ &= \exp(it_1 s_2/\hbar)U(s_1)U(t_1)V(s_2)V(t_2) \\ &= \exp(it_1 s_2/\hbar)U(s_1+t_1)V(s_2+t_2) \\ &= \exp(it_1 s_2/\hbar)W(s_1+t_1,s_2+t_2), \end{aligned}$$

and this proves that W is a projective unitary representation of $\mathbb{R}^2$. The impossibility to solve (A.14) means however that this projective representation does not arise from a true representation. □

Valentine Bargmann [4] invented a powerful criterion giving conditions under which a projective unitary representation arises from a true representation, a simple proof of which can be found in Simms' book [76]. Interestingly, a key step in the proof is to "produce differentiability", and this is done by the argument of Kuranishi used above. There are two (and only two) possible types of obstructions why a projective representation may not arise from a true representation. These are obstructions of an algebraic nature (as in the case of the previous example) and obstructions of a topological nature (as in the case of the group $SO(3)$ of Chapter 4). A consequence of Bargmann's criterion is that the projective representations of the Poincaré group $\mathcal{P}$ all arise from true representations of its double cover $\mathcal{P}^*$.

A.2 Continuous Projective Unitary Representations

The naive approach we have been using in the main text toward continuity of projective representations is not the mathematically fruitful method to consider this question. In the present section we describe the correct method. This point can be considered as a subtle and advanced topic in topology. Do not worry if you find it confusing, it is used nowhere else in the book. The single result of this section, Lemma A.2.1 (which will be used in the next section) is a very simple fact.

Let us first recall that the natural topology on the unitary group $\mathcal{U}(\mathcal{H})$ is the so-called *strong topology*. It is the topology corresponding to the notion of strong continuity of Definition 2.13.1. A basis of neighborhoods of a unitary operator U consists of the sets of unitary operators of the type

$$\mathcal{W} := \{V \in \mathcal{U}(\mathcal{H})\,;\, \forall k \le n\,,\, \|Vx_k - Ux_k\| < \eta\} \tag{A.15}$$

where $\eta > 0$, n is arbitrary and $x_1, \dots, x_n \in \mathcal{H}$.

Let us consider the map Φ from the unitary group $\mathcal{U}(\mathcal{H})$ into its quotient $\mathcal{U}_p(\mathcal{H})$ by the group consisting of multiples of identity. A map $V : \mathbb{R} \to \mathcal{U}(\mathcal{H})$ is a projective representation (as in Definition 2.10.1) if and only if the map $a \mapsto \Phi(V(a))$ is a group homomorphism. It is this quotient map that is the natural object, so the natural definition of continuity should refer to the continuity of this map $a \mapsto \Phi(V(a))$ when $\mathcal{U}_p(\mathcal{H})$ is provided with the natural topology, the quotient topology induced by the strong topology on $\mathcal{U}(\mathcal{H})$. We will describe this situation by the sentence "the map $a \mapsto \Phi(V(a))$ is continuous for the quotient topology". This quotient topology is defined as the strongest topology on $\mathcal{U}_p(\mathcal{H})$ such that the map $U \mapsto \Phi(U)$ is continuous when $\mathcal{U}(\mathcal{H})$ is provided with the strong topology. A set $\Omega \subset \mathcal{U}_p(\mathcal{H})$ is open for this topology if and only if $\Phi^{-1}(\Omega)$ is open for the strong topology. A set $\Omega \subset \mathcal{U}_p(\mathcal{H})$ is a neighborhood of $\Phi(U)$ if and only if $\Phi^{-1}(\Omega)$ is a neighborhood of U, or, equivalently, if $\Phi(\mathcal{W}) \subset \Omega$ for a certain neighborhood $\mathcal{W}$ of U. The map $a \mapsto V(a)$ is continuous for the quotient topology if and only if given any $a \in \mathbb{R}$ and any neighborhood $\mathcal{W}$ of $V(a)$ for the strong topology then $V(b) \in \Phi^{-1}\Phi(\mathcal{W})$ for b close enough to a, i.e.

$$\exists \varepsilon > 0 \,,\; |b-a| \le \varepsilon \Rightarrow \exists \lambda \in \mathbb{C}\,;\; |\lambda| = 1 \,,\; \lambda V(b) \in \mathcal{W}. \tag{A.16}$$

The nit-picking issue here is that if the map $a \mapsto V(a)$ is strongly continuous, i.e. satisfies

$$\exists \varepsilon > 0 \,,\; |b-a| \le \varepsilon \Rightarrow V(b) \in \mathcal{W},$$

then the map $a \mapsto \Phi(V(a))$ is continuous for the quotient topology, but that the converse is not true. It is a weaker and more natural hypothesis to assume that the map $a \mapsto \Phi(V(a))$ is continuous for the quotient topology than to assume that the map $a \mapsto V(a)$ is strongly continuous. The following lemma however shows that when the map $a \mapsto \Phi(V(a))$ is continuous for the quotient topology, we can redefine the map $a \mapsto V(a)$ to make it strongly continuous in a neighborhood of the identity.

Lemma A.2.1 *Consider a projective representation $V : \mathbb{R} \to \mathcal{U}(\mathcal{H})$ such that the map $a \mapsto \Phi(V(a))$ is continuous for the quotient topology. Then we can find $\varepsilon_0 > 0$ and for $|a| \le \varepsilon_0$ numbers $\mu(a)$ of modulus 1 such that the map $a \mapsto \mu(a)V(a)$ is strongly continuous for $|a| \le \varepsilon_0$.*

We state the result for $\mathbb{R}$ in order not to increase the psychological difficulty, but the very same argument works for a general topological group.

Proof Since $V(0)$ is a multiple of the identity we can assume without loss of generality that $V(0) = 1$. By the very definition of the quotient topology, saying that the map $a \mapsto \Phi(V(a))$ is continuous means that the following condition is satisfied (as follows combining (A.15) and (A.16))

$$\forall a \in \mathbb{R} \,,\; \forall \eta > 0 \,,\; \forall n \in \mathbb{N} \,,\; \forall x_1, \ldots, x_n \in \mathcal{H} \,,\; \exists \varepsilon > 0 \,,\; \forall b \in \mathbb{R} \,,\; |b-a| < \varepsilon \Rightarrow$$
$$\exists \lambda \in \mathbb{C} \,,\; |\lambda| = 1 \,,\; \forall k \le n \,,\; \|\lambda V(b)x_k - V(a)x_k\| < \eta. \tag{A.17}$$

Let us fix x_0 of norm 1, so that using (A.17) for $a = 0$ (and since $V(0) = 1$) we see that for a certain $\varepsilon_0 > 0$, for all $|b| < \varepsilon_0$, we can find λ of modulus 1 with $\|\lambda V(b)x_0 - x_0\| < 1/2$. Consequently $|(x_0, V(b)x_0)| = |(x_0, \lambda V(b)x_0)| > 1/2$. Consequently, we can find $\mu(b)$ with $|\mu(b)| = 1$ such $(x_0, \mu(b)V(b)x_0) \in \mathbb{R}^+$ and $(x_0, \mu(b)V(b)x_0) \ge 1$. Replacing $V(b)$ by $\mu(b)V(b)$ we may assume that for $|b| < \varepsilon_0$ we have $(x_0, V(b)x_0) \in \mathbb{R}^+$ and $(x_0, V(b)x_0) \ge 1/2$. This redefinition of V does not change the fact that it is continuous for the quotient topology, because this continuity property involves only the map $a \mapsto \Phi(V(a))$ which has not changed under this redefinition. We prove that the map $a \mapsto V(a)$ is now strongly continuous in the neighborhood $|a| < \varepsilon_0$ of 0. Indeed, by (A.17), given $x_1, \ldots, x_n \in \mathcal{H}$, $a \in \mathbb{R}$ and $\eta > 0$ we can find $\varepsilon > 0$ such that

$$|b-a| < \varepsilon \Rightarrow \exists \lambda \,,\; |\lambda| = 1 \,,\; \|\lambda V(b)x_0 - V(a)x_0\| \le \eta \,,\; \forall k \le n \,,\; \|\lambda V(b)x_k - V(a)x_k\| \le \eta.$$

In particular we have

$$|\lambda(x_0, V(b)x_0) - (x_0, V(a)x_0)| \le \eta. \tag{A.18}$$

Assuming now that $|a|, |b| < \varepsilon_0$ both $(x_0, V(a)x_0)$ and $(x_0, V(b)x_0)$ are positive numbers $\geq 1/2$. Dividing both sides of (A.18) by $(x_0, V(a)x_0)$ implies that $|\lambda - \alpha| < 2\eta$ where $\alpha \in \mathbb{R}^+$. Since λ is a complex number of modulus 1 we then have $|\lambda - 1| < 4\eta$, so that $\|\lambda V(b)x_k - V(b)x_k\| \leq 4\eta\|V(b)x_k\| = 4\eta\|x_k\|$. Consequently, if $|b - a| < \varepsilon$, then $\|V(b)x_k - V(a)x_k\| \leq \eta(1 + 4\|x_k\|)$, concluding the argument. □

Exercise A.2.2 Combine the previous results to prove the following improved version of Theorem 2.13.2: A projective unitary representation of $\mathbb{R}$ which is continuous for the quotient topology arises from a true representation.

A.3 Projective Finite-dimensional Representations

In the present section we prove Theorem 8.6.1 in the case of finite-dimensional representations. The proof is related to the ideas of the previous section.

Theorem A.3.1 *Consider a compact group G with the following properties.*
(a) Every element a of G can be connected to the identity by a continuous curve. Moreover, given a neighborhood W of the identity we can find a neighborhood V of the identity such that if $a \in V$ we can find such a curve entirely contained in W.
(b) G is simply connected. That is, every continuous loop passing through the unit of G can be continuously deformed into the constant loop at the unit of G.
Then a finite dimensional continuous projective representation of G arises from a true representation.

Our prime example of a compact group satisfying these conditions is the group $SU(2)$ of Chapter 8, which is topologically the unit sphere of $\mathbb{R}^4$. Application to this case shows that "the projective representations of $SO(3)$ come from true representations of $SU(2)$" in the following sense.

Corollary A.3.2 *Consider a finite-dimensional continuous unitary projective representation π' of $SO(3)$ in a finite-dimensional space. Then there exists a true unitary representation π of $SU(2)$ and for $A \in SU(2)$ a complex number $\lambda(A)$ of modulus 1 such that*

$$\forall A \in SU(2)\,;\ \pi'(\kappa(A)) = \lambda(A)\pi(A). \tag{A.19}$$

Proof We apply Theorem A.3.1 to the case of $G = SU(2)$ and to the continuous projective unitary representation $A \mapsto \pi'(\kappa(A))$ of G. □

The proof of Theorem A.3.1 is self-contained. This result is an immediate consequence Propositions A.3.3 and A.3.4.

Proposition A.3.3 *Consider a compact group G and a finite-dimensional projective continuous unitary representation π' of G. Then we can find a neighborhood W of the identity and a continuous function $a \mapsto \lambda(a)$ on W, where $\lambda(a)$ is a complex number of modulus 1 such the quantity $\pi(a) = \lambda(a)\pi'(a)$ satisfies*

$$\pi(ab) = \pi(a)\pi(b) \tag{A.20}$$

whenever $a, b, ab \in W$.

Proof As explained at the end of the previous section, the correct meaning of "continuous" is for a certain quotient topology, but we proved in Lemma A.2.1 that if we restrict π' to a suitable neighborhood of the identity we may assume that the map $a \mapsto \pi'(a)$ is strongly continuous. (Lemma A.2.1 was proved in the case of the group $\mathbb{R}$ in order not to confuse the reader but the argument is very general.) As the representation is finite-dimensional, the map $a \mapsto \det(\pi'(a))$ is then continuous. We may assume that $\pi'(e) = 1$ where e is the unit of G. We may also assume that W is small enough so that $|\det \pi'(a) - 1| < 1$ for $a \in W$. Let n be the dimension of the representation. The quantity $\lambda(a) := (\det \pi'(a))^{-1/n}$ is then well defined and of modulus 1. Let us set

$$\pi(a) := \lambda(a)\pi'(a) = (\det \pi'(a))^{-1/n}\pi'(a).$$

Then $\det \pi(a) = 1$. Since π' is a projective representation, for $a, b, ab \in W$ we have

$$\pi(ab) = r(a,b)\pi(a)\pi(b)$$

for a certain number $r(a,b)$. Taking determinants shows that $r(a,b)^n = 1$. Moreover, if W is small enough, by continuity we have $|r(a,b) - 1| < 1/2n$, so that in fact $r(a,b) = 1$. □

Proposition A.3.4 *Consider a compact group G which satisfies the conditions of Theorem A.3.1, a neighborhood W of the identity and a continuous map π satisfying (A.20) which associates to each point of W an operator π(a). Then we can find a true representation π_1 of G such that $\pi_1(a) = \pi(a)$ when a is close to the identity. Moreover, if π′ denotes a projective representation of G such that $\pi'(a) = \lambda(a)\pi(a)$ for $a \in W$ (where λ(a) is a complex number of modulus 1) then π′ arises from π_1.*

Proof Given an element a of G by hypothesis there exists a continuous map φ from $[0,1]$ to G such that $\varphi(0) = e$ (the unit of G) and $\varphi(1) = a$. Using this map we will define an operator $\pi(a,\varphi)$. We will then show that in fact this operator does not depend on φ but only on a. This will be the operator $\pi_1(a)$ we are looking for. All the other properties will be rather obvious.

To define the operator $\pi(a,\varphi)$ we consider points $0 = t_0 < t_1 < \ldots < t_k = 1$ such that for $t_\ell \le s, t \le t_{\ell+1}$ we have $\varphi(s)\varphi(t)^{-1} \in W$. This is possible by continuity of φ. Consider then the quantity

$$\pi(\varphi(t_k)\varphi(t_{k-1})^{-1})\cdots\pi(\varphi(t_\ell)\varphi(t_{\ell-1})^{-1})\cdots\pi(\varphi(t_1)\varphi(t_0)^{-1}). \tag{A.21}$$

It follows from (A.20) that for $t_{\ell-1} < s < t_\ell$ we have

$$\pi(\varphi(t_\ell)\varphi(t_{\ell-1})^{-1}) = \pi(\varphi(t_\ell)\varphi(s)^{-1})\pi(\varphi(s)\varphi(t_{\ell-1})^{-1}).$$

Thus the quantity (A.20) does not change if we add some extra points t_i. Consequently it depends only on φ and not on the specific choice of the points t_ℓ. Indeed if we consider two different finite families the values of the corresponding quantities (A.21) are equal to the value of this quantity corresponding to the union of these two families. In this manner we define the operator $\pi(a,\varphi)$.

Suppose that we have a continuous map $\psi\colon [0,1] \times [0,1] \to G$ such that $\psi(0,u) = e$ and $\psi(1,u) = a$ for each $u \in [0,1]$. We may then consider the curves φ_u from e to a given by $\varphi_u(t) = \psi(t,u)$. We now prove the crucial fact: $\pi(a,\varphi_u)$ is independent of u. For this we consider a small neighborhood W' of the identity such that $b, bc, bcd \in W$ and $b^{-1} \in W$ whenever $b, c, d \in W'$. For u and v close enough we have

$$\forall t\,,\ 0 \le t \le 1\,,\ \varphi_u(t)\varphi_v(t)^{-1} \in W'. \tag{A.22}$$

It suffices to prove that under this condition we have $\pi(a,\varphi_u) = \pi(a,\varphi_v)$. Let us consider $0 = t_0 < t_1 < \ldots < t_k = 1$ such that for $t_\ell \le s, t \le t_{\ell+1}$ we have $\varphi_u(s)\varphi_u(t)^{-1} \in W'$ and $\varphi_v(s)\varphi_v(t)^{-1} \in W'$. Then $\pi(a,\varphi_u)$ and $\pi(a,\varphi_v)$ are respectively equal to

$$\pi(\varphi_u(t_k)\varphi_u(t_{k-1})^{-1})\cdots\pi(\varphi_u(t_\ell)\varphi_u(t_{\ell-1})^{-1})\cdots\pi(\varphi_u(t_1)\varphi_u(t_0)^{-1}) \tag{A.23}$$

and

$$\pi(\varphi_v(t_k)\varphi_v(t_{k-1})^{-1})\cdots\pi(\varphi_v(t_\ell)\varphi_v(t_{\ell-1})^{-1})\cdots\pi(\varphi_v(t_1)\varphi_v(t_0)^{-1}). \tag{A.24}$$

To prove that these quantities are equal to each other, we write for each ℓ

$$\varphi_u(t_\ell)\varphi_u(t_{\ell-1})^{-1} = b_\ell c_\ell d_\ell$$

where

$$b_\ell = \varphi_u(t_\ell)\varphi_v(t_\ell)^{-1} \;;\; c_\ell = \varphi_v(t_\ell)\varphi_v(t_{\ell-1})^{-1} \;;\; d_\ell = \varphi_v(t_{\ell-1})\varphi_u(t_{\ell-1})^{-1}$$

all belong to W'. Therefore (A.20) implies

$$\pi(\varphi_u(t_\ell)\varphi_u(t_{\ell-1})^{-1}) = \pi(b_\ell)\pi(c_\ell)\pi(d_\ell).$$

We substitute this equality in each term of (A.23), and we notice that $d_\ell = b_{\ell-1}^{-1} \in W$, so that $\pi(d_\ell) = \pi(b_{\ell-1})^{-1}$. Since $b_k = d_1 = e$ all the terms $\pi(b_\ell)$ and $\pi(d_\ell)$ cancel, leaving the product (A.24). Thus we have proved that $\pi(b,\varphi_u)$ does not depend on u.

Consider now the case $a = e$. A continuous curve φ from e to a is then a loop. The assumption that G is simply connected means that we can continuously contract this loop to the origin, that is we can find a continuous map $\psi\colon [0,1]\times[0,1] \to G$ such that $\psi(t,0) = \varphi(t)$ and $\psi(t,1) = e$, whereas $\psi(0,u) = \psi(1,u) = e$. Writing as before $\varphi_u(t) = \psi(t,u)$ it is obvious that $\pi(e,\varphi_1) = 1$. Since $\pi(e,\varphi_u)$ does not depend on u we then have

$$\pi(e,\varphi) = 1. \tag{A.25}$$

From this point on everything is easy.

Let us then consider a continuous curve φ from e to a and a continuous curve φ' from e to another point b. We can define a continuous curve ψ from e to ab by $\psi(t) = \varphi'(2t)$ for $0 \le t \le 1/2$ and $\psi(t) = \varphi(2t-1)b$ for $1/2 \le t \le 1$. It is straightforward to check from the definition that

$$\pi(ab,\psi) = \pi(a,\varphi)\pi(b,\varphi'). \tag{A.26}$$

Also, it is equally straightforward that if we define $\hat\varphi(t) := \varphi(1-t)$ then

$$\pi(a,\varphi)^{-1} = \pi(a^{-1},\hat\varphi a^{-1}). \tag{A.27}$$

Considering two continuous maps φ_1 and φ_2 from e to a certain element a of G, let us prove that $\pi(a,\varphi_1) = \pi(a,\varphi_2)$. Consider the loop given by $\varphi(t) = \varphi_1(2t)$ for $0 \le t \le 1/2$ and $\varphi(t) = \varphi_2(2-2t)$ for $1/2 \le t \le 1$, so that $\pi(e,\varphi) = 1$ by (A.25). Combining with (A.26) for $b = a^{-1}$ and $\varphi'(t) = \varphi_2(1-t)a^{-1}$ yields $1 = \pi(a,\varphi_1)\pi(a^{-1},\varphi')$. Since $\pi(a^{-1},\varphi') = \pi(a,\varphi_2)^{-1}$ by (A.27) we conclude as desired that $\pi(a,\varphi_1) = \pi(a,\varphi_2)$. So we have proved that $\pi_1(a) := \pi(a,\varphi)$ depends only on a. By hypothesis there is a neighborhood V of the identity such that φ is valued entirely inside W when $a \in V$. It is then obvious that $\pi_1(a) = \pi(a)$. It follows from (A.26) that π_1 is a representation.

Consider now a projective representation π' and assume that $\pi'(a) = \lambda(a)\pi(a)$ for $a \in W$, where as usual $|\lambda(a)| = 1$. Therefore

$$\pi'(a) = \lambda(a)\pi_1(a) \tag{A.28}$$

for $a \in V$. In the expression (A.21) we may as well assume that each term $\varphi(t_\ell)\varphi(t_{\ell-1})^{-1}$ belongs to V. Applying (A.28) to each such term, and using that π' is a projective representation yields that the quantity (A.21) equals $\lambda\pi'(\varphi(1))$ where $|\lambda| = 1$. But this quantity also equals $\pi_1(\varphi(1))$ by the definition of π_1. Thus, the projective representation π' arises from π_1. □

A.4 Induced Representations for Finite Groups

In this section all groups are finite. Consider a finite group G and a representation W of G, that is a map W from G into the group of invertible linear transformations of a linear space $\mathcal{G}$ with $W(AB) = W(A)W(B)$.[5] Consider a subgroup H of G. It may happen that there exists a subspace $\mathcal{H}$ of $\mathcal{G}$ such

[5] In this section there are no inner products on the linear spaces involved, and we are not concerned about unitary representations, only about representations.

that $W(A)(\mathcal{H}) = \mathcal{H}$ for $A \in H$. In this manner we obtain a representation U of H on $\mathcal{H}$: $U(A)$ is the restriction of $W(A)$ to $\mathcal{H}$. We call U the *restriction of* W *to* H *and* $\mathcal{H}$, and we will say that the representation U of H is a *subrepresentation of* W. We will say that this subrepresentation *spans* W if $\mathcal{G}$ is spanned by the union of the spaces $W(A)(\mathcal{H})$ for $A \in G$.

We say that two representations W, W' of a group G into spaces $\mathcal{G}, \mathcal{G}'$ are *equivalent* if there is an isomorphism (= invertible linear map) T from $\mathcal{G}$ to $\mathcal{G}'$ such that $TW(A) = W'T(A)$ for each $A \in G$. We say that a representation W of G in $\mathcal{G}$ is *irreducible* is there are no proper subspaces of $\mathcal{G}$ which are invariant under all the transformations $W(A)$ for $A \in G$.

Suppose now that we are given the representation U of H. What can we say of the representations of G which are spanned by a subrepresentation equivalent to U? We will prove the following facts.

Theorem A.4.1 *(a) Every representation of G which is spanned by a subrepresentation equivalent to U has a dimension $\leq \dim \mathcal{H} \times \operatorname{card} G/\operatorname{card} H$. Up to equivalence there is a unique such representation of this dimension, denoted $\operatorname{Ind}_H^G U$.*
(b) Every representation of G which is spanned by a subrepresentation equivalent to U is in a natural sense (which will be made precise in Lemma A.4.3) a quotient of $\operatorname{Ind}_H^G U$.
(c) When $\operatorname{Ind}_H^G U$ is irreducible, every representation of G which is spanned by a subrepresentation equivalent to U is equivalent to $\operatorname{Ind}_H^G U$.

Let us stress that up to equivalence, $\operatorname{Ind}_H^G U$ depends only on G, H and U.

We will start by analyzing the representations of G which are spanned by a subrepresentation equivalent of U even though we have not proved yet that there is any such representation. We begin with some very basic observations. The binary relation $A\mathcal{R}B$ on G defined by $A\mathcal{R}B$ if and only if $B^{-1}A \in H$ is an equivalence relation. All the equivalence classes have cardinality $\operatorname{card} H$. The set of equivalence classes is denoted by G/H. Given $A \in G$ if $B, C \in G$ are such that $B\mathcal{R}C$ then $AB\mathcal{R}AC$. In this manner for $A \in G$ and $w \in G/H$ we define $A \cdot w$ as being the equivalence class consisting of the elements AB for B in the equivalence class w. This defines an action of G on G/H, that is for $A, B \in G$ and $w \in G/H$ we have $A \cdot (B \cdot w) = (AB) \cdot w$. This formula will be of constant use. Obviously $\operatorname{card} G/H = \operatorname{card} G/\operatorname{card} H$. We denote by w^* the class of H in G/H. Thus $A \cdot w^* = w^*$ if and only if $A \in H$.

The following simple lemma goes a long way toward explaining the structure of representations which are spanned by a given subrepresentation.

Lemma A.4.2 *Consider a representation W of G on a space $\mathcal{G}$. Consider a subspace $\mathcal{H}$ of $\mathcal{G}$ and assume that $W(A)(\mathcal{H}) = \mathcal{H}$ for $A \in H$. Then for $A \in G$ the space $W(A)(\mathcal{H})$ depends only on $w := A \cdot w^*$, the class of A in G/H. We denote it by $\mathcal{H}_w$. Furthermore $W(A)(\mathcal{H}_w) = \mathcal{H}_{A \cdot w}$.*

Proof If $A \cdot w^* = B \cdot w^*$ we have $B^{-1}A \in H$. Since $A = B(B^{-1}A)$ we have $W(A) = W(B)W(B^{-1}A)$. Since $W(B^{-1}A)(\mathcal{H}) = \mathcal{H}$ we have $W(A)(\mathcal{H}) = W(B)(\mathcal{H})$. To prove the last property we notice that if $w = B \cdot w^*$ then $\mathcal{H}_w = W(B)(\mathcal{H})$ and $W(A)(\mathcal{H}_w) = W(AB)(\mathcal{H}) = \mathcal{H}_{AB \cdot w^*} = \mathcal{H}_{A \cdot w}$. □

In particular if $\mathcal{G}$ is spanned by the spaces $W(A)(\mathcal{H})$ it is spanned by the spaces $\mathcal{H}_w$. Each of these has dimension $\dim \mathcal{H}$ and there are $\operatorname{card} G/\operatorname{card} H$ of them, so that $\dim \mathcal{G} \leq \dim \mathcal{H} \times \operatorname{card} G/\operatorname{card} H$. When there is equality, $\mathcal{G}$ is the direct sum of the spaces $\mathcal{H}_w$, and the action of an operator $W(A)$ shuffles these spaces around. We may then write uniquely every element x of $\mathcal{G}$ as a sum $x = \sum_{w \in G/H} x_w$ where $x_w \in \mathcal{H}_w$. Then

$$W(A)(x) = \sum_{w \in G/H} W(A)(x_w) = \sum_{w \in G/H} W(A)(x_{A^{-1} \cdot w}),$$

where we have used in the last equality that $w \mapsto A^{-1} \cdot w$ is a permutation of G/H. The point of this last expression is that $W(A)(x_{A^{-1} \cdot w}) \in \mathcal{H}_w$ because $A \cdot (A^{-1} \cdot w) = w$. If we identify a family $(x_w)_{w \in G/H}$ with a function $f \colon G/H \to \mathcal{G}$ with $f(w) \in \mathcal{H}_w$ we have discovered the formula (A.29).

Lemma A.4.3 *Under the conditions of Lemma A.4.2 consider the space $\mathcal{Z}$ of functions $f: G/H \to \mathcal{G}$ such that $f(w) \in \mathcal{H}_w$ for each $w \in G/H$. Then the formula*

$$Z(A)(f)(w) = W(A)f(A^{-1} \cdot w) \tag{A.29}$$

defines a representation Z of G in $\mathcal{Z}$. This representation is spanned by the space $\mathcal{H}' = \{f \in \mathcal{Z}; w \neq w^* \Rightarrow f(w) = 0\}$. Its restriction to H and $\mathcal{H}'$ is equivalent to the restriction U of W to H and $\mathcal{H}$. The map $S: \mathcal{Z} \to \mathcal{G}$ given by $S(f) = \sum_{w \in G/H} f(w)$ satisfies

$$SZ(A) = W(A)S \tag{A.30}$$

for each $A \in G$. Furthermore if U spans the representation W this map is onto.

$$\begin{array}{ccc} \mathcal{Z} & \xrightarrow{Z(A)} & \mathcal{Z} \\ \downarrow S & & S \downarrow \\ \mathcal{G} & \xrightarrow{W(A)} & \mathcal{G} \end{array}$$

W as a quotient of Z.

In words, if a representation W of G in a space $\mathcal{G}$ is spanned by a subrepresentation equivalent to U, it is in a natural sense a quotient of the representation Z of (A.29). The remarkable fact is that the representation Z of (A.29), which seems to depend on the whole representation W, up to equivalence depends only on the restriction U of W to H and $\mathcal{H}$. This is not apparent yet, but is the key to everything.

Proof of Lemma A.4.3 We note that $Z(A)(f)(w) = W(A)f(A^{-1} \cdot w) \in \mathcal{H}_w$ because $f(A^{-1} \cdot w) \in \mathcal{H}_{A^{-1} \cdot w}$ and $W(A)(\mathcal{H}_{A^{-1} \cdot w}) = \mathcal{H}_w$. It is then straightforward to check that the formula (A.29) defines a representation. The rest is obvious. □

Corollary A.4.4 *Assume either that* $\dim \mathcal{G} = \dim \mathcal{H} \times \operatorname{card} G/\operatorname{card} H$ *or that the representation Z of (A.29) is irreducible. Then S is an isomorphism and the representations Z and W are equivalent.*

Proof Since $\dim \mathcal{Z} = \dim \mathcal{H} \times \operatorname{card} G/\operatorname{card} H$, when $\dim \mathcal{G} = \dim \mathcal{H} \times \operatorname{card} G/\operatorname{card} H$, then $\dim \mathcal{Z} = \dim \mathcal{G}$. Then since S is onto it is an isomorphism, and (A.30) witnesses the equivalence of the representations Z and W. Furthermore, according to (A.30), the space $\ker S$ is invariant by all the operators $Z(A)$ so that when Z is irreducible this space must be reduced to $\{0\}$ and S is again an isomorphism. □

Next, let us try to understand better the representation Z of (A.29). There are many natural maps between $\mathcal{H}$ and $\mathcal{H}_w$, namely all the maps $W(A)$ where $A \cdot w^* = w$. It clarifies matters to distinguish such a map, even though we have to do it in an arbitrary manner. So, for any $w \in G/H$ let us fix $D_w \in G$ with $D_w \cdot w^* = w$.

Lemma A.4.5 *For each $f \in \mathcal{Z}$ the function $T(f)$ on G/H given by $T(f)(w) = W(D_w^{-1})(f(w))$ is valued in $\mathcal{H}$. Furthermore*

$$T(Z(A)(f)) = X(A)(T(f)) \tag{A.31}$$

where for a function $\varphi: G/H \to \mathcal{H}$ we define $X(A)(\varphi)$ by

$$X(A)(\varphi)(w) = W(D_w^{-1} A D_{A^{-1} \cdot w})(\varphi(A^{-1} \cdot w)). \tag{A.32}$$

Proof Since $f(w) \in \mathcal{H}_w$, we have $T(f)(w) = W(D_w^{-1})(f(w)) \in \mathcal{H}_{D_w^{-1}\cdot w} = \mathcal{H}$. Next, using in the last equality that $f(w) = W(D_w)(T(f)(w))$ for $A^{-1} \cdot w$ rather than w we obtain

$$T(Z(A)(f))(w) = W(D_w^{-1}A)(f(A^{-1}\cdot w)) = W(D_w^{-1}AD_{A^{-1}\cdot w})(T(f)(A^{-1}\cdot w))$$

and the right-hand side is just $X(A)(T(f))(w)$. □

Lemma A.4.6 *For each $w \in G/H$ we have $D_w^{-1}AD_{A^{-1}\cdot w} \in H$.*

Proof Let us recall that $A \in H$ if and only if $A \cdot w^* = w^*$. But then

$$(D_w^{-1}AD_{A^{-1}\cdot w})\cdot w^* = (D_w^{-1}A)\cdot(A^{-1}\cdot w) = D_w^{-1}\cdot w = w^*. \qquad \square$$

As a consequence the right-hand side of (A.32) depends only on the restriction U of W to $\mathcal{H}$ and H. Since it is obvious that T is invertible, (A.31) shows that this is also the case of the representation Z of (A.29).

Now that we have discovered the formula (A.32), it is easy to actually construct a representation of G spanned by a subrepresentation equivalent to U. We assume for simplicity that $D_{w^*} = e$, the unit of G.

Lemma A.4.7 *Consider the space $\mathcal{X}$ of functions f from G/H to $\mathcal{H}$. Then the formula*

$$X(A)(f)(w) = U(D_w^{-1}AD_{A^{-1}\cdot w})(f(A^{-1}\cdot w)) \tag{A.33}$$

defines a representation of G in $\mathcal{X}$, which is spanned by a subrepresentation equivalent to U. Furthermore $\dim \mathcal{X} = \dim \mathcal{H} \times \dim G/\dim H$. Finally the representation Z of (A.29) is equivalent to the representation X.

Proof Since $D_w^{-1}AD_{A^{-1}\cdot w} \in H$ by Lemma A.4.6 the formula (A.33) makes sense. It requires only straightforward algebra to check that this formula defines a representation. The space $\mathcal{H}'$ of functions $f \in \mathcal{X}$ such that $f(w) = 0$ for $w \neq w^*$ is invariant under the transformations $X(B)$ for $B \in H$ and the restriction of X to H and $\mathcal{H}'$ is equivalent to U because for $B \in H$ we have $D_{B\cdot w^*}BD_{w^*} = B$ since D_{w^*} is the unit of G. That the spaces $X(A)(\mathcal{H}')$ span $\mathcal{X}$ is a consequence of the fact that every point of G/H is of the type $A \cdot w^*$ for some $A \in G$. The claim concerning the dimension of $\mathcal{X}$ is obvious and the last claim has been proved in (A.31). □

Proof of Theorem A.4.1 We have proved that if a representation W of G is spanned by a subrepresentation equivalent to U, then it is a quotient of the representation Z of (A.29), and we have proved in Lemma A.4.7 that up to equivalence this representation Z depends only on U. The rest follows from Corollary (A.4.4). □

Finally, we give a far more elegant (but also more mysterious) construction of $\mathrm{Ind}_H^G U$.

Lemma A.4.8 *Consider a group G, a subgroup H of G and a representation U of H in a space $\mathcal{H}$. Consider the space $\mathcal{G}$ of functions φ from G to $\mathcal{H}$ which satisfy the condition*

$$\forall A \in G\,,\ \forall C \in H\,,\ \varphi(AC) = U(C^{-1})(\varphi(A)). \tag{A.34}$$

Then the formula

$$R(A)(\varphi)(B) := \varphi(A^{-1}B) \tag{A.35}$$

defines a representation of G in $\mathcal{G}$. This representation is equivalent to $\mathrm{Ind}_H^G U$.

Proof For $x \in \mathcal{H}$ let us define the function $T(x)$ on G by $T(x)(A) = 0$ if $A \notin H$ and $T(x)(A) = U(A^{-1})(x)$ if $A \in H$. It is straightforward to check that $T(x) \in \mathcal{G}$. For $C, A \in H$ and $x \in \mathcal{H}$ we have

$$\begin{aligned} T(U(C)(x))(A) &= U(A^{-1})(U(C)(x)) = U(A^{-1}C)(x) \\ &= U((C^{-1}A)^{-1})(x) = T(x)(C^{-1}A) = R(C)(T(x))(A) \end{aligned}$$

and thus $TU(C) = R(C)T$, so that the restriction of R to $T(\mathcal{H})$ is equivalent to U. It is obvious that $\mathcal{G}$ is of dimension $\dim \mathcal{H} \times \operatorname{card} G/\operatorname{card} H$ so that by Theorem A.4.1 the representation R is equivalent to $\operatorname{Ind}_H^G U$. □

If it is not obvious to you that $\mathcal{G}$ is of dimension $\dim \mathcal{H} \times \operatorname{card} G/\operatorname{card} H$ this is a consequence of the fact that the map T below is an isomorphism between $\mathcal{G}$ and $\mathcal{X}$.

Exercise A.4.9 Give a direct proof that the representations of Lemmas A.4.7 and A.4.8 are equivalent. Hint: For $\varphi \in \mathcal{G}$ define $T(\varphi) \in \mathcal{X}$ by $T(\varphi)(w) = \varphi(D_w)$.

A.5 Representations of Finite Semidirect Products

A.5.1 Semidirect Products

As we mentioned at the end of Section 9.3, the method of induced representations describes all the irreducible unitary representations of the group $\mathcal{P}^*$. The full proof of this theorem involves an amount of technicality which does not fit with the spirit of the present book. On the other hand, it is interesting and not difficult to understand why the theorem can be true by considering a simpler situation where there are no technicalities. In this section, we describe all the irreducible unitary representations of a semidirect product $N \rtimes H$ where N and H are finite groups and N is commutative.

First let us explain what is a semidirect product. Since N is commutative, we will denote its operation by addition (to bring forward the similarity with the Poincaré group). The basic structure to construct the semidirect product is a homomorphism κ from H into the group of automorphisms[6] of N. That is, for each $A \in H$, $\kappa(A)$ is an automorphism of N i.e.

$$\forall a, b \in N\ ,\ \kappa(A)(a+b) = \kappa(A)(a) + \kappa(A)(b),$$

and for $A, B \in H$ we have

$$\kappa(A)\kappa(B) = \kappa(AB).$$

Definition A.5.1 The semidirect product $N \rtimes H$ is the set $N \times H$ endowed with the operation

$$(a, A)(b, B) = (a + \kappa(A)(b), AB). \tag{A.36}$$

It is straightforward to check that (A.36) defines a group operation. Please observe that the map κ does not appear in the notation $N \rtimes H$. When $\kappa(A)$ is the identity of N for each $A \in H$, $N \rtimes H$ is simply the direct product of N and H. When $N = \mathbb{R}^{1,3}$, $H = SL(2,\mathbb{C})$ and when for $A \in SL(2,\mathbb{C})$, $\kappa(A) \in SO^{\uparrow}(1,3)$ is given by (8.21) then $N \rtimes H = \mathcal{P}^*$.

We now assume that N and H are finite, and we first note a few simple facts.

Definition A.5.2 A character w of N is a group homomorphism from N into the complex numbers of modulus 1. The dual group $\widehat{N}$ of N is the set of characters provided with pointwise multiplication, i.e. $ww'(a) = w(a)w'(a)$.

Let us observe in particular that a character satisfies

$$w(-a) = w(a)^{-1} = w(a)^*. \tag{A.37}$$

As the reader will prove in Exercise A.5.10, $\widehat{N}$ is finite and $\operatorname{card} \widehat{N} = \operatorname{card} N$. The following obvious result explains how κ induces an operation $\widehat{\kappa}$ of H on $\widehat{N}$.

6 An automorphism of a group is an isomorphism of the group into itself.

Lemma A.5.3 *For $A \in H$ and $w \in \widehat{N}$ we define $\widehat{\kappa}(A)(w) \in \widehat{N}$ by*

$$\widehat{\kappa}(A)(w)(a) = w(\kappa(A^{-1})(a)).$$

Then for $A, B \in H$ we have $\widehat{\kappa}(A)\widehat{\kappa}(B) = \widehat{\kappa}(AB)$.

A.5.2 Finding All Irreducible Unitary Representations of Finite Semidirect Products

Our goal is to describe all the irreducible unitary representations of $N \rtimes H$. We use first, as is proved in the next section, that such an irreducible representation is finite-dimensional. So let $\pi(a, A)$ be such an irreducible representation of $N \rtimes H$ in a finite-dimensional Hilbert space $\mathcal{H}$. We first consider the representation Π of N given by $\Pi(a) = \pi(a, e)$ where e denotes the unit of H. For $w \in \widehat{N}$ we denote

$$\mathcal{V}_w = \{x \in \mathcal{H} \,;\, \forall a \in N \,;\, \Pi(a)(x) = w(a)x\}.$$

This is a subspace of $\mathcal{H}$.

Lemma A.5.4 *If $w \neq w'$ then $\mathcal{V}_w$ and $\mathcal{V}_{w'}$ are orthogonal.*

Proof Let us consider $a \in N$ with $w(a) \neq w'(a)$, and $x \in \mathcal{V}_w, y \in \mathcal{V}_{w'}$. On the one hand, we have

$$(x, \Pi(a)y) = w'(a)(x, y)$$

and on the other hand, using that $\Pi(a)$ is unitary, and recalling (A.37),

$$\begin{aligned}(x, \Pi(a)y) &= (\Pi(a)^\dagger x, y) = (\Pi(a)^{-1}x, y) = (\Pi(-a)x, y) = (w(-a)x, y)\\ &= (w(a)^* x, y) = w(a)(x, y),\end{aligned}$$

so that indeed $(x, y) = 0$. □

Lemma A.5.5 *There exists $w^\sharp \in \widehat{N}$ such that $\mathcal{V}_{w^\sharp}$ is not reduced to 0.*

Proof Each subspace of $\mathcal{H}$ (and in particular $\mathcal{H}$ itself) which is invariant under all the $\Pi(a)$ contains a minimal non-trivial such subspace $\mathcal{H}'$, which constitutes an irreducible representation of N. Since N is commutative, this representation is of dimension 1 by Schur's lemma (Lemma 4.5.7). Considering $x \neq 0 \in \mathcal{H}'$, for $a \in N$ we then have $\Pi(a)(x) = w^\sharp(a)x$ for a certain complex number $w^\sharp(a)$ of modulus 1, and thus obviously $w^\sharp$ is a character for which $x \in \mathcal{V}_{w^\sharp}$. □

Let us recall that 0 denotes the unit of N, and for $A \in H$ let us write $\Xi(A) = \pi(0, A)$. We observe the relations $(a, A) = (a, e)(0, A)$ and $(a, A) = (0, A)(\kappa(A^{-1})(a), e)$ and their consequences

$$\begin{aligned}\pi(a, A) &= \pi(a, e)\pi(0, A) = \Pi(a)\Xi(A)\\ \pi(a, A) &= \pi(0, A)\pi(\kappa(A^{-1})(a), e) = \Xi(A)\Pi((\kappa(A^{-1})(a)),\end{aligned} \tag{A.38}$$

which we will use many times.

Lemma A.5.6 *For $A \in H$ we have $\Xi(A)(\mathcal{V}_w) = \mathcal{V}_{\widehat{\kappa}(A)(w)}$.*

Proof Consider $x \in \mathcal{V}_w$. Then, using (A.38), and using in the last equality that $\Pi(\kappa(A^{-1})(a))(x) = w(\kappa(A^{-1})(a))x$ since $x \in \mathcal{V}_w$,

$$\Pi(a)(\Xi(A)(x)) = \pi(a, A)(x) = \Xi(A)\Pi(\kappa(A^{-1})(a))(x) = w(\kappa(A^{-1})(a))\Xi(A)(x),$$

and since $w(\kappa(A^{-1})(a)) = \widehat{\kappa}(A)(w)(a)$ this implies that $\Xi(A)(x) \in \mathcal{V}_{\widehat{\kappa}(A)(w)}$. Thus $\Xi(A)(\mathcal{V}_w) \subset \mathcal{V}_{\widehat{\kappa}(A)(w)}$. The other direction follows by replacing A by A^{-1} and w by $\widehat{\kappa}(A)(w)$. □

Consider then

$$\mathcal{O} = \{\widehat{\kappa}(A)(w^\sharp) \; ; \; A \in H\}, \tag{A.39}$$

the orbit of $w^\sharp$ under the action $\widehat{\kappa}$ of H on $\widehat{N}$. It follows from Lemma A.5.6 that for $w \in \mathcal{O}$ the space $\mathcal{V}_w$ is not reduced to 0. Furthermore this lemma implies that the linear span of the spaces $\mathcal{V}_w$ for $w \in \mathcal{O}$ is invariant under all the transformations $\pi(a, A)$, and since π is irreducible this space is $\mathcal{H}$ itself. Thus $\mathcal{H}$ is the orthogonal sum of the spaces $\mathcal{V}_w$ as w varies in $\mathcal{O}$. In particular $x \in \mathcal{H}$ can be written in a unique manner as a sum $x = \sum_{w \in \mathcal{O}} x_w$ where $x_w \in \mathcal{V}_w$, and the square of the norm of x is the sum of the squares of the norms of the x_w. All the spaces $\mathcal{V}_w$ have the same dimension and each $\pi(a, A)$ just shuffles them around. In fact, we have already a complete description of π, as the following lemma shows.

Lemma A.5.7 *We have*

$$\pi(a, A)\Big(\sum_{w \in \mathcal{O}} x_w\Big) = \sum_{w \in \mathcal{O}} w(a)\Xi(A)(x_{\widehat{\kappa}(A^{-1})(w)}), \tag{A.40}$$

and

$$\Xi(A)(x_{\widehat{\kappa}(A^{-1})(w)}) \in \mathcal{V}_w. \tag{A.41}$$

Proof We first observe that (A.41) follows from Lemma A.5.6. Using in the second equality that $w \mapsto \widehat{\kappa}(A^{-1})(w)$ is a permutation of $\mathcal{O}$ we obtain using (A.38),

$$\pi(a, A)\Big(\sum_{w \in \mathcal{O}} x_w\Big) = \sum_{w \in \mathcal{O}} \Pi(a)\Xi(A)(x_w) = \sum_{w \in \mathcal{O}} \Pi(a)\Xi(A)(x_{\widehat{\kappa}(A^{-1})(w)}) \tag{A.42}$$

and the result by (A.41), and since $\Pi(a)(y) = w(a)y$ for $y \in \mathcal{V}_w$. □

There are obvious relations between the present situation and the theory of induced representations discussed in the previous section, but we will discuss these in Exercise A.5.12 and we give self-contained arguments.

For each $w \in \mathcal{O}$ let us pick an element $D_w \in H$ with $\widehat{\kappa}(D_w)(w^\sharp) = w$. To each function φ from $\mathcal{O}$ to $\mathcal{V} := \mathcal{V}_{w^\sharp}$ we associate the element $T(\varphi) := \sum_{w \in \mathcal{O}} \Xi(D_w)(\varphi(w))$ of $\mathcal{H}$. Thus, if $x = T(\varphi)$ then $x = \sum_{w \in \mathcal{O}} x_w$ where $x_w = \Xi(D_w)(\varphi(w)) \in \mathcal{V}_w$. Then we obtain

$$\pi(a, A)(T(\varphi)) = \sum_{w \in \mathcal{O}} w(a)\Xi(AD_{\widehat{\kappa}(A^{-1})(w)})(\varphi(\widehat{\kappa}(A^{-1})(w))).$$

Furthermore if $x = \sum_{w \in \mathcal{O}} x_w \in \mathcal{H}$ then $T^{-1}(x)(w) = \Xi(D_w^{-1})(x_w)$, so that we obtain

$$T^{-1}\pi(a, A)T(\varphi)(w) = w(a)\Xi(D_w^{-1}AD_{\widehat{\kappa}(A^{-1})(w)})(\varphi(\widehat{\kappa}(A^{-1})(w))). \tag{A.43}$$

Let us define the little group $H_{w^\sharp} := \{A \in H \; ; \; \widehat{\kappa}(A)(w^\sharp) = w^\sharp\}$. Then for $A \in H_{w^\sharp}$ we have (recalling that $\mathcal{V} = \mathcal{V}_{w^\sharp}$)

$$\Xi(A)(\mathcal{V}) = \mathcal{V},$$

so that the restriction U of Ξ to $H_{w^\sharp}$ and $\mathcal{V}$ is unitary. The remarkable feature of the right-hand side of (A.43) is that $D_w^{-1}AD_{\widehat{\kappa}(A^{-1})(w)} \in H_{w^\sharp}$ (as in Lemma A.4.6) so that this right-hand side depends only on U. Furthermore, the map T is unitary between $\ell^2(\mathcal{O}, \mathcal{V})$ and $\mathcal{H}$. We may then state the main result of the present section.

Proposition A.5.8 *(a) Given a character $w^\sharp$ on N, consider its orbit O under the action of H and a unitary representation U of the little group $H_{w^\sharp}$ of $w^\sharp$ in a finite-dimensional Hilbert space $\mathcal{V}$. For each $w \in O$ fix $D_w \in H$ such that $\hat{\kappa}(D_w)(w^\sharp) = w$. Then the formula*

$$\pi'(a,A)(y)(w) = w(a)U(D_w^{-1}AD_{\widehat{\kappa}(A^{-1})(w)})[y(\widehat{\kappa}(A^{-1})(w))] \tag{A.44}$$

defines a unitary representation of $N \rtimes H$ in the space $\ell^2(O,\mathcal{V})$.
(b) Every irreducible unitary representation of $N \rtimes H$ is unitarily equivalent to a representation of the previous type.

The formula (A.44) should be compared to the formula (9.7).

Proof (a) It is straightforward algebra to check that the formula (A.44) defines a unitary representation of $N \rtimes H$, following the same path as in the proof of Theorem 9.4.2.
(b) This follows from (A.43). □

Proposition A.5.9 *The representation of $N \rtimes H$ constructed in Proposition A.5.8 (a) is irreducible if and only if U is irreducible.*

Proof If a subspace $\mathcal{G}$ of $\mathcal{V}$ is invariant under all the operators $U(A)$, $A \in H_{w^\sharp}$, then $\ell^2(O,\mathcal{G})$ is invariant under all the operators $\pi'(a,A)$. Consequently U must be irreducible if π' is irreducible. The converse is the object of the next exercise. □

Exercise A.5.10 (a) Prove that if $w \neq w' \in \widehat{N}$, then $\sum_{a\in N} w'(a)^* w(a) = 0$. As a consequence, the elements of $\widehat{N}$ are linearly independent: If a sum $\sum_{w\in\widehat{N}} \alpha_w w$ is such that $\sum \alpha_w w(a) = 0$ for each $a \in N$ then $\alpha_w = 0$ for each $w \in \widehat{N}$.
(b) Prove that card N = card $\widehat{N}$.
(c) Consider the space $\mathcal{F}$ of functions $f : H \to \mathcal{V}$ with the property that for $A \in H, C \in H_{w^\sharp}$ we have $f(AC) = U(C^{-1})f(B)$, provided with the norm $\|f\|^2 = \sum_{A\in H} \|f(A)\|^2$. Prove that the formula

$$\lambda(a,A)(f)(B) = \widehat{\kappa}(B)(w^\sharp)(a)f(A^{-1}B)$$

defines a unitary representation of $N \rtimes H$ on $\mathcal{F}$ and that this representation is unitarily equivalent to $\pi(a,A)$.
(d) Prove that the previous representation is irreducible when U is irreducible.

Exercise A.5.11 To better connect the ideas of the present section with those of Section A.4 describe a natural map R from $H/H_{w^\sharp}$ to O such that for $A \in G$ and $v \in H/H_{w^\sharp}$ we have $R(A \cdot v) = \widehat{\kappa}(A)(R(v))$.

Exercise A.5.12 We recall the notion of induced representations from the previous section.
(a) Prove that the representation Ξ is equivalent to $\mathrm{Ind}_{H_{w^\sharp}}^{H}(U)$.
(b) Prove that the representation π is equivalent to the representation $\mathrm{Ind}_{N\rtimes H_{w^\sharp}}^{N\rtimes H}(U')$ where U' is the restriction of π to $N \rtimes H_{w^\sharp}$ and $\mathcal{V}$.

A.5.3 Classifying All Unitary Representations of Finite Semidirect Products

We have proved that an irreducible unitary representation of $N \rtimes H$ is always equivalent to a representation given by the formula (A.44), and that conversely this formula (A.44) produces irreducible representations. The remaining question, then, is how to recognize when two representations given by the formula (A.44) are equivalent. A first observation is that the representation does not depend on the choice of D_w such that $\widehat{\kappa}(D_w)(w^\sharp) = w$. The simple proof is hinted at in Exercise 9.4.3. Another simple observation is that in (A.39) the point $w^\sharp$ is after all an *arbitrary* point of its orbit O. It is the orbit O, not the specific point $w^\sharp$ of this orbit which is important. If we choose another point of this orbit we will describe the same representation in another manner. This simple result is described in Proposition 9.4.5. (It is important to observe that the little groups of the points in the same orbit are all isomorphic.) Besides these obvious cases, the formula (A.44) produces different representations, as is expressed by the following, where we call the orbit O *the orbit associated to the representation.*

Theorem A.5.13 *Consider two irreducible unitary representations of $N \rtimes H$ produced by the formula (A.44). If the two orbits associated to these representations are disjoint, the representations are not unitarily equivalent. If the two orbits associated to these representations coincide, and if we use the same point $w^\sharp$ of these orbits to describe the two representations by the formula (A.44), the representations are unitarily equivalent if and only if the corresponding representations of the little group of $w^\sharp$ are unitarily equivalent.*

As we have no use for this result, which is an easy consequence of arguments given earlier in this section we will not give the proof.

A.5.4 Formal Application to the Poincaré Group.

Now we describe the corresponding procedure in the situation of $\mathcal{P}^* = \mathbb{R}^{1,3} \rtimes SL(2,\mathbb{C})$. We recall that what we denoted in this case as $A(x)$ for $x \in \mathbb{R}^{1,3}$ and $A \in SL(2,\mathbb{C})$ was in fact $\kappa(A)(x)$, where $\kappa(A)$ is given by (8.21). The group $\widehat{\mathbb{R}}^{1,3}$ of (continuous) characters identifies to $\mathbb{R}^{1,3}$, a point $p \in \mathbb{R}^{1,3}$ corresponding to the character $\chi_p : x \mapsto \exp(\mathrm{i}(x,p)/\hbar)$. The action $\widehat{\kappa}(A)$ of $A \in SL(2,\mathbb{C})$ on $\widehat{\mathbb{R}}^{1,3} = \mathbb{R}^{1,3}$ is given by

$$\begin{aligned}\widehat{\kappa}(A)(\chi_p)(x) &= \chi_p(\widehat{\kappa}(A)^{-1}(x))\\ &= \exp(\mathrm{i}(\kappa(A^{-1})(x),p)/\hbar)\\ &= \exp(\mathrm{i}(x,\kappa(A)(p)/\hbar)\\ &= \chi_{\kappa(A)(p)}(x), \end{aligned} \tag{A.45}$$

so that the action of $SL(2,\mathbb{C})$ on the group of characters is just its natural action on $\mathbb{R}^{1,3}$.

This explains the importance of the orbits in $\mathbb{R}^{1,3}$ under the action of $SL(2,\mathbb{C})$. We will leave the determination of these orbits as an easy exercise to the reader. These are the sets $X_m = \{p; p^2 = m^2c^2, p^0 > 0\}$; $-X_m = \{p; p^2 = m^2c^2, p^0 < 0\}$; $X_0 \setminus \{0\} = \{p; p^2 = 0, p^0 > 0\}$; $\{p; p^2 = 0, p^0 < 0\}$; $\{p; p^2 = -m^2c^2\}$, where $m > 0$, and, finally, $\{0\}$. Given such an orbit O the space $\ell^2(O,\mathcal{V})$ of Proposition A.5.8 has to be replaced by the space $L^2(O,\mathrm{d}\lambda,\mathcal{V})$ where $\mathrm{d}\lambda$ is a measure on O invariant under the action of $SL(2,\mathbb{C})$. By analogy with the case of finite groups, it is not that surprising that an irreducible representation of $\mathcal{P}^*$ always arises in this manner. Not all the orbits however are physically relevant. Those which are relevant are orbits of points p with $p^2 \ge 0$ and $p^0 \ge 0$. These orbits are exactly the sets X_m for $m > 0$ and the set $X_0 \setminus \{0\}$. The natural choice of an invariant measure on O is then $\mathrm{d}\lambda = \mathrm{d}\lambda_m$. Given a specific choice p^* in X_m, in the formula (A.44) one has to replace the value of the character w on a by $\exp(\mathrm{i}(a,p^*)/\hbar)$, and this formula is then just the formula of induced representations (9.7), describing an irreducible representation of $\mathcal{P}^*$ in the space $L^2(X_m,\mathrm{d}\lambda_m,\mathcal{V})$. The choice of p^* in X_m is up to us. When the orbit is X_m, $m > 0$ a natural choice is $p^* = (mc,0,0,0)$. A further advantage of this choice is that given p in X_m there is a canonical choice of $D_p \in SL(2,\mathbb{C})$ such that $\kappa(D_p)(p^*) = p$, namely a pure boost.

A.6 Representations of Compact Groups

In the present section we prove the following fundamental theorem. In Section D.3 we apply it to the definition of angular momentum.

Theorem A.6.1 *A unitary strongly continuous representation of a compact group has a finite-dimensional invariant subspace.*

The proof will start in about a page, after that of the following corollary.

Corollary A.6.2 *A unitary strongly continuous representation of a compact group in a separable Hilbert space is a direct sum of finite-dimensional representations.*

That is, the Hilbert space is a direct sum of finite-dimensional orthogonal subspaces $\mathcal{H}_n$, each of which is invariant under all the operators $\pi(A)$.

Proof Mathematicians often appeal to Zorn's lemma for this proof, which I find absurd, so I give a constructive proof. Consider an orthonormal basis $(e_\ell)_{\ell\geq 1}$ of $\mathcal{H}$. For a finite-dimensional subspace $\mathcal{F}$ of $\mathcal{H}$, let us define

$$a(\mathcal{F}) = \sum_{\ell\geq 1} 2^{-\ell}\|P_{\mathcal{F}}(e_\ell)\|^2,$$

where $P_{\mathcal{F}}$ is the orthogonal projection on $\mathcal{F}$. Let us denote by $\mathcal{I}$ the class of finite-dimensional invariant subspaces of $\mathcal{H}$. We construct by induction a sequence $\mathcal{H}_k \in \mathcal{I}$ of mutually orthogonal spaces, as follows. Having constructed $\mathcal{H}_1,\ldots,\mathcal{H}_k$, the orthogonal complement $\mathcal{G}_k$ of $\bigoplus_{\ell\leq k}\mathcal{H}_\ell$ is invariant under T, so that by Theorem A.6.1 it contains a least one space $\mathcal{F}\in\mathcal{I}$. We then chose $\mathcal{H}_{k+1}$ with

$$a(\mathcal{H}_{k+1}) \geq \frac{1}{2}\sup\{a(\mathcal{F})\ ;\ \mathcal{F}\subset\mathcal{G}_k\ ,\ \mathcal{F}\in\mathcal{I}\}. \tag{A.46}$$

We prove now that the closed span $\mathcal{G}$ of the spaces $\mathcal{H}_k$ equals $\mathcal{H}$. Otherwise, for some ℓ, the projection e'_ℓ of e_ℓ on the orthogonal complement $\mathcal{G}'$ of $\mathcal{G}$ is not zero. Now, the closed linear span of the spaces $\mathcal{F}\in\mathcal{I},\mathcal{F}\subset\mathcal{G}'$ is all of $\mathcal{G}'$ (for otherwise the orthogonal complement of this space in $\mathcal{G}'$ would violate Theorem A.6.1). Consequently we can find $\mathcal{F}\in\mathcal{I},\mathcal{F}\subset\mathcal{G}'$ with $\|P_{\mathcal{F}}(e'_\ell)\|^2 > 0$. Since $\|P_{\mathcal{F}}(e_\ell)\|^2 = \|P_{\mathcal{F}}(e'_\ell)\|^2$ this shows that $a(\mathcal{F}) > 0$. Since $\mathcal{F}\subset\mathcal{G}_k$ we deduce from (A.46) that for each k we have

$$a(\mathcal{H}_{k+1}) \geq \frac{1}{2}a(\mathcal{F}).$$

But since the spaces $\mathcal{H}_k$ are orthogonal, for each ℓ we have $\sum_k \|P_{\mathcal{H}_k}(e_\ell)\|^2 \leq \|e_\ell\|^2 = 1$, so that $\sum_k a(\mathcal{H}_k)\leq 1$, which contradicts the previous condition. □

In the rest of the section we give a self-contained proof of Theorem A.6.1. The reader who is afraid of compact groups may assume that the group is $SU(2)$. The fundamental ingredient of the proof is that on a compact group G there exists a translation invariant probability measure $\mathrm{d}a$ called the *Haar measure*. (It is known to be unique.) Thus, for any continuous function f on G and any $b\in G$ we have

$$\int f(a)\mathrm{d}a = \int f(ba)\mathrm{d}a = \int f(ab)\mathrm{d}a. \tag{A.47}$$

Exercise A.6.3 We have seen from Lemma 8.1.7 that $SU(2)$ naturally identifies with the sphere of $\mathbb{R}^4$. There is a natural probability measure on this sphere. Prove that it is the Haar measure. Hint: Prove (A.47).

Let us then fix a strongly continuous unitary representation π of a compact group G. Our goal is to construct a finite-dimensional invariant subspace. (This will imply Theorem A.6.1.) We fix once and for all a unit vector u of $\mathcal{H}$ and for $x\in\mathcal{H}$ we define

$$T(x) = \int (\pi(a)(u),x)\pi(a)(u)\mathrm{d}a \in \mathcal{H}. \tag{A.48}$$

Lemma A.6.4 *The operator T is defined everywhere, of norm ≤ 1, symmetric and not zero. It commutes with all the operators $\pi(a)$. Moreover, if (x_n) is a bounded sequence in $\mathcal{H}$,*

$$\forall y \in \mathcal{H}, \lim_{n\to\infty} (y, x_n) = 0 \Rightarrow \lim_{n\to\infty} \|T(x_n)\| = 0. \tag{A.49}$$

Proof Since $\|\pi(a)(u)\| \leq 1$ it is obvious that $\|T(x)\| \leq \|x\|$. We write

$$\begin{aligned}(y, T(x)) &= \int (\pi(a)(u), x)(y, \pi(a)(u))\mathrm{d}a \\ &= \int (x, \pi(a)(u))^*(\pi(a)(u), y)^*\mathrm{d}a \\ &= \Big(\int (\pi(a)(u), y)(x, \pi(a)(u))\mathrm{d}a\Big)^* \\ &= (x, T(y))^* = (T(y), x),\end{aligned} \tag{A.50}$$

proving that T is symmetric. Now,

$$(x, T(x)) = \int |(\pi(a)(u), x)|^2\mathrm{d}a.$$

For $x = u$ the integrand is 1 when a is the unit of G. The map $a \mapsto \pi(a)(u)$ is continuous since we assume π to be strongly continuous. Since a neighborhood of the identity has positive Haar measure, the right-hand side is positive. Thus $T(u) \neq 0$.

Next we show that T commutes with $\pi(b)$. We have, using (A.47) in the second equality, using in the third equality that $(\pi(ba)(u), \pi(b)(y)) = (\pi(a)(u), y)$ since π is unitary, and using that $\pi(ba) = \pi(b)\pi(a)$ in the fourth equality,

$$\begin{aligned}T(\pi(b)(y)) &= \int (\pi(a)(u), \pi(b)(y))\pi(a)(u)\mathrm{d}a \\ &= \int (\pi(ba)(u), \pi(b)(y))\pi(ba)(u)\mathrm{d}a \\ &= \int (\pi(a)(u), y)\pi(ba)(u)\mathrm{d}a \\ &= \pi(b)(T(y)).\end{aligned} \tag{A.51}$$

Finally we turn to the proof of (A.49). First we note that the set $K = \{\pi(a)(u); a \in G\}$ is compact because G is compact and $a \mapsto \pi(a)(u)$ is continuous. Next, since $\|\pi(a)(u)\| \leq 1$ it holds that

$$\|T(x)\| \leq \int |(\pi(a)(u), x)|\|\pi(a)(u)\|\mathrm{d}a \leq \sup_{y\in K} |(y, x)|.$$

Since K is compact, given $\varepsilon > 0$ we can find a finite subset F of K such that each point of K is within distance ε of F, and then

$$\sup_{y\in K} |(y, x)| \leq \sup_{y\in F} |(y, x)| + \varepsilon\|x\|,$$

and this inequality should make the result obvious. □

If one knows spectral theory, Theorem A.6.1 is obvious at this stage, because there exist projections which are "functions" of the non-trivial symmetric operator T and therefore commute with every $\pi(a)$, and one can use (A.49) to show that these projections must have a finite-dimensional range. Rather than using spectral theory, we prove Theorem A.6.1 by a simple self-contained argument.

Lemma A.6.5 *There exists a unit vector x with $\|T(x)\| = \lambda$ where $\lambda = \|T\| > 0$. Furthermore $T^2(x) = \lambda^2 x$.*

Proof Consider a sequence (y_n) of unit vectors of $\mathcal{H}$ with $\lim_{n\to\infty} \|T(y_n)\| = \lambda = \|T\|$. Fixing an orthonormal basis (e_k) we may assume by taking a subsequence that for each k the limit $w_k := \lim_{n\to\infty}(e_k, y_n)$ exists. Since for each n we have $\sum_k |(e_k, y_n)|^2 = 1$ it follows that $\sum_k |w_k|^2 \le 1$. Let $x = \sum_k w_k e_k$, so that $\|x\| \le 1$. Consider $x_n = y_n - x$. Then $\|x_n\| \le 2$, and for each k we have $\lim_{n\to\infty}(e_k, x_n) = 0$. It follows that for each $z \in \mathcal{H}$ we have $\lim_{n\to\infty}(z, x_n) = 0$, by approximating in norm such a vector z by a linear combination of finitely many basis vectors. Then (A.49) implies that $\lim_{n\to\infty} \|T(x_n)\| = 0 = \lim_{n\to\infty} \|T(y_n) - T(x)\|$, and thus that $\lim_{n\to\infty} \|T(y_n)\| = \|T(x)\|$. Consequently $\|T(x)\| = \lambda = \|T\|$ and since $\|x\| \le 1$ it follows that x is a unit vector. Then $\|T^2(x)\| \le \|T\|^2 = \lambda^2$. On the other hand, since T is symmetric,

$$(T^2(x), x) = (T(x), T(x)) = \|T(x)\|^2 = \lambda^2.$$

Thus the vector $y := T^2(x)/\lambda^2$ satisfies $\|y\| \le 1$ and $(y, x) = 1$. Since x is a unit vector we have $y = x$ i.e. $T^2(x) = \lambda^2 x$. □

Proof of Theorem A.6.1 The eigenspace $\mathcal{H}'$ of T^2 corresponding to the eigenvalue λ^2 is invariant by π because each $\pi(a)$ commutes with T^2. Assume then that $\mathcal{H}'$ is infinite-dimensional, and consider an orthogonal basis (e_n) of this space. Then $\|T^2(e_n)\| = \|\lambda^2 e_n\| = \lambda^2$ whereas $\lim_{n\to\infty}(y, e_n) = 0$ for all $y \in \mathcal{H}$. But this contradicts (A.49) since $\|T^2(e_n)\| \le \|T\|\|T(e_n)\| \to 0$. Therefore $\mathcal{H}'$ is a finite-dimensional invariant subspace. □

Appendix B

End of Proof of Stone's Theorem

In this appendix we lighten notation by taking a system of units where $\hbar = 1$. Our goal, given a self-adjoint operator A, is to construct the one-parameter unitary group $U(t)$ of which it is the infinitesimal generator, that is

$$\forall x \in \mathcal{D}(A)\,,\ A(t) = \lim_{t\to 0} \frac{1}{\mathrm{i}t}(U(t)x - x).$$

The proof follows the lines of the proof of the Hille–Yosida theorem on semigroups of contractions in a Banach space, with a little twist as we need to construct a group rather than a semigroup. The construction is based on the Green operator $G(z) = (z1 - A)^{-1}$ of Definition J.1. To learn about the Green operator, the reader is referred to Appendix J, but we now make a list of the properties of this operator we shall need. The Green operator $G(z)$ exists for $\mathrm{Im}\, z \neq 0$. It is a bounded operator-valued in $\mathcal{D}(A)$ which satisfies $\|G(z)\| \leq |\mathrm{Im}\, z|^{-1}$ and

$$\forall x \in \mathcal{D}(A), G(z)(z1 - A)(x) = x\ ;\ \forall x \in \mathcal{H}, (z1 - A)G(z)(x) = x,$$

which we simply write as

$$G(z)(z1 - A) = (z1 - A)G(z) = 1.$$

Furthermore if $\mathrm{Im}\, z, \mathrm{Im}\, z' \neq 0$, then

$$G(z)G(z') = G(z')G(z). \tag{B.1}$$

Let us first explain the strategy. We rewrite the equality $G(z)(z1 - A) = 1$ as

$$zG(z) - 1 = G(z)A. \tag{B.2}$$

We specialize to the case $z = \mathrm{i}\lambda$ where $\lambda \in \mathbb{R}$ and lighten notation by writing

$$V(\lambda) = \mathrm{i}\lambda G(\mathrm{i}\lambda).$$

Informally we have

$$G(z) = \frac{1}{z1 - A}$$

so that

$$V(\lambda) = \frac{\mathrm{i}\lambda}{\mathrm{i}\lambda 1 - A},$$

and we should expect that

$$\lim_{\lambda\to\pm\infty} V(\lambda) = 1. \tag{B.3}$$

A fundamental property is that

$$\|V(\lambda)\| \le |\lambda|\|G(\mathrm{i}\lambda)\| \le |\lambda||\mathrm{Im}\,(\mathrm{i}\lambda)|^{-1} = 1. \tag{B.4}$$

Now, from (B.2) it holds that

$$\lambda(V(\lambda) - 1) = -\mathrm{i}V(\lambda)A \tag{B.5}$$

and we should expect from (B.3) that the bounded operator $\lambda(V(\lambda)-1)$ approaches $-\mathrm{i}A$ as $\lambda \to \pm\infty$. We then should have

$$U(t) = \exp(\mathrm{i}tA) = \lim_{\lambda\to\infty} \exp(-t\lambda(V(\lambda) - 1)).$$

Lemma B.1 *The limit (B.3) holds in the strong sense.*

Proof Consider $x \in \mathcal{D}(A)$. Multiplying the identity $\mathrm{i}\lambda x - (\mathrm{i}\lambda - A)(x) = A(x)$ to the left by $G(\mathrm{i}\lambda)$ we get

$$V(\lambda)(x) - x = G(\mathrm{i}\lambda)A(x),$$

and the second inequality in (B.4) implies that this quantity goes to 0 as $\lambda \to \infty$. The set of x for which $\lim_{\lambda\to\pm\infty} V(\lambda)(x) = x$ therefore contains $\mathcal{D}(A)$ so that it is dense. Since $\|V(\lambda)\| \le 1$, this set is closed, and this concludes the proof. □

Let us then define

$$U_\lambda(t) := \exp(-t\lambda(V(\lambda) - 1)) = \exp(t\lambda)\sum_{n\ge 0}\frac{(-t\lambda)^n}{n!}V(\lambda)^n. \tag{B.6}$$

Consequently, using (B.4)

$$\|U_\lambda(t)\| \le \exp(t\lambda)\sum_{n\ge 0}\frac{|t\lambda|^n}{n!} = \exp(t\lambda + |t\lambda|). \tag{B.7}$$

This bound is useful only when t and λ have opposite signs. This creates a technical difficulty.

Let us observe that all the $U_\lambda(t)$ commute with each other because, as follows from (B.1), the operators $V(\lambda)$ commute with each other. Let us also observe that for $x \in \mathcal{D}(A)$ we have, using (B.5) in the second equality,

$$\lim_{t\to 0}\frac{1}{\mathrm{i}t}(U_\lambda(t)(x) - x) = \frac{-1}{\mathrm{i}}\lambda(V(\lambda) - 1)(x) = V(\lambda)A(x). \tag{B.8}$$

Lemma B.2 *For $t \ge 0$ the limit $U(t) = \lim_{\lambda\to-\infty} U_\lambda(t)$ exists and $\|U_\lambda(t)\| \le 1$. Moreover for $s,t \ge 0$ we have $U(s)U(t) = U(s+t)$ and for $x \in \mathcal{D}(A)$ we have*

$$\lim_{t\to 0+}\frac{1}{\mathrm{i}t}(U(t)(x) - x) = A(x). \tag{B.9}$$

Furthermore for $x \in \mathcal{D}(A)$ we have $U(t)(x) \in \mathcal{D}(A)$ and $AU(t)(x) = U(t)A(x)$.

Proof The crucial point is that for $t \ge 0$ and $\lambda \le 0$ we have $\|U_\lambda(t)\| \le 1$ by (B.7).

For two commuting operators S, T of norm ≤ 1 it holds that

$$S^n - T^n = (S^{n-1} + S^{n-2}T + \cdots + T^{n-1})(S - T)$$

and therefore

$$\|(S^n - T^n)(x)\| \le n\|(S - T)(x)\|.$$

Consider $x \in \mathcal{D}(A)$, $\mu \le 0$ and let us use the previous inequality for $S = U_\lambda(t/n)$ and $T = U_\mu(t/n)$. Then

$$\|(U_\lambda(t) - U_\mu(t))(x)\| \le n\|U_\lambda(t/n)(x) - U_\mu(t/n)(x)\|.$$

Writing the right-hand side as

$$t\left\|\frac{U_\lambda(t/n)(x)-x}{t/n}-\frac{U_\mu(t/n)(x)-x}{t/n}\right\|,$$

letting $n \to \infty$ and using (B.8) we obtain

$$\|(U_\lambda(t)-U_\mu(t))(x)\| \le t\|(V(\lambda)-V(\mu))A(x)\|. \tag{B.10}$$

It then follows from Lemma B.1 that the limit $\lim_{\lambda\to-\infty} U_\lambda(t)(x)$ exists for all $t > 0$ and all $x \in \mathcal{D}(A)$, so that it exists for all x using again that $\|U_\lambda(t)\| \le 1$. The property $U(s)U(t) = U(s+t)$ should be obvious. Given $x \in \mathcal{D}(A)$, (B.8) implies

$$\frac{\mathrm{d}}{\mathrm{d}t}U_\lambda(t)(x) = \mathrm{i}U_\lambda(t)V(\lambda)A(x). \tag{B.11}$$

We write

$$\begin{aligned}
&\|U_\lambda(t)V(\lambda)A(x) - U(t)A(x)\|\\
&\le \|(U_\lambda(t)-U(t))V(\lambda)A(x)\| + \|U_\lambda(t)(V(\lambda)-1)A(x)\|\\
&\le \|V(\lambda)\|\|(U_\lambda(t)-U(t))A(x)\| + \|U_\lambda(t)\|\|(V(\lambda)-1)A(x)\|\\
&\le \|(U_\lambda(t)-U(t))A(x)\| + \|(V(\lambda)-1)A(x)\|.
\end{aligned} \tag{B.12}$$

As $\lambda \to -\infty$, $U_\lambda(t) \to U(t)$ and $V(\lambda) \to 1$ in the strong sense so that the right-hand side of (B.11) converges to $\mathrm{i}U(t)A(x)$ uniformly over each bounded interval in t, and the derivative of the limit as $\lambda \to \infty$ is the limit as $\lambda \to \infty$ of the derivative i.e.

$$\frac{\mathrm{d}}{\mathrm{d}t}U(t)(x) = \mathrm{i}U(t)A(x).$$

To prove the last point of the lemma let us first observe a general principle. Let us say that an operator W commutes with A if $W(x) \in \mathcal{D}(A)$ for $x \in \mathcal{D}(A)$ and $AW(x) = WA(x)$. If the operators W_λ commute with A, and if the strong limit $\lim_{\lambda\to\infty} W_\lambda$ exists then this limit commutes with A. This is because for $x \in \mathcal{D}(A)$

$$(W_\lambda(x), W_\lambda A(x)) = (W_\lambda(x), AW_\lambda(x))$$

is a point in the graph of A, and this graph is closed by Lemma 2.5.14, because A is an adjoint (since it is self-adjoint). Consequently $(W(x), WA(x))$ belongs to the graph of A, i.e. $W(x) \in \mathcal{D}(A)$ and $WA(x) = AW(x)$. Now, the operators $V(\lambda)$ commute with A, and using the previous principle, taking the limit over the partial sums which define $U_\lambda(t)$ we obtain that $U_\lambda(t)$ also commutes with A. Taking $\lambda \to \infty$ and using the same principle again shows that $U(t)$ commutes with A. □

Lemma B.3 *For $t \le 0$ the limit $W(t) = \lim_{\lambda\to\infty} U_\lambda(t)$ exists and $\|W(t)\| \le 1$. Moreover for $s,t \le 0$ we have $W(s)W(t) = W(s+t)$ and for $x \in \mathcal{D}(A)$ we have*

$$\lim_{t\to 0-}\frac{1}{\mathrm{i}t}(W(t)(x)-x) = A(x). \tag{B.13}$$

Furthermore the operators $W(s)$ and $U(t)$ commute. Finally for $x \in \mathcal{D}(A)$ we have $W(s)(x) \in \mathcal{D}(A)$ and $AW(s)(x) = WA(s)$.

The proof is entirely similar to the proof of Lemma B.2.

To complete the proof of Stone's theorem we define $U(t) = W(t)$ for $t < 0$. It remains only to prove that in this manner we define a one-parameter group of operators, that is we have to prove the identity $U(t)U(s) = U(t+s)$. We already know that this property holds for s and t of the same sign, so let us consider without loss of generality the case $s+t \ge 0$ and $t \ge 0 \ge s$. Changing s into $-s$, it suffices to prove that $U(t)W(-s) = U(t-s)$ for $0 \le s \le t$. For this we fix t, we consider $x \in \mathcal{D}(A)$, and the function $\varphi(s) = U(t)W(-s)(x) - U(t-s)(x)$. For $s < t$ and $0 \le u \le t - s$ we write

$$\varphi(s+u) - \varphi(s) = U(t)(W(-u)W(-s)(x) - W(-s)(x)) + U(t-s-u)(U(u)(x) - x).$$

Then, using (B.9) and (B.13) together with the fact that A commutes with $U(t)$ and $W(s)$ we obtain that $\mathrm{d}\varphi(s)/\mathrm{d}s = -\mathrm{i}A\varphi(s)$. Since A is symmetric,

$$\frac{\mathrm{d}}{\mathrm{d}s}\|\varphi(s)\|^2 = \frac{\mathrm{d}}{\mathrm{d}s}(\varphi(s),\varphi(s)) = (-\mathrm{i}A\varphi(s),\varphi(s)) + (\varphi(s), -\mathrm{i}A\varphi(s)) = 0.$$

Consequently, and since $\varphi(0) = 0$, we obtain that $\varphi(s) = 0$ for each $s \le t$, i.e. $U(t)W(-s)(x) = U(t-s)(x)$ for all x in $\mathcal{D}(A)$ so that $U(t)W(-s) = U(t-s)$ since $\mathcal{D}(A)$ is dense. □

Appendix C

Canonical Commutation Relations

Let us recall from Section 2.14 the following one-parameter unitary groups of $L^2 = L^2(\mathbb{R})$:

$$U(t) = \exp(\mathrm{i}tX/\hbar)\ ;\ V(s) = \exp(\mathrm{i}sP/\hbar), \tag{C.1}$$

so that

$$U(t)(f)(x) = \exp(\mathrm{i}tx/\hbar)f(x)\ ;\ V(s)(f)(x) = f(x+s). \tag{C.2}$$

They satisfy the relation (2.74)

$$V(s)U(t) = \exp(\mathrm{i}ts/\hbar)U(t)V(s). \tag{C.3}$$

As we explained in Section 2.14 the previous relation is the fruitful form of the canonical commutation relation $[X,P] = \mathrm{i}\hbar 1$. We will call it the 1-ICCR, integrated canonical commutation relation, the "1" referring to the dimension. We start to learn how to compute with this relation in Section C.1. In Section C.2 we explain the coherent states for the harmonic oscillator and their remarkable time-evolution. In Section C.3 we state and prove (for $n = 1$) the foundational Stone–von Neumann theorem. In Section C.4 we explain in great detail why and how this theorem fails in the case "$n = \infty$". Finally, in Section C.5 we try to explain how this question is treated in physics textbooks.

C.1 First Manipulations

Definition C.1.1 For s,t in $\mathbb{R}$ we set

$$S(s,t) = \exp(\mathrm{i}st/2\hbar)U(t)V(s) = \exp(-\mathrm{i}ts/2\hbar)V(s)U(t). \tag{C.4}$$

This makes sense because the last two terms are equal thanks to (C.3). The reason behind this definition will become clear very soon, in Lemma C.1.3. The following is straightforward using (C.3) and (C.4).

Lemma C.1.2 *We have*

$$S(s,t)S(s',t') = \exp(\mathrm{i}(st'-s't)/2\hbar)S(s+s',t+t'). \tag{C.5}$$

This formula encompasses at the same time the commutation relation (C.3) and the fact that $U(t)$ and $V(s)$ are one-parameter groups. From now on, we mostly forget about $U(t)$ and $V(s)$ and think in terms of $S(s,t)$.

Lemma C.1.3 *For $s,t \in \mathbb{R}$ we have*

$$S(s,t) = \exp(\mathrm{i}(tX+sP)/\hbar). \tag{C.6}$$

While $tX+sP$ is symmetric and defined, say, on the Schwartz functions, it is by no means obvious that this is a self-adjoint operator. As the proof will show, there exists a self-adjoint operator equal to $tX+sP$ on the space $\mathcal{S}$ of Schwartz functions.

Proof Fixing s and t for $r \in \mathbb{R}$ define $T(r) = S(rs, rt)$. As a consequence of (C.5), we obtain $T(r)T(r') = T(r+r')$. Thus T is a one-parameter group of unitary operators. Moreover, for $f \in \mathcal{S}$ it holds that

$$\lim_{r\to 0} \frac{\hbar}{ir}(T(r)(f) - f) = tX(f) + sP(f).$$

This is obtained in a straightforward manner by writing the value of $T(r)(f)(x)$ using (C.4) and taking the derivative. Consequently, by Stone's theorem, there exists a self-adjoint operator Y which coincides with $tX + sP$ on $\mathcal{D}(X) \cap \mathcal{D}(P)$, and the associated one-parameter group is given by the formula (C.6). □

After the previous argument, it would be a waste not to make a small detour to prove a useful formula connected to (C.3). It is much abused in physics textbooks which routinely apply it to operators which do not satisfy the required hypothesis or even for which the meaning of the exponential is obscure.

Proposition C.1.4 *Consider two bounded operators A and B which both commute with* $[A, B]$. *Then*

$$\exp(A + B) = \exp(-[A, B]/2) \exp A \exp B. \tag{C.7}$$

For bounded operators the exponential is given by the sum of the usual power series. A fundamental property is that $\exp(A + B) = \exp A \exp B$ when A and B commute. This special case of (C.7) is used throughout the proof. Also used is the fact that $(\exp tA)' = A \exp tA = (\exp tA)A$.

Proof Considering

$$\varphi(t) = \exp(-t^2[A, B]/2) \exp tA \exp tB,$$

it is obvious that $\varphi'(0) = A + B$. The plan is to prove that $\varphi(s)\varphi(t) = \varphi(s + t)$. Indeed, then $\varphi'(t) = \varphi'(0)\varphi(t)$ and thus $\varphi(t) = \exp t\varphi'(0) = \exp t(A+B)$. Straightforward algebra proves that the relation $\varphi(s)\varphi(t) = \varphi(s + t)$ is equivalent to the relation

$$\exp sB \exp tA \exp(-sB) = \exp(-st[A, B]) \exp tA. \tag{C.8}$$

Given s, both the right-hand and the left-hand sides are functions $\psi(t)$ which satisfy $\psi(t)\psi(t') = \psi(t + t')$ so that by the previous argument it suffices to prove that their derivatives at $t = 0$ are equal, i.e. $\exp(sB)A \exp(-sB) = -s[A, B] + A$, which is in turn obvious by taking the derivative in s, and since $[A, B]$ commutes with $\exp(sB)$. □

The following exercise explains how the commutation relations are often presented in textbooks.[1]

Exercise C.1.5 [The Heisenberg Group] For $(s, t, r) \in \mathbb{R}^3$ we define the unitary operator

$$T(s, t, r) = \exp(ir/\hbar)S(s, t). \tag{C.9}$$

(a) Prove that

$$T(s, t, r)T(s', t', r') = T(s + s', t + t', r + r' + (st' - s't)/2). \tag{C.10}$$

(b) Prove that $\mathbb{R}^3$ provided with the composition law

$$(s, t, r) * (s', t', r') = (s + s', t + t', r + r' + (st' - s't)/2)$$

forms a group, called the Heisenberg group.

[1] As to whether this presentation helps at an elementary level is left to the judgment of the reader.

(c) Reformulate (C.10) by saying that T defines a unitary representation of the Heisenberg group. This representation is called the *Schrödinger representation of the Heisenberg group.*

Let us then learn how to differentiate (C.3). In order not to worry about domain issues, we think of all equalities between operators below as equalities valid when applied to Schwartz functions. Differentiating both sides in t implies

$$V(s)U(t)X = (s\mathbf{1} + X)\exp(its/\hbar)U(t)V(s),$$

which is straightforward to rewrite as

$$S(s,t)XS(s,t)^{-1} = X + s\mathbf{1}. \tag{C.11}$$

Effectively, conjugation by the unitary map $S(s,t)$ transforms X into $X+s\mathbf{1}$, a remarkable property. Next

$$S(s,t)X^2S(s,t)^{-1} = (S(s,t)XS(s,t)^{-1})^2 = X^2 + 2sX + s^2\mathbf{1}.$$

Consequently, using the result of Exercise 2.5.20, given a state $|z\rangle$ the state $|z_{s,t}\rangle := S(s,t)^{-1}|z\rangle$ satisfies

$$\langle z_{s,t}|X|z_{s,t}\rangle = s + \langle z|X|z\rangle \tag{C.12}$$

and

$$\langle z_{s,t}|X^2|z_{s,t}\rangle = s^2 + 2s\langle z|X|z\rangle + \langle z|X^2|z\rangle$$

so that in particular if for an operator V we define $\Delta_z^2 V = \langle z|V^2|z\rangle - \langle z|V|z\rangle^2$ we have

$$\Delta_{z_{s,t}}^2 X = \Delta_z^2 X. \tag{C.13}$$

Thus we obtain the remarkable result that when going from $|z\rangle$ to $|z_{s,t}\rangle$ the average position is shifted by s, but the spread around this average position is unchanged. A similar computation (watching for a treacherous sign) yields

$$S(s,t)PS(s,t)^{-1} = P - t\mathbf{1}, \tag{C.14}$$

from which one deduces

$$\langle z_{s,t}|P|z_{s,t}\rangle = -t + \langle z|P|z\rangle, \tag{C.15}$$

and

$$\Delta_{z_{s,t}}^2 P = \Delta_z^2 P. \tag{C.16}$$

C.2 Coherent States for the Harmonic Oscillator

The present section is a side story within this appendix. The main story starts in the next section.

We start with an obvious consequence of Stone's theorem.

Lemma C.2.1 *If Y is a self-adjoint operator and W is unitary, then $W^{-1}YW$ is self-adjoint with domain $W^{-1}\mathcal{D}(Y)$ and* $\exp(\mathrm{i}W^{-1}YW/\hbar) = W^{-1}\exp(\mathrm{i}Y/\hbar)W$.

The reader may now review Section 2.18 devoted to the harmonic oscillator with Hamiltonian $H = (P^2 + \omega^2 m^2 X^2)/2m$. Let $W(t) = \exp(-\mathrm{i}tH/\hbar)$ be the time-evolution. We recall that the time-evolution (in the Heisenberg picture) of an operator Z is the operator $W(t)^{-1}ZW(t)$. It is then very simple in principle to compute the time-evolution of an operator $\exp(\mathrm{i}rX + \mathrm{i}sP)$, since by Lemma C.2.1 it is given by

$$W(t)^{-1}\exp((\mathrm{i}rX + \mathrm{i}sP)/\hbar)W(t) = \exp\Big(\mathrm{i}rW(t)^{-1}XW(t)/\hbar + \mathrm{i}sW(t)^{-1}PW(t)/\hbar\Big). \tag{C.17}$$

However, it is not P and X which have a nice time-evolution, but rather it is

$$a = \alpha X + \mathrm{i}\beta P \; ; \; a^\dagger = \alpha X - \mathrm{i}\beta P$$

for $\alpha = \sqrt{\omega m/2\hbar}$ and $\beta = \sqrt{1/2\omega m\hbar}$. This is expressed by (2.104) and (2.105) i.e.

$$W(t)^{-1} a W(t) = \exp(-\mathrm{i}\omega t)a \; ; \; W(t)^{-1} a^\dagger W(t) = \exp(\mathrm{i}\omega t)a^\dagger . \tag{C.18}$$

Therefore to compute time-evolution it is easier to express combinations of X and P as combinations of a and $a^\dagger$ instead. For this let us consider a complex number $\gamma = u + \mathrm{i}v$, $u, v \in \mathbb{R}$. Then

$$\gamma a^\dagger - \gamma^* a = 2\mathrm{i}v\alpha X - 2\mathrm{i}u\beta P, \tag{C.19}$$

and we define the operator[2]

$$A(\gamma) := \exp((\gamma a^\dagger - \gamma^* a)/\hbar) = S(-2u\beta, 2v\alpha).$$

where we have used (C.6) in the second equality. Combining with (C.17) and (C.18) we obtain the magic formula

$$W(t)^{-1} A(\gamma) W(t) = A(\gamma \exp(\mathrm{i}\omega t)). \tag{C.20}$$

Consider now the ground state $|0\rangle$ of H (that is, the eigenvector with smallest eigenvalue), so that $H|0\rangle = (\omega\hbar/2)|0\rangle$ and thus $W(t)|0\rangle = \exp(-\mathrm{i}\omega t/2)|0\rangle$. Let us define the state $|\gamma\rangle := A(\gamma)|0\rangle$. Applying (C.20) to the state $|0\rangle$ we get $\exp(-\mathrm{i}\omega t/2) W(t)^{-1}|\gamma\rangle = |\gamma \exp(\mathrm{i}\omega t)\rangle$. Changing t into $-t$ we obtain the time-evolution of $|\gamma\rangle$:

$$W(t)|\gamma\rangle = \exp(-\mathrm{i}\omega t/2)|\gamma \exp(-\mathrm{i}\omega t)\rangle. \tag{C.21}$$

The states $|\gamma\rangle$ are called *coherent states*. A remarkably simple formula describes them. To discover it, let us pretend that we are permitted to apply (C.7) to the operators $A = \gamma a^\dagger/\hbar$ and $B = \gamma^* a/\hbar$. Then (without worrying about what is really meant by the operator[3] $\exp(\gamma a^\dagger/\hbar)$) we obtain

$$A(\gamma) = \exp\Big(-\frac{|\gamma|^2}{2\hbar^2}\Big) \exp(\gamma a^\dagger/\hbar) \exp(\gamma^* a/\hbar),$$

and since $a|0\rangle = 0$ this yields the formula

$$|\gamma\rangle = A(\gamma)|0\rangle = \exp\Big(-\frac{|\gamma|^2}{2\hbar^2}\Big) \exp(\gamma a^\dagger/\hbar)|0\rangle.$$

This argument, found in many textbooks, is at best heuristic, but as the following shows, it led us to the correct formula when one interprets the quantity $\exp(\gamma a^\dagger/\hbar)|0\rangle$ in the obvious manner.

Proposition C.2.2 *We have*

$$|\gamma\rangle = \exp\Big(-\frac{|\gamma|^2}{2\hbar^2}\Big) \sum_{n\ge 0} \frac{1}{n!}\Big(\frac{\gamma a^\dagger}{\hbar}\Big)^n |0\rangle. \tag{C.22}$$

Proof We recall from our analysis of the harmonic oscillator that the elements $(e_n)_{n\ge 0}$ defined recursively by $e_0 = |0\rangle$ and $e_{n+1} = a^\dagger(e_n)/\sqrt{n+1}$ form an orthonormal basis, so that in particular $(a^\dagger)^n|0\rangle = \sqrt{n!}e_n$ is of norm $\sqrt{n!}$. This implies the convergence of the series in (C.22) and justifies the way we will manipulate series below. Combining (C.11) and (C.14) shows that

$$S(s,t) a^\dagger S(s,t)^{-1} = a^\dagger + (\alpha s + \mathrm{i}\beta t)1.$$

[2] Not be confused with an annihilation operator.

[3] We know how to define $\exp(B)$ when B is bounded, by the usual power series. We also know how to define $\exp(\mathrm{i}B)$ when B is self-adjoint by Stone's theorem, but here neither case applies.

Now, for $\gamma = u + \mathrm{i}v$ we have $A(\gamma) = S(-2u\beta, 2v\alpha)$. Since $\alpha\beta = 1/(2\hbar)$ the preceding equality yields

$$A(\gamma)a^\dagger A(\gamma)^{-1} = a^\dagger - (\gamma^*/\hbar)\mathbf{1}$$

and

$$A(\gamma)a^\dagger = (a^\dagger - \gamma^*/\hbar)A(\gamma). \tag{C.23}$$

Consider then the function $\varphi(t) = A(t\gamma)|0\rangle$ so that $\varphi(1) = |\gamma\rangle$, and using (C.23) in the last equality,

$$\hbar\varphi'(t) = A(t\gamma)(\gamma a^\dagger - \gamma^* a)|0\rangle = \gamma A(t\gamma)a^\dagger|0\rangle = \gamma(a^\dagger - t\gamma^*/\hbar)\varphi(t).$$

Consequently the function $\psi(t) = \exp(t^2|\gamma|^2/2\hbar^2)\varphi(t)$ satisfies $\hbar\psi'(t) = \gamma a^\dagger\psi(t)$ and $\psi(0) = |0\rangle$. Since this is also the case for the function $\sum_{n\geq 0}(\gamma t a^\dagger/\hbar)^n|0\rangle/n!$ these functions are equal and taking $t = 1$ proves the result. □

Exercise C.2.3 Prove directly from the formula (C.22) that $\langle\gamma|\gamma\rangle = 1$.

Exercise C.2.4 Copy the proof of (C.23) to show that $A(\gamma)a = (a - \gamma/\hbar)A(\gamma)$. Prove that $a|\gamma\rangle = \gamma|\gamma\rangle/\hbar$. Prove that this property characterizes $|\gamma\rangle$ up to a multiplicative constant.

The special case $\gamma = t \in \mathbb{R}$ of (C.22) reads

$$\exp(-2\mathrm{i}\beta tP/\hbar)|0\rangle = \exp\Big(-\frac{t^2}{2\hbar^2}\Big)\sum_{n\geq 0}\frac{1}{n!}\Big(\frac{ta^\dagger}{\hbar}\Big)^n|0\rangle. \tag{C.24}$$

As we learned in Section 2.18, the state $|0\rangle$ corresponds to the function

$$\varphi_0(x) = \theta\exp(-\omega m x^2/2\hbar) = \theta\exp(-\alpha^2x^2), \tag{C.25}$$

where $\theta = (m\omega/\pi\hbar)^{1/4}$ is the normalizing factor, whereas for a certain polynomial P_n we have $(a^\dagger)^n|0\rangle = P_n(\alpha x)\varphi_0(x)$. Let us recall from (C.1) and (C.2) that

$$\exp(\mathrm{i}uP/\hbar)(f)(x) = f(x+u).$$

Using this for $f = \varphi_0$ and $u = -2\beta t$, (C.24) yields the identity

$$\exp(-\alpha^2(x - 2\beta t)^2) = \exp\Big(-\frac{t^2}{2\hbar^2}\Big)\sum_{n\geq 0}\frac{t^nP_n(\alpha x)}{\hbar^n n!}\exp(-\alpha^2x^2).$$

Since $\alpha\beta = 1/2\hbar$, setting $y = \sqrt{2}\alpha x$ and $u = t/\hbar\sqrt{2}$ we have proved the following.

Proposition C.2.5 *The polynomials* $H_n(y) := 2^{n/2}P_n(y/\sqrt{2})$ *satisfy the identity*

$$\exp(2uy - u^2) = \sum_{n\geq 0}\frac{u^nH_n(y)}{n!}. \tag{C.26}$$

The polynomials H_n are classically known as the *Hermite polynomials*[4] and (C.26) is described by the expression "$\exp(2uy - u^2)$ is the generating function of the Hermite polynomials".

Exercise C.2.6 Try to deduce from (C.26) that

$$H_n(y) = (-1)^n\exp(y^2)\frac{\mathrm{d}^n}{\mathrm{d}y^n}\exp(-y^2) = \exp(y^2/2)\Big(y - \frac{\mathrm{d}}{\mathrm{d}y}\Big)^n\exp(-y^2/2). \tag{C.27}$$

How does this relate to the formula (2.97)?

[4] According to Wikipedia, there are two versions of these polynomials, one in probability, one in physics. Here it is the "physics version".

Exercise C.2.7 Prove that the functions e_n of Section 2.18 are given by

$$e_n(x) = \left(\frac{m\omega}{\pi\hbar}\right)^{1/4} \frac{1}{\sqrt{2^n n!}} H_n(x\sqrt{m\omega/\hbar}) \exp(-m\omega x^2/2\hbar).$$

The following properties relate coherent states with the macroscopic behavior of the harmonic oscillator.[5]

Exercise C.2.8 (a) Using (C.13) and (C.16) convince yourself that for a coherent state the uncertainty on momentum and position is as small as permitted by (2.12). (Even though we did not discuss (2.12) in the context used here, it can be shown to be true in this context.)
(b) Convince yourself using (C.11) and (C.14) that the average position and momentum of the state $|\gamma \exp(\omega t)\rangle$ correspond to a trajectory of the classical oscillator (2.95).

C.3 The Stone–von Neumann Theorem

From this point on we use a system of units in which $\hbar = 1$. Consider the space $L^2(\mathbb{R}^n)$, thought of as a position space of a particle, the position operators X_i given by $X_i(f)(x) = x_i f(x)$ for $x = (x_i)_{i \le n}$ and the momentum operators $P_i = -\mathrm{i}\partial/\partial x_i$. These self-adjoint operators satisfy the canonical commutation relations

$$[X_i, P_i] = \mathrm{i}1, \tag{C.28}$$

with all other commutators being 0. To avoid domain issues, as in the case $n = 1$ we consider a stronger version of these relations. For $t = (t_i)_{i \le n}$ let us define the operator $U(t)$ by

$$U(t)(f)(x) = \exp(\mathrm{i}t \cdot x) f(x), \tag{C.29}$$

where $t \cdot x$ is the dot product $\sum_{i \le n} t_i x_i$ of t and x. For $s = (s_i)_{i \le n}$ let us define the operator $V(s)$ by

$$V(s)(f)(x) = f(x+s). \tag{C.30}$$

Then, recalling (2.74) (or proceeding directly) we have the following

$$U(t)U(t') = U(t+t')\ ;\ V(t)V(t') = V(t+t') \tag{C.31}$$

and

$$V(s)U(t) = \exp(\mathrm{i}t \cdot s)U(t)V(s), \tag{C.32}$$

which are called the *integrated version* of the canonical commutation relations (C.28), and are a multidimensional version of (2.74).

We say that two strongly continuous families $V(s), U(t)$ for $s, t \in \mathbb{R}^n$ of unitary operators on a Hilbert space $\mathcal{H}$ *constitute a realization of n integrated canonical commutation relations* (n-ICCR) if they satisfy the relations (C.31) and (C.32). We say that this realization is *irreducible* if no non-trivial subspace of $\mathcal{H}$ is invariant under all the maps $U(t)V(s)$.

Theorem C.3.1 (The Stone–von Neumann Theorem) *Two non-trivial irreducible realizations of the n-ICCR are unitarily equivalent.*

That is, if $U'(t)$ and $V'(s)$ constitute another non-trivial irreducible realization of the n-ICCR on a Hilbert space $\mathcal{H}'$ then there exists a one-to-one unitary map $W : \mathcal{H}' \to \mathcal{H}$ such that

$$U(t)V(s) = WU'(t)V'(s)W^{-1}. \tag{C.33}$$

[5] It is typically a difficult problem to use Quantum Mechanics to derive the macroscopic properties of matter. Here we brilliantly succeed in the case of the harmonic oscillator.

This theorem is foundational, as it shows the uniqueness of the model implied by the relations (C.28). A very readable proof can be found in [40]. Nonetheless, since the proof in Hall's textbook [40] is presented within the study of Weyl quantization, a topic which we do not cover, we present in this section a self-contained proof in the simplest possible case $n = 1$.

We denote by $U_0(t), V_0(s)$ the operators (C.2) and $S_0(s,t)$ the corresponding quantity (C.4). We consider another irreducible realization $U(t)$, $V(s)$ of the 1-ICCR, and $S(s,t)$ the corresponding quantity (C.4). We proceed to prove that these two realizations are unitarily equivalent, that is to construct an operator W such that $WS_0(s,t)W^{-1} = S(s,t)$. The relation (C.5), both for S and S_0 will be of constant use, for the reason explained below (C.5). The following simple fact is left as an exercise, as we have often used this type of argument.

Exercise C.3.2 Prove that the realization $S_0(s,t)$ of the 1-ICCR is irreducible.

First we recall the following standard result (whose proof can be found in Appendix N.)

Lemma C.3.3 *For any $\alpha > 0$ and any $a \in \mathbb{C}$ we have*

$$\int \exp(-\alpha x^2 + ax)\mathrm{d}x = \sqrt{\frac{\pi}{\alpha}} \exp\left(\frac{a^2}{4\alpha}\right). \tag{C.34}$$

Then the norm-1 function

$$\tau_0(x) = (\pi)^{-1/4} \exp(-x^2/2) \tag{C.35}$$

has the property that

$$(\tau_0, S_0(s,t)(\tau_0)) = \exp(-(s^2 + t^2)/4), \tag{C.36}$$

as follows from (C.34) in a straightforward manner.[6]

The importance of (C.36) is that, thanks to the 1-ICCR, it entirely determines all the inner products

$$(S_0(s',t')(\tau_0), S_0(s,t)(\tau_0)),$$

and therefore all the structure of the operators $S_0(s,t)$ (as will be detailed at the end of the proof).

Suppose now that you are given a realization of the 1-ICCR which is unitarily equivalent to S_0, that is, for a certain unitary operator, we have $S(s,t) = WS_0(s,t)W^{-1}$. Then, obviously, the function $\tau = W(\tau_0)$ satisfies

$$(\tau, S(s,t)(\tau)) = \exp(-(s^2 + t^2)/4). \tag{C.37}$$

Consequently, for the Stone–von Neumann theorem to be true, we must be able to find such a function τ in any realization of the 1-ICCR. Moreover, since the relation (C.37) will also encode all the structure of the $S(s,t)$, finding such a function will be the key step of the proof.

The natural idea to find such a function τ is to "write a formula for τ_0" in terms of the operators $S_0(s,t)$, hoping that the very same formula, but using instead the operators $S(s,t)$ will let us find τ. To "write a formula for τ_0" we describe the orthogonal projector P_0 on the span of τ_0 as a mixture of the operators $S_0(s,t)$, i.e.

$$P_0 = \int g(s,t)S_0(s,t)\mathrm{d}s\mathrm{d}t.$$

Therefore the function $g(s,t)$ should be such that for each (reasonable) function f one has

$$\int g(s,t)\exp(\mathrm{i}st/2)\exp(\mathrm{i}tx)f(x+s)\mathrm{d}s\mathrm{d}t = P_0(f)(x) = \tau_0(x)(\tau_0, f)$$
$$= (\pi)^{-1/2}\exp(-x^2/2)\int \exp(-y^2/2)f(y)\mathrm{d}y. \tag{C.38}$$

[6] If you wonder what kind of magic is going on here, let me reassure you that any resemblance between (C.35) and (2.103) is certainly *not* a coincidence, although some choice of parameters has been made to obtain clean formulas.

It is then easy to discover the correct formula for g,

$$g(s,t) = \frac{1}{2\pi} \exp\Big(-\frac{s^2+t^2}{4}\Big), \tag{C.39}$$

or to check that it satisfies (C.38) by integrating first in t using (C.34) and making the change of variables $y = x + s$.

Lemma C.3.4 *The map* $(s,t) \mapsto S(s,t) = \exp(\mathrm{i}st/2)U(t)V(s)$ *is strongly continuous.*

Proof It suffices to prove that for each $x \in \mathcal{H}$ the map $(s,t) \to U(t)V(s)(x)$ is continuous. For simplicity of notation we prove continuity at $s = t = 0$. Since the maps $t \mapsto U(t)$ and $s \mapsto V(s)$ are strongly continuous by hypothesis, given $\varepsilon > 0$ we can find $\alpha > 0$ such that $\|U(t)(x) - x\| \le \varepsilon/2$ and $\|V(s)(x) - x\| \le \varepsilon/2$ for $|t|, |s| \le \alpha$. Then for such s,t, we have

$$\begin{aligned}\|U(t)V(s)(x) - x\| &\le \|U(t)V(s)(x) - U(t)(x)\| + \|U(t)(x) - x\| \\ &\le \|V(s)(x) - x\| + \|U(t)(x) - x\| \le \varepsilon\end{aligned}$$

since $U(t)$ is an isometry. □

Lemma C.3.5 *In any realization of the 1-ICCR, if* $h \in \mathcal{S}^2 = \mathcal{S}(\mathbb{R}^2)$ *is such that*

$$\int h(s,t)S(s,t)\mathrm{d}s\mathrm{d}t = 0 \tag{C.40}$$

then $h = 0$.

Proof It is straightforward from (C.5) that

$$S(-s', -t')S(s,t)S(s',t') = \exp(\mathrm{i}(st' - s't))S(s,t). \tag{C.41}$$

We denote by $\mathcal{C}$ the class of functions h which satisfy (C.40). Applying $S(-s', -t')$ to the left of (C.40) and $S(s',t')$ to the right, and using (C.41) we get that if $h \in \mathcal{C}$ then $(s,t) \mapsto \exp(\mathrm{i}(st' - s't))h(s,t) \in \mathcal{C}$. Consequently for any $g \in \mathcal{S}^2$ we have $gh \in \mathcal{C}$ because g can be written as an average of the functions $\exp(\mathrm{i}(st' - s't))$, using the Fourier transform. Assume for contradiction that $h(s_0,t_0) \ne 0$. Consider a unit vector x. Then by Lemma C.3.4 there exists $\alpha > 0$ such that $\|S(s,t)(x) - x\| \le 1/2$ for $|s - s_0|, |t - t_0| \le \alpha$. Consider then $g \in \mathcal{S}^2$ such that $g(s,t)h(s,t) = 0$ unless $|s - s_0|, |t - t_0| \le \alpha$ and $gh \ge 0, \int g(s,t)h(s,t)\mathrm{d}s\mathrm{d}t = 1$. Then clearly

$$\Big\|\int g(s,t)h(s,t)S(s,t)(x)\mathrm{d}s\mathrm{d}t - x\Big\| \le \int g(s,t)h(s,t)\|S(s,t)(x) - x\|\mathrm{d}s\mathrm{d}t \le 1/2,$$

so that $\|\int g(s,t)h(s,t)S(s,t)(x)\mathrm{d}s\mathrm{d}t\| \ge 1/2$, contradicting the fact that $gh \in \mathcal{C}$. □

The heart of the proof of the Stone–von Neumann theorem is then the following, where g is the function (C.39).

Proposition C.3.6 *In any realization of the 1-ICCR, the operator*

$$P = \int g(s,t)S(s,t)\mathrm{d}s\mathrm{d}t \tag{C.42}$$

is self-adjoint and satisfies

$$PS(a,b)P = \exp\Big(-\frac{a^2+b^2}{4}\Big)P. \tag{C.43}$$

Proof We observe from (C.5) that $S(s,t)^\dagger = S(s,t)^{-1} = S(-s, -t)$ and then

$$P^\dagger = \int g(s,t)^* S(-s,-t)\mathrm{d}s\mathrm{d}t = \int g(-s,-t)^* S(s,t)\mathrm{d}s\mathrm{d}t,$$

so that $P^\dagger = P$ because $g(-s,-t)^* = g(s,t)$ from (C.39). To prove (C.43), let us compute $PS(a,b)P$. Using (C.42), and the consequence

$$S(s,t)S(a,b)S(s',t') = RS(s+s'+a,t+t'+b)$$

for $R := \exp(\mathrm{i}(sb+st'-ta-ts'+at'-bs')/2)$ from the relation (C.5), we obtain

$$PS(a,b)P = \int h(s,t,s',t',a,b)S(s+s'+a,t+t'+b)\mathrm{d}s\mathrm{d}t\mathrm{d}s'\mathrm{d}t'$$

where

$$h(s,t,s',t',a,b) = \exp(\mathrm{i}(sb+st'-ta-ts'+at'-bs')/2)g(s,t)g(s',t'). \tag{C.44}$$

Let us make a change of variables $t' = x-t-b$ and $s' = y-s-a$ so that

$$PS(a,b)P = \int h(s,t,y-s-a,x-t-b,a,b)S(x,y)\mathrm{d}s\mathrm{d}t\mathrm{d}x\mathrm{d}y.$$

Therefore it suffices to show that

$$\int h(s,t,y-s-b,x-t-a,a,b)\mathrm{d}s\mathrm{d}t = \exp(-(a^2+b^2)/4)g(x,y). \tag{C.45}$$

Recalling (C.44), this amounts to computing an elementary integral. This computation is not very appealing but a magic argument (which I learned in Hall's book [40]) bypasses it. According to Lemma C.3.5, to prove that two functions of x,y in $\mathcal{S}(\mathbb{R}^2)$ are equal, it suffices to check that they have the same integral when multiplied by $S_0(x,y)$. We use this for the two functions of x,y given by the left- and right-hand sides of (C.45), thinking of a and b as parameters. Performing the same manipulations as previously in the reverse order, the equality (C.45) is equivalent to the equality

$$P_0S_0(a,b)P_0 = \exp(-(a^2+b^2)/4)P_0$$

which is an obvious consequence of (C.36) and the fact that $P_0(f) = \tau_0(\tau_0,f)$. □

We are at last ready to prove the Stone–von Neumann theorem in the case $n=1$. Consider the operator P given by (C.42). Then, using (C.43) for $a=b=0$ we obtain that $P^2 = P$. We also know from Lemma C.3.5 that $P \neq 0$. Consider then a unit vector τ in the range of P. Then $\tau = P(y)$ for some $y \in \mathcal{H}$ and since $P^2 = P$ we have $\tau = P(\tau)$. The rest of the proof amounts to showing (as will be used later) that "τ entirely determines W". Using that P is self-adjoint in the second equality and (C.43) in the third equality, together with the fact that τ is a unit vector, we get

$$(\tau,S(s,t)(\tau)) = (P(\tau),S(s,t)P(\tau)) = (\tau,PS(s,t)P(\tau)) = \exp\Big(-\frac{s^2+t^2}{4}\Big).$$

The exact form of the right-hand side is not relevant, what matters is that this is the right-hand side of (C.36), and thus

$$(\tau,S(s,t)(\tau)) = (\tau_0,S_0(s,t)(\tau_0)). \tag{C.46}$$

Now we observe, using (C.46), (C.5) (and also $S(s,t)^\dagger = S(-s,-t)$), that

$$\begin{aligned}(S(s,t)(\tau),S(s',t')(\tau)) &= (\tau,S(-s,-t)S(s',t')(\tau))\\ &= \exp(-\mathrm{i}(st'-s't))(\tau,S(s'-s,t'-t)(\tau)),\end{aligned}$$

and a similar relation for S_0. Consequently (C.46) implies the all-important relation that for all s,t,s',t' we have

$$(S(s,t)(\tau),S(s',t')(\tau)) = (S_0(s,t)(\tau_0),S_0(s',t')(\tau_0)). \tag{C.47}$$

The span of the vectors of the type $S_0(s,t)(\tau_0)$ is dense in L^2 because this realization of the 1-ICCR is irreducible as previously checked. The span of the vectors of the type $S(s,t)(\tau)$ is dense in $\mathcal{H}$

because we assume that the realization is irreducible. Moreover it follows from (C.47) that the linear combinations $\sum_i \alpha_i S(s_i,t_i)(\tau)$ and $\sum_i \alpha_i S_0(s_i,t_i)(\tau_0)$ have the same norm. Consequently there exists a linear isometry $W: L^2 \to \mathcal{H}$ such that for each s,t

$$WS_0(s,t)(\tau_0) = S(s,t)(\tau).$$

Now,

$$\begin{aligned} WS_0(a,b)S_0(s,t)(\tau_0) &= \exp(\mathrm{i}(at-bs)/2)WS_0(a+s,b+t)(\tau_0) \\ &= \exp(\mathrm{i}(at-bs)/2)S(a+s,b+t)(\tau) \\ &= S(a,b)S(s,t)(\tau) \\ &= S(a,b)WS_0(s,t)(\tau_0), \end{aligned}$$

so that $WS_0(a,b)(h) = S(a,b)W(h)$ for each element h of the type $S_0(s,t)(\tau_0)$ and since these elements span L^2 we have $WS_0(a,b) = S(a,b)W$. The proof is complete.

Exercise C.3.7 The Stone–von Neumann theorem is not as trivial as you might think. To convince yourself of this consider real numbers $\alpha,\beta,\gamma,\delta,u,v$ with $\alpha\delta - \beta\gamma = 1$, and

$$P' = \alpha P + \beta X + u1 \ ; \ X' = \gamma P + \delta X + v1.$$

In as many cases as you can, find explicitly a unitary map with $W^{-1}P'W = P$ and $W^{-1}X'W = X$. (E.g. try the case $\alpha = \delta = 1, \beta = \gamma = 0$.)

Exercise C.3.8 (a) Convince yourself that neither the exact form of (C.36) nor the actual value of g is really used in the proof (although computing g might be the fastest way to prove that such a function exists). (b) Prove the Stone–von Neumann theorem for general n by a simple extension of the previous arguments.

We have proved that there is no non-trivial subspace invariant under all the operators $S(s,t)$. If we were working with finite-dimensional spaces, we could conclude from Schur's lemma that every operator commuting with these is a multiple of identity. This argument does not work here because we are not in a finite-dimensional space. Using the spectral theory for unitary operators one may show in full generality that a unitary operator commuting with all the operators of an irreducible unitary representation is a multiple of the identity, see [48, Proposition 3.4.17]. We give an elementary self-contained proof in the present case. This simple argument will be useful later.

Theorem C.3.9 *Two irreducible realizations of the 1-ICCR are unitarily equivalent in an essentially unique way, that is the intertwining operator (C.33) is unique up to multiplication by a phase (i.e. a complex number of modulus 1).*

Proof By Theorem C.3.1 we reduce to the case where these representations are both S_0, so it suffices to prove that if a unitary operator W on $L^2(\mathbb{R})$ is such that

$$\forall s,t \in \mathbb{R} \ ; \ WS_0(s,t)W^{-1} = S_0(s,t)$$

then W is a multiple of the identity. The map W is entirely determined by $\tau := W(\tau_0)$ since $WS_0(s,t)(\tau_0) = S_0(s,t)(\tau)$ and since the elements $S_0(s,t)(\tau_0)$ are dense in $\mathcal{H}$. Furthermore τ satisfies

$$\forall s,t \ , \ (\tau, S_0(s,t)(\tau)) = \exp(-(s^2+t^2)/4), \tag{C.48}$$

which can be written explicitly as

$$\int \tau(x)^* \exp(\mathrm{i}st/2)\exp(\mathrm{i}tx)\tau(x+s)\mathrm{d}x = \exp(-(s^2+t^2)/4). \tag{C.49}$$

Taking $s = 0$ implies that $|\tau|^2$ is the inverse Fourier transform of $\exp(-t^2/4)$, so that $|\tau|^2$ and hence $|\tau|$ decreases fast. Taking $t = 0$ implies

$$\int \tau(x)^* \tau(x+s)\mathrm{d}x = \exp(-s^2/4).$$

Using Plancherel's formula, this yields

$$\int \hat{\tau}(p)^* \hat{\tau}(p) \exp(\mathrm{i}ps) \frac{\mathrm{d}p}{2\pi} = \exp(-s^2/4),$$

so that $|\hat{\tau}|^2$ is the Fourier transform of $\exp(-s^2/4)$ and hence $\hat{\tau}$ decreases fast. Using that τ is the inverse Fourier transform of $\hat{\tau}$ (and the formulas corresponding to (1.23) for the inverse Fourier transform) shows that τ is infinitely differentiable and that all its derivatives are bounded. This justifies the differentiation operations we perform in the sequel. Taking $s = 0$ in (C.49) and differentiating twice in t at $t = 0$ we obtain

$$\int \tau(x)^* \tau(x) x^2 \mathrm{d}x = \frac{1}{2}.$$

Taking $t = 0$ and differentiating twice in s at $s = 0$ we obtain, using also integration by parts,

$$\int \tau'(x)^* \tau'(x)\mathrm{d}x = -\int \tau(x)^* \tau''(x)\mathrm{d}x = \frac{1}{2}.$$

Differentiating (C.49) in t at $t = 0$ we obtain

$$\int \tau(x)^* \Big(\frac{\mathrm{i}s}{2} + \mathrm{i}x\Big) \tau(x+s)\mathrm{d}x = 0$$

and differentiating this at $s = 0$ yields

$$\int \tau(x)^* \Big(\frac{\tau(x)}{2} + x\tau'(x)\Big)\mathrm{d}x = 0$$

and therefore since $(\tau, \tau) = 1$, as follows from (C.49) for $s = t = 0$,

$$\int x\tau'(x)^* \tau(x)\mathrm{d}x = \int x\tau(x)^* \tau'(x)\mathrm{d}x = -\frac{1}{2}.$$

Combining these we obtain

$$\int |x\tau(x) + \tau'(x)|^2 \mathrm{d}x = 0.$$

Therefore $\tau'(x) = -x\tau(x)$, from which it follows that $\tau = \lambda\tau_0$. Moreover $|\lambda| = 1$ since these are both unit vectors. □

Let us summarize the situation.

Corollary C.3.10 *To each irreducible representation $S(s,t)$ of the 1-ICCR corresponds an element τ of the state space such that $(\tau, S(s,t)(\tau)) = \exp(-(s^2+t^2)/4)$. This element is unique up to a phase (i.e. multiplication by a complex number of modulus 1).*

Exercise C.3.11 Prove the generalization of Theorem C.3.9 to the n-ICCR. Hint: Use exactly the same argument to show that $\partial_k \tau = -x_k \tau$ for each $k \le n$.

C.4 Non-equivalent Unitary Representations

To get some concrete feeling, let us consider U and V as in (C.29) and (C.30), and a number $\rho > 0$. Let us then define $U^\rho(t) = U(\rho t)$ and $V^\rho(s) = V(s/\rho)$, which obviously satisfy the n-ICCR. It is quite straightforward to see that (C.33) is satisfied for the unitary map

$$W(f)(x) = \rho^{-n/2} f(x/\rho).$$

The map W however does not make sense for $n = \infty$. In imprecise terms we are going to show that indeed as soon as one considers an infinite number of relations (C.28) then the Stone–von Neumann theorem fails. As it makes no sense to consider $L^2(\mathbb{R}^n)$ for $n = \infty$ we first prove the following.

Lemma C.4.1 *Consider on $\mathbb{R}$ the probability measure μ_1 of density*

$$(\pi)^{-1/2} \exp(-x^2).$$

The operators on $L^2(\mu_1)$ given by

$$U(t)(f)(x) = \exp(\mathrm{i}tx) f(x) \tag{C.50}$$

$$V(s)(f)(x) = \exp\Big(-sx - \frac{s^2}{2}\Big) f(x+s) \tag{C.51}$$

are unitary and satisfy (C.32). Moreover if τ_0 denotes the function of constant value 1, and if

$$S_0(s,t) = \exp(\mathrm{i}st/2) U(t) V(s) \tag{C.52}$$

then

$$(\tau_0, S_0(s,t)(\tau_0)) = \exp(-(s^2+t^2)/4). \tag{C.53}$$

Proof It would be straightforward to check these claims directly, but this is not necessary as they are obtained from the corresponding formulas in $L^2(\mathbb{R})$ by transporting these formulas to $L^2(\mathrm{d}\mu_1)$ using the unitary map W from $L^2(\mathbb{R})$ to $L^2(\mathrm{d}\mu_1)$ given by $W(f)(x) = (\pi)^{1/4}\exp(x^2/2) f(x)$. □

The operators X and P now take the form

$$Xf(x) = xf(x)\ ;\ Pf(x) = -\mathrm{i}f'(x) + \mathrm{i}xf(x).$$

The operator P "does not look symmetric" because of the last term, but here the base measure is not Lebesgue's measure.

Exercise C.4.2 Convince yourself using integration by parts that P is symmetric.

To state our example, we need the fact that there exists a probability measure μ on $\mathbb{R}^{\mathbb{N}}$ which is the product measure when one puts μ_1 on each factor. As the reader may not be familiar with this object, we list the properties which one needs to understand the rest of the section. First, consider a function f on $\mathbb{R}^{\mathbb{N}}$, which depends only on the first n coordinates of $x \in \mathbb{R}^{\mathbb{N}}$. In other words, $f(x) = g(x_1, x_2, \ldots, x_n)$ for a certain function g on $\mathbb{R}^n$. Then one has the formula

$$\int f(x)\mathrm{d}\mu(x) = (\pi)^{-n/2} \int \ldots \int \exp\Big(-\sum_{k\le n} x_k^2\Big) g(x_1,\ldots,x_n)\mathrm{d}x_1 \cdots \mathrm{d}x_n.$$

In the space $L^2(\mathrm{d}\mu)$ the subspaces L_n^2 of functions depending only on the first n coordinates form an increasing sequence:

$$L_0^2 \subset \cdots \subset L_n^2 \subset L_{n+1}^2 \subset \cdots \subset L^2(\mathrm{d}\mu).$$

A fundamental fact is that $\bigcup_n L^2_n$ is dense in $L^2(\mathrm{d}\mu)$. Equivalently, denoting by Π_n the orthogonal projection of $L^2(\mathrm{d}\mu)$ on L^2_n, we have

$$\forall f \in L^2(\mathrm{d}\mu)\ ;\ \lim_{n\to\infty} \|f - \Pi_n f\| = 0. \tag{C.54}$$

Let us denote by $\mathbb{R}^{(\mathbb{N})}$ the set of sequences $s = (s_i)_{i\geq 1}$ such that only finitely many of the terms s_i are not zero. Then for $s,t \in \mathbb{R}^{(\mathbb{N})}$ the quantity $t \cdot x = \sum_{k\geq 1} t_k x_k$ is well defined for $x \in \mathbb{R}^{\mathbb{N}}$, and the quantity $\|s\|^2 = \sum_{k\geq 1} s_k^2$ is also well defined. We can define the following operators on $L^2(\mathrm{d}\mu)$:

$$U(t)(f)(x) = \exp(\mathrm{i}t \cdot x) f(x) \tag{C.55}$$

$$V(s)(f)(x) = \exp\Big(-s\cdot x - \frac{\|s\|^2}{2}\Big) f(x+s). \tag{C.56}$$

These are unitary and satisfy the relations (C.31) and (C.32). This can be checked directly, or deduced from Lemma C.4.1. These are the integrated versions of the canonical commutation relations, but now for an infinite sequence of operators rather than for just n of them. The corresponding self-adjoint operators X_k, P_k for $k \geq 1$ are given by

$$X_k f(x) = x_k f(x)$$

and

$$P_k f(x) = -\mathrm{i}\frac{\partial f}{\partial x_k}(x) + \mathrm{i}x_k f(x).$$

These operators satisfy

$$[X_k, P_k] = \mathrm{i}1$$

all the other commutators being 0. That is, there is now a whole sequence of operators satisfying the relations (C.28). The relations (C.31) and (C.32) can also be reformulated as

$$S(s,t)S(s',t') = \exp(\mathrm{i}(s\cdot t' - s'\cdot t)/2)S(s+s', t+t') \tag{C.57}$$

for $S(s,t) = \exp(\mathrm{i}s\cdot t/2)U(t)V(s)$. In the rest of this appendix, such a family of operators will be simply called a realization of the ICCR. The particular realization described above will be called the canonical realization of the ICCR. On the other hand, the case $n = 1$ will still be called the 1-ICCR. This terminology is by no means standard or to be remembered beyond this appendix, its sole purpose is purely local. The following is better left as an exercise.

Lemma C.4.3 *The canonical realization of the ICCR is irreducible, in the sense that there is no non-trivial subspace of $L^2(\mathrm{d}\mu)$ which is invariant under all maps $U(t)V(s)$.*

We now investigate whether another realization (which without loss of generality we may assume to take place on the same space $L^2(\mathrm{d}\mu)$) of the ICCR is unitarily equivalent to the canonical realization. There is a simple criterion to decide this. We adorn with a $'$ the operators relative to this new realization.

Theorem C.4.4 *A realization $S'(s,t)$ of the ICCR on $L^2(\mathrm{d}\mu)$ is unitarily equivalent to the canonical realization if and only if it is irreducible and there exists a function $\tau \in L^2(\mathrm{d}\mu)$ such that*

$$\forall s,t \in \mathbb{R}^{(\mathbb{N})}\ ;\ (\tau, S'(s,t)\tau) = \exp(-(\|s\|^2 + \|t\|^2)/4). \tag{C.58}$$

The most important part for our purpose is the "only if" one: If no such function τ exists, the realization $S'(s,t)$ is not unitarily equivalent to the canonical representation.

Proof If there exists a unitary map W such that

$$S'(s,t)W = WS(s,t),$$

then $\tau = W\tau_0$ satisfies

$$(\tau, S'(s,t)\tau) = (W\tau_0, S'(s,t)W\tau_0) = (W\tau_0, WS(s,t)\tau_0) = (\tau_0, S(s,t)\tau_0)$$

and this is $\exp(-(\|s\|^2 + \|r\|^2)/4)$ by using (C.53) on each coordinate. To prove the converse, we proceed exactly as in the proof of Theorem C.3.1 (the straightforward arguments are given after (C.46)). □

We first use this criterion in a simple situation.

Theorem C.4.5 *For $\rho > 0$ consider $U^\rho(t) = U(\rho t)$ and $V^\rho(s) = V(s/\rho)$. These operators are an irreducible realization of the commutation relations (C.31) and (C.32). Yet for two different values of ρ these representations of the commutations relations are not unitarily equivalent.*[7]

Proof We prove here only that the case $\rho = 1$ is not unitarily equivalent to the case $\rho \neq 1$. Otherwise, according to (C.58) there would exist a unit vector τ such that (among other conditions)

$$(\tau, V^\rho(s_n)(\tau)) = \exp(-1/4), \tag{C.59}$$

where $s_n \in \mathbb{R}^{(\mathbb{N})}$ has all its components equal to zero except the component of rank n which is equal to 1. Consider $\varepsilon > 0$. Then, according to (C.54), for some m large enough, $\tau_m := \Pi_m \tau$ satisfies $\|\tau - \tau_m\| \leq \varepsilon$. Since $V^\rho(s)$ is unitary, the most brutish estimates yield

$$|(\tau_m, V^\rho(s_n)(\tau_m)) - (\tau, V^\rho(s_n)(\tau))| \leq 3\varepsilon. \tag{C.60}$$

The key observation is that since τ_m does not depend on the $(m+1)$th coordinate, one has

$$(\tau_m, V^\rho(s_{m+1})(\tau_m)) = (\tau_m, V(s_{m+1}/\rho)(\tau_m)) = \exp(-1/(4\rho^2))\|\tau_m\|^2, \tag{C.61}$$

as would follow from (C.53) if it were true that τ_m is constant. To see this, when computing the inner product $(\tau_m, V(s_{m+1}/\rho)(\tau_m))$ we integrate first in the $(m+1)$th coordinate at a given value of the first m coordinates, using (C.53), and then integrate in these afterwards. But then (C.59) and (C.61) contradict (C.60) if ε is small enough. The irreducibility is proved as before. □

We now analyze far more precisely some general situations of interest. All our arguments are rigorous, but we gloss over some purely technical points which would belong to a course on measure theory but not to the present text.[8]

Among the elements of $L^2(d\mu)$, some remarkable ones are the functions which are simply a product of functions depending only on one coordinate. To bring this idea forward, for a sequence of functions w_k of $L^2(d\mu_1)$, such that $w_k \equiv 1$ for k large enough, we define the function $\bigotimes_{k\geq 1} w_k$ by

$$\bigotimes_{k\geq 1} w_k(x) = \prod_{k\geq 1} w_k(x_k),$$

where the product makes sense since all terms for k large enough are equal to 1. The canonical representation of the ICCR has the remarkable property that (recalling the operator S_0 of (C.52))

$$S(s,t)\bigotimes_{k\geq 1} w_k = \bigotimes_{k\geq 1} S_0(s_k,t_k)w_k.$$

[7] It is essential for this result to be in "infinite dimensions" as is shown by the first equation in this section.
[8] As this author spent far too many years studying this type of question, you are probably safe to trust him for these matters.

We now investigate in detail the situation where the representation S' of the ICCR has a similar property

$$S'(s,t)\bigotimes_{k\geq 1} w_k = \bigotimes_{k\geq 1} S_k(s_k,t_k)w_k, \tag{C.62}$$

where $S_k(s,t)$ are unitary maps on $L^2(\mathrm{d}\mu_1)$ which are a realization of the 1-ICCR. According to Corollary C.3.10, there exists a function $\tau_k \in L^2(\mathrm{d}\mu_1)$ such that

$$\forall s,t \in \mathbb{R}\,,\ (\tau_k, S_k(s,t)\tau_k) = \exp(-(s^2+t^2)/4). \tag{C.63}$$

To get a clean statement we assume that $\int \tau_k \mathrm{d}\mu_1 \neq 0$ for each k. Since τ_k is defined only up to a phase, we may then assume that

$$\int \tau_k \mathrm{d}\mu_1 > 0.$$

As τ_k is of norm 1, the Cauchy-Schwarz inequality then implies

$$0 < A_k := \int \tau_k \mathrm{d}\mu_1 \leq 1. \tag{C.64}$$

We are going to prove the very clean following result:

Theorem C.4.6 *A realization $S'(s,t)$ of the ICCR satisfying the previous conditions is unitarily equivalent to the canonical realization if and only if*

$$\prod_{k\geq 1} A_k = \prod_{k\geq 1}\int \tau_k \mathrm{d}\mu_1 > 0. \tag{C.65}$$

Since $0 < A_k \leq 1$ the infinite product in (C.65) is well defined as

$$\prod_{k\geq 1} A_k := \lim_{m\to\infty} \prod_{1\leq k\leq m} A_k.$$

Condition (C.65) is rather stringent. In particular it always fails when S_k is independent of k (except in the trivial case where $S' = S$). Let us note for further use the following consequence of (C.65):

$$\lim_{m\to\infty} \prod_{k\geq m} A_k = 1. \tag{C.66}$$

To prove Theorem C.4.6, the strategy is to use Theorem C.4.4. The obvious candidate for the function τ there is given by

$$\tau(x) = \prod_{k\geq 1} \tau_k(x_k). \tag{C.67}$$

Lemma C.4.7 *The formula (C.67) defines a bona fide function of $L^2(\mathrm{d}\mu)$ if and only if (C.65) holds.*

Proof Consider the function g_m defined by

$$g_m(x) = \prod_{1\leq k\leq m} \tau_k(x_k).$$

We observe that if $n \geq m$ we have $\int g_n^* g_m \mathrm{d}\mu = \int |g_m|^2 \mathrm{d}\mu \int \prod_{m<k\leq n} \tau_k^* \mathrm{d}\mu = \prod_{m<k\leq n} A_k$ so that

$$\|g_m - g_n\|^2 = 2\Big(1 - \prod_{m<k\leq n} A_k\Big),$$

so that as a consequence of (C.66) the sequence (g_m) is a Cauchy sequence and its limit is what we mean by (C.67). Conversely, (staying a bit informal for these measure-theoretic technicalities) if the formula (C.67) makes sense, for a function $h \in L^2_m$ we should have

$$(h,\tau) = (h,g_m)\prod_{k>m} A_k \le \|h\| \prod_{k>m} A_k, \tag{C.68}$$

as is seen by integrating first in all the variables x_k for $k > m$. But from (C.54) we can find m large enough that $(\Pi_m\tau,\tau) > 0$ and using the previous inequality for $h = \Pi_m\tau$ proves (C.65). □

"Half" of Theorem C.4.6 is now obvious: Under (C.65), the realization of the ICCR is equivalent to the canonical realization, because the function (C.67) satisfies (C.58). This is an obvious consequence of (C.62) and (C.63) because (C.58) has to be checked only for $s,t \in \mathbb{R}^{(\mathbb{N})}$, so that s,t have only finitely many non-zero coordinates.

To prove the converse, the strategy is to prove first that the only choice for a function τ in (C.58) is given by (C.67). But then this function can exist only under (C.65). The main step of the program is to prove that for each k the function τ has to be of the type $\tau(x) = \tau_k(x_k)g(x)$ where $g(x)$ is independent of x_k. It is obtained through an extension of the argument of Theorem C.3.9, which we formulate now.

Lemma C.4.8 *Consider a function $\tau(x,\lambda)$, where $x \in \mathbb{R}$ and λ belongs to some parameter space, on which there is a probability measure ν. Consider a realization $\bar{S}$ of the 1-ICCR on $L^2(\mathrm{d}\mu_1)$. Assume that for each $s,t \in \mathbb{R}$ one has*

$$\int (\tau(\cdot,\lambda), \bar{S}(s,t)\tau(\cdot,\lambda))\mathrm{d}\nu(\lambda) = \exp(-(s^2+t^2)/4), \tag{C.69}$$

where $\tau(\cdot,\lambda)$ denotes the element of $L^2(\mathrm{d}\mu_1)$ given by $x \mapsto \tau(x,\lambda)$. Then τ is of the form $\tau(x,\lambda) = \tau_0(x)g(\lambda)$ where $\int \mathrm{d}\nu(\lambda)|g(\lambda)|^2 = 1$ and $\int |\tau_0(x)|^2\mathrm{d}\mu_1(x) = 1$.

Proof Let us first consider a function $\tau(x,\lambda)$ such that for each $s,t \in \mathbb{R}$ one has

$$\int \tau(x,\lambda)^* \exp(ist/2)\exp(itx)\tau(x+s,\lambda)\mathrm{d}x\mathrm{d}\nu(\lambda) = \exp(-(s^2+t^2)/4). \tag{C.70}$$

We will leave to the reader as a challenge to prove that it is actually permitted to differentiate in (C.70) (which is more cumbersome to prove than in the case of Theorem C.3.9). Given this, it is rather amazing that the same argument goes through, as we obtain now (denoting by τ' the derivative in x)

$$\int \tau(x,\lambda)^*\tau(x,\lambda)x^2\mathrm{d}x\mathrm{d}\nu(\lambda) = \frac{1}{2}$$

$$\int \tau'(x,\lambda)^*\tau'(x,\lambda)\mathrm{d}x\mathrm{d}\nu(\lambda) = \frac{1}{2}$$

$$\int x\tau'(x,\lambda)^*\tau(x,\lambda)\mathrm{d}x\mathrm{d}\nu(\lambda) = \int x\tau(x,\lambda)^*\tau'(x,\lambda)\mathrm{d}x\mathrm{d}\nu(\lambda) = -\frac{1}{2},$$

which can be combined into

$$\int |\tau'(x,\lambda) + x\tau(x,\lambda)|^2\mathrm{d}x\mathrm{d}\nu(\lambda) = 0.$$

Therefore $\tau'(x,\lambda) = -x\tau(x,\lambda)$ so that $\tau(x,\lambda) = \tau_0(x)g(\lambda)$. Taking $s = t = 0$ in (C.69) shows that $\int |\tau(x,\lambda)|^2\mathrm{d}x\mathrm{d}\nu(\lambda) = 1$ so that we may assume $\int |g(\lambda)|^2\mathrm{d}\lambda = \int |\tau_0(x)|^2\mathrm{d}x = 1$.

To prove the lemma, we use that $\bar{S}$ is equivalent to S_0. Thus there exists a unitary operator $W: L^2(\mathrm{d}\mu_1) \to L^2(\mathbb{R})$ such that for all $s,t \in \mathbb{R}$ we have

$$(f,\bar{S}(s,t)f) = (Wf, S_0(s,t)Wf) = \int (Wf)(x)^* \exp(ist/2)\exp(itx)(Wf)(x+t)\mathrm{d}x.$$

Thus (C.69) implies that (C.70) holds for the function $\bar{\tau}$ such that $\bar{\tau}(\cdot,\lambda) = W(\tau(\cdot,\lambda))$ instead of τ. We then obtain that $\tau(x,\lambda) = g(\lambda)W^{-1}(\tau_0)(x)$. □

Proof of Theorem C.4.6 Let us assume that the representation $S'(s,t)$ is unitarily equivalent to the canonical realization, and consider the function τ of (C.58). Considering $k \geq 1$ we consider this function as a function of x_k and λ, where $\lambda = (x_1,\dots,x_{k-1},x_{k+1},\dots)$. Thus (C.69) holds for $\bar{S} = S_k$, and we obtain that $\tau(x)$ is of the type $\tau_k(x_k)g(\lambda)$. By iteration of this principle, for each m the function τ must be of the type $\tau(x) = \prod_{k\leq m}\tau_k(x_k)g_m(x)$ where g_m does not depend on $x_1,\dots,x_m$. Computing the L^2 norm shows that $\|g_m\| = 1$ so that $\int |g_m(x)|\mathrm{d}\mu(x) \leq 1$ by the Cauchy-Schwarz inequality, and then

$$\int |\tau(x)|\mathrm{d}\mu(x) \leq \int \prod_{k\leq m}\tau_k(x_k)\mathrm{d}\mu(x) = \prod_{k\leq m} A_k. \tag{C.71}$$

This proves (C.65) because the left-hand side of (C.71) is positive since τ is of norm 1. □

C.5 Orthogonal Ground States!

There is little doubt that most physicists would consider the previous material as extremely bizarre, as they certainly do not think in these terms.[9] Let us go back to the space $L^2(\mathrm{d}\mu)$. Then, rather than the operators X_k and P_k let us consider

$$a_k = \frac{1}{\sqrt{2}}(X_k + \mathrm{i}P_k)\ ;\ a_k^\dagger = \frac{1}{\sqrt{2}}(X_k - \mathrm{i}P_k), \tag{C.72}$$

which satisfy the commutation relations

$$[a_k, a_k^\dagger] = 1, \tag{C.73}$$

with all the other commutators equal to zero. This is what we call "the canonical commutation relations" in the rest of this appendix. As a physicist never uses an explicit state space, these commutation relations describe all the information available. She asserts that there should exist a distinguished element, $|0\rangle$, which is sent to 0 by all the annihilation operators a_k:

$$\forall k \geq 1\ ,\ a_k|0\rangle = 0. \tag{C.74}$$

The interpretation here is simply that a_k "removes particles of type k" and that $|0\rangle$ is "the vacuum", that is the state where there are no particles of any type to be removed. The "boson Fock space" $\mathcal{H}$ is then obtained by repeated applications of the creation operators to the vacuum, i.e. considering $a_{k_1}^\dagger a_{k_2}^\dagger \cdots a_{k_n}^\dagger|0\rangle$. With our choice $L^2(\mathrm{d}\mu)$ as a state space, the vacuum $|0\rangle$ is simply the constant function 1. Applying $a_k^\dagger$ to this function one gets the function $x \mapsto x_k/\sqrt{2}$. The functions of this type span the subspace $\mathcal{H}_0$ of $L^2(\mathrm{d}\mu)$ consisting of functions of the type $x \mapsto \sum_{n\geq 1}\alpha_n x_n$. We encourage the reader to convince herself that applying the creation operators to the vacuum generates the whole space $L^2(\mathrm{d}\mu)$. It is worth stressing the picture here:

$L^2(\mathrm{d}\mu)$ can be viewed in this manner as a concrete realization of the boson Fock space of $\mathcal{H}_0$. (C.75)

Consider now another sequence a'_k of operators which satisfies the same commutation relations (C.73). Two examples often considered in physics textbooks are

$$a'_k = \cosh s a_k + \sinh s a_k^\dagger\ ;\ a_k'^\dagger = \cosh s a_k^\dagger + \sinh s a_k, \tag{C.76}$$

9 They might describe $L^2(\mathrm{d}\mu)$ as "the state space of a sequence of independent harmonic oscillators".

for some $s \in \mathbb{R}$, or, for $\gamma_k \in \mathbb{C}$,

$$a'_k = a_k + \gamma_k 1 \; ; \; a'^\dagger_k = a^\dagger_k + \gamma^*_k 1. \tag{C.77}$$

These will be analyzed in exercises below. Corresponding to these new operators is a new vacuum $|0\rangle'$ which satisfies

$$\forall k \geq 1 \, , a'_k|0\rangle' = 0. \tag{C.78}$$

The first questions that come to the mind of a mathematician here are whether such an object exists and is unique but physicists are trained differently, and worry about such matters only if ignoring them creates disaster. The operator W satisfying

$$W(a^\dagger_{k_1} a^\dagger_{k_2} \cdots a^\dagger_{k_n}|0\rangle) = a'^\dagger_{k_1} a'^\dagger_{k_2} \cdots a'^\dagger_{k_n}|0\rangle' \tag{C.79}$$

is then a natural candidate for an operator for which $W^{-1}a^\dagger_k W = a'^\dagger_k$ (or, equivalently, if W is unitary, $Wa_kW^{-1} = a'_k$, i.e. W transforms a_k into a'_k). At least at the formal level, this is *exactly* what we do in Theorem C.4.4. The function τ there corresponds to the state $|0\rangle'$, and completely determines W. (In the setting of this theorem, the operators X'_k and P'_k are obtained in the obvious manner by differentiation of $S'(s,t)$ and the operators a'_k and $a'^\dagger_k$ are defined as the suitable linear combinations.)

What, then, could go wrong in this beautifully simple picture? Explicit computation in some cases yields

$$\langle 0|0\rangle' = 0, \tag{C.80}$$

at which point many physics textbooks conclude without further comment that "one says in this case that the representations of the canonical commutation relations are not unitarily equivalent". I cannot fathom which mental image they form to reach this conclusion. What is certain however is that (C.80) is obtained by showing that a certain infinite product of quantities $0 < A_k \leq 1$ is zero. The same argument (which in our language is contained in (C.68)) shows that actually

$$\forall k_1, \ldots, k_n \; ; \; \langle 0|a_{k_1} \cdots a_{k_n}|0\rangle' = 0, \tag{C.81}$$

and *this* spells trouble, as it implies that the unit vector $|0\rangle'$ is orthogonal to each element $a^\dagger_{k_1} \cdots a^\dagger_{k_n}|0\rangle$ of the boson Fock space. In physics flowery language, the vacuum $|0\rangle'$ "is outside of the Fock space".[10] In the setting of Theorem C.4.6 the condition $\langle 0|0\rangle' = 0$ means simply $\int \tau \mathrm{d}\mu = \prod_{k\geq 1} A_k = 0$, so it is a less formal but nonetheless correct way to express the same conclusions as our analysis.

Exercise C.5.1 Show that the situation of (C.76) corresponds to the situation of Theorem C.4.5 for $\rho = \exp s$.

Exercise C.5.2 The purpose of this exercise is to show that, as a consequence of Theorem C.4.6, the representation (C.77) of the canonical commutation relations is unitarily equivalent to the canonical representation if and only if $\sum_{k\geq 1} |\gamma_k|^2 < \infty$.
(a) Defining $A_k(\gamma) = \exp(\gamma a^\dagger_k - \gamma^* a_k)$ prove that

$$A_k(\gamma)a_kA_k(\gamma)^{-1} = a_k - \gamma 1. \tag{C.82}$$

(b) Prove that this corresponds to the situation where in (C.63) one has $\tau_k = A_k(\gamma_k)\tau_0$ where here τ_0 is the constant function 1 in $L^2(\mathrm{d}\mu_1)$.
(c) Prove that $\int \tau_k \mathrm{d}\mu_1 = \exp(-|\gamma_k|^2/2)$.
(d) Conclude.

[10] In the same line of flowery language, this also reads sometimes as two vectors "lie in different Hilbert spaces".

We discussed the "discrete setting" of a single sequence of operators for mathematical clarity, but in physics one needs the "continuous setting". This problem is addressed in the following exercise. It is probably best to delay studying this exercise until after reading Chapter 5.

Exercise C.5.3 We consider the space $\mathcal{H} = L^2(\mathrm{d}^3\boldsymbol{p}/(2\pi\hbar)^3)$ and the corresponding boson Fock space $\mathcal{B}$. Consider the operators $b(\boldsymbol{p})$ and $b^\dagger(\boldsymbol{p})$ as in (3.59). Consider a function $\gamma(\boldsymbol{p})$, which is assumed to be infinitely differentiable for simplicity. Define formally the operators

$$a(\boldsymbol{p}) = b(\boldsymbol{p}) + \gamma(\boldsymbol{p})1 \; ; \; a^\dagger(\boldsymbol{p}) = b^\dagger(\boldsymbol{p}) + \gamma(\boldsymbol{p})^* 1. \tag{C.83}$$

The problem is to know whether there is a unitary operator W on $\mathcal{B}$ such that

$$Wb(\boldsymbol{p})W^{-1} = a(\boldsymbol{p}) \; ; \; Wb^\dagger(\boldsymbol{p})W^{-1} = a^\dagger(\boldsymbol{p}). \tag{C.84}$$

(a) Prove that this can be the case only if $\gamma \in \mathcal{H}$ i.e. if γ is square-integrable. Hint: Reduce to the previous exercise by considering operators $B(\varphi_k)$ where the φ_k are appropriately chosen, in particular with disjoint supports.

The goal is now to prove the converse.

(b) Prove the following abstract result, in the setting of Theorem 3.4.2. If ψ is a continuous linear functional on $\mathcal{H}$ the family of operators on $\mathcal{B}$ given by

$$A(\xi) = B(\xi) + \psi(\xi)1 \; ; \; A^\dagger(\eta) = B^\dagger(\eta) + \psi(\eta)^* 1,$$

is unitarily equivalent to the family $B(\xi), B^\dagger(\eta)$. Hint: Choose an appropriate basis.

(c) Prove that W exists as in (C.84) when $\gamma \in \mathcal{H}$.

Appendix D

A Crash Course on Lie Algebras

Even though in our treatment Lie Algebras play a minimum role, it is of importance to learn the basic ideas underlying this fundamental tool, at least at a very elementary level, if only to understand the relationship between some of the material we present and what is found in physics textbooks. Excellent sources are available, e.g. [39]. Our limited goal here is to provide a self-contained treatment in mathematical language of the material found at the beginning of any treatise on Quantum Field Theory. This also does bring light to some topics we have covered. Our treatment is at the same level of detail as the rest of the book, which is not as high a level as achieved in Hall's book [39].

We denote by $Gl(n,\mathbb{C})$ the group of invertible $n \times n$ complex matrices.

Definition D.1 A Lie group of matrices is a subgroup of $Gl(n,\mathbb{C})$ which is closed for the natural topology.[1]

Observe that the definition does not prevent the Lie group from consisting only of matrices with real entries.

The only examples which really concern us here are $O(3)$, the group of 3×3 real orthogonal matrices, the Lorentz group $O(1,3)$ and two groups of 2×2 matrices, the groups $SL(2,\mathbb{C})$ and $SU(2)$.[2] Thus the largest matrices you will have to think about are 4×4. Since we are not using any other Lie group than groups of matrices, we shall simply say "group" rather than "Lie group of matrices".

The two basic ideas of the theory are:

- Instead of studying the Lie group G, a complicated set, try to study its Lie algebra, a finite-dimensional vector space over $\mathbb{R}$.
- Instead of studying the product on G, a complicated operation, try to study a simpler operation, the Lie bracket on its Lie algebra.

D.1 Basic Properties and $\mathfrak{so}(3)$

The essential tool is the exponential of matrices. We recall that for a square matrix X its exponential is defined by the series

$$\exp X = \sum_{n \geq 0} \frac{X^n}{n!} \tag{D.1}$$

[1] $Gl(n,\mathbb{C})$ is a subset of $\mathbb{C}^{n^2}$ on which the natural topology is induced, say, by the usual distance.

[2] These two groups of 2×2 matrices were introduced in Section 8.1. Do not worry if you have not read it yet.

where $X^0 = 1$.[3] Thus

$$XY = YX \Rightarrow \exp(X+Y) = \exp X \exp Y, \tag{D.2}$$

because the term of degree n in the right-hand side is

$$\sum_{k \le n} \frac{X^k}{k!}\frac{Y^{n-k}}{(n-k)!} = \frac{(X+Y)^n}{n!}.$$

Let us stress that in general $\exp(X+Y) \ne \exp X \exp Y$. In fact, it is not easy at all to relate these two quantities (see however Proposition C.1.4). The following should also be obvious:

$$\exp X^\dagger = (\exp X)^\dagger, \tag{D.3}$$

as well as a similar formula for the transpose X^T. Again obvious is that

$$A(\exp X)A^{-1} = \exp(AXA^{-1}). \tag{D.4}$$

As a consequence of (D.2), $\exp(t+u)X = \exp tX \exp uX$, where to lighten notation we write $\exp tX$ rather than $\exp(tX)$. Given X, the matrices of the type $\exp tX$ form a group. Since $\exp(t+s)X - \exp tX = (\exp sX - 1)\exp tX = \exp tX(\exp sX - 1)$ it follows that

$$\frac{\mathrm{d}}{\mathrm{d}t}\exp tX = X \exp tX = (\exp tX)X. \tag{D.5}$$

Definition D.1.1 The Lie algebra $\mathfrak{g}$ of a group G of $n \times n$ matrices is the set of $n \times n$ matrices X such that

$$\forall t \in \mathbb{R}\,;\ \exp tX \in G.$$

Mathematicians and physicists differ again here. Physicists would insert a factor i in the exponential. Here we will follow mathematician's practice.[4]

In the previous definition, it is the small values of t which matter, as the following shows.

Exercise D.1.2 Assume that for some sequence $\varepsilon_k \to 0$ we have $\exp \varepsilon_k X \in G$. Prove that then $X \in \mathfrak{g}$. Hint: Use that G is closed.

Let us consider the case where $G = O(3)$, the group of real orthogonal matrices. Thus, $A \in G$ if and only if A is a 3×3 matrix with real entries which satisfies $AA^T = 1$ where T is the transpose of A. In particular for $A \in O(3)$ we have $1 = \det A \det A^T = (\det A)^2$, so that $\det A = \pm 1$. Consider X in the Lie algebra of G. First, since for all t the matrix $\exp tX \in O(3)$ has real entries this is also the case for X by (D.5). Furthermore $\exp tX(\exp tX)^T = 1$ for each $t \in \mathbb{R}$. Taking the derivative at $t = 0$ yields $X + X^T = 0$, i.e. $X^T = -X$. Such a matrix is called *skew-symmetric*. Conversely, if X is skew-symmetric, $\exp tX$ satisfies

$$\exp tX(\exp tX)^T = \exp tX \exp(tX^T) = \exp tX \exp -tX = 1,$$

so that $\exp tX \in O(3)$. Thus the Lie algebra of $O(3)$ is exactly the set of skew-symmetric 3×3 matrices with real entries. We observe now that the function $t \mapsto \det(\exp tX)$ is continuous. When X is skew-symmetric, this function is valued in $\{1, -1\}$, and for $t = 0$ it takes the value 1. It thus takes the value 1 everywhere. Consequently, $\exp X \in SO(3)$, the subgroup of $O(3)$ consisting of matrices of determinant 1. In particular $O(3)$ and $SO(3)$ have the same Lie algebra (the space of skew-symmetric matrices) and this Lie algebra is denoted by $\mathfrak{so}(3)$. Thus we have proved:

[3] This series converges in operator norm because $\|X^n\| \le \|X\|^n$.

[4] This practice has the distinct advantage of being less confusing when one works both with real and complex Lie algebras, which we will not do here. Generally speaking physicists often do not distinguish clearly which numbers are real and which can be complex, and this can be very confusing, even to themselves.

Proposition D.1.3 *The Lie algebra $\mathfrak{so}(3)$ of the groups $O(3)$ and $SO(3)$ is the set of real 3×3 skew-symmetric (i.e. $X = -X^T$) matrices.*

We have just seen in the case $G = O(3)$ that the exponential map from $\mathfrak{g}$ to G is in general not onto. As will be soon apparent it is in general not one-to-one either, so the basic idea of replacing the study of G by that of $\mathfrak{g}$ will at best work approximately.

Definition D.1.4 The Lie bracket $[X,Y]$ of two $n \times n$ matrices is defined as

$$[X,Y] = XY - YX. \tag{D.6}$$

The Lie bracket is anti-symmetric, $[Y,X] = -[X,Y]$. It is straightforward that the Lie bracket satisfies the Jacobi identity

$$[X,[Y,Z]] + [Y,[Z,X]] + [Z,[X,Y]] = 0. \tag{D.7}$$

This identity is fundamental in the theory of Lie algebras, although this will not become apparent during our short excursion there.

The Lie bracket is closely connected to the group structure of G. The reason why this group structure is hard to understand is that in general $AB \neq BA$. This lack of commutativity is measured by the fact that $ABA^{-1}B^{-1}$ is usually not the identity matrix. When $A = \exp tX$, $B = \exp tY$, for t small this matrix differs from the identity by about $t^2[X,Y]$:

$$\exp tX \exp tY (\exp tX)^{-1} (\exp tY)^{-1} = 1 + t^2[X,Y] + O(t^3). \tag{D.8}$$

In the right-hand side the addition is the ordinary addition of matrices. The physicist's notation $O(t^3)$ in (D.8) denotes a matrix of the form $t^3 R(t)$ where $R(t)$ is a matrix that stays bounded (for example in the sense that the sum of the absolute values of its entries stays bounded) as t stays bounded. The relation (D.8) is proved simply by expanding the exponentials and computing the coefficients of t and t^2. It shows that the Lie bracket should capture some of the "non-commutativity" of the group. When G is commutative, the left-hand side of (D.8) is 1 so that the bracket is identically zero on its Lie algebra.

Theorem D.1.5 *The Lie algebra $\mathfrak{g}$ of a matrix group G is a vector space over the* real *numbers. It is closed under the bracket (D.6), i.e. if $X, Y \in \mathfrak{g}$ then $[X,Y] \in \mathfrak{g}$. Moreover, if $X \in \mathfrak{g}$ and $A \in G$ then $AXA^{-1} \in \mathfrak{g}$.*

As this theorem is obvious by inspection in all the specific cases we consider, we refer to Hall's textbook [39] for the proof for a general matrix group (which takes only a few lines).

We will at times need the following result.

Lemma D.1.6 *There exists a neighborhood W of 0 in $\mathfrak{g}$ such that the map $X \mapsto \exp X$ is one-to-one from W to a neighborhood of unity in G.*

Proof of Lemma D.1.6 for $G = SO(3)$ The difficult part of the proof is to show that $V := \exp W$ is a neighborhood of unity. For this we have to find a neighborhood V' of the unity in G such that $A \in V'$ is of the type $\exp X$ for $X \in W$. The obvious choice is $X = \log A$. The function log is well defined in a neighborhood of unity by the formula

$$\log A = \sum_{n \geq 1} (-1)^n \frac{(A-1)^n}{n}.$$

What is not obvious though is that $\log A \in \mathfrak{g}$. It is true that $\exp(\log A) = A \in G$ but this does not suffice.[5] In the present case of $SO(3)$ since $A^T = A^{-1}$ we have

$$(\log A)^T = \log(A^T) = \log A^{-1} = -\log A$$

[5] To prove that $\log A \in \mathfrak{g}$ we have to prove $\exp t \log A \in G$ for all t.

so that $\log A \in \mathsf{so}(3)$. Clearly $\log A \in W$ if V' is small enough, and Lemma D.1.6 is proved in that case. □

It will be equally obvious how to prove this result in all the specific cases we need. For the general case, we are unable to improve upon the (simple) arguments in Hall's book [39, Theorem 3.42], to which we refer the reader.

We define now what should be called a *real* Lie algebra of matrices. As we shall not explicitly consider complex Lie algebras[6] we will simply call this a Lie algebra.

Definition D.1.7 A Lie algebra h is a set of square matrices of given dimension which is a vector space over the real numbers and which is closed under the Lie bracket, $X, Y \in \mathsf{h} \Rightarrow [X,Y] \in \mathsf{h}$.

Thus our terminology makes sense: Theorem D.1.5 shows that the Lie algebra g of a group G is a Lie algebra in the sense of Definition D.1.7. Again, the idea is that the structure of the Lie bracket on g encodes much of the structure of G.

For the rest of this section we concentrate on the study of $\mathsf{so}(3)$ and of the properties of the exponential map from $\mathsf{so}(3)$ to $SO(3)$. Let us stress the following obvious but fundamental general fact. Since the Lie bracket $[X,Y]$ is bilinear it is completely determined by the values it takes for X, Y in a basis of the Lie algebra (where "basis" means "basis as a vector space"). The traditional basis of $\mathsf{so}(3)$ consists of the matrices

$$J_1 = \begin{pmatrix} 0 & 0 & 0 \\ 0 & 0 & -1 \\ 0 & 1 & 0 \end{pmatrix} \; ; \; J_2 = \begin{pmatrix} 0 & 0 & 1 \\ 0 & 0 & 0 \\ -1 & 0 & 0 \end{pmatrix} \; ; \; J_3 = \begin{pmatrix} 0 & -1 & 0 \\ 1 & 0 & 0 \\ 0 & 0 & 0 \end{pmatrix}. \tag{D.9}$$

One then calculates

$$[J_1, J_2] = J_3 \; ; \; [J_2, J_3] = J_1 \; ; \; [J_3, J_1] = J_2. \tag{D.10}$$

The relations (D.10) completely determine the structure of the Lie algebra $\mathsf{so}(3)$. When exploring a new Lie algebra the first task is often to compute the value of the Lie bracket on a suitable basis.

Often people use the notation $\mathbf{J} = (J_1, J_2, J_3)$ and $\boldsymbol{u} \cdot \mathbf{J} = \sum_i u_i J_i$. Let us denote by $\boldsymbol{v} \wedge \boldsymbol{u}$ the cross product of the vectors $\boldsymbol{v}$ and $\boldsymbol{u}$. Writing down what this means, one obtains the following very useful fact.

Lemma D.1.8 *The matrix of the operator* $\boldsymbol{x} \mapsto \boldsymbol{u} \wedge \boldsymbol{x}$ *is* $\boldsymbol{u} \cdot \mathbf{J}$.

In the present case the nature of the exponential map is clarified by the following result.

Proposition D.1.9 *Consider a unit vector* $\boldsymbol{u}$ *and* $\theta \in \mathbb{R}$. *Then*

$$R(\boldsymbol{u}, \theta) := \exp \theta \boldsymbol{u} \cdot \mathbf{J} \tag{D.11}$$

is the matrix of a rotation of axis $\boldsymbol{u}$ *and angle* θ.

Proof We have

$$\frac{\mathrm{d}}{\mathrm{d}\theta} R(\boldsymbol{u}, \theta) = \boldsymbol{u} \cdot \mathbf{J} R(\boldsymbol{u}, \theta),$$

so that (identifying a matrix with the corresponding operator on $\mathbb{R}^3$) we deduce from Lemma D.1.8 that

$$\frac{\mathrm{d}}{\mathrm{d}\theta} R(\boldsymbol{u}, \theta)(\boldsymbol{x}) = \boldsymbol{u} \cdot \mathbf{J} R(\boldsymbol{u}, \theta)(\boldsymbol{x}) = \boldsymbol{u} \wedge R(\boldsymbol{u}, \theta)(\boldsymbol{x}). \tag{D.12}$$

6 A complex Lie algebras is simply a Lie algebra which is also a complex vector space, e.g. the set $\mathsf{gl}_{\mathbb{C}}(m)$ of all $m \times m$ complex matrices.

Denoting now by $S(\theta)$ the rotation of angle θ around the axis determined by a unit vector $\boldsymbol{u}$, you will convince yourself using high-school geometry that for any vector $\boldsymbol{x}$ we have the relation

$$\frac{\mathrm{d}}{\mathrm{d}\theta}S(\theta)(\boldsymbol{x}) = \boldsymbol{u} \wedge (S(\theta)(\boldsymbol{x})).$$

Comparing with (D.12) and using the uniqueness of solutions of ordinary differential equations this completes the proof. □

Exercise D.1.10 Use (D.10) to prove that for $\boldsymbol{u}, \boldsymbol{v} \in \mathbb{R}^3$ one has

$$[\boldsymbol{u} \cdot \mathbf{J}, \boldsymbol{v} \cdot \mathbf{J}] = (\boldsymbol{u} \wedge \boldsymbol{v}) \cdot \mathbf{J}. \tag{D.13}$$

Exercise D.1.11 Prove that every element X of $SO(3)$ is of the type $\exp \theta \boldsymbol{u} \cdot \mathbf{J}$. Hint: Use the fact proved in Section 4.2 that the operator of matrix X has 1 as an eigenvalue.

Exercise D.1.12 In this exercise we identify a matrix A and the corresponding linear operator on $\mathbb{R}^3$. Prove that for $A \in SO(3)$ and $\boldsymbol{v} \in \mathbb{R}^3$ we have

$$A(\boldsymbol{v} \cdot \mathbf{J})A^{-1} = A(\boldsymbol{v}) \cdot \mathbf{J}. \tag{D.14}$$

This property (D.14) is often expressed by saying that $\mathbf{J}$ "transforms as a vector under rotations". Hint: $A = \exp \boldsymbol{u} \cdot \mathbf{J}$ for a certain $\boldsymbol{u}$. For $A(t) = \exp t\boldsymbol{u} \cdot \mathbf{J}$ compute the derivative of the function $t \mapsto A(t)^{-1}\big((A(t)(\boldsymbol{v})) \cdot \mathbf{J}\big)A(t)$ using (D.13) and (D.12).

D.2 Group Representations and Lie Algebra Representations

In this section we explain why Lie algebras are such a convenient method to study (finite-dimensional) group representations. In the present appendix we are *not* particularly interested in unitary representations. Thus in contrast with the main text, where "representation" means by default "unitary representation" in the present appendix a *representation of G is a* continuous *group homomorphism from G to the group of invertible linear operators on a finite-dimensional real or complex vector space*. Most of the time this finite-dimensional space will be complex, and we then take it to be $\mathbb{C}^m$. (Obvious adaptations are needed when the space is real rather than complex vector space.) The group of invertible linear operators on $\mathbb{C}^m$ is then the linear group $Gl(m,\mathbb{C})$ and a representation of G is a continuous group homomorphism from G to $Gl(m,\mathbb{C})$. The Lie algebra $\mathfrak{gl}_{\mathbb{C}}(m)$ of $Gl(m,\mathbb{C})$ is the set of all $m \times m$ complex matrices. Let us stress here that with this point of view, we do not distinguish between a linear operator on $\mathbb{C}^m$ and its matrix in the canonical basis of $\mathbb{C}^m$. The Lie algebra $\mathfrak{gl}_{\mathbb{C}}(m)$ has a distinguished feature: it is naturally a vector space over the complex numbers, not only over the real numbers.

As we stressed, the essential structure in a Lie algebra is the Lie bracket (D.6). This motivates the following definition.

Definition D.2.1 A Lie Algebra homomorphism π' from a Lie algebra $\mathfrak{g}$ to a Lie algebra $\mathfrak{g}'$ is a real linear map from $\mathfrak{g}$ to $\mathfrak{g}'$ which respects the Lie bracket,

$$\pi'([X,Y]) = [\pi'(X), \pi'(Y)]. \tag{D.15}$$

We can now state and prove the following fundamental result.

Theorem D.2.2 *For any continuous group homomorphism $\pi : G \to Gl(m,\mathbb{C})$ and any X in $\mathfrak{g}$ the derivative*

$$\pi'(X) := \frac{\mathrm{d}}{\mathrm{d}t}\pi(\exp tX)|_{t=0} \tag{D.16}$$

exists. The map $\pi' : \mathfrak{g} \to \mathfrak{gl}_{\mathbb{C}}(m)$ is a Lie algebra homomorphism. Moreover

$$\pi(\exp X) = \exp \pi'(X). \tag{D.17}$$

Please note that while π is a group homomorphism (multiplicative) then π' is linear. A Lie algebra homomorphism from $\mathfrak{g}$ to $\mathfrak{gl}_{\mathbb{C}}(m)$ is sometimes called a *representation* of $\mathfrak{g}$. We will use both terminologies interchangeably. Let us stress the content of Theorem D.2.2.

- The representation π of G determines a representation π' of $\mathfrak{g}$.
- This representation π' is a linear map. It is a simpler object than the group homomorphism π.
- Moreover the knowledge of π' determines the value of π on the elements of G of the type $\exp X$, so probably it basically determines π.

The following principle will be of constant use.

Lemma D.2.3 *Consider a continuous one-parameter group $V(t)$ of linear operators on a finite-dimensional space. Then the map $t \mapsto V(t)$ is differentiable and $V(t) = \exp tV'(0)$.*

Here of course the exponential is defined by the power series (D.2).

Proof The argument of Lemma 2.14.5 proves that the derivative exists, and that $V'(t) = V'(0)V(t)$. □

Proof of (D.17) It is straightforward that $V(t) = \pi(\exp tX)$ is continuous and satisfies $V(s+t) = V(s)V(t)$. Thus $\pi'(X) = V'(0)$ exists and Lemma D.2.3 implies (D.17). □

Lemma D.2.4 *If $A = 1 + O(t^k)$ then $\pi(A) = 1 + O(t^k)$.*

Proof We use Lemma D.1.6 and its proof. If A is close to unity,

$$\pi(A) = \pi(\exp \log A) = \exp \pi'(\log A).$$

Then we obtain successively that $\log A = O(t^k)$, $\pi'(\log A) = O(t^k)$ and $\pi(A) = 1 + O(t^k)$. □

Proof of Theorem D.2.2 It remains to prove that π' is a Lie algebra homomorphism. We first prove that π' is linear. It is obvious from the definition that $\pi'(aX) = a\pi'(X)$ for $a \in \mathbb{R}$ so we prove that π' is additive. Given $X, Y \in \mathfrak{g}$ we define $S(t) \in G$ by

$$S(t) := \exp t(X+Y)(\exp tY)^{-1}(\exp tX)^{-1}. \tag{D.18}$$

Applying π to this identity and using (D.17) we obtain

$$\exp t\pi'(X+Y) = \pi(S(t)) \exp t\pi'(X) \exp t\pi'(Y). \tag{D.19}$$

It follows from (D.18) (replacing each exponential by the corresponding power series) that, in the physicists notation, $S(t) = 1 + O(t^2)$ so that the desired equality $\pi'(X+Y) = \pi'(X) + \pi'(Y)$ follows by using Lemma D.2.4 for $k = 2$ and looking at the terms of order t in (D.19).

Next we prove that π' respects the Lie bracket. Given again X, Y in $\mathfrak{g}$ we define now $S(t) \in G$ by

$$\exp tX \exp tY (\exp tX)^{-1}(\exp tY)^{-1} := \exp(t^2[X,Y])S(t). \tag{D.20}$$

It follows from (D.8) that $S(t) = 1 + O(t^3)$, so that $\pi(S(t)) = 1 + O(t^3)$ by using Lemma D.2.4 for $k = 3$. Applying π to (D.20) and using (D.17) we get

$$\exp t\pi'(X) \exp t\pi'(Y)(\exp t\pi'(X))^{-1}(\exp t\pi'(Y))^{-1} = \exp(t^2\pi'([X,Y]))(1 + O(t^3)).$$

The required equality follows by equating the coefficients of t^2 in the left- and right-hand sides. □

As a first example, consider the group homomorphism π from $Gl(m,\mathbb{C})$ to $Gl(1,\mathbb{C}) = \mathbb{C}\backslash\{0\}$ given by $\pi(A) = \det A$. Then obviously $\det\exp tX = \det(1+tX+O(t^2))) = 1+t\mathrm{tr}X+O(t^2)$, where $\mathrm{tr}X$ is the trace of X, that is the sum of its diagonal coefficients. Thus (D.16) implies that $\pi'(A) = \mathrm{tr}A$. Then (D.17) takes the form

$$\det(\exp X) = \exp(\mathrm{tr}X). \tag{D.21}$$

Furthermore tr is a Lie algebra homomorphism, $\mathrm{tr}[A,B] = [\mathrm{tr}A,\mathrm{tr}B] = 0$ i.e. $\mathrm{tr}AB = \mathrm{tr}BA$, which of course can be simply proved directly.

The reader should watch for the tricky notation we use now. To each group homomorphism π we associate a Lie algebra homomorphism π', but the notation π' also stands for a generic Lie algebra homomorphism. The fundamental question then is whether Theorem D.2.2 has a converse: Given a Lie algebra homomorphism π' from $\mathfrak{g}$ to $\mathfrak{gl}_{\mathbb{C}}(m)$ can we find a representation π which satisfies (D.17)? This will be discussed shortly. In the meantime we show that deciding whether the representation π is irreducible can be done by studying a similar property for π'.

Definition D.2.5 A Lie algebra homomorphism $\pi'\colon \mathfrak{g} \to \mathfrak{gl}_{\mathbb{C}}(m)$ is called irreducible if there is no proper subspace $\mathcal{H}$ of $\mathbb{C}^m$ which is invariant under all the operators $\pi'(X)$, i.e. for which $\pi'(X)(\mathcal{H}) \subset \mathcal{H}$ for all $X \in \mathfrak{g}$.

Theorem D.2.6 *Assume that the group G is generated by the elements* $\exp X$ *for* $X \in \mathfrak{g}$. *Then the group homomorphism $\pi : G \to Gl(m,\mathbb{C})$ is an irreducible representation of G if and only if the Lie algebra homomorphism $\pi'\colon \mathfrak{g} \to \mathfrak{gl}_{\mathbb{C}}(m)$ is irreducible.*

Proof If π is not irreducible, there is a proper subspace $\mathcal{H}$ of $\mathbb{C}^m$ such that $\pi(A)(\mathcal{H}) \subset \mathcal{H}$ for each A in G and it should be obvious that this implies that $\pi'(X)(\mathcal{H}) \subset \mathcal{H}$ for each $X \in \mathfrak{g}$. Conversely if the space $\mathcal{H}$ has this property then (D.17) and (D.1) imply that $\pi(A)(\mathcal{H}) \subset \mathcal{H}$ whenever A is of the type $\exp X$ for $X \in \mathfrak{g}$ and since these elements generate G by hypothesis this is true for each $A \in G$, so π is not irreducible. □

As we have seen (in Exercise D.1.11) $SO(3)$ is generated by the elements of the type $\exp X$, but this is not the case for $O(3)$. More generally it is possible to show that G is generated by the elements $\exp X$ if and only if it is connected (in the topological sense), see [39].

Let us also note the following trivial fact.

Proposition D.2.7 *If the group homomorphism π is valued in a Lie group $G' \subset Gl(m,\mathbb{C})$ then π' is valued in the Lie algebra $\mathfrak{g}'$ of G', so is a Lie algebra homomorphism from $\mathfrak{g}$ to $\mathfrak{g}'$.*

D.3 Angular Momentum

Consider first a unitary representation π of $SO(3)$ in a Hilbert space $\mathcal{H}$ (possibly of infinite dimension). Given a non-zero vector $\boldsymbol{u}$ we recall from (D.11) that $\exp(\theta\boldsymbol{u}\cdot\mathbf{J})$ is a rotation of angle $\theta\|\boldsymbol{u}\|$ and axis $\boldsymbol{u}$. Let us then consider

$$U(\theta) = \pi(\exp\theta\boldsymbol{u}\cdot\mathbf{J}). \tag{D.22}$$

This is a one-parameter semigroup of unitary operators, so that by Stone's theorem (and using a unit system where $\hbar = 1$) we have

$$U(\theta) = \exp(-\mathrm{i}\theta J'_{\boldsymbol{u}}) \tag{D.23}$$

for a certain self-adjoint operator $J'_{\boldsymbol{u}}$. When $\boldsymbol{u}$ is a unit vector this operator is called the *operator of angular momentum with respect to* $\boldsymbol{u}$. But how do we relate these operators as $\boldsymbol{u}$ varies? It is not even clear that they are commonly defined on a dense subspace. But Corollary A.6.2 allows one to reduce to the case where $\mathcal{H}$ is finite-dimensional. Then we know from (D.17) that $\pi(\exp\theta\boldsymbol{u}\cdot\mathbf{J}) = \exp\theta\pi'(\boldsymbol{u}\cdot\mathbf{J})$

and comparison with (D.23) shows that $J'_{\boldsymbol{u}} = \mathrm{i}\pi'(\boldsymbol{u}\cdot\mathbf{J})$. Writing $J'_k := J'_{e_k} = \mathrm{i}\pi'(J_k)$ where e_1, e_2, e_3 are the standard basis vectors, we then have

$$J'_{\boldsymbol{u}} = \mathrm{i}\sum_k u_k \pi'(J_k) = \sum_k u_k J'_k,$$

whereas the commutation relation (D.10) together with the fact that π' respects the Lie bracket yields the relations

$$[J'_1, J'_2] = \mathrm{i}J'_3\ ;\ [J'_2, J'_3] = \mathrm{i}J'_1\ ;\ [J'_3, J'_1] = \mathrm{i}J'_2.$$

D.4 su(2) = so(3)!

Let us determine the Lie algebra su(2) of the group $SU(2)$ of 2×2 complex unitary matrices with determinant 1, so $A \in SU(2)$ if and only if $AA^\dagger = I$ and $\det A = 1$. Using (D.3), a matrix X is in su(2) if and only if $\exp tX \exp tX^\dagger = I$ and $\det\exp tX = 1$ for each t. Taking the derivative at $t = 0$ of the first condition gives $X + X^\dagger = 0$ i.e. $X = -X^\dagger$. Such a matrix is called *skew-adjoint*. It follows from (D.21) that the condition $\det\exp tX = 1$ is equivalent to $\mathrm{tr}\,X = 0$. A matrix of trace zero is called *traceless* and we have proved the following.

Proposition D.4.1 *The Lie algebra* su(2) *of* $SU(2)$ *is the space of skew-adjoint traceless* 2×2 *complex matrices.*

To study this Lie algebra we look for a nice basis. To find it, let us recall the Pauli matrices

$$\sigma_1 = \begin{pmatrix} 0 & 1 \\ 1 & 0 \end{pmatrix}\ ;\ \sigma_2 = \begin{pmatrix} 0 & -\mathrm{i} \\ \mathrm{i} & 0 \end{pmatrix}\ ;\ \sigma_3 = \begin{pmatrix} 1 & 0 \\ 0 & -1 \end{pmatrix}. \tag{D.24}$$

These are traceless, but self-adjoint (=Hermitian) rather than skew-adjoint. One checks by computation that they satisfy the identities

$$\sigma_1\sigma_2 - \sigma_2\sigma_1 = 2\mathrm{i}\sigma_3\ ;\ \sigma_2\sigma_3 - \sigma_3\sigma_2 = 2\mathrm{i}\sigma_1\ ;\ \sigma_3\sigma_1 - \sigma_1\sigma_3 = 2\mathrm{i}\sigma_2. \tag{D.25}$$

The matrices

$$X_j = -\frac{\mathrm{i}\sigma_j}{2} \tag{D.26}$$

are traceless skew-adjoint. They form a basis of su(2), and (D.25) implies that they satisfy the relations

$$[X_1, X_2] = X_3\ ;\ [X_2, X_3] = X_1\ ;\ [X_3, X_1] = X_2. \tag{D.27}$$

These are the same relations as in (D.10), and this shows that the Lie algebras so(3) and su(2) are isomorphic.[7] The corresponding groups $SO(3)$ and $SU(2)$ are however rather different. What happens here is a special feature of a general fact: The Lie algebra of a group captures only the "local" structure of a group, where "local" means in a neighborhood of the identity, but it does not capture its global structure.

We now turn to the construction of a fundamental two-to-one homomorphism $\tilde{\kappa}$ from $SU(2)$ to $SO(3)$. (This map is simply the restriction of the map κ of Section 8.4 to $SU(2)$ but we perform the construction from scratch again.) First, if X is self-adjoint traceless, this is also the case for AXA^{-1} for any $A \in SU(2)$. The Pauli matrices form a basis of the space of self-adjoint traceless matrices, so each such matrix is of the form

$$\boldsymbol{x}\cdot\boldsymbol{\sigma} := \sum_{j\le 3} x_j\sigma_j = \begin{pmatrix} x_3 & x_1 - \mathrm{i}x_2 \\ x_1 + \mathrm{i}x_2 & -x_3 \end{pmatrix}, \tag{D.28}$$

[7] Specifically, the unique linear map $T\colon$ so(3) $\to$ su(2) such that $T(J_i) = X_i$ is a Lie algebra isomorphism.

for a certain $x \in \mathbb{R}^3$. This is similar to (8.19), with the difference that we are here considering only traceless matrices. Since the matrix $A(x \cdot \sigma)A^{-1} = A(x \cdot \sigma)A^\dagger$ is traceless Hermitian, it is of the type $(\tilde{\kappa}(A)(x)) \cdot \sigma$ for a certain vector $\tilde{\kappa}(A)(x) \in \mathbb{R}^3$. Thus we have constructed a group homomorphism $\tilde{\kappa}$ from $SU(2)$ into the group $Gl(3,\mathbb{R})$ of linear transformations of $\mathbb{R}^3$, these transformations being identified with their corresponding matrices. The homomorphism $\tilde{\kappa}$ is valued in $O(3)$ because

$$\|x\|^2 = -\det x \cdot \sigma = -\det A(x \cdot \sigma)A^{-1} = -\det(\tilde{\kappa}(A)(x)) \cdot \sigma = \|\tilde{\kappa}(A)(x)\|^2.$$

When $\tilde{\kappa}(A) = I$ the matrix A commutes with all the matrices $x \cdot \sigma$. Just writing what it means that A commutes with the matrices σ_1 and σ_2 implies that A is then diagonal, so that $A = \pm I$ since $\det(A) = 1$. Since $\tilde{\kappa}$ is a group homomorphism this shows that $\tilde{\kappa}$ is two-to-one.

The map $\tilde{\kappa}$ is continuous, since for any $x \in \mathbb{R}^3$ the components of $\tilde{\kappa}(A)(x)$ are second-degree polynomials in the entries of A. Since $SU(2)$ is connected its image is contained in the connected component of the identity in $O(3)$ i.e. in $SO(3)$. It then follows from Proposition D.2.7 that the map $\tilde{\kappa}'$ is a Lie algebra homomorphism from su(2) to so(3). To compute it we consider $Y \in$ su(2) and we differentiate the equality

$$(\tilde{\kappa}(\exp tY)(x)) \cdot \sigma = (\exp tY)(x \cdot \sigma)\exp(-tY)$$

at $t = 0$, getting

$$(\tilde{\kappa}'(Y)(x)) \cdot \sigma = [Y, x \cdot \sigma]. \tag{D.29}$$

The safest attitude here is to take a pencil and check that for X_j given by (D.26) this implies that

$$\tilde{\kappa}'(X_j) = J_j, \tag{D.30}$$

so that $\tilde{\kappa}'$ is just the natural isomorphism between su(2) and so(3) that we had described earlier. A far fancier way is to use your pencil to check the property of Exercise D.1.10 and proceed as follows. For $u \in \mathbb{R}^3$ define $u \cdot \mathbf{X} = \sum_{j\le 3} u_j X_j$, so that $u \cdot \mathbf{X} = -\mathrm{i}u \cdot \sigma/2$ and (D.29) becomes

$$\tilde{\kappa}'(Y)(x) \cdot \mathbf{X} = [Y, x \cdot \mathbf{X}].$$

In the case where $Y = y \cdot \mathbf{X}$ the right-hand side is $(y \wedge x) \cdot \mathbf{X}$. Consequently

$$\tilde{\kappa}'(y \cdot \mathbf{X})(x) = y \wedge x,$$

and Lemma D.1.8 implies that the matrix of the operator $\tilde{\kappa}'(y \cdot \mathbf{X})$ is therefore $y \cdot \mathbf{J}$, implying (D.30).

Having calculated $\tilde{\kappa}'$ we may now use (D.17) to write

$$\tilde{\kappa}(\exp u \cdot \mathbf{X}) = \exp(u \cdot \mathbf{J}). \tag{D.31}$$

This may give you the illusion that you understand $\tilde{\kappa}$ very well. The problem however is that to use this formula to calculate $\tilde{\kappa}(A)$ for $A \in SU(2)$ you must write A as an exponential $\exp u \cdot \mathbf{X}$ and this is not a trivial task.

Let us spell out the following consequence of (D.31): If u is a unit vector, the image by $\tilde{\kappa}$ of

$$\exp\Big(-\frac{\mathrm{i}}{2}\sum_{j\le 3}\theta u_j \sigma_j\Big) = \exp(-\mathrm{i}\theta u \cdot \sigma/2)$$

is the rotation of angle θ around the axis u, see (D.11).

Now, in the reverse direction, does there exist a group homomorphism λ from $SO(3)$ to $SU(2)$ such that λ': so(3) $\to$ su(2) is the Lie algebra isomorphism given by $u \cdot \mathbf{J} \to u \cdot \mathbf{X}$? Such a homomorphism would satisfy

$$\lambda(\exp(u \cdot \mathbf{J})) = \exp(u \cdot \mathbf{X}). \tag{D.32}$$

In particular one would have

$$\lambda(\exp(\theta J_3)) = \exp(-\mathrm{i}\theta\frac{\sigma_3}{2}).$$

The obvious problem with this formula is that for $\theta = 2\pi$, $\exp(\theta J_3)$ is a rotation of angle 2π, i.e. the identity I, while the right-hand side is $-I$, so that λ cannot be a group homomorphism.

This does not stop physicists from "defining" a representation of $SO(3)$ on $\mathbb{C}^2$ precisely by the formula (D.32). They acknowledge of course that there is a "sign ambiguity" in this definition, in the sense that the right-hand side is "defined up to a sign".[8] Thus, in our language, they construct a "double-valued representation".

D.5 From Lie Algebra Homomorphisms to Lie Group Homomorphisms

Although this is not obvious at all, the root of the problem with $SO(3)$ is that this group is not nice topologically. As we discuss in Section 8.5 it is not simply connected. For simply connected groups (i.e. in which all loops can be continuously contracted to one point), we have the following:

Theorem D.5.1 *If G is a compact simply connected Lie group of matrices and φ a Lie algebra homomorphism from $\mathfrak{g}$ to $\mathfrak{gl}_{\mathbb{C}}(m)$, then there is a group homomorphism Φ from G to $Gl(m, \mathbb{C})$ such that $\Phi(\exp X) = \exp \varphi(X)$ for $X \in \mathfrak{g}$.*

This is a rather deep result, which we shall not use. As I wanted to understand this result, I provide an elementary and essentially self-contained proof. (The hypothesis that G is compact is not really needed.)

Let us consider a neighborhood W of 0 in $\mathfrak{g}$ as in Lemma D.1.6. Then each element of $V := \exp W$ is of the type $\exp X$ for a unique $X \in \mathfrak{g}$. We may then define a map $\Pi : V \to Gl(m, \mathbb{C})$ by

$$\Pi(\exp X) = \exp \varphi(X). \tag{D.33}$$

The main step of the proof is as follows.

Theorem D.5.2 *There exists a neighborhood V' of the unit in G such that whenever $A, B, AB \in V'$ then*

$$\Pi(AB) = \Pi(A)\Pi(B). \tag{D.34}$$

Proof of Theorem D.5.1 It follows from (the first part of) Proposition A.3.4 that there exists a group homomorphism Φ from G to $Gl(m, \mathbb{C})$ such that $\Phi(A) = \Pi(A)$ for A close to unity. Then $\Phi(\exp X) = \exp \varphi(X)$ for $X \in \mathfrak{g}$ close to 0. Thus $\varphi = \Phi'$ and $\Phi(\exp X) = \exp \varphi(X)$ for each X. □

The main tool for the proof of Theorem D.5.2 is the following.

Proposition D.5.3 *For square matrices X, Y we have*

$$\frac{\mathrm{d}}{\mathrm{d}t} \exp(tX + Y)\Big|_{t=0} = \Psi(X, Y) \exp Y, \tag{D.35}$$

where $\Psi(X, Y)$ is given by the converging series

$$\Psi(X, Y) = X + \frac{1}{2!}[Y, X] + \frac{1}{3!}[Y, [Y, X]] + \frac{1}{4!}[Y, [Y, [Y, X]]] + \cdots. \tag{D.36}$$

More generally, for a differentiable matrix-valued function $U(t)$ we have

$$\frac{\mathrm{d}}{\mathrm{d}t} \exp U(t) = \Psi(U'(t), U(t)) \exp U(t). \tag{D.37}$$

[8] Then they point out that the sign ambiguity does not matter for Quantum Mechanics so that everything is fine.

During our analysis of this result, we will see that it boils down to a clever algebraic identity. We could not care less about the exact form of the function Ψ. What matters is that since φ is a Lie algebra homomorphism, for $X, Y \in \mathfrak{g}$ we have

$$\Psi(\varphi(X), \varphi(Y)) = \varphi(\Psi(X, Y)). \tag{D.38}$$

Proof of Theorem D.5.2 We choose V' small enough that for $\exp X$, $\exp Y \in V'$ and $0 \leq t \leq 1$ we have $\exp tX \exp Y \in V = \exp W$. Fixing such $X, Y \in \mathfrak{g}$ we may then define $U(t)$ by $\exp U(t) = \exp tX \exp Y$. First, differentiating both sides and using (D.37) we obtain

$$\Psi(U'(t), U(t)) \exp U(t) = X \exp U(t),$$

and hence the identity

$$\Psi(U'(t), U(t)) = X. \tag{D.39}$$

Combining with (D.38) we then get

$$\Psi\big((\varphi(U'(t)), \varphi(U(t))\big) = \varphi(X). \tag{D.40}$$

Consider then the function

$$A(t) := \exp \varphi(U(t)).$$

Then, using (D.37) and (D.40) we obtain

$$A'(t) = \Psi\big(\varphi(U'(t)), \varphi(U(t))\big) A(t) = \varphi(X) A(t).$$

The function $B(t) = \exp t\varphi(X) \exp \varphi(Y)$ satisfies the same differential equation. Since $A(0) = \exp \varphi(Y) = B(0)$ we then have $A(t) = B(t)$ for each $0 \leq t \leq 1$. For $t = 1$ this means

$$\Pi(\exp X \exp Y) = \exp \varphi(U(1)) = \exp \varphi(X) \exp \varphi(Y) = \Pi(\exp X) \Pi(\exp Y).$$ □

Next, we turn to the proof of Proposition D.5.3. We will prove only (D.35) as the proof of (D.37) is entirely similar. We fix X and Y. We define $Z_0 = X$ and inductively $Z_\ell = [Y, Z_{\ell-1}]$ so that

$$\Psi(X, Y) = \sum_{\ell \geq 0} \frac{Z_\ell}{(\ell + 1)!}.$$

We first reduce the proof of Proposition D.5.3 to the proof of the following clever identity.

Lemma D.5.4 *For each k we have*

$$S_k := Y^k X + Y^{k-1} X Y + \cdots + X Y^k = \sum_{0 \leq \ell \leq k} \binom{k+1}{\ell+1} Z_\ell Y^{k-\ell}. \tag{D.41}$$

Proof of Proposition D.5.3 It is straightforward from the series expansion that the left-hand side of (D.35) equals

$$\sum_{k \geq 0} \frac{S_k}{(k+1)!}. \tag{D.42}$$

On the other hand, using the explicit expression for $\binom{k+1}{\ell+1}$ one may rewrite (D.41) as

$$\frac{S_k}{(k+1)!} = \sum_{0 \leq \ell \leq k} \frac{Z_\ell}{(\ell+1)!} \frac{Y^{k-\ell}}{(k-\ell)!}.$$

Substitution in (D.42) and regrouping the terms completes the proof. □

One way to discover (D.41) is to use the commutation relations to push all the factors X to the left. Thus, since $YX = XY + Z_1$, we have

$$Y^k X = Y^{k-1} X Y + Y^{k-1} Z_1.$$

Next, using also $YZ_1 = Z_1 Y + Z_2$, we obtain

$$Y^k X = Y^{k-2} X Y^2 + 2Y^{k-2} Z_1 Y + Y^{k-2} Z_2,$$

and by induction over n we obtain that for $1 \leq n \leq k$ one has

$$Y^k X = \sum_{0 \leq \ell \leq n} \binom{n}{\ell} Y^{k-n} Z_\ell Y^{k-\ell}.$$

In particular for $n = k$ we have finished pushing the X factors to the left:

$$Y^k X = \sum_{0 \leq \ell \leq k} \binom{k}{\ell} Z_\ell Y^{k-\ell}. \tag{D.43}$$

As the following proof shows, this is all there is to (D.41).

Proof of Lemma D.5.4 The proof goes by induction over k. For the step from k to $k+1$, using that $S_{k+1} = S_k Y + Y^{k+1} X$ we have to check that

$$\sum_{0 \leq \ell \leq k} \binom{k+1}{\ell+1} Z_\ell Y^{k+1-\ell} + Y^{k+1} X = \sum_{0 \leq \ell \leq k+1} \binom{k+2}{\ell+1} Z_\ell Y^{k+1-\ell}.$$

Using that

$$\binom{k+2}{\ell+1} = \binom{k+1}{\ell+1} + \binom{k+1}{\ell},$$

this coincides with (D.43). □

D.6 Irreducible Representations of $SU(2)$

Our goal is to classify the finite-dimensional irreducible representations of $SU(2)$. All results will be easy consequences of the following central fact.

Proposition D.6.1 *Consider a finite-dimensional complex vector space $\mathcal{H}$ and three operators Z, A^+, A^- on $\mathcal{H}$ which satisfy the relations*

$$[Z, A^+] = 2A^+ \; ; \; [Z, A^-] = -2A^- \tag{D.44}$$

$$[A^+, A^-] = Z. \tag{D.45}$$

Then the largest eigenvalue of Z is an integer m. Consider further an eigenvector e_0 corresponding to this eigenvalue, and define recursively the vectors

$$e_{k+1} = A^-(e_k). \tag{D.46}$$

Then (remembering that m is the largest eigenvalue of Z),

$$e_m \neq 0 \; ; \; e_{m+1} = 0. \tag{D.47}$$

Furthermore,

$$Z(e_k) = (m-2k)e_k, \tag{D.48}$$

$$A^+(e_0) = 0\,;\ k > 0 \Rightarrow A^+(e_k) = k(m-k+1)e_{k-1}, \tag{D.49}$$

the vectors $e_0, e_1, \ldots, e_m$ are linearly independent and their span $\mathcal{G}$ is invariant under the operators $Z, A^{\pm}$. In particular if no proper subspace of $\mathcal{H}$ is invariant under these operators then $\mathcal{H} = \mathcal{G}$. Moreover, no proper subspace of $\mathcal{G}$ is invariant under Z and $A^{\pm}$.

The relations (D.46) and (D.48) to (D.49) completely describe the operators Z, A^+, A^- in the basis $e_0, \ldots, e_m$ of $\mathcal{G}$. Conversely, when $\mathcal{G} = \mathcal{H}$ it is straightforward that the operators determined by these relations satisfy the required commutation relations (D.44) and (D.45).

Proposition D.6.1 relates to representations of $SU(2)$ as follows.

Lemma D.6.2 *Consider a representation π of $SU(2)$ in $\mathbb{C}^k$ and the map π' as provided by (D.17). Consider the elements X_j of $\mathfrak{su}(2)$ given by (D.26), and the elements $Z_j = \pi'(X_j)$. Then the operators*

$$Z = 2iZ_3\,;\ A^+ = iZ_1 - Z_2\,;\ A^- = iZ_1 + Z_2 \tag{D.50}$$

satisfy the relations (D.44) and (D.45).

Proof The elements X_j satisfy the commutation relations (D.27) and therefore the elements $Z_j = \pi'(X_j)$ satisfy the same commutation relations. The relations (D.44) and (D.45) then follow by straight computation. For example,

$$[A^+, A^-] = [iZ_1 - Z_2, iZ_1 + Z_2] = i([Z_1, Z_2] - [Z_2, Z_1]) = 2iZ_3 = Z$$

and

$$[Z, A^+] = [2iZ_3, iZ_1 - Z_2] = -2[Z_3, Z_1] - 2i[Z_3, Z_2] = -2Z_2 + 2iZ_1 = 2A^+. \qquad \square$$

Proof of Proposition D.6.1 The fundamental observation is as follows. Assume that we are given a vector e with $Z(e) = \lambda e$. Applying the identity $ZA^- - A^-Z = -2A^-$ to e we obtain $ZA^-(e) - \lambda A^-(e) = -2A^-(e)$. The vector $f = A^-(e)$ satisfies $Z(f) = (\lambda - 2)f$. A similar result holds for A^+.

Since we are working with complex vector spaces, the operator Z has an eigenvalue. Let us consider the eigenvalue λ of Z with the largest real part, and a corresponding eigenvector e_0. For $k \geq 0$ we define recursively $e_{k+1} = A^-(e_k)$. As we previously showed,

$$Z(e_k) = (\lambda - 2k)e_k. \tag{D.51}$$

Let us next define $e_{-1} = 0$ and prove by induction over $k \geq 0$ that

$$A^+(e_k) = k(\lambda - k + 1)e_{k-1}. \tag{D.52}$$

For $k = 0$ this means $A^+(e_0) = 0$. This holds because $ZA^+(e_0) = (\lambda + 2)A^+(e_0)$ and we have assumed that λ is the eigenvalue of Z with the largest possible real part. To go from k to $k+1$ we write, using (D.45) in the second equality,

$$A^+(e_{k+1}) = A^+A^-(e_k) = A^-A^+(e_k) + Z(e_k).$$

We then use the induction hypothesis and (D.51) to obtain

$$A^+(e_{k+1}) = A^-(k(\lambda - k + 1)e_{k-1}) + (\lambda - 2k)e_k = (k+1)(\lambda - k)e_k,$$

completing the proof of (D.52).

Now, since $\mathcal{H}$ is finite-dimensional, (D.51) implies that the vectors e_k cannot be all non-zero, for then they would be linearly independent. Thus there exists exactly one m with $e_m \neq 0$ and $e_{m+1} = 0$.

Using (D.52) for $k = m+1$ yields $\lambda = m$. The span of $e_0, \dots, e_m$ is invariant under the action of Z, A^+, A^-. According to (D.48) these vectors are independent, so that they form a basis of $\mathcal{G}$.

It remains to prove that no proper subspace of $\mathcal{G}$ is invariant under Z and $A^{\pm}$. Consider such a subspace $\mathcal{G}'$ and $x \in \mathcal{G}', x \neq 0$. Then $x = \sum_{0 \le k \le m} a_k e_k$, where not all the coefficients a_k are zero. Let ℓ be the largest such that $a_\ell \neq 0$. Then $A^{+\ell}(x) \in \mathcal{G}'$, and (D.49) proves that $A^{+\ell}(x)$ is a non-zero multiple of e_0. Thus $e_0 \in \mathcal{G}'$ and $\mathcal{G}' = \mathcal{G}$ since by (D.46) $\mathcal{G}'$ contains all vectors e_k. □

Lemma D.6.3 *Consider an irreducible representation π of $SU(2)$ in $\mathbb{C}^k$. Then the vectors $e_0, \dots, e_m$ provided by Proposition D.6.1 span $\mathbb{C}^k$ (so that $k = m+1$).*

Proof Indeed their span $\mathcal{G}$ is invariant under the operators Z, A^+, A^-. According to (D.50) every operator $\pi'(X_j)$ is a linear combination of these, so that $\mathcal{G}$ is invariant under all the operators $\pi'(X)$. Since we assume that the representation is irreducible the space $\mathcal{G}$ equals $\mathbb{C}^k$ by Theorem D.2.6, and since every element of $SU(2)$ is of the type $\exp X$ for $X \in \mathsf{su}(2)$, as is obvious using an orthonormal basis where this element is diagonal. □

Let us recall that we say that two representations π_1, π_2 of a group in vector spaces $\mathcal{H}_1$ and $\mathcal{H}_2$ are *equivalent* if there is an invertible linear operator $V : \mathcal{H}_1 \to \mathcal{H}_2$ such that $V\pi_1(g) = \pi_2(g)V$ for each element g of the group.

Theorem D.6.4 *Two irreducible representations of $SU(2)$ in spaces of the same dimension are equivalent.*

Proof Let us consider an irreducible representation π of $SU(2)$ in $\mathbb{C}^k$, and recall the operators Z, A^+, A^- of (D.50). According to Lemma D.6.3, Proposition D.6.1 then completely describes the operators Z, A^+, A^- on $\mathbb{C}^k$ in the distinguished basis $e_0, \dots, e_{k-1}$ of $\mathbb{C}^k$. According to (D.50) each operator $\pi'(X_j)$ $(j \le 3)$ is a linear combination of these, so that this proposition completely describes the operators $\pi'(X_j)$ on this basis. Consider now two irreducible representations π_1 and π_2 of $SU(2)$ in $\mathbb{C}^k$ with corresponding distinguished basis $e_0^1, \dots, e_{k-1}^1$ and $e_0^2, \dots, e_{k-1}^2$ respectively. Then for $j \le 3$ the description of $\pi_1'(X_j)$ in the basis (e_ℓ^1) is the same as the description of $\pi_2'(X_j)$ in the basis (e_ℓ^2). The linear operator V of $\mathbb{C}^k$ such that $V(e_\ell^1) = e_\ell^2$ for $0 \le \ell \le k-1$ therefore satisfies $V\pi_1'(X_j) = \pi_2'(X_j)V$, and since the X_j form a basis of $\mathsf{su}(2)$, it satisfies $V\pi_1'(X) = \pi_2'(X)V$ for every element X of $\mathsf{su}(2)$. Using that $\pi(\exp X) = \exp \pi'(X)$ this operator then witnesses that the representations are equivalent. □

One good point of this result is that the proof is constructive, allowing us (in principle) to build the required intertwining operator.

In order to deal with *unitary* representations, we prove now the simple fact that there is essentially *at most* one Hilbertian norm that makes a given irreducible representation unitary.

Proposition D.6.5 *Consider a finite-dimensional irreducible unitary representation π of a group G and another Hilbertian norm $N(\cdot)$ for which this representation is also unitary. Then N is proportional to the original norm.*

Proof Saying that the norm N is Hilbertian means that $N(x)^2 = B(x,x)$ where $B(x,y)$ is a sesquilinear form which satisfies $B(y,x) = B(x,y)^*$. Defining the operator A by $B(x,y) = (x, A(y))$, we obtain

$$(A(y),x) = (x,A(y))^* = B(x,y)^* = B(y,x) = (y,A(x))$$

so that A is Hermitian. We can therefore diagonalize it in an orthonormal basis. That is, if n is the dimension of the underlying space, we can find an orthonormal basis $(e_i)_{i \le n}$ for the original norm such that $N(\sum_i x_i e_i)^2 = \sum_{i \le n} \alpha_i |x_i|^2$ where the sequence α_i is non-increasing. Assume that k is such that $\alpha_1 = \alpha_2 = \dots = \alpha_k > \alpha_{k+1}$. Then the span of $e_1, \dots, e_k$ consists exactly of the vectors x

such that $\sqrt{\alpha_1}\|x\| = N(x)$ so it is stable under all the maps $\pi(C)$ for $C \in G$ since these are isometries for both norms. Since π is irreducible we have $k = n$. □

Corollary D.6.6 *Consider two irreducible unitary representations π_1 and π_2 of $SU(2)$ on Hilbert spaces $\mathcal{H}_1, \mathcal{H}_2$ and assume that for some operator $W\colon \mathcal{H}_1 \to \mathcal{H}_2$ one has $W\pi_1(A) = \pi_2(A)W$ for each $A \in SU(2)$. Then either $W = 0$ or there is $\lambda > 0$ such that λW is an isometry. In particular if π_1 and π_2 are equivalent they are unitary equivalent.*

Proof Use Proposition D.6.5 for the norm N on $\mathcal{H}_1$ given by $N(x) = \|W(x)\|$. □

Although we do not need this fact, for each finite-dimensional representation of $SU(2)$ (or, in fact of a compact group) one can find an equivalent norm which makes it unitary. (Proving this is the goal of the next exercise.) This explains why the study of unitary representations is not different from the study of general representations.

Exercise D.6.7 This exercise uses the standard fact that a compact group has a Haar measure, that is a measure $d\mu$ which is invariant under translations. That is, if f is a continuous function on G, then for each $h \in G$ it holds that

$$\int f(g)d\mu(g) = \int f(gh)d\mu(g) = \int f(hg)d\mu(g).$$

Consider now an irreducible representation V of G on a Hilbert space $\mathcal{H}$ with norm $\|\cdot\|$. Defining the norm N by

$$N(x)^2 = \int \|V(g)(x)\|^2 d\mu(g),$$

prove that V is unitary when $\mathcal{H}$ is provided with the norm N. Prove (at least when $\mathcal{H}$ is finite-dimensional) that N is equivalent to the original norm and that it is Hilbertian.

Theorem D.6.8 *Two irreducible unitary representations of $SU(2)$ of the same dimension are unitarily equivalent.*

Proof This follows from Theorem D.6.4 and Corollary D.6.6. □

Let us also note the following.

Proposition D.6.9 *A representation π of $SU(2)$ in a space $\mathcal{H}$ of dimension $m+1$ is irreducible if and only if the operator $Z = 2i\pi'(X_3)$ has m as an eigenvalue.*

Proof It should be obvious that the condition is necessary. To prove that it is sufficient we recall with the notation of Proposition D.6.1 that we proved there that the largest eigenvalue m' of Z is an integer and that $\mathcal{G}$ is of dimension $m'+1$. Thus if m is an eigenvalue of Z then $m' \geq m$ and thus $\dim \mathcal{G} = m'+1 \geq m+1 = \dim \mathcal{H}$. Thus $\mathcal{G}$ coincides with $\mathcal{H}$, and irreducibility follows from the last assertion of this proposition. □

It is good to keep in mind for further use the following simple fact.

Lemma D.6.10 *For an irreducible representation π of $SU(2)$ of dimension $m+1$ the eigenvalues of Z are exactly the elements of the set*

$$A_m := \{m, m-2, \ldots, -m\}. \tag{D.53}$$

Exercise D.6.11 With the notation of the proof of Theorem D.6.4, find explicitly the operators Z, A^+, A^- in the case of the representation π_j of Definition 8.2.3. Find an eigenvector of Z with eigenvalue j.

Exercise D.6.12 Combine the previous exercise with the last argument of Proposition D.6.1 to obtain a self-contained proof that π_j is irreducible. Convince yourself that this proof is identical to the second proof we gave of this property.

D.7 Decomposition of Unitary Representations of $SU(2)$ into Irreducibles

A finite-dimensional unitary representation π of $SU(2)$ on a finite-dimensional space $\mathcal{H}$ decomposes in an orthogonal sum of irreducible representations. (Let us recall that this is an almost obvious consequence of the fact that the orthogonal complement of an invariant subspace is invariant.) How may one determine these? The answer turns out to be very simple, as explained in this section. To lighten terminology, a subspace $\mathcal{G}$ of $\mathcal{H}$ which is invariant under π and such that the restriction of π to $\mathcal{G}$ is irreducible will be called "an irreducible". Let us consider a decomposition of $\mathcal{H}$ as an orthogonal sum $\mathcal{H} = \bigoplus_{1\le k\le N} \mathcal{H}_k$ of irreducibles. According to Theorem D.6.8 the restriction of π to $\mathcal{H}_k$ is equivalent to the representation π_{j_k} of Definition 8.2.3 for a certain $j_k \ge 0$ (and $\dim \mathcal{H}_k = j_k + 1$). We want to determine the sequence $(j_k)_{1\le k\le N}$, which without loss of generality we may assume to be non-increasing (i.e. $j_{k+1} \le j_k$).

Definition D.7.1 For $\ell \in \mathbb{Z}$ we define m_ℓ as the dimension of the eigenspace $\mathcal{G}_\ell$ of $Z = 2i\pi'(X_3)$ with eigenvalue ℓ.[9]

Lemma D.7.2 *The sequence $(j_k)_{1\le k\le N}$ determines the sequence $(m_\ell)_{\ell\in\mathbb{Z}}$ and conversely, the sequence $(m_\ell)_{\ell\in\mathbb{Z}}$ determines N and the sequence $(j_k)_{1\le k\le N}$. Moreover $m_\ell = m_{-\ell}$ so that the sequence $(m_\ell)_{\ell\ge 0}$ determines N and the sequence $(j_k)_{1\le k\le N}$.*

Proof According to Lemma D.6.10, in $\mathcal{H}_k$ the operator Z has each element of the set A_{j_k} as an eigenvalue. As k varies, the corresponding eigenvectors are orthogonal. We then get one dimension for $\mathcal{G}_\ell$ for each value of k such that $\ell \in A_{j_k}$. Thus

$$m_\ell \ge \operatorname{card}\{1 \le k \le N \,;\, \ell \in A_{j_k}\}.$$

On the other hand,

$$\sum_{\ell\in\mathbb{Z}} m_\ell = \sum_{\ell\in\mathbb{Z}} \dim \mathcal{G}_\ell \le \dim \mathcal{H}$$

whereas

$$\begin{aligned}\sum_{\ell\in\mathbb{Z}} \operatorname{card}\{1 \le k \le N \,;\, \ell \in A_{j_k}\} &= \operatorname{card}\{(\ell,k) \,;\, \ell \in \mathbb{Z},\ 1 \le k \le N,\ \ell \in A_{j_k}\} \\ &= \sum_{1\le k\le N} \operatorname{card} A_{j_k} = \sum_{1\le k\le N} (j_k + 1) \\ &= \sum_{1\le k\le N} \dim \mathcal{H}_k = \dim \mathcal{H} \qquad \text{(D.54)}\end{aligned}$$

and consequently for each $\ell \in \mathbb{Z}$,

$$m_\ell = \operatorname{card}\{1 \le k \le N \,;\, \ell \in A_{j_k}\}. \qquad \text{(D.55)}$$

This proves that the sequence $(j_k)_{1\le k\le N}$ determines the sequence $(m_\ell)_{\ell\in\mathbb{Z}}$. Since by the definition of A_j we have $\ell \in A_j$ if and only if $-\ell \in A_j$ this also proves that $m_\ell = m_{-\ell}$.

So we have proved that the sequence $(j_k)_{k\le N}$ determines the sequence $(m_\ell)_{\ell\in\mathbb{Z}}$ by the formula (D.55). We prove that conversely the sequence $(m_\ell)_{\ell\in\mathbb{Z}}$ given by (D.55) determines the sequence $(j_k)_{1\le k\le N}$. For this consider another sequence $(j'_k)_{1\le k\le N'}$ and the sequence $(m'_\ell)_{\ell\in\mathbb{Z}}$ given by

$$m'_\ell = \operatorname{card}\{1 \le k \le N \,;\, \ell \in A_{j'_k}\}.$$

Assuming that the sequences $(j_k)_{1\le j\le N}$ and $(j'_k)_{1\le j\le N'}$ are different we will prove that the sequences $(m_\ell)_{\ell\in\mathbb{Z}}$ and $(m'_\ell)_{\ell\in\mathbb{Z}}$ are different. Assume first that there exists $k \le \min(N, N')$ for

[9] And $m_\ell = 0$ if ℓ is not an eigenvalue.

which $j_k \neq j'_k$, and consider the smallest such k. Assume without loss of generality that $j'_k < j_k$. Then $j_k \notin A_{j'_s}$ for $s \geq k$ because $j_k > j'_k \geq j'_s$. Consequently

$$m'_{j_k} = \text{card}\,\{1 \leq s < k\,;\, j_k \in A_{j'_s}\}.$$

On the other hand, $j'_s = j_s$ for $s < k$, and $j_k \in A_{j_k}$, so that $m_{j_k} \geq m'_{j_k} + 1$. Thus the sequences (m_j) and (m'_j) are different. The case where $j_k = j'_k$ for $k \leq \min(N, N')$ and $N \neq N'$ is similar. □

As the sequence (m_j) is defined independently of any decomposition of $\mathcal{H}$ as a sum of irreducibles we obtain the following.

Corollary D.7.3 *The sequence $(j_k)_{1 \leq k \leq N}$ does not depend on the decomposition of $\mathcal{H}$ as an orthogonal sum of irreducibles.*

This does *not* say that the decomposition $\mathcal{H} = \bigoplus \mathcal{H}_k$ is unique, a question we investigate later in Exercises D.7.5 and D.7.9. The trouble comes from the different values of k which have the same value of j_k. On the other hand, when all the terms of the sequence (j_k) are different, the decomposition of $\mathcal{H}$ as an orthogonal sum of irreducibles is *unique* and when this is the case we will talk of "the decomposition of $\mathcal{H}$ into irreducibles".

Corollary D.7.4 *If Z has an eigenvalue $\geq \ell \geq 0$ then $\mathcal{H}$ contains an irreducible of dimension $\geq \ell + 1$.*

Proof Consider an eigenvalue $\ell' \geq \ell$ of Z. Then $m_{\ell'} \geq 1$, and by (D.55) we can find k for which $\ell' \in A_{j_k}$, so that $j_k \geq \ell' \geq \ell$, and $\dim \mathcal{H}_k = j_k + 1 \geq \ell + 1$. □

Exercise D.7.5 Consider two integers $n, n' \geq 0$ and assume that $j_k = n + n' - 2k$ for $0 \leq k \leq N = \min(n, n')$, so that the sequence $(j_k)_{1 \leq k \leq N}$ is the sequence

$$n + n', n + n' - 2, \ldots, |n - n'|.$$

Consider the sequence (m_ℓ) given by (D.55). Prove that $m_\ell = 0$ unless $\ell = n + n' - 2r$ for some $r \in \mathbb{N}$. Prove that moreover when $\ell \geq 0$ is such that $\ell = n + n' - 2r$ it holds that

$$m_\ell = 1 + \min(r, n, n') = \text{card}\,\{(k, k')\,;\, 0 \leq k \leq n\,;\, 0 \leq k' \leq n'\,;\, k + k' = r\}. \tag{D.56}$$

Let us define the representation $\pi = \pi_n \otimes \pi_{n'}$ by

$$\pi_n \otimes \pi_{n'}(C)(x \otimes x') = \pi_n(C)(x) \otimes \pi_{n'}(C)(x') \tag{D.57}$$

for $x \in \mathcal{H}_n$, $x' \in \mathcal{H}_{n'}$. As a sample application we prove the following standard result of Quantum Mechanics (which is known there under the name of "addition of angular momenta").

Proposition D.7.6 *For the representation $\pi_n \otimes \pi_{n'}$ the sequence (j_k) is given by $n + n', n + n' - 2, \ldots, |n - n'|$.*

Proof Copying the proof of the Leibniz's rule, the definition of π' and (D.57) imply

$$\pi'(X_3)(x \otimes x') = \pi'_n(X_3)(x) \otimes x' + x \otimes \pi'_{n'}(X_3)(x'). \tag{D.58}$$

Now there is a basis $e_0, \ldots, e_n$ of $\mathcal{H}_n$ with $2i\pi'_n(X_3)(e_k) = (n - 2k)e_k$ for $0 \leq k \leq n$, and a basis $e'_0, \ldots, e'_{n'}$ of $\mathcal{H}_{n'}$ with $2i\pi'_{n'}(X_3)(e'_{k'}) = (n' - 2k')e'_{k'}$ for $0 \leq k' \leq n'$. Therefore from (D.58), letting $Z = 2i\pi'(X_3)$, we have

$$Z(e_k \otimes e'_{k'}) = (n + n' - 2k - 2k')(e_k \otimes e'_{k'}).$$

In this manner we obtain $(n + 1)(n' + 1) = \dim \mathcal{H}_n \otimes \mathcal{H}_{n'}$ linearly independent eigenvectors of Z, so we have found a basis of eigenvectors. Thus for $r \in \mathbb{N}$ we know that the dimension m_ℓ of the eigenspace of Z with eigenvalue $\ell = n + n' - 2r$ is exactly the number of possible choices for

$0 \le k \le n$ and $0 \le k' \le n'$ with $k + k' = r$. Now Lemma D.7.2 shows that the sequence (m_ℓ) and the sequence (j_k) determine each other, and in Exercise D.7.5 we computed the sequence (j_k) corresponding to the particular sequence (m_ℓ) here. □

Proposition D.7.7 *Consider a finite-dimensional unitary representation π of $SU(2)$. Then:*
(a) For any irreducible $\mathcal{G}$ and any eigenvector e of $Z = 2i\pi'(X_3)$ with eigenvalue $> \dim \mathcal{G} - 1$, e is orthogonal to $\mathcal{G}$.
(b) Two irreducibles $\mathcal{G}, \mathcal{G}'$ of different dimensions are orthogonal.

Proof We first show that, as a consequence of the representation being unitary, Z is self-adjoint. For this we copy the argument with which we determined $\mathfrak{su}(2)$. For each t, $\pi(\exp tX_3) = \exp t\pi'(X_3)$ is unitary, so that $\exp t\pi'(X_3)^\dagger \exp t\pi'(X_3) = 1$ and taking the derivative at $t = 0$ shows that $\pi'(X_3)$ is skew-adjoint so that Z is self-adjoint. Consequently, eigenvectors of Z with different eigenvalues are orthogonal. The rest of the argument relies again on the analysis of Proposition D.6.1. Let $m = \dim \mathcal{G} - 1$. Then $\mathcal{G}$ is spanned by $e_0, e_1, \ldots, e_m$ which are eigenvectors of Z with eigenvalues $\le m$. Thus if e is an eigenvector of eigenvalue $> m$ it is orthogonal to $e_0, e_1, \ldots$ and hence to $\mathcal{G}$, and this proves (a). To prove (b) assume without loss of generality that $m' = \dim \mathcal{G}' - 1 > m$. Then we know that $\mathcal{G}'$ contains an eigenvector e'_0 of $Z = 2i\pi'(X_3)$ of eigenvalue m', so that e'_0 is orthogonal to $\mathcal{G}$ by (a). Thus e'_0 is contained in the orthogonal complement $\mathcal{G}^\perp$ of $\mathcal{G}$. Since $\mathcal{G}^\perp$ is invariant under π and $\mathcal{G}'$ is irreducible then $\mathcal{G}' \subset \mathcal{G}^\perp$. □

Exercise D.7.8 Prove that the conclusion of Proposition D.7.7 does not hold if $\dim \mathcal{G} = \dim \mathcal{G}'$. Hint: Consider an irreducible representation π of $SU(2)$ on $\mathcal{H}$ and the representation Π on $\mathcal{H} \oplus \mathcal{H}$ given by $\Pi(X)(x, y) = (\pi(X)(x), \pi(X)(y))$. For $\alpha \in \mathbb{C}$ consider the subspace $\mathcal{H}_\alpha$ of $\mathcal{H} \oplus \mathcal{H}$ given by $\{(x, y)\,;\, x = \alpha y\}$. Show that $\mathcal{H}_\alpha$ is an irreducible.

Exercise D.7.9 Prove that in the decomposition $\mathcal{H} = \bigoplus \mathcal{H}_k$ in an orthogonal sum of irreducibles, the space

$$\mathcal{G}_n = \bigoplus_{\dim \mathcal{H}_k = n} \mathcal{H}_k$$

is independent of the decomposition. Hint: Use Proposition D.7.7.

D.8 Spherical Harmonics

The material of the present section is not used at all in the main text, and is provided simply because it is standard in many physics textbooks and requires very little extra effort at this stage. We have already met the fundamental representation of $SO(3)$ in $L^2(\mathbb{R}^3, d^3\boldsymbol{x})$ given by

$$U(R)(f)(\boldsymbol{x}) = f(R^{-1}(\boldsymbol{x})). \tag{D.59}$$

This representation is very far from being irreducible because if $r_1 < r_2$ the space of functions f which are zero unless $r_1 \le \|\boldsymbol{x}\| \le r_2$ is invariant. To avoid this phenomenon, we consider the sphere $\mathbb{S}^2$ of $\mathbb{R}^3$ provided with its canonical probability measure $d\mu$, which is invariant under rotations. We can then define a unitary representation V of $SO(3)$ in $L^2(\mathbb{S}^2, d\mu)$ by

$$V(R)(f)(y) = f(R^{-1}(y)) \tag{D.60}$$

for $y \in \mathbb{S}^2$. We will prove that this representation decomposes as an orthogonal sum of finite-dimensional irreducibles which we will determine. (Generally speaking, we know that a *finite-dimensional* unitary representation decomposes as a sum of irreducibles. Furthermore, as proved in Section A.6, every unitary representation of a *compact* group decomposes as an orthogonal sum of finite-dimensional irreducibles.)

We first note that the space $\mathcal{H}_j$ of restrictions of homogeneous polynomials of degree j to $\mathbb{S}^2$ is invariant. The dimension of this space is the number of triples a,b,c of elements of $\mathbb{N}$ with sum equal to j. There are $(j-a+1)$ such triples where a is fixed, so the total number of triples is $(j+1)+j+\cdots+1=(j+1)(j+2)/2$. Next the map which restricts a homogeneous polynomial of degree j to $\mathbb{S}^2$ is injective, because obviously this restriction determines the polynomial (using that the polynomial is homogeneous of degree j). Consequently

$$\dim \mathcal{H}_j = \frac{(j+1)(j+2)}{2}. \tag{D.61}$$

When $f \in \mathcal{H}_{j-2}$ is the restriction to $\mathbb{S}^2$ of the homogeneous polynomial Q of degree $j-2$, it is also the restriction of the homogeneous polynomial $(x_1^2+x_2^2+x_3^2)Q$ of degree j. Consequently $\mathcal{H}_j \supset \mathcal{H}_{j-2} \supset \mathcal{H}_{j-4} \supset \ldots$. The restriction of V to $\mathcal{H}_j$ is not irreducible because its subspace $\mathcal{H}_{j-2}$ is invariant. Let us denote by $\mathcal{G}_j$ the orthogonal complement of $\mathcal{H}_{j-2}$ in $\mathcal{H}_j$. Then

$$\dim \mathcal{G}_j = \dim \mathcal{H}_j - \dim \mathcal{H}_{j-2} = \frac{(j+1)(j+2)}{2} - \frac{(j-1)j}{2} = 2j+1, \tag{D.62}$$

and $\mathcal{G}_j$ is invariant because both $\mathcal{H}_j$ and $\mathcal{H}_{j-2}$ are invariant.

Theorem D.8.1 *The restriction of V to $\mathcal{G}_j$ is irreducible. The restriction f_j to $\mathbb{S}^2$ of the polynomial $P_j=(x_1+\mathrm{i}x_2)^j$ belongs to $\mathcal{G}_j$ and is an eigenvector of eigenvalue $2j$ of $Z = 2\mathrm{i}V'(J_3)$ (where J_3 is given in (D.9)). The spaces $\mathcal{G}_j$ are orthogonal to each other and they span $L^2(\mathbb{S}^2,\mathrm{d}\mu)$.*

As a preparation for the proof, let us compute Z in the most elementary way. If $R(\theta)=\exp(\theta J_3)$ is the rotation of angle θ around the third axis, then $R(\theta)$ sends the point (x_1,x_2,x_3) to the point $(x_1\cos\theta - x_2\sin\theta, x_1\sin\theta + x_2\cos\theta, x_3)$, and thus for a differentiable function P on $\mathbb{R}^3$,

$$U(R(\theta))(P)(\boldsymbol{x}) = P(R(\theta)^{-1}(\boldsymbol{x})) = P(x_1\cos\theta + x_2\sin\theta,\ -x_1\sin\theta + x_2\cos\theta, x_3),$$

so that

$$U'(J_3)(P)(\boldsymbol{x}) := \frac{\mathrm{d}U(R(\theta))(P)(\boldsymbol{x})}{\mathrm{d}\theta}\Big|_{\theta=0} = x_2\frac{\partial P}{\partial x_1}(\boldsymbol{x}) - x_1\frac{\partial P}{\partial x_2}(\boldsymbol{x}). \tag{D.63}$$

For a homogeneous polynomial P on $\mathbb{R}^3$ let us denote by TP its restriction to $\mathbb{S}^2$. It should be obvious from the previous formula that

$$V'(J_3)(TP) = T\Big(x_2\frac{\partial P}{\partial x_1} - x_1\frac{\partial P}{\partial x_2}\Big), \tag{D.64}$$

from which it is obvious that $f_j = TP_j$ is an eigenvector of $Z = 2\mathrm{i}V'(J_3)$ of eigenvalue $2j$.

Another observation is that results such as Proposition D.7.7 remain true for a unitary representation π of $SO(3)$ rather than $SU(2)$ (and when then $Z = 2\pi'(X_3)$ is replaced by $Z = 2\pi'(J_3)$). This is seen by applying Proposition D.7.7 (etc.) to the representation $A \mapsto \pi(\tilde{\kappa}(A))$ where $\tilde{\kappa}$ is the two-to-one homomorphism $SU(2) \to SO(3)$ constructed in Section D.4.

Proof of Theorem D.8.1 We prove by induction over j that the irreducibles of $\mathcal{H}_j$ are $\mathcal{G}_j,\mathcal{G}_{j-2},\ldots$ where the last term is $\mathcal{G}_1$ when j is odd and $\mathcal{G}_0$ when j is even. This is obvious for $j=0$ and $j=1$. Since $\mathcal{H}_j$ contains f_j, which is an eigenvector of Z with eigenvalue $2j$, by Corollary D.7.4 (in the version described before the start of this proof) it contains an irreducible $\mathcal{G}'_j$ of dimension $\geq 2j+1$, and this irreducible is orthogonal to $\mathcal{H}_{j-2}$ by the induction hypothesis and Proposition D.7.7, (b). Thus $\mathcal{G}'_j \subset \mathcal{G}_j$ and then $\mathcal{G}'_j = \mathcal{G}_j$ because $\dim \mathcal{G}_j = 2j+1$. This completes the proof that $\mathcal{H}_j$ decomposes into the irreducibles $\mathcal{G}_j,\mathcal{G}_{j-2},\ldots$. Next, it follows from Proposition D.7.7 (a) that f_j is orthogonal to $\mathcal{G}_{j-2},\mathcal{G}_{j-4},\ldots$ and therefore $f_j \in \mathcal{G}_j$.

It remains only to prove that the spaces $\mathcal{G}_j$ span $L^2(\mathbb{S}^2, d\mu)$, or equivalently that the union of the spaces $\mathcal{H}_j$ is dense in this space. This is really easy, since a continuous function on $\mathbb{S}^2$ can be uniformly approximated by a polynomial. □

The next result provides another standard characterization of $\mathcal{G}_j$, see e.g. [40, Theorem 17.12]. We recall that on $\mathbb{R}^3$ the Laplacian Δ is the operator $\sum_{j\le 3} \partial^2/\partial x_j^2$, and that a function is *harmonic* if its Laplacian is zero. A fundamental property of the Laplacian is that it commutes with the operators $U(R)$,

$$\Delta U(R) f = U(R) \Delta f. \tag{D.65}$$

The proof is straightforward using the chain rule.

Exercise D.8.2 Write down the proof.

Proposition D.8.3 *$\mathcal{G}_j$ coincides with the space $\widehat{\mathcal{G}}_j$ of restrictions to $\mathbb{S}^2$ of harmonic homogeneous polynomials of degree j.*

Proof According to (D.65) the space $\widehat{\mathcal{G}}_j$ is invariant under the representation V. It is obvious that $f_j \in \widehat{\mathcal{G}}_j$ so that $\mathcal{G}_j \subset \widehat{\mathcal{G}}_j$. To prove that $\mathcal{G}_j = \widehat{\mathcal{G}}_j$ it then suffices to prove that they have the same dimension $2j+1$. Consider the space $\mathcal{F}_j$ of homogeneous polynomials of degree j. Obviously $\Delta(\mathcal{F}_j) \subset \mathcal{F}_{j-2}$. We will prove that $\Delta(\mathcal{F}_j) = \mathcal{F}_{j-2}$ so that $\dim \widehat{\mathcal{G}}_j = \dim \ker \Delta = \dim \mathcal{F}_j - \dim \Delta(\mathcal{F}_j) = \dim \mathcal{F}_j - \dim \mathcal{F}_{j-2} = 2j+1$. For this we note that

$$\Delta(x_1^{n_1} x_2^{n_2} x_3^{n_3}) = n_1(n_1-1)x_1^{n_1-2} x_2^{n_2} x_3^{n_3} + n_2(n_2-1)x_1^{n_1} x_2^{n_2-2} x_3^{n_3} + n_3(n_3-1)x_1^{n_1} x_2^{n_2} x_3^{n_3-2}.$$

Taking $n_1 = n_2 = 0, n_3 = j$ proves that $x_3^{j-2} \in \Delta(\mathcal{F}_j)$, and then by induction on $q \le j-2$ that $x_1^{n_1} x_2^{n_2} x_3^{j-q-2} \in \Delta(\mathcal{F}_j)$ for $n_1 + n_2 = q$. □

According to Proposition D.6.1 a basis of $\mathcal{G}_j$ is obtained by the functions $(A^-)^k f_j$ for $0 \le k \le 2j$, where $A^- = \mathrm{i}V'(J_1) + V'(J_2)$. These functions go under the name of *spherical harmonics* and are useful in physics.

D.9 $\mathfrak{so}(1,3) = \mathfrak{sl}_{\mathbb{C}}(2)$!

Let us recall the matrix Ξ of (4.10), and that a matrix A belongs to the Lorentz group $O(1,3)$ if and only if $A^T \Xi A = \Xi$. Denoting by $\mathfrak{so}(1,3)$ the Lie algebra of the Lorentz group, and computing the differential at zero of the map $t \mapsto (\exp tX)^T \Xi \exp tX$ we obtain that if $X \in \mathfrak{so}(1,3)$ then

$$X^T \Xi + \Xi X = 0. \tag{D.66}$$

Conversely, if $\Xi^{-1} X^T \Xi = -X$ then for each t, $\Xi^{-1}(\exp tX)^T \Xi = \exp(-tX)$ so that $\exp tX \in O(1,3)$ and $X \in \mathfrak{so}(1,3)$. The reason for the notation $\mathfrak{so}(1,3)$ is that this is also the Lie algebra of the group $SO^{\uparrow}(1,3)$. Indeed for $t \in \mathbb{R}$ and $X \in \mathfrak{so}(1,3)$ we have $\exp tX \in SO^{\uparrow}(1,3)$ by continuity of the map $t \mapsto \exp tX$.

Since the matrix Ξ is diagonal the condition (D.66) is far simpler than it looks. The diagonal elements of X must be 0, and the elements above the diagonal determine those below the diagonal so that $\mathfrak{so}(1,3)$ is of dimension 6. The traditional choice of a convenient basis is

$$J_1 = \begin{pmatrix} 0 & 0 & 0 & 0 \\ 0 & 0 & 0 & 0 \\ 0 & 0 & 0 & -1 \\ 0 & 0 & 1 & 0 \end{pmatrix}; \; J_2 = \begin{pmatrix} 0 & 0 & 0 & 0 \\ 0 & 0 & 0 & 1 \\ 0 & 0 & 0 & 0 \\ 0 & -1 & 0 & 0 \end{pmatrix}; \; J_3 = \begin{pmatrix} 0 & 0 & 0 & 0 \\ 0 & 0 & -1 & 0 \\ 0 & 1 & 0 & 0 \\ 0 & 0 & 0 & 0 \end{pmatrix} \tag{D.67}$$

and

$$K_1 = \begin{pmatrix} 0 & 1 & 0 & 0 \\ 1 & 0 & 0 & 0 \\ 0 & 0 & 0 & 0 \\ 0 & 0 & 0 & 0 \end{pmatrix} ; K_2 = \begin{pmatrix} 0 & 0 & 1 & 0 \\ 0 & 0 & 0 & 0 \\ 1 & 0 & 0 & 0 \\ 0 & 0 & 0 & 0 \end{pmatrix} ; K_3 = \begin{pmatrix} 0 & 0 & 0 & 1 \\ 0 & 0 & 0 & 0 \\ 0 & 0 & 0 & 0 \\ 1 & 0 & 0 & 0 \end{pmatrix}. \tag{D.68}$$

One then has to compute the commutation relations between these objects. Get ready, this requires patience. The good news is that the computation is done once and for all:

$$[J_1, J_2] = J_3 \ ; \ [J_2, J_3] = J_1 \ ; \ [J_3, J_1] = J_2 \tag{D.69}$$

$$[K_1, K_2] = -J_3 \ ; \ [K_2, K_3] = -J_1 \ ; \ [K_3, K_1] = -J_2 \tag{D.70}$$

$$[K_1, J_1] = 0 \ ; \ [K_1, J_2] = K_3 \ ; \ [K_1, J_3] = -K_2 \tag{D.71}$$

$$[K_2, J_1] = -K_3 \ ; \ [K_2, J_2] = 0 \ ; \ [K_2, J_3] = K_1 \tag{D.72}$$

$$[K_3, J_1] = K_2 \ ; \ [K_3, J_2] = -K_1 \ ; \ [K_3, J_3] = 0. \tag{D.73}$$

To write these in a compact, intelligible form, for $\boldsymbol{u}, \boldsymbol{v} \in \mathbb{R}^3$ we introduce the notation

$$\boldsymbol{u} \cdot \mathbf{J} := \sum_{j \le 3} u_j J_j \ ; \ \boldsymbol{v} \cdot \mathbf{K} := \sum_{j \le 3} v_j K_j.$$

Then (D.69) and (D.70) become respectively[10]

$$[\boldsymbol{u} \cdot \mathbf{J}, \boldsymbol{u}' \cdot \mathbf{J}] = (\boldsymbol{u} \wedge \boldsymbol{u}') \cdot \mathbf{J} \ ; \ [\boldsymbol{v} \cdot \mathbf{K}, \boldsymbol{v}' \cdot \mathbf{K}] = -(\boldsymbol{v} \wedge \boldsymbol{v}') \cdot \mathbf{J}, \tag{D.74}$$

while the other three relations become

$$[\boldsymbol{u} \cdot \mathbf{J}, \boldsymbol{v} \cdot \mathbf{K}] = (\boldsymbol{u} \wedge \boldsymbol{v}) \cdot \mathbf{K}. \tag{D.75}$$

While studying the Lorentz group we have introduced "rotations" and "pure boosts". After our study of $SO(3)$ it should be obvious that "rotations" are exactly the elements of the form $\exp \boldsymbol{u} \cdot \mathbf{J}$, and we turn to the investigation of pure boosts.

We first prove a remarkable consequence of (D.75). To state it we note that $A = \exp \boldsymbol{u} \cdot \mathbf{J}$ is a transformation of $\mathbb{R}^4$ which leaves e_0 and the span of e_1, e_2, e_3 invariant. It thus induces a transformation $\hat{A}$ of $\mathbb{R}^3$.

Lemma D.9.1 *If* $A = \exp \boldsymbol{u} \cdot \mathbf{J}$ *then*

$$A(\boldsymbol{v} \cdot \mathbf{K})A^{-1} = \hat{A}(\boldsymbol{v}) \cdot \mathbf{K}. \tag{D.76}$$

In physics, this property is described by the sentence "**K** transforms like a vector under rotations".

Proof The simplest-minded proof is for $A(t) = \exp(t\boldsymbol{u} \cdot \mathbf{J})$ to consider the function

$$t \mapsto A(t)^{-1}\big(\hat{A}(t)(\boldsymbol{v}) \cdot \mathbf{K}\big)A(t)$$

and to show that its derivative is zero. This derivative is

$$A(t)^{-1}\Big(-[\boldsymbol{u} \cdot \mathbf{J}, \hat{A}(t)(\boldsymbol{v}) \cdot \mathbf{K}] + (\hat{A}'(t)(\boldsymbol{v})) \cdot \mathbf{K}\Big)A(t),$$

whereas, using (D.75) in the first equality and (D.12) in the second equality,

$$[\boldsymbol{u} \cdot \mathbf{J}, \hat{A}(t)(\boldsymbol{v}) \cdot \mathbf{K}] = (\boldsymbol{u} \wedge \hat{A}(t)(\boldsymbol{v})) \cdot \mathbf{K} = (\hat{A}'(t)(\boldsymbol{v})) \cdot \mathbf{K}. \qquad \square$$

[10] The second relation below, showing that rotations may arise from boosts has a remarkable consequence in relativity called Thomas precession.

Proposition D.9.2 *Given a unit vector $\boldsymbol{v} \in \mathbb{R}^3$ and $s \in \mathbb{R}^+$, we have*

$$\exp(s\boldsymbol{v} \cdot \mathbf{K}) = B_p, \tag{D.77}$$

where B_p is the pure boost sending $(1,0,0,0)$ *to the point* $p = (\cosh s, \sinh s\boldsymbol{v}) \in \mathbb{R}^4$.

Proof Combining (D.4) and (D.76) we obtain that if $A = \exp(\boldsymbol{u} \cdot \mathbf{J})$ then

$$A \exp(\boldsymbol{p} \cdot \mathbf{K}) A^{-1} = \exp(\hat{A}(\boldsymbol{p}) \cdot \mathbf{K}).$$

Observing that $A(p^0, \boldsymbol{p}) = (p^0, \hat{A}(\boldsymbol{p}))$ and comparing to (4.23) it suffices to prove the result when $\boldsymbol{v}$ is parallel to the third axis. The proof is then straightforward by explicit computation. □

Let us then turn to the study of the Lie algebra $\mathfrak{sl}_{\mathbb{C}}(2)$ of $SL(2,\mathbb{C})$. It is obvious that this is the set of traceless 2×2 complex matrices. As a real vector space, it has dimension 6. For a nice basis, besides the matrices X_j of (D.26) we use the matrices

$$Y_j = \frac{\sigma_j}{2} = \mathrm{i}X_j. \tag{D.78}$$

With obvious notation we may write the commutation relations as

$$[\boldsymbol{u} \cdot \mathbf{X}, \boldsymbol{u}' \cdot \mathbf{X}] = (\boldsymbol{u} \wedge \boldsymbol{u}') \cdot \mathbf{X} \; ; \; [\boldsymbol{v} \cdot \mathbf{Y}, \boldsymbol{v}' \cdot \mathbf{Y}] = -(\boldsymbol{v} \wedge \boldsymbol{v}') \cdot \mathbf{X} \; ; \; [\boldsymbol{u} \cdot \mathbf{X}, \boldsymbol{v} \cdot \mathbf{Y}] = (\boldsymbol{u} \wedge \boldsymbol{v}) \cdot \mathbf{Y}. \tag{D.79}$$

The first relation is the form (D.13) of (D.25) and the other two relations follow from this one and the fact that $\mathbf{Y} = \mathrm{i}\mathbf{X}$. These are the same relations as in the case of $\mathfrak{so}(1,3)$: Sending X_j to J_j and Y_j to K_j witnesses the isomorphism of these algebras.

Exercise D.9.3 Prove that the homomorphism κ of (8.21) from $SL(2,\mathbb{C})$ to $SO^{\uparrow}(1,3)$ satisfies

$$\kappa'(X_j) = J_j \; ; \; \kappa'(Y_j) = K_j.$$

Exercise D.9.4 Prove that (with obvious notation) the matrix $\exp(\boldsymbol{v} \cdot \mathbf{Y})$ is *positive* Hermitian. Hint: Use the result of Lemma D.9.1 to reduce to the case $v_1 = v_2 = 0$.

D.10 Irreducible Representations of $SL(2,\mathbb{C})$

In this section we prove that all irreducible representations of $SL(2,\mathbb{C})$ are equivalent to a representation of the type (j,k), as introduced in Definition 8.3.2. As we use only these representations for $0 \le j,k \le 1$, the material of this section is presented just for fun. The point is again that much of the work has already been done.

To study irreducible representations π of $\mathfrak{su}(2)$ we made the clever combinations (D.50) of the operators $\pi'(X_i) \in \mathfrak{gl}_{\mathbb{C}}(m)$ satisfying simpler commutation relations, which could then be analyzed through Proposition D.6.1. Our strategy here will be exactly the same. Given an irreducible representation π of $\mathfrak{sl}_{\mathbb{C}}(2)$ in $\mathbb{C}^m$ we consider the operators $L_j := \pi'(X_j)$ and $M_j = \pi'(Y_j)$ which satisfy commutation relations as in (D.69) to (D.73). We set

$$\mathbf{A} = \frac{1}{2}(\mathbf{L} - \mathrm{i}\mathbf{M}) \; ; \; \mathbf{B} = \frac{1}{2}(\mathbf{L} + \mathrm{i}\mathbf{M}), \tag{D.80}$$

and we observe the relations

$$[\boldsymbol{u} \cdot \mathbf{A}, \boldsymbol{v} \cdot \mathbf{A}] = (\boldsymbol{u} \wedge \boldsymbol{v}) \cdot \mathbf{A} \; ; \; [\boldsymbol{u} \cdot \mathbf{B}, \boldsymbol{v} \cdot \mathbf{B}] = (\boldsymbol{u} \wedge \boldsymbol{v}) \cdot \mathbf{B} \; ; \; [\boldsymbol{u} \cdot \mathbf{A}, \boldsymbol{v} \cdot \mathbf{B}] = 0.$$

Thus, we have found two sets of operators, each with the same commutation structure as in the case of $\mathfrak{su}(2)$, and which commute with each other. The joint analysis of two sets of operators commuting with each other just cannot be much more difficult than the analysis of the two sets separately (which we already performed). Before doing this, however, let us build our intuition by investigating the simple situation of the representation $\pi = (1,0)$, which, now that we identify operators with the corresponding matrices, is simply the identity from $SL(2,\mathbb{C})$ to itself. Thus

$$L_j = X_j = -\frac{i\sigma_j}{2} \; ; \; M_j = Y_j = \frac{\sigma_j}{2} = iX_j$$

and in that case we have $B_j = 0$ and $A_j = X_j$.

Physicists "define" the action of a Lorentz transformation $\exp(\boldsymbol{u} \cdot \mathbf{J} + \boldsymbol{v} \cdot \mathbf{K})$ on a column vector $C \in \mathbb{C}^2$ by the formula

$$\exp(\boldsymbol{u} \cdot \mathbf{J} + \boldsymbol{v} \cdot \mathbf{K})C := \exp((\boldsymbol{u} + i\boldsymbol{v}) \cdot \mathbf{X})C. \tag{D.81}$$

As in the case of $SO(3)$, this is really a "double-valued representation".

In the case of the representation $(0,1)$, which is the map $A \to A^{\dagger -1}$, then $\pi'(X) = -X^\dagger$. Thus

$$L_j = X_j = -\frac{i\sigma_j}{2} \; ; \; M_j = -Y_j = -\frac{\sigma_j}{2} = -iX_j$$

and in that case we have $A_j = 0$. The formula (D.81) is modified in the obvious manner $\exp(\boldsymbol{u} \cdot \mathbf{J} + \boldsymbol{v} \cdot \mathbf{K})C = \exp((\boldsymbol{u} - i\boldsymbol{v}) \cdot \mathbf{X})C$.

Let us now move to the key point. Looking back at (D.50), we set

$$Z = 2iA_3, A^\pm = iA_1 \mp A_2 \; ; \; W = 2iB_3, C^\pm = iB_1 \mp B_2.$$

We turn to the analysis of these relations following the pattern of Proposition D.6.1.

Proposition D.10.1 *Consider a finite-dimensional complex vector space $\mathcal{H}$. Consider six operators Z, A^+, A^- and W, C^+, C^- on $\mathcal{H}$. Assume that each of the first three operators commute with each of the last three. Moreover assume that (D.44) and (D.45) hold, as well as the corresponding relations for the last three operators:*

$$[W, C^+] = 2C^+ \; ; \; [W, C^-] = -2C^- \tag{D.82}$$

$$[C^+, C^-] = W. \tag{D.83}$$

Assume moreover that there is no proper subspace of $\mathcal{H}$ which is invariant by $A^\pm$, Z, $C^\pm$, W. Then we can find integers m, n and a basis $e_{k,\ell}$, $0 \le k \le m, 0 \le \ell \le n$ of $\mathcal{H}$ with the following properties:

$$Z(e_{k,\ell}) = (m - 2k)e_{k,\ell}, \tag{D.84}$$

$$W(e_{k,\ell}) = (n - 2\ell)e_{k,\ell}, \tag{D.85}$$

$$A^-(e_{m,\ell}) = 0 \; ; \; k < m \Rightarrow A^-(e_{k,\ell}) = e_{k+1,\ell}, \tag{D.86}$$

$$C^-(e_{k,n}) = 0 \; ; \; \ell < n \Rightarrow C^-(e_{k,\ell}) = e_{k,\ell+1}, \tag{D.87}$$

$$A^+(e_{0,\ell}) = 0 \; ; \; k > 0 \Rightarrow A^+(e_{k,\ell}) = k(m - k + 1)e_{k-1,\ell}, \tag{D.88}$$

$$C^+(e_{k,0}) = 0 \; ; \; \ell > 0 \Rightarrow C^+(e_{k,\ell}) = \ell(n - \ell + 1)e_{k,\ell-1}. \tag{D.89}$$

When $Z = A^+ = A^- = 0$ the preceding means that we simply construct a sequence $(e_\ell)_{\ell \le m}$ which satisfies (D.85), (D.87) and (D.89).

Proof The operators Z and W commute, so there exist vectors which are eigenvectors for both these operators. We select such a vector $e_{0,0}$, for which the sum of the real parts of the corresponding eigenvalues λ and μ is as large as possible. We then set $e_{k,\ell} = (A^-)^k(C^-)^\ell(e_{0,0})$, and we compute $A^+(e_{k,\ell})$ by induction on k and $C^+(e_{k,\ell})$ by induction on ℓ, and using in an essential manner that Z, A^+, A^- commute with W, C^+, C^-. We then find that $\lambda = m$, where m is the largest integer for which $e_{m,0} \neq 0$, while $\mu = n$ where n is the largest integer with $e_{0,n} \neq 0$. The proof goes very much as in Proposition D.6.1 so the details are left to the reader. □

So it follows that the irreducible representations of $SL(2,\mathbb{C})$ are classified by a family depending on two integers m and n, and it remains to show that these are in fact the representations (m,n) of Definition 8.3.2, a task we leave to the reader in the next two exercises.

Exercise D.10.2 Consider a basis $(e_{k,\ell})_{0\leq k\leq m, 0\leq \ell\leq n}$ and operators Z, A^+, A^-, W, C^+ and C^- as in (D.84) to (D.89). Prove that there is no proper subspace that is invariant by all the six operators.

Exercise D.10.3 Since we have not proved yet that the representation (k,ℓ) is irreducible we cannot use Proposition D.10.1 to study it. Show directly that the structure described by this proposition exists in this representation by exhibiting it. Conclude using the previous exercise that this representation is irreducible. Prove then that every irreducible representation of $SL(2,\mathbb{C})$ is equivalent to a representation (m,n).

Exercise D.10.4 Consider an irreducible representation π of $SL(2,\mathbb{C})$. Prove that it is of the type $(m,0)$ if and only if $\pi'(X_i) = \mathrm{i}\pi'(Y_i)$ for $i \leq 3$, and of the type $(0,n)$ if and only if $\pi'(X_i) = -\mathrm{i}\pi'(Y_i)$ for $i \leq 3$.

D.11 QFT Is Not for the Meek

The purpose of this section is to explain how the Lie algebra $\mathfrak{so}(1,3)$ comes about in some physics textbooks, e.g. [77, page 32] or [24, page 173]. We have brought the notations closer to ours, but we reproduce their approach and try to make the link with mathematicians' view. You may skip this section if you are not interested in reading physics textbooks.[11] It requires some familiarity with the mechanism of lowering and raising indices described at the end of Section 4.1. We will write matrices in $\mathfrak{so}(1,3)$ the same way we write Lorentz matrices, using two indices that can be lowered or raised like any other Lorentz indices. The first step is to notice that the relation (D.66) means

$$X_{\mu,\nu} = -X_{\nu,\mu}. \tag{D.90}$$

To see this we observe that taking the product ΞX amounts to changing the sign of X^μ_ν when $\mu \geq 1$ i.e. to consider $X_{\mu,\nu}$.

Consequently X depends linearly on the six independent parameters $X_{\mu,\nu}$ for $0 \leq \mu < \nu \leq 3$, so that we can write

$$X = \sum_{0\leq\mu<\nu\leq 3} X_{\mu,\nu} J^{\mu,\nu}, \tag{D.91}$$

where $J^{\mu,\nu}$ is "the 4×4 matrix of the 'coefficients' of $X_{\mu,\nu}$". Defining now $J^{\mu,\nu} = -J^{\nu,\mu}$ for $\mu > \nu$ and using (D.90) one can rewrite (D.91) in the much nicer form

$$X = \frac{1}{2} X_{\mu,\nu} J^{\mu,\nu}, \tag{D.92}$$

[11] In all fairness it must be said that the sample presented here has not been chosen in an unbiased manner, but rather to illustrate again the near un-bridgeable gap of cultures.

where repeated indices are summed as usual. This is an implicit way to introduce the matrices (D.67) and (D.68). Furthermore, if we have operators $A^{\mu,\nu}$ such that

$$\forall X \in \mathfrak{so}(1,3)\ ;\ X_{\mu,\nu}A^{\mu,\nu} = 0, \tag{D.93}$$

then

$$\forall \mu,\nu\ ,\ A^{\mu,\nu} = A^{\nu,\mu}, \tag{D.94}$$

which a physicist would express by saying that "the anti-symmetric part of $A^{\mu,\nu}$ is zero". To prove this we simply observe that

$$X_{\mu,\nu}A^{\mu,\nu} = \sum_{0\le\mu<\nu\le 3} X_{\mu,\nu}(A^{\mu,\nu} - A^{\nu,\mu})$$

and that the variables $X_{\mu,\nu}$ in the right-hand side can take any real values.

Let us consider a representation π of $SO^{\uparrow}(1,3)$. For $A, B \in SO^{\uparrow}(1,3)$ it holds that

$$\pi(ABA^{-1}) = \pi(A)\pi(B)\pi(A)^{-1}. \tag{D.95}$$

We think of a matrix $X \in \mathfrak{so}(1,3)$ as having the property that for ε infinitesimal, $1+\varepsilon X$ is an "infinitesimal Lorentz transformation", but rather than εX physicists use δX, an "infinitesimal matrix". They consider as self-evident that

$$\pi(1+\delta X) \simeq 1 + \frac{1}{2}\delta X_{\mu,\nu}M^{\mu,\nu}, \tag{D.96}$$

for certain operators $M^{\mu,\nu}$, where repeated indices are summed, and where $\delta X_{\mu,\nu}$ are the entries of the matrix δX.[12] Of course, once (D.92) has been spelled out, (D.96) is simply the linearity of the derivative, and $M^{\mu,\nu} = \pi'(J^{\mu,\nu})$.

Using (D.95) for $B = 1 + \delta X$ and also (D.96) one obtains

$$\pi(1 + A\delta XA^{-1}) \simeq 1 + \frac{1}{2}\delta X_{\mu,\nu}\pi(A)M^{\mu,\nu}\pi(A)^{-1}, \tag{D.97}$$

which is simply another way to obtain the mathematician's relation

$$\pi'(AXA^{-1}) = \pi(A)\pi'(X)\pi(A)^{-1}. \tag{D.98}$$

We now want to use (D.96) again, this time for the infinitely small matrix $A\delta XA^{-1}$ rather than δX. We have to figure out the entries of this matrix. Generally speaking for matrices A,B,C we have $(ABC)_{\mu,\nu} = A_{\mu,\alpha}B^{\alpha}{}_{\beta}C^{\beta}{}_{\nu}$. Now, as is shown at the end of Section 4.1, for $C = A^{-1}$ we have $C^{\beta}{}_{\nu} = A_{\nu}{}^{\beta}$, so that $(ABA^{-1})_{\mu,\nu} = A_{\mu,\alpha}B^{\alpha}{}_{\beta}A_{\nu}{}^{\beta} = A_{\mu}{}^{\alpha}A_{\nu}{}^{\beta}B_{\alpha,\beta}$. Using this for $B = \delta X$, and then (D.96) for the matrix $A\delta XA^{-1}$ we obtain

$$\pi(1 + A\delta XA^{-1}) \simeq 1 + \frac{1}{2}A_{\mu}{}^{\alpha}A_{\nu}{}^{\beta}\delta X_{\alpha,\beta}M^{\mu,\nu},$$

and comparing with (D.97) this yields

$$\frac{1}{2}A_{\mu}{}^{\alpha}A_{\nu}{}^{\beta}\delta X_{\alpha,\beta}M^{\mu,\nu} \simeq \frac{1}{2}\delta X_{\mu,\nu}\pi(A)M^{\mu,\nu}\pi(A)^{-1}.$$

Since the coefficients of $\delta X_{\mu,\nu}$ are anti-symmetric they must be equal (as implied by (D.94)),

$$\pi(A)M^{\mu,\nu}\pi(A)^{-1} = A_{\alpha}{}^{\mu}A_{\beta}{}^{\nu}M^{\alpha,\beta}, \tag{D.99}$$

which the physicists express by saying that the quantities $M^{\mu,\nu}$ "transform as tensors" (see more about this in Section D.12). This is simply a way to rewrite (D.98) in another language, and brings no

[12] This atrocious notation is unfortunately rather common.

extra information. We can now use (D.99) itself in the case $A = 1 + \delta X$. Comparing the "first-order terms" in (D.99) we then obtain

$$\frac{1}{2}\delta X_{\alpha,\beta}[M^{\alpha,\beta}, M^{\mu,\nu}] = \delta X_{\alpha}^{\ \mu} M^{\alpha,\nu} + \delta X_{\beta}^{\ \nu} M^{\mu,\beta}$$
$$= \delta X_{\alpha,\lambda}\eta^{\lambda,\mu} M^{\alpha,\nu} + \delta X_{\beta,\gamma}\eta^{\gamma,\nu} M^{\mu,\beta}, \qquad \text{(D.100)}$$

which we rewrite as

$$\delta X_{\alpha,\beta}\Big(\frac{1}{2}[M^{\alpha,\beta}, M^{\mu,\nu}] - \eta^{\beta,\mu} M^{\alpha,\nu} + \eta^{\alpha,\nu} M^{\mu,\beta}\Big) = 0,$$

and then expressing as in (D.93), (D.94) that "the anti-symmetric part of the coefficient of $X_{\alpha,\beta}$ is 0" we obtain the relation

$$[M^{\alpha,\beta}, M^{\mu,\nu}] = \eta^{\beta,\mu} M^{\alpha,\nu} - \eta^{\alpha,\nu} M^{\mu,\beta} - \eta^{\alpha,\mu} M^{\beta,\nu} - \eta^{\beta,\nu} M^{\mu,\alpha} \qquad \text{(D.101)}$$

In the case $M_{\mu,\nu} = J_{\mu,\nu}$ (corresponding to the case where π is the identity) one can check that (D.101) is a (particularly cryptic) way to write the relations (D.69) to (D.73), and the general case of these relations is equivalent to this special case since $M_{\mu,\nu} = \pi'(J_{\mu,\nu})$ and π' is a Lie algebra homomorphism. Physicists sometimes call the relations (D.101) "the Lie algebra of the Lorentz group".[13] The Lie algebra relations for the Poincaré group can be obtained in a similar manner.

D.12 Some Tensor Representations of $SO^{\uparrow}(1,3)$

In this section we enjoy certain concrete non-trivial representations of $SO^{\uparrow}(1,3)$.

Proposition D.12.1 *If $j + \ell$ is even there exists a representation $\Pi_{j,\ell}$ of $SO^{\uparrow}(1,3)$ such that the representation $A \mapsto \Pi_{j,\ell}(\kappa(A))$ of $SL(2,\mathbb{C})$ is equivalent to the representation (j,ℓ) of Definition 8.3.2. Furthermore a representation U of $SO^{\uparrow}(1,3)$ is equivalent to $\Pi_{j,\ell}$ if and only if the representation $A \mapsto U(\kappa(A))$ of $SL(2,\mathbb{C})$ is equivalent to the representation (j,ℓ).*

Proof For clarity we denote by $\pi_{j,\ell}$ the (j,ℓ) representation of $SL(2,\mathbb{C})$. The key point is that (as is obvious from the definition of this representation)

$$\pi_{j,\ell}(-A) = (-1)^{j+\ell}\pi_{j,\ell}(A).$$

Thus when $j + \ell$ is even (and only in that case) we have $\pi_{j,\ell}(A) = \pi_{j,\ell}(-A)$, so that we can define $\Pi_{j,\ell}(B)$ for $B \in SO^{\uparrow}(1,3)$ by setting $\Pi_{j,\ell}(B) = \pi_{j,\ell}(A)$ where A is such that $\kappa(A) = B$. This does not depend on the choice of A, since the only other possible choice is $-A$ which gives the same value $\pi_{j,\ell}(-A)$. It is then straightforward to check that $\Pi_{j,\ell}$ is a representation. The last assertion is also straightforward. □

The goal of this section is to investigate some of the representations $\Pi_{j,\ell}$ of $SO^{\uparrow}(1,3)$. In physics they are called *tensor representations* as opposed to the case $j + \ell$ odd, where the representations (j,ℓ) (seen as projective representations of $SO^{\uparrow}(1,3)$) are called *spinor representations*.

We will first show in Exercise D.12.4 that $\Pi_{1,1}$ is the "defining representation" of $SO^{\uparrow}(1,3)$ obtained by letting a matrix operate on column vectors of $\mathbb{R}^4$ by matrix multiplication. (What else indeed could it be?) The main step toward this result is as follows.

Proposition D.12.2 *The representation $(1,1)$ of $SL(2,\mathbb{C})$ is equivalent to the representation θ of $SL(2,\mathbb{C})$ on the space $\mathcal{M}$ of 2×2 matrices such that $\theta(A)$ is the operator $M \mapsto AMA^{\dagger}$.*

[13] Several authors present the previous arguments in a far more concise manner, and let the reader try and figure out the meaning of (D.101) entirely on her own. This motivates the title of this section.

Proof Since the representations $A^{\dagger}$ and A^{*-1} are equivalent, it should be clear that θ is equivalent to the representation r given by

$$r(A) : M \mapsto AMA^{*-1} = AMB^T, \tag{D.102}$$

where $B = A^{\dagger -1}$. Denoting by $m_{i,j}$ the entries of the matrix M, the entries $t_{n,k}$ of $r(A)(M)$ are given by

$$t_{n,k} = \sum_{i,j=1,2} A_{n,i} m_{i,j} B_{k,j} = \sum_{i,j=1,2} A_{n,i} B_{k,j} m_{i,j}. \tag{D.103}$$

Comparing with the formula describing the action of A on a tensor $(x_{i,j})$ in the representation $(1,1)$ proves the result. □

Exercise D.12.3 Give a complete proof of the "it should be clear that".

Exercise D.12.4 If you have not already done so, solve Exercise 8.4.4. Hint: Use Proposition D.12.2. Using the last assertion of Proposition D.12.1 proves that $\Pi_{1,1}$ is equivalent to the defining representation of $SO^{\uparrow}(1,3)$.

In the rest of the present section we investigate concrete representations which are equivalent to $\Pi_{2,0}$ and $\Pi_{0,2}$. To follow the computations the reader should be familiar with the technique of raising and lowering indices as explained at the end of Section 4.1. The starting point is the defining representation of $SO^{\uparrow}(1,3)$ on $\mathbb{R}^4$. Taking the tensor product of this representation with itself, we consider tensors (x^{μ_1,μ_2}) where $0 \le \mu_1, \mu_2 \le 3$. This space of tensors is of dimension 16, and $L \in SO^{\uparrow}(1,3)$ operates on it by

$$x = (x^{\mu_1,\mu_2}) \mapsto U(L)(x) = (L^{\mu_1}_{\nu_1} L^{\mu_2}_{\nu_2} x^{\nu_1,\nu_2}). \tag{D.104}$$

Here as usual repeated indices are summed over. This representation is not irreducible because the space of anti-symmetric tensors, i.e. tensors satisfying $x^{\mu_1,\mu_2} = -x^{\mu_2,\mu_1}$ is invariant. This space, of dimension 6, is not irreducible either. There exists an operator T which commutes with each $U(L)$ and satisfies $T^2 = -1$. The two eigenspaces corresponding to the eigenvalues $\pm \mathrm{i}$ are therefore invariant. To describe T we consider the unique anti-symmetric tensor $(\varepsilon^{\mu_1,\mu_2,\mu_3,\mu_4})$ with $\varepsilon^{0,1,2,3} = 1$ (so that e.g. $\varepsilon^{0,0,1,2} = 0$ and $\varepsilon^{0,1,3,2} = -1$). For $x = (x^{\mu_1,\mu_2})$ we define $T(x)$ by

$$T(x)^{\mu_1,\mu_2} = \frac{1}{2}\varepsilon^{\mu_1,\mu_2,\mu_3,\mu_4} x_{\mu_3,\mu_4},$$

where in the right-hand side we have lowered the indices of x. The right-hand side represents a sum in which there are only two terms that are not zero and are equal, so we get very simple relations such as $T(x)^{0,1} = x_{2,3}$. To prove that $T^2 = -1$ we observe that

$$T(x)_{\mu_3,\mu_4} = \frac{1}{2}\varepsilon_{\mu_3,\mu_4,\mu_5,\mu_6} x^{\mu_5,\mu_6},$$

so that

$$T^2(x)^{\mu_1,\mu_2} = \frac{1}{4}\varepsilon^{\mu_1,\mu_2,\mu_3,\mu_4}\varepsilon_{\mu_3,\mu_4,\mu_5,\mu_6} x^{\mu_5,\mu_6}.$$

Observing that $\varepsilon_{\mu_3,\mu_4,\mu_5,\mu_6} = -\varepsilon^{\mu_3,\mu_4,\mu_5,\mu_6}$ we check that

$$\varepsilon^{\mu_1,\mu_2,\mu_3,\mu_4}\varepsilon_{\mu_3,\mu_4,\mu_5,\mu_6}$$

equals -2 if $(\mu_1,\mu_2) = (\mu_5,\mu_6)$, equals 2 if $(\mu_1,\mu_2) = (\mu_6,\mu_5)$ and equals 0 in the other cases, proving that $T^2 = -1$. Next we prove that T commutes with $U(L)$. The relation $TU(L) = U(L)T$ is equivalent to

$$\varepsilon^{\mu_1,\mu_2,\mu_3,\mu_4} L^{\nu_3}_{\mu_3} L^{\nu_4}_{\mu_4} = L^{\mu_1}_{\nu_1} L^{\mu_2}_{\nu_2} \varepsilon^{\nu_1,\nu_2,\nu_3,\nu_4}. \tag{D.105}$$

Recalling that by (4.18) the matrices $L^{\mu}_{\ \nu}$ and $L_{\nu}^{\ \mu}$ are inverse of each other, multiplying both sides by $L_{\mu_1}^{\lambda_1} L_{\mu_2}^{\lambda_2}$, summing over μ_1, μ_2 and changing ν_3, ν_4 to λ_3, λ_4 this is equivalent to

$$\varepsilon^{\mu_1,\mu_2,\mu_3,\mu_4} L_{\mu_1}^{\lambda_1} L_{\mu_2}^{\lambda_2} L_{\mu_3}^{\lambda_3} L_{\mu_4}^{\lambda_4} = \varepsilon^{\lambda_1,\lambda_2,\lambda_3,\lambda_4}, \tag{D.106}$$

which simply says that the tensor $\varepsilon^{\mu_1,\mu_2,\mu_3,\mu_4}$ is Lorentz invariant. To prove this we observe that both sides are zero if the indices λ_j are not all distinct. Using anti-symmetry it then suffices to consider the case where $\lambda_j = j - 1$ in which case the left-hand side is simply the determinant of L, which we know to be equal to 1.

A trained physicist will think of the previous situation in a rather different way. In physics, a tensor (x^{μ_1,μ_2}) is "a quantity which transforms under the rule (D.104) under a Lorentz transformation", with obvious adaptation of this rule for more indices including lower indices. Then (D.106) means that "$\varepsilon^{\mu_1,\mu_2,\mu_3,\mu_4}$ is a tensor". Furthermore, it is a general fact that "contracting some upper indices with some lower indices" in a tensor one still gets a tensor. An example of such a contraction is the tensor $\tilde{x}^{\mu_1,\mu_2} = (1/2)\varepsilon^{\mu_1,\mu_2,\mu_3,\mu_4} x_{\mu_3,\mu_4}$, sometimes called the dual of x^{μ_1,μ_2}. The property that one obtains a tensor by contracting tensor indices is a consequence of the fact that by (4.18) the matrices $L^{\mu}_{\ \nu}$ and $L_{\nu}^{\ \mu}$ are inverse of each other. With this point of view, it is obvious to a physicist that $U(L)$ commutes with T, because she has learned to use the elements of the previous proof in a systematic way.

We now know that U induces a representation of $SO^{\uparrow}(1,3)$ on the spaces of anti-symmetric tensors T that satisfy either the condition $T(x) = \mathrm{i}x$ or $T(x) = -\mathrm{i}x$. Each of these spaces is of dimension 3. Since $\Pi_{j,\ell}$ is of dimension $(j+1)(\ell+1)$, the only representations of this type which are of dimension 3 are $\Pi_{2,0}$ and $\Pi_{0,2}$. But which is which? To answer this question, we will look at it another way. Let us recall the representation θ of Proposition D.12.2, which we view as a representation on the space $\mathcal{M}$ of tensors $(m_{ij})_{i,j=1,2}$. Then we may define a representation $\theta \otimes \theta$ on $\mathcal{M} \otimes \mathcal{M}$ such that $\theta \otimes \theta(A) = \theta(A) \otimes \theta(A)$. We may identify $\mathcal{M} \otimes \mathcal{M}$ with the space of tensors $m = (m_{i_1,j_1,i_2,j_2})$ with all indices equal to 1 or 2.

We recall the Pauli matrices σ_ν. The following asserts that the representations $A \mapsto U(\kappa(A))$ and $A \mapsto \theta \otimes \theta(A)$ are equivalent.

Lemma D.12.5 *Consider the map $V : \mathbb{R}^4 \otimes \mathbb{R}^4 \to \mathcal{M} \otimes \mathcal{M}$ given for $x = (x^{\mu_1,\mu_2})$ by*

$$V(x) = x^{\mu_1,\mu_2} \sigma_{\mu_1} \otimes \sigma_{\mu_2}.$$

Then for $A \in SL(2,\mathbb{C})$ we have

$$V(U(\kappa(A))(x)) = \theta \otimes \theta(A)(V(x)). \tag{D.107}$$

Proof Recalling that $M(x) = x^{\mu}\sigma_{\mu}$ the definition $M(\kappa(A)(x)) = AM(x)A^{\dagger}$ of $\kappa(A)$ reads $\kappa(A)^{\mu}_{\ \nu} x^{\nu}\sigma_{\mu} = x^{\mu} A\sigma_{\mu}A^{\dagger}$ so that $\kappa(A)^{\mu}_{\ \nu}\sigma_{\mu} = A\sigma_{\nu}A^{\dagger} = \theta(A)(\sigma_{\nu})$. Now,

$$U(\kappa(A))(x)^{\mu_1,\mu_2} = \kappa(A)^{\mu_1}_{\nu_1}\kappa(A)^{\mu_2}_{\nu_2} x^{\nu_1,\nu_2},$$

and hence

$$\begin{aligned} V(U(\kappa(A))(x)) &= \kappa(A)^{\mu_1}_{\nu_1}\kappa(A)^{\mu_2}_{\nu_2} x^{\nu_1,\nu_2}\sigma_{\mu_1} \otimes \sigma_{\mu_2} \\ &= x^{\nu_1,\nu_2}\theta(A)(\sigma_{\nu_1}) \otimes \theta(A)(\sigma_{\nu_2}) \\ &= x^{\nu_1,\nu_2}(\theta \otimes \theta)(A)(\sigma_{\nu_1} \otimes \sigma_{\nu_2}) \\ &= (\theta \otimes \theta)(A)(x^{\nu_1,\nu_2}\sigma_{\nu_1} \otimes \sigma_{\nu_2}) \end{aligned}$$ □

The invariant subspaces of $\theta \otimes \theta$ are easy to analyze.

Proposition D.12.6 *The following subspaces of $\mathcal{M} \otimes \mathcal{M}$ are invariant under $\theta \otimes \theta$:*

$$\{(m_{i_1,j_1,i_2,j_2})\ ;\ \forall i_1,i_2,j_1,j_2\ ,\ m_{i_1,j_1,i_2,j_2} = m_{i_2,j_1,i_1,j_2}\ ,\ m_{i_1,j_1,i_2,j_2} = m_{i_1,j_2,i_2,j_1}\}. \tag{D.108}$$

$$\{(m_{i_1,j_1,i_2,j_2})\,;\ \forall i_1,i_2,j_1,j_2\ ,\ m_{i_1,j_1,i_2,j_2} = m_{i_2,j_1,i_1,j_2}\ ,\ m_{i_1,j_1,i_2,j_2} = -m_{i_1,j_2,i_2,j_1}\}. \quad \text{(D.109)}$$

$$\{(m_{i_1,j_1,i_2,j_2})\,;\ \forall i_1,i_2,j_1,j_2\ ,\ m_{i_1,j_1,i_2,j_2} = -m_{i_2,j_1,i_1,j_2}\ ,\ m_{i_1,j_1,i_2,j_2} = m_{i_1,j_2,i_2,j_1}\}. \quad \text{(D.110)}$$

$$\{(m_{i_1,j_1,i_2,j_2})\,;\ \forall i_1,i_2,j_1,j_2, m_{i_1,j_1,i_2,j_2} = -m_{i_2,j_1,i_1,j_2}, m_{i_1,j_1,i_2,j_2} = -m_{i_1,j_2,i_2,j_1}\}. \quad \text{(D.111)}$$

These spaces are respectively of dimensions 9,3,3,1. The restriction of $\theta\otimes\theta$ to these spaces are respectively equivalent to the representations $(2,2)$, $(2,0)$, $(0,2)$, $(0,0)$.

Proof Let us write the action of $\theta\otimes\theta(A)$ on a tensor m:

$$\theta\otimes\theta(A)(m)_{i_1,j_1,i_2,j_2} = \sum_{k_1,\ell_1,k_2,\ell_2} A_{i_1,k_1} B_{j_1,\ell_1} A_{i_2,k_2} B_{j_2,\ell_2} m_{k_1,\ell_1,k_2,\ell_2},$$

where $B = A^*$. It is then obvious that the subspaces above are invariant under this action. The statements about the dimension of these spaces are elementary. To prove that the restriction to the space (D.109) is equivalent to $(2,0)$ we observe that $m_{i_1,1,i_2,1} = m_{i_1,2,i_2,2} = 0$ and $m_{i_1,2,i_2,1} = -m_{i_1,1,i_2,2}$, so that using that $\det B = 1$ we obtain

$$\theta\otimes\theta(A)(m)_{i_1,1,i_2,2} = \sum_{k_1,k_2} A_{i_1,k_1} A_{i_2,k_2} m_{k_1,1,k_2,2}.$$

The case of the space (D.110) is similar, and the case of the space (D.108) is much easier. □

Corollary D.12.7 *Consider a subspace $\mathcal{G}$ of $\mathcal{M}\otimes\mathcal{M}$ and assume that $\mathcal{G}$ is invariant under $\theta\otimes\theta$, and that the restriction of $\theta\otimes\theta$ to $\mathcal{G}$ is equivalent to the representation (j,ℓ). Then $V^{-1}(\mathcal{G})$ is invariant under U and the restriction of U to $V^{-1}(\mathcal{G})$ is equivalent to $\Pi_{j,\ell}$.*

Proof The first statement is obvious. The last one follows from Proposition D.12.1. □

Proposition D.12.8 *The image under V of the space of anti-symmetric tensors x that satisfy the relation $T(x) = \mathrm{i}x$ is the space (D.109). Consequently the restriction of U to this space is $\Pi_{2,0}$.*

Proof The second statement is obvious. As for the first statement, I wish I knew a conceptual proof. This image is an invariant space of dimension 3, so that it can be only one of the two spaces (D.109) or (D.110) and the statement can be painfully checked by hand. □

Exercise D.12.9 Show that the inverse image by V of the space (D.108) is the space of symmetric traceless tensors, i.e. $x^{\mu}_{\ \mu} = 0$. Conclude that the restriction of U to this space is $\Pi_{2,2}$.

Appendix E

Special Relativity

E.1 Energy–Momentum

A good method to describe a curve in $\mathbb{R}^3$ is to parameterize it as the collection of points $\boldsymbol{x} = \boldsymbol{x}(\theta)$ where the parameter θ runs in a certain interval. Many different parameterizations describe the same curve. However when the curve is nice, one learns (or, depending on one's age, one teaches) that among all the parameterizations there is a distinguished one, $\boldsymbol{x} = \boldsymbol{x}(s)$, the parameterization by *arc length* which (once a specific point of the curve and the direction in which one moves are given) is determined by

$$\left\|\frac{\mathrm{d}\boldsymbol{x}}{\mathrm{d}s}\right\|^2 = 1.$$

This parameterization is intrinsic, in the sense that it does not depend on the coordinate system but only on the Euclidean norm.

Suppose now that we want to describe the trajectory of a massive point in space-time. I will propose a parameterization

$$x(t) = (ct, \boldsymbol{x}(t)). \tag{E.1}$$

This however has no intrinsic value because this definition is not invariant under Lorentz transformations.[1] To find an intrinsic parameterization, it is the idea of parameterization by arc length which succeeds, but the "Euclidean norm has to be replaced by the Lorentz norm". To describe a curve in space-time, we want to use a parameter τ such that

$$\left(\frac{\mathrm{d}x}{\mathrm{d}\tau}, \frac{\mathrm{d}x}{\mathrm{d}\tau}\right) = c^2. \tag{E.2}$$

The factor c^2 is motivated by the fact that the left-hand side has the dimension of the square of a speed. It is not always possible to parameterize a curve this way, because the Lorentz "norm" of a non-zero vector can be zero. For example, whatever the parameterization $x(\theta)$ of the trajectory of a light ray the vector $\mathrm{d}x/\mathrm{d}\theta$ has Lorentz seminorm zero. When we consider however the curve $x(\theta)$ which describes the trajectory in space-time of a point of mass $m > 0$, the condition that this point moves at a speed less than the speed of light is expressed by

[1] More specifically, what I call "time" is for other observers an unremarkable mixture of what they call "time" and "space".

$$\left(\frac{\mathrm{d}x}{\mathrm{d}\theta},\frac{\mathrm{d}x}{\mathrm{d}\theta}\right) > 0,$$

allowing to make a change of parameter in order to reach (E.2).

Definition E.1.1 The parameter τ is called the *proper time* along the trajectory, and the vectors

$$\frac{\mathrm{d}x}{\mathrm{d}\tau} \text{ and } m\frac{\mathrm{d}x}{\mathrm{d}\tau}$$

are called respectively the *four-velocity* and the *four-momentum* of the moving particle.[2]

It should be clear that the proper time is invariant under Lorentz transformation, and that the four-momentum is a "four-vector", i.e. it transforms as expected under a change of coordinates in space-time given by a Lorentz transformation.[3] By construction, the energy–momentum has Lorentz "norm" mc, i.e. (4.29) holds.

Let us examine the meaning of these concepts in the case where the curve is given by (E.1) in my coordinate system. Then, using the notation $\boldsymbol{v}(t) = \mathrm{d}\boldsymbol{x}(t)/\mathrm{d}t$, we have $\mathrm{d}x/\mathrm{d}t = (c,\boldsymbol{v}(t))$ and $(\mathrm{d}x/\mathrm{d}t,\mathrm{d}x/\mathrm{d}t) = c^2 - \boldsymbol{v}(t)^2$. Since

$$\frac{\mathrm{d}x}{\mathrm{d}t} = \frac{\mathrm{d}\tau}{\mathrm{d}t}\frac{\mathrm{d}x}{\mathrm{d}\tau},$$

using (E.2) we obtain $c^2(\mathrm{d}\tau/\mathrm{d}t)^2 = c^2 - \boldsymbol{v}(t)^2$ so that

$$\frac{\mathrm{d}\tau}{\mathrm{d}t} = \sqrt{1-\frac{\boldsymbol{v}(t)^2}{c^2}} := \alpha. \tag{E.3}$$

When $\boldsymbol{v}(t)$ is constant an observer moving with the point is an inertial observer, legitimate to compute the proper time. In his own reference frame he will measure his own position as $x(t') = (ct',0,0,0)$, where t' is time as he measures it and will find by the previous calculation that $\tau = t'$.[4] the case where $\boldsymbol{v}(t)$ is constant, the previous example together with Lorentz invariance shows that the proper time is the time as measured by an observer moving with this point. Then α is the "time contraction factor". In one second of my time I feel the relativistic traveler has aged only α seconds. In my coordinates, the four-momentum is then given by

$$\left(\frac{cm}{\sqrt{1-\boldsymbol{v}(t)^2/c^2}},\frac{m\boldsymbol{v}(t)}{\sqrt{1-\boldsymbol{v}(t)^2/c^2}}\right) = (cm', m'\boldsymbol{v}(t)), \tag{E.4}$$

where $m' = m/\alpha$, a formula which in popular books is described by saying that as a result of the motion the mass has apparently increased from m to m'. Observe also from the previous formula that, as used just below (12.79), writing $(p^0,\boldsymbol{p})$ the four-momentum of the particle, one has the relation

$$\boldsymbol{v} = \frac{c\boldsymbol{p}}{p^0}. \tag{E.5}$$

At non-relativistic velocity $|\boldsymbol{v}(t)| \ll c$, (E.4) shows that the four-momentum is about

$$\left(mc + \frac{m\boldsymbol{v}(t)^2}{2c}, m\boldsymbol{v}(t)\right),$$

giving as expected the energy $E = mc^2 + m\boldsymbol{v}(t)^2/2$.

[2] The four-momentum is also called the *energy–momentum*.

[3] That is, if your coordinates y are related to my coordinates x by $y = L(x)$, and if I measure four-momentum p you will measure four-momentum $L(p)$.

[4] Provided of course that we ask that $\tau = 0$ for $t = 0$.

E.2 Electromagnetism

This section assumes that the reader is familiar with Section 4.1. It might also benefit the reader to study Section 10.12 in parallel with the present one. In Special Relativity the electrical field $\boldsymbol{E}$ and the magnetic field $\boldsymbol{B}$ are components of a single object, the electromagnetic tensor[5]

$$(F_{\mu\nu}) = \begin{pmatrix} 0 & E_x/c & E_y/c & E_z/c \\ -E_x/c & 0 & -B_z & B_y \\ -E_y/c & B_z & 0 & -B_x \\ -E_z/c & -B_y & B_x & 0 \end{pmatrix}.$$

Here we use physics-type terminology: "Tensor" means "tensor field", a map which assigns the tensor $F_{\mu\nu}(x)$ to each point of a certain domain Ω. Contained in the name "electromagnetic tensor" is the assertion that it "transforms as a tensor" under a Lorentz transformation L, i.e. the tensor field $F_{\mu\nu}(x)$ becomes

$$L_\nu^{\ \lambda} L_\mu^{\ \alpha} F_{\lambda\alpha}(L^{-1}(x)). \tag{E.6}$$

Here we raise and lower indices as explained at the end of Section 4.1 to define the coefficients $L_\nu^{\ \lambda}$. So the tensor $F^{\mu\nu}$ transforms as $L^\mu_{\ \lambda} L^\nu_{\ \alpha} F^{\lambda\alpha}(L^{-1}(x))$.

The components of the electromagnetic tensor *mix together* under Lorentz transformations, which is not surprising when one remembers that a fixed electrical charge produces only an electrical field, but a moving charge produces a magnetic field too. The electromagnetic tensor obeys the Maxwell equations, which, in the absence of electrical charge, take the form[6]

$$\forall \kappa, \mu, \nu\ ,\ \partial_\kappa F_{\mu\nu} + \partial_\mu F_{\nu\kappa} + \partial_\nu F_{\kappa\mu} = 0, \tag{E.7}$$

$$\forall \nu\ ,\ \partial^\mu F_{\mu\nu} = 0, \tag{E.8}$$

where of course there is a summation over the repeated index in (E.8).

Exercise E.2.1 Convince yourself by writing the proof in complete detail that the Maxwell equations are invariant under Lorentz transformations.

The electromagnetic tensor can in turn (when the domain Ω is sufficiently nice) be described by a four-vector A called the *electromagnetic four-potential* via the formula

$$F_{\mu\nu} = \partial_\mu A_\nu - \partial_\nu A_\mu. \tag{E.9}$$

Here $A = A(x)$ is defined at every point of the domain Ω and of course "four-vector" means that under a Lorentz transformation L it transforms according to the rule $A_\nu(x) \to L_\nu^{\ \lambda} A_\lambda(L^{-1}(x))$.

Exercise E.2.2 Write all details to convince yourself that if A is a four-vector, then (E.9) is a tensor.

The existence of the four-potential is actually a consequence of the equation (E.7). The equation (E.7) is then automatically satisfied while the equation (E.8) becomes

$$\partial^\mu F_{\mu\nu} = (\partial^\mu \partial_\mu) A_\nu - \partial_\nu(\partial^\mu A_\mu) = 0. \tag{E.10}$$

Let us stress however that it is the components of the electromagnetic tensor which are physically measurable, not those of the four-potential. The components $F_{\mu\nu}$ do not determine A. Indeed they are invariant under the change $A_\mu \to A_\mu + \partial_\mu \chi$ for any smooth function χ on $\mathbb{R}^{1,3}$. Such a transformation is called a *gauge transformation* (an extremely successful idea). All the physically measurable quantities should be invariant under such a gauge transformation.

5 We are using SI units.
6 Let us recall the tricky fact that $\partial_\mu = \partial/(\partial x^\mu)$ and $\partial^\mu = \partial/(\partial x_\mu)$.

Since the possibility of a gauge transformation leaves us much choice about A, we may like to restrict this choice by imposing further conditions on A. The *Lorenz gauge* (named after Ludvig Lorenz and not after Hendrik Lorentz) imposes the Lorentz invariant condition

$$\partial_\mu A^\mu = \partial^\mu A_\mu = 0. \tag{E.11}$$

To see that this is possible, starting with any potential A one replaces A_μ by $A_\mu + \partial_\mu \chi$ where χ satisfies the equation $\partial^\mu \partial_\mu \chi = -\partial^\mu A_\mu$.[7] In the Lorenz gauge, the Maxwell equation (E.10) becomes

$$(\partial^\mu \partial_\mu) A_\nu = 0. \tag{E.12}$$

That is, each component C of the potential satisfies the "wave equation" $\partial^\mu \partial_\mu C = 0$ but these components are not independent, as they satisfy (E.11).

The electromagnetic tensor and the four-potential are *real-valued*. This does not stop physicists from writing formulas involving complex numbers, but what they mean is the *real part* of what they write, and this is how the forthcoming formulas must be understood. Using the Fourier transform, it is rather easy to solve the equations (E.12) and (E.11) (or at least to find all the well-behaved solutions). These solutions are mixtures of solutions of the type $A(x) = a \exp(\mathrm{i}(x,p)/\hbar)$ where $p^2 = 0$ (to satisfy (E.12)) and the four-vector a satisfies $(a,p) = 0$ (to satisfy (E.11)). In other words, $A_\mu(x) = a_\mu \exp(\mathrm{i}(x,p)/\hbar)$ with $a_\mu p^\mu = p_\mu p^\mu = 0$. These solutions are called *plane waves*. One may think at first that, besides p, a plane wave depends on three parameters (the degrees of freedom in the choice of a left by the condition $(a,p) = 0$), but this is not the case: When two vectors a differ by a multiple of p the corresponding plane waves are related by a gauge transformation and represent the same physical process, as is shown by the identity

$$(a_\mu + \lambda p_\mu) \exp(\mathrm{i}(x,p)/\hbar) = a_\mu \exp(\mathrm{i}(x,p)/\hbar) + \partial_\mu \chi \tag{E.13}$$

where $\chi(x) = -\mathrm{i}\lambda\hbar \exp(\mathrm{i}(x,p)/\hbar)$. Thus there are actually only two parameters that determine a plane wave. It is this fact which physically justifies taking the quotient by $\mathcal{H}_{\mathrm{null}}$ in Proposition 9.13.6.

In (E.13) one may take $\lambda = -a_0/p_0$ to reduce to the case $A_0 = 0$ and $A_i(x) = a_i \exp(\mathrm{i}(x,p)/\hbar)$ for $i = 1,2,3$.[8] Let us write $p = (p_0, \boldsymbol{p})$ (so that $p_0 = \pm|\boldsymbol{p}|$ since $p^2 = 0$). Then $\boldsymbol{a} = (a_1, a_2, a_3)$ satisfies $\boldsymbol{a} \cdot \boldsymbol{p} = 0$, which is expressed by the fact that electromagnetic waves are *transverse* to the direction of motion, and makes it transparent that the plane wave depends only on two parameters. In elementary textbooks one likes to consider two orthonormal vectors $\boldsymbol{e}_1(\boldsymbol{p}), \boldsymbol{e}_2(\boldsymbol{p})$ which are orthogonal to $\boldsymbol{p}$ and which define so-called directions of polarization, but this cannot be done in an intrinsic manner.

What is more intrinsic is to consider potentials of the type

$$A^{\pm}(x) = (\boldsymbol{e}_1(\boldsymbol{p}) \pm \mathrm{i}\boldsymbol{e}_2(\boldsymbol{p})) \exp(\mathrm{i}(x,p)/\hbar), \tag{E.14}$$

which have the property that rotating around the direction of motion by an angle θ multiplies them by $\exp(\mp \mathrm{i}\theta)$.

Exercise E.2.3 Check the previous claim when $\boldsymbol{p}$ is parallel to the third axis. Prove then that it holds for any direction of motion by using first a rotation to bring $\boldsymbol{p}$ in the direction of the third axis.

Plane waves as in (E.14) are said to have *circular polarization*. A plane wave with a given $\boldsymbol{p}$ is a mixture of two waves with circular polarization. This is closely connected to the fact that photons are mixtures of photons of helicity ± 2, see Section 9.13.

7 It is not a very difficult fact that under sufficient regularity of f one may solve the equation $\partial^\mu \partial_\mu \chi = f$. In the related case of the Laplace equation, this is briefly explained in Section N.2.

8 The procedure is general. The Lorenz gauge does not determine the potential since one can still replace A_μ by $A_\mu + \partial_\mu \chi$ where $\partial^\mu \partial_\mu \chi = 0$. One may choose a χ such that $\partial_0 \chi = -A_0$, and the new potential satisfies $A_0 = 0$. It also satisfies the condition $\sum_{1 \le i \le 3} \partial^i A_i = 0$, a condition known as the Coulomb gauge, and usually written as $\nabla \cdot \mathbf{A} = 0$.

Appendix F

Does a Position Operator Exist?

In this section we go back to the setting of Proposition 4.9.2. We interpret $\mathcal{H} = L^2(X_m, d\lambda_m)$ as the state space of a particle of mass m with the Hamiltonian and momentum operator as in that proposition. Is it possible to define a position operator? On general grounds, it is difficult to imagine that one could locate precisely a particle, because by the Heisenberg uncertainty relations, this particle would have large momentum and hence large energy, and would therefore create new particles. For this reason, it seems impossible to construct a consistent relativistic one-particle theory. The considerations of the present appendix are nonetheless of interest. We show that under very reasonable conditions there is essentially only one possible candidate for a position operator, but that interpreting this operator as an *exact* position operator implies that the particle is permitted faster-than-light travel. It seems better to reject that possibility even at the microscopic level. (On the other hand there is no doubt that even relativistic particles display some degree of localization!) Our treatment is a transcription of pages 12–16 of Sydney Coleman's flamboyant course [13]. It is written in more mathematical language but *does not constitute a mathematical proof.* More mathematical arguments can be found in Wightman's paper [89] but they are considerably less entertaining.

We work in the Schrödinger picture, where the states evolve but the operators do not. As we will use space translations and rotations, but never boosts, it is more convenient to work with the state space $\mathcal{H}' = L^2(\mathbb{R}^3, d^3\boldsymbol{p}/(2\pi\hbar)^3)$ rather than $\mathcal{H}$. We will need the canonical unitary map J from $\mathcal{H}'$ to $\mathcal{H}$ given by $J(\varphi)(p) = \sqrt{2\omega_{\boldsymbol{p}}}\varphi(\boldsymbol{p})$ where as usual $p = (\omega_{\boldsymbol{p}}, \boldsymbol{p})$.

A position operator consists of three commuting self-adjoint operators which would measure the position of the particle along each coordinate. It is customary to use vector notation $\boldsymbol{X}$ to denote this triplet of operators. The obvious candidate is the standard position operator $\boldsymbol{X}'$ from non-relativistic Quantum Mechanics with components X'_1, X'_2, X'_3 given by

$$X'_i = i\hbar \frac{\partial}{\partial p_i}. \tag{F.1}$$

Can it be interpreted as a position operator? Let us consider the space $\mathcal{H}'' = L^2(\mathbb{R}^3, d^3\boldsymbol{x})$ and the Fourier transform $\mathcal{F}\colon \mathcal{H}'' \to \mathcal{H}'$. Then the operator $\mathcal{F}^{-1}X'_i\mathcal{F}$ on $\mathcal{H}''$ is simply "multiplication by x_i". Consequently, if we believe that $\boldsymbol{X}'$ is a position operator, it seems impossible to escape the conclusion[1] that if $\varphi \in \mathcal{H}'$ is of norm 1, then its inverse Fourier transform $f \in \mathcal{H}''$ should represent the probability density of finding the particle at a given point. In particular, if we consider a function $f \in \mathcal{S}(\mathbb{R}^3) \subset \mathcal{H}''$, which is zero outside a region A, its Fourier transform $\varphi \in \mathcal{H}'$ should represent the state of a particle that is localized in A. Consider now another function $f' \in \mathcal{S}(\mathbb{R}^3) \subset \mathcal{H}''$ which is zero outside a region B, so that its Fourier transform $\varphi' \in \mathcal{H}'$ should represent the state of a particle which is localized in B. Now if we let the system evolve for a time t, the state φ evolves to

[1] Remember that in this appendix, we do *not* prove anything: we *argue as physicists!*

$\exp(-ic\omega_{\boldsymbol{p}}t/\hbar)\varphi$. As is explained at the end of Section 2.1, the probability of observing φ' after time t is the square of the amplitude

$$\begin{aligned}\theta(t) := (\varphi', \exp(-ic\omega_{\boldsymbol{p}}t/\hbar)\varphi) &= \int \frac{d^3\boldsymbol{p}}{(2\pi\hbar)^3}\varphi'(\boldsymbol{p})^*\varphi(\boldsymbol{p})\exp(-ict\omega_{\boldsymbol{p}}/\hbar) \\ &= \int d^3\boldsymbol{x}d^3\boldsymbol{y}f'(\boldsymbol{x})^* f(\boldsymbol{y})\mathcal{I}_t(\boldsymbol{x}-\boldsymbol{y}), \end{aligned} \tag{F.2}$$

where the distribution $\mathcal{I}_t(\boldsymbol{x})$ is formally given by

$$\mathcal{I}_t(\boldsymbol{x}) = \int \frac{d^3\boldsymbol{p}}{(2\pi\hbar)^3}\exp(i(-ct\omega_{\boldsymbol{p}} + (\boldsymbol{x}\cdot\boldsymbol{p})/\hbar).$$

Now if A and B satisfy the condition

$$\boldsymbol{x} \in A\,,\; \boldsymbol{y} \in B \Rightarrow (\boldsymbol{x}-\boldsymbol{y})^2 > c^2t^2$$

the amplitude must be zero, for otherwise this would mean that the particle has traveled faster than light. This would require that $\mathcal{I}_t(\boldsymbol{x}) = 0$ for $\boldsymbol{x}^2 > c^2t^2$, but this is not true. (In Lemma 5.1.7 we compute a function whose derivative in t equals $\mathcal{I}_t(\boldsymbol{x})$ in the domain $\boldsymbol{x}^2 > c^2t^2$.) Thus $\boldsymbol{X}'$ cannot be interpreted as an exact position operator. Given $\varphi \in \mathcal{H}'$, calling its inverse Fourier transform $\mathcal{F}^{-1}(\varphi) \in L^2(\mathbb{R}^3, d^3\boldsymbol{x})$ "the wave function in position state space" as certain elementary textbooks do is therefore a risky proposition.

If one writes $\boldsymbol{X}'$ as an operator on $\mathcal{H}$, i.e. one considers the operator $J\boldsymbol{X}'J^{-1}$, a straightforward calculation yields that the ith component of this operator is given by

$$-i\hbar\left(\frac{\partial}{\partial p^i} + \frac{p^i}{2\omega_{\boldsymbol{p}}^2}\right).$$

This operator is known as the Newton–Wigner operator, see formula (11) of [59].

Let us go back to the consideration of a general position operator $\boldsymbol{X}$, the goal being to show that it basically has to be the operator (F.1). We expect that $\boldsymbol{X}$ behaves well with respect to space translations and rotations. We will show that these requirements essentially force that $\boldsymbol{X}$ equals $\boldsymbol{X}'$. For $\boldsymbol{a} \in \mathbb{R}^3$ and $R \in SO(3)$ let us define the action $V(\boldsymbol{a}, R)$ on $\mathcal{H}'$ by

$$V(\boldsymbol{a}, R)(\varphi)(\boldsymbol{p}) = \exp(-i\boldsymbol{a}\cdot\boldsymbol{p}/\hbar)\varphi(R^{-1}(\boldsymbol{p})). \tag{F.3}$$

We must first show that when dealing with space translations and rotations the previous action is just another way to write the action $U(c, C)$ of the Poincaré group on $\mathcal{H}$ given by (4.61). Using the canonical unitary map J from $\mathcal{H}'$ to $\mathcal{H}$, this means that $U(a, R)J = JV(\boldsymbol{a}, R)$ if $a = (0, \boldsymbol{a})$ and R is seen as an element of $SO^{\uparrow}(1,3)$. It is straightforward to check this.

Next, we argue that the position operator should satisfy the relation

$$\boldsymbol{X}V(\boldsymbol{a}, R) = V(\boldsymbol{a}, R)(\boldsymbol{a}1 + R\boldsymbol{X}) \tag{F.4}$$

whenever $\boldsymbol{a} \in \mathbb{R}^3$ and $R \in SO(3)$. Here the notation $R\boldsymbol{X}$ represents a triplet of operators, product of a matrix and a column vector, that is if $S_{i,j}$ denotes the matrix of S, $S\boldsymbol{X}$ is the triplet of operators $(\sum_{j\le 3} S_{i,j}X_j)_{i\le 3}$, and (F.4) represents three different equalities between operators. The relation (F.4) expresses that if we translate the system by $\boldsymbol{a}$ and apply the position operator, the value it returns has been translated by $\boldsymbol{a}$, and that if we rotate the system by a rotation R, so does the value given by our position operator.[2] To understand this, consider the ideal case where φ is an eigenvalue of $\boldsymbol{X}$, $\boldsymbol{X}\varphi = \boldsymbol{b}\varphi$, so that (F.3) expresses as desired that $\psi := V(\boldsymbol{a}, R)\varphi$ satisfies $\boldsymbol{X}\psi = \boldsymbol{X}V(\boldsymbol{a}, R)\varphi = V(\boldsymbol{a}, R)(\boldsymbol{a}1 + R\boldsymbol{X})\varphi = V(\boldsymbol{a}, R)(\boldsymbol{a} + R\boldsymbol{b})\varphi = (\boldsymbol{a} + R\boldsymbol{b})V(\boldsymbol{a}, R)\varphi = (\boldsymbol{a} + R\boldsymbol{b})\psi$, where we have used linearity in the third equality.

2 The term $\varphi(R^{-1}(\boldsymbol{p}))$ in (F.3) means that we have rotated the system by a rotation R.

Next we prove that the standard position operator $\boldsymbol{X}'$ satisfies (F.4). Let $\psi(\boldsymbol{p}) := V(\boldsymbol{a}, R)(\varphi)(\boldsymbol{p}) = \exp(-\mathrm{i}\boldsymbol{a}\cdot\boldsymbol{p}/\hbar)\varphi(R^{-1}(\boldsymbol{p}))$. We compute $\boldsymbol{X}'(\psi)(\boldsymbol{p})$. The term $\exp(-\mathrm{i}\boldsymbol{a}\cdot\boldsymbol{p}/\hbar)$ on the right creates a term $\boldsymbol{a}\psi(\boldsymbol{p})$, and the computation of the second term is obtained through the chain rule

$$\frac{\partial}{\partial p_i}\varphi(R^{-1}(\boldsymbol{p})) = \sum_j R^{-1}_{j,i}\frac{\partial\varphi}{\partial p_j}(R^{-1}(\boldsymbol{p})) = \sum_j R_{i,j}\frac{\partial\varphi}{\partial p_j}(R^{-1}(\boldsymbol{p})),$$

using also that R is a rotation so that its matrix $R_{i,j}$ is orthogonal. Therefore, using linearity in the last line,

$$\begin{aligned}\boldsymbol{X}'(\psi)(\boldsymbol{p}) &= \exp(-\mathrm{i}\boldsymbol{a}\cdot\boldsymbol{p}/\hbar)\big(\boldsymbol{a}\varphi(R^{-1}(\boldsymbol{p})) + R\boldsymbol{X}'(\varphi)(R^{-1}(\boldsymbol{p}))\big)\\ &= V(\boldsymbol{a},R)(\boldsymbol{a}\varphi(\boldsymbol{p}) + R\boldsymbol{X}'(\varphi)(\boldsymbol{p})),\end{aligned}$$

proving that $\boldsymbol{X}'$ satisfies (F.4). The difference $\boldsymbol{X}'' = \boldsymbol{X} - \boldsymbol{X}'$ hence satisfies

$$\boldsymbol{X}''V(\boldsymbol{a},R) = V(\boldsymbol{a},R)R\boldsymbol{X}''. \tag{F.5}$$

In full, for any φ in the domain of $\boldsymbol{X}''$ and any $\boldsymbol{a}\in\mathbb{R}^3$,

$$\boldsymbol{X}''(\exp(-\mathrm{i}\boldsymbol{a}\cdot\boldsymbol{p}/\hbar)\varphi(R^{-1}(\boldsymbol{p})) = \exp(-\mathrm{i}\boldsymbol{a}\cdot\boldsymbol{p}/\hbar)R\boldsymbol{X}''(\varphi)(R^{-1}(\boldsymbol{p})). \tag{F.6}$$

Then, presumably, taking mixtures in (F.6), for any $\psi\in\mathcal{S}(\mathbb{R}^3)$, any $R\in SO(3)$, and any φ in the domain of $\boldsymbol{X}''$,

$$\boldsymbol{X}''(\psi(\boldsymbol{p})\varphi(R^{-1}(\boldsymbol{p})) = \psi(\boldsymbol{p})R\boldsymbol{X}''(\varphi)(R^{-1}(\boldsymbol{p})). \tag{F.7}$$

Here let us take a leap of faith. Let us pretend that $\boldsymbol{X}''$ is self-adjoint. (There is no problem to show that $\boldsymbol{X}''$ is symmetric, but claiming that this operator is self-adjoint means that it is defined on a big domain, while, for all we know it could be defined only on the zero function…). If we knew more about self-adjoint operators, we could switch back to a more rigorous continuation, but let us proceed with analogies: In finite dimension an operator which commutes with all functions of a diagonal operator is itself a function of this diagonal operator, so here, just using (F.7) in the case R the identity, we may believe that $\boldsymbol{X}''$ is of the type "multiplication by a function $\boldsymbol{F}(\boldsymbol{p})$ of $\boldsymbol{p}$". That is, $\boldsymbol{X}''(\varphi)(\boldsymbol{p}) = \boldsymbol{F}(\boldsymbol{p})\varphi(\boldsymbol{p})$. From (F.7) this function satisfies $\boldsymbol{F}(\boldsymbol{p}) = R\boldsymbol{F}(R^{-1}(\boldsymbol{p}))$ for any rotation R. Considering first the case where R leaves $\boldsymbol{p}$ invariant we obtain that $\boldsymbol{F}(\boldsymbol{p})$ must be a multiple of $\boldsymbol{p}$, and this multiple can depend only on $\boldsymbol{p}^2$, so that $\boldsymbol{F}(\boldsymbol{p}) = F_1(\boldsymbol{p}^2)\boldsymbol{p}$ for a certain function F_1, and F_1 is real-valued since $\boldsymbol{X}''$ is self-adjoint. This relation implies that $\boldsymbol{F}$ is the gradient of a real-valued function V, $\boldsymbol{F} = \nabla V$, and we have proved that

$$\boldsymbol{X} = \boldsymbol{X}' + \nabla V,$$

where the last term means multiplication by ∇V. Consider now the function $W = \exp(\mathrm{i}V/\hbar)$, which is of modulus 1. Then

$$\begin{aligned}\boldsymbol{X}(W\varphi) &= \boldsymbol{X}'(W\varphi) + (\nabla V)W\varphi\\ &= W\boldsymbol{X}'(\varphi) + (\mathrm{i}\hbar\nabla W + (\nabla V)W)\varphi\\ &= W\boldsymbol{X}'(\varphi),\end{aligned} \tag{F.8}$$

where on the second line we use the simple computation of $\boldsymbol{X}'(W\varphi)$ from the explicit form of $\boldsymbol{X}'$ and on the third line that $\nabla W = (\mathrm{i}\hbar^{-1}\nabla V)W$. Therefore if J' is the unitary operator on $\mathcal{H}'$ given by $J'(\varphi) = W^{-1}\varphi$ we have $J'\boldsymbol{X}J'^{-1} = \boldsymbol{X}'$. Now, J' commutes with the Hamiltonian (multiplication by the function $\mathrm{i}c\omega_{\boldsymbol{p}}$), so that we can make the "change of basis" in $\mathcal{H}'$ to reduce to the case where $\boldsymbol{X} = \boldsymbol{X}'$. Thus, up to this change of basis, the standard position operator (F.1) is the only candidate as a position operator. As this candidate was unsatisfactory, we have argued that there is no exact position operator at all.

Appendix G

More on the Representations of the Poincaré Group

G.1 A Fun Formula

In Section 9.3 we start using a specific element D_p which sends p^* to p, and the reader might want to see explicit formulas for such an element. In this section we give such a formula for the canonical choice of D_p in the case $m > 0$.

Proposition G.1.1 *The unique positive Hermitian matrix $V \in SL(2,\mathbb{C})$ such that $\kappa(V)$ is the pure boost which sends $e_0 = (1,0,0,0)$ to $p \in X_1$ (i.e. $p^2 = 1$) is given by the formula*

$$V = \frac{1}{\sqrt{2(1+p^0)}}\begin{pmatrix} p^0+1+p^3 & p^1 - ip^2 \\ p^1 + ip^2 & p^0+1-p^3 \end{pmatrix} = \frac{1}{\sqrt{2(1+p^0)}}(I + M(p)). \tag{G.1}$$

Proof Since V is positive Hermitian it is of the type $V = UWU^{-1}$ where U is unitary and W is diagonal. Since $V \in SL(2,\mathbb{C})$, the eigenvalues of W are of the type e^s, e^{-s}, that is, using (8.3) V is of the type

$$\begin{aligned} V &= \begin{pmatrix} \alpha & \beta \\ -\beta^* & \alpha^* \end{pmatrix}\begin{pmatrix} e^s & 0 \\ 0 & e^{-s} \end{pmatrix}\begin{pmatrix} \alpha^* & -\beta \\ \beta^* & \alpha \end{pmatrix} \\ &= \begin{pmatrix} \cosh s + a\sinh s & b\sinh s \\ b^*\sinh s & \cosh s - a\sinh s \end{pmatrix}, \end{aligned} \tag{G.2}$$

where the second line is obtained through computation, and where $a = \alpha\alpha^* - \beta\beta^*$, $b = -2\alpha\beta$. Since $M(e_0)$ is the identity matrix,

$$M(\kappa(V)(e_0)) = VV^\dagger = UW^2U^\dagger = \begin{pmatrix} \alpha & \beta \\ -\beta^* & \alpha^* \end{pmatrix}\begin{pmatrix} e^{2s} & 0 \\ 0 & e^{-2s} \end{pmatrix}\begin{pmatrix} \alpha^* & -\beta \\ \beta^* & \alpha \end{pmatrix},$$

whose value is given by the formula (G.2) (for $2s$ rather than s). Now, to find V with $\kappa(V)(e_0) = p$ it suffices that the previous matrix is $M(p)$, yielding the relations $\cosh 2s \pm a\sinh 2s = p^0 \pm p^3$ and $b\sinh 2s = p^1 - ip^2$, $b^*\sinh 2s = p^1 + ip^2$, i.e.

$$\cosh 2s = p^0\ ;\ a\sinh 2s = p^3\ ;\ b\sinh 2s = p^1 - ip^2. \tag{G.3}$$

Finally $\cosh s = \sqrt{(p^0+1)/2}$ and $\sinh s = \sqrt{(p^0-1)/2}$, and since $\sinh 2s = 2\sinh s\cosh s$, using this in (G.2) yields (G.1). □

Exercise G.1.2 In the case $p^1 = p^2 = 0$, $p^0 = \cosh s$, $p^3 = \sinh s$ convince yourself that (G.1) reduces to

$$V = \begin{pmatrix} e^{s/2} & 0 \\ 0 & e^{-s/2} \end{pmatrix}.$$

Corollary G.1.3 *The unique positive Hermitian matrix D_p such that $\kappa(D_p)$ sends $p^* = (mc,0,0,0)$ to $p \in X_m$ is given by*

$$D_p = \frac{1}{\sqrt{2mc(mc+p^0)}}(mcI + M(p)). \tag{G.4}$$

In view of (9.36) one has

$$D_p^2 = D_p D_p^\dagger = \frac{1}{mc} M(p),$$

and some physics textbooks (such as [64, equation (3.50)]) write formulas such as $D_p = \sqrt{M(p)/mc}$.

Exercise G.1.4 Check the formula (10.98) i.e. $D_{P(p)} = D_p^{-1}$ from (G.4).

G.2 Higher Spin: Bargmann–Wigner and Rarita–Schwinger

The present section illustrates the methods of Section 9.5. Even though up to unitary equivalence there is only one representation of $\mathcal{P}^*$ describing a particle of mass $m > 0$ and given spin j, we may present this representation in many different ways, and we explain in this section two ingenious ways this has been done, trying to present these results in a less mysterious way than is usually done.

Let us first review the basic construction, a version of Theorem 9.5.3 including parity. We start with a representation V of $SL^+(2,\mathbb{C})$ on a finite-dimensional space $\mathcal{H}_0$, and a subspace $\mathcal{G}$ of $\mathcal{H}_0$ such that the restriction of V to $\mathcal{G}$ and $SU(2)^+$ is unitary and irreducible. Considering the space $\mathcal{G}_p = V(D_p)(\mathcal{G})$, the desired representation of $\mathcal{P}^{*+}$ lives on the space $\mathcal{H}$ of functions $\varphi\colon X_m \to \mathcal{H}_0$ such that $\varphi(p) \in \mathcal{G}_p$, provided with the norm

$$\|\varphi\|^2 = \int \|V(D_p)^{-1}(\varphi(p))\|^2 \mathrm{d}\lambda_m(p). \tag{G.5}$$

It is given by the formula

$$U(a,A)(\varphi)(p) = \exp(\mathrm{i}(a,p))V(A)[\varphi(A^{-1}(p))]. \tag{9.24}$$

There are two objectives in such a construction.[1]

- Get a nice description of $\mathcal{H}$ as a set of functions satisfying certain equations.
- Get a pleasant formula for the norm (G.5).

Both constructions we present will use the representation S of $SL^+(2,\mathbb{C})$ on $\mathbb{C}^4$ considered in Section 8.10 and the γ matrices (8.49) which the reader should review now. In both constructions $V(P')$ will be the identity on $\mathcal{G}$, so the restriction of V to $\mathcal{G}$ and $SU(2)^+$ is determined by the restriction of V to $\mathcal{G}$ and $SU(2)$, and this restriction will be unitarily equivalent to the representation π_j, as expected for a particle of spin $j/2$.

G.2.1 Bargmann–Wigner

Our first construction is a generalization of Theorem 9.10.4 so that the reader should review Section 9.10 now. We consider for $\mathcal{H}_0$ the space of symmetric tensors $x = (x_{n_1,n_2,\ldots,n_j})_{n_1,n_2,\ldots,n_j \in \{1,2,3,4\}}$, and the representation $V = S^{\otimes j}$ of $SL^+(2,\mathbb{C})$ in $\mathcal{H}_0$ given by

$$V(A)(x)_{n_1,\ldots,n_j} = \sum_{k_1,\ldots,k_j \in \{1,2,3,4\}} S(A)_{n_1,k_1} \cdots S(A)_{n_j,k_j} x_{k_1,\ldots,k_j}.$$

[1] One may argue that these objectives are somewhat illusory. A simpler-looking formula need not necessarily provide deeper understanding.

Consider the two-dimensional subspace W of $\mathbb{C}^4$ consisting of vectors u such that $\gamma_0(u) = u$, or, equivalently, $\gamma(p^*)(u) = mcu$. This space is invariant by all the operators $S(A)$ for $A \in SU(2)^+$. This is because if $A \in SU(2)^+$ then $S(A)$ commutes with γ_0 (as is straightforward to check). In fact the action of $S(P')$ on W is trivial (as is also straightforward to check). Consider the space $\mathcal{G} = W_s^{\otimes j} \subset \mathcal{H}_0$ consisting of symmetric tensors in $W^{\otimes j}$. Then the restriction of V to $SU(2)^+$ and $\mathcal{G}$ is unitary. According to Proposition 8.3.1, the restriction of V to $SU(2)$ and $\mathcal{G}$ is π_j. To understand how to describe $\mathcal{G}_p$ it is useful to first prove the following simple fact about tensor products.

Lemma G.2.1 *Consider a finite-dimensional space X and a linear operator U on X. Then for another finite-dimensional space Y we have*

$$\ker(U \otimes 1) = \ker U \otimes Y.$$

Proof Since $(U \otimes 1)(x \otimes y) = U(x) \otimes y$ is zero whenever $U(x) = 0$, it is obvious that $\ker U \otimes Y \subset \ker(U \otimes 1)$. Equality follows as these spaces have the same dimension, since

$$\begin{aligned}\dim(\ker U \otimes 1) &= \dim X \dim Y - \operatorname{rank}(U \otimes 1)\\ &= (\dim X - \operatorname{rank} U)\dim Y\\ &= \dim(\ker U)\dim Y\\ &= \dim \ker U \otimes Y.\end{aligned}$$

□

Lemma G.2.2 *Consider a finite-dimensional space X, a linear operator U on X and $Z = \ker U$. For $k \le j$ consider the operator $U_k = \bigotimes_{\ell \le j} U_{\ell,k}$ where $U_{k,k} = U$ and $U_{\ell,k} = 1$ if $\ell \ne k$. Then for $x \in X^{\otimes j}$ we have $x \in Z^{\otimes j}$ if and only if for all $k \le j$ it holds that $U_k(x) = 0$. In particular if x is a symmetric tensor then $x \in Z^{\otimes j}$ (or, equivalently, $x \in Z_s^{\otimes j}$) if and only if $U_1(x) = 0$.*

Proof The non-trivial statement is that if $x \in X^{\otimes j}$ is such that $U_k(x) = 0$ for each $k \le j$ then $x \in Z^{\otimes j}$. It is proved by induction over j as a consequence of the preceding lemma. Since $U_1(x) = 0$ one has $x = z \otimes y$ with $z \in \ker U$ and one uses to induction hypothesis on y. □

We are now ready to identify $\mathcal{G}_p$. Since $\mathcal{G} = W_s^{\otimes j}$, then $\mathcal{G}_p = V(D_p)(\mathcal{G}) = (S(D_p)(W))_s^{\otimes j}$, the space of symmetric tensors in $(S(D_p)(W))^{\otimes j}$. As explained just after (9.69) the space $S(D_p)(W)$ is described by the equation $\gamma(p)(u) = mcu$, i.e. $S(D_p)(W) = \ker U$ where $U(u) = \gamma(p)(u) - mcu$. According to Lemma G.2.2 the space $\mathcal{G}_p$ is described by the equation $\gamma(p) \otimes 1 \otimes \cdots \otimes 1(x) = mcx$.

On the space of functions $\varphi\colon X_m \to \mathcal{H}_0$, for $k \le j$ let us define the operator $\widehat{D}_k(\varphi)$ by

$$\widehat{D}_k(\varphi)(p) = \Big(\bigotimes_{\ell \le j} U_{\ell,k}\Big)(\varphi(p)),$$

where $U_{k,k} = \gamma(p)$ and $U_{\ell,k} = 1$ for $\ell \ne k$. Then the space $\mathcal{H}$ of functions φ such that $\varphi(p) \in \mathcal{G}_p$ is described by the equations

$$\forall k \le j\,,\ \widehat{D}_k(\varphi) = mc\varphi, \tag{G.6}$$

which are equivalent to the first one of them, $\widehat{D}_1(\varphi) = mc\varphi$. These are called the *Bargmann–Wigner equations*, see [5]. The norm $\|\cdot\|_p$ on $\mathcal{G}$ can be calculated just as in the case $j = 1$ of Section 9.10. For this we consider the sesqui-linear form B on $(\mathbb{C}^4)^{\otimes j} \otimes (\mathbb{C}^4)^{\otimes j}$ such that

$$B(y_1 \otimes \cdots \otimes y_j, z_1 \otimes \cdots \otimes z_j) = \prod_{k \le j} y_k^\dagger z_k,$$

where $y_k^\dagger$ is the row vector which is the conjugate-transpose of y_k. Using an orthonormal basis one checks that

$$B(x,x) = \|x\|^2. \tag{G.7}$$

We then prove that $C(x,y) := B(x, \gamma_0^{\otimes j} y)$ satisfies $C(V(A)x, V(A)y) = C(x,y)$, so that for $x \in \mathcal{G}_p$ we have $\|x\|_p^2 = C(x,x)$ because this holds for $p = p^*$ by (G.7). No new ideas are required compared to the case $j = 1$, so we simply state the result.

Theorem G.2.3 *The space of functions φ which associate to each point a symmetric tensor in $(\mathbb{C}^4)^{\otimes j}$ and which satisfy the Bargmann–Wigner equation $\widehat{D}_1(\varphi) = mc\varphi$ defines a unitary representation of $\mathcal{P}^{*+}$ when the action of the group is given by (9.24) and the norm is given by*

$$\|\varphi\|^2 = \int C(\varphi(p), \varphi(p))\mathrm{d}\lambda_m(p).$$

This representation corresponds to a particle of mass m and spin $j/2$.

G.2.2 Rarita–Schwinger

Our second construction can be used to describe particles of spin $n + 1/2$, although we consider only the case $n = 1$.[2] Here the space $\mathcal{H}_0$ is $\mathbb{C}^4 \otimes \mathbb{C}^4$ and for $A \in SL^+(2,\mathbb{C})$ we set

$$V(A) = S(A) \otimes \kappa(A), \tag{G.8}$$

where $\kappa(A)$ is considered as an operator on $\mathbb{C}^4$. The coordinates in this copy of $\mathbb{C}^4$ are numbered from 0 to 3. We will denote a point x of $\mathcal{H}_0$ by $x = (x_n^\mu)$ where the *spinor index* n ranges from 1 to 4 and the *Lorentz index* μ ranges from 0 to 3. When there are both spinor and Lorentz indices, it is customary not to write the spinor indices. Thus x^μ denotes a four-dimensional column vector (i.e. a 4×1 matrix, which we will call a spinor), and $x \in \mathcal{H}_0$ consists of four spinors x^0, x^1, x^2, x^3. Thus $\gamma_\nu x^\mu$, the product of a 4×4 matrix and 4×1 matrix is also a 4×1 matrix, i.e. a spinor. In this manner, given $x \in \mathcal{H}_0$ we define the spinor $T(x) = \gamma_\mu x^\mu$ where as usual repeated Lorentz indices are summed. The following lemma contains much of the magic of this construction.

Lemma G.2.4 *For each $A \in SL^+(2,\mathbb{C})$ we have*

$$TV(A) = S(A)T. \tag{G.9}$$

Proof Writing for once both spinor and Lorentz indices, $V(A)(x)$ is given by its components $V(A)(x)_n^\mu$:

$$V(A)(x)_n^\mu = \sum_{k\le 4}\sum_{0\le\nu\le 3} S(A)_{n,k}\kappa(A)^\mu{}_\nu x_k^\nu,$$

or, writing only Lorentz indices,

$$V(A)(x)^\mu = \kappa(A)^\mu{}_\nu S(A)x^\nu, \tag{G.10}$$

so that, using $\gamma_\mu\kappa(A)^\mu{}_\nu = S(A)\gamma_\nu S(A)^{-1}$ by (8.51),

$$TV(A)(x) = \gamma_\mu\kappa(A)^\mu{}_\nu S(A)x^\nu = S(A)\gamma_\nu x^\nu = S(A)T(x).$$ □

[2] No elementary particle of spin $\ge 3/2$ has yet been observed, but the consideration of such particles is of theoretical interest. In particular in certain theories there exists a particle of spin 3/2, the gravitino.

Our next goal is to identify a subspace $\mathcal{G}'$ of $\mathcal{H}_0$ of dimension 6 on which the restriction of V to $SU(2)$ is equivalent to $\pi_1 \otimes \pi_2$. Then, as shown by Proposition D.7.6, this representation of dimension 6 decomposes in two irreducible spaces, respectively of dimension 2 and 4, corresponding to representations of spin 1/2 and 3/2 respectively. We will then identify the space $\mathcal{G}$ of dimension 4 such that the restriction of V to $SU(2)$ and $\mathcal{G}$ is equivalent to π_3.

Lemma G.2.5 *The following subspaces of $\mathcal{H}_0$ are invariant under the restriction of V to $SU(2)^+$:*

$$\mathcal{G}_1 = \{x\ ;\ \forall \mu \le 3, \gamma_0 x^\mu = x^\mu\},$$

$$\mathcal{G}_2 = \{x\ ;\ x^0 = 0\},$$

$$\mathcal{G}_3 = \{x\ ;\ T(x) = 0\}.$$

Proof The fact that $\mathcal{G}_3$ is invariant is an obvious consequence of Lemma G.2.4.

To prove that $\mathcal{G}_1$ is invariant we recall that the subspace $W = \{y; \gamma_0 y = y\}$ of $\mathbb{C}^4$ is invariant under $S(A)$ for $A \in SU(2)^+$ so that $\mathcal{G}_1 = W \otimes \mathbb{C}^4$ is invariant under $V(A) = S(A) \otimes \kappa(A)$. Another way to see this is to write the definition of $\mathcal{G}_1$ as

$$\mathcal{G}_1 = \{x\ ;\ (\gamma_0 \otimes 1)x = x\},$$

and to observe that $\gamma_0 \otimes 1$ commutes with $V(A)$ for $A \in SU(2)$.

To prove that $\mathcal{G}_2$ is invariant we observe that $W' = \{y \in \mathbb{C}^4; y^0 = 0\}$ is invariant under $\kappa(A)$ for $A \in SU(2)^+$ so that $\mathcal{G}_2 = \mathbb{C}^4 \otimes W'$ is invariant under $V(A) = S(A) \otimes \kappa(A)$. □

To get a more concrete description of these subspaces, we decompose each spinor x^μ in two parts y^μ and z^μ composed respectively of the first two and the last two components. The space $\mathcal{G}_1$ consists simply of the tensors x which satisfy

$$\forall \mu\ ;\ y^\mu = z^\mu. \tag{G.11}$$

Using the explicit definition (8.49) of the matrices γ_μ, the condition $T(x) = 0$ is equivalent to

$$y^0 - \sigma_1 y^1 - \sigma_2 y^2 - \sigma_3 y^3 = 0\ ;\ z^0 + \sigma_1 z^1 + \sigma_2 z^2 + \sigma_3 z^3 = 0.$$

For $x \in \mathcal{G}_1$ this means

$$y^0 = 0\ ;\ \sigma_1 y^1 + \sigma_2 y^2 + \sigma_3 y^3 = 0. \tag{G.12}$$

This shows that $\mathcal{G} := \mathcal{G}_1 \cap \mathcal{G}_3$ is of dimension 4, as the second part of (G.12) gives two relations between the six components of y^1, y^2, y^3. Moreover $\mathcal{G} \subset \mathcal{G}' := \mathcal{G}_1 \cap \mathcal{G}_2$.

Lemma G.2.6 *The space $\mathcal{G}' = \mathcal{G}_1 \cap \mathcal{G}_2$ is of dimension 6 and the restriction of V to $SU(2)$ and $\mathcal{G}'$ is unitarily equivalent to $\pi_1 \otimes \pi_2$.*

Proof A point x of $\mathcal{G}'$ is determined by the components x_k^μ for $k = 1,2$ and $1 \le \mu \le 3$, so that $\mathcal{G}' = \mathbb{C}^2 \otimes \mathbb{C}^3$. To prove the last claim one has to go back to the definition of V and convince oneself that for $A \in SU(2)$ the restriction of $V(A)$ to $\mathcal{G}'$ identifies with $S(A)' \otimes \kappa(A)'$ where $S(A)'$ is the restriction of $S(A)$ to the space $\{x \in \mathbb{C}^4; \gamma_0 x = x\}$ and $\kappa(A)'$ is the restriction of $\kappa(A)$ to $\{x \in \mathbb{C}^4; x^0 = 0\}$, and to use Exercise 8.4.5. □

Since the subspace $\mathcal{G}$ of $\mathcal{G}'$, of dimension 4, is invariant under the restriction of V to $SU(2)$ and $\mathcal{G}'$, the restriction of V to $SU(2)$ and this subspace has to be equivalent to π_3. This can also be checked directly but it is not necessary to do so.

The reader may check using previous arguments that the space $\mathcal{G}_p := V(D_p)(\mathcal{G})$ is described by the equations

$$\forall \mu\ ,\ \gamma(p)x^\mu = mcx^\mu\ ;\ \gamma_\mu x^\mu = 0. \tag{G.13}$$

The space $\mathcal{H}$ of functions $\varphi\colon X_m \to \mathcal{H}_0$ which satisfy $\varphi(p) \in \mathcal{G}_p$ for each p is therefore described by the equations

$$\forall\mu\ ,\ \widehat{D}(\varphi^\mu) = mc\varphi^\mu\ ;\ \gamma_\mu\varphi^\mu = 0, \tag{G.14}$$

which are called the *Rarita–Schwinger equations* (here in momentum form). Furthermore as we show now there is a simple formula to compute the norm $\|\cdot\|_p$.

Lemma G.2.7 *For $x \in \mathcal{H}_0$ the quantity*

$$B(x) = -\eta_{\mu,\nu}(x^\mu)^\dagger\gamma_0 x^\nu = -x_\mu^\dagger\gamma_0 x^\mu = -(x^0)^\dagger\gamma_0 x^0 + \sum_{1\le\mu\le 3}(x^\mu)^\dagger\gamma_0 x^\mu \tag{G.15}$$

satisfies $B(V(A)(x)) = B(x)$ and

$$x \in \mathcal{G} \Rightarrow B(x) = \|x\|^2\Big(=\sum_{k,\mu}|x_k^\mu|^2\Big). \tag{G.16}$$

Proof Using (G.15), we obtain, using that $\kappa(A)^\mu_{\ \lambda}$ is a real number

$$\begin{aligned} B(V(A)(x)) &= -\eta_{\mu,\nu}(\kappa(A)^\mu_{\ \lambda}S(A)x^\lambda)^\dagger\gamma_0\kappa(A)^\nu_{\ \eta}S(A)x^\eta \\ &= -\eta_{\mu,\nu}\kappa(A)^\mu_{\ \lambda}\kappa(A)^\nu_{\ \eta}(S(A)x^\lambda)^\dagger\gamma_0 S(A)x^\eta. \end{aligned} \tag{G.17}$$

Now, $\eta_{\mu,\nu}\kappa(A)^\mu_{\ \lambda}\kappa(A)^\nu_{\ \eta} = \eta_{\lambda,\eta}$ because $\kappa(A) \in SO^\uparrow(1,3)$ while

$$(S(A)x^\lambda)^\dagger\gamma_0 S(A)x^\eta = (x^\lambda)^\dagger\gamma_0 x^\eta$$

by (9.72). This proves that $B(V(A)(x)) = B(x)$. Next, for $x \in \mathcal{G}$ we have $x^0 = 0$ and $\gamma_0 x^\mu = x^\mu$, so that $(x^\mu)^\dagger\gamma_0 x^\mu = (x^\mu)^\dagger x^\mu = \sum_{k\le 4}|x_k^\mu|^2$ and this proves (G.16). □

We may then state the following:

Theorem G.2.8 *The space of functions φ from X_m to $\mathbb{C}^4\otimes\mathbb{C}^4$ satisfying the Rarita–Schwinger equations (G.14), provided with the norm*

$$\|\varphi\|^2 = -\int(\varphi_\mu(p))^\dagger\gamma_0\varphi(p)^\mu \mathrm{d}\lambda_m(p)$$

gives a unitary representation of $\mathcal{P}^{*+}$ when the action of the group is given by the formula (9.24). This representation models a particle of mass m and spin $3/2$.

Appendix H

Hamiltonian Formalism for Classical Fields

We first investigate how one might construct a Hamiltonian for the Lagrangian with Lagrangian density (10.57). We then present complements to Sections 6.4 and 6.5, and we try to explain in mathematical terms some potentially confusing considerations which are ubiquitous in elementary physics textbooks, see e.g. [35, page 532]. The reader needs to be fluent with Section 10.12.

H.1 Hamiltonian for the Massive Vector Field

Let us recall that the massive vector field has a Lagrangian density given by (10.57),

$$\mathcal{L} = -\frac{\hbar^2 c^2}{4} F_{\nu\mu} F^{\nu\mu} + \frac{m^2 c^4}{2} A_\nu A^\nu, \tag{10.57}$$

where $F_{\nu\mu} = \partial_\nu A_\mu - \partial_\mu A_\nu$. In this formula, the variables are the A_ν and we are trying to write equations of motion for $(A_\nu)_{0\le\nu\le3}$. As we pointed out, the variables A_ν are not independent, and we proceed to show that A_1, A_2, A_3 can be identified as "a proper set of independent variables". More precisely, assuming that A_1, A_2, A_3 are independent variables, we will guess from the general method (6.36) (or more exactly its extension to the case of classical fields) what the Hamiltonian should be, and we will check that this Hamiltonian gives the correct equations of motion. We recall the notation $\partial_t = \partial/\partial t = c\partial_0$. We think of the Lagrangian density as a function of the "independent variables" A_1, A_2, A_3 and their derivatives $\partial_\nu A_k$. For each variable A_k, $k = 1,2,3$ we compute its "conjugate momentum"

$$\pi_k = \frac{\partial \mathcal{L}}{\partial(\partial_t A_k)} = -c\hbar^2 F^{0k} = c\hbar^2 F_{0k}. \tag{H.1}$$

and the tentative Hamiltonian density is given as[1]

$$\mathcal{H} = \sum_{k\le3} \pi_k \partial_t A_k - \mathcal{L} = c\sum_{k\le3} \pi_k \partial_0 A_k - \mathcal{L}. \tag{H.2}$$

This is exactly the procedure we used to obtain (6.50), but in (6.50) there was a single independent variable, while here there are three of them. For things to make sense (H.2) has to be expressed as a function of the π_k, the A_k and their derivatives $\partial_\nu A_k$, and $\partial_\nu \pi_k$ for $\nu \ge 1$. The goal is then to eliminate the quantities $\partial_0 A_k$ and A_0 from the expression (H.2) (using somewhat ad hoc manipulations) as follows. First we rewrite the Lagrangian density (10.57) as

$$\mathcal{L} = \frac{1}{2\hbar^2}\sum_{k\le3} \pi_k^2 - \frac{\hbar^2 c^2}{4}\sum_{k,\ell\le3} F_{k\ell}^2 + \frac{m^2 c^4}{2}(A_0^2 - \sum_{k\le3} A_k^2), \tag{H.3}$$

[1] Let us recall the convention that the Latin indexes range from 1 to 3, so that below $\sum_{k\le3}$ means $\sum_{1\le k\le3}$.

Next, using the definition of F_{0k} and then (H.1),

$$c\partial_0 A_k = cF_{0k} + c\partial_k A_0 = \frac{1}{\hbar^2}\pi_k + c\partial_k A_0,$$

so that

$$c\pi_k \partial_0 A_k = \frac{1}{\hbar^2}\pi_k^2 + c\pi_k \partial_k A_0 = \frac{1}{\hbar^2}\pi_k^2 + c\partial_k(\pi_k A_0) - cA_0\partial_k \pi_k. \tag{H.4}$$

Furthermore, by (10.62) we have $m^2c^2 A_0 + \hbar^2 \partial^\nu F_{\nu 0} = 0$ i.e., using (H.1),

$$m^2c^2 A_0 = -\hbar^2 \sum_{k\le 3} \partial_k F_{0k} = -\frac{1}{c}\sum_{k\le 3}\partial_k \pi_k, \tag{H.5}$$

so that $-c\sum_{k\le 3} A_0 \partial_k \pi_k = m^2c^4 A_0^2$. Consequently, by summation of the relations (H.4),

$$\sum_{k\le 3} c\pi_k \partial_0 A_k = \frac{1}{\hbar^2}\sum_{k\le 3}\pi_k^2 + m^2c^4 A_0^2 + c\sum_{k\le 3}\partial_k(\pi_k A_0),$$

and, using algebra in the first line and again (H.5) in the form $A_0 = -1/(m^2c^3)\sum_{k\le 3}\partial_k \pi_k$ in the second line,

$$\begin{aligned}\mathcal{H} &= \frac{1}{2\hbar^2}\sum_{k\le 3}\pi_k^2 + \frac{\hbar^2 c^2}{4}\sum_{k,\ell\le 3} F_{k\ell}^2 + \frac{m^2c^4}{2}(A_0^2 + \sum_{k\le 3} A_k^2) + \mathcal{R} \\ &= \frac{1}{2\hbar^2}\sum_{k\le 3}\pi_k^2 + \frac{\hbar^2 c^2}{4}\sum_{k,\ell\le 3} F_{k\ell}^2 + \frac{m^2c^4}{2}\sum_{k\le 3} A_k^2 + \frac{1}{2m^2c^2}\Big(\sum_{k\le 3}\partial_k \pi_k\Big)^2 + \mathcal{R},\end{aligned} \tag{H.6}$$

where $\mathcal{R}$ is a sum of partial derivatives. Now we recall that the Hamiltonian is the integral of the Hamiltonian density. Since $\mathcal{R}$ is of integral zero because it is a sum of derivatives (and assuming of course that everything goes to zero fast enough at infinity) it does not contribute to the Hamiltonian. In Section H.4 we prove that this Hamiltonian density does provide the correct equations of motion.

H.2 From Hamiltonians to Lagrangians

In Section 6.5 we have explained how in favorable circumstances one may go from Lagrangian formalism to Hamiltonian formalism. Here we first briefly explain that conversely we may also go in the reverse direction. To describe a mechanical system with n degrees of freedom, say a massive point in dimension n, it is possible to start directly from a Hamiltonian, a function $H(x,p)$ on phase space $\mathbb{R}^n \times \mathbb{R}^n$. The equations of motion are Hamilton's equations of motion (6.39) and (6.40)

$$\dot{p}_k = -\frac{\partial H}{\partial x_k}\ ;\ \dot{x}_k = \frac{\partial H}{\partial p_k}. \tag{H.7}$$

Let us set

$$v_k = \frac{\partial H}{\partial p_k}, \tag{H.8}$$

and $v = (v_k)_{k\le n}$. In good situations, we can compute p as a function $p(x,v)$ of x and v and then H as a function $H(x,p(x,v))$ of x and v where now x and v are viewed as independent variables. We may then define the Lagrangian

$$\mathcal{L}(x,v) = v\cdot p(x,v) - H(x,p(x,v)).$$

Then a solution $x(t), p(t)$ of the equations of motion (H.7) is also a solution of the Euler-Lagrange equations. Indeed,

$$\frac{\partial \mathcal{L}}{\partial x_k} = -\frac{\partial H}{\partial x_k} \; ; \; \frac{\partial \mathcal{L}}{\partial v_k} = p_k$$

because the other terms eliminate thanks to (H.8), so that

$$\frac{\mathrm{d}}{\mathrm{d}t}\frac{\partial \mathcal{L}}{\partial v_k} = \dot{p}_k = -\frac{\partial H}{\partial x_k} = \frac{\partial \mathcal{L}}{\partial x_k}.$$

The method to go from the function $L(x,v)$ to the function $H(x,p)$ and back is called *the Legendre transform.*

Let us examine our favorite example

$$H(x,p) = \frac{1}{2m} p \cdot p + V(x).$$

Then $v_k = p_k/m$ and $p(x,v) = mv$ so that

$$\mathcal{L}(x,v) = p \cdot v - H(v, p(x,v)) = \frac{m}{2} v \cdot v - V(x).$$

H.3 Functional Derivatives

At the end of Section 6.4 we met no particular difficulties in extending the Lagrangian approach from the case of n degrees of freedom to the case of a continuous number of degrees of freedom. We now start the same program in the case of the Hamiltonian formalism: how to extend the equations of motion (H.7) to the continuous case. For simplicity we will discuss a specific case, the generalizations of which are obvious. The main issue is that the equations (H.7) are defined in terms of the partial derivatives of a function $H(x,p)$, when $x, p \in \mathbb{R}^n$. In the continuous case $H(x,p)$ is replaced by $H(u,\pi)$, a *functional* of two functions u, π on $\mathbb{R}^k$. We follow here the usual terminology to call a function a functional if its arguments are themselves functions. Then, as we already explained, u is a family of numbers $(u(x))_{x\in\mathbb{R}^k}$ which is the continuous version of the family $(x_k)_{k\le n}$. As the contents of the rest of this appendix are very formal, it is better not to specify what is meant by "function" although probably it would be a good idea to consider only elements of $\mathcal{S}^4$. We want to define "the partial derivative of H with respect to $u(x)$". This is the fearsome *functional derivative* of physicists, who rarely formulate it in a mathematician-friendly language. Functional derivative is an extension of the idea of partial derivative when a finite number of variables is replaced by a continuous family of variables. To understand the idea of the forthcoming definition, let us observe the following formula, in the case of n degrees of freedom:

$$\lim_{\varepsilon\to 0} \varepsilon^{-1}(H(x+\varepsilon y, p) - H(x,p)) = \sum_{k\le n} y_k \frac{\partial H}{\partial x_k},$$

which determines the numbers $\partial H/\partial x_k$. In the continuous case one simply replaces the discrete sum by an integral.

Definition H.3.1 Given a functional $H(u,\pi)$ and two specific choices u and π of the arguments, a distribution ξ in the variable x is called a functional derivative and is denoted[2]

$$\frac{\delta H}{\delta u(x)}$$

[2] I find this traditional notation horrendous. I think $\delta H/\delta u$ would be much better.

at these arguments u, π if for each test function $\varphi \in \mathcal{S}^k$ we have

$$\int \varphi(x) \mathrm{d}\xi(x) := \xi(\varphi) = \lim_{\varepsilon \to 0} \varepsilon^{-1}(H(u + \varepsilon\varphi, \pi) - H(u, \pi)). \tag{H.9}$$

We are going to give many concrete examples below. The horrendous notation does not help. Again, given u and π the functional derivative $\delta H/\delta u(x)$ is a distribution in the variable x. In good cases, $\delta H/\delta u(x)$ will be a concrete expression depending on x, u, π. Then, given x, $\delta H/\delta u(x)$ is a new functional of u and π, and this is what motivates the notation. One difficulty here is that one should resist attempting to directly define in a general setting the value of $\delta H/\delta u(x)$ for a given x, because as the first example below shows this does not make much sense. (On the other hand as soon as one moves to the specific situations we will consider below what $\delta H/\delta u(x)$ means becomes clear.) We define $\delta H/\delta \pi(x)$ in a similar manner. As a functional derivative itself depends on the two functions u and π, the procedure can be iterated.

There is little doubt that the functional derivative, as well as certain other tricky notions that will be explained below, can be rigorously defined in a suitable space of distributions. The purpose of this appendix is however not to build a rigorous theory but to explain the meaning of what you may read in a physics textbook.

Example H.3.2 Consider a given point y of $\mathbb{R}^k$ and the case where $H(u, \pi) = u(y)$. Then it is pretty obvious from (H.9) that

$$\frac{\delta H}{\delta u(x)} = \delta_y(x).$$

Example H.3.3 Consider the case where $H(u, \pi) = \int u(x)\varphi(x)\mathrm{d}^k x$ for a certain test function φ. Then

$$\frac{\delta H}{\delta u(x)} = \varphi(x).$$

Example H.3.4 Consider the case where

$$H(u, \pi) = \int u(x)^{2n} \mathrm{d}^k x. \tag{H.10}$$

It is assumed that the integral exists so that the formula makes sense. It should then be pretty obvious that

$$\frac{\delta H}{\delta u(x)} = 2nu(x)^{2n-1}.$$

Example H.3.5 Consider the case where

$$H(u, \pi) = \int \Big(\frac{\partial u}{\partial x_j}\Big)^2 (x) \mathrm{d}^k x \tag{H.11}$$

Then

$$\lim_{\varepsilon \to 0} \varepsilon^{-1}(H(u + \varepsilon\varphi, \pi) - H(u, \pi)) = 2 \int \frac{\partial u}{\partial x_j}(x) \frac{\partial \varphi}{\partial x_j}(x) \mathrm{d}^k x$$

and integration by part shows that

$$\frac{\delta H}{\delta u(x)} = -2 \frac{\partial^2 u}{\partial x_j^2}(x). \tag{H.12}$$

This means exactly that, in the sense of distributions, $\delta H/\delta u(x)$ is the partial derivative with respect to x_j of the function $-2\partial u/\partial x_j$. If we are lucky (H.12) might be an equality of functions.

Example H.3.6 More generally, we may consider expressions of the type

$$H(u,\pi) = \int \psi(x,u(x),\pi(x),\partial_1 u(x),\ldots,\partial_k u(x))\mathrm{d}^k x, \tag{H.13}$$

or even more complicated ones. In this case, if all goes well, and pretending in our notation that all the arguments of ψ are independent variables,

$$\frac{\delta H}{\delta u(x)} = \frac{\partial \psi}{\partial u} - \sum_{i\le k} \partial_i \frac{\partial \psi}{\partial_i u}, \tag{H.14}$$

where on the right-hand side the arguments are the same as in (H.13).

Definition H.3.7 For a Hamiltonian $H(u,\pi)$ the equations

$$\dot\pi = -\frac{\delta H}{\delta u(x)}\ ;\ \dot u = \frac{\delta H}{\delta \pi(x)} \tag{H.15}$$

are called Hamilton's equations of motion.

In this compact notation we assume that u (and π) is a time-dependent function $u(x,t)$ of x, and we write $\dot u(x,t) = \partial u(x,t)/\partial t$. That is, $\dot u$ is the time-derivative of u seen as a point in a function space. Then the first equation above means

$$\dot\pi(x,t) = -\frac{\delta H}{\delta u(x)}(u,\pi),$$

where the right-hand side is the value of $\delta H/\delta u(x)$ computed at the functions u,π on $\mathbb{R}^k$ given by $u\colon y \mapsto u(y,t)$ and $\pi\colon y \mapsto \pi(y,t)$. If we are lucky, just as in the discrete case, given a Hamiltonian H the equation of motion determines the functions $u(x,t)$ and $\pi(x,t)$ given the "initial conditions" $u(x,0)$ and $\pi(x,0)$.

H.4 Two Examples

In this section we compute the equations of motion for two specific Hamiltonians. As a first illustration we consider the case where

$$H(u,\pi) = \int \mathcal{H}(u,\pi)(\boldsymbol{x})\mathrm{d}^3\boldsymbol{x}, \tag{H.16}$$

where the Hamiltonian density $\mathcal{H}(u,\pi)$ is given by the formula (6.50):

$$\mathcal{H}(u,\pi) = \frac{1}{2\hbar^2}\pi^2 + \frac{1}{2}\hbar^2 c^2 \sum_{1\le\nu\le 3} (\partial_\nu u)^2 + \frac{1}{2} m^2 c^4 u^2. \tag{H.17}$$

Then

$$\frac{\delta H}{\delta \pi(\boldsymbol{x})} = \frac{1}{\hbar^2}\pi(\boldsymbol{x})\ ;\ \frac{\delta H}{\delta u(\boldsymbol{x})} = -\hbar^2 c^2 \sum_{1\le\nu\le 3} \partial_\nu^2 u(\boldsymbol{x}) + m^2 c^4 u(\boldsymbol{x}), \tag{H.18}$$

so that the equations of motion are

$$\pi = \hbar^2 \dot u\ ;\ \dot\pi = \hbar^2 c^2 \sum_{1\le\nu\le 3} \partial_\nu^2 u - m^2 c^4 u,$$

and we recover the Klein-Gordon equation in the form

$$\hbar^2 \ddot u - \hbar^2 c^2 \sum_{1\le\nu\le 3} \partial_\nu^2 u + m^2 c^4 u = 0.$$

We turn to our second and more challenging example: the Hamiltonian (H.16) whose density $\mathcal{H}$ is given by (H.6), i.e.

$$\mathcal{H} = \frac{1}{2\hbar^2}\sum_{k\le 3}\pi_k^2 + \frac{\hbar^2c^2}{4}\sum_{k,\ell\le 3}F_{k\ell}^2 + \frac{m^2c^4}{2}\sum_{k\le 3}A_k^2 + \frac{1}{2m^2c^2}\Big(\sum_{k\le 3}\partial_k\pi_k\Big)^2 \qquad (H.6),$$

where $F_{k\ell} = \partial_k A_\ell - \partial_\ell A_k$. That is, here we have three variables A_1, A_2, A_3 and three "conjugate variables" π_k, each of which is a function on $\mathbb{R}^3$. Our goal is to show that Hamilton's equations of motion, which here become

$$\dot{\pi}_k = -\frac{\delta H}{\delta A_k(\boldsymbol{x})}\ ;\ \dot{A}_k = \frac{\delta H}{\delta \pi_k(\boldsymbol{x})}$$

allow us to recover the equations of motion (10.62). To lighten notation we define A_0 as in (H.5) i.e.

$$\sum_{k\le 3}\partial_k\pi_k = -m^2c^3A_0. \qquad (H.19)$$

Then the reader will prove the following formulas (whose proof follows from the various situations we have already explained):

$$\frac{\delta H}{\delta \pi_k(\boldsymbol{x})} = \frac{1}{\hbar^2}\pi_k + c\partial_k A_0\ ;\ \frac{\delta H}{\delta A_k(\boldsymbol{x})} = m^2c^4A_k - \hbar^2c^2\sum_{\ell\le 3}\partial_\ell F_{\ell k}.$$

For example to prove the first equality one uses that $\mathcal{H}$ depends on π_k through both the first and the last term of (H.6) and one uses (H.19). This gives the equations of motion

$$\dot{A}_k = \frac{1}{\hbar^2}\pi_k + c\partial_k A_0\ ;\ \dot{\pi}_k = -m^2c^4A_k + \hbar^2c^2\sum_{\ell\le 3}\partial_\ell F_{\ell k}. \qquad (H.20)$$

Let us now recall that $F_{0k} = \partial_0 A_k - \partial_k A_0 = c^{-1}\dot{A}_k - \partial_k A_0$, so that the first equation of motion above means

$$\pi_k = \hbar^2 c F_{0k}$$

and the second equation of motion becomes

$$\hbar^2 c\dot{F}_{0k} = -m^2c^4A_k + \hbar^2c^2\sum_{\ell\le 3}\partial_\ell F_{\ell k} \qquad (H.21)$$

i.e.

$$m^2c^4A_k + \hbar^2c^2\Big(\partial_0 F_{0k} - \sum_{\ell\le 3}\partial_\ell F_{\ell k}\Big) = 0, \qquad (H.22)$$

while (H.19) becomes

$$m^2c^4A_0 - \hbar^2c^2\sum_{k\le 3}\partial_k F_{k0} = 0. \qquad (H.23)$$

Thus (H.22) and (H.23) recover the equations of motion (10.62).

H.5 Poisson Brackets

Let us go back to the space $\mathbb{R}^n \times \mathbb{R}^n$, a point of which is denoted by (x, p). We recall that this space is called the phase space. Given two (smooth) functions f, g on phase space we define a new function $\{f, g\}$ on phase space called their *Poisson bracket* by the formula

$$\{f,g\} = \sum_{k\le n} \frac{\partial f}{\partial x_k}\frac{\partial g}{\partial p_k} - \frac{\partial g}{\partial x_k}\frac{\partial f}{\partial p_k}. \tag{H.24}$$

This operation satisfies remarkable properties. First it is anti-symmetric

$$\{f,g\} = -\{g,f\}. \tag{H.25}$$

Second, it satisfies the Leibniz's rule

$$\{f,gg'\} = g\{f,g'\} + g'\{f,g\}. \tag{H.26}$$

Finally, it satisfies the Jacobi identity

$$\{f,\{g,h\}\} + \{g,\{h,f\}\} + \{h,\{f,g\}\} = 0. \tag{H.27}$$

The verification of these is straightforward, but a bit tedious in the case of the Jacobi identity. From the definition we have the following

$$\{x_k,x_j\} = \{p_k,p_j\} = 0\ ;\ \{x_k,p_j\} = \delta_k^j. \tag{H.28}$$

Here of course x_k is seen as a function on phase space, the function $(x,p) \mapsto x_k$, etc. Moreover, Hamilton's equation of motion (H.7) can be written in the form

$$\dot{p}_k = \{p_k,H\}\ ;\ \dot{x}_k = \{x_k,H\}. \tag{H.29}$$

More generally, consider a function f on phase space, and think of $f = f(x(t),p(t))$ as a function of time, where $x(t)$ and $p(t)$ describe the motion of the system. Then

$$\dot{f} = \sum_{k\le n} \dot{x}_k\frac{\partial f}{\partial x_k} + \dot{p}_k\frac{\partial f}{\partial p_k}.$$

Substitution of the values of $\dot{x}_k$ and $\dot{p}_k$ given by the equations of motion (H.7) yields the following generalization of (H.29):

$$\dot{f} = \{f,H\}. \tag{H.30}$$

What precedes is not only a nice set of notations. It brings forward the Poisson bracket as the fundamental structure, paving the way to the use of transformations of phase space which preserve this bracket, the canonical transformations which we already mentioned very briefly, and for which we refer to Arnol'd's textbook [3], a really enjoyable book.

Poisson brackets being such an important and fruitful idea, it is rather natural to try to extend it to the setting of classical fields. In the rest of this appendix we explain how this is done in elementary physics texts.[3] It is now the set of pairs of functions (u,π) on $\mathbb{R}^n$ which play the role of the phase space. Given two functionals F,G on the phase space we would like to define a third functional, their Poisson bracket. A look at (H.24) suggests the formula

$$\{F,G\} = \int\Big(\frac{\delta F}{\delta u(x)}\frac{\delta G}{\delta\pi(x)} - \frac{\delta G}{\delta u(x)}\frac{\delta F}{\delta\pi(x)}\Big)\mathrm{d}^k x. \tag{H.31}$$

If we are lucky, all the functional derivatives occurring in this expression will be functions of x depending upon the variables u,π, the integral will make sense, and will define a new functional. When F,G are given by formulas such as in (H.13), (H.31) will often make sense. Using now (H.31) in the case where $F(u,\pi) = u(x)$, we find

$$\{u,H\} = \frac{\delta H}{\delta\pi(x)},$$

[3] Thus do not be surprised that our considerations sound like fairy tales: it is because they are!

and this means that the second equation of motion (H.15) takes the same form

$$\dot{u} = \{u, H\} \tag{H.32}$$

as in the classical case, and this is also the case of the first equation. Let us now look at (H.31) in the case where

$$F(u,\pi) = \int \varphi(x)u(x)\mathrm{d}^k x \; ; \; G(u,\pi) = \int \psi(x)\pi(x)\mathrm{d}^k x,$$

where φ, ψ are two given test functions. Then

$$\{F,G\} = \int \varphi(x)\psi(x)\mathrm{d}^k x. \tag{H.33}$$

The meaning of this formula is that the functional of u, π on the left-hand side (which we think of as a function rather than a distribution) equals the constant number on the right-hand side. On the other hand, we would like our Poisson bracket to behave well with respect to integrals,

$$\left\{ \int \varphi(x)u(x)\mathrm{d}^k x, \int \psi(y)\pi(y)\mathrm{d}^k y \right\} = \iint \varphi(x)\psi(y)\{u(x),\pi(y)\}\mathrm{d}^k x\mathrm{d}^k y, \tag{H.34}$$

and comparing with (H.33) one reaches the formula

$$\{u(x),\pi(y)\} = \delta(x-y), \tag{H.35}$$

which would have been hard to reach directly from (H.31), as the functional derivatives in the integrand are both distributions and cannot be easily multiplied. The same approach yields

$$\{u(x),u(y)\} = \{\pi(x),\pi(y)\} = 0. \tag{H.36}$$

Another approach is to declare from the very start that the continuous version of (H.28) must be (H.36) and (H.35). Combining several layers of formal manipulations as in (H.34), or by differentiation of the relations (H.35) one may "compute" $\{F, G\}$ at least when F, G are of the type (H.13). In particular, one may "compute" the right-hand side of (H.32) and "derive" the equations of motion (H.15) for the classical field given the Hamiltonian. This approach is found in many textbooks.[4]

[4] I probably simply lack the imagination to understand what this formalism brings more than declaring from the onset that (H.15) are the correct equations of motion.

Appendix I

Quantization of the Electromagnetic Field through the Gupta–Bleuler Approach

This appendix assumes that the reader is familiar with Sections 4.1 and E.2, and to follow the details of the construction she should master the material between (9.93) and Proposition 9.13.6. We cover quite fascinating material, but this material is exotic and will not be used elsewhere. The overall goal is to construct for $\nu = 0, 1, 2, 3$ operators $A_\nu(x)$ (which are actually some kinds of operator-valued distributions) which represent the quantization of the electromagnetic four-potential.

To preserve Lorentz invariance we would like to quantize the electromagnetic field in the Lorenz gauge (which is recalled in (I.1)). The "standard pattern" one should try first follows a road we have traveled several times: finding a Lagrangian that describes the correct equations of motion, and imposing the canonical commutation relations between the "dynamic variables" and their conjugates. One can look e.g. in Schweber's book [75, Chapter 9, section b], for an account of such a trial and why it fails.

Physicists have developed an intriguing method to deal with these problems, the so-called Gupta–Bleuler formalism. It is not so obvious what they mean by statements such as "the Hilbert space of the photon is endowed with an indefinite metric". In the rest of this section we describe the formalism in more mathematical terms. Our treatment is an expansion of the rather concise treatment of Dimock [23].

The spectacular part of the Gupta–Bleuler formalism involves using a "state space" $\mathcal{B}$ provided with a sesqui-linear form $(\cdot,\cdot)_\eta$ which is *not definite positive*. The expected value of an operator B in the state $|\varphi\rangle$ is then computed as $\langle\varphi|B|\varphi\rangle_\eta$, which is to be understood as $(|\varphi\rangle, B|\varphi\rangle)_\eta$. The whole interpretation of Quantum Mechanics is in danger with the approach, because one may run into meaningless "negative probabilities". This is avoided by declaring that the only meaningful computations will be those done using states in a certain subspace of the "state space" $\mathcal{B}$, the state $\mathcal{B}_{\mathrm{phys}}$ of "physical states".[1]

In Proposition 9.13.6 we met exactly the same situation at the level of a single particle. Forgetting about the operation of taking quotients, we constructed there a representation of the Poincaré group $\mathcal{P}$ which takes place in a Hilbert subspace of a space "with an indefinite metric". Starting with this representation, the approach will then follow the standard path: We construct the "boson Fock space" and define the quantum field using appropriate creation and annihilation operators, with each of these constructions modified in the appropriate manner. What is certainly not obvious at the beginning however is that in this manner we will obtain a useful result.

Let us start with some general considerations which will later tie nicely with the specific properties of the construction. We expect that the "operators" $A_\nu(x)$ will in some sense satisfy both the wave equations $(\partial^\mu\partial_\mu)A_\nu = 0$ and the Lorenz gauge condition

$$\partial^\mu A_\mu = 0, \tag{I.1}$$

[1] This appendix is for entertainment, and is an homage to the physicists' inventiveness, but it might be confusing. Read at your own risk.

because in the Lorenz gauge the classical four-potential satisfies these relations. We further expect that as usual we will get non-trivial equal-time commutation relations. That is, if the time components of x and x' coincide, the commutators $[A_\mu(x), A_\nu(x')]$ will be zero for $\mu \neq \nu$ but not be identically zero for $\mu = \nu$. On the other hand the relation (I.1) implies, using in the last equality that $[A_\mu(x), A_\nu(x')] = 0$ for $\mu \neq \nu$:

$$0 = [\partial^\mu A_\mu(x), A_\nu(x')] = \frac{\partial}{\partial x_\mu}[A_\mu(x), A_\nu(x')] = \frac{\partial}{\partial x_\nu}[A_\nu(x), A_\nu(x')]. \tag{I.2}$$

The derivative with respect to x_ν of the complicated quantity $[A_\nu(x), A_\nu(x')]$ is unlikely to be identically zero. Therefore the identity (I.1) cannot hold as an equality between operators and has to be weakened. To explain the correct condition, we note first that we expect that A_μ will consist of an annihilation part A_μ^+ and a creation part A_μ^-. Gupta and Bleuler then impose the condition

$$\forall\,|\varphi\rangle \in \mathcal{B}_{\rm phys}\,,\ \partial^\mu A_\mu^+(x)|\varphi\rangle = 0. \tag{I.3}$$

Let us then assume (as will be the case in the explicit construction below) that the creation part A_ν^- of A_ν is the adjoint of the annihilation part A_ν^+ with respect to the sesqui-linear form $(\cdot,\cdot)_\eta$ (as expressed precisely in (I.6)). Then assuming (I.3), for $\varphi, \varphi' \in \mathcal{B}_{\rm phys}$ we then have $\langle\varphi'|\partial^\mu A_\mu^+|\varphi\rangle = 0$ (obviously) and $\langle\varphi'|\partial^\mu A_\mu^-|\varphi\rangle = 0$ (by (I.6)) so that $\langle\varphi'|\partial^\mu A_\mu|\varphi\rangle_\eta = 0$. This says that as long as we consider only physical states, the operator $\partial^\mu A_\mu$ appears to vanish. This is the proper way to understand (I.1).

We now start the detailed mathematical construction. We consider the space $\mathcal{H} = L^2(X_0, d\lambda_0, \mathbb{C}^4)$. We recall the space $\mathcal{B}_0$, the algebraic sum of the spaces $\mathcal{H}_{n,s}$ and its completion, the boson Fock space $\mathcal{B}$. On $\mathcal{H}$ we consider the isometry η given by

$$\varphi = (\varphi_\mu) = (\varphi_0, \varphi_1, \varphi_2, \varphi_3) \mapsto \eta\varphi := (-\varphi_0, \varphi_1, \varphi_2, \varphi_3) = (-\varphi^\mu).$$

Thus, for $\varphi, \psi \in \mathcal{H}$ we have

$$(\varphi, \eta\psi) = \int (\varphi(p), \psi(p))_\eta d\lambda_0(p),$$

where $(\cdot,\cdot)_\eta$ denotes the sesqui-linear form on $\mathbb{C}^4$ as in (9.93) i.e.

$$(x,y)_\eta := -\eta_{\mu\nu}(x^\mu)^* y^\nu = -x^{0*}y^0 + \sum_{i\le 3} x^{i*}y^i. \tag{9.93}$$

The unitary operator η has a canonical extension $\eta_{\mathcal{B}}$ to $\mathcal{B}$, and on $\mathcal{B}\times\mathcal{B}$ we may define the sesqui-linear form $(x,y)_\eta := (x, \eta_{\mathcal{B}} y)$. Let us note that

$$|(x,x)_\eta| = |(x, \eta_{\mathcal{B}} x)| \le \|x\|^2. \tag{I.4}$$

Theorem I.1 *For ξ in $\mathcal{H}$ we can define operators $B(\xi)$ and $B^\dagger(\xi)$ with the properties of Theorem 3.4.2, except that everywhere the inner product $(\cdot,\cdot)$ is replaced by the sesqui-linear form $(\cdot,\cdot)_\eta$.*

These operators are not defined everywhere, but they are defined on $\mathcal{B}_0$ and they send $\mathcal{B}_0$ into itself. So relations such as

$$\forall \xi, \zeta \in \mathcal{H}\,,\ [B(\xi), B^\dagger(\zeta)] = (\xi, \zeta)_\eta 1 \tag{I.5}$$

make perfectly good sense as operators on $\mathcal{B}_0$. The previous relation is what replaces (3.33), and (3.32) is replaced by

$$\forall \xi \in \mathcal{H}\,,\ \forall x, y \in \mathcal{B}_0\,,\ (B^\dagger(\xi)(x), y)_\eta = (x, B(\xi)(y))_\eta, \tag{I.6}$$

which is what is meant when one says that $B^\dagger(\xi)$ is the adjoint of $B(\xi)$ with respect to the sesqui-linear form $(\cdot,\cdot)_\eta$.

In essence the proof of Theorem I.1 is very similar to that of Theorem 3.4.2. One has however to write the proof using abstract tensor products. For this we need to find proper substitutes for (3.24) and (3.25) as follows. Given an element $\xi \in \mathcal{H}$, there exists an operator $C_n(\xi)$ from $\mathcal{H}_n$ to $\mathcal{H}_{n-1}$ such that

$$C_n(\xi)(\zeta_1 \otimes \zeta_2 \otimes \cdots \otimes \zeta_n) = \sqrt{n}(\xi, \zeta_n)_\eta(\zeta_1 \otimes \cdots \otimes \zeta_{n-1}) \tag{I.7}$$

and we define $B(\xi)$ by the property that for each n its restriction to $\mathcal{H}_{n,s}$ coincides with the restriction of $C_n(\xi)$ to $\mathcal{H}_{n,s}$. This provides the desired substitute for (3.24). We then replace (3.25) by the condition

$$B^\dagger(\zeta)(\zeta_1 \otimes \zeta_2 \otimes \cdots \otimes \zeta_n) = \sqrt{n+1}S_{n+1}(\zeta \otimes \zeta_1 \otimes \cdots \otimes \zeta_n),$$

where $\zeta_k \in \mathcal{H}$, and where S_n denotes the canonical projection (symmetrization) from $\mathcal{H}_n$ to $\mathcal{H}_{n,s}$. We leave checking the straightforward details to the interested reader.

We now define the operators A_ν. We denote by e_ν the canonical basis vectors of $\mathbb{C}^4$. For a function $f \in \mathcal{S}^4_{\mathbb{R}}$ the function $\hat{f}e_\nu$ is an element of $\mathcal{H}$. We then define the operator-valued distribution A^+_ν, A^-_ν by

$$A^-_\mu(f) = B^\dagger(\hat{f}e_\nu)\,;\ A^+_\mu(f) = B(\hat{f}e_\nu). \tag{I.8}$$

and we set

$$A_\nu(f) = A^-_\mu(f) + A^+_\mu(f) = B^\dagger(\hat{f}e_\nu) + B(\hat{f}e_\nu). \tag{I.9}$$

Next we study the properties of this construction with respect to the action of the Poincaré group $\mathcal{P}$, as there is no need to use $\mathcal{P}^*$ here. We observe first that $C \in SO^\uparrow(1,3)$ defines an operator on $\mathbb{C}^4$ (the one with the same matrix). For $(c,C) \in \mathcal{P}$ we consider first the action $U(c,C)$ on $\mathcal{H} = L^2(X_0, \mathrm{d}\lambda_0, \mathbb{C}^4)$ given for $p \in X_0$ and $\varphi \in \mathcal{H}$ by

$$U(c,C)(\varphi)(p) = \exp(\mathrm{i}(c,p)/\hbar)C(\varphi(C^{-1}(p))). \tag{I.10}$$

This representation satisfies

$$(U(c,C)(\varphi), U(c,C)(\psi))_\eta = (\varphi,\psi)_\eta, \tag{I.11}$$

as follows from (9.94). We note that $U(c,C)$ is a bounded operator on $\mathcal{H}$, so that it extends to $\mathcal{B}_0$.[2] We abuse notation by denoting this extension $U_{\mathcal{B}}(c,C)$. Let us also denote by $V(c,C)$ the usual action on the functions of $f \in \mathcal{S}^4_{\mathbb{R}}$,

$$V(c,C)(f)(x) = f(C^{-1}(x-c)).$$

The invariance properties of the fields A_ν are described as follows.

Proposition I.2 *As operators on $\mathcal{B}_0$ we have*

$$U_{\mathcal{B}}(c,C) \circ A_\nu(f) \circ U_{\mathcal{B}}(c,C)^{-1} = C^\mu{}_\nu A_\mu(V(c,C)f). \tag{I.12}$$

If we remember that, as proved at the end of Section 4.1, it holds that $C^\mu_\nu = (C^{-1})_\nu{}^\mu$, we may rewrite (I.12) as

$$U_{\mathcal{B}}(c,C) \circ A^\nu(f) \circ U_{\mathcal{B}}(c,C)^{-1} = (C^{-1})^\nu{}_\mu A^\mu(V(c,C)f).$$

This is exactly the covariance condition (10.6) for the representation S of $SO^\uparrow(1,3)$ where each element acts on $\mathbb{C}^4$ by the operator associated with the matrix of this element.

2 However this operator is not unitary, and it does NOT extend to the boson Fock space $\mathcal{B}$; but all our operators are considered as operators from $\mathcal{B}_0$ to itself.

Proof As in Proposition 3.5.2 we obtain

$$U_{\mathcal{B}}(c,C)\circ A_\nu(f)\circ U_{\mathcal{B}}(c,C)^{-1} = B(U(c,C)\hat{f}e_\nu) + B^\dagger(U(c,C)\hat{f}e_\nu). \tag{I.13}$$

Now, using (5.4) in the last equality, and since $C(e_\nu) = C^\mu{}_\nu e_\nu$,

$$U(c,C)(\hat{f}e_\nu)(p) = \exp(\mathrm{i}(c,p)/\hbar)C(e_\nu)\hat{f}(C^{-1}(p)) = C^\mu_\nu e_\mu \widehat{V(c,C)(f)}(p).$$

Combining with (I.9) and (I.13), and using that the numbers C^μ_ν are real implies (I.12). □

The following expresses the commutation relations satisfied by the fields A_μ. It is a consequence of (I.5) as in the case of the scalar free field. The distribution Δ_0 is given by (5.11) for $m = 0$.

Proposition I.3 *We have*

$$[A_\mu(x), A_\nu(y)] = -\eta_{\mu\nu}(\Delta_0(y-x) - \Delta_0(x-y))1. \tag{I.14}$$

In particular $-\eta_{0,0} = -1$, the opposite of what one might expect.

Now comes the key step: Let us try to discover a natural way to satisfy the condition (I.3). From (I.9) the annihilation part A_ν^+ of A_ν is such that $A_\nu^+(f) = B(\hat{f}e_\nu)$. Thus $\partial^\nu A_\nu^+(f) = A_\nu^+(-\partial^\nu f) = B(-\widehat{\partial^\nu f}e_\nu) = -B(\mathrm{i}p^\nu \hat{f}e_\nu)$ and (I.3) means that for $f \in \mathcal{S}^4_{\mathbb{R}}$ and each physical state $|\varphi\rangle$ we should have

$$B(\hat{f}p^\nu e_\nu)|\varphi\rangle = 0. \tag{I.15}$$

Since the restriction of $B(\xi)$ to $\mathcal{H}_{n,s}$ satisfies (I.7), when $|\varphi\rangle$ is a one-particle state, i.e. is given by an element of $\zeta \in \mathcal{H}$, (I.15) means that $(\hat{f}p^\nu e_\nu, \zeta)_\eta = 0$, i.e.

$$\int \mathrm{d}\lambda_0(p)\hat{f}^*(p)(p,\zeta(p))_\eta = (\hat{f}p^\nu e_\nu, \zeta)_\eta = 0.$$

Since f is arbitrary this implies $(p,\zeta(p))_\eta = 0$ a.e. so that $|\varphi\rangle = \zeta \in \mathcal{H}'$ where

$$\mathcal{H}' = \{\zeta \in \mathcal{H}\,;\, \forall p \in X_0\,;\, (p,\zeta(p))_\eta = 0\}. \tag{I.16}$$

Thus, when $|\varphi\rangle = \zeta$ is a one-particle state, the natural condition (I.3) is equivalent to the equally natural condition that $\zeta(p)$ belongs to the space $\mathcal{V}_p = \{x \in \mathbb{C}^4; (x,p) = 0\}$.

Having described "the one-particle spaces $\mathcal{H}'$, we construct now the corresponding "boson Fock space" $\mathcal{B}'$: we define $\mathcal{B}'$ as the closure (for the usual topology of $\mathcal{B}$) of the algebraic sum $\mathcal{B}'_0$ of the spaces $\mathcal{H}'_{n,s}$. Thus "$\mathcal{B}'$ is the boson Fock space of $\mathcal{H}'$", with the small proviso that the norm is not really positive definite on $\mathcal{H}'$.[3] It is not hard to show using (I.7) that (I.15) holds for $|\varphi\rangle \in \mathcal{B}'$. Furthermore, as follows from Lemma 9.13.4 we have $(x,x)_\eta \geq 0$ for $x \in \mathcal{B}'$. For $x \in \mathcal{B}'$ we may then define

$$\|x\|_\eta = \sqrt{(x,x)_\eta},$$

and we have now avoided the dreadful "negative probabilities" since $\|x\|_\eta \geq 0$.

On $\mathcal{B}'$ the "norm" $\|\cdot\|_\eta$ is positive but it is not positive definite. We would like our state space to be a true Hilbert space, and for this it suffices to take a suitable quotient. We define

$$\mathcal{B}_{\mathrm{null}} := \{u \in \mathcal{B}'\,;\, \|u\|_\eta = 0\},$$

and

$$\mathcal{B}_{\mathrm{phys}} = \mathcal{B}'/\mathcal{B}_{\mathrm{null}}.$$

The image in $\mathcal{B}_{\mathrm{phys}}$ of $x \in \mathcal{B}'$ is denoted by $[x]$. We provide $\mathcal{B}_{\mathrm{phys}}$ with the sesqui-linear form defined by $([x],[y])_\eta = (x,y)_\eta$. It is positive definite. We will show later that $\mathcal{B}_{\mathrm{phys}}$ is complete for this norm.

[3] This will be solved by taking a quotient.

We denote by $\mathcal{B}_{\mathrm{phys},0}$ the image of $\mathcal{B}'_0$ in $\mathcal{B}_{\mathrm{phys}}$. It is a dense subspace of $\mathcal{B}_{\mathrm{phys}}$. The point of the construction is as follows:

Proposition I.4 *(a) The representation $U_{\mathcal{B}}(c,C)$ of the Poincaré group induces a representation on $\mathcal{B}_{\mathrm{phys}}$ (denoted also $U_{\mathcal{B}}(c,C)$).*
(b) If we consider functions f_μ with $\partial^\mu f_\mu = 0$ then $A_\mu(f^\mu)$ acts on $\mathcal{B}_{\mathrm{phys},0}$.
(c) Furthermore if the functions f_μ are as in (b) we have

$$U_{\mathcal{B}}(c,C) \circ A_\mu(f^\mu) \circ U_{\mathcal{B}}(c,C)^{-1} = C^\mu{}_\nu A_\mu(U(c,C)f^\nu). \tag{I.17}$$

This result does not quantize the four-potential A.[4] This is not a real problem because this potential is not an observable. The observables are the electrical and magnetic fields. Quantized versions of the components of these are given by the quantities $F_{\mu\nu}(x) = \partial_\mu A_\nu(x) - \partial_\nu A_\mu(x)$. As an important consequence of (b), these operators (or more precisely a smeared version of them) act on $\mathcal{B}_{\mathrm{phys},0}$. To see this, we may assume that $\mu \neq \nu$. Then for $f \in \mathcal{S}^4_{\mathbb{R}}$ we have

$$F_{\mu\nu}(f) = \partial_\mu A_\nu(f) - \partial_\nu A_\mu(f) = A_\nu(-\partial_\mu f) - A_\mu(-\partial_\nu f).$$

The right-hand side is $A_\rho(g^\rho)$ where $g^\rho \in \mathcal{S}^4_{\mathbb{R}}$ is given by $g^\nu = -\partial_\mu f$, $g^\mu = \partial_\nu f$, and $g^\rho = 0$ if $\rho \notin \{\mu,\nu\}$. Consequently, $\partial^\rho g_\rho = \partial_\rho g^\rho = -\partial_\nu\partial_\mu f + \partial_\mu\partial_\nu f = 0$. Thus $F_{\mu\nu}(f) = A_\rho(g^\rho)$ where $\partial^\rho g_\rho = 0$. Therefore by (b) of the proposition $F_{\mu\nu}(f)$ acts on $\mathcal{B}_{\mathrm{phys},0}$.

That the operators $F_{\mu\nu}(f)$ act on $\mathcal{B}_{\mathrm{phys},0}$ means it makes sense to consider the quantities $F_{\mu\nu}(f)|\varphi\rangle$ for $|\varphi\rangle \in \mathcal{B}_{\mathrm{phys},0}$, so that we can compute quantities related to the electrical and magnetic fields, such as their average value in a physical state.

We prepare for the proof of Proposition I.4 with the following.

Lemma I.5 *For $g \in \mathcal{H}'$ the operators $B(g)$ and $B^\dagger(g)$ naturally induce operators on $\mathcal{B}_{\mathrm{phys},0}$.*

Proof First we prove that $B(g)$ and $B^\dagger(g)$ define operators on $\mathcal{B}'_0$. It suffices for this to show that if x is a tensor $\xi_1 \otimes \xi_2 \otimes \cdots \otimes \xi_n$ with $\xi_1,\ldots,\xi_n \in \mathcal{H}'$ when $g \in \mathcal{H}'$then $B(g)(x) \in \mathcal{H}'_{n-1,s}$ and $B^\dagger(g)(x) \in \mathcal{H}'_{n+1,s}$, but this should be obvious.

Next we have to show that we can go to quotients by $\mathcal{B}_{\mathrm{null}}$, that is for $g \in \mathcal{H}'$ we must prove that

$$u \in \mathcal{B}_{\mathrm{null}} \cap \mathcal{B}_0 \Rightarrow B(g)(u), B^\dagger(g)(u) \in \mathcal{B}_{\mathrm{null}}.$$

In other words, we have to prove that $\|u\|_\eta = 0 \Rightarrow \|B(g)(u)\|_\eta = 0$ and similarly for $B^\dagger(g)$. We observe that on $\mathcal{B}'_0$ the sesqui-linear form $(x,y)_\eta$ is positive, so that it satisfies Cauchy's inequality $|(x,y)_\eta| \leq \|x\|_\eta \|y\|_\eta$, and thus $(x,u)_\eta = 0$ when $\|u\|_\eta = 0$. Thus by (I.6), if $\|u\|_\eta = 0$ then $(x, B(g)(u))_\eta = (B^\dagger(g)(x), u)_\eta = 0$, and taking $x = B(g)(u)$ we obtain $\|B(g)(u)\|_\eta = 0$. The same argument works for $B^\dagger(g)(u)$. □

Proof of Proposition I.4 (a) This simply follows from the fact that (with obvious notation) $\mathcal{B}'_0$ and $\mathcal{B}_{\mathrm{null},0}$ are stable under each map $U(c,C)$ as is straightforward to check.
(b) By definition of $A_\mu(f) = B(\widehat{f e_\mu}) + B^\dagger(\widehat{f e_\mu})$ we have $A_\mu(f^\mu) = B(g) + B^\dagger(g)$ where $g \in \mathcal{H}$ is given by $g(p) = (\widehat{f^0}(p), \widehat{f^1}(p), \widehat{f^2}(p), \widehat{f^3}(p))$. Now, the condition $\partial^\mu f_\mu = 0$ implies $p^\mu \widehat{f_\mu} = 0 = p_\nu \widehat{f^\nu}$. This means that $(g(p), p)_\eta = 0$ for each p and thus $g \in \mathcal{H}'$. The conclusion then follows from Lemma I.5.
(c) Straightforward from (I.12). □

[4] For this we would need to have the individual operators $A_\mu(f)$ acting on $\mathcal{B}_{\mathrm{phys},0}$, whereas we know only that the combinations $A_\mu(f_\mu)$ act on this space.

Let us now consider the space $\mathcal{H}_{\rm null} := \{u \in \mathcal{H}' \,;\, \|u\|_\eta = 0\}$ and the space $\mathcal{H}_{\rm phys} = \mathcal{H}'/\mathcal{H}_{\rm null}$. The canonical map $\mathcal{H}' \to \mathcal{H}_{\rm phys}$ is denoted by $\varphi \to [\varphi]$. The nature of the space $\mathcal{B}_{\rm phys}$ is clarified by the following.

Proposition I.6 *The space $\mathcal{B}_{\rm phys}$ naturally identifies with the boson Fock space of $\mathcal{H}_{\rm phys}$.*

Proof This proof is mathematical and contains no physical idea. Throughout, we denote by (x,y) the standard Hermitian product on $\mathbb{C}^4$.

The first part of the proof consists in showing that the boson Fock space of $\mathcal{H}_{\rm phys}$ identifies with the boson Fock space of a concrete space $\mathcal{M}$ which we construct now. Consider for each $p \in X_0$ the subspaces $\mathcal{M}_p \subset \mathcal{V}_p \subset \mathbb{C}^4$ defined by

$$\mathcal{V}_p = \{(x) \in \mathbb{C}^4 \,;\, (p,x)_\eta = 0\} \,;\, \mathcal{M}_p = \{(x) \in \mathbb{C}^4 \,;\, (p,x)_\eta = 0, x_0 = 0\}.$$

Since $(p,p)_\eta = 0$ for all $p \in X_0$ we have $\mathbb{C}p \subset \mathcal{V}_p$. Moreover, for $x \in \mathcal{M}_p$ we have $(p,x)_\eta = 0$, and also $(x,p) = 0$. That is, $\mathcal{V}_p$ is the direct sum of $\mathcal{M}_p$ and $\mathbb{C}p$ and these two spaces are orthogonal, both for the product $(\cdot,\cdot)_\eta$ and the standard inner product. Furthermore for $x,y \in \mathcal{M}_p$ we have $(x,y) = (x,y)_\eta$. The definition (I.16) of $\mathcal{H}'$ now becomes

$$\mathcal{H}' = \{\xi \in \mathcal{H} \,;\, \forall p\,,\, \xi(p) \in \mathcal{V}_p\},$$

so that

$$\mathcal{M} := \{\xi \in \mathcal{H} \,;\, \forall p\,,\, \xi(p) \in \mathcal{M}_p\} \subset \mathcal{H}'.$$

Also, it follows from Lemma 9.13.4 that

$$\mathcal{H}_{\rm null} = \{\xi \in \mathcal{H} \,;\, \forall p\,,\, \xi(p) \in \mathbb{C}p\}.$$

Since $\mathcal{V}_p$ is the direct sum of $\mathcal{M}_p$ and $\mathbb{C}p$, $\mathcal{H}'$ is the direct sum of $\mathcal{M}$ and $\mathcal{H}$. On $\mathcal{M}$ we have $(\xi,\xi) = (\xi,\xi)_\eta$ and the map $\varphi \to [\varphi]$ from $\mathcal{M}$ to $\mathcal{H}_{\rm phys}$ is an isomorphism. Therefore the boson Fock space $\mathcal{B}_{\mathcal{M}}$ of $\mathcal{M}$ identifies canonically with the boson Fock space of $\mathcal{H}_{\rm phys}$.

The second part of the proof is to show that we can canonically identify $\mathcal{B}_{\mathcal{M}}$ and $\mathcal{B}_{\rm phys}$. This will be done by finding an isometry between these spaces.

Let us consider the orthogonal projection P from $\mathcal{H}'$ onto $\mathcal{M}$ (for the standard Hilbert structure). It has a natural extension $P_{\mathcal{B}}$ from $\mathcal{B}'$ to the boson Fock space $\mathcal{B}_{\mathcal{M}}$ of $\mathcal{M}$. We will show that this is the isometry we are looking for. The crucial property is that

$$\forall x \in \mathcal{B}'\,,\ \|x - P_{\mathcal{B}}(x)\|_\eta = 0. \tag{I.18}$$

We first prove this when $x \in \mathcal{H}'_{n,s}$ is of the type $\xi_1 \otimes \cdots \otimes \xi_n$ where $\xi_k \in \mathcal{H}'$. We decompose each element ξ_k as a sum of an element of $\mathcal{M}$ and an element of $\mathcal{H}_{\rm null}$. Then $x - P_{\mathcal{B}}(x)$ is a sum of elements of the type $\zeta_1 \otimes \cdots \otimes \zeta_n$ where at least one of the elements belongs to $\mathcal{H}_{\rm null}$. Then

$$\|\zeta_1 \otimes \cdots \otimes \zeta_n\|_\eta = \prod_{k \le n} \|\zeta_k\|_\eta$$

is zero as soon as one of the ζ_k belongs to $\mathcal{H}_{\rm null}$. It then follows that (I.18) holds when x is of the type $\xi_1 \otimes \cdots \otimes \xi_n$ where $\xi_k \in \mathcal{H}'$, and then also when x is a linear combination of such elements. This set of linear combinations is dense in $\mathcal{B}'$, and we will finish the proof of (I.18) using a continuity argument. For $x \in \mathcal{B}'_0$ by (I.4) we have $\|x\|_\eta \le \|x\|$. The map $x \mapsto a(x) := \|x - P_{\mathcal{B}}(x)\|_\eta$ satisfies

$$|a(x) - a(y)| \le \|x - y\|_\eta + \|P_{\mathcal{B}}(x) - P_{\mathcal{B}}(y)\|_\eta \le \|x - y\| + \|P_{\mathcal{B}}(x) - P_{\mathcal{B}}(y)\| \le 2\|x - y\|,$$

so that it is continuous. Since $a(x) = 0$ on a dense set it is zero everywhere and we have proved (I.18). (The same continuity argument is actually needed to see that the norm $\|\cdot\|_\eta$ is defined on all of $\mathcal{B}'$.)

It follows from (I.18) that for $x \in \mathcal{B}'$ it holds that

$$\|x\|_\eta = \|P_{\mathcal{B}}x\|_\eta = \|P_{\mathcal{B}}x\|$$

because $\|\cdot\|_\eta$ and $\|\cdot\|$ coincide on $\mathcal{M}$. Consequently $P_{\mathcal{B}}$ induces an isometry from $\mathcal{B}_{\mathrm{phys}} = \mathcal{B}'/\mathcal{B}_{\mathrm{null}}$ to $\mathcal{B}_{\mathcal{M}}$. □

To describe the situation using a physicist's words we recall from Section E.2 that a plane wave satisfying the Maxwell equations is described by a pair a, p of four-vectors with $p^2 = 0$ and $(a, p) = 0$. This description uses redundant parameters. Using the Coulomb gauge removes these redundant parameters. Then a is of the form $a = (\boldsymbol{a}, 0)$ where "$\boldsymbol{a}$ is transversal" i.e. $\boldsymbol{a} \cdot \boldsymbol{p} = 0$. When not using the Coulomb gauge but using a general point $a \in \mathbb{R}^{1,3}$ to describe a plane wave, the contribution of a^0 might be called "a time-like photon", and the contribution of the part of $\boldsymbol{a}$ which is parallel to $\boldsymbol{p}$ a "longitudinal photon". Both time-like photons and longitudinal photons are unphysical, but in any calculation involving only physical states their contributions exactly cancel, leaving only the physical contribution of the "transverse photons". The simplest level at which this mechanism can be explained is when computing $(a,a)_\eta$ for $a \in \mathbb{R}^{1,3} \subset \mathbb{C}^4$. Given $p \in X_0$ we may write

$$a = a_{\mathrm{time}} + a_{\mathrm{long}} + a_{\mathrm{trans}}$$

where a_{time} has no space component, the other two have no time components, and their space components are respectively perpendicular and parallel to that of $\boldsymbol{p}$. Then we have

$$(a, p)_\eta = 0 \Rightarrow (a,a)_\eta = \|a_{\mathrm{trans}}\|^2.$$

So, in the computation of $(a,a)_\eta$, only the transversal part of a matters.

Computing the norm $(\zeta,\zeta)_\eta = \int \mathrm{d}\lambda_0(p)(\zeta(p),\zeta(p))_\eta$ of the single-particle state $\zeta \in \mathcal{H}$ and using the previous method to compute each quantity $(\zeta(p),\zeta(p))_\eta$ shows that only "the transversal part" of ζ contributes to this norm, as the contribution of the "time-like photon" part is negative and exactly cancels the contribution of the "longitudinal photon". The same phenomenon occurs when computing physical quantities of the type $\langle\varphi|H|\varphi\rangle_\eta$ for $|\varphi\rangle \in \mathcal{B}_{\mathrm{phys}}$.

Appendix J

Lippmann–Schwinger Equations and Scattering States

This appendix complements Section 12.2. It is not needed to follow the main story. Its purpose is to explain to a non-physicist some related material found in many textbooks that students of the topic are bound to read.

Our first result is the actual starting point of the rigorous study of self-adjoint operators. The reader should review Section 2.5 now.

Theorem J.1 *Consider a self-adjoint operator H and a complex number z with $\mathrm{Im}\, z \neq 0$. Then there exists a bounded operator $(H+z\mathbf{1})^{-1}$ from $\mathcal{H}$ to $\mathcal{D}(H)$ such that $(H+z\mathbf{1})(H+z\mathbf{1})^{-1} = \mathbf{1}$ and $(H+z\mathbf{1})^{-1}(H+z\mathbf{1})(x) = x$ for $x \in \mathcal{D}(H)$. The operator norm of $(H+z\mathbf{1})^{-1}$ is $\leq |\mathrm{Im}\, z|^{-1}$.*

For simplicity we will simply say that "$(H+z\mathbf{1})^{-1}$ is the inverse of $H+z\mathbf{1}$".

To get an intuitive meaning of this theorem, think of the case where H is the "multiplication by x" operator on L^2. Then $H+z\mathbf{1}$ is the "multiplication by $x+z$" operator, the inverse of which is the "multiplication by $(x+z)^{-1}$ operator". The point of course is that $|x+z|^{-1} \leq |\mathrm{Im}\, z|^{-1}$ for x real.

We start with a few very simple facts.

Lemma J.2 *Consider an operator A on a Hilbert space $\mathcal{H}$. Then*

$$(\mathrm{range}\ A)^{\perp} = \ker A^{\dagger}.$$

Proof If $x \in (\mathrm{range}\ A)^{\perp}$ then $(x, A(y)) = 0 = (0, y)$ for each $y \in \mathcal{D}(A)$ and by definition of $A^{\dagger}$ this implies that $x \in \mathcal{D}(A^{\dagger})$ and $A^{\dagger}(x) = 0$. Thus $(\mathrm{range}\ A)^{\perp} \subset \ker A^{\dagger}$, and the converse is similar. □

Lemma J.3 *Consider a symmetric operator A on a Hilbert space. Then for $x \in \mathcal{D}(A)$ and $z \in \mathbb{C}$ we have*

$$\|(A+z\mathbf{1})(x)\| \geq |\mathrm{Im}\, z|\|x\|. \tag{J.1}$$

Proof Since $a\mathbf{1}$ is symmetric for $a \in \mathbb{R}$ we may absorb $\mathrm{Re}\, z\mathbf{1}$ into A and assume that $z = ib$ for $b \in \mathbb{R}$. Then, using that $(A(x), x) = (x, A(x))$ in the second line

$$\begin{aligned}(A(x)+ibx, A(x)+ibx) &= (A(x), A(x)) - ib(x, A(x)) + ib(A(x), x) + b^2(x,x)\\ &= (A(x), A(x)) + b^2(x,x)\\ &\geq b^2(x,x),\end{aligned}$$

which is the desired result. □

Proof of Theorem J.1 The operator $B = H + z\mathbf{1}$ satisfies

$$\|B(x)\| \geq |\mathrm{Im}\, z|\|x\| \tag{J.2}$$

by (J.1). Next we prove that $B\colon \mathcal{D}(H) \to \mathcal{H}$ is onto. First, using again (J.1) but for $B^{\dagger}$ rather than B we obtain that $\ker B^{\dagger} = \{0\}$ so that by Lemma J.2 the range of B is dense. Consider then $y \in \mathcal{H}$, so that y is the limit of a sequence $y_n = B(x_n)$. In particular (y_n) is a Cauchy sequence and thus by (J.2) this is also the case for (x_n) which therefore converges to a certain $x \in \mathcal{H}$. Moreover the relation $y_n = B(x_n)$ implies $y_n = zx_n + H(x_n)$, and using Lemma 2.5.14 for $H = (H^\dagger)^\dagger$ proves that $y = zx + H(x)$ i.e. $y = B(x)$. Thus B is onto.

Consequently, given $y \in \mathcal{H}$ there is a unique $x \in \mathcal{D}(B)$ with $B(x) = y$. We define $(H + z\mathbb{1})^{-1}(y) = x$. It is straightforward to check that this operator has the required properties. □

Definition J.1 For $\mathrm{Im}\, z \neq 0$ the operator

$$G(z) = (z\mathbb{1} - H)^{-1}$$

is called the Green operator associated to H.

For further use let us prove the following.

Lemma J.4 *For $u \in \mathcal{D}(H)$ we have*

$$G(z)H(u) = HG(z)(u). \tag{J.3}$$

Moreover for any z, z' with $\mathrm{Im}\, z, \mathrm{Im}\, z' \neq 0$ it holds that

$$G(z)G(z') = G(z')G(z). \tag{J.4}$$

Proof Let $G(z)(u) = v$ so that $u = zv - H(v)$. Since $u \in \mathcal{D}(H)$ by hypothesis and $v \in \mathcal{D}(H)$ because $v = G(z)(u)$ and $G(z)$ is valued in $\mathcal{D}(H)$, it holds that $w := H(v) \in \mathcal{D}(H)$. Applying H to the relation $u = zv - w$ gives $H(u) = zw - H(w)$ i.e. $G(z)H(u) = w = H(v) = HG(z)(u)$.

To prove the second statement consider u and $v' = G(z')(u)$ so that $z'v' - H(v') = u$. Applying $G(z)$ to both sides and using (J.3) gives $z'G(z)(v') - H(G(z)(v')) = G(z)(u)$ i.e. $G(z')G(z)(u) = G(z)(v') = G(z)G(z')(u)$. □

Lemma J.5 *If $\mathrm{Im}\, z > 0$ it holds that*

$$G(z) = -\mathrm{i}\int_{-\infty}^{0} \exp(-\mathrm{i}t(z\mathbb{1} - H))\mathrm{d}t. \tag{J.5}$$

In this expression and in the rest of this section the operator $\exp(-\mathrm{i}t(z\mathbb{1} - H))$ is defined as $\exp(-\mathrm{i}tz)\exp(\mathrm{i}tH)$ where the existence of the second term follows from the converse of Stone's theorem as proved in Appendix B.

Proof For $a < b$ we have

$$\int_a^b (-\mathrm{i}(z\mathbb{1} - H))\exp(-\mathrm{i}t(z\mathbb{1} - H))\mathrm{d}t = \exp(-\mathrm{i}b(z\mathbb{1} - H)) - \exp(-\mathrm{i}a(z\mathbb{1} - H)), \tag{J.6}$$

where the equality is in the sense of operators on $\mathcal{D}(H)$, i.e. when both sides of (J.6) are applied to an element of the domain of H. Using that $\|\exp(-\mathrm{i}t(z\mathbb{1} - H))\| = |\exp(-\mathrm{i}tz)| = \exp(t\mathrm{Im}\, z)$, taking $b = 0$ and letting $a \to -\infty$ we get

$$(z\mathbb{1} - H)(-\mathrm{i})\int_{-\infty}^{0} \exp(\mathrm{i}t(z\mathbb{1} - H))\mathrm{d}t = 1.$$

We multiply this equality by $G(z)$ and we obtain that the two bounded operators of (J.5) coincide on each element of $\mathcal{D}(H)$ and hence everywhere. □

Let us now go back to the setting of Section 12.2, where we considered the Hamiltonians H_0 and $H = H_0 + V$ where

$$H_0 = \frac{\boldsymbol{P}^2}{2m} := -\frac{1}{2m}\sum_{i \le 3}\frac{\partial^2}{\partial x_i^2}, \tag{12.2}$$

We denote by $G_0 = G_0(z)$ and $G = G(z)$ the corresponding Green operators. We expect that G will be very difficult to study, while it will be possible to evaluate G_0 (as we will do later). Thus it will be useful to relate G and G_0, with the intention of expressing as far as possible all results in terms of G_0 rather than G.

Lemma J.6 *We have*

$$G = G_0 + G_0 V G \,;\; G_0 = G + G V G_0. \tag{J.7}$$

Proof We note that $V = H - H_0 = B_0 - B$ where $B = z\mathbf{1} - H$, $B_0 = z\mathbf{1} - H_0$. Then, since $BG = \mathbf{1}$, and since $G_0 B_0$ is the identity on the domain of H and hence on the range of G we have

$$G_0 V G = G_0(B_0 - B)G = G_0 B_0 G - G_0 B G = G - G_0,$$

proving the first part of J.7. The second part is similar. □

Let us recall from Section 12.2 the unitary operators $U_0(t) = \exp(-\mathrm{i}tH_0)$ and $U(t) = \exp(-\mathrm{i}tH)$ and

$$\Omega_+(\varphi) = \lim_{t\to-\infty} U(t)^{-1} U_0(t)(\varphi) \tag{12.9}$$

Lemma J.7 *Let us assume that the potential V satisfies the condition (12.8), i.e.* $\int V(\boldsymbol{x})^2 \mathrm{d}^3\boldsymbol{x} < \infty$. *Then for each* $\alpha > 0$ *and* $\boldsymbol{a} \in \mathbb{R}^3$ *it holds that*

$$\begin{aligned}\Omega_+(\varphi_{\alpha,\boldsymbol{a}}) &= \varphi_{\alpha,\boldsymbol{a}} - \mathrm{i}\int_{-\infty}^{0} U(t)^{-1} V U_0(t)(\varphi_{\alpha,\boldsymbol{a}})\\ &= \varphi_{\alpha,\boldsymbol{a}} - \mathrm{i}\lim_{\varepsilon\to0^+}\int_{-\infty}^{0} \exp(t\varepsilon) U(t)^{-1} V U_0(t)(\varphi_{\alpha,\boldsymbol{a}})\mathrm{d}t.\end{aligned} \tag{J.8}$$

where $\varphi_{\alpha,\boldsymbol{a}} = \exp(-(\boldsymbol{x}-\boldsymbol{a})^2/2\alpha)$ *as in (12.3).*

Proof In the proof of Proposition 12.2.6 we have shown the convergence of the integral $\int_0^\infty \|U(t)^{-1} V U_0(t)(\varphi_{\alpha,\boldsymbol{a}})\|\mathrm{d}t$. Together with (12.13) this proves the first equality in (J.8) and the second one is a consequence of Lebesgue's convergence theorem. □

From this point on our arguments become heuristic. The reader may like to review Section 2.8 as we now return to Dirac's notation: we write $|\alpha,\boldsymbol{a}\rangle$ instead of $\varphi_{\alpha,\boldsymbol{a}}$. Then

$$U(t)^{-1} V U_0(t)|\alpha,\boldsymbol{a}\rangle = \int U(t)^{-1} V U_0(t)|\boldsymbol{p}\rangle\langle\boldsymbol{p}|\alpha,\boldsymbol{a}\rangle \frac{\mathrm{d}^3\boldsymbol{p}}{(2\pi)^3}. \tag{J.9}$$

Now, setting $E_{\boldsymbol{p}} = \boldsymbol{p}^2/2m$, we have $H_0|\boldsymbol{p}\rangle = E_{\boldsymbol{p}}|\boldsymbol{p}\rangle$ and therefore

$$U_0(t)|\boldsymbol{p}\rangle = \exp(-\mathrm{i}tE_{\boldsymbol{p}})|\boldsymbol{p}\rangle.$$

Consequently, $U(t)^{-1} V U_0(t)|\boldsymbol{p}\rangle = \exp(-\mathrm{i}tE_{\boldsymbol{p}})\exp(\mathrm{i}tH)V|\boldsymbol{p}\rangle$ and (J.9) yields

$$\exp(\varepsilon t)U(t)^{-1} V U_0(t)(\varphi_{\alpha,\boldsymbol{a}}) = \int \exp(-\mathrm{i}t(E_{\boldsymbol{p}} + \mathrm{i}\varepsilon - H))V|\boldsymbol{p}\rangle\langle\boldsymbol{p}|\alpha,\boldsymbol{a}\rangle \frac{\mathrm{d}^3\boldsymbol{p}}{(2\pi)^3}. \tag{J.10}$$

From (J.5) we have the perfectly rigorous formula

$$G(E_{\boldsymbol{p}} + \mathrm{i}\varepsilon) = -\mathrm{i}\int_{-\infty}^{0} \exp(-\mathrm{i}t(E_{\boldsymbol{p}} + \mathrm{i}\varepsilon - H))\mathrm{d}t.$$

This is an equality between operators on $\mathcal{H}$. Let us pretend that it still makes sense when applied to the improper vector $V|\boldsymbol{p}\rangle$, and that we can integrate in t inside the integral in (J.10), so that

$$-\mathrm{i}\int_{-\infty}^{0} \exp(\varepsilon t)U(t)^{-1} V U_0(t)|\alpha,\boldsymbol{a}\rangle\mathrm{d}t = \int G(E_{\boldsymbol{p}} + \mathrm{i}\varepsilon)V|\boldsymbol{p}\rangle\langle\boldsymbol{p}|\alpha,\boldsymbol{a}\rangle \frac{\mathrm{d}^3\boldsymbol{p}}{(2\pi)^3}.$$

Then, combining with (J.8),

$$\Omega_+(\varphi_{\alpha,\boldsymbol{a}}) = |\alpha,\boldsymbol{a}\rangle + \lim_{\varepsilon\to 0+}\int G(E_{\boldsymbol{p}}+\mathrm{i}\varepsilon)V|\boldsymbol{p}\rangle\langle\boldsymbol{p}|\alpha,\boldsymbol{a}\rangle\frac{\mathrm{d}^3\boldsymbol{p}}{(2\pi)^3}$$

$$= \lim_{\varepsilon\to 0+}\int (|\boldsymbol{p}\rangle + G(E_{\boldsymbol{p}}+\mathrm{i}\varepsilon)V|\boldsymbol{p}\rangle)\langle\boldsymbol{p}|\alpha,\boldsymbol{a}\rangle\frac{\mathrm{d}^3\boldsymbol{p}}{(2\pi)^3}. \tag{J.11}$$

Since

$$\Omega_+|\alpha,\boldsymbol{a}\rangle = \int \Omega_+|\boldsymbol{p}\rangle\langle\boldsymbol{p}|\alpha,\boldsymbol{a}\rangle\frac{\mathrm{d}^3\boldsymbol{p}}{(2\pi)^3},$$

and since there are sufficiently many elements $|\alpha,\boldsymbol{a}\rangle$ we may expect that in some sense

$$|\boldsymbol{p}\rangle_+ := \Omega_+|\boldsymbol{p}\rangle = |\boldsymbol{p}\rangle + \lim_{\varepsilon\to 0^+} G(E_{\boldsymbol{p}}+\mathrm{i}\varepsilon)V|\boldsymbol{p}\rangle,$$

a formula which it is customary to abbreviate as

$$|\boldsymbol{p}\rangle_+ = |\boldsymbol{p}\rangle + G(E_{\boldsymbol{p}}+\mathrm{i}0_+)V|\boldsymbol{p}\rangle. \tag{J.12}$$

The notation Ω_+ was designed to make the plus signs match in this formula. To make the formula useful, it remains to replace G by G_0. For this we use the first part of (J.7):

$$GV = G_0V(1+GV)$$

so that using (J.12) again,

$$G(E_{\boldsymbol{p}}+\mathrm{i}0_+)V|\boldsymbol{p}\rangle = G_0(E_{\boldsymbol{p}}+\mathrm{i}0_+)V\big(|\boldsymbol{p}\rangle + G(E_{\boldsymbol{p}}+\mathrm{i}0_+)V|\boldsymbol{p}\rangle\big)$$

$$= G_0(E_{\boldsymbol{p}}+\mathrm{i}0_+)V|\boldsymbol{p}\rangle_+, \tag{J.13}$$

and therefore we obtain the *Lippmann–Schwinger equations*

$$|\boldsymbol{p}\rangle_+ = |\boldsymbol{p}\rangle + G_0(E_{\boldsymbol{p}}+\mathrm{i}0_+)V|\boldsymbol{p}\rangle_+. \tag{J.14}$$

The states $|\boldsymbol{p}\rangle_+$ are called the *scattering states*. As we already mentioned, in some sense they are eigenvalues of H, so that decomposition of an element along this continuous basis allows easily to determine its time-evolution.

All of this is pretty formal, and to make it really useful one has to find conditions on V such that these equations have solutions given by true functions $\psi_{\boldsymbol{p}}(\boldsymbol{x}) := \langle\boldsymbol{x}|\boldsymbol{p}\rangle_+$, so we end this topic by computing $G_0(E_{\boldsymbol{p}}+\mathrm{i}0_+)$ and writing (J.14) as a true integral equation. Applying $\langle\boldsymbol{x}|$ to the left of (J.14) we obtain

$$\psi_{\boldsymbol{p}}(\boldsymbol{x}) = \exp(\mathrm{i}\boldsymbol{x}\cdot\boldsymbol{p}) + \langle\boldsymbol{x}|G_0(E_{\boldsymbol{p}}+\mathrm{i}0_+)V|\boldsymbol{p}\rangle_+$$

$$= \exp(\mathrm{i}\boldsymbol{x}\cdot\boldsymbol{p}) + \int \langle\boldsymbol{x}|G_0(E_{\boldsymbol{p}}+\mathrm{i}0_+)V|\boldsymbol{y}\rangle\langle\boldsymbol{y}|\boldsymbol{p}\rangle_+\mathrm{d}^3\boldsymbol{y}$$

$$= \exp(\mathrm{i}\boldsymbol{x}\cdot\boldsymbol{p}) + \lim_{\varepsilon\to 0+}\int \langle\boldsymbol{x}|G_0(E_{\boldsymbol{p}}+\mathrm{i}\varepsilon)|\boldsymbol{y}\rangle V(\boldsymbol{y})\psi_{\boldsymbol{p}}(\boldsymbol{y})\mathrm{d}^3\boldsymbol{y}, \tag{J.15}$$

where we have used that $V|\boldsymbol{y}\rangle = V(\boldsymbol{y})|\boldsymbol{y}\rangle$ since in position state space V is "multiplication by the function $V(\boldsymbol{y})$". Let us now compute the Green operator $G_0(z)$. By definition it satisfies

$$(z1 - H_0)G_0(z) = 1, \tag{J.16}$$

which means that for any function f in $L^2(\mathbb{R}^3)$ the differential operator $z1 - H_0$ applied to $G_0(z)(f)$ gives f. In Dirac formalism one simply writes

$$G_0(z)(f)(\boldsymbol{x}) = \langle\boldsymbol{x}|G_0(z)(f)\rangle = \int \langle\boldsymbol{x}|G_0(z)|\boldsymbol{y}\rangle\langle\boldsymbol{y}|f\rangle\mathrm{d}^3\boldsymbol{y},$$

and, recalling the value of H_0, (J.16) becomes

$$\left(z + \frac{1}{2m} \sum_{i \le 3} \frac{\partial^2}{\partial x_i^2}\right) \langle \boldsymbol{x} | G_0(z) | \boldsymbol{y} \rangle = \delta_{\boldsymbol{y}}^{(3)},$$

so that the kernel $\langle \boldsymbol{x} | G_0(z) | \boldsymbol{y} \rangle$ is the fundamental solution of a certain differential equation (see Section N.2 if needed for a very brief review). These are traditionally called Green functions, and this motivates the name "Green operator". A mathematician computes these objects using the Fourier transform, but the Fourier transform is part of Dirac formalism, by writing

$$\langle \boldsymbol{x} | G_0(z) | \boldsymbol{y} \rangle = \int \langle \boldsymbol{x} | \boldsymbol{p}' \rangle \langle \boldsymbol{p}' | G_0(z) | \boldsymbol{p} \rangle \langle \boldsymbol{p} | \boldsymbol{y} \rangle \frac{\mathrm{d}^3 \boldsymbol{p}}{(2\pi)^3} \frac{\mathrm{d}^3 \boldsymbol{p}'}{(2\pi)^3}.$$

In momentum state space, $z1 - H_0$ is simply multiplication by $z - E_{\boldsymbol{p}}$, so that $G_0(z)$ is multiplication by $(z - E_{\boldsymbol{p}})^{-1}$ i.e. $G_0(z)|\boldsymbol{p}\rangle = (z - E_{\boldsymbol{p}})^{-1}|\boldsymbol{p}\rangle$ and thus $\langle \boldsymbol{p}' | G_0(z) | \boldsymbol{p} \rangle = (2\pi)^3 \delta^{(3)}(\boldsymbol{p}' - \boldsymbol{p})(z - E_{\boldsymbol{p}})^{-1}$. Hence

$$\langle \boldsymbol{x} | G_0(z) | \boldsymbol{y} \rangle = \int \frac{\exp(\mathrm{i}\boldsymbol{p} \cdot (\boldsymbol{x} - \boldsymbol{y}))}{z - E_{\boldsymbol{p}}} \frac{\mathrm{d}^3 \boldsymbol{p}}{(2\pi)^3}.$$

By rotational invariance, this quantity depends only on $s := |\boldsymbol{x} - \boldsymbol{y}|$, and to compute it one may assume that $\boldsymbol{x} - \boldsymbol{y} = s e_3$, where e_3 is the third coordinate unit vector. Going to spherical coordinates one can perform two integrations to get

$$\langle \boldsymbol{x} | G_0(z) | \boldsymbol{y} \rangle = \frac{-i}{4\pi^2 s} \int_0^\infty r \frac{\exp(irs) - \exp(-irs)}{z - r^2/2m} \mathrm{d}r = \frac{im}{2\pi^2 s} \int_{-\infty}^\infty \frac{r \exp irs}{r^2 - 2mz} \mathrm{d}r.$$

The integral can be evaluated using a contour integral (a topic that is reviewed in Section N.1). The appropriate contour consists of the line segment $(-R, 0), (R, 0)$ together with a half-circle of radius R in the upper half-plane. The only pole of the integrand (seen as a function of the complex variable r) inside of the contour is the unique number r_z with $r_z^2 = 2mz$ and $\mathrm{Im}\, r_z > 0$, so that by Cauchy's Theorem

$$\langle \boldsymbol{x} | G_0(z) | \boldsymbol{y} \rangle = -\frac{m}{2\pi s} \exp(i r_z s).$$

Now, for $z = E_{\boldsymbol{p}} + \mathrm{i}\varepsilon$, as $\varepsilon \to 0+$ we have $r_z \to |\boldsymbol{p}|$. Substitution of this result in (J.15) yields the Lippmann–Schwinger equations in concrete form:

$$\psi_{\boldsymbol{p}}(\boldsymbol{x}) = \exp(\mathrm{i}\boldsymbol{x} \cdot \boldsymbol{p}) - \frac{m}{2\pi} \int \frac{\exp(\mathrm{i}|\boldsymbol{p}||\boldsymbol{x} - \boldsymbol{y}|)}{|\boldsymbol{x} - \boldsymbol{y}|} V(\boldsymbol{y}) \psi_{\boldsymbol{p}}(\boldsymbol{y}) \mathrm{d}^3 \boldsymbol{y}.$$

The interest of the abstract formulation (J.14) is of course that it remains the same in far more general cases.

Appendix K

Functions on Surfaces and Distributions

The technical study of distributions starts in the next appendix. Our goal in the present appendix is to make some remarks at the purely heuristic level.

The power of the theory of distributions is that it can describe a wide variety of objects in a single unified setting. This at times can be overwhelming. In particular certain objects we use throughout the book are described through distributions, which might make them appear more mysterious than what they actually are. Describing a function as a distribution (as in (1.8)) is certainly not the simplest way to think about a function. Many of the objects we consider are hardly more complicated than functions, *they are functions defined on a specific surface.* Then in the expression of these objects as distributions there is a factor which is a delta function (or a product of delta functions), and expresses that we are not concerned about what happens outside the specific surface, and another factor which describes what happens on the surface. Among the simplest possible examples, if I want to describe a function on the diagonal $x = y$ of $\mathbb{R}^2$, I will describe it by a distribution of the type $\delta(x-y)f(x)$, of $\delta(x-y)g(x,y)$. In the second formula only the values of g on the diagonal matter.

A more elaborate example of the same situation occurs in (12.32), where we consider the distribution $\delta(\boldsymbol{p}_1^2 - \boldsymbol{p}_2^2)\Xi(\boldsymbol{p}_1, \boldsymbol{p}_2)$. Here we are just studying, in the language of distributions, the function $\Xi(\boldsymbol{p}_1, \boldsymbol{p}_2)$ on the manifold $\boldsymbol{p}_1^2 = \boldsymbol{p}_2^2$. The delta function term $\delta(\boldsymbol{p}_1^2 - \boldsymbol{p}_2^2)$ carries the information that we are only interested in the case $\boldsymbol{p}_1^2 = \boldsymbol{p}_2^2$. The reason for this is conservation of momentum, the outgoing particle has the same momentum as the incoming particle. One may think of $\delta(\boldsymbol{p}_1^2 - \boldsymbol{p}_2^2)$ as describing a certain surface measure $\mathrm{d}\mu$ on the manifold $\boldsymbol{p}_1^2 = \boldsymbol{p}_2^2$, and the distribution can be thought of as integration of the test function against the measure $\Xi \mathrm{d}\mu$.

Another ubiquitous use of distributions is to express the value of Feynman diagrams. If, say, there are three particles involved, this value will often be of the type $(2\pi)^4\delta^{(4)}(p_1 + p_2 + p_3)f(p_1, p_2, p_3)$ for a nice function f. This expression contains two pieces of information: One, the delta function asserts that we must have $p_1 + p_2 + p_3 = 0$. Second, given that $p_1 + p_2 + p_3 = 0$, say $p_3 = -p_1 - p_2$, it is the function $(p_1, p_2) \mapsto f(p_1, p_2, -p_2 - p_3)$ which describes the process of interest.

Appendix L

What Is a Tempered Distribution Really?

"Informal distribution theory" as we practiced liberally has its limits when one tries to actually prove results, as we will do in Appendix M. In the present appendix we give the rigorous definition of tempered distributions, and we prove a few sample results. This appendix is a required prerequisite to Appendix M.

L.1 Test Functions

Given an infinitely differentiable complex-valued function f on $\mathbb{R}^n$ and a multiindex $\alpha = (\alpha_1, \ldots, \alpha_n)$ with $\alpha_j \in \mathbb{N}$ we define $|\alpha| := \sum_{j \le n} \alpha_j$ and

$$D^\alpha f = \frac{\partial^{|\alpha|}}{\partial x_1^{\alpha_1} \cdots \partial x_n^{\alpha_n}} f.$$

For $x \in \mathbb{R}^n$ we denote $|x| = \sum_{j \le n} |x_j|$ and for each integer k we denote

$$\|f\|_k = \sup_{x \in \mathbb{R}^n} (1 + |x|)^k \sum_{|\alpha| \le k} |D^\alpha f(x)|.$$

The Schwartz space $\mathcal{S}^n = \mathcal{S}(\mathbb{R}^n)$ consists of all the functions f for which the seminorms $\|f\|_k$ are finite for each k. In words a function belongs to $\mathcal{S}^n$ if and only if all its derivatives decrease faster at infinity than the reciprocal of any polynomial. Given a multiindex β we write $x^\beta = x_1^{\beta_1} \ldots x_n^{\beta_n}$. To avoid confusion of notation we denote by $N(f)$ the L^2 norm of $f \in L^2(\mathbb{R}^n)$.

Lemma L.1.1 *(a) For a test function $f \in \mathcal{S}^n$ and two multiindices α and β we have $x^\beta D^\alpha f \in \mathcal{S}^n$.*

(b) For each n there exists a constant C such that for each $f \in \mathcal{S}^n$ it holds that

$$N(f) \le C\|f\|_n. \tag{L.1}$$

(c) If $f \in L^2(\mathbb{R}^n)$ is infinitely differentiable, for some integer k and a constant L it holds that

$$\sup_x |f(x)| \le L \max_{|\alpha|+|\beta| \le k} N(x^\beta D^\alpha(f)). \tag{L.2}$$

Proof The claim (a) is obvious. To prove (b) we observe that for each k and each $x \in \mathbb{R}^n$ it holds that $|f(x)|(1+|x|)^k \le \|f\|_k$ so that $|N(f)| \le C\|f\|_k$ as soon at the function $(1+|x|)^{-k}$ is square-integrable, which occurs in particular for $k = n$ (with room to spare).[1]

[1] In dimension n the function $(1+|x|)^{-k}$ is integrable whenever $k > n$. This is because its integral on the set where $2^p \le |x| \le 2^{p+1}$ is of order $2^{p(n-k)}$.

Since the proof of (c) is a bit messy we give the details only for $n = 1$. Since

$$f(x) - f(0) = \int_0^x f'(y)\mathrm{d}y = \int_0^x \frac{1}{1+|y|}(1+|y|)f'(y)\mathrm{d}y$$

and since the function $1/(1+|y|)$ is square-integrable, the Cauchy-Schwarz inequality implies that for some constant C independent of x it holds that

$$|f(x) - f(0)| \le CN((1+|y|)f') \le C \max_{|\alpha|+|\beta|\le 2} N(y^\beta D^\alpha f).$$

Since $f \in L^2$ there are values of x for which $f(x)$ is arbitrarily small. Using the inequality $|f(0)| \le |f(x)| + |f(x) - f(0)|$ for these values of x we obtain $|f(0)| \le C \max_{|\alpha|+|\beta|\le 2} N(x^\beta D^\alpha f)$ and the result follows since $|f(x)| \le |f(x) - f(0)| + |f(0)|$. □

Given a test function $f \in \mathcal{S}^n$, we denote by $\mathcal{F}_m(f)$ its mathematician's Fourier transform as in (1.22).

Proposition L.1.2 *If $f \in \mathcal{S}^n$ then $\mathcal{F}_m(f) \in \mathcal{S}^n$. Moreover the map $f \to \mathcal{F}_m(f)$ is continuous in the sense that for each k there exists a k' and a constant C such that $\|\mathcal{F}_m(f)\|_k \le C\|f\|_{k'}$.*

Proof It suffices to prove that given multiindices α and β there is a constant C and an integer k such that for any $f \in \mathcal{S}^n$ we get

$$\sup_x |x^\beta D^\alpha(\mathcal{F}_m(f))(x)| \le C\|f\|_k. \tag{L.3}$$

According to (L.2) it suffices to obtain an estimate

$$N(x^\beta D^\alpha(\mathcal{F}_m(f))(x)) \le C\|f\|_k. \tag{L.4}$$

But it follows from Plancherel's theorem and the suitable n-dimensional version of (1.23) that

$$N(x^\beta D^\alpha(\mathcal{F}_m(f))(x)) = N(\mathcal{F}_m(D^\beta(x^\alpha f))) = N(D^\beta(x^\alpha f)),$$

and the conclusion follows from (L.1). □

L.2 Tempered Distributions

Definition L.2.1 A tempered distribution Φ on $\mathbb{R}^n$ is a linear functional on $\mathcal{S}^n$ which is continuous for one of the seminorms $\|\cdot\|_k$ i.e. for some constant C and all $f \in \mathcal{S}^n$,

$$|\Phi(f)| \le C\|f\|_k. \tag{L.5}$$

Tempered distributions are sometimes called *generalized functions.*

Exercise L.2.2 Consider a (measurable) function h on $\mathbb{R}^n$ and assume that for some k we have $|h(x)| \le 1 + |x|^k$. Prove that the formula $\Phi(f) = \int f(x)h(x)\mathrm{d}^n x$ defines a tempered distribution.

We recall that the *support* of a function is the closure of the set of points where it is different from zero.

Definition L.2.3 We say that a tempered distribution Φ is supported by a closed subset F of $\mathbb{R}^n$ if $\Phi(f) = 0$ whenever $f \in \mathcal{S}^n$ has a compact support which does not intersect F.

Lemma L.2.4 *When a tempered distribution is supported by two closed sets, it is supported by their intersection.*

Proof Consider two closed subsets F_1, F_2 each supporting a tempered distribution Φ. Consider $f \in \mathcal{S}^n$ whose support K does not meet $F_1 \cap F_2$. The argument relies on the existence of "partitions of unity" of the following form: There exist $h_1, h_2 \in \mathcal{S}^n$ such that the support of h_j does not meet F_j while $h_1 + h_2 = 1$ on K. Then $f = fh_1 + fh_2$ while $\Phi(fh_j) = 0$ because the support of fh_j does not meet F_j. □

Exercise L.2.5 Prove that the intersection of all the closed sets which support a tempered distribution also supports this distribution. This set is called the support of the distribution.

Exercise L.2.6 Prove that the support of the delta function given by $\delta(f) = f(0)$ is the set $\{0\}$.

Exercise L.2.7 Prove that $\Phi(f)$ need not be zero when f is zero on the support of Φ. Hint: Take $n = 1$ and $\Phi = \delta'$, the derivative of the delta function in the sense of distributions.

Lemma L.2.8 *Consider a tempered distribution Φ and $u \in \mathcal{S}^n$. Assume that for each $v \in \mathcal{S}^n$ with compact support, one has $\Phi(uv) = 0$. Then $\Phi(u) = 0$.*

Proof According to (L.5), we have $|\Phi(u)| = |\Phi(u - uw)| \le C\|u - uw\|_k$, so that it suffices to find $w \in \mathcal{S}^n$ with compact support such that $\|u(1 - w)\|_k$ is arbitrarily small. For this let us fix v with compact support, equal to 1 in a neighborhood of the origin, and define $v_a(x) = v(ax)$. The idea is to let $a \to 0$ so that v_a equals 1 on a very large set. For any test function w one has $\lim_{a\to 0} \sup_x |w(x)(1 - v_a(x))| = 0$ and for any $\alpha \neq 0$ one has $\lim_{a\to 0} \sup_x |w(x)D^\alpha v_a(x))| = 0$ by straightforward estimates. Expanding the derivatives $D^\alpha(u(1 - v_a))$ according to Leibniz's rule proves that $\lim_{a\to 0} \|u(1 - v_a)\|_k = 0$. □

Exercise L.2.9 Prove that the support of a non-zero tempered distribution is not empty.

The following result illustrates how the continuity condition (L.5) might be used in an actual proof.

Proposition L.2.10 *If the tempered distribution Φ on $\mathcal{S}^n$ is supported by the origin it is of the type*

$$\Phi(u) = \sum_\alpha c_\alpha D^\alpha u(0)$$

for a finite sum over multiindices α.

We fix the tempered distribution Φ as above and integer k for which (L.5) holds and start the preparations for the proof.

Lemma L.2.11 *If two functions $u, w \in \mathcal{S}^n$ coincide in a neighborhood of the origin then $\Phi(u) = \Phi(w)$.*

Proof Given any $v \in \mathcal{S}^n$ with compact support, the function $v(u - w)$ has a compact support which does not include the origin. Since Φ is supported by the origin, we have $\Phi((u - w)v) = 0$. By Lemma L.2.8, this implies $\Phi(u - w) = 0$. □

Lemma L.2.12 *Consider a function $w \in \mathcal{S}^n$ and assume that $D^\gamma(w)(0) = 0$ whenever $|\gamma| \le k$. Then*

$$|w(x)| \le C|x|^{k+1}, \tag{L.6}$$

where C is independent of x.

Proof Apply Taylor's formula to the function $t \to w(tx)$. □

Let us now fix a function $v \in \mathcal{S}^n$, with compact support, equal to 1 in a neighborhood of the origin. We define the function v_a by $v_a(x) = v(ax)$. We are interested in the case $a \to \infty$, so that the support of v_a shrinks to 0. From now on $C_1, C_2, \dots$ denote constants independent of a and x. The following is obvious by change of variables.

Lemma L.2.13 *We have* $|D^\gamma v_a(x)| \le C_1 a^{|\gamma|}$.

The key to the proof of Proposition L.2.10 is the following estimate.

Lemma L.2.14 *Assume that* $D^\alpha(u)(0) = 0$ *whenever* $|\alpha| \le k$. *Then for* $a \ge 1$ *it holds that* $\|uv_a\|_k \le C/a$.

Proof Consider α with $|\alpha| \le k$. Leibniz's rule gives us $D^\alpha(uv_a)$ as a sum of products $D^\beta u D^\gamma v_a$, where $\beta + \gamma = \alpha$, so that $|\beta| + |\gamma| = |\alpha| \le k$. By hypothesis $D^\delta D^\beta u(0) = 0$ for $|\delta| \le k - |\beta|$, so that by Lemma L.2.12 we have $|D^\beta u(x)| \le C_2|x|^{k-|\beta|+1}$. Thus

$$|D^\beta u(x) D^\gamma v_a(x)| \le C_3 |x|^{k-|\beta|+1} a^{|\gamma|}. \tag{L.7}$$

Since v has a compact support, we have $|x| \le C_4$ for $v(x) \ne 0$ so that $|x| \le C_4/a$ for $v_a(x) \ne 0$, and then $|x|^{k-|\beta|+1} a^{|\gamma|} \le C_5 a^{-k+|\beta|-1+|\gamma|} \le C_5/a$. When $a \ge 1$ and $v_a(x) \ne 0$ it holds that $|x| \le C_4/a \le C_4$ and thus $(1+|x|)^k \le C_6$ so that

$$\sup_{x \in \mathbb{R}^n} (1+|x|^k)|D^\alpha(uv_a)(x)| \le C/a.$$

□

Corollary L.2.15 *Assume that* $D^\alpha(u)(0) = 0$ *for* $|\alpha| \le k$. *Then* $\Phi(u) = 0$.

Proof According to (L.5) and Lemma L.2.14 it holds that $\lim_{a\to\infty} \Phi(uv_a) = 0$. But $\Phi(uv_a) = \Phi(u)$ according to Corollary L.2.11. □

Proof of Proposition L.2.10 Consider the polynomial w_α such that $D^\beta w_\alpha = \delta_\alpha^\beta$. Since $v = 1$ in a neighborhood of zero, $\tilde{u} := \sum_{|\alpha| \le k} v w_\alpha D^\alpha u(0)$ is a test function with $D^\alpha \tilde{u}(0) = D^\alpha u(0)$ for $|\alpha| \le k$. We apply Corollary L.2.15 to the function $u - \tilde{u}$ to see that $\Phi(u) = \Phi(\tilde{u})$ and we obtain the result for $c_\alpha = \Phi(w_\alpha v)$. □

L.3 Adding and Removing Variables

A function f of one variable x can also be considered as a function g of two variables x, y by setting $g(x,y) = f(x)$. The function g has the property that $g(x,y) = g(x, y+a)$ for all a. Conversely a function $g(x,y)$ with this latter property "does not depend on the second variable and thus arises from a function of one variable". We will show that the same property holds for distributions, although it requires a little proving.

Definition L.3.1 A tempered distribution Φ is invariant under translations along the nth coordinate if for each $f \in \mathcal{S}^n$ and each $a \in \mathbb{R}$ it holds that $\Phi(f_a) = \Phi(f)$ where $f_a(x_1, \dots, x_n) = f(x_1, \dots, x_{n-1}, x_n + a)$.

We first show how to associate to a distribution on $\mathbb{R}^{n-1}$ a distribution on $\mathbb{R}^n$. The following result is basically obvious.

Proposition L.3.2 *Consider a tempered distribution* Ψ *on* $\mathbb{R}^{n-1}$. *Then there exists a tempered distribution* Φ *on* $\mathbb{R}^n$ *such that for all* $f \in \mathcal{S}^n$ *we have*

$$\Phi(f) = \Psi(G(f)), \tag{L.8}$$

where $G(f) \in \mathcal{S}^{n-1}$ is given by

$$G(f)(x_1,\ldots,x_{n-1}) = \int \mathrm{d}y f(x_1,\ldots,x_{n-1},y). \tag{L.9}$$

Moreover Φ is invariant under translations along the nth coordinate.

Exercise L.3.3 (a) Show that to prove that Φ is a distribution, it suffices to prove that the map $f \mapsto G(g)$ from $\mathcal{S}^n$ to $\mathcal{S}^{n-1}$ is continuous in the sense that for each k there exists k' for which $\|G(f)\|_k \le \|f\|_{k'}$.
(b) Prove this inequality.

It is the converse that deserves a little proof. This converse has been formulated in a way that its application can be easily iterated.

Proposition L.3.4 *Consider a tempered distribution Φ on $\mathbb{R}^n$ and assume that Φ is invariant by translation along the last coordinate. Then there is a tempered distribution Ψ on $\mathbb{R}^{n-1}$ such that for each $f \in \mathcal{S}^n$ we have, using the notation (L.9),*

$$\Phi(f) = \Psi(G(f)).$$

Moreover if Φ is invariant under translations along the kth coordinate, so is Ψ.

Proof The key point of the proof is to show that

$$G(f) = 0 \Rightarrow \Phi(f) = 0. \tag{L.10}$$

To prove this we observe that if $f \in \mathcal{S}^n$ is such that $G(f) = 0$ then we can define a function $h \in \mathcal{S}^n$ by

$$h(x_1,\ldots,x_n) = \int_{-\infty}^{x_n} \mathrm{d}y f(x_1,\ldots,x_{n-1},y).$$

It is tedious but straightforward that $h \in \mathcal{S}^n$. The point of the condition $G(f) = 0$ is that otherwise it is not even true that h approaches zero as the variables go to infinity. Next, one checks that for each k

$$\lim_{a\to 0}\left\|\frac{1}{a}(h_a - h) - f\right\|_k = 0$$

where $h_a(x_1,\ldots,x_n) = h(x_1,\ldots,x_{n-1},x_n + a)$ and thus

$$\Phi(f) = \lim_{a\to 0} a^{-1}(\Phi(h_a) - \Phi(h)) = 0$$

because the right-hand side is identically zero since $\Phi(h_a) = \Phi(h)$. This proves (L.10).

The rest is easy. Let us consider $\theta \in \mathcal{S}^1$ of integral 1. For $g \in \mathcal{S}^{n-1}$ let us define $C(g) \in \mathcal{S}^n$ by

$$C(g)(x_1,\ldots,x_n) = g(x_1,\ldots,x_{n-1})\theta(x_n)$$

and let us define the tempered distribution Ψ on $\mathbb{R}^{n-1}$ by

$$\Psi(g) = \Phi(C(g)).$$

We observe that $G(w) = 0$ where $w = f - C(G(f))$, so that by (L.10) it holds that $\Phi(w) = 0$ i.e. $\Phi(f) = \Phi(C(G(f)) = \Psi(G(f))$ and this proves (L.8). The rest is obvious. □

L.4 Fourier Transforms of Distributions

The subject of Fourier transforms of distributions is confusing because the definition used by mathematicians does not exactly coincide with the definition required here. Mathematicians define the Fourier transform of a distribution Φ by

$$\mathcal{F}_m(\Phi)(f) = \Phi(\mathcal{F}_m(f)) \tag{L.11}$$

for a test function f. In our version however, this definition has to be changed, because the Fourier transform sends functions on position space to functions on momentum space, and these are not the same, see (1.30) and (1.31). In order to keep track of which is which, we denote by Latin letters functions on position space and by Greek letters functions on momentum space.

Given a distribution Φ on position space we then define its Fourier transform $\hat{\Phi}$ (on momentum space) by

$$\hat{\Phi}(\xi) = \Phi(\xi^{\sharp\check{}}), \tag{L.12}$$

where $\xi^\sharp$ is defined by

$$\xi^\sharp(p) = \xi(-p), \tag{L.13}$$

and where $\xi^{\sharp\check{}}$ denotes the inverse Fourier transform of $\xi^\sharp$. Let us observe that if we identify position and momentum space, this coincides with the definition (L.11), because then it is immediate that $\mathcal{F}_m^{-1}(\xi^\sharp) = \mathcal{F}_m(\xi)$.

The point of this definition is to make it coincide with the usual Fourier transform when Φ is actually a function and

$$\Phi(f) = \int d^n x \, \Phi(x) f(x). \tag{L.14}$$

We check that this is the case in the remainder of the appendix. The first point is that

$$\widehat{f^{*\sharp}} = (\hat{f})^*,$$

where of course $f^*(x) = f(x)^*$, as it is straightforward to see from the definition of $\hat{f}$. To lighten notation we denote by $(\cdot,\cdot)$ the inner product of two functions, either on position or on momentum space. Thus (L.14) becomes $\Phi(f) = (\Phi^*, f)$. Then, using Plancherel's formula in the third equality, and the change of variable $p \to -p$ in the fourth one,

$$\hat{\Phi}(\xi) = \Phi(\xi^{\sharp\check{}}) = (\Phi^*, \xi^{\sharp\check{}}) = (\widehat{\Phi^*}, \xi^\sharp) = (\widehat{\Phi^{*\sharp}}, \xi) = ((\hat{\Phi})^*, \xi), \tag{L.15}$$

which exactly says that the Fourier transform of the function Φ seen as a distribution coincides with its Fourier transform seen as a function.

Appendix M

Wightman Axioms and Haag's Theorem

In this appendix we introduce the Wightman axioms, one of the most successful approaches to rigorous results in Quantum Field Theory, in the simplest possible case of a Hermitian scalar field. Our purpose is twofold. First, we want to describe some of the methods used in this rigorous approach. Second, we prove two significant results. We prove part of Haag's theorem, enough to clearly demonstrate the difficulties inherent with the "interaction picture" in Quantum Field Theory. We also prove a rigorous version of the Källén–Lehmann representation of the dressed propagator, which is fundamental to relate the properties of this propagator to the physical mass of the particle.

The two main mathematical tools required in developing the Wightman axioms are distribution theory and the theory of analytic functions of several variables. Therefore Appendix L is a prerequisite here. On the other hand, we have managed to entirely avoid the use of analytic functions and the corresponding deep results, such as the "edge of the wedge" theorem, and this allows for a reasonably compact presentation. This appendix is far more mathematical than physical, and the relevance of the present material to the physical world is a matter of debate (a point which is even more acute when one uses sophisticated results about analytic functions). Nonetheless, the mathematics are certainly appealing. The results of this appendix are fully rigorous. We have even at a few places been fastidious enough to actually work with distributions rather than pretending they are functions. While we hope that the sample of results we present here is more accessible than Bogoliubov et al.'s [9] or even Streater and Wightman's books [79], there is no doubt that it is still a significant undertaking.

We use a system of units such that $\hbar = 1$.

M.1 The Wightman Axioms

In this setting, the states of the theory are described by the points (or, more accurately, the unitary rays) of a separable Hilbert space $\mathcal{H}$. The Poincaré group $\mathcal{P}$ acts on $\mathcal{H}$ by a unitary representation $U(c,C)$. (Since we do not consider particles with spin there is no need to use the group $\mathcal{P}^*$.) We assume that the map $c \mapsto U(c,1)$ is strongly continuous. The representation $U(c,1)$ is assumed to satisfy a fundamental property, which physically expresses that there are no particles of negative mass. Unfortunately the standard way to express this property appeals to the spectral theory of operators, a topic we have not covered. We will briefly explain this property using the language of spectral theory, and then give an equivalent formulation which does not appeal to this theory and which is the one we will use. This point is by far the most difficult part of the axioms to understand, and the reader should not be discouraged at this stage.

The operators $U(c,1)$ constitute a unitary representation of $\mathbb{R}^4$, and $\mathbb{R}^4$ is a commutative group. We will admit the fact that they have a common spectral resolution, that is we have a formula

$$U(c,1) = \int \exp(\mathrm{i}(c,p))\mathrm{d}E(p), \tag{M.1}$$

where E is a *projection-valued measure*. The next two paragraphs explain what this means and how to understand the integral above.

A projection-valued measure on $\mathbb{R}^4$ associates to any Borel subset A of $\mathbb{R}^4$ a projection $E(A)$ in $\mathcal{H}$. That $E(A)$ is a projection means that there is an orthogonal decomposition $\mathcal{H} = \mathcal{H}_1 + \mathcal{H}_2$ of $\mathcal{H}$, so that each $x \in \mathcal{H}$ can be written in a unique manner $x = x_1 + x_2$ with $x_1 \in \mathcal{H}_1$ and $x_2 \in \mathcal{H}_2$, and $E(A)(x) = x_1$. Thus $\mathcal{H}_1$ is the range $E(A)(\mathcal{H})$ of $E(A)$. The projection-valued measure E is additive in the sense that if A_1 and A_2 are disjoint, then the ranges of the projectors $E(A_1)$ and $E(A_2)$ are orthogonal to each other. It is also σ-additive in the sense that the range of $E(\cup A_n)$ is the closure of the union of the ranges of $E(A_n)$ for an increasing sequence A_n. This sounds very complicated at first, but the rather trivial example M.1.1 is nonetheless almost generic.

Given a projection-valued measure on $\mathbb{R}^4$ and a nice (say continuous and bounded) real-valued function ψ on $\mathbb{R}^4$ it does not take much imagination to define the integral $\int \psi(p)\mathrm{d}E(p)$ as in (M.1), by considering first the case where ψ takes only a finite number of values.[1] The value of this integral is an operator.

Example M.1.1 Consider a positive measure $\mathrm{d}\mu$ on $\mathbb{R}^n$, and assume for simplicity that each bounded set has a finite measure. (If you do not like this generality, please assume that $\mathrm{d}\mu$ has a density with respect to the volume measure.) Let $\mathcal{H} = L^2(\mathbb{R}^n, \mathrm{d}\mu)$. Consider the operator $E(A)$ given by multiplication of a function by the indicator of A. That is, $E(A)(f)(x) = f(x)$ for $x \in A$ and $E(A)(f)(x) = 0$ otherwise. This defines a projection-valued measure. When $\mu(A) = 0$ then $E(A) = 0$. Furthermore if the function f takes finitely many values, the integral $\int f(x)\mathrm{d}E(x)$ is simply the operator "multiplication by the function f", and this remains true when f is well behaved, say bounded continuous.

Example M.1.2 Consider now $m > 0$ and the measure $\mathrm{d}\lambda_m$, seen as a positive measure on $\mathbb{R}^4$, and the space $\mathcal{H} = L^2(X_m, \mathrm{d}\lambda_m)$, which we may identify with $L^2(\mathbb{R}^4, \mathrm{d}\lambda_m)$. To each set $A \subset \mathbb{R}^4$ we associate the operator $E_0(A)$ on $\mathcal{H}$, the multiplication by the indicator function $\mathbf{1}_A$. The operator $\int \exp(\mathrm{i}(c,p))\mathrm{d}E_0(p)$ is then simply the operator "multiplication by $\exp(\mathrm{i}(c,p))$", that is the action of space-time translations as given by (4.61). Consider the boson Fock space $\mathcal{B}$ of $\mathcal{H}$. The operator $E_0(A)$ has a canonical extension $E(A)$ to $\mathcal{B}$. To visualize this, we recall that $\mathcal{B}$ is the Hilbert sum of the spaces $L^2_{n,s}$ of symmetric functions on X_m^n (or $(\mathbb{R}^4)^n$), and for such a function ϕ, $E(A)(\phi)$ is given by

$$E(A)(\phi)(p_1, \ldots, p_n) = \mathbf{1}_A(p_1) \ldots \mathbf{1}_A(p_n)\phi(p_1, \ldots, p_n).$$

Writing what this means, there is no difficulty to check that E defines a projection-valued measure, and that moreover the action of space-time translations on $\mathcal{B}$ is described by the formula (M.1). In this case, the projection-valued measure E is supported by X_m, in the sense that $E(A) = 0$ whenever $A \cap X_m = \emptyset$.

Definition M.1.3 The interior[2] of the forward light cone is the set

$$\bar{\mathcal{V}} := \{p = (p^0, \boldsymbol{p}) \in \mathbb{R}^4 \; ; \; p^0 \geq \|\boldsymbol{p}\|\}. \tag{M.2}$$

We then make the following requirement on the projection-valued measure E of (M.1).

Axiom 1 We have

$$A \cap \bar{\mathcal{V}} = \emptyset \Rightarrow E(A) = 0. \tag{M.3}$$

[1] That is, if ψ takes only the values $y_1, \ldots, y_n$, the value of this integral is $\sum_{i \leq n} y_i E(\psi^{-1}(y_i))$.
[2] This is standard terminology, but one must beware that despite its name this set is topologically a *closed* set. The correct name for this set should be "the closure of the interior of the forward light cone".

We have already seen in Example M.1.2 that this holds in the case of the action of the Poincaré group "on the state space of the free scalar field". There, we even had the stronger conclusion that $E(A) = 0$ when $A \cap X_m = \emptyset$, reflecting the fact that this free field corresponds to a particle of mass exactly m.

Our next goal is to explain why Axiom 1 means that "there are no negative masses" (even though as we are going to see soon, the *Wightman axioms describe a quantum field without any reference to particles*). For this we have to recall the relation between the momenta operators and the action of the Poincaré group as explained on page 126, namely (using relativistic notation) that the operator P^μ of momentum along the direction of coordinate $\mu = 1,2,3$ should be such that $\exp(-\mathrm{i}sP^\mu)$ is the action of a translation of s in the same direction, with a similar interpretation in the case $\mu = 0$ with the Hamiltonian instead of the momentum operator. Thus (M.1) implies that (for $\mu \geq 1$) we have

$$\exp(-\mathrm{i}sP^\mu) = \int \exp(-\mathrm{i}sp^\mu)\mathrm{d}E(p), \tag{M.4}$$

so that (taking the differential in s at $s = 0$)

$$P^\mu = \int p^\mu \mathrm{d}E(p), \tag{M.5}$$

and a similar argument also works for $\mu = 0$. Now comes the heuristic physical argument: The formula $m^2 = p^\mu p_\mu$ for a particle indicates that the operator $Q := P^\mu P_\mu$ should measure "the square of the mass". Furthermore differentiating twice (M.4) at $s = 0$ implies that this operator is given by $\int p^\mu p_\mu \mathrm{d}E(p)$. Axiom 1 implies that the projection-valued measure $\mathrm{d}E(p)$ lives on the set where $p^\mu p_\mu$ is ≥ 0. Technically this is expressed by the fact that "the operator Q has a positive spectrum" and this means that the observable corresponding to the operator Q, the square of the mass, is always ≥ 0.

Since we do not want to go into the innards of spectral theory, we reformulate Axiom 1 as follows.

Axiom 1'. If the function $f \in \mathcal{S}^4$ is such that its Fourier transform $\hat{f}$ has a compact support which does not meet $\bar{\mathcal{V}}$ then

$$\int f(c)U(c,1)\mathrm{d}^4c = 0. \tag{M.6}$$

At least formally, this is a consequence of (M.1) and Axiom 1, since

$$\begin{aligned}\int f(c)U(c,1)\mathrm{d}^4c &= \int f(c)\Big(\int \exp(\mathrm{i}(c,p))\mathrm{d}E(p)\Big)\mathrm{d}^4c \\ &= \int\Big(\int f(c)\exp(\mathrm{i}(c,p))\mathrm{d}^4c\Big)\mathrm{d}E(p) \\ &= \int \hat{f}(p)\mathrm{d}E(p) = 0,\end{aligned}$$

where the last integral should be zero as an easy consequence of Axiom 1 once one has properly defined the integral with respect to a projection measure. This formal argument can be made rigorous if one knows spectral theory.

We stated Axiom 1 only to make informally the link with the standard way to formulate matters, and we next reformulate Axiom 1' in the exact form we will actually use. This reformulation is a consequence of the fact that a bounded operator W vanishes if and only if $(\xi, W\eta) = 0$ whenever $\xi,\eta \in \mathcal{H}$. (We do not use Dirac's notation in this appendix.)

Axiom 1*. Given $\xi,\eta \in \mathcal{H}$,[3] and a function $f \in \mathcal{S}^4$ such that its Fourier transform has a compact support which does not intersect $\bar{V}$ then

$$\int f(c)(\xi, U(c,1)\eta)\mathrm{d}^4c = 0.$$

The main idea now is to define a *quantum field* φ as an operator-valued distribution. That is, for $f \in \mathcal{S}^4$ the quantity $\varphi(f)$ is an unbounded operator on the state space $\mathcal{H}$. We assume that as f varies these operators have a common domain $\mathcal{D}$ of definition, which is also invariant under the action of the Poincaré group. This is the content of the next two axioms, where we recall that $\mathcal{S}^4_{\mathbb{R}}$ denotes the space of real-valued test functions.

Axiom 2 There exists a dense subspace $\mathcal{D}$ of $\mathcal{H}$ and a linear map φ from $\mathcal{S}^4_{\mathbb{R}}$ into the space of (unbounded) operators on $\mathcal{H}$ with the following properties.

$$f \in \mathcal{S}^4_{\mathbb{R}} \Rightarrow \varphi(f) \text{ is defined on } \mathcal{D} \text{ and } \varphi(f)(\mathcal{D}) \subset \mathcal{D}. \tag{M.7}$$

$$\xi,\eta \in \mathcal{D}, f \in \mathcal{S}^4_{\mathbb{R}} \Rightarrow (\varphi(f)\xi,\eta) = (\xi,\varphi(f)\eta). \tag{M.8}$$

$$\forall (c,C) \in \mathcal{P}\,,\ U(c,C)\mathcal{D} \subset \mathcal{D}. \tag{M.9}$$

Condition (M.8) means that we consider only symmetric operators. This is for simplicity, and the axiomatic setting itself does not require this. As a consequence of (M.8), when we allow $f \in \mathcal{S}^4$ to take complex values we have

$$(\varphi(f^*)\xi,\eta) = (\xi,\varphi(f)\eta). \tag{M.10}$$

Axiom 3 Given $\xi,\eta \in \mathcal{D}$ the map $f \mapsto (\xi,\varphi(f)\eta)$ is a tempered distribution.

In the sequel we will actually have to consider also a more general version where there are several fields φ_j, $j \le k$. In that case the proper formulation of Axiom 3 involves all products $\varphi_1(f_1)\cdots\varphi_n(f_n)$.

The next two axioms should be very familiar. We have first met these properties when constructing the free scalar field. They express respectively that the quantum field transforms as expected under the Poincaré group, and microcausality.

Axiom 4 Given $(c,C) \in \mathcal{P}$ and $f \in \mathcal{S}^4$ we have

$$U(c,C)\varphi(f)U(c,C)^{-1} = \varphi(V(c,C)f), \tag{M.11}$$

where $V(c,C)f(x) = f(C^{-1}(x-c))$.

Axiom 5 If $f,g \in \mathcal{S}^4$ are such that their supports are causally separated (i.e. $(x-y)^2 < 0$ for x in the support of f and y in the support of g) then $[\varphi(f),\varphi(g)] = 0$ as an operator defined on $\mathcal{D}$.

The next axiom concerns "existence and uniqueness of the vacuum state".

Axiom 6 There exists an element Ω of $\mathcal{D} \subset \mathcal{H}$ (the "vacuum state") such that $U(c,1)\Omega = \Omega$ and every element with this invariance property is a multiple of Ω.[4]

The last axiom expresses that there are sufficiently many fields $\varphi(f)$.

3 We have tried to consistently denote functions on momentum space by Greek letters. There is unfortunately a shortage of such letters, so ξ,η are also used to denote points in the state space $\mathcal{H}$. On the other hand, the letter φ is used only for the quantum field.

4 In Streater and Wightman's book [79] this axiom is stated in the apparently weaker form that Ω is unique (up to a phase) under the stronger condition that $U(c,C)\Omega = \Omega$ for all $(c,C) \in \mathcal{P}$. I cannot prove that this implies Axiom 6. However on page 141, line 8 from bottom, these authors seem to use the strong form of Axiom 6 as stated here.

Axiom 7 $\mathcal{H}$ is the closed linear span of the vectors $\varphi(f_1)\varphi(f_2)\cdots\varphi(f_n)\Omega$ for $n \geq 0, f_j \in \mathcal{S}^4$.

These axioms have been the basis of a large body of rigorous work. It is a really simple matter to show that they are satisfied by the free scalar field we constructed in Chapter 4. Only proving that Axiom 6 holds needs any work at all and this will be explained in Lemma M.3.3. Unfortunately, to this day there is essentially no example (in four space-time dimensions) of situations really different from a free field where these axioms are satisfied. This sharply decreases their appeal. Another reservation one might have is that, while statements like Axioms 4 and 5 are completely natural, why should one use tempered distributions in our model?

Still, we shall try in the following few pages to describe the type of methods which can be used to prove theorems under these assumptions. This will bring forward the fact that some of these assumptions are far more powerful than is immediately apparent.

We first show that the vacuum state must be invariant under the action of the Poincaré group.

Proposition M.1.4 *For each* $(c, C) \in \mathcal{P}$ *we have* $U(c, C)\Omega = \Omega$.

Proof Since $U(c, C) = U(c, 1)U(0, C)$ and since by Axiom 6, $U(c, 1)\Omega = \Omega$ it suffices to prove this when $c = 0$. Since $U(c, 1)U(0, C) = U(0, C)U(C^{-1}(c), 1)$, the vector $U(0, C)\Omega$ is invariant under all the transformations $U(c, 1)$. Thus by hypothesis of Axiom 6, $U(0, C)\Omega = A(C)\Omega$ for a certain number $A(C)$ of modulus 1. Then A defines a one-dimensional representation of the Lorentz group, and therefore is trivial, $A(C) = 1$ for all C. (The proof of this fact is similar to the proof we gave in the case of the rotation group in Lemma 8.5.2.) □

Our next result shows how Axiom 1* might be used.

Proposition M.1.5 *Consider a bounded operator B on $\mathcal{H}$ and assume that*

$$\xi, \eta \in \mathcal{D}\,,\ f \in \mathcal{S}^4_{\mathbb{R}} \Rightarrow (\xi, B\varphi(f)\eta) = (\varphi(f)\xi, B\eta). \tag{M.12}$$

Then B is a multiple of the identity.

If we knew that $\varphi(f)$ is defined on each vector $B\eta$ then (M.12) would simply mean that B and $\varphi(f)$ commute. Thus we may heuristically rephrase Proposition M.1.5 as saying "every operator B which commutes with all the $\varphi(f)$ is a multiple of the identity".

In the sequel we need Proposition M.1.5 only for the free scalar field. In that case we provide later a direct proof. So, the general argument which follows, while not really more complicated than the direct proof, is not required reading. Its point is essentially to illustrate the power of Axiom 1*.

Proof Iteration of (M.12) shows that for each n, any real-valued test functions $f_1, \ldots, f_n$ and $\xi, \eta \in \mathcal{D}$ we have

$$(\xi, B\varphi(f_1)\cdots\varphi(f_n)\eta) = (\varphi(f_n)\cdots\varphi(f_1)\xi, B\eta). \tag{M.13}$$

Let us specialize this relation to the case $\xi = \eta = \Omega$ and replace f_j by $V(c, 1)f_j$. Then (M.11) for $f = f_j$ means exactly that $\varphi(V(c, 1)f_j) = U(c, 1)\varphi(f_j)U(c, 1)^{-1}$. Together with the fact that $U(c, 1)^{-1}\Omega = \Omega$ we obtain

$$\begin{aligned}(\Omega, BU(c, 1)\varphi(f_1)\cdots\varphi(f_n)\Omega) &= (U(c, 1)\varphi(f_n)\cdots\varphi(f_1)\Omega, B\Omega)\\ &= (\varphi(f_n)\cdots\varphi(f_1)\Omega, U(-c, 1)B\Omega),\end{aligned} \tag{M.14}$$

using in the last equality that $U(c, 1)^\dagger = U(-c, 1)$. Consider now the function $w(c)$ which is defined either by the left-hand side or the right-hand side of the above equality. This function is bounded because B is a bounded operator, and it is continuous because the map $c \mapsto U(c, 1)$ is strongly continuous. Let us write the left-hand side of (M.14) as

$$(\xi, U(c, 1)\eta) := (B^\dagger\Omega, U(c, 1)\varphi(f_1)\cdots\varphi(f_n)\Omega) = w(c).$$

Axiom 1* implies that whenever the Fourier transform of f has a compact support which does not intersect $\bar{V}$, then $\int f(c)w(c)\mathrm{d}^4c = 0$. Denoting by $\check{\psi}$ the inverse Fourier transform of $\psi \in \mathcal{S}^4$, the formula

$$\Psi(\psi) = \int \check{\psi}(c)w(c)\mathrm{d}^4c \tag{M.15}$$

defines a distribution Ψ on $\mathbb{R}^4$. Since the Fourier transform of $\check{\psi}$ is ψ the previous property implies that Ψ is supported by $\bar{V}$.

Using now the right-hand side of (M.14) we also have

$$(\xi', U(-c,1)\eta') := (\varphi(f_n)\cdots\varphi(f_1)\Omega, U(-c,1)B\Omega) = w(c)$$

and using again Axiom 1* shows that Ψ is supported by $-\bar{V}$. Lemma L.2.4 then asserts that the distribution Ψ is supported by $\bar{V}\cap-\bar{V}$ which is reduced to the origin. Thus we may apply Proposition L.2.10 to conclude that Ψ must be of the type described in this proposition. Let us then prove that $w(c)$ must be a polynomial. For this it suffices to show that we can find a polynomial which satisfies (M.15) because knowing all the integrals $\int \check{\psi}(c)w(c)\mathrm{d}^4c$ determines w. But that we can find such a w follows from the extension of certain formulas of Section 1.5, namely the fact that $\int \check{\psi}(c)\mathrm{d}^4c = \psi(0)$ and formulas such as the first part of (1.23). Now since w is a polynomial and is bounded it must be constant. Thus for $c, c' \in \mathbb{R}^4$ we have $(\xi', U(c,1)B\Omega) = (\xi', U(c',1)B\Omega)$ whenever ξ' is of the type $\varphi(f_1)\cdots\varphi(f_n)\Omega$. As these vectors are dense in $\mathcal{H}$ by Axiom 7, this means that $U(c,1)B\Omega$ is independent of c. It then follows from Axiom 6 that $B\Omega$ is a multiple of Ω, $B\Omega = a\Omega$. Going back to (M.13) in the case $\eta = \Omega$ we obtain

$$(\xi, B\varphi(f_1)\cdots\varphi(f_n)\Omega) = a(\varphi(f_n)\cdots\varphi(f_1)\xi, \Omega) = a(\xi, \varphi(f_1)\cdots\varphi(f_n)\Omega). \tag{M.16}$$

Since both the vectors of the type $\xi \in \mathcal{D}$ and $\varphi(f_1)\cdots\varphi(f_n)\Omega$ are dense in $\mathcal{H}$ this implies that $B = a\mathbf{1}$. □

We now present a direct proof of Proposition M.1.5 in the case of the free field. It does not use distribution theory. The state space is the boson Fock space $\mathcal{B}$ of $L^2(X_m, \mathrm{d}\lambda_m)$. It is the Hilbert sum of the spaces $\mathcal{H}_{n,s}$, where $\mathcal{H}_{n,s}$ is the space of square-integrable functions on X_m^n which are symmetric in their arguments. The state Ω is a basis of the space $\mathcal{H}_{0,s} = \mathbb{C}$.

Lemma M.1.6 *Given any $\psi \in \mathcal{S}^4$ there is $f \in \mathcal{S}^4_{\mathbb{R}}$ such that $\hat{f}$ coincides with ψ on X_m.*

Proof We simply find $\eta \in \mathcal{S}^4$ with $\eta(p) = \eta(-p)^*$ and $\eta(p) = \psi(p)$ for $p \in X_m$. For this we find a neighborhood U of X_m such that U and $-U$ are disjoint. We find a function $\gamma \in \mathcal{S}^4$ with support contained in U such that $\gamma(p) = \psi(p)$ if $p \in X_m$. We define $\eta(p) = \gamma(p)$ if $p \in U$, $\eta(p) = \gamma(-p)^*$ if $p \in -U$ and $\eta(p) = 0$ otherwise. Then $f = \check{\eta}$ satisfies $f^* = f$ so that it is real-valued and $\hat{f}(p) = \eta(p) = \psi(p)$ for $p \in X_m$. □

Lemma M.1.7 *If W is a linear function of $\xi \in \mathcal{S}^4$ and W' is an anti-linear function of $\xi \in \mathcal{S}^4$ such that $W(\xi) + W'(\xi) \equiv 0$ then $W(\xi) \equiv 0$ and $W'(\xi) \equiv 0$.*

Proof Changing ξ into $\mathrm{i}\xi$ in the relation $W(\xi) + W'(\xi) = 0$ yields $\mathrm{i}W(\xi) - \mathrm{i}W'(\xi) = 0$. □

Proof of Proposition M.1.5 for the free field The plan is to deduce from (M.13) that $B\Omega$ is a multiple of Ω, and to conclude as previously with (M.16) that $B = a\mathbf{1}$. Using (M.13) in the case where $\xi = \eta = \Omega$ we obtain that for any real-valued test functions $f_1, \ldots, f_n$ we have

$$(B^\dagger\Omega, \varphi(f_1)\cdots\varphi(f_n)\Omega) = (\varphi(f_n)\cdots\varphi(f_1)\Omega, B\Omega). \tag{M.17}$$

We recall that the free field is given by the formula $\varphi(f) = A(\hat{f}) + A^\dagger(\hat{f})$, where A and $A^\dagger$ are as in Theorem 3.4.2. We then conclude from Lemma M.1.6 that for any n and any test functions $\xi_1, \ldots, \xi_n$ we have

$$(B^\dagger\Omega, (A^\dagger(\xi_1)+A(\xi_1))\cdots(A^\dagger(\xi_n)+A(\xi_n))\Omega)$$
$$= ((A^\dagger(\xi_n)+A(\xi_n))\cdots(A^\dagger(\xi_1)+A(\xi_1))\Omega, B\Omega). \tag{M.18}$$

Thinking of $\xi_1,\ldots,\xi_{n-1}$ as fixed, both the right-hand side and the left-hand are sums of a part which is linear in ξ_n and a part which is anti-linear. The anti-linear part of the left-hand side is $(B^\dagger\Omega, (A^\dagger(\xi_1)+A(\xi_1))\cdots A(\xi_n)\Omega)$ which is zero since $A(\xi_n)\Omega = 0$. According to Lemma M.1.7 the anti-linear part of the right-hand side is also zero, that is

$$(A^\dagger(\xi_n)(A^\dagger(\xi_{n-1})+A(\xi_{n-1}))\cdots(A^\dagger(\xi_1)+A(\xi_1))\Omega, B\Omega) = 0.$$

Thinking now of $\xi_1,\ldots,\xi_{n-2},\xi_n$ as fixed, as a function of ξ_{n-1} the previous quantity is the sum of a linear and an anti-linear part. Expressing that both are zero we obtain

$$(A^\dagger(\xi_n)A^\dagger(\xi_{n-1})(A^\dagger(\xi_{n-2})+A(\xi_{n-2}))\cdots(A^\dagger(\xi_1)+A(\xi_1))\Omega, B\Omega) = 0.$$

Continuing in this fashion we obtain that $(A^\dagger(\xi_n)\cdots A^\dagger(\xi_1)\Omega, B\Omega) = 0$. Since the closed linear span of the elements of the type $A^\dagger(\xi_n)\cdots A^\dagger(\xi_1)\Omega$ is $\mathcal{H}_{n,s}$, and since this holds for each n this implies that $B\Omega$ is a multiple of Ω. □

Let us now introduce some vocabulary.

Definition M.1.8 A set $\mathcal{A}$ of operators defined on $\mathcal{D}$ is irreducible if any bounded operator B which satisfies

$$(\xi, BA\eta) = (A^\dagger\xi, B\eta)$$

for each $A \in \mathcal{A}$ and each $\xi,\eta \in \mathcal{D}$ is a multiple of the identity.

Thus Proposition M.1.5 expresses that the set of operators of the form $\varphi(f)$ for $f \in \mathcal{S}^4_{\mathbb{R}}$ is irreducible.

M.2 Statement of Haag's Theorem

Consider a field φ which satisfies the previous axioms. Furthermore assume that given $t \in \mathbb{R}$ and $g \in \mathcal{S}^3 = \mathcal{S}(\mathbb{R}^3)$ we can define the operators $\varphi_t(g)$ in a way that

$$g \mapsto (\xi, \varphi_t(g)\eta)$$

are tempered distributions. Assume that for each $f \in \mathcal{S}^4_{\mathbb{R}}$ we have

$$(\xi, \varphi(f)\eta) = \int (\xi, \varphi_t(f_t)\eta)\mathrm{d}t$$

where $f_t \in \mathcal{S}^3_{\mathbb{R}}$ is given by $f_t(\boldsymbol{x}) = f(t,\boldsymbol{x})$. Assume that this decomposition of φ is done in a way consistent with Axiom 4, namely that, writing

$$W(\boldsymbol{a}) = U((0,\boldsymbol{a}),1), \tag{M.19}$$

we have

$$W(\boldsymbol{a})\varphi_t(g)W(\boldsymbol{a})^{-1} = \varphi_t(g_{\boldsymbol{a}}),$$

where $g_{\boldsymbol{a}}$ is given by $g_{\boldsymbol{a}}(\boldsymbol{x}) = g(\boldsymbol{x}-\boldsymbol{a})$. In particular we have

$$W(\boldsymbol{a})\varphi_0(g)W(\boldsymbol{a})^{-1} = \varphi_0(g_{\boldsymbol{a}}). \tag{M.20}$$

Assume moreover that the derivatives $\dot\varphi_t$ of φ_t with respect to t exist, so that we also have

$$W(\boldsymbol{a})\dot\varphi_0(g)W(\boldsymbol{a})^{-1} = \dot\varphi_0(g_{\boldsymbol{a}}). \tag{M.21}$$

Consider now a free scalar field $\varphi^{\rm free}$ of mass $m > 0$, defined on another Hilbert space $\mathcal{H}^{\rm free}$. As we have shown in Section 6.9 (and as will be recalled below) we know how to define the operators $\varphi_t^{\rm free}$ and $\dot\varphi_t^{\rm free}$.

Theorem M.2.1 (Haag's theorem) *Assume that there exists a unitary operator* $V : \mathcal{H} \to \mathcal{H}^{\rm free}$ *such that*

$$\forall g \in \mathcal{S}^3_{\mathbb{R}}\ ,\ V\varphi_0(g)V^{-1} = \varphi_0^{\rm free}(g)\ ;\ V\dot\varphi_0(g)V^{-1} = \dot\varphi_0^{\rm free}(g). \tag{M.22}$$

Then φ *is also a free field of mass* m.

Implicit in (M.22) is the fact that the domains of definition $\mathcal{D}$ and $\mathcal{D}^{\rm free}$ of the operators φ are exchanged to each other by V. The last statement means that φ is unitarily equivalent to the free field $\varphi^{\rm free}$ of mass m.

The importance of Haag's theorem was explained in Section 13.4, where we argued that in the interaction picture an interacting field must satisfy (M.22). Something goes wrong: It just cannot happen that every interacting field is a free field. The trouble lies with the interaction picture (which was never rigorously defined): Haag's theorem shows that this interaction picture *is not mathematically consistent.*

In the rest of the appendix, we analyze the situation (M.22). While we do not fully prove Haag's theorem we obtain enough information about a field φ which satisfies this condition to make clear that it cannot be an interacting field of interest.

M.3 Easy Steps

Our approach starts with two easy results concerning the free field.

Proposition M.3.1 *The set of operators* $\varphi_0^{\rm free}(g)$, $\dot\varphi_0^{\rm free}(g)$ *for* $g \in \mathcal{S}^3_{\mathbb{R}}$ *is irreducible.*

Proof It follows from Lemma M.3.2 that the linear span of this set of operators contains the set of operators of the type $\varphi^{\rm free}(f)$ for $f \in \mathcal{S}^4_{\mathbb{R}}$, which is irreducible according to Proposition M.1.5. □

Lemma M.3.2 *Given* $f \in \mathcal{S}^4_{\mathbb{R}}$ *there exist* g_1 *and* $g_2 \in \mathcal{S}^3_{\mathbb{R}}$ *such that* $\varphi^{\rm free}(f) = \varphi_0^{\rm free}(g_1) + \dot\varphi_0^{\rm free}(g_2)$.

Proof We recall from (6.66) that the operators $\varphi_t^{\rm free}$ are given by the formula

$$\varphi_t^{\rm free}(g) = A(\mathcal{F}_t(g)) + A^\dagger(\mathcal{F}_t(g)), \tag{M.23}$$

where A and $A^\dagger$ denote the annihilation and creation operators on the space $\mathcal{H}^{\rm free}$, which is the boson Fock space of the space $L^2(X_m, \mathrm{d}\lambda_m)$ and where for $g \in \mathcal{S}^3_{\mathbb{R}}$,

$$\mathcal{F}_t(g)(p) := \exp(\mathrm{i}tp^0)\mathcal{F}(g)(\boldsymbol{p})\ ;\ \mathcal{F}(g)(\boldsymbol{p}) := \int \exp(-\mathrm{i}\boldsymbol{x}\cdot\boldsymbol{p})g(\boldsymbol{x})\mathrm{d}^3\boldsymbol{x}. \tag{M.24}$$

Consequently,

$$\varphi_0^{\rm free}(g) = A(\mathcal{F}(g)) + A^\dagger(\mathcal{F}(g)) \tag{M.25}$$

$$\dot\varphi_0^{\rm free}(g) = A(\mathcal{G}(g)) + A^\dagger(\mathcal{G}(g)), \tag{M.26}$$

where

$$\mathcal{F}(g)(p) = \mathcal{F}(g)(\boldsymbol{p})\ ;\ \mathcal{G}(g)(p) = \mathrm{i}p^0\mathcal{F}(g)(\boldsymbol{p}). \tag{M.27}$$

Looking at (M.25) and (M.26), this means that given $f \in \mathcal{S}^4_{\mathbb{R}}$ it suffices to find $g_1, g_2 \in \mathcal{S}^3_{\mathbb{R}}$ such that $\hat{f} = \mathcal{F}(g_1) + \mathcal{G}(g_2)$ as elements of $L^2(X_m, \mathrm{d}\lambda_m)$. That is, it suffices that this equality holds for any point $p \in X_m$, or, equivalently, that for $\boldsymbol{p} \in \mathbb{R}^3$ we have

$$\hat{f}(p) = \mathcal{F}(g_1)(\boldsymbol{p}) + \mathrm{i}\omega_{\boldsymbol{p}}\mathcal{F}(g_2)(\boldsymbol{p}), \tag{M.28}$$

where $\omega_{\boldsymbol{p}} = \sqrt{m^2 + \boldsymbol{p}^2}$ and $p = (\omega_{\boldsymbol{p}}, \boldsymbol{p})$. Let us try to guess from this formula what g_1 and g_2 should be. Let us first observe that a function g on $\mathbb{R}^3$ is real-valued if and only if for each $\boldsymbol{p}$ we have

$$\mathcal{F}(g)(\boldsymbol{p})^* = \mathcal{F}(g)(-\boldsymbol{p}). \tag{M.29}$$

Consider then $Pp = (\omega_{\boldsymbol{p}}, -\boldsymbol{p})$. We then deduce that if (M.28) holds then

$$\begin{aligned}\hat{f}(Pp)^* &= \mathcal{F}(g_1)(-\boldsymbol{p})^* - \mathrm{i}\omega_{\boldsymbol{p}}\mathcal{F}(g_2)(-\boldsymbol{p})^* \\ &= \mathcal{F}(g_1)(\boldsymbol{p}) - \mathrm{i}\omega_{\boldsymbol{p}}\mathcal{F}(g_2)(\boldsymbol{p}).\end{aligned}$$

Therefore we must have

$$2\mathcal{F}(g_1)(\boldsymbol{p}) = \hat{f}(p) + \hat{f}(Pp)^* \; ; \; 2\mathrm{i}\omega_{\boldsymbol{p}}\mathcal{F}(g_2)(\boldsymbol{p}) = \hat{f}(p) - \hat{f}(Pp)^*.$$

Since the function $p \mapsto \hat{f}(p) \pm \hat{f}(Pp)^*$ is a test function, we may find test functions g_1 and g_2 which satisfy the above relations. These are real-valued from (M.29), and it is obvious that (M.28) holds. □

As a consequence of Lemma M.3.2 one may say that "φ^{free} is determined by $\varphi^{\mathrm{free}}_0$ and $\dot{\varphi}^{\mathrm{free}}_0$".[5] Denoting by U^{free} the action of the Poincaré group on $\mathcal{H}^{\mathrm{free}}$, we now write

$$W^{\mathrm{free}}(\boldsymbol{a}) = U^{\mathrm{free}}((0, \boldsymbol{a}), 1).$$

Lemma M.3.3 *Consider an element $\xi \in \mathcal{H}^{\mathrm{free}}$ and assume that*

$$\forall \boldsymbol{a} \in \mathbb{R}^3 \; ; \; W^{\mathrm{free}}(\boldsymbol{a})\xi = c(\boldsymbol{a})\xi \tag{M.30}$$

for a certain number $c(\boldsymbol{a})$. Then $\xi = c\Omega^{\mathrm{free}}$ for a number c.

Proof We recall that $\mathcal{H}^{\mathrm{free}}$ is the boson Fock space based on $L^2 = L^2(X_m, \mathrm{d}\lambda_m)$. It is a direct sum of the spaces $\mathcal{H}_{n,s}$ of symmetric functions on X_m^n, each of which is invariant under the maps $W^{\mathrm{free}}(\boldsymbol{a})$. Therefore it suffices to prove that if $\xi \in \mathcal{H}_{n,s}$ satisfies (M.30) then $\xi = 0$ if $n \geq 1$. Now, the action of $W^{\mathrm{free}}(\boldsymbol{a})$ on $\mathcal{H}_{n,s}$ is multiplication by $\exp(\mathrm{i}\boldsymbol{a} \cdot (\sum_{j \leq n} \boldsymbol{p}_j))$, so that ξ can be transformed into a multiple of itself by such multiplication only if it is (almost everywhere) zero except on a set where this function is constant. These sets are of measure zero for $\lambda_m^{\otimes n}$ so that this is possible only if $\xi = 0$. □

We are now ready to take a first step in the proof of Haag's theorem.

Lemma M.3.4 *Under (M.22) we have*

$$V\Omega = c\Omega^{\mathrm{free}} \tag{M.31}$$

for a number c of modulus 1.

[5] In physics this is often presented as a self-evident consequence of the fact that φ^{free} satisfies the Klein-Gordon equation (as shown in Proposition 6.1.3). The idea is that a function $\varphi(t, \boldsymbol{x})$ satisfying this equation is indeed determined by $\varphi(0, \boldsymbol{x})$ and $\dot{\varphi}(0, \boldsymbol{x})$. Here however φ^{free} is not a function but an operator-valued distribution, so one may feel a bit unsafe to use this argument. It is however essentially correct: Our rigorous argument uses only that the Fourier transform of φ vanishes on test functions which vanish on X_m. i.e. only that φ satisfies the Klein-Gordon equation in the sense of distributions.

Proof Let us note that as in (M.20) we have

$$W^{\text{free}}(\boldsymbol{a})\varphi_0^{\text{free}}(f)W^{\text{free}}(\boldsymbol{a})^{-1} = \varphi_0^{\text{free}}(f_a). \tag{M.32}$$

Then, using in succession (M.22), (M.20), (M.22), and (M.32),

$$\begin{aligned} W^{\text{free}}(\boldsymbol{a})^{-1}VW(\boldsymbol{a})V^{-1}\varphi_0^{\text{free}}(f) &= W^{\text{free}}(\boldsymbol{a})^{-1}VW(\boldsymbol{a})\varphi_0(f)V^{-1} \\ &= W^{\text{free}}(\boldsymbol{a})^{-1}V\varphi_0(f_a)W(\boldsymbol{a})V^{-1} \\ &= W^{\text{free}}(\boldsymbol{a})^{-1}\varphi_0^{\text{free}}(f_a)VW(\boldsymbol{a})V^{-1} \\ &= \varphi_0^{\text{free}}(f)W^{\text{free}}(\boldsymbol{a})^{-1}VW(\boldsymbol{a})V^{-1}. \end{aligned}$$

Therefore the operator $S := W^{\text{free}}(\boldsymbol{a})^{-1}VW(\boldsymbol{a})V^{-1}$ commutes with every operator $\varphi_0^{\text{free}}(f)$, in the sense that for each $\xi \in \mathcal{D}^{\text{free}}$ we have

$$S\varphi_0^{\text{free}}(f)\xi = \varphi_0^{\text{free}}(f)S\xi.$$

In a similar manner we prove the commutation relation

$$S\dot{\varphi}_0^{\text{free}}(f)\xi = \dot{\varphi}_0^{\text{free}}(f)S\xi.$$

It then follows from Proposition M.3.1 that S is a multiple of the identity, i.e.

$$W^{\text{free}}(\boldsymbol{a})^{-1}VW(\boldsymbol{a})V^{-1} = c(\boldsymbol{a})1,$$

so that

$$VW(\boldsymbol{a}) = c(\boldsymbol{a})W^{\text{free}}(\boldsymbol{a})V$$

and hence

$$V\Omega = c(\boldsymbol{a})W^{\text{free}}(\boldsymbol{a})V\Omega.$$

It then follows from Lemma M.3.3 that $V\Omega = c\Omega^{\text{free}}$. Moreover $|c| = 1$ because V is unitary and both Ω and Ω^{free} are unitary vectors.[6] □

This result will be used only through the following corollary.

Corollary M.3.5 *Under (M.22) for $f_1, \ldots, f_n \in \mathcal{S}^3_{\mathbb{R}}$ we have*

$$(\Omega, \varphi_0(f_1)\cdots\varphi_0(f_n)\Omega) = (\Omega^{\text{free}}, \varphi_0^{\text{free}}(f_1)\cdots\varphi_0^{\text{free}}(f_n)\Omega^{\text{free}}). \tag{M.33}$$

Proof Replace $\varphi_0(f)$ by $V^{-1}\varphi_0^{\text{free}}(f)V$ in the left-hand side and use (M.31). □

Corollary M.3.6 *If φ is a free field of mass $m' \neq m$ then (M.33) does not hold.*

Proof The most direct method is to appeal to (M.24) to obtain

$$\begin{aligned} (\Omega^{\text{free}}, \varphi_0^{\text{free}}(f_1)\varphi_0^{\text{free}}(f_2)\Omega^{\text{free}}) &= (\varphi_0^{\text{free}}(f_1)\Omega^{\text{free}}, \varphi_0^{\text{free}}(f_2)\Omega^{\text{free}}) \\ &= (A^\dagger(\mathcal{F}(f_1))\Omega^{\text{free}}, A^\dagger(\mathcal{F}(f_2))\Omega^{\text{free}}). \end{aligned}$$

By construction, this quantity is just the inner product of $\mathcal{F}(f_1)$ and $\mathcal{F}(f_2)$, i.e. that quantity

$$\int \mathcal{F}(f_1)(\boldsymbol{p})^*\mathcal{F}(f_2)(\boldsymbol{p})\frac{\mathrm{d}^3\boldsymbol{p}}{\sqrt{2\omega_{m,\boldsymbol{p}}}},$$

where $\omega^2_{m,\boldsymbol{p}} = m^2 + \boldsymbol{p}^2$ certainly depends on m, so that (M.33) cannot hold if φ is a free field of mass $m' \neq m$. □

[6] Recalling that states are described by unitary vectors of the state space.

M.4 Wightman Functions

It follows from Axioms 2 and 3 (using also that $\Omega \in \mathcal{D}$) that for each n the quantity

$$(\Omega, \varphi(f_1) \cdots \varphi(f_n)\Omega)$$

is a tempered distribution in each f_j when all the other functions f_k are fixed. It then follows from the Schwartz Kernel Theorem [33] (which we briefly met in Section 12.3 to define the scattering matrix) that there exists a tempered distribution Ξ on $\mathbb{R}^{4n}$ such that

$$(\Omega, \varphi(f_1) \cdots \varphi(f_n)\Omega) = \Xi(f_1 \otimes \cdots \otimes f_n), \tag{M.34}$$

where $f_1 \otimes \cdots \otimes f_n(x_1, \ldots, x_n) = f_1(x_1) \cdots f_n(x_n)$. Writing this distribution as a function $\mathcal{W}(x_1, \ldots, x_n)$ the definition becomes

$$\mathcal{W}(x_1, \ldots, x_n) := (\Omega, \varphi(x_1) \cdots \varphi(x_n)\Omega). \tag{M.35}$$

These are called the *Wightman functions* and play an essential role in the approach based on the Wightman axioms.

The purpose of the present section is to study in detail the Wightman function for $n = 2$. The basic idea is that, pretending that distributions are functions, we have translation invariance: the function $\mathcal{W}(x_1, x_2) = (\Omega, \varphi(x_1)\varphi(x_2)\Omega)$ satisfies $\mathcal{W}(x_1, x_2) = \mathcal{W}(x_1 + a, x_2 + a)$ for any a. Thus $\mathcal{W}(x_1, x_2)$ depends only of $x_1 - x_2$, so that it is function of $x_1 - x_2$,

$$\mathcal{W}(x_1, x_2) = (\Omega, \varphi(x_1)\varphi(x_2)\Omega) = W(x_1 - x_2). \tag{M.36}$$

For once, we will however not pretend that distributions are functions, and we will give a completely rigorous version of the previous argument. We will prove that the distribution W is rather regular, because (essentially) its Fourier transform is a positive measure supported by $\bar{\mathrm{V}}$. We will obtain in Proposition M.4.11 a precise description of W, and this enables us to perform a further step in the proof of Haag's theorem. Namely, we will show in Theorem M.4.12 that for $n = 2$ the Wightman function of a field satisfying the hypothesis of Haag's theorem coincides with the corresponding function of a free field.

We first spell out, in the case $n = 2$, some simple properties of the distribution Ξ.

Lemma M.4.1 *For each $(c, C) \in \mathcal{P}$ and $f \in \mathcal{S}^8$ we have*

$$\Xi(f) = \Xi(V(c, C)f), \tag{M.37}$$

where $V(c, C)f(x_1, x_2) = f(C^{-1}(x_1 - c), C^{-1}(x_2 - c))$. Moreover, if $v \in \mathcal{S}^4$ is such that its Fourier transform has a compact support which does not intersect $\bar{\mathrm{V}}$ then for $f_1, f_2 \in \mathcal{S}^4$ it holds that $\Xi(f_1 \otimes f_2') = 0$ where $f_2'(x) = \int \mathrm{d}^4 c f_2(x - c)v(c)$. Finally, for each $h \in \mathcal{S}^4$ we have

$$\Xi(h^* \otimes h) \geq 0. \tag{M.38}$$

Proof Let us recall that by definition of Ξ for $f_1, f_2 \in \mathcal{S}^4$ it holds that

$$\Xi(f_1 \otimes f_2) = (\Omega, \varphi(f_1)\varphi(f_2)\Omega). \tag{M.39}$$

To prove (M.38) we write, using (M.10) in the second equality,

$$\Xi(h^* \otimes h) = (\Omega, \varphi(h^*)\varphi(h)\Omega) = (\varphi(h)\Omega, \varphi(h)\Omega) \geq 0.$$

It suffices to prove (M.37) when $f = f_1 \otimes f_2$ for $f_1, f_2 \in \mathcal{S}^4$ because Ξ is determined by its values on such functions, as follows from Schwartz's kernel theorem. Using (M.11) it holds that $U(c, C)\varphi(f_1)\varphi(f_2)U(c, C)^{-1} = \varphi(V(c, C)f_1)\varphi(V(c, C)f_2)$ and recalling that $U(c, C)^{-1}\Omega = \Omega$ by Proposition M.1.4 we obtain

$$(\Omega, \varphi(f_1)\varphi(f_2)\Omega) = (\Omega, \varphi(V(c, C)f_1)\varphi(V(c, C)f_2)\Omega).$$

Combining with (M.39) completes the proof of (M.37).

Next, consider $v \in \mathcal{S}^4$ such that its Fourier transform has a compact support which does not intersect $\bar{\mathcal{V}}$. Then by Axiom 1* we have

$$\int v(c)(\varphi(f_1)\Omega, U(c,1)\varphi(f_2)\Omega)\mathrm{d}^4c = 0. \tag{M.40}$$

Using again (M.11) for $C = 1$ we obtain $U(c,1)\varphi(f_2)U(-c,1) = \varphi(f_{2,c})$ where $f_{2,c}(x) = f_2(x-c)$ and, using (M.34) in the last equality

$$(\varphi(f_1)\Omega, U(c,1)\varphi(f_2)\Omega) = (\varphi(f_1)\Omega, \varphi(f_{2,c})\Omega) = (\Omega, \varphi(f_1)\varphi(f_{2,c})\Omega) = \Xi(f_1 \otimes f_{2,c}).$$

Therefore, combining with (M.40),

$$0 = \int v(c)\Xi(f_1 \otimes f_{2,c})\mathrm{d}^4c = \Xi(f_1 \otimes f_2'),$$

where the second equality is justified by the fact that for any k the function f_2' can be arbitrarily well approximated in the seminorm $\|\cdot\|_k$ by a linear combination of functions $f_{2,c}$; a tedious exercise. □

When working with distributions there is a constant pressure by need of clarity to pretend that they are functions, and the classical book by Streater and Wightman [79] yields to this pressure at the present stage of the argument. While it is not very difficult to write a fully correct argument without using this shortcut, this is not completely trivial either. As we try to exemplify in this appendix how to be really rigorous while using distributions, we give a rigorous version of (M.36). This is reserved for uncompromising readers, whereas others should jump to Lemma M.4.3.

Proposition M.4.2 *Under the Wightman axioms there exists a tempered distribution W on $\mathbb{R}^4$ such that for any function $f \in \mathcal{S}^8$ it holds that*

$$\Xi(f) = W(H(f)) \tag{M.41}$$

where the function $H(f) \in \mathcal{S}^4$ is given by

$$H(f)(x) = \int f(x+z,z)\mathrm{d}^4z. \tag{M.42}$$

The equality (M.41) formulates in a rigorous way, at the level of distributions, the idea that $\mathcal{W}(x_1,x_2) = W(x_1 - x_2)$. Indeed, for functions, making the change of variable $x_1 = x_2 + y$ we have

$$\begin{aligned}\iint \mathcal{W}(x_1,x_2)f(x_1,x_2)\mathrm{d}^4x_1\mathrm{d}^4x_2 &= \iint W(x_1-x_2)f(x_1,x_2)\mathrm{d}^4x_1\mathrm{d}^4x_2 \\ &= \iint W(y)f(x_2+y,x_2)\mathrm{d}^4y\mathrm{d}^4x_2 \\ &= \int W(y)H(f)(y)\mathrm{d}^4y, \end{aligned}\tag{M.43}$$

so that if we pretend that distributions are functions this coincides with (M.41).

Proof The idea is to use Proposition L.3.4, but first we have to make a change of variables for this, using $x_1 - x_2$ and x_2 as new variables, after which change of variable our distribution "will not depend on x_2". First, as a consequence of (M.37) in the case $C = 1$ and $c = -a$, we have

$$\Xi(f) = \Xi(f^a), \tag{M.44}$$

where for $x_1, x_2 \in \mathbb{R}^4$ we define

$$f^a(x_1,x_2) := f(x_1+a, x_2+a)\,;\ f_a(x_1,x_2) := f(x_1, x_2+a). \tag{M.45}$$

Given a function $w \in \mathcal{S}^8$ consider the function $F(w)(x_1,x_2) = w(x_1 - x_2, x_2)$. Then

$$F^a(w)(x_1,x_2) = F(w)(x_1 + a, x_2 + a) = w(x_1 - x_2, x_2 + a)$$
$$= w_a(x_1 - x_2, x_2) = F(w_a)(x_1,x_2), \tag{M.46}$$

and thus $F(w_a) = F^a(w)$. Consider moreover the distribution Φ on $\mathbb{R}^8$ given for $w \in \mathcal{S}^8$ by $\Phi(w) = \Xi(F(w))$. Then, using also (M.44), we obtain

$$\Phi(w_a) = \Xi(F(w_a)) = \Xi(F^a(w)) = \Xi(F(w)) = \Phi(w).$$

Consequently, Proposition L.3.4 used four times shows that there exists a tempered distribution Ψ on $\mathbb{R}^4$ such that $\Phi(w) = \Psi(G(w))$, where

$$G(w)(x) = \int w(x,a)\mathrm{d}^4a.$$

Now consider $f \in \mathcal{S}^8$ and the function $Z(f)$ given by $Z(f)(x_1,x_2) = f(x_1 + x_2, x_2)$, so that $F(Z(f)) = f$. Consequently,

$$\Xi(f) = \Xi(F(Z(f)) = \Phi(Z(f))) = \Psi(G(Z(f))),$$

and recalling that $G(Z(f)) = H(f)$ by (M.42), this proves (M.41) for $W = \Psi$. □

We now start the study of the distribution W.

Lemma M.4.3 *The distribution W of Proposition M.4.2 is Lorentz invariant in the sense that $W(w) = W(V(0,C)w)$ for $C \in SO^{\uparrow}(1,3)$ and $w \in \mathcal{S}^4$.*

Proof Consider a function $\xi \in \mathcal{S}^4$ such that $\int \xi(x)\mathrm{d}^4x = 1$. Then the function f given by $f(x,y) = w(x - y)\xi(y)$ is such that $f \in \mathcal{S}^8$ and $w = H(f)$. It is straightforward that $H(V(0,C)f) = V(0,C)H(f) = V(0,C)w$ because Lebesgue's measure d^4x is Lorentz invariant. Then, using (M.37) in the third equality,

$$W(V(0,C)w) = W(H(V(0,C)f)) = \Xi(V(0,C)f) = \Xi(f) = W(H(f)) = W(w).$$

□

Lemma M.4.4 *The distribution W of Proposition M.4.2 has the following property: For each $w \in \mathcal{S}^4$, each $v \in \mathcal{S}^4$ whose Fourier transform has a compact support which does not meet $\bar{\mathrm{V}}$, setting $u(x) = \int w(x + a)v(a)\mathrm{d}^4a$ we have $W(u) = 0$.*

Proof Consider two functions $w, f \in \mathcal{S}^4$. It is proved in Lemma M.4.1 that $\Xi(w \otimes f') = 0$ where $f'(x) = \int \mathrm{d}^4c f(x - c)v(c)$. Then according to (M.41) it holds that $W(H(w \otimes f')) = 0$. Now

$$H(w \otimes f')(x) = \int \mathrm{d}^4z w(x + z)f'(z) = \iint \mathrm{d}^4z\mathrm{d}^4c w(x + z)f(z - c)v(c)$$
$$= \iint \mathrm{d}^4z\mathrm{d}^4c w(x + z + c)f(z)v(c) = \int \mathrm{d}^4z u(x + z)f(z).$$

To prove that $W(u) = 0$ it suffices to prove that given k one may find f such that $\|u - \int \mathrm{d}^4z u(x + z)f(z)\|_k$ is arbitrarily small. For this one takes $f \geq 0$ of integral 1 and peaked around the origin together with routine estimates. □

We recall that for two functions $f, g \in \mathcal{S}^4$ their convolution $f * g$ is defined by

$$f * g(x) = \int f(x - y)g(y)\mathrm{d}^4y. \tag{M.47}$$

We will use the classical formula

$$\widehat{f * g} = \hat{f}\hat{g}, \tag{M.48}$$

which can be checked in a few lines in the spirit of (1.25). In the forthcoming computations unpleasant minus signs occur in the arguments of the variables.[7] To minimize this annoyance, we introduce the following notation. Given a function $f \in \mathcal{S}^4$ we define the function $f^\sharp$ by

$$f^\sharp(x) = f(-x).$$

Thus the formula $u(x) = \int w(x+a)v(a)\mathrm{d}^4a$ of Lemma M.4.4 means that $u = w * v^\sharp$, as is seen by changing a into $-a$, so that this lemma asserts that $W(w * v^\sharp) = 0$ whenever $\hat{v}$ has a compact support which does not meet $\bar{\mathcal{V}}$. Let us also observe the straightforward fact that

$$\widehat{f^\sharp} = (\hat{f})^\sharp, \tag{M.49}$$

where for a function ξ on momentum space we also define $\xi^\sharp$ by $\xi^\sharp(p) = \xi(-p)$.

Lemma M.4.5 *We have $W(f) = 0$ whenever the Fourier transform of f has a compact support which does not meet $-\bar{\mathcal{V}}$.*

Proof Consider a function $\xi \in \mathcal{S}^4$, which is equal to 1 on the support of $\hat{f}$, and which has a compact support which does not meet $-\bar{\mathcal{V}}$. Then $\hat{f} = \hat{f}\xi$, so that, taking the inverse Fourier transform of both sides, by (M.48) we have $f = f * \check{\xi}$. Let $g = (\check{\xi})^\sharp$ so that $f = f * g^\sharp$ while $\hat{g} = \xi^\sharp$ has a compact support which does not meet $\bar{\mathcal{V}}$. Consequently $W(f) = W(f * g^\sharp) = 0$ by the consequence of Lemma M.4.4 explained just above (M.49). □

Definition M.4.6 We define the distribution $\hat{W}$ by the formula[8]

$$\hat{W}(\xi) = W(\check{\xi}^\sharp). \tag{M.50}$$

It will turn out that $\hat{W}$ is a very well behaved distribution: It is a positive measure.

Exercise M.4.7 Show that the formula

$$\hat{W}(p) = \int \exp(\mathrm{i}(p,x))W(x)\mathrm{d}^4x$$

is an informal way to write (M.50).

Lemma M.4.8 *The distribution $\hat{W}$ is supported by $\bar{\mathcal{V}}$.*

Proof This is the content of Lemma M.4.5. Indeed, if ψ has a compact support which does not meet $\bar{\mathcal{V}}$, the function $g = \check{\psi}^\sharp$ is such that its Fourier transform, i.e. $\psi^\sharp$, has a compact support which does not meet $-\bar{\mathcal{V}}$, so that $\hat{W}(\psi) = W(g) = 0$. □

Now we come to the crucial point.

Lemma M.4.9 *The distribution $\hat{W}$ is positive, i.e. $\hat{W}(\theta) \geq 0$ if $\theta \in \mathcal{S}^4$ is ≥ 0 at each point.*

Proof We first prove that $\hat{W}(\eta^2) \geq 0$ where $\eta \in \mathcal{S}^4_{\mathbb{R}}$ has a compact support. To lighten notation, let $f = \check{\eta}$. As a consequence of (M.48), $\xi = \eta^2$ satisfies $\check{\xi} = f * f$, so that

$$\check{\xi}^\sharp(x) = \int f(-x-a)f(a)\mathrm{d}^4a = H(f^\sharp \otimes f)(x).$$

Now, since η is real-valued, it holds that $f(-x) = f(x)^*$, i.e. $f^\sharp = f^*$ so that

$$\hat{W}(\eta^2) = \hat{W}(\xi) = W(\check{\xi}^\sharp) = W(H(f^* \otimes f)) = \Xi(f^* \otimes f) \geq 0$$

by (M.38).

7 These are related to the problem of the proper definition of the Fourier transform of a distribution in the present setting, see Section L.4, but we do not want to be more confusing than necessary by addressing this issue here.

8 According to Definition L.12, $\hat{W}$ is the Fourier transform of W as a distribution.

Consider then $\theta \geq 0 \in \mathcal{S}^4$ with compact support. Consider $\xi \in \mathcal{S}^4$ with compact support, such that $\xi = 1$ on a neighborhood of the support of θ. Then for $\alpha > 0$ the function $\eta := \sqrt{\alpha\xi^2 + \theta}$ belongs to $\mathcal{S}^4$ and has compact support. (Observe that this function is infinitely differentiable in the interior of the support of ξ, and also in the complement of the support of θ because there it equals $\sqrt{\alpha}\xi$.) Consequently

$$\hat{W}(\eta^2) = \hat{W}(\theta) + \alpha\hat{W}(\xi^2) \geq 0$$

for each $\alpha > 0$ and therefore $\hat{W}(\theta) \geq 0$. □

Proposition M.4.10 *There exists a positive measure μ on $\bar{V}$ such that for $\psi \in \mathcal{S}^4$ it holds that*

$$\hat{W}(\psi) = \int \psi \mathrm{d}\mu. \tag{M.51}$$

The integral exists because μ has the following property: There exist numbers $k > 0$ and $C > 0$ such that the mass for the measure μ of the Euclidean ball of center 0 and radius R is at most $C(1+R)^k$. Moreover μ is Lorentz invariant.

If the concept of positive measure turns you off, please be patient, a formula in the language of physicists will be given in equation (M.53). If you like to think in terms of distributions, a crucial point is that the distribution $\hat{W}$ satisfies a much stronger inequality than (L.5) namely that for some positive integer $k > 0$ we have

$$|\hat{W}(\psi)| \leq C \sup_p (1 + \|p\|)^k |\psi(p)|.$$

As the proof below shows, this stronger inequality is a simple consequence of positivity. The measure μ is simply a mixture of invariant measures on the mass shells X_m as m varies.

Proof Consider $\theta \in \mathcal{S}^4$ with compact support, $\theta \geq 0$ and such that $\theta = 1$ on the Euclidean unit ball of $\mathbb{R}^4$ centered at the origin. For $a \geq 1$ define θ_a by $\theta_a(p) = \theta(p/a)$. We first show that, as a consequence of W being a tempered distribution it holds that

$$0 \leq \hat{W}(\theta_a) \leq Ca^k \tag{M.52}$$

for a certain integer k and a constant C independent of a. Indeed since W is a tempered distribution it satisfies an inequality $|W(f)| \leq C\|f\|_k$ for a certain k. Thus s $|\hat{W}(\theta_a)| = |W(f)| \leq C\|f\|_k$ where $f = \check{\psi}^\sharp$ for $\psi = \theta_a$. It is straightforward to check that $f(x) = a^4\check{\theta}^\sharp(ax)$ and that this function satisfies $\|f\|_k \leq C'a^{k+4}$. Replacing k by $k+4$, (M.52) follows (for a new constant C) since $\hat{W}(\theta_a) \geq 0$ by Lemma M.4.9.

For a function $\eta \in \mathcal{S}^4_{\mathbb{R}}$ we define its uniform norm $\|\eta\| := \sup_p |\eta(p)|$. Assume that η has a compact support and satisfies $\|\eta\| \leq 1$. Consider a large enough that the support of η is contained in the set where $\theta_a = 1$. Then $|\hat{W}(\eta)| \leq \hat{W}(\theta_a)$ because $\theta_a \pm \eta \geq 0$ and hence $\hat{W}(\theta_a) \pm \hat{W}(\eta) \geq 0$. Consequently for any $\eta \in \mathcal{S}^4_{\mathbb{R}}$ whose support is contained in the set where $\theta_a = 1$ we have

$$|\hat{W}(\eta)| \leq \hat{W}(\theta_a)\|\eta\|.$$

This is the key relation. It implies that W extends to a positive linear functional on the space of continuous functions with compact support, simply because a continuous function with compact support is the uniform limit of functions in $\mathcal{S}^4$ with compact support. This is exactly what is meant by saying that $\hat{W}$ is a positive measure. Moreover the measure of the balls centered at the origin is controlled by (M.52). The Lorentz invariance follows from Lemma M.4.3 and the good behavior of the Fourier transform with respect to Lorentz transformation, as in Lemma 4.8.1. □

The set $\bar{V}$ is the disjoint union of the sets X_m for $m \geq 0$. An invariant measure on $\bar{V}$ is a mixture of invariant measures on the sets X_m. This is seen by looking at the restriction of the measure to the sets $m^2 \leq p^2 \leq m^2 + \varepsilon$, scaling properly and letting $\varepsilon \to 0$. We do not give the details, which are not surprising. One has to be cautious about the case $m = 0$ because the Lorentz group does not act transitively on X_0 since the origin is a fixed point, so that a Lorentz invariant measure of X_0 has two parts, a point mass at the origin and a part proportional to $d\lambda_0$. As a consequence, there exists an increasing function ρ on $\mathbb{R}^+$ such that

$$\hat{W}(\psi) = c\psi(0) + \int_{m\geq 0} d\rho(m) \int \psi d\lambda_m \tag{M.53}$$

(all numerical factors are absorbed by ρ in this formula) and furthermore ρ is of polynomial growth from (M.52). Since $W(f) = \hat{W}(\hat{f}^{\sharp})$ by (M.50), substitution of the value

$$\hat{f}^{\sharp}(p) = (2\pi)^{-4} \int \exp(-i(x,p)) f(x) d^4 x$$

in (M.50), writing W as a function, we obtain the following.

Proposition M.4.11 (The Källén–Lehmann representation of W**)** *For some polynomially bounded function ρ it holds*

$$W(x) = c + \iint \exp(-i(x,p)) d\rho(m) d\lambda_m(p). \tag{M.54}$$

Our goal in the rest of this section is to argue that the following holds.

Theorem M.4.12 *Under (M.33) we have in fact*

$$\forall x_1, x_2\ ;\ \mathcal{W}(x_1,x_2) = \mathcal{W}^{\text{free}}(x_1,x_2), \tag{M.55}$$

or, equivalently,

$$(\Omega, \varphi(f_1)\varphi(f_2)\Omega) = (\Omega^{\text{free}}, \varphi^{\text{free}}(f_1)\varphi^{\text{free}}(f_2)\Omega^{\text{free}}).$$

It seems quite obvious that no interacting quantum field of interest is going to satisfy the very strong requirement (M.55) that the two-point Green function coincides with that of a free field. Therefore Theorem M.4.12 demonstrates clearly the problems inherent to the interaction picture. Very significant work remains to be done to fully prove Haag's theorem. This work is the content e.g. of Streater and Wightman's book [79, Theorem 4-15]. The analytic properties of the Wightman functions (which we managed to skirt in our arguments) seemingly would be required in the proof, so we will not attempt to go beyond Theorem M.4.12.

Unfortunately, giving a truly complete mathematical proof of Theorem M.4.12 would require more work than the reader is likely to accept at this stage, so that we will only give supporting arguments in favor of this theorem.[9]

Lemma M.4.13 *The formula (M.54) defines a true function in the domain $x^2 < 0$. The value $W(x)$ is given by the formula*

$$W(x) = c + \int m^2 I(-mx) d\rho(m) \tag{M.56}$$

where $I(x)$ is given by (5.13) for $m = 1$, i.e. $I(x) = \int \exp(i(x,p)) d\lambda_1(x)$.

Supporting argument. The first claim has been proved in Lemma 5.1.7. For the second assertion we use that $\int d\lambda_m(p) \exp((-i(x,p)) = m^2 I(-mx)$ as is shown by the change of variable $p \to mp$, using e.g. (4.36) to get the factor m^2.

The following exercise will help you to understand the argument of the next lemma.

9 These supporting arguments are considerably more convincing than what we called supporting arguments in Chapter 14 .

Exercise M.4.14 Assume that for all $x > 0$ we have the identity

$$\exp(-x) = \int_0^\infty \exp(-mx)\mathrm{d}\rho(m).$$

Prove that $\mathrm{d}\rho(m)$ consists of a unit mass at $m = 1$.

Lemma M.4.15 *Assume that on the domain $x^2 < 0$ the function (M.56) satisfies*

$$W(x) = m_0^2 I(-m_0 x). \tag{M.57}$$

Then $c = 0$ and $\rho(m) = 0$ for $m < m_0$ while $\rho(m) = 1$ for $m \geq m_0$, so that $\mathrm{d}\rho(m) = \delta(m - m_0)$. Then (M.56) (which holds for all x) becomes $W(x) = m_0^2 I(-m_0 x)$.

In particular, if the function W coincides at each point x where $x^2 < 0$ with the corresponding function W^{free} for the free scalar field of mass m_0, then it coincides everywhere with this function.[10]

Supporting argument. Let us write $J(a) = I(0,0,0,-a)$. We "computed" this function in the proof of Lemma 5.1.7: it is given by the right-hand side of (5.21) for $b = a$. Our argument needs simply that $J(a)$ decreases very fast as $a \to \infty$, which simply follows from successive integrations by parts of the right-hand side of (5.21). The equality (M.57) implies

$$m_0^2 J(m_0 a) = c + \int m^2 J(ma)\mathrm{d}\rho(m) := V(a),$$

and the argument consists of comparing the rates of decay of both sides as $a \to \infty$. First one sees that $c = 0$ and that ρ is constant for $m < m_0$ because otherwise the right-hand side $V(a)$ would not decay as fast as the left-hand side at infinity.[11] Next, we must have $\lim_{a\to\infty} V(a)/m_0^2 J(am_0) = 1$ and this forces ρ to have a jump of 1 at the point m_0 because the contributions of the larger values of m have a faster decay than $J(am_0)$. And then removing the contribution of this jump to $V(a)$ we see that ρ must be constant for $m > m_0$.

Supporting argument for Theorem M.4.12 Since W is a true function on the domain $x^2 < 0$, this is also the case of $\mathcal{W}$ on the domain $(x_1 - x_2)^2 < 0$ and then $\mathcal{W}(x_1, x_2) = W(x_1 - x_2)$. As a consequence of (M.33) we have

$$\mathcal{W}(x_1, x_2) = \mathcal{W}^{\mathrm{free}}(x_1, x_2)$$

whenever both x_1 and x_2 have their time-coordinate equal to zero. Consequently, since $W(x) = \mathcal{W}(x, 0)$ we get

$$W(x) = W^{\mathrm{free}}(x)$$

whenever x has its time coordinate equal to zero. By Lorentz invariance this equality remains true whenever $x^2 < 0$ and the result then follows from Lemma M.4.15. □

[10] More generally it seems true that the value of $W(x)$ for $x^2 < 0$ determines the values of c and of ρ, but I do not see an easy argument for this.

[11] The gap in mathematical rigor is that it does not seem to be true that $J \geq 0$ so that in principle there could be cancellations in the right-hand side.

Appendix N

Feynman Propagator and Klein-Gordon Equation

In this appendix we explain how physics textbooks reach the formula (13.55), and we use the opportunity to review a few basic fact, which, while definitely in the category "undergraduate mathematics", might have faded from the reader's mind.

N.1 Contour Integrals

Here is a very brief review of this nearly miraculous method. If V is an open set in $\mathbb{C}$, and a a point of V, a function f which is holomorphic in $V \setminus \{a\}$ is said to have a *simple pole at a* if $(z-a)f(z)$ can be extended to a holomorphic function on V. In other words, $f(z) = g(z)/(z-a)$ where g is holomorphic in V.[1] The *residue* $\mathrm{res}_a f$ is then defined as $g(a) = \lim_{z\to a}(z-a)f(z)$. Consider a function f which is holomorphic in $\mathbb{C}$ except maybe at a finite number of points where it has simple poles. Then if $\mathcal{C}$ is a *contour*, that is, a closed simple (i.e. has no double points) curve in $\mathbb{R}^2$ oriented counterclockwise, Cauchy's theorem asserts that

$$\int_{\mathcal{C}} f(z)\mathrm{d}z = 2\mathrm{i}\pi \sum_a \mathrm{res}_a f, \tag{N.1}$$

where the sum is taken over the poles which are inside the closed curve $\mathcal{C}$.

As a first example consider the function $f(z) = 1/(1+z^2) = 1/(z-\mathrm{i})(z+\mathrm{i})$, which has a simple pole at $z = \mathrm{i}$ with residue $1/(2\mathrm{i})$. Consider the contour $\mathcal{C}$ consisting of the line segment $\mathcal{C}_1$ from $(-R,0)$ to $(R,0)$, and the half-circle $\mathcal{C}_2$ of radius R centered at the origin in the upper half-plane. Then Cauchy's theorem implies

$$\int_{\mathcal{C}_1} f(z)\mathrm{d}z + \int_{\mathcal{C}_2} f(z)\mathrm{d}z = \pi. \tag{N.2}$$

Letting $R \to \infty$ it is straightforward to see that the second term goes to 0, yielding the classical formula

$$\int_{-\infty}^{\infty} \frac{\mathrm{d}x}{1+x^2} = \pi. \tag{N.3}$$

More generally, consider a complex number a with $\mathrm{Re}\, a > 0$, and

$$f(z) = \frac{\exp(\mathrm{i}tz)}{z^2 + a^2},$$

where $t > 0$ is a real number. Considering the same contour, and using that $|\exp(\mathrm{i}tz)| = \exp(-t\mathrm{Im}\, z)$, on $\mathcal{C}_2$ we have $|\exp(\mathrm{i}tz)| \le 1$ and it is straightforward to prove that the contribution of $\int_{\mathcal{C}_2}$ vanishes

[1] One usually calls a a pole only when $g(a) \neq 0$, but this makes no difference to us.

as $R \to \infty$, while the only pole inside the contour is for $z = ia$, with a residue $\exp(-at)/2ia$, so that the right-hand side of (N.2) is now

$$2i\pi\frac{\exp(-at)}{2ia} = \pi\frac{\exp(-at)}{a},$$

yielding the formula

$$\int \frac{\exp(itx)}{a^2+x^2}\mathrm{d}x = \pi\frac{\exp(-at)}{a}. \tag{N.4}$$

When $t < 0$, change of variable $x \to -x$ shows that the right-hand side has to be changed to $\pi\exp(at)/a$. It is however far more instructive to obtain this result by a contour integral. One replaces $\mathcal{C}_2$ by a half-circle in the *lower* half-plane. The resulting contour is now oriented *clockwise*, which has to be compensated by a minus sign in Cauchy's formula. On this contour we now have $|\exp(itz)| \le 1$, and the contribution from $\mathcal{C}_2$ vanishes as $R \to \infty$. The pole inside the contour is at $z = -ia$, and the residue there is $-\exp(at)/2ia$. This technique of closing the contour either in the upper or the lower half-plane is of constant use.

Considering the same contour, but now $f(z) = (\exp(\mathrm{i}z) - 1)/z$, which is holomorphic in the entire complex plane, so that $\int_{\mathcal{C}_1} f(z)\mathrm{d}z + \int_{\mathcal{C}_2} f(z)\mathrm{d}z = 0$. Again the contribution of $\int_{\mathcal{C}_2}\exp(\mathrm{i}z)\mathrm{d}z/z$ vanishes as $R \to \infty$ (although it takes now a little proof) while $\int_{\mathcal{C}_2} -\mathrm{d}z/z = -\mathrm{i}\pi$. Therefore

$$\int_{-\infty}^{\infty}\frac{\exp(\mathrm{i}x)-1}{x}\mathrm{d}x = \mathrm{i}\pi.$$

In a similar manner one obtains

$$\int_{-\infty}^{\infty}\frac{\exp(-\mathrm{i}x)-1}{x}\mathrm{d}x = -\mathrm{i}\pi,$$

but now the half-circle closing the contour has to be taken in the lower half-plane. Consequently, since $\sin x = (\exp(\mathrm{i}x) - \exp(-\mathrm{i}x))/2\mathrm{i}$, we obtain

$$\int_{-\infty}^{\infty}\frac{\sin x}{x}\mathrm{d}x = \pi. \tag{N.5}$$

In a similar manner,

$$\int_{-\infty}^{\infty}\frac{\exp(2\mathrm{i}x)-2\mathrm{i}x-1}{x^2}\mathrm{d}x = -2\pi.$$

Changing x into $-x$ and adding we get

$$\int_{-\infty}^{\infty}\frac{\sin^2 x}{x^2}\mathrm{d}x = \pi. \tag{N.6}$$

As another example of this method, let us prove the formula (C.34). Completing the square, it suffices to prove that for any complex number c one has

$$\int \exp(-\alpha(x+c)^2)\mathrm{d}x = \sqrt{\frac{\pi}{\alpha}}.$$

Obviously one may assume that c is purely imaginary, $c = id$. The result is certainly true for $d = 0$, so it suffices to show that the integral is independent of d. This follows from the fact that the integral of $\exp(-az^2)$ on the contour which is the boundary of the rectangle of vertices $(-R,0),(R,0),(R,d),(-R,d)$ is zero, and letting $R \to \infty$.

N.2 Fundamental Solutions of Differential Equations

Consider a linear differential operator D with constant coefficients. Our fundamental example is the Klein-Gordon operator

$$D(u) = \partial_t^2 u - \sum_{k\leq 3} \partial_k^2 u + m^2 u, \tag{N.7}$$

and the toy version

$$D^0(u) = \partial_t^2 u + m^2 u, \tag{N.8}$$

where the three dimensions of physical space have been removed. A *fundamental solution* of the operator D is a function G such that in the sense of distributions one has

$$D(G) = \delta^{(4)}. \tag{N.9}$$

The reason why it is fundamental is that once you have computed it you know how to solve the equation $D(u) = f$. Provided things make sense, a solution is given by

$$u(x) = \int G(x-y) f(y) \mathrm{d}^4 y.$$

The fundamental solution is by no way unique. It is determined only up to a solution of the homogeneous equation $D(u) = 0$, and to specify it one has to fix "boundary conditions". To see how this works, let us look at the toy version (N.8). Then the following functions u satisfy $D^0(u) = \delta$:

$$u(t) = 0 \text{ if } t \leq 0\ ;\ u(t) = \frac{1}{m}\sin(mt) \text{ if } t \geq 0. \tag{N.10}$$

$$u(t) = -\frac{1}{m}\sin(mt) \text{ if } t \leq 0\ ;\ u(t) = 0 \text{ if } t \geq 0. \tag{N.11}$$

$$\forall t \in \mathbb{R}\ ;\ u(t) = \frac{\mathrm{i}\exp(-\mathrm{i}|t|m)}{2m}. \tag{N.12}$$

The verification is straightforward. For example, in the case (N.12),

$$u'(t) = \frac{1}{2}\operatorname{sign} t \exp(-\mathrm{i}|t|m)\ ;\ u''(t) = \delta(t) - m^2 u(t). \tag{N.13}$$

The fundamental solutions (N.10) and (N.11) are completely determined by the condition $u = 0$ for $t < 0$ (respectively $u = 0$ for $t > 0$).

Theorem N.2.1 *The Feynman propagator*

$$\Delta_F(x) = \Delta_F(t,\boldsymbol{x}) = \mathrm{i}\int \frac{\mathrm{d}^3\boldsymbol{p}}{(2\pi)^3 2\omega_{\boldsymbol{p}}} \exp(-\mathrm{i}|t|\omega_{\boldsymbol{p}} + \mathrm{i}\boldsymbol{p}\cdot\boldsymbol{x}) \tag{13.55}$$

is a fundamental solution of the Klein-Gordon operator.

Proof Here is a somewhat formal proof: Using that the function $t \mapsto \mathrm{i}\exp(-\mathrm{i}|t|\omega_{\boldsymbol{p}})/2\omega_{\boldsymbol{p}}$ satisfies (N.13) for $m = \omega_{\boldsymbol{p}}$, we get

$$\partial_t^2 \Delta_F(t,\boldsymbol{x}) = \int \delta(t)\exp(\mathrm{i}\boldsymbol{x}\cdot\boldsymbol{p})\frac{\mathrm{d}^3\boldsymbol{p}}{(2\pi)^3} - \mathrm{i}\int \omega_{\boldsymbol{p}}^2 \frac{\exp(-\mathrm{i}|t|\omega_{\boldsymbol{p}} + \mathrm{i}\boldsymbol{x}\cdot\boldsymbol{p})}{2\omega_{\boldsymbol{p}}}\frac{\mathrm{d}^3\boldsymbol{p}}{(2\pi)^3}. \tag{N.14}$$

The first term on the right is $\delta(t)\delta^{(3)}(\boldsymbol{x}) = \delta^{(4)}(x)$, and since $\omega_{\boldsymbol{p}}^2 = m^2 + \boldsymbol{p}^2$ the second term on the right is $(-m^2 + \sum_{j\leq 3}\partial_j^2)\Delta_F(x)$. □

Among the fundamental solutions of the Klein-Gordon operator, the easiest one to picture is the solution which is zero outside the forward light cone. It somewhat corresponds to what happens when you throw a stone in a quiet pond and watch the waves die on the shores.[2] This is the situation (N.10). Running the movie of this event backward, one obtains another physically possible situation, this one much harder to realize: The waves are born on the shore and concentrate at a point where they suddenly annihilate each other. This corresponds to the situation (N.11). The Feynman propagator corresponds to neither case, but to a mixture of them as in (N.12).

Let us now try to find a method to discover such fundamental solutions, starting with the toy example of D^0. If $\partial_t^2 u + m^2 u = \delta$, taking the Fourier transform and using $\widehat{\partial_t u} = \mathrm{i}\omega\hat{u}$ yields $-\omega^2\hat{u}(\omega) + m^2\hat{u} = 1$, so that

$$\hat{u}(\omega) = \frac{1}{-\omega^2 + m^2},$$

and taking the inverse Fourier transform gives the tentative formula

$$u(t) = \int \frac{\exp(\mathrm{i}t\omega)}{-\omega^2 + m^2}\frac{\mathrm{d}\omega}{2\pi}. \tag{N.15}$$

The integral does not quite make sense, but let us consider it as a contour integral, using the methods of Section N.1. We can make sense of the integral if we avoid the poles $\omega = \pm m$ by making a little detour outside the real axis to bypass them. This detour can be made either above or below the real axis. We observe also that for $t > 0$ we must close the contour by a half-circle in the upper half-plane, while for $t < 0$ this must be done in the lower half-plane. If we bypass both poles from below, e.g. by replacing ω by $\omega - \mathrm{i}\varepsilon$ for $\varepsilon > 0$, then for $t > 0$ it is straightforward to obtain

$$\int \frac{\exp(\mathrm{i}t\omega)}{-(\omega - \mathrm{i}\varepsilon)^2 + m^2}\frac{\mathrm{d}\omega}{2\pi} = \frac{\sin(mt)}{m}$$

through the residue formula, while the integral is zero if $t < 0$. This recovers the formula (N.10), while the formula (N.11) is recovered by bypassing both poles from above. Yet another way to bypass the poles is to replace (N.15) by

$$u(t) = \int \frac{\exp(\mathrm{i}t\omega)}{-\omega^2 + m^2 - \mathrm{i}\varepsilon}\frac{\mathrm{d}\omega}{2\pi} \tag{N.16}$$

for a vanishingly small $\varepsilon > 0$, and this time one obtains the solution (N.12).

Let us now go back to the case of the Klein-Gordon equation, where, using again the Fourier transform, the fundamental solution should be given by (if this would make sense)

$$u(x) = \int \frac{\exp(-\mathrm{i}(x,p))}{-p^2 + m^2}\frac{\mathrm{d}^4 p}{(2\pi)^4}. \tag{N.17}$$

Since $-p^2 = -p_0^2 + \boldsymbol{p}^2$, the problem is with the ill-defined integration in p_0. This is the same problem as we have considered in the toy case, replacing m^2 by $m^2 + \boldsymbol{p}^2$, and we can fix it by the same device of setting

$$u(x) = \lim_{\varepsilon \to 0^+} \int \frac{\exp(\mathrm{i}(x,p))}{-p^2 + m^2 - \mathrm{i}\varepsilon}\frac{\mathrm{d}^4 p}{(2\pi)^4}, \tag{N.18}$$

rediscovering the formula for the Feynman propagator.

Exercise N.2.2 The point of this exercise is to discover a fundamental solution for the Laplace equation $\nabla^2 f = \delta^{(3)}$, a very important computation.

[2] The waves on the pond do not satisfy the Klein-Gordon equation, but another differential equation.

(a) Using the Fourier transform, show that $\boldsymbol{k}^2 \hat{f}(\boldsymbol{k}) = -1$.

(b) Express f as the inverse Fourier transform of $-1/\boldsymbol{k}^2$. Show by rotational symmetry that $f(\boldsymbol{x})$ depends only on $\|\boldsymbol{x}\|$. Assuming that $\boldsymbol{x}$ is in the z-direction, perform the integral in cylindrical coordinates to obtain

$$f(\boldsymbol{x}) = -\frac{1}{(2\pi)^2} \int_0^\infty r\mathrm{d}r \int \frac{\exp(ikx_3)}{r^2 + k^2} \mathrm{d}k.$$

(c) Use (N.4) to show that the last integral is $\pi \exp(-x_3 r)/r$ and conclude that

$$f(\boldsymbol{x}) = -\frac{1}{4\pi \|\boldsymbol{x}\|}.$$

Appendix O

Yukawa Potential

In this appendix we present two variations on the model of Section 13.6, and the reader is assumed to have mastered that section. These models go one small step closer to realistic models. We still consider only bosons, but we no longer assume that the particle is its own anti-particle.

In the first model three particles exist. First, the a-particle, a massive spinless particle which is different from its anti-particle $\bar{a}$. We denote by a and $a^\dagger$ the annihilation and creation operators of this particle, and by $\bar{a}$ and $\bar{a}^\dagger$ those of the anti-particle. The free field relative to the a-particle is no longer self-adjoint, and is given by

$$\varphi_a(x) = \int \big(\exp(-\mathrm{i}(x,p))a(\boldsymbol{p}) + \exp(\mathrm{i}(x,p))\bar{a}^\dagger(\boldsymbol{p})\big)\frac{\mathrm{d}^3\boldsymbol{p}}{(2\pi)^3\sqrt{2\omega_{\boldsymbol{p}}}}, \tag{O.1}$$

$$\varphi_a^\dagger(x) = \int \big(\exp(-\mathrm{i}(x,p))\bar{a}(\boldsymbol{p}) + \exp(\mathrm{i}(x,p))a^\dagger(\boldsymbol{p})\big)\frac{\mathrm{d}^3\boldsymbol{p}}{(2\pi)^3\sqrt{2\omega_{\boldsymbol{p}}}}, \tag{O.2}$$

so that we have formulas such as

$$\langle 0|a(p)\varphi_a(x)|0\rangle = 0\ ;\ \langle 0|a(p)\varphi_a^\dagger(x)|0\rangle = \exp(\mathrm{i}(x,p)) \tag{O.3}$$

$$\langle 0|\varphi_a(x)a^\dagger(p)|0\rangle = \exp(-\mathrm{i}(x,p))\ ;\ \langle 0|\varphi_a^\dagger(x)a^\dagger(p)|0\rangle = 0 \tag{O.4}$$

$$\langle 0|\bar{a}(p)\varphi_a^\dagger(x)|0\rangle = 0\ ;\ \langle 0|\bar{a}(p)\varphi_a(x)|0\rangle = \exp(\mathrm{i}(x,p)) \tag{O.5}$$

$$\langle 0|\varphi_a^\dagger(x)\bar{a}^\dagger(p)|0\rangle = \exp(-\mathrm{i}(x,p))\ ;\ \langle 0|\varphi_a(x)\bar{a}^\dagger(p)|0\rangle = 0. \tag{O.6}$$

There exists another particle, the c-particle, which as previously is spinless, massive and its own anti-particle (and for which the free field is given by the usual formulas discovered in Chapter 5). The interacting Hamiltonian density is given by

$$\mathcal{H}(x) = g\ {:}\varphi_a(x)\varphi_a^\dagger(x)\varphi_c(x){:}\,. \tag{O.7}$$

Our goal is to compute some elements of the S matrix at the second order of perturbation theory. These computations proceed very much as in Chapter 13 and are better left as exercises. Many of the amplitudes one may write are zero, and to see this it helps to formulate in words the content of the formulas (O.3) to (O.6) in words: To produce contractions that are not zero, a term $\varphi_a(x)$ has to be contracted with an incoming a-particle or an out-going $\bar{a}$-particle whereas a term $\varphi_a^\dagger(x)$ has to be contracted with an incoming $\bar{a}$-particle or an outgoing a-particle.

Exercise O.1 Prove that at the first order of perturbation theory one has

$$\langle 0|a(p_1)a(p_2)S_1c^\dagger(q)|0\rangle = 0 \tag{O.8}$$

and

$$\langle 0|a(p_1)\bar{a}(p_2)S_1 c^\dagger(q)|0\rangle = -\mathrm{i}g(2\pi)^4\delta^{(4)}(p_1 + p_2 - q).$$

In words, a c-particle may decompose into an a-particle and an $\bar{a}$-particle but not into two a-particles.

Exercise O.2 Consider now particles $(b_j)_{j\le 4}$ such that each b_j is either a or $\bar{a}$. We are interested in computing at the second order of perturbation theory the amplitude $\langle 0|b_4(p_4)b_3(p_3) S_2 b_2^\dagger(p_2)b_1^\dagger(p_1)|0\rangle$. Assuming the incoming momenta of the particles are all different from their outgoing momenta, prove that the amplitude is zero unless the number of incoming a-particle plus the number of out-going $\bar{a}$ particles is 2. For example $\langle 0|\bar{a}(p_4)a(p_3)S_2 a^\dagger(p_1)a^\dagger(p_2)|0\rangle = 0$.

Exercise O.3 Assuming the incoming momenta of the particles are all different from their outgoing momenta, show that when computing the quantity

$$\langle 0|a(p_4)a(p_3)S_2\, a^\dagger(p_1)a^\dagger(p_2)|0\rangle, \tag{O.9}$$

the two relevant diagrams are those at the left of Figure O.1. Compute then the values of these diagrams.

To simplify the analysis we consider now a different model where an interaction similar to the one in the previous exercise occurs through a single channel. In this model we have five particles, of types a, b and c, and the anti-particles $\bar{a}$ and $\bar{b}$ of the particles a and b respectively. The corresponding free fields are given by (O.1) and (O.2) for a and similar formulas for b. The c-particles are their own anti-particles. The interacting Hamiltonian density is then given by

$$\mathcal{H}(x) = g\ {:}\varphi_a(x)\varphi_a^\dagger(x)\varphi_c(x){:} + g\ {:}\varphi_b(x)\varphi_b^\dagger(x)\varphi_c(x){:}\ .$$

Exercise O.4 Assuming as usual that the momenta of the incoming and outgoing particles are all different, prove that in this model the computation at the second order of perturbation theory of the amplitude

$$\langle 0|b(p_4)a(p_3)S_2 a^\dagger(p_2)b^\dagger(p_1)|0\rangle \tag{O.10}$$

involves the diagram (b) of Figure O.1. Show that its value is

$$(-\mathrm{i})^3 g^2(2\pi)^4\frac{\delta^{(4)}(p_4 + p_3 - p_1 - p_2)}{m_c^2 - (p_1 - p_3)^2}. \tag{O.11}$$

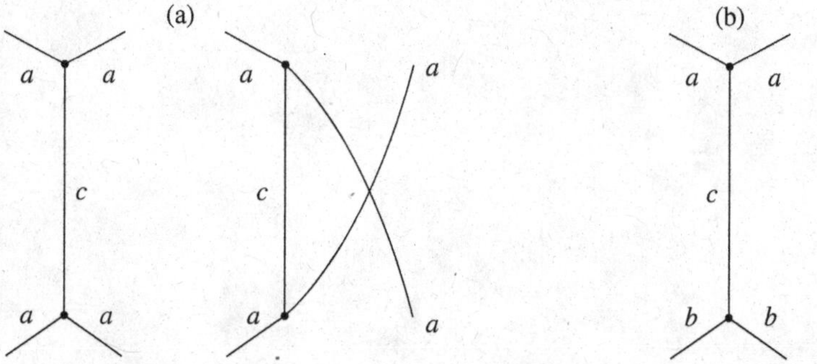

Figure O.1 (a) The two relevant diagrams for (O.9). (b) The diagram relevant to (O.10).

Performing the non-relativistic approximation of (O.11) as in (13.88) we obtain

$$ig^2(2\pi)^4 \frac{\delta^{(4)}(p_4 + p_3 - p_1 - p_2)}{m_c^2 + (\boldsymbol{p}_1 - \boldsymbol{p}_3)^2}. \tag{O.12}$$

Comparing (O.12) with the Born approximation (13.8) shows that the potential V such that

$$\widehat{V}(\boldsymbol{p}) = -\frac{1}{m_c^2 + \boldsymbol{p}^2} \tag{O.13}$$

is involved here. It is a simple computation (once you know the result!) to check using spherical coordinates that the potential

$$V(\boldsymbol{x}) = -\frac{\exp(-m_c|\boldsymbol{x}|)}{4\pi|\boldsymbol{x}|} \tag{O.14}$$

has precisely the property (O.13). This potential is called the Yukawa potential. In the limit $m_c \to 0$ it is the electrostatic potential.

Appendix P

Principal Values and Delta Functions

In this appendix we present a more abstract version of Lemma 18.1.1. As we have shown in this lemma, for a smooth function φ the limit

$$\lim_{\varepsilon\to 0^+}\int_{-1}^{1}\frac{\varphi(x)}{x-\mathrm{i}\varepsilon}\mathrm{d}x$$

exists. This means that the function $1/(x-\mathrm{i}\varepsilon)$ has a limit in the sense of distributions. This limit can be computed explicitly. This is not difficult although the result looks a bit scary at first, because it involves the "principal value", a distribution which may not be familiar.[1]

Proposition P.1 *In the sense of distributions,*

$$\lim_{\varepsilon\to 0^+}\frac{1}{x+\mathrm{i}\varepsilon}=\mathcal{P}\left(\frac{1}{x}\right)-\mathrm{i}\pi\delta_0, \tag{P.1}$$

where $\mathcal{P}(1/x)$ denotes the principal value,

$$\mathcal{P}\left(\frac{1}{x}\right)(\varphi):=\lim_{\alpha\to 0^+}\left(\int_{-\infty}^{-\alpha}\frac{\varphi(x)}{x}\mathrm{d}x+\int_{\alpha}^{\infty}\frac{\varphi(x)}{x}\mathrm{d}x\right). \tag{P.2}$$

In a similar manner,

$$\lim_{\varepsilon\to 0^+}\frac{1}{x-\mathrm{i}\varepsilon}=\mathcal{P}\left(\frac{1}{x}\right)+\mathrm{i}\pi\delta_0. \tag{P.3}$$

That the limit exists in (P.2) is part of the result. The meaning of (P.1) is simply that for every test function $\varphi\in\mathcal{S}$,

$$\lim_{\varepsilon\to 0^+}\int\frac{\varphi(x)}{x+\mathrm{i}\varepsilon}\mathrm{d}x=\mathcal{P}\left(\frac{1}{x}\right)(\varphi)-\mathrm{i}\pi\varphi(0).$$

Proof We write

$$\frac{1}{x+\mathrm{i}\varepsilon}=\frac{x}{x^2+\varepsilon^2}-\mathrm{i}\frac{\varepsilon}{x^2+\varepsilon^2}. \tag{P.4}$$

Taking (N.3) into account we obtain (with the change of variable $x=\varepsilon y$)

$$\lim_{\varepsilon\to 0^+}\int\frac{\varepsilon\varphi(x)}{x^2+\varepsilon^2}\mathrm{d}x=\pi\varphi(0),$$

[1] We state this result to help the reader connect with the literature which overwhelmingly states Proposition P.1. However the existence of the limit in Lemma 18.1.1 is arguably far more useful than the exact value of this limit.

so that it suffices to prove that

$$\lim_{\varepsilon\to 0^+}\int \frac{x\varphi(x)}{x^2+\varepsilon^2}\mathrm{d}x = \mathcal{P}\left(\frac{1}{x}\right)(\varphi). \tag{P.5}$$

First, if φ is constant on the interval $[-a,a]$, then for each $0<\alpha\le a$ one has

$$\int \frac{x\varphi(x)}{x^2+\varepsilon^2}\mathrm{d}x = \int_{-\infty}^{-\alpha}\frac{x\varphi(x)}{x^2+\varepsilon^2}\mathrm{d}x + \int_{\alpha}^{\infty}\frac{x\varphi(x)}{x^2+\varepsilon^2}\mathrm{d}x,$$

because the contribution in the interval $(-\alpha,\alpha)$ is zero by symmetry. As $\varepsilon\to 0^+$ the limit of this is

$$\begin{aligned}\int_{-\infty}^{-a}\frac{\varphi(x)}{x}\mathrm{d}x + \int_{a}^{\infty}\frac{\varphi(x)}{x}\mathrm{d}x &= \lim_{\alpha\to 0^+}\int_{-\infty}^{-\alpha}\frac{\varphi(x)}{x}\mathrm{d}x + \int_{\alpha}^{\infty}\frac{\varphi(x)}{x}\mathrm{d}x\\ &= \mathcal{P}\left(\frac{1}{x}\right)(\varphi).\end{aligned}$$

Next, if $\varphi(0)=0$, then it is much easier to understand what $\mathcal{P}(1/x)(\varphi)$ is:

$$\mathcal{P}\left(\frac{1}{x}\right)(\varphi) = \int \frac{\varphi(x)}{x}\mathrm{d}x,$$

and (P.5) is basically obvious by dominated convergence. Finally any function $\varphi\in\mathcal{S}$ is the sum of two functions in $\mathcal{S}$, one of which is constant in a neighborhood of the origin, the other one being zero at the origin. This completes the proof of (P.5). □

Solutions to Selected Exercises

Solutions to selected exercises are available for download from the book's catalog page www.cambridge.org/9781316510278.

I am indebted to Krzysztof Smutek for having suggested that I write this chapter. He worked out many of these solutions, and I am grateful for his permission to reproduce them. There are indications of how to solve almost every exercise, except those which are embarrassingly easy, or those whose solutions can be found in the given references. Please be aware that I did not work out the exercises with the same dedication as I wrote the main text, and brace yourself for a higher density of typos and possibly a positive density of plain nonsense. After all, the point of the exercises is to make the reader (not the author) work.

Reading Suggestions

As detailed in the introduction this book would not exist if I had found expositions of the topic in which I could have learned easily. This is not my topic and I have a limited knowledge of the literature.[1] Nonetheless, I painfully experienced that it is very difficult to learn from a single book, and there is no reason why this one should be an exception. So, here is some advice (for what it is worth).

- For Quantum Mechanics, try the book of Brian Hall [40], and the treatise by Claude Cohen-Tannoudji et al. [11]. The book [61] by Kalyanapuram Parthasarathy could also be helpful. Peter Woit's work [91] has some simple material at the beginning which is well explained and hard to find elsewhere. You may also find Cresser's text [17] useful.
- I am fully aware that I do not say enough about Special Relativity that you could survive if you really have never heard of it. This is an easy topic and there is plenty of material available. I have enjoyed Taylor and Wheeler's book [81] and the simpler book [41] by Harris.
- When you have the impression that you understand something about a topic in the present book, check in the much more compact book of Gerald Folland [31] if you understand his treatment.
- If you want to learn how to read physicists' work, when you have the impression that you understand something about a topic in the present book, check in the book by Tom Lancaster and Stephen Blundell [50] if you understand their treatment. Their book makes an unparalleled effort at being accessible; but it is very definitely not written with a mathematically oriented reader in mind.
- There is a whole series of books by Walter Greiner and coauthors (such as [35]) with elementary and very detailed presentations. They can however be extremely confusing to a mathematician.
- The book [24] by Anthony Duncan is a real challenge, but is of great potential interest for a mathematician.
- The references comprise a number of papers and books which I found of interest.

[1] You must have guessed too that I feel that some published books have a *strong negative* value, and you will forgive me for not naming them.

References

[1] Anselmi, D. 2019. Renormalization. `http://renormalization.com/pdf/14B1.pdf` (Cited on pages 5 and 496.)

[2] Araki, H. 2009. *Mathematical Theory of Quantum Fields.* International Series of Monographs on Physics **101**. Oxford University Press, Oxford. (Cited on pages 208 and 420.)

[3] Arnol'd, V. 1989. *Mathematical Methods of Classical Mechanics.* Graduate Texts in Mathematics **60**. Springer-Verlag, New York. (Cited on page 683.)

[4] Bargmann, V. 1954. On unitary ray representations of continuous groups. *Ann. of Math. (2)* **59**, pp. 1–46. (Cited on pages 199, 212, and 596.)

[5] Bargmann, V.; Wigner, E. P. 1948. Group theoretical discussion of relativistic wave equations. *Proc. Nat. Acad. Sci. U.S.A.* **34**, pp. 211–223. (Cited on page 673.)

[6] Barut, A. O.; Raczka, R. 1986. *Theory of Group Representations and Applications.* Second edition. World Scientific, Singapore.

[7] Bergère, M. C.; Zuber, J. B. 1974. Renormalization of Feynman amplitudes and parametric integral representation. *Comm. Math. Phys.* **35**, pp. 113–140. (Cited on pages 471 and 498.)

[8] Blasone, M.; Vitiello, G.; Jizba, P. 2011. *Quantum Field Theory and Its Macroscopic Manifestations.* Imperial College Press, London.

[9] Bogoliubov, N. N.; Logunov, A. A.; Oksak, A. I.; Todorov, I. T. 1990. *General Principles of Quantum Field Theory.* Kluwer Academic Publishers, Dordrecht.(Cited on pages 2, 420, and 704.)

[10] Berndt, R. 2007. *Representations of Linear Groups: An Introduction Based on Examples from Physics and Number Theory.* Vieweg, Wiesbaden.

[11] Cohen-Tannoudji, C.; Diu, B.; Laloë, F. 1988–1992. *Mécanique quantique.* 1988-1992. Hermann, Paris. (Cited on pages 21 and 732.)

[12] Cohen-Tannoudji, C.; Diu, B.; Laloë, F. 1991. *Quantum Mechanics.* Vol. 1 and 2. Wiley, New York.

[13] Coleman, S. Notes from Sidney Coleman's Physics 253a. `https://arxiv.org/abs/1110.5013` (Cited on pages 364, 463, and 668.)

[14] Coleman, S. 2019. *Quantum Field Theory.* World Scientific, Singapore.

[15] Collins, J. 1984. *Renormalization: An Introduction to Renormalization, the Renormalization Group, and the Operator-Product Expansion.* Cambridge Monographs on Mathematical Physics. Cambridge University Press, Cambridge. (Cited on pages 5, 496, and 537.)

[16] Connes, A.; Marcolli, M. 2008. *Noncommutative Geometry, Quantum Fields and Motives.* American Mathematical Society Colloquium Publications **55**. American Mathematical Society, Providence, RI.

[17] Cresser, J. Quantum physics notes. `http://physics.mq.edu.au/~jcresser/Phys304/Handouts/QuantumPhysicsNotes.pdf` (Cited on page 732.)

[18] Deligne, P. (Ed.) 1999. *Quantum Fields and Strings: A Course for Mathematicians.* American Mathematical Society, Providence, RI. (Cited on page 3.)

[19] Derezinski, J.; Gérard, C. 2013. *Mathematics of Quantization and Quantum Fields.* Cambridge Monographs on Mathematical Physics. Cambridge University Press, Cambridge. (Cited on page 420.)

[20] d'Espagnat, B. 1976. *Conceptual Foundations of Quantum Mechanics.* Second revised edition. Mathematical Physics Monograph **20**. W. A. Benjamin, Reading, MA. (Cited on page 27.)

[21] Dirac, P. A. M. 1947. *The Principles of Quantum Mechanics.* Third edition Clarendon Press, Oxford.

[22] Dirac, P. A. M. 1950. Generalized Hamiltonian dynamics. *Canad. J. Math.* **2**, pp. 129-148. (Cited on page 290.)

[23] Dimock, J. 2011. *Quantum Mechanics and Quantum Field Theory: A Mathematical Primer.* Cambridge University Press, Cambridge. (Cited on pages 4, 74, 140, and 685.)

[24] Duncan, A. 2012. *The Conceptual Framework of Quantum Field Theory.* Oxford University Press, Oxford. (Cited on pages 31, 262, 265, 424, 463, 496, 535, 658, and 732.)

[25] Dütsch, M. 2019. *From Classical Field Theory to Perturbative Quantum Field Theory.* Progress in Mathematical Physics **74**. Birkhäuser/Springer, Cham. (Cited on page 496.)

[26] Dyson, F. 1949. The *S* matrix in quantum electrodynamics. *Phys. Review* **75**, no. 11, pp. 1736–1755. (Cited on page 481.)

[27] Earman, J.; Fraser, D. 2005. Haag's theorem and its implications for the foundations of quantum field theory. `http://philsci-archive.pitt.edu/2673/`

[28] Epstein, H.; Glaser, V. 1973. The role of locality in perturbation theory. *Ann. Inst. H. Poincaré Sect. A* **19**, pp. 211–295. (Cited on pages 360 and 510.)

[29] Fetter, A.; Walecka, J. 1971. *Quantum Theory of Many-Particle Systems.* Dover, New York. (Cited on page 431.)

[30] Frankel, T. 2004. *The Geometry of Physics: An Introduction.* Second edition. Cambridge University Press, Cambridge.

[31] Folland, G. 2008. *Quantum Field Theory: A Tourist Guide for Mathematicians.* Mathematical Surveys and Monographs **149**. American Mathematical Society, Providence, RI. (Cited on pages 2, 4, 5, 9, 143, 149, 155, 208, 239, 320, 322, 372, 446, 508, 549, and 732.)

[32] Friedman, A.; Susskind, L. 2014. *Quantum Physics: The Theoretical Minimum.* Basic Books, New York.

[33] Gask, H. 1960. A proof of Schwartz's Kernel Theorem. *Math. Scand.* **8**, pp. 327–332. (Cited on pages 330 and 714.)

[34] Glimm, J.; Jaffe, A. 1987. *Quantum Physics: A Functional Integral Point of View.* Springer-Verlag, New York.

[35] Greiner, W.; Reinhardt, J. 1996. *Field Quantization.* Springer-Verlag, New York. (Cited on pages 170, 287, 677, and 732.)

[36] Gross, L. 2011. Some physics for mathematicians. `http://pi.math.cornell.edu/~gross/712notes2011.pdf`

[37] Gustafson, S.; Sigal, I. M. 2011. *Mathematical Concepts of Quantum Mechanics.* Second edition. Universitext. Springer, Heidelberg.

[38] Hairer, M. 2018. An analyst's take on the BPHZ theorem. *Computation and Combinatorics in Dynamics, Stochastics and Control.* Abel Symposium, 13, Springer, Cham, pp. 42–476.

[39] Hall, B. 2003. *Lie Groups, Lie Algebras, and Representations: An Elementary Introduction.* Graduate Texts in Mathematics **222**. Springer-Verlag, New York. (Cited on pages 118, 326, 635, 637, 638, and 641.)

[40] Hall, B. 2013. *Quantum Theory for Mathematicians.* Graduate Texts in Mathematics **267**. Springer, New York. (Cited on pages 2, 21, 35, 36, 40, 622, 624, 654, and 732.)

[41] Harris, P. Special relativity, available at `http://sro.sussex.ac.uk/id/eprint/101172/1/Relativ.pdf`. (Cited on page 732.)

[42] Itzykson, C.; Zuber, J.-B. 1980. *Quantum Field Theory.* International Series in Pure and Applied Physics. McGraw-Hill, New York.

[43] Isham, C. J. 1995. *Lectures on Quantum Theory: Mathematical and Structural Foundations.* Imperial College Press, London. (Cited on pages 26 and 75.)

[44] Jaffe, R. 2005. The Casimir effect and quantum vacuum. `https://arxiv.org/pdf/hep-th/0503158v1.pdf` (Cited on page 87.)

[45] Jost, R. 1965. *The General Theory of Quantized Fields.* American Mathematical Society, Providence, RI. (Cited on pages 420 and 463.)

[46] Kato, T. 1972. Schrödinger operators with singular potentials. Proceedings of the International Symposium on Partial Differential Equations and the Geometry of Normed Linear Spaces (Jerusalem, 1972). *Israel J. Math.* **13**, pp. 135–148. (Cited on page 327.)

[47] Kester, D. 1961. *Introduction à l'électrodynamique quantique.* Travaux et Recherches Mathématiques **VI**, Dunod, Paris.

[48] Kowalski, E. Representation theory. `https://people.math.ethz.ch/~kowalski/representation-theory.pdf` (Cited on page 625.)

[49] Kuranishi, M. 1950. On Euclidean local groups satisfying certain conditions. *Proc. Amer. Math. Soc.* **1**, pp. 372–380. (Cited on page 595.)

[50] Lancaster, T.; Blundell, S. 2014. *Quantum Field Theory for the Gifted Amateur.* Oxford University Press, Oxford. (Cited on pages 395 and 732.)

[51] McMahon, D. 2008. *Quantum Field Theory Demystified.* McGraw-Hill, New York.

[52] Manoukian, E. 1983. *Renormalization.* Pure and Applied Mathematics **106**. Academic Press [Harcourt Brace Jovanovich], New York. (Cited on pages 5 and 496.)

[53] McComb, W. 2004. *Renormalization Methods: A Guide for Beginners.* The Clarendon Press, Oxford University Press, Oxford.

[54] Messiah, A. 1959. *Mécanique quantique.* 2 vols. Dunod, Paris. (Cited on pages 309 and 312.)

[55] Molinari, L. 2007. Another proof of Gell-Mann and Low's theorem. *J. Math. Phys.* **48**.

[56] Moretti, V. 2013. *Spectral Theory and Quantum Mechanics: With an Introduction to the Algebraic Formulation.* Unitext **64**. La Matematica per il 3+2. Springer, Milan.

[57] Munkres, J. 1975. *Topology: A First Course.* Prentice Hall, Englewood Cliffs, NJ. (Cited on page 186.)

[58] Muta, T. 2010. *Foundations of Quantum Chromodynamics.* World Scientific Lecture Notes in Physics **78**. World Scientific, Singapore. (Cited on page 5.)

[59] Newton, T.; Wigner, E. Localized states for elementary systems. *Rev. Mod. Phys.* **21**, pp. 400–406. (Cited on page 669.)

[60] Nenciu, G.; Rasche, G., 1989. Adiabatic theorem and Gell-Mann & Low formula. *Helv. Phys. Acta* **62**, pp. 372–388. (Cited on page 433.)

[61] Parthasarathy, K. R. 1992. *An Introduction to Quantum Stochastic Calculus.* Modern Birkhäuser Classics. Birkhäuser/Springer Basel AG, Basel. (Cited on page 732.)

[62] Paugam, F. 2014. *Towards the Mathematics of Quantum Field Theory.* Springer-Verlag, New York. (Cited on page 3.)

[63] Penrose, R. 2005. *The Road to Reality: A Complete Guide to the Laws of the Universe.* Alfred A. Knopf, New York. (Cited on page 5.)

[64] Peskin, M.; Schroeder, V. 1995. *An Introduction to Quantum Field Theory.* Addison-Wesley, Advanced Book Program, Reading, MA. (Cited on pages 141, 373, 423, 428, 430, 446, 447, 450, 463, and 672.)

[65] Petz, D. 1990. *An Invitation to the Algebra of Canonical Commutation Relations.* Leuven Notes in Mathematical and Theoretical Physics. Series A: Mathematical Physics **2**. Leuven University Press, Leuven.

[66] Polyak, M. Feynman diagrams for pedestrians and mathematicians. `https://arxiv.org/abs/math/0406251`

[67] Preskill, J. Functional integration. `www.theory.caltech.edu/~preskill/ph205/205Chapter4-Page1-63.pdf` (Cited on page 539.)

[68] Reed, M.; Simon, B. 1978. *Methods of Modern Mathematical Physics. IV. Analysis of Operators.* Academic Press, New York/London. (Cited on page 36.)

[69] Reed, M.; Simon, B. 1979. *Methods of Modern Mathematical Physics. III. Scattering Theory.* Academic Press, New York/London. (Cited on page 327.)

[70] Rivasseau, V. 1991. *From Perturbative to Constructive Renormalization.* Princeton Series in Physics. Princeton University Press, Princeton, NJ. (Cited on pages 5, 420, 472, and 496.)

[71] Robertson, H. 1929. The uncertainty principle. *Phys. Review* **34**, no. 1, pp. 16–164. (Cited on page 29.)

[72] Sakurai, J.; Napolitano, J. 1994. *Modern Quantum Mechanics.* Addison-Wesley, San Francisco. (Cited on page 322.)

[73] Sénéchal, D. Course PHQ 430. Université de Sherbrooke. `www.physique.usherbrooke.ca/pages/node/8600` (Cited on page 45.)

[74] Serre, J.-P. 1978. *Représentations linéaires des groupes finis.* Third revised edition. Hermann, Paris. (Cited on page 115.)

[75] Schweber, S. 1961. *An Introduction to Relativistic Quantum Field Theory.* Row, Peterson and Company, Evanston, IL/Elmsford, NY. (Cited on page 685.)

[76] Simms, D. J. 1971. A short proof of Bargmann's criterion for the lifting of projective representations of Lie groups. *Rep. Math. Phys.* **2**, no. 4, pp. 283–287. (Cited on page 596.)

[77] Srednicki, M. 2007. *Quantum Field Theory.* Cambridge University Press, Cambridge. (Cited on pages 447, 538, and 658.)

[78] Sternberg, S. 1994. *Group Theory and Physics.* Cambridge University Press, Cambridge.

[79] Streater, R.; Wightman, A. 1964. *PCT, Spin, Statistics, and All That.* Benjamin, New York. (Cited on pages 420, 704, 707, 715, and 719.)

[80] Schwartz, M. 2014. *Quantum Field Theory and the Standard Model.* Cambridge University Press, Cambridge. (Cited on pages 250, 404, and 426.)

[81] Taylor, E.; Wheeler, J. 1992. *Spacetime Physics.* Second edition. W. H. Freeman, New York. (Cited on pages 102 and 732.)

[82] Taylor, J. 1972. *Scattering Theory: The Quantum Theory of Nonrelativistic Collisions*. John Wiley and Sons, New York. (Cited on pages 322 and 339.)

[83] Teller, P. 1995. *An Interpretive Introduction to Quantum Field Theory.* Princeton University Press, Princeton, NJ.

[84] Wald, R. 1984. *General Relativity.* University of Chicago Press, Chicago.

[85] Ticciati, R. 1999. *Quantum Field Theory for Mathematicians.* Cambridge University Press, Cambridge. (Cited on page 463.)

[86] Weinberg, S. 1960. High energy behavior in Quantum Field Theory. *Phys. Review* **110**, no. 3, pp. 838–849. (Cited on page 481.)

[87] Weinberg, S. 2005. *The Quantum Theory of Fields. Vol. I. Foundations.* Cambridge University Press, Cambridge. (Cited on pages 2, 4, 132, 250, 252, and 463.)

[88] Weinberg, S. 2013. *Lectures on Quantum Mechanics.* Cambridge University Press, Cambridge. (Cited on pages 26 and 311.)

[89] Wightman, A. 1962. On localizability of Quantum Mechanical systems, *Rev. Mod. Phys.* **34**, no. 4, pp. 845–872. (Cited on page 668.)

[90] Wigner, E. 1939. On unitary representations of the inhomogeneous Lorentz group. *Ann. of Math.* **40**, no. 11, pp. 149–204. (Cited on page 208.)

[91] Woit, P. 2017. *Quantum Theory, Groups and Representations: An Introduction.* Springer, Cham. (Cited on page 732.)

[92] Wigner, E. 1959. *Group Theory and Its Application to the Quantum Mechanics of Atomic Spectra.* Pure and Applied Physics **5**. Academic Press, New York/London. (Cited on page 46.)

[93] Zee, A. 2010. *Quantum Field Theory in a Nutshell.* Second edition. Princeton University Press, Princeton, NJ. (Cited on pages 99, 176, and 496.)

[94] Zimmermann, W. 1968. The power counting theorem for minkowski metric, *Commun. Math. Phys.* **11**, pp. 1–8. (Cited on page 552.)

[95] Zimmermann, W. 1969. Convergence of Bogoliubov's method of renormalization in momentum space. *Comm. Math. Phys.* **15**, pp. 208–234. (Cited on page 563.)

[96] The sum of all numbers is -1/12. `www.youtube.com/watch?v=w-I6XTVZXww` (Cited on page 179.)

Index

Printed in the United States
by Baker & Taylor Publisher Services

Printed in the United States
by Baker & Taylor Publisher Services